Virus-Like Particl

Virus-Like Particl

Virus-Like Particles
A Comprehensive Guide

Paul Pumpens, Dr. habil. biol.
Professor of Molecular Biology
University of Latvia
Riga, Latvia

Peter Pushko, PhD
President and Chief Scientific Officer
Medigen, Inc.
Frederick, Maryland, USA

with cartoons of viruses
by
Philippe Le Mercier, PhD
Team Leader, Swiss Institute of Bioinformatics
Geneve, Switzerland

CRC Press
Taylor & Francis Group
Boca Raton   London

CRC Press is an imprint of the
Taylor & Francis Group, an **informa** business

The cover image is generated by *Chimera* software (Pettersen EF, Goddard TD, Huang CC, Couch GS, Greenblatt DM, Meng EC, Ferrin TE. UCSF Chimera - a visualization system for exploratory research and analysis. *J Comput Chem.* 2004;25:1605 -1612) and demonstrates 3D reconstruction of the chimeric hepatitis B virus core particles carrying a model foreign epitope inserted at the major immunodominant region.

CRC Press
Boca Raton and London

First edition published 2022
by CRC Press
6000 Broken Sound Parkway NW, Suite 300, Boca Raton, FL 33487–2742

and by CRC Press
4 Park Square, Milton Park, Abingdon, Oxon, OX14 4RN

CRC Press is an imprint of Taylor & Francis Group, LLC

© 2022 Taylor & Francis Group, LLC

Library of Congress Cataloging-in-Publication Data
Names: Pumpens, Paul, author. | Pushko, Peter, author.
Title: Virus-like particles : a comprehensive guide / Paul Pumpens, Peter Pushko.
Description: First edition. | Boca Raton : CRC Press, 2022. | Includes bibliographical references and indexes. | Summary: "This book represents the first complete and systematic guide to the virus-like particles (VLPs) and their applications as vaccines, therapeutic tools, nanomaterials, and nanodevices. The grouping of the VLPs follows the most recent virus taxonomy and the traditional Baltimore classification of viruses, which are based on the genome structure and mechanism of mRNA synthesis. Within each of the seven Baltimore classes, the order taxon serves as a framework of the chapter's arrangement. The term "VLP" is used as a universal designation for the virus–, core–, or capsid-like structures, which became an important part of the modern molecular virology. The 3D structures, expression systems, and nanotechnological applications are described for VLPs in the context of the original viruses and uncover their evolving potential as novel vaccines and medical interventions"— Provided by publisher.
Identifiers: LCCN 2021051808 (print) | LCCN 2021051809 (ebook) | ISBN 9780367476779 (hardback) | ISBN 9781032246734 (paperback) | ISBN 9781003279716 (ebook)
Subjects: MESH: Viral Structural Proteins—classification | Viral Structural Proteins—therapeutic use | Vaccines, Virus-Like Particle
Classification: LCC QR189 (print) | LCC QR189 (ebook) | NLM QW 15 | DDC 615.3/72—dc23/eng/20220204
LC record available at https://lccn.loc.gov/2021051808
LC ebook record available at https://lccn.loc.gov/2021051809

ISBN: 978-0-367-47677-9 (hbk)
ISBN: 978-1-032-24673-4 (pbk)
ISBN: 978-1-003-27971-6 (ebk)

DOI: 10.1201/b22819

Typeset in Times LT Std
by Apex CoVantage, LLC

Access References online in Support Materials: www.routledge.com/9780367476779

What changes, endures

Rainis, *The Golden Horse*

Contents

SECTION II Single-Stranded DNA Viruses

SECTION III Double-Stranded RNA Viruses

SECTION IV   Positive Single-Stranded RNA Viruses

SECTION V Negative Single-Stranded RNA Viruses

SECTION VI Single-Stranded RNA Viruses Using Reverse Transcription

SECTION VII Double-Stranded DNA Viruses Using Reverse Transcription

References (Available online in Support Materials: www.routledge.com/9780367476779)

Preface

Knowing is not enough; we must apply. Willing is not enough; we must do.

Johann Wolfgang von Goethe

As people are walking all the time, in the same spot, a path appears.

John Locke

This book is the first attempt to write the complete systematic guide for the classification of various virus-like particles—or VLPs—by using modern taxonomical categories. The VLPs can be both natural or synthetic structures and building blocks, which represent a considerable part of the current nanobiotechnology. In addition to describing taxonomical classification of various VLPs, the book focuses on the generation of, and manipulation with, the VLPs from all viral taxons, as potential vaccines, gene therapy, and diagnostic tools. We also describe VLP-based, emerging nanomaterials, nanodevices, and nanomachines, together with the nanoscale methodologies for new materials design and fabrication. Finally, the history of the VLP discovery, as well as continuous improvement in the VLP manufacturing by using recombinant technologies, genetic and protein engineering techniques, and chemical synthesis methodologies, represent some of the central ideas of our book.

We believe that this overview based on the current taxonomical classification comes at the right time, since the current situation in the VLP technologies, which were born more than 30 years ago in the middle of the 1980s, could be characterized by Arthur Schopenhauer's sentence: "Just remember, once you're over the hill you begin to pick up speed." We realize that a comprehensive guide to the VLPs is an ambitious goal and cannot possibly mention every effort in the VLP field; we often remembered Antoine de Saint-Exupery's quote: "The one thing that matters is the effort."

At the beginning, when we were looking for the most appropriate ways of VLP classification, we started with the optimistic Lewis Carroll quote: "If you don't know where you are going, any road will get you there." Eventually, we decided to follow the current official taxonomy developed by the International Committee on Taxonomy of Viruses, or ICTV. To make the story understandable in the classic terms of viral genomes, we complemented the ICTV taxonomy with the traditional Baltimore classification, which is based on the genome structure and mechanism of the viral mRNA synthesis. Thus, according to Baltimore classification, the VLPs are described in this book in seven main sections, in the following order:

(i) VLPs for double-stranded DNA, or dsDNA, viruses

(ii) Single-stranded DNA (ssDNA) viruses

(iii) Double-stranded RNA (dsRNA) viruses

(iv) Positive single-stranded RNA (plus-ssRNA) viruses with plus-strand or sense RNA

(v) Negative single-stranded RNA (minus-ssRNA) viruses with minus-strand or antisense RNA

(vi) Single-stranded RNA viruses using reverse transcription (ssRNA-RT) with plus-strand or sense RNA and DNA intermediate in life cycle

(vii) Double-stranded DNA viruses using reverse transcription (dsDNA-RT) with intermediate pregenomic RNA in life cycle

Within each of the seven Baltimore classes, the VLPs are arranged in most cases in the alphabetical order of their parent viruses in the hierarchy of the ICTV taxons: realm, order (in the case when the first two higher taxons are assigned), family, genus, and species. The taxon *order* is chosen as the backbone of the book because of the familiar traditional nomenclature, together with relatively small number of *orders*, currently 59, in contrast to the traditionally used taxon *family*. The number of *families* currently stands at 189.

It should be noted from the very beginning that we are using the term *VLP* as a broad concept for all variants of so-called virus-, core-, and capsid-like structures, including particles resulting from the separate expression of cores and envelopes of enveloped viruses or preparation of monolayered particles from the naturally multilayered viruses. The main argument in favor of such terminology is the fact that the VLPs resemble viruses in their spatial composition, although not necessarily the specific parent virus that served as the source of a VLP. Unfortunately, space limitations did not allow inclusion of the other symmetric particles or VLPs, which were formed by nonviral proteins using synthetic methods.

Each chapter begins with the general taxonomic introduction. Then, the following information is provided for each VLP: (i) the spatial 3D structure of both the parent virus and genetically engineered particles, whenever possible and (ii) genes used for VLP expression, as well as the recombinant gene expression systems. Then, the VLP examples are provided as (i) natural—or nonmodified—and (ii) chimeric—or genetically or chemically modified particles—if the latter have been prepared. The different types of the chimeric VLP are described according to their molecular composition: uniform, mosaic, or multilayered. Preparation of chemically modified VLPs by performing chemical reactions with VLPs and/or whole virions as substrates is also described, in parallel with the preparation of VLPs using genetic engineering techniques. Finally, packaging, targeting, and visualization approaches are described when these have been applied for the VLPs.

The guide aims to describe briefly the practical VLP applications including, first of all, display/exposure of foreign immunological epitopes to prepare experimental vaccines, packaging of genetic materials for improving immunogenicity of vaccines and/or for gene delivery, targeting to the specific cells by attaching the appropriate tags/addresses, and generation of novel bionanomaterials, e.g. therapeutic and imaging tools. These specific applications are outlined separately for the natural—or original—VLPs, as well as for the chimeric derivatives of the original VLPs. The application of the engineered natural and/or chimeric VLPs is described further in the following order:

(i) Vaccines
(ii) Delivery vehicles/vectors
(iii) Novel nanomaterials, including diagnostic tools

However, we did not focus specifically on the types and clinical applications of approved VLP-based vaccines, although clinical trials are described in the literature.

The guide describes currently available information not only for the VLPs, but also for live, infectious viruses, when the latter have been modified using genetic and/or chemical engineering. We briefly described but did not intend to provide detailed information on (i) phage display technique based on filamentous single-stranded DNA phages; (ii) DNA vaccination; and (iii) vector technologies and gene therapy aspects, which utilize viruses as expression vectors for the inserted foreign genes.

The guide is primarily recommended for both professionals and students in various fields from classic virology to nanotechnology. Furthermore, the book can be of interest to specialists in life sciences including history of science, molecular biology, microbiology, genetics, immunology, biochemistry, environmental science, evolutionary research, and vaccinology. This book can also draw attention from readers without specialized biological or scientific background who could be interested in an informative introduction to the virology and nanotechnology, with applications to the modern medicine and healthcare. Finally, the book is intended to expand knowledge horizons of anyone interested in biotechnology and science and to improve the situation characterized by Albert Einstein: "Technological progress is like an axe in the hands of a pathological criminal."

This book would not have been possible without our long-term work on VLPs in Riga. First and foremost, we thank our Chief, Professor Elmārs Grēns, who conveyed a continuous spirit of adventure regarding the VLP research.

We are highly indebted to our colleagues Dr. habil. biol. Tatyana Kozlovska, Dr. Andris Dišlers, Dr. habil. biol. Velta Ose, Dr. Regīna Renhofa, Dr. Indulis Cielēns, Dr. Dace Skrastiņa, Dr. Vadim Bichko (now in San Diego), and Dr. habil. biol. Alexander Tsimanis (now in Rehovot), who have been extremely supportive and who established the RNA phage and hepatitis B virus VLP fields in Riga and shared readily their unpublished data described in this book. We are happy that Professor Kaspars Tārs, Dr. Andris Zeltiņš, and Dr. Andris Kazāks continue the good VLP traditions in Riga.

We are grateful to all of those with whom we have had the pleasure to work on the VLPs in Riga: Ms. Ināra Akopjana, Dr. Ina Baļķe, Mr. Jānis Bogāns, Dr. Olga Borschukova (now in Boston), Dr. Maija Bundule, Professor Yury Dekhtyar, Dr. Dzidra Dreiliņa, Dr. Anda Dreimane, Dr. Jānis Freivalds, Dr. Edith Grēne Carron (now in Leonardtown, MD), Dr. Esther Grinstein (now in Berlin), Mr. Indulis Gusārs, Dr. Juris Jansons, Dr. Ieva Kalnciema, Dr. Gints Kalniņš, Dr. Vera Krieviņa, Dr. Ilva Liekniņa, Dr. Marija Mihailova, Dr. Guna Mežule, Dr. Baiba Niedre-Otomere, Dr. Ligita Orna, Dr. Ivars Petrovskis, Dr. Jānis Rūmnieks, Dr. Dagnija Sniķere, Dr. Irina Sominskaya, Dr. Anna Strelnikova, Dr. Arnis Strods, Mr. Valdis Tauriņš, Dr. Vladimir Tsibinogin, Dr. Inta Vasiljeva, and Dr. Tatjana Voronkova-Kazāka.

We also thank colleagues and VLP enthusiasts in the EU, United States, and worldwide. We would especially like to thank Professor Martin F. Bachmann (Zürich—Bern—Oxford), who led the VLP studies from basic science to the real nanotechnological tools by starting many amazing innovative projects and remaining active and successful in the VLP field up to now. In the context of the RNA phage VLP nanotechnology in Riga, we are highly indebted to Dr. Alain Tissot (Penzberg), Dr. Gary Jennings (Zürich), and Dr. Stephan Oehen (now in Basel). Warm gratitude to the Nobelist Rolf M. Zinkernagel (Zürich) for his friendly attitude and guidance on the in-depth immunological studies of the hepatitis B core and RNA phage VLPs.

We are highly indebted to Albuquerque's Professors David S. Peabody and Bryce Chackerian, who developed science and applications of the RNA phage VLPs.

We are very thankful to Professor R. Anthony Crowther (Cambridge) for his collaboration and constant guidance in 3D structures of RNA phage and hepatitis B core VLPs and to Professor David J. Rowlands (Leeds) for long-term collaboration on hepatitis B core VLP technologies. Many thanks to Professor David Klatzmann (Paris) who initiated joint biotechnological projects on the hepatitis B core and RNA phage VLPs. We are grateful to Professor Dr. Rüdiger Schmitt (Regensburg) for providing unique expertise in the generation of the RNA phage AP205 VLPs.

Many thanks to the pioneers of the VLP field Professor David R. Milich (San Diego), Professor Michael Nassal (Freiburg), Professor Adam Zlotnick (Bloomington), Professor Michael Roggendorf (Essen), Professor George P. Lomonossoff (Norwich), Professor Trevor Douglas (Bloomington), Professor Polly Roy (London), Professor Bogdan Dragnea (Bloomington), Professor Bettina Böttcher (Würzburg), Professor Reinhold Schirmbeck (Ulm), Dr. John A. Berriman (Charmouth), and Dr. Alan M. Roseman (Manchester).

A warm thank you to our colleagues and friends Professor Indriķis Muižnieks, Professor Kęstutis Sasnauskas (Vilnius), Professor Matti Sällberg (Stockholm), Professor Henrick Garoff (Stockholm), Professor Britta Wahren (Stockholm), Dr. Nikolai Granovski (Dnipro), Dr. Lionel

Crawford, F.R.S. (Cambridge, UK), Dr. Jonathan Smith, Dr. Michael Parker, Dr. Pamela Glass, Dr. Robin Robinson, Dr. Rick Bright, Dr. Vittoria Cioce, Dr. Gale Smith, Dr. Terrence Tumpey, Dr. Rahul Singhvi, Thomas Kort, Dr. Irina Tretyakova, Professor Lars Magnius, Professor Sang-Moo Kang, and Professor Kenneth Lundstrom (Lutry). We are very thankful to the Essen's Professors Sergei Viazov and Stefan Roß and to Professor Michael Kann (Göteborg). Our warm thanks to Dr. Rainer G. Ulrich (Greifswald-Insel Riems), Dr. Helga Meisel (Berlin), and Professor Detlev H. Krüger (Berlin); to the Tartu Professors Richard Villems and Mart Ustav; and to Professor Mart Saarma (Helsinki), Professor Lars Magnius (Stockholm), Professor Joseph Holoshitz (Ann Arbor), and Professor Mikhail Mikhailov (Moscow).

We are thankful to the architects of the novel viral taxonomy Professors Eugene Koonin (Bethesda), Mart Krupovic (Paris), Vadim Agol (Moscow), David Prangishvili (Paris and Tbilisi), Colin Hill (Cork), Elliot J. Lefkowitz (Birmingham), Alexander Gorbalenya (Leiden), Jens Kuhn (Frederick), and many other ICTV enthusiasts for the timely general revision of the ICTV files, which enabled current classification of VLP in this Guide.

Many thanks to the Nobelist George P. Smith for his friendly attitude and a photograph and to Professor Lee Makowski (Boston) for the excellent image of the phage M13. We are grateful to Dr. Vijay S. Reddy (San Diego) for the constant friendly support and 3D images of VLPs at the excellent VIPERdb. We are deeply grateful to Professor Rozanne Sandri-Goldin (Irvine) for the assistance with the illustrations from *Journal of Virology*. Our thanks to Dr. Pavel Zayakin for his professional help and advice on IT applications.

A cordial thanks to Professors Gunārs Duburs, Alexander Rappoport, Katrīna Ērenpreisa, Jāzeps Keišs, Andrejs Ērglis, Ludmila Vīksna, and Uga Dumpis and to Drs. Valentīna Sondore, Rūta Brūvere, and Uldis Bērziņš for constant friendship and support.

We would like to thank Riga enthusiasts Dr. Venta Kocere, director of the Latvian Academic Library grounded in 1524; Professor Jānis Krastiņš, a well-known expert in art history; Mr. Eugene Gomberg, a great supporter of Riga Art Nouveau, and Mr. Alexander Kiselev / YL2RR, as well as Dr. Tamara Pererva from Kyiv for the true friendship and creative discussions.

We are highly indebted to our older colleagues and friends Professor Viesturs Baumanis, Professor Valdis Bērziņš, Dr. Galina Borisova, Professor Rita Kukaine, Mr. Juris Ozols, Dr. Oleg Plotnikov, Dr. Anatoly Sharipo, Dr. Zinaida Shomstein, Professor Eva Stankeviča, Professors Hans-Alfred and Sinaida Rosenthal (Berlin), Professor Lev Kisselev (Moscow), and Professor Nikolai Kiselev (Moscow), who will never see this book. We are thankful to Professor Peter H. Hofschneider (München), who not only discovered RNA phages but also cloned the full-length hepatitis B virus genome and initiated the story of the hepatitis B core VLPs.

Last but not least, we would like to express the deepest appreciation to Professor Wolfram H. Gerlich (Giessen) for his constant friendly support and advice on our VLP studies and many special thanks to Dr. Yury Khudyakov (Atlanta) who encouraged our collaboration with Taylor & Francis and to Dr. Philippe Le Mercier (Geneve) who generated the beautiful cartoons of viruses.

Finally, we wish to thank Dr. Charles "Chuck" Crumly, Dr. Daina Habdankaitė, Ms. Linda Leggio, Mrs. Kate Fornadel, Mr. Christian Munoz (cover designer), and the entire CRC Press/Taylor & Francis team for publishing this work.

A complete and searchable Reference List is available in eResources for Free Download at: www.routledge.com/9780367476779

Paul Pumpens
Peter Pushko

Authors

Paul Pumpens, Dr. habil. biol., graduated from the Chemical Department of the University of Latvia in 1970 and earned his PhD in molecular biology from the Latvian Academy of Sciences, Riga, and DSc from the Institute of Molecular Biology of the USSR Academy of Sciences, Moscow, USSR. Dr. Pumpens started his research career as a research fellow at the Institute of Organic Synthesis, where he conducted research from 1973–1989. He served as head of the Laboratory of Protein Engineering at the Institute of Organic Synthesis (1989–1990), as head of the Department of Protein Engineering at the Institute of Molecular Biology of the Latvian Academy of Sciences (since 1993, the Biomedical Research and Study Centre) in Riga (1990–2002), and as scientific director of the Biomedical Research and Study Centre (2002–2014). He served as a professor of the Biological Department of the University of Latvia from 1999 until 2013. Dr. Pumpens pioneered genetic engineering research in Latvia. His group was one of the first in the world to perform successful cloning of the hepatitis B virus genome and the expression of hepatitis B virus genes in bacterial cells. His major scientific interests are in designing novel recombinant vaccines and diagnostic reagents and development of tools for gene therapy based on virus-like particles. Dr. Pumpens is an author of more than 300 scientific papers and issues or pending patents. He has edited, together with Dr. Yury E. Khudyakov, the CRC Press/Taylor & Francis book *Viral Nanotechnology* (2015) and authored the CRC Press/Taylor & Francis book *Single-Stranded RNA Phages: From Molecular Biology to Nanotechnology* (2020).

Peter Pushko, PhD, graduated in 1984 from the University of Latvia, Riga, USSR and received his doctorate degree in molecular biology in 1990 from the University of Tartu (Tartu, Estonia) under mentorship of Dr. Paul Pumpens. He worked at the Institute of Molecular Biology (Riga, Latvia) on chimeric virus-like particles derived from bacteriophage fr and from hepatitis B virus. He completed postdoctoral studies at the Institute of Virology (Berlin, Germany), State Bacteriological Laboratory (Stockholm, Sweden) and Imperial Cancer Research Fund (Department of Pathology, University of Cambridge, UK) From 1994 to 2000, Dr. Pushko worked at the U.S. Army Medical Research Institute for Infectious Diseases (USAMRIID, Fort Detrick, Maryland, USA) as the National Research Council fellow. During this period, he developed RNA vector system from Venezuelan equine encephalitis (VEE) virus replicon and prepared experimental vaccines against Ebola, Marburg, and Lassa hemorrhagic fever viruses. In 2000–2010, Dr. Pushko worked at Novavax, Inc., (Rockville, MD), where he developed recombinant virus-like particles (VLPs) as vaccines against influenza. Since 2010, Dr. Pushko has served as President and Chief Scientific Officer at Medigen Inc. (Frederick, Maryland), a biopharmaceutical company developing innovative vaccines for emerging infectious diseases and cancer. His scientific interests include vaccine development for emerging viruses, such as pandemic influenza and alpha-, arena-, and flaviviruses. He developed several novel vaccine technologies including VEE replicon vector, influenza VLPs, and DNA-launched live-attenuated vaccines. Dr. Pushko is the author of over 50 patents and 70 publications.

Prologue

Every dogma must have its day.

H.G. Wells

I am a cage, in search of a bird.

Franz Kafka

THE CONCEPT OF VLPs

As already mentioned in the Preface, we use the term *virus-like particles* (VLPs) as a universal designation for the virus-, core-, and capsid-like structures. VLPs include particles resulting from the expression of cores and envelopes of enveloped and nonenveloped viruses, preparations of monolayered particles from naturally multilayered viruses. Symmetric particles formed by nonviral proteins or artificially designed symmetric particles also can be considered VLPs; however, these deserve a special book and are not described in this guide. The size of particles in the range of less than 20 and more than 100 nm in diameter could be considered as a typical size for most VLPs. The term VLP is used therefore in the sense that such structures are intended to resemble or mimic the basic elements of viral architecture—such as size, symmetry, or self-assembly—but do not necessarily copy the exact structure of a specific virus, including the source of structural proteins. Remarkably, Hyman et al. (2021) recently raised the question of different meanings of the term *VLP* in different disciplines.

The early understanding of VLPs as virus-resembling and immunologically and practically useful structures came in the early 1980s due to rapid progress of the genetic engineering methodology and development of efficient heterologous platforms for the powerful expression of mammalian genes. As supramolecular structures built symmetrically from hundreds of subunits of one or more types, the VLPs were viewed primarily as exceptional candidates for vaccines. First, VLPs resembled the immunogenic properties of their viral parent but were noninfectious since they did not contain any infectious genetic material. Second, the VLP production and purification seemed feasible and easily achievable, because of affordable bacterial, insect, and/or yeast expression platforms. Third, the genetically cloned and expressed viral structural genes allowed genetic manipulations that could result in knowledge-based reconstructions of the natural VLPs.

The two structural genes of hepatitis B virus (HBV), namely core (HBc) and surface (HBs), were among the first VLP candidates prepared in heterologous expression platforms that are described in detail in Chapters 37 and 38. These early efforts resulted in the first recombinant human vaccine. The prophylactic HBV vaccine was prepared in yeast, purified in the form of the HBs VLPs, and has been used successfully since 1986 in human healthcare.

Next, the cervical cancer vaccine was produced in the form of human papillomavirus (HPV) L1 VLPs in the yeast or baculovirus expression systems, which went on the market in 2006 and 2007, respectively, as described in Chapter 7.

After global success of both vaccines, the VLP-based hepatitis E vaccine was approved in 2011. Then, numerous animal vaccines followed, where, for example, nonchimeric circovirus and parvovirus VLPs were approved as vaccines against infections in pigs and dogs. Other animal vaccines were generated against calicivirus (RHDV), papillomavirus (BPV and CRPV), reovirus (BTV), birnavirus (AHSV), and other viruses by using the appropriate natural VLPs. It is important to emphasize that all previously mentioned vaccines were based on the expression of the natural, unmodified, or nonchimeric viral genes and could be therefore identified as the *natural* or *native* VLPs. These and other numerous vaccines and vaccine candidates are described in the corresponding chapters.

The historic Figure P.1 classifies many parent viruses for the VLPs by size and morphology. The figure shows nonenveloped viruses resolved by x-ray crystallography, as well as enveloped viruses displayed as electron micrographs. Notably, the scale of the bottom part of the figure is twice of that of the upper part. Needless to say, many viruses are not shown, including recent successful VLP platforms, such as popular RNA phage AP205 and cucumber mosaic virus (CuMV), which are described later. These recent platforms are described in the appropriate chapters of this book.

CHIMERIC VLPs

In parallel with this continuing work with the natural VLPs, a strong interest emerged from the early 1980s to the *chimeric* VLPs expressed by using engineered—or chimeric—carrier genes and carrying foreign structural elements on the external or internal surface of the final chimeric particles to address the needs of numerous viral nanotechnology applications, in contrast to the solely immunological purposes. The term *chimeric VLPs* (Ulrich et al. 1996; Krüger et al. 1999) originated from *Chimaera*, a fire-breathing monster of Lycia in Asia Minor, composed of the parts of more than one animal, usually depicted as a lion, with the head of a goat arising from its back and a tail that might end with a snake's head.

The historic Figure P.2 demonstrates the way the natural VLPs could be converted into the chimeric ones, which would display foreign sequences on their outer surface.

Generally, the idea of the chimeric VLPs was triggered at that time by the successful genetic engineering of proteins, which flourished in the early 1980s. Alan R. Fersht and his

DOI: 10.1201/b22819-1

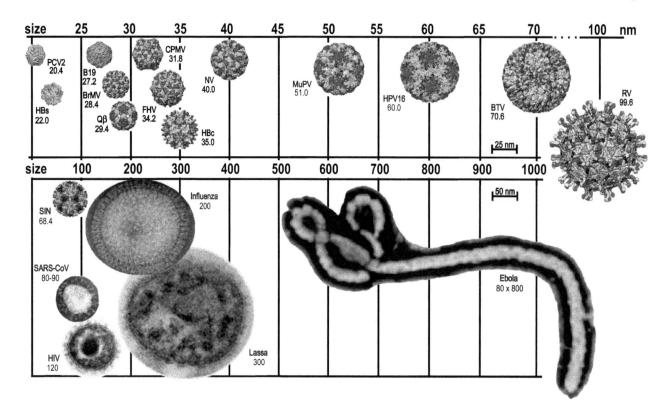

FIGURE P.1 The size-scaled presentation of the crystal structures of nonenveloped viruses and the electron micrographs of enveloped viruses that appeared as popular sources of VLPs. The 3D models of the structurally resolved particles are from the VIPERdb database (VIrus Particle ExploreR), *http://viperdb.scripps.edu/* (Carrillo-Tripp et al. 2009). The structure of the HBs particle with octagonal symmetry is from Gilbert et al. (2005). The electron micrographs of the enveloped viruses are extracted from the publicly available resources. Abbreviations: B19 (human parvovirus B19); BrMV (brome mosaic virus); BTV (bluetongue virus); CPMV (cowpea mosaic virus); Ebola (Ebola virus); FHV (Flock house virus); HBc (hepatitis B virus core protein); HBs (hepatitis B virus surface protein); HIV (human immunodeficiency virus); HPV16 (human papillomavirus type 16); Influenza (influenza A virus); Lassa (Lassa virus); MuPV (murine polyomavirus); NV (Norwalk virus); PCV2 (porcine circovirus type 2); Qβ (RNA phage Qβ); RV (rotavirus); SARS-CoV (severe acute respiratory syndrome-related coronavirus); SIN (sindbis virus). (Adapted with permission of S. Karger AG, Basel from Pushko P, Pumpens P, Grens E. *Intervirology.* 2013;56:141–165.)

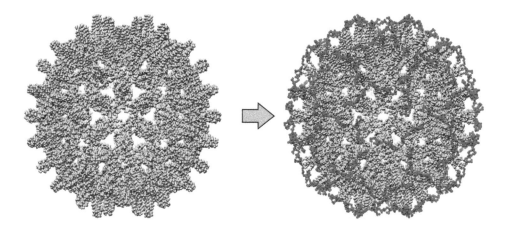

FIGURE P.2 The 3D model demonstrating how foreign sequences are exposed on the VLPs. Hepatitis B core (HBc) VLPs are used as an example in this case. The red peptides are epitopes exposed on the outer surface of the HBc VLPs after insertion of the corresponding foreign fragment into the major immunodominant region (MIR) of the HBc protein. The structural data is compiled from the VIPERdb (http://viperdb.scripps.edu) database (Carrillo-Tripp et al. 2009). The 3D reconstruction is performed by a comparative modeling program 3D-JIGSAW, *https://bmm.crick.ac.uk/* (Bates et al. 2001) and visualized using Chimera software (Pettersen et al. 2004).

colleagues conceived of the idea of the specialized branch of genetic engineering, named protein engineering, meaning mutational intervention into the structure of proteins, which was based on the spatial and functional knowledge of enzymes. In fact, the early protein engineering was oriented toward the creation of artificially improved proteins with desired functions (Winter et al. 1982; Wilkinson et al. 1983, 1984; Leatherbarrow and Fersht 1986; Arnold 1988). Since then, the protein engineering has achieved enormous success, resulting in the new forms of enzymes and their inhibitors and elucidating the basic rules of protein folding and changing their specific activities (Arnold 1990; Fersht and Winter 1992).

The successful protein engineering of enzymes, i.e., relatively small monomeric or oligomeric proteins, eventually led to the demand for the preparation of artificial multimeric structures based on the VLPs, which could be considered a primary tool to enhance immunological activity of specific immunological epitopes from any protein of interest. The success in the novel field, namely protein engineering of VLPs, was ensured by the two incredibly important directions and methodologies set in motion simultaneously in the early 1980s.

The first direction led to the clear comprehension of the significance of specific epitopes as antigenic regions, which were necessary and sufficient for (i) recognition of antigens by monoclonal or polyclonal antibodies and (ii) induction of antigen-specific immune response. The idea gained general acceptance, and many epitope sequences were successfully mapped. The fine mapping of the immunological epitopes resulted, first, in the determination of a minimal epitope sequence, which could be defined as a shortest stretch within the protein molecule, which is still capable of inducing a specific immunological response and binding to the appropriate antibody. The next level of recognition within the epitope was defined as an antigenic determinant, namely the exact side groups of amino acid residues involved in the precise epitope recognition by the corresponding paratope sequences of the corresponding antibody. The mapping techniques, based in general on the examination of libraries of short overlapping synthetic peptides or bacteria-produced fusion proteins, revealed the existence of linear and conformational forms of the epitopes. The linear epitopes were defined as a complex of antigenic determinants that were localized in a short stretch of amino acid residues, usually from four to ten, and their recognition was possible within the polypeptides deprived of specific folding, for example, within the short synthetic peptides or within the SDS-denatured fusion proteins. In contrast, the conformational epitopes had their antigenic determinants distributed in the more distant polypeptide stretches and were strongly dependent on the specific protein folding within the correct 3D structure. In 1980s, the peak of the epitope mapping techniques was achieved thanks to the genetic engineering that offered experimental resources to construct DNA copies of both linear and conformational epitopes, to define their length and composition at a single amino acid resolution, to combine them in a different order, and to introduce them into carrier genes, preferably into the genes encoding the self-assembly competent monomers capable of assembling into the VLPs.

The increasing manipulations with the VLPs were further encouraged by the simultaneous development of the second of the two major directions mentioned previously. This direction was motivated by rapid development of structural knowledge about viral structural units using high-resolution techniques. Thus, the x-ray crystallography and, later, electron cryomicroscopy, revealed in 1980–1990 the high-resolution 3D structures of the first experimental VLP carriers, such as rhinovirus (Rossmann et al. 1985), poliovirus (Hogle et al. 1985), tobacco mosaic virus (TMV; Namba and Stubbs 1986; Namba et al. 1989), bluetongue virus (BTV; Grimes et al. 1997), RNA phages (Valegård et al. 1990; Liljas et al. 1994; Golmohammadi et al. 1996; Tars et al. 1997, 2000), and the already mentioned HBV core antigen HBc (Crowther et al. 1994; Böttcher et al. 1997; Conway et al. 1997; Wynne et al. 1999). Some of these structures are shown in Figure P.1. The 3D structures of the VLPs enabled the knowledge-based engineering of the most efficient regular arrangement of the foreign chains in the desired positions of the outer or inner surface of the 3D-resolved VLP carriers.

In contrast to the monomeric and oligomeric protein carriers, the VLP carriers were able to provide not only a high density of introduced foreign oligopeptides per particle but also a distinctive 3D conformation, which could be especially important for the presentation of the conformational epitopes. Therefore, the regular repetitive pattern and correct conformation of inserted epitopes remained the factors that encouraged continuing work on the functional activity of the engineered chimeric VLPs in inducing immunologically protective response and, later, by other nanotechnological applications.

PROTEIN ENGINEERING OF VLPs

The monomeric constituents of viral capsids, envelopes, and rods of viral origin and later the artificial self-assembling units all became targets of the VLP protein engineering and construction of the engineered chimeric VLPs.

The success in structural research of the candidate carrier VLP scaffolds and of the insertion-intended sequences, as well as the development of protein structure prediction techniques, have led to recognition of the protein engineering of VLPs as a specific branch of molecular biology and biotechnology until 2000, although the first thorough review article on the VLPs as an epitope carrier tool was published in the late 1980s (Gren and Pumpens 1988). Amazingly, the historic Figure P.3 reproduced from the more than 30-year-old paper describes well the classical methodology of the VLP protein engineering, when a single gene is engineered by insertion of a foreign epitope, with or without deletion of the host epitopes. These are therefore the two types of VLP vectors to distinguish, i.e. the insertion and replacement

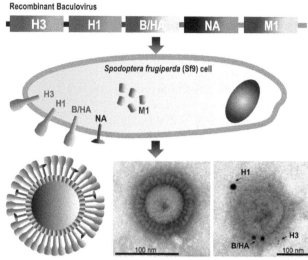

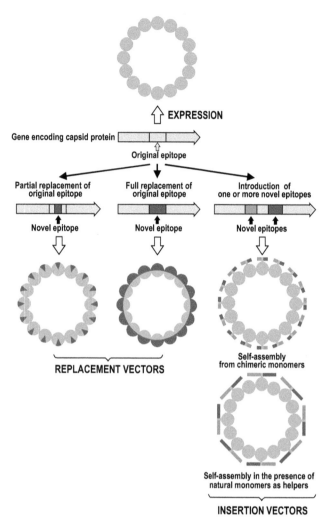

FIGURE P.3 The historical scheme presenting the generation strategies for the chimeric VLPs. (Adopted in color with kind permission of *Russian Chemical Journal* from Gren EJ, Pumpen PP. Recombinant viral capsids as a new age of immunogenic proteins and vaccines. *J All-Union Mendeleyev's Chem Soc.* 1988;33:531–536.)

ones. Then, there are mosaic particles consisting of both natural and chimeric subunits when the chimeric subunits are unable to self-assemble into VLPs and the natural non-chimeric units are needed as helpers to rescue correct VLP structures.

Protein engineering of VLPs appears even more diverse when it involves complex enveloped viruses. The historic Figure P.4 gives a first rough idea how the influenza VLP platform works when it involves for example different hemagglutinin (HA) subtypes derived from three distinct influenza virus strains (Pushko et al. 2011). To be self-assembled, the VLPs are combined in this case from independently expressed HA genes of three seasonal influenza viruses together with neuraminidase (NA) and matrix (M1) genes in a single insect cell. The assembly of the independently synthesized proteins within the VLPs somewhat resembles the phenotypic mixing that is observed with structural proteins of live viruses. Such structures consisting of the

FIGURE P.4 The preparation and electron microscopy of a triple-subtype VLP vaccine containing seasonal influenza A H1 and H3 subtypes and influenza B virus HA. The recombinant baculovirus (rBV) for the expression of the triple-HA VLPs in *Sf*9 cells contained influenza HA gene sequences derived from A/New Caledonia/20/1999 (H1N1), A/New York/55/2004 (H3N2) and B/Shanghai/361/2002. NA and M1 gene sequences were from the A/Indonesia/05/2005 (H5N1) virus (Pushko et al. 2011). Negative stain transmission electron microscopy (bottom middle) was performed by staining with 1% phosphotungstic acid. Immunoelectron microscopy (bottom right) was performed using sucrose gradient-purified VLPs. VLPs were probed with a mixture of primary antibodies specific for H1, H3, and type B influenza from rabbit, mouse, and guinea pigs, respectively. Secondary antibodies were donkey antirabbit labeled with 18-nm gold particles, antimouse labeled with 6-nm gold particles, and antiguinea pig antibodies labeled with 12-nm gold particles. Bars: 100 nm. (Reprinted with permission of S. Karger AG, Basel from Pushko P, Pumpens P, Grens E. *Intervirology.* 2013;56:141–165.)

desired full-length proteins could be named *hybrid* or *reassortant* VLPs rather than *chimeric* or *mosaic* ones.

Furthermore, Figure P.5 illustrates the next and more complex situation when structural genes of different viruses, members of the *Articulavirales* and *Nidovirales* orders are coexpressed, and one of the genes is chimeric. To avoid increase of specific terms, the same designation *hybrid* VLPs could also be applicable in this case. The details of this approach applied to members of the *Articulavirales* order are described in Chapter 33.

As a whole, the VLP protein engineering was defined at that time as a discipline working on the knowledge-based approaches for the theoretical design and experimental preparation of recombinant genomes and genes that might enable the efficient synthesis and correct self-assembly of the chimeric VLPs with the programmed structural and functional properties (Pumpens and Grens 2002).

According to later definition by Pushko et al. (2013), the VLPs are nanodimensional structures that are (i) built from one or several viral structural constituents in the form of recombinant proteins synthesized in efficient homologous or

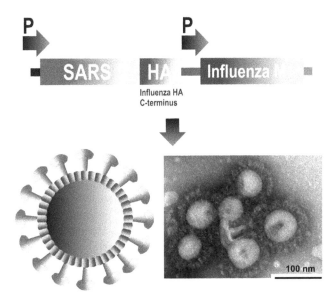

Influenza HA
C-terminus

FIGURE P.5 The preparation of hybrid/chimeric influenza-SARS-CoV VLPs using influenza M1 protein and chimeric SARS-CoV S protein (GenBank AAP13441), containing a carboxy-terminal TM sequence derived from influenza HA. The TM helices were predicted using TransMem based upon a Hidden Markov Model. The TM domain was derived from influenza A/Indonesia/5/2005 HA (aa 531–568). The chimeric SARS-CoV S and influenza M1 proteins (ABI36004) were coexpressed from recombinant baculovirus in *Spodoptera frugiperda* (*Sf*9) cells (Liu et al. 2011). An electron photomicrograph of hybrid/chimeric SARS-CoV VLPs is shown along with the schematic model. (Reprinted with permission of S. Karger AG, Basel from Pushko P, Pumpens P, Grens E. *Intervirology.* 2013;56:141–165.)

primarily heterologous expression systems (bacteria, yeast, or eukaryotic cell culture); (ii) identical or closely related by their 3D architecture and immunochemical characteristics to naturally occurring viral structures; and (iii) lacking viral genomes and therefore not infectious. The term *chimeric* VLPs was accepted therefore for the structures, where the original structural proteins were covalently modified by the addition of new stretches and/or substitution of original sequences by foreign polypeptide stretches with desired functional properties, such as immunological epitopes, cell-targeting, or cargo-packaging signals. The covalent integration into the chimeric VLPs could be achieved either by the expression of the chimeric VLP monomer genes containing the appropriate insertions encoding the desired protein sequences and generated by the genetic engineering methodology (Gren and Pumpen 1988) or by the chemical coupling of peptides, proteins, glycoproteins, carbohydrates, or other functional molecules to the original natural or chimeric VLPs (Bachmann and Dyer 2004). Remarkably, despite the different ways to achieve the result, both generic engineering and chemical coupling will finally result in the covalent integration of the desired sequence and demonstrate the pattern similar to that in Figure P.2.

The protein engineering of VLPs targeted the three main functional applications: (i) presentation of foreign epitopes

leading to novel immunological content and subsequently to the creation of novel vaccines; (ii) encapsidation of various therapeutic or diagnostic agents, such as nucleic acids as adjuvants for vaccines or gene therapy tools, proteins or mRNAs for diagnostic or therapeutic purposes, or low molecular-mass drugs to be delivered to specific cells; and (iii) specific targeting of desired organs, tissues, or cells.

The latest advances in the protein engineering of VLPs expanded their functionality by including functional motifs, other than immunological epitopes, such as DNA or RNA binding and packaging sites, receptors and receptor binding sequences, immunoglobulins, elements recognizing low molecular mass substrates, carbohydrates, etc. This inevitably broadened the VLP field from the conventional area of vaccines and diagnostic tools to the advanced viral nanotechnology applications by preparing complex and highly specific cell-targeted VLP nanocages, etc.

The DNA copies of the foreign oligopeptide sequences, which could be selected for the insertion into the VLP carrier genes, comprised therefore the essential components of the VLP protein engineering technology. Since these sequences might differ in length from a few to several hundred amino acid residues, the appropriate DNA copies were prepared by PCR multiplication with the use of specific synthetic primers. These DNA copies were adjusted at that time to specific insertion sites on carrier genes, which were selected and proven computationally and experimentally for each VLP candidate and fitted then with synthetic polylinker sequences. Nowadays, not only the DNA copies of desired inserts but also full-length-designed chimeric genes are generated preferentially by DNA chemical synthesis.

The next most crucial steps in the successful implementation of the protein engineering of VLPs involved the choice of the suitable expression platform among prokaryotic bacterial cells and eukaryotic yeast, plant, insect, or mammalian cells. The choice of the expression system determined the nature of the regulatory elements, which needed to be added to the natural or chimeric genes for the optimal expression. These regulatory elements (promoters, ribosome binding sites, transcription terminators, enhancers, etc.) were prepared most often by the chemical synthesis and comprised, after their introduction into the appropriate positions within the expression cartridge, the essential parts of the sophisticated vector arrangement that could be responsible for the desired high-level production of the complete VLPs. Some distinct groups of the VLP candidates, first and foremost the single-stranded RNA phages of the former *Leviviridae* family (the class *Leviviricetes* by the current taxonomy), provided many functional elements necessary for the high-level production in the bacterial expression platform. For simple and easy purification of VLPs, the latter were supplied frequently with genetically fused specific tags that facilitated the purification process.

Nowadays, the methodological tools for VLP protein engineering are broadened by chemical functionalization of the parent VLPs. The chemical functionalization uses reactivity of the initial amino acid residues—such as lysine, cysteine, tyrosine, or tryptophan—together with point mutations in the parent VLP genes, as well as introduction of unnatural amino acid residues. Both methods allow complex and targeted introduction of chemically active groups on the specific outer and interior surfaces of the engineered VLPs. The "green chemistry" technologies not only empower the capabilities of the chemical coupling technique but also combine the latter with the genetic manipulations and erase boundaries between two classic techniques of the VLP protein engineering. Furthermore, the global functionalization approach is combined with such innovative techniques as bacterial superglue, nanoglue, or plug-and-display, which is described on the appropriates pages of Chapters 25, 26, 37, and 38.

Special attention is devoted in the guide to the various imaging techniques, when the VLPs were provided with fluorescent markers and fluorescent proteins—not only by covalent fusion—but also by encapsulation as a cargo into the inner space of the VLPs. The imaging abilities were combined with other aims of the VLP protein engineering, such as gene and drug delivery. After vaccines, diagnostics, and gene and drug delivery, the VLPs are now used for generation of novel nanomaterials, nanodevices, and nanomachines. In this case, the VLPs are combined widely with other organic and inorganic nanoparticles, such as ferromagnetic and gold and many others.

The modern reengineering and functionalization of the VLPs was described aptly as the *stick, click, and glue* technology (Brune and Howarth 2018). Generally, the current scientific developments have led to a novel discipline that could be classified as a complex *physical, chemical,* and *synthetic virology,* with a special goal to create reprogrammed VLPs as controllable nanodevices, as reported in a thoughtful review by Chen et al. (2019).

FIRST VLP PLATFORMS

The first candidates for the VLP protein engineering were described in the mid-1980s and included at once the three main structural forms of the experimental VLP candidates. First, the filamentous phage f1 came from the *Inoviridae* family of the *Tubulavirales* order covering the rod-shaped phages (Smith 1985), which are described in Chapter 11. This groundbreaking paper paved the way to the enormous field of the phage display methodology. It is special pride of the whole VLP protein engineering field that George Pearson Smith was rewarded with the Nobel Prize 2018 in Chemistry, together with Sir Gregory P. Winter "for the phage display of peptides and antibodies" and Frances H. Arnold "for the directed evolution of enzymes." As stated by the Nobel Committee, George Smith developed in 1985

an elegant method known as phage display, where a bacteriophage . . . can be used to evolve new proteins. Gregory Winter used phage display for the directed evolution of

antibodies, with the aim of producing new pharmaceuticals. The first one based on this method, adalimumab, was approved in 2002 and is used for rheumatoid arthritis, psoriasis, and inflammatory bowel diseases. Since then, phage display has produced antibodies that can neutralize toxins, counteract autoimmune diseases and cure metastatic cancer.

(www.nobelprize.org/prizes/chemistry/
2018/press-release/)

The rod-shaped viruses were represented in the early works also by TMV (Haynes et al. 1986), a member of the *Tobamovirus* genus from the family *Virgaviridae*, which is described in Chapter 19. Then, the outer envelope, surface antigens, or HBs, proteins of the previously mentioned HBV from the *Hepadnaviridae* family were used to expose foreign sequences on the so-called 22-nm particles that fit our VLP definition as the HBs VLPs (Valenzuela et al. 1985; Delpeyroux et al. 1986; see Chapter 37).

At last, the regular icosahedral particles were prepared as VLPs—first, the RNA phage coats (Borisova et al. 1987; Gren et al. 1987; Kozlovskaia et al. 1988; Kozlovska et al. 1993) of the former *Leviviridae* family from the former *Levivirales* order, as described in Chapter 25. The order *Levivirales* that always remained extremely important for the VLP technologies was taxonomically elevated in 2021 to the class *Leviviricetes*, but the RNA phages in question represent now the huge novel family *Fiersviridae* from the novel order *Norzivirales*. Second, the icosahedral carriers were represented by the HBV cores HBc (Borisova et al. 1987; Clarke et al. 1987; Newton et al. 1987) of the *Hepadnaviridae* family, the sole family of the novel order *Blubervirales*, which is presented in Chapter 38. Third, the VLPs driven by yeast retrotransposon Ty1 (Adams et al. 1987) of the *Ortervirales* order were developed (Chapter 35). At last, the nonenveloped live virions capsids (but not VLPs) of picornaviruses (Burke et al. 1988) from the *Picornavirales* order were used to construct live chimeric viruses, as described in Chapter 27.

The first VLPs based on the two major HBV products, HBs and HBc proteins, were reviewed by us at that time and later (Pumpens and Grens 1999, 2001, 2002, 2016; Pumpens et al. 2008), as well as the first VLPs based on the RNA phages (Pumpens et al. 2016).

The envelope-based VLPs from complex enveloped viruses, such as influenza viruses of the *Orthomyxoviridae* family from the order *Articulavirales* (Pushko et al. 2008; Pushko and Tumpey 2015)—described in Chapter 33 or Ebola virus of the family *Filoviridae* from the *Mononegavirales* order (Warfield et al. 2003)—described in Chapter 31—were successfully introduced into the chimeric VLP category in the early 2000s.

The first attempts to construct the chimeric VLP were extremely productive not only from the viewpoint of the VLP symmetries and composition but also in laying down the course of development of the two main directions of the chimera engineering based on the live infectious virion and the noninfectious VLP structures. Whereas the latter did not contain any genomic material capable of productive infection or replication as represented by the previously mentioned RNA phage coats, HBs, HBc, and Ty1 VLPs (Chapter 35) and formed therefore the basis of a long list of further replication-noncompetent models, the chimeric derivatives of the filamentous DNA phage f1 and poliovirus initiated the line of the replication-competent viruses used as the models to construct chimeras. Thus, the phage f1, together with its close taxonomic neighbors fd and M13, initiated preparation of the phage-display technology, described in brief in Chapter 11. Furthermore, poliovirus, rhinovirus, and many plant viruses were developed as the productive replication-competent models. The latter could be divided again into the two main categories. First, there were replication-competent vectors, which carried full-length infectious viral genomes with hybrid genes, allowing production of the chimeric virus progeny. The second type represented the replication-non-competent vectors, which were infectious but gave rise to the chimeric progeny only in special replication conditions, e.g., in the presence of special helpers. Therefore, the genomes enveloped by such chimeric progeny were unable to generate the chimeric particles.

Concerning technology, the full-length genomes and long genome fragments were usually either obtained at that time by PCR amplification using specific synthetic oligonucleotides as primers, or the individual carrier genes were generated by total chemical synthesis. Thus, the first fully synthetic genes of the VLP models were constructed, namely the tobacco mosaic virus coat gene constructed by Joel R. Haynes et al. (1986) and the HBc gene synthesized by Michael Nassal (1988). The advantage of the synthetic gene approach was that it allowed optimization of codon usage for a particular expression system by high-level production of VLPs.

LINKAGE TO VIRAL TAXONOMY

A special feature of this guide is in the close linkage of the VLP models to the current viral classification and taxonomy. Although we always adhered to the principle of Blaise Pascal that "All generalizations are false, including this one," we tried to rely on the current and well-supported viral classification delivered recently by the International Committee on Taxonomy of Viruses (ICTV 2020) as the ICTV Taxonomy Release #35, upgraded recently to #36. Moreover, we planned not only to demonstrate application of each taxonomic unit to the VLP protein engineering but also to convey the considerable progress in the current viral classification.

In contrast to the previous efforts of virus taxonomists to focus on the grouping of relatively closely related viruses, the novel approach also intended to include the basic evolutionary structural relationships among distantly related viruses (ICTV 2020). Thus, the ICTV has changed its code from

5-rank to the 15-rank classification hierarchy that closely aligns with the Linnaean taxonomic system and may accommodate the entire spectrum of genetic divergence in the virosphere. In brief, the viral taxonomy moved from the genera and families predominantly to the whole world consisting of the eight principal ranks including four that were already in use (*order*, *family*, *genus*, and *species*) and four that are new: *realm*, *kingdom*, *phylum*, and *class*, all of which are above the *order* rank (ICTV 2020). Formally, the voting on the novel taxonomy was closed on 21 March 2020 (Walker et al. 2020) but recently updated (Walker et al. 2021).

The rank *class* in this series is not to be confused with the genomic classes described by David Baltimore (1974–1975), the 1975 Nobelist. The Baltimore classes were based on the genome structure and mechanism of mRNA production by the viruses. It should be emphasized that the actuality and logic of the Baltimore classes was recently confirmed by Koonin et al. (2021), who demonstrated that this early classification covered the diversity of virus genome expression schemes.

As just noted in the Preface, our guide uses the Baltimore classes to set up the general scaffold of seven sections devoted to the seven genomic classes in the following order of viral genomes: (i) dsDNA, (ii) ssDNA; (iii) dsRNA; (iv) positive ssRNA; (v) negative ssRNA (vi) ssRNA using reverse transcription (ssRNA-RT) with DNA intermediate in life cycle; (vii) dsDNA using reverse transcription (dsDNA-RT) with pregenomic RNA intermediate in life cycle. Further, the viruses are arranged in accordance with their current taxonomical affiliation within each of the seven sections. To structure our guide as systematically as possible, we choose the rank *order* as a central unit to organize the chapters. This is because the *order* is the lowest rank that still allows one to follow easily to the list of the taxons, since there are 59 *orders* only, in contrast to the numerous *families*, currently 189 in total, including 136 subfamilies, 2,224 genera, 70 subgenera, and 9,110 species, according to the current Taxonomy EC52 release #36 (Walker et al. 2021). Moreover, the orders allow grouping of viruses into clear associations with certainly visible common features. Figure P.6 demonstrates how the 59 orders in question are involved in the higher ranks that currently consist in total of 6 *realms*, 10 *kingdoms*, 17 *phyla*, 2 *subphyla*, and 39 *classes*, in accordance with the recent Master Species List #36.

Figure P.7 illustrates the actual content of the guide chapters and demonstrates how the new realms cover the entire scale of virus divergence and reveal the virus relationships by going through numerous RNA viruses of the *Riboviria* realm and closing the circle at the border of dsDNA and dsDNA-RT viruses. Remarkably, the unique *Riboviria* realm includes all viruses of the five Baltimore classes III, IV, V, VI, and VII. It is evident also that not all 59 orders have their own chapters. The remaining orders, which are poorly developed by viral nanotechnology at this time, are grouped into the "other" virus chapters in accordance with the Baltimore

World of viruses as VLP sources

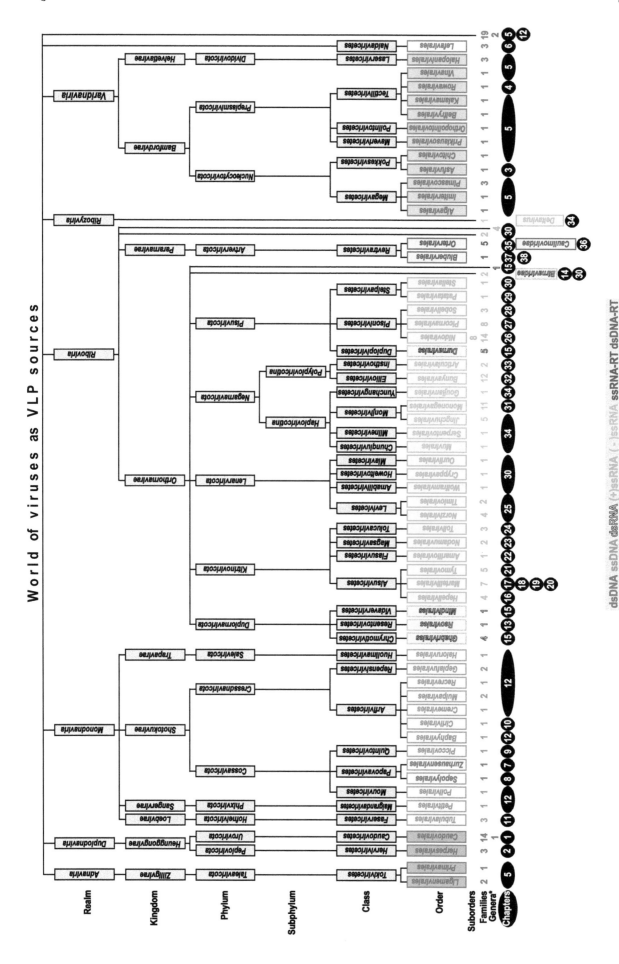

FIGURE P.6 The representation of the virus world in the present guide, in accordance with the current ICTV taxonomy. The orders and order-independent families that are subjects of the corresponding chapters are marked by the appropriate colors, according to the Baltimore class, shown at the bottom. The number of suborders and families within the orders is indicated. The number of genera (marked by star) indicates the family- or other higher-taxon independent genera only and not the total number of the genera within the order.

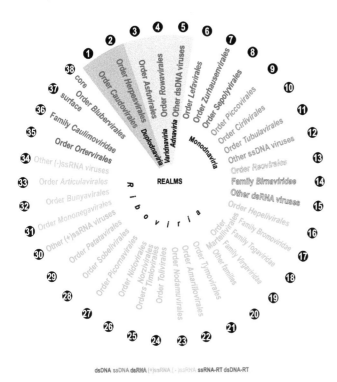

FIGURE P.7   The circular presentation of the guide contents. The colors correspond to the Baltimore classes, shown at the bottom.

classification. In fact, the viruses from these orders represent potential novel sources for the VLP applications.

As with any ground-breaking concept, this novel viral taxonomy can be described by Niccolò Machiavelli's phrase from *The Prince*: "There is nothing more difficult to take in hand, more perilous to conduct or more uncertain in its success than to take the lead in the introduction of a new order of things." Anyway, the new classification has been absolutely instrumental in providing the backbone to our guide and is extremely useful for in-depth understanding of the viral universe.

Section I

Double-Stranded DNA Viruses

1 Order *Caudovirales*

I choose a block of marble and chop off everything
I don't need.

François-Auguste Rodin

ESSENTIALS AND GENOME

The tailed bacteriophages forming the order *Caudovirales*
have a unique position in biology, representing an absolute
majority of all organisms in the biosphere, and the genetic
diversity of their population is extremely high (Hatfull and
Hendrix 2011). In contrast to the currently well-defined
realm *Riboviria*, which occupies Chapter 13 to Chapter 38
in this guide, the taxonomy of the DNA phages remains
now at the crossroads and is going through dramatic but
timely changes. This is convincingly demonstrated by the
well-developed roadmap for the genome-based phage taxonomy, which was recently proposed by Turner et al. (2021).
The reason for some taxonomic concern comes from the
fact that the classification of the DNA phages remains yet
the most archaic among other DNA viruses. This new roadmap calls for the abandoning the historically known DNA
phage families *Myoviridae*, *Podoviridae*, and *Siphoviridae*
and of the huge *Caudovirales* order itself by requiring the
genome-based differentiation of species, genus, subfamily and family-level ranks of the tailed phages. In fact, the
Bacterial and Archaeal Viruses Subcommittee of the ICTV
regularly highlights the creation of new—and reorganization of the old—taxonomic units. The last official ICTV
summary of these activities was recently published by
Adriaenssens et al. (2020).

According to the current ICTV (2020) situation and considering the latest corrections, the huge *Caudovirales* order
of the tailed DNA phages currently involves 14 families,
73 subfamilies, 927 genera, and 2,814 species altogether.
The order *Caudovirales* is a single member of the class
Caudoviricetes, the only member of the *Uroviricota* phylum.
The latter, together with the relatively small *Peploviricota*
phylum, the only order of which, namely *Herpesvirales*,
is presented in the neighboring Chapter 2, form the
Heunggongvirae kingdom of the realm *Duplodnaviria*.

As described in the ninth official ICTV report by Lavigne
et al. (2012), the *Caudovirales* order originally consisted of
the three huge families of the tailed bacteriophages infecting bacteria and archaea, which were classified by the
structure of their tails: *Myoviridae* with the long contractile
tails, *Siphoviridae* with the long noncontractile tails, and
Podoviridae with the short noncontractile tails. Figure 1.1
demonstrates typical images (we will call them portraits)
of the representatives of these three great classical families.

As reviewed by Lavigne et al. (2012), the virions of the
Caudovirales order members have no envelope and consist
of two parts, the head and the tail. The head contains a
single linear double-stranded DNA molecule of 18 to >500
kb, while the tail is a protein tube whose distal end binds
the surface receptors on susceptible bacterial cells. Then,
the DNA is injected through the tail tube into the cell. The
heads of 45–170 nm in diameter possess icosahedral symmetry of the true icosahedrons or their prolates with known
triangulation numbers of T = 4, 7, 12, 13, 16, and 52. The
prolate heads are derived from icosahedra by addition of
equatorial belts of capsomers and can be up to 230 nm long.
The tail shafts have 6-fold or, rarely, 3-fold symmetry and
are helical or stacks of disks of subunits from 3 and 825
nm in length. There are 7 to 49 different virion structural
proteins, where the typical head shells are made up of 60T
molecules of a single main building block coat and 12 molecules of portal protein through which DNA enters and
leaves, but they can also contain varied numbers of proteins that plug the portal hole, proteins to which tails bind,
proteins that bind to the outside of the coat shell, and other
proteins. The noncontractile tails are made of one major
shaft or tube protein, and the contractile tails have a second major protein, the sheath protein that forms a cylinder
around the central tube. The tails also have small numbers
of varied specific proteins at both ends. Those at the end
distal from the head form a structure called the tail tip in
siphoviruses or baseplate in myoviruses, to which the tail
fibers are attached (Lavigne et al. 2012).

As reviewed by Lavigne and Ceyssens (2020), the
Myoviridae tails are contractile, more-or-less rigid, long,
and relatively thick (80–455 × 16–20 nm). They consist of
a central core built of stacked rings of six subunits and surrounded by a helical contractile sheath, which is separated
from the head by a neck. The heads and tails are assembled in separate pathways. The *Podoviridae* virions have
short, noncontractile tails about 20 × 8 nm (Lavigne and
Kropinski 2020). The heads are assembled first, and tail
parts are added to them sequentially. The *Siphoviridae* virions have long, noncontractile, thin tails of 65–570 × 7–10
nm in size, which are often flexible and built of stacked
disks of six subunits (Hendrix et al. 2012).

The dsDNA genome of the *Myoviridae* family members
is about 33–244 kb, encoding for 40–415 proteins, while
the *Podoviridae* genome of 40–42 kb encodes for 55 genes,
and the *Siphoviridae* genome is ~50 kb and contains about
70 genes. The last systematic description of the three classical families was published in the ninth report of the ICTV
(Hendrix et al. 2012; Lavigne and Ceyssens 2020; Lavigne
and Kropinski 2020).

As of 2021, six novel families of the *Caudovirales*
order have been officially ratified, namely the three new
families of myoviruses: *Ackermannviridae*, *Chaseviridae*,

DOI: 10.1201/b22819-3

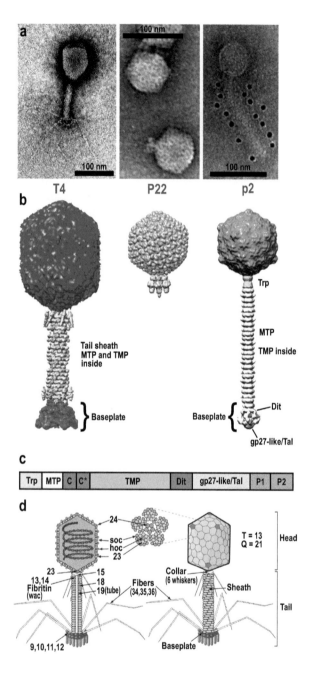

activity of the lactococcal abortive infection mechanism AbiT. *Appl Environ Microbiol.* 2012;78:6890–6899.) (b) The cartoons of the corresponding phages. (c) The schematic representation of the typical genome organization within the *Siphoviridae* tail morphogenesis module (this organization is also observed for several myophages with some adaptations). Trp, tail terminator; MTP, major tail protein; C and C*, tail chaperones; TMP, tape measure protein; Dit, distal tail protein; gp27-like/Tal (tail-associated lysozyme or tail fiber), the presence of a C-terminal domain depends on the phage considered; P1 and P2, baseplate/tip peripheral proteins (their number varies among phages). (The (b) and (c) are reprinted from Veesler D and Cambillau C. A common evolutionary origin for tailed-bacteriophage functional modules and bacterial machineries. *Microbiol Mol Biol Rev.* 2011;75:423–433.) (d) A fine schematic model of the phage T4 taken with a kind permission from the ViralZone, Swiss Institute of Bioinformatics (Hulo C et al. 2011). (Courtesy Philippe Le Mercier.)

FIGURE 1.1 The three classical *Caudovirales* families. (a) From left to right are the *Myoviridae* (*Escherichia* phage T4), *Podoviridae* (*Salmonella* phage P22), and *Siphoviridae* (*Lactococcus* phage p2) representatives. The electron micrographs are from the following resources: T4 (reprinted from *Virus Taxonomy, Classification and Nomenclature of Viruses. Ninth Report of the International Committee on Taxonomy of Viruses*, King AMQ, Lefkowitz E, Adams MJ, Carstens EB (Eds), Lavigne R, Ceyssens PJ, Family—*Myoviridae*. 46–62, Copyright 2012, with permission from Elsevier.), P22, courtesy Sherwood Casjens (reprinted from *Virus Taxonomy, Classification and Nomenclature of Viruses. Ninth Report of the International Committee on Taxonomy of Viruses*, King AMQ, Lefkowitz E, Adams MJ, Carstens EB (Eds), Lavigne R, Kropinski AM, Family—*Podoviridae*. 63–85, Copyright 2012, with permission from Elsevier.), and p2 with antibodies specific for the major tail protein (reprinted from Labrie SJ et al. Involvement of the major capsid protein and two early expressed phage genes in the

Herelleviridae; two for the siphoviruses: *Demerecviridae*, and *Drexlerviridae*; and one of podoviruses, namely *Autographiviridae* (Turner et al. 2021). Currently, five more families are added to the list.

FAMILY *AUTOGRAPHIVIRIDAE*

SUBFAMILY *STUDIERVIRINAE*

Genus *Teseptimavirus*: Phage T7

The phage T7 of the *Escherichia virus T7* species is an icosahedral T = 7l shell with a diameter of ~51 nm resolved first to ~10 Å resolution by electron cryomicroscopy (Agirrezabala et al. 2007) and a tail of ~19 nm in diameter and 28.5 nm in length. Later, the electron cryomicroscopy allowed resolution to 4.6, 3.5 and 3.6 Å, respectively, the so-called capsid I, a DNA-free form, the capsid II that, with minor changes, becomes the mature capsid of the T7 phage composed of 420 copies of gp10 protein (Guo et al. 2014). Recently, Whitford et al. (2020) used this structure and focused on the fine mechanisms of the capsid assembly. Figure 1.2 presents the appropriate structures of the T7 head from this excellent study.

Unlike its far relative and more complicated T4 phage, the phage T7 progressed rapidly to a favorite and widely used phage-display technology tool. The first commercial T7 phage-display system was announced by Rosenberg et al. (1996). Danner and Belasco (2001) demonstrated that the T7 phage display can be used to clone rapidly and selectively any protein that binds a known RNA regulatory element, including those that bind with low affinity or that must compete for binding with other proteins. By creating the T7 phage display libraries, the foreign sequences were fused to the C-terminus of the minor capsid protein 10B (398 aa in length), which was due to a translational shift to the -1 frame of the major capsid protein 10A mRNA and constituted approximately one tenth of the total number of capsid proteins (Danner and Belasco 2001). The T7 phage display system is commercially available now and demonstrates

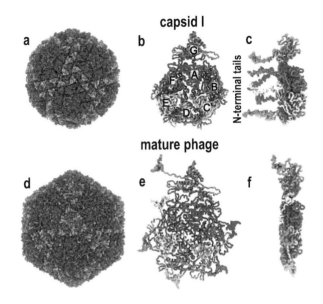

capsid I

a b c

G

F A

B

E

D C

N-terminal tails

mature phage

d e f

FIGURE 1.2 The expansion during phage maturation involves ordering of the N-terminal tails of gp10. (a) The electron cryomicroscopy structure of T7 capsid I (PDB:3J7V; Guo et al. 2014). (b) The exterior-shell view of asymmetric subunit in capsid I conformation. Monomer positions labeled A–F. (c) The side view of panel (b). The modeled gp10 N-terminal tails (vdW representation) are on the interior of the shell. (d) the electron cryomicroscopy structure of the mature phage (PDB:3J7X; Guo et al. 2014). (e) The exterior-shell view of asymmetric subunit. (f) The side view of the panel (e). All structural representations are generated with the VMD molecular graphics (Humphrey et al. 1996). (Reprinted from Whitford et al. Simulations of phage T7 capsid expansion reveal the role of molecular sterics on dynamics. *Viruses.* 2020;12:1273.)

clear advantages over the previous filamentous phage display technologies, since a more diverse aa sequence repertoire can be displayed due to differing processes of viral morphogenesis, as reviewed, e.g., by Krumpe et al. (2006) or later by Teesalu et al. (2012). The applications of the chimeric lytic T7 phages, which belong ideologically to the subject of this guide, are presented later, but there are not many papers using the T7 phage display technology for the preparation of different specialized peptide libraries. It should be noted that Krumpe and Mori (2014) presented detailed methodology of the T7-driven phage display.

Concerning putative vaccines and/or diagnostics, the following sequences were displayed on the T7 capsids: an ALT-2 antigen from *Brugia malayi* (Gnanasekar et al. 2004); the latent membrane protein 1 (LMP1) encoded by Epstein-Barr virus (EBV) and invented as a tumor prevention tool (Gao J et al. 2011); the 40-aa VP1 immunodominant epitope 131–170 of foot-and-mouth disease virus (FMDV) as a diagnostic tool (Wong et al. 2013); the aa 111–156 fragment of HBsAg of hepatitis B virus (HBV; Tan et al. 2005); an immunodominant H-2k(d)-restricted CTL epitope derived from the rat HER2/neu oncoprotein (Pouyanfard et al. 2012); matrix protein 2 ectodomain (M2e) of influenza

A virus (Hashemi et al. 2012); M2e peptides of avian influenza A virus (Xu et al. 2013); a Tsp10 polypeptide encoded by a cDNA fragment of *Trichinella spiralis* intestinal infective larvae (Cui et al. 2013); the Ep15 epitope derived from the West Nile virus (WNV) E protein DIII domain for an ELISA assay (Herrmann et al. 2007). Shadidi et al. (2008) used a proteomics-based approach to identify a set of tumor antigens recognized by serum antibodies from patients with breast cancer, expressed them on the surface of the T7 phage, and found specific immune responses in mice following oral immunization. Moreover, these immune responses inhibited tumor growth and metastasis of the 4T1 mammary adenocarcinoma cell line (Shadidi et al. 2008). It is worth mentioning that Pleiko et al. (2021) recently developed the differential binding-based high throughput sequencing in vivo T7 phage display, which demonstrated the brain-specific differential binding of brain homing phage and resulted in identification of novel lung- and brain-specific homing peptides.

Concerning the specific targeting of the T7-derived chimeras, the appropriate specificity was demonstrated in the case of hepatocytes by displaying aa 60–108 fragment of the preS1 region of HBV (Tang et al. 2009) and to the appropriate receptors by the Cry1Ac toxin from *Bacillus thuringiensis* (Pacheco et al. 2006). It should be noted that the T7 phage was found to enhance accumulation of an anticancer drug at the tumor site when administered with a tumor-penetrating peptide RGD (Sugahara et al. 2010).

As to the T7 application by the generation of the novel advanced nanomaterials, the phage T7 was used by attaching and screening libraries of synthetic compounds (Woiwode et al. 2003). Then, the phage T7 particles were used as a nanocontainer and filled with a fluorescent europium complex (Liu CM et al. 2005) or metallic cobalt (Liu CM et al. 2006). Next, the biotinylation of the T7 virions was performed with further conjugation to the streptavidin-coated quantum dots (Edgar et al. 2006). Similarly, Caberoy et al. (2010) engineered a modified T7 phage for the detection of Tubby-like protein 1. The authors expressed both 3C protease cleavage sites and biotinylation tags onto the capsid; upon selection of bacteriophages bound to streptavidin, the authors cleaved the 3C site to generate a biotinylated complementary DNA library.

A stable copper complex of the T7 virions was engineered and demonstrated prospective anticancer activity (Dasa et al. 2012). The biosensing of the T7 virions was achieved by the binding of the carboxymethyl chitosan capped gold nanoparticles (Kannan et al. 2014).

At last, the phage T7 display model was used as a visualization tool. Thus, the mosaic T7 phage carrying B10 protein with enhanced yellow fluorescent protein (eYFP) inserted as a result of translation frameshift were generated (Slootweg et al. 2006). Tsuboyama and Maeda (2013) constructed fluorescent nanobioprobes on the T7 phage scaffold. The T7 applications in the biomaterial field were reviewed soon after (Pokorski and Steinmetz 2011; Lee et al. 2013; Farr et al. 2014).

FAMILY *DREXLERVIRIDAE*

Genus *Vilniusvirus*: Phage NBD2

Špakova et al. (2019) produced the recombinant tail tube protein gp39 of the siphovirus vB_EcoS_NBD2 (NBD2), a single member of the *Escherichia virus NBD2* species, which is a single member of the genus *Vilniusvirus* within the novel *Drexlerviridae* family, to approximately 33% and 27% of the total cell protein level in *Escherichia coli* and *Saccharomyces cerevisiae* expression systems, respectively. The gp39 protein was self-assembled in vivo into well-ordered tubular structures (polytubes) in the absence of any other phage proteins. The diameter of these structures was the same as the diameter of the tail of phage NBD2, approximately 12 nm, and the length of these structures varied from 0.1 μm to >3.95 μm, which was 23-fold the normal NBD2 tail length (Špakova et al. 2019). Špakova et al. (2020b) reported that the recombinant polytubes were immunogenic in mice, as well as allowed addition of the His_6 tag at their C-terminus. At last, Špakova et al. (2020a) clearly demonstrated the great suitability of the NBD2 phage-originated polytubes as a scaffold for the presentation of foreign sequences in the *S. cerevisiae* expression system. While the insertion of the foreign sequences at the N-terminus of the gp39 had a dramatic effect on its ability to self-assemble, the C-terminal fusions resulted in the polytube formation. Moreover, the modified polytubes could display the inserts of different origin and size on the outer exterior of the nanotubes. The insertion of the entire 238 aalong eGFP resulted in the short fluorescent polytubes that retained eGFP functional activity (Špakova et al. 2020a).

FAMILY *MYOVIRIDAE*

Subfamily *Peduovirinae*

Genus *Peduovirus*: Phages P2 and P4

The phages P2 and satellite phage P4 are members of the *Escherichia virus P2* species. The phage P4 is dependent on the structural proteins, which are supplied by the helper phage P2 to assemble infectious virions. Normally, the phage P2 forms an icosahedral capsid with the icosahedral T = 7d symmetry, but in the presence of P4, however, the same structural proteins are assembled into a smaller capsid with T = 4 symmetry (Dearborn et al. 2012). The P4 genome encodes replication and regulatory genes but lacks structural genes and uses P2 as a helper to obtain the tail and capsid components. The P4-size procapsids can be produced by coexpression of the two proteins only: the P2-encoded gpN and the P4-encoded Sid (Dokland et al. 2002). The P2 and P4 structures were resolved by electron microscopy to 9.5 Å, and the 40-nm diameter was attributed to the P4 capsid (Dearborn et al. 2012). The corresponding structures are depicted in Figure 1.3. The P4 capsid can be assembled in vitro in the absence of any other gene products (Wang S et al. 2000). These features were

highly promising for the development of the P4 phage as a prospective VLP model. However, the only attempt in this direction in the P2-P4 phage system was performed 25 years ago (Lindqvist and Naderi 1995). The P4 capsid decoration component Psu of 20 kDa in size, which is nonessential for the P4 growth but enhances the stability of the P4 capsid by binding to its exterior, was used as a carrier protein. A synthetic oligonucleotide encoding a 10-aa peptide, whose sequence was part of the human p62[c-myc] protein, was inserted at the third aa position of the Psu protein. The chimeric Psu showed full capsid binding activity and reacted with monoclonal antibodies directed against the c-myc peptide (Lindqvist and Naderi 1995). Remarkably, Ranjan et al. (2013) found later that the P4-decorating Psu protein moonlighted as a transcription antiterminator of the Rho-dependent termination.

Subfamily *Tevenvirinae*

Genus *Tequatrovirus*: Phage T4

The phage T4 from the *Escherichia virus T4* species, a representative member of the relatively populated *Tequatrovirus* genus (75 species in total), is a large, tailed virus with a head in the form of the 120-nm-long and 86-nm-wide prolate icosahedron. Generally, the phage T4 remains one of the best-known assemblies of more than 40 different protein building blocks, which functioned as a natural supramolecular nanomachine (Leiman et al. 2003). The capsid has a unique "portal" vertex to which a ~140-nm-long contractile tail is attached, which is terminated by a baseplate to which are bound six long tail fibers (for references see Chen Z et al. 2017). Remarkably, the T4 head, tail, and fibers assemble by separate pathways (Leiman et al. 2003). As reviewed briefly by Chen Z et al. (2017), the capsid assembly proceeds via formation of a prohead, the formation of which is initiated by the dodecameric protein gene product (gp) 20. The latter nucleates the assembly of the inner scaffolding core made by the major core protein gp22, the prohead protease gp21 (24), and other proteins. The major capsid protein gp23 coassembles with the core to form the outer shell of the prohead, while the pentamers of the vertex protein, gp24, form the other 11 vertices of the prohead. When the prohead assembly has been completed, the gp21 protease cleaves the inner core proteins and removes 65 N-terminal aa residues from the major capsid protein (gp23) and 10 N-terminal aa from the vertex protein, gp24, producing the gp23* and gp24*, respectively. The outer surface of the capsid is decorated by the small outer capsid (Soc) protein and the highly immunogenic outer capsid (Hoc) protein, where Soc helps to stabilize the capsid against extremes of pH and temperature and Hoc probably assists in finding and infecting the bacterial host. After the completion of genome packaging, the portal vertex is sealed by the assembly of gp13–gp14 complex (i.e., "neck"), which also creates the attachment site for the independently assembled phage tail, as condensed by Chen Z et al. (2017).

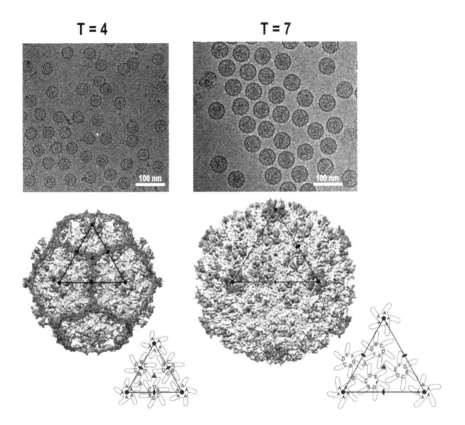

FIGURE 1.3 The P4 (T = 4) and P2 (T = 7) procapsids. The electron cryomicrographs and the isosurface representations of the P4 and P2 procapsid reconstructions, viewed down a 2-fold axis of symmetry. The large triangles correspond to one icosahedral face, delimited by three 5-fold axes of symmetry (pentagons). Three 2-fold symmetry axes (ovals) are located along the edges of the triangle. The four nonequivalent gpN subunits in the asymmetric unit in P4 and the seven subunits in P2 are labeled A through G. The same triangle is represented schematically below the corresponding reconstructions, showing the subunit arrangement in the T = 4 and T = 7 *dextro* lattices. The icosahedral (type 1) threefold axis is indicated by a filled triangle, while the local type 2 and type 3 3-fold axes are indicated by open triangles. (Reprinted from *J Struct Biol.* 178, Dearborn AD, Laurinmaki P, Chandramouli P, Rodenburg CM, Wang S, Butcher SJ, Dokland T, Structure and size determination of bacteriophage P2 and P4 procapsids: Function of size responsiveness mutations, 215–224, Copyright 2012, with permission from Elsevier.)

The structure of the mature T4 head was studied by using electron cryomicroscopy (Iwasaki et al. 2000; Olson et al. 2001; Fokine et al. 2004; Sun et al. 2015). The resulting structure is demonstrated in Figure 1.4a, c. The best resolution that had been achieved for the prolate head was approximately 10 Å (Sun et al. 2015). These earlier studies showed that the major capsid protein gp23* is organized into a hexagonal lattice characterized by triangulation numbers $T_{cap} = 13l$ for the icosahedral caps and $T_{mid} = 20$ for the elongated midsection (Fokine et al. 2004).

The Hoc monomers bound at the centers of the gp23* hexameric capsomers, whereas Soc molecules bound at the interface between adjacent gp23* hexamers and clamped neighboring capsomers together (Qin et al. 2010). As finally summarized by Chen Z et al. (2017), the T4 head contains 155 hexamers of gp23*, 12 copies of gp20 (61 kDa) at the special portal vertex, 11 pentamers of gp24* (44 kDa) at the remaining 5-fold vertices, 155 copies of Hoc (40.4 kDa), and 870 copies of Soc (9.1 kDa). Chen Z et al. (2017) used for the atomic structural studies a mutant T4 phage with a mutation in the major capsid protein gp23, which

produced a predominantly empty isometric but not prolate heads. These isometric particles, which are presented in Figure 1.4b, d, had icosahedral symmetry except for the unique portal vertex occupied by the dodecameric portal protein gp20. The gp23* protein in these isometric heads formed a hexagonal lattice characterized by the triangulation number T = 13l, the structure of which was resolved to 3.3 Å. The capsid was stabilized by 660 copies of the outer capsid protein, Soc, which clamped adjacent gp23 hexamers. Remarkably, the Soc molecules reinforced the structure where there was the greatest strain in the gp23 hexagonal lattice. The electron density map established the atomic-resolution structures of the gp23*, gp24*, and Soc molecules in the capsid environment (Chen Z et al. 2017).

In spite of its large size and extremely complicated structure, the phage T4 grew up as a powerful tool for the phage display. It means that, unlike the classical VLP approach, the chimeric T4 phages remained infectious. The T4-based phage display employed the decorative proteins Hoc and Soc, as well as the minor T4 fibrous protein fibritin encoded by the *wac* gene, which withstands a lengthening

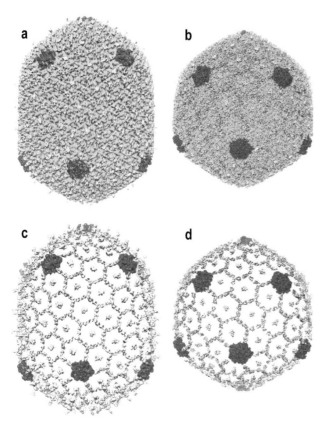

FIGURE 1.4 The structure of the T4 prolate (a and c) and isometric (b and d) heads. Gp23*, gp24*, Hoc, and Soc are colored blue, magenta, yellow, and white, respectively. Note that Soc is absent around the gp24* pentamers. In (c) and (d), the gp23* protein has been removed for clarity. (Reprinted from Chen Z et al. Cryo-EM structure of the bacteriophage T4 isometric head at 3.3-Å resolution and its relevance to the assembly of icosahedral viruses. *Proc Natl Acad Sci U S A.* 2017;114:E8184-E8193.)

at the C-terminus without impairing its folding or binding to the phage particle and forms moreover a trimeric foldon domain.

Both Hoc and Soc proteins are nonessential and can be deleted without affecting phage viability or infectivity. Therefore, the first studies used the in vivo display, where short pathogen peptides were fused to the Hoc or Soc, expressed in *E. coli* cells either from recombinant phage genome or a recombinant plasmid, and displayed on the phage particles produced in the same cells by *hoc⁻soc⁻* mutant infection (Ren ZJ et al. 1996; Jiang et al. 1997; Sathaliyawala et al. 2006; Li et al. 2007). These successful attempts were explained by the structure determinations of Soc and Hoc, showing that the N- and C-termini of Soc and the N-terminus of Hoc were accessible to solvent and could be effectively used for molecular display (Qin et al. 2010; Fokine et al. 2011). Since the in vivo phage display had strong fundamental limitations, the defined in vitro phage assembly system was elaborated (Sathaliyawala et al. 2006; Shivachandra et al. 2006; Li et al. 2007; Tao et al. 2013a). The Soc- or Hoc-fused antigens with an affinity tag at the N- or C-termini were expressed in *E. coli* from a

strong promoter such as the phage T7 promoter and purified by a single-step affinity chromatography. Moreover, this could be done also using a mammalian expression system, especially for antigens requiring specific post-translational modifications. The purified proteins were then functionally and/or antigenically characterized and displayed on phage by simple mixing the recombinant protein with the purified T4 *hoc⁻soc⁻* phage. As argued in detail in the exhaustive review of Tao et al. (2019), this approach has many advantages and opened the way for the generation of powerful vaccine candidates.

The vaccine candidates, which have been based on the Hoc and Soc display options, are listed here in the alphabetical order of the application target and provided with some technical details for clarity: anthrax antigens: 83-kDa protective antigen (PA) from *Bacillus anthracis* fused to the N-terminus of Hoc (Shivachandra et al. 2006); protective antigen (PA), lethal factor (LF), and edema factor (EF) fused to Hoc with an N-terminal His tag (Shivachandra et al. 2007); tripartite anthrax toxin completed of PA63 hexamers, LFn, and EF via Hoc and Soc fusions (Li Q et al. 2006); full-length or structural domains of anthrax toxins, PA and LF via Soc fusion (Li Q et al. 2007); a high electron cryomicroscopy resolution of the decorated T4 particles displaying the complex of PA63 heptamers with LFn-Soc on the phage surface, which is shown in Figure 1.5 (Fokine et al. 2007); combination with a novel adjuvant and testing in monkey model (Rao et al. 2011); anticancer vaccine displaying mouse VEGFR2 via Soc fusion (Ren S et al. 2011); antigenic determinant cluster mE2 (123 aa) and full-length primary antigen E2 (371 aa) of classical swine fever virus (CSFV) via Soc and Hoc fusions, respectively (Wu et al. 2007); full-length capsid precursor polyprotein (P1, 755 aa), as well as sub-full-length capsid structural proteins, and proteinase 3C (213 aa) of FMDV separately displayed on different T4 phage particle surfaces via Soc fusion (Ren ZJ et al. 2008); HIV-1 antigens: the 43-residue V3 loop domain of gp120 via C-terminal fusion to Soc (Ren ZJ et al. 1996); the p24 Gag, Nef, and an engineered gp41 C-peptide trimer via Hoc fusions (Sathaliyawala et al. 2006); display of gp41 trimers via Soc fusion (Gao G et al. 2013) and the adjuvanting of particles by liposomes containing both glucosyl ceramide and MPLA (Alving et al. 2011); major protein VP2 from the infectious bursal disease virus (IBDV) via Soc fusion (Cao et al. 2005); 36-aa PorA peptide from *Neisseria meningitidis* as fusions at the N terminus of Hoc or Soc (Jiang et al. 1997); plaque vaccine, where mutated (mut) capsular antigen F1 and the low-calcium-response V antigen of the type 3 secretion system (F1mutV) protein (56 kDa) of *Yersinia pestis* was arrayed on T4 particles via Soc fusion (Tao et al. 2013a, b, 2016); VP1 capsid protein (312 aa) of poliovirus (PV) as a C-terminal fusion to Soc (Ren ZJ et al. 1996); detailed elucidation of Hoc as a display vector (Ren ZJ et al. 1997; Sathaliyawala et al. 2010); and recognition of the Hoc N-terminus as the best position for effective presentation on the phage (Ceglarek et al. 2013).

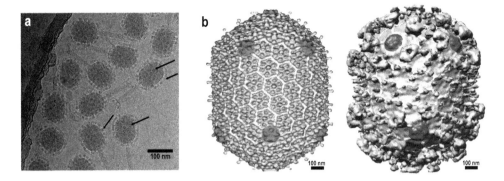

FIGURE 1.5   The fine structure of the chimeric T4 heads. (a) The electron cryomicroscopy image of the T4 phage virions decorated with anthrax proteins. Note the position of the PA63 heptamers covering all parts of the capsid (see arrows). (b) The electron cryomicroscopy reconstructions: *left*, the wild-type T4 capsid at 22 Å resolution (figure from Fokine et al. 2004); *right*, the T4 capsid with attached (PA63)$_7$ and LFn-Soc molecules at 35 Å resolution. The gp23* hexamers are shown in blue, the gp24* pentamers in magenta, the Hoc molecules in yellow, and the Soc molecules in white. Protrusions corresponding to the anthrax toxin proteins are colored in pink. (Reprinted from *Virology*. 367, Fokine A, Bowman VD, Battisti AJ, Li Q, Chipman PR, Rao VB, Rossmann MG, Cryo-electron microscopy study of bacteriophage T4 displaying anthrax toxin proteins, 422–427, Copyright 2007, with permission from Elsevier.)

Tao et al. (2018) engineered a dual vaccine against anthrax and plaque. Using the in vitro assembly system, the T4 heads were arrayed with anthrax and plague antigens fused to the Soc. The antigens included the anthrax PA of 83 kDa and the F1mutV of 56 kDa. These nanoparticles elicited robust anthrax- and plague-specific immune responses and provided complete protection against inhalational anthrax and/or pneumonic plague in three animal challenge models, namely mice, rats, and rabbits. The protection was demonstrated even when the animals were simultaneously challenged with lethal doses of both anthrax lethal toxin and *Y. pestis* CO92 bacteria (Tao et al. 2018).

Recently, Zhu et al. (2021) developed a "universal" T4-based vaccine design platform that can rapidly generate multiplex vaccine candidates. Using severe acute respiratory syndrome coronavirus 2 (SARS-CoV-2) pandemic virus as a model, the authors employed the CRISPR engineering of the T4 phage. The appropriate pipeline of vaccine candidates was engineered by incorporating various SARS-CoV-2 components into the compartments of the phage nanoparticle. These included expressible spike genes in genome, spike and envelope epitopes as surface decorations, and nucleocapsid proteins in packaged core. The phage decorated with spike trimers was found to be the most potent vaccine candidate in mouse and rabbit models. Without any adjuvant, this vaccine stimulated robust immune responses—both Th1 and Th2 IgG subclasses—blocked virus-receptor interactions, neutralized viral infection, and conferred complete protection against viral challenge (Zhu et al. 2021).

In addition to the great machine-resembling T4 head, the foldon domain of fibritin was chosen as a scaffold that efficiently presented the target polypeptides. The trimer structure of the 53-nm long fibritin is stabilized by the foldon domain with 30 residues at the C-terminus, as demonstrated in Figure 1.6. The first fibritin modifications were performed when the 45-aa fragment of the preS2 region

of hepatitis B virus (HBV) was C-terminally fused to the fibritin molecule (Efimov et al. 1995). The foldon was used to stabilize the protein trimer of envelope glycoprotein in HIV-1 composed of uncleaved external domains of gp120 and gp41 (Yang X et al. 2002; Du SX et al. 2009; Ringe et al. 2015; Shrivastava et al. 2018; Li T et al. 2019). The trimer structures of collagen, the gp26 fiber from the phage P22, and the gp5 needle protein from the phage T4 were stabilized by fusing them to the foldon domain (Frank et al. 2001; Bhardwaj et al. 2008; Yokoi et al. 2010). The specific function of foldon was applied to construct the vaccine candidates, which involved the trimeric subunits of targets from respiratory syncytial virus (RSV), namely fusion F glycoprotein (McLellan et al. 2013; Stewart-Jones et al. 2015), hemagglutinin (HA) domains of influenza virus (Du L et al. 2011; Krammer et al. 2012; Eggink et al. 2014; Lu et al. 2014, Yu et al. 2015), receptor-binding domain RBD of middle east respiratory syndrome coronavirus (MERS-CoV; Tai et al. 2016), and, last, the protein S1 of both MERS-CoV and novel SARS-CoV-2 (Kim et al. 2020).

The T4 as a successful scaffold in vaccine design was quickly and exhaustively reviewed (Gamkrelidze and Dąbrowska 2014; Bao et al. 2019; Chen Y et al. 2019; Tao et al. 2019; El-Gazzar and Enan 2020; Fralick and Clark 2021; Nguyen et al. 2021).

Concerning the specific targeting, the T4 scaffold was employed first by attachment of a 271-aa heavy and light chain fused IgG anti-egg-white lysozyme antibody to the Soc C-terminus (Ren ZJ and Black 1998). The GST and His affinity tags were added by Oślizło et al. (2011). Tao et al. (2013b) performed specific delivery of the chimeric T4 derivatives to antigen-presenting dendritic cells via a MAbs or a CD40 ligand.

The T4 scaffold was successfully used for the packaging of foreign proteins. The protein expression, packaging, and processing (PEPP) platform was elaborated, which was originally employed for the following proteins:

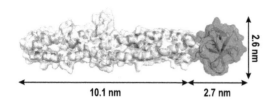

FIGURE 1.6 The model structure of fibritin E (from Glu368 to Ala486 of fibritin) from PDB ID: 1AA0. The foldon domain at the C-terminus is presented in red with dimensions of 2.7 × 2.6 nm. The structure was rebuilt by PyMOL. (Reprinted from Nguyen et al. The versatile manipulations of self-assembled proteins in vaccine design. *Int J Mol Sci.* 2021;22:1934.)

β-galactosidase and β-globin (Hong and Black 1993), GFP (Mullaney and Black 1998a; Mullaney et al. 2000), protease of HIV-1 (Mullaney and Black 1998b), luciferase (Mullaney and Black 1996), staphylococcal nuclease (Mullaney and Black 1998a), T7 polymerase (Hong et al. 1995). Later, Mullaney and Black (2014) published a full and detailed description of the PEPP platform, which enabled production of human HIV-1 protease, micrococcal endonuclease from *Staphylococcus aureus*, restriction endonuclease EcoRI, luciferase, human granulocyte colony stimulating factor (GCSF), GFP, and the 99 aa C-terminus of amyloid precursor protein (APP).

The T4 scaffold was used by the generation of novel bionanomaterials, first by immobilization onto a gold platform for the detection of bacteria by outwards orientated tail fibers (Arya et al. 2011); next by the introduction of biotin binding peptide and cellulose binding module via Soc fusion for biosensor applications (Minikh et al. 2010; Tolba et al. 2010); then by the conjugation of fluorescent dyes for cellular imaging and flow cytometry (Robertson et al. 2011). Boss and Lieberman (2009) decorated T4 with magnetic nanoparticles for the detection of bacteria. The T4 virions were used as a scaffold for the construction of 3D monodisperse metal-particle arrays (Hou et al. 2010). The T4 capsid-only constructs without tail fibers were immobilized onto sensor platforms (Archer and Liu 2009; Liu JL et al. 2012). The T4-based nanomaterials were quickly and thoroughly reviewed (Chung et al. 2011; Lee et al. 2013; Farr et al. 2014). Kaźmierczak et al. (2021) generated the red fluorescent protein (RFP)-labeled T4 phages as a new tool for the phage detection in living cells and tissues.

FAMILY *PODOVIRIDAE*

GENUS *LEDERBERGVIRUS*: PHAGE P22

The phage P22, a member of the *Salmonella virus P22* species, infects *Salmonella typhimurium* when intact tail fibers are present. The early experiments (Prevelige et al. 1988) demonstrated that the coat and scaffolding subunits derived from the procapsids of the phage P22 coassembled rapidly and efficiently into icosahedral shells in vitro under native conditions, and the in vitro reaction exhibited the regulated

features observed in vivo, while neither coat nor scaffolding subunits alone self-assembled into large structures. Upon simply mixing the subunits together, they polymerized into procapsid-like shells with the in vivo coat and scaffolding protein composition. The subunits in the purified coat protein preparations are monomeric. These results confirmed that the P22 procapsid formation did not proceed through the assembly of a core of scaffolding, which then organized the coat but required copolymerization of coat and scaffolding (Prevelige et al. 1988). Furthermore, due to the long-standing efforts of the famous Trevor Douglas team by the highly advanced P22-based protein cargo tools, this phage became a definite grand of the viral nanotechnology. Thanks to their ingenious solutions, the P22 VLPs paved a way for further multidisciplinary applications.

The atomic models for the P22 procapsids and infectious virions were resolved from electron cryomicroscopy density maps at 3.8 and 4.0 Å resolution, respectively, with the diameter of the T = 7 procapsid of 61 nm and virion of 71 nm (Chen DH et al. 2011). Both procapsid and virion structures are depicted in Figure 1.7.

At first, Kang et al. (2006) introduced cysteine residue in the middle of the coat loop region by substitution T182C, which did not alter procapsid assembly or capsid integrity, while Kang et al. (2008) demonstrated that the cysteine residues were selectively and covalently modified with cysteine reactive reagents. Moreover, the inherent structural flexibility of viral capsids was taken into account and finished with the generation of morphologically different types of viral nanoplatforms, so-called wiffle-balls or expanded forms, which were missing scaffolding proteins but presented reactive thiol groups in each protein of their 420-subunit cage after Ser → Cys exchange on the interior surface of the protein cage (Kang et al. 2010).

As to the efficient expression of the phage heads, the P22 coat and scaffold proteins were expressed in *E. coli*, purified, and assembled in vitro, and the purified P22 VLPs were heated at 65°C to remove the scaffold protein (Kang et al. 2010). Schwarz et al. (2015a) used the trimeric decoration protein (Dec) from bacteriophage L as a means of controlled exterior presentation on the mature P22 VLPs, to which it binds with high affinity. Through genetic fusion to the C-terminus of the Dec protein, either the 17 kDa soluble region of murine CD40L or a minimal peptide designed from the binding region of the "self-marker" CD47 was independently presented on the P22 VLP capsid exterior. Both candidates retained function when presented as a Dec-fusion. The Dec-mediated presentation offered a robust, modular means of decorating the exposed exterior of the P22 capsid in order to further orchestrate responses to internally functionalized VLPs within biological systems (Schwarz et al. 2015a).

The canonic version of the P22-driven cargo experiments was based on the fact that the phage P22 assembled from 420 copies of the coat-protein gp5 subunit into an icosahedral capsid with the assistance of 100–330 copies of a scaffold protein, the C-terminus of which interacted with the

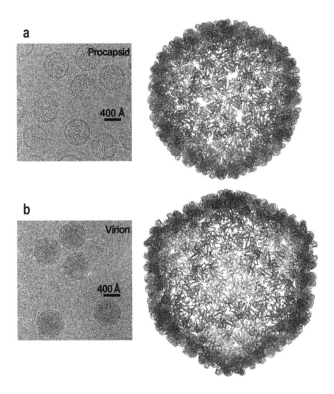

FIGURE 1.7 The icosahedral T = 7 structure of the P22 procapsids and virions. (a) The typical electron cryomicroscopy image of the P22 procapsid and the Cα backbone model of gp5 (residues 10–425) for the entire procapsid at 3.8-Å resolution. (b) The typical electron microscopy image of the P22 virion and the Cα backbone model of gp5 (residues 1–425) for the virion at 4.0-Å resolution. (Chen DH et al. Structural basis for scaffolding-mediated assembly and maturation of a dsDNA virus. *Proc Natl Acad Sci U S A.* 108:1355–1360, Copyright 2011 National Academy of Sciences, U.S.A.)

coat protein and was necessary for self-assembly, whereas the N-terminus could be severely truncated or mutated with little to no effect on assembly (O'Neil et al. 2011). Therefore, the system genetically fused the cargo proteins to the truncated form of the P22 scaffold protein, which acted further as a template for capsid assembly as well as a specific encapsulation signal for the cargo. In this way, the additional space within the capsid from the truncation of the SP enabled the cargo to be assembled in vivo with high packaging efficiency. This method did not alter the coat protein in any way; eliminated the need for affinity tags, chemical linkers, or bridging ligands; and perhaps avoided the misassembly of VLPs. Instead, the fusion of the scaffold protein to the cargo naturally provided the necessary affinity to the interior of the capsid, and assembly was templated around the cargo. Thus, a vector containing the gene for P22 coat protein and a truncated variant of the wild-type scaffold protein encoding for amino acids 141–303 (SP141) was designed. This vector enabled the insertion of a series of fluorescent-protein genes, including eGFP and mCherry, for the creation of N-terminal fluorescent protein—SP141 fusions. In these protein fusions, the thrombin-cleavage site was located between SP141 and the fluorescent protein and acted as both a flexible linker sequence and an accessible cleavage site for the release of the fluorescent protein from the SP141 as needed. The fluorescent procapsids were indistinguishable from the wild-type P22 procapsids during the expression and purification process except for their brilliant color, which indicated the expression, encapsulation, and proper folding of the fluorescent protein (Figure 1.7b). On average, the expression yields were 150 mg of P22 with encapsulated fluorescent cargo per liter of *E. coli* culture after a single purification step by sucrose-cushion ultracentrifugation (O'Neil et al. 2011).

Summarizing the numerous encapsulation experiments and generation of novel bionanomaterials, the further studies should be mentioned first: the addition of biotin linkers and encapsulation of streptavidin (Kang et al. 2010); the encapsulation of the alcohol dehydrogenase D enzyme from the hyperthermophile *Pyrococcus furiosus* (Patterson et al. 2012) and an enzyme complex for the formation of a segregated metabolon (Patterson et al. 2014); packaging of Fe_2O_3 (Reichhardt et al. 2011); the cargos for the intermolecular communication between the coencapsulated proteins (O'Neil et al. 2012); the hybrid materials as contrast agents for magnetic resonance imaging (Qazi et al. 2014) including conjugation of Gd^{III}-chelating agent complexes (Min et al. 2013; Qazi et al. 2013); the metal-ligand coordination bond to the direct packing of guest molecules (Uchida et al. 2012); the encapsulation of the cross-linked poly(2-aminoethyl methacrylate) polymer as a scaffold for the attachment of small functional molecules (Lucon et al. 2012); the semiconducting nanocrystals of CdS confined inside the VLPs (Kale et al. 2013; Zhou et al. 2014); the sequestration and solubilization of recombinant proteins, which were usually trafficked to insoluble inclusion bodies, into the VLP cages (Patterson et al. 2013a).

Recently, the P22 cages were used as a model to create synthetic protein-based protocells that can confine smaller functionalized proto-organelles and additional macromolecules to support a range of biochemical reactions. Thus, Waghwani et al. (2020) demonstrated a strategy for controlled copackaging of subcompartments, ferritin cages, inside the P22 VLPs, together with active cellobiose-hydrolyzing β-glycosidase enzyme macromolecules, where stoichiometry of both components was modulated to control the degree of compartmentalization. Remarkably, the coencapsulated enzyme showed catalytic activity even when packaged at high total macromolecular concentrations comparable to an intracellular environment (Waghwani et al. 2020). Wang Y et al. (2020) designed active nanoreactors capable of catalyzing the partial or complete pathway for biosynthesis of glutathione, which was realized by encapsulating essential enzymes of the pathway inside the P22 VLPs. The two enzymes were selected as biocatalyst cargos: the glutamate cysteine ligase from *S. cerevisiae*—which catalyzes the rate-limiting step of glutathione biosynthesis—and the glutathione synthetase from *Pasteurella multocida*, which is capable of the complete glutathione biosynthesis pathway. Using the previously described SP fusion strategy, where the enzyme was genetically fused to the N-terminus of the truncated SP of aa 239–303, the multiple copies of these enzymes were encapsulated inside the P22 capsids by coexpression and self-assembly with the coat protein. In fact, these nanoreactors were the first examples of nanocages specifically designed for the biosynthesis of oligomeric biomolecules and were suggested as a potential liver-tropic nanocarrier in vivo, as a promising tool for a novel treatment for glutathione-deficient hepatic diseases (Wang Y et al. 2020).

At last, Díaz-Barriga et al. (2021) used the P22 VLPs to encapsulate asparaginase II from *E. coli* to overcome such undesirable side effects by treating of acute lymphoblastic leukemia, such as short lifetimes, susceptibility to proteases, and remarkable immunogenicity. Such asparaginase-P22 nanoreactors were PEGylated, evaluated in a human leukemic cell line, and suggested as a novel alternative for the cancer treatment.

Selivanovitch et al. (2021) investigated the porosity of the P22 VLPs with encapsulated alcohol dehydrogenase-D and developed a suitable methodology to study diffusion across porous barriers. Thus, the activity of the encapsulated enzyme was used to probe the access of different synthetic substrates of varying size across the porous barrier of the P22 VLPs. Furthermore, Kraj et al. (2021) proposed predictive models to prepare the polymer-coated P22 VLPs at very low ionic strength and evaluated the effect of coating on the activity of the encapsulated enzyme.

After the enzyme nanomachines and nanomaterial packaging, the P22 model was used for the engineering of some putative vaccine candidates. Thus, Patterson et al. (2013b) performed encapsulation and sequestration of the conserved nucleoprotein from influenza on the interior of the P22 VLPs. A novel vector category was created for further capsid modifications by addition of His$_6$ tag to the extended C-terminus of coat protein (Servid et al. 2013). Richert et al. (2012) described the pretreatment of mice with P22 VLPs in order to enhance immunologic response to an ovalbumin model, when the latter was chemically coupled to the nonviral small heat shock protein VLPs. An early attempt, which was not connected to the Trevor Douglas team investigations, should be also mentioned, when the aa 134–156 epitope of the FMDV VP1 was fused to the C-terminus of the P22 tailspike protein and abled to reconstitute infectious viruses by in vitro association with tailless particles (Carbonell and Villaverde 1996).

The specific targeting of the P22 VLPs was achieved by the chemical attachment of catechol ligands to the interior surface of capsids for encapsulation of an anticancer drug bortezomib and targeted delivery to hepatocellular carcinoma cells by chemically conjugated targeting peptide SFSIIHTPILPL to the exterior surface of the capsid nanocomposites (Min et al. 2014). The polyglutamate domains of the VLP nanocages were specifically bound to hydroxyapatite-containing materials (Culpepper et al. 2013).

The successful use of the P22 nanocages was quickly reviewed (Bruckman et al. 2013; Koudelka et al. 2015; Schwarz et al. 2015b; Maassen et al. 2016; Raeeszadeh-Sarmazdeh et al. 2016; Zhang Y et al. 2016).

GENUS *RAUCHVIRUS*: PHAGE BPP1

The phage BPP1, a member of the *Bordetella virus BPP1* species, infects *Bordetella bronchiseptica* bacteria that cause whooping cough. The electron cryomicroscopy structure of the T = 7*l* procapsid with ~67 nm diameter revealed folds and localization of the two major structural proteins: cement protein and major capsid protein, the termini of which are ideally positioned for the diversity generating retroelement (DGR)-based phage-display engineering (Zhang X et al. 2013). The DGR is a natural instrument to alter tail-fiber protein of the BPP1 phage and provokes synthesis of self-made phage libraries (SMPLs) by introducing DNA mutations into the gene encoding the major tropism determinant (Mtd) protein on the tail fibers. The first attempt to involve the phage BPP1 into the VLP-based technologies identified four sites amenable to the insertion of <19-aa heterologous peptides within the variable region of the Mtd protein and resulted in heterologous SMPLs employed for selections targeting anti-FLAG antibody, immobilized metal affinity chromatography microtiter plates, and HIV-1 gp41 (Overstreet et al. 2012). Further experiments were directed to engineering of Mtd variants with high affinity to a model protein, namely T4 lysozyme (Yuan et al. 2013).

FAMILY *SIPHOVIRIDAE*

GENUS *BYRNIEVIRUS*: PHAGE HK97

The phage HK97, a member of the *Escherichia virus HK97* species, is a long-tailed phage, which provided a highly productive experimental system for investigating how virus capsids are assembled from their protein components and how the assembled capsids mature to their final form (Hendrix and Johnson 2012). The icosahedral T = 7*l* procapsid of HK97, 54.2 nm in diameter and composed of 415 copies of the capsid protein (gp5), was the first described by the high-resolution crystallographic structure and triggered the line of virus folds called HK97-like capsid protein fold (Wikoff et al. 2006). Unlike in the case of papillomaviruses (Chapter 7) and polyomaviruses (Chapter 8), in which the capsid protein was preformed into pentamers, the capsid protein subunits in the HK97-like fold have to switch between hexameric and pentameric clustering on the T = 7 lattice, as described in substantial detail by Hendrix and Johnson (2012) and Prasad and Schmid (2012). Remarkably, when evaluated by x-ray crystallography, the HK97 phage capsid protein demonstrated a polypeptide fold similar to that of the T4 protein gp24 that formed the pentameric vertices of the T4 capsid (Fokine et al. 2005). The full HK97 VLP assembly and maturation pathway is depicted in Figure 1.8.

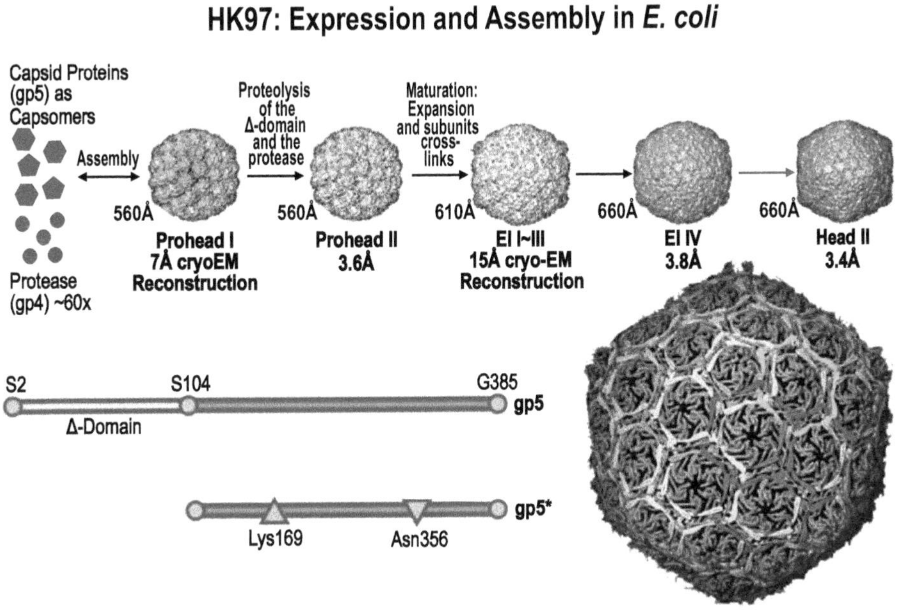

HK97: Expression and Assembly in *E. coli*

FIGURE 1.8 The HK97 VLP assembly and maturation pathway that is followed when only the coat protein and protease are coexpressed in *E. coli*. At the top are all the intermediates that have been characterized by crystallography and/or electron cryomicroscopy. Below is shown the processing that occurs to the capsid protein (gp5) and the residues that form the autocatalytic cross-link. At right is an enlargement of the final mature particle indicating that the cross-linked gp5 proteins mechanically chain-link the particle together. Each ring of the same color corresponds to five or six subunits chain-linked together by the isopeptide bond formed by the side chains of Asn356 and Lys169. Assembly and maturation: the capsid protein (gp5) immediately assembles into skewed hexamers and pentamers (generically called capsomers) shown in blue. The protease (gp4) is shown in red. The capsomers (60 hexamers and 12 pentamers) and approximately 60 copies of the protease coassemble to form Prohead I, a roughly spherical particle, 51 nm in diameter. The hexamers in this particle are not symmetric but "skewed" and correspond roughly to two shifted trapezoids (each composed of three subunits) related by 2-fold symmetry. Under normal circumstances, Prohead I is transient. The protease becomes active upon assembly and cleaves residues 2–103 (the Δ-domain) from all the gp5 subunits, creating gp5* as shown. The protease autodigests, and all of the fragments diffuse from openings in the capsid. This creates Prohead II (51 nm), which is morphologically closely similar to Prohead I (including the skewed hexamers and virtually the same diameter); however, the mass of this particle is 13 MDa, compared to 17 MDa for Prohead I. Prohead I can be stabilized for study by not expressing the protease or by coexpressing a mutant, inactive protease. Either of these Prohead I particles can be disassembled and reassembled under mild conditions as indicated by the arrow from capsomers to Prohead I. Prohead II is a metastable particle that can only transition to EI-I. Conditions that disassemble Prohead I either cause nothing to happen to Prohead II or cause the transition to EI-I. There are many conditions that cause the transition, but dropping the pH from 7 to 4.0 was used for most of the in vitro studies of maturation. The transition to EI-I (56.0 nm), triggered by the pH drop, has a half-life of ~3 min and is stochastic, without populated intermediates. EI-I hexamers are sixfold symmetric, and the particle is cross-link competent, with cross-link initiation commencing virtually immediately after this particle is formed. Approximately 60% of the cross-links form before the morphology of EI-I changes (EI-II, III have essentially the same morphology as EI-I with increasing numbers of cross-links) to EI-IV. The process resembles a Brownian ratchet (a process by which thermal energy is captured by driving a process in only one direction) in which the loop containing Lys169 fluctuates until the covalent bond with Asn356 forms, locking it down and incrementally raising the energy of the particle. When a sufficient number form (~60%), the particle crosses the energy barrier and transitions to EI-IV, again without populated intermediates, a round (~62.5 nm), thin-shelled particle eventually forming all but 60 cross-links. This particle is the end point at pH 4.0 and was studied by crystallography and cryo-EM. These studies showed that subunits in the pentamers had not formed their final cross-link and that they were dynamic, fluctuating by 1.4 nm along the fivefold particle axes. Neutralizing EI-IV completes the cross-links with pentamer subunits, forming the fully mature, ~65 nm, faceted particle shown in the lower right. (Hendrix RW and Johnson JE: Bacteriophage HK97 capsid assembly and maturation. *Adv Exp Med Biol.* 2012.726.351–363. Copyright Wiley-VCH Verlag GmbH & Co. KGaA. Reproduced with permission.)

The collection of switching structures from the procapsid to the mature capsid head II was among the most robust VLPs characterized (Gertsman et al. 2010). These VLPs were used as versatile platforms for the engineering of multifunctional nanoparticles (Huang et al. 2011). The first practical example in this direction consisted in the engineering of the HK97 viral nanoparticles for the tumor cell-specific targeting (Huang et al. 2011). In this study, the noninfectious HK97 VLPs, which did not contain the viral genome, were produced in *E. coli* by coexpression of the gp5 and the virally encoded protease gp4, which was required for the maturation of the HK97 head. The introduction of functional thiol groups to the capsid surface was achieved by Cys exchanges and chemical coupling of transferrin and resulted in the specific delivery of the VLPs to the cancer cells in vitro (Huang et al. 2011). Karimi et al. (2016) referred to the transferrin-mediated targeting of the phage HK97 VLPs to tumor cells by their evaluation of the phage-inspired nanocarriers for targeted delivery of therapeutic cargos.

Genus *Lambdavirus*: Phage Lambda

The phage lambda, or λ, a member of the *Escherichia virus Lambda* species and the near relative of the phage HK97, has served for more than 50 years as a major priority model in gene engineering and molecular biology, especially by the phenomenon of lysogeny and deciphering of numerous regulatory mechanisms at the molecular level, as reviewed by Węgrzyn et al. (2012). In the phage λ, both the capsid and the tail have been engineered to deliver foreign molecules (Zanghi et al. 2007). Unlike HK97, the whole structure of the phage λ capsid was studied less efficiently: no one high-density resolution of the full lambda capsid is available in the VIPERdb database, although the structure of the λ major capsid protein gpD was resolved in 2000 (Yang F et al. 2000). Later, a pseudo-atomic model of gpD in the context of the assembled procapsid shell was constructed by a combination of chemical cross-linking/mass spectrometry and computational modeling (Singh et al. 2013). Regarding the VLP ideology, the phage λ is an accepted carrier for the phage display technology, where both N- and C-termini of the gpD protein were employed, as reviewed by Beghetto and Gargano (2013) and Nicastro et al. (2014). The λ phage-based vaccine approaches were developed in the two main directions by (i) display of the desired epitopes on the capsid surface and (ii) delivery of the DNA cargo to antigen-presenting cells, with or without introduction of any specific targeting elements. Combination of both approaches is welcome (Nicastro et al. 2014).

Concerning putative vaccine candidates, the general principles of the peptide and protein display were described by Maruyama et al. (1994) and Sternberg and Hoess 1995). The gene delivery was performed with genes encoding following proteins, in chronological order: hepatitis B surface (HBs) protein of hepatitis B virus (Clark and March 2004; March et al. 2004; Clark et al. 2011); the *Mycoplasma mycoides* genes selected by phage display (March et al.

2006); oncoprotein E7 of human papillomavirus (HPV; Ghaemi etal. 2010, 2011); ompA, the major outer membrane protein, of *Chlamydophila abortus* (Ling et al. 2011); core protein of hepatitis C virus (HCV; Saeedi et al. 2014). The following epitopes were displayed on the λ VLPs: four immunodominant regions of porcine circovirus 2 (PCV2) capsid protein (Gamage et al. 2009; Hayes et al. 2010); gp140 spikes of HIV-1 (Mattiacio et al. 2011); hemagglutinin H5 of influenza virus (Domm et al. 2014). Thomas et al. (2012) presented a model to compare the efficiency of protein and gene immunization.

The phage λ-based display was used for the cell targeting. Thus, the cancer cells were targeted by the display of high-affinity nanobody against HER2 (Shoae-Hassani et al. 2013). The human 293T cell line was targeted by the display of chemically coupled human holotransferrin, and delivery of the GFP-encoding gene was realized (Khalaj-Kondori et al. 2011). The mammalian cell lines were also targeted by display of the following addressing tools: the RGD motif (Dunn 1996); TAT transduction domain of HIV-1 (Eguchi et al. 2001; Wadia et al. 2013); single-chain antibodies (Gupta et al. 2003) and single-chain anti-CEA antibody (Vaccaro et al. 2006); the full-length adenoviral penton base or its central domain of aa 286–393 (Piersanti et al. 2004); αvβ3 integrin-binding peptide (Zanghi et al. 2005) together with delivery of luciferase gene (Lankes et al. 2007); ubiquitinylation and CD-40 binding motifs (Zanghi et al. 2007); FcγRI (Sapinoro et al. 2008). The Cry1Ac toxin was displayed to address the VLPs to the receptor of *Bacillus thuringiensis* insecticidal toxin (Vílchez et al. 2004).

The visualization strategy of the phage λ VLPs was achieved either by the delivery of GFP reporter gene to *E. coli* cells for sensitive detecting of bacteria (Funatsu et al. 2002) or by the GFP display (Sokolenko et al. 2012; Nicastro et al. 2013). The diagnostic use of the phage λ VLPs was ensured by simultaneous display of anti-CEA single-chain antibody fragment and GFP or alkaline phosphatase on gpD and gpV lambda proteins (Pavoni et al. 2013) or by the display of IgG-binding domains of the *Staphylococcus aureus* protein A, β-lactamase, or β-galactosidase (Mikawa et al. 1996).

Concerning the bionanomaterial applications, the CdSe/ZnS quantum dots (QD) were attached to the phage λ and used for the highly sensitive bacterial detection (Yim et al. 2009).

2 Order *Herpesvirales*

> You can't depend on your eyes when your imagination is out of focus.
>
> **Mark Twain**
> *A Connecticut Yankee in King Arthur's Court*

ESSENTIALS AND GENOME

The *Herpesvirales* is an order of enveloped, spherical to pleomorphic, dsDNA viruses 150–200 nm in diameter. As reviewed by the ICTV reports (Davison et al. 2009; Pellett et al. 2012), the range of host species is very wide, and it is not excluded that all vertebrates carry multiple herpesvirus species, and a herpesvirus has also been identified in invertebrates, namely, mollusks. The order members are medically important and cause such human infections as cold sores, genital herpes, chickenpox, shingles, and glandular fever. According to the present ICTV (2020) situation, the *Herpesvirales* order currently includes 3 families, 3 subfamilies, 23 genera, and 130 species altogether. The order *Herpesvirales* is a sole member of the class *Herviviricetes*, the sole member of the *Peploviricota* phylum. The latter, together with the huge *Uroviricota* phylum, the sole order of which, namely, *Caudovirales* uniting the tailed DNA phages, is described in the preceding Chapter 1, belong to the *Heunggongvirae* kingdom of the realm *Duplodnaviria*. The central great family of the *Herpesvirales* is *Herpesviridae*, which involves 3 subfamilies, 17 genera, and 115 species infecting mammals, birds, and reptiles, while two other families, *Alloherpesviridae* (4 genera and 13 species) infecting frogs and fish and *Malacoherpesviridae* (2 genera and 2 species) infecting mollusks, are pretty small yet.

Figure 2.1 demonstrates typical images and schematic cartoons of the herpesviruses. As reviewed in the ninth official ICTV report by Pellett et al. (2012), the virions of the *Herpesvirales* members are comprised of the core, capsid, tegument, and envelope. The mature capsid is a T = 16 icosahedron. For example, in virions of human herpesvirus 1 (HHV-1), the 160-nm-thick protein shell has an external diameter of 125 nm. The 11 pentons and 150 hexons (a total of 161 capsomers) are composed primarily of five and six copies, respectively, of the same protein and are joined by masses, termed *triplexes*, which are made of two smaller proteins present in a 2:1 ratio. The twelfth pentonal position is occupied by a ring-like structure consisting of 12 copies of the capsid portal protein. The capsid is embedded in a proteinaceous matrix called the tegument, which in its turn is enclosed by a glycoprotein-containing lipid envelope. The complicated structure of the tegument is recently well-defined at atomic resolutions, as described later. There are inner (capsid-associated) and outer (envelope-associated) tegument layers, and the capsids may be nonsymmetrically situated within

the envelope. Remarkably, the enveloped tegument structures lacking capsids can assemble and are released from cells along with virions. The envelope is a lipid bilayer that is intimately associated with the outer surface of the tegument and contains a number of different integral viral glycoproteins that form a network of closely spaced spikes of at least three distinct morphologies. Generally, the mature capsid is composed of four major and several minor proteins, while the tegument contains at least 15 different polypeptides, many of which are dispensable in vitro and are therefore not required for virion morphogenesis (Pellett et al. 2012).

The linear dsDNA genomes of 125 to 295 kBp in size are packed in a liquid-crystalline array that fills the entire internal volume of the capsid. The number of ORFs contained within herpesvirus genomes that potentially encode proteins ranges from about 70 to more than 200. In addition to proteins, the herpesvirus genomes also harbor varying numbers of microRNA genes and express numerous putatively nontranslated transcripts of unknown function, while a subset of about 40 protein-coding genes is conserved among the *Herpesviridae* family members (Pellett et al. 2012).

The usual prototype of the order is herpes simplex virus 1 (HSV-1), a member of the *Human alphaherpesvirus 1* species from the genus *Simplexvirus*, subfamily *Alphaherpesvirinae* of the family *Herpesviridae*. In HSV-1, the core contains 162 capsomers each including either six (for hexons) or five (for pentons) molecules of the major capsid protein VP5 (pUL19), where the VP5 of hexons (but not pentons) bind one molecule of VP26 each, as reviewed by Everett (2014). As described earlier, the core is surrounded by an amorphous tegument and embedded into the lipid-containing envelope. The capsids C are mature; the capsids A do not contain viral DNA and are likely to result from abortive packaging events (Everett 2014). In 2018–2020, an impressive set of atomic structures of the herpesvirus cores consisting each of more than 3,000 proteins was successfully resolved by electron cryomicroscopy.

FAMILY *HERPESVIRIDAE*

Subfamily *Alphaherpesvirinae*

Genus *Simplexvirus*: Herpes Simplex Viruses 1 and 2

The first structures of whole virions (Zhou et al. 1999) and cores (Zhou et al. 2000) of HSV-1 that causes cold sores were resolved by electron cryomicroscopy to the relatively low resolution. At last, Dai and Zhou (2018) published an atomic model of the HSV-1 capsid with capsid-associated tegument complexes (CATC), comprising multiple conformers of the capsid proteins VP5, VP19c, VP23, and VP26 and tegument proteins pUL17, pUL25, and pUL36. This structure is

DOI: 10.1201/b22819-4

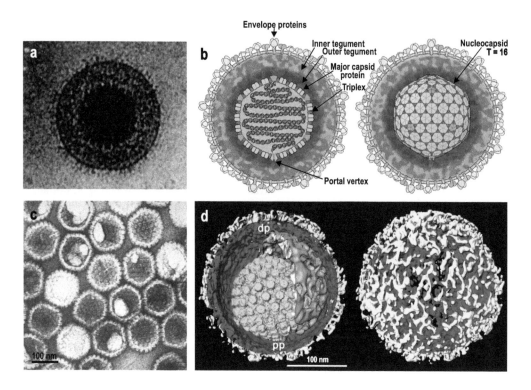

FIGURE 2.1   The images and schematic cartoons of typical representatives of the order *Herpesvirales*. (a) The electron micrograph of a frozen hydrated herpes simplex virus type 1 (HSV-1) virion, courtesy Wah Chiu (Reprinted from *Semin Virol.* 4, Rixon FJ, Structure and assembly of herpesviruses, 135–144, Copyright 1993, with permission from Elsevier.) (b) The cartoons of the *Herpesvirales* representatives are taken with kind permission from the ViralZone, Swiss Institute of Bioinformatics, http://viralzone.expasy.org (Hulo C et al. 2011). Courtesy Philippe Le Mercier. (c) The numerous herpes simplex virions in electron micrograph of 1975, provided by Fred Murphy and Sylvia Whitfield to the public CDC collection (ID 10230). (d) The segmented surface rendering of a single virion tomogram after denoising (left). Outer surface showing the distribution of glycoprotein spikes (yellow) protruding from the membrane (blue) (right). Cutaway view of the virion interior, showing the capsid (light blue) and the tegument "cap" (orange) inside the envelope (blue and yellow). pp, proximal pole; dp, distal pole. (From Grünewald K, Desai P, Winkler DC, Heymann JB, Belnap DM, Baumeister W, Steven AC. Three-dimensional structure of herpes simplex virus from cryo-electron tomography. *Science.* 2003;302:1396–1398. Reprinted with permission of AAAS.)

presented in Figure 2.2. Crowning every capsid vertex are five copies of heteropentameric CATC. The pUL17 monomer in each CATC bridges over triplexes Ta and Tc on the capsid surface and supports a coiled-coil helix bundle of a pUL25 dimer and a pUL36 dimer, thus positioning their flexible domains for potential involvement in nuclear egress and axonal transport of the capsid. It should be noted that the definite architectural similarities were found between herpesvirus triplex proteins and the corresponding proteins of the phages λ and HK97 that are described in Chapter 1, indicating therefore the commonality between tailed bacteriophages and herpesviruses (Dai and Zhou 2018).

In parallel, Yuan et al. (2018) resolved to 3.1 Å the atomic structure of herpes simplex virus type 2 (HSV-2), a member of the *Human alphaherpesvirus 2* species, which causes genital herpes. This structure is shown in Figure 2.3 in comparison with that of typical representatives of two other subfamilies of the *Herpesviridae* family, namely, *Betaherpesvirinae* and *Gammaherpesvirinae*. Again, it was found that both hexons and pentons contained the major capsid protein, VP5, while hexons also contained a small capsid protein, VP26, and triplexes comprised VP23

and VP19C. Acting as core organizers, the VP5 proteins formed extensive intermolecular networks, involving multiple disulfide bonds, about 1,500 in total, and noncovalent interactions, with VP26 proteins and triplexes (Yuan et al. 2018).

Thomsen et al. (1994) used recombinant baculoviruses to express the six HSV-1 capsid genes UL18 (VP23), UL19 (VP5), UL26 (VP21 and VP24), UL26.5 (VP22a), UL35 (VP26), and UL38 (VP19C) in insect cells. The *Sf*9 cells infected with a mixture of the six recombinant baculoviruses produced the HSV-1-like capsids, which had the same size and appearance as the authentic HSV-1 B capsids, and the protein composition of these capsids was nearly identical to that of the B capsids isolated from the HSV-1-infected Vero cells. By infections in which single capsid genes were left out, it was found that the UL18, UL19, UL38, and either the UL26 or the UL26.5 genes were required for assembly of 100-nm capsids, where the protein VP22a was shown to form the inner core of the B capsid, but the protein VP26 was not required for the assembly of the 100-nm capsids, although assembly of the B capsids was more efficient when it was present. It was concluded therefore that the products

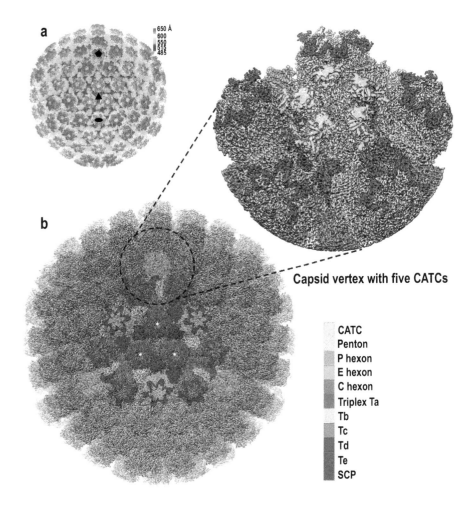

Capsid vertex with five CATCs

CATC
Penton
P hexon
E hexon
C hexon
Triplex Ta
Tb
Tc
Td
Te
SCP

FIGURE 2.2 The electron cryomicroscopy reconstruction and atomic modeling of the HSV-1 capsid. (a) The radially colored cryo-EM density map of the HSV-1 capsid viewed along a threefold axis. Fivefold, threefold, and twofold axes are denoted by a pentagon, triangle, and oval, respectively. (b) The structure of the HSV-1 capsid with capsid-associated tegument proteins. Surface view of a 4.2-Å resolution map of the icosahedral capsid, with a single facet shown in color. The structure of the vertex region (magnified view) was improved to 3.5-Å resolution by subparticle refinement. P, peripentonal; C, center; E, edge; Ta to Te, heterotrimeric triplexes composed of Tri1, Tri2A, and Tri2B. (From Dai X, Zhou ZH. Structure of the herpes simplex virus 1 capsid with associated tegument protein complexes. *Science.* 2018;360: eaao7298. Reprinted with permission of AAAS.)

of the UL18, UL19, UL35, and UL38 genes self-assemble into structures that form the outer surface (icosahedral shell) of the capsid, but the products of the UL26 and/or UL26.5 genes are required as scaffolds for assembly of 100-nm capsids and the interaction of the outer surface of the capsid with the scaffolding proteins requires the product of the UL18 gene (Thomsen et al. 1994). Newcomb et al. (1994) infected *Sf*9 cells separately with each of the six recombinant baculoviruses described earlier, prepared cell extracts from each, mixed them, and demonstrated the spontaneous self-assembly of the HSV-1 capsids in the in vitro cell-free system. Trus et al. (1995) confirmed the structural identity of the natural and baculovirus-expressed HSV-1 capsids by electron cryomicroscopy and 3D image reconstruction at ~27 Å. Furthermore, Newcomb et al. (1996) presented the first detailed description of how the assembly of the HSV-1 capsid occurs and characterized all intermediates observed during the cell-free capsid formation.

In contrast to the previously presented putative application of the capsid-based structures, Pardoe and Dargan (2002) recommended so-called L-particles (L as light) of HSV-1 as a promising candidate to the VLP vaccines. The L-particles were observed earlier by Szilágyi and Cunningham (1991) and superficially resembled the HSV-1 virions but lacked the capsid and were not infectious. Ibiricu et al. (2013) evaluated the L-particle assembly and egress pathways in cultured neurons by electron cryotomography, compared the 3D ultrastructure of the intracellular and extracellular L-particles, and quantified their diameters. The intracellular L-particles were nearly spherical, while the isolated L-particles were more heterogeneous in size and shape. Thus, only extracellular L-particles with a relatively spherical shape were considered, while the average diameter of intracellular and extracellular L-particles was 180 nm and 177 nm, respectively.

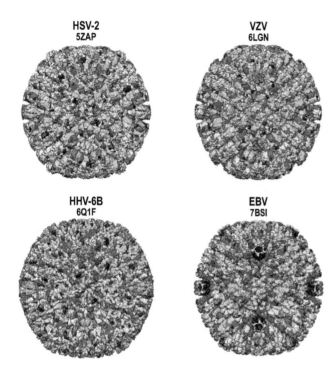

FIGURE 2.3 The cryo-EM structure of capsids of the *Herpesvirales* order members: herpes simplex virus 2 (HSV-2), species *Human alphaherpesvirus 2*, genus *Simplexvirus*, subfamily *Alphaherpesvirinae*, family *Herpesviridae*, 3.10 Å resolution, outer diameter 1296 Å (Yuan et al. 2018); varizella zoster virus (VZV), species *Human alphaherpesvirus 3*, genus *Varicellovirus*, subfamily *Alphaherpesvirinae*, family *Herpesviridae*, 5.30 Å, 1286 Å (Wang et al. 2020); human herpesvirus 6B (HHV-6B), species *Human betaherpesvirus 6B*, genus *Roseolovirus*, subfamily *Betaherpesvirinae*, family *Herpesviridae*, 9.0 Å, 1314 Å (Zhang et al. 2019); Epstein Barr virus (EBV), species *Human gammaherpesvirus 4*, genus *Lymphocryptovirus*, subfamily *Gammaherpesvirinae*, family *Herpesviridae*, 4.10 Å, 1284 Å (Li et al. 2020). The 3D structures are taken from the VIPERdb (http://viperdb.scripps.edu) database (Carrillo-Tripp et al. 2009). The size of particles is in scale. All structures possess the icosahedral T = 16 symmetry. The corresponding protein data bank (PDB) ID numbers are given under the appropriate virus names.

Dolter et al. (1993) constructed the first two HSV-1 chimeras by inserting the human CD4 gene into the HSV-1 genome between the gC promoter and the gC structural gene. These viruses synthesized a significant quantity of CD4 that was expressed on the surface of infected cells, while a small but detectable quantity of CD4 was incorporated into the chimeric virions. These results suggested that the specific virion-incorporation signals were not strictly required for inclusion of glycoproteins into HSV-1 virions (Dolter et al. 1993). Spear et al. (2003) concentrated on the surface attachment proteins and paved the way to the numerous chimeric HSV-1 virions. Thus, the HSV-1 amplicon plasmid was constructed, which carried the gene encoding gC, a major constituent of HSV-1 virions and primary attachment protein, which was provided with unique restriction sites flanking the heparan sulfate (HS) binding domain HSBD at aa residues 33–176 to allow rapid, high efficiency

substitution with foreign peptide domains. The system was tested with a His tag by transfection of Vero cells in the presence of an HSV-1 helper virus, where the His-modified gC was demonstrated. Generally, the generated amplicon system provided means to rapidly and efficiently generate HSV-1 amplicon and viral vector expressing surface attachment proteins modified with different peptide epitopes.

Hu et al. (2013) engineered a fusion of the gC by hijacking the cis-acting signals to direct incorporation of the chimeric protein into the virion. The purified virions, printed on glass slides, formed a high-density Virion Display (VirD) Array, and this VirD approach was proposed as a platform for profiling functional membrane proteins and screening putative ligands and drugs.

The visualization of the HSV-1 virions was achieved by fusion of GFP to structural genes: UL35 (VP26) (Desai and Person 1998), UL25 (Cockrell et al. 2009; Conway et al. 2010), UL37, UL46 (VP11/12), UL47 (VP13/14), UL49 (VP22), UL36 (VP1/2), and UL48 (VP16) (Antinone and Smith 2010).

It should be noted that HSV-1 demonstrated high potential as a gene therapy vector (Goins et al. 2014). Suffice it to say that the replication-competent oncolytic HSV-1 vector (T-VEC; talimogene laherparepvec; OncoVex GM-CSF) passed phase III trial to treat stage III/IV malignant melanoma (Andtbacka et al. 2013) and had the chance to be the second approved gene therapy product, after Glybera using adeno-associated virus (AAV) to express lipoprotein lipase to treat the rare autosomal recessive lipoprotein lipase deficiency (Ylä-Herttuala 2012). Moreover, the HSV-1 amplicon vectors appeared as excellent genetic vaccines, as described in detail by Laimbacher and Fraefel 2014, against foot-and-mouth (D'Antuono et al. 2010) and rotavirus (Laimbacher et al. 2012) infection.

Genus *Varicellovirus*

Bovine Herpesvirus

The bovine herpesvirus 1 (BHV-1) of the *Bovine alphaherpesvirus 1* species causes rhinotracheitis and infectious pustular vulvovaginitis in cattle. Kühnle et al. (1998) engineered the chimeric BHV-1 virions by integration of the membrane glycoprotein G of bovine respiratory syncytial virus (BRSV), a member of the *Mononegavirales* order (Chapter 31), into the BHV-1 envelope in the correct orientation. Moreover, Keil (2000) demonstrated that the aa stretch 1–71 of the BRSV glycoprotein G, which encompassed the cytoplasmic domain and the membrane anchor, was sufficient to target proteins into the envelope of BHV-1 without notably affecting the in vitro replication of BHV-1. Using this approach, the GFP molecule was incorporated into the chimeric BHV-1 virions, and the possibility of the construction of virions with altered biological properties was raised for potential use in vaccine development (Keil 2000). Furthermore, Keil et al. (2005) reported the next efficient methodology where a second furin cleavage site was introduced into the gB protein of BHV-1, together with intervening polypeptides, and the viable BHV-1 chimeras were isolated, which expressed the GFP and bovine alpha interferon as the furin-excisable proteins, secreting from infected cells. This original expression strategy that used therefore the BHV-1 gB as transporter for a cargo protein embedded in the chimeric gB as a furin-excisable polypeptide was described in detail by Keil (2009) and developed further by Keil et al. (2010).

Equid Herpesvirus

The small capsid protein VP26 of equid herpesvirus type 1 (EHV-1), a representative of the *Equid alphaherpesvirus 1* species and major pathogen of horses, was visualized by fusion with the monomeric red fluorescent protein mRFP1 (Frampton et al. 2010). The infection with the fluorescent chimeric virus allowed exhaustive investigation of the role of microtubules and dynein as a microtubule motor protein by the intracellular herpesvirus trafficking to the nucleus for viral replication.

Pseudorabies Virus

The visualization of pseudorabies virus (PrV), a member of the *Suid alphaherpesvirus 1* species, causing Aujeszky's disease in swine, was achieved by numerous approaches. First, Smith et al. (2001) realized the chimeric PrV virions by the N-terminal addition of GFP to the capsid protein VP26. As a result, the authors tracked the individual chimeric PrV capsids as they moved in axons away from infected neuronal cell bodies in culture. Next, Luxton et al. (2005) engineered the dual-fluorescent and FLAG-tagged viruses derived from the infectious PrV clone, where the dual-fluorescent PrV expressed capsids fused to the mRFP1 and a tegument protein fused to the GFP. Moreover, Bohannon et al. (2012) described the fusion of the fluorescent proteins to the three PrV proteins that allowed imaging of capsids in both in vitro and in vivo infection models. The PrV strains expressing the fluorescent pUL25 and pUL36 (VP1/2) preserved wild-type properties better than the traditional fluorescent pUL35 (VP26) isolates in assays of plaque size and virulence in mice.

The simultaneous expression of the PrV pUL31 and pUL34 genes in stably transfected rabbit kidney cells resulted in the formation of vesicles in the perinuclear space that resembled primary envelopes without a nucleocapsid (Klupp et al. 2007). These particles contained both pUL31 and pUL34 products as shown by immunolabeling and were derived from the nuclear envelope. Thus, the coexpression of only two conserved herpesvirus envelope proteins without any other viral factor was sufficient to induce the formation of vesicles from the nuclear membrane.

Varicella Zoster Virus

The fine structure of the capsids of varicella zoster virus (VZV) belonging to the *Human alphaherpesvirus 3* species and causing chickenpox and shingles in humans of all ages was determined recently by electron cryomicroscopy at near-atomic resolution (Sun et al. 2020; Wang et al. 2020). An example of such structures is demonstrated in Figure 2.3, in comparison to other herpesviruses. Sun et al. (2020) resolved to 3.7 Å the 3D structure of the VZV A-capsids that are empty protein shells resulting from abortive DNA packaging. Overall, the VZV capsid had a similar architecture to that of other known herpesviruses. The major capsid protein assembled into pentons and hexons, forming extensive intra- and inter-capsomer interaction networks that were further secured by the small capsid protein and the heterotriplex. Wang et al. (2020) reported the electron cryomicroscopy structures of the purified VZV A-capsid and of the C-capsid, which is the full capsid particle, as well as of the DNA-containing capsid inside the virion. The corresponding atomic models showed that, despite enclosing a genome that is substantially smaller than those of other human herpesviruses, VZV had a similarly sized capsid, consisting of 955 major capsid protein (MCP), 900 small capsid protein (SCP), 640 triplex dimer (Tri2), and 320 triplex monomer (Tri1) subunits. The VZV capsid had high thermal stability, although with relatively fewer intra- and inter-capsid protein interactions and less stably associated tegument proteins compared with other human herpesviruses (Wang et al. 2020).

By analogy with HSV-1, VZV produced the L-particles, i.e., envelopes without capsids, which contained glycoproteins gE, gI, and gB in similar numbers as in the complete virions, demonstrating therefore no difference in the glycoprotein content of the L-particles and suggesting further nanotechnological applications (Carpenter et al. 2008).

SUBFAMILY *BETAHERPESVIRINAE*

Genus *Cytomegalovirus*: Human Cytomegalovirus

The 235-kilobase genome of human cytomegalovirus (HCMV) of the *Human betaherpesvirus 5* species, which is frequently associated with the salivary glands' infections, is by far the largest of any herpesvirus, yet it has been unclear how its capsid, which is similar in size to those of other herpesviruses, is stabilized. Yu et al. (2017) reported the atomic structure of HCMV, which described the herpesvirus-conserved capsid proteins MCP, Tri1, Tri2, and SCP and the HCMV-specific tegument protein pp150-totaling ~4000 molecules and 62 different conformers. The MCPs manifested as a complex of insertions around the phage HK97 gp5-like domain, which gave rise to the three classes of capsid floor-defining interactions; triplexes, composed of two "embracing" Tri2 conformers and a "third-wheeling" Tri1, fasten the capsid floor. The HCMV-specific strategies included using hexon channels to accommodate the genome and pp150 helix bundles to secure the capsid via cysteine tetrad-to-SCP interactions. Remarkably, the authors proposed the clear impact of the structure for the rational design of countermeasures against HCMV, other herpesviruses, and even HIV/AIDS (Yu et al. 2017).

Talbot and Almeida (1977) were the first ones to purify the so-called HCMV dense bodies, the structures antigenically related to, but morphologically distinct from, the enveloped virion, which consisted of tegument proteins with a complete viral envelope with all glycoproteins. Weisel et al. (2010) characterized these dense bodies, termed them VLPs, and used them as an immunizing antigen to generate and study the HCMV-specific memory B cells. Schneider-Ohrum et al. (2016) performed an exhaustive study on these dense-body VLPs and provided a real foundation for their future development as an HCMV-based VLP vaccine.

It should also be noted here that Wen et al. (2014) compared the immunogenicity of the HCMV replicon particles (VRPs) expressing the protein gB, the gH/gL complex, and the pentameric gH/gL/UL128/UL130/UL131A complex, administered in the presence or absence of the MF59 adjuvant, where the pentameric complex elicited in mice significantly higher levels of neutralizing antibodies than the gH/gL complex and that MF59 significantly increased the potency of each complex.

Concerning the baculovirus expression system, Vicente et al. (2014) reported the purification steps of the so-called HCMV reVLP that was generated using the patented Redbiotec's rePAX proprietary coexpression baculovirus system. As indicated by the authors, the HCMV reVLP was composed of more than two different HCMV proteins: one tegument protein of ~65 kDa—against which a mouse monoclonal antibody (antitegument antibody) was available for quality control—and at least three envelope proteins, including one of ~80 kDa, against which a second mouse monoclonal antibody (antisurface antibody) was available for quality control.

The growing role of the VLP-based candidates for the future HCMV vaccines was reviewed in detail (Gomes et al. 2019; Perotti and Perez 2019).

To visualize HCMV particles in living cells, Sampaio et al. (2005) generated the chimeric HCMV expressing enhanced GFP fused to the C-terminus of the capsid-associated tegument protein pUL32 (pp150). The UL32-EGFP-HCMV replicated similarly to wild-type virus in fibroblast cultures and fluorescent virions were released from infected cells. Tavalai et al. (2008) fused enhanced yellow fluorescent protein (eYFP) to the N-terminus of the UL82-encoded tegument protein pp71. Intriguingly, the insertion of the eYFP-UL82 coding sequence into the HCMV genome gave rise to a chimeric virus that replicated to significantly higher titers than the wild-type one.

Genus *Muromegalovirus:* Murine Cytomegalovirus

Liu et al. (2019) used electron cryomicroscopy with subparticle reconstruction method to obtain the first atomic structure of the murine cytomegalovirus (MCMV) capsid with the associated tegument protein pp150. Surprisingly, the capsid-binding patterns of the pp150 differed between HCMV and MCMV despite their highly similar capsid structures. In MCMV, the pp150 was absent on triplex Tc and existed as a "Λ"-shaped dimer on other triplexes, leading to only 260 groups of two pp150 subunits per capsid in contrast to 320 groups of three pp150 subunits each in a "Δ"-shaped fortifying configuration. Many more aa residues

contributed to the pp150-pp150 interactions in MCMV than in HCMV, making MCMV pp150 dimer inflexible and thus incompatible to instigate triplex Tc-binding as observed in HCMV. While the pp150 was essential in HCMV, the pp150-deletion mutant of MCMV remained viable, though with attenuated infectivity and exhibiting defects in retaining viral genome (Liu et al. 2019).

Regarding the chimeric derivatives of MCMV, the CTL epitope 43–54 of the nucleoprotein from Zaire Ebolavirus (EBOV) was inserted into the MCMV vector, namely by the C-terminal addition of the EBOV epitope to the IE2, a nonessential MCMV protein, which conferred full protection of mice against a lethal challenge of animals by Zaire EBOV (Tsuda et al. 2011).

Genus *Roseolovirus:* Human Herpesvirus 6B

The 3D structure of human herpesvirus 6B (HHV-6B), a member of the *Human betaherpesvirus 6B* species and the cause of the common childhood illness exanthema subitum, or roseola infantum, was resolved to 9.0 Å by Zhang et al. (2019). This structure is shown in Figure 2.3, as a representative for β-herpesviruses. Compared to other β-herpesviruses, the HHV-6B exhibited high similarity in capsid structure but organizational differences in its CATC (pU11 tetramer). The 180 "VΛ"-shaped CATCs were observed in HHV-6B, distinguishing from the 255 "Λ"-shaped dimeric CATCs observed in MCMV and the 310 "Δ"-shaped CATCs in HCMV. This trend in CATC quantity correlated with the increasing genome sizes of these β-herpesviruses (Zhang et al. 2019).

SUBFAMILY *GAMMAHERPESVIRINAE*

Genus *Lymphocryptovirus:* Epstein-Barr Virus

As the first discovered human cancer virus, Epstein-Barr virus (EBV), a representative of the *Human gammaherpesvirus 4* species, causes Burkitt's lymphoma and nasopharyngeal carcinoma, as well as infectious mononucleosis. The complete atomic model of the EBV capsid was presented in parallel by Liu et al. (2020) and Li et al. (2020). The latter is presented in Figure 2.3, as a representative of the γ-herpesviruses. The structure resolved by Li et al. (2020) included the icosahedral capsid, the dodecameric portal, and the CATC. In this case, the in situ portal from the tegumented capsid adopted a closed conformation with its channel valve holding the terminal viral DNA and with its crown region firmly engaged by three layers of ring-like dsDNA, which, together with the penton flexibility, effectively alleviated the capsid inner pressure placed on the portal cap. In contrast, the CATCs, through binding to the flexible penton vertices in a stoichiometric manner, accurately increased the inner capsid pressure to facilitate the pressure-driven genome delivery. This explained the mechanism by which the EBV capsid, portal, packaged genome and the CATCs coordinately achieved a pressure balance to simultaneously benefit both viral genome retention and ejection (Li et al. 2020). Liu et al. (2020) accentuated the structural plasticity among the 20 conformers of the MCP, 2 conformers of the SCP, 4 conformers of the triplex monomer proteins and 2 conformers of the triplex dimer proteins. The plasticity reached the greatest level at the capsid-tegument interfaces involving SCP and CATC: SCPs crowned pentons/hexons and mediated tegument protein binding, and CATCs bound and rotated all five periportal triplexes but notably only about one peri-penton triplex. These results offered insights into a mechanism for recruiting cell-regulating factors into the tegument compartment as a "cargo" and should contribute further nanotechnological applications (Liu et al. 2020).

Delecluse et al. (1999) were the first ones to generate a virus-free packaging cell line that allowed encapsidation of plasmids into the EBV VLPs. This cell line harbored an EBV mutant whose packaging signals were deleted. The gene vectors, which could encompass very large, contiguous pieces of foreign DNA, carried all cis-acting elements involved in amplification and encapsidation into VLPs as well as those essential for extrachromosomal maintenance in the recipient cell. In fact, this approach opened the way to delivery and stable maintenance of any transgene in human B cells (Delecluse et al. 1999). The EBV VLPs lacking DNA were nevertheless able to bind to the EBV target cells (Feederle et al. 2005). Adhikary et al. (2008) used the EBV VLPs by the advanced immunological studies of the EBV infection.

Ruis et al. (2011) presented the EBV VLPs as the first safe and efficient vaccine candidate by engineering of a dedicated producer cell line for the EBV-derived VLPs. This cell line contained a genetically modified EBV genome, which was devoid of all potential viral oncogenes but provided viral proteins essential for the assembly and release of VLPs. The human B cells readily took up the EBV-based VLPs and presented viral epitopes to T cells. Consequently, the EBV-based VLPs were highly immunogenic and elicited humoral and strong CD8$^+$ and CD4$^+$ T cell responses in vitro and in a preclinical murine model in vivo (Ruis et al. 2011). Moreover, Pavlova et al. (2013) constructed a series of novel EBV mutants that lacked proteins involved in maturation and assessed their ability to produce viral DNA-free VLPs. These mutants fulfilled all criteria of efficacy and safety expected from a preventative vaccine.

Recently, Kühne et al. (2020) studied the potency of the EBV VLPs to induce the virus-specific T cell responses in vitro and presented a novel strategy for vaccination of immunocompromised renal transplant recipients to prevent EBV reactivation especially under drug-based immunosuppression. The growing role of the VLP-based vaccination against EBV was reviewed by Rühl et al. (2020).

Genus *Rhadinovirus*: Kaposi Sarcoma-Associated Herpesvirus

Kaposi sarcoma-associated herpesvirus (KSHV), a member of the *Human gammaherpesvirus 8* species, is the etiologic agent of Kaposi sarcoma, an endothelial cell tumor (Chang et al. 1994). Dai et al. (2018) reported a 4.2 Å resolution the KSHV capsid structure, determined by electron cryo-microscopy and containing 46 unique conformers of the MCP, the SCP, and the triplex proteins Tri1 and Tri2. The structure and mutagenesis results revealed a groove in the upper domain of the MCP that contained hydrophobic residues that interacted with the SCP, which in turn cross-linked with neighboring MCPs in the same hexon to stabilize the capsid (Dai et al. 2018).

It should be mentioned here that the promising VLP-based anti-KSHV vaccine candidate (Barasa et al. 2017) was elaborated on the scaffold of Newcastle disease virus (NDV) from the *Mononegavirales* order in the same way as the analogous EBV vaccine (Perez et al. 2017), as described in Chapter 31.

3 Order *Asfuvirales*

Absence of proof is not proof of absence.

Michael Crichton

ESSENTIALS AND GENOME

According to the present ICTV (2020) situation, the extremely small *Asfuvirales* order currently includes a sole family *Asfarviridae* with the sole genus *Asfivirus* involving the only recognized *African swine fever virus* species, well known because of its ecological and economical danger. The order *Asfuvirales*, together with the well-known *Chitovirales* order including the famous *Poxviridae* family, which is touched on briefly in Chapter 5, forms the class *Pokkesviricetes*. The latter, together with the class *Megaviricetes*, forms the *Nucleocytoviricota* phylum. The latter, together with the *Preplasmiviricota* phylum, forms the *Bamfordvirae* kingdom, while the latter, together with the kingdom *Helvetiavirae*, belongs in turn to the great realm *Varidnaviria*.

As reviewed by the actual ICTV reports (Dixon et al. 2012; Alonso et al. 2018), African swine fever virus (ASFV) is the only virus with the double-stranded DNA genome, which is transmitted by arthropods. The infection of domestic pigs and wild boar results in hemorrhagic fever and, for virulent isolates, death in 5–15 days. The virus is endemic in Africa, where warthogs and bush pigs act as reservoirs. The disease was first spread from West Africa into Europe through Portugal in 1957, was endemic in parts of the Iberian Peninsula from 1960 until 1995, and was finally eradicated from Europe except for the island of Sardinia. In 2007, the virus again spread out of Africa through Georgia in the Trans-Caucasus to Europe and is currently causing outbreaks in the Russian Federation and several neighboring countries such as the Baltic Republics, Poland, Czech Republic, Moldova, and Romania (Alonso et al. 2018).

Figure 3.1 demonstrates some typical images and schematic cartoons of ASFV. As reviewed in the recent official ICTV report by Alonso et al. (2018), the ASFV virions are multilayered and comprised of a nucleoprotein core structure, 70–100 nm in diameter, surrounded by an internal lipid layer and an icosahedral capsid, 170–190 nm in diameter, which in turn is eventually surrounded by an external lipid-containing envelope, which is dispensable for infection. The extracellular enveloped virions have a diameter of 175–215 nm. The capsid exhibits icosahedral symmetry (T = 189–217) corresponding to 1,892–2,172 capsomers, where each capsomer is 13 nm in diameter and appears as a hexagonal prism with a central hole, and the intercapsomeric distance is 7.4–8.1 nm. In total, the ASFV virion contains more than 50 different proteins, including a number of enzymes and factors needed for early mRNA transcription and processing. The polyprotein processing by a viral protease yields multiple subunit structural proteins. The latter include the CD2v protein (EP402R), p24 (pKP177R), pP2 (pO61R) on the outer envelope; the major capsid protein p72 (pB646L), p49 (pB438L) and pE120R on the capsid shell; the p17 (pD117L), p54 or j13L (pE183L), and probably j18L (pE199L) and j5R (pH108R) in the internal envelope. The products of the pp220 polyprotein (pCP2475L) that is cleaved to give four structural proteins (p150, p37, p14 and p34) and the products of the pp62 polyprotein (pCP530R) that is cleaved to give two structural proteins (p35 and p15) are localized in the matrix or inner core shell (Alonso et al. 2018).

The ASFV genome consists of a single molecule of linear, covalently close-ended, dsDNA of 170–194 kbp and encodes 151–167 genes. The end sequences are present as two flip-flop forms that are inverted and complementary with respect to each other, and adjacent to both termini are identical arrays of directly repeated 2.1 kbp units. These ORFs are closely spaced with intergenic distances generally less than 200 nt and read from both DNA strands. The complete nucleotide sequences include the tissue culture-adapted Ba71V isolate (ASFV-BA71V) and numerous field isolates from Africa and Europe (Alonso et al. 2018).

3D STRUCTURE

Salas and Andrés (2013) published an exhaustive review on the general structure and morphogenesis of the large (an average diameter of 200 nm) and terrifyingly complex ASFV particles. At last, in 2019–2020, the ASFV structure was determined by electron cryomicroscopy at near-atomic resolution in parallel by three great teams. Figure 3.2 demonstrates a brief summary of these progressive structural investigations. Wang et al. (2019) used an optimized image reconstruction strategy and solved the ASFV capsid structure up to 4.1 Å. The capsid was found to consist of 17,280 proteins, including one major (MCP) p72 and four minor capsid proteins M1249L, p17, p49 and H240R, organized into pentasymmetrons and trisymmetrons. The minor capsid proteins formed a complicated network below the outer capsid shell, stabilizing the capsid by holding adjacent capsomers together. Acting as core organizers, 100-nm-long M1249L proteins run along each edge of trisymmetrons bridging two neighboring pentasymmtrons and form extensive intermolecular networks with other capsid proteins, driving the formation of the capsid framework (Wang et al. 2019). In parallel, Liu et al. (2019) solved the ASFV structure to 4.6 Å and found 8,280 copies of the MCP p72, 60 copies of the penton protein, and at least 8,340 minor capsid proteins, of which there might be 3 different types. Andrés

DOI: 10.1201/b22819-5

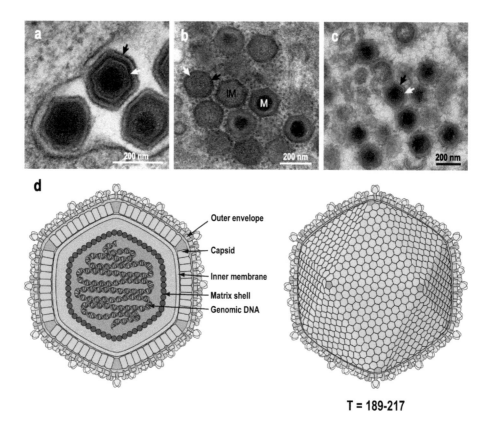

T = 189-217

FIGURE 3.1 The portraits and schematic cartoons of African swine fever virus (ASFV). The electron micrographs of the: (a) extracellular virions, black arrow is outer envelope, white arrow is virus membrane; (b) intracellular virions, IM is immature virion, M is mature virion, black arrow is capsid protein, white arrow is virus membrane; (c) intracellular virions, black arrow is capsid protein, white arrow is virus membrane. (Courtesy Pippa Hawes and Linda Dixon.) (Reprinted from *Virus Taxonomy, Classification and Nomenclature of Viruses. Ninth Report of the International Committee on Taxonomy of Viruses*, King AMQ, Lefkowitz E, Adams MJ, Carstens EB (Eds), Dixon LK, Alonso C, Escribano JM, Martins C, Revilla Y, Salas ML, Takamatsu H, Family—*Asfarviridae*. 153–162, Copyright 2012, with permission from Elsevier.) (d) The cartoons of ASFV are taken with kind permission from the ViralZone, Swiss Institute of Bioinformatics, http://viralzone.expasy.org (Hulo C et al. 2011). (Courtesy Philippe Le Mercier.)

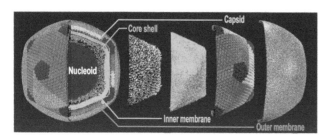

FIGURE 3.2 The multilayer architecture of African swine fever virus (ASFV). The five layers, including the genome core, nucleocapsid, inner membrane, capsid, and outer membrane, are colored in red, orange, green, blue, and gray, respectively. For a better observation on the capsid, the outer membrane has been depicted with 70% transparency. The capsid surface was colored pentasymmetrons (dark blue) and trisymmetrons (lighter blue colors including cyan, dodger blue, and cornflower blue). (Reprinted from *Trends Biochem Sci.* 45, Xian Y, Xiao C, The structure of ASFV: Advances the fight against the disease, 276–278, Copyright 2020, with permission from Elsevier.)

et al. (2020) reported the complete structure of the ASFV virion, resolving the inner capsid built from polyprotein components pp220 and pp62 in fine structural detail. This study concluded that the ASFV virion possessed a radial diameter of ~2,080 Å, enclosed a genome-containing nucleoid surrounded by two distinct icosahedral protein capsids and two lipoprotein membranes. The outer capsid formed a hexagonal lattice with triangulation number T = 277 and was composed of 8,280 copies of the double jelly-roll MCP p72—arranged in trimers displaying a pseudo-hexameric morphology—and of 60 copies of a penton protein at the vertices. The inner protein layer, organized as a T = 19 capsid, confined the core shell, and it was composed of the mature products derived from the ASFV polyproteins pp220 and pp62, namely, from the protein components p35, p15, and p8 derived from the pp62 and the protein components p5, p14, p34, p37, and p150 derived from the pp220. This structure is presented in Figure 3.3. Also, an icosahedral membrane lies between the two protein layers, whereas

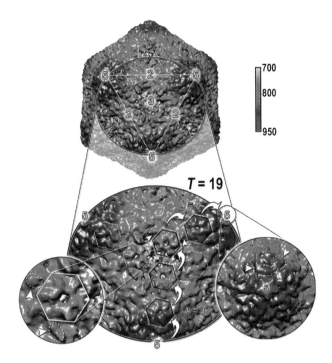

FIGURE 3.3 The ASFV inner capsid organization. Surface representation of the icosahedral inner capsid colored by radial distance (radial color scale bar in Å) with the 5-, 3-, and 2-fold icosahedral symmetry axes labeled with cardinal numbers and with a white triangle marking a facet (the volume was resymmetrized in XMIPP (de la Rosa-Trevín et al. 2013), Gaussian filtered with a width of 2.94 Å, and displayed in Chimera at 0.083 threshold). Inset below, enlarged view of a facet with pseudo-hexameric capsomers marked by black hexagons and with curved white arrows marking the *h* and *k* vectors of the hexagonal lattice with triangulation number T = 19; white hexagons mark other capsomers within the facet. Left insets, enlarged views of (i) a capsomer with the appearance of six petals of a daisy and of (ii) the penton and peripentonal capsomers viewed along the 5-fold axis; additional density in between capsomers is marked by white arrowheads and dotted lines. (Reprinted from *J. Biol Chem.* 295, Andrés G, Charro D, Matamoros T, Dillard RS, Abrescia NGA, The cryo-EM structure of African swine fever virus unravels a unique architecture comprising two icosahedral protein capsids and two lipoprotein membranes, 2020; 1–12.)

a pleomorphic envelope wraps the outer capsid (Andrés et al. 2020). The groundbreaking near-atomic ASFV structures were quickly reviewed and commented by Gallagher and Harris (2020) and Xian and Xiao (2020).

Meanwhile, Li et al. (2020) presented the crystal structure of the ASFV protein component p35 of 304 aa residues and explored its potential role in the core shell assembly. The full-length p35 was expressed in *Drosophila* S2 cells with a His$_6$ tag at its C-terminus and purified by immobilized metal ion affinity chromatography and size exclusion chromatography. The ASFV p35 existed in solution as a monomer. Remarkably, the recombinant p35 adopted a novel fold structure, which shared no homology with any reported structure.

Remarkably, Epifano et al. (2006) demonstrated earlier that the generation of filamentous instead of icosahedral particles occurred, when the structural ASFV protein B438L, encoded by the B438L gene, was absent. The mutant ASFV virions lacking B438L changed viral morphogenesis from icosahedral to filamentous particles covered by a capsid-like layer that, although containing the major capsid protein p72, did not acquire icosahedral morphology. Figure 3.4 demonstrates the filamentous ASFV variants.

EXPRESSION

The full-length ASFV MCP protein p72 was synthesized in vitro by the rabbit reticulocyte lysate system, as well as in vivo in *E. coli* cells (Cistué and Tabarés 1992). Freije et al. (1993) obtained a high expression level of the intact p72 in *E. coli*, but the product appeared in the insoluble fraction of the bacterial extracts. Nevertheless, the purified polypeptide showed an intense reaction with the antibodies present in the sera of ASFV-infected animals and was used to develop a serological test of the disease. Alcaraz et al. (1995) elaborated a Western blot test based on the protein p54, one of the most antigenic ASFV structural proteins, expressed in *E. coli* as a fusion to the N-terminus of polymerase of the RNA phage MS2 from the class *Leviviricetes* (Chapter 25). This (at the time) modern technology was reviewed in detail by Pumpens (2020).

In the baculovirus expression system, the first produced ASFV protein was hemagglutinin (Ruiz-Gonzalvo et al. 1996). Oviedo et al. (1997) described the baculovirus-driven expression of the two ASFV genes E183L and CP204L encoding the two of the most antigenic ASFV-induced proteins, p54 and p30. These proteins reacted in the serological tests in use for ASFV, improving the antibody detection in sera from experimentally inoculated pigs with attenuated viruses. Barderas et al. (2001) engineered a chimera of the p54 and p30 and produced the chimeric protein p54/30 in insect cells and in *Trichoplusia ni* larvae. The chimera retained antigenic determinants present in both proteins and reacted with a collection of sera from inapparent ASFV carrier pigs. The pigs immunized with the chimeric protein developed neutralizing antibodies and survived the challenge with a virulent ASFV. Later, Neilan et al. (2004) immunized pigs with the baculovirus-expressed ASFV proteins p30, p54, p72, and p22 and found that the neutralizing antibodies to these ASFV proteins were not sufficient for antibody-mediated protection. Furthermore, Gallardo et al. (2006) synthesized the full-length polyprotein pp62 in insect cells and used it in diagnostic tests. Argilaguet et al. (2013) constructed a BacMam vector, namely, the baculovirus-based vector for gene transfer into mammalian cells, expressing a fusion protein comprising three in tandem ASFV antigens: p54, p30, and the extracellular domain of the viral hemagglutinin, under the control of the human cytomegalovirus immediate early promoter. The immunization of pigs with this vector induced specific

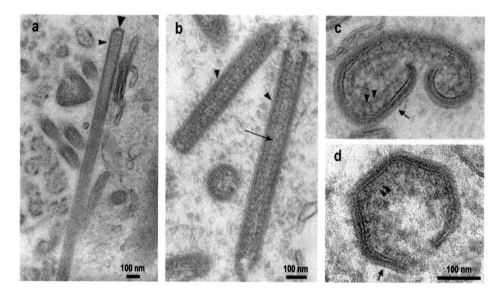

FIGURE 3.4 The electron microscopy of the ultrathin sections of the ASFV-mutant vB438Li-infected Vero cells. The panels (a) and (b) show higher magnifications of the tubular structures. The outer electron-dense layer and the apparent central channel of these structures are indicated by arrowheads and an arrow, respectively. The closed end of the tubule in panel (a) is indicated by a large arrowhead. The panel (c) shows a higher magnification of a bilobulate structure. The outer and inner domains of these structures are indicated by an arrow and arrowheads, respectively. These layers resemble the capsid and the inner envelope (arrow) and the core shell domain (arrowheads) of a normal assembling particle (d). (Reprinted with permission of American Society for Microbiology from Epifano C et al. *J Virol.* 2006;80:11456–11466.)

T-cell responses, although no specific antibody responses were detectable prior to ASFV challenge. Nevertheless, the capability of the BacMam-derived vector immunization to protect pigs against a sublethal homologous challenge with ASFV in the absence of antibodies was proved.

Concerning the putative expression of the ASFV proteins in yeast, the three recombinant *Saccharomyces cerevisiae* were constructed, which expressed the chimeric ASFV proteins fused to porcine Ig heavy chains (Chen et al. 2021). Thus, the P30-Fcγ and P54-Fcα fusion proteins were displayed on the surface of *S. cerevisiae* cells by fusion of the Fc fragment of porcine immunoglobulin IgG1 or IgA1 with the p30 or p54 encoding ASFV genes, respectively and proposed as a cheap and safe oral *S. cerevisiae*-vectored vaccine. Needless to say, all expression variants did not pretend to produce any kind of the ASFV VLPs.

CHIMERIC VIRIONS

Brun et al. (1999) were the first ones to demonstrate the feasibility of incorporating foreign aa sequences, up to 18 residues, into a protein component of the ASFV virion without affecting virus viability. Thus, the authors developed the ASFV transfer vectors carrying the gene encoding for the highly antigenic structural ASFV protein p54 encoded by the E183L gene, in which foreign sequences were introduced. The oligodeoxyribonucleotide sequences encoding continuous linear epitopes, namely, the antigenic site A from foot-and-mouth disease virus (FMDV) VP1 protein and the DA3 antigenic determinant from transmissible

gastroenteritis coronavirus (TGEV) nucleoprotein N, were separately cloned into the p54 gene, in a region encoding a nonessential domain of the protein. The chimeric p54 genes were inserted by homologous recombination into the thymidine kinase locus of the ASFV genome. The resulting chimeric viruses efficiently expressed both chimeric proteins under transcriptional control of the p54 promoter, and the chimeric gene products were recognized by antibodies to both p54 and foreign epitopes. The modified p54 proteins were also found in the viral particles and complemented the function of the wild-type p54, since deletion of the p54 gene from recombinant viruses did not affect virus replication in Vero cells (Brun et al. 1999). Moreover, Hernaez et al. (2006) presented the chimeric ASFV virions, in which the single copy of the gene E183L was exchanged with a chimeric gene encoding p54-eGFP. The chimeric virus was fully viable and exhibited growth kinetics similar to those of its parental virus. The p54-eGFP was incorporated into the virus particle with the same efficiency as the original p54. The presence of p54-eGFP in the virion resulted in fluorescent particles, which were readily visualized with a fluorescence microscope, allowing therefore the visualization of the p54-eGFP within live cells (Hernaez et al. 2006).

It should be noted that the CRISPR/Cas9 system was used to reconstruct the ASFV genes and engineer the most suitable vaccine candidates (Woźniakowski et al. 2020). Generally, the perspectives of the vaccination of pigs against the dangerous African swine fever disease were exhaustively reviewed (Escribano et al. 2013; Blome et al. 2020).

4 Order *Rowavirales*

Big results require big ambitions.

Heraclitus

ESSENTIALS AND GENOME

The *Rowavirales* is a compact order with a sole family *Adenoviridae* involving nonenveloped icosahedral pseudoT = 25 double-stranded DNA viruses that have been shown to be excellent models of both viral nanotechnology and cancer oncolytic virotherapy. According to the modern taxonomy (ICTV 2020), the family *Adenoviridae* currently involves 6 genera and 86 species altogether. Generally, the order *Rowavirales* is one of the four orders belonging to the class *Tectiliviricetes*, together with the three small orders *Belfryvirales*, *Kalamavirales*, *and Vinavirales* and one nonordered family *Autolykiviridae*, which are briefly touched on in Chapter 5. The *Tectiliviricetes* class belongs to the *Preplasmiviricota* phylum from the kingdom *Bamfordvirae*, realm *Varidnaviria*.

The members of the *Adenoviridae* family have a broad range of vertebrate hosts. Owing to their genetic heterogeneity, diverse tropism, and infections of a variety of organs and tissues, the numerous human adenoviruses (HAdVs) are causing a wide range of illnesses, from highly contagious ocular infections and common cold in young children to severe and potentially lethal diarrhea in the pediatric setting, fatal courses of pneumonia or myocarditis, and life-threatening multiorgan disease in people with a weakened immune system (Lion 2019). As reviewed by Lion (2019), the number of HAdVs is steadily expanding and currently includes more than 100 different types, which are divided into seven species termed from A to G. Although adenoviruses have been demonstrated to display oncogenic potential in experimental mammalian models, it is not possible thus far to provide clear indications for a causal relationship between HAdV infection and tumorigenicity in humans.

Figure 4.1 presents a typical portrait and schematic cartoons of the family members. As stated in the ninth official ICTV report by Harrach et al. (2012), the adenovirus virions are 70–90 nm in diameter and consist of 720 hexon subunits arranged as 240 trimer nonvertex capsomers (hexons) 8–10 nm in diameter and 12 vertex capsomers (penton bases), each with a fiber protruding from the virion surface giving the characteristic morphology, where the penton base and fiber together make up the penton. The length of fibers typically ranges between 9 and 77.5 nm. The human adenoviruses 40 and 41 have fibers of two different lengths that occur alternately on the vertexes. The members of the genus *Aviadenovirus* have two fiber proteins per vertex. The 240 hexons are formed by the interaction of three identical polypeptides (designated II) and consist of two distinct parts:

a triangular top with three "towers" and a pseudohexagonal base with a central cavity. The hexon bases are tightly packed, forming a protein shell that protects the inner components. In members of the genus *Mastadenovirus*, 12 copies of polypeptide IX are found between 9 hexons in the center of each facet. The polypeptide IX is not present in the other four genera. Two monomers of IIIa are located underneath the vertex region. The multiple copies of protein VI form a ring underneath the peripentonal hexons. The polypeptide VIII has been assigned to the inner surface of the hexon capsid (Harrach et al. 2012). Summarizing, the icosahedral pT = 25 capsids of adenoviruses consist of three major capsid proteins: hexon, penton base, and fiber and four minor, or cement, proteins IIIa, VI, VIII, and IX.

The adenovirus genome, as digested by Harrach et al. (2012), is represented by nonsegmented, linear dsDNA usually of 35–36 kb, while the size of genomes can range between 26 and 48 kb. The genome has terminally redundant sequences that have inverted terminal repetitions (ITR) of 36 to 371 bp. The terminal protein (TP) is covalently attached to each end of the genome. The central part of the genome is well conserved throughout the family, whereas the two ends show large variations in length and gene content. Altogether, about 40 different polypeptides are produced, mostly via complex splicing mechanisms, where almost a third compose the virion, including a virus-encoded cysteine protease (23 kDa), which is necessary for the processing of some precursor proteins. Apart from proteins V and IX, the other structural proteins are well conserved in every genus of the *Adenoviridae* family (Harrach et al. 2012).

Currently, the *Adenoviridae* family members are favorite gene and cancer therapy vectors for a transgene expression of a given antigen and for a direct oncolytic virotherapy, covering a huge number of publications and excellent specialized reviews (Matthews QL 2011; Beatty and Curiel 2012; Chillón and Bosch 2014; Crystal 2014; Majhen et al. 2016; Barry et al. 2020; Bullard et al. 2020; Hemminki et al. 2020; Neukirch et al. 2020; Chaurasiya et al. 2021; Hasanpourghadi et al. 2021; Lundstrom 2021).

3D STRUCTURE

VIRIONS

The early crystallographic and electron cryomicroscopy and 3D reconstruction studies were concentrated on the human adenoviruses from the genus *Mastadenovirus* and enabled location of the major and some of the minor capsid proteins (Stewart et al. 1991, 1993). The near-atomic resolution data were made available for the hexon, a major capsid protein, of HAdV-2 (Roberts MM et al. 1986) and HAdV-5

DOI: 10.1201/b22819-6

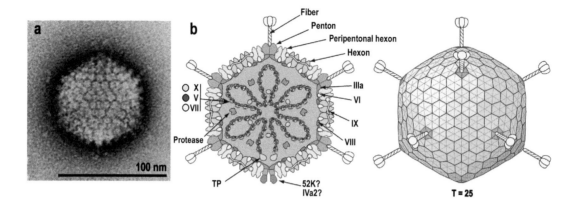

FIGURE 4.1 The portrait and schematic cartoons of a typical representative of the order *Rowavirales*. (a) The particle of fowl adenovirus 9 of the genus *Aviadenovirus*, negatively stained with uranyl acetate and showing the characteristic double fibers of fowl adenoviruses. (With kind permission from Springer Science+Business Media: *Arch Virol*, The fibers of fowl adenoviruses, 72, 1982, 289–298, Gelderblom H, Maichle-Lauppe I.) (b) The cartoons of the *Rowavirales* representatives are taken with kind permission from the ViralZone, Swiss Institute of Bioinformatics, http://viralzone.expasy.org (Hulo C et al. 2011). (Courtesy Philippe Le Mercier.)

(Rux and Burnett 2000), the fiber head domains of HAdV-5 (Xia et al. 1994), HAdV-12 (Bewley et al. 1999), HAdV-3 (Durmort et al. 2001), HAdV-37 (Burmeister et al. 2004), and the HAdV-2 fiber head and part of its shaft (van Raaij et al. 1999). At last, Zubieta et al. (2005) resolved the first crystal structure of the HAdV-2 penton. As a result, Fabry et al. (2005) published the first full quasi-atomic model of the human HAdV-5 capsid at the 10.00 Å resolution, where the particle diameter was determined as 948 Å.

Reddy et al. (2010) reported the x-ray structure of the HAdV-5 at 3.5 Å resolution and described in detail the interactions between the hexon and several accessory molecules that stabilized the capsid. Moreover, the structure revealed an altered association between the penton base and the trimeric fiber protein—perhaps reflecting an early event in cell entry—and provided a substantial advance toward understanding the assembly and cell entry mechanisms of a large double-stranded DNA, markedly improving the rapidly developing adenovirus-mediated gene transfer techniques. In parallel, Liu H et al. (2010) published a 3.6 Å resolution structure of the HAdV-5 by electron cryomicroscopy single-particle analysis, enabling to construct atomic models for the minor capsid proteins and to resolve critical regions of the hexon and penton base not seen by x-ray crystallography. Reddy and Nemerow (2014) presented a highly advanced structure of the HAdV-5 displaying a short and flexible fiber from Ad35, termed Ad5F35, with the structures of all of the adenoviral cement proteins IIIa, VI, VIII, and IX. Dai et al. (2017) revealed the specific place of the minor proteins VI and VII within the near-atomic capsid structure of the HAdV-5. Yu et al. (2017) resolved the electron cryomicroscopy structure of the HAdV-D26 to 3.70 Å and revealed the general conservation of structural organization among human adenoviruses. Finally, the Vijay S. Reddy team published the revised crystal structure of the Ad5F35 at the 3.80 Å resolution and determined its diameter as 940 Å (Kundhavai Natchiar et al. 2018). This

structure is shown as a prototype in Figure 4.2. In parallel with the Ad5F35 structure; the picture demonstrates the electron cryomicroscopy structures of bovine adenovirus type 3 (BAdV-3) (Cheng et al. 2014), human adenovirus F41, a leading cause of diarrhea in children (Rafie et al. 2021), as well as the first structurally resolved representative of the other *Mastadenovirus* genus, namely, lizard atadenovirus type 2 (LAdV-2) from the *Atadenovirus* genus, the structure of which showed, however, close similarity to that of the previously described *Mastadenovirus* members (Marabini et al. 2021).

DODECAHEDRONS

In terms of the VLP methodology, a special structural attention should be devoted to the penton base protein that was able to form the dodecahedral T = 1 particles of ~27–29 nm in diameter. The small dodecahedral particles, an association of 12 pentons, were seen in the HAdV-3 preparations by electron microscopy as early as 1964 (Norrby 1964). According to the exhaustive review of Fender et al. (1997), native penton-dodecahedra have been observed in the extracts of Ad3-, Ad4-, Ad7-, Ad9-, Ad11-, and Ad15-infected cells. Gelderblom et al. (1967) described the Ad3 dodecahedron as having a regular array of 12 morphological subunits arranged according to icosahedral symmetry.

When the two HAdV-3 proteins, penton base and fiber, were coexpressed in the baculovirus expression system, the dodecahedral particles were formed spontaneously (Fender et al. 1997). No such particles were observed when the Ad2 penton base was used instead of the Ad3 penton base; dodecahedra were not formed (Fender et al. 1997).

Schoehn et al. (1996) presented the first electron cryomicroscopy and 3D reconstructions of the dodecahedron of the baculovirus-expressed Ad3 penton both with and without the fiber and found a remarkable difference in

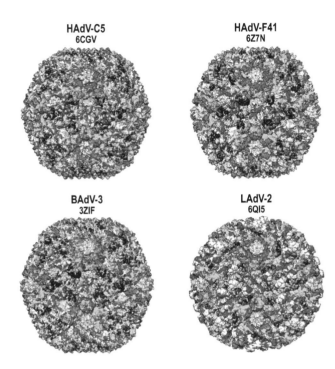

HAdV-C5
6CGV

HAdV-F41
6Z7N

BAdV-3
3ZIF

LAdV-2
6QI5

FIGURE 4.2 The near-atomic structure of the *Adenoviridae* family members: human adenovirus type 5 (HAdV-C5), referred to in the paper as Ad5F35, species *Human mastadenovirus C*, genus *Mastadenovirus*, x-ray diffraction at 3.80 Å resolution, outer diameter 940 Å (Kundhavai Natchiar et al. 2018); HAdV-F41, species *Human mastadenovirus F*, genus *Mastadenovirus*, electron cryomicroscopy at 3.77 Å resolution, outer diameter 932 Å (Rafie et al. 2021); bovine adenovirus type 3 (BAdV-3), species *Bovine mastadenovirus B*, genus *Mastadenovirus*, electron cryomicroscopy at 4.50 Å, outer diameter 934 Å (Cheng et al. 2014); lizard atadenovirus type 2 (LAdV-2), species *Lizard atadenovirus A*, genus *Atadenovirus*, electron cryomicroscopy at 3.40 Å resolution, outer diameter 936 Å (Marabini et al. 2021). The 3D structures are taken from the VIPERdb (http://viperdb.scripps.edu) database (Carrillo-Tripp et al. 2009). The size of particles is in scale. All structures possess the icosahedral pT = 25 symmetry. The corresponding protein data bank (PDB) ID numbers are given under the appropriate virus names.

the penton base structure with and without the fiber. The five small protuberances on the outer surface of each base moved away from the 5-fold axis by ~15 Å when the fiber was present. These protuberances were of relatively low density and represented a flexible loop containing the RGD site involved in the receptor binding. The fiber was apparently bound to the outer surface of the penton base, rather than inserted into it. The fiber was flexible, and the shaft contained two distinct globular regions 26 Å in diameter. The inner cavity was ~8.7 nm in diameter, and so the inner volume of the dodecahedron was calculated as 350 ± 100 nm³ (Schoehn et al. 1996).

Later, Zubieta et al. (2005) expressed the full-length HAdV-2 penton base of the 571 aa residues in the baculovirus system and reported the 3.3 Å crystal structure of the dodecahedral form, which provided therefore the last remaining major piece of the adenovirus capsid puzzle at

that time. The penton base monomer had a basal jelly-roll domain and a distal irregular domain formed by two long insertions, a similar topology to the adenovirus hexon. The Arg-Gly-Asp (RGD) motif, required for interactions with cellular integrins, occurred on a flexible surface loop. In parallel, the authors resolved to 3.5 Å the crystal structure of the penton base in complex with the interacting N-terminal fiber peptide, showing the binding of the fiber motif FNPVYPY to the insertion domain of the penton base and giving insight into the symmetry mismatch from pentameric penton base and trimeric fiber. The corresponding dodecahedra structures are shown in Figure 4.3. Zubieta et al. (2006) replaced the hypervariable loop of the Ad-2 penton base with the corresponding—but much shorter—loop of the Ad-12 and reported the 3.6 Å crystal structure of the Ad-2/12 penton base chimera crystallized as a dodecamer. The structure was generally similar to the Ad-2 penton base, with the main differences localized to the fiber protein-binding site.

Fuschiotti et al. (2006b) used electron cryomicroscopy to resolve the structure of the fibreless Ad-3 dodecahedron to 9 Å and described the internal cavity of the dodecahedron of ~80 Å in diameter, where the interior surface was accessible to solvent through perforations of ~20 Å between the pentamer towers. The intact N-terminal domain appeared to interfere with pentamer-pentamer interactions, and its absence by mutation or proteolysis was essential for dodecamer assembly. Remarkably, the differences between the 9 Å dodecahedron structure and the crystallographic Ad-2 model correlated closely with the sequence differences. These 3D data allowed the authors to propose the adenovirus dodecahedron for targeted drug delivery. Later, Szolajska et al. (2012) resolved the crystal structure of the Ad-3 penton base dodecamer to 3.8 Å and found that the dodecahedral structure was stabilized by strand-swapping between neighboring penton base molecules. Such unique N-terminal strand-swapping did not occur in the Ad2 case, a serotype that did not form dodecahedrons under physiological conditions. The evidence was provided that the distal N-terminal residues were externally exposed and available for attaching the putative cargo. Fender (2014) described in detail the methodology of the penton-dodecahedra devoid of genetic information and composed of 12 pentamers of penton base and 12 trimers of fiber protein and showed how the dodecahedron is expressed in the baculovirus system in two forms: a fiber-devoid dodecahedron made only of 12 penton bases (called base-dodecahedron: Bs-Dd) and the fiber-containing dodecahedron (called penton dodecahedron: Pt-Dd). At last, Vragniau et al. (2019) used the Ad-3 penton base protein as a starting point and engineered the synthetic self-assembling ADDomer platform for facile and rapid custom DNA insertion, ensuring easy display of the genetically fused multiepitope stretches. To preserve structural integrity of the penton base protomer, the restriction sites were designed in silico to be located outside of and adjacent to known secondary structure elements in the penton base protein. The resulting ADDomer gene was codon-optimized for the baculovirus expression system and served as a platform for the display of foreign epitopes, as described later (Vragniau et al. 2019). The 3D structure of the ADDomer dodecahedron is shown in Figure 4.3. The lower part of the Figure 4.3 shows electron micrographs of the recombinant HAdV-3 dodecahedrons pseudotyped with various fibers of different length.

The full story of the expression and structural deciphering of the adenovirus dodecahedrons, including the fishing of the long elusive adenovirus receptor, desmoglein-2, and the potential use of the dodecahedron in therapeutic development, was recently presented in the exhaustive review of Besson et al. (2020). Apart from the recombinant dodecahedrons, the full chimeric live adenoviruses of different serotypes and belonging therefore to the different species were engineered by the elaboration of the numerous novel vaccine and gene therapy vectors. The following typical examples of both approaches are cited later, where their affiliation to the various HAdV serotypes belonging to the different HAdV species is not always indicated.

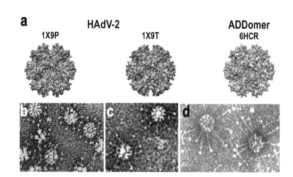

FIGURE 4.3 The adenovirus dodecahedrons. (a) The crystal structure of the adenovirus penton base dodecahedrons: HAdV type 2 (HAdV-2), species *Human mastadenovirus C*, genus *Mastadenovirus*, (left) original dodecahedrons at 3.30 Å resolution, outer diameter 290 Å, (right) the dodecahedrons cocrystallized with the 21-aa N-terminal stretch 1-MKRARPSED-TFNPVYPYDTEC-21 of the HAdV-2 fiber protein at 3.50 Å resolution, outer diameter 274 Å (Zubieta et al. 2005); ADDomer, an adenovirus-derived multimeric protein-based self-assembling nanoparticle scaffold, at 3.50 Å resolution, outer diameter 294 Å (Vragniau et al. 2019). The 3D structures are taken from the VIPERdb (http://viperdb.scripps.edu) database (Carrillo-Tripp et al. 2009). The size of particles is in scale with the full adenovirus particles in Figure 4.2. All structures possess the icosahedral T = 1 symmetry. The corresponding protein data bank (PDB) ID numbers are given under the appropriate subject names. (b, c, d) The electron micrographs of the recombinant HAdV3 dodecahedrons pseudotyped with different fibers. Base Dodecahedron (Bs-Dd) coexpressed with its corresponding HAdV3 fibers (b), the enteric HAdV41 with short fiber (c), coincubated with the HAdV2 fiber (d) observed by negative staining at the same scale. (Reprinted from Besson et al. The adenovirus dodecahedron: beyond the platonic story. *Viruses.* 2020;12:718.)

CHIMERIC VIRIONS

REPLACEMENTS

Wu et al. (2002) constructed a chimeric adenovirus vector, Ad-5/H3, by replacing the Ad-5 hexon gene with the hexon gene of Ad-3. The chimeric viruses were successfully rescued in 293 cells. Compared to that for the control Ad-5/H5, the growth rate of Ad-5/H3 was significantly slower, and the final yield was about 1 log order less. These data indicated that the Ad-3 hexon was able to encapsidate the Ad-5 genome but with less efficiency than the original Ad-5 hexon. Most important, the preimmunization of mice with one of the two types of viruses also did not prevent subsequent infection of the other type.

Nanda et al. (2005) constructed the chimeric HAdV-35 vectors carrying the Ad-5 fiber knob (rAd35k5) and compared the immunogenicities of rAd5, rAd35k5, and rAd35 vectors expressing simian immunodeficiency virus (SIV) Gag and HIV-1 Env in mice and rhesus monkeys. The in vivo studies showed that rAd35k5 vectors were more immunogenic than the rAd35 vectors in both animals. These data suggested that the Ad-5 fiber knob contributed substantially to the immunogenicity of the adenovirus vectors and paved the way for the further combination of the beneficial immunologic and serologic properties of the different adenovirus serotypes.

Roberts DM et al. (2006) constructed chimeric HAdV-5 vectors in which the seven short hypervariable regions (HVRs) on the surface of the hexon protein were replaced with the corresponding HVRs from the rare adenovirus serotype HAdV-48. These HVR-chimeric vectors were produced at high titers and were stable through serial passages in vitro. In the presence of high levels of preexisting anti-HAdV-5 immunity, the immunogenicity of HVR-chimeric vectors was not detectably suppressed, whereas the immunogenicity of the parental HAdV-5 vectors was abrogated. These data demonstrated also that the functionally relevant adenovirus-specific neutralizing antibodies are focused on epitopes located within the hexon HVRs, as described in more detail later.

HEXON MODIFICATIONS

The hexon as the most abundant structural protein was used incredibly early to display the foreign epitopes on the surface of the chimeric adenovirus virions (Crompton et al. 1994). As the trimeric hexon constituted 240 out of the 252 capsomers of the virus, the foreign epitope was repeated 720 times on the virion surface. The 8-aa residues epitope 93-EQPTTRAQ-100 from the major antigenic site in the VP1 capsid protein of poliovirus type 3 (PV3) was engineered into the two regions of the HAdV-2 hexon. The two loop regions chosen to accommodate the foreign sequences were exposed on the surface of the virion, showed sequence variation between serotypes, and were the sites of interaction with neutralizing antibodies. The chimeric adenoviruses with the PV3 epitope within one of the chosen loops were recognized and efficiently neutralized by anti-sera specific for the poliovirus sequence. The antiserum raised against the chimeric adenovirus specifically recognized the poliovirus VP1 capsid protein (Crompton et al. 1994). Then, the short sequences attaining adenovirus vector retargeting were successfully inserted into the hexon (Vigne et al. 1999). Therefore, the way was generally paved for the further insertion of the foreign epitopes into the hexon stretches, against which an antibody response was desired, namely, into the poorly conserved domains that were termed hypervariable regions (HVRs) and mapped to small surface loops of the hexon towers. Because the HVRs did not appear to be involved in maintaining the structural integrity of hexon, Rux and Burnett (2000) consented to the idea that the HVRs could be changed without affecting the viability of the virus. The idea was further approved by the insertion of the His_6 tag into the different HVRs (Wu et al. 2005). Later, Zhong et al. (2012) inserted the model His_6, FLAG (aa sequence DYKDDDDK), and His_6-FLAG tags into the HVRs.

Worgall et al. (2005) modified the hexon protein by insertion of a 14-aa epitope 8 of *P. aeruginosa* OprF (Epi8) in the loop 1 of the HVR5 of the HAdV-5 hexon. The immunization of mice with the chimeric virions resulted in detectable serum anti-*P. aeruginosa* and anti-OprF humoral responses. Worgall et al. (2007) focused on the development of a new clinical vaccine candidate that was based on the previously described chimeric virions that appeared as more effective immunogens compared to a comparable wild-type adenovirus capsid. Sharma et al. (2013) compared the vaccine candidates that carried the same 14-aa epitope Epi8 epitope within the adenovirus hexon or fiber, as described later.

McConnell et al. (2006) used a B-cell epitope from *Bacillus anthracis* protective antigen (PA) as a model antigen to characterize HVR5 of the HAdV-5 hexon as a site for peptide insertion. The authors engineered chimeric virus genomes, in which the sequence encoding the 13 aa residues of the HVR5 (aa 269 to 281) was replaced with sequences encoding increasingly larger fragments of the *B. anthracis* PA or eGFP. The HVR5 accommodated a peptide of up to 36 aa without adversely affecting virus infectivity, growth, or stability. The viruses containing chimeric hexons elicited antibodies against PA in mice, both IgG1 and IgG2a subtypes, suggesting that Th1 and Th2 immunity had been stimulated. Remarkably, the coinjection of wild-type adenovirus and a synthetic peptide from PA produced no detectable antibodies, indicating that incorporation of the epitope into the capsid was crucial for immune stimulation (McConnell et al. 2006).

Krause et al. (2006) incorporated the common epitope YPYDVPDYA of the hemagglutinin (HA) protein of the influenza A virus into the loop 1 of HVR5 between aa residues 268 and 269, resulting in 720 HA epitopes per capsid, 3 for each of the 240 hexon trimers. As presented later, the HA sequences in this study were incorporated in parallel into the penton base, fiber knob, or protein IX. This systematic incorporation of the same HA epitope into the different

adenovirus capsid proteins therefore provided clear insights into the correlation between epitope position and antiepitope immunity.

Matthews QL et al. (2008) demonstrated that the HVR2 or HVR5 of the HAdV-5 hexon can accommodate large heterologous polypeptides. In order to assess the vector capacity, the authors incorporated incrementally increasing fragments of the RGD-containing loop of the penton base, while the fragments were engineered to contain the RGD motif in the middle, flanked by penton base-derived sequences of equal lengths on both sides. The length of each flanking sequence in the shortest construct was 15 aa residues and this was increased by 10-aa increments in succeeding constructs. Thus, a total of six fragments encoding the penton base protein ranging in size from 33, 43, 53, 63, 73, and 83 aa were amplified and incorporated into the HAdV-5 hexon HVR2 or HVR5 region (Matthews QL et al. 2008).

Shiratsuchi et al. (2010) constructed an adenovirus-based malaria vaccine by inserting a B cell epitope derived from a *Plasmodium yoelii* circumsporozoite (CS) protein into the hypervariable regions within a recombinant adenovirus expressing both *P. yoelii* CS protein and GFP as transgenes.

Concerning the putative human immunodeficiency virus (HIV) vaccine candidates, Abe et al. (2009) exposed fragments of the V3 loop of the HIV gp120 within the HAdV-5 hexon. Gu et al. (2013) were the first ones to report the display of dual antigens within the same adenovirus virion. These vectors utilized HVR1 as an incorporation site for a 7-aa epitope ELDKWAS of the HIV gp41, in combination with the His_6 tag incorporation, within the HVR2 or HVR5. These multivalent antigen vectors were viable, presented the gp41 epitope as well as His_6 tag within the same virion, and elicited both HIV- and His_6-specific humoral immune response in mice. Furthermore, Gu et al. (2014) inserted the short 318-RGPGRAFVTI-327 version of the V3 loop into the hexon and found that some spacers ensured the display of the V3 epitopes similar to the natural V3 structure within the native HIV virions.

Tian et al. (2012) constructed the HAdV-3-based EV71 vaccine vectors by incorporating a neutralizing epitope SP70 containing 15 aa residues derived from the capsid protein VP1 of enterovirus 71 (EV71) within the different surface-exposed domains of the hexon of Ad3-eGFP, a recombinant HAdV-3 expressing enhanced green fluorescence protein. The thermostability and growth kinetic assays suggested that the SP70 epitope incorporation into the HVR1, HVR2, or HVR7 of the hexon did not affect virus fitness. The SP70 epitopes were exposed on all hexon-modified intact virion surfaces. Remarkably, the immunization of mother mice with the chimeric virions conferred protection in vivo to neonatal mice against the lethal EV71 challenge, and the chimeric virions-immunized mice serum also conferred passive protection against the lethal challenge in newborn mice. Moreover, Xue et al. (2014) succeeded by a multivalent HAdV-3 vaccine that displayed two neutralizing epitopes from EV71 within the hexon. No differences were found between the viruses with two epitopes incorporated into the HVR1 and HVR2 of the hexon, and both epitopes were exposed on the virion surface. The immunization of mice with the chimeric virions elicited higher IgG titers and higher neutralization titers against EV71 in vitro than immunization with the chimeric adenovirus with only one epitope incorporated into the HVR1 (Xue et al. 2014).

FIBER MODIFICATIONS

Wickham et al. (1997) were the first who constructed the two adenovirus vectors that contained modifications to the fiber protein that redirected virus binding to either α_v-integrin or heparan sulfate cellular receptors. Next, the so-called HI loop of the fiber knob domain was identified as a promising site for the incorporation of targeting moieties (Dmitriev et al. 1998; Krasnykh et al. 1998). Moreover, it was hypothesized that the structural properties of this loop would allow for the insertion of a wide variety of ligands, including large polypeptide molecules. Thus, Belousova et al. (2002) tested this hypothesis by deriving a family of adenovirus vectors whose fibers contained polypeptide inserts of incrementally increasing lengths. By assessing the levels of productivity and infectivity and the receptor specificities of the resultant virions, it was shown that the polypeptide sequences exceeding the size of the knob domain by 50% could be incorporated into the fiber with only marginal negative consequences on these key properties of the vectors. The study has also revealed a negative correlation between the size of the ligand used for vector modification and the infectivity and yield of the resultant virions, thereby predicting the limits beyond which further enlargement of the fiber knob would not be compatible with the virion's integrity (Belousova et al. 2002). The numerous tropism-changing modifications were engineered, some examples of which are described in the *Specific Targeting* paragraph. Later, Matsui et al. (2009) developed exempli gratia the chimeric fiber HAdV-F35 vectors containing the foreign RGD- and FLAG-carrying peptides in the fiber knob and confirmed that the HI loop was the most suitable for the insertion of the foreign peptides, while the fiber-mutant vectors were proposed as a platform for targeted gene delivery systems.

In the previously described great comparative study, in the case of the fiber knob-modified vector, Krause et al. (2006) placed the hemagglutinin (HA) sequence into the HI loop of fiber knob between residues 543 and 544, resulting in 36 HA epitopes per capsid, against the 720 HA epitopes per capsid in the case of the hexon modification. Nevertheless, the highest primary IgM and secondary IgG anti-HA humoral and cellular CD4 γ-interferon and interleukin-4 responses against HA in mice were always achieved with the adenovirus vector carrying the HA epitope in the fiber knob. Remarkably, these observations suggested that the immune response against an epitope inserted into the adenovirus capsid proteins was not necessarily dependent on the capsid protein number (Krause et al. 2006). Sharma et al. (2013) expressed the previously described

14-aa *P. aeruginosa* OprF epitope Epi8 in five distinct sites of the HAdV-5 fiber, namely, four loops and C-terminus and, together with the hexon-derived construct, compared their capacity to elicit anti-*P. aeruginosa* immunity to the recombinant adenovirus expressing the entire OprF protein. The most efficient inducers of specific antibodies and protectors against the respiratory infection with *P. aeruginosa* were found among the fiber-based chimeras. Strikingly, the following three doses of the two fiber-derived chimera variants elicited immunity surpassing that induced by the OprF-expressing adenovirus (Sharma et al. 2013).

Seregin et al. (2010) constructed several novel Ad5-based vectors displaying the complement-regulatory peptide (COMPinh) as fiber or pIX fusion proteins. These novel vectors dramatically minimize adenovirus-dependent activation of the human and nonhuman primate complement systems and broadened the safe use of adenovirus vectors in putative gene therapy applications.

PENTON BASE MODIFICATIONS

Wickham et al. (1996) engineered a new vector, AdFLAG, which incorporated the FLAG peptide DYKDDDDK into the penton base protein. As described later, this vector was used by the specific retargeting of the adenovirus chimeras to the α_v-integrin receptors by Einfeld et al. (1999). For this purpose, Einfeld et al. (1999) introduced the hemagglutinin (HA) epitope into the penton base or into the HI loop of the fiber protein, enabling therefore the transduction of the pseudoreceptor expressing cells by the chimeric virions.

Later, Krause et al. (2006), as mentioned earlier, compared the different adenovirus proteins as the display targets for the foreign epitopes. For the penton base-modified vector, the HA epitope was inserted into the RGD loop of the penton base between residues 301 and 302, resulting in 60 copies per capsid.

PROTEIN IX MODIFICATIONS

Rosa-Calatrava et al. (2001) performed the fine structural analysis of the HAdV-5 protein pIX encoded by the gene IX, which was known to actively participate in the stability of the viral icosahedron, acting as a capsid cement and contributing also as a transcriptional activator of several viral and cellular TATA-containing promoters. By extensive mutagenesis, the authors delineated the functional domains involved in each of the pIX properties, where aa residues 22 to 26 of the highly conserved N-terminal domain were crucial for incorporation of the protein into the virion, but specific residues of the C-terminal leucine repeat were responsible for the pIX interactions with itself.

Dmitriev et al. (2002) were the first ones who used the pIX protein for the genetic incorporation of the foreign sequences, namely targeting ligands, into its C-terminus. The appropriate chimeric vectors contained the modified pIX carrying a C-terminal FLAG epitope along with a heparan sulfate binding motif consisting of either eight consecutive lysines or a polylysine sequence. Using an anti-FLAG antibody, it was shown that modified pIXs were incorporated into virions and displayed the FLAG-containing C-terminal sequences on the capsid surface and both lysine octapeptide and polylysine ligands were accessible for binding to heparin-coated beads. Parks (2005) narrated at that time the full story of the pIX protein as a versatile platform for the presentation of polypeptides on the surface of the viral capsid, including ligands for virus retargeting and fluorescent proteins for visualizing the virus in vitro and in vivo.

Furthermore, when Krause et al. (2006) compared different adenovirus capsid proteins for the display of foreign epitopes, as stated earlier, the HA gene was added to the C-terminus of protein IX, resulting therefore in 240 HA copies per capsid. The previously described complement-regulatory peptide COMPinh was inserted not only into the fiber but also at the C-terminus of the protein pIX (Seregin et al. 2010).

TARGETING

The specific adenovirus retargeting was provoked by a major hurdle to adenovirus-mediated gene transfer when the target tissue lacked sufficient levels of receptors to mediate vector attachment via the adenovirus fiber protein. Thus, the endothelial and smooth muscle cells that were primary targets in gene therapy approaches to prevent restenosis following angioplasty or to promote or inhibit angiogenesis were poorly recognized by the traditional adenovirus vectors. As mentioned earlier, Wickham et al. (1996) decided to overcome the problem by targeting the adenovirus binding to α_v-integrin receptors that were sufficiently expressed by these cells. To target the α_v-integrins, a bispecific antibody that comprised a monoclonal antibody to the FLAG peptide epitope, DYKDDDDK, and a monoclonal antibody to α_v-integrins was constructed. In conjunction with the bispecific antibody, the previously mentioned vector AdFLAG, which incorporated the FLAG peptide epitope into its penton base protein, was engineered. The complexing of the AdFLAG virions with the bispecific antibody increased the model gene transduction to the target cells via α_v-integrin receptors (Wickham et al. 1996).

Wickham et al. (1997) provided fiber protein with one of the targeting motifs that was known to interact with high affinity to the α_v-integrins or to heparin sulfate proteoglycans. One motif contained an RGD integrin-binding sequence, ACDCRGDCFCG, while the other was a string of seven lysine residues KKKKKKK that contained multiple overlapping consensus motifs that allowed high-affinity binding to heparin and heparan sulfate. The use of the RGD motif within the fiber protein was optimized by Reynolds et al. (1999) and later by Lavilla-Alonso et al. (2010). It should be noted here that the introduction of the RGD motif into the fiber protein was successfully used to enhance the mucosal immunogenicity of the nonhuman primate-based AdV-C7 platform (Krause et al. 2013).

Developing further the retargeting idea, Einfeld et al. (1999) have constructed a membrane-anchored single-chain antibody that recognized a linear hemagglutinin (HA) epitope, while the HA epitope was incorporated into the HI loop of the fiber protein or into the penton base. This enabled the chimeric HA-tagged virions to transduce the pseudoreceptor expressing cells.

Jullienne et al. (2009) provided the chimeric adenovirus vectors with a new entry pathway into tumors, by insertion of the so-called NGR peptide, namely CNGRCVSGCAGRC, into either fiber or hexon proteins. This strategy allowed a very efficient entry pathway in both endothelial cells and tumor cells, and both vectors were efficient tools to deliver angiostatin K1–5 cDNA into the cells, thus leading to a dramatic inhibition of their proliferation and increased cell death.

Matsui et al. (2009) have developed the chimeric fiber HAdV-F35 vectors containing the foreign RGD- and DYKDDDDK (FLAG epitope)-carrying peptides in the fiber knob and confirmed the idea that the HI loop was suitable for the insertion of the foreign peptides.

Ballard et al. (2012) used a set of peptide motifs originally isolated using phage display technology that evince tumor specificity in vivo. The selected peptides were incorporated into the HI loop of the fiber protein, and the corresponding targeted adenovirus vectors were generated. Remarkably, the incorporation of relatively low affinity peptide ligands into the fiber knob, while effective in vitro, had only minimal targeting efficacy in vivo and highlighted the importance of high affinity ligand:receptor interactions to achieve tumor targeting (Ballard et al. 2012).

Yamamoto et al. (2014) elaborated an efficient method to construct an adenovirus library displaying random peptides on the fiber knob and proposed a fiber-modified plasmid library. Behr et al. (2014) established the cell type-specific entry of HAdV-5-based vectors by the ligand insertion into a chimeric fiber with shaft and knob domains of the short HAdV-41 fiber. This fiber format ablated transduction in vitro and biodistribution to the liver in vivo. When a YSA peptide, which binds to the pan-cancer marker EphA2, was inserted into three positions of the chimeric fiber, the strong transduction of EphA2-positive but not EphA2-negative cells of human melanoma biopsies and of tumor xenografts after intratumoral injection was observed (Behr et al. 2014).

Krasnykh et al. (2001) presented an original T4 phage-involving strategy to derive an adenovirus vector with enhanced targeting potential by a radical replacement of the fiber protein in the virion with a chimeric molecule containing a heterologous trimerization motif and a receptor-binding ligand. This approach, which capitalized upon the overall structural similarity between the human HAdV-5 fiber and bacteriophage T4 fibritin proteins, has resulted in the generation of the chimeric Ad5 virions incorporating the chimeric fiber-fibritin proteins targeted to artificial receptor molecules. The following gene transfer studies employing this novel viral vector have demonstrated its capacity to efficiently deliver a transgene payload to the target cells in a receptor-specific manner (Krasnykh et al. 2001). Based on the first generation fiber-fibritin molecule, several new chimeric fibers containing variable amounts of fibritin and the Ad5 fiber shaft were analyzed by Noureddini et al. (2006). Moreover, the incorporation of the Fc-binding domain of the *Staphylococcus aureus* protein A at the C-terminus of this chimeric fiber facilitated targeting of the vector to a variety of cellular receptors by means of coupling with monoclonal antibodies (Noureddini et al. 2006). Hedley et al. (2006) incorporated a stable scFv into a deknobbed, fibritin-foldon trimerized adenovirus fiber and demonstrated selective targeting to the cognate epitope expressed on the membrane surface of cells. The deknobbed, fibritin-foldon trimerized fiber was used for the incorporation of antihuman carcinoembryonic antigen (hCEA) single variable domains derived from the heavy chain (VHH) camelid family of antibodies (Kaliberov et al. 2014).

As mentioned earlier, Dmitriev et al. (2002) used protein pIX to modify the adenovirus tropism. Remarkably, in contrast to virus bearing lysine octapeptide fused to the pIX ectodomain, the adenovirus vector displaying a polylysine could recognize cellular heparan sulfate receptors. Thus, the incorporation of a polylysine motif into the pIX resulted in a significant augmentation of the adenovirus fiber knob-independent infection of cell types that were deficient in the natural coxsackie and adenovirus receptor (CAR). The pIX ectodomain appeared therefore as an alternative to the fiber knob, penton base, and hexon proteins for incorporation of targeting ligands for the purpose of the adenovirus tropism modification (Dmitriev et al. 2002).

Concerning genetic modifications in the hexon protein, Di et al. (2012) proposed a method for the rapid generation of the adenovirus vectors with the chimeric hexons. As a result, the targeting NGR, RGD or Tat PTD peptides were inserted into the hexon HVR5, and the transduction efficiency of the Tat PTD-modified virions was significantly enhanced in the specific cell lines.

Packaging and Drug Delivery

First, the eGFP molecules were fused to the protein IX and the infectious fluorescent virions were developed, which possessed full virus function and utility in the tracking of infection as well as the detection of viral biodistribution by the in vivo investigations (Le et al. 2004; Meulenbroek et al. 2004). Second, the ability to track the adenovirus core and genome initiated a potential detection strategy distinct from capsid labeling, namely, the C-terminal fusions of eGFP to the native core proteins Mu, pV, and pVII (Matthews DA; Lee et al. 2003, 2004). Moreover, Le et al. (2006) achieved the genetic labeling of the adenovirus core through the use of a chimeric expression system of both the native and fluorescent fusion core proteins. Thus, the adenovirus core labeling offered a unique way to follow the adenovirus core with potential for studying adenovirus infection and biology as well as tracking adenoviral vectors in gene therapy applications (Le et al. 2006).

Tresilwised et al. (2010) prepared the adenovirus virions surrounded by magnetic nanoparticles. The oncolytic adenovirus dl520, a derivative of HAdV-5, which did not replicate in normal cells (Haley et al. 1984; Wong and Ziff 1994), was chosen for this study, in order to boost the efficacy of the oncolytic effect by associating it with magnetic nanoparticles and magnetic-field-guided infection in multidrug-resistant (MDR) cancer cells in vitro and upon intratumoral injection in vivo. The virions were complexed by self-assembly with core-shell nanoparticles having a magnetite core of about 10 nm and stabilized by a shell containing lithium 3-[2-(perfluoroalkyl)ethylthio]propionate) and 25 kDa branched polyethylenimine. The ultrastructural analysis by electron and atomic force microscopy showed structurally intact virions surrounded by magnetic particles that occasionally bridged several virus particles and the viral uptake into cells was enhanced tenfold compared to nonmagnetic virus when infections were carried out under the influence of a magnetic field. The increased virion internalization resulted in a tenfold enhancement of the oncolytic potency in terms of the IC_{50} dose (Tresilwised et al. 2010).

Liu L et al. (2014) incorporated metallothionein (MT), a low-molecular-mass metal-binding protein, as a fusion to pIX and investigated different fusions of MT within pIX to optimize functional display. As a result, the authors identified a dimeric MT construct that showed a significant increase in ^{99m}Tc-radiolabeling activity to a pIX-MT fusion and characterized metal binding and noninvasively observed the adenovirus biodistribution and the kinetics of the virion uptake in vivo on a whole-body level in mice, where most of the radioactivity was found in the liver, spleen, kidneys, and bladder, with significant differences between the groups observed in the liver (Liu L et al. 2014).

CHIMERIC DODECAHEDRONS

Vaccines

The adenoviral penton base dodecahedron (Dd, or Bs-Dd) VLPs were used as a platform for the display of the influenza epitopes derived from the influenza M1 protein for induction of cell-mediated immunity (Szurgot et al. 2013). The vaccine bearing the two M1 epitopes 40–57 and 55–72 was constructed, produced in the baculovirus system, and purified. Remarkably, the study confirmed the previously mentioned observation that the expression in bacteria yielded free pentameric penton bases only. It was important that both cellular and humoral immune responses were elicited in the absence of an adjuvant upon chicken vaccination with the Dd-M1 VLPs.

Vragniau et al. (2019) used the previously described synthetic dodecahedron ADDomer to generate a vaccine against Chikungunya virus (CHIKV), a representative of the *Togaviridae* family, order *Martellivirales* (Chapter 18). The authors have chosen the major CHIKV neutralizing epitope E2EP3 that is located at the very N terminus of the viral E2 glycoprotein, comprising the first 18 aa residues and inserted the E2EP3 into the ADDomer. To recapitulate the exposed conformation of the E2EP3 in Chikungunya virus, a site for a highly specific protease, namely tobacco etch virus (TEV) protease, was inserted right in front of the epitope, which, upon cleavage, generated a native-like conformation comprising an exposed N-terminal serine. The electron micrographs confirmed that the ADDomer dodecahedron remained intact notwithstanding multiple (60-fold) polypeptide backbone cuts, as demonstrated in Figure 4.4. The immunization experiments compellingly validated this approach. The uncut ADDomer-CHIK did not yield E2EP3-specific immunoglobulin titers, while the quantitatively cleaved VLPs exposing the native-like E2EP3 epitope elicited very strong specific IgG response concomitant with barely detectable IgM, in excellent agreement with the immune reaction to Chikungunya viral infection (Vragniau et al. 2019).

TARGETING

The Bs-Dd VLPs interact with cellular heparan sulfate proteoglycans (HSPGs) and facilitate in turn the particle binding to integrins that is a prerequisite for targeting and entry (Vivès et al. 2004; Fender et al. 2008). Gout et al. (2010) designed a set of mutants to study the respective roles of the RGD sequence (an RGE mutant) and of a basic sequence located just downstream by the HSPG recognition. It was found that the RGE mutant binding to the heparan sulfate deficient CHO cells was abolished, and, unexpectedly,

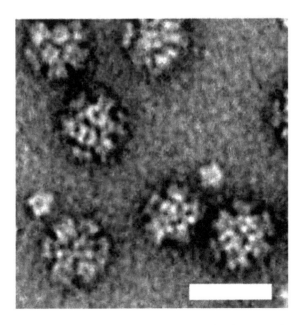

FIGURE 4.4 A representative negative-stain electron microscopy image of the TEV-cleaved ADDomer-CHIK VLPs reveals stable dodecamers. Scale bar, 30 nm. (Reprinted from Vragniau et al. Synthetic self-assembling ADDomer platform for highly efficient vaccination by genetically encoded multiepitope display. *Sci Adv.* 2019;5:eaaw2853.)

mutation of the basic sequence (KQKR to AQAS) dramatically decreased integrin recognition by the VLPs (Gout et al. 2010).

PACKAGING AND DRUG DELIVERY

Schoehn et al. (1996) were the first ones to propose the Dd VLPs as the putative tools for viral nanotechnology, although their inner volume was rather small and precluded the packaging of the sufficient genetic information. The authors predicted the possibility of linking the gene to the Dd surface in the hope that it would be internalized with the dodecahedron VLPs. Fender et al. (1997) emphasized the idea that the Dd retained many of the advantages of adenovirus for gene transfer such as efficiency of entry, efficient release of DNA from endosomes, and wide range of cell and tissue targets. Because it consisted of only one or two adenovirus proteins instead of many contained in an adenovirus virion and it did not contain the viral genome, it seemed potentially a safer alternative to the chimeric viable adenovirus. Thus, the uptake of the Dd particles was demonstrated in HeLa cells. Then, a luciferase reporter plasmid was attached to the outside of the Dd. Since in the penton the fiber protein was attached to the penton base through its N-terminus, the authors synthesized a bifunctional peptide composed of 20 aa residues corresponding to the N-terminus of the Ad-3 fiber with a C-terminal extension of 20 lysines. After mixing Dd, peptide, and DNA, the bifunctional peptide attached with the fiber-like N-terminus to the penton base, fixed and condensed DNA with the C-terminal polylysines, and ensured the appropriate level of the gene transfection (Fender et al. 1997).

Fender et al. (2003) showed that the dodecahedrons efficiently penetrated human cells when delivered large multimeric proteins such as immunoglobulins. The complexes of the Bs-Dd with an anti-Dd MAb were obtained by incubation of purified Dds with purified MAbs, and the latter were delivered to HeLa cells with the Bs-Dd as a vector. Since the two proline-proline-x-tyrosine (PPxY) motifs were found in the penton base sequence (Galinier et al. 2002), it was proposed to bind foreign proteins to the Dd VLPs via interaction with the protein structural domains called WW, a fragment of 23–35 aa residues flanked by two tryptophans (Fender et al. 2003). Taking advantage of the Dd interaction with the WW domains, Garcel et al. (2006) elaborated a universal adaptor to attach a protein of interest to the Dd vector. Thus, the tandem of three WW structural domains derived from the Nedd4 protein, a protein belonging to the ubiquitin E3-ligase family, were used as a universal protein adaptor to recognize the PPxY motif present at the N-terminal extremity of the penton base protein. The protein of interest, namely, maltose binding protein of *E. coli*, was fused to the triple WW linker and delivered by the Bs-Dd in 100% of cells in culture with on average more than ten million molecules per cell (Garcel et al. 2006).

Naskalska et al. (2009) proposed a novel influenza vaccine composed of the Dd as delivery platform carrying an internal influenza matrix protein M1 that was provided with the WW domains. Then, Naskalska et al. (2013) constructed a novel flu vaccine version when prepared with six different constructs of hemagglutinin (HA), A/swan/Poland/467/2006(H5N1), by fusion with the tandem of three WW adaptor domains. The best behaving variant was successfully delivered into human cells in vitro.

Villegas-Mendez et al. (2010) engineered WW-ovalbumin, which was successfully delivered to the dendritic cells by the Pt-Dd VLPs. Thus, the immunization with the WW-OVA/Pt-Dd resulted in 90% protection against B16-OVA melanoma implantation in syngeneic mice. This high level of protection correlated with the development of the OVA-specific CD8[+] T cells. Moreover, vaccination with the WW-OVA Pt-Dd induced robust humoral responses in mice and resulted in complete tumor regression in 100% of cases (Villegas-Mendez et al. 2010).

Fuschiotti et al. (2006a) demonstrated passive encapsidation as a cargo of the marker gold nanoparticles that had a 14-Å core of gold atoms surrounded by an organic shell of triphenylphosphines with an entire diameter of ~27 Å. The encapsidation into the Dd was ensured by shifting buffer conditions from disassembly to assembly after incubation with a marker molecule and subsequent imaging by electron cryomicroscopy to allow visualization of internal density (Fuschiotti et al. 2006a).

Zochowska et al. (2009) used the Dd VLPs as a vector for delivery of the lipophilic, nonpermeant and labile glycopeptide anticancer antibiotic bleomycin (BLM), which resulted in significant increase in BLM bioavailability. In fact, the BLM is an extremely cytotoxic agent once inside the cell nucleus, where it cleaves the DNA, but its cytotoxicity is limited by its inability to freely diffuse through membranes and by its cleavage by intracellular proteases. Therefore, the Dd display was able to resolve the problem. The BLM molecules were attached to the Dd carrier chemically, which resulted in a Dd-BLM conjugate with each penton base monomer carrying on the average one BLM moiety attached and 60 BLM molecules per particle (Zochowska et al. 2009).

Sumarheni et al. (2016) observed the intracellular trafficking of the Pt-Dd VLPs by direct conjugation with the two fluorophore dyes, Cy5 and Alexa Fluor 555, having a different wavelength of excitation/emission in ~650/670 nm and ~555/565 nm, respectively, on fixed and live cell lines. Then, the authors tried to examine the enzymatic activity and intracellular fate of the protein linked to the Pt-Dd VLPs after internalization by using horseradish peroxidase (HRP), a reporter molecule which has an enzymatic activity. In this case, the activated HRP molecules were linked

to the carrier by chemical coupling over the primary amines of the penton base protein but not over the WW mechanism (Sumarheni et al. 2016).

CHEMICAL COUPLING

The introduction of Cys residue into the hexon protein with further covalent coupling of polyethylene glycol (PEG), transferrin, or dextran onto the surface of adenovirus virions was presented by Prill et al. (2011). This approach permitted both detargeting from and targeting to hepatocytes with evasion from Kupffer cell scavenging. Thus, the virions modified with small PEG moieties specifically on hexon exhibited decreased transduction of hepatocytes by shielding from blood coagulation factor binding (Prill et al. 2011).

Hicks et al. (2011) developed a novel strategy using disrupted adenovirus for the development of immunity to cocaine. It was supposed that covalently linking a cocaine analog to Ad capsid proteins would elicit high-affinity, high-titer antibodies against cocaine sufficient to sequester systemically administered drug and prevent access to the brain, thus suppressing cocaine-induced behavior. To circumvent any risk of using an infectious virus, the virions were disrupted, resulting in a vaccine comprised of capsid proteins that retained the immunologic adjuvant properties of the original adenovirus. The putative vaccine dAd5GNE was prepared using a cocaine hapten termed GNE, a stable cocaine analog that was covalently linked to the capsid proteins of the disrupted HAdV-5 mutant. Koob et al. (2011) summarized quickly the development and assessment of the dAd5GNE vaccine in murine and rat models challenged with cocaine. Maoz et al. (2013) employed the positron emission tomography and a radiotracer to measure cocaine occupancy of the dopamine transporter in rhesus macaques. The repeat administration of the dAd5GNE vaccine induced high anticocaine titers, and the vaccinated animals significantly reduced cocaine occupancy (Maoz et al. 2013). Hicks et al. (2014) showed that the levels of cocaine in the blood of vaccinated animals rapidly decreased, suggesting that, while the antibody limited access of the drug and its active metabolites to the brain and sensitive organs of the periphery, it did not prolong drug levels in the blood compartment. Moreover, the gross and histopathology of major organs found no vaccine-mediated untoward effects.

In parallel, an antismoking vaccine was engineered by a similar way, when the nicotine hapten AM1 was chemically coupled in this case to the purified HAdV-5 hexon protein (Rosenberg et al. 2013). The immunized mice demonstrated high antinicotine antibody titers sufficient to inhibit nicotine-induced behavior of the nicotine-sensitized animals.

5 Other Double-Stranded DNA Viruses

See that the imagination of nature is far, far greater than the imagination of man.

Richard P. Feynman

ESSENTIALS

Koonin et al. (2020) were the first ones to follow systematically the readily traceable hierarchy of the evolutionary relationships in the great world of DNA viruses. The authors outlined this entire hierarchy and presented it within a global sketch of a megataxonomy, which was proposed to the ICTV in a series of the three formal taxonomic proposals (Koonin et al. 2019a, b, c) that were approved by the ICTV Executive Committee in December 2019 and convincingly described by Koonin et al. (2020). This hierarchy resulted in the three novel DNA virus realms, *Monodnaviria*, *Varidnaviria*, and *Duplodnaviria*, which were followed shortly by the latest realm *Adnaviria* (Krupovic et al. 2021). The realm *Duplodnaviria* that unifies viruses with dsDNA genomes that produce virions with icosahedral capsids formed by major capsid proteins with the HK97 fold was the subject of Chapter 1 and Chapter 2. The great realm *Monodnaviria* that comprises viruses with predominantly ssDNA genomes encoding rolling-circle replication initiation endonucleases of the HUH superfamily is encompassed by Chapters 7–12. The contribution of the members of the latest realm *Adnaviria*, which encompasses archaeal filamentous viruses with dsDNA genomes that adopt the A-form conformation within their virions, to the VLP story is a subject of the first paragraphs of the present chapter. The next paragraphs are devoted to the great realm *Varidnaviria* that comprises dsDNA viruses with icosahedral capsids built from major capsid proteins with double jelly-roll fold. The two orders of the *Varidnaviria* realm, which played a remarkable role by the VLP technologies, namely *Asfuvirales* and *Rowavirales*, were considered in Chapter 3 and Chapter 4, respectively. Noteworthy, the authors of the novel megataxonomy emphasized that the hierarchy is an open system in which the emergence of new high-rank taxa, including realms, can be anticipated (Koonin et al. 2020).

REALM *ADNAVIRIA*

ESSENTIALS

Recently, Krupovic et al. (2021) announced the creation of a new realm, *Adnaviria*, which unified archaeal filamentous viruses with linear A-form double-stranded DNA genomes and characteristic major capsid proteins unrelated to those encoded by other known viruses. Historically, the

archaeal viruses that formed filamentous virions were classified into four families: *Clavaviridae*, *Lipothrixviridae*, *Rudiviridae*, and *Tristromaviridae* (Prangishvili et al. 2017). Based on the shared gene content, structural similarity of their virions, and homology of the major capsid proteins, the families *Lipothrixviridae* and *Rudiviridae* were included into the order *Ligamenvirales* (Prangishvili and Krupovic 2012).

As reviewed by Krupovic et al. (2021), clavavirids (Prangishvili et al. 2019a) and tristromavirids (Prangishvili et al. 2019b) did not encode proteins with recognizable sequence similarity to proteins of ligamenvirals. Hence, the two families were not included in the *Ligamenvirales* order.

However, based on the structural similarity between members of *Ligamenvirales* and *Tristromaviridae*, the latter was assigned to a new order, *Primavirales*, which, along with the *Ligamenvirales* order, was included in a new class, *Tokiviricetes*.

High-resolution structures are now available for virions of six distinct tokiviricetes (Baquero et al. 2020; Wang F et al. 2020). Nevertheless, there is no detectable sequence similarity among the capsid proteins of viruses from different tokiviricete families. To bridge the gap between the class and realm taxa, the intermediate kingdom and phylum taxa, named *Zilligvirae* and *Taleaviricota*, respectively, were established (Krupovic et al. 2021).

As commented by Krupovic et al. (2021), the structural characterization of the clavavirid Aeropyrum pernix bacilliform virus 1 (APBV1) particles by electron cryomicroscopy confirmed that the fold of its major capsid protein and its overall virion organization are unrelated to that of ligamenviral, tristromavirid, or other characterized virus particles (Ptchelkine et al. 2017). Therefore, notwithstanding their filamentous virions, clavavirids were not included into the realm *Adnaviria* and likely represent another realm to be established in the future (Krupovic et al. 2021).

ORDER *LIGAMENVIRALES*

Figure 5.1 presents portraits and cartoons of the typical representatives of the two families, *Lipothrixviridae* and *Rudiviridae*, included currently into the *Ligamenvirales* order. As reviewed by Prangishvili (2012e), the *Lipothrixviridae* virions are flexible filaments that vary from 410 to 2,200 nm in length and 24 to 38 nm in diameter. They are enveloped, and the envelope consists of viral proteins and host-derived lipids. The helical nucleoprotein core contains a single molecule of linear dsDNA, 15.9–56 kb long. The virion ends have specific structures that vary between four genera: *Alpha-*, *Beta-*, *Gamma-*, and *Deltalipothrixvirus* (Prangishvili 2012e).

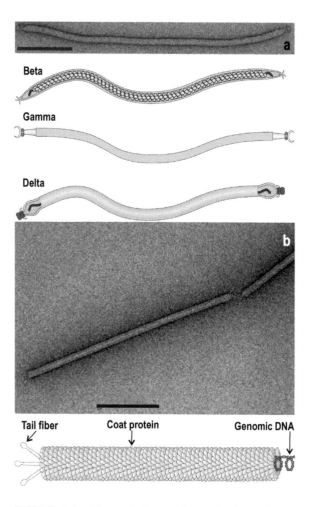

FIGURE 5.1 The typical portraits and schematic cartoons of the representatives of the *Ligamenvirales* order. (a) Family *Lipothrixviridae*: a full virion of an archaeal lipothrixvirid *Acidianus* filamentous virus 2 (AFV2) (top). (Reprinted with permission of American Society for Microbiology from Häring M et al. *J Bacteriol*. 2005b; 187:3855–3858); the cartoon of the family members [bottom] taken with kind permission from the ViralZone, Swiss Institute of Bioinformatics, http://viralzone. expasy.org; Hulo C et al. 2011.) (Courtesy Philippe Le Mercier.) (b) Family *Rudiviridae*: a full virion particle of an archaeal rudivirid, Stygiolobus rod-shaped virus (SRV), with a discontinuous central line along the virion, scale bar, 200 nm (top). (Reprinted with permission of American Society for Microbiology from Vestergaard G et al. *J Bacteriol*. 2008;190:6837–6845); the cartoon of the family members [bottom] taken with kind permission from the ViralZone.) (Courtesy Philippe Le Mercier.)

The *Rudiviridae* virion has a stiff rod shape and measures about 600–900 × 23 nm (Prangishvili 2012f). It is not enveloped and consists of a tube-like superhelix formed by dsDNA of 24.6–35.5 kb and multiple copies of a major capsid protein. At each end, the tube carries plugs, about 50 × 6 nm, to which three tail fibers are anchored. These tail fibers appear to be involved in adsorption onto the host cell surface. The length of the virions is proportional to the size of the packaged viral DNA (Prangishvili 2012f).

The major capsid proteins (MCPs) of ligamenviral particles have a unique α-helical fold first found in the MCP of rudivirid Sulfolobus islandicus rod-shaped virus 2 (SIRV2; DiMaio et al. 2015). This structure is shown in Figure 5.2. The lipothrixvirids and rudivirids share a characteristic feature in that the interaction between the MCP dimer and the linear dsDNA genome maintains the DNA in the A form. Consequently, the entire genome adopts the A-form in virions (DiMaio et al. 2015; Liu et al. 2018; Krupovic et al. 2021).

Vestergaard et al. (2008) expressed the major capsid protein of rudivirid Acidianus rod-shaped virus 1 (ARV1) in *E. coli*, and a His-tagged protein was purified to homogeneity on an Ni^{2+}-NTA-agarose column. The protein was shown by transmission electron microscopy to self-assemble and produce filamentous structures of uniform widths and different lengths. The optimal conditions for the assembly, 75°C and pH 3, were close to those of the natural environment, and no additional energy source was required for this process. The self-assembled filaments had structural parameters similar to those of the native virions, with a diameter of 21 ± 3 nm and a periodicity of (4.2 nm)$^{-1}$, suggesting that the single major capsid protein alone can generate the body of the virion (Vestergaard et al. 2008). Figure 5.3 demonstrates the self-assembled ARV1 VLPs.

Quax et al. (2011) demonstrated that the rudivirus SIRV2 encodes other autonomous structures in addition to the capsid: the virus-associated pyramids (VAPs) on the host cell surface, which are also able to self-assemble. The VAPs were produced in *Sulfolobus acidocaldarius* and *E. coli* by heterologous expression of the SIRV2 gene P98 encoding PVAP, the 10-kDa membrane protein that forms VAPs, without the need for any other cellular component. The results confirmed that the PVAP is the only constituent of the VAPs and demonstrated that no other viral protein was involved in the assembly of pyramids. Thus, while the major capsid protein self-assembled into the filament VLPs with the same diameter as the native linear virion (Vestergaard et al. 2008), the PVAP self-assembled into pyramids. As commented by Quax et al. (2011), these two autonomous structures have different functions. Whereas the capsid protein functions in DNA packaging and virus particle formation, the VAPs are specifically designed to release virus particles from the host cell.

To identify which parts of the PVAP are required for the VAP assembly, the truncated mutants lacking the last 10, 20, 30, 40, or 70 C-terminal residues were constructed (Daum et al. 2014). The electron microscopy analysis of *E. coli* cells transfected with these constructs revealed VAPs only in case of the ΔC10 mutant. The VAPs did not form after truncation of 20–70 C-terminal residues corresponding to one to three C-terminal α-helical PVAP segments, and the expression of these constructs resulted mostly in protein aggregates. As a result, the requirements for the formation of the VAP, a specific archaeoviral egress structure that takes the shape of a large, sevenfold pyramid in the host

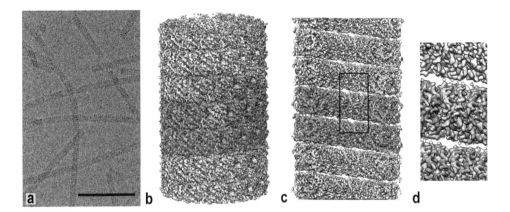

FIGURE 5.2 The electron cryomicroscopy and 3D reconstruction of rudivirid Sulfolobus islandicus rod-shaped virus 2 (SIRV2). (a) The micrograph showing SIRV2 virions in vitreous ice. Scale bar, 1000 Å. (b) The side view of the reconstructed virion with a ribbon model for the protein (magenta). The asymmetric unit in the virion contains a protein dimer, and one is shown with one chain in yellow and the other in green. (c) The cutaway view showing the hollow lumen with the all α-helical protein segments that line the lumen. These α helices wrap around the dsDNA (blue) and encapsidate it. (d) The close-up view of the region shown within the rectangle in (c). (From DiMaio F, Yu X, Rensen E, Krupovic M, Prangishvili D, Egelman EH. A virus that infects a hyperthermophile encapsidates A-form DNA. *Science*. 2015;348:914–917. Reprinted with permission of AAAS.)

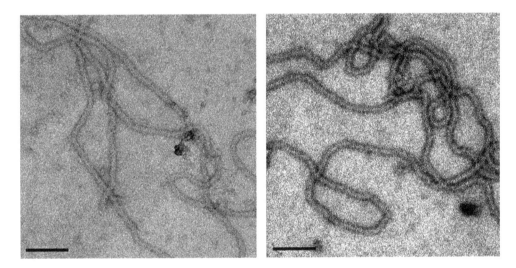

FIGURE 5.3 Electron micrograph images of the self-assembled major coat protein of ORF134 from ARV1 after negative staining with 3% uranyl acetate. Bar, 100 nm (Reprinted with permission of American Society for Microbiology from Vestergaard G et al. *J Bacteriol*. 2008;190:6837–6845.)

membrane and has no parallel in biology, were considered (Daum et al. 2014).

ORDER *PRIMAVIRALES*

The sole family of the order *Primavirales* is *Tristromaviridae*, involving two small genera *Alphatristromavirus* and *Betatristromavirus* (Prangishvili et al. 2017, 2019b; Krupovic et al. 2021). Figure 5.4 presents a portrait and cartoon of a typical representative of the family.

The structure of particles produced by tristromavirid Pyrobaculum filamentous virus 2 (PFV2) were recently resolved by electron cryomicroscopy (Wang F et al. 2020). As reviewed by Krupovic et al. (2021), the two nucleocapsid

proteins of tristromavirids were found to be structurally related to those of ligamenvirals, and the virion organizations were discovered to be remarkably similar, including the A-form conformation of the genomic dsDNA. The virions were filamentous—$400 \pm 20 \times 32 \pm 3$ nm in size—and contained a lipid envelope and an inner core consisting of a rod-shaped helical nucleocapsid formed by two major virion proteins VP1 and VP2, each with a molecular mass of 14 kDa, and a nucleocapsid-encompassing protein sheath composed of a single virion protein VP3 of 18 kDa. The sheath layer was sandwiched between the nucleocapsid and the lipid envelope, akin to the matrix protein layer found in some eukaryotic, negative-sense RNA viruses. The virions also contained at least five minor proteins with molecular masses in the range of 11–30 kDa and

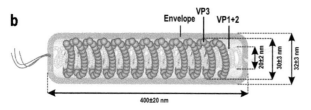

FIGURE 5.4 The typical portrait and schematic cartoon of a representatives of the *Primavirales* order. (a) Electron micrograph of a virion of Pyrobaculum filamentous virus 1 (PFV1) negatively stained with 2% uranyl acetate. (b) Schematic of the virion organization with the position of the three major capsid proteins indicated. Modified with permission from (Rensen et al. 2016; reprinted from Prangishvili D et al. ICTV virus taxonomy profile: *Tristromaviridae. J Gen Virol.* 2019b;100:135–136.)

the double-stranded DNA genomes of 16–18 kb. For example, the linear dsDNA genome of Pyrobaculum filamentous virus 1 (PFV1) was 17 714 bp—including 60 bp terminal inverted repeats—and was predicted to encode 39 proteins, most of which do not show similarities to the sequences in public databases (Krupovic et al. 2018a, 2021).

REALM *VARIDNAVIRIA*

CLASS *MEGAVIRICETES*

Essentials

The class *Megaviricetes* belongs, together with the following described class *Pokkesviricetes*, to the phylum *Nucleocytoviricota* that unites so-called giant viruses—or the nucleocytoplasmic large DNA viruses. The *Megaviricetes* members are known for typically being much larger than other viruses in two aspects. First, their genomes are much larger than the typical virus, encoding many genes involved in DNA repair, DNA replication, transcription, translation, and other processes that most viruses usually lack. Second, the *Megaviricetes* members are physically much larger in size than the typical virus, sometimes exceeding the size of typical bacteria. The class *Megaviricetes* includes the three orders that are characterized in brief later, 5 families, 2 subfamilies, 18 genera, and 65 species altogether. The related class *Pokkesviricetes* consists in turn of 2 orders, 2 families, 2 subfamilies, 23 genera, and 84 species. The phylum *Nucleocytoviricota*, together with the phylum *Preplasmiviricota*, forms the kingdom *Bamfordvirae* of the realm *Varidnaviria*.

Order *Algavirales*

As reviewed by the most recent ICTV report (Wilson et al. 2012), the virions of the *Phycodnaviridae* family, a sole

member of the order *Algavirales*, which covers 6 genera, were assumed to be large icosahedral structures of 120–220 nm in diameter with a multilaminate shell surrounding an electron-dense core and lacking an external membrane, as shown in Figure 5.5. In addition, one of the vertices in the representative Paramecium bursaria chlorella virus 1 (PBCV-1) of the genus *Chlorovirus* has a cylindrical spike or tail, 250 Å long and 50 Å wide. Generally, PBCV-1 has a 330-kb genome that encodes 416 predicted proteins and 11 tRNA molecules and involves 149 different proteins in the mature virion (Dunigan et al. 2012).

According to the first electron cryomicroscopy investigations of PBCV-1 (Yan et al. 2000; Simpson et al. 2003), the external PBCV-1 capsid was built of capsomers that were arranged into tri- and pentasymmetrons. Each tri- and pentasymmetron contained 66 and 31 capsomers (30 pseudo-hexameric capsomers and a pentameric capsomer), respectively. The atomic structure of the major capsid protein (MCP), Vp54, was determined by x-ray crystallography and consisted of two sequential jelly-roll folds with four N-linked glycans (Nandhagopal et al. 2002; De Castro et al. 2018). Each pseudo-hexameric capsomer consisted of three copies of the MCP arranged with pseudo-6-fold symmetry relating the six jelly-roll domains. The overall arrangement of the capsomers in the icosahedral lattice followed T = 169d triangulation. Zhang X et al. (2011) reached the 8.5 Å resolution by the electron cryomicroscopy of PBCV-1 virion and determined its diameter as 190 nm. At last, Fang et al. (2019) extended the resolution of the PBCV-1 structure to 3.5 Å that resulted in the identification of 13 different minor capsid proteins and an atomic model of the viral capsid consisting of 6,900 polypeptide chains. This was the largest viral capsid structure that has been determined to near-atomic resolution and the first near-atomic description of a capsid of the so-called nucleocytoplasmic large DNA viruses (NCLDVs), including phycodnaviruses, mimiviruses, iridoviruses, ascoviruses, marseilleviruses, and poxviruses, which all are presented in the current chapter, as well as asfarviruses described in Chapter 3.

Order *Imitervirales*

The order includes the sole *Mimiviridae* family with two genera, *Cafeteriavirus* and *Mimivirus*, and two species altogether (Claverie and Abergel 2012, 2018). As commented by Claverie and Abergel (2018), since 1998, when Jim van Etten's team initiated characterization of the previously described PBCV-1 from the order *Algavirales*, it was the largest known DNA virus, both in terms of particle size and genome complexity. However, in 2003, the Acanthamoeba polyphaga mimivirus (APMV) from the *Mimivirus* genus (La Scola et al. 2003; Raoult et al. 2004, 2007) of the order *Imitervirales* superseded PBCV-1, opening the era of giant viruses, i.e., with virions large enough to be visible by light microscopy and with genomes encoding more proteins than many bacteria (Claverie and Abergel 2018).

As reviewed by Claverie and Abergel (2018), the *Mimiviridae* family exhibited the broadest distribution of

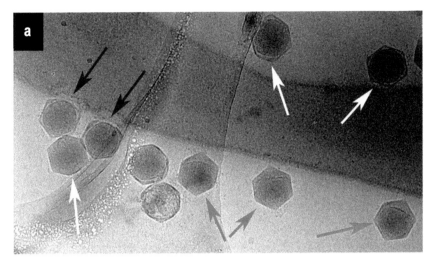

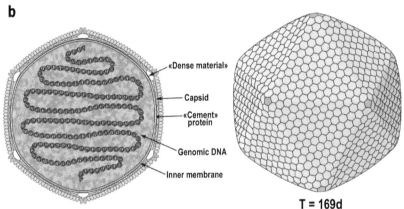

T = 169d

FIGURE 5.5   The typical portrait and schematic cartoon of a representative of the family *Phycodnaviridae* from the *Algavirales* order. (a) The electron cryomicroscopy image shows some particles of Paramecium bursaria chlorella virus 1 (PBCV-1) attached to the cell wall. The virus particles marked with white arrows are either too far away from the cell wall or overlap with the cell wall in the projection to be considered as definitely having recognized the cell wall. The orientation of the unique vertex (recognizable by the pocket under the vertex) can be seen for the virus particles marked with red arrows. The virus particles marked by black arrows have approached the cell wall such that the unique vertex is no longer recognized, and the virion has presumably partially disassembled. (Zhang X et al. Three-dimensional structure and function of the Paramecium bursaria chlorella virus capsid. *Proc Natl Acad Sci U S A*. 108:14837–14842, Copyright 2011 National Academy of Sciences, USA.) (b) The cartoon of the family members is taken with kind permission from the ViralZone, Swiss Institute of Bioinformatics, http://viralzone.expasy.org (Hulo C et al. 2011). (Courtesy Philippe Le Mercier.)

genome sizes, from 370 kb for Aureococcus anophageffe-rens virus (AaV) to 1.51 Mb for Tupanvirus deep Ocean (TupanDO, or TPV-DO), as well as of particle sizes (from 750–140 nm for icosahedral virions, up to 2.3 µm for the tailed tupanviruses). About tupanviruses and the proposed genus *Tupanvirus* see Rodrigues et al. (2019). Despite these huge differences, all members of the family demon-strated similar architecture of particles, including an inter-nal lipid membrane. Figure 5.6. demonstrates a portrait of Acanthamoeba polyphaga mimivirus (APMV), a reference strain of the *Mimivirus* genus. However, the technical bar-riers have prevented high-resolution reconstructions of APMV (Xiao et al. 2009).

Xiao et al. (2017) presented the reconstruction of the capsid of Cafeteria roenbergensis virus (CroV) from the

Cafeteriavirus genus, one of the largest viruses analyzed by the electron cryomicroscopy to date. This reconstruction is shown in Figure 5.7. The CroV capsid has a diameter of 300 nm and a triangulation number T = 499. Unlike APMV from Figure 5.6, the CroV capsid was not decorated with glycosylated surface fibers but featured 30 Å-long surface protrusions that were formed by loops of the major capsid protein. It was proposed that the capsids of CroV and related giant viruses are assembled by a newly conceived assembly pathway that initiates at a 5-fold vertex and continuously proceeds outwards in a spiraling fashion. The latter is dem-onstrated in Figure 5.7 for CroV and two other giant viruses, the previously described PBCV-1 from the *Phycodnaviridae* family, order *Algavirales*, and Chilo iridescent virus (CIV) from the *Iridoviridae* family, order *Pimascovirales*.

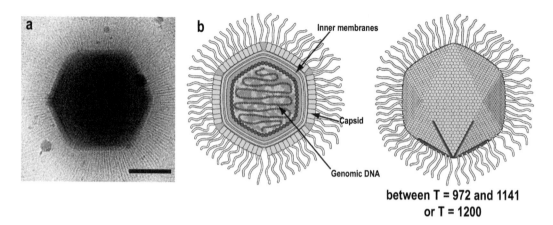

FIGURE 5.6   The portrait and schematic cartoon of Acanthamoeba polyphaga mimivirus (APMV) from the *Mimivirus* genus. (a) Single APMV particle. Scale bar, 200 nm (Reprinted from Xiao C et al. *Sci Rep.* 2017;7:5484.) (b) The cartoon is taken with kind permission from the ViralZone, Swiss Institute of Bioinformatics, http://viralzone.expasy.org (Hulo C et al. 2011). Courtesy Philippe Le Mercier. The capsid is of 400 nm in diameter with 125 nm-long closely packed fibers projecting out from the capsid surface. A 5-fold axe displays a starfish-shaped structure.

Order *Pimascovirales*

This next order of giant viruses included 3 families, namely *Ascoviridae, Iridoviridae,* and *Marseilleviridae;* 2 subfamilies; 10 genera; and 30 species altogether. The virions of the members of the *Iridoviridae* family, as reviewed by the latest ICTV report (Chinchar et al. 2017), are nonenveloped, 150–200 nm in diameter; the principal component of their capsids is the major capsid protein of 48 kDa. The genome is represented by a linear, double-stranded, circularly permuted, terminally redundant DNA of 103–220 kb, encoding 92–211 proteins. Chilo iridescent virus (CIV) from the *Invertebrate iridescent virus 6* species of the *Iridovirus* genus serves as a reference strain of the *Iridoviridae* family. The CIV virion structure is shown in Figure 5.8 in comparison with that of the previously described PBCV-1 and Melbournevirus (MelV) from the *Marseilleviridae* family. The first electron cryomicroscopy structure of CIV was presented by Yan et al. (2000) in parallel with the PBCV-1 structure. Later, Yan et al. (2009) resolved the CIV structure to 13 Å, when a homology model of P50, the CIV major capsid protein (MCP), was built based on its aa sequence and the structure of the homologous PBCV-1 Vp54 MCP.

Remarkably, the authoritative review of Dedeo et al. (2011) referenced the first attempt to use giant NCLDV virions by nanotechnological applications. Thus, with the goal of making materials with desirable optical properties, Radloff et al. (2005) approved CIV as a useful core substrate in the fabrication of metallodielectric, plasmonic nanostructures and reported its use as a scaffold to position gold nanocrystals and subsequently form continuous nanoshells using electroless deposition techniques. A gold shell was assembled around the wild-type viral core by attaching small, negatively charged 2–5-nm gold nanoparticles to the virus surface through electrostatic interactions. Then, reduction of gold[III] salts from solution afforded a thin

shell of gold surrounding the entire biotemplate (Radloff et al. 2005).

The first representative of the *Marseilleviridae* family, namely Acanthamoeba polyphaga marseillevirus (APMaV), was isolated in 2007 from water collected from a cooling tower in Paris (Boyer et al. 2009). Thus, APMaV was described five years after the discovery of the first giant virus APMV and was found to be smaller than APMV with respect to the sizes of the capsid and the genome. Nonetheless, with a capsid diameter of approximately 250 nm and a genome composed of 368,454 bp encoding 457 genes, APMaV was large enough to represent the giant NCLDV virions. The arguments in favor of the novel *Marseilleviridae* family were published quickly by Colson et al. (2013) and Aherfi et al. (2016).

Okamoto et al. (2018) resolved the electron cryomicroscopy structure of the Melbournevirus (MelV) particle, which still remains an unclassified marseillevirus. to 26.3 Å. The MelV virion that is shown in Figure 5.8 possessed the triangulation number T = 309 and was constructed by 3080 pseudohexagonal capsomers. The most distinct feature of the particle was a large and dense body (LDB) of 30 nm consistently found inside all particles and proposed to be a genome/protein complex, setting MelV apart from the other NCLDVs (Okamoto et al. 2018).

At last, the members of the *Ascoviridae* family demonstrated the most unusual structural appearance. As summarized by the official ICTV report (Asgari et al. 2017), the virions of ascoviruses are either bacilliform, ovoidal, or allantoid in shape, and, depending on the species, have complex symmetry, involve at least 20 proteins, and measure ~130 nm in diameter by 200–400 nm in length. The ascovirus genome is a single molecule of circular dsDNA ranging from 100 to 200 kb and encoding from 117 to 180 genes, of which 40 are common among the family members (Asgari et al. 2017).

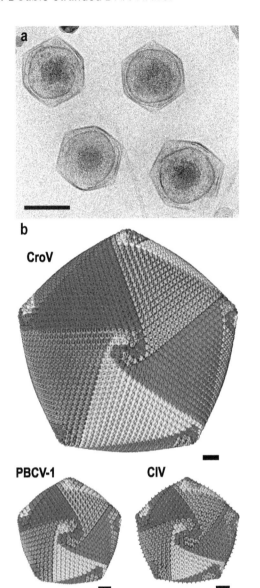

FIGURE 5.7 The portrait and 3D reconstruction of the capsid of Cafeteria roenbergensis virus (CroV) from the *Cafeteriavirus* genus. (a) The electron cryomicrograph of four CroV particles. Scale bar, 200 nm. (b) The proposed spiral assembly pathway of large icosahedral capsids of CroV, Paramecium bursaria chlorella virus 1 (PBCV-1) from the *Phycodnaviridae* family, order *Algavirales*, and Chilo iridescent virus (CIV) from the *Iridoviridae* family, order *Pimascovirales*. The isosurfaces are centered on the 5-fold axis. The capsomers in all three panels are colored based on their orientation in red, blue, green, cyan, and orange. The pentameric capsomers are depicted as purple stars. (Reprinted from Xiao C et al. *Sci Rep.* 2017;7:5484.)

Class *Pokkesviricetes*

Order *Chitovirales*

After the ultrasmall *Asfuvirales* order that is presented in Chapter 3, the class *Pokkesviricetes* includes the famous *Chitovirales* order with the sole *Poxviridae* family, 2 subfamilies, 22 genera, and 83 species altogether. A typical portrait and the schematic cartoon of poxviruses is presented in

Figure 5.9. According to the latest ICTV review (Skinner et al. 2012), the poxvirus virions are pleomorphic, generally brick-shaped (220–450 nm long × 140–260 nm wide × 140–260 nm thick) with a lipoprotein surface membrane displaying tubular or globular units (10–40 nm). They can also be ovoid (250–300 nm long × 160–190 nm diameter) with a surface membrane possessing a regular spiral filament (10–20 nm in diameter). The poxvirus virions are known as the mature virion (MV) or intracellular mature virus (IMV) and intracellular enveloped virus (IEV) or wrapped virions (WV). Some MV is wrapped by an additional double layer of intracellular membranes (derived from the trans-Golgi or endosomes) to form wrapped virions (WV; also known as intracellular enveloped virus, IEV). The latter are externalized, losing the outermost of the additional membrane layers via fusion with the cell membrane, to form extracellular virions (EV). The EV are antigenically distinct from MV, due to the presence of envelope-specific proteins, and they can be released into the extracellular medium in a form of the extracellular enveloped virus (EEV). The poxvirus genome is a single, linear molecule of covalently closed dsDNA, 130–375 kb in length, encoding 150–300 proteins, about 100 of which can be presented in virions (Skinner et al. 2012).

Cyrklaff et al. (2005) resolved the structure of the intracellular mature vaccinia virus (VV) of the *Vaccinia virus* species from the genus *Orthopoxvirus*, a classical representative of poxviruses, to 40–60 Å by the combination of electron cryomicroscopy and tomographic reconstruction. This surface-rendering representation revealed brick-shaped viral particles with slightly rounded edges and dimensions of ~360 × 270 × 250 nm. The outer layer was consistent with a lipid membrane (5–6 nm thick), below which usually two lateral bodies were found, built up by a heterogeneous material without apparent ordering or repetitive features. The internal core presented an inner cavity with electron dense coils of presumptive DNA—protein complexes, together with areas of very low density. The core was surrounded by two layers comprising an overall thickness of ~18–19 nm; the inner layer was consistent with a lipid membrane. The outer layer was discontinuous, formed by a periodic palisade built by the side interaction of T-shaped protein spikes that were anchored in the lower membrane and were arranged into small hexagonal crystallites (Cyrklaff et al. 2005).

The brilliant history of the vaccinia virus stretches back to 1796, when the breakthrough discovery by Edward Jenner paved the way to prevent and finally eradicate smallpox (Jenner 1799, 1800). The attenuated VV variants, such as, exempli gratia, modified vaccinia Ankara (MVA), remain favorite expression vectors for the viral genes and valuable boosters by vaccination nowadays. Although the great structural complexity prevented the generation of the VV-protein based VLPs, the VV vectors were frequently used to produce VLPs. Some typical and mostly recent examples of the VV-driven VLP production are listed later. Thus, at the very beginning, Karacostas et al. (1989)

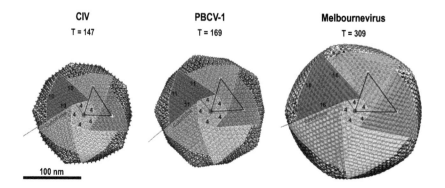

CIV
T = 147

PBCV-1
T = 169

Melbournevirus
T = 309

100 nm

FIGURE 5.8 The trisymmetron and pentasymmetron organization of Chilo iridescent virus (CIV), PBCV-1, and Melbournevirus (MelV) particles. The numbers of capsomers on each edge are shown in the figure. The yellow triangles indicate 3-fold axes and a red pentagon a 5-fold axis. The pentasymmetrons with 4-capsomer-long edges are common to all virus species. (Reprinted from *Virology*. 516, Okamoto K, Miyazaki N, Reddy HKN, Hantke MF, Maia FRNC, Larsson DSD, Abergel C, Claverie JM, Hajdu J, Murata K, Svenda M, Cryo-EM structure of a *Marseilleviridae* virus particle reveals a large internal microassembly, 239–245, Copyright 2018, with permission from Elsevier.)

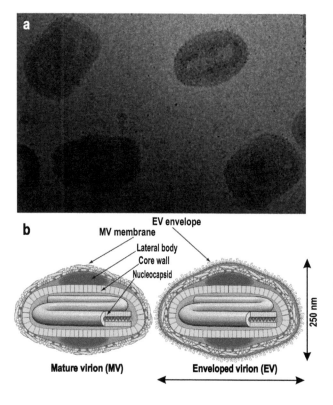

FIGURE 5.9 The portrait and schematic cartoon of a member of the *Poxviridae* family. (a) The electron cryomicrograph of vaccinia virus (VV). A typical projection image comprising several particles at different orientations is shown. (Cyrklaff M et al. Cryo-electron tomography of vaccinia virus. *Proc Natl Acad Sci U S A*. 102: 2772–2777, Copyright 2005 National Academy of Sciences, U.S.A.) (b) The cartoon is taken with kind permission from the ViralZone, Swiss Institute of Bioinformatics, http://viralzone.expasy.org (Hulo C et al. 2011). (Courtesy Philippe Le Mercier.)

designed a recombinant VV to express the entire gag-pol precursor protein of the human immunodeficiency virus type 1 (HIV-1; Chapter 35). As a result, the electron micrographs revealed immature retrovirus-like particles budding from the plasma membrane and extracellular particles with morphological characteristics of immature and mature HIV-1. Recently, Perdiguero et al. (2019) described the generation and preclinical evaluation of the MVA-based candidates expressing the HIV-1 clade C membrane-bound gp145 trimeric protein and/or the Gag-Pol-Nef polyprotein that was processed to form the Gag-induced VLPs. The VLP particles as well as purified MVA virions containing Env and Gag were visualized by immunoelectron microscopy (Perdiguero et al. 2019). A prospective MVA-based HIV-1 vaccine was presented by van Diepen et al. (2019). Overall, these and other numerous results supported the consideration of the MVA-based HIV-1 VLPs as efficacious potential vaccine candidates.

Schweneker et al. (2017) constructed a MVA coexpressing VP40 and GP of Ebola virus (EBOV) Mayinga and the nucleoprotein of Taï Forest virus (TAFV) to launch noninfectious EBOV VLPs (Chapter 31). Lázaro-Frías et al. (2018) used the same approach to generate an MVA-based vaccine against Zaire Ebolavirus (EBOV) and Sudan Ebolavirus (SUDV). Malherbe et al. (2020) constructed the MVA-driven vaccine expressing Marburg virus (MARV) VLPs (Chapter 31) from the MARV envelope glycoprotein GP and the matrix protein VP40. The electron microscopy confirmed self-assembly and budding of the MARV VLPs from infected cells.

Pérez et al. (2018) developed a novel Zika virus (ZIKV) vaccine based on the highly attenuated MVA expressing the ZIKV pre-membrane (prM) and envelope (E) structural genes and ensuring production of the ZIKV VLPs (Chapter 22). The immunization of mice with MVA-ZIKV elicited antibodies that were able to neutralize ZIKV and induced potent and polyfunctional ZIKV-specific CD8[+] T cell responses (Pérez et al. 2018). Jasperse et al. (2021) developed several MVA-based ZIKV VLP vaccine candidates, which expressed the ZIKV prM and E genes.

Reuschel et al. (2019) used recombinant vaccinia viruses to produce hepatitis B virus (HBV) surface protein (HBs) particles (Chapter 37) in HuH-7 hepatoma cells.

Salvato et al. (2019) constructed the MVA-based vaccine expressing the glycoprotein precursor (GPC) and zinc-binding matrix protein (Z) from the prototype Josiah strain lineage IV of Lassa virus (LASV) (Chapter 32). When expressed together, GP and Z formed the LASV VLPs in cell culture. This MVA-driven vaccine protected mice against challenge with a lethal dose of a Mopeia/Lassa reassortant virus delivered directly into the brain. It should be emphasized that this was the first report showing that a single dose of a replication-deficient MVA vector can confer full protection against a lethal challenge with LASV (Salvato et al. 2019).

The role of the MVA vectors by the development of the Covid-19 vaccine candidates (Chapter 26) was thoroughly reviewed by Lundstrom (2021). The vaccine-driven and potentially oncolytic function of the VV-derived vectors was referenced in a huge number of reviews, which are represented here by only some limited examples (Badrinath et al. 2016; Sasso et al. 2020; Zhang Z et al. 2021).

CLASS *MAVERIVIRICETES*

Order *Priklausovirales*

This order is the sole representative of the small class *Maveriviricetes* and involves a sole family *Lavidaviridae*, 2 genera, and 3 species altogether. The novel fascinating *Lavidaviridae* family, including two genera, *Sputnikvirus* and *Mavirus*, was proposed by Krupovic et al. (2016).

As recently reviewed by Fischer (2021), the members of the family *Lavidaviridae*, commonly known as virophages, depend on a coinfecting giant dsDNA virus of the *Mimiviridae* for their propagation. Instead of replicating in the nucleus, the virophages multiply in the cytoplasmic virion factory of a coinfecting giant virus inside a phototrophic or heterotrophic protistal host cell. Therefore, the virophages are parasites of giant viruses and can inhibit their replication, which may lead to increased survival rates of the infected host cell population. Only few virophage representatives, namely Sputnik of the *Mimivirus-dependent virus Sputnik* species, mavirus of the *Cafeteriavirus-dependent mavirus* species, and Zamilon of the *Mimivirus-dependent virus Zamilon* species, have been isolated in laboratory culture and are thus amenable to structural studies. The virophage particles that have been examined so far are 50–75 nm in diameter and possess icosahedral symmetry. Figure 5.10 demonstrates the two typical portraits of virophages, as well as the electron cryomicroscopy resolution

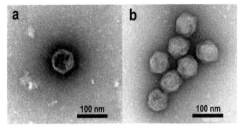
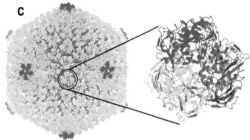

FIGURE 5.10 The structure of lavidavirus capsids. (a) The negative stain electron micrograph of a Sputnik particle (courtesy of M. Gaia and B. La Scola, Univ. Aix-Marseille, France). (b) The negative stain EM image of CsCl gradient-purified mavirus particles (U. Mersdorf and M. Fischer, Max Planck Institute for Medical Research, Germany). (c) The electron cryomicroscopy reconstruction of the Sputnik virion at 3.5 Å resolution with a magnified major capsid protein trimer (PDB entry 3J26). (Modified from Zhang X et al. 2012.) (Reprinted from Fischer MG. The virophage family *Lavidaviridae*. *Curr Issues Mol Biol.* 2021;40:1–24.)

of Sputnik at 3.5 Å, which was achieved by Zhang X et al. (2012). As summarized by Fischer (2021), the Sputnik capsid is composed of the major capsid protein (MCP) encoded by the V20 gene, and the minor capsid protein (mCP) encoded by the V18/V19 gene. The MCP contains a double jelly-roll fold and forms trimeric capsomers (hexons) with pseudohexagonal symmetry that build the 20 faces of the icosahedral particle, whereas the mCP is a single jelly-roll protein that forms pentameric capsomers (pentons) occupying the 12 vertices. The mature capsid consists of 260 hexons and 12 pentons that are arranged in a lattice with a triangulation number T = 27 (Sun et al. 2010; Zhang et al. 2012). The genomes of virophages are 17–33 kb long and encode 16–34 proteins.

Born et al. (2018) evaluated assembly and maturation of mavirus, a marine virophage, by combining structural and stability studies on capsomers, VLPs, and native virions. The mavirus protease was found to process the double jelly-roll MCP at multiple C-terminal sites and that these sites were conserved among virophages. The mavirus MCP assembled in *E. coli* in the absence and presence of penton protein, forming VLPs with defined size of 60–75 nm and shape. While quantifying VLPs in *E. coli* lysates, the full-length rather than the processed MCP was the competent state for the capsid assembly. The authors proposed that the MCP C-terminal domain served as a scaffolding domain by adding strain on MCP to confer assembly competence. The mavirus protease processed MCP more efficiently after capsid assembly, which provided a regulation mechanism for timing capsid maturation (Born et al. 2018).

Class *Polintoviricetes*

Order *Orthopolintovirales*

The novel class *Polintoviricetes* includes the sole *Orthopolintovirales* order involving the sole family *Adintoviridae*, which were proposed to classify Polintons (also known as Mavericks) into the global virus taxonomy. The Polintons were first discovered in the mid-2000s as a class of large eukaryotic transposons of 15–20 kb encoding up to ten individual proteins and named for their hallmark type B DNA *pol*ymerase and retrovirus-like *int*egrase genes (Pritham et al. 2007). Fischer and Suttle (2011) were the first ones to discover an evolutionary link between the previously described mavirus virophage from the *Lavidaviridae* family and the Polinton/Maverick transposons and hypothesized that the latter might have originated from ancient relatives of Mavirus. Yutin et al. (2013) performed strong phylogenomic analysis of the virophages and related genetic elements and concluded that the Polintons evolved from a Mavirus-like ancestor.

Krupovic et al. (2014a) showed that both Polintons and the so-called Tlr elements of *Tetrahymena thermophila* encode two key virion proteins, the major capsid protein (MCP) with the double jelly-roll fold and the minor capsid protein, known as the penton, with the single jelly-roll topology. This observation along with others strongly suggested that the Polintons and Tlr elements combined features of bona fide viruses and transposons. The authors proposed the name polintoviruses to denote these putative viruses that could have played a central role in the evolution of several groups of DNA viruses of eukaryotes (Krupovic et al. 2014a). Krupovic and Koonin (2015) delineated the evolutionary relationships among putative polintoviruses and bacterial tectiviruses, adenoviruses, virophages, and large and giant DNA viruses of eukaryotes and hypothesized that the Polintons were the first group of eukaryotic dsDNA viruses to evolve from bacteriophages and that they gave rise to most large DNA viruses of eukaryotes and various other selfish genetic elements.

At last, Tisza et al. (2020) proposed to create the new class *Polintoviricetes*, one new order *Orthopolintovirales*, one new family *Adintoviridae*, two genera, and two species in the phylum *Preplasmaviricota*. The proposal was approved by the ICTV, but the full scientific background of the proposed classification was convincingly argued by Starrett et al. (2020), who used a data-mining approach to investigate the distribution and gene content of the Polinton-like elements and related DNA viruses in animal genomic and metagenomic sequence datasets. The family name *Adintoviridae* was proposed, which connoted similarities to adenovirus virion proteins and the presence of a retrovirus-like integrase gene. The sequences resembling adintovirus virion proteins and accessory genes appeared to be restricted to animals, although degraded adintovirus sequences were endogenized into the germlines of a wide range of animals, including humans. Therefore, the newly described and classified adintovirus capsid proteins are waiting to go into the VLP applications.

Class *Tectiliviricetes*

Essentials

The great class *Tectiliviricetes* involves 4 orders, 5 families, 15 genera, and 105 species altogether. The largest and most studied order *Rowavirales* was described in Chapter 4. The class *Tectiliviricetes*, together with the two other previously described classes, *Maveriviricetes* and *Polintoviricetes*, forms the phylum *Preplasmaviricota* from the *Bamfordvirae* kingdom of the realm *Varidnaviria*.

Order *Belfryvirales*

The small *Belfryvirales* order involves the sole *Turriviridae* family including the sole genus *Alphaturrivirus* with the two species *Sulfolobus turreted icosahedral virus 1* and *Sulfolobus turreted icosahedral virus 2* altogether. As reviewed by Iranzo et al. (2016), the turriviruses are known to infect hyperthermophilic crenarchaea of the order *Sulfolobales*, but proviruses encoding homologous MCPs and genome-packaging ATPases have also been described in organisms from other orders of the *Crenarchaeota* and the phylum *Euryarchaeota*. The linear dsDNA genome is of 17 kb and encodes 36 ORFs. The well-studied Sulfolobus turreted icosahedral virus (STIV) is an archaeal virus isolated from an acidic hot spring (pH 2–4, 72–92°C) in Yellowstone National Park.

The deep structural analysis of the STIV virions was performed by the famous John E. Johnson's team. Thus, Khayat et al. (2005) reported the crystal structure of the STIV major capsid protein (MCP). The structure was nearly identical to the MCP structures of the eukaryotic PBCV and the bacteriophage PRD1, and it showed a common fold with the mammalian adenovirus. The structural analysis of the capsid architecture, determined by fitting the subunit into the electron cryomicroscopy reconstruction of the virus, identified a number of key interactions that were akin to those observed in adenovirus and PRD1. The similar capsid proteins and capsid architectures strongly suggest that these viral capsids originated and evolved from a common ancestor (Khayat et al. 2005).

In parallel, Brumfield et al. (2009) investigated the ultrastructural changes of STIV using both scanning and transmission electron microscopy by a near-synchronous STIV infection. The authors noticed assembly of the STIV particles—including particles lacking DNA—and formation of pyramid-like projections from the cell surface prior to cell lysis. These projections appeared to be caused by the protrusion of the cell membrane beyond the bordering surface protein layer (S-layer) and were thought to be sites at which progeny virus particles were released from infected cells.

Fu et al. (2010) applied the whole-cell electron cryoto-mography to the STIV virions and presented the vivid images of the STIV assembly, maturation, and particle distribution within infected archaeon *Sulfolobus* cells, including the intrapyramidal bodies that often occupied the volume of the pyramids. The mature virions, procapsids without genome cores, and partially assembled particles were identified, suggesting that the capsid and inner membrane coassemble in the cytoplasm to form a procapsid. Remarkably, the virions tended to form tightly packed clusters or quasicrystalline arrays while procapsids mostly scattered outside or on the edges of the clusters (Fu et al. 2010). Khayat et al. (2010) resolved the STIV particles at 12.5 Å by the electron cryomicroscopy, described the three viral morphologies that corresponded to biochemical disassembly states of STIV, identified the location of multiple proteins forming the large turret-like appendages at the icosahedral vertices, observed heterogeneous glycosylation of the capsid shell, and located mobile MCP C-terminal arms responsible for tethering and releasing the underlying viral membrane to and from the capsid shell. Wang K et al. (2011) presented the electron cryotomograms of the *Sulfolobus* cells infected with STIV at ~60 Å resolution, and the maximum likelihood algorithm was employed to compute reconstructions of virions within the cell. At this stage, Fu and Johnson (2012) summarized the structural investigations of STIV as one of the well-established model systems to study archaeal virus replication and viral-host interactions. The common capsid ancestry with the adenoviruses, the chlorella virus PBCV, and the phage PRD1 was emphasized. The viral-induced pyramid-like protrusions on cell surfaces were characterized as a novel viral release mechanism and previously uncharacterized functions in viral replication.

At last, the single-particle electron cryomicroscopy and x-ray crystallography techniques allowed resolution of the structure of STIV at 4.4 Å (Veesler et al. 2013a, b), as shown in Figure 5.11. As described by Veesler et al. (2013b), the STIV virions had a total radius of 480 Å comprising an icosahedral capsid with a radius of 365 Å decorated with a turret structure at each fivefold vertex that extended 115 Å above the capsid. The virus architecture was based on a pseudo T = 31d capsid symmetry with each icosahedral asymmetric unit encompassing 15 copies of the coat subunit (B345), one copy of the A223 penton base protein, one copy of the C381 turret protein, and one copy of the A55 membrane protein (A, B, and C denote different reading frames in the viral genome and the number corresponds to the number of amino acid residues in the ORF). The 17.6-kb circular dsDNA genome was enclosed within the viral membrane and filled the volume defined by it (Veesler et al. 2013b).

It should be emphasized that that the STIV MCP structure remains of special interest for the VLP nanotechnology to provide insights into the stabilizing forces required for extracellular hyperthermophilic proteins to tolerate high-temperature hot springs. Furthermore, the turrivirus pyramids could be regarded as a promising model for the future development of a novel class of the VLPs.

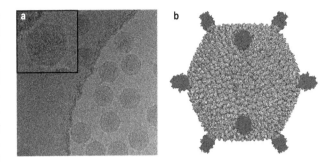

FIGURE 5.11 The near-atomic resolution reconstruction of the mature Sulfolobus turreted icosahedral virus (STIV) virion using a FEI TF20 Twin electron microscope and a Gatan K2 Summit direct detector. (a) A micrograph of ice-embedded STIV virions. (b) The overall virus reconstruction is displayed with the different protein components individually colored (coat subunit/B345, light blue; penton base/A223, light pink; turret protein/C381, purple; reprinted from *J Struct Biol.* 184, Veesler D, Campbell MG, Cheng A, Fu CY, Murez Z, Johnson JE, Potter CS, Carragher B, Maximizing the potential of electron cryomicroscopy data collected using direct detectors, 193–202, Copyright 2013, with permission from Elsevier.)

Order *Kalamavirales*

The *Kalamavirales* order involves the sole *Tectiviridae* family including 5 genera, 10 species altogether. The tectiviruses are lipid-containing bacteriophages infecting different bacteria but have no characteristic head-tail structures, although they can form the producing tail-like tubes of ~ 60 ×10 nm upon adsorption or after chloroform treatment. As reviewed by the latest ICTV report (Oksanen and Bamford 2012), the tectivirus virions have no external envelope, measure ~66 nm from facet to facet, and have flexible spikes extending about 20 nm from the virion vertices. The capsid of a typical representative, namely the phage PRD1 from the *Pseudomonas virus PRD1* species, genus *Alphatectivirus*, is constructed of 240 major capsid protein P3 trimers that form a pseudo T = 25 lattice, as shown in Figure 5.12. The protein P3 contains two β-barrels and forms very tight trimers. Underneath the capsomers, the minor coat protein P30 dimers stretch out from one vertex to another cementing the P3 facets together. The spikes are formed by the two proteins P2 and P5 extending from the penton protein P31 and are used for receptor recognition. One virion vertex is different and used for the phage DNA packaging. The packaging vertex contains packaging ATPase P9 and three other proteins P6, P20, and P22. The capsid encloses an inner membrane formed of approximately equal amounts of virus-encoded proteins and lipids derived from host-cell plasma membrane. During the process of infection, tectiviruses produce a membranous tube derived from the inner membrane of the viral particle, lined from the inside by proteins P7, P11, and P32. This tube is used to inject the viral dsDNA into the bacterial host. The single linear dsDNA molecule of about 15 kb encodes 32 proteins. Remarkably, the virions contain about 15% lipids

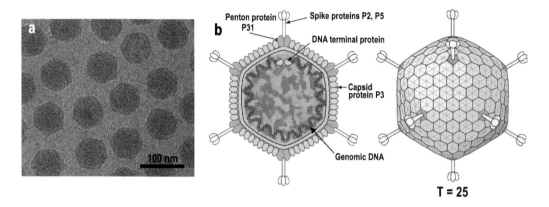

FIGURE 5.12 The portrait and schematic cartoon of a typical representative of the *Tectiviridae* family. (a) A typical micrograph of the PRD1 virions from the *Pseudomonas virus PRD1* species, genus *Alphatectivirus*. (Reprinted from Hong C et al. A structural model of the genome packaging process in a membrane-containing double stranded DNA virus. *PLoS Biol.* 2014;12:e1002024.) (b) The cartoon is taken with kind permission from the ViralZone, Swiss Institute of Bioinformatics, http://viralzone.expasy.org (Hulo C et al. 2011). (Courtesy Philippe Le Mercier.)

by mass, forming an internal membrane with virus-specific proteins and constituting about 60% of the inner membrane mass (Oksanen and Bamford 2012).

The x-ray crystal structure of the PRD1 protein P3, which was obtained from the purified virions, was resolved to 1.65 Å by Benson et al. (2001) and demonstrated close similarity to that of the adenovirus hexon protein. San Martín et al. (2001) fitted the P3 atomic structure into the electron cryomicroscopy reconstructions for three types of PRD1 particles: the wild-type virion, a packaging mutant without DNA, and a P3-shell lacking the membrane and the vertices and achieved capsids at 20–28 Å resolution. The striking architectural similarities between the phage PRD1 and the mammalian adenovirus were demonstrated. San Martín et al. (2002) provided fine details of the molecular interactions within the particles at 13.5 Å and demonstrated that the N- and C-termini of the P3 acted as conformational switches bridging capsid to membrane and linking the P3 trimers. Abrescia et al. (2004) improved the PRD1 resolution to 4.2 Å. At last, Hong et al. (2014) used single particle electron cryomicroscopy and symmetry-free image reconstruction to determine the fine structures of PRD1 virion, procapsid, and packaging deficient mutant particles. Comparing these structures allowed to propose an assembly pathway for the genome packaging apparatus in the PRD1 virion. The structure and functioning of the phage PRD1 were recently reviewed by Oksanen and Abrescia (2019).

Reddy et al. (2019) published the electron cryomicroscopy structure of the phage PR772, a close relative of the PRD1. It should be noted that PR772 was used as a model by the introduction of novel single-particle imaging techniques using hard x-ray free-electron lasers (XFELs), which were different from electron cryomicroscopy, where single biological particles have to be preserved at liquid nitrogen temperatures to determine their structure. Thus, the XFELs provided another dimension for the study of biological

systems with a definite contribution of the phage PR772 (Rose et al. 2018; Assalauova et al. 2020; Li et al. 2020).

The close structural similarity of tecti- and adenoviruses allowed to use the phage PR772 as a potential surrogate for adenovirus in the ecological investigations, such as water control and disinfection (Gall et al. 2016). It is noteworthy that tectiviruses are infecting wine bacterial spoilers. Recently, Chaïb et al. (2020) isolated and characterized by electron cryomicroscopy the phage GC1 from the *Gluconobacter virus GC1* species of the novel genus *Gammatectivirus*, which infects the grape-associated acetic acid bacterium *Gluconobacter cerinus*.

Order *Vinavirales*

The *Vinavirales* order involves the sole *Corticoviridae* family including the sole *Corticovirus* genus, and 2 species, *Pseudoalteromonas virus Cr39582* and *Pseudoalteromonas virus PM2*, altogether. As reviewed by the most recent ICTV report (Oksanen et al. 2017), the *Corticoviridae* is a family of icosahedral, internal membrane-containing phages of ~57 nm diameter between facets with highly supercoiled, double-stranded circular DNA genomes of approximately 10 kb containing 17 genes and 4 additional ORFs, where the phage PM2 from the *Pseudoalteromonas virus PM2* species remains the reference strain. The outer protein capsid is an icosahedron with a pseudo T = 21 symmetry and an internal protein-rich membrane enclosing the genome. The phage PM2 infects gram-negative *Pseudoaltermonas* bacteria and was the first phage in which the presence of lipids within the virion has been demonstrated. The viral lipids, mainly phosphatidyl ethanolamine and phosphatidyl glycerol, are acquired selectively during virion assembly from the host cytoplasmic membrane. Currently, PM2 does not share significant sequence similarity to any other virus.

Abrescia et al. (2008) reported the x-ray crystal structure at 7 Å resolution of the intact PM2 virion and higher-resolution of 2.5 Å or better of the isolated major capsid

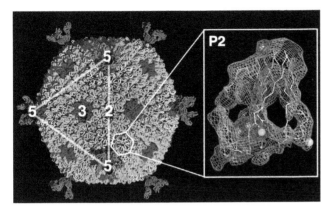

FIGURE 5.13 Structural components of PM2. (Left) Major capsid protein P2 trimers within the icosahedral asymmetric unit color coded in yellow, green, cyan, and blue. P1 is colored blue. The white triangle delineates a virus facet, and numbers indicate icosahedral symmetry axes. (Right) C_α trace of P2 (yellow) atomic structure fitted in the virus map (1σ, blue), and SeMet difference Fourier (4σ, red). (Reprinted from *Moll Cell*. 31, Abrescia NG, Grimes JM, Kivelä HM, Assenberg R, Sutton GC, Butcher SJ, Bamford JK, Bamford DH, Stuart DI, Insights into virus evolution and membrane biogenesis from the structure of the marine lipid-containing bacteriophage PM2, 749–761, Copyright 2008, with permission from Elsevier.)

protein P2 and much of the receptor-binding protein P1, as shown in Figure 5.13.

Family *Autolykiviridae*

The order-unassigned family *Autolykiviridae* involves the 2 genera, *Livvievirus* and *Paulavirus*, and 5 species altogether. Kauffman et al. (2018) characterized a group of marine dsDNA nontailed viruses with 10-kb genomes isolated during a study that quantified the diversity of viruses infecting *Vibrionaceae* bacteria. The authors proposed the name *Autolykiviridae* for the novel family within the ancient lineage of double jelly-roll (DJR) capsid viruses. Remarkably, the DJR viruses were identified in the genomes of diverse major bacterial and archaeal phyla and in marine water column and sediment metagenomes.

CLASS *LASERVIRICETES*

Order *Halopanivirales*

The relatively small class *Laserviricetes* includes the sole *Halopanivirales* order involving the 3 families, namely *Matshushitaviridae*, *Simuloviridae*, and *Sphaerolipoviridae*; 3 genera; and 9 species altogether. The members of the most-studied family *Sphaerolipoviridae*, which was approved in 2014, include both archaeal viruses and bacteriophages that possess a tailless icosahedral T = 28d capsid with an internal membrane (Iranzo et al. 2016; Demina et al. 2017). The putative genera *Alphasphaerolipovirus* and *Betasphaerolipovirus* comprise viruses that infect halophilic euryarchaea, whereas the phages P23–77 and IN93 of thermophilic *Thermus thermophilus* bacteria belong to

the putative genus *Gammasphaerolipovirus*. Demina et al. (2017) concentrated on a novel species, namely *Haloarcula virus HCIV1*, from the genus *Alphasphaerolipovirus*; described it in detail; and are awaiting now the ICTV approval of the three proposed genera.

Rissanen et al. (2013) developed a capsid model for the phage P23–77, in which capsomers were formed by various building blocks of the two major capsid proteins VP16 and VP17, and the 22-kDa minor capsid protein VP11, suggesting a novel assembly mode not observed in other icosahedral viruses. Pawlowski et al. (2015) showed that a mixture of VP16, VP17, and VP11, which were expressed in *E. coli* and purified separately, led to the formation of complexes in vitro, which could be regarded as a prelude to viral assembly and formation of VLPs. Zhai et al. (2021) assessed for the first time the expression of ORF13 and ORF14 of the phage IN93, the homologues of the P23–77 VP16 and VP17, in *E. coli* and generated therefore a novel VLP platform. Thus, the coexpressed truncated versions of ORF13 and ORF14 formed oval structures that look like VLPs, but their sizes of ~75 nm—100 nm in diameter were less than those of an authentic IN93 virion of ~130 nm (Zhai et al. 2021).

UNASSIGNED FAMILIES

The members of the archaeal virus family *Ampullaviridae* demonstrate the most unusual morphologies. As reviewed by the official ICTV reports (Prangishvili 2012a; Prangishvili et al. 2017, 2018a), the champagne-bottle-like shape of their virions is determined by the extraordinary way the linear dsDNA genome is condensed by the capsid proteins into a cone-shaped inner core and encased with the lipid-containing envelope, as shown in Figure 5.14a (Häring et al. 2005a). As summarized by Prangishvili (2012a), the virion has an overall length of about 230 nm and a width varying from about 75 nm at the broad end to 4 nm at the pointed end. The broad end of the virion exhibits 20 ± 2 thin rigid filaments, 20 nm long and 3 nm in width, which appear to be interconnected at their bases and regularly distributed around—and inserted into—a disc or ring. The 9-nm-thick envelope encases a funnel-shaped core formed by a torroidally supercoiled nucleoprotein filament, 7 nm in width. The core structure shows striations running perpendicularly to the long axis, with periodicities of 13 nm^{-1} and 4.3 nm^{-1}, indicative of helical arrangement of subunits. The virions contain six major proteins in the size range 15 to 80 kDa. The genome is represented by one dsDNA molecule of ~23.9 kb encoding 57 predicted proteins and containing two large noncoding regions of about 600 and 300 bp (Prangishvili 2012a).

As commented by Prangishvili et al. (2017), some of the most widespread and abundant archaea-specific viruses have spindle-shaped virions. The viruses that have this morphology belong to the *Bicaudaviridae* or the *Fuselloviridae* families (Krupovic et al. 2014b) and are presented in Figure 5.14b and c, respectively. The members of these two families share little sequence similarity, and their

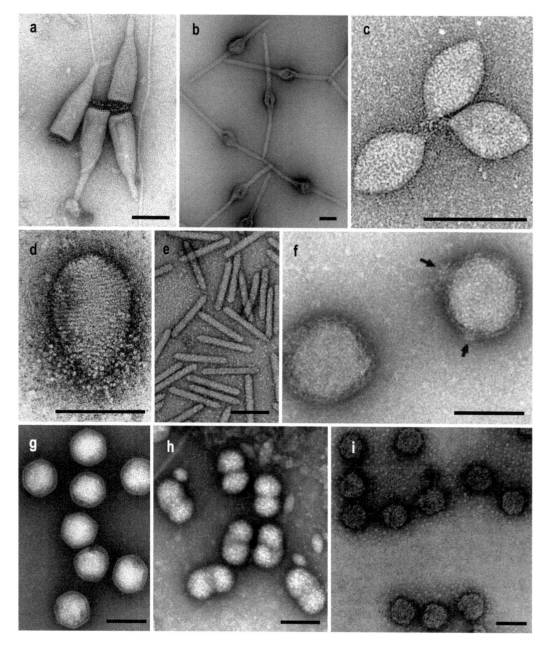

FIGURE 5.14   The electron microscopy portraits of the typical representatives of the order-unassigned dsDNA virus families. (a) *Ampullaviridae*: Acidianus bottle-shaped virus (ABV) particles attached to each other with their thin filaments at the broad end. (Reprinted with permission of American Society for Microbiology from Häring M et al. *J Virol*. 2005a;79: 9904–9911.) (b) *Bicaudaviridae*: Acidianus two-tailed virus (ATV). (Reprinted from *J Mol Biol*. 359, Prangishvili D, Vestergaard G, Häring M, Aramayo R, Basta T, Rachel R, Garrett RA, Structural and genomic properties of the hyperthermophilic archaeal virus ATV with an extracellular stage of the reproductive cycle, 1203–1216, Copyright 2006, with permission from Elsevier.) (c) *Fuselloviridae*: Sulfolobus spindle-shaped virus 1 (SSV-1). (With kind permission from Springer Science+Business Media: *The Springer Index of Viruses*, Tidona C, Darai G. (Eds), Fusellovirus. *Fuselloviridae*, 2011, 561–566, Stedman K.) (d) *Guttaviridae*: Sulfolobus neozealandicus droplet-shaped virus (SNDV.) (Reprinted from *Virology*. 272, Arnold HP, Ziese U, Zillig W, SNDV, a novel virus of the extremely thermophilic and acidophilic archaeon Sulfolobus, 409–416, Copyright 2000, with permission from Elsevier.) (e) *Clavaviridae*: Aeropyrum pernix bacilliform virus 1 (APBV1). (Reprinted from *Virology*. 402, Mochizuki T, Yoshida T, Tanaka R, Forterre P, Sako Y, Prangishvili D, Diversity of viruses of the hyperthermophilic archaeal genus *Aeropyrum*, and isolation of the Aeropyrum pernix bacilliform virus 1, APBV1, the first representative of the family *Clavaviridae*, 347–354, Copyright 2010, with permission from Elsevier.) (f) *Globuloviridae*: Pyrobaculum spherical virus (PSV), spherical protrusions are marked by arrows. (Reprinted from *Virology*. 323, Häring M, Peng X, Brügger K, Rachel R, Stetter KO, Garrett RA, Prangishvili D, Morphology and genome organization of the virus PSV of the hyperthermophilic archaeal genera *Pyrobaculum* and *Thermoproteus*: a novel virus family, the *Globuloviridae*, 233–242, Copyright 2004, with permission from Elsevier.) (g) *Portogloboviridae*: Sulfolobus polyhedral virus 1 (SPV1). (Reprinted with permission of American Society for Microbiology from Liu Y et al. *J Virol*. 2017;91:e00589–17.) (h) *Ovaliviridae*: Sulfolobus ellipsoid virus 1 (SEV1). (Reprinted with permission of American Society for Microbiology from Wang H et al. *J Virol*. 2018;92:e01727–17.) (i) *Plasmaviridae*: Acholeplasma virus L2 (AVL2.) (Reprinted with permission of S. Karger AG, Basel from Poddar SK et al. *Intervirology*. 1985;23:208–221.) Scale bar, 100 nm.

virions consist of unrelated major capsid proteins (MCPs), suggesting that their shared morphology is the result of convergent evolution. The *Bicaudaviridae* members are generally highly pleomorphic and have tails of different lengths at one or both pointed ends of the spindle-shaped body. For the type species of the *Bicaudaviridae*, Acidianus two-tailed virus (ATV), the tails, which were shown to develop extracellularly, consist of helically arranged globular subunits. The virions of the *Fuselloviridae* have a bundle of filaments at one of the two pointed ends and show a certain degree of pleomorphicity (Prangishvili et al. 2017).

As summarized by the actual ICTV reports (Prangishvili 2012b; Prangishvili et al. 2018b), the members of the *Bicaudaviridae* that infect hyperthermophilic archaea in the genus *Acidianus* are spindle-shaped upon release with a size of 120 × 80 nm, subsequently developing two tails each of up to 400 nm in length. The tails have a tube-like structure with walls that are approximately 6 nm thick. The virions are built by 11 major structural proteins. The circular dsDNA genome of ATV consists of 62.7 kb and encodes 72 predicted proteins and carries four putative transposable elements (Prangishvili et al. 2018b).

The *Fuselloviridae* virions are lemon-shaped with short tail fibers attached to one pole and slightly heterogeneous in size with 55–60 nm in their short dimension and 80–100 nm in their long dimension. A small fraction of the Sulfolobus spindle-shaped virus 1 (SSV-1) of the *Alphafusellovirus* genus, up to 1%, is larger, with a particle length of about 300 nm. The main constituents of the SSV-1 viral envelope are two basic proteins VP1 and VP3 of 73 and 92 aa, respectively, as deduced from the DNA sequence and from the N-terminal protein sequencing. The genome is a circular dsDNA, from 14.8 to 17.8 kb, encoding at least 34 genes (Prangishvili 2012c).

The virions like that of the *Fuselloviridae* members were identified to the sole *Salterprovirus His1* species from the genus *Salterprovirus* (Bath et al. 2006). Despite this morphological similarity, there was no genetic relationship, and the replication strategies of viruses were entirely different. Thus, the *Salterprovirus* genus was recently classified as the sole member of the novel family *Halspiviridae*. In brief, the enveloped His1 virus infects extremely halophilic archaea *Haloarcula hispanica*, has lemon-shaped or spindle-shaped morphology and possesses linear dsDNA genome of 14.5 kb encoding 35 ORFs, four of which have been shown to encode structural proteins of the virion with the protein VP21 as the major capsid protein (MCP; Pietilä et al. 2013). Hong et al. (2015) used electron cryotomography and symmetry-free and model-free subtomogram averaging to reveal the tail organization of the lemon-shaped virus His1, with a central tail hub and six surrounding tail spikes.

As commented by Prangishvili et al. (2017), another unusual virion morphology was observed in members of the family *Guttaviridae* infecting hyperthermophilic archaea. Thus, the enveloped guttavirus virions resemble spindles in which one of the two pointed ends has been rounded, rendering them droplet shaped (Arnold et al. 2000; Mochizuki et al. 2011), as shown in Figure 5.14d. The droplet-shaped morphology is unprecedented among viruses of bacteria and eukaryotes and represents a group of archaea-specific virion morphotypes (Prangishvili et al. 2017). Notably, the guttaviruses share several genes in common with fuselloviruses, suggesting that the two families of viruses are evolutionarily related (Iranzo et al. 2016; Prangishvili et al. 2017).

Generally, the *Guttaviridae* family includes two genera, *Alphaguttavirus* and *Betaguttavirus*, each with a single species. The guttavirus virions are ovoid or droplet-shaped, with a diameter of 55–80 nm and a length of 75–130 nm. The genome is a circular dsDNA molecule of around 14–20 kb, and the genome of a type species, Sulfolobus neozealandicus droplet-shaped virus (SNDV), contains 21 ORFs that could encode proteins of more than 56 aa residues (Prangishvili et al. 2018c).

The bacilliform *Clavaviridae* virions are nonenveloped and contain a single MCP; the ends of the virion are asymmetric—one is rounded and the other is slightly pointed, as seen in Figure 5.14e (Mochizuki et al. 2010). The virion does not have any terminal filaments, and it is unclear how it interacts with the host cell (Prangishvili et al. 2017). As summarized by the official ICTV report (Prangishvili et al. 2019a), the family *Clavaviridae* includes viruses that replicate in hyperthermophilic archaea from the genus *Aeropyrum*. The nonenveloped rigid virions are rod shaped, about 143 × 16 nm, with terminal cap structures, one of which is pointed and carries short fibers, while the other one is rounded. The virion displays helical symmetry and is constructed from a single major α-helical protein—which is heavily glycosylated—and several minor capsid proteins. The 5,278 bp, circular, double-stranded DNA genome of Aeropyrum pernix bacilliform virus 1 (APBV1) of the *Clavavirus* genus is packed inside the virion as a left-handed superhelix and held in place by interactions with the molecules of the major capsid protein. The genome contains 14 ORFs larger than 40 codons, all of which are located on the same DNA strand (Mochizuki et al. 2010). None of the putative gene products share similarities with sequences in public databases (Prangishvili et al. 2019a).

As summarized by Prangishvili et al. (2017), the spherical archaea-specific viruses are classified into the families *Globuloviridae* and *Portogloboviridae*. The virions of the *Globuloviridae* have a lipid-containing envelope and a superhelical nucleoprotein core that contains linear dsDNA. In the virions of the *Portogloboviridae*, the circular dsDNA genome is condensed by capsid proteins into a spherical nucleoprotein coil, which forms an inner core of the virion. The portraits of both family members are shown in Figure 5.14f and g, respectively. The spherical inner core is surrounded by a lipid membrane and is further encased by an outer icosahedral protein shell. Specifically, the *Globuloviridae* virions are spherical, 70–100 nm in diameter, with spherical protrusions that are about 15 nm in diameter and carry a lipid-containing envelope that encases a superhelical core, consisting of linear dsDNA and nucleoproteins. The virions contain three major structural proteins, VP1, VP2, and VP3, with molecular masses of about 16, 20, and 33 kDa, respectively. The genome is a single molecule of linear dsDNA, comprising 28,337 bp for Pyrobaculum spherical virus (PSV) and ~21.6 kb for Thermoproteus tenax spherical virus 1 (TTSV1), both from the genus *Alphaglobulovirus*. The PSV and TTSV1 genomes have 48 and 38 ORFs, respectively, of which only 15 are shared between the two viruses (Prangishvili 2012d; Prangishvili et al. 2018c).

The *Portogloboviridae* virions are icosahedral, 87 nm in diameter from vertex to vertex and 83 nm from facet to facet and consist of three structural units: the outer icosahedral protein shell, the subcapsomer proteins and lipid membrane, and a circular nucleoprotein wrapped into a spherical core (Prangishvili et al. 2021). The electron cryomicroscopy and 3D imaging allowed to resolve the atomic structure of Sulfolobus polyhedral virus 1 (SPV1) from the *Sulfolobus alphaportoglobovirus 1* species, genus *Alphaportoglobovirus*, to 3.7 Å (Wang F et al. 2019). The two capsid proteins VP4 and VP10 with variant single jelly-roll folds formed pentamers and hexamers that assembled into a T = 43 icosahedral shell. The circular dsDNA genomes of SPV1 and Sulfolobus polyhedral virus 2 (SPV2) were 20,222 bp and 20,424 bp long and encoded 45 and 46 ORFs, respectively. Rematkably, the genomic SPV1 DNA was in an A-form, as it was described earlier for the filamentous archaeal viruses of the novel realm *Adnaviria*.

The family *Ovaliviridae* includes the sole genus *Alphaovalivirus* with the sole *Sulfolobus ellipsoid virus 1* species represented by the single archaeal Sulfolobus ellipsoid virus 1 (SEV1) yet, which was isolated from an acidic hot spring in Costa Rica (Wang H et al. 2018). As summarized by the recent official ICTV statement (Huang et al. 2021), the SEV1 virions, 115 nm long and 78 nm wide, contain a protein capsid with 16 regularly spaced striations and are enveloped with a lipid membrane. The unique shape and architecture of the ovalivirus particles have not been observed among bacterial or eukaryotic viruses and are specific to archaeal viruses, as shown in Figure 5.14h. According to the electron cryomicroscopy investigation (Wang H et al. 2018), the 16 helical striations aligned perpendicular to the longitudinal axis of the particle with a periodicity of about 5 nm were clearly visible on the capsid. Each electron-dense stripe was about 2.8 nm wide. The longitudinal section of the 3D electron cryotomography of a virion revealed a tube-like structure of about 8 nm in diameter at the center of the capsid. As shown in the 2D sectioned slices from the electron cryotomography of the capsid, the apparent striations were also evident in the interior of the particle, while the capsid was probably formed by coiling of a nucleoprotein filament (Wang H et al. 2018). The 6-fold symmetrical virus-associated pyramids appear on the surface of SEV1-infected cells, which are ruptured to form a hexagonal opening for subsequent release of progeny

virus particles. The genome is a linear dsDNA of 23,219 bp with 172 bp inverted terminal repeats, probably in the form of a nucleoprotein filament, which wraps around the longitudinal axis of the virion in a plane to form a multilayered, disk-like structure with a central hole, and 16 of these structures are stacked to generate a spool-like capsid. The genome encodes 38 proteins, where four genes—*vp1*, *vp2*, *vp3*, and *vp4*—encode the structural proteins of the virus. It should be emphasized that no related viruses are known yet (Huang et al. 2021).

The family *Plasmaviridae* includes the sole genus *Plasmavirus* with the single species *Acholeplasma virus L2*, which infects *Acholeplasma* species, a wall-less bacteria of the class *Mollicutes*, and is released by budding through the cell membrane without causing host-cell lysis (Gourlay 1971; Poddar et al. 1985). As summarized by the regular ICTV reports (Maniloff 2012; Krupovic et al. 2018b), the enveloped virions are pseudospherical and slightly pleomorphic, as seen in Figure 5.14i, with a diameter of about 80 nm in the range of 50–125 nm. At least three distinct virion forms are produced during infection: ~75% of virions are 70–80 nm, ~20% are 80–90 nm, and ~5% are 110–120 nm (Poddar et al. 1985). The absence of a regular capsid structure suggests that plasmavirus virions consist of a condensed nucleoprotein bounded by a proteinaceous lipid vesicle. At least four major proteins of about 64, 61, 58, and 19 kDa were identified. The plasmavirus genome is circular, supercoiled dsDNA of ~12 kb encoding 15 ORFs, all of which are encoded on the same strand (Krupovic et al. 2018b).

The family *Polydnaviridae* includes two genera—*Bracovirus* and *Ichnovirus*—and forms a symbiotic relationship with parasitoid wasps, where ichnoviruses occur in ichneumonid wasp species and bracoviruses in braconid wasps. Surprisingly, these wasps are themselves parasitic on the *Lepidoptera* moths and butterflies. As narrated in the exhaustive reviews (Bézier et al. 2009; Strand and Drezen 2012; Strand and Burke 2015), there is little or no sequence homology between ichnoviruses and bracoviruses, suggesting that the two genera evolved independently for a long time. The virions have a complex construction, consisting of a nucleocapsid and a single- or double-layer envelope. They are enveloped, contain at least 20–30 polypeptides with sizes ranging from 10 to 200 kDa, and demonstrate either prolate ellipsoid form by ichnovirus or cylindrical structure by bracovirus. Thus, the ichnovirus virions consist of nucleocapsids of uniform size of ~85 × 330 nm, having the form of a prolate ellipsoid, surrounded by two unit membrane envelopes. The inner envelope appears to be assembled de novo in the nuclei of calyx cells, while the outer envelope is acquired by budding through the plasma membrane of calyx cells. The bracovirus virions were described as enveloped cylindrical electron-dense nucleocapsids of uniform diameter 34–40 nm but of variable 8–150 nm length, which may contain one or more nucleocapsids within a single envelope assembled de novo in the nuclei of calyx cells and possess long, unipolar, tail-like appendages. The encapsidated genomes of polydnaviruses consist of segmented supercoiled multiple dsDNAs of variable size, ranging from approximately 190 to more than 500 kb (Strand and Drezen 2012).

Cui et al. (2018) used the electron cryomicroscopy to reveal the near-native morphology of the two nucleocapsid-containing model bracoviruses: Microplitis bicoloratus bracovirus (MbBV) and Microplitis mediator bracovirus (MmBV). The inconsistent nucleocapsid diameters are 34–69.9 nm in MbBV and 46–69.9 nm in MmBV, and the largest observed cylindrical area length was expanded to 126 nm.

Concerning the gene-engineering applications, it should be mentioned that Deng and Webb (1999) cloned a His-tagged structural gene of a polydnavirus in *E. coli*, while Asgari et al. (2003) tried to characterize the polydnavirus-connected VLPs in wasps.

The family *Thaspiviridae* includes the sole *Nitmarvirus* genus with the single *Nitmarvirus NSV1* species. As summarized by the current ICTV report (Kim et al. 2021), the virions are spindle-shaped, measuring 64 ± 3 nm in diameter and 112 ± 6 nm in length, with short fibers at one pole, similar therefore to that of members of the families *Fuselloviridae* and *Halspiviridae*, while the linear dsDNA genomes are of 27 to 29 kbp and encode 48 or 51 ORFs by different isolates.

UNASSIGNED GENERA

The HcDNAV, a large dsDNA virus of the single species *Heterocapsa circularisquama DNA virus 01* from the unassigned genus *Dinodnavirus*, was isolated from Japanese coastal waters in August 1999 during a *H. circularisquama* bloom (Tarutani et al. 2001). The virus was icosahedral, lacking a tail, approximately 180–210 nm in diameter and contained an electron-dense core. Its genome size was estimated to be about 356 kb. In fact, this was the first report that has been isolated and maintained in culture of a virus infecting dinoflagellates. Later, Takano et al. (2018) performed detailed observation of the HcDNAV particle, and its infection process was conducted via field emission scanning electron microscopy and epifluorescence microscopy. Each 5-fold vertex of the icosahedral virion was decorated with a protrusion, which may be related to the entry process of HcDNAV into the host. The transverse groove of host cells is proposed to be the main virus-entry site.

The unassigned genus *Rhizidiovirus* contains the sole *Rhizidiomyces virus* species. As reviewed by the most recent ICTV report (Ghabrial 2012), the virions are nonenveloped, about 60 nm in diameter, with icosahedral, round, or isometric geometries. The genome is nonsegmented, linear dsDNA of ~25.5 kb, which encodes at least 14 proteins with sizes in the range of 26–84.5 kDa. Remarkably, no research has been done on this dsDNA virus system since 1983 (Ghabrial 2012).

6 Order *Lefavirales*

Just remember, once you're over the hill you begin to pick up speed.

Arthur Schopenhauer

ESSENTIALS

After a very long time of the unassigned status of the *Baculoviridae* family within the viral word, the order *Lefavirales* was established as a sole member of the novel class *Naldaviricetes*, in full accordance with the recent proposal of Harrison et al. (2020b). Nevertheless, the class *Naldaviricetes* is not assigned to any higher taxonomic unit and still remains independent of the great dsDNA virus realms. Along with the *Lefavirales* order, the class *Naldaviricetes* includes a nonordered family *Nimaviridae*, which is described in this chapter. The *Lefavirales* order, in turn, currently involves 3 families, namely, *Baculoviridae*, *Hytrosaviridae*, and *Nudiviridae*, 10 genera, and 98 species altogether. The members of the *Baculoviridae* family infect exclusively larvae-stage insects from the orders *Lepidoptera*, *Hymenoptera*, and *Diptera* (Harrison et al. 2018), were isolated from more than 600 host insect species, and play an important ecological role regulating the size of insect populations (Slack and Arif 2007). Thus, some baculoviruses have been developed as biopesticides for targeted biocontrol agents against forestry and agriculture pests. Moreover, the baculoviruses played an enormously great role as vectors by the expression of a huge number of recombinant proteins including the successful production of numerous VLPs, which are described in virtually all the chapters of the present VLP guide. Thus, the baculovirus-driven expression in insect cells and larvae generated a set of widely used and highly efficient methods to produce both natural and chimeric VLPs, as proved by numerous reviews (Kost and Condreay 1999; Fernandes et al. 2013; Liu et al. 2013; Rohrmann 2013, 2019; Lin SY et al. 2014; Yamaji 2014; Chambers et al. 2018; Gopal and Schneemann 2018; Gupta et al. 2019; Possee et al. 2020; Puente-Massaguer et al. 2020). As direct subjects of the VLP ideology, baculoviruses may be considered first in the role of the live carriers displaying foreign proteins on the surface of infectious pseudotyped virions.

FAMILY *BACULOVIRIDAE*

ESSENTIALS AND GENOME

Figure 6.1 demonstrates a typical portrait and schematic cartoons of the order representative of the insect-specific baculoviruses. The typical member of the latter is Autographa californica multiple nucleopolyhedrovirus C6

(L22858), AcMNPV for short, of the species *Autographa californica multiple nucleopolyhedrovirus* from the genus *Alphabaculovirus*. Historically, this virus was named in parallel Autographa californica nucleopolyhedrosis virus (AcNPV). For this reason, we have kept this past term by the papers using this term.

As summarized by the official ICTV reports (Herniou et al. 2012; Harrison et al. 2018), the one or two virion phenotypes are involved in baculovirus infections. The infection is initiated by virions contained within a crystalline protein occlusion body, which may be polyhedral in shape containing many virions (members of the genera *Alphabaculovirus*, *Gammabaculovirus* and *Deltabaculovirus*) or ovocylindrical containing only one, rarely two or more, virions (members of the genus *Betabaculovirus*). The virions within occlusions, referred to as occlusion-derived virus (ODV), consist of one or more rod-shaped nucleocapsids that have a distinct structural polarity and are enclosed within an envelope. The nucleocapsids are 30–60 nm in diameter and 250–300 nm in length. The occlusion bodies measure 0.15–5 μm and consist of a matrix composed of a single viral protein expressed at high levels during infection. The virions of the second phenotype, which is termed budded virus (BV), are generated when nucleocapsids bud through the plasma membrane at the surface of infected cells. The BVs typically contain a single nucleocapsid, while their envelopes are derived from the cellular plasma membrane and appear as a loose-fitting membrane that contains a major envelope glycoprotein gp64, which forms peplomers usually observed at one end of the virion. The virion may contain as few as 23 and as many as 73 different polypeptides. The nucleocapsids from both ODV and BV contain a major capsid protein VP39, a basic DNA-binding protein P6.9 complexed with the viral genome, and at least 2–3 additional proteins. The glycoprotein gp64 is present in a group of alphabaculoviruses that include AcMNPV and close relatives. The genome of the *Baculoviridae* is a circular supercoiled dsDNA ranging from 80 to 180 kb and encoding 100–200 proteins (Harrison et al. 2018).

3D STRUCTURE

Braunagel et al. (2003) performed an extensive analysis of the protein composition of the AcMNPV ODV virion and produced a list of 44 ODV-associated viral proteins. The nonatomic but nevertheless compelling structures of both BV and ODV virions were generated by Slack and Arif (2007). It was established that the 21 proteins of these 44 AcMNPV-specific ODV proteins are conserved among all baculovirus genomes. As to the electron microscopic characterization of baculoviruses, due to the fragility and

DOI: 10.1201/b22819-8

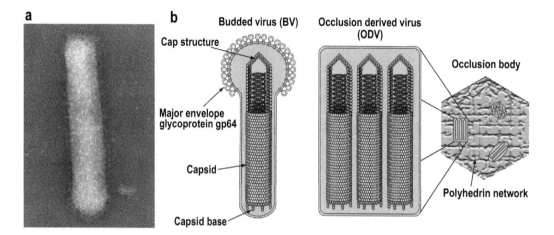

FIGURE 6.1 The typical portrait and schematic cartoons of the representatives of the *Baculoviridae* family. (a) The negative stain electron micrograph of Autographa californica nucleopolyhedrovirus (AcNPV). (With kind permission from Springer Science+Business Media: *The Springer Index of Viruses*, Tidona C, Darai G. (Eds), *Alphabaculovirus. Baculoviridae*, 72, 2011, 105–117, Carstens EB.) (b) The cartoon of the family members is taken with kind permission from the ViralZone, Swiss Institute of Bioinformatics, http://viralzone.expasy.org (Hulo C et al. 2011). (Courtesy Philippe Le Mercier.)

flexibility of the BV particles, the procedure of negative-staining and subsequent drying severely impaired the structural integrity of the viral envelope and caused virions to collapse. The intact BV particles were therefore rare and usually have lost part of their envelope. Wang Q et al. (2016) were the first ones who applied electron cryomicroscopy to reveal the near-native morphology of the two intensively studied baculoviruses, AcMNPV and Spodoptera exigua MNPV (SeMNPV), as representative models for BVs carrying the proteins gp64 and F as the envelope fusion proteins on the surface. The obtained near-native baculovirus structures are presented in Figure 6.2. Thus, the well-preserved AcMNPV and SeMNPV BV particles had a remarkable elongated, ovoid shape leaving a large, lateral space between nucleocapsid and envelope. The nucleocapsid demonstrated a distinctive cap and base structure interacting tightly with the envelope. The electron cryomicroscopy also revealed that the viral envelope contained two layers with a total thickness of ~6–7 nm, which is significantly thicker than a usual biological membrane (<4 nm) as measured by x-ray scanning. The most spikes were densely clustered at the two apical ends of the virion, although some envelope proteins were also found more sparsely on the lateral regions. Remarkably, the spikes on the surface of the AcMNPV BVs appeared distinctly different from those of SeMNPV (Wang Q et al. 2016).

CHIMERIC VIRIONS

Kondo and Maeda (1991) isolated hybrid baculoviruses of AcNPV and Bombyx mori nuclear polyhedrosis virus (BmNPV), both belonging to the genus *Alphabaculovirus*, capable of replicating in both BmN (not susceptible to AcNPV) and *Sf*21 (not susceptible to BmNPV) cells. Maeda et al. (1993) localized a fragment that could alone

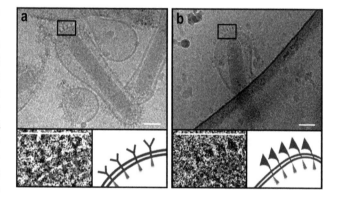

FIGURE 6.2 The ultrastructural characterization of baculovirus envelope proteins. (a, b) The electron cryomicroscopy images of envelope proteins on AcMNPV (a) and SeMNPV (b) BVs. The squared regions on the viral envelope were further focused in the lower panels showing more detailed structural patterns. Scale bars represent 50 nm. (Reprinted from *J Invertebr Pathol*. 134, Wang Q, Bosch BJ, Vlak JM, van Oers MM, Rottier PJ, van Lent JWM, Budded baculovirus particle structure revisited, 15–22, Copyright 2016, with permission from Elsevier.)

expand the host range of AcNPV and indicated that the expanded host-range characteristics of the chimeric virions were solely the result of recombination within the coding region of a putative DNA helicase gene.

Boublik et al. (1995) were the first ones to develop AcNPV as a vector for the display of distinct proteins on the viral surface in a manner that was analogous to the established bacterial "phage display" systems. As a model system, the marker gene encoding the 26 kDa protein glutathione-S-transferase (GST) was used to construct several fusions with the major envelope glycoprotein gp64 gene. Following expression in *Sf*9 cells, the yield and cellular distribution of each GST-gp64 protein was assessed. One fusion, in which GST was inserted between the leader peptide and the nature protein, was found to be efficiently secreted into the cell medium. In the context of expression of the full length gp64, the hybrid GST-gp64 was shown by immunogold labeling to be incorporated onto the virion surface. In addition, the affinity purification of the soluble transmembrane gp64-GST fusion protein resulted in the copurification of wild-type gp64, suggesting that cooligomerization of the GST-tagged fusion and the wild-type molecule was the basis for the formation of mosaic virions. The authors efficiently displayed the major surface glycoprotein gp120 of human immunodeficiency virus type 1 (HIV-1) in functional form on the viral surface following fusion to the N-terminus of gp64. Moreover, a general expression vector was constructed, in which the multiple cloning sites were positioned in-phase between the gp64 signal sequence and the sequence encoding the mature protein, aa 38 and 39 of the unprocessed gp64, under the control of the polyhedrin promoter (Boublik et al. 1995). Grabherr et al. (1997) chose the glycosylated envelope protein gp41 of HIV-1 to display on the baculovirus. The gp41 ectodomain was fused in this case to the entire gp64 and to the membrane anchor sequence of gp64, and two different promoters were compared. Furthermore, Mottershead et al. (1997) displayed more foreign proteins on the AcNPV virion surface by the fusion of proteins to the N-terminus of gp64. Such fusion proteins were incorporated into the baculoviral virion and displayed the model FLAG epitope tag. Furthermore, the chimeric baculoviruses displaying the green fluorescent protein (GFP) and the envelope proteins E1 and E2 of rubella virus were engineered. In fact, the chimeric viruses possessed two copies of the gp64 coding sequence, one being the wild-type gp64, essential for virus infectivity, and the second being the gp64 fusion protein, under the control of the polyhedrin gene promoter. Therefore, the chimeric AcNPV virions produced in *Sf*21 cells and displaying foreign sequences were mosaic (Mottershead et al. 1997).

Tami et al. (2000) constructed the chimeric AcNPV, displaying on their surface and in the membrane of infected cells the small, immunodominant antigenic site—the so-called site A—or the large polyprotein P1 coding for the four structural proteins of foot-and-mouth disease virus (FMDV). The coding sequences were inserted at the N-terminus of gp64. Although both mosaic viruses were able to compete with the FMDV-specific MAbs, their patterns of reactivity were different. Furthermore, Tami et al. (2004) demonstrated that the mice, vaccinated intraperitoneally with gp64-P1 immunogens, showed a low-antibody response and a variable degree of protection. However, when mice received recombinant baculoviruses or infected insect cells expressing the fusion protein gp64-site A, high immune response and strong protection against challenge with virulent FMDV were elicited (Tami et al. 2004).

Yang et al. (2007) constructed an influenza vaccine, where the histidine-tagged hemagglutinin of the H5N2 subtype was linked to the cytoplasmic domain derived

gp64. The product was efficiently incorporated into baculovirus and demonstrated strong immunogenicity. In fact, this was the first report demonstrating the potential of the hemagglutinin-pseudotyped baculovirus as an avian influenza vaccine. Tang et al. (2010) engineered another influenza vaccines, where chimeric hemagglutinins were provided with different segments of gp64. The signal peptide and cytoplasmic tail domains of gp64 enhanced the hemagglutinin display the baculovirus surface, while the transmembrane domain of gp64 impaired the display. The baculovirus simultaneously displaying four hemagglutinins derived from four subclades of H5N1 influenza viruses was constructed. This tetravalent H5N1 vaccine provided 100% protection against lethal doses of homologous H5N1 viruses (Tang et al. 2010). Lin W et al. (2011) combined the characteristics of baculovirus as a gene delivery vehicle (which is described later) and surface display system and constructed a *Baculovirus Dual Expression System* (BV-Dual), which was capable of displaying the H9N2 hemagglutinin on the surface of the viral envelope and expressing it upon transduction in mammalian cells. The results showed that stronger immune responses were induced in a mouse model immunized with BV-Dual-HA than in those vaccinated with a DNA vaccine encoding the same antigen. Moreover, the BV-Dual-HA vaccine candidate ensured complete protection of mice against lethal challenge with H9N2 virus (Lin W et al. 2011).

The same approach was used by the generation of the baculovirus display-based malaria vaccine candidates. Thus, Yoshida et al. (2003) generated a chimeric baculovirus that displayed rodent malaria *Plasmodium berghei* circumsporozoite protein (PbCSP) on the virion surface as a fusion protein with the gp64. The PbCSP-gp64 fusion protein was incorporated and oligomerized on the virion surface and ensured 60% protection of mice against sporozoite challenge. To prepare a second generation baculovirus-based malaria vaccine, Yoshida et al. (2009) used the previously mentioned BV-Dual system, which not only displayed PbCSP on the viral envelope but also expressed the PbCSP gene upon transduction of mammalian cells. This vaccine induced strong protective immune responses against pre-erythrocytic parasites. Next, Yoshida et al. (2010) developed a new vaccine comprising a 19-kDa C-terminus of merozoite surface protein 1 of *P. yoelii*, which was fused to the N-terminus of the gp64 and displayed therefore on the surface of the virus envelope in its native 3D structure.

Concerning the human malaria *P. falciparum* vaccines, Strauss et al. (2007) produced and tested three baculovirus-based vaccine variants that displayed the PfCSP on the baculovirus surface as a fusion with gp64, expressed the the PfCSP gene in antigen-presenting cells, and both displayed and expressed the PfCSP. The latter variant was superior in inducing the anti-PfCSP immune responses (Strauss et al. 2007). Mlambo et al. (2010) generated a potent *P. falciparum* transmission-blocking vaccine candidate, where the chimeric baculovirus virions displayed Pfs25, a sexual-stage antigen.

Mizutani et al. (2014) presented a multistage *P. vivax* vaccine that simultaneously expressed the PvCSP and Pvs25 as a fusion protein linked to the baculovirus protein gp64. Again, the BV-Dual system was evaluated and demonstrated great efficiency. This vaccine induced high levels of antibodies to Pvs25 and PvCSP and elicited high protection levels against transgenic *P. berghei* parasites expressing the corresponding *P. vivax* antigens in mice. Remarkably, Emran et al. (2018) demonstrated further that the baculovirus itself effectively induced fast-acting, innate immune responses that provided powerful first lines of both defensive and offensive attacks against preerythrocytic parasites and killed liver-stage *Plasmodium* in mice.

The baculovirus display via the gp64 fusion strategy was applied also for the N- and C-terminal domains of the p67 antigen of *Theileria parva*, an intracellular protozoan parasite that causes East Coast fever, a severe lymphoproliferative disease in cattle (Kaba et al. 2003); the spike protein S of SARS-associated coronavirus (SARS-CoV) (Feng et al. 2006); the glycoprotein D of bovine herpesvirus-1 (BHV-1) (Peralta et al. 2007); the envelope proteins E^{rns} (Xu et al. 2008a) and E2 (Xu et al. 2008b) and nonstructural protein NS3 (Xu et al. 2009) of classical swine fever virus (CSFV); the σC and σB proteins of avian reovirus (ARV; Lin YH et al. 2008); the envelope GP3 protein of porcine reproductive and respiratory syndrome virus (PRRSV; Wang ZS et al. 2011); the envelope glycoprotein E of Japanese encephalitis virus (JEV; Xu et al. 2011b); the ovalbumin fragment acting as an adjuvant and promoting specific CD4 and CD8 responses against OVA and strong enough to reject a challenge with OVA-expressing melanoma cells in mice (Molinari et al. 2011); the VP1 protein of enterovirus 71 (EV71; Meng et al. 2011; Kolpe et al. 2012); the VP2 protein of infectious bursal disease virus (IBDV; Xu et al. 2011a); the capsid protein of porcine circovirus type 2 (PCV2; Ye et al. 2013); the glycoprotein S1 of infectious bronchitis virus (IBV; Zhang et al. 2014); and the hemagglutinin protein of *peste des petits ruminants virus* (PPRV; Zhao et al. 2020);

Peralta et al. (2013) subjected the protein VP1 of foot-and-mouth disease virus (FMDV) to be displayed on the surface of baculoviruses with either polar, when fused to gp64, or nonpolar distribution, when fused to anchor membrane from vesicular stomatitis virus (VSV) protein G. The insect cells infected with the different recombinant baculoviruses expressed VP1 fusion protein at high levels. However, the recombinant VP1 protein was not carried by budded virions. The subcellular localization of the FMDV VP1 revealed that the trafficking of the fusion protein to the cell plasma membrane was impaired. It was concluded that the FMDV VP1 may contain cryptic domains that interfere with protein secretion and subsequent incorporation into budded baculoviruses (Peralta et al. 2013).

After the previously described studies on the AcMNPV or AcNPV, the analogous attempts were performed with *Bombyx mori* nucleopolyhedrovirus (BmNPV) from the same *Alphabaculovirus* genus. Thus, Rahman et al. (2003)

used the BmNPV gp64 to display the immunodominant ectodomains of fusion glycoprotein F of PPRV and the hemagglutinin protein of rinderpest virus (RPV) on the budded virions as well as the surface of the infected host cells. The chimeric virions induced immune response in mice against PPRV or RPV. Furthermore, Jin et al. (2008) constructed the chimeric BmNPV virions by the gp64 fusion with the H5N1 influenza virus hemagglutinin and displayed therefore the latter on the surface of the baculovirus. The immunization with this vaccine protected monkeys against influenza virus infection.

It should be noted that the baculovirus-infected insect cells were developed as a display platform for class II MHC (MHCII) molecules covalently bound to a library of potential peptide mimotopes (Crawford et al. 2004). The peptide mimotope/MHC complexes that bound to the soluble receptors and stimulated T cells bearing the same receptors were gained by the "fishing" in this library with soluble fluorescent T cell receptors. Next, this approach was adapted for use with MHCI (Wang Yibing et al. 2005) and extensively reviewed by Crawford et al. (2006).

Vaccine Vectors

The use of baculovirus as a vaccination vector for foreign genes, without any fusions to the baculovirus genes, was first achieved by Aoki et al. (1999), who demonstrated that inoculating mice intranasally or intramuscularly with the recombinant baculovirus expressing pseudorabies virus (PrV) glycoprotein B elicited a measurable humoral response directed against the target PrV glycoprotein. Facciabene et al. (2004) demonstrated the induction in mice of antigen-specific immune response mediated by the baculovirus vectors expressing the E2 glycoprotein of hepatitis C virus, the carcinoembryonic antigen, or the glycoprotein G of VSV.

The numerous vaccine candidates that were able to induce strong protection from the lethal influenza virus challenge in experimental animals were engineered by the insertion of the influenza hemagglutinin gene into the baculovirus genome. This technique allowed the hemagglutinin display on the baculovirus surface, while the baculovirus particles served as an adjuvant. Thus, the hemagglutinin genes of different influenza subtypes were tested: H1N1 (Abe et al. 2003; Prabakaran et al. 2010, 2011; Choi et al. 2013), H5N1 (Lu et al. 2007; Prabakaran et al. 2008, 2010; Wu et al. 2009), H6 (Syed Musthaq et al. 2014), H7N9 (Prabarakan et al. 2014). Recently, a baculovirus vaccine expressing the hemagglutinin of H7N9 strain A/Chicken/Jiaxing/148/2014 was prepared (Hu et al. 2019). The baculovirus generated in this study showed favorable growth characteristics in insect cells, good safety profile, and induced high-level hemagglutination inhibition antibody titer. Moreover, this vaccine demonstrated better efficacy than inactivated whole-virus vaccine JX148, provided complete protection of chickens against challenge with the H7N9 virus, and effectively inhibited viral shedding.

The AcMNPV pseudotyped with the glycoprotein G of VSV on the envelope was used as a vector to express the murine telomerase reverse transcriptase (mTERT) in order to induce antitumor immunity (Kim CH et al. 2007); the GP5 and M genes of PRRSV (Wang S et al. 2007); capsid protein of PCV2 (Fan et al. 2008); the envelope glycoprotein E of JEV (Li et al. 2009); major immunodominant surface antigen SAG1 of *Toxoplasma gondii* (Fang et al. 2010); glycoprotein G of rabies virus (RABV; Huang et al. 2011).

Syed Musthaq et al. (2009) pseudotyped baculovirus with the major envelope protein VP28 of white spot syndrome virus (WSSV) from the related *Nimaviridae* family that belongs to the same class *Naldaviricetes*—like the *Lefavirales* order including baculoviruses—but is not a member of the order, as mentioned earlier and described later. The chimeric baculovirus successfully acquired the VP28 protein from the insect cell membrane via the budding process, displayed VP28, and was used as a vaccine against WSSV in *Penaeus monodon* shrimps (Syed Musthaq et al. 2009; Syed Musthaq and Kwang 2011).

The AcMNPV pseudotyped with the envelope protein of human endogenous retrovirus (HERV) was used to express the protein L1 of human papillomavirus 16 (HPV16; Lee et al. 2010, 2012; Cho et al. 2014). Recently, Cho et al. (2021) constructed the HERV-pseudotyped baculovirus as a DNA vaccine vector for delivery of the human coronavirus genes: full S, S1 subunit, or receptor-binding domain (RBD) of MERS-CoV; full S or S1 of SARS-CoV-2; and the fusion of the MERS-CoV N-terminal domain (NTD) with the SARS-CoV-2 RBD. These DNA vaccines showed complete protection against MERS-CoV or SARS-CoV-2, where the Covid 19-S vaccine provided the greatest protection against SARS-CoV-2 challenge.

Moreover, as mentioned in Chapter 26 by nidoviruses, the full-length protein S and the S1 subunit of another coronavirus, namely porcine epidemic diarrhea virus (PEDV), was successfully displayed on the chimeric AcMNPV (Chang et al. 2018; Hsu et al. 2021).

CHIMERIC VLPs

Rao et al. (2018) presented a strategy for controllable assembly of flexible bio-nanotubes consisting of the *E. coli* expressed capsid protein of baculovirus Helicoverpa armigera nucleopolyhedrovirus (HearNPV) in vitro. These protein-only nanotubes were studied as a new structural platform for high-density presentation of multiple active molecules on the exterior surface by direct fusion of the protein of interest to the N-terminus of HearNPV capsid protein (HaCP). The structural characterization by electron cryomicroscopy demonstrated that the HaCP assembled into the two closely related but structurally distinct tube types, narrow and wide ones referred to as N-tubes and W-tubes, respectively. The bio-nanotubes are shown in Figure 6.3. The authors concluded that the tunable HaCP interaction network was the major contributor to the flexibility of the HaCP nanotubes, and the flexible nanotubes could tolerate larger molecular modifications compared with the classical tobacco mosaic virus (TMV)-based templates, described in Chapter 19 and could be used as promising candidates for versatile molecular loading applications.

TARGETING

The baculovirus targeting was improved or changed to restrict the putative gene therapy vector homing and transgene expression only to the cells of interest. The goal was achieved via incorporation of specific targeting moieties onto the baculovirus surface. Thus, the chimeric baculoviruses carrying the gp64 fusion with the CSP protein of *P. berghei*, which were constructed by Yoshida et al. (2003) and mentioned earlier, demonstrated a twelvefold increase in the binding activity of the chimeric virions to HepG2 cells. These data demonstrated that the pseudotyped baculoviruses may ensure specific targeting and, in this case, a liver-directed gene delivery.

The specific delivery of the baculovirus vectors was ensured by the pseudotyping of the virions with the

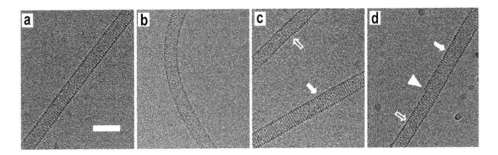

FIGURE 6.3 The analysis of 2D class averaging from the electron cryomicroscopy images of HaCP nanotubes. (a—d) The electron cryomicroscopy images of selective HaCP nanotubes embedded in a thin layer of vitreous ice, recorded at a defocus range of 1–2.5 μm: (a) a straight tube; (b) a bent tube; (c) an N-tube (unfilled arrow) and a W-tube (solid arrow); (d) a tube consisting of a narrow domain (unfilled arrow) and a wide domain (solid arrow), with a transition region indicated by a white triangle. Scale bar: 50 nm. (Reprinted with permission from Rao G et al. Controllable assembly of flexible protein nanotubes for loading multifunctional modules. *ACS Appl Mater Interfaces*, 25135–25145. Copyright 2018, American Chemical Society.)

glycoprotein G of VSV (Barsoum et al. 1997; Mangor et al. 2001; Kitagawa et al. 2005; Facciabene et al. 2004; Kim CH et al. 2007; Wang S et al. 2007; Fan et al. 2008; Li et al. 2009; Wu et al. 2009; Fang et al. 2010; Pan et al. 2010; Huang et al. 2011), the measles virus receptors CD46 or SLAM (Kitagawa et al. 2005), the envelope protein of human endogenous retrovirus (HERV; Lee et al. 2010, 2012; Choi et al. 2013; Cho et al. 2014, 2021; Gwon et al. 2016; Cho et al. 2021), or the glycoprotein G of rabies virus (RABV; Wu et al. 2014).

The gp64 fusion technique was used to improve or change the specific baculovirus targeting by the display of a functional single chain antibody fragment (scFv) or the IgG binding domains ZZ derived from protein A of *Staphylococcus aureus* (Mottershead et al. 2000; Ojala et al. 2001, 2004), avidin (Räty et al. 2004), tumor-homing peptides (Mäkelä et al. 2006, 2008a), the RGD motifs (Ernst et al. 2006; Matilainen et al. 2006), and human immunoglobulin Fc (Martyn et al. 2009). Dojima et al. (2010) used the gp64 fusion technique to display the ZZ domain on the surface of the BmNPV virions.

It should be noted that the gp64 fusion was used also to display GFP on the surface of the AcNPMV virions (Toivola et al. 2002). Next, Kukkonen et al. (2003) displayed enhanced GFP on the baculovirus surface by the fusions to the N-terminus or C-terminus of the major capsid protein vp39 without compromising the viral titer or functionality. The GFP and eGFP display on the BmNPV virions by the same gp64 fusion technique was achieved by Rahman and Gopinathan (2003).

Mäkelä et al. (2008b) evaluated the functionality of the ODV envelope protein p74 for the display of the IgG-binding Z domains.

Kaikkonen et al. (2008) developed the metabolically biotinylated baculoviral vectors by displaying a small biotin acceptor peptide (BAP) fused either to different sites in the baculovirus glycoprotein gp64 or to the transmembrane anchor of the VSV protein G. The baculovirus virions were biotinylated during vector production by coexpression of E. coli biotin ligase. The insertion of BAP at aa position 283 of gp64 resulted in the most efficient biotin display. These vectors showed improved transduction when retargeted to transferrin, epidermal growth factor, and CD46 receptors overexpressed on rat glioma and human ovarian carcinoma cells (Kaikkonen et al. 2008).

Remarkably, the human (pro)renin receptor lacking the transmembrane domain was found localized on the budded virions of AcMNPV (Kato et al. 2009) and BmNPV (Kato et al. 2011, 2012).

NANOMATERIALS

This approach started with the baculovirus display of the specifically inorganic-binding peptides. Thus, Song et al. (2010a) engineered the AcNPV virions with a His_6-tagged ZnO binding peptide fused to the N-terminus of the viral capsid protein vp39. The binding of nanosized ZnO

powders to the virus capsid was visualized by transmission electron microscopy and served as the first report of the display of the inorganic-binding peptide on the baculovirus capsid, which could be useful by the preparation of novel functional nanodevices. In parallel, Song et al. (2010b) fused the ZnO binding peptide to the N-terminus of gp64. The inorganic-specific peptide was successfully displayed on the surface of the virions and visualized by transmission electron microscopy. As by the vp39-forced display, the chimeric virions maintained both the viral infectivity and the specific binding activity of the displayed peptide.

It should be noted in this context that the baculovirus virions were coated with poly(ethylene glycol) (PEG) (Kim YK et al. 2006) or folate-PEG (Kim YK et al. 2007) to obtain efficiency and specificity of gene delivery.

POLYHEDRIN

Besides their use as the previously described display tools, the baculoviruses served as the source of polyhedrin, which could be regarded as a special class of highly structured carriers. As mentioned briefly earlier, the polyhedrin is synthesized in massive amounts and forms a protective crystal around the viruses, which allows them to remain viable for many years outside the insect larvae (Rohrmann 1986). The x-ray crystallographic examination of polyhedrins started in early 1980s (Rohrmann 1986). The chimeric AcMNPV polyhedrin proteins carrying an influenza hemagglutinin epitope not only showed influenza-specific antigenic and immunogenic properties but also presented the foreign epitope on the surface of the chimeric occlusion bodies (McLinden et al. 1992). Furthermore, chimeras of the AcMNPV polyhedrin and Trichoplusia ni granulosis virus (TnGV) granulin were constructed (Eason et al. 1998). The results clearly showed that the size and structure of occlusions was affected by the primary structure of the product, although the involvement of other viral proteins in the virion occlusion body assembly and shape complicated interpretation and further application of this system (Chen et al. 2013).

FAMILY *HYTROSAVIRIDAE*

The salivary gland hypertrophy viruses (SGHVs) of the *Hytrosaviridae* family are similar to baculoviruses and share at least 12 core genes with the *Baculoviridae* members (Jehle et al. 2013). The hytrosaviruses infect the hematophagous tsetse flies and the filth-feeding houseflies and are classified into the two genera, *Glossinavirus* and *Muscavirus*, each with a single species, *Glossina hytrovirus* and *Musca hytrovirus*, respectively. As summarized further by the recent ICTV report (Kariithi et al. 2019), the *Hytrosaviridae* virions are nonoccluded rod-shaped particles of 50–80 nm in diameter and 500–1,000 nm in length with a bilayer lipid envelope and rounded and/or conical ends. The dense internal capsid has a helical organization, and the envelope contains virus-encoded proteins with

7 Order *Zurhausenvirales*

There is one thing stronger than all the armies in the world, and that is an idea whose time has come.

Victor Hugo

And fact is the most stubborn thing in the world.

Mikhail Bulgakov
The Master and Margarita

ESSENTIALS

In contrast to other double-stranded DNA viruses from the corresponding Baltimore's class, papillomaviruses of the freshly established order *Zurhausenvirales*, as well as polyomaviruses of the order *Sepolyvirales* that is the subject of Chapter 8, belong to the realm *Monodnaviria* in accordance with the modern viral taxonomy (ICTV 2020), where they are united together with the classical single-stranded DNA viruses. The reason is that the realm *Monodnaviria*, which comprises viruses with predominantly single-stranded DNA genomes, involves the members encoding rolling-circle replication initiation endonucleases of the HUH superfamily, in contrast to the "true" dsDNA viruses of the realms *Adnaviria*, *Duplodnaviria*, and *Varidnaviria* (Krupovic et al. 2021), as well as the realm-unassigned order *Lefavirales* that were described before in Chapters 1 to 6.

The order *Zurhausenvirales* currently involves the sole family *Papillomaviridae*, 2 subfamilies, 53 genera, and 133 species altogether and forms the class *Papovaviricetes*, together with the previously mentioned and closely related order *Sepolyvirales*. The *Papovaviricetes* class, together with the two classes, *Mouviricetes* and *Quintoviricetes*, forms the phylum *Cossaviricota* of the kingdom *Shotokuvirae*, realm *Monodnaviria*.

The numerous papillomavirus genera of the most studied *Firstpapillomavirinae* subfamily are termed alphabetically from *Alphapapillomavirus* till *Zetapapillomavirus*. The small *Secondpapillomavirinae* subfamily contains the sole *Alefpapillomavirus* genus with the sole *Alefpapillomavirus 1* species. The epidemiological and biological data are primarily available for the human papillomaviruses (HPV), mostly of the genus *Alphapapillomavirus*, and specifically those viruses are associated with cervical cancer. According to the current WHO information (www.who.int/news-room/fact-sheets/detail/human-papillomavirus-(hpv)-and-cervical-cancer), at least 14 papillomaviruses are cancer-causing, where the two HPV types 16 (HPV16) and 18 (HPV18) cause 70% of cervical cancers and precancerous cervical lesions, linking HPV with cancers of the anus, vulva, vagina, penis, and oropharynx. According to Tumban (2019) and Yadav et al. (2020), 42 HPV types can be transmitted sexually via anogenital to anogenital sex or anogenital to oral sex, out of which ~19 types called high-risk types (oncogenic types 16, 18, 26, 31, 33, 35, 39, 45, 51–53, 56, 58, 59, 66, 68, 70, 73, and 82) are associated with cancers. The remaining types, known as low-risk HPV types (6, 11, 40–44, 54, 61, 72, 81, etc.), are associated with genital warts and recurrent respiratory papillomatosis. Generally, the cervical cancer is the fourth most common cancer among women globally, with an estimated 570,000 new cases in 2018. Remarkably, nearly 90% of the 311,000 deaths worldwide in 2018 occurred in low- and middle-income countries. Briefly, HPV is spread by skin-to-skin contact, and infections with genital HPV remain the most common sexually transmitted infection. Therefore, the great interest in papillomaviruses was based not only on their favorable structural properties but also on the need for a vaccine to prevent, first, genital tract HPV infections, especially those types associated with genital tract malignancy. There was also a special interest in the development of a therapeutic vaccine, which, unlike a prophylactic vaccine, could be used to treat individuals who are already infected.

The great contribution of the VLP technique consisted therefore in the fact that the vaccination against HPV was possible since 2006–2007 and the VLP vaccines that protect against HPV16 and HPV18 were approved and recommended by WHO for use in many countries. The numerous clinical trials and postmarketing surveillance have shown that the HPV vaccines were safe and effective in preventing HPV infections, according to the previously cited WHO digest. Therefore, the *Papillomaviridae* family presented the second, after the family *Hepadnaviridae* with the recombinant VLP-based hepatitis B vaccine that is described in Chapter 37, successful and world-widely accepted VLP-based vaccine—as reviewed by Schiller and Lowy (2012)—and triggered therefore a huge number of medical investigations, which are not referenced here.

The current taxonomy (Bernard et al. 2012; Van Doorslaer et al. 2018) clearly established the relationship among the classical HPV serotypes and the modern classification. Thus, the species *Alphapapillomavirus 1* is represented by HPV32; the dangerous HPV18, together with HPV39, HPV45, HPV59, and others belongs to the species *Alphapapillomavirus 7*; the other oncogenic HPV16, together with HPV31, HPV33, HPV35, HPV52, HPV58, and HPV67 are members of the species *Alphapapillomavirus 9*; the non-oncogenic HPV6 and HPV11, together with some others, belong to the species *Alphapapillomavirus 10*.

As summarized by the official ICTV reports (Bernard et al. 2012; Van Doorslaer et al. 2018), the members of the *Papillomaviridae* family primarily infect mucosal and keratinized epithelia and have been isolated from fish, reptiles,

DOI: 10.1201/b22819-9

birds, and mammals. Despite a long coevolutionary history with their hosts, some papillomaviruses remain pathogens of their natural host species.

Figure 7.1 presents papillomaviruses as nonenveloped capsids of ~55 nm in diameter, possessing the icosahedral T = 7d symmetry. As summarized by Van Doorslaer et al. (2018), the capsids consist of 360 copies of the major capsid protein L1—which is arranged as 72—pentamers, and ~12 molecules of the minor capsid protein L2. The expression of the L1 gene, with or without L2, allows for self-assembly of the VLPs, as described in detail later. The structural features of the papillomavirus structural proteins L1 and L2 are thoroughly reviewed by Buck et al. (2013) and Wang JW and Roden (2013), respectively. The general contribution of papillomaviruses alone and together with polyomaviruses to the VLP story was assessed many times, for example in the *Viral Nanotechnology* book (Dalianis 2015; Kazaks and Voronkova 2015; Suchanová et al. 2015).

GENOME

Figure 7.2 shows the genomic structure of the *Papillomaviridae* members. Each capsid packages a single copy of the viral circular dsDNA. The packaged viral DNA is associated with core histone proteins. The size of the genome ranges from 5,748 bp to 8,607 bp, and a typical papillomavirus encodes six to nine proteins. However, the ancestral papillomavirus may have only contained a core set of four major proteins E1, E2, L1, and L2. The papillomavirus gene expression is tightly regulated at the level of transcription and RNA processing, including alternative mRNA polyadenylation and splicing, while the temporal expression of the viral genome is associated with tissue differentiation. The early region encodes viral proteins involved in transcription, replication, and manipulation of the cellular milieu. The late region encodes the capsid proteins L1 and L2. The putative oncogenes E6 and E7 have been shown to play key roles in usurping the cellular environment to allow for replication. The E6 protein contains two zinc-binding motifs essential for its function, while the N-terminus of E7 contains regions of similarity to conserved regions CR1 and CR2 of the mastadenovirus E1A protein, and the polyomavirus (PyV) large T antigen and the C-terminus of E7 contains a single zinc-binding motif homologous to the E6 motifs. The E6 and E7 proteins appear to be essential for members of the genus *Alphapapillomavirus*, but they are not encoded by all papillomaviruses (Van Doorslaer et al. 2018).

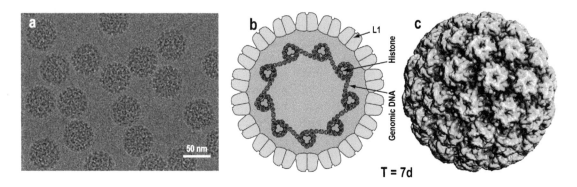

FIGURE 7.1   A portrait, cartoon, and the 3D structure of typical representatives of the *Papillomaviridae* family. (a) The representative region of the electron cryomicrographs of the human papillomavirus 16 (HPV16) particle capsids. (Reprinted from *Structure*. 25, Guan J, Bywaters SM, Brendle SA, Ashley RE, Makhov AM, Conway JF, Christensen ND, Hafenstein S, Cryoelectron microscopy maps of human papillomavirus 16 reveal L2 densities and heparin binding site, 253–263, Copyright 2017, with permission from Elsevier.) (b) The cartoon taken with kind permission from the ViralZone, Swiss Institute of Bioinformatics, http://viralzone.expasy.org (Hulo C et al. 2011). (Courtesy Philippe Le Mercier.) (c) The near-atomic structure of human papilloma virus 16, PDB ID: 5KEQ, electron cryomicroscopy at 4.30 Å resolution, outer diameter 586 Å (Guan et al. 2017), presented as rendered surface and taken from the VIPERdb (http://viperdb.scripps.edu) database (Carrillo-Tripp et al. 2009). The structure possesses the icosahedral T = 7d symmetry.

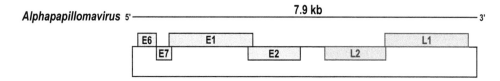

FIGURE 7.2   The genomic structure of the *Papillomaviridae* family represented by the genome of human papillomavirus 32 (HPV32) from the species *Alphapapillomavirus 1* of the genus *Alphapapillomavirus*. The structural genes are colored dark pink. The circular character of the genome is displayed by the lower connecting bracket, while the linearization is performed with *ori* as the opening site.

GENUS *ALPHAPAPILLOMAVIRUS*

3D Structure

The human papillomavirus type 16 (HPV16) belonging to the *Alphapapillomavirus 9* species is a reference strain of the *Alphapapillomavirus* genus. The members of this genus preferentially infect the oral or anogenital mucosa in humans and primates, while specific members of certain species (e.g. *Alphapapillomavirus 7* and *Alphapapillomavirus 9*) are considered as oncogenic in view of their regular presence in malignant tissue. All members of this genus code for a hydrophobic E5 protein, located at the 3′-end of the early region (Van Doorslaer et al. 2018).

The T = 7d structure of the HPV capsids was visualized by the electron cryomicroscopy reconstructions (Baker et al. 1991; Buck et al. 2008; Cardone et al. 2014; Lee et al. 2014; Guan et al. 2015). These studies reached the 9-Å resolution, from which a pseudoatomic model was developed using the major structural protein L1 (Cardone et al. 2014). Therefore, the 55–60-nm HPV capsid was found to be composed of 360 copies of L1 arranged into 72 capsomers, each composed of five copies of L1. Twelve capsomers lay on the icosahedral 5-fold vertices and were referred to as pentavalent capsomers, whereas the remaining 60 capsomers were each surrounded by 6 other capsomers and referred to as hexavalent capsomers. An extension of the L1 C-terminus—or C-terminal arm—linked capsomers together and provided interactions leading to the formation of disulfide bonds between Cys175 and Cys428 from two neighboring L1 proteins. Besides the major capsid protein L1, there was an uncertain number of the L2 minor structural proteins that coassembled with L1 to form infectious virus capsids. Specifically, the N-terminal residues of L2 (1–88) were exposed on the surface of the capsid and were bound by the anti-L2 antibody, RG-1, which recognized the aa residues 17–36 of the L2 protein (Okun et al. 2001; Day et al. 2008a, b). The naturally produced virions averaged approximately 24–36 L2 molecules per capsid, as summarized by Guan et al. (2017). At last, Guan et al. (2017) presented the near-atomic resolution structures of the HPV16 alone and the HPV16 interacting with the heparin receptor molecules at 4 Å resolutions. The outcome of this structural investigation is shown in Figure 7.3. It was concluded that the HPV capsids are heterogeneous in nature but contained subpopulations of capsids with consistent diameters. As a result, the calculation of the L1-only capsid was used to identify the location of L2 densities in the maps, which was possible by solving the L1 asymmetric unit. Remarkably, the binding of heparin induced no conformational changes in the capsid that included local changes in L1, but there were alterations to putative L2 densities. Generally, these accelerated 3D studies allowed to propose a realistic model for the virus entry mechanism (Guan et al. 2017).

The truncation of 10 N-terminal aa residues of the HPV16 protein L1 led to formation of the T = 1 particles consisting of the 12 L1 pentamers (Chen XS et al. 2000) presented

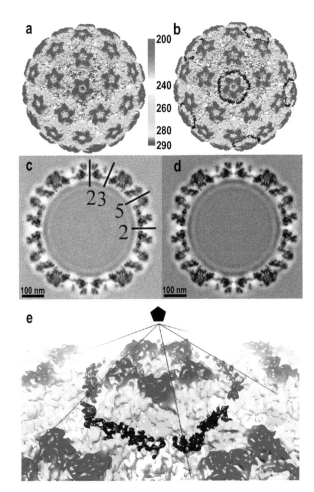

FIGURE 7.3 The electron cryomicroscopy reconstruction of HPV16 and HPV16-heparin complex. (a, b) The surface rendered 3D maps of HPV16 (a) and capsid-heparin complex (b) were radially colored according to the distance from the center (color key in angstroms) of the capsid and surface rendered at 1σ. The heparin difference density (black, difference map rendered at 4.2σ) was found at positions surrounding only the 5-fold vertices of the capsid. (c, d) The central sections through the electron cryomicroscopy density maps show the quality of the reconstructions. The capsids were cut vertically through the 2-, 3-, and 5-fold icosahedral symmetry axes (black lines), with the central 2-fold axis appearing at the 12 o'clock position. (e) The zoomed-in view of the HPV16-heparin complex shows the detailed features of bound heparin molecules. The 5-fold symmetric axis was represented by black pentagon. (Reprinted from *Structure*. 25, Guan J, Bywaters SM, Brendle SA, Ashley RE, Makhov AM, Conway JF, Christensen ND, Hafenstein S, Cryoelectron microscopy maps of human papillomavirus 16 reveal L2 densities and heparin binding site, 253–263, Copyright 2017, with permission from Elsevier.)

in Figure 7.4. The x-ray crystallographic analysis of these particles at 3.50 Å showed that the L1 protein closely resembled the VP1 protein from PyVs described in Chapter 8. The surface loops contained the sites of sequence variation among HPV types and the locations of dominant neutralizing epitopes. The ease with which the small T = 1 VLPs might be obtained from the L1 protein expressed in *E. coli*

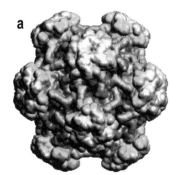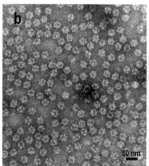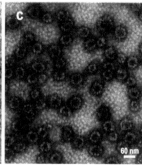

FIGURE 7.4   The T = 1 particles of human papillomavirus. (a) The near-atomic structure of the T = 1 particles, PDB ID: 1DZL, x-ray diffraction at 3.50 Å resolution, outer diameter 320 Å (Chen XS et al. 2000). The 3D structure is taken from the VIPERdb (http://viperdb.scripps.edu) database (Carrillo-Tripp et al. 2009). (b, c) The assembly of virus-like particles from the HPV16 L1 pentamers, monitored by negative stain electron microscopy: (b) the T = 1 VLPs derived from crystals of the small, 12-pentamer VLPs assembled at pH 5.2 from L1 ΔN-10; (c) the particles that resemble 72-pentamer virions assemble at pH 5.2 from full-length L1. (Reprinted from *Mol Cell.* 5, Chen XS, Garcea RL, Goldberg I, Casini G, Harrison SC, Structure of small virus-like particles assembled from the L1 protein of human papillomavirus 16, 557–567, Copyright 2000, with permission from Elsevier.)

made them attractive candidates for the VLP nanotechnologies (Chen XS et al. 2000). Moreover, the near-atomic T = 1 structures played a critical role by the 3D reconstruction of the T = 7d particles of human papillomavirus. Thus, Bishop et al. (2007) fitted the crystal L1 structure from the T = 1 particle into the electron cryomicroscopy reconstruction of bovine papillomavirus (BPV; Trus et al. 1997), performed a set of structure-guided internal deletions of the HPV L1 helices, and found regions that were indispensable for the assembly of T = 7d particles.

EXPRESSION

An inability to efficiently propagate the papillomaviruses in cultured cells stimulated construction of the native-like VLP vaccine candidates. First, the production of noninfectious HPV virion-like particles was achieved by the coexpression of the capsid protein genes L1 and L2 (but not either protein alone) of HPV16 using vaccinia virus in mammalian cells, where VLPs consisted of capsomeres similar to HPV and contained major capsid protein L1 in glycosylated form (Zhou et al. 1991). The corresponding HPV VLPs of ~40 nm in diameter were observed and purified by sucrose gradient centrifugation from the nuclei of cells synthesizing both L1 and L2, but not from the cells synthesizing either protein alone. It is remarkable that Cason et al. (1993) were the first reviewers to evaluate the vaccinia virus-driven production of the HPV16 VLPs as a possible source of the prospective HPV vaccine.

Hagensee et al. (1993) used vaccinia virus vectors to express the L1 and L2 proteins of HPV1 (which is classified now to the *Mupapillomavirus*, but not to the *Alphapapillomavirus* genus), where all constructs expressed the appropriate-sized HPV proteins, and both L1 and L2, singly or in combination, were localized to the nucleus. The VLPs were purified by cesium chloride density gradient centrifugation from the nuclei of cells infected with a vaccinia virus encoding

the L1 or both L1 and L2 genes but not from the L2 alone expressing cells. The electron microscopy showed that the particles were ~55 nm in diameter and had icosahedral symmetry. The immunogold-labeled antibodies confirmed the presence of the L1 and L2 proteins in the HPV1 VLPs. The VLPs containing L1 alone were fewer and more variable in size and shape than the VLPs containing the L1 and L2 proteins. Moreover, the latter were indistinguishable in appearance from the HPV1 virions obtained from plantar warts (Hagensee et al. 1993). The x-ray crystallography of the papillomavirus VLPs was not undertaken, but the 3D reconstructions of the HPV1 L1 alone or both L1 and L2 VLPs were generated by electron cryomicroscopy and image analysis techniques (Hagensee et al. 1994). According to this analysis, the HPV VLPs carried the same 72 pentameric capsomeres arranged on a T = 7d icosahedral lattice. Each particle, ~ 60 nm in diameter, consisted of an approximately 2-nm-thick shell of protein with a radius of 22 nm with capsomeres that extended ~6 nm from the shell. At the low 35 Å resolution, the VLPs appeared identical to the capsid structure of the native HPV (Baker et al. 1991).

The efficient and correct self-assembly of the HPV16 L1 and L2 VLPs was achieved by the Semliki Forest virus (SFV) expression system (Heino et al. 1995), which allowed for the pseudotyping of papillomaviruses, i.e., Bos taurus papillomavirus 1 (BPV1) genomes of the *Deltapapillomavirus 4* species (see later) coated by HPV16 L1 and L2 (Roden et al. 1996). Moreover, the HPV33 VLPs carrying packed copies of a marker plasmid have been generated in Cos-7 cells (Unckell et al. 1997), and the suggestion was formulated for the HPV16 L1 (Touze and Coursaget 1998) and BPV1 L1 and L2 (Zhao et al. 1998) VLPs as promising vehicles to package and deliver unrelated plasmid DNA. Furthermore, a poliovirus recombinant vector was used to express the HPV16 L1, a replication-competent poliovirus was obtained after transfection of the recombinant RNA into tissue culture cells, and the HPV16 VLPs were identified (van Kuppeveld et al. 2002). Remarkably, when susceptible transgenic mice carrying the poliovirus receptor were infected with the recombinant poliovirus, all animals demonstrated a modest but consistent immune response against HPV16 infection.

Iyengar et al. (1996) demonstrated the generation of the virion-like HP16 L1 VLPs by the cloning of the L1 gene in an expression plasmid that was transcribed and translated in vitro in a rabbit reticulocyte lysate. Although the amounts of the obtained L1 protein were small, the ability to produce the correct VLPs in vitro seemed useful in the studies of virus assembly and immunological properties.

The famous John T. Schiller's team was the first to report the self-assembly of the protein L1 of human HPV16 or bovine BPV1 papillomaviruses into the VLPs by the baculovirus-driven expression system in insect cells (Kirnbauer et al. 1992). The protein L1 alone was sufficient for assembly of the VLPs that were morphologically similar to native virions. Moreover, these VLPs induced high-titer neutralizing antisera in rabbits, which were able to prevent the papillomavirus infection in vitro (Kirnbauer et al. 1992). To accelerate the success, Kirnbauer et al. (1993) expressed the L1 genes of two HPV16 isolates, both of which showed the yield of VLPs three orders of magnitude higher than what has been obtained previously with L1 derived from the prototype HPV16. Remarkably, the DNA sequence comparison identified a single nonconserved aa residue change to be responsible for the inefficient self-assembly of the prototype L1. The VLPs were also obtained by expressing L1 of HPV6, HPV11, and cottontail rabbit papillomavirus (CRPV), indicating that the L1 from a variety of papillomaviruses had the intrinsic capacity to self-assemble into the VLPs. Furthermore, the coexpression of the HPV16 L1 plus L2 proteins by using a baculovirus double-expression vector also resulted in the efficient VLP self-assembly, and the average particle yield increased about fourfold in comparison to when L1 only was expressed (Kirnbauer et al. 1993). In fact, this study paved the way for the efficient production of the first preparative amounts of HPV16 L1 and L1/L2 VLPs, which was used for the development of serological assays useful by the immunoprophylaxis against HPV16 infection and further for the generation of HPV vaccines.

In parallel, Rose et al. (1993) purified the HPV11 protein L1 from the insect *Sf*9 cells and spontaneously assembled it in vitro into various aggregates, including the VLPs appearing similar to empty virions. Rose et al. (1994a) prepared the insect cell-produced VLPs of HPV11, HPV16, and HPV18, immunized rabbits with them, and found that the resulting antibodies reacted only with the specific VLP type used as the immunogen. These results clearly suggested the antigenic distinction of the HPV types and provoked future studies of immune responses to HPV. Rose et al. (1994b) demonstrated that the HPV11 VLPs can be used to elicit a neutralizing antibody response in rabbits and can substitute faithfully for native virions in the development of the serodiagnostic HPV immunoassays. The correct antigenic

structure of the insect cell-produced HPV6 and HPV11 VLPs was proven by Christensen et al. (1994a, b). The insect cell-produced HPV11 VLPs induced the systemic virus-neutralizing antibodies after oral vaccination (Rose et al. 1999).

Next, the HPV16 VLPs were produced in *Sf*21 cells by Le Cann et al. (1994), who also confirmed the presence of the native conformational epitopes on the particles, while Dupuy et al. (1997) demonstrated the T-cell response characterized by a Th1 profile in mice against the HPV16 protein L1-formed VLPs. Gerber et al. (2001) showed that the HPV16 and HPV18 VLPs were immunogenic in mice when administered orally and that oral coadministration of these antigens with *E. coli* heat-labile enterotoxin or CpG oligonucleotide can significantly improve anti-HPV humoral responses in peripheral blood and in genital mucosal secretions.

Meanwhile, Volpers et al. (1994) reported high-level production of the VLPs of HPV33, another human papillomavirus associated with malignant genital lesions, by the expression of the gene L1 alone or together with the L2 gene in insect cells. The spherical VLPs of 50–60 nm diameter and tubular structures of either 25–30 nm or 50–60 nm diameter and variable length were extracted from nuclei of the recombinant baculovirus-infected insect cells. However, the icosahedral 50- to 60-nm particles were found predominantly in the cell culture medium during long-term infections. The L2 protein seemed to be modified and was shown to be tightly associated with L1 using density gradient and sedimentation analysis. Approximately 50% of the L1 molecules were cross-linked by intermolecular disulfide bonds (Volpers et al. 1994). Readily, this was the first study that characterized in detail the HPV VLPs containing both the major and the minor capsid protein using the baculovirus expression system. As stated earlier by Kirnbauer et al. (1993), the outcome of the HPV VLPs was markedly increased during coexpression of the L1 with the L2 (Volpers et al. 1994). Sapp et al. (1994) used the HPV33 VLPs for the identification of type-dependent virus-neutralizing epitopes. Benincasa et al. (1996) accelerated the production cycle of the HPV11 and CRPV VLPs. Touzé et al. (1996) produced the HPV33 VLPs in *Sf*9 cells, allowing evaluation of serological reactivities between HPV types 6, 11, 16 and 45 (Touzé et al. 1998). The exhaustive review of the insect cells-driven production of the HPV VLPs was published at that time by Kirnbauer (1996).

Reddy et al. (2004) described the efficient baculovirus-driven production of the HPV16 VLPs in *Sf*9 cells, as a control to the engineered non-VLPs peptide immunogens, so-called hypervariable epitope constructs, which represented sequence variants of the HPV L1 immunodominant B-cell epitopes and were able to overcome the restriction of the type-specific immunity. Ryding et al. (2007) produced in the *Sf*9 insect cells the HPV16 VLPs devoid of the H16.V5 epitope, a major neutralizing epitope defined by the monoclonal antibody H16.V5, by site-directed mutagenesis of ten nonconserved, surface-exposed residues. Remarkably, the lack of the H16.V5 epitope had only a marginal effect on antigenic reactivity with antibodies in sera from infected subjects but affected immunogenicity in experimental immunization of mice, with reduced induction of both antibody responses and CTL responses (Ryding et al. 2007).

Later, Senger et al. (2009) employed the MultiBac, a modified baculovirus-based expression system, which permitted the substantially improved VLP production of several HPV types up to 40-fold. The highest VLP yields were achieved when two copies of the L1 gene were expressed from independently controlled cassettes. The production of the HPV57 L1 VLPs were evaluated in more detail, where the level of the HPV57 L1 protein was only slightly increased, but the strongly enhanced yield of the HPV57 VLPs was observed, suggesting necessity of a critical concentration of L1 within the producer cell for the efficient VLP assembly (Senger et al. 2009). Thönes et al. (2008) performed a direct comparison of the insect cells-produced HPV16 L1 VLPs and capsomeres and showed that the latter demonstrated a lower immunogenicity in mice than the VLPs, with respect to the induced antibody response.

After the insect cells, the HPV VLPs were extensively produced in bacteria. At the very beginning, a number of studies had reported successful expression of the HPV protein L1 in *E. coli* (Banks et al. 1987; Doobar and Gallimore 1987; Li CC et al. 1987; Thompson and Roman 1987; Jenison et al. 1988; Kelsall and Kulski 1995), but the products remained incorrectly folded, did not form VLPs, and demonstrated therefore considerably less immunogenicity than the properly assembled particles. As follows from Chapter 8, the related polyomaviruses (PyVs) preceded papillomaviruses by the expression and VLP formation in *E. coli*, when the PyV major capsid protein VP1 was self-assembled into the corresponding VLPs after purification from *E. coli* (Salunke et al. 1986). This study served as a prototype for the successful production of the HPV11 protein L1 VLPs in *E. coli*, which was performed by Li M et al. (1997). However, the truncated L1 protein missing 86 C-terminal aa residues revealed a domain structure that differed from that of the PyV VP1, when formed pentamers but was unable to assemble further into the virion-like VLPs. Thus, this first report on the HPV L1 expression in *E. coli* did not go beyond the statement that the L1 protein was purified to near homogeneity and appeared in electron microscopy as pentameric capsomeres that self-assembled into the VLPs (Li M et al. 1997). Furthermore, Zhang W et al. (1998) produced the protein L1 of HPV16 in *E. coli* as a fusion of the coding sequence at the N-terminus to 31 aa residues including a 24-aa leader sequence, pelB, and at the C-terminus to His_6 tag. The resulting protein formed insoluble inclusion bodies, and the yield was more than 10% of total cell proteins. The inclusion bodies were isolated and solubilized with 8 M urea, and the L1 proteins were purified by chromatographic separation. Following removal of the urea by gradual dialysis, the denatured L1 protein spontaneously renatured and subsequently assembled into polymorphic aggregations in vitro. After separation from the pool of

polymorphic structures by sucrose gradient sedimentation, the correctly formed virion-like VLPs were recognized by a HPV16 type-specific, conformational-dependent monoclonal antibody (Zhang W et al. 1998).

As mentioned earlier, the *E. coli*-produced HPV16 protein L1 missing 10 N-terminal aa residues formed the T = 1 particles consisting of the 12 L1 pentamers (Chen XS et al. 2000). Next, Chen XS et al. (2001) described the highly efficient expression of the L1 genes of HPV11 and HPV16 in *E. coli* in the form of fusions with glutathione-S-transferase and purified the chimeric products to near homogeneity as pentamers, after thrombin cleavage from the GST moiety. Then, the sequences at the N- and C-termini contributing to the formation of L1 pentamers and to the in vitro capsid assembly were identified by deletion analysis. For both HPV11 and HPV16 L1, up to at least 10 residues could be deleted from the N-terminus and 30 residues from the C-terminus without affecting pentamer formation. The HPV16 pentamers assembled into the virion-like VLPs at low pH, with the exception of the N-terminally truncated HPV16 L1, which assembled into the small T = 1 VLPs (Chen XS et al. 2001). Schädlich et al. (2009) were the first ones to measure the endotoxin—a lipopolysaccharide (LPS)—concentration at various stages of the purification of the HPV16 L1 from *E. coli* and to determine how it enhanced the immunogenicity of the HPV16 VLPs. A suitable protocol using Triton X-114 phase separation for the removal of LPS was elaborated by the VLP purification. Remarkably, the LPS-free capsomeres purified from *E. coli* induced neutralizing L1-specific antibodies (Schädlich et al. 2009).

It should be emphasized, however, that, unlike *E. coli*, the first clear demonstration of the papillomavirus VLP self-assembly in prokaryotes was achieved via expression of the HPV16 L1 gene in *Salmonella typhimurium* (Nardelli-Haefliger et al. 1997). The efficient formation of the HPV VLPs in *Salmonella* opened the door for the idea of the possible use of oral and nasal routes for immunization (Nardelli-Haefliger et al. 1997) and led to a set of the advanced studies and discussions on the advantages of mucosal over parenteral immunization in mice (Balmelli et al. 1998; Benyacoub et al. 1999; Dupuy et al. 1999; Nardelli-Haefliger et al. 1999; Revaz et al. 2001). Moreover, it was found that the immunogenicity against the HPV16 VLPs could be strongly enhanced by the specific PhoPc phenotype in *Salmonella* (Baud et al. 2004a) and by using a codon-optimized version of the L1 gene (Baud et al. 2004b). At last, Fraillery et al. (2007) proposed the *Salmonella enterica* serovar Typhi Ty21a expressing HPV16 L1 as a potential live vaccine against cervical cancer and typhoid fever.

Meanwhile, the true virion-like HPV16 L1 VLPs were produced in bacteria *Shigella flexneri* (Yang et al. 2005), *Lactobacillus casei* (Aires et al. 2006), *Lactococcus lactis* (Cho et al. 2007; Cortes-Perez et al. 2009), and *Listeria monocytogenes* (Mustafa et al. 2009). Baek et al. (2012) produced the HPV33 L1 VLPs in *Bacillus subtilis*.

Nevertheless, the bacterial expression systems have proven to be quite limited in producing economically significant quantities of the HPV VLPs (Lai and Middelberg 2002).

The successful story of the yeast-supported production of the HPV VLPs started in early 1990s when Sasagawa et al. (1995) demonstrated the self-assembly of the HPV6 and HPV16 VLPs from the expressed L1 alone and both L1 and L2 together, in *Schizosaccharomyces pombe*, although L2 was not incorporated into the VLPs. Remarkably, the two HPV16 L1 variants, isolated from benign cervical samples, produced many more (68- and 14-fold) VLPs than the prototype L1 derived from cervical carcinoma. A bit later, the HPV6a L1 or the L1 together with L2 ORFs were expressed in the yeast *Saccharomyces cerevisiae*, and the self-assembly of the L1 protein into the VLPs was demonstrated both in the L1 as well as L1 + L2 coexpressing yeast strains (Hofmann et al. 1995). In this case the copurification of the L1 and L2 proteins showed complex formation of the L1 and L2 proteins in the yeast-derived VLPs of the coexpressing strains. Next, the *S. cerevisiae* yeast was used to get VLPs by the expression of the HPV6/11 hybrid L1 (Neeper et al. 1996; Cook et al. 1999; Brown et al. 2001), HPV16 L1 (Koutsky et al. 2002; Kim SN et al. 2007; Park et al. 2008; Kim HJ et al. 2012), and HPV18 L1 (Woo et al. 2008) genes. Bazan et al. (2009) performed the expression of the codon-optimized HPV16 L1 gene and characterization of the corresponding VLPs produced by the methylotrophic yeast *Pichia pastoris*. The expression of the codon-optimized L1 genes of HPV16 and HPV18 in *Pichia pastoris* was presented by Hanumantha Rao et al. (2011). Furthermore, the L1 genes of HPV16, 18, and 33 (Coimbra et al. 2011) and HPV58 (Jiang et al. 2011) were expressed in *Pichia pastoris*.

The rapid success of the HPV VLPs in plants was directly incited by the idea of edible vaccines. Thus, Warzecha et al. (2003) synthesized a plant codon-optimized version of the HPV11 L1 gene and introduced it into tobacco and potato. The full-length L1 protein was expressed and localized in plant cell nuclei, and the expression of L1 in plants was enhanced by removal of the C-terminal nuclear localization signal sequence. The plant-expressed L1 self-assembled into the empty VLPs with immunological properties comparable to those of the native HPV virions. Moreover, the ingestion of transgenic L1 potato was associated with activation of an anti-VLP immune response in mice that was qualitatively similar to that induced by the parenteral administration of VLPs, and this response was enhanced significantly by subsequent oral boosting with the purified insect cell-derived VLPs (Warzecha et al. 2003). At the same time, Biemelt et al. (2003) described the generation of transgenic tobacco and potato plants carrying the HPV16 gene L1 under the control of the cauliflower mosaic virus 35S promoter. The feeding of tubers from transgenic potatoes to mice induced an anti-L1 antibody response in 3 out of 24 mice, although an anti-L1 response was primed in about half of the 24 animals.

In parallel, the HPV16 L1 gene, with and without nuclear localization signals, was expressed in *Nicotiana tabacum*

cv. Xanthi plants (Varsani et al. 2003b). The VLP yield was estimated to be 2–4 µg per kg of fresh leaf material, and the rabbits immunized with small doses of the plant-produced VLPs elicited a specific anti-L1 immune response. The L1 yield was increased further to 20–37 µg/kg of fresh leaf material by the expression of the HPV16 gene L1 in *N. benthamiana* using an infectious tobamovirus vector (Varsani et al. 2006). The expression of the HPV16 L1 gene in transgenic tobacco plants was also reported by Liu HL et al. (2005).

Kohl et al. (2007) achieved expression of the HPV11 gene L1, but only with the C-terminal nuclear localization signal-encoding region removed and not with the full-length gene, in two different plant species, namely *N. tabacum* and *Arabidopsis thaliana*, and increased yields of HPV11 L1 protein by between 500- and 1,000-fold compared to the previous report of Warzecha et al. (2003). Maclean et al. (2007) optimized the expression of the HPV L1 VLPs in *N. benthamiana* leaves, concluded that the L1 gene resynthesized to reflect human codon usage expressed better than the native gene, which expressed better than a plant-optimized gene, and reached the highest levels of the HPV L1 expression reported for plants. However, Fernández-San Millán et al. (2008) improved further the yield to ~3 mg L1/g fresh mass and thus equivalent to 24% of total soluble protein by expression of the HPV16 gene L1 in tobacco chloroplasts.

The VLPs of HPV8, a member of the *Betapapillomavirus 1* species from the *Betapapillomavirus* genus were obtained in *N. benthamiana* by Matić et al. (2012). Remarkably, a fourfold yield increase in the L1 gene expression was obtained when 22 C-terminal aa residues were deleted, eliminating therefore the nuclear localization signal. The plant-made HPV8 L1 proteins assembled in the appropriate VLPs of T = 1 or T = 7d symmetry and revealed their accumulation in the cytoplasm in the form of VLPs or paracrystalline arrays. The full story of the HPV VLPs in plants was revealed at that time by Thuenemann et al. (2013). Noris (2018) published a method for the efficient expression of the HPV16 L1 gene in *N. benthamiana* and the production of highly purified HPV16 L1 VLP preparations. At last, Naupu et al. (2020) presented the *N. benthamiana*-produced L1 VLPs of eight high-risk (HPV 16, 18, 31, 33, 35, 45, 52, and 58) and two low-risk (HPV 6 and 34) HPV types. The immunogenicity studies were conducted in mice utilizing HPV35, 52, and 58 and showed that the type-specific L1-specific antibodies were produced, which were able to successfully neutralize homologous HPV pseudovirions in pseudovirion-based neutralization assays. This work confirmed the great potential for using plant-based transient expression systems to produce affordable and immunogenic HPV vaccines, particularly for developing countries (Naupu et al. 2020).

NATURAL VLPS AS VACCINES

From the very beginning, the HPV VLPs were proven as a highly promising vaccination tool against cervical cancer, the third most common cancer in women worldwide. Thus, Rudolf et al. (1999) showed that the HPV16 L1 VLPs stimulated an MHC class I restricted CTL response with human peripheral blood lymphocytes (PBL) in vitro. The immunization of patients carrying genital warts with the HPV6b VLPs induced immunity to the L1 epitopes recognized during natural infection and accelerated regression of warts (Zhang LF et al. 2000). The insect cell-produced HPV11 L1 VLP candidate vaccine was measured in phase I trials in humans and found safe (Evans et al. 2001; Emeny et al. 2002). Koutsky et al. (2002) demonstrated that the administration of the HPV16 VLPs significantly lowers infection with HPV16 and leads to a reduction of HPV-16-associated disease. The exhaustive review on the HPV VLPs as the putative prophylactic human vaccines was published at that time by Lowy and Frazer (2003).

Hu et al. (2014) described an open-label phase-I clinical trial in Jiangsu province, China, of the *E. coli*-expressed bivalent HPV16 and HPV18 vaccine candidate, which appeared safe and well tolerated in healthy women and supported further immunogenicity and efficacy studies for this novel HPV VLP vaccine candidate.

Xu et al. (2008) suggested to improve immunogenicity of the HPV VLPs by genetic fusion of the modified adjuvant, mLTK63, to the C-terminus of HPV16 L2 protein. The coexpression of the HPV16 L1 and L2-mLTK63 genes in insect cells led to the efficient assembly of the chimeric HPV16 L1/L2-mLTK63 VLPs, which combined therefore the antigen and adjuvant as a unit. The intramuscular immunization of mice with the chimeric VLPs induced higher titers of the HPV16-specific long-lasting neutralizing antibodies and stronger T cell response (Xu et al. 2008).

As to the combination of the proteins L1 and L2 from different HPV types, Combelas et al. (2010) produced in insect cells the HPV58 L1 VLPs carrying the HPV31 L2 and demonstrated high level of cross-neutralizing antibodies in mice.

Gardasil™ and Cervarix™

The HPV vaccine was the second great VLP and anticancer vaccine, after the hepatitis B virus (HBV) vaccine described in Chapter 37, which was introduced into the extensive clinical practice. For about 15 years, the two popular prophylactic HPV vaccines have been on the market, namely Gardasil™ (Merck, now Sanofi Pasteur MSD), established in 2006 and Cervarix™ (GlaxoSmithKline), from 2007. The FDA approval of the HPV vaccine is described in the official CDC (2010a, b) documents. The numerous studies have shown that both vaccines are safe, well tolerated, and highly immunogenic. First, the Merck company developed the quadrivalent HPV vaccine Gardasil™, which included the HPV6, 11, 16 and 18 L1 VLPs and was produced in yeast *S. cerevisiae*. The Gardasil™ vaccine was approved by two large Phase III trials and licensed for use in the United States in June 2006 (Wright et al. 2006) and was shown to be effective for five years for prevention of infection by HPV

types 6, 11, 16 and 18 (Villa et al. 2006; Schiller et al. 2008; Palmer et al. 2009). McCormack (2014) summarized the Gardasil™ vaccination data and concluded that the vaccine provided high-level protection against infection or disease caused by the vaccine HPV types over 2–4 years in females aged 15–45 years who were negative for the vaccine HPV types and provided a degree of cross-protection against certain nonvaccine HPV types. Moreover, the vaccine also provided high-level protection against persistent infection, anogenital precancerous lesions, and genital warts caused by the vaccine HPV types over 3 years in susceptible males aged 16–26 years. The protection has been demonstrated for up to 8 years. In subjects who were negative for the vaccine HPV types, the high seroconversion rates and high levels of anti-HPV antibodies were observed in females of all age ranges from 9–45 years and in males aged 9–26 years. The vaccine was generally well tolerated and was usually predicted to be cost effective in girls and young women. Therefore, the HPV vaccine offered an effective means to substantially reduce the burden of HPV-related anogenital disease in females and males, particularly cervical cancer and genital warts (McCormack 2014).

Second, the bivalent Cervarix™, containing the HPV16 and HPV18 L1 VLPs produced in baculovirus-infected insect cells and adjuvanted with AS04, has also shown sustained efficacy for up to 4.5 years (Harper et al. 2006; Schiller et al. 2008). Szarewski (2012) summarized the Cervarix™ vaccination data and stressed the significant cross-protection against some HPV types not included in the vaccine, where protection against HPV45 was particularly important, as this HPV type was relatively more common in adenocarcinoma. Moreover, the vaccine's antibody response profile suggested a long duration of immunity.

Recently, Nicoli et al. (2020) published a comprehensive six-year surveillance study following immunization with Cervarix™ and Gardasil™ in adolescent and young adult women in Italy. The humoral responses against HPV16 and HPV18 (and HPV6 and HPV11 for Gardasil™) were high in both vaccines up to six years from the third dose. Remarkably, Cervarix™ induced significantly higher and more persistent antibody responses, while the two vaccines were rather equivalent in inducing memory B cells against HPV16 and HPV18. Moreover, the percentage of subjects with vaccine-specific memory B cells was even superior among Gardasil™ vaccinees, and, conversely, Cervarix™-vaccinated individuals with circulating antibodies but undetectable memory B cells were found. Finally, a higher proportion of Cervarix™-vaccinated subjects displayed cross-neutralizing responses against nonvaccine types HPV31 and HPV45. It was concluded that Gardasil™ and Cervarix™ may differently affect long-lasting humoral immunity from both the quantitative and qualitative point of view (Nicoli et al. 2020).

The most recent approved HPV vaccine, Gardasil-9, an upgrade of the Gardasil-4, is a nonavalent vaccine and protects against seven HPV types 16, 18, 31, 45, 33, 52, 58 associated with ~90% of cervical cancer and against two

HPV types 6 and 11 associated with ~90% genital warts with little cross-protection against nonvaccine HPV types, as reviewed by Zhai L and Tumban (2016).

Chimeric VLPs

Genetic Fusions

HPV Vaccines

Müller et al. (1997) were the first ones to produce the chimeric HPV VLPs. The latter were engineered by replacing the 34 C-terminal aa residues of the HPV16 L1 protein with various parts of the HPV16 E7 protein up 60 aa and producing the chimeric VLPs in insect cells by the baculovirus expression system. The chimeric VLPs were able to induce a neutralizing antibody response in mice, recognized the cellular receptor, and were proposed to deliver peptides into mammalian cells in vitro and in vivo, possibly reaching the pathway for MHC class I presentation (Müller et al. 1997). The chimeric VLPs formed by the protein L1, where 50 C-terminal aa were replaced by 55 or 60 N-terminal aa of E7, induced an E7-specific CTL response in the absence of an adjuvant and prevented outgrowth of E7-expressing tumor cells in mice, even if inoculation of cells was performed two weeks before vaccination (Schäfer et al. 1999). The further examination led to the clear conclusion that the chimeric HPV L1-E7 VLPs provided not only protection against tumor growth but also a therapeutic effect on preexisting tumors (Jochmus et al. 1999). It should be noted that the E7 moiety was not exposed on the surface of the chimeric VLPs (Müller et al. 1997). Kaufmann et al. (2001) found that the HPV L1-E7 VLPs induced in mice a primary T cell response and activated CTL to recognize MHC-restricted HPV16 L1 and E7 peptide epitopes. As a result, these CTLs were cytolytic toward HPV16-positive cervical cancer cells. Finally, Kaufmann et al. (2007) subjected the HPV L1-E7 VLPs to a randomized, double blind, placebo-controlled clinical trial in HPV16 mono-infected, high-grade cervical intraepithelial neoplasia (CIN) patients. The outcome demonstrated evidence for safety and a nonsignificant trend for the clinical efficacy of the vaccine candidate (Kaufmann et al. 2007). Meanwhile, Freyschmidt et al. (2004) found that unmethylated CpG motifs (CpG ODN) as well as lipopolysaccharides and sorbitol enhanced the stimulation of mouse-bone-marrow-derived dendritic cells that were induced by the HPV L1-E7 VLPs.

Sharma et al. (2012) were the first to produce the HPV L1-E7 VLPs in *E. coli*. The authors expressed and purified recombinant HPV16 L1(ΔN26)-E7(ΔC38) protein in *E. coli*, which was assembled into the chimeric VLPs in vitro. These VLPs induced neutralizing antibodies and triggered cell-mediated immune response in a murine model of cervical cancer, exhibiting antitumor efficacy.

An alternative approach to generating chimeric HPV16 VLPs was carried out by the fusion of the entire E7 or E2 proteins to the minor capsid protein L2 (Greenstone et al. 1998). The 98-aa E7 protein was of HPV16, while the 391-aa E2 gene was taken from CRPV and fused to the C-terminal aa of the HPV16 L2 gene. Such chimeric L1/L2-E7/E2 VLPs were indistinguishable from the parental VLPs in their morphology, although they carried the chimeric L2 protein with a ratio of approximately 1:30 to the protein L1 and elicited high titers of neutralizing antibodies. Moreover, the VLPs carrying the E7 oncoprotein protected mice from tumor challenge, and this protection was mediated by the MHC class I restricted cytotoxic lymphocytes (Greenstone et al. 1998). The effect of preexisting neutralizing antibodies on the antitumor immune response induced by the chimeric L1/L2-E7 (Da Silva et al. 2001) and the role of heterologous boosting (Da Silva et al. 2003) were studied in mice. Nevertheless, Wakabayashi et al. (2002) found that the HPV16 L1-E7 VLPs were superior to the HPV16 L1/L2-E7 VLPs in eliciting tumor protection at equivalent doses, although both types of particles were able to protect mice from tumor challenge. In both cases, the first 57 aa of E7 were used to overcome the size limitation and limited VLP production imposed by inserting polypeptides into the chimeric L1 VLPs.

In parallel, Xu et al. (2007) engineered and produced in insect cells the chimeric HPV6b VLPs carrying the protein L2 with the C-terminally added E7 that was in this case an artificial synthetic gene designed by codon modification, point mutation, and gene shuffling. Remarkably, the fusion of the 158 aa HPV16 E7 protein to L2 did not prevent the correct self-assembly of 55-nm particles similar to the non-chimeric HPV-6b L1/L2 VLPs, but the E7 moiety had an internal location within the VLPs (Xu et al. 2007).

Christensen et al. (2001) prepared the chimeric HPV VLPs using complementary regions of L1 from HPV11 and HPV16. The two small, noncontiguous hypervariable regions of the HPV16 L1, when replaced into the HPV11 L1 backbone, produced an assembly positive hybrid L1, which was recognized by the type-specific, conformationally dependent HPV16 neutralizing monoclonal antibody. When a set of the chimeric VLPs was tested as immunogens in rabbits, antibodies to both HPV11 and HPV16 L1 VLPs were obtained. One of the chimeric VLPs containing hypervariable FG and HI loops of the HPV16 L1 replaced into an HPV11 L1 background provoked neutralizing activity against both HPV11 and HPV16 (Christensen et al. 2001).

Another set of promising HPV vaccine candidates involved the HPV L2 epitopes. In contrast to the protein L1 that is not conserved among HPV types, the L2 is highly conserved and was regarded therefore for decades as a promising alternative target antigen to develop a broadly protective HPV vaccine. Thus, Varsani et al. (2003a) replaced regions of the HPV16 L1 with the cross-neutralizing HPV16 L2 peptide defined by aa 108-LVEETSFIDAGAP-120 and produced the chimeric VLPs by the baculovirus expression system. Kondo et al. (2008) inserted the L2 sequences of aa 18–38, 56–75, or 96–115 between the L1 aa residues 430 and 433. When expressed from the recombinant baculovirus in insect *Sf*9 cells, the three chimeric VLPs were shown to present the L2 peptides on their surface. By immunizing rabbits with the VLPs, the antibodies neutralizing HPV16 and cross-neutralizing the infectious HPV18, 31, 52, and 58 pseudovirions were induced (Kondo et al. 2008). McGrath et al. (2013) replaced the region downstream of the L1 aa 413 with the HPV16 L2 epitopes at aa 108–120, 56–81, and 17–36 and demonstrated production of the appropriate VLPs in insect cells. Schellenbacher et al. (2013) concentrated on the HPV16 L2 aa stretch 17–36 known as an RG1 epitope, which was inserted within the DE-surface loop of the HPV16 L1. The immunization of rabbits with the RG1-VLPs adjuvanted with human-applicable alum-MPL (aluminum hydroxide plus 3-O-desacyl-4'-monophosphoryl lipid A) induced robust L2 antibodies, which (cross-)neutralized mucosal high-risk HPV16/18/45/37/33/52/58/35/39/51/59/68/73/26/69/34/70, low-risk HPV6/11/32/40, and cutaneous HPV2/27/3/76. Moreover, in vivo, mice were efficiently protected against experimental vaginal challenge with mucosal high-risk pseudovirion types HPV16/18/45/31/33/52/58/35/39/51/59/68/56/73/26/53/66/34 and low-risk HPV6/43/44. Therefore, the RG1 epitope appeared as a highly promising next-generation vaccine component with broad efficacy against all relevant mucosal and also cutaneous HPV types. Recently, Huber et al. (2021) published a comprehensive review about the RG1-carrying vaccine candidates ready for a first-in-human clinical study, including the promising candidates based on the RNA phage VLPs reviewed by Yadav et al. (2019) and described in detail in Chapter 25.

The plant expression-based chimeric HPV vaccine candidates deserve to be highlighted. Paz De la Rosa et al. (2009) reported for the first time the expression in plants of a chimeric particle containing the HPV16 L1 sequence and a string of T-cell epitopes from HPV16 E6 and E7 fused to the L1 C-terminus. The resulting tomato-produced VLPs induced a significant antibody and CTL response in mice. Monroy-García et al. (2011) elaborated a novel ELISA assay and demonstrated that the tomato-produced chimeric VLPs carrying the E6 and E7 epitopes enabled much better detection of IgG antibodies in the sera of cervical patients positive for HPV16 infection than those obtained with VLPs containing only the HPV16 L1 protein. Furthermore, Monroy-García et al. (2014) provided evidence that these VLPs induced persistent IgG antibodies for over 12 months and ensured efficient protection and inhibition of tumor growth in mice. Remarkably, the significant tumor reduction of 57% was observed in mice after immunization with the chimeric VLPs in comparison with the HPV16 L1 VLPs or in tumor control.

Other Vaccines

Continuing the strong success of the first chimeric HPV VLPs described earlier, Nieland et al. (1999) evaluated an antitumor potential and linked the P1A epitope LPYLGWLVF derived from the murine P815 tumor-associated antigen to the L1 protein, resulting in the synthesis of the chimeric HPV16 L1-P1A VLPs in insect cells. The immunotherapy

with the VLPs suppressed the growth of established progressor P815 tumors and led to a significant survival advantage of treated mice compared with the nonchimeric VLP-treated control mice against a lethal tumor challenge (Nieland et al. 1999).

Slupetzky et al. (2001) engineered the 8-aa B-cell epitope LELDKWAS from the human immunodeficiency virus type 1 (HIV-1) gp41 envelope glycoprotein into several hypervariable regions of the L1 protein of HPV16 and BPV1 and produced the appropriate chimeric VLPs by the baculovirus expression system. For the HPV16 L1, the peptide was inserted between aa residues 286/287, 281/287 (replacing aa 282–286) or fused to the C-terminus. Dale et al. (2002) have chosen simian-human immunodeficiency virus (SHIV) epitopes for the insertion into the HPV6b VLPs produced in insect *Sf*9 cells. SHIV was a chimeric primate lentivirus virus utilizing an SIV backbone with the HIV-1 *env*, *tat*, and *rev* genes replacing the homologous SIV genes. The chimeric VLPs carrying the fragments of SIV Gag p27, HIV-1 Tat, and HIV-1 Rev proteins were administered to macaques both systemically and mucosally. Although HPV L1 antibodies were induced in all immunized macaques, weak antibody or T cell responses to the chimeric SHIV antigens were detected only in animals receiving the DNA prime/HPV-SHIV VLP boost vaccine regimen. A significant but partial protection from a virulent mucosal SHIV challenge was also detected only in the prime/boosted macaques and not in animals receiving the HPV-SHIV VLP vaccines alone (Dale et al. 2002).

Sadeyen et al. (2003) introduced the immunodominant loop epitope 78-DPASRE-83 of the core protein (HBc) of hepatitis B virus (HBV), which is described in detail in Chapter 38, within the different loops of the HPV16 L1 protein at positions 56/57, 140/141, 179/180, 266/267, 283/284, or 352/353. All these chimeric L1 proteins were capable of self-assembly into the VLPs. Although the antigenicity and immunogenicity of some of these VLPs were reduced compared to the levels observed with wild-type VLPs, all were nevertheless able to induce HPV-neutralizing antibodies, and all except the one with insertion at position 56/57 were also able to induce anti-HBc antibodies, thus suggesting exposure of the HBc epitope on the VLP surface. Next, Carpentier et al. (2005) engineered six other L1 protein mutants by either insertion or substitution of six aa residues corresponding to the HBc epitope within the FG surface loop of the HPV16 L1 protein. The chimeric products retained their ability to self-assemble by the baculovirus-driven expression in *Sf*21 cells. Continuing the HBV epitope story, Joshi et al. (2013) used the HPV16 T = 1 platform to generate in silico the chimeric L1 VLPs, where one of the highly immunogenic epitopes of HPV, the EF loop at aa residues 170–189 was computationally replaced with the preS1 epitope of HBV, namely, the 37-NSNNPDWDF-45 stretch that was found to be responsible for eliciting an immune response against HBV virions (Liang 2009).

Zhu et al. (2010) inserted the Epstein-Barr virus (EBV) multiepitope derived from latency membrane protein 2

(LMP2) and containing T- and B-cell epitope-rich peptides into the C-terminus of the HPV6b L1. The chimeric structure was expressed in mammalian Cos-7 cells, while the intramuscular administration of the chimeric gene in mice was able to elicit not only antibodies against HPV6b L1 VLPs and EBV LMP2 but also a CTL response against the EBV LMP2 epitopes.

Matić et al. (2011) used the HPV16 L1 as a carrier of the two epitopes from influenza A virus, namely M2e2–24, ectodomain of the M2 protein (M2e) of 24 aa residues, which is highly conserved among all influenza A isolates— or M2e2–9, a shorter 9-aa version of M2e containing the N-terminal highly conserved epitope, which was common for both M1 and M2 influenza proteins. The synthetic HPV16 L1 gene optimized with human codon usage was used as a backbone gene to design four chimeric sequences containing either the M2e2–24 or the M2e2–9 epitopes in two predicted surface-exposed L1 positions: helix 4 and the coil region connecting helix 4 and the β-J sheet. All chimeric constructs were transiently expressed in plants *N. benthamiana* using a cowpea mosaic virus-derived expression vector. The chimeric proteins made in plants spontaneously assembled into the T = 1, T = 7 VLPs, or capsomeres (Matić et al. 2011). In fact, this was not only the first report of the transient expression and the self-assembly of a chimeric HPV16 L1 bearing the M2e influenza epitope in plants but also the first record of a successful expression of the chimeric HPV-16 L1 carrying an epitope of a heterologous virus in plants, as reviewed by Thuenemann et al. (2013).

It should be noted that Murata et al. (2009) proposed the HPV16 L1 helix 4 (h4) region as a novel antigen display site within the L1 pentamer but not VLPs. Thus, the HPV L1 monomers bearing one of the two conserved neutralizing epitope of the human respiratory syncytial virus (RSV) fusion protein F, aa residues 255–278 and 423–436, instead of the deleted L1 h4 region, were produced in insect *Trichoplusia ni* cells and self-assembled into pentameric ring-like capsomeres that elicited antibody response against the encoded foreign epitopes. When injected into mice, each of the capsomere derivatives was immunogenic with respect to L1 protein and resulted in RSV nonneutralizing antisera that recognized purified RSV F protein in immunoblots (Murata et al. 2009).

Chemical Coupling

The conjugation idea offered uncoupling of the VLP scaffold production from the preparation of the target to be displayed by the covalent or noncovalent allocation of the desired antigen onto the VLP surface. In the case of HPV, after the first study on the BPV1 as a scaffold (Chackerian et al. 2001; described later), the two novel scaffolds, namely HPV16 and human polyomavirus BK (see Chapter 8), were involved (Chackerian et al. 2002). The authors used the noncovalent coupling approach to evaluate the structural correlates of autoantibody induction, using a system in which mice were vaccinated with a fusion protein containing self (TNF-α) and foreign (streptavidin) components,

conjugated to the biotinylated VLPs. Remarkably, similar titers of autoantibodies to TNF-α were elicited using conjugated papillomavirus and polyomavirus VLPs, indicating that acute activation of dendritic cells by the antigen is not required. Moreover, the strong autoantibody responses were also induced by conjugated papillomavirus capsid pentamers, indicating that this order of self-assembly was sufficient. However, a reduction of self-antigen density on the VLP surfaces dramatically reduced the efficiency of the autoantibody induction (Chackerian et al. 2002). The same approach was used by Chackerian et al. (2006) to prepare VLP immunogens for Alzheimer's disease, which induced specific antibody responses against amyloid-β (Aβ) without concomitant T cell responses. The HPV16 VLPs were produced in this study by expression in mammalian 293TT cells in accordance with Buck et al. (2004, 2005b) and used in parallel with the RNA phage Qβ VLPs, the subject of Chapter 25. The purified HPV16 VLPs were biotinylated, and the Aβ conjugated VLPs were generated by linking biotinylated Aβ peptides of different length to the biotinylated VLPs via streptavidin. The immunization with the Aβ peptide conjugated to the HPV16 VLPs or to the RNA phage Qβ elicited anti-Aβ antibody responses at low doses and without the use of adjuvants. Remarkably, both HPV16 VLP- and Qβ phage-based Aβ vaccines induced weak or negligible T cell responses against Aβ, while the T cell responses were largely directed against linked viral epitopes (Chackerian et al. 2006). Furthermore, Chackerian et al. (2008) used the biotinylated HPV16 VLPs to link via streptavidin the hen egg lysozyme (HEL) as a neo-self-antigen in a B-cell receptor transgenic mouse model. The VLP-conjugated HEL was more potent than trivalent HEL in both in vivo and in vitro studies.

Ionescu et al. (2006) were the first ones to present a real HPV VLP-based vaccine candidate that was constructed by chemical coupling to the capsid surface. This approach avoided therefore the potential technical problems of the genetic fusion and purification of chimeric VLPs, as well as of the biotinylation procedure. Compared to the latter, the chemical coupling allowed much higher peptide loads per VLP. Thus, the extracellular peptide fragment of influenza type A M2 protein was coupled to the HPV16 VLPs, produced in *S. cerevisiae* according to Tobery et al. (2003), as the result of the reaction between a C-terminal cysteine residue on the peptide and the maleimide-activated capsids. The conjugates carried ~4000 copies of the antigenic peptide per VLP, had an average particle size slightly larger than the carrier, were highly immunogenic, and conferred good protection against lethal challenge of influenza virus in mice (Ionescu et al. 2006).

PACKAGING AND DELIVERY

Unckell et al. (1997) were the first ones to show that the Cos-7 cells-produced HPV33 VLPs packaged the original plasmid and anticipated the development of the packaging and delivery methodologies. The conditions for the quantitative disassembly and reassembly of *Trichoplusia ni* (HighFive) cell-produced and purified HPV11 L1 VLPs to the level of capsomeres were elaborated after demonstration that disulfide bonds alone were essential to maintaining capsid structure, which simplified packaging of selected exogenous compounds within the reassembled VLPs (McCarthy et al. 1998).

In parallel, Stauffer et al. (1998) demonstrated that the DNAs ranged in size from 5.4 to 7.9 kb were encapsidated into the HPV18 L1/L2 VLPs produced in mammalian 293T cells. When encapsidated plasmids contained either the β-galactosidase gene or the puromycin-resistance gene, the pseudovirions were shown to be infectious in that they could transfer β-galactosidase activity or confer resistance to puromycin to a number of cell types. The DNA encapsidation was independent of the HPV DNA sequences and of the plasmid DNA replication. However, the L1 protein alone was unable to encapsidate DNA, although it was able to form VLPs (Stauffer et al. 1998).

Touze and Coursaget (1998) reported that the insect cell-produced HPV16 L1 VLPs can package unrelated plasmid DNA in vitro and then deliver this foreign DNA in the form of pseudovirions to eukaryotic cells with the subsequent expression of the encoded gene. The results indicated higher gene transfer than with DNA alone or with liposome. Therefore, the HPV VLPs were proclaimed as a very promising vehicle for delivering genetic material into target cells. Moreover, the preparation of the HPV VLP gene transfer vehicle was relatively easy (Touze and Coursaget 1998). Looking for new vehicles, Combita et al. (2001) approved the ability of nine insect cell-produced HPV VLPs, namely, HPV16, 18, 31, 33, 39, 45, 58, 59, and 68, to package heterologous DNA and serve for the gene transfer into Cos-7 cells.

Kawana et al. (1998) engineered the HPV16 L1/L2 pseudovirions by disassembling of the insect cells-produced VLPs with further incorporation of plasmid DNAs into the reassembled L1/L2 capsids. After the baculovirus expression system, the assembly of the *S. cerevisiae* yeast-produced HPV16 L1 pseudovirions was performed by Rossi et al. (2000). Buck et al. (2005a) presented a comprehensive methodology for the generation of HPV pseudovirions in mammalian cells.

Using the insect cell-produced HPV16 L1 VLPs, Shi et al. (2001) engineered pseudovirions with packaged DNA that encoded the well-studied CTL epitope gp33, namely, aa 33-KAVYNFATC-41, of lymphocytic choriomeningitis virus (LCMV). This famous epitope is characterized in more detail in Chapter 25. Such pseudovirions induced a stronger CTL response than a DNA plasmid vaccine encoding the same epitope given systemically. Oh et al. (2004) demonstrated the enhanced mucosal and systemic immunogenicity of HPV16 L1 VLPs encapsidating DNA encoding the cytokine IL2 gene as a genetic adjuvant.

Malboeuf et al. (2007) evaluated the VLP-mediated delivery and expression of a GFP reporter construct in vitro, which was found to be highly dependent upon the presence of full-length L2 protein within the insect cell-produced

HPV16 VLPs. Similarly, the expression of GFP and luciferase reporter plasmids in vivo was strongly enhanced by coadministration of the L1/L2 VLPs. Remarkably, the GFP expression was detected in migrating antigen presenting cells (APCs) recovered from mice inoculated with the VLP-packaged GFP plasmid but not in APCs recovered from mice inoculated with the plasmid alone (Malboeuf et al. 2007).

The HPV31 pseudovirions were used for the vectorization of a DNA vaccine against hepatitis E virus (Renoux et al. 2008). The HEV genes ORF2(112–660) and ORF2(112–608) were optimized for expression in mammalian cells and inserted in a baculovirus-derived vector for expression in insect cells. When expressed in the *Sf*21 insect cells, the HEV ORF2(112–660) led to the production of irregular 15-nm particles that accumulated in the cytoplasm of the cells, whereas ORF2(112–608) induced the production of 18-nm particles that were present in both the cell culture medium and the cell cytoplasm, as described in Chapter 16. The delivery into mice of two HEV ORF2 genes via the HPV VLPs was very effective in the induction of anti-HEV antibodies. In addition, an effective immune response HPV occurred. These engineered pseudovirions were thus demonstrated to induce immune responses to both HEV and HPV when they were administered to mice intramuscularly (Renoux et al. 2008).

Peng et al. (2010) confirmed the idea that the DNA vaccines delivered using HPV pseudovirions represent an efficient delivery system that can potentially affect the field of DNA vaccine delivery. The authors generated the HPV16-OVA pseudovirions encapsidating a DNA vaccine encoding the gene of the model antigen, namely ovalbumin (OVA). The subcutaneous vaccination of mice with the HPV16-OVA pseudovirions elicited significantly stronger OVA-specific CD8+ T-cell immune responses compared with OVA DNA vaccination via gene gun in a dose-dependent manner. Furthermore, a subset of the infected cells in draining lymph nodes became labeled on vaccination with fluorescein isothiocyanate-labeled HPV16-OVA pseudovirions in injected mice (Peng et al. 2010).

Bousarghin et al. (2009) performed the silencing of E6 and E7 expression in cervical carcinoma cells by the HPV pseudovirions that were packaged with plasmids coding for a short hairpin RNA (shRNA) for RNA interference. Such pseudovirions were acquired by the disassembly/reassembly of the insect cell-produced HPV31 VLPs together with the plasmid encoding the shRNA sequences. The results indicated the degradation of E6 and E7 mRNAs when the shRNAs against E6 or E7 were initiated by the HPV pseudovirions in the HPV-positive cells in vitro. Moreover, the treatment of mice with the HPV VLP vectors coding for the E7 shRNA sequence resulted in dramatic inhibition of tumor growth (Bousarghin et al. 2009).

Concerning the specific targeting of the HPV VLPs, Wang H et al. (2009) tested the hypothesis that the DNA-mimotope peptide DWEYSVWLSN can target anti-dsDNA antibody-producing cells and decrease the number of these cells. Thus, the chimeric HPV16 L1 VLPs were achieved by insertion of the targeting DWEYSVWLSN peptide or a mismatched irrelevant WSLDYWNEVS control. The chimeric VLPs were packaged with plasmids encoding diphtheria toxin A (DTA), cocultured with target cells, and lupus-prone mice were vaccinated to assess the killing efficiency in vivo. The results showed that the chimeric VLPs carrying the targeting peptide possessed great potential of killing anti-dsDNA antibody-producing B cells with high efficiency (Wang H et al. 2009).

VISUALIZATION

Bergsdorf et al. (2003) investigated whether HPV VLPs may serve as an efficient carrier of low molecular mass compounds, e.g. hormones, vitamins, peptides etc., into the eukaryotic cells. Thus, the Cos-7 cells were incubated with the insect cell-produced HPV16 L1/L2 VLPs labeled with the fluorescence dye carboxyfluorescein diacetate succinimidyl ester (CFDA-SE) that is only activated after cell entry. It appeared that the labeled VLPs were specifically bound to the cell surface followed by their complete internalization and could be therefore recognized as promising vehicles for highly efficient delivery of low-molecular-mass compounds into cells. In parallel, Drobni et al. (2003) performed the same CFDA-SE labeling of the HPV6 L1, HPV6 L1/L2, and HPV16 L1 VLPs produced in insect *Sf*21 cells and suggested a primary interaction between HPV and cell-surface heparan sulfate.

Windram et al. (2008) fused green fluorescent protein (GFP) to the N- or C-terminus of both L1 and L2, with L2 chimeras being coexpressed with native L1 in the insect *Sf*21 cells. The purified chimeric VLPs were comparable in size of ~55 nm to the native HPV VLPs and retained their antigenicity. The VLPs were acceptable for gene transfer and encapsidated DNA in the range of 6–8 kb. As indicated earlier, the visualization of cells occurred also after expression of the packaged DNA that encoded the GFP gene (Malboeuf et al. 2007).

GENUS *DELTAPAPILLOMAVIRUS*

3D STRUCTURE

The *Deltapapillomavirus* genus includes seven species, the natural hosts of which are ruminants. Although European elk papillomavirus (EEPV) from the *Deltapapillomavirus 1* species remains a reference strain, the structural and VLP studies were performed with BPV1 of the *Deltapapillomavirus 4* species.

The T = 7d structure of the BPV-1 capsids of 60 nm in diameter was characterized first, in parallel with the previously described HPV1 structure, by electron cryomicroscopy and 3D image reconstruction at 25 Å resolution (Baker et al. 1991). The five-coordinate (pentavalent) and six-coordinate (hexavalent) capsomeres both exhibited distinct 5-fold axial symmetry as was observed for SV40 and polyoma viruses, as described in Chapter 8. Then, the 3D

structure of BPV1 was improved by electron cryomicroscopy to 9-Å resolution, and the location of the L2 minor capsid protein in the center of the pentavalent capsomeres was settled (Trus et al. 1997). Finally, the structure of BPV1 was solved by electron cryomicrosopy at ~3.6 Å resolution (Wolf et al. 2010). The structure shown in Figure 7.5 demonstrates how the compact capsids are assembled from 72 pentamers of the protein L1 and reveals how the N- and C-terminal arms of a subunit (extensions from its β-jellyroll core) associate with a neighboring pentamer. The critical contacts come from the C-terminal arm, which loops out from the core of the subunit, forms contacts (including a disulfide) with two subunits in a neighboring pentamer, and reinserts into the pentamer from which it emanates.

EXPRESSION

As described earlier, John T. Schiller's team was at the origin of the papillomavirus VLPs, reporting the

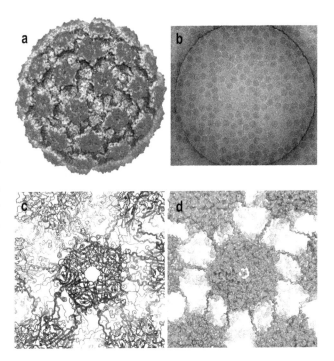

FIGURE 7.5 Near-atomic structure of BPV1. (a) Surface representation of the BPV 3D image reconstruction. Radial coloring emphasizes the isolated connections formed by the C-terminal loops (red layer). (b) Representative electron micrograph with BPV particles suspended in amorphous ice. The diameter of the circular hole in the carbon film is 1.0 μm. (c, d) Partial radial section of the particle viewed from the outside along the axis of a six-coordinated pentamer. Colors as in (a). (c) Ribbon representation of a pentamer and part of its surrounding pentamers. The L1 C-terminal arms are emphasized as worms, and the interpentamer disulfides are highlighted as yellow spheres. (d) Semitransparent, depth cued density map (contoured at 2.5σ above the average density), with embedded model. (Wolf et al. Subunit interactions in bovine papillomavirus. Proc Natl Acad Sci U S A. 107:6298–6303, Copyright 2010 National Academy of Sciences, U.S.A.)

self-assembly of the protein L1 of BPV1 and HPV16 into the virion-like VLPs by the baculovirus-driven expression in insect cells (Kirnbauer et al. 1992). Moreover, the protein L1 alone was sufficient for the VLP assembly, and the VLPs induced high-titer neutralizing antisera in rabbits, which could prevent papillomavirus infection in vitro (Kirnbauer et al. 1992). Kirnbauer et al. (1996) produced in insect cells the VLPs containing either the L1 and L2 or only the L1 of BPV4, and both VLP preparations proved to be extremely effective prophylactic vaccines, as described later. Shafti-Keramat et al. (2009) expressed the L1 gene of BPV1 and Bos taurus papillomavirus 2 (BPV2) of the same *Deltapapillomavirus 4* species in *Sf*9 cells. The antisera induced in rabbits by either VLP vaccine were able to robustly cross-neutralize heterologous as well as homologous types, indicating that BPV1 and BPV2 are closely related serotypes and suggesting that a monovalent BPV1 or BPV2 VLP vaccine may potentially protect against both infections (Shafti-Keramat et al. 2009). To identify areas within the BPV1 L1 that are important for virus assembly, Paintsil et al. (1996) constructed a set of 24 baculovirus recombinants expressing the BPV1 L1 deletion mutants that span the entire L1 open reading frame. The C-terminal truncation mutant, lacking the last 24 aa residues, formed VLPs (threefold more efficiently than the original BPV1 L1), while most of the mutants screened formed capsomeres and aggregates.

Zhou et al. (1993) produced the BPV1 pseudovirions by the vaccinia virus (VV)-driven expression, where the BPV1 L1 and L2 genes were expressed. The VLPs were observed in the nucleus of mammalian cells infected with a VV recombinant for the BPV1 L1 protein, and greater numbers of similar particles were seen in the nuclei of cells infected with a VV double recombinant for L1 and L2. Moreover, the presence of L2 was necessary for the BPV1 DNA packaging (Zhou et al. 1993).

Zhao and Frazer (2002a) found that yeast *S. cerevisiae* is permissive for the BPV1 replication. Thus, the *S. cerevisiae* protoplasts took up the BPV1 and allowed the viral episomal DNA to replicate after uptake. The BPV1 VLPs were assembled in the infected *S. cerevisiae* cultures from newly synthesized capsid proteins and also packaged the newly synthesized DNA, including full-length and truncated viral DNA and *S. cerevisiae*-derived DNA. Then, the yeast-produced BPV1 VLPs and virions were able to convey packaged DNA to mammalian cells. Recently, Chen J et al. (2021) elaborated a comprehensive model for the long-term BPV1 infection in *S. cerevisiae*.

In yeast *P. pastoris*, Jesus et al. (2012) expressed the codon-optimized L1 gene of BPV1, BPV2, as well as of Bos taurus papillomavirus 4 (BPV4), belonging to the species *Xipapillomavirus 1* of the *Xipapillomavirus* genus, with the C-terminal addition of the His$_6$ tag. Therefore, after methanol induction, the recombinant yeast cells were able to produce L1 proteins of the three different BPV types under bioreactor conditions, but no data concerning their ability to self-assemble were demonstrated.

In plants, Love et al. (2012) transiently expressed the BPV1 L1 in *N. benthamiana* as a prelude to producing a candidate vaccine. Remarkably, the plant codon optimization of L1 gave higher levels of expression than its non-optimized counterpart. Following protein extraction, the authors reached the yields of 183 mg/kg fresh mass leaf tissue of the relatively pure L1, which had self-assembled into the virion-like VLPs. The latter elicited a highly specific and strong immune response in rabbits (Love et al. 2012). As mentioned earlier by the plant production of the HPV16 L1 VLPs, Noris (2018) presented a plant-based papillomavirus VLP production method that had also been successfully applied not only to other HPVs but also to BPV1. Pietersen et al. (2020) were the first ones to produce the functional BPV1 pseudovirions in plants. The authors transiently expressed the BPV1 L1 and L2 and a self-replicating reporter plasmid in *N. benthamiana* to get not only VLPs but also pseudovirions. The obtained pseudovirions were able to infect mammalian cells and express their encapsidated reporter genes in vitro, and their functionality as reagents in pseudovirion-based neutralization assay was demonstrated (Pietersen et al. 2020).

Natural VLPs

Kirnbauer et al. (1996) immunized calves with the insect cell-produced BPV4 L1 or L1/L2 VLPs. Thirteen of 15 animals were refractory to experimental challenge with high doses of BPV4 and did not develop papillomas, while 9 of 10 control animals developed multiple oral papillomas. However, the VLPs were not efficient as therapeutic vaccine in calves with established papillomas, although the VLP-vaccinated animals appeared to undergo tumor regression more rapidly than nonvaccinated control animals. The antibody responses in the VLP-vaccinated calves were associated with prevention of disease but not with regression of papillomas (Kirnbauer et al. 1996). Schiller (2007) concluded that the putative BPV1 VLP vaccine was not successful in inducing regression of preexisting tumors in cattle, thus lacking therapeutic efficacy.

Hainisch et al. (2012) used the insect cell-produced BPV1 L1 VLP vaccine to vaccinate horses with different doses of the VLPs and Alum as an adjuvant. The trial was successful, since none of 12 vaccinated horses showed adverse reactions upon intramuscular vaccination, which resulted in a long-lasting antibody response against BPV1 (Hainisch et al. 2012). Hainisch et al. (2017) evaluated the protective potential of the BPV1 L1 VLP vaccine against experimental BPV1 and BPV2 challenge and studied the safety and immunogenicity of a bivalent equine papillomavirus type 2 (EcPV2)/ BPV1 L1 VLP vaccine in horses. The Equus caballus papillomavirus 2 (EcPV2) belongs to the *Dyoiotapapillomavirus 1* species of the *Dyoiotapapillomavirus* genus. The monovalent BPV1 L1 VLP vaccine proved highly effective in protecting horses from BPV1-induced pseudo-sarcoid formation. The incomplete protection from BPV2-induced tumor development conferred by the bivalent vaccine was due to the poorer immune response by immune interference or lower cross-neutralization titres to heterologous BPV2 virions (Hainisch et al. 2017). Harnacker et al. (2017) demonstrated that the immunization of horses with the BPV1 L1 VLPs induced long-lasting protection against experimental BPV1 virion-induced disease. Thus, the authors assessed the BPV1 L1 VLP vaccine-mediated long-term protection from experimental tumor formation in seven horses five years after immunization with three different doses of BPV1 L1 VLPs and three unvaccinated control animals. In vaccinated horses, the BPV1 challenge did not result in any apparent lesions irrespective of vaccine dosage and BPV1-neutralizing antibody titers that had dropped considerably over time and below the detection limit in one individual. The control horses developed pseudo-sarcoids at all inoculation sites (Harnacker et al. 2017).

Chimeric VLPs

Genetic Fusions

BPV Vaccines

In full analogy to the previously described chimeric HPV16 L1-E7 VLPs (Müller et al. 1997), a therapeutic vaccine consisting of the chimeric BPV1 L1-E7 VLPs was developed, produced, tested in horses and donkeys in a phase I clinical trial to cure of naturally occurring papillomavirus-associated tumors, and published in two accompanied papers (Ashrafi et al. 2008; Mattil-Fritz et al. 2008). Thus, the three different fragments of the BPV1 E7 gene, namely aa 1–54, 46–100, and 72–127, were inserted into the L1 truncated version lacking the 26 C-terminal aa residues and the chimeric genes were expressed in the insect HighFive cells. Only the construct carrying the E7 aa stretch 1–54 was suitable to produce the large quantities of VLPs required for the subsequent vaccination trial. Remarkably, the yield of the chimeric VLPs was comparable to that of the wild-type BPV1 L1 VLPs. The clinical trial was performed by the immunization of sarcoid-bearing horses (Mattil-Fritz et al. 2008) or donkeys (Ashrafi et al. 2008). However, the effects of the vaccine were appraised as a partially beneficial: Although sarcoid regression was observed in both trials, many sarcoids remained stationary.

In parallel with the previously described chimeras based on the HPV16 L1 and bearing different fragments of the HPV16 L2 as putative HPV vaccines, McGrath (2013) similarly constructed the HPV16 L1 chimera carrying the aa residues 1–88 of the BPV1 L2 and produced it in insect cells. However, the BPV-bearing chimera was a poor vaccine candidate due to low levels of expression with concomitant lack of immunogenicity (McGrath 2013).

HPV Vaccines

Müller et al. (1997), in parallel with the previously described chimeric HPV16 L1 derivatives, generated the first chimeric BPV1 VLPs by fusion of the fragments of the HPV16 E7 gene encoding the aa residues 1–50, 1–60, 1–98, 25–75,

40–98, 50–98 to the C-terminus of the BPV1 L1 missing the last 25 aa residues and expression by the baculovirus system. Peng et al. (1998) fused the HPV16 E7 CTL epitope, mapped at the aa stretch 49-RAHYNIVTF-57, to the C-terminally truncated BPV L1, in parallel with a CTL epitope from HIV-1 (see following description). The CTL responses induced by immunization with the BPV1L1/HPV16E7CTL VLPs protected mice against challenge with E7-transformed tumor cells. Furthermore, a high titer-specific antibody response against BPV1L1 VLPs was also induced, and this antiserum could inhibit papillomavirus-induced agglutination of mouse erythrocytes, suggesting that the antibody may recognize conformational determinates relevant to virus neutralization (Peng et al. 1998). Moreover, Liu XS et al. (1998) found that the mice immunized with the BPV1L1/HPV16E7CTL VLPs intranasally elicited cellular and humoral immunity and that the humoral immunity extended to production of the VLP-specific antibodies at mucosal surface.

Greenstone et al. (1998) fused the full-length HPV16 E7 gene coding for 98 aa to the 3' end of the full-length gene of BPV1 L2 coding for 469 aa in parallel with the previously described fusion of the E7 to the HPV16 L2 and got the virion-like VLPs by the baculovirus-driven expression of this chimeric L2 gene together with the BPV L1.

Slupetzky et al. (2007) generated the chimeric BPV1 L1 VLPs, in which HPV16 L2 neutralization epitopes comprising L2 aa residues 69–81 or 108–120 were inserted within an immunodominant surface loop, between aa 133/134, of the BPV1 L1. The immunization of rabbits with the chimeric VLPs induced L2-specific antibodies with high titers.

Schellenbacher et al. (2009) inserted the peptides comprising aa 2–22, 13–107, 18–31, 17–36, 35–75, 75–112, 115–154, 149–175, and 172–200 of the HPV16 L2 into the DE surface loop of the BPV1 L1 protein. Except for chimeras 35–75 and 13–107, the fusion proteins assembled into the VLPs. Remarkably, the vaccination of rabbits with Freund's adjuvanted VLPs induced higher HPV L2-specific antibody titers than vaccination with corresponding SDS-denatured proteins in both rabbits and mice. The immune sera to epitopes within aa residues 13 to 154 neutralized HPV16 in the pseudo-virion neutralization assays, whereas chimera 17–36 induced additional cross-neutralization to divergent high-risk HPV18, 31, 45, 52, and 58, as well as to low-risk HPVs.

Other Vaccines

The generation of the first chimeric BPV1 VLPs was performed simultaneously with that of the development of the previously described chimeric HPV VLP carriers. Thus, Peng et al. (1998) fused the H-2^d-restricted CTL epitope corresponding to aa residues 318-RGPGRAFVTI-327 from the HIV IIIB gp160 protein, to the C-terminally truncated BPV L1, in parallel with the HPV16 E7 CTL epitope described earlier. The mice immunized with these hybrid VLPs mounted strong CTL responses against the relevant target cells in the absence of any adjuvants (Peng et al. 1998).

This study clearly demonstrated the ability of the BPV1 carrier to present foreign CTL epitopes. Next, Slupetzky et al. (2001) designed the BPV1 L1 chimeras that incorporated the 8-aa B-cell epitope LELDKWAS from the HIV-1 gp41 envelope glycoprotein in three different hypervariable regions, between aa residues 133/134, 282/286 (replacing aa 283–285), or 351/355 (replacing aa 352–354). Remarkably, the electron micrographs demonstrated pentamers and full-size or smaller VLPs, indicating that the HIV-1 peptide insertion did not prevent the L1 self-assembly into higher-ordered structures in the baculovirus expression system (Slupetzky et al. 2001).

Liu WJ et al. (2000) constructed the putative chimeric BPV1-HIV VLPs by replacing the 23-aa C-terminal aa stretch of the BPV1 L1 with an artificial "polytope" minigene, containing known CTL epitopes of human HPV16 E7 protein, HIV IIIB gp120 P18 (a shortened form of the V3 loop), Nef, reverse transcriptase proteins, and an HPV16 E7 linear B epitope. The chimeric L1 protein assembled into VLPs when expressed in Sf9 cells by recombinant baculovirus and the purified VLPs induced in mice, in the absence of any adjuvant, serum antibodies which reacted with both polytope VLPs and wild-type BPV1 L1 VLPs. The CTL precursors specific for the polytope CTLs were also detected in the spleen of immunized mice. Thus, the polytope VLPs were able to deliver multiple B and T epitopes as immunogens to the MHC class I and class II pathways, extending the utility of VLPs as self-adjuvanting immunogen delivery systems (Liu WJ et al. 2000). Liu XS et al. (2002) concentrated on the BPV1 L1 VLPs carrying the immunodominant sequences of the HIV-1 gp120, either the V3 loop or the shorter peptide P18 containing the known CTL epitope. Remarkably, the denatured VLPs induced a much reduced immune response when compared with the native VLPs, and the immune responses following mucosal administration of the VLPs were generally weaker than following systemic administration (Liu XS et al. 2002).

Zhang H et al. (2004) engineered the BPV1 L1-HIV-1 gp41 fusion protein in which ELDKWA of gp41 was inserted into the N-terminus of BPV1 L1, aa residues 130 to 136. The expression of the fused gene in insect cells led to the assembly of the chimeric VLPs that had sizes similar to those of the BPV1 virions and were able to bind to the cell surface and penetrate the cell membrane. Importantly, the oral immunization of mice with the chimeric VLPs induced gp41-specific serum immunoglobulin G (IgG) and intestinal secretory IgA (Zhang H et al. 2004).

Continuing this idea, Zhai Y et al. (2013) exposed on the BPV1 VLPs the HIV-1 neutralizing epitopes from the membrane-proximal external region (MPER) of the gp41. The chimeric VLPs presenting MPER domain resembled the HIV-1 natural epitopes better than the chimeric VLPs presenting single epitopes. The oral immunization of mice with the chimeric VLPs displaying the MPER domain or single epitopes elicited epitope-specific serum IgGs and mucosal secretory IgAs, and the antibodies induced by the chimeric VLPs presenting MPER domain were able to partially neutralize HIV-1 viruses from clade B and clade C (Zhai Y et al. 2013).

The BPV1 L1 carrier was applied successfully to the induction of autoantibodies against selfantigens. Thus, Chackerian et al. (1999) inserted a peptide representing an extracellular loop of the mouse chemokine receptor CCR5, namely HYAANEWVFGNIMCKV, into the BPV1 L1 immunodominant sites, replacing the L1 stretches of aa 130–136, 275–285, or 344–350. The anti-CCR5 autoantibodies that were induced in mice not only inhibited binding of its ligand RANTES but also blocked HIV-1 infection of an indicator cell line expressing a human–mouse CCR5 chimera (Chackerian et al. 1999).

Concerning Alzheimer's disease, Zamora et al. (2006) generated a chimeric fusion protein by incorporating the N-terminal 9 aa residues of Aβ, namely 1-DAEFRHDSG-9, which contained a B cell epitope but did not include known T cell epitopes, into a hypervariable region of the BPV1 L1 between aa 133/134. Following expression and self-assembly in insect cells, the Aβ-epitope was exposed at high density (360 times) on the surface of the assembled BPV1 L1-Aβ VLPs. The latter induced an effective humoral immune response to Aβ in a mouse model (Zamora et al. 2006).

Handisurya et al. (2007) engineered the chimeric BPV1 L1 VLPs that displayed a 9-aa B-cell epitope, 144-DWEDRYYRE-152, of the murine/rat prion protein in an immunogenic capsid surface loop, between aa 133/134. The immunization with the chimeric VLPs induced high-titer antibodies to prion in rabbits and rats, without inducing overt adverse effects.

Chemical Coupling

The first progressive attempt to uncouple expression and purification of the papillomavirus VLP scaffold and the heterologous protein to be displayed, with subsequent conjugation of the desired antigen to the VLP surface, was performed by Chackerian et al. (2001). Thus, the biotinylated BPV1 L1 VLPs were decorated with a streptavidin fusion of the self-polypeptide TNF-α. This composition elicited high-titer protective autoantibodies in a mouse model for type II collagen-induced arthritis and paved the way to the further development of the VLP-based therapeutic tools against arthritis (Chackerian et al. 2001).

The same approach was used by the generation of a conjugated VLP-based HIV vaccine that induced a strong anti-CCR5 autoantibody response in pig-tailed macaques (Chackerian et al. 2004). The N-terminal domain of pig-tailed macaque CCR5 was fused to streptavidin that, when conjugated at high density to the biotinylated BPV L1 VLPs, induced high-titer IgG that bound to a macaque CCR5-expressing cell line and could block infection of CCR5-tropic simian/human immunodeficiency virus (SHIV) in vitro.

Li Q et al. (2004) engineered a potential Alzheimer's disease vaccine, where the biotinylated Aβ1–40 peptide was linked to the biotinylated insect cell-produced BPV1 VLPs using streptavidin.

Pejawar-Gaddy et al. (2010) generated a tumor vaccine candidate based on conjugation of a MUC1 peptide to polyionic BPV1 VLPs. To do this, the authors displayed a repetitive array of polyanionic docking sites, which could serve for covalent coupling of polycationic fusion proteins, on the surface of the insect cells-produced BPV1 VLPs. Thus, the polyglutamic-cysteine stretch was inserted into the BC, DE, HI loops and the H4 helix of the BPV1 L1, while the assembled VLPs were yielded in insect cells only from the HI loop insertion. The insertion in the DE loop and H4 helix resulted in partially formed VLPs and capsomeres, respectively. Then, the polyanionic sites on the surface of the BPV1 VLPs and capsomeres were decorated with a polycationic MUC1 peptide containing a polyarginine-cysteine residue fused to twenty aa residues of the MUC1 tandem repeat through electrostatic interactions and redox-induced disulfide bond formation. The MUC1 fully assembled VLP-induced robust activation of bone-marrow-derived dendritic cells, which could then present MUC1 antigen to MUC1-specific T cell hybridomas and primary naïve MUC1-specific T cells obtained from a MUC1-specific TCR transgenic mice. The immunization of human MUC1 transgenic mice, where MUC1 is a self-antigen, with the VLP vaccine induced MUC1-specific CTL, delayed the growth of MUC1 transplanted tumors and elicited complete tumor rejection in some animals (Pejawar-Gaddy et al. 2010).

PACKAGING AND DELIVERY

Zhao et al. (1998) evaluated the encapsidation of circular DNA by the BPV1 VLPs in mammalian Cos-1 cells. The BPV1 VLPs assembled from the protein L1 alone packaged little plasmid DNA, whereas VLPs assembled from BPV1 L1 and L2 packaged plasmid DNA at least 50 times more effectively. BPV-1 L1/L2 VLPs packaged a plasmid containing BPV-1 sequence 8.2 ± 3.1 times more effectively than a plasmid without BPV1 sequences. Using a series of plasmid constructs comprising a core BPV1 sequence and spacer DNA, the BPV1 pseudovirions accommodated a maximum of about 10.2 kb of plasmid DNA, and that longer closed circular DNA was truncated to produce less-dense virions with shorter plasmid sequences. Zhao et al. (2000) studied the BPV1 L1/L2 VLP packaging not only in mammalian Cos-1 but also in insect *Sf*9 cells. The packaging efficiency of the model plasmid was estimated at one plasmid per 10^4 VLPs in both Cos-1 and *Sf*9 cells. In each cell type, the expression of the BPV1 early region protein E2 in trans increased the packaging efficacy. Remarkably, the E2-mediated enhancement of packaging favored 8-kb plasmid incorporation over incorporation of shorter DNA sequences and the resultant BPV1 pseudovirions incorporated significant amounts of E2 protein. Liu Y et al. (2001) used the BPV1 pseudovirions produced in Cos-1 cells to deliver an 8-kb plasmid carrying a GFP reporter gene. Shi et al. (2001), in parallel with the previously described use

of the HPV16 L1 pseudovirions, packaged the DNA that encoded the well-studied CTL epitope gp33 of LCMV also into the insect cells-produced BPV1 VLPs.

Buck et al. (2004) published a method for efficient intracellular production of BPV1 and HPV16 gene transfer vectors carrying reporter plasmids in mammalian 293 cells. The production of these vectors required both L1 and L2. Although earlier data had suggested a potential role for the viral early protein E2, the expression of the E2 gene did not enhance the intracellular production of the BPV1 vectors. It was also possible to encapsidate reporter plasmids devoid of BPV1 DNA sequences. The BPV1 vector production efficiency was significantly influenced by the size of the target plasmid being packaged. The use of 6-kb target plasmids resulted in the BPV1 vector yields that were higher than those with target plasmids closer to the native 7.9-kb size of papillomavirus genomes (Buck et al. 2004). The methodological aspects of the production of the BPV1 pseudovirions were unveiled by Buck et al. (2005a).

Pietersen et al. (2020) suggested plants as an alternate source of the BPV1 pseudovirions. Thus, the BPV1 L1 and L2, together with a self-replicating reporter plasmid, were transiently expressed in *N. benthamiana* to produce the BPV1 pseudovirions. The latter demonstrated their ability to infect mammalian cells and express their encapsidated reporter genes in vitro, as well as their functionality as putative reagents by pseudovirion-based neutralization assay (Pietersen et al. 2020). In fact, this was the first success by the generation of the BPV pseudovirions in plants, which demonstrated great potential for the development of therapeutic veterinary vaccines in planta.

Concerning the visualization, Peng et al. (1999) were the first to construct and produce in *Sf*9 cells the chimeric fluorescent BPV1 VLPs carrying the L2 protein in which GFP was inserted into the N-terminal region at aa residue 88. The fluorescent VLPs could also be assembled from a GFP/L2 fusion protein in which part of the L2 sequence had been deleted. In vitro, the fluorescent VLPs could bind to the mammalian CV-1 cells, and this VLP/cell interaction could be analyzed by FACS assay (Peng et al. 1999). Next, the GFP-labeled BPV1 VLPs were made by cotransfecting 293TT cells with plasmids encoding BPV1 L1 and a BPV1 L2-GFP fusion protein (Lenz et al. 2003). Such BPV1-GFP VLPs were used in the previously mentioned study of Chackerian et al. (2008).

Watanabe et al. (2020) produced the chimeric BPV6 L1 VLPs displaying an entire eGFP as a model on its surface. The authors predicted the 3D structure of the BPV6 L1 and selected the putative insertion sites for the foreign protein. Only the fusion protein in which eGFP was inserted between aa residues 136 and 137 of the BPV6 L1 self-assembled into VLPs and did not exhibit hindrance of the eGFP conformation by the baculovirus-driven expression. The immunized mice demonstrated specific IgG response against both BPV6 and eGFP (Watanabe et al. 2020).

GENUS *KAPPAPAPILLOMAVIRUS*

The cottontail rabbit papilloma virus (CRPV), also known as Shope papilloma virus (SPV), belonging to the *Kappapapillomavirus 2* species, is a reference strain of the *Kappapapillomavirus* genus. CRPV infects rabbits, causing keratinous carcinomas resembling horns, typically on or near the animal's head (Bernard et al. 2012).

Kirnbauer et al. (1993) obtained the virion-like VLPs by the baculovirus-driven expression in *Sf*9 cells of the CRPV L1 and L1/L2 genes, in parallel with the corresponding genes of BPV1, HPV6, and HPV11, indicating that the capsid proteins from a variety of papillomaviruses had the intrinsic capacity to self-assemble into the VLPs. The corresponding VLPs are shown in Figure 7.6. Breitburd et al. (1995) tested the insect cells-produced CRPV L1 or L1/L2 VLPs as a vaccine in rabbits. The groups of rabbits were immunized with native or denatured VLPs from CRPV or BPV1 with Freund's or Alum adjuvant opposing flanks. In contrast to control groups immunized with native or denatured BPV1 L1/L2 or with the denatured CRPV L1/L2, the animals inoculated with the native CRPV L1 or L1/L2 VLPs developed fewer lesions, which all regressed except for those on one rabbit, and none developed cancer within one year of infection. The rabbits vaccinated with the native CRPV VLPs developed high-titer antibodies and passive transfer of serum or immunoglobulin G from the rabbits immunized with the CRPV VLPs protected against CRPV challenge (Breitburd et al. 1995).

In parallel, Christensen et al. (1996) immunized rabbits with the insect cells-produced CRPV L1 VLPs, where the HPV11 L1 VLPs were used as a negative control. At two weeks after the CRPV L1 VLP immunizations, the rabbits were completely protected against virus challenge. Moreover, the strong protection was also observed at both 6 and 12 months after the CRPV L1 VLP immunizations. Therefore, the results demonstrated the strong and long-lasting protection against experimental challenge with CRPV by the CRPV L1 VLP immunization of rabbits (Christensen et al. 1996). It should be noted that the insect cells-produced CRPV L1 VLPs from the John Schiller' lab were used by Sundaram et al. (1997), who demonstrated that the intracutaneous vaccination of rabbits with the CRPV L1 gene protects against virus challenge.

Jansen et al. (1995) produced the CRPV VLPs consisting of the capsid proteins L1 or L1/L2 in yeast *S. cerevisiae*. Moreover, the three immunizations with the Ll VLPs formulated on Alum adjuvant efficiently protected rabbits from challenge with CRPV. The sera of immunized rabbits contained high-titer CRPV-neutralizing antibodies.

Kohl et al. (2006) expressed the CRPV L1 gene transgenically by a tobacco mosaic virus (TMV) vector in plant *Nicotiana spp.* The protein did not detectably assemble into the VLPs; however, the immunoelectron microscopy showed presumptive pentamer aggregates and extracted protein reacted with conformation-specific and neutralizing monoclonal antibodies. The sera of the immunized rabbits reacted with the baculovirus produced CRPV L1; however, they did not detectably neutralize infectivity in an in vitro assay. The vaccinated rabbits were, nevertheless, protected against wart development on subsequent challenge with live virus (Kohl et al. 2006). In parallel to the plant-driven expression, Kohl et al. (2006) produced the CRPV L1 VLPs by the baculovirus expression system. Using these insect cells-produced VLPs, Govan et al. (2008) explored whether regression of the CRPV-induced papillomas could be achieved following the CRPV L1 VLP immunization of rabbits. In fact, the gained data demonstrated the therapeutic potential of the papillomavirus VLPs in a well-understood animal model with potential important implications for human therapeutic vaccination (Govan et al. 2008).

Mejia et al. (2006) developed a model using chimeric HPV capsid/CRPV genome particles to permit the direct testing of the HPV VLP vaccines in rabbits. The animals vaccinated with the CRPV, HPV16, or HPV11 VLPs were challenged with both homologous (CRPV capsid) and chimeric (HPV16 capsid) particles, and the strong type-specific protection was observed, demonstrating the potential application of this approach (Mejia et al. 2006).

GENUS *LAMBDAPAPILLOMAVIRUS*

The canine oral papillomavirus (COPV) belonging to the *Lambdapapillomavirus 2* species is a reference strain of the *Lambdapapillomavirus* genus. The members of this genus infect cats and dogs, causing mucosal and cutaneous lesions (Bernard et al. 2012).

Suzich et al. (1995) obtained the virion-like COPV L1 VLPs by the baculovirus expression system in *Sf*9 cells.

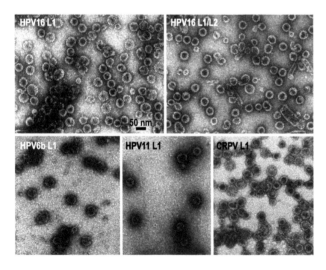

FIGURE 7.6 The purified HPV16 L1 and HPV16 L1/L2, HPV6b L1, HPV11 L1, and CRPV L1 VLPs. The particles were purified from recombinant baculovirus-infected insect cells on preparative CsCl gradients, stained with uranyl acetate, and examined by TEM. Magnification, × 36,000. Bar, 50 nm. (Adapted with permission of American Society for Microbiology from Kirnbauer R et al. *J Virol.* 1993;67:6929–6936.)

The purified VLPs were injected intradermally into the foot pad of beagles. The vaccinated animals developed circulating antibodies against COPV and became completely resistant to experimental challenge with COPV. Moreover, the serum immunoglobulins passively transferred from the COPV L1 VLP-immunized beagles to naïve beagles conferred protection from experimental infection with COPV. Using such insect cells-produced COPV L1 VLPs as a therapeutic vaccine, Kuntsi-Vaattovaara et al. (2003) resolved in a Siberian husky dog the persistent oral papillomatosis that that did not regress spontaneously and was refractory to surgical treatment over a six-month period. The regression of the papillomas was achieved by administering a series of experimental vaccinations starting at the time of the last surgery. The vaccine acted therapeutically, causing the papillomas that had regrown to shrink, and no side-effects were noted (Kuntsi-Vaattovaara et al. 2003).

The systematical analysis of the COPV L1 VLPs produced in insect cells showed that 26 C-terminal aa are abundant for self-assembly, but further truncation of the L1 C-terminus for 67 aa resulted in a capsid protein that formed VLPs but that failed to express conformational epitopes (Chen Y et al. 1998). The deletion of the first 25 N-terminal aa residues of the COPV L1 also abolished expression of conformational epitopes without altering VLP formation, but the native conformation of this deletion mutant could be restored by the addition of the N-terminus of the HPV11 L1 (Chen Y et al. 1998).

Yuan et al. (2001) expressed the COPV L1 protein as a glutathione S-transferase (GST) fusion protein in *E. coli* and assayed its immunogenic activity in an established COPV model that previously validated the efficacy of VLP vaccines. The GST-COPV L1 fusion protein formed pentamers, but these capsomere-like structures did not assemble into the VLPs. However, despite the lack of VLP formation, the GST-COPV L1 protein retained its native conformation as determined by reactivity with conformation-specific anti-COPV antibodies. Most importantly, the GST-COPV L1 pentamers completely protected dogs from high-dose viral infection of their oral mucosa (Yuan et al. 2001).

GENUS *PIPAPILLOMAVIRUS*

The *Rattus norvegicus papillomavirus* 1 (RnPV2)—a reference strain of the genus—and the popular Mus musculus, or murine, papillomavirus type 1 (MusPV, MmuPV1) that naturally infects laboratory mice both belong to the species *Pipapillomavirus 2*.

Joh et al. (2012) produced the virion-like MmuPV1 L1 VLPs by the baculovirus expression system in *Sf*9 cells and used them in ELISA screening assays. Furthermore, Joh et al. (2014) searched for the proper initiating site of the L1 protein to generate the optimal MmuPV1 VLPs. Thus, the MmuPV1 L1 gene had four Met codons at the aa positions 1, 2, 28, and 30. The three L1 genes of MmuPV1, starting from the second, third, and fourth Mets, were expressed using the baculovirus expression system and characterized for their ability to self-assemble into the VLPs. While the MmuPV1 L1 starting from the second Met expressed an L1 protein that did not fold into VLPs, the L1 starting from the third and fourth Mets generated the correct VLPs in abundant quantities (Joh et al. 2014). Joh et al. (2016) found that the passive transfer of hyperimmune sera from normal congenic mice immunized with the MmuPV1 L1 VLPs to T cell-deficient strains of mice prevented infection by virions of experimental mice.

GENUS *UPSILONPAPILLOMAVIRUS*

The members of this genus are associated with genital lesions in cetaceans. Remarkably, the genus members lack an E7 ORF (Van Doorslaer et al. 2018). In dolphins and porpoises, a high prevalence of orogenital tumors was documented with at least four distinct novel species-specific papillomavirus types detected in such lesions. Rehtanz et al. (2009) generated the corresponding VLPs to establish a serological screening test to determine the prevalence of papillomavirus infection in Atlantic bottlenose dolphins *Tursiops truncatus*. Using the baculovirus expression system, the VLPs were generated from the L1 proteins of the two virus types, Tursiops truncatus papillomavirus 1 (TtPV1) of the *Upsilonpapillomavirus 1* species, and Tursiops truncatus papillomavirus 2 (TtPV2) of the *Upsilonpapillomavirus 2* species. The polyclonal antibodies against the TtPV VLPs were produced in rabbits, and their specificity for the VLPs was confirmed. The electron microscopy and ELISA studies revealed that the generated virion-like VLPs presented the correctly folded conformational immunodominant epitopes and could be regarded as a potential TtPV vaccine (Rehtanz et al. 2009). The TtPV1 L1 VLP-based ELISA test was used to demonstrate antibody prevalence in bottlenose dolphins and led to conclusion that the papillomavirus infection is common, while the main route of virus transmission among dolphins may be horizontal and orogenital neoplasia may develop in early life stages of certain free-ranging bottlenose dolphins.

GENUS *ZETAPAPILLOMAVIRUS*

The Equus caballus, or equine, papillomavirus 1 (EcPV1) belonging to the *Zetapapillomavirus 1* species, is a reference strain of the *Zetapapillomavirus* genus. The members of this genus produce cutaneous infections in horses (Van Doorslaer et al. 2018). To produce the EcPV1 VLPs, the L1 protein was expressed in insect cells using a recombinant baculovirus vector (Ghim et al. 2004). The self-assembled EcPV1 VLPs were morphologically indistinguishable from the wild-type papillomavirus virions. The monoclonal antibodies were developed against the intact and denatured EcPV1 VLPs. When tested by ELISA, all monoclonal antibodies produced against intact and some against denatured EcPV1 VLPs reacted with the intact EcPV1 VLPs only, demonstrating that the VLPs carried type-specific conformational as well as linear epitopes on their surface.

The insect cells-produced EcPV1 VLPs were suggested as a basis for a putative noninfectious vaccine to prevent and eradicate equine cutaneous papillomatosis (Ghim et al. 2004). Schellenbacher et al. (2015) found a novel equine papillomavirus EcPV2 and engineered the EcPV2 L1 VLPs by the baculovirus expression system. The vaccination of rabbits and mice with the EcPV2 L1 VLPs using Freund's or Alum respectively as adjuvant induced high-titer neutralizing serum antibodies. Moreover, the passive transfer with the rabbit EcPV2-VLP immune sera completely protected mice from the experimental vaginal EcPV2 pseudovirion infection (Schellenbacher et al. 2015). As mentioned earlier in the BPV1 VLP presentation, Hainisch et al. (2017) used not only the monovalent BPV1 but also a bivalent EcPV2/BPV1 L1 VLP vaccine against experimental BPV1 and BPV2 challenge in horses. The immunization with the bivalent vaccine was safe, resulted in lower median day 42 antibody titers, and conferred significant yet incomplete cross-protection from BPV2-induced tumor formation, with 11/14 horses developing small, short-lived papules, while the control horses developed pseudo-sarcoids at all inoculation sites.

8 Order *Sepolyvirales*

Ideas are like rabbits. You get a couple and learn how to handle them, and pretty soon you have a dozen.

John Steinbeck

ESSENTIALS

As mentioned in Chapter 7, the order *Sepolyvirales*, together with the related order *Zurhausenvirales*, forms the class *Papovaviricetes*, which in turn is linked to the phylum *Cossaviricota*, together with the two classes *Mouviricetes* and *Quintoviricetes*, all belonging to the kingdom *Shotokuvirae* of the realm *Monodnaviria*. Therefore, like papillomaviruses from the *Zurhausenvirales* order, the polyomaviruses that are the members of the *Sepolyvirales* order are included into the realm *Monodnaviria*, together with the classical single-stranded DNA viruses after Baltimore's classification. The reason is that the realm *Monodnaviria*, which comprises viruses with predominantly single-stranded DNA genomes, involves the viruses encoding rolling-circle replication initiation endonucleases of the HUH superfamily, in contrast to the "true" dsDNA viruses of the realms *Adnaviria*, *Duplodnaviria*, and *Varidnaviria* (Krupovic et al. 2021) and the realm-unassigned order *Lefavirales* that all were described previously in Chapters 1 to 6.

The order *Sepolyvirales* currently involves the sole family *Polyomaviridae*, 6 genera, and 117 species altogether. The two genera *Alphapolyomavirus* and *Betapolyomavirus* are the most populated and well studied, while five polyomavirus species still remain genera-independent. As summarized by the official ICTV reports (Norkin et al. 2012; Moens et al. 2017), the members of the *Polyomaviridae* family infect mammals, birds, and fish and may cause different forms of neoplasia, while each family member has a restricted host range. Thus, Merkel cell polyomavirus (MCPyV) and raccoon polyomavirus (RacPyV) are associated with cancer in their host. Other members are human and veterinary pathogens causing symptomatic infection in their natural host, whereas the clinical manifestations are observed primarily in immunocompromised patients (Moens et al. 2017).

Figure 8.1 presents the polyomaviruses (PyVs) as non-enveloped icosahedral T = 7 particles of around 40–45 nm in diameter. The polyomavirus virion consists of 72 capsomers, each of which contains a pentamer of major capsid proteins VP1 surrounding one minor capsid protein, either VP2 or VP3. The minor capsid proteins are encoded by the same coding DNA sequence, but translation of VP3 is initiated downstream of VP2, forming an N-terminally truncated form of VP2 (for detailed reviews see Dalianis (2012) and Teunissen et al. (2013)). For most mammalian

polyomaviruses, VP2 and VP3 are minor capsid proteins, while bird polyomaviruses have an additional unique VP4. The capsomers are interlinked by the C-terminal arm of VP1. The capsomer contacts are further stabilized by calcium ions and disulfide bonds between the pentamers. A single copy of VP2 or VP3 binds in a hairpin manner into the cavity on the internal face of each pentamer. The VP4 of bird polyomaviruses is located between VP1 and the viral genome. Remarkably, both T = 7 *dextro* and *laevo* symmetries were found in the polyomavirus particles. In contrast to the capsids of typical representatives of polyomaviruses, such as murine polyomavirus (MPyV) and simian virus 40 (SV40), which possess T = 7d symmetry, hamster polyomavirus (HaPyV) assembles in accordance with the T = 7l symmetry (Siray et al. 1999). Generally, the early data on the PyV capsid assembly (Montross et al. 1991) suggested not only common structural features between the *Polyomaviridae* and *Papillomaviridae* families but also similar control of the intracellular site and mechanism of the self-assembly. The modern assessment of the polyomavirus evolution and taxonomy was recently presented by Ehlers et al. (2019).

GENOME

Figure 8.2 shows the genomic structure of the *Polyomaviridae* members, where each capsid packages a single copy of the viral circular dsDNA of ~5 kb encoding for 5–9 proteins and the mature viral genome is organized as a minichromosome packed with histone proteins. The genomes of members of the recognized species vary from 3,962 bp for giant guitarfish polyoma virus (GfPyV1) from the species *Rhynchobatus djiddensis polyomavirus 1* to 7,369 bp for black-sea-bass-associated polyomavirus 1 (BassPyV1) from the species *Centropristis striata polyomavirus 1*. Of the known human polyomaviruses, Merkel cell polyomavirus (MCPyV), species *Human polyomavirus 5*, has the largest genome of 5,387 bp, and St. Louis polyomavirus (STLPyV), species *Human polyomavirus 11*, has the smallest one of 4,776 bp. Among known bird polyomaviruses, budgerigar fledgling disease virus (BFDV) from the species *Aves polyomavirus 1* has a genome of 4,981 bp, and the largest so far known avian polyomavirus genome for canary polyomavirus (CaPyV) from the species *Serinus canaria polyomavirus 1* is 5,421 bp (Calvignac-Spencer et al. 2016).

The extraordinary role of polyomaviruses by the development of the VLP nanotechnologies was accented in the exhaustive reviews (Ulrich et al. 2008; Dalianis 2015; Diederich et al. 2015; Suchanová et al. 2015; Zhang W et al. 2017).

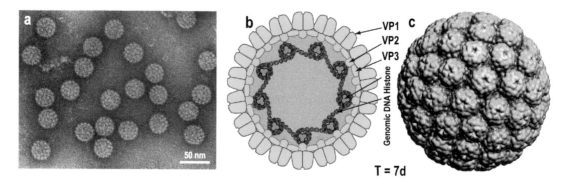

FIGURE 8.1 A portrait, cartoon, and the 3D structure of typical representatives of the *Polyomaviridae* family. (a) The electron micrograph of polyomavirus particles suspended in stain over a hole in the carbon substrate. This method ensures "two-side" images, i.e., superposition patterns of detail from the near and far sides of the virus particles. The film of stain was broken and contracted, and, although the particles are somewhat compressed, the distortion in this field is fairly isometric. (Reprinted with permission of Microbiology Society from Finch JT. *J Gen Virol*. 1974;24:359–364.) (b) The cartoon taken with kind permission from the ViralZone, Swiss Institute of Bioinformatics, http://viralzone.expasy.org (Hulo C et al. 2011). (Courtesy Philippe Le Mercier.) (c) The near-atomic structure of polyomavirus SV40, PDB ID: 1SVA, x-ray diffraction at 3.10 Å resolution, outer diameter 494 Å (Stehle et al. 1996), presented as rendered surface and taken from the VIPERdb (http://viperdb.scripps.edu) database (Carrillo-Tripp et al. 2009). The structure possesses the icosahedral T = 7d symmetry.

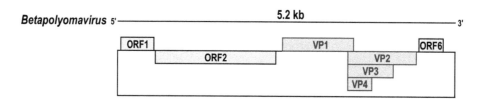

FIGURE 8.2 The genomic structure of the *Polyomaviridae* family represented by the genome of simian virus 40 (SV40) from the species *Macaca mulatta polyomavirus 1* of the genus *Betapolyomavirus*. The structural genes are colored dark pink. The circular character of the genome is displayed by the lower connecting bracket, while the linearization is performed with *ori* as the opening site.

GENUS *ALPHAPOLYOMAVIRUS*

MOUSE POLYOMAVIRUS

Although the *Human polyomavirus 9* species is mentioned as a reference strain of the genus, mouse polyomavirus (MPyV) of the species *Mus musculus polyomavirus 1* remains one of the most studied genus members. It should be noted that MPyV can potentially infect a very broad range of cells and produces a wide variety of tumors in its natural host. Even in the late 1970s, Adolph et al. (1979) resolved the crystal structure of the MPyV virion to 30 Å, the high resolution 3.65-Å and 3.8-Å structures of, respectively, MPyV (Stehle et al. 1994; Stehle and Harrison 1996) and the similar SV40 virions (Liddington et al. 1991) were achieved by x-ray crystallography more than ten years later. The polyomavirus capsid was thus comprised of 72 pentamers of 360 copies of the 42.5-kDa VP1 protein associated with the minor virion proteins VP2 and VP3. Like the pentamers of papillomaviruses, the MPyV was arranged on the T = 7d icosahedral lattice (Belnap et al. 1996). The MPyV structure is presented in Figure 8.3 and accompanied by the 3D structures of the most studied representatives of the *Polyomaviridae* family. The MPyV structure was determined in complex with model compounds for

both straight-chain and branched-chain sialoglycoconjugates, which were bound to a shallow groove on the surface of VP1. The data collected from crystals soaked at different oligosaccharide concentrations established that both receptor fragments had similar, low affinities for the virion. The MPyV structure was compared with that of the SV40 described later, which did not recognize sialylated oligosaccharides as putative receptors. The comparison of the loop structures at the surface revealed the molecular basis for the different specificities of these structurally very similar viruses (Stehle and Harrison 1996).

Brady and Consigli (1978) were the first ones to show that the treatment of the purified MPyV virions with 6 M guanidine-hydrochloride and 0.01 M β-mercaptoethanol resulted in the dissociation and separation of the capsid VP1, VP2, and VP3 proteins and histone proteins VP4–7, while the renaturation of the purified VP1 protein resulted in the formation of subunits that were morphologically, biophysically, and immunologically similar to native virion capsomeres.

The MPyV VP1 protein was expressed in *E. coli* as early as the 1980s (Leavitt et al. 1985) and appeared as dissociated pentameric capsomeres capable of the spontaneous in vitro assembly into the virion-like structures and

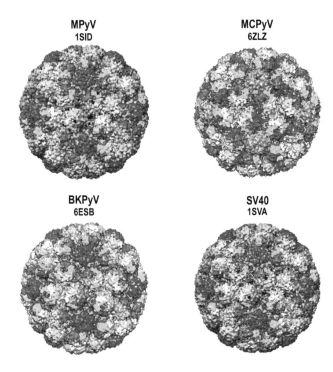

MPyV
1SID

MCPyV
6ZLZ

BKPyV
6ESB

SV40
1SVA

FIGURE 8.3 The near-atomic structure of the *Polyomaviridae* family members: mouse polyomavirus (MPyV), species *Mus musculus polyomavirus 1*, genus *Alphapolyomavirus*, x-ray diffraction at 3.65 Å resolution, outer diameter 510 Å (Stehle and Harrison 1996); Merkel cell polyomavirus (MCPyV), species *Human polyomavirus 5*, genus *Alphapolyomavirus*, x-ray diffraction at 3.52 Å resolution, outer diameter 504 Å (Bayer et al. 2020); BK polyomavirus (BKPyV), species *Human polyomavirus 1*, genus *Betapolyomavirus*, electron cryomicroscopy at 3.40 Å, outer diameter 500 Å (Hurdiss et al. 2018); simian virus 40 (SV40), species *Macaca mulatta polyomavirus*, genus *Betapolyomavirus*, x-ray diffraction at 3.10 Å resolution, outer diameter 494 Å (Stehle et al. 1996). The 3D structures are taken from the VIPERdb (http://viperdb.scripps.edu) database (Carrillo-Tripp et al. 2009). The size of particles is in scale. All structures possess the icosahedral T = 7d symmetry. The corresponding protein data bank (PDB) ID numbers are given under the appropriate virus names.

polymorphic aggregates (Salunke et al. 1986, 1989). The C-terminal truncation of the VP1 protein blocked assembly of capsomeres (Garcea et al. 1987), but its N-terminus was found to be responsible for DNA binding and nuclear localization (Moreland et al. 1991). The *E. coli*-produced proteins VP2 and VP3 of MPyV interacted with the purified VP1 in vitro, affected the biochemical properties of the VP1 capsomeres, and changed their epitope accessibility (Cai et al. 1994; Delos et al. 1995). The crystal structure of a complex of the single VP2/VP3 copy with the pentameric major capsid protein VP1 was determined at 2.2 Å resolution (Chen XS et al. 1998). Braun et al. (1999) expressed the VP1 gene in *E. coli* as a fusion with the completely removable N-terminal His$_6$ tag. The pentameric morphology of the recombinant VP1 protein was confirmed by electron microscopy after affinity chromatography and factor Xa cleavage under conditions of low ionic strength. The

self-assembly of the VP1 VLPs was induced by increasing the ionic strength with $(NH_4)_2SO_4$. These VP1 VLPs were packed in vitro with antisense oligonucleotides and plasmid DNA (Braun et al. 1999). Furthermore, Yang and Chen (2000) generated the *E. coli*-produced VP1-based MPyV pseudovirions, both in vitro and in vivo, which were able to transfer the exogenous DNA with following expression of the encoded reporter genes to mammalian cells or to the livers of Wistar rats. Stubenrauch et al. (2000) inserted a polyionic sequence of eight glutamic acid residues into the exposed VP1 loop to improve purification properties of the MPyV VP1 VLPs in *E. coli*.

In the early 2000s, the Anton P.J. Middelberg's team increased outcome of the MPyV VP1 by an order of magnitude above yields reported before and demonstrated expression levels of up to 180 mg L^{-1} (Chuan et al. 2008b). These yields were achieved from low-cell density *E. coli* cultures using a simple and low cost strategy and considering the effects of host strain, plasmid, inducer concentration, pre-induction temperature, and cell density at induction with design of experiment. Remarkably, these data supported the notion that full codon optimization might be unnecessary to improve expression of viral genes rich in *E. coli* rare codons, while using a strategically modified host cell could provide a simpler and cheaper alternative. The study of Chuan et al. (2008b) justified further research of the cell-free in vitro assembly route for VLPs, in both manufacturing and preparative settings. Figure 8.4 demonstrates the excellent *E. coli*-produced MPyV VLPs. Furthermore, Liew et al. (2010) optimized the biotechnological MPyV VP1 process in *E. coli* by the production of pentamers, obtained as glutathione-S-transferase (GST) fusion proteins GST-VP1, and their purification by affinity chromatography followed by the proteolytic detachment of GST by thrombin. Further purification of capsomeres by size exclusion chromatography yielded capsomeres which were subsequently self-assembled into the VLPs in a cell-free reactor by direct dilution, a promising method for large-scale self-assembly (Liew et al. 2010).

Montross et al. (1991) found the virion-like VLPs in the nucleus of the *Sf*9 insect cells expressing the MPyV protein VP1. The cytoplasmic VP1 assembled after the cells were treated with the calcium ionophore ionomycin. Thus, the MPyV VP1 assembled in vivo into the T = 7 capsids independent of the minor VP2 and VP3 or viral DNA, while the nuclear assembly could result from increased available calcium in this subcellular compartment (Montross et al. 1991). When coinfected with the VP1 expressing baculovirus, both VP2 and VP3 became predominantly localized to the nucleus in association with the virion-like structures (Delos et al. 1993). Although VP2 and VP3 alone could not generate recombinant particles, they became correctly incorporated into the virion-like VLPs when expressed with VP1 in *Sf*9 cells (An et al. 1999a). The recombinant particles with different MPyV structural proteins were obtained by using different combined expression of these proteins in *Sf*9 cells. Moreover, the cellular DNA of 5 kb in

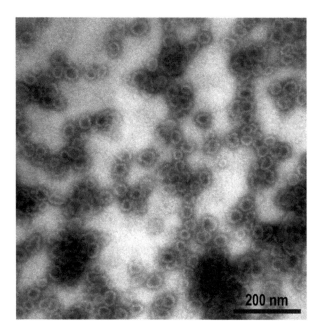

FIGURE 8.4 The transmission electron micrograph of MPyV VLPs assembled from VP1 protein expressed in *E. coli*. (Reprinted from *J Biotechnol*. 134, Chuan YP, Lua LH, Middelberg APJ, High-level expression of soluble viral structural protein in *Escherichia coli*, 64–71, Copyright 2008b, with permission from Elsevier.)

size was packaged in all of the recombinant particles, which showed the same diameter as that of native virions. Thus, as in the case of the *E. coli* expression, the VLPs obtained by the baculovirus-driven expression in insect cells were found to be capable of packing DNA and were suggested as promising delivery tools for human gene therapy (Forstová et al. 1993, 1995; Gillock et al. 1997). The "pseudoinfection" with the insect cells-produced MPyV VLPs was shown to be efficient enough for the transfer of up to 7.2 kbp plasmid DNA (Soeda et al. 1998).

The insect cells-produced MPyV VP1 VLPs were used by a comparative immunological study of Szomolanyi-Tsuda et al. (1998). The results suggested that the highly organized, repetitive nature of the VLPs was insufficient to account for their ability to elicit T-cell independent IgG response and that signals generated by live polyomavirus infection might be essential for the switch to IgG production in the absence of T cells (Szomolanyi-Tsuda et al. 1998). Heidari et al. (2002) produced in insect cells the mutant MPyV VLPs, the VP1 of which carried a deletion that removed the aa stretch 146-TDTVNTK-152. To facilitate large-scale production of VLPs free from cellular contaminants, Ng et al. (2007) used the stable *Drosophila* cell lines expressing either wild-type protein or VP1 tagged with a secretion signal for targeting to the extracellular medium. Both wild-type and tagged VP1 expressed at 2–4 mg L⁻¹ VP1 per of culture. While the wild-type protein self-assembled into the VLPs, the tagged VP1 was efficiently secreted to the extracellular medium but was also glycosylated, unlike the wild-type VP1. Despite this fact, a small

fraction of the recombinant secreted protein assembled into the VLP-like structures that had altered disulphide bonding (Ng et al. 2007).

The insect cells-produced MPyV VP1 VLPs not only prevented MPyV infection in a T-cell immune deficient experimental mouse model (Vlastos et al. 2003) but also conferred protection against some polyomavirus-induced tumors (Franzén et al. 2005).

In yeast *S. cerevisiae*, the correct and highly purified MPyV VP1 VLPs were obtained, in parallel with a set of other polyomavirus VLPs, by the Kęstutis Sasnauskas team in Vilnius (Sasnauskas et al. 2002). Simon et al. (2013) used the nonconventional yeast *Kluyveromyces lactis* to produce the in-vivo-assembled MPyV VP1 VLPs. These highly purified VLPs showed a homogeneous species of 45-nm particles and a high resistance against proteolysis compared to the conventional in-vitro-assembled VLPs. The reasons of this superior stability were unveiled by Simon et al. (2014). Thus, the intermolecular disulfide linkages involving five of the six cysteines of VP1 were found, indicating a highly coordinated disulfide network within the in-vivo-assembled VLPs involving the N-terminal region of VP1. The electron cryomicroscopy revealed structured termini not resolved in the published crystal structure of the bacterially expressed VLP that appeared to clamp the pentameric subunits together (Simon et al. 2014).

The genetic reconstruction of the MPyV VLPs as carriers for foreign epitopes started as late as 1999. At first, all cysteine residues of the MPyV VP1 produced in *E. coli* were replaced by serines, and a new unique cysteine residue was introduced for the attachment of the fluorescence marker, which would be helpful for labeling virions to study virus–cell interactions preceding gene delivery (Schmidt et al. 1999). In parallel, the dihydrofolate reductase from *E. coli* was introduced as a model protein into position 293/294 of the MPyV VP1 (Gleiter et al. 1999). The formation of pentameric capsomeres and their ability to assemble into capsids were not influenced by the insertion, and the inserted dihydrofolate reductase, though less stable than the wild-type form, proved to be a fully active enzyme (Gleiter et al. 1999).

To construct the chimeric MPyV VLPs as putative vaccine models or candidates, the foreign sequences were inserted first into the protein VP1. By the *E. coli*-driven expression, Neugebauer et al. (2006) inserted model B cell epitopes of 12 and 14 aa in length into the BC2 loop of VP1, and the chimeric pentamers and VLPs carrying the surface exposed epitopes induced a strong and long-lasting humoral immune response against VP1 and the inserted foreign epitope. Remarkably, the epitope-specific antibody response was only moderately decreased when VP1 pentamers were used instead of VLPs (Neugebauer et al. 2006). Brinkman et al. (2004, 2005) achieved a nearly complete melanoma tumor remission in mice by fusion of ovalbumin CTL epitope 257–264 or tyrosinase-related protein 2 (TRP2) CTL epitope 180–188 to the C-terminus of VP1 and producing the corresponding chimeric VLPs in *E. coli*.

The full-length murine tumor antigen survivin, an inhibitor of apoptosis protein, of 16.4 kDa was fused to the C-terminus of the MPyV VP1 by Schumacher et al. (2007). The *E. coli*-produced correctly sized VLPs of ~45 nm in diameter were characterized by irregularly shaped outer rims, which indicated that the large C-terminal fusion interfered with the correct assembly. The mice that had received the VP1-survivin VLPs showed significantly increased survival after inoculation of melanoma cells (Schumacher et al. 2007).

The two insertion sites on the surface of the MPyV VP1 were exploited for the presentation of the M2e antigen from influenza and the J8 peptide from Group A *Streptococcus* (GAS; Middelberg et al. 2011). Remarkably, the VLPs were self-adjuvating, and the antibodies raised against the GAS J8 peptide were bactericidal against a GAS reference strain. Furthermore, the MPyV VLPs displaying one or two GAS J8i epitopes induced significant titers of J8i-specific IgG and IgA antibodies, indicating significant systemic and mucosal responses when delivered intranasally to outbred mice without adjuvant. The GAS colonization in the throats of mice challenged intranasally was reduced in these immunized mice, and protection against lethal challenge was observed (Chuan et al. 2013; Rivera-Hernandez et al. 2013). Continuing this story, Chuan et al. (2014) inserted into the VP1 three heterologous N-terminal peptides (GAS1, GAS2, and GAS3) from the GAS surface M-protein and clearly demonstrated the innovative ways to design GAS vaccines.

To engineer an influenza vaccine, a helical peptide antigen element, helix 190 (H190) from the influenza hemagglutinin (HA) receptor binding region was displayed on the *E. coli*-produced MPyV VLPs, using two strategies aimed to promote H190 helicity on the VLPs (Anggraeni et al. 2013). In the first strategy, H190 was flanked by GCN4 structure-promoting elements within the antigen module; in the second, dual H190 copies were arrayed as tandem repeats in the module. The molecular dynamics simulation predicted that tandem repeat arraying would minimize secondary structural deviation of modularized H190 from its native conformation. The in vivo testing supported this finding, showing that, although both modularization strategies conferred high H190-specific immunogenicity, the tandem repeat arraying of H190 led to a strikingly higher immune response quality in mice, as measured by ability to generate antibodies recognizing a recombinant HA domain and split influenza virion (Anggraeni et al. 2013). Wibowo et al. (2013) applied an innovative strategy of inserting multiple antigenic modules, up to 45 M2e modules, in a single capsomere. This approach was used further by the generation of the putative malaria vaccine, where the chimeric MyPV VLPs and capsomeres each incorporated defined CD8+ and CD4+ T cell or B cell repeat epitopes derived from *Plasmodium yoelii* circumsporozoite protein (Pattinson et al. 2020). According to this recent study, both chimeric VLP and capsomere vaccine platforms induced robust CD8+ T cell responses at similar levels, while the

capsomere platform was, however, more efficient at inducing CD4+ T cell responses and less efficient at inducing antigen-specific antibody responses.

The MPyV VLPs were used by the Middelberg's team as brilliant models by quantitative analysis of the VLP size and distribution by field-flow fractionation (Chuan et al. 2008a) and electrospray differential mobility analysis and transmission electron microscopy (Pease et al. 2009), far-infrared spectroscopy (Falconer et al. 2010), as well as by modeling the competition between aggregation and self-assembly (Ding et al. 2010). Moreover, the MPyV VLPs were used as a model for the theoretical VLP engineering aided by computational chemistry methods (Zhang L et al. 2013, 2015; Tekewe et al. 2016).

By the yeast *S. cerevisiae*-driven expression, Kęstutis Sasnauskas's team exposed the so-called preS1$_{phil}$ sequence from the hepatitis B virus gene S (see Chapters 37 and 38) into the HI loop between aa residues 293/295 of the MPyV VP1. The total length of the inserted stretch reached 90 aa residues, 77 of which belonged to the HBV preS1$_{phil}$ sequence. Figure 8.5 presents the obtained chimeric MPyV VLPs. The adjuvant-free immunization of mice with these particles induced high preS1-specific antibody response.

An alternate approach to generating the chimeric MPyV VLPs enabled the addition of foreign sequences to the minor MPyV capsid proteins and was used by Tina Dalianis's team. Thus, the putative insect cells-produced MPyV VLP anticancer vaccine carried human or rat breast cancer oncogene Her2/neu stretch fused to the C-terminus of the minor VP2 protein (Tegerstedt et al. 2005a, b, 2007; Andreasson et al. 2009). These VLPs protected mice from outgrowth of the Her2-expressing murine tumor cells, as well as spontaneous tumor formation in transgenic *neu* mice. Eriksson et al. (2011) engineered the MPyV VLPs carrying the entire human prostate-specific antigen that was fused to the C-terminus of the VP2/VP3. To enhance the efficient PSA-specific tumor protective immune

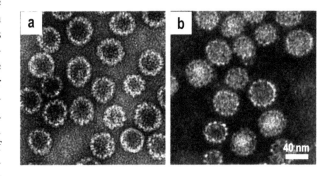

FIGURE 8.5 The electron microscopy evaluation of the MPyV VP1 VLPs. (a) The natural MPyV VP1 VLPs as a control. (b) MPyV VP1-preS1phil VLPs. (Reprinted from *Vaccine*. 26, Skrastina D, Bulavaite A, Sominskaya I, Kovalevska L, Ose V, Priede D, Pumpens P, Sasnauskas K, High immunogenicity of a hydrophilic component of the hepatitis B virus preS1 sequence exposed on the surface of three virus-like particle carriers, 1972–1981, Copyright 2008, with permission from Elsevier.)

response in mice, the PSA-MPyV VLPs were coinjected in this study with adjuvant CpG, in combination with loading them onto murine dendritic cells in vitro before immunization.

Hruskova et al. (2009) fused a fragment of the Bcr-Abl hybrid protein containing the epitope of chronic myeloid leukemia fusion region to the C-terminus of the VP3, but the Bcr-Abl breakpoint epitopes appeared to be weak immunogens and even the MPyV-VLPs did not provide sufficient adjuvant ability to support induction of immune responses specific to the Bcr-Abl fusion zone epitope.

The insect cells-produced MPyV VP1 VLPs were used to package foreign DNA and deliver it to mammalian cells (Forstová et al. 1993, 1995; Gillock et al. 1997; Krauzewicz et al. 2000a, b; Clark et al. 2001; Lipin et al. 2008).

The *E. coli*-driven expression of the MPyV VP1 VLPs for the packaging of DNA was also elaborated (Braun et al. 1999; Henke et al. 2000; Yang and Chen 2000). The specific targeting of the *E. coli*-produced DNA-packaged MPyV VLPs was achieved by coupling of a recombinant tumor-specific antibody fragment (Stubenrauch et al. 2001; May et al. 2002); addition of a Z domain (Gleiter and Lilie 2001, 2003) or WW domain recognizing proline-rich peptide sequences (Schmidt et al. 2001). Abbing et al. (2004) used the *E. coli*-produced MPyV VLPs for the encapsulation of proteins (GFP) and low-molecular-mass drugs (methotrexate) via a 49-aa stretch of VP2 as an anchor, either expressed as a fusion protein with GFP or covalently linked to methotrexate. The targeting to the urokinase plasminogen activator receptor was ensured by the insertion of fragments of urokinase plasminogen activator and baculovirus-driven expression of the chimeric DNA-carrying MPyV VLPs (Shin and Folk 2003).

As a part of a large comparative study with the evaluation of the RNA phage Qβ VLPs, described in Chapters 25, polymer-coated nanodiamonds and polymeric poly(HPMA) nanoparticles, the MPyV VLPs were tested for their ability to display low-molecular-mass inhibitors of glutamate carboxypeptidase II (GCPII), a membrane protease that is overexpressed by prostate cancer cells (Neburkova et al. 2018). Regardless of the diversity of the investigated nanosystems, they all strongly interacted with the GCPII and effectively targeted the GCPII-expressing cells.

Jitka Forstová's team engineered the baculovirus-driven vector for preparation of the MPyV VLPs for transfer of foreign peptides or proteins into cells (Bouřa et al. 2005). The pseudovirions carrying eGFP fused to the C-terminus of the VP3 minor protein entered mouse epithelial cells, fibroblasts, and human and mouse dendritic cells efficiently and were processed by both lysosomes and proteasomes. When used for intranasal immunization of mice, these VLPs induced strong anti-VP1 but not anti-eGFP antibody production, and no specific cytotoxic activities against VP1 and eGFP were detected, appearing therefore to be promising delivery and adjuvant vehicles for therapeutic proteins (Fric et al. 2008).

HAMSTER POLYOMAVIRUS

Hamster polyomavirus (HaPV or HaPyV) belongs to the *Mesocricetus auratus polyomavirus 1* species. It was originally described in 1967 by Arnold Graffi as a cause of epithelioma in Syrian hamsters *M. auratus*, as reviewed by Scherneck et al. (2001). The great success of the HaPyV VLP story was achieved due to the Rainer G. Ulrich and Kęstutis Sasnauskas teams.

First, Siray et al. (1999) reported that the virions isolated from epitheliomas of the HaPyV-infected hamsters were shown by electron microscopy to be spherical particles with the typical icosahedral structure of polyomaviruses. However, in contrast to the capsids of yV and SV40, resolved at that time, the T = 7*l* but not the T = 7*d* of the HaPyV virions was observed.

Second, the highly efficient expression and self-assembly of the authentic 384-aa (42 kDa) and an N-terminally 4-aa extended derivative (resulting from an in-frame-situated AUG codon) HaPyV VP1 was shown in yeast *S. cerevisiae* (Sasnauskas et al. 1999). The nuclear localization of HaPyV VP1 suggested the functional activity of its nuclear localization signal in yeast cells, with the VP1 derivative being N-terminally extended by 4 aa residues. At the same time, the HaPyV VP1 VLPs were obtained by the expression in the plasmid-mediated *Drosophila* Schneider (S2) cells, as referred by Siray et al. (2000) at that time and described later in detail in the paper of Voronkova et al. (2007). In parallel with the HaPyV VLPs, Sasnauskas et al. (2002) published the high-yield production in *S. cerevisiae* of an impressing set of other polyomavirus VP1 VLPs that are described later, namely human JC and BK polyomaviruses, SV40, and budgerigar fledgling disease virus (BFDV) were expressed. In all cases, the empty VLPs formed by all yeast-produced VP1 proteins were dissociated into pentamers and reassociated into the VLPs by defined ion and pH conditions. Gedvilaite et al. (2006b) compared the polyomavirus VLPs immunologically and found that the exposure of the human dendritic cells (hDCs) to the highly immunogenic HaPyV and MPyV VLPs induced hDC maturation, in contrast to the exposure to the BK, JC, and SV40 VLPs, which induced the hDC maturation only marginally.

The ability of the yeast-expressed VP1 variants to form VLPs strongly depended on the size and position of the VP1 truncation (Gedvilaite et al. 2006a). Thus, the VP1 variants lacking 21, 69, and 79 aa residues at the C-terminus efficiently formed VLPs similar to those formed by the unmodified VP1, while the C-terminal truncations of 35 to 56 aa residues failed to form VLPs. The VP1 mutants with a single A336G exchange or internal deletions of aa 335 to aa 346 and aa 335 to aa 363 resulted in the formation of VLPs of a smaller size with 20 nm diameter (Gedvilaite et al. 2006a).

Third, the highly efficient expression of HaPyV VP1 gene in *E. coli* carried out by Tatyana Voronkova and Rainer G. Ulrich in 1998 was of special interest since this highly productive bacterial model allowed the first practical application of the HaPyV1 VLPs as an epitope carrier and packager of foreign DNA (Voronkova et al. 2007). Thus, the establishment of the in vitro disassembly/reassembly system allowed the encapsidation of plasmid DNA into the VLPs. When transfected with the HaPyV VP1 VLPs carrying a plasmid encoding the eGFP gene, the mammalian Cos-7 and CHO cells demonstrated the appropriate eGFP synthesis. Remarkably, the CsCl gradient centrifugation allowed the separation of the VLPs with encapsidated DNA from the empty VLPs (Voronkova et al. 2007). It should be mentioned that earlier the HaPyV proteins including the capsid proteins VP1 and VP2/3 were produced in *E. coli* as β-galactosidase-, TrpE- and dihydrofolate reductase-fusion proteins (Ulrich et al. 1996). Remarkably, the efficient expression in *E. coli* of the HaPyV VP1 gene fragments (Voronkova et al. 2007) was achieved by use of the RNA phage fr-based vectors that are described in Chapter 25. Later, the *E. coli*-produced HaPyV VLPs were used to establish the IgA antibody production by intrarectal immunization of mice (Messerschmidt et al. 2012).

At last, using the established 3D structure of the related polyomaviruses MPyV and SV40, the superficially exposed regions of the HaPyV VP1 were predicted to be located at aa positions 81–88, 222/223, 244–246, 289–294, and at the C-terminus, while the 81–88 epitope and the C-terminal part were found immunodominant by epitope mapping (Siray et al. 1999, 2000).

The popular model epitope DPAFR was taken from the hepatitis B virus (HBV) preS1 stretch of the gene S, the story of which was used to map the putative insertion sites for the further generation of the planned chimeric HaPyV VLPs (Gedvilaite et al. 2000). The linear DPAFR epitope is recognized by the classical anti-preS1 monoclonal antibody MA18/7 generated by the famous Wolfram H. Gerlich's team, and the full story of the epitope usage is presented in Chapters 25 and 38. As the C-terminal part of VP1 was responsible for the pentamer–pentamer interactions, the model epitope DPAFR was introduced between aa 80/89, 221/224, 243/247, and 288/295. All chimeras were capable of being self-assembled into the VLPs and induce the DPAFR-specific antibody response in mice (Gedvilaite et al. 2000).

The prospective HaPyV model was used for the generation of numerous chimeric VLPs in *S. cerevisiae*. Thus, Gedvilaite et al. (2004) demonstrated that the insertion of 45 or 120 aa-long segments from the N-terminus of Puumala hantavirus (PUUV) nucleocapsid protein into the sites aa 80/89 (loop BC) and aa 288/295 (loop HI) of VP1 allowed the highly efficient formation of the chimeric VLPs. In contrast, the expression level and assembly capacity of fusions into the aa 222/225 and aa 243/247 were drastically reduced. The immunization of mice with the chimeric VLPs induced a high-titered antibody response against the hantavirus nucleocapsid protein, even in the absence of any adjuvant, while the strongest response was observed in the mice immunized with the VLPs harboring 120 aa of the PUUV nucleocapsid protein (Gedvilaite et al. 2004).

The chimeric HaPyV VLPs carrying the 120-aa sequence of PUUV nucleocapsid protein promoted the generation of five monoclonal antibodies of IgG isotype specific to PUUV nucleocapsid (Zvirbliene et al. 2006).

Aleksaitė and Gedvilaitė (2006) inserted the three different human tumor-associated antigen (TAA) epitopes, namely TRP—tyrosinase-related protein-2 epitope FVWLHYYSV, MAGE—the MAGE A family protein epitope LVHPLLLKY, and hTERT—human telomerase reverse transcriptase epitope ILAKFLHWT, into the VP1 positions 81/88, 244/246, and 289/294 separately and into the 289/294 as one fused insert produced from all three TAAs.

The 9-aa CTL epitope STAPPVHNV of human mucin 1 (MUC1) with possible anticancer activity was flanked with GSSG linkers and inserted simultaneously at aa positions 80–89 and 288–295 of a single VP1 molecule (Zvirbliene et al. 2006). The chimeric VLPs elicited a strong epitope-specific humoral immune response in mice and promoted the production of MUC1-specific mAbs. Furthermore, Dorn et al. (2008) demonstrated that these MUC1 mAbs reacted specifically with human tumor cells and the cocultivation of the chimeric VLP-primed hDCs with autologous peripheral blood leukocytes resulted in the activation of the MUC1 epitope-specific CD8+ T cells, indicating therefore a potential of the chimeric HaPyV VP1 VLPs as a delivery vehicle for immunotherapeutic targets. Lawatscheck et al. (2007) inserted the arcinoembryonic antigen-derived T cell epitope CAP-1-6D (CEA) YLSGADLNL between the VP1 aa residues 80–89 or 288–295 or simultaneously at both positions and *S. cerevisiae*-expressed VP1 carrying the CEA insertion assembled into chimeric VLPs, independently from additional flanking linkers. The mice immunized with adjuvant-free VLPs developed VP1- and epitope-specific antibodies, while the level of the CEA-specific antibody response was determined by the insertion site, the number of inserts, and the flanking linker. The strongest CEA-specific antibody response was observed in mice immunized with VP1 proteins harboring the CEA insert at site 80–89 (Lawatscheck et al. 2007). Mazeike et al. (2012) inserted into the same two VP1 positions the popular model CTL epitope gp33 from the surface glycoprotein of lymphocytic choriomeningitis virus (LCMV), namely the 33-KAVYNFATM-41 stretch. More details about the history of the gp33 model are found in Chapters 25 and 38. The chimeric VP1-gp33 VLPs were effectively processed by antigen presenting cells in vitro and in vivo and induced antigen-specific CD8+ T cell proliferation. The mice immunized only once with the VLPs without adjuvant developed an effective gp33-specific memory T cell response: 70% were fully and 30% partially protected from LCMV infection. Moreover, the aggressive growth of tumors expressing gp33 was significantly delayed in these mice in vivo (Mazeike et al. 2012). Furthermore, Zvirbliene et al. (2014) inserted into the same two VP1 sites a 99-aa segment of the PUUV Gc protein. The chimeric product self-assembled into the VLPs that induced an efficient insert-specific antibody response in mice and allowed generation of MAbs of IgG isotype specific to hantavirus Gc glycoprotein.

An alternate *S. cerevisiae* expression approach implemented the generation of the pseudotyped HaPyV VLPs by coexpression of the natural VP1 with a VP2 fusion. Thus, Pleckaityte et al. (2011a, b) fused the single-chain variable fragment (scFv) protein neutralizing the cytolytic activity of vaginolysin—the main virulence factor of *Gardnerella vaginalis*—and derived from the appropriate hybridoma cell line with the human IgG1 Fc domain, then added the recombinant scFv-Fc fragment to the N-terminus of VP2 and coexpressed in yeast with VP1. The recombinant scFv-Fc molecules were displayed on the surface of the VLPs and neutralized the vaginolysin-mediated lysis of human erythrocytes and HeLa cells with high potency comparable to that of the full-length antibody (Pleckaityte et al. (2011b). By analogy, Lasickienė et al. (2012) designed the pseudotyped HaPyV VLPs consisting of an intact VP1 protein and VP2 protein fused N-terminally with the target antigen cellular marker p16^{INK4A}. Both genes coexpressed in *S. cerevisiae* were self-assembled to the VLPs carrying the inserted antigen on the surface. The pseudotyped VLPs were used for the generation of antibodies against p16^{INK4A} that represented a potential biomarker for cells transformed by high-risk human papillomavirus (HPV) and induced a strong immune response against the target antigen VLPs in mice. These antisera showed specific immunostaining of p16^{INK4A} protein in malignant cervical tissue (Lasickienė et al. 2012).

Next, Pleckaityte et al. (2015) generated the hepatitis B virus (HBV) markers-targeted pseudotyped HaPyV VLPs. Thus, the VLPs harbored the surface-exposed functionally active neutralizing antibody specific to the HBV surface (HBs) protein. In fact, the two types of the pseudotyped VLPs were generated carrying either the scFv or Fc-engineered Fc-scFv on the VLP surface. Both types of the pseudotyped VLPs were functionally active and showed a potent HBV-neutralizing activity on the HBV-susceptible primary hepatocytes from *Tupaia belangeri*, which was comparable to that of the parental monoclonal antibody. The VP2-fused scFv molecules were incorporated into the VLPs with higher efficiency as compared to the VP2-fused Fc-scFv. However, the pseudotyped VLPs with displayed VP2-fused Fc-scFv molecule showed higher antigen-binding activity and HBV-neutralizing capacity that might be explained by a better accessibility of the Fc-engineered scFv of the VLP surface (Pleckaityte et al. 2015).

Developing further the pseudotyped HaPyV VLP approach, Eiden et al. (2021) engineered a prospective vaccine against bovine spongiform encephalopathy (BSE) in cattle, a disease like scrapie in sheep or Creutzfeldt-Jakob disease in humans that are fatal neurodegenerative diseases characterized by the conformational conversion of the normal, mainly α-helical cellular prion protein (PrP^C) into the abnormal β-sheet rich infectious isoform PrP^{Sc}. Since the immunization against prions is hampered by the self-tolerance to PrP^C in mammalian species, the authors presented the nine different PrP variants onto the VP1/

VP2-pseudotyped VLPs. When the immunized mice were subsequently challenged intraperitoneally with the mouse scrapie strain, the increased mean survival time was observed (Eiden et al. 2021).

The exhaustive inventory of the long-term HaPyV VLP studies performed by the two great laboratories was recently published by Jandrig et al. (2021).

MERKEL CELL POLYOMAVIRUS

Merkel cell polyomavirus (MCV, or MCPyV), a member of the *Human polyomavirus 5* species and the first human viral pathogen discovered using metagenomics (Feng et al. 2008), is suspected to cause Merkel cell carcinoma (MCC), a relatively uncommon but highly lethal form of skin cancer (Rotondo et al. 2017). The 3D structure of MCPyV resolved by Bayer et al. (2020) is presented in Figure 8.3.

The plasmids carrying codon-modified versions of the MCPyV VP1 and VP2 genes of the isolate 339 were used to produce the appropriate VLPs by transfecting human embryonic kidney-derived 293TT cells (Pastrana et al. 2009). For the initial optimization experiments, the VP1 and VP2 expression plasmids were cotransfected with a reporter plasmid encoding GFP. The transfected cells produced high yields of capsids with a VP1:VP2 ratio of about 6:1 and a fraction of the particles encapsidated the GFP reporter plasmid. The authors found that MCPyV, in contrast to other polyomaviruses, did not encode a functional VP3 protein. Remarkably, the VP1 from another isolate, MCV350, was rapidly degraded to undetectable levels in 293TT cell lysates and seemed structurally defective VP1 protein, possibly due to mutations arising during tumorigenesis. The purified MCPyV VP1/VP2 VLPs were highly immunogenic in rabbits and mice (Pastrana et al. 2009).

Kean et al. (2009) expressed the MCPyV isolate 339 VP1, in parallel with that of a long set of other polyomaviruses, in *E. coli* in the form of GST fusions that were affinity purified as soluble proteins but not suspected to self-assemble into VLPs.

Touzé et al. (2010) expressed three different MCPyV VP1 in insect cells using recombinant baculoviruses. The VLPs were obtained with only one of the three VP1 genes, namely of the VP1 derived from the MKT-21 clone isolated from a French patient with MCC, but only protein aggregates were detected for VP1 proteins derived from clones MCC350 and MKT-26. The high-titer antibodies against the VP1 VLPs were detected in the immunized mice with MCV VLPs, and limited cross-reactivity was observed with BK polyomavirus (BKPyV) and lymphotropic polyomavirus (LPV). Viscidi et al. (2011) produced in insect Sf9 cells the VP1 VLPs of the previously mentioned 339 strain and used to evaluate the association between exposure to MCPyV and MCC, with the BKPyV and JCPyV VLPs as controls. Kumar et al. (2012) used the MCPyV and Trichodysplasia spinulosa-associated polyomavirus (TSPyV) VLPs to evaluate correlation between virus-specific Th-cell and antibody responses in patients. Šroller et al. (2014) assessed the occurrence of serum antibodies against MCPyV, BKPyV, and JCPyV in a healthy population of the Czech Republic, using the corresponding VLPs produced in insect cells. Li TC et al. (2015) detected in insect cells the two different sizes of the released MCPyV VLPs, where DNA molecules of 1.5- to 5-kb, which were derived from host insect cells, were packaged in large, ~50-nm spherical particles but not in small, ~25-nm particles. The electron cryomicroscopy showed that the large MCPyV VLPs were composed of 72 pentameric capsomeres arranged in the T = 7 icosahedral surface lattice and were 48 nm in diameter. The MCPyV VLPs did not share antigenic determinants with the BKPyV and JCPyV VLPs. The VLP-based enzyme immunoassay was applied to investigate age-specific prevalence of MCPyV infection in the general Japanese population aged 1–70 years (Li TC et al. 2015).

Norkiene et al. (2015) produced in yeast *S. cerevisiae* the VP1 VLPs from MCPyV and ten other newly identified human polyomaviruses including the later-described alphapolyomaviruses HPyV9, TSPyV, and New Jersey polyomavirus (NJPyV), as well as a rare betapolyomavirus—namely Karolinska Institutet polyomavirus (KIPyV)—and deltapolyomaviruses HPyV7, MWPyV (or HPyV10), and the St. Louis polyomavirus (STLPyV) described later. The obtained VLPs varied in size with diameters ranging from 20 to 60 nm. The smaller-sized VLPs of 25–35 nm in diameter predominated in preparations from Washington University polyomavirus (WUPyV)—a betapolyomavirus—and HPyV6—a deltapolyomavirus. The expression of the VP1 of HPyV12, an unclassified polyomavirus, resulted in a 364-aa-long VP1 protein, which efficiently self-assembled into the typical polyomavirus VLPs (Norkiene et al. 2015).

OTHER ALPHAPOLYOMAVIRUSES

Concerning human polyomavirus 9 (HPyV9) from the *Human polyomavirus 9* species, the corresponding VLPs were produced by the baculovirus expression system (Nicol et al. 2012), in parallel with the VLPs of human polyomavirus 6 (HPyV6) and 7 (HPyV7; Nicol et al. 2013), both from the genus *Deltapolyomavirus*, and TSPyV, all described later. As mentioned previously, the HPyV9 VLPs were produced in *S. cerevisiae* by Norkiene et al. (2015).

Nicol et al. (2014) produced in insect HighFive cells the VP1 VLPs of orangutan polyomavirus (OraPyV1) from the species *Pongo abelii polyomavirus 1* or *Pongo pygmaeus polyomavirus 1*, as well as of chimpanzee polyomaviruses PtvPyV1 and PtvPyV2 from the *Pan troglodytes polyomavirus 1* species. The study indicated serological identity between the two chimpanzee polyomaviruses and a high level of cross-reactivity with MCPyV, while cross-reactivity was not observed between OraPyV1 and TSPyV. Earlier, Zielonka et al. (2011) produced the VP1 VLPs of chimpanzee polyomavirus, ~45 nm in diameter, in yeast *S. cerevisiae* cells.

The VP1 VLPs of Trichodysplasia spinulosa-associated polyomavirus (TSPyV) of the *Human polyomavirus 8* species were produced in insect cells (Chen T et al. 2011; Kumar et al. 2012; Nicol et al. 2013, Nicol et al. 2014). As mentioned earlier, Norkiene et al. (2015) produced the TSPyV VLPs in *S. cerevisiae*. Moreover, Gedvilaite et al. (2015) exploited the TSPyV VP1 protein as a carrier for construction of the chimeric VLPs harboring selected B and T cell-specific epitopes and evaluated the novel model in comparison to the previously described yeast-produced HaPyV VP1 VLPs. Thus, the chimeric TSPyV VLPs were engineered, which exposed on the surface the model HBV preS1 epitope DPAFR or a universal T cell-specific epitope AKFVAAWTLKAAA at the HI or BC loop and produced in *S. cerevisiae*. The VLPs induced a strong immune response in mice, activated dendritic cells and T cells in spleen cell cultures and appeared therefore as a novel promising candidate of the chimeric VLPs (Gedvilaite et al. 2015). Zaveckas et al. (2018) developed a highly efficient and scalable purification procedure for the TSPyV VP1 VLPs based on two chromatographic steps, ion-exchange monolith and core-bead chromatography, which allowed recovery of 42% of TSPyV VP1 with a purity of 93% and served as an alternative to the conventional ultracentrifugation-based purification methods.

Yamaguchi et al. (2014) demonstrated the VP1 VLPs of vervet monkey polyomavirus 1 (VmPyV), a member of the *Chlorocebus pygerythrus polyomavirus 1* species. Thus, the full-length VmPyV VP1 was subcloned into a mammalian expression plasmid and expressed in human embryonic kidney 293T (HEK293T) cells. The VmPyV VLPs were purified and demonstrated a diameter of ~50 nm. The C-terminal truncation of 116 aa residues did not prevent the VLP formation, although it reduced their number.

The VP1 VLPs of New Jersey polyomavirus (NJPyV), a member of the *Human polyomavirus 13* species, were produced in *S. cerevisiae* by Norkiene et al. (2015).

GENUS *BETAPOLYOMAVIRUS*

BK POLYOMAVIRUS

BK polyomavirus (BKPyV), a member of the *Human polyomavirus 1* species, is usually associated with post-transplant interstitial nephritis. The near-atomic structure of the BKPyV presented in Figure 8.3, together with other polyomavirus structures, was resolved by electron cryomicroscopy (Hurdiss et al. 2016). Moreover, Hurdiss et al. (2018) presented the 3.4-Å electron cryomicroscopy structure of the native infectious BKPyV in complex with the receptor fragment of a ganglioside.

The BKPyV VP1 VLPs were produced in insect cells (Touzé et al. 2001, 2010; Li TC et al. 2003; Viscidi et al. 2011; Šroller et al. 2014). Nilsson et al. (2005) explored the assembly process of the insect cells-produced VP1, when compared the 3D structure of the two VLPs formed in a calcium-dependent manner. It was demonstrated for the first time in polyomaviruses that the small VP1-formed particle, 26.4 nm in diameter, was built by the pentameric capsomeres on icosahedral T = 1 surface lattice with meeting densities at the 3-fold axes that interlinked three capsomers. In the larger particle, 50.6 nm in diameter, the capsomeres formed the T = 7 icosahedral shell with three unique contacts (Nilsson et al. 2005).

Kęstutis Sasnauskas's team produced the VP1 VLPs of the two antigenic variants of BKPyV, namely, strains SB and AS, as well as JC polyomavirus (JCPyV) described later, in *S. cerevisiae* (Hale et al. (2002). The VP1s of BKPyV/AS and JCPyV were of expected molecular mass, whereas that of the BKPyV/SB appeared to be smaller than anticipated. However, all VP1s self-assembled into the VLPs retaining sialic acid-binding and antigenic properties of the native virions. The full story of the early period in the generation of the human and nonhuman polyomavirus VLPs in yeast was narrated at that time by Sasnauskas et al. (2002).

Concerning packaging and delivery, the insect cells-produced VLPs were able to transfer DNA to mammalian cells by (Touzé et al. 2001).

Chackerian et al. (2002) used the insect cells-produced BKPyV VLPs in the noncovalent coupling approach, using the idea described in Chapter 7, when mice were vaccinated with a fusion protein containing self (TNF-α) and foreign (streptavidin) components, conjugated to the biotinylated VLPs.

JC POLYOMAVIRUS

JC polyomavirus (JCPyV), a member of the species *Human polyomavirus 2*, is associated with progressive multifocal leucoencephalopathy, a fatal neurological disease. The JCPyV VLPs were produced in mammalian Cos-7 cells by the expression of the VP1 and VP2/VP3 genes (Shishido et al. 1997). The VLP assembly was observed in the nucleus of the expressing cells.

In insect cells, the efficient production of the virion-like JCPyV VP1 VLPs was demonstrated (Chang D et al. 1997; Goldmann et al. 1999; Viscidi et al. 2011; Šroller et al. 2014). In *E. coli*, Ou et al. (1999) demonstrated that the JCPyV VP1 self-assembled into the VLPs represented by the virion-like pseudovirion and empty capsid-like populations. The pseudovirions contained DNA and RNA molecules, but the pseudocapsids did not contain any nucleic acid. Moreover, the pseudovirions were able to package and deliver exogenous DNA into human fetal kidney epithelial cells, acting as a potent human gene transfer vector (Ou et al. 1999). To investigate the minimal sequences on the JCPyV VP1 required for the assembly, Ou et al. (2001) truncated it by the first 12 and 19 aa at the N-terminus and the last 16, 17, and 31 aa at the C-terminus. The ΔN12 and ΔC16 VP1 variants were self-assembly competent, while the ΔN19, ΔC17, and ΔC31 formed a pentameric capsomere structure only. It should be noted that the *E. coli* chaperone DnaK is incredibly important for the correct VP1 assembly and reduction of protein aggregates (Saccardo et al. 2014).

As mentioned earlier, the JCPyV VP1 VLPs were produced in *S. cerevisiae* by Hale et al. (2002). A bit earlier, Chen PL et al. (2001) published the similar expression process in *S. cerevisiae* with an aim to investigate the role of disulfide bonds in the VLP structure. The disulfide bonds were found in the VLPs, which caused dimeric and trimeric VP1 linkages, although the VLPs remained intact when disulfide bonds were reduced by dithiothreitol. Remarkably, the VLPs without disulfide bonds were disassembled into capsomeres by EGTA alone, while those with disulfide bonds could not be disassembled by EGTA. The capsomeres were reassembled into VLPs in the presence of calcium ions. However, the capsomeres formed irregular aggregations instead of VLPs when treated with diamide to reconstitute the disulfide bonds (Chen PL et al. 2001).

The JCPyV V1 VLPs were enthusiastically exploited as a packaging and delivery vector, as it was suggested first by Ou et al. (1999). Wang M et al. (2004) used the yeast-produced JCPyV VLPs to package and deliver an antisense oligodeoxynucleotide against SV40 large tumor antigen into the SV40-transformed human cells in order to inhibit expression of the oncoprotein and demonstrated apoptosis of the treated cells. The yeast-produced JCPyV VLPs were shown for the first time as a nontoxic tool to deliver RNAi in both murine macrophage cells and mice for gene therapy purposes (Chou et al. 2010). Citkowicz et al. (2008) presented the scalable gene delivery system by reassembling the insect cells-produced JCPyV VP1 pentamers in the presence of the desired DNA. The efficient gene transfer using the *E. coli*-produced JCPyV VLPs that inhibited human colon adenocarcinoma growth in a nude mouse model was published by Chen LS et al. (2010). Fang et al. (2012) showed that the JCPyV VLPs were able to package plasmid DNA in *E. coli* up to at least 9.4 kb and deliver it successfully into human neuroblastoma cells.

The widespread development of the JCPyV VLP-based gene delivery system, which appeared to be easily generated in large quantities and at low cost, was comprehensively reviewed at that time by Chang CF et al. (2011). Lin et al. (2014) used the *E. coli*-produced JCPyV VLPs to target the BKPyV-infected human kidney cells and block the BK virus replication by the delivery of the BK large tumor antigen-specific shRNA.

Goldmann et al. (2000) were the first to use the JCPyV VLPs as putative carriers for pharmaceutical substances of low molecular mass. Thus, propidium iodide (PI) was packaged into the VLP as a fluorescent marker, and the PI-containing VLPs were followed directly by flow cytometry. Qu et al. (2004) conjugated the *E. coli*-produced JCPyV VLPs with fluorescein isothiocyanate (FITC) and packaged the fluorescent dye Cy3 into the FITC-labeled VLPs. Ohtake et al. (2008) improved the uptake of the FITC-labeled VLPs by display of sialic acid. Ohtake et al. (2010) went to the direct incorporation of the visualizing proteins into the JCPyV VLPs, fused GFP to the N terminus of VP2, and coexpressed the fusion together with VP1 in *E. coli*. The expressed VP1 and GFP-VP2 associated with each other and formed the GFP-incorporated VLPs inside *E. coli*. Furthermore, the GFP-VP2/VP1 VLPs were provided with the His$_6$ tag that was added N-terminally to the GFP-VP2, and the resultant VLPs were packed with nitrilotriacetic acid-sulforhodamine (NTA-SR), which contained both a His$_6$-tag-targeting NTA segment and a fluorescent sulforhodamine SR segment. The cellular uptake and release of encapsulated NTA-SR was examined in NIH3T3 cells, where encapsulation of the NTA-SR markedly enhanced the cellular uptake compared to the NTA-SR alone (Ohtake et al. 2010). In fact, this was the first report of small molecule encapsulation within VLPs using specific peptide-tag interaction and of the controlled release in an environment-responsive manner.

Niikura et al. (2013) presented the JCPyV VLPs capable of glutathione-triggered release of drug molecules encapsulated via disulfide bonds. To do this, the authors engineered the VLPs coupled with cyclodextrins as hydrophobic pockets through disulfide bonds inside the VLPs, where hydrophobic drugs could be incorporated and reported the intracellular delivery of hydrophobic dyes or drugs encapsulated in VLPs through cyclodextrins with high efficiency and their subsequent release in cells in response to glutathione. As a model anticancer drug, paclitaxel (PTX)-cyclodextrin complexes were encapsulated inside VLPs, and the PTX-loaded VLPs exhibited a dose-dependent cytotoxic effect in NIH3T3 cells, suggesting a highly promising approach to deliver hydrophobic drugs without chemical modification of them (Niikura et al. 2013).

Since the JCPyV VLPs bound to sialic acid, Niikura et al. (2009) performed the conjugation of sialic-acid-linked gold particles with the VLPs enabling the spatial arrangement of the gold particles on the VLP surface. This structure produced a red shift in the absorption spectrum due to plasmon coupling between adjacent gold particles, leading to the construction of an optical virus detection system (Niikura et al. 2009).

MOUSE PNEUMOTROPIC VIRUS

Mouse pneumotropic virus (MPtV), previously called Kilham polyomavirus, the second identified murine polyomavirus that was isolated by Kilham and Murphy (1953), belongs to the *Mus musculus polyomavirus 2* species. The MPtV VLPs were studied by Tina Dalianis's team. Thus, Tegerstedt et al. (2003) obtained the MPtV VP1 VLPs by the baculovirus-driven expression and demonstrated that the MPtV VLPs could potentially complement the MPyV VP1 VLPs as vectors for prime-boost gene therapy. The MPtV VLPs bound to different cells, independent of MHC class I antigen expression, and sialic acid receptors and did not cause hemagglutination of red blood cells. Moreover, the MPtV VLPs were shown to transduce foreign DNA in vitro and in vivo (Tegerstedt et al. 2003). The usefulness of the MPtV VLPs for gene therapy, immune therapy and as vaccines, in the prime-boost combination with the MPyV VLPs, was summarized by Tegerstedt et al. (2005a).

After the MPyV VLPs carrying human Her2/neu prevented the outgrowth of a human Her2/neu expressing tumor in a transplantable tumor model as well as outgrowth of spontaneous rat Her2/neu carcinomas in mice, as described earlier, Andreasson et al. (2009) demonstrated the same prophylactic and therapeutic protection with the MPtV VLPs carrying human or rat Her2/neu that was fused to the 48 C-terminal aa of the VP2/3. Remarkably, the therapeutic immunization with human Her2/neu VLPs together with CpG given up to six days after challenge protected against tumor. Andreasson et al. (2010) demonstrated that the immunization of mice with the Her2/neu VLPs was efficient in stimulating several compartments of the immune system and induced an efficient immune response including long-term memory. In addition, when depleting mice of isolated cellular compartments, the tumor protection was not as efficiently abolished as when depleting several immune compartments together (Andreasson et al. 2010).

SIMIAN VIRUS 40

SV40 belonging to the *Macaca mulatta polyomavirus* species remains the most studied representative of the genus *Betapolyomavirus*. The high-resolution crystal structure of the SV40 virions was achieved at 3.8 Å by Liddington et al. (1991) and refined further to 3.1 Å (Stehle et al. 1996). The structures of SV40 and of the similar previously described MPyV were compared in detail (Stehle and Harrison 1996; Yan et al. 1996). This near-atomic structure is presented in Figure 8.3, together with other polyomavirus structures.

The SV40 genes encoding VP1, VP2, and VP3 were expressed in insect cells by the baculovirus system (Kosukegawa et al. 1996). When the VP1 gene was expressed alone or coexpressed with the VP2 and VP3, the virion-like VLPs were produced, while the VP2 and VP3 were incorporated into the VLPs. Sandalon and Oppenheim (1997) not only expressed the SV40 VP1, VP2, and VP3, in *Sf*9 cells with the three recombinant baculoviruses at equal multiplicities and demonstrated the presence of the SV40-like structures and heterogeneous aggregates of variable size, mostly 20–45 nm, in nuclear extracts of the infected cells but also characterized their properties. Thus, the SV40 VLPs were disrupted by treatment with the reducing agent dithiothreitol and the calcium chelator EGTA, and the VLPs were found to be significantly less stable than the SV40 virions but were partially stabilized by calcium ions (Sandalon and Oppenheim 1997). Sandalon et al. (1997) engineered the SV40 pseudovirions when the nuclear insect cell extracts containing the three proteins were allowed to interact with purified SV40 DNA or with plasmid DNA produced and purified from *E. coli*, significantly larger than the SV40 DNA. Thus, the intact supercoiled DNA was packaged and transmitted further into the target cells, while the packaging process was not dependent on the SV40 packaging signal (Sandalon et al. 1997). Sandalon and Oppenheim (2001) presented a detailed methodology for the scalable preparation of the SV40 VLPs and pseudovirions. To unveil the role of the disulfide linkage and calcium ion-mediated interactions by the VP1-formed VLP assembly/disassembly, Ishizu et al. (2001) performed a comprehensive mutational analysis of the VP1 produced by the baculovirus expression system. Yokoyama et al. (2007) showed that the progressive deletions from the C-terminus of the insect cells-produced VP1, up to 34 aa residues, caused size and shape variations in the resulting VLPs, including tubular formation, whereas deletions beyond 34 aa simply blocked the VP1 self-assembly. The mutants carrying point mutations in the critical 301–312 aa region formed small VLPs resembling T = 1 symmetry. The chimeric VP1, in which this region was substituted with the homologous regions from VP1 of other polyomaviruses, assembled only into the small T = 1 VLPs. The latter were recommended as a novel vector to develop drug delivery systems using the different SV40 VLPs (Yokoyama et al. 2007).

In *E. coli*, the SV40 VLPs were assembled from His$_6$-tagged VP1 that was efficiently produced in a strain deficient in the GroELS chaperone machine but not in the wild-type strain (Wróbel et al. 2000).

In yeast *S. cerevisiae*, Sasnauskas et al. (2002) obtained the SV40 VP1 VLPs at high yields, where the empty SV40 VLPs were formed and dissociated into pentamers and reassociated back into the VLPs by defined ion and pH conditions.

Concerning the putative chimeric VLPs, Takahashi et al. (2008) identified the two sites within the VP1 surface loops DE and HI that accommodated foreign peptides in such a way that they were displayed on the surface of the VLPs without interfering with the assembly or the packaging of viral DNA. The insertion of the Flag tags but not RGD integrin-binding motifs at these sites strongly inhibited cell attachment of the VLPs. Instead, the VLPs carrying the RGD motifs bound to integrin in vitro and to the cell surface in an RGD-dependent manner (Takahashi et al. 2008).

Kawano et al. (2014b) developed the first SV40-based vaccine candidate when produced in the *Sf*9 insect cells the SV40 VP1 VLPs with an HLA-A*02:01-restricted CTL epitope corresponding to the influenza A virus matrix peptide 58-GILGFVFTL-66 (FMP58–66) inserted into the DE loop or the HI loop. The chimeric SV40 VLPs effectively induced the influenza-specific CTLs and hetero-subtypic protection against influenza A viruses without the need of adjuvants. The clear advantages of the SV40 VLPs as a vaccine platform were substantially reviewed and summarized at that time by Kawano et al. (2014a) and later by Kawano et al. (2018). In parallel, Kawano et al. (2015) showed that the insect cells-produced and purified VP1 pentamers covered polystyrene beads measuring 100, 200, and 500 nm in diameter, as well as silica beads. In addition to covering spherical beads, the VP1 pentamers covered cubic magnetite beads, as well as the distorted surface structures of liposomes. These findings indicated that the VP1 pentamers could coat artificial beads of various shapes and sizes larger than the natural capsid and would be useful in providing a VLP-like surface for enclosed materials, enhancing their

stability and cellular uptake for drug delivery systems (Kawano et al. 2015).

Inoue et al. (2008) engineered the chimeric SV40 VLPs where the heterologous proteins were fused to VP2/3 and efficiently incorporated into the VP1 VLPs. Using eGFP as a model protein, the fusion to the C-terminus of VP2/3 appeared preferable and the C-terminal VP1-interaction domain of VP2/3 was found sufficient for the incorporation into the VLPs. The eGFP-carrying VLPs retained the ability to attach to the cell surface and enter the cells. Using this system, yeast cytosine deaminase, a prodrug-modifying enzyme that converts 5-fluorocytosine to 5-fluorouracil, was attached to the VLPs. When CV-1 cells were challenged with the resulting VLPs, they became sensitive to 5-fluorocytosine-induced cell death (Inoue et al. 2008).

As to the application of chemical coupling to the SV40 VLPs, Kitai et al. (2011) conjugated human EGF to the N138C variant of the VP1 VLPs and showed that the resultant VLPs acquired selectivity for the cells overexpressing the EGF receptor.

As described earlier, the great DNA packaging and delivery potential of the SV40 VLPs was established first in late 1990s (Sandalon et al. 1997). The in vitro packaging methodology that was elaborated by Kimchi-Sarfaty et al. (2003) overcame restrictions of the other SV40 systems such as the requirement for SV40 sequences and the limitation in size of DNA that can be packaged. The in vitro packaging system used the four SV40 proteins VP1, VP2, VP3, and agno or VP1 only. The packaged plasmids ranged in size from 4.2–17.6 kb, did not require any SV40 sequence, and only slightly affected particle size. Furthermore, Kimchi-Sarfaty et al. (2005) demonstrated the first use of the SV40 pseudovirions to deliver into human cells both principal types of RNAi effector molecules: plasmid-expressed short hairpin RNAs (shRNAs) and synthetic siRNAs. Kimchi-Sarfaty et al. (2006) delivered a toxin to treat human adenocarcinomas in mice when a truncated *Pseudomonas* exotoxin gene was transferred by the SV40 pseudovirions. Later, Lund et al. (2010) and Kimchi-Sarfaty and Gottesman (2012) published the detailed procedure on how to generate the SV40 pseudovirions and deliver the packaged genes.

Mukherjee et al. (2010) examined the synergism between the SV40 VP1 pentamer-pentamer interaction and the pentamer-DNA interaction using a minimal system of purified VP1 and a linear dsDNA 600-mer. Thus, at low VP1/DNA ratios, large tubes were observed. As the VP1 concentration increased, the tubes were replaced by particles of 20 nm. At high VP1/DNA ratios, a progressively larger fraction of particles was similar to the 45 nm virions. Paradoxically, the stable complexes appeared only at high ratios of VP1 to DNA (Mukherjee et al. 2010). In parallel, Enomoto et al. (2011) reconstituted in vitro the SV40 pseudovirions from all three proteins VP1, VP2, and VP3, when packaged them with a ~5-kb circular nucleosomal DNA with hyperacetylated histones. Remarkably, when inoculated into mammalian cells, the VP1/2/3 pseudovirions containing nucleosomal DNA with a reporter gene yielded a significantly higher level of gene expression than the VP1-only particles containing the corresponding naked DNA.

The SV40 VLPs were intensively studied as putative nanocages to deliver different nanomaterials, such as quantum dots (Li F et al. 2009, 2010; Gao et al. 2013), gold nanoparticles (Wang T et al. 2011), gold nanoparticles and quantum dots (Li F et al. 2011, 2012), and magnetic nanoparticles (Enomoto et al. 2013). Zhang W et al. (2017) summarized the data on the encapsulation of inorganic nanomaterials into the SV40 VLPs on the background of other popular VLP models.

Recently, Xu et al. (2020) eliminated the polymorphism of the SV40 VLPs when engineering 15 VP1 mutants through the substitution of double cysteines at the VP1 pentamer interfaces, generating a group of VLPs with altered size distributions. One of the mutants specifically formed homogenous T = 1 VLPs of 24 ± 3 nm in diameter. The homogeneous self-assembly and stability enhancement of these VLPs was attributed to the new disulfide bonds contributed by Cys102 and Cys300. The novel SV40 T = 1 VLPs were proposed as an evolved version of this popular VLPs in future studies and applications (Xu et al. 2020).

GENUS *GAMMAPOLYOMAVIRUS*

The first known avian polyomavirus, namely budgerigar fledgling disease polyomavirus (BFDV), isolated in the 1980s, a member of the *Aves polyomavirus 1* species (known also as simply avian polyomavirus (APV)), caused a fatal, multiorgan disease among several bird species, and was first discovered in budgerigars. The APV structure was not only resolved by electron cryomicroscopy to 11.30 Å but also was compared to that of mammalian polyomaviruses, particularly JCPyV and SV40, as shown in Figure 8.6 (Shen et al. 2011). The treatment of APV with 250 mM L-arginine (APV+R) markedly reduced particle aggregation and substantially narrowed the distribution of the APV particle sizes to a range that was comparable to that observed for the SV40 particles under mild buffer conditions. As a result, the structure of the pentameric VP1 was mostly conserved, although the APV VP1 had a unique, truncated C-terminus that eliminated an intercapsomere-connecting β-hairpin observed in other polyomaviruses. It was postulated that the terminal β-hairpin locked other polyomavirus capsids in a stable conformation and that the absence of the hairpin led to the observed capsid size variation in APV. The plug-like density features were observed at the base of the VP1 pentamers, consistent with the known location of VP2 and VP3. However, the plug density was more prominent in APV and might include VP4, a minor capsid protein unique to bird polyomaviruses (Shen et al. 2011).

Rodgers et al. (1994) expressed the BFDV VP1 gene in *E. coli*, and the product was purified to near homogeneity by immunoaffinity chromatography and appeared as pentameric capsomeres. The virion-like VLPs were formed in vitro from the purified VP1 capsomeres with the addition of Ca2+ ions and the removal of chelating and reducing

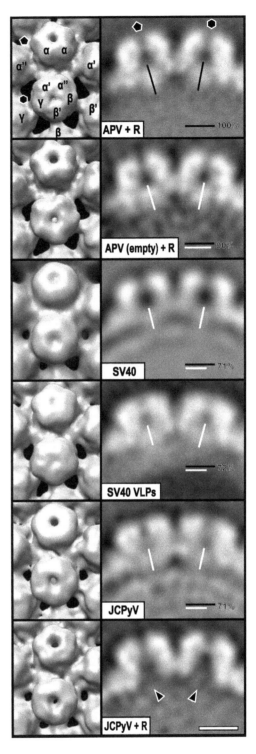

FIGURE 8.6 The views of pentavalent and hexavalent capsomeres based on electron cryomicroscopy reconstructions of APV+R, APV+R empty capsids, SV40, SV40 VLPs, JCPyV, and JCPyV+R all rendered at 25 Å resolution to allow for direct comparison. Left column, the surface rendering views of 5-fold symmetric pentavalent (top, filled pentagon) and nonsymmetric hexavalent (bottom, empty hexagon) capsomeres and intercapsomere connections rendered at 1σ density level. Top, VP1 monomers labeled according to Liddington et al. (1991). Right column, cross-sectional density slices of pentavalent (left) and hexavalent (right) capsomeres. Plug densities were observed at the
(Continued)

VP1 pentamer-dsDNA interface but not when JCPyV was treated with arginine (black arrowheads). Black (APV+R) or white lines overlaid against plug densities show the measured length of the plug-like density. Lines at the bottom right corner of each panel compare the average pentavalent and hexavalent plug length against the APV+R plug length, with percentage given. Scale bar (striped), 5 nm, for right column only. (Reprinted from *Virology*. 411, Shen PS, Enderlein D, Nelson CD, Carter WS, Kawano M, Xing L, Swenson RD, Olson NH, Baker TS, Cheng RH, Atwood WJ, Johne R, Belnap DM, The structure of avian polyomavirus reveals variably sized capsids, non-conserved inter-capsomere interactions, and a possible location of the minor capsid protein VP4, 142–152, Copyright 2011, with permission from Elsevier.)

---

agents. Rodgers and Consigli (1996) made the two deletions in the VP1 protein to identify the calcium binding domain and to further characterize the role of calcium ions in the capsid assembly process. Thus, the recombinant VP1 lacking a putative calcium binding domain D237-D248 failed to bind radioactive $^{45}Ca^{2+}$ yet associated into capsomeres. These capsomeres were similar in shape to the wild-type VP1 but were unable to assemble into the VLPs. Likewise, the recombinant VP1 lacking ten C-terminal aa E334-R343 also formed capsomeres that were unable to assemble into the VLPs (Rodgers and Consigli 1996).

In insect cells, when the cells were coinfected with the BFDV VP1 and either VP2 or VP3 recombinant baculoviruses, the VP1 protein was transported to the nucleus by both the VP2 and VP3 minor proteins (An et al. 1999b). However, the minor proteins could not be coisolated with VP1 protein by immunoaffinity chromatography using a monoclonal anti-VP1 serum. Nevertheless, the VP1 capsomeres were isolated by immunoaffinity chromatography and assembled in vitro into the VLPs. The electron microscopic observation of thin-sectioned *Sf*9 cells, which were coinfected with VP1, VP2 and VP3 recombinant baculoviruses, demonstrated capsomere-like structures in the nucleus, but the virion-like particles were not observed (An et al. 1999b).

Sasnauskas et al. (2002) reported the highly efficient production of the BFDV VLPs in yeast *S. cerevisiae*. The correct formation of the VP1-based VLPs of other avian polyomaviruses was achieved in *S. cerevisiae*, namely for crow polyomavirus (CPyV) of the *Corvus monedula polyomavirus 1* species (Zielonka et al. 2012), finch polyomavirus (FPyV) of the *Pyrrhula pyrrhula polyomavirus 1* species (Zielonka et al. 2012), and goose hemorrhagic polyomavirus (GHPyV) of the *Anser anser polyomavirus 1* species (Zielonka et al. 2006, 2012). The GHPyV VP1 VLPs were obtained earlier in insect cells (Zielonka et al. 2006). A single injection of goslings with the insect cells-produced GHPyV VP1 VLPs induced 95% protection against challenge with a virulent polyomavirus isolate, while the booster vaccination regimen ensured 100% protection (Mató et al. 2009).

GENUS *DELTAPOLYOMAVIRUS*

As mentioned earlier by the *Alphapolyomavirus* genus, the VLPs of human polyomavirus 6 (HPyV6) of the *Human polyomavirus 6* species and human polyomavirus 7 (HPyV7) of the *Human polyomavirus 7* species were produced in parallel with the HPyV9 VLPs by the baculovirus expression system (Nicol et al. 2012, 2013).

Earlier, the VP1 VLPs of HPyV6 and HPyV7 were produced in mammalian cells, in parallel with the previously described MCPyV VLPs (Schowalter et al. 2010). At last, Norkiene et al. (2015) produced the HPyV7, VLPs in *S. cerevisiae*.

The VP1 VLPs of Malawi polyomavirus (MWPyV), or HPyV10, a member of the *Human polyomavirus 10* species, were produced in insect cells by Nicol et al. (2013) and in *S. cerevisiae* by Norkiene et al. (2015), in parallel with a set of other human polyomaviruses.

As mentioned earlier, Norkiene et al. (2015) produced in yeast *S. cerevisiae* and characterized in detail the VP1 VLPs of St. Louis polyomavirus (STLPyV), a member of the *Human polyomavirus 11* species.

UNCLASSIFIED POLYOMAVIRUSES

The popular B-lymphotropic African green monkey polyomavirus (AGMPyV or LPV) remains as a related strain that is unclassified as of yet. Pawlita et al. (1996) obtained the virion-like LPV VP1 VLPs that were assembled spontaneously in the nuclei of insect cells by the baculovirus-driven expression. The LPV VP1 was expressed in large amounts and unspecifically encapsidated linear, double-stranded with a predominant size of 4.5 kb. The fraction of DNA-containing VP1 particles increased with time and dose of baculovirus vector infection. The encapsidated DNA consisted of insect cell and baculoviral sequences with no apparent strong homology to the LPV sequences. In parallel with other early polyomavirus data, these results suggested that the LPV VP1 alone can be sufficient to encapsidate linear DNA in a sequence-independent manner (Pawlita et al. 1996). Later, Touzé et al. (2010) prepared the LPV VP1 VLPs in insect cells to study the cross-reactivity with the insect cells-produced MCPyV and BKPyV VLPs.

To change the receptor binding specificity of the LPV VLPs, Langner et al. (2004) have replaced sets of three aa in the three predicted VP1 surface loops by RGD; the ten mutants gave rise to the expected 40 kDa VP1 protein upon the baculovirus-driven expression, and the five of the VP1 mutants representing all three surface loops have retained the ability to spontaneously assemble to the VLPs in the nuclei of the insect cells. As expected, all mutant VLPs had lost specific binding to the LPV receptor but showed specific binding to $\alpha v\beta 3$ integrin (Langner et al. 2004).

As mentioned earlier, the expression of the VP1 gene of HPyV12, an unclassified polyomavirus, in *S. cerevisiae* resulted in a 364-aa-long VP1 protein, which efficiently self-assembled into the typical polyomavirus VLPs (Norkiene et al. 2015).

Section II

Single-Stranded DNA Viruses

9 Order *Piccovirales*

As the builders say, the larger stones do not lie well without the lesser.

Plato

The simplest things are often the truest.

Richard Bach

ESSENTIALS

The *Piccovirales* is an order of resilient, nonenveloped T = 1 icosahedral viruses, 21–22 nm in diameter, with the virion consisting of 60 copies of VP proteins and belonging, therefore, together with circoviruses (Chapter 10), to the smallest known viruses. The order *Piccovirales* currently involves the sole family *Parvoviridae*, 3 subfamilies, namely the two traditional *Densovirinae* and *Parvovirinae* and a newly established *Hamaparvovirinae*; 26 genera; and 126 species altogether and forms the class *Quintoviricetes* as a single member. As summarized by the official ICTV reports (Tijssen et al. 2012; Cotmore et al. 2019; Pénzes et al. 2020), the two historically initial subfamilies *Parvovirinae* and *Densovirinae* were distinguished primarily by the respective ability of their member viruses to infect vertebrates, including humans, versus invertebrates. The *Hamaparvovirinae* subfamily was split in 2019 from the *Densovirinae* subfamily that served as a specific melting pot for all invertebrate-infecting parvoviruses. Some order members cause diseases, which range from subclinical to lethal. A few, such as adeno-associated virus (AAV) of the genus *Dependoparvovirus* from the subfamily *Parvovirinae*, require coinfection with helper viruses from other families.

The class *Quintoviricetes* is a part of the *Cossaviricota* phylum that also includes the great class *Papovaviricetes* involving the orders *Zurhausenvirales* and *Sepolyvirales* described in Chapters 7 and 8, respectively, as well as the class *Mouviricetes* with the sole order *Polivirales*, which is characterized briefly in Chapter 12. The phylum *Cossaviricota*, together with the large *Cressdnaviricota* phylum, form the kingdom *Shotokuvirae* of the realm *Monodnaviria*.

The *Parvoviridae* family has been advanced to the forefront of the VLPs at the very beginning of the protein engineering era for the two major reasons. They owned the old traditions of structural investigations, and their capsids were flexible enough to allow high-level production in the form of VLPs in insect cells and to retain self-assembly properties after foreign insertions. In fact, the most promoted was the *Parvovirinae* subfamily, where numerous representatives of the genera *Erythroparvovirus* and *Protoparvovirus* played the great role by the development of the classical VLP carriers in 1990s, as substantially reviewed at that time (Casal 1999; Casal et al. 1999). The AAV of the *Dependoparvovirus*

genus, in turn, occupied a place of the favorite gene therapy vector; for the first historical reviews see Büeler (1999) and Grimm and Kleinschmidt (1999).

Figure 9.1 presents some pictures of the numerous *Parvoviridae* x-ray and electron microscopy structures available at the PDB and VPERdb collections. As reviewed extensively by Cotmore et al. (2019), the T = 1 virions are 23–28 nm in diameter and exceptionally rugged, often remaining infectious in the environment for months or years. The capsids are assembled from a nested set of 60 viral protein (VP) proteins, typically encoded on the single structural VP gene that includes the entire coding sequence for VP1 (generally 75–100 kDa), while one or more smaller forms (VP2–5) share a common C-terminal sequence but have N-terminal truncations of different length. Thus, the capsids of the important *Erythroparvovirus* and *Protoparvovirus* genera are composed of 60 copies of the major capsid protein VP2 and 6 to 10 copies of the minor capsid protein VP1, where VP1 is identical to VP2 but contains an additional N-terminal stretch of 227 aa. A portion of that region of VP1 is external to the capsid, but the VP1 itself is not required for capsid formation. The VP monomers interdigitate extensively to create 60 asymmetric units per particle, which coordinate at 12 5-fold, 20 3-fold and 20 2-fold axes. Each VP chain forms a core β-barrel structure of at least 8 strands, while individual β-strands are linked by loops of variable length, sequence, and conformation, most of which project toward the outer surface of the capsid and give individual viruses their unique surface topology.

GENOME

Figure 9.2 presents some examples of the *Parvoviridae* genomes, according to the current ICTV report (Cotmore et al. 2019). They are linear nonpermuted ssDNA molecules of 4–6 kb in which a long coding region is bracketed by short, 116 to ~550 nucleotide, imperfect palindromes that fold into dynamic hairpin telomeres. The viruses possess two major gene cassettes; a nonstructural replication initiator gene (NS) located in the 3′ (by convention the "left") half of the negative-sense strand and a single capsid sequence (VP) located in the right half. Most genomes also encode a small number of ancillary proteins in alternate and/or overlapping open reading frames, which are not shown in the simplified Figure 9.2. Some parvoviruses preferentially excise and encapsidate ssDNA of negative polarity, e.g. minute virus of mice (MVM) of the *Protoparvovirus* genus, while others encapsidate strands of either polarity in equivalent—e.g. adeno-associated virus 2 (AAV2) of the *Dependoparvovirus* genus—or different proportions, e.g. bovine parvovirus 1 (BPaV1) of the *Bocaparvovirus* genus (Cotmore et al. 2019).

DOI: 10.1201/b22819-12

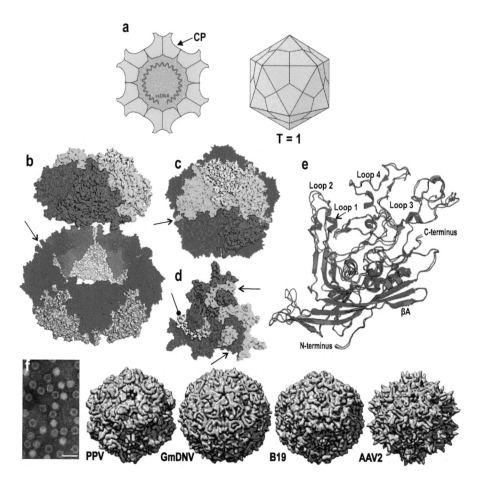

FIGURE 9.1 Morphology of parvoviruses. (a) The general cartoon of the *Parvoviridae* representatives taken with a kind permission from the ViralZone, Swiss Institute of Bioinformatics, http://viralzone.expasy.org (Hulo C et al. 2011). Courtesy Philippe Le Mercier. (b) Side view, at a resolution of 3.5 Å, of a tilted model of a porcine parvovirus (PPV) capsid with the top five trimers translated 120 Å along its 5-fold axis from the middle body of ten trimers (five shown in dark blue and five in light blue); the bottom five trimers are not shown. The 5-fold axes are located at the intersection of five trimers (e.g. at arrow: green, magenta, two dark blue and one light blue colored trimer). (c) Top view of model shown in (b), without light-blue trimers, along the 5-fold axis. The channels at the 5-fold axes are clearly visible (arrow indicates same 5-fold axis as shown by arrow in (b)). (d) Structure of a PPV trimer. The arrows indicate the intertwining of the GH-loop (between β strands G and H) in the counter-clockwise located proteins from near the 3-fold axis toward the 2-fold axis (arrows). The GH-loop actually consists of two loops of which one (loop 3, yellow), running to the 2-fold axis, is partly covered by loop 4 (cyan) near the 3-fold axis (oval arrow). (e) Parvoviruses may have remarkably similar structures despite low sequence identities. This figure shows the alignment of minute virus of mice (MVM, in red; 549 amino acid residues, PDB 1MVM) and PPV (in blue; 542 amino acid residues, PDB 1K3V) structural proteins that have only 52% sequence identity. Nevertheless, 528 C_{α}s (97%) of the residue pairs occupy the same position in the capsid (root-mean-square error of 1.0 Å). (f) Negative contrast electron micrograph of empty and full PPV particles (bar 50 nm) and space-filling models of the capsid structures of PPV, Galleria mellonella densovirus (GmDNV; PDB 1DNV), human B19 virus (PDB 1S58), and adeno-associated virus 2 (AAV2; PDB 1LP3) shown at a resolution of 4 Å. In each case, the view is down a 2-fold axis at the center of the virus, with 3-fold axes left and right of center and 5-fold axes above and below. Models (b-e) have been rendered by PyMOL and the space-filling models by Chimera (Multiscale Models). (Reprinted from *Virus Taxonomy, Classification and Nomenclature of Viruses. Ninth Report of the International Committee on Taxonomy of Viruses*, King AMQ, Lefkowitz E, Adams MJ, Carstens EB (Eds), Tijssen P, Agbandje-McKenna M, Almendral JM et al., Family—*Parvoviridae*. 405–425, Copyright 2012, with permission from Elsevier.)

SUBFAMILY *PARVOVIRINAE*

GENUS *DEPENDOPARVOVIRUS*

Adeno-Associated Virus

The primate AAVs of the *Adeno-associated dependoparvovirus A* and *B* species, where serotypes 1–4 and 6–13 belong to *Adeno-associated dependoparvovirus A* and AAV5 to

Adeno-associated dependoparvovirus B (Cotmore et al. 2019), are the most famous representatives of *Parvoviridae* family, since they are the extraordinarily successful gene therapy vectors and remain a "gold standard" of the current gene therapy, first, for inherited forms of blindness and hemophilia B. The causes of the great gene therapy interest lie in the facts that the AAVs are not currently known

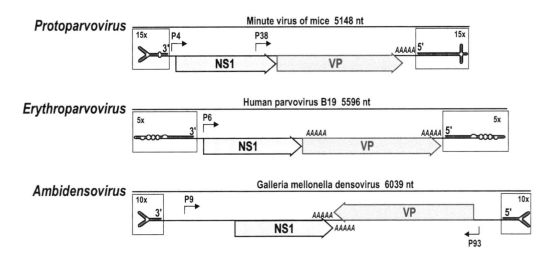

FIGURE 9.2 The genus-specific organization of the *Parvoviridae* genomes. The terminal hairpins are magnified relative to the coding region to show predicted secondary structures. The ORFs are indicated by arrowed boxes, and the structural genes are colored dark pink. The angled arrows indicate transcriptional promoters, and AAAAA indicates polyadenylation sites. (The structures are redrawn from the official ICTV paper, and simplified from Cotmore SF et al. ICTV virus taxonomy profile: *Parvoviridae. J Gen Virol.* 2019;100:367–368.)

to cause disease, have limited capacity to induce immune responses in humans, and are infecting both dividing and quiescent cells. Moreover, the AAV vectors can persist in an extrachromosomal state without integrating into the genome of the host cell, although the native AAVs do integrate into the host genome. The subject of the AAV-based gene therapy is not only too huge (suffice it to say that there are more than 12,000 entries in Medline) but also stands a bit apart from the VLP nanotechnology mainstream to be referenced in the present book. It should be mentioned, however, that the first success of the AAV2 gene therapy vectors resulted in the transduction of nondividing cells in the brain (Kaplitt et al. 1994), muscle (Kessler et al. 1996), liver (Snyder et al. 1997), and eye (Flannery et al. 1997). The recombinant AAV vectors were typically produced in tissue-cultured mammalian cells by providing DNA plasmids that contained the therapeutic gene flanked by the origin of AAV replication, genes for AAV replication proteins, genes of the structural proteins VP1, VP2, and VP3, and a plasmid containing early genes from adenovirus (Grimm et al. 1998; Grimm and Kleinschmidt 1999; Salvetti et al. 1998; Xiao X et al. 1998). At last, some milestone reviews concerning the continual development of the AAV gene therapy vectors could be recommended (Carter 2004; Grieger and Samulski 2005; Daya and Berns 2008; Agbandje-McKenna and Kleinschmidt 2011; Zinn and Vandenberghe 2014; Naso et al. 2017; Castle 2019; Frederick et al. 2020). As a compromise, the following AAV story is confined to the examples, which are closer to the VLP ideology and demonstrate, first, strong and successful parallelism of the functional gene therapy vector and high-resolution structural electron cryomicroscopy and x-ray crystallography investigations in the AAV field. Second, the typical examples of how to improve the cell targeting properties of the AAV vectors by their

surface modification are included, since this is the typical goal of the VLP approach.

The three AAV capsid proteins VP1, VP2, and VP3 are encoded by overlapping sequences of the same open reading frame by the initiation at the internal initiation codons. The first high-resolution AAV structures were resolved by electron cryomicroscopy and image reconstruction for the mammalian cells-produced AAV2 VP1/VP2/VP3 VLPs to 1.05 nm (Kronenberg et al. 2001) and for the virions of AAV5 to 16 Å (Walters et al. 2004) and AAV4 to 13 Å (Padron et al. 2005). Then, the insect cells-produced VLPs were resolved by x-ray crystallography for the AAV5 VP1/VP2/VP3 VLPs to 3.2 Å (DiMattia et al. 2005; Govindasamy et al. 2013), AAV1 VP1/VP2/VP3 VLPs to 4.0 Å (Miller et al. 2006), AAV8 VP1/VP2/VP3 VLPs to 2.6 Å resolution (Nam et al. 2007), and AAV9 VP1/VP2/VP3 VLPs to 2.8 Å (Mitchell et al. 2009; DiMattia et al. 2012).

The x-ray crystallography was used at the early time to resolve the first AAV gene therapy vector structures: AA2 to 3.0 Å resolution (Xie et al. 2002, 2003), AA6 to 3.2 Å (Xie et al. 2008). Later, the atomic structures were resolved for the great number of the AAV gene therapy vectors: AAV1 (Huang et al. 2016; Zhang R et al. 2019b), AAV2 (McCraw et al. 2012; Drouin et al. 2016; Tan et al. 2018; Meyer et al. 2019; Zhang R et al. 2019a), AAV2.5 (Burg et al. 2018), AAV2.7m8 (Bennett et al. 2020), AAV 3B (Lerch et al. 2010), AAV4 (Govindasamy et al. 2006), AAV5 (Govindasamy et al. 2013; Zhang R et al. 2019b; Silveria et al. 2020), AAV6 (Ng et al. 2010; Xie et al. 2011; Huang et al. 2016), AAV7 (Nam et al. 2011), AAV8 (Nam et al. 2007; Rumachik et al. 2020; Mietzsch et al. 2020a), AAV-DJ, a chimeric mix of serotypes 2, 8 and 9 (Lerch et al. 2010; Xie et al. 2013, 2017, 2020), AAV9 (Penzes et al. 2021). At last, Mietzsch et al. (2021) completed the AAV structural atlas and reported the remaining capsid structures of AAV7, AAV11, AAV12, and AAV13 determined by electron cryomicroscopy and 3D image reconstruction to 2.96, 2.86, 2.54, and 2.76 Å resolution—respectively—and localized the structures to the specific clades. Thus, the AAV7 represented the first clade D capsid structure; AAV11 and AAV12 were of a currently unassigned clade that would include AAV4; and AAV13 represented the first AAV2-AAV3 hybrid clade C capsid structure. The obtained atlas of surface loop configurations compatible with capsid assembly paved a way for the future vector engineering efforts with respect to specific tissue targeting, transduction efficiency, antigenicity, or receptor retargeting (Mietzsch et al. 2021).

The rhesus macaques' isolates possessing lower seroprevalence in human populations compared to the standard AAV2 and AAV8 vectors were resolved: AAVrh32.33 (Mikals et al. 2014), AAVrh.8 (Halder et al. 2015), AAVrh.10, and AAVrh.39 (Mietzsch et al. 2020a).

Guenther et al. (2019) developed a protease-activatable AAV9 vector named provector that responded to elevated extracellular protease activity commonly found in diseased tissue microenvironments and resolved its structure by electron cryomicroscopy. In the in vivo model of myocardial infarction, the provector was able to deliver transgenes site-specifically to high-protease activity regions of the damaged heart, with concomitant decreased delivery to many off-target organs, including the liver (Guenther et al. 2019).

Concerning the first production examples of the AAV VLPs in insect cells, the separate expression of the AAV2 VP1, VP2, and VP3 genes by recombinant baculoviruses in insect *Sf*9 cells was achieved by mutation of the internal translation initiation codons (Ruffing et al. 1992). The coexpression of VP1 and VP2, VP2 and VP3, and all three capsid proteins and the expression of VP2 alone resulted in the production of VLPs resembling empty capsids generated during infection of HeLa cells with AAV2 and adenovirus, suggesting the requirement for VP2 in the formation of the VLPs. Figure 9.3 demonstrates the first AAV2 VLPs obtained by the baculovirus-driven expression. In parallel, the individual and combined expression of the structural AAV2 genes was performed in mammalian HeLa cells (Ruffing et al. 1992). Steinbach et al. (1997) separately expressed the AAV2 VP1, VP2, and VP3 by recombinant baculoviruses, purified the products under denaturing conditions, renatured them in the presence of 0.5 M arginine, and recovered VLPs in vitro, however, only in the presence of a HeLa extract. Next, the baculovirus expression system was used to produce the AAV2 VLPs as a gene therapy vector (Urabe et al. 2002; Aucoin et al. 2006, 2007). Kohlbrenner et al. (2005) extended the baculovirus platform to other AAV serotype gene therapy vectors, as well as produced the first pseudotyped vectors, where the AAV5 or AAV8 vectors incorporated the AAV2 VP1 in a mosaic vector particle. Rayaprolu et al. (2013) generated in insect cells the VP1/VP2/VP3 VLPs of AAV1, AAV2, AAV5, and AAV8 and compared stability and dynamics of the

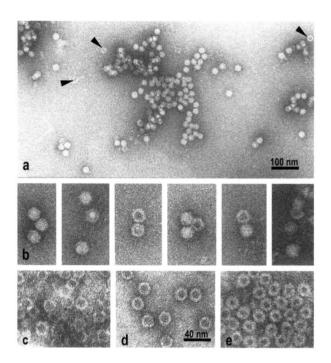

FIGURE 9.3 The comparison of reconstituted and wild-type capsid structures of AAV2. The different images of negatively stained AAV2 capsid particles present in an AAV-2 virus preparation (a, b) were compared with empty capsids formed in *Sf*9 cells after coinfection with recombinant baculoviruses expressing VP1 and VP2 (c), VP2 and VP3 (d), and VP1, VP2, and VP3 (e). The arrows in panel (a) indicate ringlike structures, which probably represent empty capsids, some of which are shown in detail in panel (b). The specimens were negatively stained with uranyl acetate. Bars correspond to 100 nm in panel (a) and 40 nm in panels (b) through (e). (Reprinted with permission of American Society for Microbiology from Ruffing M et al. *J Virol.* 1992;66:6922–6930.)

obtained capsids. Rumachik et al. (2020) directly compared the insect *Sf*9 and mammalian HEK293 cells to produce the AAV VLPs and resolved the unexplained differences in vector performance, which have been seen clinically and preclinically. First, the VLPs had different posttranslational modifications, including glycosylation, acetylation, phosphorylation, and methylation. Second, the AAV genomes were differently methylated. As a result, the human cell-produced AAVs were more potent than insect-cells vectors in various cell types in vitro, in various mouse tissues in vivo, and in human liver in vivo (Rumachik et al. 2020).

In yeast *S. cerevisiae*, Backovic et al. (2012) demonstrated that the AAV2 VLPs can be produced when at least two out of three AAV structural proteins, VP1 and VP3, are simultaneously synthesized in yeast cells, and their intracellular stoichiometry has to resemble the one found in the particles derived from mammalian or insect cells.

The presentation of foreign epitopes on the AAV2 VLPs was started with the Flag tag fusion to the N-terminus of VP3 and formation of the mosaic VLPs from wild-type VP2 and modified VP3 in insect cells, where the chimeric VLPs, however, were not recognized by anti-Flag antibody (Hoque et al. 1999). Furthermore, the relatively large proteins were incorporated into the AAV virions via N-terminal fusion to VP2 without compromising viral infectivity or affecting viral tropism (Warrington et al. 2004; Lux et al. 2005; Asokan et al. 2008; Münch et al. 2013). Rybniker et al. (2012) displayed a target antigen Ag85A of *Mycobacterium tuberculosis*, without its mycobacterial leader peptide sequence, by fusion to the N-terminus of the AAV2 VP2 protein and expression in mammalian cells. This study clearly demonstrated that the combination of antigen incorporation with antigen overexpression after cell transduction can dramatically enhance the antigenic potential of AAV-based vaccines (Rybniker et al. 2012). The human papillomavirus (HPV) epitopes were displayed on the AAV2 VLPs and demonstrated protective immunity after vaccination of mice and rabbits with the HPV16 and HPV31 L2 aa 17–36 stretch inserted at positions 587 and 453 of the AAV2 VP3, respectively (Nieto et al. 2012; Jagu et al. 2015). The exhaustive evaluation of the antigen-displaying AAV VLPs as future candidates for personalized cancer vaccination was recently performed by Neukirch (2021).

The great AAV virion and VLP targeting story is presented later by the few typical examples given in chronological order. Thus, at the very beginning, the preferential infectivity for the CD34+ human myoleukemia cell line was ensured by the packaging of all three wild-type AAV2 capsid proteins in mammalian cells with the addition of a single N-terminal fusion of a single-chain antibody sFv against human CD34 to either VP1, or VP2, or VP3 proteins of the AAV2 vector produced in mammalian cells (Yang Q et al. 1998). Girod et al. (1999) inserted the 14 aa peptide QAGTFALRGDNPQG containing the RGD motif into different six putative loops of the AAV2 VP protein and got the AAV particles with extended tropism to the cells possessing integrin receptors. Wu P et al. (2000) performed exhaustive mutational analysis of the AAV2 VP gene and finally engineered a vector with altered tropism by N-terminal insertion of serpin receptor ligand into the VP1 and VP2.

Furthermore, the random peptide libraries were displayed on the AAV2 vector to select for targeted gene therapy vectors on specific cells (Müller et al. 2003; Waterkamp et al. 2006; Naumer et al. 2012a, b). Adachi and Nakai (2010) engineered a novel peptide display library platform based on the AAV1 virion with aa 445–568 being replaced with those of AAV9 in order to impair infectivity and delay blood clearance. The insect cells-produced AAV2 VLPs were coated with polyethyleneimine to ensure efficient siRNA delivery in breast cancer therapy (Shao et al. 2012). Lee NC et al. (2012) improved axial muscle tropism and decreased liver tropism by insertion of acidic six aspartic acids D_6 oligopeptide into heparan sulfate proteoglycan binding region within the AAV2 VP1. Kienle et al. (2012) generated the synthetic AAV vectors by DNA shuffling. The targeting to the cancer cell-surface marker EGFR was improved by fusion of two modular targeting molecules (DARPin or Affibody) to N-terminus of the AAV2 VP2 (Hagen et al. 2014). Pandya et al. (2014) ensured the specific

targeting to dendritic cells by mutational modification of the surface of the AAV6 vector. The visualization of the AAV vectors was achieved by replacement of VP2 by the VP2 with N-terminally fused eGFP protein (Lux et al. 2005).

Nonprimate Adeno-Associated Viruses

Muscovy duck (MDPV) and goose (GPV) parvoviruses, both members of the *Anseriform dependoparvovirus 1* species, contributed to the VLP technologies. The VLPs were achieved in insect cells by the baculovirus-driven synthesis of the MDPV VP2 and VP3 proteins (Le Gall-Reculé et al. 1996) or of the GPV VP1, VP2, and VP3 (Ju et al. 2011) or GPV VP2 alone (Chen Z et al. 2012). Wang CY et al. (2005) obtained in *E. coli* two structural proteins of MDPV and GPV, namely VP3 and a truncated form of VP1 that was designated VP1N and contained sequences present in VP1 and VP2, but not in VP3. They used them for diagnostic purposes, but no attempts were undertaken to obtain VLPs in bacteria. The GPV VLPs, primarily VP2 VLPs, appeared to be good candidates for the vaccination of goslings against Derzsy's disease, goose hepatitis, or gosling plague (Ju et al. 2011; Chen Z et al. 2012; Wang Q et al. 2014).

Xiao S et al. (2021) developed a VLP vaccine candidate against duckling short beak and dwarfism syndrome virus (SBDSV), an emerging goose parvovirus, which has caused short beak and dwarfism syndrome in Chinese duck flocks since 2015, by the engineering of the SBDSV VP2 VLPs in insect *Sf*9 cells by the baculovirus-driven system. A single dose of SBDSV was able to demonstrate high protective efficacy (Xiao S et al. 2021).

Bossis and Chiorini (2003) were first to clone the genome of avian adeno-associated virus (AAAV), a member of the *Avian dependoparvovirus 1* species, to generate the AAAV particles carrying a lacZ reporter gene in the mammalian 293T cells and demonstrate their transduction efficiency in both chicken primary cells and several cell lines. Later, Wang AP et al. (2017) established the scalable AAAV production method by using the baculovirus expression system in insect *Sf*9 cells. Thus, the three recombinant baculoviruses expressing the AAAV gene *rep* and modified VP gene and the inverted terminal repeats-flanked GFP gene were generated. The rAAAV-GFP virions were produced by triple infection of insect cells or triple transfection of HEK293 cells for comparison purpose, and the formation of typical AAAV particles was revealed by electron microscopy, while the insect cell-produced rAAAV yield was almost twenty-five-fold higher than that produced by HEK293 cells. It should be noted that such rAAAV vectors ensured protection of ducklings against duck hepatitis A virus type 1 (DHAV-1) challenge, when the major DHAV-1 capsid VP1 gene was used instead of the model GFP gene (Wang AP et al. 2019).

Mietzsch et al. (2020b) reported the first capsid structure of a nonprimate AAV, which was isolated from bats. The capsid structure of this bat adeno-associated virus (BtAAV) was determined by electron cryomicroscopy and

3D reconstruction to 3.03 Å resolution. The comparison of the mammalian cells-produced empty and genome-containing capsids showed that the capsid structures were almost identical except for an ordered nucleotide in a previously described nucleotide-binding pocket, the density in the 5-fold channel, and several amino acids with altered side chain conformations. Compared to other dependoparvoviruses, for example AAV2 and AAV5, the BtAAV displayed unique structural features including insertions and deletions in capsid surface loops. The BtAAV capsids were capable of packaging AAV2 vector genomes and thus were proposed as gene delivery vectors, since a screen with human sera showed lack of recognition by the BtAAV capsid, rendering it "invisible" to potential preexisting neutralizing human anti-AAV antibodies (Mietzsch et al. 2020b).

Genus Erythroparvovirus

The human parvovirus B19 (B19V) belonging to the *Primate erythroparvovirus 1* species played an outstanding role by the early development of the VLP technologies. The first described production of the B19V VP1 and VP2 was achieved in Chinese hamster ovary cells—where they formed virion-like VLPs (Kajigaya et al. 1989)—and in Cos-7 cells by the replication of hybrid B19V-SV40 origin vectors (Beard et al. 1989). Meanwhile, the B19V VLPs were produced in mammalian Cos-7 cells (Cohen et al. 1995). The mammalian cells-produced B19V proteins self-assembled into the VLPs that were morphologically and antigenically similar to the native B19V virions and could substitute for native antigen in a B19V IgM assay (Cohen et al. 1995). Later, Zhi et al. (2010) reported the mammalian cell type-specific production of the B19V VLPs and greatly increased outcome of both VP1 and VP2 in nonpermissive 293T and HeLa cells by codon optimization, which appeared to be a key factor in the capsid protein production in mammalian cells.

After only a year after the first mammalian expression, both B19V VP1 and VP2 were expressed in insect cells by the baculovirus vectors (Brown et al. 1990), where VP2 alone was able to assemble into the VLPs, but stoichiometry of the capsids containing both VP1 and VP2 was like that previously observed in the B19V-infected cells (Brown et al. 1991; Kajigaya et al. 1991; Tsao EI et al. 1996). The immunization of rabbits with such VLPs resulted in the production of neutralizing antibodies (Kajigaya et al. 1991). The B19V VLPs, consisting of either VP2 alone or of both VP1 and VP2, appeared therefore to be similar to native virions in size and appearance and were thought to be an excellent source of antigen, first of all, for immunodiagnostic kits (Salimans et al. 1992). Then, Michel et al. (2008) engineered polyhistidine-tagged versions of the B19V VP1 and VP2 and produced the corresponding VLPs by the baculovirus expression system, allowing thus the VLP purification by immobilized metal-ion affinity chromatography. The downstream processing (Ladd Effio et al. 2015) and high-throughput characterization (Ladd Effio et al. 2016) of the insect cells-produced B19V VLPs was elaborated later.

The crystal structure of the insect cells-produced B19V VLPs was resolved at first to 8 Å (Agbandje et al. 1991), and then the structure of the full virions and empty capsids of FPV was resolved to at least 3.3 Å (Agbandje et al. 1993). Further work on the B19V structure did not improve the 8 Å resolution, but it established the significant difference of the surface structure of B19V, canine parvovirus (CPV), and feline panleukopenia virus (FPV), both from the *Protoparvovirus* genus described later, where B19V lacked prominent spikes on the 3-fold icosahedral axes (Agbandje et al. 1994). Further, electron cryomicroscopy of the B19V VP2 capsids complexed with their cellular receptor, globoside, were used to confirm the x-ray data (Chipman et al. 1996). At last, Kaufmann et al. (2004) determined the B19V structure to approximately 3.5-Å resolution by x-ray crystallography and achieved the first near-atomic structure of an erythroparvovirus. The polypeptide fold of the major capsid protein VP2 was a "jelly roll" with a β-barrel motif similar to that found in many icosahedral viruses. The large loops connecting the strands of the β-barrel formed surface features that differentiated B19V from other parvoviruses. Although B19V VP2 had only 26% sequence identity to the AAV VP3, 72% of the C_α atoms were aligned structurally with a rms deviation of 1.8 Å. Remarkably, both viruses required an integrin as a coreceptor (Kaufmann et al. 2004). After 15 years, Sun et al. (2019) presented the first electron cryomicroscopy structure of the B19V VLPs complexed with the Fab fragment of a human neutralizing antibody at 3.2 Å resolution, and the antigenic residues on the surface of the capsid were identified. Remarkably, the neutralizing antibody bridged the B19V capsid by binding to a quaternary structure epitope formed by residues from three neighboring VP2 capsid proteins, as evident from Figure 9.4 (Sun et al. 2019).

In yeast *S. cerevisiae*, the B19V VLPs were formed by VP2 in the absence of VP1 and resembled the insect cells-produced B19V VLPs with respect to size, molecular mass, and antigenicity as shown by antigen-capture ELISA and T-cell proliferation tests (Lowin et al. 2005).

In *E. coli*, Rayment et al. (1990) expressed a number of polypeptides derived from the B19V capsid proteins, including native VP1 and VP2 and also fusions to β-galactosidase containing differing amounts of the N-terminus of the VP1/2 polypeptide. Although each of these was expressed at high levels and the majority were produced as full-length proteins, only one, namely a β-galactosidase fusion, was soluble. Later, Sánchez-Rodríguez et al. (2012) reported the efficient in vitro self-assembly of the B19V VLPs derived from the VP2 produced in *E. coli* and the critical effects of pH and ionic strength on the assembly process. Thus, at neutral pH, the homogeneous VLPs assembled, while at acidic and basic pHs, with low ionic strength, the major assemblies were small intermediates. The in vitro self-assembled VLPs were highly stable at 37°C, and a significant fraction of particles remained correctly assembled after 30 min at 80°C (Sánchez-Rodríguez et al. 2012).

The external location and special immunological importance of the B19 VP1 was confirmed by direct immune

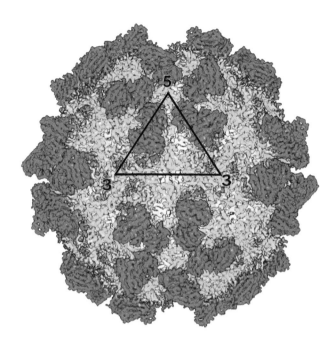

FIGURE 9.4 The structure of B19 VLP complexed with 860–55D Fab molecules. The surface-rendered electron cryomicroscopy map of the B19-Fab complex at 3.2-Å resolution. Yellow (100 Å), green (125 Å), and red (140 Å) coloring indicate increasing distances from the center of mass. (Reprinted with permission of American Society for Microbiology from Sun Y et al. *J Virol.* 2019;93:e01732–18.)

electron microscopy investigation (Rosenfeld et al. 1992). The VP1 was shown to be dispensable for the encapsidation of virus ssDNA but responsible for virus entry, subsequent to cell binding and prior to the initiation of DNA replication (Tullis et al. 1993). The B19 VLPs composed of only VP2 and VP2 containing varying amounts of VP1 protein were evaluated in mice, guinea pigs, and rabbits, and the importance of VP1 for high levels of neutralizing activity was demonstrated (Bansal et al. 1993). The epitope mapping revealed the unique region of the B19 VP1 at the N-terminal portion that was responsible for neutralizing activity (Saikawa et al. 1993; Rosenfeld et al. 1994).

Although the minor capsid protein VP1, longer by 227 aa than VP2, alone did not form empty capsids in insect cells, Wong et al. (1994) found that the VLPs were formed in *Sf*9 cells by truncated versions of VP1. Thus, severely shortened VP1, extended beyond the VP2 core sequence by about 70 aa of the unique region, formed the VLPs normal in appearance, while longer versions of VP1 also formed capsids but did so progressively less efficiently and produced capsids of more markedly dysmorphic appearance as the VP1-unique region was lengthened. Thus, the unique N-terminal part of the VP1 appeared as an ingenuous target for foreign insertions (Wong et al. 1994). The progressive truncation of the unique region of the B19 VP1 with the addition of the Flag tag DYKDDDK at the N terminus led to the formation of VLPs not only from the mosaic VP2-VP1 particles but also from the truncated Flag-VP1 proteins only, and most

of the VP1 unique region was confirmed to be external to the capsid and accessible to antibody binding (Kawase et al. 1995b). The N-terminal truncation of the B19V VP2 showed that aa position 25 was critical for the self-assembly, and VP2 truncated to aa 26 to 30 failed to self-assemble but did participate with normal VP2 in the capsid structure, whereas truncations beyond aa 30 were incompatible with either self-assembly or coassembly (Kawase et al. 1995a).

The insect cells-produced B19V vaccine candidate against erythema infectiosum, transient aplastic crisis in individuals with underlying hemolytic disorders and hydropsfetalis in humans, was preliminarily elucidated by Bansal et al. (1993). The empty B19V VLPs composed of only VP2 and VP2 capsids containing 4%, 25%, 35%, or 41% VP1 protein were evaluated in mice, guinea pigs, and rabbits. The VP2 capsids devoid of VP1 protein—or consisting of only 4% VP1, the composition of naturally occurring virions—were generally poor at eliciting high levels of virus-neutralizing activity, while the capsids consisting of more than 25% VP1 protein efficiently and consistently provoked vigorous B19V virus neutralizing responses (Bansal et al. 1993). Franssila et al. (2001) used the B19V vaccine candidate to induce the T helper cell-mediated in vitro response of recently and remotely infected subjects. Franssila and Hedman (2004) concluded that, whereas VP1 contained important B-cell epitopes, VP2 appeared to provide the major target for B19-specific Th-cells years or decades after natural infection. The safety and immunogenicity trials of the insect cells-produced B19V vaccine in healthy adults were successful (Bostic et al. 1999; Ballou et al. 2003; Bernstein et al. 2011).

Chandramouli et al. (2013) developed new VLP-based B19V vaccine candidates, produced by coexpressing VP2 and either wild-type VP1 or phospholipase-negative VP1 in a regulated ratio from a single plasmid in *S. cerevisiae.* These VLPs were expressed efficiently and were homogeneous and easy to purify. Although VP2 alone formed VLPs, in mouse immunizations, VP1 and the adjuvant MF59 were required to elicit a neutralizing response, where both wild-type VLPs and those with phospholipase-negative VP1 were equivalently potent (Chandramouli et al. 2013). Penkert et al. (2017) performed a preclinical trial of the yeast-produced B19V vaccine candidate in mice.

The generation of the empty chimeric B19 VLPs broke out after the construction and expression in insect cells of the VP2 with an epitope VII consisting of aa 9–21 from the envelope glycoprotein gD of human herpes simplex virus type 1 (HSV1) and an epitope A from the spike protein of murine hepatitis virus (MHV) strain A59 inserted at the N terminus and at a predicted surface region (Brown et al. 1994). The immune electron microscopy indicated that the epitopes inserted into the loop were exposed on the surface of the chimeric particles. The chimeric capsids were not only immunogenic in mice but they also induced partial protection against a lethal challenge infection with either MHV or HSV (Brown et al. 1994). In the other approach, the 227-aa N-terminal portion of the B19V VP1 was substituted by a sequence encoding the 147 aa of hen egg-white lysozyme (HEL) with variable amounts of retained VP1 sequence joined to the VP2 backbone (Miyamura et al. 1994). The authors demonstrated the external presentation of HEL on the VLPs, as well as its enzymatic activity and immunogenicity in rabbits.

As to other putative vaccine candidates, Ogasawara et al. (2006) engineered an anthrax vaccine where the B19V VLPs displayed the small-loop peptide and the full-length domain 4 of protective antigen (PA) of *Bacillus anthracis*. The epitopes were added in different combinations to the N-terminus of VP2 retaining short fragments of VP1, and the chimeric VP2 were self-assembled by the baculovirus-driven expression in *Sf*9 cells. The chimeric VLPs retained the immunogenicity of the displayed microbial PA-epitopes, elicited robust levels of anti-PA antibody titers, and showed potential for the prevention of lethal toxin-induced mortality of mouse-macrophage cells (Ogasawara et al. 2006).

To produce a dengue vaccine candidate, Amexis and Young (2006) produced in the *Sf*9 cells the empty B19V VLPs that carried the two dengue 2-specific epitopes comprising the aa 352–368 and 386–397 of domain BIII of the envelope glycoprotein, which were added to the VP2 N-terminus. The chimeric VLPs induced high antidengue 2 titers in mice, as well as robust 50%-plaque-reduction neutralization test (Amexis and Young 2006).

Del Carmen Morán-García et al. (2016) composed the chimeric B19 VLPs from the VP2 protein fused, at its N-terminus, with two peptides derived from the fusion glycoprotein F of the respiratory syncytial virus (RSV) and produced in *E. coli*. The chimeric VLPs elicited a humoral immune response in mice against RSV, although these antibodies did not cross-react with RSV in ELISA tests.

The in vitro encapsulation of heterologous dsDNA fragments of different sizes between 120 and 551 bp into the B19V VLPs was achieved, where the DNA and denatured VP2 protein were coincubated, and the assembly process was conducted by one dialysis step (Sánchez-Rodríguez et al. 2015).

As to the visualization of the chimeric B19V VLP derivatives, Toivola et al. (2004) labeled the B19V VLPs with the fluorescent dye Oregon Green 488. Moreover, Gilbert et al. (2005) inserted eGFP at the VP2 N-terminus and demonstrated the assembly of the chimeric VLPs in insect cells. Remarkably, the fluorescent VLPs were very similar in size to wild-type B19 virions, while the eGFP was displayed on the surface of the particles. Moreover, the fluorescent VLPs were able to efficiently enter cancer cells and traffic to the nucleus via the microtubulus network (Gilbert et al. 2005). The disassembly of the B19V eGFP-VP2 VLPs by urea and SDS was elucidated, in parallel with the analogous CPV-based construct (Toivola et al. 2005). Ylihärsilä et al. (2015) biotinylated the insect cells-produced B19V VP2 VLPs.

Bustos-Jaimes et al. (2017) have chosen the flexible loop 301-EGDSSSTGAGKAL-313 of the B19V VP2 as a target for the insertions of proteins to be folded, namely eGFP and a lipase from *B. pumilus*. As a result, the model proteins were inserted at the tip of this loop, between residues Thr307 and Gly308, and the corresponding VLPs were obtained by the expression in *E. coli*.

Cayetano-Cruz et al. (2018) used the advanced SpyTag/SpyCatcher, which is described in detail in Chapter 25, to expand the repertoire of the B19V VLPs. Thus, the α-glucosidase Ima1p of *S. cerevisiae* was attached to the surface of the in vitro assembled B19 VLPs. The latter demonstrated a noticeable increase in size compared to the nondecorated VLPs and were proposed for the further development as part of a therapy of lysosomal storage diseases derived from defects in the human acid α-glucosidase.

Salazar-González et al. (2019) used the elaborated *E. coli* expression system to engineer a putative vaccine candidate against breast cancer. Thus, the two chimeric B19V VP2 were engineered by addition of the epitopes from the insulin-like growth factor-1 receptor (IGF-1R), namely epitopes 249 (GDLTNRCTMEEKPMEK) and P8 (RQPQDGYLYRHNYCSK), to the VP2 N-terminus. This resulted in the multiepitope anticancer VLP vaccine, which prevented and delayed tumor growth when used in a prophylactic scheme of four weekly immunizations prior to cancer cell inoculation in mice and led to idea of a library of chimeric VP2-fused epitopes for further assembly in a designed and personalized epitope delivery system (Salazar-González et al. 2019). To realize this idea, a set of neoepitopes, namely Tmtc2, Gprc5a, and Qars, were selected from experimental studies of next-generation sequencing of the breast cancer 4T1 cell line—as well as an epitope of surviving—and were added to the VP2 N-terminus, and the obtained VLPs were evaluated in a breast cancer model (Jiménez-Chávez et al. 2019).

After the famous B19V, another member of the *Primate erythroparvovirus 1* species, namely an isolate V9, was used to produce the corresponding VLPs in insect cells (Heegaard et al. 2002). Ekman et al. (2007) obtained the insect cells-produced VP2, alone or together with VP1, VLPs of the A6-, LaLi-, and V9-like isolates that nevertheless belong to the same B19V species.

Genus *Protoparvovirus*

Carnivore Parvoviruses

The carnivore-infecting parvoviruses include members of the *Carnivore protoparvovirus 1* species, such as canine parvovirus (CPV), feline panleukopenia virus (FPV), and mink enteritis virus (MEV). As reviewed by the current ICTV report (Cotmore et al. 2019), the behavior of the carnivore parvoviruses is in sharp contrast to that of the relatively mild human parvoviruses. Thus, CPV replicates in dogs in intestinal and bone marrow cell populations that continue dividing throughout the animals' lifespan and can reach mortality rates around 90% if untreated (Kailasan et al. 2015a).

The CPV virion structure was the subject of the first x-ray study performed on parvoviruses (Luo et al. 1988).

The infectious virions of CPV were measured by 25.7 nm in diameter and contained 60 copies of VP2 within the T = 1 icosahedral capsids, and the central structural motif of VP2 had the same topology, namely an eight-stranded antiparallel β-barrel, as has been found at that time in many other icosahedral viruses (Tsao J et al. 1991). The structure demonstrated a 22-Å-long protrusion on the 3-fold axes where major antigenic regions were found, a 15-Å-deep canyon circling around each of the five cylindrical structures at the 5-fold axes, and a 15-Å-deep depression at the twofold axes (Tsao J et al. 1991). The structure of the empty CPV capsids revealed some conformational differences between the full and empty viruses in the region where some ordered DNA has been observed to bind in the CPV-full particles (Wu Hao and Rossmann 1993; Wu Hao et al. 1993). Finally, the structure of the CPV capsid was resolved at 2.9 Å (Xie and Chapman 1996). Remarkably, it was documented that approximately 87% of the VP2 have N-termini on the inside of the capsid, but for approximately 13% the polypeptide starts on the outside and runs through one of the pores surrounding each 5-fold axis, explaining apparently conflicting antigenic data (Xie and Chapman 1996). Llamas-Saiz et al. (1996) resolved the structure of a single aa Ala/Asp CPV mutant that demonstrated a loss of canine host range and altered antigenic properties. Based on the CPV model, Agbandje et al. (1993) resolved the atomic FPV structure to 3.3 Å. Furthermore, Simpson et al. (2000) and Govindasamy et al. (2003) compared the CPV and FPV structures in order to localize elements responsible for the recognition of canine transferrin receptor and for the canine cell infection. Later, Organtini et al. (2015) resolved the crystal structure of a CPV2a strain, characterized by four VP2 mutations, that emerged in 1978, spread worldwide within two years, and completely replaced the original strain CPV2. The CPV2a structure showed that each mutation conferred small local changes. In parallel, the electron cryomicroscopy technique was used to characterize the CPV complex with a virus-neutralizing Fab molecule to near-atomic 4 Å resolution (Organtini et al. 2016).

In parallel with the structural investigations, the direct epitope mapping of the CPV VP2 revealed the N-terminal portions as responsible for neutralizing activity (Casal et al. 1995), by analogy with the previously described B19V.

Saliki et al. (1992) expressed the VP2 genes of CPV and of a recombinant consisting of CPV and FPV sequences by the baculovirus system and obtained the virion-like VLPs. The immunization of dogs with the lysates of the baculovirus-infected cells protected them from clinical disease upon challenge with a virulent CPV isolate. López de Turiso et al. (1992) obtained the CPV VP2 VLPs in insect cells and used them to immunize dogs in different doses and combinations of adjuvants and demonstrated a good protective response, higher than that with a commercially available inactivated vaccine. The efficiency of the insect cells-produced CPV VP2 VLPs as a vaccine in minks against MEV infection was demonstrated by Langeveld et al. (1995). Choi et al. (2000) achieved high-level production of the CPV VP2

VLPs in *Bombyx mori* larvae. Later, Feng et al. (2014) produced the CPV VP2 VLPs in insect cells and pupae and investigated their immunogenicity in mice and dogs. When immunized intramuscularly with purified VLPs, in the absence of an adjuvant the mice elicited CD4+ and CD8+ T cell and neutralizing antibody responses, while the oral administration of raw homogenates containing VLPs to the dogs resulted in a systemic immune response and long-lasting immunity (Feng et al. 2014). Jin et al. (2016) achieved the codon-optimized baculovirus-driven expression of the VP2 gene from the currently prevalent CPV2a strain and demonstrated high antibody titers after intramuscular or oral immunization of dogs.

Wu Hongchao et al. (2020) produced the MEV VP2 VLPs in the insect *Sf*9 cells, using and a single-dose injection of the VLPs resulted in complete protection of mink against virulent MEV challenge for at least 180 days. Jiao et al. (2020) used the baculovirus expression system to obtain the tiger FPV VP2 VLPs, where the VP2 gene was taken from an infected Siberian tiger *Panthera tigris altaica*. Remarkably, the key aa of this gene was the same as that of FPV, whereas the 101st aa position was the same as that of CPV. The antibody production was induced in cats after subcutaneous immunization and persisted for at least 12 months. It should be noted that the insect cells-produced CPV VP2 VLPs were employed by immunization of hens to obtain specific chicken anti-CPV IgY single chain fragment variables (scFv; Ge et al. 2000).

In *E. coli*, Xu J et al. (2014) used the small ubiquitin-like modifier (SUMO) fusion motif to express the whole natural CPV VP2. After the cleavage of the fusion motif, the CPV VP2 protein self-assembled into the VLPs that resemble the authentic virions by size and shape. However, the self-assembly efficiency of VLPs was affected by different pH levels and ionic strengths. The bacterial CPV VLPs did not differ from the natural CPV virions by induction of a specific anti-CPV antibody response in mice. Nan et al. (2018) produced the soluble CPV VP2 in *E. coli* by coexpression with chaperone trigger factor (Tf16) and assembled the CPV VLPs, where NaCl concentration and pH were critical. The immunization of guinea pigs with the VLPs induced high-titer virus-neutralizing antibodies.

Concerning the development of the putative chimeric CPV VLPs, Cortes et al. (1993) were the first to perform the thorough topographical analysis of the CPV virions and VLPs. Based on these results, it was hypothesized that the N-terminus of VP2 was barely or not at all exposed on the surface of the native virions but became accessible after some virion steric change. The CPV VP2 allowed for N-terminal truncation of 9 and 14 but not of 24 aa, and it allowed for the deletion of loop 2 but not of the other three loops of the protein, proposing this specific loop 2 as a possible target for foreign insertions (Hurtado et al. 1996). Casal et al. (1999) described the methodological approaches on how to insert a foreign epitope, namely, C3:B poliovirus epitope comprising aa 93–104 of VP1, into the CPV VP2 N-terminus and all four loops. Rueda et al. (1999a,

b) evaluated in detail the insertion of this epitope into the four loops and at the C-terminus of the VP2 protein. Later, Feng et al. (2011) produced in silkworm pupae the CPV VP2 VLPs that carried the 23-aa virus-neutralizing and/or 8-aa CTL antigen epitopes of rabies virus (RV), which were inserted into the loop 2 of VP2 with deletion of aa 226–233 of VP2 and a T cell epitope added to the N-terminus of VP2. Wang C et al. (2017) engineered a vaccine consisting of the chimeric CPV VP2 VLPs carrying the receptor binding domain (RBD) of Middle East respiratory syndrome coronavirus (MERS-CoV). The stretch spanning the spike glycoprotein S aa residues 367–606 of the MERS-CoV RBD was fused N-terminally to the CPV VP2. The chimeric virion-like VLPs induced the RBD-specific humoral and cellular immune responses in mice, and antisera from these animals were able to prevent pseudotyped MERS-CoV entry into susceptible cells, demonstrating a promise as a prophylactic vaccine candidate against MERS-CoV in a potential outbreak situation (Wang C et al. 2017).

The visualization of the CPV VP2 VLPs was achieved by the insertion of eGFP at the N-terminus of VP2 without altering assembly of the latter (Gilbert et al. 2004). Thus, the fluorescent VLPs were abundantly produced in insect cells and displayed a very similar size and appearance when compared to the original CPV VP2 VLPs. The feeding of mammalian cells susceptible to CPV infection with these fluorescent VLPs indicated that entry and intracellular trafficking could be observed (Gilbert et al. 2004). Gilbert et al. (2006) fused the eGFP to the N-terminally truncated VP2 lacking 14, 23, and 40 aa residues and produced the constructs in insect cells. While the nontruncated eGFP-VP2 fusion protein and the eGFP-VP2 constructs truncated by 23 and by as much as 40 aa were able to form VLPs, the construct containing eGFP-VP2 truncated by 14 aa was not able to assemble into the VLP-resembling structures. The disassembly of the CPV eGFP-VP2 by urea and SDS was elucidated, in parallel with the analogous previously mentioned B19V eGFP-VP2 VLPs (Toivola et al. 2005).

The natural tropism of the insect cells-produced CPV VP2 VLPs was used to develop tumor targeting and drug delivery strategy by the chemical conjugation of small molecules and further evaluation of the VLP binding to transferrin receptors over-expressed by a variety of cancer cells of canine as well as of human origin (Singh et al. 2006; Singh 2009). The structural modeling suggested that six lysines per VP2 subunit were presumably addressable for bioconjugation on the CPV capsid exterior. Therefore, between 45 and 100 of the possible 360 lysines/particle could be routinely derivatized with dye molecules depending on the conjugation conditions. The dye conjugation also demonstrated that the CPV VLPs could withstand conditions for chemical modification on lysines. The attachment of fluorescent dyes neither impaired binding to the transferrin receptors nor affected internalization of the VLPs into several human tumor cell lines, exhibiting therefore highly favorable characteristics for development as a novel nanomaterial for tumor targeting (Singh et al. 2006). These

studies were comprehensively reviewed and commented by Manchester and Singh (2006) and later by Li K et al. (2010) and by Dedeo et al. (2011).

Yan et al. (2015) regarded the natural CPV tropism to the transferrin receptor as a tool to increase the targeting ability of quantum dots (QDs) in cell imaging. The CPV VP2 VLPs produced in *E. coli* according to the previously described approach of Xu J et al. (2014) were encapsulated with the QDs and delivered to cells expressing transferrin receptor. As a result, the CPV-VLPs-QDs significantly reduced the cytotoxicity of QDs and selectively labeled the cells with high-level transferrin receptors. It was confirmed therefore that the CPV VP2 VLPs can significantly promote the biocompatibility of nanomaterials and could expand the application of VLPs in bionanomedicine. Again, the CPV VLPs were suggested as promising carriers to facilitate the targeted delivery of encapsulated nanomaterials into cells via receptor-mediated pathways (Yan et al. 2015).

Porcine Parvovirus

Porcine parvovirus (PPV), a member of the *Ungulate protoparvovirus 1* species, causes reproductive failure of swine characterized by embryonic and fetal infection and death, usually in the absence of outward maternal clinical signs. Martínez et al. (1992) were the first to produce the PPV VP2 VLPs in the *Sf*9 cells by the baculovirus-driven expression. These first PPV VLPs were used to immunize two pigs. The latter elicited an immune response that was identical to that of a commercial vaccine. The amount of recombinant antigen needed in a vaccine dose was only 3 μg in a primary dose and 1.5 μg in the booster (Martínez et al. 1992). José Ignacio Casal (1996) presented the vaccine as highly immunogenic and protecting sows against reproductive failure in PPV challenge experiments. Rueda et al. (2000) elaborated an efficient protocol for the safe purification of the insect cells-produced PPV VP2 VLPs, first to avoid dangerous contamination by recombinant baculoviruses. As a result, the insect cells-produced PPV VP2 VLPs were subjected to the efficient large-scale production (Maranga et al. 2002, 2003, 2004) and converted into the prospective vaccine preventing the reproductive failure in gilts (Antonis et al. 2006). To apply the Ni-NTA affinity column chromatography for the PPV VLP purification, Zhou et al. (2010) provided the VP2 protein with the N-terminally added His$_6$ tag.

The near-atomic structure of the insect cells-produced PPV VP2 VLPs was solved using x-ray crystallography to 3.5 Å and was found to be similar to the related CPV and minute virus of mice (MVM; Simpson et al. 2002). The arguments for this similarity are presented in Figure 9.5. It should be remembered, however, that the PPV VP2 had only 57% and 49% aa sequence identity with CPV and MVM respectively, but the degree of conservation of surface-exposed residues was lower than average. Consequently, most of the structural differences appeared on the surface and were the probable cause of the known variability in antigenicity and host range. Remarkably, the strains of PPV demonstrated distinct tissue tropisms and pathogenicity,

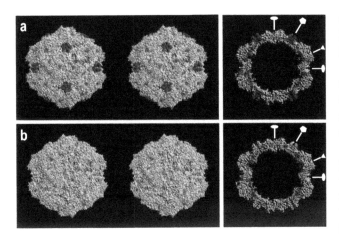

FIGURE 9.5 The conservation of surface features of PPV relative to FPV and MVM. (a) PPV compared to FPV; (b) PPV compared to MVM, with red being the most conserved and blue being the most variable. Left stereo figures show surface views and right mono figures show central cross-sections. Symmetry axes are shown in white. (Reprinted from *J Mol Biol.* 315, Simpson AA, Hébert B, Sullivan GM, Parrish CR, Zádori Z, Tijssen P, Rossmann MG, The structure of porcine parvovirus: comparison with related viruses, 1189–1198, Copyright 2002, with permission from Elsevier.)

which were mediated by one or more of the aa residues 381, 386, and 436, located on or near the surface of the virion (Simpson et al. 2002).

Xu Y and Li (2007) used *Lactobacillus casei* as a delivery vehicle for oral administration by the expression of the PPV VP2, but no data about the possible assembly of the produced VP2 within the cells was mentioned in the paper. Qi and Cui (2009) expressed the codon-optimized PPV VP2 gene, provided with the His tag at the N-terminus, in *E. coli*, however, without any self-assembly indications.

The Kęstutis Sasnauskas team in Vilnius obtained the virion-like PPV VP2 VLPs in *S. cerevisiae*, generated nine MAbs against the VP2, and developed a highly sensitive indirect IgG ELISA test for detection of PPV-specific antibodies in swine sera (Tamošiūnas et al. 2014).

Guo C et al. (2014) expressed the PPV VP2 gene in yeast *Pichia pastoris* but without any self-assembly indications. At last, Yang D et al. (2021) presented the large-scale production and purification of the PPV VP2 VLPs from yeast *Kluyveromyces marxianus*, which was positioned as the highest production of the PPV VLPs achieved to date.

In plants, Rymerson et al. (2003) observed in planta some self-assembled particles after expression of the PPV VP2 gene in low alkaloid transgenic tobacco.

Chen Y et al. (2011) adopted a pseudorabies virus (PRV) system to express the PPV VP2 gene and proposed a bivalent vaccine against both PRV and PPV infection. As a result, the immunization of piglets with recombinant virus elicited PRV- and PPV-specific humoral immune responses and provided complete protection against a lethal dose of PRV challenges. However, the gilts immunized with recombinant virus induced PPV-specific antibodies and

significantly reduced the mortality rate following virulent PPV challenge compared with the control (Chen Y et al. 2011).

Concerning the PPV VP2-derived chimeric VLPs, the foreign sequences were introduced into the N- and C-termini of the PPV VP2, and the fusion of the poliovirus (PV) VP1 T- and B-cell epitopes to the VP2 N-terminus did not alter the self-assembly into the VLPs (Sedlik et al. 1995). Moreover, the VLPs containing C3:T epitope of the poliovirus VP1 were able to induce a T-cell response in vivo, while the VLPs containing the C3:B epitope did not induce any peptide-specific antibody response. Therefore, the N-terminus of the PPV VP2 was declared to be located in an internal position, useful for the T-cell epitope presentation but inadequate for the insertion of B-cell epitopes (Sedlik et al. 1995).

The real breakthrough in the development of the PPV VLPs was offered by the addition of the standard CTL epitope 118-RPQASGVYMGNLTAQ-132 from the lymphocytic choriomeningitis virus (LCMV) nucleoprotein to the N-terminus of VP2 (Sedlik et al. 1997). The immunization of mice with these VLPs carrying a single viral CTL epitope, without any adjuvant, induced a strong CTL response and complete protection of mice against a lethal LCMV infection (Sedlik et al. 1997, 2000). Further detailed immunological investigations unveiled mechanisms of how the PPV VLPs can be presented by MHC class I and class II molecules (Lo-Man et al. 1998). Thus, the strong CTL responses and neutralizing antibodies against the LCMV were achieved after intranasal (but not oral) immunization of mice, with the PPV VLPs carrying the single viral CTL epitope (Lo-Man et al. 1998). The insertion of the poliovirus C3:B epitope into the four loops and at the C-terminus of the CPV VP2 allowed for the recovery of capsids in all of the mutants but only the insertion at aa 225 of loop 2 was able to elicit a significant antipeptide antibody response, not poliovirus-neutralizing antibodies (Sedlik et al. 1999). To fine-modulate this insertion site in loop 2, the authors reinserted the epitope into adjacent positions 226, 227, and 228 (Rueda et al. 1999a). Surprisingly, these chimeric VLPs were able to elicit a strong neutralizing antibody response against poliovirus, demonstrating that the minor displacements in the insertion place may cause dramatic changes in the accessibility of the epitope and the induction of antibody responses. The potential of combining different types of epitopes in different positions of canine and porcine parvovirus VLPs to stimulate different branches of the immune system paves the way to the elaboration of novel vaccine candidates (Rueda et al. 1999b).

In parallel with the previously described LCMV CTL epitope 118–132, Lo-Man et al. (1998) engineered the PPV VLPs with the T-helper cell, MHC class II restricted preS epitope 120-MQWNSTTFHQTL-132 (corresponding to aa positions 1–12 of the preS2 region) from the HBs envelope of hepatitis B virus (HBV) referred to as PreS:T. The results demonstrated how the chimeric PPV VLPs can be presented by MHC class I and class II molecules and under-scored the wide potency of the PPV VLP system to deliver foreign antigens for vaccine design (Lo-Man et al. 1998).

The popular CTL epitope at aa 257-SIINFEKL-264 of chicken ovalbumin (OVA) was inserted into different positions of the PPV VP2 with a special emphasis on the functional influence of the epitope-flanking sequences (Rueda et al. 2004). Thus, the presentation of the OVA epitope was considerably improved by insertion of short natural flanking sequences, which indicated the relevance of the flanking sequences on the processing of the PPV VP2 VLPs. Only PPV VLPs carrying two copies of the OVA epitope linked by two glycines were able to be properly processed, suggesting that the introduction of flexible residues between the two consecutive OVA epitopes may be necessary for the correct presentation of these dimers by the PPV VLPs (Rueda et al. 2004). It should be noted that Boisgérault et al. (2005) established protection of mice against OVA-bearing melanoma after vaccination with the natural VP2 VLPs as an adjuvant and OVA peptide 257–264, both carried by microspheres and fulfilling the cross-presentation requirements.

To construct a malaria vaccine candidate, Rodríguez et al. (2012) generated the PPV VP2 VLPs carrying the CD8$^+$ T cell epitope SYVPSAEQI of the circumsporozoite (CS) protein from *Plasmodium yoelii* and tested it for efficacy in a rodent malaria model.

Pan et al. (2008) constructed a bivalent VLP vaccine where the PPV VP2 carried aa residues 165–200 from the porcine circovirus 2 (PCV2) nucleoprotein. This was the sole PPV-based vaccine candidate that was produced in mammalian HEK-293 but not in insect cells. The obtained PPV-PCV2 VLPs induced strong antibody responses against PCV2 in the absence of any adjuvant. Moreover, Pan et al. (2013) performed a systemic evaluation of the eGFP display on the PPV VP2 VLPs by the adenovirus vector-driven expression in mammalian cells. First, the 3D structure of the PPV capsid protein and surface loops deletion mutants were analyzed to define essential domains in PPV VP2 for the assembly of VLPs and the presence of abundant VLPs in a loop2, namely aa 212–245, deletion mutant of expected size and appropriate morphology was established. Second, the eGFP was displaced as a model in different loop constructs.

After vaccines, the PPV VP2 VLPs were used as a carrier of four copies of somatostatin fused to the N-terminal of VP2 when produced by the baculovirus system in the *Sf*9 cells (Zhang X et al. 2010). The chimeric VLPs demonstrated sufficient immunogenicity in mice, as measured by PPV-specific neutralizing antibody, somatostatin antibody, and growth hormone levels.

Primate Parvoviruses

Several members of the *Protoparvovirus* genus, capable of infecting humans, were discovered, including bufavirus (BuV), cutavirus (CuV), and tusavirus (TuV). BuV belongs to the *Primate protoparvovirus 1* species, CuV is a member of the *Primate protoparvovirus 3* species, and TuV belongs to the *Primate protoparvovirus 4* species.

In the case of bufaviruses, the VP2 VLPs were produced in insect cells by the baculovirus-driven expression for BuV1, BuV2, and BuV3; the current known genotypes and the structure of these VLPs was resolved by electron cryomicroscopy and image reconstruction to 2.84, 3.79, and 3.25 Å, respectively (Ilyas et al. 2018). The structure conserved the β-barrel (βB-βI) and α-helix A observed for all parvovirus structures. However, unlike the animal protoparvoviruses, the bufavirus capsid contained three separated protrusions surrounding the 3-fold axes, as has been reported for members of the *Dependoparvovirus* genus, human parvovirus B19, and human bocaviruses. The structure comparison among the BuV strains and to other available protoparvovirus structures revealed capsid surface variations in previously defined common variable regions associated with serotype and strain-specific functions (Ilyas et al. 2018).

Mietzsch et al. (2020c) produced in insect cells the VP2 VLPs of CuV and TuV, resolved their structure to ~2.9 Å by electron cryomicroscopy and image reconstruction, and compared the structures to that of BuV and other protoparvoviruses. The CuV and TuV, which displayed low sequence identities, shared common capsid features with other parvoviruses, and the overall VP2 structure topologies were similar, with the core eight-stranded β-barrel superposable. Again, the major differences were localized within the surface loops, in previously defined variable regions involved in receptor binding, cellular trafficking, transcription, and antigenic reactivity (Mietzsch et al. 2020c).

Rodent Parvoviruses

The popular parvoviruses—minute virus of mice (MVM), H-1 parvovirus (H-1PV), LuIII, mouse parvovirus, and tumor virus X—belong to the *Rodent protoparvovirus 1* species. The MVM, H-1PV, and LuIII viruses are pathogenic to rodents and selectively replicate in and kill human cancer cells (Chen YQ et al. 1986; Guetta et al. 1986; Geletneky et al. 2010; Paglino and Tattersall 2011). Importantly, a Phase I/IIa clinical trial has extended these observations to patients in which H-1PV was used to treat recurrent glioblastoma multiforme tumors (Geletneky et al. 2012, 2015, 2017). Some exhaustive reviews on the tumor-suppressing activity of H-1PV could be recommended (Rommelaere and Cornelis 1991; Geletneky et al. 2005; Rommelaere et al. 2010; Moehler et al. 2014).

The 3D structure of the MVM virions similar to the CPV and FPV structures was resolved to 3.5 Å for both full and empty virions by Llamas-Saiz et al. (1997) and analyzed in detail by Agbandje-McKenna et al. (1998). Simpson et al. (2002) unveiled homologies of the 3D structures of MVM, FPV, and PPV, as shown in Figure 9.5. Hernando et al. (2000) not only produced the MVM VP2 VLPs in insect cells but also reported the first crystals of parvovirus-like particles that were self-assembled in a heterologous system and diffracted x-rays to high resolution. Livingston et al. (2002) used the insect cells-produced MVM VLPs for ELISA serodiagnosis of MVM and mouse parvovirus infection. The direct structural comparison of the native wild-type empty MVM capsids containing VP1 and VP2 and of the insect cells-expressed MVM VP2 VLPs was achieved at 3.75 and 3.25 Å resolutions, respectively (Kontou et al. 2005). The crystal structure of H-1PV was resolved to 2.7 Å (Halder et al. 2012, 2013). Pittman et al. (2017) used single-particle electron cryomicroscopy and image reconstruction to resolve the LuIII structure to 3.17 Å and compare it to MVM and H-1PV.

Carreira et al. (2004) have chosen the insect cells-produced MVM VP2 VLPs as a model for the in vitro disassembly of the capsid and for the stability evaluation of the chimeric VLPs carrying heterologous epitope insertions. The two viral antigenic peptides that contained well-characterized continuous B-cell epitopes were selected as models: the B-C loop of VP1 of poliovirus type 3—previously used to test the immunogenicity of the CPV chimeras (Rueda et al. 1999a), as described earlier—and the G-H loop of VP1 of FMDV type C. The previously described 3D structure of the MVM capsid was inspected for sites of insertion that could have the least disruptive effects on capsid stability. The positions chosen were located close to the tips of some loops highly exposed to solvent on the capsid surface and involved in few or no interactions with residues in neighboring loops, especially those in other subunits (Carreira et al. 2004).

It is noteworthy that Callaway et al. (2019) reconstructed the three ancient endogenous capsid gene remnants found in rodent genomes within the so-called parvovirus-derived endogenous viral elements (EVEs). The reconstructed VP2 genes were expressed by mammalian and baculovirus expression systems. The VP2 sequence from Algerian mouse *Mus spretus* assembled into capsids, which had high thermal stability and entered murine L cells. The 3.89-Å structure of the *M. spretus* VLPs, determined by electron cryomicroscopy, showed similarities to rodent and porcine parvovirus capsids. The repaired VP2 sequences from brown rat *Rattus norvegicus* and wood mouse *Apodemus sylvaticus* did not assemble, but chimeras combining capsid surface loops from *R. norvegicus* with CPV assembled (Callaway et al. 2019).

GENUS *AMDOPARVOVIRUS*

A DNA fragment encoding the structural proteins VP1 and VP2 of Aleutian mink disease parvovirus (ADV) from the *Carnivore amdoparvovirus 1* species was expressed first in mammalian cells by a vaccinia virus vector, when both proteins were found in the empty icosahedral 27-nm VLPs that demonstrated ADV-specific response by the immunization of mice (Clemens et al. 1992).

Christensen et al. (1993) inserted the ADV VP1-VP2, and VP2 alone into a recombinant baculovirus and expressed them in *Sf*9 insect cells. The ADV VP1 and VP2 showed nuclear localization in the cells and formed VLPs that were indistinguishable, by electron microscopy, from wild-type virus. Christensen et al. (1994) obtained the purified VP2 VLPs from insect cells and demonstrated strong protection of the immunized minks upon challenge with a virulent isolate of virus. The 3D structure of the insect cell-produced ADV VP2 VLPs was resolved by electron cryomicroscopy and image reconstruction to 22 Å (McKenna et al. 1999) due to the sequence alignment with the structurally explored VP2 of CPV. The ADV VP2 VLPs demonstrated a mean diameter of 25.6 nm and possessed several surface features like those found in B19V, CPV, FPV, and MVM.

In parallel, Wu WH et al. (1994) obtained the similar ADV VLPs, 23–25 nm in diameter, consisting of both VP1 and VP2 proteins in the same *Sf*9 cells by the baculovirus-driven expression and suggested to use the VLPs as diagnostic antigens. Later, Knuuttila et al. (2009) used the insect cells-produced ADV VP2 VLPs of a Finnish wild-type strain to construct a sensitive and specific high-throughput ELISA test for identifying ADV antibodies from mink.

The preliminary optimism by the VLP vaccination of mink was shaken by the thorough trial of Aasted et al. (1998), who found that the immunization of minks with the insect cells-produced VP1/VP2 VLPs resulted in the progress of disease by the dangerous antibody-dependent enhancement mechanism.

GENUS *BOCAPARVOVIRUS*

As reviewed by Cotmore et al. (2019) concerning bocaviruses in humans, there is a significant proportion of babies that become infected with human bocavirus 1 (HBoV1) of the *Primate bocaparvovirus 1* species in the early months of life, developing their own antibody responses as soon as waning maternal IgG levels render them vulnerable to infection from their environment. However, while parvoviruses can induce a broad range of pathology in humans (Qiu et al. 2017), in most people these are resulting in respiratory and gastrointestinal infections and are rarely life threatening.

The numerous near-atomic bocavirus structures are available at the VIPERdb, namely those of HBoV1 (Mietzsch et al. 2017; Luo Mengxiao et al. 2021); HBoV2, species *Primate bocaparvovirus 2* (Luo Mengxiao et al. 2021); both HBoV3, species *Primate bocaparvovirus 1*, and HBoV4, species *Primate bocaparvovirus 2* (Mietzsch et al. 2017), as well as of bovine parvovirus 1 (BPaV1) of the *Ungulate bocaparvovirus 1* species (Kailasan et al. 2015b) and gorilla bocavirus 1 (GBoV1) (Yu et al. 2021).

The VP2 VLPs of HBoV1 of 21–25 nm in diameter were produced in insect cells and applied in seroepidemiological tests in different countries (Kahn et al. 2008; Lin et al. 2008; Lindner et al. 2008a, b; Cecchini et al. 2009; Söderlund-Venermo et al. 2009; Hedman et al. 2010; Guido et al. 2012) in parallel with HBoV2, HBoV3, and HBoV4 (Kantola et al. 2011; Guo L et al. 2012; Fang et al. 2014) The structure of the HBoV1 VP2 VLPs was determined by electron cryomicroscopy and image reconstruction at 7.9 Å resolution (Gurda et al. 2010). The immune response elicited by the HBoV1 VP2 VLPs by comparison with that against the VLPs of B19V (Kumar et al. 2011) or HBoV2 (Deng et al. 2014).

Kęstutis Sasnauskas's team produced the VP2 VLPs of all four HBoV by the expression of the appropriate genes in yeast *S. cerevisiae* (Tamošiūnas et al. 2016). Moreover, the mosaic VLPs composed of both VP2 and VP1 capsid proteins of HBoV1 were generated. All HBoV VLPs were similar to native HBoV virions in size and morphology. The purified VLPs were successfully used in ELISA tests to detect prevalence of HBoV infection in groups of Lithuanian patients with clinical symptoms of respiratory tract infection (Tamošiūnas et al. 2016).

The VP2 gene of HBoV1 was directly cloned, and those of HBoV2 to -4 were synthesized after codon optimization, and all were expressed by the baculovirus-driven system in *Sf*9 cells (Kailasan et al. 2016).

The GBoV1 VLPs were produced using recombinant baculovirus expressing either of GBoV1 VP1, VP2, and VP3 (GBoV1 WT) or only VP3 (GBoV1 VP3-only) in *Sf*9 insect cells (Yu et al. 2021). In fact, the purified GBoV1 WT sample carried the VP1, VP2, and VP3 molecules with a corresponding molecular mass of approximately 60, 65 and 80 kDa. For the GBoV1 VP3-only, only one band of 60 kDa was present. The GBoV1 VLP structure was resolved to 2.76 Å resolution using electron cryomicroscopy, its strong relationship to other parvoviruses was confirmed, and recognition by mouse monoclonal antibodies and human sera was mapped (Yu et al. 2021).

As reviewed by Yu et al. (2021), the HBoV1 and GBoV1, along with HBoV2–4, were proposed as gene therapy delivery vectors. The interest in HBoV1 was raised by its specific tropism for the apical side of polarized human airway epithelia (pHAE), which is optimal for the treatment of cystic fibrosis, a classical field of the AAV-based vectors. In addition to the favorable tropism of HBoV1, the expanded genome capacity of bocaviruses (5.5 kb), compared to AAV, allows the packaging of the full-length 4.7 kb CFTR gene. Due to these advantages, various studies have aimed to optimize a recombinant rAAV2/HBoV1 pseudotyped vector, as reviewed in detail by Yu et al. (2021).

The VP2 VLPs of BPaV1 were produced in insect cells and applied in seroepidemiological tests in parallel with human bocaviruses (Kantola et al. 2011). Chang et al. (2019) checked ten sites on the N-terminus and different surface variable regions of the BPaV1 VP2 for insertion of the type O foot-and-mouth disease virus (FMDV) conserved neutralizing epitope 8E8, namely, the ARGDLQVLTPKA stretch. The ten chimeric BPaV VP2 capsid proteins carrying the 8E8 epitope formed the virion-like VLPs when expressed in *Sf*9 cells, and each of the ten chimeric VLPs reacted with both anti-BPaV serum and antitype O FMDV mAb 8E8, demonstrating lack of interference therefore with the immunoreactivity of VP2 or VLP formation and correct display of the 8E8 epitope on the VLP surface. Moreover, the chimeric VLPs induced FMDV-specific antibodies in mice, where epitope insertions into positions 391/392 and

395/396 ensured robust neutralizing capacity of the induced antibodies (Chang et al. 2019).

GENUS *TETRAPARVOVIRUS*

Human parvovirus 4 (PARV4) of the *Primate tetraparvovirus 1* species was identified in 2005 in nuclease-digested plasma from patients with acute viral infection syndrome and detected, with widely varying frequencies, in plasma samples worldwide, but its clinical associations remained uncertain (Cotmore et al. 2019).

Lahtinen et al. (2011) obtained the PARV4 VP2 VLPs by the baculovirus expression system and used them by serodiagnosis of primary infections in Finland. Maple et al. (2013) produced the virion-like PARV4 VP2 VLPs in the insect *Sf*9 cells, applied them to prepare a PARV4 IgG time-resolved fluorescence immunoassay (TRFIA), and first examined the utility of the TRFIA with a panel of sera from people who inject drugs, a high-prevalence population for PARV4 infection, in England.

Kęstutis Sasnauskas's team expressed the PARV4 VP2 in yeast *S. cerevisiae*, and the purified VLPs were used for serological detection of virus-specific IgG and IgM antibodies in the sera of patients with acute respiratory diseases in Lithuania (Tamošiūnas et al. 2013). Figure 9.6 demonstrates the yeast-produced virion-like PARV4 VP2 VLPs. Lazutka et al. (2021) used these VLPs to generate mAbs in mice and mapped therefore at least three distinct antigenic sites on the VP2 surface. These antigenic sites corresponding to the

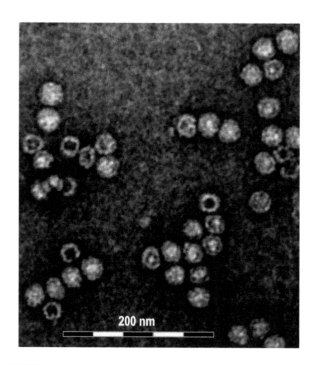

FIGURE 9.6 The electron micrograph of the CsCl-purified PARV4 VP2 VLPs. (Reprinted with permission of S. Karger AG, Basel from Tamošiūnas PL, Simutis K, Kodzė I, Firantienė R, Emužytė R, Petraitytė-Burneikienė R, Zvirblienė A, Sasnauskas K. *Intervirology.* 2013;56:271–277.)

loops at aa 144–158, 248–262, and 507–513 were replaced with a model His₇ peptide IHHHHHHHD tag, and the corresponding VLPs were found to expose the His tag on the VLP surface. However, the obtained VLPs were 35–50 nm in diameter compared to the 25–28 nm in diameter of the original nonmodified VP2 VLPs (Lazutka et al. 2021). Thus, the first chimeric derivatives of the PARV VLPs were engineered and purified.

SUBFAMILY *DENSOVIRINAE*

GENUS *ITERADENSOVIRUS*

Bombyx mori densovirus 1 (BmDV1)—a silkworm pathogen and member of the *Lepidopteran iteradensovirus 1* species—and four other species of the *Iteradensovirus* genes infect insects from the order *Lepidoptera*.

Kaufmann et al. (2011) produced the BmDV1 VP3 VLPs in insect cells, since the VP3 was the dominant protein among the capsid proteins of BmDV1. The near-atomic crystal structure of the empty VP3 VLPs was resolved to 3.1 Å, and this first 3D structure within the *Iteradensovirus* genus was compared to the structures of a set of parvoviruses. The particles consisted of 60 copies of the 55-kDa VP3 coat protein. The VP3 had a β-barrel "jelly-roll" fold similar to that found in many diverse icosahedral viruses, as well as other parvoviruses. Most of the surface loops had little structural resemblance to other known parvovirus capsid proteins. In contrast to vertebrate parvoviruses, the N-terminal β-strand of the BmDV1 VP3 was positioned relative to the neighboring 2-fold related subunit in a "domain-swapped" conformation, similar to findings for other invertebrate parvoviruses, suggesting that domain swapping was an evolutionarily conserved structural feature of the *Densovirinae*. Thus, the N-terminal β-strands of 2-fold related VP3 subunits were in positions similar to those in Galleria mellonella densovirus (GmDNV) and Penaeus stylirostris penstyldensovirus 1 (PstDV1) described later. Their positions were swapped relative to the N-terminal β-strands in vertebrate parvoviruses. The BmDV1 structure was of special interest as a putative basis for the design of capsid-binding antiviral compounds to protect silkworms against BmDV1 infections.

GENUS *PEFUAMBIDENSOVIRUS*

The x-ray crystal structure of the mature cricket parvovirus, or Acheta domesticus densovirus (AdDNV), a member of the *Blattodean pefuambidensovirus 1* species, was resolved to 3.5 Å (Meng et al. 2013). This study confirmed the observation that the vertebrate and invertebrate parvoviruses have evolved independently, although there were common structural features among all parvovirus capsid proteins. It was shown that raising the temperature of the AdDNV particles caused a loss of their genomes. The structure of these emptied particles was determined by electron cryomicroscopy to 5.5 Å resolution, and the empty AdDNV

VLP structure was found to be the same as that for the full, mature virus. The VP1 N-termini could be externalized without significant damage to the capsid, while in vitro this externalization of the VP1 N-termini was accompanied by the release of the viral genome (Meng et al. 2013).

GENUS *PROTOAMBIDENSOVIRUS*

The most structurally defined representatives, Galleria mellonella densovirus (GmDNV) and Junonia coenia densovirus (JcDV), belong to the species *Lepidopteran protoambidensovirus 1*.

The timely information for the structural comparison of parvoviruses was obtained from the first fine 3D structure of the more distantly related insect parvovirus of the *Protoambidensovirus* genus, namely, GmDNV, where the same motif of externalization of the N-terminal region of the unique VP1 sequence was found as by other parvoviruses analyzed at that time (Simpson et al. 1998). The full-length GmDNV VP and its truncated forms were expressed by the baculovirus system (Tijssen et al. 2003). The expression of the VP1 yielded the same four proteins as for the virus, with similar relative amounts, and these proteins assembled into the VLPs that could not be distinguished from virus particles. The context of the candidate AUGs for VP2 and VP3 initiation was improved by changing them to the context found for VP4. The size of the proteins that were obtained corresponded to that obtained for proteins from the virus and the electron microscopy revealed that all expression experiments yielded VLPs (Tijssen et al. 2003).

Croizier et al. (2000) expressed in the insect *Sf*9 cells the coding sequences of four overlapping polypeptides starting at four different in-frame AUG codons and coterminating at the stop codon of the cap gene of JcDV to generate the VP1 (88 kDa), VP2 (58 kDa), VP3 (52 kDa), and VP4 (47 kDa) expression variants. In all cases, the baculovirus-driven resulted in the production of the VLPs of 22–25 nm in diameter. The VLPs produced by the three VP2, VP3 and VP4 were abundant and contained three, two and one polypeptides, respectively. The VP4, the shortest polypeptide, thus appeared to be sufficient for the assembly of VLPs morphologically similar to those formed with two to four polypeptides. The ratio of VPs did not appear to be critical for assembly of the particles. The polypeptide starting at the first AUG immediately downstream from the promoter was always the most abundantly expressed in infected cells, regardless of the construct (Croizier et al. 2000). Bruemmer et al. (2005) presented the structure of native JcDV at 8.7 Å resolution and of the two insect cells-produced VLPs formed essentially from VP2 and VP4 at 17 Å resolution, as determined by electron cryomicroscopy. The capsid displayed a remarkably smooth surface, with only two very small spikes that defined a pentagonal plateau on the 5-fold axes, and the JcDV capsid was very closely related to that of the previously described GmDNV. The difference imaging revealed that the 21 disordered aa residues at the N-terminus of the VP4 were located inside the capsid at the 5-fold axis, but the additional 94 aa residue extension of VP2 was not visible, suggesting that it was highly disordered (Bruemmer et al. 2005).

SUBFAMILY *HAMAPARVOVIRINAE*

The penstyldensoviruses, members of the *Penstylhamaparvovirus* genus and highly pathogenic shrimp viruses causing significant damage to farmed and wild shrimp populations, were known previously as infectious hypodermal and hematopoietic necrosis viruses (IHHNV) of prawns and shrimp. Therefore, the earlier-described IHHNV is in fact Penaeus stylirostris penstyldensovirus 1 (PstDV1) belonging to the single *Decapod penstylhamaparvovirus 1* species of the *Penstylhamaparvovirus* genus.

Hou et al. (2009) expressed the IHHNV capsid gene in *E. coli* and found the product to self-assemble into the VLPs with homogeneous size and shape similar to native IHHNV particles. Furthermore, the IHHNV VLPs encapsidated RNA and DNA with a predominant size of 0.5 kb. The stability experiments revealed that the VLPs could not be disassembled by EDTA, but their structure was completely disrupted by dithiothreitol, suggesting that intra-capsid disulfide bonds were essential for maintaining the structural integrity of the capsid. Moreover, the VLPs efficiently entered primary hemocytes of the shrimp *Litopenaeus vannamei*, which implied that the IHHNV VLPs would be promising vehicles for the delivery of antiviral agents (Hou et al. 2009).

Kaufmann et al. (2010) established that PstDV1 has only one type of 37.5-kDa capsid protein—expressed the corresponding gene by the baculovirus system—and resolved the structure of the VLPs, composed of 60 copies of the coat protein and the smallest parvoviral capsid protein reported, to 2.5 Å resolution by x-ray crystallography. The capsid protein had the typical β-barrel "jelly-roll" motif similar to that found other parvoviruses. The N-terminal portion of the PstDNV coat protein adopted a "domain-swapped" conformation relative to its 2-fold-related neighbor similar to the previously described insect parvovirus GmDNV but in stark contrast to vertebrate parvoviruses. Again, most of the surface loops had little structural resemblance to any of the known parvoviral capsid proteins (Kaufmann et al. 2010). The putative application of the PstDV1 VLPs by the drug delivery to shrimps was first reviewed and commented by Itsathitphaisarn et al. (2017). Sinnuengnong et al. (2018) developed a single step strategy to form PstDV1 VLPs and encapsulate dsRNA of yellow head virus (YHV) in *E. coli*, through the coexpression of the shrimp PstDV1 capsid protein with the YHV dsRNA termed dsRNA-YHV-Pro in the same *E. coli* cells. These dsRNA-YHV-Pro-PstDV1 VLPs were then purified and delivered to shrimp for the determination of their anti-YHV property and stability. In fact, the administration of these VLPs gave higher levels of YHV suppression and a greater reduction in shrimp mortality than the delivery of naked dsRNA-YHV-Pro and provided therefore remarkable protection against YHV in shrimp (Sinnuengnong et al. 2018). Furthermore, Weerachatyanukul et al. (2021) encapsulated VP37 and VP28 dsRNA into the PstDV1—or IHHNV in this paper—VLPs and investigated protection against white spot syndrome virus (WSSV). It was apparent that the coencapsulation of dual dsRNA showed a superior WSSV silencing ability than the single dsRNA counterpart.

Li Z et al. (2008) expressed by the baculovirus system the various lengths of VPs from Aedes albopictus C6/36 cell densovirus (C6/36DNV), which was presented by the authors as possessing the simplest and most compact capsid in brevidensovirus and could be regarded by the modern taxonomy as a member of species *Dipteran brevihamaparvovirus 1* or *2* of the *Brevihamaparvovirus* genus. The result showed that the N-terminal 23-GGSG-26 sequence—a highly conserved glycine-rich region in *Parvoviridae*—and the C-terminal 344-GTGGVVTCMP-353 sequence were essential for the VLP assembly, while the N-terminal nuclear localization signal 15-GTKRKR-20 was nonessential for the assembly of the VLPs but did affect the formation of crystalline arrays in the infected *Sf*9 cells. The 3D structure of capsid of C6/36DNV was resolved earlier to 14 Å by electron cryomicroscopy and image reconstruction (Cheng et al. 2004, 2007).

10 Order *Cirlivirales*

One never notices what has been done; one can only see what remains to be done.

Marie Curie

ESSENTIALS

The *Cirlivirales* is an order of nonenveloped T = 1 icosahedral viruses, 15–25 nm in diameter, with the covalently closed ssDNA genome and the capsid consisting of 60 copies of VP protein, including the smallest known viral pathogens of animals and belonging, therefore, together with parvoviruses (Chapter 9), to the smallest known viruses. They replicate in the nucleus of infected cells, using the host polymerase for genome amplification. The order *Cirlivirales* currently involves the sole family *Circoviridae* with 2 genera—*Circovirus* and *Cyclovirus*—and 101 species altogether and forms the class *Arfiviricetes*, together with 4 other orders that are briefly presented in Chapter 12. It should be noted that the former genus *Gyrovirus* of the *Circoviridae* family, including the popular chicken anemia virus (CAV) as a reference strain, was transferred to the order-unassigned family *Anelloviridae* (Breitbart et al. 2017; Rosario et al. 2017), which is briefly described in Chapter 12. The class *Arfiviricetes* is a part of the large *Cressdnaviricota* phylum, which forms the kingdom *Shotokuvirae* of the realm *Monodnaviria*, together with another large phylum *Cossaviricota*, which includes the great VLP-generating orders *Zurhausenvirales* (Chapter 7), *Sepolyvirales* (Chapter 8), and *Piccovirales* (Chapter 9).

As commented by the current ICTV report (Breitbart et al. 2017), all available biological information has been gathered from a few members of the genus *Circovirus*, mainly beak and feather disease virus (BFDV) and porcine circoviruses (PCV1 and PCV2), and there is no biological data regarding the infectivity, transmission, or host range of members, as well as no virion structure or VLP engineering data concerning the genus *Cyclovirus*.

Figure 10.1 presents an introductory historical picture to the *Circoviridae* family, when CAV was still regarded as a part of the family, and R. Anthony Crowther et al. (2003) were the first to compare the fine structures of three early discovered circoviruses BFDV and PCV2 with that of the former "circovirus" CAV. This first structural analysis based on the 3D reconstruction of electron cryomicroscopy-derived data revealed that the circovirus virions have similar appearance with 60 capsid protein subunits arranged in 12 pentameric clusters (Crowther et al., 2003). As reviewed by the official ICTV reports (Biagini et al. 2012; Breitbart et al. 2017; Rosario et al. 2017), the virion morphology is only known for some members of the genus *Circovirus*, which have an icosahedral T = 1 symmetry, while the virions

were not observed for members of the genus *Cyclovirus*. The BFDV virions are associated with up to three proteins, 26.3, 23.7, and 15.9 kDa. The virions of PCV1 and PCV2 are each comprised of one structural protein Cap, for which approximate masses of 36 and 30 kDa have been estimated, respectively (Breitbart et al. 2017).

GENOME

Figure 10.2 presents the representative genome of the *Circovirus* genus. The circovirus virions contain covalently closed circular ssDNA in size from 1.7–2.1 kb and encode the replication-associated protein Rep and capsid protein Cap on different strands of the dsDNA replicative form. The genomes of PCV1 and PCV2, 1,759 and 1,768 bases, respectively, are the smallest viruses shown to replicate autonomously in mammalian cells. All virus genomes contain a putative *ori* marked by the conserved nonanucleotide (T/n)A(G/t)TATTAC, where lowercase nucleotides are observed at low frequency and "n" represents any nucleotide, at the apex of a potential stem-loop structure (Breitbart et al. 2017; Rosario et al. 2017). The members of the genera *Circovirus* and *Cyclovirus* are distinguished by the location of the *ori* relative to the coding regions and the length of the intergenic regions. Thus, the *Circovirus* members have the *ori* on the same strand as the *rep* ORF, whereas the genus *Cyclovirus* members have the putative *ori* on the same strand as the *cap* ORF (Breitbart et al. 2017).

GENUS *CIRCOVIRUS*

BEAK AND FEATHER DISEASE VIRUS

Beak and feather disease virus (BFDV) from the *Beak and feather disease virus* species causes severe disease characterized by irreversible feather disorders and serious immunosuppression in over 40 species of psittacine birds (Breitbart et al. 2017).

As mentioned previously and depicted in Figure 10.1, the first 3D insights into circoviruses were published by the pioneering work of Crowther et al. (2003). The next structural efforts related to the VLP structure. Although the first attempts to express the full-length BFDV Cap-encoding gene in *E. coli* did not lead to the VLPs (Johne et al. 2004; Patterson et al. 2013), Sarker et al. (2015) presented overexpression, production, and purification of milligram quantities of highly pure, soluble Cap in *E. coli*, which not only possessed high specific antigenicity but was also amenable for the crystallization followed by further serious structural studies. At least, Sarker et al. (2016) presented three high-resolution x-ray crystallographic structures for distinct macromolecular assemblies of the BFDV Cap,

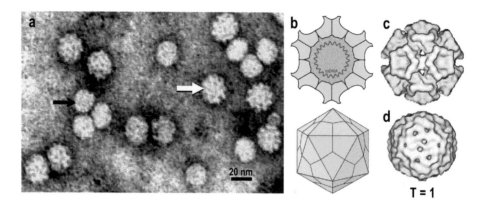

FIGURE 10.1 The cartoon, portraits, and 3D images of beak and feather disease virus (BFDV) and porcine circovirus-2 (PCV2), typical representatives of the order *Cirlivirales*, accompanied with those of chicken anemia virus (CAV), a member of the former *Gyrovirus* genus of the *Circoviridae* family, which is moved now to the order-unassigned *Anelloviridae* family described in Chapter 12. (a) The negative contrast electron microscopy of particles of an isolate of BFDV (white arrow) and CAV (black arrow), stained with uranyl acetate. (b) The cartoon taken with kind permission from the ViralZone, Swiss Institute of Bioinformatics, http://viralzone. expasy.org (Hulo C et al. 2011). (Courtesy Philippe Le Mercier.) (c) The electron cryomicroscopy image of a particle of an isolate of CAV. A structural model comprising 60 subunits (T = 1) arranged in 12 trumpet-shaped pentameric rings has been proposed by Crowther et al. (2003). (d) The electron cryomicroscopy image of a particle of an isolate of PCV2. A structural model comprising 60 subunits (T = 1) arranged in 12 flat pentameric morphological units has been proposed by Crowther et al. 2003. (Reprinted from *Virus Taxonomy, Classification and Nomenclature of Viruses. Ninth Report of the International Committee on Taxonomy of Viruses*, King AMQ, Lefkowitz E, Adams MJ, Carstens EB (Eds), Biagini P, Bendinelli M, Hino S et al, Family—*Circoviridae*. 342–349, Copyright 2012, with permission from Elsevier.)

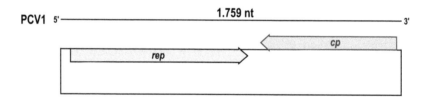

FIGURE 10.2 The genomic structure of a typical *Circoviridae* representative. The ORFs are indicated by arrowed boxes and the structural genes are colored dark pink. The circular character of the genome is displayed by the lower connecting bracket, while the linearization is performed with *ori* as the opening site.

which included a 10-nm assembly—resolved to 2.0 Å and comprised of two face-to-face pentamers of the Cap molecules—and two 17-nm assemblies comprised of 60 Cap monomers, arranged as 12 pentamers, in the absence and presence of ssDNA, determined to 2.5 and 2.3 Å resolution, respectively. These assemblies, shown in Figure 10.3, exhibited distinct monomeric and pentameric units and unique inverted morphologies that reversed the accessibility of the DNA binding and nuclear localization signal domains present within the N-terminal arginine rich motif (ARM), important for viral assembly (Sarker et al. 2016). The impressing structural studies on BFDV and PCV were recently reviewed by Nath et al. (2021).

Concerning the use of the baculovirus expression system, Heath et al. (2006) were the first to synthesize the BFDV proteins Rep and Cap in the insect *Sf*21 cells. The intracellular localization of the BFDV Cap was shown to be directed by three partially overlapping bipartite nuclear localization signals (NLSs) situated between aa residues 16 and 56 at the N-terminus of the Cap. Moreover, a DNA

binding region was also mapped to the N-terminus of the Cap and appeared within the region containing the three putative NLSs. Meanwhile, the deletion of portions of the Cap N-terminus resulted in the increased expression of the truncated Cap over the full-length Cap. Remarkably, it was established that the nuclear localization of Rep in insect cells is Cap-dependent, strongly implying that Rep and Cap of BFDV do interact. However, no indications on the presence of the putative BFDV Cap VLPs in the insect cells were revealed. Stewart et al. (2007) expressed the entire coding region of the BFDV Cap in the *Sf*9 insect cells using the baculovirus expression system and demonstrated that the recombinant Cap self-assembled to produce the virion-like VLPs. The BFDV VLPs also possessed hemagglutinating activity, which provided further evidence that the self-assembled BFDV VLPs retained receptor-mediated biological activity and that the determinants for the BFDV hemagglutination activity relied solely on the Cap protein. The BFDV VLPs reacted with anti-BFDV sera from naturally immune parrots and cockatoo and from

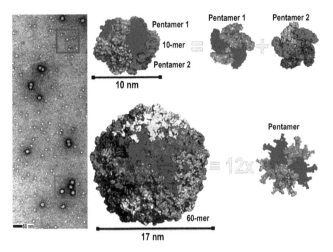

FIGURE 10.3 The structural characterization of two distinct BFDV-Cap complexes. Left panel, negatively stained electron micrograph of the BFDV Cap protein shows two populations corresponding to VLPs (red box), and a smaller assembly of ~10 nm in diameter (blue box). Right panel, x-ray crystal structures allow modeling of the two complexes to 2.0 Å (10 nm, top), and 2.5 Å (17 nm, bottom). The smaller complex is comprised of 10 Cap molecules arranged as two interlocking discs, with each disc containing five Cap molecules. The larger VLP is comprised of 12 pentamers arranged with T = 1 icosahedral symmetry. (Sarker S et al. Structural insights into the assembly and regulation of distinct viral capsid complexes. *Nat Commun.* 2016;7:13014.)

chickens experimentally inoculated with native BFDV in both Western blots and hemagglutination inhibition assay and were used as a suitable replacement antigen for serological detection of antibodies against BFDV (Stewart et al. 2007). Bonne et al. (2009) provided strong evidence that the insect cells-produced BFDV VLPs are immunogenic in hand-raised long-billed corellas *Cacatua tenurostris* collected in Australia and might be a suitable candidate vaccine to prevent beak and feather disease in psittacine birds. Thus, the vaccinated birds did not develop feather lesions, had only transient PCR-detectable viraemia, and had no evidence of persistent infection 270 days postchallenge, whereas the nonvaccinated control corellas developed transient feather lesions and had all necessary test results consistent with the psittacine beak and feather disease (Bonne et al. 2009).

In plants, the full-length BFDV Cap and a truncated Cap (ΔN40) were transiently expressed in tobacco *Nicotiana benthamiana* as fusions to elastin-like polypeptide (Duvenage et al. 2013). These two proteins were C-terminally fused to the elastin-like polypeptides of different lengths, 140 or 255 aa residues, in order to increase expression levels and to provide a simple means of purification, while the self-assembly of the products into VLPs, which could be potentially affected by long fusions, was not assessed. Regnard et al. (2017) published the first report of the plant-made full-length BFDV Cap assembling into the VLPs, and the putative pseudovirions were suggested to further efficacy

studies as putative vaccines. Thus, the virion-like Cap VLPs were successfully produced in *N. benthamiana*, albeit at relatively low yield. Moreover, the VLPs spontaneously incorporated amplicon DNA produced from the replicating bean yellow dwarf virus (BeYDV) plant vector (Regnard et al. 2017).

In yeast, Sariya et al. (2014) expressed the BFDV Cap gene in *Pichia pastoris*, where the recombinant protein appeared in inclusion bodies. Although the obtained Cap was purified by affinity binding with Ni-NTA resin, no further self-assembly into higher aggregates was addressed.

The first chimeric BFDV VLPs were obtained by Chen JK et al. (2020) in *E. coli*. Thus, the authors engineered the BFDV VLPs including pure Cap proteins, mutant Cap proteins CapΔNLS$_{54}$, Cap ΔNLS$_{62}$, Cap C$_{228}$S, and Cap ΔNES—where NES was a nuclear export sequence—and chimeric Cap proteins carrying the aa 64–70 epitope of the replication-associated protein Rep. All of the aforementioned VLPs were observed via transmission electron microscopy and verified through immunogold labeling. Moreover, the study was the first to describe the precise position of the NLS of the BFDV Cap protein between aa residues 55–62 and the interaction of Cap protein with importin α and importin β in vitro (Chen JK et al. 2020).

Porcine Circovirus

PCV2 of the species *Porcine circovirus 2* appears to be restricted to pigs—including various commercial pig breeds and wild boars—and causes so-called porcine circovirus associated disease or postweaning multisystemic wasting syndrome (PMWS). PCV1 from the *Porcine circovirus 1* species readily infects but is not known to cause disease in swine. The coat proteins of PCV1 and PCV2 are antigenically distinct (Mahé et al. 2000) and both, presumably PCV2, are antigenically distinct from BFDV (Todd et al. 1991).

Since PCV2 is the one of the most economically important pathogens of pigs, a valuable breakthrough in the circovirus VLP story consisted in the elaboration of the commercial VLP vaccines, such as Circoflex by Boehringer Ingelheim, Circumvent by Intervet/Merck, and Porcillis PCV by Schering-Plough/Merck, which were based on the capsid gene expressed in insect cells by the baculovirus system. The problem of the VLP vaccination against PCV2 was extensively reviewed (Chae 2012; Ellis 2014; Afghah et al. 2017; Nath et al. 2021).

Nawagitgul et al. (2000) were the first to express the PCV2 ORF2 in the insect *Sf*9 cells by the baculovirus system, to demonstrate that the 30-kDa product self-assembled into the virion-like VLPs, and to characterize therefore the PCV2 ORF2 as the structural *cap* gene. Kim Y et al. (2002) confirmed the correct self-assembly of the PCV2 Cap into the VLPs by the baculovirus-driven expression. Liu LJ et al. (2008) achieved the highly efficient production of the PCV2 VLPs by a recombinant baculovirus. Thus, the *cap* gene was expressed by using insect *Tn*5 cells, and a large amount of 28-kDa protein was released into the culture medium and self-assembled into the virion-like PCV2 VLPs, 20 nm in diameter, yielding 1 mg of purified particles per 10^7 $Tn5$ cells. The PCV2 VLPs were used to develop a sensitive method for detecting PCV2-specific IgG antibodies and recommended as the most promising PCV2 vaccine candidate by virtue of their potent immunogenicity (Liu LJ et al. 2008).

Fachinger et al. (2008) performed the extensive study of the efficacy of vaccination in suffering pigs by the insect cells-produced VLPs against PCV2 in the control of the PCV2-associated porcine respiratory disease complex. The vaccine was designated as Ingelvac® CircoFLEX™, Boehringer Ingelheim Vetmedica GmbH and the single intramuscular injection reduced the mean PCV2 viral load by 55–83% and the mean duration of viraemia by 50%. During a period of study (from 3–25 weeks of age), the vaccinated animals exhibited a reduced mortality rate, an improved average daily mass gain, as well as a reduced time to market (Fachinger et al. 2008).

In parallel, the insect cells-produced PCV2 VLPs were extensively studied and always regarded as a safe vaccine against PMWS in pigs (Fan et al. 2005, 2007; Fort et al. 2008, 2009; Pérez-Martín et al. 2010; Martelli et al. 2011; Park C et al. 2014; Zhang H et al. 2015). Kekarainen et al. (2008) demonstrated, however, that the insect cells-produced PCV2 VLPs differ from the natural PCV2 virions by the fine mechanisms of the immune response induction. Recently, Kim K and Hahn (2021) successfully evaluated the novel insect cells-produced vaccine of the PCV type 2d (PCV2d) in pigs naturally infected with PCV2d.

Meanwhile, López-Vidal et al. (2015) improved the insect cell production of the PCV2 VLPs by the involvement of a novel baculovirus expression cassette, while Liu Y et al. (2015) enhanced the VLP production in $Sf9$ cells by translational enhancers. Masuda et al. (2018) used the silkworm-baculovirus expression vector system (silkworm-BEVS) for mass production of the PCV2 VLPs and established a simple three-step protocol for its purification from pupae, including extraction by detergent, ammonium sulfate precipitation, and anion exchange column chromatography. Cao et al. (2019) developed a simple and high-yielding fed-batch process to produce the PCV2 VLPs in the insect $Sf9$ cells. He et al. (2020) developed another high-efficiency method for the optimized production and purification of the PCV2 VLPs. By this method, the Cap was provided with the His$_6$ tag and produced in baculovirus-infected silkworm larvae, as well as in parallel in the efficient $E.\ coli$ BL21 (DE3) prokaryotic expression system. Remarkably, the PCV2 Cap proteins purified from silkworm larvae self-assembled into the VLPs in vitro, while the Cap proteins purified from bacteria were unable to self-assemble. The process of the baculovirus-driven expression was recommended for the development of a low-cost and efficient vaccine (He et al. 2020).

Kim K et al. (2020) used the baculovirus system to express complete Cap(1–233) and mutant Cap(Δ169–180), in which

the so-called decoy epitope 169-STIDYFQPNNKR-180 was deleted and evaluated the immune response to these in mice. The immunization with mutant Cap(Δ169–180) protein, which formed a very low level of VLPs, elicited significantly lower levels of the PCV2 CP-specific IgG antibodies and a slightly lower neutralizing activity than immunization with the complete Cap(1–233) protein. This suggested that the complete Cap was important for the efficient VLP assembly and induction of the PCV2-specific IgG antibodies and neutralizing antibodies in mice (Kim K et al. 2020).

Concerning expression in $E.\ coli$, Marcekova et al. (2009) expressed the codon-optimized full-length PCV2 cap gene and purified the Cap protein to greater than 95% homogeneity, but the product was unable to self-assemble into the VLPs. Yin et al. (2010) were the first to report the production of the PCV2 Cap VLPs in $E.\ coli$. In this case, the small ubiquitin-like modifier (SUMO) expression system was used to successfully produce the whole native Cap protein by making it highly soluble in $E.\ coli$, while the SUMO tag was subjected to cleavage with SUMO protease, and the whole Cap protein simultaneously self-assembled into the VLPs in vitro (Yin et al. 2010).

The $E.\ coli$ cells-produced PCV2 VLPs provided the first crystal structure and fine atomic description for the $Circovirus$ family (Khayat et al. 2011). The choice of the PCV2 consensus sequence was made in this study in a desire to pursue a crystal structure for what would be the most representative of the PCV2 population. Such N-terminally truncated PCV2^{CS} Cap was purified from $E.\ coli$ in the form of monomeric subunits yet assembled to form VLPs during the crystallization trials. As a result, the authors reported the 2.3-Å crystal structure of the PCV2^{CS} VLPs. This pioneering structure is presented in Figure 10.4. Then, an electron cryomicroscopy image reconstruction was performed to the PCV2 VLPs derived from a native PCV2 strain N12 (PCV2^{N12}) and expressed in the baculovirus system to affirm that the PCV2^{CS} subunits assembled into

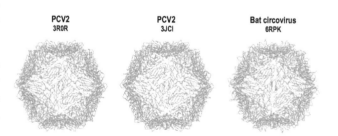

FIGURE 10.4 The near-atomic structure of the $Circoviridae$ family members: PCV2, crystal structure, PDB ID 3R0R, 2.35 Å, 204 Å (Khayat et al. 2011); PCV2, electron cryomicroscopy, PDB ID 3JCI, 2.90 Å, 204 Å (Liu Z et al. 2016); bat circovirus, electron cryomicroscopy, PDB ID 6RPK, 2.84 Å, 204 Å (Nath et al. 2021). The 3D structures are taken from the VIPERdb (http://viperdb. scripps.edu) database (Carrillo-Tripp et al. 2009). The size of particles is in scale. All structures possess the icosahedral T = 1 symmetry. The corresponding protein data bank (PDB) ID numbers are given under the appropriate virus names.

the biologically relevant VLPs. In contrast to the PCV2^{CS} VLPs, the V2^{N12} VLPs autoassembled in vivo and thus were likely to reflect fully the structure of the infectious virions. The PCV2^{N12} image reconstruction at 9.6 Å agreed with the PCV2^{CS} crystal structure and displayed density corresponding to the packaged nucleic acid and the N-terminus of the capsid protein, since both were absent in the PCV2^{CS} VLP crystal structure. Briefly, the crystal structure revealed that the capsid protein fold was a canonical viral jelly roll. The loops connecting the strands of the jelly roll defined the limited features of the surface. The sulfate ions interacting with the surface and electrostatic potential calculations strongly suggested a heparan sulfate binding site that allowed PCV2 to gain entry into the cell. The crystal structure also allowed the previously determined epitopes of the capsid to be visualized. The electron cryomicroscopy image reconstruction showed that the location of the N terminus, absent in the crystal structure, was inside the capsid (Khayat et al. 2011). Liu Z et al. (2016) resolved the near-atomic structure of the insect cells-produced PCV2 VLPs to 2.90 Å by the electron cryomicroscopy. This pioneering structure is also presented in Figure 10.4. At last, Khayat et al. (2019) used electron cryomicroscopy to identify differences among the VLPs of PCV2d and PCV2a, b, and d genotypes that accompanied the emergence of PCV2b from PCV2a, and PCV2d from PCV2b, as shown in Figure 10.5. These differences indicated that the sequence analysis of genotypes was insufficient and that it was important to determine the PCV2 capsid structure as the virus evolved. The structure-based sequence comparison demonstrated that each genotype possessed a unique combination of aa residues located on the surface of the capsid that underwent substitution (Khayat et al. 2019). It should be noted that the VLPs of the different PCV2 variants were produced in this case in mammalian cells, rightly believing that the VLP expression using mammalian cells provided an analogous system to that utilized by the natural virus infection. It should be noted that a bit earlier Chi et al. (2014) demonstrated the virion-like PCV2 VLPs by the mammalian cell expression system using a recombinant pseudorabies virus (PRV) that was gE gene deficient, a widely used PRV marker vaccine. The produced VLPs induced significant response of specific antibodies in mice and guinea pigs (Chi et al. 2014).

Meanwhile, Wu et al. (2012) produced a large amount of Cap1–233—the full-length Cap protein of 233 aa residues without any tags—and got virion-like VLPs when using the *E. coli* expression system and optimized rare codons. The N-terminal deletion mutant Cap51–233 without nuclear localization signal domain and an internal deletion mutant CapΔ51–103 without the dimerization domain failed to form VLPs, suggesting that both regions were essential for the Cap self-assembly. The Cap1–233-immunized pigs demonstrated specific antibody immune responses and were prevented from PCV2 challenge (Wu et al. 2012). Furthermore, Wu et al. (2016) presented an efficient scheme for the *E. coli* expression and large-scale purification of the virion-like PCV-2 VLPs. Later, Mo et al.

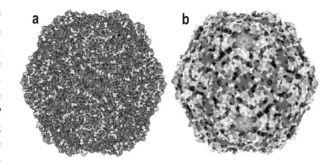

FIGURE 10.5 The sequence comparison of 1,278 PCV2 capsid protein entries plotted on the PCV2d atomic coordinates. (a) The sequence diversity and variation plotted on the PCV2d atomic coordinates. The red-to-blue color gradient represents diversity (sequence identity). The size of the atoms and tube represent variation (AL2CO: sequence entropy), with smaller atoms/tubes representing lower entropy/variation and larger atoms/tubes representing greater entropy/variation. Lower entropy indicates fewer different aa present in the alignment at a position, and larger entropy indicates more different aa present at a position. The plotting of variation allows one to appreciate the frequency of different amino acids at each position. The image was made with UCSF Chimera (Pettersen et al. 2004). (b) Surface representation of the PCV2d atomic coordinates with the three conserved patches colored in green (Tyr55, Thr56, Asp70, Met71, Arg73, Asp127), blue (Pro82, Gly83, Gly85) and red (Asp168, Thr170, Gln188, Thr189). The image made with UCSF ChimeraX and colored using flat lighting (Goddard et al. 2018). (Reprinted from *Virology*. 537, Khayat R, Wen K, Alimova A, Gavrilov B, Katz A, Galarza JM, Gottlieb P, Structural characterization of the PCV2d virus-like particle at 3.3 Å resolution reveals differences to PCV2a and PCV2b capsids, a tetranucleotide, and an N-terminus near the icosahedral 3-fold axes, 186–197, Copyright 2019, with permission from Elsevier.)

(2019) established that that the N-terminal truncation of the PCV2 Cap up to 27 aa residues still enabled the formation of VLPs, while truncation of more than 30 aa residues prevented self-assembly. Moreover, the crystal structure of the *E. coli*-produced PCV2 VLPs that were assembled from the N-terminal NLS-truncated PCV2 Cap was resolved to 2.8 Å resolution, while electron cryomicroscopy was used to resolve the structure of the PCV2 VLPs assembled from full-length PCV2 capsid protein at 4.1 Å resolution. Remarkably, the NLS-truncated PCV2 Cap only formed instable VLPs, which were easily disassembled in solution, whereas the full-length PCV2 Cap formed stable VLPs due to interaction between 15-PRSHLGQILRRRP-27 (α-helix) and 33-RHRYRWRRKN-42 (NLS-B) in a repeated manner. Moreover, the first fine structure of the PCV2 VLPs in complex with a specific Fab of virus-neutralizing antibody was resolved to 15 Å by electron cryomicroscopy (Mo et al. 2019). Zhan et al. (2020) demonstrated the critical role of the C-terminus of the PCV2 Cap by the VLP assembly, cell entry, and propagation, when the removal of the C-terminal loop led to abrogation of the in vitro Cap self-assembly into the VLPs.

The *E. coli*-produced PCV2 VLPs were employed in ELISA tests that were engineered for large-scale PCV2 antibody monitoring to exclude the PCV1 antibody interference after PCV2 vaccination (Han et al. 2016; Zhang Y et al. 2016). Xi et al. (2016) characterized the *E. coli*-produced PCV2 VLPs by electron microscopy single particle resolution at 4.5 Å and demonstrated its protective efficacy against PCV2 challenge on three-week-old piglets.

Fang et al. (2015) synthesized the gene coding for the PCV2b Cap, expressed it in *E. coli* in a soluble form, dialyzed it into three different buffers, and assembled it into VLPs. Remarkably, the self-assembly process was followed not only by transmission electron microscopy but also by fluorescence spectroscopic analysis (FSA), where the assembled VLPs showed a distinct FSA curve with a peak at 320 nm, providing a simple solution for monitoring VLP assembly during production of the VLP-based vaccines (Fang et al. 2015).

Li Y et al. (2020a) suggested a combination of the *E. coli*-produced PCV2 VLPs with the elastin-like polypeptides (ELP) that can undergo temperature-dependent inverse phase transition and simplify purification of the ELPylated proteins by inverse transition cycling (ITC). Thus, the PCV2b Cap, together with the virus-neutralizing epitopes of PCV2a, PCV2d, and PCV2e, was expressed in *E. coli* as an ELPylated soluble protein and purified by ITC in the presence of mild detergents. In parallel, the Cap protein was also expressed as a His-tagged protein in the form of insoluble inclusion bodies and purified by nickel affinity chromatography in denaturing conditions. Nevertheless, the two purified fusion proteins assembled into the VLPs with similar morphology. However, the ELPylated VLPs were more immunogenic in mice than the His-tagged VLPs (Li Y et al. 2020a, b). Liu X et al. (2020) engineered the PCV2 VLPs in *E. coli* by fusing a truncated form of flagellin FliC (TFlg$_{85-111}$) to the Cap. The addition of this TFlg to the C-terminus of Cap protein did not affect the formation of VLPs and boosted both humoral and cellular immune responses in mice. After a challenge with PCV2, in the Cap-TFlg vaccinated group, viremia was milder and viral loads were lower as compared with those in the native Cap-vaccinated group (Liu X et al. 2020).

Concerning other than *E. coli* bacterial expression systems, the putative live oral PCV2 vaccines delivering the PCV2 Cap were engineered by using *Lactococcus lactis* (Wang K et al. 2008), *Bordetella bronchiseptica* aroA mutant strain (Kim T et al. 2009), and attenuated *Salmonella enterica* serovar Typhimurium (Xu et al. 2012). No one of these studies touched the problem of the VLP formation.

In yeast, Bucarey et al. (2009) expressed the codon-optimized PCV2 *cap* gene in *Saccharomyces cerevisiae* and demonstrated by electron micrography that the yeast-derived PCV2 Cap protein self-assembled into VLPs that were morphologically and antigenically similar to the insect cell-derived VLPs. Feeding the raw yeast extract containing Cap protein to mice elicited both serum- and fecal-specific antibodies against the antigen. Thus, it appeared feasible to use *S. cerevisiae* as a safe and simple system not only to produce the appropriate VLPs but also to efficiently induce anti-PCV2 antibodies in a mouse model by oral yeast-mediated antigen delivery as an alternative strategy to injectable formulations. To develop this idea further, Bucarey et al. (2014) used chitosan microparticles to base a PCV2 oral vaccine, where the yeast-produced PCV2 VLPs were loaded onto chitosan and induced anti-PCV2 cellular responses in the vaccine-fed mouse model.

Meanwhile, the PCV2 VLPs in *S. cerevisiae* were efficiently produced by Kęstutis Sasnauskas's team (Nainys et al. 2014). The obtained VLPs driven by the native PCV2 *cap* gene as well by a codon-optimized gene were used to generate a set of MAbs that did not show any cross-reactivity with PCV1-infected cells and served as a highly specific antigen for the development of an indirect IgG PCV2 Cap VLP-based ELISA for the detection of virus-specific IgG antibodies in swine sera collected in Lithuania. Zaveckas et al. (2015) elaborated a reliable protocol for the chromatographic purification of the PCV2 Cap VLPs from clarified *S. cerevisiae* lysate. Chen P et al. (2018) adopted the YeastFab assembly method to synthesize transcriptional units in a single tube by piecing up promoter, open reading frame, and terminator in *S. cerevisiae*. By this technique, the two yeast recombinants were successfully constructed to secrete the PCV2 VLPs of 18 nm in diameter.

To achieve high-level expression of the PCV2 Cap in methylotrophic yeast *Pichia pastoris*, the wild-type Cap gene was codon-optimized, and the intracellular soluble opti-Cap reached high 174 µg/mL without concentration in a shake flask, whereas the wt-Cap could not be detectable (Tu et al. 2013). However, the putative VLP-or-not status of the efficiently produced Cap was not addressed. Xiao et al. (2018) used yeast *Hansenula polymorpha* to express an NLS-deleted capsid protein ΔCat of PCV2b based on the capsid protein of the PCV2b strain Y-7 isolated in China. The purified ΔCat self-assembled into the VLPs with similar morphology of the VLPs formed by the Cat and induced high levels of specific antibodies in mice (Xiao et al. 2018). At last, the efficient production of the PCV2 VLPs was demonstrated in a nonconventional yeast *Kluyveromyces marxianus* (Duan et al. 2019). After codon optimization, the synthesized PCV2 Cap assembled spontaneously into VLPs and reached ~1.91 g/L of PCV2 VLP antigen in a 5-L bioreactor after high cell density fermentation for 72 h. That yield greatly exceeded to the earlier reported production by baculovirus-insect cells, *E. coli* and *P. pastoris*. By the means of two-step chromatography, 652.8 mg of PCV2 VLP antigen was obtained from 1 L of the recombinant *K. marxianus* cell culture. The PCV2 VLPs induced high levels of anti-PCV2 IgG antibody in mice and decreased the virus titers in both livers and spleens of the challenged mice (Duan et al. 2019).

In plants, the Ed Rybicki's team transiently expressed the PCV-2 Cap in *Nicotiana benthamiana* via agroinfiltration, and the PCV-2 Cap was successfully purified using sucrose gradient ultracentrifugation (Gunter et al. 2019).

The Cap self-assembled into the VLPs resembling native virions, and up to 6.5 mg of VLPs could be purified from 1 kg of leaf wet mass. The mice immunized with the plant-produced PCV2 VLPs elicited specific antibody responses to the PCV2 Cap (Gunter et al. 2019). Park Y et al. (2021) reported the transient expression of the PCV2 Cap in *N. benthamiana* and purification of the His_6 tag-provided PCV2 VLPs by affinity chromatography, with a yield of 102 mg from 1 kg plant leaves. In this case, the constructs were designed to direct expression of the Cap to either ER or chloroplasts, where one construct encoded the BiP signal peptide and the HDEL retention signal for accumulation of the Cap in the ER, with a His_6 tag for affinity purification, while the other construct encoded an N-terminal rubisco transit peptide, for accumulation of the Cap in chloroplasts, again with the His_6 tag (Park Y et al. 2021).

Concerning the live chimeric PCV2 virions, Fenaux et al. (2003, 2004) demonstrated first that a chimeric PCV (PCV1–2) with the capsid gene of PCV2 inserted into the backbone of PCV1 was infectious but attenuated in pigs, and the appropriate inactivated commercial vaccine based on the chimeric PCV1–2 appeared on the market (Gillespie et al. 2008; Segalés et al. 2009). Based on the previously constructed chimeric PCV1–2 variant (Beach et al. 2010), Beach et al. (2011) aimed to identify genomic locations that could tolerate small insertions of epitope tags and to produce an epitope-tagged vaccine virus for use as a potential tractable modified live-attenuated vaccine. Thus, four mutants were constructed, where each contained an influenza virus hemagglutinin (HA) tag YPYDVPDYA inserted in frame at the N- or C-termini of Rep and Cap ORFs. No infectious virus was detectable from cells transfected with N-HA or C-HA Rep mutants, while both N-HA and C-HA Cap insertion mutants were infectious in the appropriate cells. The C-terminus was regarded as a target of choice, since it could be displayed on the surface of the virion, and five PCV1–2 mutants were constructed and tested in vitro for infectivity, with each mutant containing a different tag inserted in frame at the C terminus of the capsid: a single HA tag, an HA tag dimer, an HA tag trimer, a glu-glu tag from mouse polyomavirus medium T antigen CEEEEYMPME, and a KT3 tag KPPTPPPEPET from simian virus 40 (SV40) large T antigen. Each of the five mutants was infectious in vitro, and each of the inserted epitope tags was properly expressed on the surface of virions, resulting in double labeling of infected cell nuclei with both anti-PCV2 capsid and antiepitope tag antibodies. The results showed therefore that the Cap C-terminus tolerated epitope tag insertions as large as 27 aa residues, as mutants with the HA tag dimer and trimer were both infectious in vitro. Moreover, the chimeric PCV1–2 vaccine viruses with epitope tags inserted at the Cap C-terminus were infectious in specific-pathogen-free pigs, inducing both antiepitope tag and anti-PCV2 neutralizing antibodies (Beach et al. 2011). In parallel, Huang L et al. (2011) engineered a stable infectious PCV2 virions, which displayed on its surface a 14-aa V5 epitope tag from simian parainfluenza virus 5 (PIV5) that was added to the

C-terminus of the PCV2 Cap. Then, the neutralizing VP1 epitope 141–160 of foot-and-mouth disease virus (FMDV) was fused to the Cap C-terminus, and the resulting chimeric virus was infectious and induced both PCV2 and FMDV VP1 epitope antibodies (Huang L et al. 2014).

Piñeyro et al. (2015) developed a bivalent vaccine candidate that could protect pigs against PCV2 and porcine reproductive and respiratory syndrome virus (PRRSV), since both PCV2 and PRRSV coinfections are very common in swineherds worldwide. Thus, the four different linear B-cell antigenic epitopes of PRRSV were inserted into the Cap C-terminus of the noninfectious PCV1–2a vaccine virus. The insertion of 12 and 14 aa residues did not impair the replication of the resulting chimeric viruses in vitro. The immunogenicity study in pigs revealed that two of the four chimeric viruses elicited neutralizing antibodies against PRRSV, as well as PCV2 (Piñeyro et al. 2015). Remarkably, Hu et al. (2016) developed in parallel a putative PCV2/PRRSV vaccine based on the PCV2 *E. coli*-produced VLP platform, where the loop CD of the PCV2 Cap tolerated insertion and surface presentation of the GP5 epitope B of PRRSV. The 3D structure of the chimeric PCV2 was simulated by homology modeling, and it appeared that the GP5 epitope B could fold as a relatively independent unit, separated from the PCV2 Cap backbone. In fact, the purified chimeric proteins self-assembled into the chimeric VLPs, which could enter the appropriate mammalian cells and induced strong virus-neutralizing antibody response against both PCV2 and PRRSV in mice (Hu et al. 2016). Wang D et al. (2018) developed further this modeling-based approach at the CD loop and designed the chimeric VLPs displaying four peptides—including two epitopes derived from structural proteins of porcine epidemic diarrhea virus (PEDV) and porcine parvovirus (PPV)—and two artificial peptides. Moreover, this substitution had no adverse effect on eliciting PCV2-neutralizing antibodies in pigs (Wang D et al. 2018). Next, Wang N et al. (2019) inserted into this location between 85G and 86S of the PCV2 Cap loop CD the B-cell epitope 228-QQITDA-233 from the structural PPV protein and the appropriate chimeric VLPs demonstrated in guinea pigs the strong protective effect against PCV2 challenge, with some protective immunity against PPV. Another bivalent PCV2/PPV vaccine solution was proposed by Liu G et al. (2020), who produced in *E. coli* the full-length PCV2 Cap and PPV VP2, which self-assembled into the VLPs of ~20 and 25 nm, respectively. By immunization of piglets, both VLPs did not antagonize each other and induced stronger humoral and cellular immune responses and provided the best protection against PPV and PCV2 coinfection (Liu G et al. 2020).

Concerning other applications of the chimeric PCV2 VLPs, Li W et al. (2013) aimed to obtain a VLP vaccine both for PCV2 prevention and growth promotion. To do this, the somatostatin gene was fused to the Cap C-terminus, the fusion protein was expressed by the baculovirus system in insect cells, and the virion-like VLPs were observed under electron microscopy. When piglets were vaccinated twice subcutaneously with the chimeric VLPs, the antibodies against both PCV2 Cap and somatostatin were induced, and the level of somatostatin concentration in the blood of pigs was significantly decreased. Moreover, the relative daily weight gain of pigs in the experimental group was obviously higher than that in the control groups. After challenge with PCV2, pigs in the vaccinated groups had no clearly clinical signs, and the weight gain was significantly higher than that in the challenge control group (Li W et al. 2013).

Zhang H et al. (2014) replaced the nuclear localization signal NLS stretch at aa 1–39 aa of the PCV2 Cap with classical swine fever virus (CSFV) T-cell epitope, aa 1446–1460, CSFV B-cell epitope, aa 693–716, and the CSFV T-cell epitope conjugated with the B-cell epitope. The chimeric proteins were expressed using the baculovirus expression system and retained ability to form VLPs. Moreover, the three chimeric Cap proteins induced efficient humoral and cellular immunity against PCV2 and CSFV in mice (Zhang H et al. 2014).

Yu et al. (2016) replaced the PCV2 Cap decoy epitope 169–180 with the Cap neutralizing B cell epitope 113–131 to improve the efficacy of the PCV2b vaccine. The chimeric Cap with the replacement was produced in *E. coli*, formed "untypical" VLPs with diameters around 20 nm instead of 17 nm for the nonchimeric control, but demonstrated nevertheless beneficial favoring induction of protective immune responses against PCV2b.

Li X et al. (2018) developed a bivalent PCV2 and FMDV vaccine candidate based on the chimeric *E. coli*-produced PCV2 Cap VLPs. Thus, the neutralizing FMDV B cell epitope region 135–160 replaced the regions aa 123–151 and aa 169–194 of the PCV2b Cap. The chimeric protein was expressed in *E. coli* and the purified product assembled into the VLPs through dialysis. The chimeric VLPs induced effective immune response against both FMDV and PCV2b in mice and guinea pigs, without inducing antibodies against the PCV2 decoy epitope (Li X et al. 2018).

Jung et al. (2019) constructed a fusion of the PCV2 Cap and Rep and expressed it by the baculovirus system. The expressed Rep-Cap fusion proteins could assemble to form VLPs on the baculovirus envelope. The mice immunized with the baculoviruses exposing Rep-Cap or Cap only successfully induced the Cap-specific immunoreaction in mice, while the humoral immune response to Bac-Rep-Cap was significantly lower than that of Bac-Cap, suggesting that Rep in the Rep-Cap fusion protein had a suppressive effect as an antigen (Jung et al. 2019). Jung et al. (2020) developed further the idea of multivalent baculovirus-based vaccines and replaced the PCV2 Cap decoy epitope 169–180 with the PRRSV GP3 epitope I, aa 61–72, PRRSV GP5 epitope IV, aa 187–200, or the PRRSV GP3 epitope I conjugated with the GP5 epitope IV. The three chimeric PCV2 variants were expressed in the *Sf*21 cells and ensured production of the corresponding VLPs, but the immunization of mice was performed with the baculovirus-containing insect cell supernatants and not with the purified VLPs. The baculovirus-immunized mice demonstrated high levels of

both PCV2 and PRRSV neutralization antibodies (Jung et al. 2020).

Ding et al. (2019) proposed a novel bivalent vaccine, fused different copy numbers of the matrix protein 2 ectodomain (M2e) from influenza A virus (IAV) to the C-terminus of the PCV2 Cap, and expressed the chimeric genes in *E. coli*. The Cap could carry at least 81 aa residues, namely three copies of M2e, at the Cap C-terminus without impairing VLP formation. The chimeric Cap-3M2e VLPs induced the highest levels of the M2e-specific immune responses, conferring protection against lethal challenge of IAVs from different species and induced specific immune responses consistent with the PCV2 commercial vaccines in mice. Moreover, the Cap-3M2e VLPs induced high levels of the M2e-specific antibodies and PCV2-specific neutralizing antibodies in pigs (Ding et al. 2019).

Tao et al. (2020) elaborated another bivalent vaccine based on the nonchimeric PCV2 VLPs and a recombinant chimera of three antigens from *Mycoplasma hyopneumoniae* that is also an economically important pathogen of swine. The recombinant proteins were efficiently produced in *E. coli*, and the combined immunization with them induced strong humoral and cellular immune responses against all four antigens in mice and piglets.

Li G et al. (2021) continued the exploitation of the PCV2 Cap loop CD and replaced the core region of Loop CD 75-LPPGGGSN-82, as well as the C-terminus 222-KDPPL-226 with the two epitopes, namely epitope B (EpB, 37-SHIQLIYNL-45) and epitope 7 (Ep7, 196-QWGRL-200), from GP5 of PRRSV. The chimeric PCV2 Cap was soluble and efficiently produced in *E. coli*, self-assembled into the chimeric VLPs with a diameter of 12–15 nm, and provided pigs with partial protection against homologous PRRSV strains (Li G et al. 2021).

Finally, it should be emphasized that the VLPs were produced for the newly emerging porcine circovirus 3, a member of the *Porcine circovirus 3* species, which may threaten to reduce the pig population dramatically worldwide. Thus, Bi et al. (2020) produced the PCV3 Cat VLPs in *E. coli*, using the previously mentioned SUMO-tag technique. The electron cryomicroscopy structure of the PCV3 VLPs was resolved to 8.5 Å, showing unique structural features of N-terminus and the CD-loop of the PCV3 Cap, as well as definite differences from the PCV2 VLPs, as shown in Figure 10.6. The PCV3 VLPs were applied by indirect ELISA to diagnose the PCV3-infected pigs (Bi et al. 2020). Wang Y et al. (2020) also reported production of the PCV3 VLPs in *E. coli*, expressed from a codon-optimized gene and provided with the His_6 tag in this case. However, the purified VLPs were of 10 nm in diameter, smaller therefore than the expected 17 nm, and this difference remained unexplained (Wang Y et al. 2020).

OTHER CIRCOVIRUSES

The infections with goose circovirus (GoCV) of the *Goose circovirus* species are associated with growth retardation

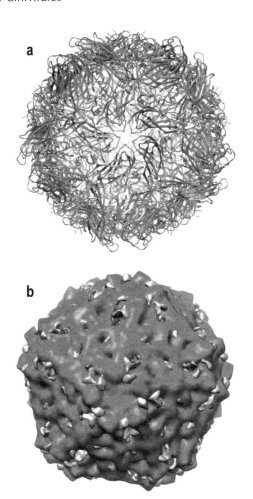

FIGURE 10.6 The PCV3 type-specific epitope mapping and structural comparison between PCV2 and PCV3. (a) The overall structural comparisons of the PCV3 VLP model and crystal structure of the PCV2 VLPs (PDB ID 5ZJU), in which the PCV2 and PCV3 VLPs are shown in pink and blue ribbon mode, respectively. (b) The structural fitting of the PCV2 and PCV3 VLPs. The PCV3 density maps are colored in red, whereas the PCV2 density maps are colored in gray. (Reprinted from Bi M et al. Structural insight into the type-specific epitope of porcine circovirus type 3. *Biosci Rep.* 2020;40:BSR20201109.)

and developmental problems in farmed geese. Scott et al. (2006) expressed the CoCV Cap in mammalian cells by the Semliki Forest virus (SFV) expression system with diagnostic purposes, but the presence of VLPs was not examined.

Pigeon circovirus (PiCV) of the *Pigeon circovirus* species is a viral agent to the development of young pigeon disease syndrome. Lai et al. (2014) expressed in a number of different *E. coli* strains the PiCV Cap that was fused with different partners including His-tag, GST-tag (glutathioine-S-transferase tag), and Trx-His-tag (thioredoxin-His tag). The Cap production was significantly increased when Cap protein was fused with either the GST-tag or the Trx-His tag rather than the His-tag. After various rare aa codons in the Cap protein were optimized, the expression level was further increased to a significant degree, and the purified GST-Cap fusion demonstrated good antigenic activity when tested against PiCV-infected pigeon sera (Lai et al. 2014). Needless to say, the VLP challenge was not addressed in the study. Gai et al. (2020) achieved the efficient baculovirus-driven production of the PiCV Cat, which self-assembled into the VLPs with a spherical morphology and diameters of 15–18 nm. The intramuscular immunization of mice with the PiCV VLPs, together with an adjuvant, induced specific antibodies against the Cap protein (Gai et al. 2020). Huang HY et al. (2021) produced the PiCV Cap in human embryonic kidney (HEK-293) cells, and the virion-like VLPs with diameters ranging from 12–26 nm were obtained. The subcutaneous immunization of pigeons with the VLPs supplemented with a water-in-oil-in-water adjuvant induced specific antibodies against PiCV and strong T cell response, while the experimentally infected pigeons that were vaccinated with the VLPs also showed no detectable viral titer (Huang HY et al. 2021).

The comprehensive VIPERdb2 database (Montiel-Garcia et al. 2021) contains to date three entries of the near-atomic structures of bat circovirus, one of which is presented in Figure 10.4. The VIPERdb2 addresses these entries to Nath et al. (2021).

11 Order *Tubulavirales*

What's done can't be undone.

William Shakespeare

ESSENTIALS

The *Tubulavirales* is a great order of nonenveloped filamentous bacteriophages. The order currently involves the three families, namely the central *Inoviridae* and two smaller *Paulinoviridae and Plectroviridae* families, 26 genera, and 34 species altogether and forms the class *Faserviricetes* as a single member. The class *Faserviricetes* is in turn a single member of the *Hofneiviricota* phylum, a single member of the *Loebvirae* kingdom that forms, together with the three other kingdoms, the realm *Monodnaviria*.

The small *Inovirus* genus of the *Inoviridae* family, which currently contains the single *Escherichia virus M13* species, shocked the world of molecular biology and served as one of the starters of the viral nanotechnologies including implicitly the VLPs. It should be emphasized that the three great members of the species, namely F-pili specific bacteriophages f1 (Loeb 1960; Loeb and Zinder 1961), M13 (Hofschneider 1963; Hofschneider and Preuss 1963), and fd (Hoffmann-Berling et al. 1963a, b; Marvin and Hoffmann-Berling 1963a, b), were discovered in close parallel and in the same line of experiments with the corresponding great F-pili specific RNA phages f2, M12, and fr, which are described in Chapter 25. The same is true for the less popular but historically important pair, namely for the filamentous DNA phage ZJ/2 discovered in parallel with the icosahedral RNA phage ZIK/1 (Bradley 1964). The phages f1, M13, and fd have about 98.5% DNA sequence identity, and for most purposes they can be considered to be almost identical (Marvin and Symmons 2014). These popular *Inovirus* members, together with phage ZJ/2, are sometimes denoted as Ff phages (for **F** specific **f**ilamentous phages) for their ability to infect bacteria bearing the F fertility factor.

After many important contributions to the classical molecular biology, the *Inovirus* members have won their remarkable place in history due to the phage display. This technique was elaborated by George P. Smith (1985) when he displayed a collection of peptides on live phage f1 libraries with following selection of one of them that was recognized by the specific antibody. The phage display was awarded with a half of the Nobel Prize in Chemistry 2018 jointly with George P. Smith and Gregory P. Winter "for the phage display of peptides and antibodies," where the other half was awarded to Frances H. Arnold "for the directed evolution of enzymes." As commented by the Nobel Committee,

Gregory Winter used phage display for the directed evolution of antibodies, with the aim of producing new

GEORGE P. SMITH, The Nobel Prize in Chemistry 2018. (Courtesy of G.P. Smith and Academic Support Center, University of Missouri.)

pharmaceuticals. The first one based on this method, adalimumab, was approved in 2002 and is used for rheumatoid arthritis, psoriasis and inflammatory bowel diseases. Since then, phage display has produced antibodies that can neutralize toxins, counteract autoimmune diseases and cure metastatic cancer.

After phage display, the filamentous phages, particularly M13, have been extensively used as cloning vectors (phagemids) in DNA sequencing, as well as in other domains of biotechnology and nanotechnology.

Figure 11.1 shows the typical structure of the inovirus M13. As reviewed by the official ICTV reports (Day 2012; Knezevic et al. 2021a), the inovirus virions are nonenveloped flexible filaments, helically organized around a positive-sense circular ssDNA, 6–10 nm in diameter and 600–2,500 nm in length, built up from 2,700 copies of the major coat protein (CoaB; p8)—with 5 copies each of p7 and p9 forming a blunt end—and of p3 (CoaA) and p6 forming a rounded end. The major coat protein is highly α-helical, and its N-terminal domains are negatively charged,

DOI: 10.1201/b22819-14

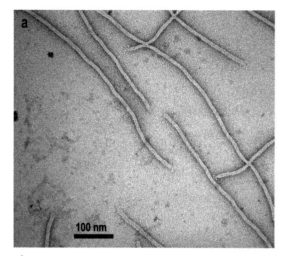

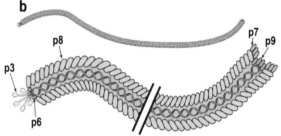

FIGURE 11.1 The typical image and schematic cartoon of bacteriophage M13 as a typical representative of the *Inoviridae* family. (a) The electron micrograph of the phage M13. Courtesy of Lee Makowski. (b) The cartoon of the *Inoviridae* family members is taken with kind permission from the ViralZone, Swiss Institute of Bioinformatics, http://viralzone.expasy.org (Hulo C et al. 2011). (Courtesy Philippe Le Mercier.)

hydrophilic, and oriented outside the virion; hydrophobic central domains stabilize subunit–subunit interactions, while C-terminal domains interact electrostatically with the DNA (Lee BY et al. 2012; Marvin et al. 2014).

In contrast to the *Inoviridae* phages, the members of the *Plectroviridae* family are rigid, asymmetric, nearly straight rods of 10–16 × 70–280 nm in size, with one end rounded and the other more variable. As commented by Knezevic et al. (2021b), the rods are shorter and wider than that of the *Inoviridae* and *Paulinoviridae* members, while the images of the negatively stained virions suggest that plectrovirus particles have 4±2 nm hollow cores.

GENOME

Figure 11.2 presents the genome of the phage M13 as the most studied and broadly applied representative of the order. As summarized by the ICTV reports (Day 2012; Knezevic et al. 2021a), the *Inoviridae* genome is a positive-sense, supercoiled, circular ssDNA molecule of 5.5–10.6 kb, encoding 7–15 proteins. The M13 genome has a DNA replication module (g2, g5, and g11), a structural module (g7, g9, g8, g3 and g6), and a morphogenesis, i.e. an assembly/extrusion module (g1, g4, and g11).

The intergenic regions contain the origin of replication, packaging signals, and promoters of various strengths. The genes are densely packed on the genome, and some of them overlap or are entirely within larger genes, being translated from internal start codons. For instance, g11 and g10 of M13 are C-terminal portions of g1 and g2, respectively.

The supercoiled, circular, ssDNA genome of the *Plectroviridae* family is about 4.5–8.3 kb and encodes 4–13 proteins. The members of this small family infect cell-wall-less bacteria from the genera *Acholeplasma* and *Spiroplasma*, adsorbing to the bacterial surface (Knezevic et al. 2021b).

PHAGE DISPLAY

In his revolutionary paper, Smith (1985) wrote that the foreign DNA fragments can be inserted into filamentous phage f1 gene g3, which is normally required for adsorption to the F-pilus, to create a fusion protein of the minor capsid protein p3 with the foreign sequence in the middle. The fusion derivative of p3 is incorporated then into the live virion, which retains infectivity and displays the foreign aa stretch in immunologically accessible form. These "fusion phages" can be enriched more than 1,000-fold over ordinary phage by affinity for antibody directed against the foreign sequence. The fusion phage may provide a simple way of cloning a gene when an antibody against the product of that gene is available (Smith 1985).

Scott and Smith (1990) described the major principles how to perform phage display and search for peptide ligands with an epitope library, where tens of millions of short peptides could be easily surveyed for tight binding to an antibody, receptor, or other binding protein. The library was presented therefore as a vast mixture of filamentous phage clones, each displaying one peptide sequence on the virion surface. The survey was accomplished by using the binding protein to affinity-purify phage that display tight-binding peptides and propagating the purified phage in *E. coli*. The aa sequences of the peptides displayed on the phage were then determined by sequencing the corresponding coding region in the viral DNAs (Scott and Smith 1990).

The phage display technique rapidly attracted great attention and was applied in numerous modifications to huge number of applications. The phage display is described by many authoritative books (Kay et al. 1996; Burton et al. 2001; Clackson and Lowman 2004; Sidhu and Geyer 2005, 2017; Hust and Lim 2017). The virus display idea was extended further to other carriers, involving DNA phages T4, T7, and lambda (Chapter 1) and RNA phages (Chapter 25). It is less known that George Pieczenik patented the phage display libraries in 1985 (Pieczenik 1999).

Generally, the inovirus phage display peptide libraries were constructed with the random peptides fused at or near the N-terminus of p3 or p8, quite often in the mosaic form, when phages contained a mixture of recombinant and wild-type coat proteins. Later, the minor structural

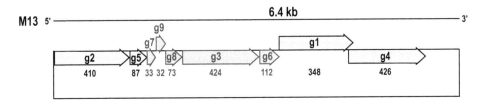

FIGURE 11.2   The linear presentation of the circular genome of M13 from the *Escherichia virus M13* species, a reference strain of the *Inovirus* genus. The genes are indicated by arrowed boxes, and the structural genes are colored dark pink. The number of aa residues is indicated under the genes. The circular character of the genome is displayed by the lower connecting bracket, while the linearization is performed with *ori* as the opening site.

proteins p7 and p9 were also employed as epitope carriers in the inovirus phage display (Løset et al. 2011a, b). Since the phage display is only marginally overlapping with the VLP technologies, the present guide is limited only by some examples of the inovirus applications, when the latter were regarded as vaccines or aimed to generate novel bionanomaterials.

The phage display libraries were able to present peptides not only in linear but also in circular form, when the epitope sequence was flanked by two cysteine residues. By building a disulfide bond, a constrained cycle was formed and presented on the phage surface (Luzzago et al. 1993). Nevertheless, the length of the inserted peptides did not exceed 6–38 aa residues, as reviewed by Hill and Stockley (1996). To overcome the capacity problem, the phagemid systems were developed, where the phagemid carried only the phage gene g3 or g8 containing the foreign sequence, and a phage with all the necessary genes for phage production, also including a copy of the wild-type coat gene, was used as a helper. Thus, both chimeric and wild-type proteins were produced and incorporated into the mosaic virions, when one of five copies of p3 and 1% to 30% of the 2700 copies of p8 were chimeric, as reviewed by Davies et al. (2000).

3D STRUCTURE

The first 3D model of the phage fd was obtained by Marvin (1966) by x-ray diffraction and electron microscope studies. The phage particles crystallized with a minimum interparticle distance of 56 Å, and the phage structure repeated regularly every 32.2 Å along its length. The proposed model was analogous to a circle of string pulled taut from opposite sides of its circumference, where the string represented an ssDNA ring enclosed in a tube of α-helices (Marvin 1966).

Marvin and Wachtel (1975) systemathized the x-ray diffraction studies of the related filamentous phages from the *Inoviridae* family, where two classes of diffraction pattern were distinguished. The class I was given by the phages fd, f1, M13, and IKe, a member of the *Salmonella virus IKe* species from the genus *Lineavirus*, while the class II pattern was given by the Pf1 and Xf strains, members of the *Pseudomonas virus Pf1* species, genus *Primolicivirus*, and a *Xanthomonas virus Xf109* from the *Xylivirus* genus, respectively. The class II pattern was simpler to interpret

and the position of diffracted intensity indicated that the protein subunits in the virus were arranged on a helix of pitch ~15 Å, with 4.4 subunits in one pitch length of the helix (Marvin and Wachtel 1975).

Furthermore, Makowski et al. (1980) resolved the Pf1 structure to 7 Å by analysis of x-ray diffraction data from partially oriented fibers of virus particles. The coat protein consisted of two α-helical segments: one, almost parallel to the particle axis, was centered at a radius of about 15 Å; the other, at about 25 Å radius, was tilted by about 25° to the particle axis. The double layer of tightly packed, intricately interlocked α-helices formed a stable, 20-Å-thick protein coat around the viral DNA (Makowski et al. 1980). Caspar and Makowski (1981) concentrated on the symmetries of the related members of the *Inoviridae* family and concluded that the difference between the symmetries of the class I and class II particles suggested that different assembly processes might have evolved to form these structures with very similar protein packing architectures.

The rapid development of the phage display demanded 3D examination of the chimeric phages in the first place. The insertions of foreign peptides were made at or near the N-terminus of p3 (Smith 1985) or p8, the major capsid protein (Ilyichev et al. 1989, 1992). Makowski (1993) described the structural constraints on the display of foreign peptides by the phage display, when successful insertion into every copy of p8 was limited by six to ten aa residues. Much larger inserts into p8 were possible using vectors containing two copies of gene g8—only one of which carried the insert—and applying therefore the mosaic particle approach. Kishchenko et al. (1994) performed the first x-ray diffraction studies of the phages containing the inserts within the p8 protein in order to determine the configuration of the displayed peptides and investigate therefore the origin of the size limit for foreign peptides.

Malik et al. (1996) showed that some large peptides were displayed at a much higher copy number than smaller ones and that some relatively small peptides were poorly displayed, if at all, in mosaic virions. The x-ray diffraction studies, together with the modeling of the epitopes of known structure, demonstrated that it was feasible to accommodate much larger structures, without perturbation of the capsid protein packing, than it has proved possible to generate in vivo. It was shown that the insertion of certain peptides greatly slowed or even prevented the processing of the p8 at the inner membrane of the *E. coli* cell (Malik et al. 1996).

Lubkowski et al. (1998) expanded the structural basis of phage display by the crystal structure of the N-terminal domains of the M13 protein p3, the main actor by the classical phage display methodology, at 1.46 Å resolution. The structure of the two N-terminal domains of p3 demonstrated that each domain consisted of either five or eight β-strands and a single α-helix. The domains were engaged in extensive interactions, resulting in a horseshoe shape with aliphatic aa residues and threonines lining the inside, delineating the likely binding site for the F-pilus. The glycine-rich linker connecting the domains was invisible in the otherwise highly ordered structure and might confer flexibility between the domains required during the infection process (Lubkowski et al. 1998). Holliger et al. (1999) reported the crystal structure of the two N-terminal domains of p3, aa residues 2–217, of the phage fd at 1.9 Å and compared it to the previously described structure of the same fragment from the M13 protein p3. Remarkably, the authors found the structure of individual domains D1 and D2 of the two phages as very similar, while there was comparatively poor agreement for the overall D1D2 structure (Holliger et al. 1999).

In parallel, Welsh et al. (1998) recorded x-ray diffraction patterns at 3.1 Å resolution from magnetically aligned fibres of the phage Pf3, a member of the *Pseudomonas virus Pf3* species from the *Tertilicivirus* genus. The patterns indicated that the Pf3 and the previously described Pf1 virions had the same helix symmetry and similar protein subunit shape. This was of particular interest, given that the primary structures of the two protein subunits and the nucleotide/protein subunit ratios were quite different. As a result, the molecular model of the Pf3 protein capsid was built on the basis of the Pf1 model. Remarkably, the Pf3 subunit appeared to be completely α-helical, beginning at the N-terminus, whereas the first few residues of the Pf1 subunit were not helical (Welsh et al. 1998).

At the same time, Papavoine et al. (1998) determined the 3D structure of the M13 protein p8, solubilized in detergent micelles, by heteronuclear multidimensional NMR and restrained molecular dynamics. The protein consisted therefore of two α-helices, running from residues 8–16 and 25–45, respectively. These two helices were connected by a flexible and distorted helical hinge region. The authors commented that the p8 structural properties resembled a flail, in which the hydrophobic helix (residues 25 to 45) was the handle and the other, amphipathic, helix was the swingle. In this metaphor, the hinge region was the connecting piece of leather (Papavoine et al. 1998).

Zeri et al. (2003) determined the structure of the fd coat in solution by solid-state NMR spectroscopy. This structure that is shown in Figure 11.3 differed from that previously determined by x-ray fiber diffraction. Most notably, the 50-residue protein was not a single curved helix but rather was a nearly ideal straight helix between residues

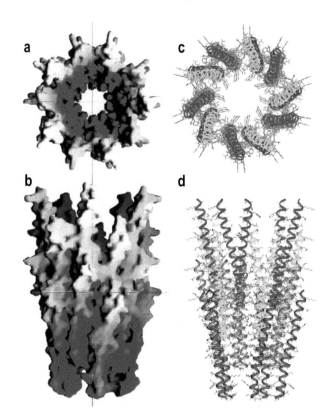

FIGURE 11.3 The model of a section of the Y21M fd filamentous bacteriophage capsid built from the coat protein subunit structure, which was determined by solid-state NMR spectroscopy. The symmetry was derived from fiber diffraction studies. (a, b) The representations of the electrostatic potential on the molecular surface of the virus, where (a) is a bottom view and (b) is a side view along the virus axis. (c, d) The views of the capsid structure showing the arrangement of the coat proteins in pentamers and further assembly of the 2-fold helical structure. (From Zeri et al. Structure of the coat protein in fd filamentous bacteriophage particles determined by solid-state NMR spectroscopy. *Proc Natl Acad Sci U S A.* 100: 6458–6463, Copyright 2003 National Academy of Sciences, U.S.A.)

7 and 38, where there was a distinct kink, and then a straight helix with a different orientation between residues 39 and 49. Residues 1–5 have been shown to be mobile and unstructured, and proline 6 terminated the helix (Zeri et al. 2003). Marvin et al. (2006) reinterpreted the NMR data and showed their consistency with the model derived from the x-ray fibre diffraction studies. In parallel, Wang YA et al. (2006) reported the first image reconstruction of the phage fd by electron cryomicroscopy. Although the thin rather featureless filaments scattered weakly, the authors achieved a nominal resolution of ~8 Å using an iterative helical reconstruction procedure. The two different conformations of the virus were found, and in both states the subunits were packed differently than in conflicting models previously proposed on the basis of x-ray fiber diffraction or solid-state NMR studies. A significant

fraction of the population of wild-type fd was either disordered or in multiple conformational states, while in the presence of the Y21M mutation this heterogeneity was greatly reduced (Wang YA et al. 2006). Moreover, the refined electron microscopy model of the phage fd closely approximated the model derived directly from x-ray fiber diffraction and solid-state NMR data (Straus et al. 2008b). The consensus structure of the x-ray and solid-state NMR data was published also for the phage Pf1 (Straus et al. 2008a, 2011). Xu J et al. (2019) resolved the structure of the phage IKe to 3.4 Å, providing therefore atomic details on the structure of the major coat protein, the symmetry of the capsid shell, and the key interactions driving its assembly. Remarkably, the phage IKe was selected for the development of the advanced 4D solid-state NMR technique (Porat et al. 2021).

VACCINES AND IMMUNOLOGY

The short repeat regions of the circumsporozoite protein (CSP) of *Plasmodium falciparum* were added to the p3 of the phage fd, displayed on the virion surface, and demonstrated the expected antigenicity, as well as immunogenicity in rabbits. Greenwood et al. (1991) displayed the CSP peptides as fusions to the phage fd major coat protein p8, in order to increase the number of the exposed copies.

Bastien et al. (1997) engineered the chimeric phage fd displaying at its surface the p3 coat protein fused to the previously identified protective B-cell epitope 173–187 from the glycoprotein G of the human respiratory syncytial virus (RSV). The mice immunized with the chimeric phage not only possessed high levels of RSV-specific antibodies but also acquired a complete resistance to RSV infection as evidenced by the lack of detectable virus particles in their lungs following intranasal challenge with live RSV (Bastien et al. 1997). In fact, this was the first report concerning the ability of the live chimeric inovirus presenting the peptide selected from random phage peptide libraries to prevent infection of immunized animals by a pathogen.

Chappel et al. (1998) presented one of the first applications of the mosaic approach, when the M13 contained two nonhomologous g8 genes, one of which, synthetic, carried N-terminal addition of a single chain antibody, namely V_H/κ polypeptide of the murine antiprogesterone antibody. The number of the chimeric p8 derivatives within the phage prepared at 25°C varied from one (34% phage) to four (1% phage), while about 50% of the phage did not display a functional antibody. Remarkably, the phage prepared at 37°C was essentially devoid of fusion (Chappel et al. 1998).

Grabowska et al. (2000) isolated the chimeric fd phages from a random phage peptide display library, which contained 15-mer peptide inserts in the p8 coat, mimicking epitopes of glycoprotein G (gG2) of herpes simplex virus 2 (HSV-2). These phages were used to immunize mice without any additional adjuvant. The phage displaying epitopes of gG2, which mapped to aa residues 551–570 and

was frequently recognized by patients infected with HSV-2, were the most immunogenic. Moreover, the mice immunized with the chimeric phage displaying a single epitope of gG2 were protected against challenge with a lethal dose of whole HSV-2 (Grabowska et al. 2000).

The short epitope display by the live mosaic filamentous phage markedly contributed to the generation of vaccine candidates against Alzheimer's disease. Thus, Frenkel et al. (2000, 2001) obtained effective autoantibodies in mice and guinea pigs, which exhibited human identity in the β-amyloid peptide (Aβ), by immunization with the chimeric fd phage displaying only four aa residues 3-EFRH-6 of the Aβ added to the fd protein p8. The effect of the immunization was long-lasting, and no toxic effect caused by autoimmune response was detected in the challenged guinea pig sections (Frenkel et al. 2001). The chimeric EFRH phage evoked effective autoimmune antibodies in the amyloid precursor protein transgenic mice that recapitulated the amyloid plaques and vascular pathology of Alzheimer's disease, and the immunization provoked a considerable reduction in the number of Aβ amyloid plaques in the brain of the transgenic mice (Frenkel et al. 2003). Remarkably, the phages expressing 300 copies of the peptide, obtained by insertion of a tandem repeat of the EFRH sequence, were more immunogenic than phages expressing 150 copies of the peptide, demonstrating that epitope density was a limiting factor within this immunization protocol (Lavie et al. 2004). Solomon (2005, 2007) developed further the fd-based EFRH vaccine. Esposito et al. (2008) further pursued Beka Solomon's strategy and compared the immunogenicity of different Aβ regions within the mosaic phage carrier system, in order to determine which region of Aβ was best suited for inclusion in a phage-based anti-Aβ vaccine. The authors promoted high epitope density by decreasing the actual epitope size, generating phages that displayed 5-aa residue epitope 2-AEFRH-6 but only had three extra aa on p8 compared to wild-type phages, and phages that displayed 7-aa residue pitope 1-DAEFRHG-7 but only had six extra aa in the chimeric p8. Other potentially active Aβ sequences were also incorporated. Interestingly, it was established that the phage displaying epitope 2–6 was more immunogenic than the phage displaying epitope 1–7 that differed only in flanking residues (Esposito et al. 2008).

A set of interesting phage display studies was connected with hepatitis B virus (HBV). For example, Wan et al. (2001) demonstrated the ability of inducing MHC class I restricted cytotoxic T lymphocytes response in vivo by the M13 phage displaying the HBV surface (HBs) epitope 28–39, when immunization of mice without any adjuvants led to specific anti-HBs CTL response eight days after injection. Ho et al. (2003) used the chimeric M13 phages that displayed random disulfide constrained heptapeptides on their p8 proteins to select for high-affinity ligands to hepatitis B core (HBc) particles. The phages bearing the amino acid sequences C-WSFFSNI-C and C-WPFWGPW-C were isolated, and a binding assay in solution showed that these phages bound tightly to full-length and truncated HBc particles (Chapter 38). Both the phages that displayed the constrained peptides were inhibited from binding to the HBc particles by a monoclonal antibody that bound specifically to their immunodominant region. Remarkably, a synthetic heptapeptide WSFFSNI derived from one of the fusion peptides inhibited the binding of large surface antigen L-HBsAg (Chapter 37) to the HBc (Ho et al. 2003). The chimeric phages displaying the anti-HBc ScFv were used as an alternative choice for the HBc/anti-HBc diagnosis in serum samples (Tan et al. 2007). Later, the HBc protein was added to the M13 phage p3, expressed on the phage surface as a whole protein, and used to develop monoclonal anti-HBc antibodies (Bahadir et al. 2011). Zhang ZC et al. (2004) generated a combinatorial library of phage-display human ScFv genes, which were derived from peripheral blood lymphocytes immunized by the HBV preS1 peptide (Chapters 37 and 38) in vitro and selected a high affinity preS1-specific single-chain antibody, which might provide a more satisfactory HBV therapy.

The phage display libraries greatly contributed to selection of mimotopes and proposed many satisfying solutions for the putative vaccines against cancer and allergy, as exhaustively reviewed at that time by Knittelfelder et al. (2009) and Rakonjac et al. (2011).

Mascolo et al. (2007) displayed the popular ovalbumin CTL epitope, aa 257–264, on the p8 and showed that this single epitope was sufficient to induce priming and sustain long-term major histocompatibility complex class I restricted cytotoxic T lymphocytes response in mice.

A novel phage display-based approach was intended by Glushkov et al. (2010) to the development of anticancer vaccines, when a benzo[a]pyrene immunomimetic peptide fused to the phage M13 protein p3 was selected.

The great potential of the phage display was unveiled in the specialized reviews devoted either to putative therapeutic applications (Bazan et al. 2012a), generation of vaccines (Bazan et al. 2012b; Prisco and De Berardinis 2012), or both (Karimi et al. 2016).

Domínguez-Romero et al. (2020) used the phage M13 display to develop a novel vaccine approach based on a new class of vaccine immunogens called variable epitope libraries (VELs). The three regions of survivin (SVN), composed of 40, 49 and 51 amino acids, were used, along with the complete SVN protein, to generate the VELs as fusions to the protein p8 within the mosaic virions. In total, eight vaccine immunogens were prepared. The mice, challenged with the aggressive and highly metastatic 4T1 cell line, were vaccinated in a therapeutic setting, and the significant tumor growth inhibition and, most importantly, strong suppression of lung metastasis was achieved after a single immunization with VEL vaccines (Domínguez-Romero et al. 2020).

Concerning the current Covid-19 era, the phage display markedly contributed toward the putative treatment and management of disease. In contrast to traditional search for vaccines, Somasundaram et al. (2020) concentrated on the passive immunotherapy and specifically on chicken egg yolk antibodies (IgY) as an alternative to mammalian antibodies that have been extensively studied against severe acute respiratory syndrome coronavirus (SARS CoV) outbreak in China. Compared to mammalian antibodies, the IgY have greater binding affinity to specific antigens, ease of extraction, and lower production costs, hence possessing remarkable pathogen-neutralizing activity of pathogens in respiratory tract and lungs. The authors argued convincingly that the monoclonal IgY scFv antibodies raised against SARS-CoV-2 spike protein S isolated from chickens by the phage display technology would be a potential model for large-scale production of high-affinity antibodies and development of rapid diagnosis and immunotherapy against SARS-CoV-2.

TARGETING AND DELIVERY

The first specific inovirus-based targeting was achieved by Hart et al. (1994). The phage fd virions that displayed an RGD motif, namely the cyclic integrin-binding peptide sequence GGCRGDMFGC, in a proportion of their g8 subunits bound to cells and were efficiently internalized. In the displayed peptide the conformation of the RGD motif was restricted within a hairpin loop formed by a disulfide bridge between the two cysteine residues. The cellular internalization of phage was demonstrated by confocal and nonconfocal immunofluorescence microscopy of tissue-cultured cells incubated with phage particles, while the cell binding and internalization was inhibited by preincubation of cells with the integrin-binding peptide GRGDSP (Hart et al. 1994). Later, Rong et al. (2008) found that the RGD-grafted phage M13 guided cell alignment and oriented the cell outgrowth along defined directions. Specifically, Chung et al. (2010) used the phage M13 with p8 displaying the RGD motifs and forming long-range-ordered liquid-crystalline matrices to dictate the direction of neuronal cell growth and assure successful regeneration of the target tissue.

Pasqualini and Ruoslahti (1996) performed a large study on the organ targeting in vivo using phage display peptide libraries by screening of random peptide sequences. The peptides capable of mediating selective localization of phage to brain and kidney blood vessels were identified and showed up to thirteenfold selectivity for these organs. One of the peptides displayed by the brain-localizing phage was synthesized and shown to specifically inhibit the localization of the homologous phage into the brain (Pasqualini and Ruoslahti 1996).

Larocca et al. (1998) investigated the general principles of the phage targeting to mammalian cell surface receptors, since the "filamentous bacteriophages represent one of nature's most elegant ways of packaging and delivering DNA." In an effort to develop novel methods for ligand discovery via phage gene delivery, the mammalian cell tropism to the phages were conferred by attaching basic fibroblast growth factor (FGF2), transferrin, or epidermal growth factor (EGF) to the phage coats and measuring CMV promoter-driven reporter gene expression in target cells. Remarkably, FGF2 was a more effective targeting agent than transferrin or EGF. The detection of green fluorescent protein or β-galactosidase activity in cells required FGF2 targeting and was phage concentration dependent. Therefore, the filamentous phages were proposed as a vehicle for targeted gene delivery to mammalian cells (Larocca et al. 1998).

Poul and Marks (1999) showed that the phage displaying the anti-ErbB2 ScFv F5 fused to p3 directly infected mammalian cells expressing the ErbB2 leading to expression of a reporter gene contained in the phage genome. This offered a strong way to discover targeting molecules for intracellular drug delivery or gene therapy by directly screening phage antibodies to identify those capable of undergoing endocytosis and delivering a gene intracellularly into the correct trafficking pathway for gene expression. The putative targeting of the phage display vectors to mammalian cells was extensively reviewed at that time by Uppala and Koivunen (2000).

Carrera et al. (2004) demonstrated efficient treatment of cocaine abuse using penetration of the central nervous system by phages displaying cocaine-sequestering antibody ScFv on the surface by fusion to the p8 peptide. This successful approach was reviewed in detail by Dickerson et al. (2005), who published a convincing illustration that is presented as Figure 11.4.

Zahid et al. (2010) incubated a cardiomyoblast cell line with a M13 phage 12-aa peptide display library and got the peptide sequence APWHLSSQYSRT, termed *cardiac targeting peptide*, which enabled transduction of

p8   ScFv

FIGURE 11.4 The enlarged region of the phage coat showing coat protein p8 displaying a single-chain antibody (ScFv; reprinted with permission of Taylor & Francis Group from Dickerson TJ et al. *Expert Opin Biol Ther.* 2005;5:773–781.)

cardiomyocytes. Rakover et al. (2010) fused the myelin oligodendrocyte glycoprotein immunodominant epitope (MOG 36–44) to the p8, enabling therefore the penetration into the central nervous system (CNS) after nasal administration. Thus, the intranasal treatment of experimental autoimmune encephalomyelitis (EAE) in mice was performed with phage carrying MOG and showed improved neuronal function, depletion of the autoantibodies against MOG, and prevention of demyelination resulting in improved clinical scores and the reduced inflammation in the CNS and periphery in the experimental mice compared to untreated sick animals (Rakover et al. 2010).

Developing further the studies on the fine mechanisms of the fusion-modified phage targeting, Wang T et al. (2011) concluded that only the binding peptide sequence matters. Thus, it was found that liposomes modified with the cancer cell MCF-7-specific phage fusion proteins, namely the MCF-7 binding peptide, DMPGTVLP, fused to the phage p8, provided a strong and specific association with target MCF-7 cancer cells but not with cocultured, nontarget cells. The substitution for the binding peptide fused to the phage p8 coat protein abolished the targeting specificity (Wang T et al. 2011).

Sartorius et al. (2011) engineered the phage fd displaying a ScFv fragment known to bind the mouse dendritic cell surface molecule DEC-205. Moreover, the DC-targeting with the virions double-displaying the anti-DEC-205 fragment on the p3 protein and the previously mentioned ovalbumin epitope 257–264 on the p8 induced potent inhibition of the growth of the B16-OVA tumor in vivo. This protection was much stronger than other immunization strategies and similar to that induced by adoptively transferred dendritic cells (Sartorius et al. 2011).

Li et al. (2010) found that genetic manipulation was not compulsory for tumor targeting of the phage virions. They involved the chemical modification of tyrosine residues of p8 to create a chemical binding site on the surface of the phage and attach the two different moieties that could be responsible for the specific targeting—such as folic acid for tumor targeting—and another, such as a fluorescent dye, as a diagnostic tool.

Displaying tissue-targeting peptides on the phage coat surface, it appeared possible to deliver therapeutic compounds into a specific cell type. Thus, Vaks and Benhar (2011) displayed an antibody on the p3 of the phage f1 and chemically conjugated that with chloramphenicol as a selected antibiotic. The conjugation took place between the amine group provided by neomycin-chloramphenicol and the carboxyl group of the phage surface. This type of conjugation process could change the properties of the phage; the immunogenicity of the phages diminished and their infectivity and toxicity was significantly reduced, whereas the half-life of the conjugated phage increased in the bloodstream. On the other hand, this experiment confirmed that the toxicity and side effects of hazardous drugs could be controlled by conjugation to the filamentous phages. Remarkably, after injection of the appropriate dose of the antibiotic-loaded phage virions, no toxic side effects of either the drug or the phage were observed (Vaks and Benhar 2011).

Rangel et al. (2012, 2013) introduced combinatorial platform technology based on so-called internalizing-phage (iPhage) libraries to identify clones that enter mammalian cells through a receptor-independent mechanism and target-specific organelles as a tool to select ligand peptides and identify their intracellular receptors. To do this, penetratin *(pen)*, an antennapedia-derived peptide, was displayed on the phage envelope and mediated receptor-independent uptake of internalizing phage into cells. It was shown that an iPhage construct displaying an established mitochondria-specific localization signal targeted mitochondria, and that an iPhage random peptide library selected for peptide motifs that localized to different intracellular compartments. As a proof of concept, it was demonstrated that one such peptide, if chemically fused to *pen*, was internalized receptor-independently, localized to mitochondria, and promoted cell death (Rangel et al. 2012). The detailed protocols of the design, cloning, construction, and production of iPhage-based vectors and libraries, along with basic ligand-receptor identification and validation methodologies for organelle receptors, was published by Rangel et al. (2013).

Ghosh et al. (2012) redesigned the M13 genome to physically separate the overlapping of the p7 and p9 genes and enabled construction of the first N-terminal genomic modification of p9 for peptide display. The refactored M13 genome was used as a platform for targeted imaging of and drug delivery to prostate cancer cells in vitro. DePorter and McNaughton (2014) engineered the prostate cancer cell-penetrating M13 phage and generated nanocarriers for the intracellular delivery of functional exogenous proteins to a human prostate cancer cell line. To do this, the cell-penetrating peptide was fused to the p3 protein and a biotin acceptor peptide (BAP) to the p9 protein. The BAP provided a biotin binding site in which a biotinylated exogenous therapeutic cargo could be attached and taken up into prostate cancer cells (DePorter and McNaughton 2014).

Jin et al. (2014) engineered M13 phages with two functional peptides, collagen mimetic peptide and streptavidin binding peptide, on their p3 and p8 coat proteins, respectively. The idea was fruitful, since collagens are overexpressed in various human cancers and subsequently degraded and denatured by proteolytic enzymes, thus making them a target for diagnostics and therapeutics. The resulting engineered phage should function as both a therapeutic and imaging tool to target degraded and denatured collagens in cancerous tissues. The ability of the engineered chimeric phages to target and label, after the conjugation with streptavidin-linked fluorescent agents, the abnormal collagens expressed on human lung adenocarcinoma cells was demonstrated experimentally (Jin et al. 2014).

Przystal et al. (2019) introduced an intravenous phage vector for dual targeting of therapeutic genes to glioblastoma, in the form of a hybrid with adeno-associated virus (AAV), described in Chapter 9, designed to deliver a recombinant AAV genome by the capsid of the phage M13. The dual tumor targeting by this vector was first achieved by phage capsid display of the RGD4C ligand that bound the $\alpha v \beta 3$ integrin receptor. Second, the genes were expressed from a tumor-activated and temozolomide (TMZ)-induced promoter of the glucose-regulated protein. The combination of TMZ and RGD4C/AAV-Grp78 targeted gene therapy exerted as a result a synergistic effect to suppress growth of orthotopic glioblastoma (Przystal et al. 2019).

FUNCTIONALIZATION AND NANOMATERIALS

By analogy with the previously described immunological and cell-targeting approaches, the filamentous inoviruses were involved into the infinite number of the pioneering investigations on the boundary of chemistry and physics, again as before in the form of complete virions but not as recombinant VLPs and mostly through the phage display methodology. Some examples of these highly innovative applications are listed later.

At the very beginning, the phage display libraries were applied for the detection of biological threats (Petrenko et al. 1996; Smith and Petrenko 1997; Petrenko and Sorokulova 2004), for example, as biosensors for the rapid detection of *Salmonella typhimurium* in solution, which was based on affinity-selected phages adsorbed to piezoelectric transducers (Olsen et al. 2006). The broad field of the nanoscale phage biosensors was extensively reviewed later by Lee JW et al. (2013).

The phage-display libraries were used to identify peptides that bound to a range of semiconductor surfaces with high specificity, depending on the crystallographic orientation and composition of the structurally similar materials (Whaley et al. 2000). As electronic devices contained structurally related materials in close proximity, such peptides were proposed for the controlled placement and assembly of a variety of practically important materials, thus broadening the scope for the so-called bottom-up fabrication approaches. Lee SW et al. (2002) fabricated the highly ordered composite semiconuctor material from the chimeric phage M13 and ZnS nanocrystals. The phage, which formed the basis of the self-ordering system, was selected to have a specific recognition moiety for ZnS crystal surfaces, coupled with ZnS solution precursors, and it spontaneously evolved a self-supporting hybrid film material. The latter was ordered at the nanoscale and at the micrometer scale into ~72-micrometer domains, which were continuous over a centimeter length scale (Lee SW et al. 2002).

Going further by the development of the reliable templates for the nucleation and orientation of semiconductor nanowires, Mao et al. (2003) selected the peptides for their ability to nucleate ZnS or CdS by using a p3 phage display library. The successful peptides were expressed then as p8 fusion proteins into the crystalline capsid of the virus, the engineered virions were exposed to semiconductor precursor solutions, and the resultant nanocrystals that were templated along the virions to form nanowires were extensively characterized by using high-resolution analytical electron microscopy and photoluminescence (Mao et al. 2003). Mao et al. (2004) used the phage M13-based scaffold for the synthesis of single-crystal ZnS, CdS, and freestanding chemically ordered CoPt and FePt nanowires, where the phage-displayed nucleating peptides exhibited control of composition, size, and phase during nanoparticle nucleation and directed the synthesis of semiconducting and magnetic materials. The removal of the viral template by means of annealing promoted the oriented aggregation-based crystal growth, forming individual crystalline nanowires. This unique ability to interchange substrate-specific peptides into the linear self-assembled filamentous construct of the phage M13 introduced a material tunability that has not been seen in previous synthetic routes (Mao et al. 2004). Reiss et al. (2004) adapted the phage display methodology to identify peptide sequences that both specifically bound to the ferromagnetic FePt materials and controlled the crystallization of FePt nanoparticles. Later, Zaman et al. (2013) demonstrated the growth of crystalline copper sulfide using the phage M13 template and $CuCl_2$ and Na_2S precursors, leading to polydisperse nanocrystals of 2–7 nm in size along the length of the viral scaffold.

Fernandes et al. (2004) involved the phage M13 into the fluorescence resonance energy transfer (FRET) methodology for the elucidation of intermolecular contacts. The phage M13 protein p8 was labeled with 7-diethylamino-3((4'iodoacetyl)amino)phenyl-4-methylcoumarin to be used as the donor in energy transfer studies, while phospholipids labeled with N-(7-nitro-2–1,3-benzoxadiazol-4-yl) were selected as the acceptors, paving the way for the protein-lipid selectivity studies (Fernandes et al. 2004).

Souza et al. (2006) established an approach for fabrication of spontaneous, biologically active molecular networks consisting of the phage directly assembled with gold nanoparticles. The spontaneous organization of these targeted networks was manipulated further by incorporation of imidazole, which induced changes in fractal structure and near-infrared optical properties, and the networks were used as labels for enhanced fluorescence and dark-field microscopy, surface-enhanced Raman scattering detection, and near-infrared photon-to-heat conversion.

Nam et al. (2006) used the phages to synthesize and assemble nanowires of cobalt oxide at room temperature. By incorporating gold-binding peptides into the filament coat, the hybrid gold-cobalt oxide wires that improved battery capacity were formed. Combining virus-templated synthesis at the peptide level and methods for controlling the 2D assembly of viruses on polyelectrolyte multilayers provided a systematic platform for integrating these nanomaterials to form thin, flexible lithium ion batteries (Nam et al. 2006). The self-assembled layer of the phage M13-templated cobalt oxide nanowires serving as the active anode material in the battery anode was formed on top of microscale islands of polyelectrolyte multilayers serving as the battery electrolyte, and this assembly was stamped onto platinum microband current collectors (Nam et al. 2008). The resulting electrode arrays exhibited full electrochemical

functionality. Next, Lee YJ et al. (2009) described fabrication of the genetically engineered high-power lithium-ion batteries from materials previously excluded because of extremely low electronic conductivity.

Zhang Z and Buitenhuis (2007) were the first to use the phage fd as a template to regulate the formation of silica nanomaterials with well-defined morphologies, including uniform silica rods, wires, and bundles on the virions. Later, Kim et al. (2013) engineered homogeneous molecular templates for forming silica coated coaxial nanocables along the mutant M13 phage displaying serine on the surface to provide hydroxyl groups for the sol-gel reaction.

Souza et al. (2010) designed a 3D tissue culture based on magnetic levitation of cells in the presence of a hydrogel consisting of gold, magnetic iron oxide nanoparticles, and the phage M13. By spatially controlling the magnetic field, the geometry of the cell mass was manipulated, and multicellular clustering of different cell types in coculture was achieved. The magnetically levitated human glioblastoma cells showed similar protein expression profiles to those observed in human tumor xenografts (Souza et al. 2010).

Developing the visualization and imaging approaches, Carrico et al. (2012) elaborated a convenient technique for the labeling of filamentous phage capsid proteins. The previous reports have shown that phage coat protein residues can be modified, but the lack of chemically distinct aa in the coat protein sequences made it difficult to attach high levels of synthetic molecules without altering the binding capabilities of the phage. To modify the phage with polymer chains, imaging groups, and other molecules, the authors developed chemistry to convert the N-terminal amines of the ~4200 coat proteins into ketone groups. These sites served as chemospecific handles for the attachment of alkoxyamine groups through oxime formation. Thus, the attachment of fluorophores and up to 3,000 molecules of 2 kDa poly(ethylene glycol) (PEG2k) to each of the phage p8 was demonstrated without significantly affecting the binding of phage-displayed ScFv fragments to EGFR and HER2. Moreover, the authors demonstrated the utility of the modified phage for the characterization of breast cancer cells using multicolor fluorescence microscopy (Carrico et al. 2012).

Hess et al. (2012) exploited bacterial sortases for the functionalization of phages with entities ranging from small molecules (e.g., fluorophores, biotin) to correctly folded proteins, such as GFP, antibodies, streptavidin, coupled in a site-specific manner. Lee JH et al. (2012) showed that p8 modification with plasmon-resonant gold nanoparticles could generate a rapid and sensitive sensor for the targeted antigen that was recognized by peptides displayed on the p3 protein.

The phage M13 as rationally designed nanocomposite with high electron mobility greatly contributed to the performance of photovoltaic devices (Dang et al. 2011). Thus, the authors reported the synthesis of single-walled carbon nanotube (SWNT)—TiO_2 nanocrystal core—shell nanocomposites using a genetically engineered M13 virions as a template. By using the nanocomposites as photoanodes in dye-sensitized solar cells, it appeared that even small fractions of nanotubes improved the power conversion efficiency by increasing the electron collection efficiency. Yi et al. (2012) showed that the genetically engineered multifunctional M13 phage can assemble fluorescent SWNTs and ligands for targeted fluorescence imaging of tumors. Nuraje et al. (2012) employed the genetically engineered M13 as effective templates to mineralize perovskite nanomaterials, particularly strontium titanate (STO) and bismuth ferrite (BFO). The phage-templated nanocrystals were small in size, highly crystalline, and showed definite photocatalytic and photovoltaic properties.

The phage fd was used for the development of biotemplated nanowire and biosensor field by deposition of gold nanoparticles after expression of the short three-methionine (MMM) aa sequence at the N-terminus of p8 to facilitate gold-sulfur interaction (Kang et al. 2010; Korkmaz 2013). Park JP et al. (2014) engineered first the M13 phage to display YEEE on the p8 and then enzymatically converted the Tyr residue to 3,4-dihydroxyl-l-phenylalanine (DOPA). The DOPA-displayed M13 phage performed two functions, namely assembly and nucleation. The engineered phage assembled various noble metals, metal oxides, and semiconducting nanoparticles into one-dimensional arrays. Furthermore, the DOPA-displayed phage triggered the nucleation and growth of gold, silver, platinum, bimetallic cobalt-platinum, and bimetallic iron-platinum nanowires. This versatile phage template enabled rapid preparation of phage-based prototype devices by eliminating the screening process (Park JP et al. 2014). Later, Korkmaz and Arslan (2019) displayed a set of Tyr containing 5-mer peptides on the fd virions for enhanced gold binding and reduction properties. Thus, the investigation of the wild-type fd with AEGDD sequence and engineered YYYYY, AYSSG, and AYGDD phages demonstrated that the presence of only one Tyr unit on five aa flexible region of p8 coat proteins increased gold binding affinities of engineered phages, while the YYYYY phages were shown to have the strongest gold surface and gold nanoparticle binding affinities. Remarkably, the recombinant phages were shown to be coated with gold clusters after one-step metallization reaction (Korkmaz and Arslan 2019).

Chen et al. (2013) used the phage M13 to build a multifunctional 3D scaffold capable of improving both electron collection and light harvesting in dye-sensitized solar cells (DSSCs). This was accomplished by binding gold nanoparticles to the virus proteins and encapsulating the gold-virus complexes in TiO_2 to produce a plasmon-enhanced and nanowire-based photoanode. The NW morphology exhibits an improved electron diffusion length compared to traditional nanoparticle-based DSSCs, and the gold nanoparticles increase the light absorption of the dye molecules through the phenomenon of localized surface plasmon resonance. Moreover, a theoretical model was proposed to predict the experimentally observed trends of plasmon enhancement (Chen et al. 2013). Later, Sokullu et al. (2020) engineered

the phage M13 to examine as a biological template to create well-defined spacing between very small gold nanoparticles of 3–13 nm and determined the effect of gold nanoparticle particle size on the enhancement of the nonlinear process of two-photon excitation fluorescence (2PEF). Such assemblies were able to clearly label *E. coli* cells and produce a 2PEF signal that was orders of magnitude higher than of isolated gold nanoparticles, providing the opportunities of such small gold nanoparticles within colloidal plasmonic assemblies for applications in biodetection or as imaging contrast agents (Sokullu et al. 2020).

The hydroxyapatite (HAP)-nucleating phages were able to self-assemble into bundles by forming β-structure between the peptides displayed on their side walls (Xu H et al. 2011). The β-structure further promoted the oriented nucleation and growth of HAP crystals within the nanofibrous phage bundles with their c-axis preferentially parallel to the bundles. The self-assembly and mineralization driven by the β-structure formation represented a new route for fabricating mineralized fibers that can serve as building blocks in forming bone repair biomaterials and mimic the basic structure of natural bones (Xu H et al. 2011). Jin et al. (2013) screened phage display libraries in a high-throughput manner to perform the phage display with inorganic crystal surfaces. Specifically, HAP was used as a model for discovery of HAP-associated proteins in bone or tooth biomineralization studies. Cao B et al. (2014) presented the bioengineering of the phage M13 to surface-display HAP-nucleating peptides derived from dentin matrix protein-1 and using the engineered phage as a biotemplate to grow HAP nanocrystals and regenerate bones.

Yoo et al. (2014) reviewed the use of the genetically engineered M13 phage as a novel tissue regeneration material by displaying a high density of cell-signaling peptides on the p8 protein for the tissue regeneration and cell therapy purposes. The long-rod shaped and monodisperse structure of the nanofibrous phage scaffolds supported cell proliferation and differentiation as well as direct orientation of the tissue growth in two or three dimensions, as shown by Merzlyak et al. (2009).

Cao J et al. (2014) chemically decorated the phage M13 viruses with phenylboronic acid moieties and demonstrated the pH responsive chiral nematic liquid crystal LC phases. Binding with biologically important diols resulted in the LC phases with microstructures that closely correlated with the molecular structure of the diols and were conveniently discerned by visual cues.

Concerning organic crystals, Cho et al. (2015) described preparation of regular, micrometer-sized, tetragonal-bipyramidal crystals of thiamethoxam (TMX), possessing high morphological uniformity, by the controlled aggregation-driven crystallization of primitive TMX crystals with the phage M13. The phage appeared to affect the supersaturation driving force for crystallization.

Juds et al. (2020) demonstrated how to combine the phage display with next-generation sequencing (NGS) for the materials sciences by a study on probing polypropylene.

Thus, the phage display biopanning with Illumina NGS was applied to reveal insights into the peptide-based adhesion domains for polypropylene. Remarkably, the single biopanning round followed by NGS selected robust polypropylene-binding peptides that were not evident through Sanger sequencing. The NGS provided a significant statistical base that enabled motif analysis, statistics on positional residue depletion/enrichment, and data analysis to suppress false-positive sequences from amplification bias (Juds et al. 2020).

In the context of the organic crystals, it should be noted that the filamentous Pf phages contributed to the self-assembly of the extracellular matrix produced by *Pseudomonas aeruginosa* into liquid crystals, which enhanced the function of bacterial biofilms by increasing adhesion and tolerance to desiccation and antibiotics (Secor et al. 2015). The authors found that the phage fd demonstrated similar biofilm-building capabilities. As found by Sawada et al. (2018), the liquid crystalline orientation of the phage M13 assemblies played an important role in the stability against heating processes, providing insight into the future use of biomolecular assemblies for reliable thermal conductive materials.

Tridgett et al. (2018) produced the multivirion assemblies of the phage M13 via chemical modification of its surface by the covalent attachment of the xanthene-based dye tetramethylrhodamine (TMR) isothiocyanate (TRITC), which induced the formation of the 3D aster-like complexes by providing adhesive action between phage particles through the formation of H-aggregates (face-to-face stacking of dye molecules). The H-aggregation of TMR was greatly enhanced by covalent attachment to M13 and was enhanced further still upon the ordered self-assembly of

M13, leading to the suggestion that M13 could be used to promote the self-assembly of dyes that form J-aggregates, a desirable arrangement of fluorescent dye, which has interesting optical properties and potential applications in the fields of medicine and light-harvesting technology (Tridgett et al. 2018).

Ohmura et al. (2019) used the phage M13 to assemble 1D biological templates into scalable, 3D structures to fabricate metal nanofoams with a variety of genetically programmable architectures and material chemistries. The nanofoam architecture was modulated by manipulating phage assembly, specifically by editing the coat protein, as well as altering templating density. These architectures were retained over a broad range of compositions including monometallic and bimetallic combinations of noble and transition metals of copper, nickel, cobalt, and gold. The phosphorous and boron incorporation was also explored (Ohmura et al. 2019).

At last, Park SM et al. (2021) investigated the drying process of the phage M13 droplet and the resultant chiral film was used to align gold nanorods, suggesting a way to use M13 as a scaffold for the multifunctional chiral structures.

The numerous classical reviews are devoted to the breakthrough applications of the inoviruses, mostly on the background of the corresponding nanoparticles from other viruses including VLPs, by the development of bioimaging and biosensing techniques (Manchester and Singh 2006; Li et al. 2010; Steinmetz 2010; Lee JW et al. 2013; Goldman and Walper 2014; Wen et al. 2015) and new materials (Lee LA et al. 2009, 2011; Chung et al. 2011; Dedeo et al. 2011; Pokorski and Steinmetz 2011; Ng et al. 2012; Wen et al. 2013, 2016; Yang et al. 2013; Farr et al. 2014; Karimi et al. 2016; Maassen et al. 2016).

12 Other Single-Stranded DNA Viruses

They are able because they think they are able.

Vergil

CLASS *MALGRANDAVIRICETES*

ORDER *PETITVIRALES*

The *Petitvirales* is an order of nonenveloped T = 1 icosahedral viruses, 25–27 nm in diameter, with the covalently closed ssDNA genome, which belong, together with parvoviruses (Chapter 9) and circoviruses (Chapter 10), to the smallest DNA viruses by the virion and genome size. The order *Petitvirales* currently involves the sole family *Microviridae* with two subfamilies, *Bullavirinae* (former genus *Microvirus*) and *Gokushovirinae*, with 7 genera and 22 species altogether. The *Bullavirinae* members infect *Enterobacteria*, while gokushoviruses are currently known to infect only obligate intracellular parasitic bacteria. Remarkably, the name *Gokushovirinae* is derived from the Japanese for "very small." The most famous representative of the *Petitvirales* order is the phage φX174 of the *Escherichia virus phiX174* species from the *Sinsheimervirus* genus belonging to the *Bullavirinae* subfamily. The phage φX174, like the famous RNA phage MS2 of the *Norzivirales* order, which is described in Chapter 25, played a unique role in the general progress of molecular biology. The full-length genomic DNA of φX174, 5,375 nucleotides in length appeared as the first sequenced DNA genome (Sanger et al. 1977). The φX174 DNA was the first genome synthesized by purified enzymes and demonstrating features of the natural virus (Goulian et al. 1967). The φX174 DNA was the first genome completely assembled in vitro from synthetic oligonucleotides (Smith et al. 2003). However, φX174, as well as other members of the current order *Petitvirales* did not play any evident role in the development of VLP technologies because of highly sophisticated assembly of their virions. Nevertheless, the breakthroughs by successful assembly of the φX174 particles in vitro (Cherwa et al. 2011) and full decompression of the overlapping φX174 genome in yeast with independent production of viral structural proteins (Jaschke et al. 2012) are giving hope to rapid progress of the *Petitvirales* members as VLP models. It is worth mentioning that the bacteriophage φX174 lysis gene E was fused with the gene L of the phage MS2, formed the fused gene E-L, and applied as a critical element in the prospective technology of bacterial ghosts by putative vaccine candidates (Harkness and Lubitz 1987; Szostak et al. 1990, 1996; Mayrhofer et al. 2005; Jawale et al. 2014).

Figure 12.1 demonstrates the structure of the φX174 virions. According to the latest ICTV review (Cherwa and Fane 2012), the virions of the subfamily *Bullavirinae* (former genus *Microvirus*) consist of 12 pentagonal trumpet-shaped pentamers (~7.1 nm wide × 3.8 nm high) and contain 60 copies each of the major capsid protein F, major spike protein G, and a small DNA-binding protein J (25–40 aa in length) and 12 copies of the DNA pilot protein H. Therefore, the 12 spikes are each composed of five G and one H proteins. The assembly of the virion uses two scaffolding proteins, the internal scaffolding protein B and external scaffolding protein D. The protein H is a multifunctional structural protein required for piloting the viral DNA into the host cell interior during the entry process. The genomes of the *Bullavirinae* members are between 5.3 and 6.1 kb in length and encode 11 genes, nine of which are essential, while several of the genes have overlapping reading frames. The φX174-like members of the *Sinsheimervirus* genus have the smallest and least variable genomes. The members of the subfamily *Gokushovirinae* have only two structural proteins: capsid proteins F (VP1) and DNA pilot protein H (VP2) and do not use scaffolding proteins. They also possess "mushroom-like" protrusions positioned at the 3-fold axes of symmetry of their icosahedral capsids. The gokushovirus genomes are in the range of 4.4–4.9 kb (Cherwa and Fane 2012).

CLASS *MOUVIRICETES*

ORDER *POLIVIRALES*

The sole *Polivirales* order of the class *Mouviricetes* includes the sole family *Bidnaviridae* with the sole genus *Bidensovirus* and the sole species *Bombyx mori bidensovirus*. The only representative of the new *Bidnaviridae* family, which was assigned by the ICTV in 2012 (Adams and Carstens 2012), namely Bombyx mori bidensovirus (BmBDV), which was previously assigned to the *Densovirinae* in the *Parvoviridae* and once termed B. mori densovirus type 2 (BmDNV-2) or B. mori parvo-like virus (BmPLV), causes fatal flacherie disease in silkworms, resulting in large losses to the sericulture industry.

As summarized by Lü et al. (2017), BmBDV is a nonenveloped T = 1 capsid of 20–24 nm in diameter and formed by 60 copies of the major capsid protein VP, although the BmBDV genome encodes two structural proteins, the major VP and the viral structural protein SP. The BmBDV virions separately pack two single-stranded linear DNA segments, viral DNA 1 (VD1, 6.6 kb) and viral DNA 2 (VD2, 6.0 kb) in capsids of the same shape and size, while an equal amount of positive and negative strands are encapsidated, so that there are four different types of full particles. The major VP is encoded by VD1-ORF3, and its molecular mass is approximately 55 kDa (Wang et al. 2007). The 133-kDa minor capsid protein SP is encoded by the VD2-ORF1.

DOI: 10.1201/b22819-15

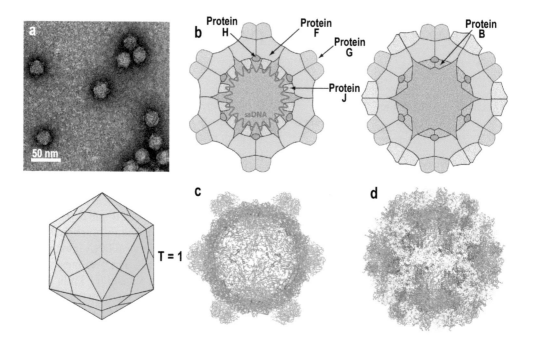

FIGURE 12.1 The portrait, cartoon, and near-atomic structure of ϕX174, a member of the *Microviridae* family. (a) The negative contrast electron micrograph of ϕX174 particles. (Reprinted from *Virus Taxonomy, Classification and Nomenclature of Viruses. Ninth Report of the International Committee on Taxonomy of Viruses*, King AMQ, Lefkowitz E, Adams MJ, Carstens EB (Eds), Cherwa JE Jr, Fane BA, Family—*Microviridae*. 385–393, Copyright 2012, with permission from Elsevier.) (b) The cartoon taken with kind permission from the ViralZone, Swiss Institute of Bioinformatics, http://viralzone.expasy.org (Hulo C et al. 2011). (Courtesy Philippe Le Mercier.) (c) ϕX174, capsid, crystal structure, PDB ID 2BPA, 3.00 Å, 342 Å (McKenna et al. 1992). (d) ϕX174, procapsid, crystal structure, PDB ID 1CD3, 3.50 Å, 352 Å (Dokland et al. 1999). The 3D structures are taken from the VIPERdb (http://viperdb.scripps.edu) database (Carrillo-Tripp et al. 2009). The size of particles is to scale. The structures possess the icosahedral T = 1 symmetry. The corresponding protein data bank (PDB) ID numbers are given under the appropriate virus names.

However, the fine evaluation of the content of the BmBDV virions from the BmBDV-infected silkworm larvae feces showed that the virion is composed of seven proteins, six of which are encoded by the *vp* gene, whereas the seventh is encoded by the *sp* gene (Lv et al. 2011).

Pan et al. (2014) were the first to get the BmBDV VLPs. Thus, the ORFs *vp1* and *vp2* of the *vp* gene, which started just upstream of the first two candidate initiation codons, were expressed in *Sf*9 cells by a baculovirus expression system. The results showed that the expressions of the *vp1* ORF yielded three proteins, namely VP1, VP1', and VP2, which were the same as the viral VPs expression in midgut of *Bombyx mori*, and the *vp2* ORF generated two VPs with the molecular mass of about 51 kDa (VP2) and 37 kDa. The electron microscopy demonstrated that these VPs can autoassemble into the VLPs that could not be distinguished from the native virions. Therefore, it was hypothesized that the BmBDV VP proteins can be expressed via a leaky scanning mechanism, which occurs in parvoviruses. However, the specific expression strategy of the VP gene remained unknown (Pan et al. 2014). To improve this understanding, Lü et al. (2017) used the novel MultiBac baculovirus expression system to further investigate the expression and assembling strategies of the BmBDV structural proteins. Thus, when the BmBDV *vp* and *sp* genes were expressed in *Sf*9 cells, the corresponding products assembled into the

VLPs of 22–24 nm in diameter. Moreover, when the second AUG codon of the VP gene was mutated to GCG, the only full-length products were produced (Lü et al. 2017), which confirmed the leaky scanning mechanism that was inferred earlier (Lv et al. 2011; Pan et al. 2014).

CLASS *ARFIVIRICETES*

The class *Arfiviricetes* involves four orders, namely *Baphyvirales* with the *Bacilladnaviridae* family, *Cremevirales* with the family *Smacoviridae*, *Mulpavirales* with families *Metaxyviridae* and *Nanoviridae*, and order *Recrevirales* with family *Redondoviridae*. The *Bacilladnaviridae* members probably have T = 1 virions. The new *Smacoviridae* family was recently introduced by Varsani and Krupovic (2018). The genomes of this family encode capsid protein, but there are no details about the putative virion structure. The *Metaxyviridae* family is represented by a single species, while the plant virus family *Nanoviridae* contains 2 well-studied genera, *Babuvirus* and *Nanovirus*, and contains 14 species altogether, as reviewed by the current ICTV reports (Vetten et al. 2018; Thomas et al. 2021). The nanovirid virions are small isometric particles of 17–19 nm in diameter, with a probable T = 1 symmetry and often displaying a hexagonal profile comprising DNA and a single species of capsid protein of about 19 kDa. No other proteins have been found associated with the nanovirid virions. Each of the 6 (genus *Babuvirus*) or 8 (genus *Nanovirus*) genomic DNAs is 0.9–1.1 kb and is separately encapsidated (Thomas et al. 2021). As reviewed recently by the ICTV report (Abbas et al. 2021), the virions of the *Redondoviridae* family members are still unknown.

CLASS *REPENSIVIRICETES*

ORDER *GEPLAFUVIRALES*

The sole order *Geplafuvirales* of the *Repensiviricetes* class involves two families, *Geminiviridae* and *Genomoviridae*, and 24 genera and 757 species altogether. According to the current ICTV reports (Brown et al. 2012; Zerbini et al. 2017), the *Geminiviridae* members are plant pathogens with DNA genomes of 2,500–5,200 bases and very interesting virions with a unique particle morphology. Thus, the geminid virions are twinned (geminate) incomplete icosahedra, T = 1, 22 × 38 nm in size, containing 110 identical coat protein molecules organized as 22 pentameric capsomers, at last in the case of maize streak virus (MSV) from the *Mastrevirus* genus.

Xu et al. (2019) used single-particle electron cryomicroscopy to determine the structure of tobacco curly shoot virus (TbCSV), genus *Begomovirus*, particle at 3.57 Å resolution; confirm the characteristic geminate architecture with single-strand DNA bound to each coat protein; and conclude that the genomic DNA plays an important role in forming a stable interface during assembly of the geminate particle. Bennett and Agbandje-McKenna (2020) summarized the present knowledge on the unique geminivirid structure that is presented in Figure 12.2.

The novel *Genomoviridae* family was presented by Krupovic et al. (2016) and recently updated from a single to 237 species by Varsani and Krupovic (2021). Remarkably, Sclerotinia sclerotiorum hypovirulence-associated DNA virus 1 (SsHADV-1), the first described genomovirus, was also the first ssDNA virus known to infect fungi. The family *Genomoviridae* includes viruses with small circular ssDNA genomes of ~1.8–2.4 kb encoding a rolling-circle replication initiation protein (Rep) and a capsid protein (CP) in an ambisense orientation. The SsHADV-1 virions are nonenveloped, isometric, 20–22 nm in diameter, and constructed from one CP (Yu et al. 2010).

CLASS *HUOLIMAVIRICETES*

The class *Huolimaviricetes* is the sole member of the *Saleviricota* phylum, which is a single member of the kingdom *Trapavirae*. The sole order *Haloruvirales* from the *Huolimaviricetes* class includes the sole family *Pleolipoviridae* with 3 genera and 15 species. As summarized by the recent ICTV report (Bamford et al. 2017), the archaeal pleolipoviruses are enveloped pseudo-spherical and pleomorphic membrane vesicles of 40–70 nm diameter with irregularly distributed spike structures, where one or

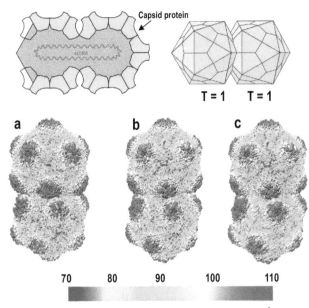

FIGURE 12.2 The structure of the family *Geminiviridae* members. (top) The cartoon taken with kind permission from the ViralZone, Swiss Institute of Bioinformatics, http://viralzone.expasy.org (Hulo C et al. 2011). Courtesy Philippe Le Mercier. (a-c) The surface representation of the geminivirus capsid based on the 110 mer capsid protein models for (a) African cassava mosaic virus (ACMV), (b) Ageratum yellow vein virus (AYVV), and (c) tobacco curly shoot virus (TbCSV). The capsids are radially colored according to the color key. The figures were generated with Chimera (Pettersen et al. 2004; Yang et al. 2012). (Reprinted from *Adv Virus Res.* 108, Bennett A, Agbandje-McKenna M, Geminivirus structure and assembly, 1–32, Copyright 2020, with permission from Elsevier.)

two types of major proteins form spikes and one or two act as internal membrane proteins. The spike and internal membrane proteins of Halorubrum pleomorphic virus 1 (HRPV1) are VP4 and VP3 respectively. Typically, virions contain a single type of transmembrane (spike) protein at the envelope and a single type of membrane protein, which is embedded in the envelope and located in the internal side of the membrane. The spike protein is anchored to the lipid membrane with a C-terminal transmembrane domain. The virions lack a capsid or nucleocapsid, while the membrane vesicle encloses different types of DNA genomes of approximately 7–16 kbp (or kilonucleotides). The ganomes can be represented by circular ssDNA of 7.0–10.7 kilonucleotides, circular dsDNA of 8.1–9.7 kbp, or linear dsDNA of 16 kbp (Bamford et al. 2017).

UNASSIGNED FAMILIES

FAMILY *ANELLOVIRIDAE*

By analogy with the *Circoviridae* family of the order *Cirlivirales* described in Chapter 10, the family *Anelloviridae* is comprised of animal viruses with circular ssDNA genomes, including well-studied chicken anemia virus (CAV) of the *Chicken anemia virus* species of the genus *Gyrovirus*. The latter was transposed recently from the *Circoviridae* family to the *Anelloviridae* family. Figure 10.1 presents the first 3D view of CAV in comparison with the true circovirus. Generally, according to the ICTV report (Biagini et al. 2012) and current ICTV taxonomy, the large *Anelloviridae* family is composed of 31 genera and 155 species altogether. The virions are nonenveloped, T = 1 icosahedra, with reported diameters of about 30 nm for torque teno viruses (TTVs, genus *Alphatorquevirus*) and torque teno mini viruses (TTMVs, genus *Betatorquevirus*). The virions contain a single molecule of circular negative-sense ssDNA, which ranges from about 2 to about 3.9 kb in size, where the ORF1 is believed to encode the putative capsid protein and replication-associated protein of human and animal anelloviruses (Biagini et al. 2012).

The urgent need for a vaccine against CAV and problems with the attenuated CAV stimulated the early VLP studies. Thus, Koch et al. (1995) cloned the CAV genes encoding the three putative proteins VPl, VP2, VP3 into a baculovirus vector and expressed in insect cells. Only lysates of insect cells that have synthesized equivalent amounts of all three proteins or cells that synthesized mainly VP1 plus VP2 induced neutralizing antibodies directed against CAV in inoculated chickens. The progeny of those chickens were protected against clinical disease after CAV challenge, while the inoculation of a mixture of lysates of cells that were separately infected with VPl-, VP2-, and VP3-recombinant baculovirus did not induce significant levels of neutralizing antibody directed against CAV, and their progeny were not protected against CAV challenge (Koch et al. 1995). Iwata et al. (1998) regarded the insect cells-produced VP2 and VP3 as putative antigens to detect anti-CAV antibodies in ELISA test. Noteborn et al. (1998) postulated the VP1 as the sole capsid protein of CAV and found that the recombinant baculovirus expressing VP1 and VP2 were a potential production system for a subunit vaccine against CAV infection. The problem of the putative VLPs and protein self-assembly was not described in these studies. Recently, Tseng et al. (2019) produced the CAV VP1 and VP2 in the *Sf*9 insect cells by the baculovirus expression system, along with chicken IL-12 as a putative biological adjuvant and demonstrated the ability of the CAV VP1 to generate the appropriate VLPs. The specific pathogen-free chickens inoculated with the VLPs and coadministered with the recombinant IL-12 produced high CAV-specific antibodies and cell-mediated immunity (Tseng et al. 2019).

Lacorte et al. (2007) synthesized the three previously described CAV VPs fused to GFP in *Nicotiana benthamiana* plants, when it was already established that the VP1 is the only structural protein; the VP2 is a protein phosphatase; and VP3, also known as apoptin, induces apoptosis. The results showed that all CAV proteins can be expressed in plant cells, though expression level of VP1 needed to be further optimized before testing its potential as an edible

subunit vaccine (Lacorte et al. 2007). The VLP problem was not raised in this study.

By the bacterial expression, Lee et al. (2011) significantly increased expression of the recombinant full-length VP1 gene of CAV by codon optimization and fusion with GST protein rather than a His-tag, followed by expression in a number of different *E. coli* strains. Lai et al. (2013) systematically assessed the CAV VPs expression in *E. coli*; separately cloned and expressed the three CAV proteins VP1, VP2 and VP3; and evaluated the putative suitability of the products for diagnostic purposes. In parallel, Moeini et al. (2011) displayed the CAV VP1 protein, in the form of the fusion with the AcmA binding domains of *Lactococcus lactis*, on *Lactobacillus acidophilus*, in order to produce an edible vaccine. By the bacterial expression, as in the case of the plant expression, the VLP problem was not touched in the discussions of these studies.

Concerning Torque teno virus (TTV) proteins, Mueller et al. (2008) expressed the ORF1, the ORF1 splace variants ORF1/1 and ORF1/2, ORF2, ORF2/2, ORF3, and ORF4 of human TTV, where ORF1 was known to encode capsid protein, in human hepatocellular carcinoma Huh7 cells. The expression of the ORF1 protein and its splice variant ORF1/1 in cell culture was detected by an ORF1-specific antiserum. Kakkola et al. (2008) expressed the six ORFs of TV genotype 6, in bacteria and insect cells. The expression of the ORF1/1-encoded protein was inefficient, while expression of the others was successful, with ORF1 and ORF1/2 as arginine-rich region depleted. Huang et al. (2011) expressed the ORF1 of porcine TTV species Torque teno sus virus 2 (TTSuV2) in *E. coli*, solubilized and purified the originally insoluble product, and used it for the development of Western blot and indirect ELISA to detect TTSuV2-specific IgG antibodies in pig sera. All studies did not touch the appearance of the putative VLPs.

OTHER UNASSIGNED FAMILIES

The family *Alphasatellitidae*, as reviewed by Briddon et al. (2018) and demonstrated by the current ICTV taxonomy, includes three subfamilies, two of which, *Geminialphasatellitinae* and *Nanoalphasatellitinae*, were established to respectively accommodate the geminivirus- and nanovirus-associated alphasatellites. The latter are circular ssDNA viruses that infect many plant species around the world but have no specific distinctive virions because the viral genomes are encapsidated within the coat protein of the helper virus.

The small family *Spiraviridae*, as stated by the recent ICTV report (Prangishvili et al. 2020), involves the sole *Alphaspiravirus* genus with the sole species *Aeropyrum coil-shaped virus* and includes viruses, such as Aeropyrum coil-shaped virus (ACV), that replicate in hyperthermophilic archaea from the genus *Aeropyrum*. As discovered by Mochizuki et al. (2012), the ACV virions demonstrate an exceptional virion architecture and possess the largest

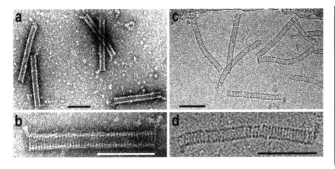

FIGURE 12.3   The electron micrographs of the Aeropyrum coil-shaped virus (ACV) virions. (a, b) Negatively stained with 2% (wt/vol) uranyl acetate. (c, d) Sample embedded in vitreous ice. Scale bars, 100 nm. (From Mochizuki et al. Archaeal virus with exceptional virion architecture and the largest single-stranded DNA genome. *Proc Natl Acad Sci U S A*. 109:13386–13391, Copyright 2012 National Academy of Sciences, U.S.A.)

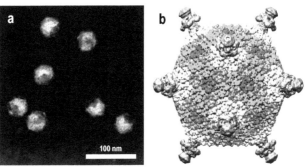

FIGURE 12.4   The structure of the Flavobacterium phage FLiP virion. (a) The virions are negatively stained with 2% phosphotungstic acid (pH 8.5) and visualized under transmission electron microscopy. (b) The electron cryomicroscopic reconstruction (Laanto et al. 2017). (Reprinted from Mäntynen S et al. ICTV virus taxonomy profile: *Finnlakeviridae*. *J Gen Virol*. 2020;101:894–895.)

single-stranded DNA genome of 24,893 nucleotides. The ACV genome is circular, positive-sense, ssDNA. The architectural solution used by ACV to package its circular genome is unprecedented among viruses of bacteria and eukaryotes. The nonenveloped, hollow, cylindrical virions are formed from a coiling fiber that consists of two intertwining halves of a single circular nucleoprotein filament, as shown in Figure 12.3. A short appendage of 20 ± 2 nm protrudes from each end at 45° angles to the axis of the cylindrical virion (Mochizuki et al. 2012). As commented by Prangishvili et al. (2017), the uncommon coil-like virion shape of members of the family *Spiraviridae* is determined by a special method of genome packing, when the circular nucleoprotein filament formed from ssDNA and capsid proteins is condensed into a rope-like structure that is further condensed into a higher-order helix. Generally, the spiraviruses, along with other archaeal viruses, may represent ancestral virus forms no longer

observed amongst extant prokaryotic or eukaryotic viruses (Prangishvili 2015).

The novel *Finnlakeviridae* family, as reviewed by the current ICTV report (Mäntynen et al. 2020), consists of the sole *Finnlakevirus* genus with the sole *Flavobacterium virus FLiP* species. The family members are icosahedral, internal membrane-containing bacterial viruses with circular ssDNA of 9,174 nucleotides and were therefore the first described ssDNA virus with an internal membrane. The virions are 59 nm in diameter. Laanto et al. (2017) achieved the electron cryomicroscopy structure of the virion at 4 Å resolution, as presented in Figure 12.4. The virion organization (pseudo T = 21 *dextro*) and major capsid protein fold (double-β-barrel) resembled those of *Pseudoalteromonas* phage PM2 of the family *Corticoviridae*, which has a dsDNA genome. A similar major capsid protein fold was also found in other dsDNA viruses in the kingdom *Bamfordvirae* (Mäntynen et al. 2020).

Section III

Double-Stranded RNA Viruses

13 Order *Reovirales*

The secret of success is constancy to purpose.

Benjamin Disraeli

ESSENTIALS

The members of the order *Reovirales* are the most "nano-technological" group of the double-stranded RNA viruses. Many order representatives played a crucial role by the development of the VLP techniques. According to the recent detailed taxonomy (ICTV 2020), the order *Reovirales* remains the only order of the class *Resentoviricetes* of the phylum *Duplornaviricota*, kingdom *Orthornavirae*, realm *Riboviria*. The order *Reovirales* includes a sole family *Reoviridae*, consisting of two subfamilies, namely *Sedoreovirinae* and *Spinareovirinae*; 15 genera; and 97 species in total. The subfamily *Spinareovirinae* contains viruses that have relatively large spikes or turrets situated at the 12 icosahedral vertices of either the virus or core particle. The subfamily *Sedoreovirinae* includes viruses that do not have large surface projections on their virions or core particles. After the unique role in the VLP technologies, the order is of great medicinal and veterinarian importance. There are such well-known pathogens as members of the species *African horse sickness virus* and *Bluetongue virus* from the *Orbivirus* genus; members of the species *Rice dwarf virus* of the *Phytoreovirus* genus, and nine rotavirus variants A to I from the *Rotavirus* genus, all from the subfamily *Sedoreovirinae*. Then, there are insects-infecting cypoviruses, short for cytoplasmic polyhedrosis virus, members of the *Cypovirus* genus from the *Spinareovirinae* subfamily. In many aspects, but not in the genome origin, cypoviruses are similar to double-stranded DNA nucleopolyhedroviruses from the *Baculoviridae* family described in Chapter 6.

Figure 13.1 presents portraits of the three most studied genera from both subfamilies. Generally, the protein capsid is organized as one, two, or three concentric layers of capsid proteins, which surround the linear dsRNA segments of the viral genome, with an overall diameter of 60–80 nm. The outer capsid of reoviruses has a T = 13*laevo* (T = 13*l*) icosahedral symmetry, the inner capsid is of a T = 2 (or designated as T = 2*) or T = 1 of 60 dimers, icosahedral symmetry (Hulo et al. 2011; Attoui et al. 2012).

In fact, the transcriptionally active core particle of the spiked viruses of the *Spinareovirinae* subfamily appears to contain only a single complete capsid layer, which has been interpreted as having T = 1 or T = 2 symmetry, to which the projecting spikes or turrets are attached. In most cases, the core is surrounded in the complete virion by an incomplete protein layer with T = 13*l* symmetry that forms the outer capsid, which is penetrated by the projections on the core

surface. These virus particles are therefore usually regarded as double-shelled.

In contrast, the virions of the nonspiked viruses of the *Sedoreovirinae* subfamily have an inner protein layer, which may be relatively fragile, having structural similarities to the innermost shell of the spiked viruses, interpreted as having T = 2* symmetry. However, in transcriptionally active core particles, the subcore is surrounded and reinforced by a complete core-surface layer, which has T = 13*l* symmetry. These double-layered cores have no surface spikes and are surrounded by a further outer capsid shell in intact virions, giving rise to three-layered virus particles that are equivalent to the two-layered particles of members of the *Spinareovirinae* subfamily, as summarized thoroughly by Attoui et al. (2012).

Globally, the members of the *Reovirales* order played an outstanding role by the development of the VLP ideology, thanks, first, to the pioneering works on the orbivirus VLPs performed by the well-known Polly Roy team. Then, the cheap and safe vaccines against rotaviruses, a leading cause of severe infantile gastroenteritis worldwide, appeared as being of prime medicinal importance, stimulating therefore the appropriate VLP studies.

GENOME

Figure 13.2 shows the segmented double-stranded RNA genomes of the representative members of the *Reovirales* order. Thus, the reovirus particles can contain 9, 10, 11, or 12 segments of linear dsRNA, depending on the genus, within the innermost protein layer of internal diameter of approximately 50–60 nm. The individual molecular mass of these RNA molecules ranges from 0.2 to 3.0×10^6, while the total molecular mass of the genome is 12–20×10^6, as reviewed by Attoui et al. (2012).

The viral RNA species are mostly monocistronic, although some segments have second functional, in-frame initiation codons or additional protein-coding ORFs. Proteins are encoded on one strand only of each duplex. Many cypoviruses also form polyhedra, which are large crystalline protein matrices that occlude virus particles and which appear to be involved in transmission between individual insect hosts.

ORBIVIRUSES

BLUETONGUE VIRUS

Expression

The pioneering advances in orbiviruses as the worldwide-known VLP carriers have been made due to the systematic

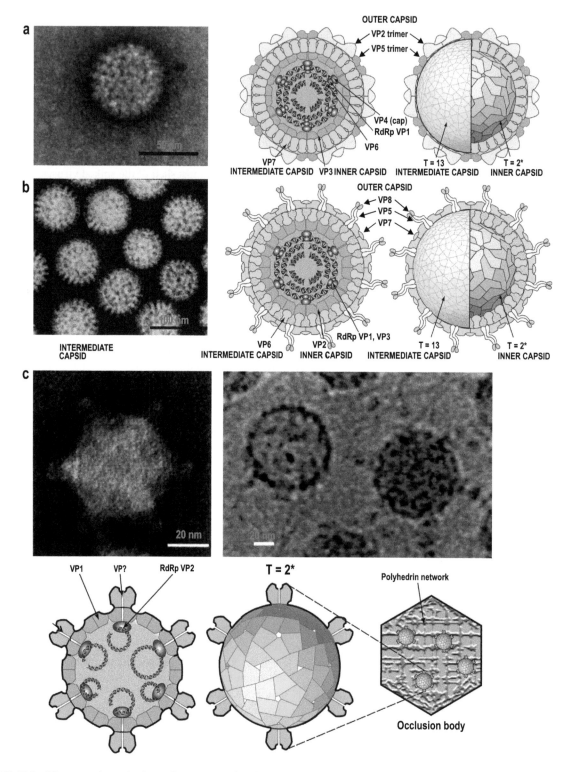

FIGURE 13.1 The portraits and schematic cartoons of typical representatives of the order *Reovirales*. (a) bluetongue virus (BTV). The negatively stained micrograph of BTV (Reprinted from Toussaint JF et al. Bluetongue in Belgium, 2006. *Emerg Infect Dis.* 2007;13:614–616.) (b) a rotavirus, courtesy of BVV Prasad; (c) a cypovirus, negative contrast electron micrograph of a nonoccluded virion of Orgyia pseudosugata cypovirus 5. (Courtesy of CL Hill.) The electron micrographs are taken from the report of the International Committee on Taxonomy of Viruses (ICTV) on reoviruses, https://talk.ictvonline.org/ictv-reports/ictv_9th_report/dsrna-viruses-2011/w/dsrna_viruses/190/reoviridae-figures. The cartoons of the *Reovirales* representatives are taken with kind permission from the ViralZone, Swiss Institute of Bioinformatics, http://viralzone.expasy.org (Hulo C et al. 2011). (Courtesy Philippe Le Mercier.)

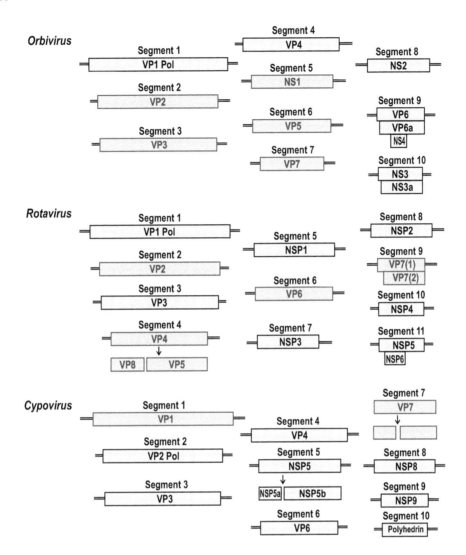

FIGURE 13.2   The genomic structure of the most representative genera of the order *Reovirales*, derived basically from the ViralZone, Swiss Institute of Bioinformatics, http://viralzone.expasy.org (Hulo et al. 2011). The structural genes are colored dark pink.

studies of Polly Roy and her group on such models as blue-tongue (BTV) and African horse sickness (AHSV) viruses, as reviewed originally by Roy (1996) and Roy et al. (1997) and later by Roy and Noad (2009) and Patel and Roy (2015). The main prototype orbivirus, BTV, is the etiological agent of a ruminant disease that can reach epidemic proportions among sheep and cattle. The double shell of the BTV is built up by four major structural proteins, where VP3 and VP7 form the inner shell, while VP2 and VP5 form the outer shell, as shown in Figure 13.1. Besides its unique property of being self-assembled in insect cells into the single- and double-shelled potentially chimeric VLPs, BTV offered another particulate candidate, a nonstructural NS1 protein, which generated tubules by expression in mammalian and insect cells (Urakawa and Roy 1988).

Table 13.1 summarizes the gained data on the formation of the VLPs of the order *Reovirales* in the used expression systems.

Since each BTV protein is encoded by a single RNA species, there was no alternative to constructing DNA clones of all ten RNA species and expressing them in baculovirus vectors at high levels. Then, the novel vectors were developed, which allowed coexpression of three, four, or five BTV genes from single recombinant vectors, and coexpressed VP3 and VP7 were shown to form BTV core-like particles (CLPs), while coexpressed VP2, VP5, VP7, and VP3 formed the true BTV VLPs (Roy et al. 1997). Historically, this work was started with the expression of the VP3 and VP7 and led to formation of large amounts of the BTV CLPs—similar to authentic BTV cores in terms of size, stoichiometric arrangement of VP3 to VP7 with a ratio of 2:15, and the predominance of VP7 on the surface of the particles—in baculovirus-driven insect cells (French and Roy 1990). These CLPs were observed as paracrystalline arrays in the infected insect cells. The next logical step consisted in assembling the true triple-layered double-shelled BTV VLPs that had the same size and appearance as authentic BTV virions, by the simultaneous expression of four structural proteins and with the addition of the outer capsid proteins VP2 and VP5 (French et al. 1990).

TABLE 13.1

Formation of CLPs and VLPs by the Combination of the Structural Proteins Derived from the *Reovirales* Order Members

Virus, Species	Genes	Expression System	References	Comments
Subfamily: *Sedoreovirinae* Genus: *Orbivirus*				
African horse sickness virus	VP3; VP7 (VP3/7)	*Spodoptera frugiperda Sf*9 insect cells	Maree S et al. 1998a, b	Low yields of CLPs.
	VP3; VP7; VP2; VP5	*Spodoptera frugiperda Sf*9 insect cells	Maree S et al. 2016	Efficient synthesis of CLPs and VLPs by different protein combinations.
	VP3; VP7; VP2; VP5 (VP2/3/5/7)	Plants *Nicotiana benthamiana*	Dennis et al. 2018a, b	The first production of the AHSV VLPs in plants.
Bluetongue virus	VP3; VP7 (VP3/7)	Rabbit reticulocyte translation system in vitro	Liu et al. 1992	The outer capsid proteins bound to the preformed CLPs.
	VP3; VP7 (VP3/7)	*Spodoptera frugiperda Sf*9 insect cells	French and Roy 1990	The BTV CLPs.
	VP3; VP7; VP2; VP5 (VP2/3/5/7)	*Spodoptera frugiperda Sf*9 insect cells	French and Roy 1990	The BTV VLPs.
	VP3; VP7; VP2; VP5 (VP2/3/5/7)	Plants *Nicotiana benthamiana*	Thuenemann et al. 2013a, b; van Zyl et al. 2015	The first plant-produced BTV VLPs.
	NS1	*Spodoptera frugiperda Sf*9 insect cells	Urakawa and Roy 1988	The tubules of 68 nm in diameter and mean lengths of about 800 nm.
Broadhaven virus	VP2; VP7 (VP2/7)	*Spodoptera frugiperda Sf*9 insect cells	Moss and Nuttall 1994	The BRDV CLPs.
Epizootic hemorrhagic disease virus	NS1	*Spodoptera frugiperda Sf*9 insect cells	Nel and Huismans 1991	The tubules of 50 nm in diameter.
Subfamily: *Sedoreovirinae* genus: *Phytoreovirus*				
Rice dwarf virus	P3	*Spodoptera frugiperda Sf*9 insect cells	Hagiwara et al. 2003	The RDV CLPs.
	P3; P7; P8	*Spodoptera frugiperda Sf*9 insect cells	Hagiwara et al. 2003	The RDV VLPs.
	P3; P8	Transgenic rice plants	Zheng et al. 2000	The RDV VLPs.
Subfamily: *Sedoreovirinae* Genus: *Rotavirus*				
Bovine rotavirus (*Rotavirus A*)	VP2	*Spodoptera frugiperda Sf*9 insect cells	Labbé et al. 1991	Single-layered empty CLPs of 52 nm in diameter.
	VP4; VP6; VP7 (VP4/6/7)	*Spodoptera frugiperda Sf*9 insect cells	Redmond et al. 1993	Double-shelled VLPs.
	VP2	High-Five cells	Shoja et al. 2013	
	VP2; VP6$_{NG}$ (VP2/6$_{NG}$)	Transgenic milk of rabbits	Soler et al. 2005	The first transgenic mammal bioreactors.
	VP2Δ; VP6; VP7 (VP2Δ/6/7)	*Saccharomyces cerevisiae*	Rodríguez-Limas et al. 2011	The use of yeast as a platform for the first time.
Bovine rotavirus (VP2) (*Rotavirus A*) and porcine rotavirus, Cowden strain (VP6) (*Rotavirus C*)	VP2; VP6 (VP2/6)	*Spodoptera frugiperda Sf*9 insect cells	Tosser et al. 1992	Double-layered single-shelled CLPs.

Virus, Species	Genes	Expression System	References	Comments
Human rotavirus (*Rotavirus A*)	VP2; VP6 (VP2/6)	*Drosophila melanogaster* Schneider 2 (S2) cells	Lee et al. 2011	Both proteins were provided with a His tag.
	VP2; VP6 (VP2/6)	African green monkey kidney cells CV-1	González SA and Affranchino 1995	
	VP6	*Escherichia coli*	Zhao et al. 2011	The appearance of CLPs after in vitro renaturation.
Murine rotavirus (*Rotavirus A*)	VP6	plants *Nicotiana benthamiana*	O'Brien et al. 2000	Paracrystalline sheets and tubes.
	VP6	plants *Solanum tuberosum*	Yu and Langridge 2003	Ability to form trimers.
	VP2; VP6 (VP2/6)	plants *Lycopersicon esculentum*	Saldaña et al. 2006	Only a small proportion of the VP2/VP6 assembled into CLPs.
	VP2; VP6; VP7 (VP2/6/7)	plants *Nicotiana tabacum*	Yang et al. 2011	The VP2/6/7 VLPs were produced and used for immunization of mice.
Simian rotavirus (*Rotavirus A*)	VP2; VP6 (VP2/6)	Baby hamster kidney cells BHK-21	Nilsson et al. 1998	No evidence of intracellular assembly was found.
	VP2; VP6; VP7 (VP2/6/7)	African green monkey kidney epithelium cells Vero	Laimbacher et al. 2012	The efficient HSV-1 vector-driven production of the VLPs.
	VP2; VP6; VP4; VP7 (VP2/4/6/7)	*Spodoptera frugiperda Sf9* insect cells	Crawford et al. 1994	Triple-layered double-shelled VLPs.
	VP2; VP6 (VP2/6)	*Spodoptera frugiperda* larvae	Molinari et al. 2008	Cost-reducing alternative for vaccine manufacturing.
	VP2; VP6 (VP2/6)	*Escherichia coli*	Li et al. 2014	The highly efficient process of the renaturation.

The minor proteins, VP1, a component of the viral RNA-directed RNA polymerase, appeared as specifically incorporated into the CLPs and VLPs when expressed together with the structural proteins, while VP4, a guanylyl transferase, and VP6 did not appear in the recombinant BTV particles (Loudon and Roy 1991; Le Blois et al. 1992).

In order to facilitate the insertion of three or four BTV genes, multiple gene transfer vectors were developed for the simultaneous expression of the appropriate genes (Belyaev and Roy 1993; Roy et al. 1997). The VLPs were formed also by combination of six VP2 from different BTV serotypes (Loudon et al. 1991).

Later, the baculovirus-driven expression of the BTV genes was improved markedly to produce milligram quantities of the BTV VLPs. Thus, a method was generated, which combined the lambda red and bacteriophage P1 Cre-recombinase systems to efficiently generate baculoviruses, where protein complexes were expressed from multiple, single-locus insertions of foreign genes (Noad et al. 2009). The new baculovirus expression strategy allowed preintegration of the genes encoding the BTV inner capsid proteins at one baculovirus locus and those encoding the outer capsid proteins at a different locus (Stewart et al. 2010). The utility of this approach was demonstrated by the production of BTV VLPs to a number of serotypes.

Concerning expression systems other than insect cells, the BTV CLPs were formed successfully in the in vitro rabbit reticulocyte translation system, and each outer capsid protein synthesized in vitro had the capacity to bind to a preformed CLP (Liu et al. 1992).

The correct BTV VLPs were produced and assembled in *Nicotiana benthamiana* after coexpression of VP2, VP3, VP5, and VP7, using the cowpea mosaic virus-based *HyperTrans* (CPMV-HT) and associated pEAQ plant transient expression vector system (Thuenemann et al. 2013a, b). Van Zyl et al. (2015) compared different vectors, which targeted recombinant proteins to the cytosol, apoplast, or chloroplast, in order to improve the efficiency of the plant-driven production of the BTV VLPs. The in situ localization technique using transmission electron microscopy allowed detection of a mixed population of CLPs—consisting of VP3 and VP7—and full VLPs that were observed as paracrystalline arrays in the cytoplasm of plant cells coexpressing all four capsid proteins.

3D Structure

The 3D structure of the single-shelled BTV virions was determined first to a resolution of 3 nm by using electron cryomicroscopy (Prasad et al. 1992). Such virion cores had T = 13 icosahedral symmetry in a left-handed configuration

T = 13*l*, in which 260 capsomeres in the outer layer were proposed to be made up of trimers of the VP7, whereas the inner layer seemed to be composed of the VP3 (Prasad et al. 1992). The outer shell of the whole BTV particle was shown to consist of 120 globular regions of VP5, located on each of the six-membered rings of VP7 trimers and of the sail-shaped spikes of the VP2 (Hewat et al. 1992b).

The electron cryomicroscopic examination of the BTV VLPs, generated by coexpression of the VP2, VP3, VP5, and VP7 in insect cells, revealed the same T = 13*l* organization and confirmed essentially the same 3D features as for the native virions (Hewat et al. 1992a, 1994). The lack of the five VP7 trimers in the recombinant CLPs (Hewat et al. 1992a) was explained by the necessity of the presence of the VP2 and VP5 for adhesion of these VP7 trimers around the 5-fold axes (Hewar et al. 1994).

In parallel with the electron cryomicroscopy, the x-ray crystallographic studies were carried on the BTV VP7 and CLPs (Basak et al. 1992; Burroughs et al. 1995; Grimes et al. 1995). The combination of the electron cryomicroscopy and x-ray crystallography data led to an atomic resolution map of the core, which consisted of 780 VP7 molecules organized into the 260 trimers on a T = 3*l* icosahedral lattice (Basak et al. 1997; Grimes et al. 1997). The map showed that the β-barrel domains of VP7 were external to the core and interacted with protein in the outer layer of the mature virion, whereas the lower, α-helical domains of VP7 interacted with the VP3 molecules, which formed the inner layer of the BTV core (Basak et al. 1997). The inner layer of the BTV cores demonstrated T = 2* organization and contained 120 copies of large, about 100 kDa, VP3 molecules arranged as 60 approximate dimers (Stuart et al. 1998).

Finally, the atomic structure of the BTV CLPs isolated from the virus-infected mammalian BHK-21 cells was determined by x-ray crystallography at a resolution approaching 3.5 Å (Grimes et al. 1998). This original 3D structure is shown in Figure 13.3.

Concerning the size of particles, the single-shelled BTV virions demonstrated diameter of 69 nm (Prasad et al. 1992). The insect cells-produced CLPs and VLPs had a diameter of 72.5 nm (Hewat et al. 1992a) and 86 nm (Hewat et al. 1994), respectively. Grimes et al. (1998) estimated diameter of the crystallography-resolved BTV CLPs as 70 nm.

At last, Zhang et al. (2016) recorded the electron cryomicroscopy images of intact BTV virions with a direct electron detector operated at super-resolution counting mode and obtained a 3.5 Å-resolution structure of the virion by single-particle analysis. Figure 13.4 demonstrates the obtained structure. Thus, the BTV virion contained an outer layer of 60 VP2 trimers and 120 VP5 trimers; a middle layer with 260 VP7 trimers; and an inner layer formed by 120 VP3 monomers. Each VP2 trimer bound atop four VP7 trimers. Situated at a six-coordinated position of the icosahedral lattice, each VP5 trimer bridged the channel formed by six surrounding VP7 trimers (Zhang et al. 2016). Remarkably, the structural features of the middle and inner layers, including aa side chains, matched the atomic

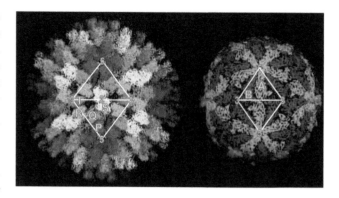

FIGURE 13.3 The molecular architecture of the BTV core. (a) The icosahedral asymmetric unit (which is a triangular area defined as marked by the symmetry axes of the icosahedron) contains 13 copies of VP7 (T = 13) arranged as 5 trimers, P, Q, R, S, and T, colored red, orange, green, yellow, and blue, respectively. Trimer T sits on the icosahedral 3-fold axis and thus contributes a monomer to the unique portion. The inner layer of VP3 (T = 2) molecules is colored red and green. Helices are shown as rods and β-strands as arrows. (b) The VP3 (T = 2) scaffold. The icosahedrally unique molecules A and B are colored in green and red, respectively. Note the completely different structural environment of the A and B molecules. (With kind permission from Springer Science+Business Media: *Nature*, The atomic structure of the bluetongue virus core, 395, 1998, 470–478, Grimes JM, Burroughs JN, Gouet P, Diprose JM, Malby R, Ziéntara S, Mertens PP, Stuart DI.)

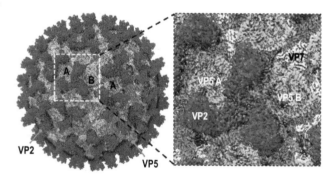

FIGURE 13.4 The electron microscopy reconstruction of the BTV virion. The electron microscopy density map is shown as radially colored surface representation, with a close-up view of the boxed area containing an asymmetric unit (With kind permission from Springer Science+Business Media: *Nat Struct Mol Biol*, Atomic model of a nonenveloped virus reveals pH sensors for a coordinated process of cell entry, 23, 2016, 74–80, Zhang X, Patel A, Celma CC, Yu X, Roy P, Zhou ZH.)

structures of core proteins VP3 and VP7 solved previously by x-ray crystallography (Grimes et al. 1998) and presented in Figure 13.3.

Vaccine

Although the BTV CLPs and VLPs did not contain viral or any other RNA and/or DNA required for replication or adjuvanting, unlike live virus vaccines, they were more immunogenic than killed viruses—or subunit vaccines—and were highly efficient in eliciting humoral, cell-mediated, and mucosal immunities, as well as long-lasting protective response against challenges of animals with infectious viruses (Roy et al. 1992, 1994, 1997; Pearson and Roy 1993; Roy 1996).

Thus, protection of sheep by low doses of VLPs consisting of VP2, VP3, VP5, and VP7 against BTV challenge was documented by Roy et al. (1992). Roy et al. (1994) demonstrated long-lasting and partial protection of VLPs against challenge with homologous or heterologous BTV serotype, respectively. Furthermore, the protection of sheep by the insect cells-produced VLPs of different serotypes was demonstrated (Stewart et al. 2010, 2012, 2013; Pérez de Diego et al. 2011). Thus, Stewart et al. (2010) generated the BTV VLPs to a number of serotypes and the VLPs demonstrated to be safe and highly effective immunogens in sheep, reducing postchallenge viraemia to levels below the threshold detection limit of quantitative RT-PCR when vaccinated animals were challenged with virulent virus.

The BTV VLPs consisting of the VP2, VP3, VP5, and VP7 proteins and produced in *N. benthamiana* plants (Thuenemann et al. 2013a, b) were shown to elicit a strong antibody response in sheep. Furthermore, they provided protective immunity against a challenge with a South African BTV-8 field isolate. These results had real implications beyond the realm of veterinary vaccines and appeared as applicable to the production of VLPs for human use in plants (Thuenemann et al. 2013a, b).

However, the trials were not always fully successive. Thus, no protective effect of the insect cells-produced VLPs consisting of the VP2, VP3, VP5, and VP7 proteins of the BTV-2 strain was demonstrated against challenge with Italian field BTV-2 isolate in cows, sheep, and goats (Savini et al. 2007). Stewart et al. (2012) found failure of full protection by the BTV CLPs but only a mitigation of the severity of disease manifestation and viraemia.

Chimeric VLPs

The BTV CLPs and VLPs demonstrated a remarkable ability to exchange proteins and produce mixed particles containing homologous subunits of other BTV subtypes (Loudon et al. 1991) and other orbiviruses such as epizootic hemorrhagic disease virus (EHDV; Le Blois et al. 1991), but not Broadhaven virus (BRDV; Moss and Nuttall 1994).

The adaptation of the BTV VLPs to the role of a molecular carrier candidate started with the deletion, point, and domain switching analyses of regions of the core proteins VP3 and VP7, which were responsible for intermolecular contacts and self-assembly (Tanaka and Roy 1994; Roy et al. 1997). The first chimeric BTV CLPs were constructed by the addition of the HBV preS2 fragment 1–48 to the N terminus of the VP7 (Belyaev and Roy 1992). The chimeric VP7-preS2 protein participated in the BTV CLPs only when the insect cells were also coinfected with a recombinant baculovirus that expressed unmodified VP7 and VP3 of BTV and mosaic particles were formed (Belyaev and

Roy 1992). No CLPs were formed therefore by the preS2-VP7 and VP3 alone. The immunoelectron microscopy of the chimeric particles indicated that the preS2 epitope was exposed on the surface of the CLPs (Belyaev and Roy 1992).

The extension of the 11-aa rabies virus sequence added to the N-terminus of VP7 was also found to allow for the CLP formation, and several mutations critical for self-assembly of VP7 were mapped (Le Blois and Roy 1993). The major immunological VP7 epitopes mapped at the N terminus of VP7 appeared as the most exposed on the BTV surface (Wang et al. 1996).

Then, a series of four potent insertion vectors was constructed on the basis of the VP3 gene (Tanaka et al. 1995) in accordance with the deletion mapping data (Le Blois et al. 1991). The insertion of 12 N-terminal aa residues of the phage T7 capsid protein into the internal VP3 regions or the addition of 13 aa of the binding domain of the protein to cellular receptors from bovine leukemia virus (BLV) glycoprotein gp51, which corresponded to aa 155–167 of the BLV gp51, to the C-terminus of VP3 did not prevent the formation of the BTV CLPs, if the VP7 was provided (Tanaka et al. 1995).

Then, a 15-aa-long C-terminal CD4[+] T-cell epitope of the influenza virus matrix protein M1 was added to the VP7, found to be included into CLPs in the presence of VP3 and able to induce an insertion-specific proliferative response (Monastyrskaya et al. 1997).

A C-terminal 15-aa T-helper cell epitope of matrix protein from influenza virus was inserted at aa 145 within the VP7 region exposed on the surface of VLPs (Adler et al. 1998).

Tan et al. (2001) introduced point mutations into the Arg-Gly-Asp (RGD) tripeptide within the VP7 protein located at aa 168–170 and generated three recombinant baculoviruses, each expressing a mutant derivative of VP7 (VP7-AGD, VP7-ADL, and VP7-AGQ). Each mutant VP7 protein was used to generate empty CLPs that were tested in an in vitro arthropod *Culicoides* cell binding assay. The mutants showed reduced binding to cells compared to wild-type CLPs and indicated the biological role of the RGD motif present on the BTV VP7 protein in the binding activity to the cells of *Culicoides*, which serves as a natural transmitter of BTV, an arthropod-borne virus.

Du et al. (2014) constructed a chimeric BTV virion by manipulation with the segmented BTV genome. The authors introduced biarsenical-tetracycteine (TC) tag, namely aa CCPGCC motif, at positions 94–95, 352–353, and 420–421 of the BTV VP2 protein followed by labeling with biarsenical dyes. Such virions were fluorescently visualized in infected cells. The positions were selected due to the VP2 structure that was resolved previously to 7 Å and appeared as nondisrupting for the 3D structure of the protein (Zhang et al. 2010). Remarkably, this was the first report on the successful tagging of a structural protein for any nonenveloped virus. The study resulted in clarification of the BTV entry pathway in mammalian cells and showed, for the first time, that the two outer capsid proteins are separated from each other during the early stages of virus entry (Du et al. 2014).

At last, the famous George P. Lomonossoff's team engineered the chimeric BTV CLPs carrying the EDIII ectodomain of flaviviruses (Ponndorf et al. 2020). Thus, the EDIII of Dengue viruses DENV1, DENV4, and Zika virus (ZIKV) was displayed on the inner surface of BTV CLPs by N-terminal fusion of the antigen to VP3 and the appropriate products were produced in plants *N. benthamiana*. The integration of the EDIII did not prevent self-assembly of the chimeric CLPs. However, the immunogenicity assays in mice revealed that the BTV-based CLPs induced lower antibody response than the DENV1 VLPs produced in parallel that are described in Chapter 22.

NS1 Tubules as a Carrier

The orbiviruses produced large numbers of tubular structures of unknown function in infected cells (Murphy et al. 1971). The BTV tubules were formed as a polymerization product of a 64 kDa nonstructural protein, NS1 (Urakawa and Roy 1988). The identical tubules were synthesized by baculovirus expression system, and they were expressed in excess of 300 mg/l of insect cells (Urakawa and Roy 1988). The size of the tubules was estimated as 68 nm in diameter and mean lengths of about 800 nm and occasionally as long as 4 microns, by baculovirus-driven expression in insect cells (Marshall et al. 1990). Hewat et al. (1992c) estimated the NS1 size on average to be 52.3 nm in diameter and up to 1,000 nm in length. Therefore, the NS1 protein was proposed as a promising carrier of foreign epitopes.

First, a series of deletion and site-directed mutants of the NS1 protein was generated, and the regions critical for its self-assembly, including both the N and C termini, were mapped (Monastyrskaya et al. 1994). By using a panel of monoclonal antibodies, it was established that the NS1 antigenic site was located near the C-terminus of the protein, which was exposed on the surface of the tubules (Monastyrskaya et al. 1995; Mikhailov et al. 1996).

Next, the epitopes corresponding to aa 131–140, 155–167, and 177–192 from envelope glycoprotein gp51 of BLV were added C-terminally, and the particles were formed by baculovirus-driven expression in insect cells (Monastyrskaya et al. 1995). Moreover, the addition of foreign sequences ranging from 44 to 116 aa in length and representing a 44-aa-long sequence from *Clostridium difficile* toxin A, the aa 1–48 sequence of the hepatitis B virus preS2 region, or of the complete 109-aa matrix protein p15 of BLV to the NS1 C-terminus resulted in all cases in the formation of tubules—or mixed tubules—when insect cells were coinfected with the three appropriate baculoviruses (Mikhailov et al. 1996). The chimeric tubules exposed foreign epitopes on their surface, demonstrated high immunogenicity, and therefore served as an illustration of the successful presentation of multiple epitopes on the NS1 (Mikhailov et al. 1996).

The successful exposition of foreign T-cell epitopes on the NS1 tubules by the C-terminal insertions followed. Thus, Ghosh et al. (2002a) added C-terminally a T-helper cell epitope aa 135–144 from VP1 of foot-and-mouth-disease virus (FMDV) and demonstrated protection of mice against FMDV infection. In parallel, when the other T-helper cell epitope, a peptide of influenza A virus, aa 186–205 of hemagglutinin (HA), was added, the chimeric NS1-HA protein was effectively assembled into tubules, easily purified, and capable of eliciting strong immune responses in mice (Ghosh et al. 2002a).

Ghosh et al. (2002b) inserted a CTL epitope of aa 118–132 from the nucleoprotein of lymphocytic choriomeningitis virus (LCMV), which was determined earlier by the famous Rolf Zinkernagel's team (Aichele et al. 1990). The protection of mice against the lethal LCMV challenge was established (Ghosh et al. 2002b).

Ghosh et al. (2002c) tested induction of HLA-A2-restricted CTL responses in HLA-A2 transgenic mice by the NS1 tubules carrying human melanoma epitopes, namely NS1-Mela1 and NS1-Mela2. The strong CTL responses specific for GnT-V/NA 17-A and gp100 (154–162) epitopes were generated in HLA-A2 transgenic mice immunized by the construct NS1-Mela1 carrying these two epitopes. The second construct NS1-Mela 2 carrying both Tyrosinase (369–377Da) and Melan-A/Mart-1(27–35) epitopes induced a weak Tyrosinase-specific CTL response in mice but failed to induce specific CTL responses against the Melan-A/Mart-1(27–35) epitope in the tested mice. As a result, the authors suggested that the NS1-derived tubular structures carrying multiple tumoral epitopes may lead to new strategies for the induction of strong tumor-specific CTL responses in cancer patients (Ghosh et al. 2002c).

In parallel, Ghosh et al. (2002b) fused NS1 with full-length green fluorescent protein (GFP) and showed that the chimeric protein retained the capacity to form tubules when expressed in insect cells. Moreover, the purified chimeric tubules were demonstrated to be internalized by macrophages and dendritic cells in mice.

Larke et al. (2005) constructed a T-cell vaccine NS1-HIVA, which delivered the HIV-1 clade A consensus-derived immunogen HIVA as large as 527 aa, consisting of the Gag p24/p17 sequences and a string of CTL epitopes, on the NS1 surface, without losing the self-assembly capability. When injected into BALB/c mice by several routes, the chimeric NS1-HIVA tubules induced HIV-1-specific MHC class I-restricted T cells.

At last, the NS1 tubules were used to construct a vaccine candidate by chemical conjugation and not by genetic fusion as before. Thus, Andrade et al. (2013) generated a vaccine against obesity by coupling of a hormone ghrelin to the insect cells-produced NS1 tubules, using a heterobifunctional cross linker EDC. In fact, the immunized mice demonstrated increasing titers of anti-ghrelin antibodies, while their cumulative food intake significantly decreased and energy expenditure was significantly enhanced, although there were no significant changes in body weight (Andrade et al. 2013).

Concerning the fine structure of the NS1 tubules, the electron cryomicroscopy revealed an average diameter of 52.3 nm and length of up to 100 nm (Hewat et al. 1992c), as mentioned earlier. The structure of their helical surface lattice was determined using computer image processing to a resolution of 40 Å. The NS1 protein was about 5.3 nm in diameter and formed a dimer-like structure, so that the tubules are composed of helically coiled ribbons of NS1 "dimers," with 21 or 22 dimers per turn. The surface lattice displayed P2 symmetry and formed a one-start helix with a pitch of 9.1 nm (Hewat et al. 1992c).

Packaging and Delivery

The BTV-derived VLPs accommodated the entire GFP genetically fused to the N-terminus of VP3 (Kar et al. 2005; Brillault et al. 2017; Thuenemann and Lomonossoff 2018). The GFP was displayed within the internal cavity of the particles and thus allowed the formation of the CLPs when coexpressed with VP7 and of the complete VLPs when coexpressed with VP7, VP5, and VP2. Moreover, Thuenemann et al. (2021) engineered constructs in which the N-terminus of the inner capsid protein VP3 was fused to the herpes simplex virus 1 thymidine kinase (HSV1-TK), the most common enzyme used in prodrug conversion therapy. The HSV1-TK is typically delivered as a gene, but in the context of the BTV VLPs it was delivered as a protein. Thus, the smaller CLPs and complete VLPs were produced, with HSV1-TK fused to the VP3. The highly efficient plant expression system allowed to get both TK-CLPs and TK-VLPs in large quantities. The TK-VLPs killed human glioblastoma cells efficiently in the presence of ganciclovir in mouse xenograft tumor models. Both BTV CLPs and VLPs containing TK were investigated for their structure and in vitro toxicity (Thuenemann et al. 2021). Therefore, the BTV VLPs paved the way to be used as nanoreactors for enzyme delivery and cancer therapy.

AFRICAN HORSE SICKNESS VIRUS

By analogy with BTV, the baculovirus-driven synthesis of structural proteins in insect cells with following formation of the appropriate particulate structures was achieved also for AHSV (Chuma et al. 1992; Maree S et al. 1998a, b). Thus, the AHSV CLPs that structurally resembled empty AHSV cores were formed after coexpression of the AHSV VP3 and VP7 proteins, but the yield of the CLPs was low (Maree S et al. 1998a, b). This fact was explained by the extremely high hydrophobicity and low solubility of the AHSV VP7 protein (Basak et al. 1996).

The artificial "chimeras" of BTV and AHSV-4 VP7 proteins were constructed on the basis of the crystal structure of the latter, but they failed to assemble into the CLPs, although they were able to form trimers (Monastyrskaya et al. 1997).

In the absence of the reliable AHSV CLPs at that time, the soluble chimeric trimers of the major capsid protein VP7 of AHSV were used as a putative vaccine delivery system by targeting some of the natural hydrophilic loops on the VP7 top domain for the insertion of foreign peptides (Rutkowska et al. 2011). Thus, the multiple clonings

sites were inserted at three different positions in the VP7 gene. These modifications inserted six aa residues at the cloning sites, and, in some cases, this converted the VP7 to a largely soluble protein without affecting the ability of the modified proteins to form trimers. The vectors were used to generate a number of soluble VP7 fusion proteins including a fusion with a 36 aa insert that overlapped immunological domains on protein VP1 of FMDV, as well as a 110 aa peptide derived from AHSV VP2. The soluble trimers of these fusion proteins were able to elicit a good insert-specific immune response in guinea pigs (Rutkowska et al. 2011).

Maree S et al. (2016) solved the problem of the reliable AHSV particles produced in insect cells. The recombinant baculovirus expression system was reported that allowed the expression of various combinations of the four major AHSV-9 structural proteins, the core proteins VP3 and VP7 and outer capsid proteins VP2 and VP5. The assembly of these proteins into architecturally complex CLPs, partial VLPs, or complete VLPs was demonstrated. Thus, the production of the fully assembled CLPs was accomplished by cosynthesis of VP3 and VP7. The outer capsid proteins could associate independently of each other with preformed cores to yield partial VLPs. The complete VLPs were synthesized, albeit with a low yield since crystalline formation of AHSV VP7 trimers impeded the high-level CLP production. The efficient assembly of CLPs composed of VP3 and more soluble but structurally similar mutant VP7 trimers was also described (Maree S et al. 2016).

At last, the famous Ed Rybicki lab produced the AHSV-5 VLPs in plants (Dennis et al. 2018a, b). Thus, the expression and complete assembly of the AHSV VLPs was achieved in the *N. benthamiana* plants using *Agrobacterium* mediated delivery of constructs encoding the four structural proteins of one of the AHSV serotypes not currently included in the live attenuated AHSV vaccines, serotype 5. The production process was fast and simple, scalable, economically viable, and guinea pig antiserum raised against the vaccine was shown to neutralize live virus in cell-based assays (Dennis et al. 2018a). Moreover, Dennis et al. (2018b) demonstrated the safety and immunogenicity of the plant-produced AHSV VLPs in horses. The recent situation in the AHSV vaccines including the pioneering research of the lab was reviewed thoroughly by Dennis et al. (2019). At last, Fearon et al. (2021) demonstrated that the insoluble AHSV-5 VP7 quasicrystals produced in *N. benthamiana* were immunogenic and induced both humoral and cell-mediated responses in guinea pigs.

Since the highly efficient production of the double-layered AHSV VLPs, like BTV VLPs, was not achieved at the turn of the century, as reviewed by Roy and Sutton (1998), Martínez-Torrecuadrada et al. (1996) analyzed the available recombinant AHSV-4 proteins, outer VP2 and VP5 and inner VP7, for their potential as candidate vaccines. The authors showed that VP2 or VP2 and VP5 in the absence of VP7 failed to induce neutralizing antibodies and to protect horses against the AHSV challenge, while the combination of the three proteins was able to confer total protection to immunized horses, which showed absence of viraemia.

The x-ray crystallographic structure was achieved for the VP7 protein of AHSV (Basak et al. 1996). The characteristics of the molecular surface of BTV and AHSV VP7 suggested why AHSV VP7 was much less soluble than BTV VP7 and indicated the possibility of attachment to the cell via attachment of the RGD motif in the top domain of VP7 to a cellular integrin for both of these orbiviruses.

Stuart et al. (1998) investigated the 3D structure of native AHSV particles, in parallel with BTV, by x-ray crystallography. As a result, a picture of the VP7 (T = 13) and VP3 (T = 2) layers for both BTV and AHSV cores was achieved. The inner layer of the BTV and AHSV cores had therefore T = 2 organization and contained 120 copies of VP3 molecules arranged as 60 approximate dimers (Stuart et al. 1998).

The tubular structures, 23 nm in diameter and up to 4 microns in length, were formed by the major nonstructural protein NS1 (Maree FF and Huismans 1997).

Epizootic Hemorrhagic Disease Virus

Le Blois et al. (1991) studied the structural EHDV proteins for the ability to form CLPs in insect cells. The inner capsid protein VP3 of EHDV-1 was able to replace the BTV VP3 in formation of the single-shelled CLPs and the double-shelled VLPs, with incorporation of all three minor core proteins VP1, VP4, and VP6 of BTV into the homologous and mixed CLPs and VLPs, indicating that the functional epitopes of the VP3 protein were conserved for the morphological events of the virus (Le Blois et al. 1991).

Nel and Huismans (1991) reported formation of tubules by the baculovirus-driven expression of the NS1-encoding gene of EHDV serotype 2 (Alberta-strain) in insect cells. It appeared that the tubules could break up into 50 nm diameter circular units, which in turn were composed of approximately 16 subunits. The circular units appeared to be hollow and stacked on top of one another, 100 units/micron tubule length, giving the tubules a segmented, ladderlike appearance. A large excess of the tubuli was found also in the appropriate baculovirus-infected *Spodoptera frugiperda* cells by electron microscopic examination of thin sections (Nel and Huismans 1991).

Great Island virus

When expressed in insect cells, the two major core proteins VP2 and VP7 of the *Great Island virus* species, which included Broadhaven virus (BRDV), a tick-borne orbivirus, formed CLPs. However, no evidence was obtained for the formation of the CLPs between the major core proteins of BRDV and those of BTV following the coexpression of BRDV VP2 and BTV VP7 or BRDV VP7 and BTV VP3, indicating that in this respect the proteins of these two orbiviruses were incompatible, unlike the situation previously described for epizootic haemorrhagic disease virus and BTV core proteins (Moss and Nuttall 1994).

PHYTOREOVIRUSES

Rice dwarf virus (RDV), a reference representative of phytoreoviruses, possesses a double-shelled particle that contains a major capsid protein P8, a major core protein P3, and several minor core proteins. Zheng et al. (2000) reported formation of the double-shelled VLPs similar to the authentic RDV particles by coexpression of the P8 and P3 genes in transgenic rice plants. The VLPs were not detected in transgenic rice plant cells expressing P8 alone.

By expression in insect cells, the single-shelled CLPs were formed by the core protein P3 in the absence of other RDV proteins, while double-shelled VLPs were observed upon mixing in vitro or coexpression in vivo of the P3 and P8 products. The core protein P7 expressed in a similar manner was incorporated into the VLPs (Hagiwara et al. 2003). When P3 of RDV was coexpressed in insect cells with P8 of rice gall dwarf virus (RGDV), another representative of the *Phytoreovirus* genus, the mixed VLPs were formed successfully (Miyazaki et al. 2005).

The first x-ray crystallographic data for the RDV virions at 6.5 Å were obtained by Mizuno et al. (1991). Later, Nakagawa et al. (2003) improved resolution of the x-ray crystallography of RDV to 3.5 Å. The first electron cryomicroscopy structure of RDV was reported with a 25 Å resolution (Lu et al. 1998). Furthermore, the resolution of the electron cryomicroscopy reconstruction reached 6.8 Å by integrating the structural analysis with bioinformatics (Zhou ZH et al. 2001). Remarkably, in the outer-shell protein, the uniquely orientated upper and lower domains were composed of similar secondary structure elements but had different relative orientations from that of BTV, while the inner-shell protein adopted a conformation similar to other members of *Reoviridae* (Zhou ZH et al. 2001).

It is noteworthy that the electron cryomicroscopy maps of the members of the order *Reovirales*, namely RDV and rotavirus, were used as examples to demonstrate applicability of the Gorgon toolkit for the 3D protein structure modeling (Baker et al. 2011).

ROTAVIRUSES

The complex molecular architecture, morphogenesis, and history of development of rotaviruses as the VLP carrier candidates are similar to those of the orbiviruses discussed earlier. However, being a leading cause of severe infantile gastroenteritis worldwide, rotaviruses provoked more interest in vaccine development. Therefore, the rotavirus VLPs were immediately brought under detailed immunological analysis as the putative human vaccine candidates, whereas construction of chimeric VLPs was set aside.

The rotaviruses are icosahedral particles consisting of a double shell—or three concentric capsid layers—enclosing a genome of 11 segments of dsRNA, as described in detail by Lawton et al. (1997). According to the electron cryomicroscopy data, the innermost layer is formed by capsid protein VP2 as an analogue of the VP3 protein in orbiviruses, which is organized in dimers of 120 quasi-equivalent molecules and acts as a scaffold for the proper assembly of the viral core (Lawton et al. 1997). The next layer of the rotavirus virion is formed by capsid protein VP6, as an analogue of the VP7 protein in orbiviruses, as shown in Figure 13.1.

EXPRESSION

The rotavirus capsid proteins were expressed at high levels in insect cells. First, the major capsid VP6 gene of simian rotavirus from the *Rotavirus A* species was expressed, and spontaneous assembly of the VP6 protein into morphologic subunits was documented (Estes et al. 1987). Second, the VP2 protein of bovine rotavirus from the same *Rotavirus A* species was shown to assemble in the cytoplasm of insect cells into the single-layered particles, or CLPs, of 52 nm in diameter (Labbé et al. 1991; Zeng et al. 1994).

The VP6 of Bowden strain from the *Rotavirus C* species itself formed trimers, whereas coexpression of the VP6 with the VP2 of bovine rotavirus resulted in the formation of the double-layered single-shelled particles (Tosser et al. 1992). The mixed CLPs were appreciated, when heterologous VP2 and VP6 of bovine and simian rotaviruses (Labbé et al. 1991) or of rotaviruses A and C (Tosser et al. 1992) were combined.

The expression of the outer membrane proteins VP4 and VP7 together with the inner protein VP6 resulted in the formation of double-shelled VP4/6/7 VLPs (Redmond et al. 1993). Moreover, the authors acknowledged formation of the VP6/7 VLPs. However, Crawford et al. (1994) opposed that coexpressing VP6 and VP7 in insect cells or mixing these proteins in vitro might lead to the VP6/7 particle-like structures but that these structures are heterogeneous and not stable.

Crawford et al. (1994) demonstrated that the use of the VP2 as a scaffolding protein permitted VLPs of different protein compositions, i.e., VP2/6, VP2/4/6, VP2/6/7, and VP2/4/6/7, to be formed by the expression of different combinations of VP2, VP4, VP6, and VP7. The recombinant VLPs were harvested from the media or cells five days postinfection, and they shared many similarities to native rotavirus particles, including the ability to be processed by routine purification procedures, excellent stability, and similar structural features. The VP2/6/7 and VP2/4/6/7 particles were 90–95% complete double-layered and triple-layered particles, respectively, as estimated by negative-stain EM, and VP2/4/6/7 particles stored at 4°C for at least one year remained intact by EM and biochemical criteria. Remarkably, the formation of the mixed particles with the VP7 protein from different virus serotypes was also permitted (Crawford et al. 1994).

Developing further the baculovirus-driven production of rotavirus VLPs in insect cells, Conner et al. (1996a) generated the VP2/4/6/7 VLPs by various combinations of VP2, VP4, VP6, and VP7 proteins from human, bovine, or simian rotaviruses. Zeng et al. (1996) achieved the rotavirus CLPs with minor proteins in all possible combinations: VP1/2/3/6, VP1/2/6, VP2/3/6, and VP2/6 by using VP1 and VP2 of bovine rotavirus and VP3 and VP6 of simian rotavirus.

Jiang et al. (1998) optimized production of the mixed VP2/6/7 VLPs derived from bovine and simian rotaviruses. Then, Jiang et al. (1999) generated the VP2/6/7 VLPs of the human rotavirus serotypes G1 and G3. Kim et al. (2002) formed the rotavirus VLPs by different combinations of the VP2, VP6, and VP7 proteins of *Rotavirus A* species with the VP6 and VP7 proteins of *Rotavirus C* species (Kim et al. 2002). In fact, this was the first report when the inner capsid VP6 of group A or group C rotavirus supported attachment of the heterologous, antigenically distinct group A or group C rotavirus outer capsid VP7 to produce hybrid VLPs in insect cells.

Vieira et al. (2005) demonstrated advantages of a tricistronic strategy over coinfection with the three monocistronic baculovirus vectors to produce the VP2/6/7 VLPs of bovine rotavirus. Next, the cost-effective purification method of the VP2/6/7 VLPs was published by Peixoto et al. (2007). Molinari et al. (2008) were first who intended *S. frugiperda* larvae for the cost-efficient production of the VP2/6 CLPs of simian rotavirus. Clark et al. (2009) described production of the VP2/6/7 or VP6/7 VLPs of human rotavirus C in insect cells.

Yao et al. (2012) constructed an efficient baculovirus-silkworm multigene expression system named Bombyx mori MultiBac, by which multiple expression cassettes were introduced into the Bombyx mori nuclear polyhedrosis virus (BmNPV) genome. The three structural genes of rotavirus and three fluorescent genes were simultaneously expressed in silkworm larvae, resulting in the formation of the VP2/6/7 VLPs of human rotaviruses from the *Rotavirus A* species and the simultaneous color change of larvae.

In parallel, an insect High-Five cell line was generated, which constitutively and stably expressed the VP2 gene of bovine rotavirus, strain RF, of the *Rotavirus A* species, leading to the formation of the VP2 CLPs (Shoja et al. 2013). After traditional *S. frugiperda* insect cells, the VP2/6 CLPs of human rotavirus of the *Rotavirus A* species were produced in stably transformed *Drosophila melanogaster* Schneider 2 (S2) cells (Lee et al. 2011).

In mammalian cells, the VP2/6 CLPs were produced by the vaccinia virus vector-driven coexpression of the appropriate human rotavirus, strain Wa, genes (González SA and Affranchino 1995) and by the Semliki Forest virus (SFV) vector-driven expression of the rotavirus A genes in primary neurons, BHK-21, and mammalian epithelial cells (Nilsson et al. 1998). Later, the VP2 CLPs of human rotavirus from the *Rotavirus A* species were obtained by the expression of the VP2 gene in a mammalian lung cell line A549 (Pourasgari et al. 2007). Laimbacher et al. (2012) achieved highly efficient production of the VP2/6/7 VLPs by herpes simplex virus type 1 HSV-1) amplicon vector-driven expression of the appropriate genes.

Meanwhile, the first report of transgenic mammal bioreactors appeared, when the VP2/6 CLPs of bovine rotavirus were produced in milk of transgenic rabbits (Soler et al. 2005). The experimental data were in the favor of an association of the VP2 and VP6$_{NG}$, a modified version referred to as "nonglycosylated," with five eliminated N-glycosylation motifs by an Asn to Gln substitution, in milk. However, the putative CLPs were not visualized due to a very large proportion of residual casein aggregates and other structured milk components (Soler et al. 2005).

In plants, the VP6 was expressed in *Nicotiana benthamiana* either as a fusion with the potato virus X (PVX) coat protein—as described in Chapter 19—or from an additional subgenomic promoter inserted to enable both VP6 and PVX coat protein to be expressed independently (O'Brien et al. 2000). Both approaches yielded VP6, which retained the ability to form trimers, while the VP6 expressed from the subgenomic promoter assembled into paracrystalline sheets and tubes. Yu and Langridge (2003) expressed the VP6 in the tuber of transformed potato *Solanum tuberosum* plants and found the VP6 trimers.

The correct plant VP2/6 CLPs of murine rotavirus were obtained for the first time in fruits of transgenic tomato *Lycopersicon esculentum*, even though only a small proportion of the VP2 and VP6 assembled into the CLPs (Saldaña et al. 2006). At last, the VP2/6/7 VLPs of human rotavirus from the *Rotavirus A* species were obtained in tobacco *Nicotiana tabacum* (Yang et al. 2011). The authors found that the plant-derived VP2, VP6, and VP7 proteins self-assembled into the VP2/6 CLPs or VP2/6/7 VLPs with a diameter of 60–80 nm.

Rodríguez-Limas et al. (2011) cloned the VP2, VP6, and VP7 genes in four *Saccharomyces cerevisiae* strains using different plasmid and promoter combinations to express one or three proteins in the same cell. To facilitate the VP2 expression, it was necessary to eliminate the coding sequence for the first N-terminal 92 aa, not necessary for VLP formation by 3D data (Lawton et al. 1997). The simultaneous production of the three proteins VP2Δ, VP6, and VP7 led to the production of the triple-layered VLPs, although yield of the latter was incredibly low in comparison with other expression platforms (Rodríguez-Limas et al. 2011).

In *E. coli*, the synthesis of VP6 was achieved with further renaturation and assembly by Zhao et al. (2011). The VP6 formed inclusion bodies in the cytoplasm, the in vitro renaturation progress of the VP6 was monitored by intrinsic fluorescence and far-UV circular dichroism spectroscopies, and diameter of the resulting CLPs was estimated as 40–90 nm.

Li et al. (2014) expressed the VP2 and VP6 genes in *E. coli* and then assembled them with high efficiency in vitro through a postpurification assembly process into the homogeneous single-layered VP6 CLPs or double-layered VP2/6 CLPs. The authors pointed out the improved thermal stability of the "bacterial" VP2/6 CLPs.

3D STRUCTURE

As its structural homolog in orbiviruses, the VP6 became the first target for studies at the atomic level by x-ray crystallography. The VP6 trimers were resolved at 2.0 Å (Petitpas et al. 1998), and a model of self-assembly of the inner capsid layer was suggested (Mathieu et al. 2001). The model assured the assembly of the 260 VP6 trimers according to T = 13*l* symmetry onto the inner T = 1 viral layer (Mathieu et al. 2001).

McClain et al. (2010) resolved the crystal structure of the bovine rotavirus VP2/6 CLP, or double-layered inner particle, at 3.8 Å resolution. Figure 13.5 presents the general 3D structure of a rotavirus. As mentioned earlier, the outer layer is formed by outer membrane proteins VP4 and VP7 that are analogues of the VP2 and VP5 in orbiviruses.

It is remarkable that earlier Ready and Sabara (1987) were the first who demonstrated in vitro assembly of the purified VP6 into a variety of structures like those observed following replication in vivo. These included tubular forms resembling those found in rotavirus-containing feces and spherical particles resembling single-shelled virus. Ready et al. (1988b) reassembled in vitro the inner capsid protein VP6 together with the outer capsid glycoprotein VP7 of bovine rotavirus into smooth VLPs resembling closely the double-shelled rotavirus particles.

Using electron cryomicroscopy and image reconstruction, Prasad et al. (1988) resolved at 40 Å the single- and double-shelled T = 13*l* icosahedral structure of simian rotavirus CLPs and VLPs produced in insect cells. In parallel, Ready et al. (1988a) demonstrated electron micrographs of tubular in vitro-assembled VP6 structures of bovine rotavirus. Lawton et al. (1997) presented an excellent electron cryomicroscopy and 3D model of CLPs formed by full-length and N-terminally truncated VP2 coexpressed with VP6 of simian rotavirus in insect cells.

Later, Libersou et al. (2008) used the electron cryomicroscopy-based quasi-atomic models to understand the molecular basis of the transcriptional machinery activation, which resulted from a structural interplay between the three capsid layers of rotavirus particles. To do this, the authors examined the two viral particles of bovine rotavirus composed of VP2, VP6, VP4, and VP7 (i.e., full triple-layered particles) and four insect cells-produced CLPs/VLPs containing various combinations of the inner VP2, inner/middle VP6, and outer VP7 layer proteins: VP6, VP2/6, VP2/6/7, and VP6/7. It was established that the absence of the VP2 layer increased the particle diameter and changed the type of quasi-equivalent icosahedral symmetry by the shift in triangulation number of the VP6 layer from T = 13 to T = 19 or more. By fitting x-ray models of VP6 into each reconstruction, the authors found that the VP6 lattices, i.e., curvature and trimer contacts, were characteristic of the particle composition. The external protein VP7 reoriented the VP6 trimers located around the 5-fold axes of the icosahedral capsid, thereby shrinking the channel through which mRNA exited the transcribing rotavirus particle. It was

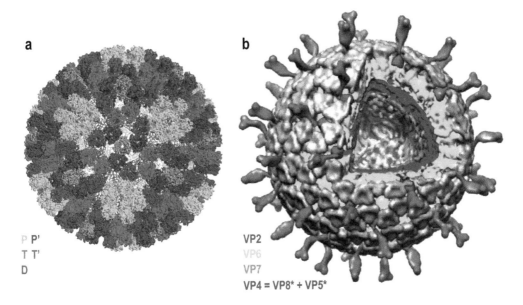

a

b

P P'
T T'
D

VP2
VP6
VP7
VP4 = VP8* + VP5*

FIGURE 13.5   3D structure of a rotavirus. (a) The VP6 shell: 260 VP6 trimers are colored relative to their positions with respect to the icosahedral symmetry axes—gold (T; on the threefold), red (T'; adjacent to T), dark blue (D; closest to the twofold), light blue (P; closest to the fivefold), and purple (P'; adjacent to P). The inner VP2 layer is shown in red and green coils. (b) Cut-away view of the complete rotavirus virion, or three-layered particle, based on the electron cryomicroscopy reconstructions and assignments of densities. The two outer layer proteins are shown in yellow (VP7) and red (VP5*/VP8*). The middle layer protein (VP6) is shown in green; the inner layer VP2 is shown in blue. (Reprinted from *J Mol Biol.* 397, McClain B, Settembre E, Temple BR, Bellamy AR, Harrison SC, X-ray crystal structure of the rotavirus inner capsid particle at 3.8 Å resolution, 587–599, Copyright 2010, with permission from Elsevier.)

concluded therefore that the constraints arising from the different geometries imposed by the external and internal layers of the rotavirus capsid constitute a potential switch regulating the transcription activity of the viral particles (Libersou et al. 2008).

VACCINES

From the very beginning, the recombinant rotavirus CLPs and VLPs were recognized as potent veterinary and human vaccines, and numerous immunological studies were performed to check their potence. First, the protective potential of bovine rotavirus proteins was tested in a murine model, and high protective efficacy of the VLPs consisting of the assembled VP6 particles together with the outer VP4 and VP7 proteins was shown, unlike with unassembled proteins (Tosser et al. 1992). An evaluation of the CLPs and VLPs from human, bovine, or simian rotavirus VP2, VP4, VP6, or VP7, as well as of mixed VLPs harboring multiple types of the outer VP7 protein, showed that they can be effective immunogens in rabbits, mice, and dairy cattle when administered parenterally (Redmond et al. 1993; Conner et al. 1996a, b; Fernandez et al. 1996; Ciarlet et al. 1998). Both VP2/6 CLPs and VP2/6/7 VLPs of simian and human rotaviruses were also shown to be protective in mice and rabbits when administered orally or intranasally (O'Neal et al. 1997, 1998) and passive immunity in newborn calves fed colostrum from VLP-immunized cows has been described (Fernandez et al. 1998). The necessity of

inducing a balanced Th1/Th2 response and the important role of adjuvants for full protectivity of VLPs was demonstrated in mice (Jiang et al. 1999).

To develop a human rotavirus vaccine, the bovine VP2 was selected as a core component, and simian VP4, VP6, and VP7 were composed on it in the mosaic VLPs (Conner et al. 1996b). Then, other compositions of bovine VP6 and simian VP7—or bovine VP2 and VP6—and simian VP7 have been studied, and the VP2/6/7 combination was found to be the most promising (Madore et al. 1999). The studies on the possible application of mixed rotavirus VLPs as a human vaccine showed that immunization with VLPs containing a different human VP7 may provide sufficient priming of the immune system to induce protective heterotypic neutralizing antibody responses, and a limited number of VP7 serotypes within the VP2/6/7 VLPs may be sufficient to provide a broadly protective subunit vaccine (Crawford et al. 1999). The protection of mice pups from mothers immunized nasally with VP2/6/7 but not VP2/6 of bovine rotavirus against rotavirus-induced diarrhea was demonstrated by Coste et al. (2000).

In total, numerous variants of the rotavirus CLP and VLP vaccine candidates in various combinations with different adjuvants were examined at the turn of the century. Thus, the protection of mice by intranasal immunization with the VP2/6 VLPs of murine rotavirus combined with a detoxified version of cholera toxin was reported (Siadat-Pajouh and Cai 2001), as well as the protection of mice and rabbits by parenteral or intranasal immunization with the

VP2/6 VLPs of simian rotavirus (Bertolotti-Ciarlet et al. 2003).

Many studies suggested that immune responses to VP4 and VP7 might be a significant component of protective immunity against rotavirus infection, although it was demonstrated earlier that high levels of protection can be achieved in mice with particles lacking VP4 and VP7 (O'Neal et al. 1997; Choi et al. 2002). Blutt et al. (2006) performed a thorough study in the murine model to examine first the role of challenge dose of the virus and the genetic background of the mouse on the induction of protection. Second, to address which viral components influence protective immunity, the authors utilized the VP2/6 CLPs administered intranasally and the VP2/6 CLPs and the VP2/6/7 and VP2/4/6/7 VLPs, inactivated viruses, and heterologous virus strains administered by the oral and intranasal routes. The conclusion sounded that the combined host, viral, and vaccine factors determined the level of protective efficacy induced by recombinant rotavirus particles.

In parallel, a neonatal gnotobiotic pig model was used to evaluate efficiency of the recombinant rotavirus vaccines. In contrary to the adult mouse or rabbit infection models, the gnotobiotic pig disease model was used to assess protection against rotavirus diarrhea. The gnotobiotic pigs presented a number of important advantages as an animal model for investigating immune responses to human rotavirus, since the pigs closely resembled humans in gastrointestinal physiology and in the development of mucosal immunity (Yuan et al. 2000). The first results indicated that the VP2/6 CLPs composed of bovine VP2 and simian or human VP6 were immunogenic in the pig model but did not confer protection against human rotavirus challenge (Yuan et al. 2000). No protection was found also by the three intranasal doses of the VP2$_{bovine}$/6$_{human}$ vaccination, but satisfactory protection was achieved by the following combination: oral inoculation of attenuated human rotavirus vaccine followed by two intranasal boosts of the VP2/6 in a gnotobiotic pig model of rotavirus disease (Yuan et al. 2001). Yuan and Saif (2002) published a review at that time about the vaccination in the gnotobiotic pig model and claimed the efficiency of the VP2/6 CLPs when administered intranasally as a booster after orally administered replicating vaccines. Such efficient oral boosting effect was demonstrated with the VP2/6 CLPs complexed with ISCOMs, immunostimulatory complexes consisting of cholesterol, phospholipid, and *Quillaja* saponin (Nguyen et al. 2003; González AM et al. 2004). Azevedo et al. (2010) evaluated the oral versus intranasal prime/boost regimen using attenuated human rotavirus together with the VP2/6 CLPs in ISCOMs.

El-Attar et al. (2009) published the first study that demonstrated some postchallenge reductions in rotavirus antigen shedding in a pig model of rotavirus disease after vaccination with the porcine rotavirus VLPs without combining with infectious rotavirus. Thus, the rotavirus antigen shedding was reduced by up to 40% after vaccination with VLPs including the neutralizing antigens VP7 and VP8* when used in combination with the adjuvant polyphosphazene poly[di(carbozylatophenoxy)phoshazene] (PCPP), while complete protection from rotavirus antigen shedding and disease was induced by vaccination with the virulent porcine rotavirus.

The studies of the recombinant human rotavirus particle vaccines in the neonatal gnotobiotic pig model of human rotavirus disease were reviewed fundamentally by Azevedo et al. (2013).

Zhou H et al. (2011) established that the prime immunization with the VLP 2/6 CLPs followed by boosting with an adenovirus expressing the VP6 gene induced protective immunization against rotavirus in mice, laying a groundwork for an alternative strategy in the rotavirus vaccine development.

Then, there were some data about the vaccination attempts with recombinant rotavirus particles that were produced by other than the traditional baculovirus-driven expression models. Thus, when orally delivered into mice with cholera toxin as an adjuvant, the previously described transgenic tobacco plant-produced VP2/6 CLPs and VP2/6/7 VLPs, without purification, as a total soluble protein extracted from plants, induced rotavirus-specific antibodies comparable with those of attenuated rotavirus vaccines (Yang et al. 2011). The *S. cerevisiae* yeast-produced VP2Δ/6/7 VLPs (Rodríguez-Limas et al. 2011), also in the form of raw yeast extracts, induced in mice an immunological response capable of reducing the replication of rotavirus after infection, when challenged with 100 50% diarrhea doses of murine rotavirus (Rodríguez-Limas et al. 2014). Remarkably, when immunizing intranasally, the protection against rotavirus infection was observed even when no increase in rotavirus-specific antibody titers was evident, suggesting that cellular responses were responsible of protection. The previously mentioned *E. coli* bacteria-produced VP2/6 CLPs conferred protection in mice (Li et al. 2014).

At last, some attempts to generate combined vaccines must be mentioned. Thus, the insect cells-produced human rotavirus VP6, which was purified in the form of tubules, was combined with norovirus GII-4 VLPs (Blazevic et al. 2011). The norovirus VLPs are described in Chapter 26. After parenteral immunization of mice, strong systemic cross-reactive and cross-blocking antibody responses were induced toward both rota- and norovirus. There was no interference of the immune response to either antigen given in combination. Rather, an adjuvant effect of the VP6 was observed on the norovirus-specific homologous and heterologous immune responses to genotypes not included in a vaccine formulation (Blazevic et al. 2011).

In this connection, Lappalainen et al. (2013) compared immunogenicity of the VP6 tubules with that of the VP2/6 CLPs. It appeared that both tubular VP6 and VP2/6 CLPs induced equally strong humoral and cellular responses against rotavirus in mice and therefore may be considered as nonlive vaccine candidates against rotavirus.

Tamminen et al. (2013) generated the trivalent combination vaccine consisting of norovirus VLPs of GI-3 and GII-4 representing the two major norovirus genogroups and the tubular VP6 (rVP6). The vaccine induced protective immune responses to the vast majority of circulating norovirus and rotavirus genotypes, major causes of viral gastroenteritis in children worldwide.

CHIMERIC VLPs

The bovine rotavirus VP6 particles were developed as a delivery system for the coupling of synthetic peptides or proteins via a "binding peptide" derived from the VP4 protein (Redmond et al. 1991). This attachment was shown to be mediated by peptide–protein interactions and did not require additional chemicals for conjugation. The resulting macromolecular structures were highly immunogenic for both the VP6 protein and the coupled peptides, as well as in the absence of any adjuvant (Redmond et al. 1991; Frenchick et al. 1992; Ijaz et al. 1995). According to this idea of the protein-peptide disulfide interaction, without any chemicals, the following peptides were coupled to the single-shelled VP6 particles of bovine rotavirus: a 25-aa peptide from VP4 of rotavirus, cyclic somatostatin, ovalbumin, aa 323–345 from gI protein of bovine herpes virus-I (BHV-1; Redmond et al. 1991; Frenchick et al. 1992), the epitopes aa 232–255 of VP4 and 275–295 of VP7 of bovine rotavirus (Ijaz et al. 1995).

On the other hand, the secreting and anchoring derivatives of the outer VP7 (Both et al. 1992) and the inner VP6 (Reddy et al. 1992) proteins were synthesized in order to present them on the surface of the cell. The expression of the chimeric proteins harboring an upstream leader sequence and a downstream membrane-spanning anchor resulted in anchoring in the cell surface membrane, with the major domains of the proteins orientated externally. Such antigens, delivered by the vaccinia virus, produced efficient stimulation of both B and T lymphocytes and may be protective in mice (Both et al. 1992; Reddy et al. 1992).

Although the rotavirus VLPs were suggested as carriers of foreign epitopes from heterologous pathogens or of drugs that need to be delivered to the gastrointestinal tract, either parenterally or orally (Estes et al. 1997), the appropriate manipulations were limited by deletion mapping of the VP6 (Affranchino and González SA 1997) and VP2 (Zeng et al. 1998). As a result, the trimerization and self-assembly domains were localized on the VP6 (Affranchino and González SA 1997), but 92 N-terminal aa residues of the VP2 were found dispensable for self-assembly. Remarkably, the truncated VP2 single- and double- (with the VP6) layered particles were unable to incorporate the VP1, a presumed RNA-dependent RNA polymerase, and lacked therefore a replicase activity (Zeng et al. 1998).

Later, Istrate et al. (2008) started a novel strategy and generated the first chimeric genetically fused VP8–2/6/7 VLPs of bovine rotavirus. These VLPs carried the aa 1–241 fragment of the VP8 protein that was fused N-terminally to the truncated VP2. A single intramuscular dose of the VP8–2/6/7 VLPs alone or combined with a PCPP adjuvant

to dams provided protective immunity against rotavirus infection to offspring in the infant mouse model.

Peralta et al. (2009) used the double-layered VP2/6 CLPs as carriers to display a 14-aa V5 epitope from P and V proteins of simian paramyxovirus 5, which was fused to the three different positions of the VP6 protein exposed on the surface of the CLPs. Although all chimeric proteins were correctly expressed in insect cells, only one of them, namely insertion at position 171↓172, resulted in spontaneous assembly of the CLPs displaying the heterologous epitope on their surface. The injection of the chimeric CLPs into mice elicited higher antibody titers than the monomeric chimeric protein (Peralta et al. 2009).

Bugli et al. (2014) developed and optimized a VP6-based biotechnological platform in *E. coli*. Thus, the three different expression protocols were compared, differing in their genetic constructs, i.e., a simple native histidine-tagged VP6 sequence, VP6 fused to thioredoxin, and VP6 obtained with the small ubiquitin-like modifier (SUMO) fusion system. The authors demonstrated that the histidine-tagged protein did not escape the accumulation in the inclusion bodies, and that the SUMO approach was largely superior to the thioredoxin-fusion tag in enhancing the expression and solubility of the VP6 protein. Moreover, the VP6 protein produced according to the SUMO fusion tag displayed well-known assembly properties, as observed in both transmission electron microscopy and atomic force microscopy images, giving rise to either VP6 trimers, 60 nm spherical CLPs, or nanotubes a few microns long (Bugli et al. 2014).

PACKAGING AND DELIVERY

The packaging, delivery, and visualization story concerning rotavirus was started by Charpilienne et al. (2001) who extended the VP2 protein with green fluorescent protein (GFP) or the DsRed protein. The two chimeric VP2 proteins were prepared, in which aa 1–92 were replaced either by the entire 238-aa GFP or by the entire 249-aa DsRed protein, both followed by a flexible linker. Such chimeric protein GFP/or DsRed-VP2 assembled perfectly well with the VP6 protein and formed the fluorescent GFP-VP2/6 or DsRed-VP2/6 CLPs when coexpressed in insect cells. The presence of GFP inside the core did not prevent the assembly of the outer capsid layer proteins VP7 and VP4 to give the VP2/4/6/7 VLPs. The electron cryomicroscopy of the purified GFP-VP2/6 CLPs showed that the GFP molecules were located at the 5-fold vertices of the core. It was possible to visualize a single fluorescent VLP in living cells by confocal fluorescent microscopy. Remarkably, the VP2/6 CLPs did not enter in vitro into permissive cells or in dendritic cells, while the fluorescent VP2/4/6/7 VLPs entered the cells and then the fluorescence signal disappeared rapidly. This allowed one to follow in real time the entry process of rotavirus and use the chimeric VLPs as "nanoboxes" carrying macromolecules to living cells (Charpilienne et al. 2001).

Cortes-Perez et al. (2010) were the first who confirmed the delivery idea both in vitro and in vivo with the same rotavirus VLP batches. Thus, the GFP-VP2/4/6/7 VLPs demonstrated marked tropism to African green monkey kidney cells MA104 or intestinal cells from healthy and 2, 4, 6-trinitrobenzene sulfonic acid (TNBS)-treated mice. The VLPs were able to enter into MA104 cells and deliver the reporter protein. Moreover, the intragastric administration of fluorescent VLPs in healthy and TNBS-treated mice resulted in the detection of GFP and viral proteins in the intestinal samples, proving the first in vivo evidence of the potential of the rotavirus VLPs as a promising safe candidate for drug delivery to intestinal cells (Cortes-Perez et al. 2010).

It is noteworthy that Fernandes et al. (2014) developed a novel baculovirus-driven expression technique that allowed great improvement in the production of the GFP-V2 CLPs.

Molinari et al. (2014) rendered the chimeric VP2/6 CLPs when replaced the first 92 aa of the VP2 protein with GFP or ovalbumin. These chimeric CLPs were efficiently uptaken by murine dendritic cells, while the heterologous sequences were able to reach the MHC-I pathway as they elicited strong and specific CTL responses.

As described earlier, Zhao et al. (2011) developed the *E. coli* bacteria-based assembling protocol to generate VP6 CLPs. The authors used their model as a targeted drug delivery system. The anticancer drug doxorubicin (DOX) was first covalently conjugated to the VP6 protein. Under appropriate protein concentration and ionic strength, the DOX-VP6 conjugates were self-assembled into the CLPs. Following that, the DOX-loaded CLPs were further linked by lactobionic acid, enabling specific targeting for hepatocytes or hepatoma cells bearing asialoglycoprotein receptors. The in vitro experiment showed that the DOX-VP6 modified by lactobionic acid were internalized specifically by hepatoma cell line HepG2 (Zhao et al. 2011).

De Lorenzo et al. (2012) applied the in vivo biotinylation technology to biotinylate the recombinant VP6 protein fused to a 15 aa-long biotin acceptor peptide (BAP), by the bacterial biotin-ligase BirA contextually coexpressed in mammalian cells. Thus, the authors constructed a stable HEK293 cell line with tetracycline-inducible expression of the VP6-BAP and constitutive expression of BirA. Upon tetracycline induction and rotavirus infection, VP6-BAP was biotinylated, recruited into viroplasms, and incorporated into newly assembled virions. The biotin molecules in the capsid allowed to use streptavidin-coated magnetic beads as a purification technique alternative to CsCl gradient ultracentrifugation. Following transfection, the double-layered particles attached to beads and were able to induce viroplasm formation and to generate infective viral progeny (De Lorenzo et al. 2012).

The packaging of RNA in vitro was demonstrated by Desselberger et al. (2013). To do this, the core shells opened with EGTA were reconstituted by the addition of di- or trivalent cations within 2 min of the opening procedure. The addition of purified, insect cells-expressed VP6 to native and reconstituted cores led to the formation of double-layered particles, which allowed packaging of heterologous nonrotavirus RNA.

NANOMATERIALS

The insect cells-produced VP6 nanotubes of simian rotavirus were used as scaffolds for constructing highly organized novel nanomaterials, namely for the synthesis of hybrid nanocomposites (Plascencia-Villa et al. 2009). The nanotubes of several micrometers in length and various diameters in the nanometer range were functionalized with Ag, Au, Pt, and Pd through strong (sodium borohydride) or mild (sodium citrate) chemical reduction. The nanocomposites obtained were characterized by electron microscopy, dynamic light scattering, and plasmon resonance. The outer surface of the VP6 nanotubes had intrinsic affinity to metal deposition that allowed in situ synthesis of nanoparticles. Furthermore, the use of preassembled recombinant protein structures resulted in highly ordered integrated materials. It was possible to obtain different extents and characteristics of the metal coverage by manipulating the reaction conditions. The electron microscopy revealed either a continuous coverage with an electrodense thin film when using sodium citrate as reductant or a discrete coverage with well-dispersed metal nanoparticles of diameters between 2 and 9 nm when using sodium borohydride and short reaction times. At long reaction times and using sodium borohydride, the metal nanoparticles coalesced and resulted in a thick metal layer. Compared to other nonrecombinant viral scaffolds used until now, the recombinant VP6 nanotubes demonstrated important advantages, including a longer axial dimension, a dynamic multifunctional hollow structure, and the possibility of producing them massively by a safe and efficient bioprocess (Plascencia-Villa et al. 2009).

To develop further the application of the VP6 nanotubes as a source of nanomaterials, Plascencia-Villa et al. (2011) elaborated an advanced method of the purification of well-standardized nanotubes after baculovirus-driven expression in insect cells.

Recently, Tamminen et al. (2020) established that the VP6 nanotubes, as well as VP6 nanospheres, are actively presented by murine bone-marrow-derived dendritic cells in vitro, supporting therefore the use of these specific VP6 nanostructures as foreign antigen delivery platforms.

CYPOVIRUSES AND OTHER *SPINAREOVIRINAE* MEMBERS

The situation with the cypovirus polyhedrin developed in the same direction as by the dsDNA nucleopolyhedrovirus polyhedrins, as described in Chapter 6. Unlike the members of the *Baculoviridae* family, the cypovirus polyhedrin crystalline inclusion bodies were resolved by x-ray powder pattern analysis at 8.2 Å (Di et al. 1991).

The expression of the polyhedrin gene in *E. coli* led to highly efficient synthesis of the product, which, however, did not form any crystalline structure in bacterial cells but accumulated in the form of insoluble inclusion bodies (Lavallee et al. 1993). The expression of Bombyx mori cytoplasmic polyhedrovirus (BmCPV) polyhedrin in an improved baculovirus expression vector and its assembly into large cubic polyhedra (Mori et al. 1993) gave onset to the mutational analysis of the putative carrier. A panel of mutated polyhedrin genes was constructed, and their expression in insect cells demonstrated clearly that N- and C-terminal mutations of the polyhedrin could turn the shape and crystallization pattern of polyhedra from hexahedral to acicular, pyramidal, or amorphous (Ikeda et al. 1998) and the localization of them from cytoplasmic to nuclear (Nakazawa et al. 1996).

Chakrabarti et al. (2010) got VLPs when expressing the Antheraea mylitta cytoplasmic polyhedrosis virus (AmCPV) ORFs S1 and S3 in insect cells via baculovirus recombinants, where the S3-encoded proteins self-assembled to form outer capsid and the VLPs maintained their stability at different pH in presence of the S1-encoded protein.

The *Spinareovirinae* family also includes a well-known mammalian orthoreovirus type 3 Dearing (MRV-3De) of the *Mammalian orthoreovirus* species, which attracted special attention as an oncolytic virus and a prospective antitumor agent (for references see Black and Morris 2012). Reolysin® (pelareorep), a clinical formulation of the latter, was evaluated, for example, to treat melanoma (Galanis et al. 2012), head and neck cancer (Kyula et al. 2012), and advanced solid tumors (Morris et al. 2013).

14 Family *Birnaviridae*

Well done is better than well said.

Benjamin Franklin

ESSENTIALS

The *Birnaviridae* is a family of double-stranded RNA viruses forming icosahedral nonenveloped single-shelled particles with a diameter of about 65 nm (Delmas et al. 2019). According to the recent taxonomy (Delmas et al. 2019; ICTV 2020), the family *Birnaviridae* is not included into any order and appears as an independent member of the kingdom *Orthornavirae*, realm *Riboviria*.

The *Birnaviridae* family currently involves 7 genera and 11 species infecting vertebrates excluding mammals and insects. The infectious pancreatic necrosis virus (IPNV) of salmonids from the *Aquabirnavirus* genus and the infectious bursal disease virus (IBDV) of poultry, the only member of the *Avibirnavirus* genus, are pathogens of major economic importance to the world's fish and poultry industries, respectively.

Figure 14.1 presents a portrait of IBDV as the most studied representative of the family, as well as a cartoon of the birnavirus virion. As in the case of the *Reovirales* order members described in Chapter 13, the birnavirus capsid follows a T = 13*laevo* (T = 13*l*) icosahedral geometry and comprises a single capsid protein, VP2, clustered in trimers and forming 260 projections at the virus surface. Inside the virus particle, the ribonucleoprotein complexes are made by the two genome segments associated with multiple copies of a ribonucleoprotein VP3 and the RNA-dependent RNA polymerase (RdRp, VP1), which do not follow the virus icosahedral symmetry, as reviewed by Delmas et al. (2019).

GENOME

Figure 14.2 shows the larger segment A of the bisegmented dsRNA genome of the two representative genera of the *Birnaviridae* family, about 6 kb of total length, where the two dsRNA segments are of 2.9–3.6 kb.

According to the recent official ICTV review (Delmas et al. 2019), the segment B encodes the viral RdRp, which is free in the virion or covalently attached to the 5'-end of the positive-sense strand of the genomic RNA segments by its N-terminal serine residue. The segment A encodes the polyprotein precursor pVP2-VP4-VP3 and a separate small protein VP5. The VP4 is a protease that cleaves its own N- and C-termini in the polyprotein, thus releasing pVP2 and VP3. Subsequent serial cleavages at the C-terminus of pVP2 yield the mature VP2 protein and peptides that remain associated with the virion.

INFECTIOUS PANCREATIC NECROSIS VIRUS

Since IPNV is an extremely dangerous pest of salmonid farming, the pure scientific interest to the IPNV VLPs was stirred up by the evident material incentive in the sense of construction of a cheap and safe vaccine. The recent situation in the development of the IPNV vaccines including recombinant ones is exhaustively reviewed by Munang'andu et al. (2014).

The first IPNV VLPs were obtained in salmonid cell culture after alphavirus-driven expression of the segment A of the INPV genome (McKenna et al. 2001). Thus, when a salmonid cell line was infected with recombinant SFV particles expressing the IPNV segment A and maintained for five days before fixing and processing for thin-section electron microscopy, the membrane-bound structures carrying isometric particles with morphology and size, diameter of 60 nm, were observed in the cytoplasm of the infected cells. These particles strongly resembled the appearance of particles found within the cytoplasm of the IPNV-infected cells (McKenna et al. 2001).

Furthermore, the IPNV VLPs were generated when the segment A of the INPV genome was expressed both in the *Spodoptera frugiperda* insect *Sf*9 cells and in cabbage looper larvae by the baculovirus-driven technique (Shivappa et al. 2005). As a result, high yields of the IPNV proteins were obtained in both models, and the structural proteins self-assembled to form VLPs of about 60 nm in diameter and morphologically similar to the IPNV virions, as indicated by qualitative electron micrographs. The immunogenicity of the VLPs was tested in immersion vaccine experiments in rainbow trout fry and by peritoneal immunization of Atlantic salmon presmolts using an oil adjuvant formulation. No indication of protection was seen by the rainbow trout immunization. The cumulative mortality rate of immunized Atlantic salmon four weeks post challenge was lower (56%) than in the control fish (77%), showing a dose-response pattern (Shivappa et al. 2005).

A new step toward the IPNV VLP vaccine was taken by Martinez-Alonso et al. (2012) who used the baculovirus-driven expression and purification of VLPs in accordance with the study of Shivappa et al. (2005). However, the authors concentrated more on the fine mechanisms of immunological responses against virus infection and immunization of fish with VLPs. It was concluded that the IPNV VLPs strongly induced nonspecific lymphocyte proliferation and specific anti-IPNV antibody production and remained therefore excellent candidates for the vaccine development.

It is noteworthy, however, that Gilmore et al. (1988) were first to express an antigenic determinant of the IPNV VP2 gene in *E. coli*, in the form of a fusion with the

DOI: 10.1201/b22819-18

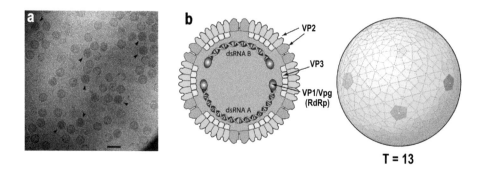

FIGURE 14.1 The typical portrait and schematic cartoon of the family *Birnaviridae* members. (a) The IBDV portrait is obtained by electron cryomicroscopy. The micrograph of the IBDV particles in an unstained, frozen, hydrated preparation. Some of the particles (arrowheads) appear uniformly dark, but most appear blotchy, with the dark ring of the capsid standing out clearly. The former are likely to be complete particles with a full complement of RNA, whereas the latter have probably lost some or all of their RNA. Bar, 100 nm. (Reprinted with permission of American Society for Microbiology from Böttcher B et al. *J Virol.* 1997;71:325–330.) (b) The cartoon of the *Birnaviridae* family members is taken with kind permission from the ViralZone, Swiss Institute of Bioinformatics (Hulo C et al. 2011). (Courtesy Philippe Le Mercier.)

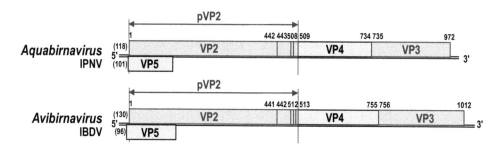

FIGURE 14.2 Schematic representation of the gene arrangement in the coding strand of the genome segment A of birnaviruses, depicting the two most studied representatives: infectious pancreatic necrosis virus (IPNV) from the *Aquabirnavirus* genus and infectious bursal disease virus (IBDV) from the *Avibirnavirus* genus. The parentheses at the 5'-ends indicate the length of the 5'-noncoding regions. The segment B of the birnavirus genome (not shown) encodes VP1 of 840 to 1050 aa. The structures are based largely on the presentations at the ViralZone, Swiss Institute of Bioinformatics, and in the official ICTV paper (Delmas et al. 2019). The structural genes are colored dark pink.

bacterial *trp*E gene and without any special intention to generate VLPs. Manning and Leong (1990) expressed the complete cDNA clone of the IPNV segment A in *E. coli* with a clear effort to develop a vaccine for IPNV in fish, and the VP2 was shown to be processed as observed in infected cells, but no indications were given to the particle formation.

Later, the subunit vaccines based on the *E. coli*-produced IPNV proteins were tested in rainbow trout and Atlantic salmon (Christie 1997; Frost and Ness 1997; Gudding et al. 1999). Thus, the bacterial lysate from *E. coli* cells expressing the segment A of the IPNV strain Sp genome induced protection against challenge with the IPNV Buhl strain, after vaccination by immersion in rainbow trout fry. By injection of Atlantic salmon parr with partly purified *E. coli*-expressed recombinant VP2, the increased resistance against IPN infection was demonstrated by virus challenge. In field trials it was shown that the vaccination of presmolt with the recombinant VP2 included in a commercial oil/glucan adjuvanted multivalent bacterial vaccine gave protection against IPN in natural outbreaks, compared

to the fish vaccinated with the same vaccine without the IPNV component.

When the VP2 gene was cloned into a yeast expression vector and expressed in *Saccharomyces cerevisiae*, the formation of ~20 nm subviral particles (SVPs) composed solely of VP2 protein was observed (Allnutt et al. 2007). Moreover, anti-IPNV antibodies were detected in rainbow trout vaccinated either by injection of the purified VP2 SVPs or by feeding recombinant yeast expressing the VP2 SVPs. When challenged with a heterologous IPNV strain, the authors demonstrated that the trout immunized with the VP2 SVPs had lower IPNV loads by both injection and orally vaccinated fish.

To evaluate the suitability of the IPNV VP2 SVPs for future development of multivalent chimeric vaccines, Dhar et al. (2010) inserted a model 10-aa epitope of a human oncogene, c-Myc, at position of aa 252 of the VP2 backbone. The model epitope was suitable, since it was well-characterized, simple, and linear, and an antibody against the epitope was commercially available. The chimeric VP2 gene was expressed in yeast *S. cerevisiae*. The transmission electron microscopy and antigenical analysis disclosed purified product as the c-Myc-VP2 SVPs of ~20 nm in diameter with the *c-myc* epitope exposed on the SVP surface. The rainbow trout immunized with the c-Myc-VP2 SVPs elicited both anti-c-Myc and anti-IPNV immune responses. When immunized fish were challenged with IPNV, the viral load in the c-Myc-VP2 SVP immunized fish was significantly lower than the sham-vaccinated controls. It was concluded that the IPNV VP2 SVPs were able to tolerate the insertion of foreign epitopes without affecting either the antigenic potential of the epitopes of the backbone protein or the inserted foreign epitope (Dhar et al. 2010).

The determination of the IPNV high-resolution structure always paralleled the corresponding studies of IBDV. For this reason, the 3D structure of both IPNV and IBDV birnaviruses is presented in the following *3D Structure* paragraph.

INFECTIOUS BURSAL DISEASE VIRUS

EXPRESSION

The baculovirus-driven expression of the IBDV structural proteins in insect cells played a critical role in deciphering the fine mechanisms of self-assembly. Vakharia et al. (1993) were the first who expressed the complete IBDV segment A encoding proteins VP2, VP3, and VP4 in the *Sf*9 cells. This resulted in the synthesis of the IBDV precursor protein, which was processed correctly by the viral protease VP4. The recombinant IBDV proteins were antigenically similar to the native IBDV proteins, but the putative IBDV-like particles were not noticed. Furthermore, the VP2 protein was synthesized individually in insect cells—again without indications on self-assembly—and used for diagnostic purposes by Dybing and Jackwood (1997).

Bentley et al. (1994) were the first team who reported production of the genuine IBDV VLPs. Thus, the expression in the insect *Sf*9 cells of a construct containing the entire coding region of the IBDV VP2, VP4, and VP3 resulted in the synthesis and correct processing of the IBDV precursor protein. When expressed in a baculovirus vector as one of a cassette, the structural proteins self-assembled into empty virions, which subsequently afforded protection in challenged chickens (see the *Vaccines* paragraph). The empty virions were characterized by electron microscopy and demonstrated correct shape and size of about 60 nm (Bentley et al. 1994).

When Kibenge et al. (1999) expressed the whole segment A in the *Sf*9 cells, they also detected the genuine IBDV VLPs of 60 nm in diameter, although no processed VP2 was observed in lysates of recombinant baculovirus-infected cells indicating the lack of the pVP2 → VP2 processing in this system and suggesting therefore that the processing is not obligatory for the capsid assembly.

Jackwood et al. (1999) were perhaps the first to express not only the polyprotein but also both the whole and truncated VP2 in the baculovirus expression system, but without

any indications on the assembly status of the gained proteins that were used for diagnostic purposes.

Next, Martinez-Torrecuadrada et al. (2000b) expressed the VPX, defined in Figure 14.1 and known now as pVP2, VP2, and VP3, singly or in different combinations, in the baculovirus expression system. The VPX, or pVP2, and especially VP3 were expressed at high levels and were simple to purify. The immunogenicity of both proteins was like that of the native virus, while the VPX was able to elicit neutralizing antibodies, but the VP3 was not. The authors found that the VPX protein formed tubule-like structures, where the capsomer structure was very similar if not identical to that within the virions, while VP3 did not form any particles.

Thus, it was concluded generally at the turn of the century that the VPX, currently pVP2, as the precursor of VP2, contained all the neutralizing domains and was probably the critical protein for the protection against IBDV. Moreover, the VPX was expressed at higher levels and was more easily purified than the polyprotein (Jackwood et al. 1999; Martinez-Torrecuadrada et al. 2000b).

Furthermore, to learn more about the processing of the polyprotein and factors affecting the correct assembly of the IBDV capsid, Martinez-Torrecuadrada et al. (2000a) made different constructs using two baculovirus transfer vectors, pFastBac and pAcYM1. The expression of the capsid proteins gave rise to different types of particles in each system. The FastBac expression led to the production of only rigid tubular structures, formed exclusively by VPX, or pVP2. These tubules revealed a hexagonal arrangement of units that were trimer clustered, similar to those observed in the IBDV virions. In contrast, the pAcYM1 expression led to the assembly of VLPs of 55 to 75 nm in diameter—as well as to the appearance of flexible tubules—and intermediate assembly products formed by icosahedral caps elongated in tubes. The presence of the processed VP2 protein correlated well with the presence of VLPs with icosahedral symmetry, while more spherical particles contained only VPX. Thus, the processing of VPX to VP2 seemed to be a crucial requirement for the proper morphogenesis and assembly of the IBDV VLPs. Another crucial factor in the morphogenetic events was the role of the VP3 protein. The latter was required for the bending of the capsomere lattice needed for the production of the VLPs as a stoichiometrically limiting factor for the capsid morphogenesis but did not participate within the tubules. It seemed that the VPX protein was processed to VP2 after interaction with VP3 and assembly into VLPs (Martinez-Torrecuadrada et al. 2000a).

Continuing these pioneering studies in the baculovirus-driven expression system, Castón et al. (2001) established that the VP2, when expressed alone, appeared as dodecahedral particles that might be assembled into larger, fragile icosahedral capsids built up by 12 dodecahedral capsids. Each dodecahedral capsid was an empty T = 1 shell of 20–30 nm in diameter and composed of 20 trimeric clusters. The structural comparison between the IBDV virions and capsids consisting of the VP2 alone allowed the determination of the major capsid protein locations and the interactions between them. Whereas VP2 formed the outer protruding trimers, the VP3 protein was found as trimers on the inner surface. Since elimination of the C-terminal region of PVX, or pVP2, correlated with the assembly of T = 1 capsids, this domain might be involved, either alone or in cooperation with VP3, in the induction of different conformations of VP2 during capsid morphogenesis (Castón et al. 2001). It must be emphasized here that the T = 1 VP2 capsids, which appeared later under the name of the IBDV subviral particles (SVPs), played an outstanding role in the further structural studies and nanotechnological applications.

Chevalier et al. (2002) were the first who reported an engineering of a morphogenesis switch to control a particular type of capsid protein assembly. First, they confirmed that the expression of the pVP2 precursor without any other polyprotein components resulted in the formation of isometric particles with a diameter of about 30 nm by the expression in insect cells. They found, however, that the genuine IBDV VLPs were generated when a large exogeneous polypeptide sequence, such as the entire GFP of 238 aa in length, was fused to the VP3 C-terminal domain. The large VLP numbers were visualized by electron microscopy, and single particles were shown to be fluorescent by standard and confocal microscopy analysis. Moreover, the final maturation process converting the pVP2 into the VP2 mature form was observed on the generated VLPs. It was concluded that the correct scaffolding of the VP3 could be artificially induced to promote the formation of the true IBDV VLPs and that the final processing of the pVP2 to VP2 is controlled by this particular assembly (Chevalier et al. 2002).

Maraver et al. (2003) showed that the VP3 protein synthesized in insect cells, either after expression of the complete polyprotein or from a VP3 gene construct, was proteolytically degraded, leading to the accumulation of product lacking the 13 C-terminal residues. This finding led to the identification of the VP3 oligomerization domain within a 24-aa stretch near the C-terminal end of the polypeptide, partially overlapping the VP1 binding domain. The inactivation of the VP3 oligomerization domain, by either proteolysis or deletion of the polyprotein gene, abolished formation of the VLPs. The formation of the VP3-VP1 complexes in cells infected with a dual recombinant baculovirus simultaneously expressing the polyprotein and VP1 prevented the VP3 proteolysis and led to the efficient VLP formation in insect cells. It seemed highly likely that fusing GFP to the VP3 C-terminal end mimicked the protective effect caused by the interaction with VP1 (Maraver et al. 2003).

Furthermore, Oña et al. (2004) reported that the 71-aa C-terminal-specific domain of the pVP2 was essential for the establishment of the VP2–VP3 interaction and that the coexpression of the pVP2 and VP3 polypeptides from independent genes resulted in the assembly of the genuine IBDV VLPs. This observation demonstrated that these two polypeptides contained the minimal information required for capsid assembly and that this process did not require the presence of the precursor polyprotein (Oña et al. 2004). In parallel, Chevalier et al. (2004) specified the last C-terminal residue of VP3, glutamic acid 257, as controlling capsid assembly.

After the classical Sf9 used in all expression studies described earlier, the structural IBDV proteins were expressed in the High-Five insect cells (Lee MS et al. 2004, 2006). When the VPXH gene, namely the VPX or pVP2, with a His-tag at the C-terminus, was expressed in the High-Five cells, the expression level was four times higher than in the Sf9 cells. Moreover, the polyprotein was efficiently processed to yield the VP2-like proteins. At least three structures of particles were observed including tubular and icosahedral particles of approximately 25 nm in diameter, namely, the SVPs (Lee MS et al. 2004).

Lee MS et al. (2006) concluded that the High-Five cells were better in terms of polyprotein processing and formation of VLPs than the Sf9 cells. However, in addition to the processing of pVP2, the VP3 protein was also degraded. With insufficient intact VP3 protein present for the formation of VLPs, the excessive VP2 formed the SVPs with a size of about 25 nm. Remarkably, the actual ratio of SVPs to VLPs was dependent on the multiplicity of infections (MOIs) used, and varied as 0.96, 16.97, and 39.29 at MOIs 0.1, 1, and 10, respectively. Remarkably, both VLPs and SVPs were found extracellularly and in large quantity (Lee MS et al. 2006).

In parallel, the same previously mentioned rVP2H protein was synthesized using insect larvae of the cabbage looper Trichoplusia ni (Lai et al. 2004). The expression level was estimated to be approximately 0.4 mg/g of larvae or 0.2 mg/larvae. The SVPs with a size of 23 nm in diameter were produced.

Jackwood (2013) generated the IBDV VLPs of 40–80 nm in diameter but mostly of approximately 60 nm by simultaneous baculovirus-driven expression of both VP2 and VP3 genes in Sf9 cells.

Chronologically, the first expression of the structural IBDV proteins was undertaken not in the insect cells but in E. coli because of the same growing interest to the novel modern IBDV vaccines. Although a live attenuated IBDV vaccine, which was first introduced in 1968, remained widely used, Azad et al. (1987) expressed the IBDV segment A encoding structural proteins VP2, VP3, and protease VP4 in E. coli, resulting in the processing of the polyprotein and establishing the VP2 protein as a major immunogen. Jagadish et al. (1988) reported the synthesis of the VP2-VP4-VP3 polyprotein in E. coli, most of which was then processed to generate constituent polypeptides.

Furthermore, Azad et al. (1991) reported that small fusions to the VP2 N-terminus led to the stable VP2 expression in E. coli, reduced the levels of inclusion body formation in comparison to the VP2 constructs with larger N-terminal fusions, and ensured appearance of some "multimeric" forms in the process of fractionation.

Next, the complete IBDV segment A was expressed in *E. coli* cells, and the correctly processed structural proteins VP2 and VP3 appeared to be assembled into the particles similar to some extent to the IBDV virions (Rogel et al. 2003). A bit later, Chen et al. (2005) cloned and expressed in *E. coli* the VP2 gene, together with its two N-terminally truncated mutants. All three proteins were provided C-terminally with His$_6$ tag and formed SVPs of icosahedral morphology and approximately 25 nm in diameter. Rong et al. (2007) reported purification of a "soluble" VP2 product from *E. coli*.

Sometime after the first bacterial expression Macreadie et al. (1990) performed the first attempt to express the structural IBDV proteins in yeast *S. cerevisiae*. A clone encoding the IBDV polyprotein but lacking the first five aa at the N-terminus was used, and resulting constructs capable of synthesizing the IBDV VP2 protein were analyzed. A substantial portion of VP2 produced in yeast was present in a high molecular mass aggregated form, and this multimeric but not the monomeric form induced a protective immune response in chickens (Macreadie et al. 1990). The synthesis of similar multimeric and highly immunogenic VP2 forms was established also in yeast *Kluyveromyces lactis* (Azad et al. 1991).

Pitcovski et al. (2003) were the first who produced the VP2 protein in yeast *P. pastoris*. Later, Taghavian et al. (2012, 2013) presented a highly efficient platform for the VP2 expression in *P. pastoris*. The spherical SVPs of 23 nm in diameter and T = 1 surface lattice were obtained up to 38 mg at 95% purity from 1 L of recombinant yeast culture in this study.

Dey et al. (2009) expressed the full-length VP2 coding gene, corresponding to 1,343 bp, in *S. cerevisiae*. The regular SVPs with the 441 aa VP2 residues and of 20 nm in diameter were purified, characterized by electron microscopy, and used as a coating antigen to generate a VP2-based enzyme linked immunosorbent assay (ELISA) detecting the IBDV-specific antibodies in chickens.

In mammalian cells, Bayliss et al. (1991) were the first who expressed the VP2 gene, namely after insertion into a fowlpox vector, in the form of a fusion with some of VP4 and β-galactosidase sequences. Therefore, this study did not indicate any assembly capabilities of the gained product.

Then, the whole IBDV segment A was expressed by a vaccinia virus system in African green monkey kidney epithelial BSC-1 cells, in which the A polyprotein was shown to be processed to give mature structural proteins VP2 and VP3 and nonstructural protein VP4, the viral protease (Fernández-Arias et al. 1998). Therefore, the authors were the first group who succeeded in mammalian cells by the production of the genuine IBDV VLPs with the correct morphology and size of approximately 60 nm in diameter. The impressive paracrystalline arrays of the IBDV-like VLPs were found in the cytoplasm of cells infected with the recombinant vaccinia virus (Fernández-Arias et al. 1998). Remarkably, the VP1 protein, the RNA-dependent RNA polymerase, formed a complex with the VP3 protein and

efficiently incorporated into the IBDV VLPs, in the absence of the IBDV RNA, after coexpressing the VP1 and IBDV ORF A in the vaccinia virus-driven system (Lombardo et al. 1999).

Concerning expression in plants, Wu H et al. (2004a) achieved efficient synthesis of the VP2 protein in *Arabidopsis thaliana*. Wu J et al. (2007) produced the VP2 protein in rice seeds. Both products were soluble and suitable for oral delivery in chickens, while no indications on the putative assembly of the VP2 protein were achieved. Marusic et al. (2021) expressed the His-tagged VP2 in *Nicotiana benthamiana* and revealed the presence of a mixed population of differently shaped particles ranging from spherical capsids, with a diameter between ~25 and ~70 nm, to tubular structures, with variable length from 100 to 400 nm. The intramuscular injection of such putative VP2 VLPs induced in pathogen-free chickens the specific anti-IBDV antibodies at titers comparable to those induced by a commercial vaccine. Moreover, all the immunized birds survived the challenge with a highly virulent IBDV strain, without major histomorphological alterations of the bursa of Fabricius, similar to what was achieved with the commercial inactivated vaccine (Marusic et al. 2021).

3D Structure

The famous Tony Crowther's team were the first to resolve the 3D map of the IBDV virion at about a 2-nm resolution by electron cryomicroscopy (Böttcher et al. 1997). The map showed that the structure of the virus was based on a T = 13 lattice and that the subunits were predominantly trimer clustered. The outer trimers seemed to correspond to the protein VP2, carrying the dominant neutralizing epitope, and the inner trimers corresponded to the VP3 protein, which had a basic C-terminal tail expected to interact with the packaged RNA (Böttcher et al. 1997).

At the same time, Granzow et al. (1997) established that the density gradient-purified IBDV preparations contained full and empty icosahedral virions, so-called type I tubules with a diameter of about 60 nm, and type II tubules of 24–26 nm in diameter.

Although the general architecture of IBDV was established clearly by electron cryomicroscopy (Böttcher et al. 1997), the precise location of the structural proteins of the virion had not been defined experimentally. Martinez-Torrecuadrada et al. (2000a) contributed strongly to the general conclusion that the IBDV shell was built up by the VP2 capsomers, while the VP3 acted to build and/or stabilize the 5-fold vertex needed to bend the capsomer lattice into icosahedral caps. In this case, the VP3 could act as a scaffolding instrument, similar to that made by the VP3 and VP7 proteins in bluetongue virus, as described in Chapter 13. The authors demonstrated the internal location of VP3 within the IBDV virion by immunoelectron microscopy of ultrathin sections (Martinez-Torrecuadrada et al. 2000a).

Castón et al. (2001) used electron cryomicroscopy and image processing techniques to determine the 3D structure of the VP2 T = 1 SVPs to a resolution of 28 Å. By comparison of both the T = 13 capsid of virions and the T = 1 SVPs, the authors clearly demonstrated the locations of the VP2 and VP3 structural proteins in the virion capsid. These results indicated that the C-terminal region of the VPX(pVP2)/VP2 capsid protein plays an important role in determining different conformations of this protein during the construction of the native IBDV capsids.

Lee CC et al. (2003) presented the first crystallization and preliminary x-ray analysis of the mosaic C-terminally His-tagged SVPs, described in their paper as 23-nm particles in diameter. These crystals diffracted x-rays to 4.5 Å resolution, but data was only collected to 6 Å. The outcome of this study was published by Lee CC et al. (2006).

Figure 14.3 presents the overall structure of the SVPs. This structure was refined further to 2.6 Å resolution (Garriga et al. 2006).

As evident from Figure 14.3, the SVPs are exclusively formed by the VP2 protein and do not contain nucleic acids or other proteins, and they are significantly different from the IBDV virion. The three VP2 subunits constituted a tight trimer, which was stabilized by a calcium ion bound to three pairs of symmetry-related Asp31 and Asp174. The treatment of the SVPs with EGTA, a Ca^{2+}-chelating reagent, indicated that the metal-ion might be important not only in maintaining highly stable quaternary structure but also in regulating the swelling and dissociation of the icosahedral particles (Lee CC et al. 2006). In parallel, Garriga et al. (2006) resolved the IBDV SVPs at 2.6 Å and also demonstrated the stabilization by calcium ions located at the 3-fold

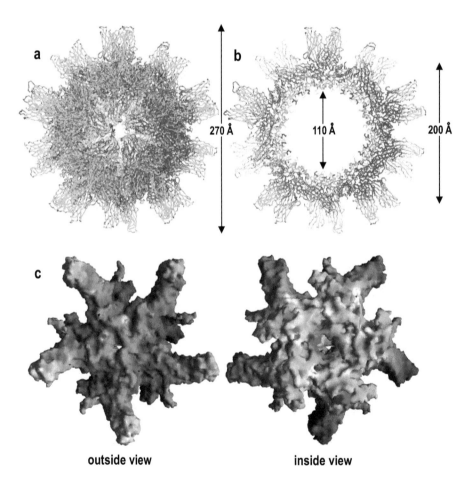

FIGURE 14.3 Overall structure of the IBDV SVPs. (a) An alpha-carbon tracing diagram of the 60-subunit SVP is viewed along an icosahedral 5-fold axis. The polypeptide chains are color-coded according to the temperature factors (or B values) of individual amino acid residues as observed in the cubic crystal. Deep blue denotes the coolest parts, and bright red shows the hottest parts. Intermediate B values are presented with ramped colors of cyan, green, yellow, and orange. (b) The same object is shown but significantly truncated in both front and back to reveal the thick shell and the large central cavity. Some relevant dimensions are indicated: the tip-to-tip distance is 270 Å, the diameter of central cavity is 110 Å, and the average surface-to-surface shell diameter is 200 Å. (c) The electrostatic surface potential for a VP2 pentamer was calculated using GRASP with a range of -15 to +15 k_BT (k_B = Boltzmann's constant; T = Kelvin temperature). It is colored with red and blue representing negative and positive charges and viewed along the fivefold axis of SVP. (Reprinted from *J Struct Biol.* 155, Lee CC, Ko TP, Chou CC, Yoshimura M, Doong SR, Wang MY, Wang AH, Crystal structure of infectious bursal disease virus VP2 subviral particle at 2.6 Å resolution: implications in virion assembly and immunogenicity, 74–86, Copyright 2006, with permission from Elsevier.)

icosahedral axes. Moreover, the structure revealed a new domain swapping that mediated interactions between adjacent trimers: a short helical segment located close to the end of the long C-terminal arm of VP2 was projected toward the 3-fold axis of a neighboring VP2 trimer, leading to a complex network of interactions that increased the stability of the T = 1 particles.

A bit earlier, Coulibaly et al. (2005) not only determined the crystal structure of the IBDV T = 1 SVPs of 260 Å in diameter at 3 Å but also resolved it at 7 Å resolution for the intact T = 13 IBDV virion of 700 Å in diameter. Figure 14.4 demonstrates the structure of the IBDV virion. Coulibaly et al. (2005) emphasized the clear relationships among different virus groups. Thus, the VP2, the only component of the IBDV icosahedral T = 13 capsid, was homologous to both T = 13 capsid protein of members of the Baltimore's dsRNA virus group belonging to the *Reovirales* order described in Chapter 13, as well as to the capsid protein of the positive-strand RNA viruses, like the T = 3 nodaviruses of the order *Nodamuvirales*, the subject of Chapter 21 (Coulibaly et al. 2005). In parallel, Pous et al. (2005) presented the T = 13 particles of both IBDV and IPNV at a resolution of approximately 15 Å by a combination of electron cryomicroscopy and 3D reconstruction. The authors showed that the VP3 and the peptides arising from the processing of pVP2→VP2 did not follow the icosahedral symmetry.

Later, Coulibaly et al. (2010) reported the 3.4-Å resolution crystal structure of the IPNV SVPs, which contained 20 VP2 trimers organized with icosahedral symmetry. As expected, the SVPs demonstrated a very similar organization

to the homologous IBDV counterparts, with VP2 exhibiting the same overall 3D fold. However, the IPNV spikes were significantly different, displaying a more compact organization with tighter packing about the molecular 3-fold axis (Coulibaly et al. 2010).

Saugar et al. (2005) solved the fine mechanism of the molecular switch. The authors showed that the molecular basis of the conformational flexibility of the pVP2 precursor was an amphipatic α helix formed by the sequence GFKDIIRAIR. The VP2 containing this α helix was able to assemble into the T = 13 capsid only when expressed as a chimeric protein with an N-terminal His tag. An amphiphilic α helix, which acted as a conformational switch, was thus responsible for the inherent structural polymorphism of the VP2 protein. The His tag mimicked the VP3 C-terminal region closely and acted as a molecular triggering factor. Using electron cryomicroscopy difference imaging at 15 Å resolution, both polypeptide elements were detected on the capsid inner surface (Saugar et al. 2005). Figure 14.5 illustrates these unique findings.

The 3D data on the viral and subviral IBDV particles was presented nicely by Navaza (2008), who explained the techniques for how the low resolution electron microscopy reconstructions were combined with the existing high resolution monomeric structures in order to produce multimeric models suitable for the molecular replacement.

VACCINES

Since the first IBDV expression experiments, the putative products were examined as putative cheap and safe vaccines

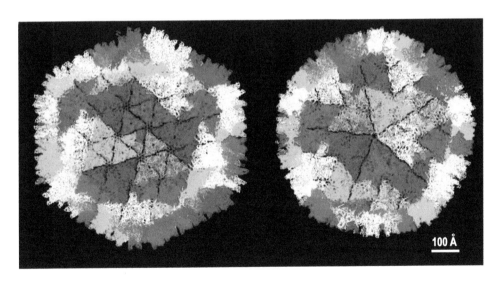

FIGURE 14.4   The IBDV virion. Half an IBDV particle viewed down the I3 (left) and the I5 (right) axes. Only the Cα carbon skeleton of the VP2 trimers fitted into the electron density is displayed. Five different colors (yellow, orange, red, green, and magenta) were used for the 60 icosahedral asymmetric units, distributed such that immediate neighbors are colored differently. A flat face of the icosahedron (perpendicular to the I3) is formed by three colors, with the I3 VP2 trimer at the center (linking three G4). For simplicity, the front half of the virion has been removed, showing the concave internal face of the particle. I3 and I5 are located where three and five colors meet, respectively. (Reprinted from *Cell* 120, Coulibaly F, Chevalier C, Gutsche I, Pous J, Navaza J, Bressanelli S, Delmas B, Rey FA, The birnavirus crystal structure reveals structural relationships among icosahedral viruses, 761–772, Copyright 2005, with permission from Elsevier.)

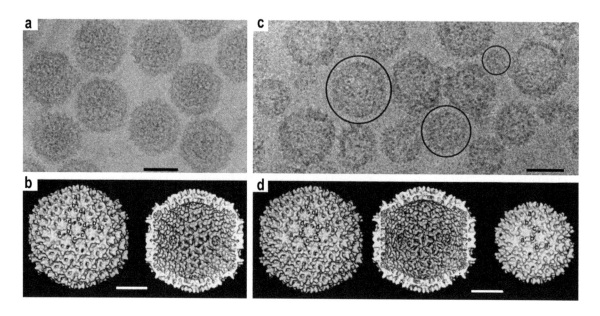

FIGURE 14.5 3D structure of the IBDV capsid. (a) The electron cryomicrograph of IBDV capsids. The scale bar represents 50 nm. (b) The surface-shaded representations of the outer (left) and inner (right) surface of the IBDV capsid viewed along a 2-fold axis of icosahedral symmetry. The surface-shaded map was contoured assuming the presence of 780 molecules of VP2 consisting of 441 aa and a value of 0.73 cm3/g as the partial specific volume of protein. To observe with clarity the pores in the shell, only the front hemisphere of the map is shown. The five types of trimeric capsomers are indicated by letters a-e. The scale bar represents 200 Å. (c) electron cryomicrograph of the HT-VP2–466, a His tagged variant of 466 aa residues. The circles enclose the three clearly discernible icosahedral assemblies with a T = 13, T = 7, and, probably, T = 1 shell. The scale bar represents 50 nm. (d) 3D structure of the HT-VP2–466 T = 13 (left and middle) and T = 7 (right) capsids. These density maps were contoured to enclose a volume for 780 (T = 13 shell) or 420 molecules (T = 7 shell) of the HT-VP2–466. The classes of the HT-VP2–466 trimers are indicated. The scale bar represents 200 Å. (Reprinted from *Structure*. 13, Saugar I, Luque D, Oña A, Rodríguez JF, Carrascosa JL, Trus BL, Castón JR, Structural polymorphism of the major capsid protein of a double-stranded RNA virus: an amphipathic alpha helix as a molecular switch, 1007–1017, Copyright 2005, with permission from Elsevier.)

against IBDV-caused Gumboro disease, a severe immunodeficiency in young chickens, which destroys the precursors of antibody-producing B cells in the bursa of Fabricius and makes chickens susceptible to other avian pathogens. Thus, Macreadie et al. (1990) were the first who demonstrated that the previously mentioned *S. cerevisiae* yeast-derived VP2, when injected into specific pathogen-free chickens, induced high titers of virus neutralizing antibodies that were capable of passively protecting young chickens from infection with IBDV. The *K. lactis* yeast-derived "highly multimeric" VP2 induced significant titers of antibodies in chickens, including virus-neutralizing ones, in contrast to the *E. coli* bacteria-derived and mostly "monomeric" VP2. The latter also produced anti-VP2 antibodies in chickens but without any significant levels of virus-neutralizing activity (Azad et al. 1991).

Bayliss et al. (1991) were the first who challenged immunized chickens with highly virulent IBDV strains after immunization and demonstrated protection against mortality but not against damage to the bursa of Fabricius. The chickens were vaccinated in this case with the previously mentioned fowlpox vector-derived β-galactosidase-VP2 fusion.

Vakharia et al. (1993, 1994) established that the insect cells-derived IBDV proteins induced a neutralizing antibody response in chickens and actively protected chickens

against virulent IBDV challenge. Moreover, Bentley et al. (1994) showed that the protective insect cell-provided IBDV antigen consisted of empty VLPs. The different approaches used for the development of recombinant IBDV vaccines by this pioneering vaccination work were reviewed at that time by Vakharia (1997).

Pitcovski et al. (1996) reported that the VP2 product that was purified from insect cells conferred protection of chickens by induction of higher antibody levels when compared with the commercial vaccine. However, no indication on the structural VP2 status was presented.

The same was true in the case of an efficient *E. coli* bacteria-derived VP2 vaccine, where authors did not notice any possible multimeric organization (Rong et al. 2005, 2007). However, Rogel et al. (2003) insisted that the protective response against their vaccine, which was also injected intramuscularly, was induced by the *E. coli*-produced IBDV-like particles.

Then, the immunization studies concentrated on the structurally well-characterized and previously described T = 1 SVPs. The pure SVP vaccines were generated in yeast *P. pastoris* (Pitcovski et al. 2003) and baculovirus-infected cells *Sf*9 (Martinez-Torrecuadrada et al. 2003). Another variant was represented by the mosaic T = 1 His-tagged SVPs produced in insect High-Five cells (Wang

MY et al. 2000; Cheng et al. 2001; Lee MS et al. 2004, 2006). The efficient SVP vaccine was produced also in the baculovirus-infected insects such as cabbage looper larvae *Trichoplusia ni* (Lai et al. 2004). These extensively purified SVPs conferred protection of chickens following parenteral administration. Recently, Hsieh et al. (2019) reported a standardization method of the SVPs by capillary zone electrophoresis, as a reliable application for the vaccine production and quality control.

Later, the insect cells-produced and previously described multivalent VLPs carrying both VP2 and VP3 proteins ensured protection against challenge by classical and variant IBDV viruses in chicken after intramuscular immunization (Jackwood 2013). Then, Lee HJ et al. (2015) generated in insect cells the genuine VLPs of 60 nm in diameter from so-called vvIBDV—or very virulent IBDV—vaccine and demonstrated 67% protection rate after intramuscular immunization of birds. Moreover, Ge et al. (2015) got up to 100% protection efficiencies when immunizing chickens with baculoviruses encoding the VP2 and VP2/4/3 cassettes.

In contrast to these injection vaccines, the plants such as *A. thaliana* (Wu H et al. 2004b) and rice (Wu J et al. 2007) produced soluble VP2 suitable for direct oral delivery without purification, although this did not achieve complete protection probably because of the inefficient assembly of the putative SVPs in plant cells. Later, Lucero et al. (2019) used intramuscular injections of the *Nicotiana benthamiana* plant-produced VP2 by elaboration of a vaccination scheme of breeder hens before the laying period.

Taghavian et al. (2013) compared the *P. pastoris* yeast-derived SVPs with a commercial nonrecombinant vaccine. The anti-IBDV antibodies were detected in chickens injected with the SVPs or fed with either purified SVPs or inactivated *P. pastoris* cells containing cell-encapsulated VP2. The challenge studies showed that intramuscular vaccination conferred full protection, achieved complete virus clearance, and prevented bursal damage and atrophy. The commercial IBDV vaccine also conferred full protection and achieved complete virus clearance, albeit with partial bursal atrophy. The oral administration of the SVPs with and without adjuvant conferred 100% protection but achieved only 60% virus clearance with adjuvant and none without it. The oral administration of *P. pastoris* VP2-containing cells resulted in 100% protection with adjuvant and 60% without and demonstrated clear dose dependence. Therefore, both oral and parenteral administration of the yeast-derived VP2 SVPs was able to induce a specific and protective immune response against IBDV without affecting the growth rate of chickens (Taghavian et al. 2013).

Using the *P. pastoris* expression system, Wang M et al. (2016) expressed the very virulent—or vvIBDV—VP2 and obtained both full-size VLPs of T = 13 of 60 nm and T = 1 SVPs of 23 nm. The immune effect of the SVPs with 100%

protection rate was significantly better than that of the full-size VLPs with protection rate of only 20%.

Finally, it must be emphasized that the IBDV SVP technique appeared efficient by the rapid generation of vaccines when novel vvIBDV strains might appear. Thus, Li G et al. (2020) constructed an efficient vaccine against an emerging vvIBDV strain, using *E. coli* bacteria-produced SVPs of 14–17 nm in diameter.

Chimeric VLPs

In spite of the tangled structural organization of particles, IBDV manifested itself as a promising carrier for the exposure of foreign epitopes. The first attempts were undertaken with tags. Thus, the first chimeric IBDV-derived particles were produced in insect cells coinfected with the recombinant baculoviruses, which encoded the VP2, VP3, and VP4 and then the VP2 alone with five histidine residues at its C-terminus (Hu et al. 1999). The purified particles including VLPs of 60 nm in diameter and SVPs of ~20 nm were mosaic and formed by native and chimeric molecules. These particles appeared immunologically identical to the wild-type IBDV strains, which contributed subunits to the chimeric VLPs and SVPs (Hu et al. 1999). The effect of the MOI ratio of both vectors on the extent of the His-tagged VP2 incorporation into the mosaic VLPs was established (Hu et al. 2001).

Furthermore, the ability of the His-tagged VP2 alone to form SVPs of approximately 20–30 nm in diameter was shown in the insect *Sf*9 cells (Wang MY et al. 2000). The addition of the six histidines to the C-terminus of the VP2 protein did not prevent formation of particles, which could be affinity-purified in one step, and assured a high level of protection against the challenge with the virulent IBDV (Wang MY et al. 2000). The well-elaborated purification protocols of the His-tagged mosaic SVPs by Ni^{2+} affinity chromatography were published (Wang MY et al. 2000; Cheng et al. 2001; Lee MS et al. 2004).

As mentioned earlier in the *Expression* paragraph, Chen et al. (2005) produced the uniform T = 1 His$_6$-tagged VLPs in *E. coli* after coexpression of the genes encoding VP2 and two its N-terminally truncated mutants, all three with C-terminally added His$_6$ tag.

Later, Jiang et al. (2016) expressed in *E. coli* the VP2 fused with a series of tags: Grifin, MBP, SUMO, thioredoxin, γ-crystallin, ArsC and PpiB, inserted after the His tag, and all the seven tags enhanced the expression and solubility of VP2 protein. After purification by Ni-NTA chromatography, the tags were excised by Tobacco etch virus (TEV) protease, and the resulting VP2 self-assembled into the SVPs of 25 nm in diameter (Jiang et al. 2016).

Concerning presentation of foreign epitopes and construction of putative vaccines against other than IBDV-caused diseases, Rémond et al. (2009) inserted a 12-aa epitope from the immunodominant GH loop of the VP1 protein of foot-and-mouth disease virus (FMDV) into the BC loop of the IBDV VP2. When expressed in insect cells, this insertion did not prevent assembly of the FMDV-VP2 fusions into the T = 1 SVPs with exposure of the FMDV epitope on the SVP surface. The mice immunized with the chimeric SVPs developed antibodies that recognized FMDV in a direct ELISA test and exhibited FMDV-neutralizing properties in vitro (Rémond et al. 2009).

Martin Caballero et al. (2012) inserted a fragment of aa 45–98 from E7 protein of human papillomavirus (HPV) at the C-terminus of VP2 or near the C-terminus at positions 436↓437 and expressed the chimeric genes in yeast *S. cerevisiae*. Both constructs formed the T = 1 SVPs of approximately 45 nm in size and expressed the E7 domain inside the particles. The therapeutic efficacy of such C-terminally modified SVPs was tested in humanized MHC Class I antigen presentation transgenic mice expressing the HLA-A2 allele, which were able to present antigens in both murine and human MHC class I molecules. It was found that the E7-VP2 vaccine elicited an IFN-γ-mediated response and was able to completely eradicate large established tumors (Martin Caballero et al. 2012).

The next attempts to construct IBDV chimeras were performed by genetic reconstruction of the live virus. The first report describing efficient expression of foreign peptides from a replication-competent IBDV demonstrating the potential of this virus as a vector was published by Upadhyay et al. (2011). Thus, by using a cDNA-based reverse genetics system, the insertions or substitutions of sequences encoding epitope tags FLAG or c-Myc and hepatitis C virus (HCV) epitopes were engineered in VP2 and nonstructural protein VP5. The attempts were made to generate chimeric IBDV that displayed foreign epitopes in the exposed loops (P_{BC} and P_{HI}) of the VP2 trimer. As a result, the chimeric IBDVs expressing c-Myc and two different virus-neutralizing epitopes of the HCV envelope glycoprotein E at the N-terminus of VP5 were obtained. The genetic analysis showed that the chimeras carrying the c-Myc/HCV epitopes maintained the foreign gene sequences and were

stable after several passages in mammalian Vero and 293T cell lines (Upadhyay et al. 2011).

Li K et al. (2014) performed the next genetic reconstruction of the live IBDV. The regions in the IBDV genome that were amenable to the introduction of a 9-aa sequence encoding the virus-neutralizing epitope of Newcastle disease virus (NDV) hemagglutinin-neuraminidase (HN) protein were identified. By using the reverse genetics approach, the insertions or substitutions of sequences encoding the NDV epitope were engineered into the exposed loops P_{BC}, P_{HI}, and $P_{AA'}$ of the VP2 and at the N-terminus of the nonstructural VP5 protein as well as the pep7a and pep7b regions of the pVP2 precursor. Three recombinant IBDVs expressing the NDV epitopes in the P_{BC}, pep7b, and VP5 regions were successfully rescued, and the expressed epitope was recognized by anti-HN antibodies. The genetic analysis showed that the chimeric IBDV carrying the NDV epitopes were stable in cell cultures and in chickens. The vaccination with the chimeric viruses generated antibody responses against both IBDV and NDV and provided 70–80% protection against IBDV and 50–60% protection against NDV (Li K et al. 2014).

FUNCTIONALIZATION AND NANOMATERIALS

As described in the *Expression* paragraph, Chevalier et al. (2002) were the first who achieved visualization of the T = 13 IBDV VLPs by fusion of GFP to the C-terminus of VP3 with inclusion of the latter into the particles by the baculovirus-driven expression in insect cells.

When eGFP-carrying VP2 was coexpressed in insect cells with the His tag-carrying VP2, the assembly yield was efficient, and the 240 copies per particle of the native eGFP structure were successfully inserted in a functional fluorescent form, while the electron cryomicroscopy showed that the eGFP molecules appeared at the inner capsid surface (Pascual et al. 2015). The immunization of mice with purified eGFP-VLPs elicited anti-eGFP antibodies.

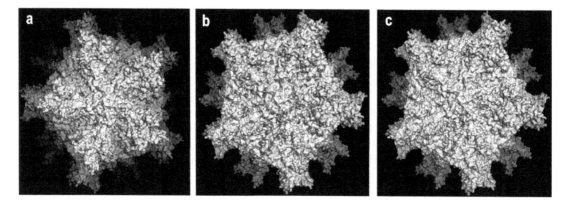

FIGURE 14.6 Structural analysis of the IBDV SVPs. (a) Surface Lys residues shown in yellow and reactive amine side chains highlighted in red. (b) Surface Glu and Asp residues shown in orange and pink, respectively. The reactive carboxyl side chains of both residues are highlighted in blue. (c) Surface Cys residues are shown in purple, but the potentially reactive thiol groups are hidden inside and not exposed to the solvent. (From Taghavian O, Mandal MK, Steinmetz NF, Rasche S, Spiegel H, Fischer R, Schillberg S. A potential nanobiotechnology platform based on infectious bursal disease subviral particles. *RSC Adv.* 2012;2:1970–1978. Reproduced by permission of the Royal Society of Chemistry.)

When hemagglutinin (HA) and matrix (M2) protein epit- opes derived from the influenza virus were exposed on the VLPs, the mice immunized with chimeric VLPs containing the HA stalk, an M2 fragment, or both antigens developed full protection against viral challenge (Pascual et al. 2015).

Taghavian et al. (2012) paved the way to the efficient functionalization of the *P. pastoris* yeast-produced T = 1 SVPs of 23 nm in diameter. The purified SVPs were able to tolerate organic solvents up to 20% concentration (etha- nol or dimethylsulfoxide); they resisted temperatures up to 65°C and remained stable over a wide pH range of 2.5–9.0. The authors achieved bioconjugation to the amine groups

of lysine residues and to the carboxyl groups of aspartic and glutamic acid residues, allowing the functionaliza- tion of the IBDV SVPs with biotin. The accessibility of surface amine groups was measured using Alexa Fluor 488 N-hydroxysuccinimide (NHS) ester, an amine-selec- tive fluorescent dye, revealing that approximately 60 dye molecules were attached to the surface of each particle. Figure 14.6 demonstrates the putative targets of biocon- jugation on the SVPs. This study opened therefore broad possibilities to use the IBDV SVPs as a robust and ver- satile nanoscaffold to display diverse functional ligands (Taghavian et al. 2012).

15 Other Double-Stranded RNA Viruses

It is not sufficient to be worthy of respect in order to be respected.

Alfred Nobel

ORDER *DURNAVIRALES*

ESSENTIALS

According to the current taxonomy (ICTV 2020), the novel order *Durnavirales* is a single member of the class *Duplopiviricetes* that belongs to the huge *Pisuviricota* phylum. The other two classes belonging to this phylum are *Pisoniviricetes* and *Stelpaviricetes*, which appear therefore as far relatives of the *Durnavirales* order members. In contrast to the small *Duplopiviricetes* class, these two classes cover many popular viruses possessing positive-sense single-stranded RNA genomes and actively involved into the VLP nanotechnologies. Thus, the huge *Pisoniviricetes* class unites such great orders as *Nidovirales* (Chapter 26) and *Picornavirales* (Chapter 27), as well as the small *Sobelivirales* order (Chapter 28), while the relatively small *Stelpaviricetes* class contains the more modest orders *Patatavirales* (Chapter 29) and *Stellavirales* (Chapter 30). The phylum *Pisuviricota* belongs to the kingdom *Orthornavirae*, realm *Riboviria*.

The order *Durnavirales* is formed altogether by 5 families, namely *Amalgaviridae*, *Curvulaviridae*, *Hypoviridae*, *Partitiviridae*, and *Picobirnaviridae*; 10 genera; and 85 species. The order members infect eukaryotes.

FAMILY *AMALGAVIRIDAE*

This family involves 2 genera, *Amalgavirus* and *Zybavirus*, and 10 species. The family members possess monosegmented double-stranded RNA genome of 3.5 kb. The nine species of the genus *Amalgavirus* are associated with plants, while the only currently accepted zybavirus species *Zygosaccharomyces bailii virus Z* was isolated from the yeast *Zygosaccharomyces bailii* (Depierreux et al. 2016). Other possible representatives could be associated with microsporidians and animals. Interestingly, the family name derives from amalgam that refers to amalgaviruses that were originally regarded as possessing typical characteristics of both partitiviruses and totiviruses (Martin et al. 2011; Krupovic et al. 2015).

The problem of a putative coat protein that could be encoded by ORF1 and produced as an icosahedral capsid-forming unit remains unclear. Krupovic et al. (2015) found some homology of the hypothetical amalgavirus coat with the nucleocapsid proteins of negative-sense RNA viruses of the genera *Phlebovirus* of the *Bunyaviridae* family and

Tenuivirus of the *Phenuiviridae* family, both from the order *Bunyavirales*. Nevertheless, Pyle et al. (2018) reasoned that the ORF1 more likely may have some other function, first because isometric VLPs have failed to be seen in amalgavirus-infected cells or purified from them. Goh et al. (2018) described the ORF1 product as a putative replication factory matrix-like protein.

FAMILY *HYPOVIRIDAE*

This small family involves a single genus *Hypovirus* and 4 species. According to the current ICTV report (Suzuki et al. 2018), the hypoviruses are capsidless viruses with positive-sense, single-stranded—but not double-stranded—nonsegmented RNA genomes of 9.1–12.7 kb that possess either a single large ORF or two ORFs. No true virions were associated with the family members. The hypoviruses have been detected in ascomycetous and basidiomycetous filamentous fungi and were considered to be replicated in host Golgi-derived lipid vesicles that contained their double-strnaded RNA as the replicative form. As a result, the pleomorphic vesicles of 50–80 nm in diameter, devoid of any detectable viral structural proteins but containing the replicative form dsRNA and polymerase activity, remain the only virus-associated particles that can be isolated from infected fungal tissue (Suzuki et al. 2018).

FAMILY *PARTITIVIRIDAE*

This relatively large family involves 5 genera and 60 species, with characteristic hosts for members of each genus: either plants or fungi for genera *Alphapartitivirus* and *Betapartitivirus*, fungi for genus *Gammapartitivirus*, plants for genus *Deltapartitivirus* and protozoa for genus *Cryspovirus* (Vainio et al. 2018). The linear bisegmented dsRNA genome of the family members is of 3–4.8 kbp. The virions are isometric, nonenveloped, and 25–43 nm in diameter. The larger dsRNA1 and smaller dsRNA2 encoding RNA dependent RNA polymerase (RdRp) and capsid protein, respectively, are separately encapsidated. Each capsid is composed of 120 copies of a single protein arranged as 60 dimers, where the dimeric surface protrusions are frequently observed on viral capsids. One or two molecules of RdRp are packaged inside each particle, as reviewed by Vainio et al. (2018).

The first 3D structure of a partitivirus, Penicillium stoloniferum virus S (PsV-S) from the genus *Gammapartitivirus*, was resolved to 7.3 Å by electron cryomicroscopy and 3D image reconstruction (Ochoa et al. 2008). The capsid, approximately 350 Å in outer diameter, contained 12 pentons, each of which was topped by 5 arched protrusions.

Each of these protrusions was, in turn, formed by a quasisymmetric dimer of coat protein, for a total of 60 such dimers per particle. This T = 1 structure usually referred to as T = 2* from the *Partitiviridae* family exhibited both similarities to and differences from the T = 2* capsids of other dsRNA viruses. Moreover, the authors reported the backbone structure of PsV-S virions, determined by homology modeling from a cryoreconstruction to ~4.5 Å resolution (Tang et al. 2010). In parallel, Pan et al. (2009) reported the crystal structure of another partitivirus, Penicillium stoloniferum virus F (PsV-F), to 3.3 Å resolution, which revealed the coat dimers related by almost-perfect local 2-fold symmetry and forming prominent surface arches. Remarkably, it appeared that the PsV-F capsid is assembled from dimers of the coat dimers, with an arrangement similar to flavivirus E glycoproteins (Chapter 22). This structure is shown in Figure 15.1. These 3D structures were thoroughly summarized and commented on by Nibert et al. (2013) and by Luque et al. (2018).

Recently, Byrne et al. (2021) utilized transient expression of VLPs in plants and high-resolution electron microscopy to determine the 3D structure of pepper cryptic virus 1 (PCV-1) from the *Deltapartitivirus* genus. This was therefore the first structure of the plant-specific partitiviruses. The overall architecture of the capsid was similar to the available structures of fungal partitivirus and the picobirnavirus (see following section) lineage of dsRNA viruses. However, the

surface topology was strikingly different with capsid protrusions emanating from the dimer interface. A disordered region was identified, which formed the outermost portion of the deltapartitivirus capsid protrusion. The disordered region was hypervariable and not required for assembly of the VLPs (Byrne et al. 2021).

FAMILIES *PICOBIRNAVIRIDAE* AND *CURVULAVIRIDAE*

These small families involve one genus each. The earlier assigned *Picobirnaviridae* includes the *Picobirnavirus* genus and 3 species grouping the picobirnaviruses (PBVs) into three genetic clusters with high sequence variability, two defined by viruses infecting vertebrates and a third with viruses found in invertebrates. As summarized by the current ICTV report (Delmas et al. 2019), the picobirnaviruses possess bisegmented (rarely nonsegmented) dsRNA genomes comprising about 4.4 kbp in total, each segment of 1.7–2.7 kbp, and produce small, nonenveloped spherical virions of 33–37 nm in diameter, with a layer of 60 symmetric capsid protein dimers.

Duquerroy et al. (2009) reported the near-atomic x-ray structure of a PBV—an animal dsRNA virus associated with diarrhea and gastroenteritis in humans. To do this, the PBV VLPs were obtained by expressing the rabbit PBV (rPBV) capsid protein gene in insect *Sf*9 cells by the baculovirus expression system. The structure was resolved to 3.4

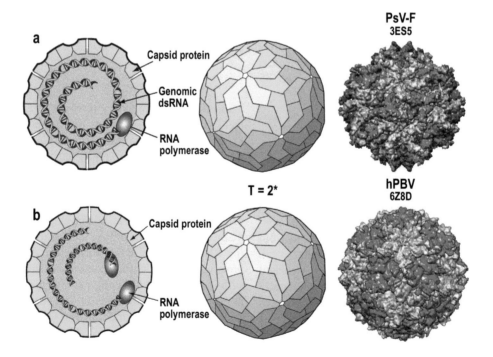

FIGURE 15.1 The schematic cartoons of the *Durnavirales* order members. (a) family *Partitiviridae*, (b) family *Picobirnaviridae*, with the atomic structures of Penicillium stoloniferum virus F (PsV-F), genus *Gammapartitivirus*, family *Partitiviridae*, 3.30 Å resolution, outer diameter 376 Å (Pan et al. 2009) and human picobirnavirus (hPBV), genus *Picobirnavirus*, family *Picobirnaviridae*, 2.63 Å, 384 Å (Ortega-Esteban et al. 2020). Both structures possess the icosahedral T = 1 (or T = 2*) symmetry. The cartoons are taken with kind permission from the ViralZone, Swiss Institute of Bioinformatics, http://viralzone.expasy.org (Hulo C et al. 2011). Courtesy Philippe Le Mercier. The 3D structures are taken from the VIPERdb (http://viperdb.scripps.edu) database (Carrillo-Tripp et al. 2009). The size of particles is in scale. The corresponding protein data bank (PDB) ID numbers are given under the appropriate virus names.

Å and showed a simple core capsid with a distinctive icosahedral arrangement, displaying 60 2-fold symmetric dimers of the coat protein with a new 3D fold. As with many non-enveloped animal viruses, the coat underwent an autoproteolytic cleavage, releasing a posttranslationally modified peptide that remained associated with nucleic acid within the capsid. These data showed that the picobirnavirus particles are capable of disrupting biological membranes in vitro, indicating that its simple capsid had evolved animal cell invasion properties (Duquerroy et al. 2009).

To obtain the atomic structure of human PBV (hPBV), Ortega-Esteban et al. (2020) designed three hPBV coat versions that differed in their N-terminal ends and could thus have different inner surfaces and properties: (i) hPBV coat containing the full-length coat of 552 aa residues, 62 kDa; (ii) hPBV Δ45-coat lacking the first 45 N-terminal aa residues (an Arg/Lys-rich segment) of 508 aa, 57 kDa; and (iii) hPBV Ht-coat, the full-length coat with an N-terminal 36 extension that included the His$_6$ tag of 588 residues, 66.2 kDa. These three coat variants were expressed in *E. coli* at high levels. All three coat constructs produced the hPBV VLPs morphologically similar to authentic virions. As a result, the structure of the *E. coli*-derived hPBV VLPs was resolved by electron cryomicroscopy to 2.6 Å (Ortega-Esteban et al. 2020). This structure is presented in Figure 15.1. Moreover, using an in vitro reversible assembly/disassembly system, the authors isolated tetramers as possible assembly intermediates and characterized the biophysical properties of the hPBV VLPs with different cargos, namely host nucleic acids or proteins.

ORDER *GHABRIVIRALES*

ESSENTIALS

The order *Ghabrivirales* is a single member of the class *Chrymotiviricetes* that forms the phylum *Duplornaviricota*, together with the class *Resentoviricetes* with the single *Reovirales* order (Chapter 13) and the class *Vidaverviricetes* with the single *Mindivirales* order that is described later. All viruses of the *Duplornaviricota* phylum are typical representatives of the Baltimore class III double-stranded RNA viruses. This phylum belongs to the kingdom *Orthornavirae*, realm *Riboviria*. The order *Ghabrivirales* contains altogether 4 families, namely *Chrysoviridae*, *Megabirnaviridae*, *Quadriviridae*, and *Totiviridae*, 9 genera, and 61 species.

FAMILY *CHRYSOVIRIDAE*

This family involves 2 genera, *Alphachrysovirus* and *Betachrysovirus*, and 31 species altogether. As summarized by the recent ICTV report (Kotta-Loizou et al. 2020), the chrysoviruses are small, nonenveloped isometric viruses of ~40 nm in diameter with multisegmented, dsRNA genomes, which typically have four segments, although some have three (collectively known as trichrysoviruses), five

(cinquechrysoviruses), or seven (settechrysoviruses) segments. The dsRNA segments are individually encapsidated in separate particles and together comprise 8.9–16.0 kbp of genomic dsRNA. The chrysoviruses infect ascomycetous or basidiomycetous fungi, plants and possibly insects.

Using the electron cryomicroscopy and 3D image reconstruction, Luque et al. (2010) resolved the structure of Penicillium chrysogenum virus (PcV) from the *Alphachrysovirus* genus to 8.0 Å. The PcV capsid, based on a T = 1 or, as mentioned earlier, so-called T = 2* lattice, contained 60 subunits of the 982-aa coat protein and remained structurally undisturbed throughout the viral cycle. The capsid protein had a high content of rod-like densities characteristic of α-helices, forming a repeated α-helical core indicative of gene duplication, despite a lack of sequence similarity between the two halves (Luque et al. 2010). Thus, in contrast to the most dsRNA virus capsids consisting of dimers of a single protein with similar folds, the PcV capsid protein had two motifs with the same fold. The spatial arrangement of the α-helical core resembled that found in the capsid protein of the L-A virus, which is discussed later. Castón et al. (2013) used the electron cryomicroscopy to solve the 3D structure of a representative of the *Cryphonectria nitschkei chrysovirus 1* species, another member of the *Alphachrysovirus* genus, in parallel with the PcV structure. At last, Luque et al. (2014) reported the 3D electron cryomicroscopy structure of the PcV capsid at near-atomic 4.1 Å resolution, which clearly presented the PcV coat as a duplication of a single domain. This structure is shown in Figure 15.2a, in parallel with the corresponding electron micrograph.

Urayama et al. (2012) expressed the structural genes of Magnaporthe oryzae Chrysovirus 1 (MoCV1) from the *Betachrysovirus* genus in *E. coli* and yeast *Saccharomyces cerevisiae*, but the putative self-assembly of the products was not covered.

FAMILY *MEGABIRNAVIRIDAE*

This extra small family contains a single genus *Megabirnavirus* with a single species *Rosellinia necatrix megabirnavirus 1*, the exemplar strain of which infects the white root rot fungus *Rosellinia necatrix* and confers hypovirulence to the host fungi. According to the current ICTV report (Sato et al. 2019), the megabirnaviruses are nonenveloped spherical viruses with segmented, 2-linear segment (each of 7.2–8.9 kbp) dsRNA genomes, comprising 16.1 kbp in total. The virions are 52 nm in diameter, where the dsRNA segments likely are individually encapsidated. Miyazaki et al. (2015) examined the Rosellinia necatrix megabirnavirus 1 (RnMBV1) isolate W779 virions by electron cryomicroscopy and 3D image reconstruction and resolved the structure to 15.7 Å. The diameter of the RnMBV1 capsid was identified as 520 Å, and the capsid was composed of 60 asymmetrical dimers in the T = 1 (or so-called T = 2*) lattice that was well conserved among many dsRNA viruses. However, RnMBV1 had putatively 120 large protrusions with a width of ~ 45 Å and a height of ~ 50 Å on the virus

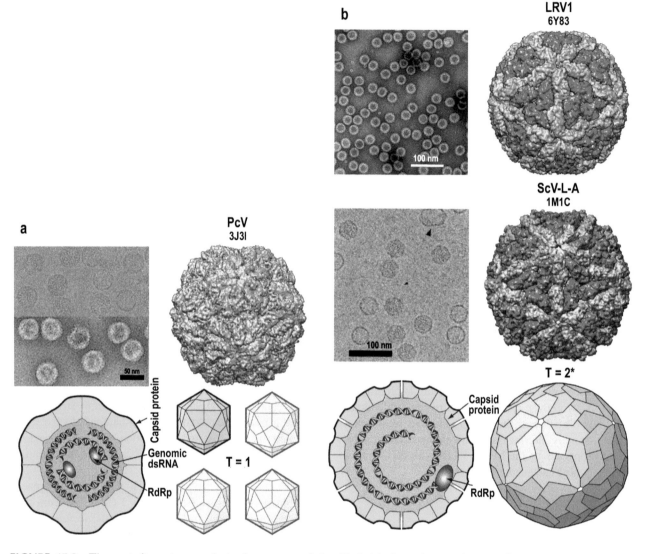

FIGURE 15.2 The portrait, cartoon, and atomic structure of the *Ghabrivirales* order members. (a) Family *Chrysoviridae* with Penicillium chrysogenum virus (PcV), genus *Alphachrysovirus*, 4.10 Å resolution, outer diameter 416 Å (Luque et al. 2014). (b) Family *Totiviridae* represented by the VLPs of Leishmania RNA Virus 1 (LRV1), genus *Leishmaniavirus*, 3.65 Å, 422 Å (Procházková et al. 2021) and Saccharomyces cerevisiae L-A virus (ScV-L-A), genus *Totivirus*, 3.40 Å, 440 Å (Naitow et al. 2002). All structures possess the icosahedral T = 1 (or T = 2*) symmetry. The cartoons are taken with kind permission from the ViralZone, Swiss Institute of Bioinformatics, http://viralzone.expasy.org (Hulo C et al. 2011). Courtesy Philippe Le Mercier. The 3D structures are taken from the VIPERdb (http://viperdb.scripps.edu) database (Carrillo-Tripp et al. 2009). The size of particles is to scale. The corresponding protein data bank (PDB) ID numbers are given under the appropriate virus names. The electron micrographs are from the ICTV book in the case of PcV (Reprinted from *Virus Taxonomy, Classification and Nomenclature of Viruses. Ninth Report of the International Committee on Taxonomy of Viruses*, King AMQ, Lefkowitz E, Adams MJ, Carstens EB (Eds), Ghabrial SA, Castón JR, Family—*Chrysoviridae*. 509–513, Copyright 2012, with permission from Elsevier) and LVR1 (Reprinted from *Virus Taxonomy, Classification and Nomenclature of Viruses. Ninth Report of the International Committee on Taxonomy of Viruses*, King AMQ, Lefkowitz E, Adams MJ, Carstens EB (Eds), Wickner RB, Ghabrial SA, Nibert ML, Patterson JL, Wang CC, Family—*Totiviridae*. 639–650, Copyright 2012, with permission from Elsevier). The electron cryomicrograph of the L-A virus is from Castón et al. (©1997 Castón et al. Originally published in *J Cell Biol*. https://doi.org/10.1083/jcb.138.5.975).

surface, making it distinguishable from the other dsRNA viruses (Miyazaki et al. 2015).

FAMILY *QUADRIVIRIDAE*

This is another extra small family involving the single genus *Quadrivirus* with a single species *Rosellinia necatrix*

quadrivirus 1. As summarized in the ICTV report by Chiba et al. (2018), the quadriviruses are nonenveloped spherical viruses with quadripartite dsRNA genomes of 3.5–5.0 kbp—comprising 16.8–17.1 kbp in total—and infect filamentous fungi, while pathogenicity of the viruses has not been reported. The quadrivirus virions of 45 nm in diameter have unique structural features when compared to those of other known dsRNA viruses. Luque et al. (2016) used the electron cryomicroscopy combined with complementary biophysical techniques to determine the structures of Rosellinia necatrix quadrivirus 1 (RnQV1) virion strains W1075 and W1118 at 8 Å resolution. The capsid was based on a single-shelled T = 1 (or T = 2*) lattice built of dimers of the P2 and P4 proteins encoded by the dsRNA2 and dsRNA4 genome fragments, respectively. The P2 and P4 proteins of the RnQV1 strains W1075 and W1118 were more than 80% identical. However, whereas the RnQV1-W1118 capsid was built of full-length coat proteins, the P2 and P4 of RnQV1-W1075 were cleaved into several polypeptides, maintaining the capsid structural organization. The RnQV1 capsid followed therefore the architectural principle for dsRNA viruses with the 120-subunit capsid but was the first T = 1 capsid with a heterodimer as an asymmetric unit reported to date (Luque et al. 2016). Despite a lack of sequence similarity between the two proteins, they had a similar α-helical domain, the structural signature shared with the lineage of the dsRNA viruses. While the P2-P4 heterodimers were organized similarly to homodimers of other spherical dsRNA viruses, the P2 and P4 had acquired domains situated on the outer surface that were hypothesized to possess actin regulatory (P2-domain) and protease (P4-domain) enzyme activities (Mata et al. 2017), as reviewed by Chiba et al. (2018).

FAMILY *TOTIVIRIDAE*

This family includes 5 genera, *Giardiavirus*, *Leishmaniavirus*, *Totivirus*, *Trichomonasvirus*, and *Victorivirus*, 28 species in total. According to Wickner et al. (2012), these viruses are associated with latent infections of their fungal or protozoan hosts. The virions are isometric of 33–40 nm in diameter, without any surface projections, contain a single coat molecule of 70–100 kDa, and possess the icosahedral T = 1 (or T = 2*) symmetry with the coat dimer as the asymmetric unit. The two highly resolved *Totiviridae* virions, together with their electron micrographs, are presented in Figure 15.2b. The *Totiviridae* genome is represented by a single molecule of linear uncapped dsRNA, 4.6–7.0 kbp in size, which contains two large, usually overlapping, ORFs: the 5'-proximal ORF encoding the coat protein called Gag in this case and the 3'-proximal ORF encoding the RdRp (Wickner et al. 2012).

The earliest and most impressing VLP-directed research in the *Totiviridae* family relates to the *Totivirus* genus. Thus, Castón et al. (1997) were the first who determined the structure of Saccharomyces cerevisiae L-A virus (ScV-L-A) from the *Totivirus* genus by the electron cryomicroscopy and 3D image reconstruction at 16 Å resolution. The yeast virus capsid was 44 nm in diameter and consisted of 60 asymmetric dimers of Gag protein of 76 kDa. Naitow et al. (2002) resolved the ScV-L-A structure to 3.4 Å by x-ray crystallogrpahy. This structure appears in Figure 15.2b.

Powilleit et al. (2007) exploited the ScV-L-A capsid for the in vivo assembly of chimeric VLPs. The authors constructed mosaic particles, using the fact that the ScV-L-A capsid has a 120-subunit structure composed of 118 Gag proteins and two copies of Gag/Pol, which extend into the interior of the capsid to ensure replication and transcription of the viral genome. The Pol was replaced by a truncated version of the immunodominant phosphoprotein pp65 (Δpp65) from human cytomegalovirus (HCMV) to modify the inner surface of the capsid. When expressed in *S. cerevisiae*, the engineered Gag/Δpp65 protein fusion of 101 kDa self-assembled via its Gag domain into the mosaic VLPs encapsulating therefore the Δpp65 as a C-terminal cargo. Furthermore, the Gag VLP-forming carrier was exploited to express and purify the GFP as a model polypeptide. For this purpose, the mosaic Gag variant was constructed encoding a 105 kDa protein fusion containing Gag at its N-terminus (to ensure the in vivo VLP assembly and GFP encapsulation), followed by an 11 amino acid T7 epitope tag (for immunological detection) and a factor Xa cleavage site to release mature GFP from Gag. The mosaic VLPs were purified, and the GFP polypeptide was successfully released by enzymatic cleavage. Then, to demonstrate the flexibility of the novel carrier for the expression of a biotechnologically relevant enzyme, the GFP moiety was replaced by the carboxylesterase EstA from *Burkholderia gladioli*. The EstA that was located inside the mosaic Gag/EstA VLPs efficiently catalyzed the release of 4-nitrophenol. Remarkably, the VLP-associated EstA was recycled and repeatedly used in multiple rounds of enzyme-catalyzed substrate conversion (Powilleit et al. 2007).

In parallel with the *Totivirus* genus, the structural research flourished with a representative of the genus *Victorivirus*. The capsid ORF of Helminthosporium victoriae virus 190S (HvV190S) from the *Victorivirus* genus was expressed in *E. coli*, resulting in the production of p88 protein (Huang and Ghabrial 1996). Next, Huang et al. (1997) used the baculovirus system to express the HvV190S capsid ORF in *Sf*9 cells. In contrast to bacteria, the expression in insect cells generated both p78 and p88, two major closely related capsid polypeptides, which successfully assembled into VLPs. Remarkably, the C-terminal but not N-terminal coat mutants retained the ability to assemble into the VLPs when expressed in insect cells. The evidence was also presented that the p78 protein was derived from the p88 via proteolytic processing at the C-terminus (Huang et al. 1997). At last, Soldevila et al. (1998) showed that the *E. coli*-produced nonphosphorylated p88 was fully competent for assembly into the empty VLPs, indicating that neither phosphorylation nor proteolytic processing of the coat protein was required for the capsid assembly.

The fine structure of the HvV190S, as the first representative of the *Victorivirus* genus, was determined by electron cryomicroscopy and 3D image reconstruction at 13.8 Å resolution by Castón et al. (2006). Later, Dunn et al. (2013) resolved the 3D structure not only to the HvV190S virions but also to the two types of VLPs (capsids lacking dsRNA and capsids lacking both dsRNA and RdRp) at estimated resolutions of 7.1, 7.5, and 7.6 Å, respectively. The HvV190S capsid was thin and smooth and contained 120 copies of coat protein arranged in the traditional T = 2* icosahedral lattice characteristic of ScV-L-A and other dsRNA viruses (Dunn et al. 2013).

Sepp et al. (1995) used the baculovirus system to express the coat gene of Giardia lamblia virus (GLV), a member of the genus *Giardiavirus*, infecting *Giardia lamblia*, the most common protozoan pathogen of the human intestine and a major agent of waterborne diarrheal disease worldwide. Although the correct product was synthesized, the authors did not report the expected VLPs. Janssen et al. (2015) presented the 6-Å electron microscopic structure of the GLV virion. Its outermost diameter was 485 Å, making it the largest totivirus capsid analyzed to date. The structural comparisons of GLV and other totiviruses highlighted the related T = 2* capsid organization and a conserved helix-rich fold in the capsid subunits.

Cadd and Patterson (1994) used the baculovirus expression system to produce the excellent VLPs of Leishmania RNA Virus 1 (LRV1), a type representative of the genus *Leishmaniavirus*. Remarkably, the species of *Leishmania* carrying LRV1 were more likely to cause severe disease and were less sensitive to treatment than those that did not contain the virus. The nice electron micrographs and sedimentation analysis indicated that the insect cells Sf9-expressed protein self-assembled into the empty VLPs of similar size and shape to authentic virions (Cadd and Patterson 1994). The baculovirus-driven expression allowed one to study the role of mutations by the LVR1 self-assembly (Cadd et al. 1994). Later, Procházková et al. (2021) published the near-atomic 3D structure of the LRV1 VLPs, which is presented

in Figure 15.2b. In this case, the excellent LRV1 VLPs were produced in and purified from *E. coli*. The electron cryomicroscopy reconstruction of the LRV1 VLPs led to a resolution of 3.65 Å. The capsid had the same T = 2* icosahedral symmetry and was formed by 120 copies of a capsid protein assembled in asymmetric dimers. Moreover, the authors presented an excellent comparison of capsids and inner capsid shells of the dsRNA viruses, which is included here as Figure 15.3.

At last, Stevens et al. (2021) determined the structure of Trichomonas vaginalis virus 2 (TVV2), a representative of the *Trichomonasvirus* genus, by electron cryomicroscopy at 3.6 Å resolution. The atomic model of TVV2 revealed the same icosahedral T = 2* composition of 60 copies of the icosahedral asymmetric unit, namely a dimer of the two coat protein conformers.

Many novel viruses infecting metazoan hosts, both arthropods (crustaceans and insects) and vertebrates (bony fish), were tentatively assigned to the family *Totiviridae*, as summarized by Janssen et al. (2015). The virions of one of these viruses, penaeid shrimp infectious myonecrosis virus (IMNV), were analyzed by the electron cryomicroscopy and 3D image reconstruction, which revealed an unusual protein-fiber complex that projected above the T = 2* capsid at the icosahedral 5-fold axes and was likely to play one or more key roles in cell entry (Tang et al. 2008). The putative role of these structures was explained in the thorough review of Nibert and Takagi (2013).

Several unassigned viruses closely related to those of the *Totiviridae* family were isolated from mosquitoes, which could cause unpredictable outbreaks of disease. One such virus, namely Omono River virus (OmRV) isolated from

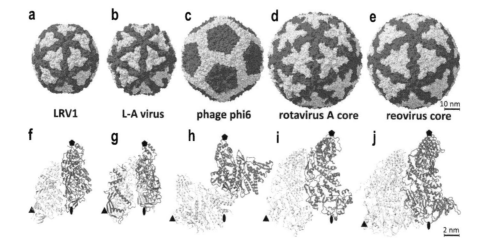

FIGURE 15.3 Comparison of capsids and inner capsid shells of viruses that replicate their double-stranded RNA genomes inside particles. (A to E) Surface representations of LRV1 (A), L-A virus (B; Naitow et al. 2002), phage phi6 (C; Sun et al. 2017), inner core of rotavirus A (D; Settembre et al. 2011), and inner core of reovirus (E; Reinisch et al. 2000). Subunits A are shown in red, and subunits B are shown in green. Scale bar, 10 nm. (F to J) Cartoon representation of proteins forming icosahedral asymmetric units of viruses. Parts of capsid proteins of L-A virus that can be superimposed onto the LRV1 structure are highlighted in cyan and magenta in subunits A and B, respectively (G). Positions of icosahedral symmetry axes are indicated by pentamers for 5-fold, triangles for 3-fold, and ovals for 2-fold axis. (Redrawn with permission of ASM from Procházková M, Füzik T, Grybchuk D, Falginella FL, Podešvová L, Yurchenko V, Vácha R, Plevka P. *J Virol*. 2021;95: e01957–20.)

Culex mosquitoes, was subjected to the high-resolution structure studies using single-particle electron cryomicroscopy (Shao et al. 2021). The final two 3D maps were resolved at 2.79 Å (full particles with genome) and 3.40 Å (empty particles without genome). The structures featured an unexpected protrusion at the 5-fold vertex of the capsid, which resulted in several conformational changes of the major capsid.

ORDER *MINDIVIRALES*

FAMILY *CYSTOVIRIDAE*

The small *Mindivirales* order possesses the single *Cystoviridae* family covering the single *Cystovirus* genus that unites 7 species in total. According to the current ICTV statement (Poranen and Bamford 2012), the cystoviruses are lytic bacteriophages that infect *Pseudomonas* bacteria. The first discovered and most studied Pseudomonas phage phi6 of the *Pseudomonas virus phi6* species typically infects plant-pathogenic *P. syringae*. The enveloped virions are spherical, about 85 nm in diameter and covered by spikes. The nucleocapsid surface shell follows T = 13 icosahedral symmetry. The virion structure is shown in Figure 15.4. The envelope surrounds an isometric nucleocapsid, about 58 nm in diameter. The turret-like extrusions of the underlying polymerase complex span the nucleocapsid surface shell layer at the 5-fold symmetry positions. The dodecahedral polymerase complex is about 50 nm in diameter. Within the polymerase complex the capsid protein dimers are arranged on a T = 1 (T = 2*) icosahedral lattice. The envelope contains three integral membrane proteins: P6, P9, and P10. The protein P8 forms the nucleocapsid surface shell, while the major capsid protein P1 (120 copies per virion) is involved in single-stranded RNA binding. Remarkably, the virions are composed of 20% lipids by mass, which are derived from host plasma membrane. The virions contain three segments of linear, double-stranded RNA: L (6.4–7.1 kb), M (3.6–4.7 kb), and S (2.6–3.2 kb). The complete genome is 12.7–15.0 kb. All the genome segments are enclosed in a single particle, and each virion contains a single copy of the genome (Poranen and Bamford 2012).

Huiskonen et al. (2006) presented the first phi6 structure by electron cryomicroscopy and 3D image reconstruction at 7.5 Å resolution. It revealed the secondary structure of the two major capsid proteins, where asymmetric P1 dimers were organized as an inner T = 1 (T = 2*) shell, and P8 trimers were organized as an outer T = 13 laevo shell. The structures of other members of the genus *Cystovirus* followed: phi8 (Jäälinoja et al. 2007; El Omari et al. 2013), phi12 (Hu et al. 2008; Wei et al. 2009), as well as novel advanced structural investigations on phi6 (Nemecek et al. 2013a, b). At last, Sun et al. (2017) reported the near-atomic structure of phi6 at the 4 Å resolution, which is demonstrated here in Figure 15.4. The structure provided a prime exemplar of bona fide domain-swapping, leading to extension of the theory of domain-swapping from the level of monomeric subunits and multimers to closed spherical

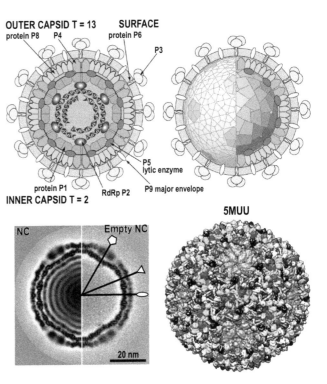

FIGURE 15.4 The structure of the *Mindivirales* order member Pseudomonas phage phi6, genus *Cystovirus*, family *Cystoviridae*. (Top) The cartoon taken with kind permission from the ViralZone, Swiss Institute of Bioinformatics, http://viralzone.expasy.org (Hulo C et al. 2011). (Courtesy Philippe Le Mercier.) (Bottom left) The halves of the central sections of the phage phi6 electron cryomicroscopy and image reconstruction for the nucleocapsid and empty nucleocapsid. The individual icosahedral 2-fold (ellipse), 3-fold (triangle), and 5-fold (pentagon) axes of symmetry are shown for the empty NC. Protein density is black. (Reprinted from *Structure* 14, Huiskonen JT, de Haas F, Bubeck D, Bamford DH, Fuller SD, Butcher SJ, Structure of the bacteriophage phi6 nucleocapsid suggests a mechanism for sequential RNA packaging, 1039–1048, Copyright 2006, with permission from Elsevier.) (Bottom right) The atomic structure at 4.00 Å resolution, outer diameter 576 Å (Sun et al. 2017). The structure possesses the icosahedral T = 13 symmetry. The 3D structure with the corresponding protein data bank (PDB) ID number is taken from the VIPERdb (http://viperdb.scripps.edu) database (Carrillo-Tripp et al. 2009).

shells, and hypothesis of a mechanism by which closed protein shells may arise in evolution (Sun et al. 2017).

FAMILY *POLYMYCOVIRIDAE*

This independent family contains a single genus *Polymycovirus* with 10 species and is classified directly under the *Riboviria* realm. The first described polymycovirus, namely Aspergillus fumigatus polymycovirus 1 (AfuPmV-1), has genome of four dsRNA segments, the smallest of which encodes a PAS-rich protein (PASrp) that apparently coats but does not encapsidate the viral genome, appearing to be an intermediate between encapsidated and capsidless dsRNA viruses (Kanhayuwa et al. 2015). Another polymycovirus, namely Colletotrichum camelliae filamentous virus 1

(CcFV-1), which was isolated from a fungal pathogen, has eight genomic dsRNA segments, ranging from 990–2,444 bp and encoding 10 putative ORFs, of which ORF4 encodes a capsid protein (Jia et al. 2017). Moreover, in contrast to the known dsRNA viruses that are mostly isometric as well as capsidless, the CcFV-1 virions are filamentous. Recently, Sato et al. (2020) concluded that the four studied members of polymycoviruses formed no true capsids, but one formed filamentous virus particles enclosing dsRNA and performed the appropriate study on a novel virus, namely Penicillium janthinellum polymycovirus 1 (PjPmV1). As a result, the authors failed to detect any PjPmV1 particles with filamentous or icosahedral structure in any virus particle-like preparations by electron microscopy, suggesting that the PASrp-associated PjPmV1 dsRNA formed a capsidless ribonucleoprotein structure, as previously proposed by Kanhayuwa et al. (2015).

GENUS *BOTYBIRNAVIRUS*

This independent genus containing one species *Botrytis porri botybirnavirus 1* is classified directly under the *Orthornavirae* kingdom, realm *Riboviria*. The genus representatives are mycoviruses, many of which were found in *Sclerotinia sclerotiorum*, an important phytopathogenic fungus. As summarized by Wang et al. (2019), the botybirnaviruses possess bipartite dsRNA genomes of 5.8–6.5 kbp for each segment. The virions have a spherical morphology with a rough surface and are 38 nm in diameter. Three specific protein bands were identified in the preparations of the Sclerotinia sclerotiorum botybirnavirus 1 (SsBRV1) virus particle (Liu et al. 2015). Then, for example, the Bipolaris maydis botybirnavirus 1 strain BdEW220 (BmBRV1-BdEW220) virions were found of 37 nm in diameter. They involved two dsRNA segments and three structural proteins of 110 kDa, 90 kDa, and 80 kDa, which were encoded by dsRNA1 and dsRNA2 ORFs (Zhai et al. 2019). Remarkably, the dsRNA genome was never completely uncoated to prevent activation of antiviral state by the cell in response to dsRNA. It should be mentioned that the structure and assembly of dsRNA mycoviruses was recently reviewed by Mata et al. (2020).

Section IV

Positive Single-Stranded RNA Viruses

16 Order *Hepelivirales*

In a trice may joy turn to sorrow, should one halt long enough over it.

Nikolai Gogol, *Dead Souls*

ESSENTIALS

The *Hepelivirales* is an order of small, single-stranded positive-sense RNA viruses having mostly nonenveloped particles of different shapes (Smith et al. 2014; Purdy et al. 2017). According to the recent taxonomy (ICTV 2020), the order *Hepelivirales* is one of the three members of the class *Alsuviricetes*, together with the two other orders: *Martellivirales* and *Tymovirales*, the subjects of the next five chapters. The *Alsuviricetes* class belongs to the *Kitrinoviricota* phylum from the kingdom *Orthornavirae*, realm *Riboviria*.

The *Hepelivirales* order currently involves 4 families, 6 genera, and 22 species. The order includes first the family *Hepeviridae* with the genera *Piscihepevirus*—whose members infect fish—and *Orthohepevirus*, whose members infect mammals and birds. The hepatitis E virus (HEV) of the *Orthohepevirus A* species is not only a typical representative of the *Orthohepevirus* genus but also the most dangerous and the most studied virus of the order. HEV is responsible for self-limiting acute hepatitis in humans and several mammalian species, where the infection may become chronic in immunocompromised individuals. Moreover, the extrahepatic manifestations of Guillain-Barré syndrome, neuralgic amyotrophy, glomerulonephritis, and pancreatitis have been described in humans in a proportion of HEV cases. The *Orthohepevirus B* species includes avian HEV that causes hepatitis-splenomegaly syndrome in chickens (Purdy et al. 2017).

The dangerous human rubella virus causing rubella, also known as German measles or three-day measles, leading to miscarriages or severe fetal defects, is a single member of the *Rubivirus* genus that is in turn the single member of the *Matonaviridae* family.

The natural hosts of the *Alphatetraviridae* family members are moths and butterflies with Nudaurelia capensis β virus (NβV) and Helicoverpa armigera stunt virus (HaSV) as reference strains.

The four species of the single *Benyvirus* genus from the *Benyviridae* family infect plants, and beet necrotic yellow vein virus is a typical representative of this family.

Figure 16.1 presents portraits of human HEV, which possesses a capsid that is composed of 180 capsid proteins—assembled into a T = 3 icosahedral particles of about 32–34 nm in diameter—and rubella virus, which is not icosahedral, about 70–80 nm in diameter and coated by envelope glycoproteins assembled in helical organization.

Remarkably, the HEV particles present in feces and bile are nonenveloped, while those in circulating blood and culture supernatants are covered with a cellular membrane, similar to enveloped viruses (Nagashima et al. 2017).

The *Alphatetraviridae* family members are composed of 240 copies of two proteins derived from the capsid precursor cleavage and assembled into a particle with the T = 4 icosahedral symmetry and about 40 nm in diameter.

The *Benyviridae* members are nonenveloped and rod shaped but not spherical. The particles are about 85–390 nm in length and 20 nm in diameter. The right-handed helix with a pitch of 2.6 nm has an axial repeat of four turns, involving 49 capsid subunits. Each capsid protein covers four nucleotides on the genomic RNA (Steven et al. 1981).

In contrast to other order members, rubella virus, the single *Matonaviridae* representative, demonstrates unique 3D structure. It is enveloped, spherical, and 70–80 nm in diameter. The capsid is not icosahedral, and envelope glycoproteins are assembled in helical organization (Mangala Prasad et al. 2013, 2017).

GENOME

Figure 16.2 shows the genomic structure of HEV and rubella virus as the most dangerous and the most studied representatives of the *Hepelivirales* order. Both genomes are monopartite and linear, where HEV genomes are of 7.2 kb and rubella virus genome of 9.7 kb in length. The virion RNAs are infectious and serve as both the genome and viral messenger RNA. The 5' ends are capped, and the 3' termini are polyadenylated. The HEV genome consists of three partially overlapping ORFs, where ORF 2 encodes the capsid protein translated by leaky scanning from the bicistronic ORF3–2 subgenomic RNA. The 72 kDa capsid protein comprises 660 aa and contains a hydrophobic stretch of 14–34 aa at the N-terminus, which functions as a signal sequence for its secretion (Jameel et al. 1996). The ORF2 has three potential glycosylation sites at the Asn positions 132, 310, and 562 (Xing et al. 2011) and is involved in virion assembly, attachment to the host cell, and immunogenicity.

The whole rubella virus genome is translated in a nonstructural polyprotein, which is processed by host and viral proteases, while the structural polyprotein is expressed through a subgenomic mRNA.

The genomes of the *Alphatetraviridae* family members are linear and monopartite in the case of the *Betatetravirus* genus or bipartite in the *Omegatetravirus* genus, with the size of about 6.5 kb. The 5' terminus is capped in omegatetraviruses but not in betatetraviruses. The 3' terminus is not polyadenylated but terminates with a distinctive tRNA-like structure.

DOI: 10.1201/b22819-21

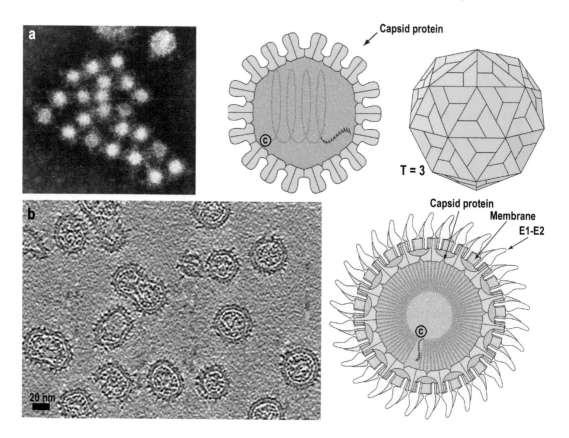

FIGURE 16.1   The typical portrait and schematic cartoon of the order *Hepelivirales* members. (a) Family *Hepeviridae*, an early electron micrograph of typical 32–34 nm enterically transmitted hepatitis non-A, non-B (now hepatitis E) virus particles recovered from stool of a case occurring in Telixtac, Mexico during a 1986 outbreak of disease. (Bradley DW. *Br Med Bull.* 1990, by permission of Oxford University Press.) (b) Family *Matonaviridae*, the section from a rubella virus tomogram showing the different morphologies of rubella virions. (Reproduced from *PLoS Pathog*, Mangala Prasad V, Klose T, Rossmann MG, Assembly, maturation and three-dimensional helical structure of the teratogenic rubella virus, 13, 2017, e1006377.) The cartoons are taken with kind permission from the ViralZone, Swiss Institute of Bioinformatics, http://viralzone.expasy.org (Hulo C et al. 2011). (Courtesy Philippe Le Mercier.)

The *Benyviridae* genomes are linear and consist of four to five segments of 1.3 to 6.7 kb in size. The genomic RNAs are capped, polyadenylated, and serve as messenger RNAs, while subgenomic RNAs are synthesized during replication for segments 2.

The short data on the genome structure is collected mostly from the ViralZone, Swiss Institute of Bioinformatics, http://viralzone.expasy.org (Hulo C et al. 2011).

FAMILY *HEPEVIRIDAE*

EXPRESSION AND VACCINES

Expression

Jameel et al. (1996) expressed the HEV ORF2 in COS-1 cells by using SV40-based expression vectors and in vitro by using a coupled transcription-translation system. It was shown that the capsid protein, encoded by the ORF2, is an 88-kDa glycoprotein that was expressed intracellularly as well as on the cell surface and had the potential to form noncovalent homodimers. This glycoprotein was synthesized as a precursor ppORF2, which was processed through signal sequence cleavage into the mature protein pORF2, which was then glycosylated to form gpORF2 (Jameel et al. 1996).

Table 1 summarizes the data concerning the putative HEV-like particles obtained by expression of various fragments of the HEV ORF2 in all classical expression systems.

The first HEV-like particles produced in insect cells were observed by Tsarev et al. (1993). Thus, the cell lysates from insect cells infected with baculovirus expressing the full-length ORF2 were fractionated by CsCl density gradient centrifugation, and the fractions were incubated with chimpanzee hyperimmune plasma. In a few immune electron microscopy experiments among many, the two kinds of HEV-like particles coated with antibody were observed in the appropriate fraction: some individual particles of ~30 nm and of ~18 nm particles that were clearly smaller than HEV. Generally, it was concluded that that the proportion of the particles driven by the full-length ORF2 in insect cells is very low.

The first reliable production of the HEV-like particles was achieved by Li TC et al. (1997) in insect cells. The capsid protein with its N terminus truncated and containing aa

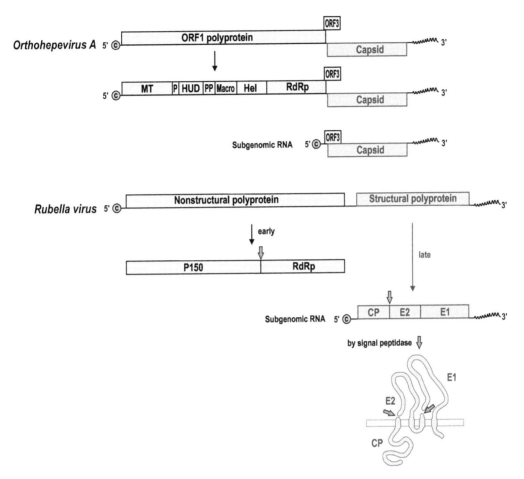

FIGURE 16.2 The genomic structures of human hepatitis E virus (HEV) from the *Orthohepevirus A* species of the *Orthohepevirus* genus, family *Hepeviridae*, and rubella virus of the *Rubivirus* genus from the *Matonaviridae* family. In the HEV case, the diagram shows a short 5′ noncoding region, a 3′ noncoding polyadenylated region, and three ORFs. ORF1 encodes nonstructural proteins including putative functional domains MT, methytransferase; P, a putative papain-like cysteine protease; HUD, *Hepeviridae* unique domain also called the Z domain; PP, a hypervariable polyproline region that is dispensable for virus infectivity; Hel, helicase; and RdRp, RNA-dependent RNA polymerase. ORF2 encodes capsid protein, and ORF3 encodes a small phosphoprotein with a multifunctional C-terminal region. ORF2 and ORF3 overlap each other, but neither overlaps ORF1. The length of the genome is about 7.2 kb. The structure is based on the official ICTV paper (Purdy et al. 2017). The structural capsid gene is colored dark pink.

residues 112–660 of ORF2 and having molecular mass of 50 kDa was synthesized in insect *Trichoplusia ni Tn*5 cells by a recombinant baculovirus. In addition to the primary translation product with a molecular mass of 58 kDa, a large amount of a further-processed molecule with a molecular mass of 50 kDa was generated and efficiently released into the culture medium. The electron microscopic observation of the culture medium revealed that the 50-kDa protein self-assembled to form empty VLPs. The diameter of VLPs was 23.7 nm, a little smaller than the 27 nm of the native HEV virions. The yield of the VLPs was 1 mg per 10^7 cells as a purified form. The particles possessed antigenicity like that of authentic HEV particles, and, consequently, they appeared to be a good antigen for the sensitive detection of HEV-specific immunoglobulins (Li TC et al. 1997).

Moreover, Li TC et al. (1997) explained why the previous attempts to get HEV VLPs did not lead to the high-yield production. Such attempts have been made earlier both in *E. coli* (Purdy et al. 1992; Li F et al. 1994; Panda et al. 1995)

and insect cells (He et al. 1993; Tsarev et al. 1993, 1997; Robinson et al. 1998). Thus, first, the N-terminal 111 aa, including the 59 distal leader sequence, should be deleted, whereas the deletion of the N-terminal 15 aa was not sufficient. However, addition of approximately 30 aa derived from the vector to the C- terminus of the truncated protein did not disrupt particle formation. In the insect *Tn*5 cells, the 50-kDa protein was released into the medium while the 58-kDa protein was tightly cell associated, although these two proteins shared the same amino acid sequences at their N-termini. These findings indicated that the truncated proteins should be further processed through a posttranslational C-terminal deletion mediated by host cellular protease(s) before or during self-assembly. The authors suggested that the viability of the infected cells was an important factor for the VLP formation, while the biochemical basis of the prolonged viability of the *Tn*5 cells compared with the *Sf*9 cells and the mechanism for particle formation specific to the *Tn*5 cells remained at that time unknown (Li TC et al. 1997).

TABLE 16.1

Expression of the ORF2 of the Species *Orthohepevirus A* by Different Expression Systems

Expression System	ORF2, aa	Molecular Mass, kDa	VLPs, Diameter	T Number	References
Human HEV					
Insect cells *Tn*5	112–660	58 and 50	50 kDa VLPs ~23 nm	T = 1	Li TC et al. 1997
	112–608		VLPs 27 nm	T = 1	Xing et al. 2010
	14–608	65.5	65.5 kDa VLPs 41 nm	T = 3	Xing et al. 2010
Insect cells *Sf*9	112–608	53	53 kDa VLPs 27 nm	T = 1	Li TC et al. 2005
Transgenic tomato plants	394–606	23	No VLPs	N/A	Ma et al. 2003
Nicotiana benthamiana	110–610	between 55 and 70 markers	His-tagged particles 100 nm	N/A	Mardanova et al. 2020
Tobacco plastids	394–606	23	No VLPs	N/A	Zhou YX et al. 2006
Transgenic potato plants	112–660 112–608	54	Limited assembly of VLPs	N/A	Maloney et al. 2005
Escherichia coli	368–606 (p239; Hecoline)	30	VLPs 20–30 nm	N/A	Li SW et al. 2005
	439–617 (p179)	20	VLPs 20–30 nm	N/A	Cao et al. 2017
	112–606 (p495)	53	VLPs 20–30 nm	N/A	Zheng et al. 2018
	394–606 (E2)	23	a dimer	N/A	Zhang JZ et al. 2001
	459–606 (E2s)	16	a dimer	N/A	Li S et al. 2009
Dromedary Camel HEV					
Insect cells *Tn*5	112–660	58 and 53	53 kDa VLPs ~24 nm	T = 1	Zhou et al. 2015
	14–660	70, 64, 53, and 40	64 kDa VLPs ~35 nm	T = 3	Zhou et al. 2015
Ferret HEV					
Insect cells *Tn*5	113–607	53	VLPs ~24 nm	T = 1	Yang T et al. 2013
Rat HEV					
Insect cells *Tn*5	101–660	58 and 53	53 kDa VLPs ~24 and 35 nm	T = 1 T = 3	Li TC et al. 2011
Swine HEV					
Nicotiana benthamiana	110–610	56	Limited assembly of VLPs	N/A	Mazalovska et al. 2017
	110–610	53	Particles 19–31 nm	N/A	Zalmanova et al. 2020
Wild Boar HEV					
Insect cells *Tn*5	112–660	58 and 53	53 kDa VLPs; ~24 nm	T = 1	Li TC et al. 2015
	14–660	71, 64, 53, and 40	64 kDa VLPs; ~35 nm	T = 3	Li TC et al. 2015

Notes: The data is acquired partly from the excellent recent review of Mazalovska and Kouokam (2020); N/A, not available.

Later, Li TC et al. (2005) investigated in detail the protein requirement for the HEV VLP formation and prepared 14 baculovirus recombinants to express the capsid proteins truncated at the N-terminus, the C-terminus, or both. In both *Sf*9 and *Tn*5 cells, the aa residues 126 to 601 were the essential elements required for the initiation of the VLP assembly. These results indicated that the cell dependence on particle formation was due to the difference between *Sf*9 and *Tn*5 cells in the modification process of the HEV ORF2 product. The electron cryomicroscopy and image processing of the VLPs produced in *Sf*9 and *Tn*5 cells indicated that they possessed the same configurations and structures.

Concerning immunogenicity of such HEV-like particles, Li TC et al. (2001) orally inoculated mice with the N-terminally truncated HEV VLPs and detected induction of specific antibodies without any adjuvant used. Because mice were not susceptible to HEV, the protection assay was performed with macaques, which were sensitive against HEV and developed acute hepatitis with biochemical, histopathological, and serological markers characteristic of the disease in humans. As a result, when cynomolgus monkeys were orally inoculated with 10 mg of the purified HEV VLPs, without any adjuvants, not only strong immune response but also protection against intravenous challenge with native HEV were achieved (Li TC et al. 2004).

Later, Li TC et al. (2011) generated in insect cells the similar HEV-like VLPs of a novel hepatitis E-like virus that was isolated from Norway rats in Germany. The N-terminally truncated ORF2 protein was synthesized in *Tn*5 and formed particles of 24 nm in diameter. Moreover, the larger rat HEV VLPs were estimated to measure 35 nm in diameter, which was similar to the size of the native rat HEV particles (Li TC et al. 2011).

In parallel with the generation of the human papillomavirus (HPV) pseudovirions vectoring a DNA vaccine against HEV, as described in Chapter 7, Renoux et al. (2008) expressed the HEV ORF2 fragment 112–660 in *Sf*21 insect cells and observed irregular 15-nm particles that accumulated in the cytoplasm of the cells, whereas the expression of the ORF2 fragment 112–608 induced the production of 18-nm particles that were present in both the cell culture medium and the cell cytoplasm. Anti-HEV immune responses were higher by the 15-nm particles of HEV(112–660) than that by the 18-nm particles of HEV(112–608; Renoux et al. 2008).

As shown in Table 1, the nice VLPs of both T = 1 and T = 3 symmetry were obtained also by the capsid gene expression in insect cells of not only of rat but also of ferret (Yang T et al. 2013), dromedary camel (Zhou X et al. 2015), and wild boar (Li TC et al. 2015) HEVs, all members of the species *Orthohepevirus A*. In all cases, the expression of the whole ORF2 did not lead to VLPs, and the N-terminal truncation of the expression variants was needed. As seen from molecular masses of products in Table 1, the C-terminal truncation of the ORF2 variants occurred during expression in insect cells.

Concerning expression in plants, when Maloney et al. (2005) expressed the HEV capsid protein gene lacking 111 aa at the N-terminus potato tubers, very limited assembly of VLPs was found, and oral immunization of mice with transgenic potatoes failed to elicit detectable anti-HEV antibody response. These obstacles suggested therefore that the weak VLP assembly was a key factor in orally delivered HEV vaccines from plants (Maloney et al. 2005). Recently, the highly efficient production of the 100-nm particles of human HEV (Mardanova et al. 2020) and 19–31-nm particles of swine HEV (Zalmanova et al. 2020) was reported, when the 110–610 aa fragment of the corresponding ORF2 was expressed in *Nicotiana benthamiana* plants.

Yang EC et al. (2010) were the first who expressed the HEV ORF2 in yeast *Pichia pastoris*. The authors selected the stretch of 293 corresponding to aa 382–674 and found it to be assembled into VLPs of 30 nm size. Later, Gupta et al. (2020) selected the classical stretch 112–608 aa of the ORF2, by precise analogy with the baculovirus-driven procedure, to express it in *P. pastoris* as an N-terminal His tag

fusion protein. The cDNA sequence encoding this stretch was fused with the α-mating factor secretion signal coding sequence for release of the fusion protein to the culture medium. The protein was secreted to the medium as an N-linked glycoprotein and formed HEV-like particles of 22 nm in diameter. The immunization of mice with these VLPs induced potent immune response as evidenced by the high ORF2-specific IgG titer and augmented splenocyte proliferation in a dose-dependent manner (Gupta et al. 2020).

Simanavicius et al. (2018) expressed full-length and truncated capsid proteins of the HEV genotype 3 and rat HEV in yeast *Saccharomyces cerevisiae*. The only yeast-expressed rat HEV capsid protein was glycosylated. The full-length rat HEV capsid protein formed round-shaped VLPs with a diameter of about 40 nm, whereas less-abundant structures of about 30 nm appeared in the case of the HEV-3 capsid protein. These structures resembled the HEV-like particles of two different sizes observed earlier in the samples of truncated HEV-1 capsid proteins synthesized in insect cells (Li TC et al. 2005, 2011; Guu et al. 2009). When both the 111 aa of the N-terminus and 52 aa (or 37 aa in the case of the rat HEV) from the C-terminus of the ORF2 of HEV-3 and rat HEV capsid proteins were removed, the abundant round-shaped VLPs of 35–40 nm diameter were observed in the case of the truncated rat HEV capsid protein, corresponding to the rat HEV aa 112–608 and suggesting that the truncation at both N- and C-termini did not affect its capacity to form VLPs. However, no higher ordered structures were observed in the samples of the truncated HEV-3 capsid protein. Thus, the only full-length yeast-expressed HEV-3 capsid protein of aa 1–660 had the capacity to self-assemble into VLPs. Using the yeast-produced VLPs, two panels of MAbs against HEV-3 and rat HEV capsid proteins were generated and used to develop novel HEV detection systems (Simanavicius et al. 2018).

Hecolin

The HEV-derived particles of ~26 nm in diameter were obtained in *E. coli* when the ORF2 fragment encoding aa 394 to 606 unable to produce VLPs was extended by the N-terminal 26 aa from the same ORF2 (Li SW et al. 2005). This putative VLP vaccine carrying the aa 368–606 stretch of the HEV genotype 1 ORF2, forming particles of 23 nm in diameter by final estimation, and named HEV239 afforded complete protection of rhesus monkeys against HEV infection. The tight homodimers were characterized as a basic scaffold of the particles (Li S et al. 2009). It is important to emphasize that the particle assembly of the HEV239 was performed after purification (Li SW et al. 2005; Yang C et al. 2013). However, in contrast to the previously described HEV-like VLPs produced in insect cells, the shape of the particles was too heterogeneous for structure determination by electron cryomicroscopy and image reconstruction, and this heterogeneity precluded the symmetry averaging. It is necessary to keep in mind that the HEV239 vaccine with 239 aa residues contained only approximately 40% of the full-length capsid ORF2 protein of 606 aa in length. The neutralizing antibodies elicited by the HEV239 showed nevertheless high neutralization titers in the in vitro cell-based models and by the in vivo challenge studies in chimpanzees, showing that the neutralizing and immunodominant epitopes are preserved on the HEV239 particles (Li SW et al. 2005; Zhao et al. 2013).

The efficacy and safety of the prophylactic HEV239 vaccine branded finally as Hecolin and produced by Xiamen Innovax Biotech (Xiamen, China) were demonstrated in randomized, double-blind, placebo-controlled, phase II (Zhang J et al. 2009) and III (Zhu et al. 2010) trials. The Hecolin vaccine was approved and licensed in China in December 2011 for prophylactic use against HEV infection and launched in China in October 2012 (Park 2012). The following research showed that the HEV vaccine could provide long-term protection, i.e., up to 4.5 years, with 86.6% efficacy (Zhang J et al. 2015). Nevertheless, the Hecolin vaccine was not placed on the priority vaccine list of the WHO prequalification. As Mazalovska and Kouokam (2020) recorded in the recent exhaustive review, in order to be recommended for global use, the two clinical trials evaluating the HEV239 vaccine are ongoing, including a phase I safety study in the USA (NCT03827395) and a phase IV trial in Bangladesh (NCT02759991) evaluating its safety and efficacy in pregnant women, who are at higher risk of acute liver failure and elevated neonatal mortality and morbidity (Zaman et al. 2020).

The story and advantages of the Hecolin vaccine, which was announced as a third great VLP vaccine, after hepatitis B (see Chapter 37) and human papilloma (see Chapter 7), were described in numerous reviews (Wu et al. 2012; Zhang J et al. 2012; Zhao et al. 2013; Zhang X et al. 2014; Li SW et al. 2015; Mazalovska and Kouokam 2020).

The shorter HEV vaccine, termed p166 and encompassing aa 452–617 of the ORF2 HEV genotype 4 protein, was able to form particles of 20 nm in diameter (Dong and Meng 2006). This protein was extended to p179 encompassing aa 439–617 (Wen et al. 2016), developed safely (Cheng et al. 2012; Wang et al. 2014), and assessed in a phase I clinical trial in China (Cao et al. 2017). It must be noted that the p179 candidate showed a difference in immunogenicity when compared with the HEV239 (or Hecolin) vaccine, possibly, due to variations in the genotype-specific neutralization epitopes (Wen et al. 2016).

At last, Zheng et al. (2018) compared the so-called p495, an ORF2 fragment of aa 112–606 lacking both its N- and C-termini of 111 and 54 aa residues, by expression in the *Tn5* insect cells, as well as in *E. coli*, after self-assembly in vitro. The characterization of the p495 particles derived from these two expression systems showed similarities in particle size, morphology, antigenicity, and immunogenicity.

NHT

Using lessons from the Hecolin vaccine, a class of novel nanoparticles was designed. This was achieved through utilization of an N-terminal hydrophobic tail (NHT), located within the HEV ORF2 aa 368–460 (Zhang X et al. 2016).

The previous studies on the Hecolin vaccine showed that the N-terminal hydrophobic leucine patch (L372, L375 and L395, among others) with this stretch of the HEV239 molecule played a key role in the particle self-assembly via hydrophobic interactions. Thus, the HEV NHT of 93 aa in length was selected as a universal tool for the generation of other particulate structures.

Following this idea, the Flu HA1, HIV gp41/gp120/p24, HBsAg and HPV16 L2 were fused with NHT, expressed in *E. coli*, and subjected to self-assembly in vitro. All the NHT-fused proteins spontaneously formed nanoparticles, as established by size-exclusion chromatography and negative electron microscopy. All of the tested particles demonstrated a globular configuration of 20–30 nm diameter, indicating that they were in a relatively regular assembly instead of a disordered aggregation. The specific antibodies elicited in mice were 2-log higher in titer than that of their parent nonassembling proteins (Zhang X et al. 2016).

3D STRUCTURE

Xing et al. (1999) were the first who solved the 3D structure of the insect cells-produced HEV-like VLPs by electron cryomicroscopy and image reconstruction and defined it at a resolution of 22 Å. The VLPs were formed with 60 copies of a 50 kDa protein arranged in the T = 1 icosahedral symmetry with protruding dimers at the icosahedral 2-fold axes. The capsid was dominated therefore by dimers that defined the 30 morphological units. The protein shell of these hollow particles extended from a radius of 50 Å outward to a radius of 135 Å (Xing et al. 1999).

Two teams published the crystal structures of the T = 1 HEV-like VLPs in the same issue of the *Proc Natl Acad Sci U S A* journal. Thus, Yamashita et al. (2009) worked with the truncated capsid protein, aa 112–608, from a genotype 3 strain, but not of genotype 1, as before (Li TC et al. 1997, 2005; Xing et al. 1999), and they solved the structure at 3.5 Å resolution. Guu et al. (2009) achieved the same 3.5 Å resolution of the T = 1 HEV-like structures built up by the same 112–608 ORF2 fragment but originated from a genotype 4 strain. Based on the T = 1 HEV-like structure, the authors modeled the full HEV T = 3 capsid by aligning the HEV structure with that of the three quasi-equivalent capsid protein molecules from tomato bushy stunt virus (see Chapter 24). Figure 16.3 presents both T = 1 and T = 3 structures, the diameters of which were estimated as 270 Å and 410 Å, respectively, in comparison with the experimental diameter 320–340 Å of the HEV virions determined earlier (Bradley et al. 1988; Bradley 1990).

It must be noticed that Li S et al. (2009) reported at the same time the crystal structure of the protruding spike (dimerization domain) of the recombinant capsid protein E2 located on the ORF2 aa 455–602 and expressed in *E. coli*, after renaturation from inclusion bodies. The dimer structure was refined up to 2.0 Å resolution.

Xing et al. (2010) found that the RNA interaction with the N-terminus of the ORF2, namely aa 14–111, led to the

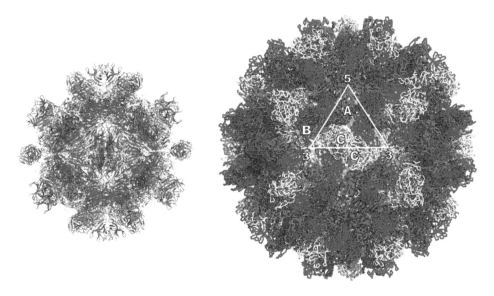

FIGURE 16.3 Structure of the hepatitis E virus-like particles. (left) The crystal T = 1 structure of HEV VLPs. The 3 domains, S, P1, and P2 are colored blue, yellow, and red, respectively. The VLP is positioned in a standard orientation with the 3 2-fold icosahedral symmetry axes aligned along the vertical, horizontal, and viewing directions, respectively. (right) The overall T = 3 HEV capsid model. The three quasi-equivalent capsid protein molecules A–C are colored blue, red, and yellow, respectively. One asymmetric unit is highlighted along with icosahedral symmetry axes. The images are to scale. (Guu et al. Structure of the hepatitis E virus-like particle suggests mechanisms for virus assembly and receptor binding. *Proc Natl Acad Sci U S A*. 106: 12992–12997, Copyright 2009 National Academy of Sciences, U.S.A.)

formation of the T = 3 virions and prevented formation of the T = 1 subviral particles.

Mori and Matsuura (2011) thoroughly reviewed the 3D structures of the HEV-like particles and spike domain protruding from the surface of the particle expressed by recombinant baculovirus or bacteria, in order to give insights into the mechanisms underlying the particle assembly, antigenicity, host cell attachment, and native virion packaging. At last, Liu et al. (2016) combined the electron cryomicroscopy and x-ray crystallography to get more details of the insect cells-produced HEV VLP structure at 3.5 Å resolution.

CHIMERIC VLPs

Niikura et al. (2002) were the first who used the HEV VLPs as a carrier for foreign antigenic epitopes by expression in insect cells. Thus, a 11 aa B cell epitope tag QPELAPEDPED from glycoprotein D of herpes simplex virus was examined for insertion into six insertion sites, including four internal positions and both the N- and the C-termini of the HEV ORF2 missing the N-terminal 111 aa, as introduced earlier by Li TC et al. (1997). Only the N- and C-terminal insertions resulted in the chimeric VLPs being released into the culture supernatant at significant amounts, although small amounts were released when the insertions were made at two of four internal positions. These results indicated that the internal insertions somehow disturbed the release of the putative VLPs into the culture supernatants. However, the N-terminal insertion also prevented formation of the reliable VLPs. Therefore, only the chimeric VLPs carrying the epitope at the C-terminus of the HEV capsid protein were

characterized further by electron microscopy as having 25 nm in diameter and morphologically appearing similar to the HEV VLPs without any tag. Moreover, the inserted epitope was exposed on the VLP surface. After oral administration, without adjuvant, the chimeric VLPs induced significant levels of specific antibodies to both the inserted epitope and HEV (Niikura et al. 2002).

Jariyapong et al. (2013) engineered the chimeric HEV VLPs carrying a 15-aa p18 epitope RIQRGPGRAFVTIGK, derived from the V3 loop of the HIV-1 Env protein gp120, at the position 485 of the antibody-binding domain P. After baculovirus-driven expression in insect cells, the T = 1 p18-VLPs resembled the tertiary and quaternary structures of the initial VLPs and appeared as a spherical projection decorated with spikes on a surface profile. The chimeric VLP reacted with an HIV-1 antibody against V3 loop but did not react with anti-HEV antibodies, due to the insertion at its antibody-binding site. Different from the initial VLPs, the chimeric VLPs were vulnerable to trypsin cleavage, suggesting that the intermolecular forces of attraction between the chimeric proteins were strong enough to maintain the icosahedral arrangement of the VLPs.

Shima et al. (2016) generated the chimeric HEV VLPs when Myc (EQKLISEEDL) or FLAG (DYKDDDDK) tags were inserted between aa residues 488 and 489, located at the exterior loop on the protruding P domain of the HEV capsid. The expression in insect cells resulted in the production of chimeric VLPs, although expression of the chimeric capsid protein carrying the HA tag YPYDVPVYA inserted at the same site failed to produce any particles. However, the coexpression with the Myc-tagged capsid protein successfully yielded a mosaic particle consisting of both recombinant capsid proteins. The immunoprecipitation analyses confirmed that the chimeric particles presented these foreign epitopes on the surface. Similar results were obtained by the expression of the chimeric capsid proteins carrying neutralizing epitopes of Japanese encephalitis virus (JEV) E protein: aa 337-IPIVSVASL-345 and 362-VATSSANS-369. Therefore, the HEV VLPs appeared to be good carriers that were able to accommodate multiple neutralizing epitopes on their surface.

Zahmanova et al. (2020) examined not only whether full-length and N- and C-terminally modified versions of the swine HEV capsid protein transiently expressed in *N. benthamiana* plants could assemble into VLPs but also assessed whether such VLPs could act as carriers of the influenza M2e peptide. The authors constructed the plant codon-optimized HEV RF2 capsid genes, in which the N-terminal, the C-terminal, or both parts of the protein were deleted. The M2e peptide was inserted into the P2 loop after the residue Gly556 of the HEV ORF2 protein by gene fusion, and three different chimeric constructs were designed. The plants expressed all versions of the HEV capsid protein up to 10% of total soluble protein, including the chimeras, but only the capsid protein consisting of aa 110–610 and chimeric M2-HEV 110–610 spontaneously assembled in higher-order structures. The chimeric VLPs assembled into particles 22–36 nm in diameter and specifically reacted with the anti-M2e antibody (Zahmanova et al. 2020).

Targeting and Delivery

To investigate whether HEV VLPs could deliver foreign genes specifically to the liver, Lee EB et al. (2019) expressed the N-terminally truncated HEV ORF2 in mammalian Huh7 cells that were transduced with recombinant baculoviruses and purified the VLPs by continuous density gradient centrifugation. The purified HEV VLPs efficiently penetrated liver-derived cell lines and the liver tissues. To evaluate the HEV VLPs as gene delivery tools, the foreign plasmids were encapsulated into the particles with a disassembly/reassembly procedure. As a result, the green fluorescence was detected at higher frequency in liver-derived Huh7 cells treated with HEV VLPs bearing GFP-encoding plasmids than in control cells. Moreover, the HEV VLPs bearing Bax-encoding plasmids induced apoptotic signatures in Huh7 cells and confirmed therefore their ability to support liver-targeted gene therapy (Lee EB et al. 2019).

FAMILY *MATONAVIRIDAE*

As stated in Figure 16.2, the glycoproteins E1 and E2 and the capsid protein of rubella virus (RV) are proteolytically cleaved from a polyprotein precursor by a natural infection. At the turn of the decade in the 1980–1990s, the genetically engineered RV structural proteins or their fragments, usually as fusion proteins, were produced in *E. coli* (Terry

et al. 1989; Wolinsky et al. 1991; Chaye et al. 1992), insect cells (Seppänen et al. 1991), and mammalian cells (Hobman et al. 1990; Baron and Forsell 1991; Sanchez and Frey 1991; Qiu et al. 1992).

The RV VLPs were obtained and released when the two envelope glycoproteins E1 and E2 and the capsid protein were expressed in a BHK cell line by using an inducible promoter (Qiu et al. 1994). The noninfectious RV VLPs contained therefore all three structural proteins and resembled RV virions in terms of their size, morphology, and antigenicity. The immunization of mice with the VLPs induced not only specific antibody responses against RV structural proteins but also generation of virus-neutralizing antibodies (Qiu et al. 1994). The BHK cell-line-produced RV VLPs were used for the greatly successful Roche Cobas Core Rubella IgG (Grangeot-Keros et al. 1995) and IgM (Grangeot-Keros and Enders 1997) assays.

In parallel, Hobman et al. (1994) developed a stably transfected Chinese hamster ovary (CHO) cell line that expressed the three structural RV proteins. The RV VLPs, which were formed by budding into the cisterna of the Golgi complex, were secreted further from the cells, and again resembled RV virions in their size and morphology and had an identical buoyant density when purified on sucrose gradients. Moreover, the VLP release was dependent upon the E1 cytoplasmic tail since deletion or substitution of this domain with the same region from vesicular stomatitis virus (VSV) G protein abrogated release of the RV proteins from transfected cells (Hobman et al. 1994).

Furthermore, the coordinated expression of the E1, E2, and capsid proteins in a stable CHO cell line, resulting in their self-assembly into the RV VLPs, as established earlier by Hobman et al. (1994), was used to study the specific fine mechanisms of rubella morphogenesis and secretion (Hobman et al. 1995; Garbutt et al. 1999a, b). The CHO-produced RV VLPs were also involved in the generation of novel diagnostic assays (Alekseev et al. 2001; Giessauf et al. 2002, 2005). Meanwhile, Lee JY et al. (1996) used the transient expression in COS cells to conclude that the dimerization of the RV capsid protein was not required for the VLP formation, while Yao and Gillam (2000) used BHK cells to map the E1 cytoplasmic domain.

Later, Claus et al. (2012) developed a particularly useful replicon *trans*-encapsidation system by transfecting a packaging cell line expressing the RV structural polyprotein with the in vitro RNA transcripts of replicons. The replicons were cDNA constructs in which the structural protein ORF was replaced with a reporter gene, and thus they were capable of replicating in transfected cells but not useful for the cell-to-cell spreading or virion formation because they did not encode the structural proteins. Such replicons were efficiently packaged with the *trans*-encapsidation system achieving secretion of the single-round infectious RV-like replicon particles (Claus et al. 2012).

The application of the RV VLPs allowed one to find out that the short self-interacting N-terminal region of the capsid protein was not essential for the VLP production but was critical for the virus infectivity, due to close cooperative actions of the capsid protein and protein p150 (Sakata et al. 2014).

FAMILY *ALPHATETRAVIRIDAE*

NUDAURELIA CAPENSIS ω VIRUS

Nudaurelia capensis ω virus (NωV), a reference species of the *Omegatetravirus* genus, remains the most-studied representative of the *Alphatetraviridae* family. The insect-infecting NωV was found in the wild infecting South African pine emperor moth larvae and other *Lepidoptera* insects, and it was not grown in any cell-culture system.

As mentioned earlier, NωV possessed the T = 4 icosahedral symmetry in contrast to the other members of the *Hepelivirales* order, and it was therefore a rare example of this symmetry, together with the hepatitis B core protein (see Chapter 38), that both appeared as resolved structures at the same time. The unique NωV structure was resolved by the famous John E. Johnson and Lars Liljas teams. Originally, it was established that each of the 240 capsid subunits contained 645 aa and the subunits assembled into a provirion. Then, an autocatalytic maturation cleavage occurred in this protein α between Asn570 and Phe571 to form proteins β and γ (Agrawal and Johnson 1992).

The expression of the capsid gene was achieved by cloning the capsid protein encoding sequences in *E. coli*, which resulted in the synthesis of a 70 kDa product (Agrawal and Johnson 1992). However, there was no cleavage of the precursor observed in protein expressed from *E. coli*, and a putative self-assembly process in bacterial cells was not achieved.

McKinney et al. (1994) inserted the NωV capsid gene into a baculovirus vector and performed expression in the insect *Sf*9 and *Sf*21 cell lines, although yield of the VLPs was rather low. Further attempts led to expression of the capsid gene in the insect cells, resulting in the formation of VLPs of a size consistent with the T = 4 quasi-symmetry observed for native virions (Agrawal and Johnson 1995). Although the initial VLP yields were rather low, an efficient one-step nondenaturing procedure involving separation on discontinuous glycerol gradients ensured quantities sufficient for further characterization. The electron microscopic observation revealed 40-nm particles that were morphologically similar to native virus. The SDS-PAGE revealed that the VLPs were composed of a 62-kDa major protein and a minor 70-kDa protein. The pulse-chase experiments revealed that the larger species was processed into the smaller one very slowly over the course of an infection. Furthermore, these particles were found to encapsidate the polyhedrin promoter-directed capsid mRNA with an apparently striking degree of specificity and selectivity. This investigation established that a specific encapsidation signal existed within the capsid coding sequences and that the components required for reconstructing most, if not all, steps in the morphogenesis of this virus could be accomplished in the baculovirus-infected cells (Agrawal and Johnson 1995). Figure 16.4a presents a portrait of the excellent NωV T = 4 VLPs produced in insect cells by the baculovirus-driven expression.

Johnson et al. (1994) published the first preliminary crystal structure of the NωV capsid at 6 Å resolution. Munshi et al. (1996) improved resolution to 2.8 Å and described finally the NωV as the T = 4 particles with a mean outer diameter of 410 Å and formed by 240 copies of a single

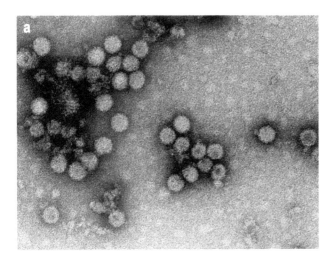 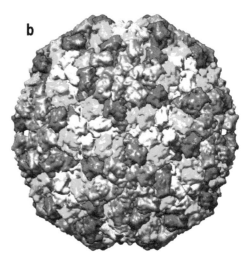

FIGURE 16.4   3D structure of the NωV capsid. (a) The electron micrograph of the NωV VLPs of 40 nm in diameter. (Reprinted from *Virology* 207, Agrawal DK, Johnson JE, Assembly of the T = 4 Nudaurelia capensis ω virus capsid protein, posttranslational cleavage, and specific encapsidation of its mRNA in a baculovirus expression system, 89–97, Copyright 1995, with permission from Elsevier.) (b) The subunit organization of the NωV VLPs, PDB ID 1OHF, according to the crystal structure at 2.8 Å resolution (Helgstrand et al. 2004). The outer and inner diameters are 42.2 and 23.4 nm, respectively. The structural data are taken from the VIPERdb (http://viperdb.scripps.edu) database (Carrillo-Tripp et al. 2009).

subunit type. The subunit was composed of a helical inner domain (where the cleavage occurred) containing residues preceding and following a canonical, viral, eight-stranded β-sandwich that formed the contiguous shell. Inserted between two strands of the shell domain were 133 residues with an immunoglobulin c-type fold. The initial gene product consisted of 644 amino acid residues and was cleaved between residues Asn570 and Phe571 in the mature particle determined in this analysis (Munshi et al. 1996), confirming therefore the early findings of Agrawal and Johnson (1992). Remarkably, a close similarity to the prefabricated pentameric helical bundles to those observed in nodaviruses, namely, black beetle and Flock House viruses (see Chapter 23), was established.

Next, the electron cryomicroscopy and image analysis allowed discovery of a procapsid of the insect cells-produced NωV VLPs (Canady et al. 2000). Furthermore, Canady et al. (2001) used the time-resolved small-angle x-ray scattering to fix the large conformational change that occurred during the NωV maturation. Taylor et al. (2002) analyzed the large-scale, pH-dependent, quaternary structure changes of the NωV VLPs. Finally, Helgstrand et al. (2004) published the refined NωV structure and explained the two dramatically different pH-dependent T = 4 assembly states of the NωV when expressed as VLPs in a baculovirus system. The procapsid was round, porous, and approximately 450 Å in diameter at pH 7. It converted, in vitro, to the capsid form at pH 5, and the capsid was sealed shut, shaped like an icosahedron with a maximum diameter of 410 Å, and underwent an autocatalytic cleavage at residue 570. The residues 571–644, the γ peptide, remained associated with the particle and were partially ordered. Thus, the refined structure allowed the description of the chemistry of molecular switching for the T = 4 capsid formation (Helgstrand et al. 2004). Figure 16.4b shows the current NωV T = 4 structure taken from the protein data bank (PDB).

Later, the detailed role of the large conformational changes and autoproteolysis in the maturation of the T = 4 VLPs was studied thoroughly (Bothner et al. 2005; Taylor and Johnson 2005; Matsui et al. 2009; Tang et al. 2009).

A first step toward generation of the chimeric NωV VLPs was made by Maree et al. (2006). Thus, a His$_6$ tag was inserted into the outer GH loop between Ala378 and Gly379 of the surface-exposed Ig-like domain of the NωV capsid protein. The His-tagged capsid protein p70 was produced in insect cells and self-assembled into His-NωV VLPs that exhibited similar morphological and RNA encapsidation properties as the wild-type NωV VLPs, whereas the tags were presented on the surface of the VLPs (Maree et al. 2006).

Concerning functionalization of the VLPs, NωV was probed by standard lysine acylation and thiol alkylation reactions (Taylor et al. 2003). This insect virus undergoes a massive conformational change upon proteolytic maturation from a 480-nm diameter procapsid to a 410-nm diameter virion (Taylor et al. 2002). Its response to both chemical reagents (Taylor et al. 2003) and proteolytic enzymes

(Bothner et al. 2005) was dramatically affected by this structural rearrangement, with the compact mature particle being much less reactive because it both exposed fewer reactive side chains on average and because it was dynamically less flexible. The NωV functionalization attempts were summarized at that time by Strable and Finn (2009).

At last, the famous George P. Lomonossoff's team demonstrated strong arguments to develop the NωV VLPs as a promising candidate of protein nanocages for oral drug delivery (Berardi et al. 2020). The authors transiently expressed the NωV VLP capsids and procapsids in plant *N. benthamiana*. The obtained capsids were highly resistant to simulated gastric fluids at pH ≥ 3. Even under the harshest conditions, which consisted of a pepsin solution at pH 1.2, the NωV capsids remained assembled as VLPs, though some digestion of the coat protein occurred. The high resistance of this protein cage to digestion and denaturation was attributed to its distinctively compact structure. The more porous form of the VLPs, the procapsid, was less stable under all conditions (Berardi et al. 2020). Furthermore, Castells-Graells et al. (2021) used expression of the NωV VLPs in *N. benthamiana* as a model to study viral maturation processes, purifying both immature procapsids and mature capsids from infiltrated leaves by varying the expression time. The electron cryomicroscopy solved the plant-produced procapsids and mature capsids to 6.6 Å and 2.7 Å, respectively, and unveiled some intrinsic maturation details, such as the ~30 Å translation-rotation of the subunits during maturation as well as conformational rearrangements in the N- and C-terminal helical regions of each subunit. Remarkably, the plant-produced mature particles were biologically active in terms of their ability to lyse membranes and had a structure that was essentially identical to authentic virus (Castells-Graells et al. 2021).

OTHER TETRAVIRUSES

Hanzlik et al. (1995) expressed the capsid protein precursor gene of the Helicoverpa armigera stunt virus (HaSV), another member of the *Omegatetravirus* genus after NωV, in *E. coli* but did not detect any VLPs. However, the expression of the HaSV p17 protein sequence in *E. coli* led to appearance of tubular structures (Hanzlik et al. 1995).

Tomasicchio et al. (2007) expressed the HaSV capsid gene in *S. cerevisiae*. The VLPs assembled as procapsids that matured spontaneously in vivo as the yeast cells began to age. The growth in the presence of hydrogen peroxide or acetic acid, which induced apoptosis or programmed cell death, resulted in the VLP maturation. These results demonstrated that the assembly dependent maturation of the tetravirus procapsids in vivo was linked to the onset of apoptosis in yeast cells and that the reduction in pH required for tetraviral maturation might be the result of cytosolic acidification, which is associated with the early onset of programmed cell death in infected cells (Tomasicchio et al. 2007).

The crystal structure of the insect cells-produced HaSV VLPs was resolved by D. Taylor and John E. Johnson to 2.5 Å and placed in the protein data bank (PDB) under the 3S6P identification code.

FAMILY *BENYVIRIDAE*

The rod-shaped VLPs of beet necrotic yellow vein virus (BNYVV) resembling the true BNYVV virions were assembled in leaves of plants *Chenopodium quinoa*, *Tetragonia expansa*, and *Beta vulgaris*, when the BNYVV capsid gene was inserted into the vectors where the expression of foreign genes was driven by the subgenomic promoter of the coat protein from zygocactus virus X (ZVX), a representative of the *Potexvirus* genus, order *Tymovirales* (Koenig et al. 2006). The formation of the rod-shaped VLPs was also observed when the capsid gene of soil-borne cereal mosaic virus (SBCMV), a representative of the *Furovirus* genus from the family *Virgaviridae*, order *Martellivirales*, was expressed by the same vector in the *C. quinoa* leaves, as described in Chapter 19.

17 Order *Martellivirales*: *Bromoviridae*

I do not seek. I find.

Pablo Picasso

ESSENTIALS

The *Martellivirales* is an order of quite different single-stranded positive-sense RNA viruses that played an outstanding role by the development of viral nanotechnology. For this reason, the story about this order is divided into four chapters, in accordance with the most studied families: *Bromoviridae* (the present chapter), *Togaviridae* (Chapter 18), *Virgaviridae* (Chapter 19) as well as Other Families (Chapter 20). According to the current taxonomy (ICTV 2020), the order *Martellivirales* is one of the three members of the class *Alsuviricetes*, together with the two other orders, *Hepelivirales* and *Tymovirales*, the subjects of the two neighboring chapters. The *Alsuviricetes* class belongs to the *Kitrinoviricota* phylum from the kingdom *Orthornavirae*, realm *Riboviria*.

The *Martellivirales* order currently involves 7 families, 25 genera, and 228 species. The order includes first the three great previously mentioned families, as well as less popular families *Closteroviridae*, *Endornaviridae*, *Kitaviridae*, and *Mayoviridae*.

The family *Bromoviridae* under the current official ICTV view (Bujarski et al. 2019; ICTV 2020) includes 6 genera and 36 species, where the genera *Alfamovirus*, *Bromovirus*, and *Cucumovirus* are of primary concern to the VLP nanotechnologies. The *Bromoviridae* members are exclusively plant viruses, whose virions are variable in morphology (i.e., spherical or bacilliform) and are transmitted between hosts mechanically, in/on the pollen and nonpersistently by insect vectors. The members of the family are responsible for major disease epidemics in fruit, vegetable, and fodder crops such as tomato, cucurbits, bananas, fruit trees, and alfalfa. The six genera are based on virus host range, genome content, and vector. The members of the genera *Alfamovirus* and *Cucumovirus* are transmitted by aphids, those of *Anulavirus* and *Ilarvirus* by thrips and/or pollen, the members of *Bromovirus* by beetles, while the transmission route for members of the genus *Oleavirus* remains unknown (Bujarski et al. 2019).

Figure 17.1 presents a portrait of a typical representative and schematic cartoons of the family members, which could be spherical or quasi-spherical of 26–35 nm in diameter (genera *Anulavirus*, *Bromovirus*, *Cucumovirus* and *Ilarvirus*) or bacilliform of 18–26 nm by 30–85 nm in size (genera *Alfamovirus*, *Ilarvirus* and *Oleavirus*) and possess predominantly the T = 3 icosahedral symmetry, while the T = 1 symmetry was also noticed.

The rapid progress of the *Bromoviridae* member-based VLP nanotechnologies was made substantial by clear recognition of the fact that plant viruses are a safer alternative to the use of bacterial and animal viruses, as reviewed originally by Porta and Lomonossoff (1998).

GENOME

Figure 17.2 shows the genomic structure of the six genera of the *Bromoviridae* family.

The tri-segmented genomes of the *Bromoviridae* members are about 8 Kb in total. The genomic RNAs are packaged in separate virions that may also contain subgenomic, defective, or satellite RNAs (Bujarski et al. 2019). The ssRNAs are provided with 5′-terminal cap structures, whereas the 3′-termini form tRNA-like or other structures that can be aminoacylated (genera *Bromovirus* and *Cucumovirus*) or not (genera *Alfamovirus*, *Anulavirus*, *Ilarvirus* and *Oleavirus*) (Bujarski et al. 2019).

The important feature of the separate virions is realized, e.g., in the case of brome mosaic virus (BMV), a reference strain of the *Bromovirus* genus, by appearance of the three separate particles that are indistinguishable by electron microscopy. The RNA1 and RNA2 are packaged into separate particles, whereas RNA3 and a subgenomic RNA4 (not shown in Figure 17.2), which is derived from RNA3 and encodes the coat protein only, are copackaged within one particle (Rao 2006; Annamalai and Rao 2007). This unusual feature permits therefore the three types of particles to package RNAs of similar total lengths. Remarkably, this is an original way of the realization of viral genome, which differs from the approach used by viruses that contain multiple RNAs within the same virions, such as reoviruses of the *Reovirales* order (Chapter 13) or influenza viruses from the order *Articulavirales* (Chapter 33).

GENUS *ALFAMOVIRUS*

ALFALFA MOSAIC VIRUS

Alfalfa mosaic virus (AMV) is the only member of the *Alfamovirus* genus. As mentioned earlier, the native AMV particles are bacilliform, which is a rare occasion among plant viruses. Bol et al. (1971) demonstrated that the N-terminus of the AMV coat of 221 aa residues is located on the surface of the virions and does not interfere with virus assembly. When the N-terminus of the AMV coat was removed by mild trypsin treatment, the particles lost their bacilliform shape and became spherical (Bol et al. 1974). If the particles were completely disrupted by salt, reassociation of the bacilliform particles was not possible in vitro (Hull 1970). In the presence of various RNA or DNA molecules, the AMV

DOI: 10.1201/b22819-22

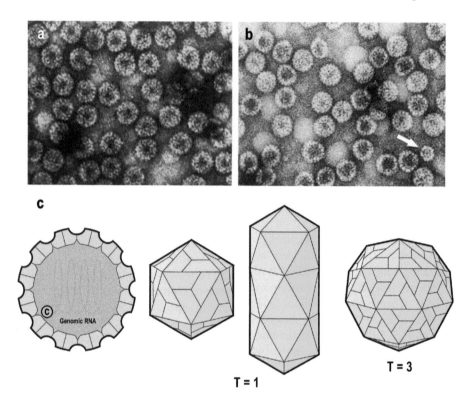

FIGURE 17.1 The portrait and schematic cartoons of the typical representatives of the *Bromoviridae* family. (a) Native brome mosaic virus (BMV) in pH 7.4 0.05 M Na-cacodylate. (b) Purified chymotrypsin-treated BMV under the same conditions. The arrow points to a small particle built on the T = 1 model, which results from the breakdown and reassembly of BMV particles, the RNA of which was already cleaved in situ: such structures form in increasing amount with increasing pH, temperature, and reaction time with chymotrypsin. Uranyl formate staining. Final magnification X 270,000. (From Pfeiffer P, Hirth L: The effect of conformational changes in brome mosaic virus upon its sensitivity to trypsin, chymotrypsin and ribonuclease. *FEBS Lett.* 1975.56.144–148. Copyright Wiley-VCH Verlag GmbH & Co. KGaA. Reproduced with permission.) (c) The general cartoons of the *Bromoviridae* representative virions are taken with a kind permission from the ViralZone, Swiss Institute of Bioinformatics, http://viralzone.expasy.org (Hulo C et al. 2011). (Courtesy Philippe Le Mercier.)

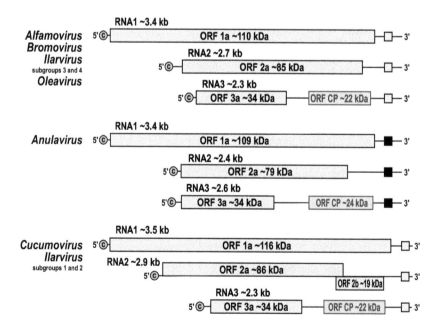

FIGURE 17.2 The genomic structure of the family *Bromoviridae* members, derived from the recent ICTV profile (Bujarski et al. 2019). The 3'-termini form either tRNA-like or complex structures and are shown as black or gray square boxes, respectively. The structural genes are colored dark pink.

coat aggregated into spherical or ovoid particles or into tubular particles with lengths characteristic of the nucleic acid (Hull 1970; Driedonks et al. 1977, 1978). In the absence of RNA, the AMV coat dimers self-assembled to form empty T = 1 icosahedral particles, which have been crystallized (Fukuyama et al. 1981) and resolved by x-ray diffraction to 4.5 Å (Fukuyama et al. 1983). This structure was reprocessed to 4.0-Å resolution, which allowed the tracing of the polypeptide chain of the coat confirming the β-sandwich fold and provided information on intersubunit interactions in the particle (Kumar et al. 1997). In this study, the particle structure was also determined by the electron cryomicroscopy and image reconstruction methods and found to be in excellent agreement with the x-ray model. Figure 17.3 presents this

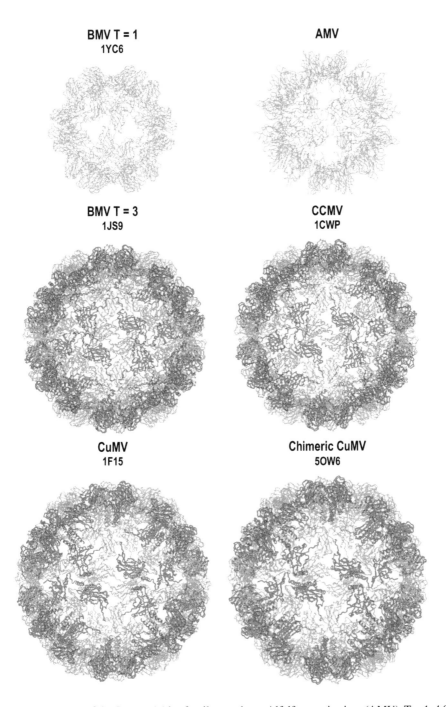

FIGURE 17.3   The crystal structure of the *Bromoviridae* family members: Alfalfa mosaic virus (AMV), T = 1, 4.0 Å resolution, outer diameter 222 Å (Kumar et al. 1997); Brome mosaic virus (BMV), T = 1, 2.90 Å, 194 Å (Larson et al. 2005); T = 3, 3.40 Å, 284 Å (Lucas et al. 2002b); CCMV, T = 3, 3.20 Å, 288 Å (Speir et al. 1995); CuMV, T = 3.3.20 Å, 302 Å (Smith et al. 2000); chimeric CuMV with tetanus-epitope, electron cryomicroscopy structure, 4.20 Å, 306 Å (Zeltins et al. 2017). The 3D structures are taken from the VIPERdb (http://viperdb.scripps.edu) database (Carrillo-Tripp et al. 2009). The size of particles is to scale. The corresponding protein data bank (PDB) ID numbers are given under the appropriate virus names.

structure, together with the 3D structures of other members of the *Bromoviridae* family.

Yusibov et al. (1996) achieved the assembly of icosahedral T = 1 particles by the recombinant AMV coat expressed in *E. coli*. The AMV coat gene was cloned and expressed in *E. coli* as a fusion protein containing a 37-aa extension with the His_6 tag for affinity purification. About half of the expressed coat was soluble upon extraction, and half remained insoluble within inclusion bodies. The empty particles were formed in vitro within six days of extensive dialysis of the Ni^{2+} resin-purified coat, were similar in diameter to those assembled from native coat (Fukuyama et al. 1981), and formed hexagonal crystals that diffracted x-rays to 5.5 Å resolution. The crystals of trypsin-treated particles were isomorphous with crystals of trypsin-treated particles of native coat protein. The preliminary analysis of the x-ray diffraction data indicated that the empty particles were similar in size and surface properties to those previously investigated by Fukuyama et al. (1983). The RNA-containing spherical particles self-assembled when the recombinant coat was combined with the in vitro transcripts of AMV RNA4, the subgenomic messenger for coat and the smallest naturally encapsidated viral RNA. The bacilliform particles that resembled native virions assembled when the recombinant coat was combined with transcripts of RNA1, the largest genomic RNA. It was concluded, first, that the N-terminal 37-aa extension did not affect the accessibility of the N-terminus for binding with the viral RNA. Second, the data proved the idea that the AMV coat can form particles of different sizes (20–60 nm) and shapes including spherical, ellipsoid, and bacilliform, depending on the length of the encapsidated RNA and reflecting the flexibility of the coat units by protein–protein interactions.

Meanwhile, the role of specific AMV coat aa residues was established for biological activity and RNA binding (Yusibov and Loesch-Fries 1998) and for the ability to form the unusually long virions (Thole et al. 1998).

The generation of the AMV-based chimeric particles was started by the famous Vidadi Yusibov's team. The AMV coat was used as a carrier to express antigenic peptides from HIV-1 and rabies virus (Yusibov et al. 1997). The selected epitopes were 47-aa V3 loop of HIV-1, isolate MN, and an engineered 40-aa antigen Drg24 that contained B-cell epitope G5–24 from rabies glycoprotein and a T cell epitope 31D from rabies nucleoprotein. The corresponding coding sequences were separately fused to the N-terminus of the full-length AMV coat gene and placed under the control of the subgenomic promoter of the tobacco mosaic virus (TMV) coat in the TMV-based plant vector (see Chapter 19). The in vitro transcripts of the recombinant viruses with sequences encoding the antigenic peptides were synthesized from DNA constructs and used to inoculate tobacco plants. The spherical AMV-like particles of the appropriate size were detected in both cases by electron microscopy. After purification, both chimeric viruses elicited specific virus-neutralizing antibodies in the immunized mice (Yusibov et al. 1997). Modelska et al.

(1998) immunized mice parenterally or orally with the spinach leaves-produced and purified or nonpurified AMV particles carrying the rabies virus epitope and demonstrated significant protection of mice against lethal challenge with rabies virus.

Next, Yusibov et al. (2002) used the same rabies virus epitope sequence, described in this paper as consisting of aa 253–275 from rabies virus glycoprotein G and aa 404–418 from nucleoprotein N. This epitope was fused again to the N-terminus of the AMV coat but expressed this time in two plant virus-based expression systems, namely *N. tabacum* providing replicative functions in trans for full-length infectious RNA3 of AMV and *N. benthamiana* and spinach *Spinacia oleracea* plants using autonomously replicating TMV vector lacking the coat gene. The chimeric virus from infected transgenic *N. tabacum* was used for parenteral immunization of mice, and the latter were protected against challenge infection. Based on the previously demonstrated efficacy of this plant virus-based experimental rabies vaccine when orally administered to mice in virus-infected unprocessed raw spinach leaves (Modelska et al. 1998), the efficacy of the spinach leaves as a vaccine was assessed in human volunteers. Three of five volunteers who had previously been immunized against rabies virus with a conventional vaccine specifically responded against the peptide antigen after ingesting spinach leaves infected with the recombinant virus. When rabies virus nonimmune individuals were fed the same material, five of nine volunteers demonstrated significant antibody responses to either rabies virus or AMV. Following a single dose of conventional rabies virus vaccine, three of these individuals showed detectable levels of rabies virus-neutralizing antibodies (Yusibov et al. 2002).

The putative vaccine against human respiratory syncytial virus (RSV), a primary cause of respiratory infection in infants, was constructed by Belanger et al. (2000). The two peptides containing aa 174–187 of the protein G, a potential virus-neutralizing epitope, of the human RSV A2 strain (NF1-RSV/172–187 and NF2-RSV/170–191) were separately engineered as N-terminal fusions with the AMV coat and individually expressed in *N. tabacum* plants through virus infection, using a newly developed AMV-based plant virus expression vector. The majority of particles recovered from the NF1-RSV- or NF2-RSV-infected plants had a spherical or ellipsoid shape, suggesting encapsidation of subgenomic RNA4 or genomic RNA3, respectively, in these virions. Long bacilliform particles suggestive of encapsidation of genomic RNA-1 or -2 were less frequently observed. Remarkably, both NF1-RSV and NF2-RSV variants were each passaged five consecutive times in new plants inoculated with an extract from systemically infected leaves of the previous passage without loss of the insert. Thus, these chimeric viruses retained the fusion protein not only during systemic movement throughout the entire plant but also on subsequent passage to uninfected plants. Moreover, high replication rate of AMV provided high levels of expression of the incorporated foreign sequences, so that on average

0.5 mg of NF1-RSV or NF2-RSV per gram of fresh tissue, equivalent to ~50 μg of pure RSV peptide, was purified. When mice were immunized intraperitoneally with three doses of the purified chimeric RSV-AMV virions, the protection against infection with RSV was achieved (Belanger et al. 2000).

Later, Yusibov et al. (2005) presented the RSV vaccine termed VMR-AMV and carrying the 21-aa epitope 170–190 of the RSV, strain A2, protein G, which was cloned in the full-length AMV RNA3-based vector VMR. The immunogenicity of the VMR-AMV vaccine was tested in vitro in human dendritic cells and in vivo in nonhuman primates, without the use of adjuvant, and significant pathogen-specific immune responses were generated in both systems.

An anthrax vaccine candidate was engineered by Brodzik et al. (2005). In this case, the small loop 15-aa epitope EGLKEVINDRYDMLN from domain-4 of the *Bacillus anthracis* protective antigen (PA-D4s) was inserted immediately after the first 25 N-terminal aa of the AMV coat, in order to retain genome activation and binding of coat to viral RNAs. The epitope was followed by 4 linker aa residues GSTA and then continued to the aa position 28 of 221 aa of the AMV coat. The chimeric virions were characterized by high yield, extended stability, systemic proliferation, and production consistency in tobacco plants. The intraperitoneal injections of mice with the chimeric virions harboring the PA-D4s epitope elicited a distinct immune response.

An AMV-based transmission-blocking malaria vaccine carrying the Pfs25 protein, which is expressed predominantly on the surface of the sexual and sporogonic stages of *Plasmodium falciparum* including gametes, zygotes, and ookinete and remains as one of the primary targets for transmission blocking malaria vaccines, was first mentioned in a conference thesis of Musivchuk et al. (2012) and reported in detail by Jones et al. (2013). The Pfs25-AMV VLPs were engineered by fusion of a stretch encompassing aa residues 23–193 of the *P. falciparum* Pfs25 to the N-terminus of AMV coat and produced in *N. benthamiana* plants using a TMV vector. To eliminate potential N-linked glycosylation sites in the Pfs25 protein, the two point mutations N112Q and N187Q were introduced. The native Pfs25 signal peptide was replaced at the N-terminus with the signal peptide of tobacco pathogenesis-related 1a protein precursor of *N. tabacum*, which was predicted to be subsequently removed during protein expression and secretion into the endoplasmic reticulum. The purified VLPs were highly consistent in size, namely, 19.3±2.4 nm in diameter, with an estimated 20–30% incorporation of Pfs25 onto the VLP surface. The immunization of mice with one or two doses of the Pfs25-AMV VLPs plus Alhydrogel induced serum antibodies with complete transmission blocking activity through the six month study period (Jones et al. 2013). Chichester et al. (2018) described production of this malaria vaccine candidate, named Pfs25 VLP-FhCMB, in *N. benthamiana* at pilot plant scale under current good manufacturing practice (cGMP) guidelines. Moreover, the results of the Phase

I clinical trials on the safety, reactogenicity, and immunogenicity assessed in healthy adult volunteers were presented. This vaccine candidate appeared therefore as the first VLP-based transmission-blocking vaccine designed for use in human clinical trials that was registered at www.ClinicalTrials.gov under reference identifier NCT02013687. However, although the vaccine induced Pfs25-specific antibodies in a clinical trial, there was limited inhibition of parasite transmission to mosquitoes, suggesting the need for improved vaccine formulations (Chichester et al. 2018). Furthermore, McLeod et al. (2019) explored the structure and function of the transmission-blocking antibodies elicited in a human vaccinated with the Pfs25 VLP-FhCMB vaccine. The authors determined the cocrystal structures of four human antibodies bound to Pfs25 and described three novel protective epitopes on Pfs25, including the most potent mAb yet described from any organism. The studies of the plasmablast lineage from which this potent transmission-blocking antibody derives revealed the evolutionary steps required for potent parasite inhibition. These insights into the maturation of potent transmission-blocking antibodies in a human were intended to guide the development of increasingly effective biomedical interventions against malaria (McLeod et al. 2019).

Shahgolzari et al. (2021) used the AMV virions that were isolated and purified from alfalfa *Medicago sativa* plant as a tool for the in situ vaccination (ISV) against 4T1, an extremely aggressive and metastatic murine triple-negative breast cancer model. The AMV used as an ISV significantly slowed down tumor progression and prolonged survival through immune mechanisms including an increase of costimulatory molecules, inflammatory cytokines, and immune effector cell infiltration and the downregulation of immune-suppressive molecules. It was concluded that AMV virions appeared to be among the more immunostimulatory plant viruses. Remarkably, they were not cytolytic and did not induce apoptosis of tumor cells directly but rather mediated their impact through stimulating the immune response against the tumor. Therefore, the AMV intratumoral treatment changed the local tumor microenvironment via induction of immune-modulating cytokines and recruitment or phenotypic change of immune cells (Shahgolzari et al. 2021).

GENUS *BROMOVIRUS*

BROME MOSAIC VIRUS

3D Structure

Bancroft et al. (1967, 1969) found that the purified coat protein of BMV self-assembled in vitro into capsids similar to those of the native virus at a pH and salt-dependent efficiency with best yields at pH 5.0 in 0.2 M NaCl. To form VLPs, the purified coat protein was simply dialyzed overnight at 6°C against the appropriate buffer, and high quality of the product was confirmed by electron microscopy. Remarkably, from the two other members of the *Bromovirus* genus, the cowpea chlorotic mottle virus (CCMV) coat formed nice VLPs, as described later, while the broad bean mottle virus (BBMV) coat did not aggregate into the recognizable organized structures.

Krol et al. (1999) were the first who clearly showed by electron cryomicroscopy that the BMV coat could assemble in vivo into two remarkably distinct capsids that selectively packaged BMV-derived RNAs in the absence of BMV RNA replication, namely by the expression in yeast cells: a 180-subunit capsid indistinguishable from the T = 3 virions produced in natural infections and a BMV capsid with 120 subunits arranged as 60 coat dimers, according to the T = 2 symmetry. Each such dimer contained two coats in distinct, nonequivalent environments, in contrast to the quasiequivalent coat environments throughout the 180-subunit capsid. This 120-subunit capsid utilized most of the coat interactions of the 180-subunit capsid plus nonequivalent coat–coat interactions. The T = 1 BMV capsids of 60 capsids were assembled earlier in vitro when the dissociated BMV virions were treated with trypsin so that the 64 N-terminal aa residues were removed and reassociation resulted in the T = 1 particles (Cuillel et al. 1981). Lucas et al. (2001) crystallized the T = 3 and T = 1 BMV particles and prepared them therefore for the x-ray crystallography studies.

Lucas et al. (2002b) determined the T = 3 structure of BMV, the type member of the *Bromoviridae* family, by x-ray crystallpgraphy at the 3.40 Å resolution. In fact, the structure was solved by molecular replacement using the model of CCMV, which closely resembled BMV and the 3D structure of which was resolved in 1995, as described later. The BMV virions were propagated in barley plants. The coat protein had the canonical "jelly-roll" β-barrel topology with extended N-terminal polypeptides as seen in other icosahedral plant viruses. Remarkably, a significant fraction of the N-terminal peptides was apparently cleaved in native BMV virions. The protein subunits forming hexameric capsomeres, particularly dimers, interacted extensively, but the subunits otherwise contacted one another sparsely about the 5-fold and quasi 3-fold axes. Thus, the virion appeared to be an assembly of loosely associated hexameric capsomeres, which could be the basis for the swelling and dissociation that occured at neutral pH and elevated salt concentration. The Mg^{2+} ion appeared to be essential for maintenance of virion stability (Lucas et al. 2002b). Figure 17.3 presents this structure, as well as the 3D structure of the T = 1 BMV particles that were resolved by x-ray crystallography to 2.90 Å by Larson et al. (2005). The T = 1 particles of BMV were created by treatment of the wild-type T = 3 virions with 1 M $CaCl_2$. The particles were composed of pentameric capsomeres from the wild-type virions, which have reoriented with respect to the original particle pentameric axes by rotations of 37° and formed tenuous interactions with one another, principally through conformationally altered C-terminal polypeptides. Otherwise, the pentamers were virtually superimposable upon those of the original T = 3 BMV particles. In fact, the T = 3 particles, in the crystals, were not perfect icosahedra but deviated slightly from exact

symmetry, possibly due to packing interactions. This suggested that the T = 1 particles were deformable, which was consistent with the loose arrangement of pentamers and latticework of holes that penetrated the surface. The T = 3 to T = 1 transition could occur by shedding of hexameric capsomeres and restructuring of remaining pentamers accompanied by direct condensation. Surprisingly, there was little resemblance between the T = 1 particles of BMV and AMV. The BMV particle, with a maximum diameter of 194 Å, was made from distinctive pentameric capsomeres with large holes along the 3-fold axis, while the AMV particle, of approximate maximum diameter 222 Å, had subunits closely packed around the 3-fold axis, large holes along the 5-fold axis, and few contacts within pentamers. In both particles crucial linkages were made about icosahedral dyads. Although residual architectural elements of the structures of the native T = 3 virions persisted in both cases, the modes and mechanisms of reassembly of the BMV and AMV particles could be quite different (Larson et al. 2005).

The specific interaction of the BMV coat with the BMV RNAs that regulated not only the RNA encapsidation and formation of VLPs but also the RNA synthesis and translation was reviewed by Kao et al. (2011).

Wang et al. (2014) integrated a number of novel technologies to resolve by electron cryomicroscopy the de novo T = 3 near-atomic structure of the BMV virions to 3.8 Å, comparable therefore with resolution of the previously described crystal structure of Lucas et al. (2002b). The diameter of the BMV virion was determined in this study as ~284 Å.

Moreover, using electron cryomicroscopy, Beren et al. (2020) depicted the RNA genome organization within the multipartite BMV virion and unveiled interaction of the genomic RNA with coat protein. The findings demonstrated that, in contrast to the monopartite viruses, like RNA bacteriophages of the *Leviviricetes* class (Chapter 25) utilizing specific protein–RNA interactions to drive virion assembly and having highly organized RNA genomes, BMV takes advantage of nonspecific electrostatic interactions between coat and RNA to ensure the separate but simultaneous packaging of two or more molecules of the multipartite genome, resulting in significantly less order in the genome (Beren et al. 2020).

Expression

By the previously described studies on the two forms of BMV capsids, Krol et al. (1999) expressed the BMV coat gene and certain BMV RNAs in yeast *Saccharomyces cerevisiae*. The coexpression of the BMV coat together with RNA2 led to formation of the BMV VLPs that were similar to the plant-derived BMV virions. When the BMV coat gene was expressed alone, the VLPs carrying coat mRNA were detected. This sort of VLP had an average diameter approximately 10% smaller than the traditional T = 3 BMV, a drastically different surface morphology, and was characterized as the putative T = 2 particles described earlier.

Choi and Rao (2000) engineered a TMV-based vector to express the BMV coat in plants *Chenopodium quinoa* via an independent subgenomic RNA. The electron microscopy revealed the presence of icosahedral VLPs with two distinct sizes: approximately 90% of VLPs measured 28 nm in diameter and were indistinguishable therefore from those of the wild-type BMV T = 3 virions, whereas the remaining 10% measured 22–24 nm in diameter and were suspected to possess T = 1 symmetry. The larger VLPs packaged the three subgenomic RNAs of the chimeric virus, while the smaller ones packaged only the two smaller subgenomic RNAs. Remarkably, the in vitro reassembly assays demonstrated that the ability of BMV capsids to display polymorphism was not dependent on the RNA size alone and appeared to be controlled by some other feature(s) of the genetically engineered RNA (Choi and Rao 2000). In these early studies, the BMV VLPs seemed to be stablized by RNA-protein interactions and could not form empty VLPs (Fox et al. 1994; Choi et al. 2000).

Gopinath et al. (2005) developed an efficient T-DNA-based gene delivery system using *Agrobacterium tumefaciens* to transiently express BMV RNAs in *N. benthamiana*. Along with other advantages, this system allowed studies of the *cis-* and *trans*-acting requirements for the BMV RNA replication in plants.

To accelerate high level production of VLPs, Moon et al. (2014) presented a technique for highly efficient expression of the synthetic BMV gene in *N. benthamiana* and production and purification of self-assembled VLPs by transient expression system using agroinfiltration. In parallel, the VLPs of cucumber mosaic virus (CuMV; described later) and maize rayado fino virus (MRFV) of the *Tymoviridae* family, order *Tymovirales* (see Chapter 21), were obtained.

New Materials

The BMV coat was characterized as a forthcoming packaging agent by the well-known Bogdan Dragnea team and emerged therefore as the first icosahedral capsid to be studied with respect to nanoparticle encapsidation (Dragnea et al. 2003; Chen et al. 2005, 2006; Dixit et al. 2006). Thus, Chen et al. (2006) described in detail the self-assembly of regular protein surfaces around nanoparticle templates, providing therefore a new class of hybrid biomaterials with potential applications in medical imaging and in bioanalytical sensing. The first example of this approach was illustrated by the VLPs having a BMV coat and a functionalized gold core. This new methodology was reported for gold nanoparticle encapsidation that provided a yield of incorporation greater than 95%. The carboxylate-terminated thiolalkylated tetraethylene glycol (TEG) was found as a ligand that satisfied both the negative charge and the stability conditions required for self-assembly, in contrast to the previously employed citrate functionalization of gold nanoparticles. After the TEG functionalization, the latter had a diameter of 16.0 ± 1.2 nm, closely corresponding to the inner diameter of the viral BMV capsid of 17–18 nm. The functionalized TEG-coated gold nanoparticles initiated

therefore the VLP assembly by mimicking the electrostatic behavior of the nucleic acid component of the native virions. The VLPs were symmetric, 26 ± 2 nm in size, with the protein stoichiometry and packaging properties indicating similarity to the icosahedral packing of the capsid, as confirmed by electron cryomicroscopy. Moreover, a pH-induced swelling transition of the VLPs was observed, in direct analogy to the native virions. It should be noted that the assembly reactions lacking gold resulted in the formation of empty VLPs with sizes varying from 20–29 nm as well as incomplete capsids (Chen et al. 2006).

Sun et al. (2007) varied the diameter of gold cores and accomplished control over the capsid structure, since the number of subunits required for a complete capsid increased with the core diameter. As shown in Figure 17.4, the VLPs of varying diameters were found to resemble to three classes, T = 1, 2, and 3, of viral particles found in cells and described earlier. The 3D reconstructions of the transmission electron micrographs of the single VLPs with 6-nm gold cores demonstrated the protein shell structure

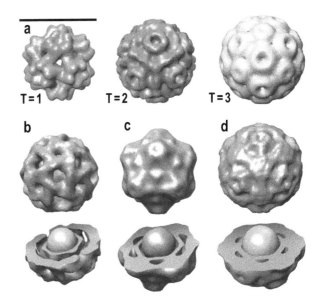

FIGURE 17.4 3D reconstructions using negative stain data for R3BMV, a control that contains packaged RNA3 and RNA4, and VLPs carrying the gold nanoparticles of an indicated size. (a) T = 1, 2, and 3 models of BMV capsids. The T = 1 and pseudoT = 2 structures were obtained from the VIPER database (Natarajan et al. 2005) and Lucas et al. (2002b). The T = 3 structure is the reconstructed image of R3BMV in this work, scale bar, 210 Å. (b) VLP$_6$ is characterized by the absence of electron density at the 3-fold symmetry axes. Its structure and diameter bring it close to a T = 1 capsid. (c) The VLP$_9$ structure is reminiscent of a pseudoT = 2. The presence of electron density at the 3-fold axes distinguishes it from the VLP$_6$ structure. (d) The VLP$_{12}$ shape resembles more the spherical shape of R3BMV although it still lacks clear evidence of hexameric capsomers. Concentric layering is a characteristic of all VLPs. (From Sun J et al. Core-controlled polymorphism in virus-like particles. Proc Natl Acad Sci U S A. 104:1354–1359, Copyright 2007 National Academy of Sciences, U.S.A.)

corresponding to the T = 1 capsids. The VLPs with 9-nm gold cores shared the same structure as pseudoT = 2 capsids, and 12-nm-diameter gold cores were encapsulated in a shell that appeared identical to the wild-type BMV, which has T = 3 capsids. As a consequence of their regularity, the VLPs formed 3D crystals under the same conditions as the wild-type virions. Such crystals represented a form of metallodielectric material that exhibited optical properties influenced by multipolar plasmonic coupling (Sun et al. 2007). The coarse-grained thermodynamic and kinetic models describing experiments in which BMV coat assembled around charge-functionalized gold nanoparticle cores were elaborated by Hagan (2009). Furthermore, Daniel et al. (2010) demonstrated a critical role of the surface charge density for the BMV VLP assembly through an analysis aimed at decoupling the surface charge and the core size. Tsvetkova et al. (2012) elaborated equilibrium models providing estimates for the thermodynamic forces driving the BMV coat assembly around gold nanoparticles, as well as the size of the critical nucleus in the case of cooperative growth.

Dixit et al. (2006) incorporated CdSe/ZnS semiconductor quantum dots (QDs) into the BMV VLPs, providing a new paradigm for the design of intracellular microscopic probes and vectors. The functionalization with HS-poly(ethylene glycol) (PEG)-COOH, a PEG with a terminal carboxylate, led to the VLPs with minimal release of photoreaction products and enhanced stability against prolonged irradiation. Remarkably, the DNA-coated QDs yielded VLPs that tended to be unstable in storage. The image reconstructions of single VLPs indicated that the protein shell structure corresponded to the pseudoT = 2 shell (Dixit et al. 2006).

Furthermore, Huang et al. (2007) reported the BMV VLPs that were formed around superparamagnetic iron oxide nanotemplates protected by carboxy-terminated PEGylated phospholipids (HOOC-PEG-PL). These VLPs were interesting for their superparamagnetic properties and possible applications such as magnetic resonance imaging and biomagnetic materials. A novel feature of the VLP assembly obtained in this work was the use of an anionic lipid micelle coat instead of a molecular layer covalently bound to the inorganic nanotemplate, as before. Remarkably, the BMV VLPs larger in size than the T = 3 virions were obtained using spherical iron oxide templates, which showed the versatility of assembling viral coat containers that could accept larger loads than the native virions (Huang et al. 2007).

Developing this issue further, Huang et al. (2011) reported formation of the BMV VLPs with cubic iron oxide nanoparticles (which were validated as MRI contrast agents) and performed a comparative study of the transit of cubic particles, 18.6 nm diameter, functionalized with PEGylated phospholipids, with and without viral coats, in *N. benthamiana* leaves. The virus capsids influenced both subcellular and long-distance transport, and the VLPs with magnetic cores were transported across long distances and

inside cells of different types, while transport of nanoparticles devoid of the BMV coats was limited (Huang et al. 2011).

Furthermore, Malyutin et al. (2015a, b) used two models, plant-derived BMV and mammals-derived hepatitis B core (HBc), which is described in Chapter 38, to explore formation of the VLPs utilizing 22–24 nm iron oxide nanoparticles as cores. To accomplish that, hydrophobic FeO/Fe$_3$O$_4$ nanoparticles prepared by thermal decomposition of iron oleate were coated with poly-(maleic acid-alt-octadecene) modified with PEG tails of different lengths and grafting densities and successfully used in MRI studies. The viral coats readily formed shells that exceeded their native size. The location of the long PEG tails upon shell formation was demonstrated by electron cryomicroscopy and small-angle x-ray scattering. The MRI studies showed unique relaxivity ratios that diminished only slightly with the gold coating. Remarkably, the PEG tails were located differently in the BMV and HBc VLPs, with the BMV VLPs preferentially entrapping the tails in the interior and the HBc VLPs allowing the tails to extend through the capsid, which highlighted the differences between intersubunit interactions in these two icosahedral viruses (Malyutin et al. 2015b). In fact, this was the first time when different packing behavior was demonstrated for PEG tails, which depended on the VLP model type.

More recently, Mieloch et al. (2018) synthesized superparamagnetic iron oxide nanoparticles Fe$_3$O$_4$ of 15 nm in diameter by thermal decomposition and functionalized with the previously mentioned COOH-PEG-PL polymer or dihexadecylphosphate and encapsulated them into the BMV VLPs.

Jung et al. (2011) were the first who encapsulated an organic near-infrared (NIR) chromophore into a plant virus for the subsequent utilization of the construct in mammalian intracellular optical imaging. Specifically, the BMV virions were disassembled and the coat protein purified and reassociated to encapsulate indocyanine green (ICG), an FDA-approved NIR chromophore. Thus, the BMV interior was doped successfully with ICG. The authors termed these nanoprobes optical viral ghosts (OVGs), emphasizing the fact that the VLPs no longer contained the genomic machinery. The constructs were highly monodispersed with standard deviation of ±3.8 nm from a mean diameter of 24.3 nm. They were physically stable and exhibit a high degree of optical stability at physiological temperature. Using human bronchial epithelial cells, the constructs were efficient for intracellular optical imaging in vitro, with greater than 90% cell viability after 3 h of incubation (Jung et al. 2011). Furthermore, Gupta et al. (2013) presented the first proof-of-principle demonstration of NIR-photoacoustic imaging with the ICG-encapsulated BMV particles. When used in appropriate amounts, the OVGs displayed a much stronger photoacoustic signal than that emitted by blood over the wavelength range of 760–820 nm. The OVGs also demonstrated superior photostability in comparison to non-encapsulated ICG in solution. Remarkably, the OVGs did not elicit an acute immunogenic response in healthy mice (Gupta et al. 2013).

Chemical Coupling

After active application of BMV in packaging techniques, Yildiz et al. (2012) were the first who explored BMV in a variety of bioconjugation chemistries. The latter were intended to display multiple copies of targeting ligands as well as therapeutic and/or imaging moieties, in addition to the encapsulated cargos. To do this, the unique cysteine side chains were introduced to the BMV surface by genetic engineering, as shown in Figure 17.5. In brief, the aa position Val168 of the BMV coat was mutated to a Cys residue using site-directed mutagenesis. The infectious cDNA clones containing the full-length genome of BMV were transformed into *A. tumefaciens* and then infiltrated into *Nicotiana benthamiana* plants to launch production of the mutant cBMV. Then, thiol-maleimide chemistry and hydrazone ligation were applied for the development of surface-modified functional cBMV nanoparticles carrying (i) fluorophores, such as OregonGreen 488 (O488); (ii) hydrophilic PEG polymers; (iii) tumor-homing proteins; (iv) chemotherapeutic moieties; and (v) cell-penetrating peptides. The dyes facilitated tracking and detection of the VLPs, while PEG reduced immunogenicity of the proteinaceous nanoparticles and enhanced

their pharmacokinetics. Moreover, the authors cleared the use of additional aa side chains on the BMV surface. Since native BMV display 1440 solvent-exposed surface Lys side chains, that is eight Lys residues per coat protein, it was confirmed that the Lys side chains are also addressable, and double labeling of BMV was performed. For example, BMV particles displaying transferrin attached to Cys residues and O488 attached to Lys side chains were successfully generated. Thus, the door for the direct application of BMV in nanomedicine was opened (Yildiz et al. 2012).

Nicole F. Steinmetz's team presented detailed protocols of the propagation, purification, and chemical conjugation methodologies for the BMV model in *N. benthamiana*, in parallel with other greatly successful models of cowpea mosaic virus (CPMV), described in Chapter 27, tobacco mosaic virus (TMV), Chapter 19, and potato virus X (PVX), Chapter 21 (Wen et al. 2012).

Recently, Nuñez-Rivera et al. (2020) loaded BMV and CCMV VLPs with the NanoOrange fluorophore and assayed them on breast tumor cells. Both BMV and CCMV VLPs were internalized into the cells but did not show any cytotoxic effect. Since the BMV VLPs were less immunogenic, they were selected as a potential carrier for the cargo delivery and loaded with small interfering RNA (siRNA) without packaging signal. First, the gene silencing was demonstrated by the BMV VLPs loaded with siGFP and tested on breast tumor cells that constitutively expressed GFP. Next, the BMV VLPs carrying siAkt1 were shown to suppress the tumor growth in mice (Nuñez-Rivera et al. 2020).

COWPEA CHLOROTIC MOTTLE VIRUS

3D Structure

It is basically significant that the CCMV VLPs, together with those of the RNA phage MS2 and Qβ from the *Leviviricetes* class (Chapter 25), were identified among the popular scaffolds by their physical stability as the most critical parameter by the VLP production, storage, and medical or industrial applications (Mateu 2011, 2013, 2016). Moreover, CCMV was the first icosahedral virus reassembled in the 1960s from purified viral protein and RNA to form infectious particles. Thus, Bancroft et al. (1967, 1968, 1969) found that the CCMV coat self-assembled in vitro with the best outcome at pH 5.0 in 0.1 M NaCl or from pH 3.5 to 5.0 in 0.2 M NaCl. However, in contrast to the previously described BMV coat, the CCMV coat protein also formed double-shelled and rosette-like particles in 0.2 M NaCl from pH 5.3 to 5.7. The narrow tubes were made at pH levels of 6.0 and higher, and the tubes were accompanied by T = 1 and T = 3 particles if ribonuclease-digested RNA was present. In the absence of NaCl in 0.01 M acetate buffer at pH 4.0 and 4.5, the CCMV coat formed laminar and plate-like aggregates (Bancroft et al. 1969). The CCMV self-assembly model was reviewed among other early putative VLP models by Fox et al. (1994). Fox et al. (1997) characterized an R26C mutant of CCMV, which displayed

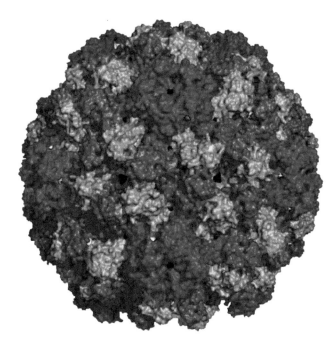

FIGURE 17.5 Structure of BMV. The 180 identical coat proteins are arranged in a T = 3 symmetry. Val168 is highlighted in yellow. The mutated particles are cBMV chimeras displaying genetically engineered Cys side chains at position 168 in the BMV coat protein. The structural information was obtained from viperdb. scripps.edu. The image was created using Pymol software. (From Yildiz I, Tsvetkova I, Wen AM, Shukla S, Masarapu MH, Dragnea B, Steinmetz NF. Engineering of Brome mosaic virus for biomedical applications. *RSC Adv.* 2012;2:3670–3677. Reproduced by permission of the Royal Society of Chemistry.)

increased virion stability and was abnormal in virion disassembly, when purified under nonreducing conditions. Remarkably, the reduced virions were infectious, whereas nonreduced virions remained noninfectious.

Furthermore, CCMV was the first bromovirus possessing a fine 3D structure. Speir et al. (1993, 1995) resolved the crystallographic CCMV structure to 3.2 Å and found it unique compared with previously determined RNA virus structures. The 3D structure of CCMV revealed specific structural features that probably explained the chemical basis for the previously mentioned formation of native and polymorphic assembly products. The structure is compared with other bromovirus 3D structures in Figure 17.3.

The CCMV capsid was made up of 180 chemically identical protein subunits that formed a 286 Å-diameter icosahedral shell with T = 3 quasisymmetry. Each protein subunit was composed of 190 aa residues taking three quasi-equivalent positions on the virus surface. The residues 1–41 in subunit A and residues 1–26 in subunits B and C were disordered. All three subunits had similar folds for equivalent residues. The residues 50–178 formed the canonical virus β-barrel motif, and residues 27–49 and 179–190 extended from the β-barrel at its N- and C-ends. Such native CCMV was stable around pH 5.0 in a compact form. When the pH of the solution was raised to 7.0 at low ionic strength in the absence of divalent cations, the viral particles underwent a 10% radial expansion at the quasi-3-fold axes and produced so-called swollen forms (Speir et al. 1995). The pseudo-atomic models of swollen CCMV were resolved from electron cryomicroscopy data to 28 Å by Liu H et al. (2003). According to this structure, the particle flexibility was accommodated primarily by changes in dimer interactions, an observation that was consistent with the flexible C-terminal polypeptide extensions that stabilized this contact in the crystal structure. In brief, the transition caused formation of 60 separate openings of ~20 Å in the protein shell.

Tang et al. (2006) found that a mutant CCMV coat lacking most of the N-terminal domain, so-called NΔ34, assembled in vitro into T = 3, T = 2, and T = 1 capsids, and a cryo-EM image of each of them was reconstructed. The T = 2 CCMV particles were similar to the previously described BMV T = 2 particles of Krol et al. (1999). In brief, the bromovirus capsid geometry appeared flexible enough to readily adapt to new requirements as the virus evolved.

It is noteworthy that the SIMNANOWORLD™ methodology searching the free-energy landscape of a nanosystem for deep minima was first applied to thermal structures of the CCMV capsid (Cheluvaraja and Ortoleva 2010).

Remarkably, the colloidal nanobubbles occuring in gas-saturated aqueous solutions triggered the self-assembly of CCMV coat in the absence of the viral genome, most likely by acting as a negatively charged template (Zhang M et al. 2020). Moreover, the nanobubble-induced self-assembly depended on protein concentration, where low coat concentrations led to assembly of 18-nm T = 1 VLPs and high concentrations led to 28-nm T = 3 VLPs.

Expression

Zhao X et al. (1995) established the assembly system to produce the whole CCMV virions from coat protein synthesized in *E. coli* and viral RNA transcribed in vitro from full-length cDNAs. When examined with electron cryomicroscopy and image reconstruction at 25 Å resolution, these particles were found indistinguishable from native virions. The mutational mapping showed that the N-terminal part of the coat protein was required for assembly of RNA-containing particles but not for the assembly of empty virions, while the C-terminus was essential for the coat protein dimer formation and particle assembly (Zhao X et al. 1995).

Fox et al. (1996) selected the CCMV mutant K42R with enhanced virion stability, the VLPs of which resisted disassembly in 1.0 M NaCl, pH 7.5, whereas the wild-type virions completely disassociated into RNA and capsid protein components. The expression of the K42R mutant coat protein in *E. coli* followed by in vitro assembly produced virions that exhibited the same salt-stable phenotype. Speir et al. (2006) resolved the crystal structure of the salt-stable K42R mutant to 2.7 Å and found an addition of 660 new intersubunit interactions per particle at the center of the 20 hexameric capsomeres, which were a direct result of the K42R mutation. The protease-based mapping experiments of intact particles demonstrated that both the swollen and closed forms of the wild-type and K42R particles had highly dynamic N-terminal regions, yet the K42R particles were more resistant to degradation (Speir et al. 2006).

Concerning other prokaryotic expression systems, Phelps et al. (2007) expressed the CCMV coat gene in *Pseudomonas fluorescens*. The coat was assembled in vivo into VLPs that were of similar size and shape to the native CCMV particles derived from plants. The CCMV VLPs were purified by PEG precipitation followed by separation on a sucrose density gradient and analyzed by electron microscopy. The DNA microarray experiments revealed that the VLPs encapsulated very large numbers of different host RNAs in a nonspecific manner (Phelps et al. 2007).

Brumfield et al. (2004) expressed the CCMV coat gene in yeast *Pichia pastoris* and got VLPs that were indistinguishable from virions produced in the natural plant host by the electron cryomicroscopy and image resonstruction analysis. In parallel, a collection of noninfectious CCMV coat mutants was expressed in the *P. pastoris* system, which could assemble into VLPs with altered architectures and function, providing thus an alternative to other heterologous expression systems, expression in which has resulted in unassembled proteins.

The formation of the CCMV VLPs in insect cells was mentioned as an unpublished observation by Hassani-Mehraban et al. (2015).

New Materials

CCMV was the first VLP that was used as a host for the synthesis of the combined regular organic-inorganic materials and appeared therefore as the first nanocontainer in the world. Trevor Douglas and Mark Young (1998) reported the pioneering mineralization of two polyoxometalate species (paratungstate and decavanadate) and the encapsulation of an anionic polymer inside the *E. coli*-produced CCMV VLPs, controlled by pH-dependent gating of the particle pores. The previously described pH-dependent swelling phenomenon allowed the virus particle to selectively entrap and release materials from within the central cavity by a reversible gating mechanism, making this a versatile strategy for materials synthesis and molecular entrapment. Moreover, this pH-dependent gating mechanism of the CCMV VLPs was coupled to a pH-dependent inorganic oligomerization reaction in order to load, crystallize, and entrap mineral particles of well-defined size. Thus, the oligomerization of aqueous molecular tungstate, WO_4^{2-}, to form paratungstate, $H_2W_{12}O_{42}^{10-}$, polyanions was induced by lowering the pH. The paratungstate polyanions had significantly lower solubility than the precursor tungstate ions. The empty virions were incubated with the inorganic precursor ions under conditions (pH>6.5) where the VLPs existed in its open swollen form and allowed all ions access to the central cavity. After incubation, the pH was lowered to a point where the virion underwent a structural transition and the pores in the protein shell closed. The VLP gating, coupled to the tungstate oligomerization, resulted in the spatially selective crystallization and entrapment of the paratungstate ion within the virion. The crystallization occurred only within the VLPs, and no bulk mineralization was observed in virion-free controls or in solutions containing only assembled VLPs without tungstate. A similar approach was used to constrain the mineralization of a vanadate $V_{10}O_{28}^{6-}$ within the CCMV VLPs, which also resulted in the formation of monodisperse nanoparticles with essentially the same size distribution as seen for the paratungstate. Then, the pH-dependent encapsulation of an anionic organic polymer (poly-antetholesulphonic acid) was undertaken. The polymer was incubated with the empty VLPs at high pH (7.5) followed by a lowering of the pH to well below the gating threshold (pH 4.5), which resulted in selective encapsulation of the polymer (Douglas and Young 1998). This revolutionary route to generate the gate-access chambers that could perform chemical reactions was reviewed by the authors (Douglas and Young 1999). Remarkably, this review mentioned the animal Norwalk virus (NV) from the *Calicivirdae* family, order *Picornavirales* (Chapter 27), as a partner model by the CCMV gate-access chamber studies. The NV VLPs formed cores virtually identical to those observed for the CCMV VLPs, although NV was not known to swell under similar conditions.

Douglas et al. (2002) genetically modified the CCMV coat by replacing 9 basic aa residues at the N-terminus with glutamic acid and prepared so-called SubE mutant. The mutant assembled readily into a cage-like architecture similar to the wild-type. This electrostatically altered viral protein cage catalyzed the rapid oxidation of Fe^{II} leading to the formation of a spatially constrained iron oxide nanoparticle within the cage. To do this, the purified *P. pastoris*-produced SubE cages were treated with aliquots of Fe^{II} at

pH 6.5 and allowed to oxidize in air. In the presence of the anionic empty SubE cages, the reaction proceeded to form a homogeneous orange solution, whereas reactions in the presence of the wild-type empty protein cages resulted in the bulk precipitation of an orange solid (Douglas et al. 2002). Later, de la Escosura et al. (2008) reported successful packaging of Prussian blue, the first modern synthetic pigment produced by oxidation of ferrous ferrocyanide salts, in the CCMV VLPs and organization of the latter into hexagonal patterns on mica and hydrophilic carbon surfaces.

Basu et al. (2003) presented the first direct evidence of metal binding to the putative metal-binding sites, originally suggested from the CCMV crystal structure. The authors used fluorescence resonance energy transfer (FRET), from tryptophan residues proximal to the putative metal-binding sites, to probe Tb^{3+} binding to the VLPs, which was investigated on the wild-type and a mutant where the RNA binding ability of the coat was removed. The Tb^{3+} binding was observed in both cases, while Ca^{2+} had about 100-fold less affinity for the binding sites. All these previously described CCMV nanocage approaches of the Douglas team were systematically reviewed by Liepold et al. (2005).

In parallel, Allen et al. (2005) demonstrated that gadolinium ions can be chelated directly into the CCMV VLPs with an apparent dissociation constant of 31 µM at the sites of the natural Ca^{2+} binding. This contributed therefore strongly to the development of novel paramagnetic nanoparticles and MRI contrast agents. Using FRET, the binding affinity of Gd^3 was characeirzed by competition with Tb^{3+}. The resulting nanoparticles exhibited high relaxivity, which was attributed to the high density of Gd^{3+} ions and the large rotational correlation time of the nanoparticle assembly. Further, this idea was used by Anderson et al. (2006) to generate RNA phage MS2-based MRI contrast agents, as described in Chapter 25, which is devoted to the class *Leviviricetes*.

Slocik et al. (2005) examined the *P. pastoris*-produced CCMV SubE, as well as the HRE-SubE variant that was engineered with the interior HRE peptide epitopes AHHAHHAAD—which were reported earlier to bind and stabilize Au^0, Ag^0, Pt^0, and Cu^0 nanoparticles—and used them as viral templates for the potentiated reduction and symmetry directed synthesis of gold nanoparticles. In the first approach, the VLPs actively potentiated the reduction of $AuCl_{4-}$ by electron transfer from surface tyrosine residues resulting in a gold nanoparticle decorated viral surface. The viral reduction appeared to be selective for gold as a collection of metal precursor substrates of Ag^+, Pt^{4+}, Pd^{4+}, and an insoluble Au^I complex were not reduced to zero-valent nanoclusters by the VLPs. Alternatively, the VLPs provided a template for the symmetry directed synthesis of Au^0 nanoparticles from a nonreducible gold precursor (Slocik et al. 2005). It is noteworthy that nanoindentation studies were performed with the SubE and wild-type capsids, both full and empty, where full capsids resisted indentation more than empty capsids, but all of the capsids were highly elastic (Michel et al. 2006).

When the gold nanoparticle cores, capped with the previously mentioned carboxylate-terminated thiolalkylated tetra(ethylene glycol) (TEG) ligands, were assembled with the N-terminally shortened CCMV NΔ34 mutants, the size distribution of nanoparticle-templated VLPs was narrower than the broad size distribution of mutant empty capsids, and its median value depended on the diameter of the core particle (Aniagyei et al. 2009).

Moving from the interior to the exterior of the VLPs by the generation of novel materials, Suci et al. (2005) performed a big study for the CCMV absorption behavior on Formvar, bare Si, Formvar-coated Si, and Si modified by aminopropyltriethoxysilane. The most appropriate modern physical methodologies were used to characterize the CCMV adsorption, while the Langmuir model was applied to provide a description of the kinetic absorption behavior.

Kostiainen et al. (2010) presented for CCMV a self-assembly and photo-triggered disassembly based on electrostatic interactions to assemble viruses into hierarchical supramolecular complexes. The assembly was achieved using photosensitive dendrons, a Newkome-type scaffold functionalized with a polyamine (spermine) surface, which bound on the CCMV surface through multivalent interactions and then acted as a molecular glue between the particles. The optical triggering induced the controlled decomposition and charge switching of dendrons, which resulted in the loss of multivalent interactions and the release of the VLPs.

Furthermore, Kostiainen et al. (2013) used nanogold to demonstrate that the CCMV capsids can guide the assembly of different materials into complex hierarchical structures. The obtained CCMV-Au binary superlattice was unique in the sense that it had no atomic or molecular counterparts and has not been observed previously with nanoscale objects. Further large hierarchically ordered structures were presented by Korpi et al. (2018) when the CCMV VLPs were cocrystallized with supercharged cationic polypeptides (SUPs) composed of 72 consecutive lysine-containing repeating units (K72) as well as GFP produced as a recombinant fusion with the same SUP tag (GFP-K72).

Brasch et al. (2011) encapsulated water-soluble zinc phthalocyanine, a very robust and versatile chromophore with numerous applications in medicine, photonics, electronics, and energy conversion, into the CCMV VLPs of the two different T = 1 and T = 3 sizes, depending on the conditions, with simply coincubation in solution. Luque et al. (2014) showed by electron cryomicroscopy an unprecedented, very high level of phthalocyanine molecule organization within both VLP classes. Further, Setaro et al. (2015) designed and synthesized negatively charged phthalocyanine dendrimers that behaved as photosensitizers for the activation of molecular oxygen into singlet oxygen and packaged them into the CCMV VLPs.

Millán et al. (2014) incorporated optically active and paramagnetic micelles of the ligand 1,4,7,10-tetraaza-1-(1-carboxymethylundecane)-4,7,10-triacetic acid cyclododecane (DOTAC10) inside the CCMV VLPs. The DOTAC10 ligand was used to complex Gd^{III}, in order to form paramagnetic micelles, as well as to encapsulate an amphiphilic zinc phthalocyanine dye that optically confirmed the encapsulation of the micelles. The incorporation of the zinc phthalocyanine molecules in the paramagnetic micelles led to high capsid loading of both Gd^{III} and phthalocyanine. The resulting protein showed good perspectives to be used as an MRI contrast agent and could be regarded as a first step toward the consecution of the CCMV cages for multimodal imaging and therapy.

Chemical Coupling

Trevor Douglas's team performed the first chemical modifications of the CCMV VLPs, which exploited the chemistry of native and engineered surface exposed functional groups for the multivalent presentation of desired ligands (Gillitzer et al. 2002). As a result, the exterior CCMV surface was chemically modified with both fluorescent molecules and small peptides. The CCMV surface modifications were obtained via coupling of fluorophores to surface-exposed amine groups of lysines, carboxylic acids of glutamates and aspartates and by engineered surface-exposed thiol groups of the cysteine Cys82 and Cys141 mutants. Moreover, a 24 aa peptide having antitumorgenesis properties was linked to the CCMV exterior through exposed lysine and cysteine (using in this case an A163C mutant) residues (Gillitzer et al. 2002).

Gillitzer et al. (2006) functionalized the exposed CCMV lysine residues with either biotin or digoxigenin ligands to generate two differentially labeled populations. These two populations were identical when examined by transmission electron microscopy, dynamic light scattering, and size-exclusion chromatography. After disassembly, the purified subunits were mixed in defined stoichiometric ratios under in vitro reassembly conditions. Again, the resulting particles were indistinguishable from initial particles (Gillitzer et al. 2006).

Klem et al. (2003) described the A163C mutant and its propagation and purification to homogeneity from yeast cells in detail. In fact, the mutant provided 180 exposed sulfhydryls on the exterior surface surrounding the pseudo 6-fold and 5-fold axes of the cage, which could be used to covalently bind the capsid to smooth Au substrate. The exposed nature of the sulfhydryls demanded the presence of a reducing agent such as 2-mercaptoethanol to prevent interparticle cross-linking via disulfide bond formation. Remarkably, breaking the symmetry of the capsid using a solid-phase approach and chemically passivating the exposed thiol groups with iodoacetic acid resulted in a capsid with exposed thiols only on one side of the particle. These symmetry-broken capsids were able to form self-assembled monolayers (SAM) on an Au surface (Klem et al. 2003).

Continuing the story of CCMV as a possible MRI contrast agent, Liepold et al. (2007) used two approaches. The first one consisted of the genetical incorporation of a 9-aa residue peptide sequence, from the Ca^{2+} binding protein calmodulin, as a genetic fusion to the N-terminus of the CCMV coat (CCMV-CAL) and the characterization of the metal binding to the genetically engineered capsid was undertaken using FRET analysis. The second approach was to covalently attach the clinically relevant contrast agent gadolinium-tetraazacyclododecane tetraacetic acid (GdDOTA) to reactive lysine residues on CCMV via an N-hydroxysuccinimide NHS) ester coupling reaction (CCMV-DOTA). Both variants indicated great potential of the CCMV capsids as high-performance contrast agents (Liepold et al. 2007).

The CCMV VLPs were functionalized by controlled integration of polymers (Comellas-Aragonès et al. 2009). In the resulting biohybrid nanomaterial, the exterior of CCMV was modified with PEG by attaching to the amine groups of lysine residues, while polystyrene sulfonate (PSS) was entrapped in the interior of the capsid after the self-assembly of the PEG-modified capsid. The PSS-CCMV-PEG VLPs were of 18 nm in diameter, possessed T = 1 symmetry, and demonstrated therefore the first example of VLPs bearing synthetic organic polymers both inside and outside the viral capsid. Earlier, Sikkema et al. (2007) reported self-assembly of the CCMV VLPs with the PSS inside into the T = 1 particles. Further, Hu et al. (2008) determined the size distributions of the VLPs formed by CCMV coat with PSS samples of increasing sizes, all comparable to or larger than the capsid sizes. Even though the size and charge of the polymer cargo increased monotonically over a broad range, only two discrete sizes of the CCMV VLPs were observed, corresponding to the two icosahedral-symmetry structures T = 2 and T = 3.

Cantin et al. (2011) conjugated influenza M2e peptides to the surface of the wild-type or two novel CCMV cysteine mutants, all expressed in *P. fluorescens*, where primary amine-directed and cysteine-directed schemes were compared. Both strategies were successful, although the cysteine-directed conjugation strategy using the CCMV cysteine mutants displayed key advantages over the primary amine-directed strategy.

Packaging and Delivery

Kaiser et al. (2007) were the first who performed biodistribution studies of the CCMV VLPs, in parallel with another protein cage nanoplatform, namely nonviral 12 nm heat shock protein (Hsp) cage. In naïve and immunized mice both nanoplatforms showed similar broad distribution and movement throughout most tissues and organs, rapid excretion, the absence of long-term persistence within mice tissue and organs, and no overt toxicity after a single injection. The two genetic variants of CCMV were used, the S102C mutant for epifluorescence studies and the K42R mutant for the biodistribution studies (Kaiser et al. 2007).

The multifunctional CCMV nanoplatform demonstrated potential for diagnosing and treating *Staphylococcus aureus* biofilm infections (Suci et al. 2007a, b). To do this, the biotynilated CCMV not only targeted protein A on *S. aureus* cells through streptavidin and biotinylated antiprotein A monoclonal antibody but also delivered MRI contrast agent, namely, GdDOTA, and imaged the extent of its diffusive transport into the *S. aureus* biofilm (Suci et al. 2007a).

When photosensitizer (Ru(bpy$_2$)phen-IA was conjugated to the biotynilated CCMV mutants S102C/K42R or S130C/K42R, the photodynamic inactivation of *S. aureus* occurred under standard light fluence conditions (Suci et al. 2007b).

Comellas-Aragonès et al. (2007) were the first who constructed a CCMV-based single-enzyme nanoreactor, when horseradish peroxidase was encapsulated as a model system and the capsid was permeable for substrate and product, while this permeability could be altered by changing pH.

Minten et al. (2009) employed the coiled-coil heterodimerization motifs to encapsulate GFP inside the CCMV VLPs in two-steps, first binding CCMV CP monomers to GFP, both modified to have the complementary motifs, followed by the VLP assembly by lowering the pH. Developing the construction of putative efficient nanoreactors, Minten et al. (2011) solved the problem of how to stabilize the CCMV VLPs at pH 7.5. As a result, the unprecedent stabilization was achieved by simple interaction of the N-terminally His-tag-modified CCMV coat with several metal ions, such as Ni^{2+} and many others.

The packaging of the CCMV VLPs with RNA molecules ranging in length from 140 to 12,000 nucleotides was evaluated by Cadena-Nava et al. (2012). Each of these RNAs was completely packaged if and only if the protein/RNA mass ratio was sufficiently high, and this critical value was the same for all of the RNAs and corresponded to equal RNA and N-terminal-protein charges in the assembly mix. For RNAs much shorter in length than the 3,000 nt of the viral RNA, two or more molecules were assembled into 24- and 26-nm-diameter capsids, whereas for much longer RNAs (>4,500 nt), a single RNA molecule was shared by two or more capsids with diameters as large as 30 nm. For intermediate lengths, a single RNA was assembled into 26-nm-diameter capsids, the size associated with the T = 3 particles (Cadena-Nava et al. 2012).

Rurup et al. (2014) reported loading of more than ten fluorescent proteins, namely teal fluorescent protein (mTFP) used as a fluorescent cargo model, into the 18-nm internal cavity of the CCMV VLPs, giving rise to a maximum efficiency of homo-FRET between the loaded proteins, as measured by fluorescence anisotropy.

Liu A et al. (2016) presented straightforward one-step conditions that were optimal for the encapsulation of gold nanoparticles of different sizes and surfactants into the CCMV cages and were applicable not only for different cargos but also for similar VLPs. The detailed protocols for the encapsulation of gold nanoparticles into the CCMV VLPs and convertion of the latter into catalytic nanoreactors were published (Liu A et al. 2018). Tagit et al. (2017) encapsulated semiconductor CdSe/ZnS quantum dots into the CCMV VLPs.

Biddlecome et al. (2019) reported a new gene-delivery platform that employed the self-amplifying mRNA replicon that was protected by the CCMV VLPs containing reporter proteins Luciferase or eYFP or the tandem-repeat model SIINFEKL epitope of ovalbumin in RNA gene form, coupled to the RNA-dependent RNA polymerase from the

Nodamura virus of the order *Nodamuvirales* (Chapter 23). The incubation of immature dendritic cells with these VLPs resulted in increased activation of maturation markers and enhanced RNA replication levels, relative to incubation with unpackaged replicon mRNA. It was concluded that the VLP protection enhanced mRNA uptake by dendritic cells.

As mentioned earlier, the NanoOrange fluorophore-loaded CCMV VLPs were used in parallel with the analogous BMV VLPs when assayed on breast tumor cells by Nuñez-Rivera et al. (2020). Although both CCMV and BMV VLPs were internalized into the cells and showed no cytotoxic effect, the BMV VLPs were less immunogenic and were used therefore for the further studies.

The CCMV VLPs were used to package CpG oligodeoxynucleotides (ODNs), namely ODN1826, which induced the phagocytic activity of macrophages by activating the Toll-like receptor 9 signaling pathway (Cai et al. 2020). The direct injection of the ODN1826-carrying CCMV VLPs into established tumors induced a robust antitumor response by increasing the phagocytic activity of tumor-associated macrophages in the tumor microenvironment. The CCMV encapsulation significantly enhanced the efficacy of ODN1826 compared to the free drug, slowing tumor growth and prolonging survival in mouse models of colon cancer and melanoma (Cai et al. 2020).

Recently, the CCMV VLPs, in parallel with the widely used VLPs of the RNA phage Qβ from the class *Leviviricetes* (Chapter 25), were used to prepare armored RNA as a positive control for the advanced Covid-19 RT-PCR tests (Chan et al. 2021a, b).

Ramos-Carreño et al. (2021) used the CCMV VLPs by the delivery of the encapsidated dsRNAi of 563 bp as antiviral therapy in shrimp against white spot syndrome virus (WSSV), a member of the nonordered family *Nimaviridae* described in Chapter 6. The WSSV glycoprotein VP28 was efficiently silenced in experimental challenges in vivo, and the shrimps were fully protected by intramuscular injection and up to 40% by oral administration (Ramos-Carreño et al. 2021). In fact, this was the first report of the oral administration of VLPs to treat shrimp against virus infection.

At last, Välimäki et al. (2021) provided the CCMV coat with a heparin-specific binding peptide and subsequently allowed to encapsulate heparin into the VLP cages. The encapsulation was specific and the resultant VLPs displayed negligible hemolytic activity, indicating proper blood compatibility and promising possibilities for heparin antidote applications.

Chimeric VLPs

Hassani-Mehraban et al. (2015) analyzed N- and C-termini and four predicted loops of the wild-type and NΔ24 CCMV coats for their potential use as target sites to foreign epitopes still allowing VLP formation by expression in *E. coli*. While insertions of the His₆ tag or M2e (7–23 aa) into the predicted external loop structures did abolish VLP formation, high yields of VLPs were obtained with all fusions of His₆ tag or various epitopes of 13 to 27 aa residues from influenza A virus proteins M2e and HA or foot-and-mouth disease virus (FMDV) proteins VP1 and 2C at the N- or C-terminal ends of CCMV CP or NΔ24-CP. The VLPs derived from the CCMV coat still encapsulated RNA, while those from the CCMV chimeras carrying a negatively charged N-terminal domain had lost this ability. Moreover, the chimeric CCMV VLPs were engineered, which contained selected sequences from the G_N and G_C glycoproteins of the recently emerged Schmallenberg orthobunyavirus (Chapter 32) at both termini of the coat (Hassani-Mehraban et al. 2015).

The well-known Martin F. Bachmann team, in collaboration with the Riga VLP group, demonstrated that the CCMV-derived VLPs of different morphology, icosahedral or rod-shaped, can be obtained by the expression in *E. coli* (Zinkhan et al. 2021). The shape of the CCMV VLPs was manipulated by adding the universal tetanus toxin (TT) epitope QYIKANSFIGITE, located at the TT aa positions 830–843, to the N- or C-terminus of the CCMV coat, by analogy with the pioneering work on the VLPs of CuMV, a representative of cucumoviruses, the evaluation of which is described later. Figure 17.6 demonstrates the electron microscopy observed outcome of these manipulations. The

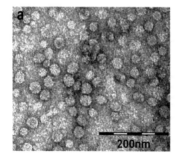
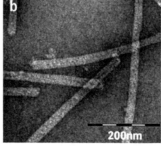
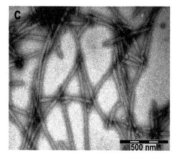

FIGURE 17.6 Directional insertion of tetanus toxin (TT) epitope in the N- or C-terminus results in round- or rod-shaped CCMV_TT VLPs. (a) Electron microscopy of round-shaped and (b, c) rod-shaped CCMV_TT VLPs, adsorbed on carbon grids and negatively stained with uranyl acetate solution, scale bars 200 nm (round) and 200 nm/500 nm (rod). Round-shaped CCMV_TT VLPs are ~30 nm in diameter, rod-shaped CCMV_TT VLPs are ~1 μm in length and ~ 30 nm in width. (Reprinted from Zinkhan S, Ogrina A, Balke I, Reseviča G, Zeltins A, de Brot S, Lipp C, Chang X, Zha L, Vogel M, Bachmann MF, Mohsen MO. The impact of size on particle drainage dynamics and antibody response. *J Control Release*. 2021;331:296–308.)

N-terminally modified CCMV$_{TT}$ VLPs formed icosahedral T = 3 particles, with a diameter of ~30 nm analogous to the parental VLPs. By incorporation of the TT epitope at the C-terminus of CCMV coat, rod-shaped VLPs, ~1 μm in length and ~ 30 nm in width, were obtained. When the draining kinetics and immunogenicity of both forms as potential B cell immunogens were evaluated, the round-shaped CCMV$_{TT}$ VLPs demonstrated greater efficiency in draining to secondary lymphoid organs to charge professional antigen-presenting cells as well as B cells. Furthermore, compared to the rod-shaped CCMV$_{TT}$ VLPs, the round-shaped ones led to more than 100-fold increased systemic IgG and IgA responses accompanied by prominent formation of splenic germinal centers. The round-shaped CCMV$_{TT}$ VLPs polarized the induced T-cell response toward Th1 (Zinkhan et al. 2021). In fact, this was the first study investigating and comparing the draining kinetics and immunogenicity of one and the same VLP monomer forming nano-sized icosahedra or rods in the micrometer size. The conclusion was that the round-shaped VLPs in the nm size range are vastly more immunogenic than the rod-shaped micron-sized particles.

GENUS *CUCUMOVIRUS*

CUCUMBER MOSAIC VIRUS

3D Structure

Smith et al. (2000) resolved the crystal structure of cucumber mosaic virus (CuMV) to 3.2 Å. This structure is presented in Figure 17.3. Despite the fact that CuMV has only 19% coat identity and 34% similarity to CCMV, the core structures of these two members of the *Bromoviridae* family are highly homologous. In CCMV, the structures of the A, B, and C subunits are initially identical except in their N termini. In contrast, the structures of two loops in subunit A of CuMV differ from those in B and C. Unlike that of CCMV, the capsid of CuMV does not undergo swelling at pH 7.0 and is stable at pH 9.0. This may be partly due to the fact that the N termini of the B and C subunits form a unique bundle of six amphipathic helices oriented down into the virion core at the threefold axes. In addition, while CCMV has a cluster of aspartic acid residues at the quasi-threefold axis that is able to bind metal in a pH-dependent manner, this cluster is replaced by complementing acids and bases in CuMV (Smith et al. 2000).

Later, the chimeric CuMV with tetanus-epitope (described later), so-called CuMV$_{TT}$ VLPs, were resolved by electron cryomicroscopy to 4.20 Å (Zeltins et al. 2017). The appropriate 3D structures of the CuMV VLPs are shown in Figure 17.3.

Expression

The synthetic gene of the CuMV coat was expressed in *E. coli*, and the corresponding VLPs were purified by Rostami et al. (2014). In parallel, Moon et al. (2014) presented a highly efficient technique to produce the CuMV VLPs in *N. benthamiana*, as described earlier for the plant-derived BMV VLPs.

The famous Bachmann's team that was earlier mentioned by the CCMV$_{TT}$ description started a series of their pioneering CuMV-based therapeutic vaccines, in collaboration with the Riga VLP group, with the cloning and expression in *E. coli* of the total RNA from CuMV-infected lily leaves collected from a private garden in Riga (Zeltins et al. 2017).

Chimeric Virions

Natilla et al. (2004) generated a chimeric CuMV, derived from the RNA3 component of the CuMV S strain, carrying the coat gene, and the RNA1 and 2 components of the CuMV D strain. This system developed mild mosaic and vein clearing in *N. tabacum* c.v. Xanthi plants three weeks after inoculation. To get chimeric virions, the coat gene was then engineered in three different internal HCV coat positions in order to expose an epitope from hepatitis C virus (HCV) of the *Amarillovirales* order (Chapter 22). The selected HCV peptide was the so-called R9 mimotope QTTVVGGSQSHTVRGLTSLFSPGASQN of 27 aa residues, a synthetic surrogate derived from a consensus profile of more than 200 hypervariable region 1 (HVR1) sequences of the HCV envelope protein E2. As a result, the chimeric virions were detected by immunoelectron microscopy, while crude plant extracts infected with the chimeric CuMV displayed a significant immunoreactivity with serum samples of chronic hepatitis C patients (Natilla et al. 2004). Piazzolla et al. (2005) demonstrated the ability of the CuMV-R9 chimeras to elicit a humoral response when parenterally administered to rabbits. Moreover, the stimulation of peripheral blood mononuclear cells (PBMC) from HCV-positive patients with the chimeras resulted in the activation of cellular immune responses in a significant percentage of patients infected with different HCV genotypes (Piazzolla et al. 2005).

Nuzzaci et al. (2007) doubled the number of the HCV R9 copies by parallel insertion of the epitope into a different position of the CuMV coat. As a result, the R9 appeared in the position between Asp176 and Ile177 in the βG-αGH region, where the first R9 had been inserted, and in the position between Ser131 and 132 in the βE-αEF, where the so-called 2R9-CuMV got the second R9. The 2R9-CuMV chimera systemically infected the host Xanthi plants, stably maintaining both inserts. Notably, it was strongly recognized by sera of HCV-infected patients and, as compared with R9-CuMV, displayed an enhanced ability to stimulate lymphocyte IFN-γ production. This stimulated the idea to develop an oral HCV vaccine (Nuzzaci et al. 2007).

Nuzzaci et al. (2009) presented excellent electron micrographs of both R9-CuMV and 2R9-CuMV virions, while Nuzzaci et al. (2010) achieved a humoral immune response in rabbits fed with the R9-CuMV-infected lettuce plants, suggesting that this system could function as a confirming tool of a bioreactor for the production of a stable edible vaccine against HCV. At last, the plant-produced R9-CuMV vaccine displayed a strong proapoptotic effect associated with activation of both caspase-8 and -9 in peripheral lymphocytes from patients with chronic hepatitis C (Piazzolla et al. 2012). It was suggested therefore that both the extrinsic and the intrinsic apoptotic pathway contributed to the downstream activation of the R9-CuMV-dependent apoptosis execution phase.

To engineer another putative vaccine, Zhao Y and Hammond (2005) inserted either a 17-aa neutralizing epitope (aa 65–81) or the fusion protein F or 8-aa neutralizing epitope (aa 346–353) of hemagglutinin-neuraminidase (HN) protein of Newcastle disease virus (NDV) from the *Mononegavirales* order (Chapter 31) into the CuMV coat and prepared chimeric infectious CuMV-NDV virions. The fusions of the F, HN or duplicated HN epitope HN_2 were made in the internal βH-βI loop within the CuMV coat. The chimeric coat gene-carrying RNA3 transcripts of the CuMV Ixora strain were inoculated on to *N. benthamiana*, together with CuMV RNA1 and 2. When the F and HN epitopes were placed in the internal motif, the chimeric virions were infectious, and the HN NDV epitope was recognized by anti-NDV sera, while a duplication of the HN epitope rendered the virus nonviable (Zhao Y and Hammond 2005).

Furthermore, Natilla et al. (2006) placed the CuMV Ixora isolate coat gene under the transcriptional control of the duplicated subgenomic coat promoter of a potato virus X (PVX)-based vector. The in vitro RNA transcripts were inoculated onto *N. benthamiana* plants and recombinant CuMV coat proteins assembled into VLPs, which were visualized by electron microscopy. The PVX/CuMV expression system was used for the transient expression of chimeric CuMV coats carrying the previously described NDV epitopes. Therefore, the CuMV-NDV chimeras were produced as authentic plant-produced VLPs (Natilla et al. 2006). Natilla and Nemchinov (2008) demonstrated by nice electron micrographs that the CuMV-NDV VLPs were homogeneous and morphologically indistinguishable from wild-type CuMV particles. Chickens immunized with the chimeric CuMV VLPs carrying the 8-aa NDV HN epitope developed antigen-specific response.

Vitti et al. (2010) attempted to produce an oral CuMV virion-based vaccine against Alzheimer's disease by insertion of Aβ-derived fragments that would stimulate mainly humoral immune responses. Six chimeric constructs, bearing the Aβ1–15 or the Aβ4–15 sequence in positions 248, 392, or 529 of the CuMV coat gene, were created. The viral products replicated well in their natural host. However, only chimeric Aβ1–15-CuMVs were detected by Aβ1–42 antiserum in Western blot analysis. The immunoelectron microscopy revealed a complete decoration of Aβ1–15-CuMV$_{248}$ and Aβ1–15-CMV$_{392}$ carrying the Aβ epitope at aa positions Gly83/Ser84 and Ser131/Ser132, respectively, following incubation with either anti-Aβ1–15 or anti-Aβ1–42 polyclonal antibodies. These two chimeric Aβ-CuMV variants appeared therefore to be endowed with features making them possible candidates for vaccination against Alzheimer's disease (Vitti et al. 2010).

Gellért et al. (2012) performed a thorough ab initio structure prediction for the CuMV particles, selected putative epitopes of the coat of porcine circovirus type 2 (PCV2) from the *Cirlivirales* order (Chapter 10), and generated a prospective live virus-based vaccine against PCV infection in pigs. The predicted PCV2 coat epitopes FNLKDPPLKP and DDNFVTKATALTYDPYVNYS were inserted at aa position 131 of the CuMV coat. *N. clevelandii* Gray and *N. tabacum* L. cv. Xanthi plants were inoculated with the recombinant and control in vitro transcripts in the presence of CuMV RNA 1 and 2 transcripts. The systemic symptoms were observed 6–8 days after the inoculation in the case of the control infection, after 8–10 days in the case of the chimera carrying 10-aa epitope, but symptoms never were observed in the case of the chimera carrying the 20-aa epitope. The chimeric virus was further propagated on *N. clevelandii* Gray plants and the yield of the purified chimeric virus was like that of the wild-type CuMV. Nevertheless, after five to six months of propagation and serial passages on the *N. clevelandii* or other tobacco plants, the complete deletion of the PCV2 epitope was observed. The chimeric virions were able to induce PCV-specific antibody response in mice, while the challenge experiment with PCV2 carried out in immunized pigs showed partial protection against the infection (Gellért et al. 2012).

Chimeric VLPs

As mentioned earlier, the famous Martin Bachmann's team, in collaboration with the Riga VLP group, generated a series of the therapeutic vaccines based on the CuMV$_{TT}$ VLPs carrying a powerful T-cell-stimulatory epitope derived from tetanus toxoid (Zeltins et al. 2017). The replacement of the first 12 N-terminal aa residues of the CuMV coat with the TT epitope QYIKANSKFIGITE, located at the TT aa positions 830–843, resulted in well-assembled VLPs with a size of approximately 30–40 nm in diameter, as detected by electron microscopy and dynamic light scattering, preserving the native T = 3 icosahedral structure. Since Th cell memory to tetanus was near universal in humans, a range of the CMV$_{TT}$-based vaccines was tested against chronic inflammatory conditions (model: psoriasis, antigen: interleukin-17), neurodegeneration (Alzheimer's, β-amyloid), and allergic disease (cat allergy, Fel-d1). The vaccine responses were uniformly strong, selective, efficient in vivo, observed even in old mice, and employed low vaccine doses. The CuMV$_{TT}$ VLP platform seemed therefore adaptable to almost any antigen, and its features and performance were ideally suited for the design of vaccines delivering enhanced responsiveness in aging populations (Zeltins et al. 2017). Since the CuMV$_{TT}$ VLPs retained capacity to package RNA, the immune response was driven to Th1 and IgG2a routes in a TLR7/8-dependent manner, as described in detail by Krueger et al. (2019) in the case of the VLPs of the RNA phage Qβ from the class *Leviviricetes* described in Chapter 25. Figure 17.7 presents the 3D structure of the CuMV$_{TT}$ VLP platform, which was resolved by electron

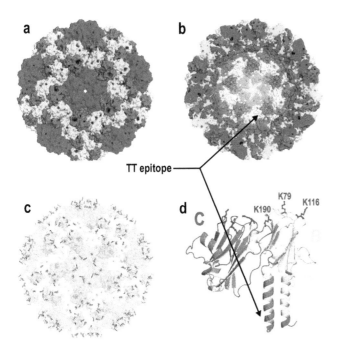

FIGURE 17.7 Cucumber mosaic virus (CuMV)-derived VLP fused to universal tetanus toxoid (TT) T-cell epitope. General structural features of CuMV$_{TT}$. (a) Surface representation of viral capsid. Its T = 3 symmetry is formed by pentamers of subunit A (blue) and hexamers of subunits B (dark cyan) and C (light cyan). (b) Cross section of the particle as shown in (a). The N-termini of both subunits B and C are colored orange to indicate the points of insertion of the tetanus toxoid epitope. (c) The same view as in (a) but with the surface lysine residues liable to conjugation highlighted in red. (d) Cartoon representation of a dimer formed by subunits B and C. All images generated from PDB 1F15. (Reprinted from Mohsen MO, Augusto G, Bachmann MF. The 3Ds in virus-like particle based-vaccines: "Design, Delivery and Dynamics." *Immunol Rev.* 2020;296:155–168.)

cryomicroscopy, as mentioned earlier, to 4.20 Å (Zeltins et al. 2017).

To test the performance of the CuMV$_{TT}$-based anti-IL17A vaccine for psoriasis, the vaccine prototype was generated by chemical conjugation of full-length, dimeric murine IL-17A to the CMV$_{TT}$ VLPs. The mice immunized with different doses of the IL-17 CMV$_{TT}$ vaccine developed high anti-IL-17A IgG levels after as low a dose as 0.5 μg. The efficacy of the vaccine was confirmed in the in vivo IL17-dependent imiquimod model of the psoriatic disease (Zeltins et al. 2017).

By a vaccine against Alzheimer's disease, Zeltins et al. (2017) coupled Aβ1–6, i.e., the N-terminus of Aβ1–42 to CuMV$_{TT}$ (Aβ1–6-CuMV$_{TT}$) or VLPs without the tetanus epitope (Aβ1–6-CuMV$_{WT}$). The control vaccine as such was analogous to the previously generated CAD106 vaccine based on the VLPs of the previously mentioned RNA phage Qβ (Chapter 25). The CuMV$_{TT}$-based vaccine demonstrated enhanced immunogenicity in mice, including older animals.

To generate a vaccine against cat allergy, the recombinant Fel d 1 protein was coupled to the CuMV$_{TT}$ VLPs via a small Cys-containing linker. The Fel d 1 coupled to the CMV$_{TT}$ VLPs was highly immunogenic and induced a 10–100-fold-stronger antibody response than free Fel d 1 or Fel d 1 mixed with VLPs, demonstrating potent immunogenicity depended on the presentation by the CMV$_{TT}$ VLPs (Zeltins et al. 2017). Furthermore, Thoms et al. (2019) developed a new strategy to treat Fel d 1-induced allergy in human subjects by immunizing cats—but not allergic cat owners—against their own major allergen, Fel d 1. The CMV$_{TT}$ VLPs-based vaccine was well tolerated and had no overt toxic effect. All cats induced high-titer and long-lasting antibody responses. The antibodies were able to neutralize the allergen both in vitro and in vivo and led to the reduction of the endogenous allergen level and a reduced allergenicity of tear samples (Thoms et al. 2019). Thoms et al. (2020) reported the positive outcome of the first field trial with ten cat-allergic participants living together with their cats, where the cats were vaccinated with the Fel-CuMV vaccine termed HypoCat™. It is noteworthy that a similar technique was developed earlier by the Bachmann team, when they displayed recombinant Fel d 1 on the surface of the VLPs of the previously mentioned RNA phage Qβ (Chapter 25). Remarkably, in both cases, the allergens displayed on the VLPs were more immunogenic than their free counterparts.

Bachmann et al. (2018) used the CuMV$_{TT}$ platform to generate a vaccine against atopic dermatitis, the most common allergic skin disease in dogs. The authors coupled the canine IL-31 the CuMV$_{TT}$ VLPs, where the free SH groups on IL-31 were introduced using N-succinimidyl S-acetylthioacetate and joined to surface Lys of the VLPs using succinimidyl 6-((beta-maleimidopropionamido)hexanoate). The current IL-31-blocking therapies are using two different drugs, namely Apoquel and an mAb, directed against IL-31. However, the CuMV$_{TT}$-based vaccine demonstrated a clear therapeutic breakthrough for the treatment of dogs (Bachmann et al. 2018).

Summarizing these first steps of application, the CuMV$_{TT}$ platform together with chemical coupling was successfully used to create the clinically efficacious therapeutic vaccines targeting IL-31 in dogs (Bachmann et al. 2018) and horses (Olomski et al. 2020), IL-5 in horses (Fettelschoss-Gabriel et al. 2018, 2019; Jonsdottir et al. 2020), Fel d 1 in cats (Zeltins et al. 2017; Thoms et al. 2019, 2020), and pain in mice (von Loga et al. 2019).

Developing further the line of the therapeutic allergy vaccines, Storni et al. (2020) coupled to the CuMV$_{TT}$ VLPs the extracts of roasted peanut (Ara R) or the single allergens Ara h 1 or Ara h 2. The model mice were sensitized intraperitoneally with peanut extract absorbed to alum, while the immunotherapy consisted of a single subcutaneous injection of CuMV$_{TT}$ coupled to the Ara R, Ara h 1, or Ara h 2. All three variants protected peanut-sensitized mice against anaphylaxis after intravenous challenge with the whole peanut extract but did not cause allergic reactions in sensitized mice. Moreover, the CuMV$_{TT}$-Ara h 1 was able to induce specific IgG antibodies, diminish local reactions after skin prick tests, and reduce the infiltration of the gastrointestinal tract by eosinophils and mast cells after oral challenge with a peanut (Storni et al. 2020).

After the impressing success with the therapeutic vaccines, a set of prophylactic vaccine candidates against infectious diseases was created by the Bachmann's team. Thus, the CuMV$_{TT}$ platform was used to generate a malaria vaccine (Cabral-Miranda et al. 2018). In this case, the thrombospondin-related adhesive protein (TRAP) of *P. falciparum* was chemically coupled to the CuMV$_{TT}$ VLPs. In mice, the vaccine formulations with the chimeric VLPs increased humoral and T-cell immunogenicity for TRAP compared to the antigen alone. The display on VLPs, in particular if formulated with an adjuvant dioleoyl phosphatidylserine (DOPS)—but not alum—induced the strongest and most protective immune response.

Next, Cabral-Miranda et al. (2019) displayed Zika-virus (ZIKV)-derived EDIII fragments on the CuMV$_{TT}$ VLPs and formulated them with DOPS adjuvant. This vaccine was able to induce antibodies efficiently in mice and neutralize the ZIKV without predisposing for antibody-dependent enhancement (ADE) with Dengue virus (DENV) infections. It was highly important, since ZIKV and DENV are flaviviruses that circulate mostly in the same environmental niche, and avoidance of disease enhancing antibodies is of critical importance (Cabral-Miranda et al. 2019).

At last, the highly promising CuMV$_{TT}$ platform was used to generate a straightforward Covid-19 vaccine candidate (Zha et al. 2021; Mohsen et al. 2021). To do this, the receptor binding domain (RBD) of the spike protein S of SARS-CoV-2 was expressed as a fusion to an Fc molecule for better production. The protein, which bound efficiently to the viral receptor ACE2, was chemically coupled to the CuMV$_{TT}$ VLPs and induced virus-neutralizing antibodies in mice.

The clear success story of the CuMV$_{TT}$ platform was quickly reviewed by Bachmann et al. (2020), with an emphasis on the allergy vaccine candidates, by Mohsen et al. (2020), accenting the 3D aspects, while Balke and Zeltins (2019, 2020) touched the novel platform against the total background of plant virus-derived VLPs as prospective vaccine candidates.

Packaging and Delivery

The intrinsic role of RNA packaged within the CuMV$_{TT}$ VLPs as putative vaccines was disclosed previously. Earlier, Lu et al. (2012) used the CuMV VLPs as a protein cage and demonstrated that the purified CuMV coat formed VLPs in the presence of both ssDNA and dsDNA oligonucleotides, with a lower size limit of 20 nt. The single-stranded 15-mers failed to produce any VLPs. The VLPs assembled from the CuMV coat and RNA exhibited a level of stability like that of virions purified from plants, while the VLPs from CuMV coat and a 20-mer oligonucleotide exhibited

comparable or greater stability. The fluorescent labeling of VLPs was achieved by the encapsidation of a fluorophore Alexa Fluor 488-labeled 45-mer oligonucleotide. Using ssDNA as a nucleating factor, encapsidation of fluorescently labeled streptavidin of 53 kDa conjugated to a biotinylated oligonucleotide was demonstrated.

Zeng et al. (2013) used the CuMV virions as a drug delivery vehicle of doxorubicin for cancer therapy. The folic acid (FA) was anchored as a targeting moiety on the CuMV capsid and loaded significant amount of doxorubicin into the interior cavity of CuMV through the formation of Dox-RNA conjugate. As a result, the FA-doxorubicin-carrying assemblies were comparable in size and morphology to the native CuMV particles. The doxorubicin-loaded particles exhibited sustained in vitro drug release profile over five days at physiological pH but were liberated from the conjugates by the addition of an elevated level of RNase. Such folate receptor-targeted assemblies demonstrated expected effects in vitro and in vivo; the cargo-transporting particles significantly decreased the accumulation of doxorubicin in the nuclei of mouse myocardial cells and improved the uptake of the drug in the ovarian cancer, leading to less cardiotoxicity and enhanced antitumor effect.

The methodology of how to use the plant virus-derived VLPs for packaging and in vivo delivery was exhaustively reviewed by Shahgolzari et al. (2020).

OTHER CUCUMOVIRUSES

In parallel with the structure of BMV (Lucas et al. 2000b), Lucas et al. (2002a) solved the 3D structure of tomato aspermy virus (TAV), a member of the *Cucumovirus* genus, to 4 Å. The TAV particle of 28 nm in diameter had the canonical β-barrel topology with a distinctive N-terminal α-helix directed into the interior of the virus where it interacted with encapsidated RNA. The N-terminal helices were joined to the β-barrels of protein subunits by extended polypeptides of six aa residues, which served as flexible hinges allowing movement of the helices in response to local RNA distribution. The side chains of Cys64 and Cys106 formed the first disulfide observed in a cucumovirus, including a unique cysteine, 106, in a region otherwise conserved. A positive ion, putatively modeled as a Mg^{2+} ion, lay on the quasi-threefold axis surrounded by three quasi-symmetric Glu175 side chains (Lucas et al. 2002a).

Llamas et al. (2006) engineered hybrids of the TAV and CuMV coats by exchange of aa portions 1–59, 60–148, and 149–219 in all possible combinations within TAV RNA3. The seven possible chimeras were able to replicate in tobacco protoplasts to similar levels, but only those having residues 1–59 or 60–148 from CuMV were infectious to tobacco plants (a common host for CuMV and TAV), and formed stable particles. No hybrid was able to infect cucumber plants, a host for CuMV and not for TAV.

Bashir et al. (2004) demonstrated that the expression of the peanut stunt virus (PSV) coat gene was essential and sufficient for production of host-dependent ribbon-like inclusions in infected plants. These inclusions appeared as long, thin, densely staining sheets, which were prevalent within the cytoplasm, accumulating most commonly near vacuoles. The formation of these ribbon-like inclusions appeared to be subgroup-specific because infection of tobacco with subgroup II PSV strains (but not subgroup I strains) induced the inclusions. Furthermore, the inclusion formation was shown to be host-specific because the inclusions were not detected in either of two leguminous host species infected with PSV subgroup II strains.

REVIEWS

The unique role of bromoviruses in the VLP nanotechnologies was described in numerous reviews, where data on the BMV and CCMV applications neighbored with those on another greatly popular icosahedral model of plant origin, namely cowpea mosaic virus (CPMV) of the *Secoviridae* family, order *Picornavirales*, described in Chapter 27; as well as RNA phage MS2, order *Norzivirales*, Chapter 25; rod-like tobacco mosaic virus (TMV), family *Virgaviridae* from the *Martellivirales* order, Chapter 19; and last such prospective but nonviral protein cages as lumazine synthase

of 15 nm, ferritin of 12 nm, small heat shock protein of 12 nm, and DNA binding protein from starved cells of 9 nm in diameter. The reviews that distinguished BMV as a nanotechnological model and analyzed it in detail were published by Aniagyei et al. (2008), Young et al. (2008), Dedeo et al. (2011), Pokorski and Steinmetz (2011), Liu Z et al. (2012), Glasgow and Tullman-Ercek (2014), Koudelka et al. (2015), Shukla and Steinmetz (2015), Wen et al. (2015), Wen and Steinmetz (2016), and Zhang W et al. (2017), while reviews of Flenniken et al. (2009), Cardinale et al. (2012), Samanta and Medintz (2016), Almeida-Marrero et al. (2018), and Wege and Koch (2020) were directed to CCMV. Mohsen et al. (2020) concentrated on both CCMV and CuMV models. A set of reviews was oriented onto both CCMV and CPMV models (Manchester and Singh 2006; Uchida et al. 2007; Lee LA et al. 2009, 2011; Steinmetz et al. 2009; Li et al. 2010; Soto and Ratna 2010; Steinmetz 2010; Ma et al. 2012; Bittner et al. 2013; Bruckman et al. 2013; Wen et al. 2013; Lee EJ et al. 2016; Maassen et al. 2016; Zhang Y et al. 2016; Shahgolzari et al. 2020; Shukla et al. 2020). The tremendous reviews unveiling the novel CuMV$_{TT}$ platform must be specifically emphasized (Balke and Zeltins 2019, 2020; Mohsen et al. 2020).

18 Order *Martellivirales*: *Togaviridae*

The more perfect a thing is, the more susceptible to good and bad treatment it is.

Dante Alighieri

ESSENTIALS

The family *Togaviridae* is a substancious and specific part of the order *Martellivirales* that was presented in Chapter 17. According to the current taxonomy (Chen R et al. 2018; ICTV 2020), the *Togaviridae* family contains a single genus *Alphavirus* with 32 species. Before April 2019, the family also contained the genus *Rubivirus*, with a single species *Rubella virus*, which has now been moved to the family *Matonaviridae*, order *Hepelivirales* (Chapter 16).

The *Alphavirus* members are mosquito-borne viruses. Most alphaviruses are cytopathic to vertebrates such as humans, nonhuman primates, equids, birds, amphibians, reptiles, rodents, and pigs and cause in humans a short febrile illness that can lead to prolonged arthritis or encephalitis but is rarely fatal. There are two aquatic alphaviruses, southern elephant seal virus and salmon pancreas disease virus, often referred to as salmonid alphavirus (SAV), infecting sea mammals and fish, respectively.

Therefore, many alphaviruses are important human and veterinary pathogens, e.g., chikungunya virus (CHIKV), eastern equine encephalitis (EEEV), western equine encephalitis (WEEV), and Venezuelan equine encephalitis (VEEV) viruses. The alphaviruses are divided into seven complexes. The most studied of them is the so-called Semliki Forest virus complex with 7 viruses and popular CHIKV, Ross River virus (RRV), and Semliki Forest virus (SFV) among them. The other important complexes are eastern equine encephalitis, Venezuelan equine encephalitis, and western equine encephalitis complexes. Some alphaviruses, including the previously mentioned SAV, are not classified within the complexes.

Figure 18.1 presents a portrait of a typical representative and a schematic cartoon of the alphavirus virions, which are enveloped, 65–70 nm spherical particles of regular structure with a single capsid protein and three envelope glycoproteins, possessing the T = 4 icosahedral symmetry. The particle consists of a nucleocapsid core surrounded by a lipid bilayer that is embedded with glycoprotein spikes. The nucleocapsid core comprises 240 copies of capsid protein. The capsid protein consists of two domains, a highly charged, N-terminal domain that interacts with the viral RNA in the interior of the nucleocapsid core and a C-terminal domain that has a chymotrypsin-like fold. The lipid bilayer is host-derived from the site of budding.

The 80 trimeric glycoprotein spikes cover the surface of alphavirions. Each spike is composed of three E1-E2 heterodimers. However, in some alphaviruses, the E3 glycoprotein remains noncovalently associated to these spikes. The precursor of the envelope E2 and E3 proteins (pE2 or p62) forms heterodimers with the protein E1. The E1 has been shown to undergo disulfide rearrangement during assembly, and the same is predicted for the protein E2. The E3 stabilizes the E2/E1 heterodimers and trimers during spike assembly by preventing dissociation of the complex as it transits through the secretory pathway to the plasma membrane. The E1 protein is a class II fusion protein that mediates fusion between the virus membrane and the host cell membrane in the endosome. The spikes formed by the E1 and E2 proteins are responsible for viral entry. The E2 protein binds to the host-cell receptor and interacts with the capsid protein (for references see Chen R et al. 2018).

Because of the high medicinal and veterinary importance of the *Togaviridae* family members, the prevalent goals of the VLP studies relate to putative vaccines. In that sense, as well as by the structure of envelopes and by the engineering of self-replicating vectors together with packaging cell cultures, the alphaviruses have a lot in common with flaviviruses from the *Amarillovirales* order, which are described in Chapter 22. The point is, the alphaviruses, like flaviviruses, as well as also measles viruses and rhabdoviruses of the *Mononegavirales* order with negative sense RNA genome (see Chapter 31), possess the capacity of highly efficient self-amplification of RNA in host cells, which makes them attractive vehicles not only for vaccine development but also as putative anticancer agents, as summarized recently in an exhaustive review written by Kenneth Lundstrom (2020). Such vaccines are based on the single-round infectious VLPs, or sriVLPs, often called virus-like replicon particles (VRPs), which are described in more detail later, as well as in Chapter 22 and Chapter 31. In fact, the alphaviruses and flaviviruses are these that remain the best candidates to engineer and administer the sriVLPs.

The recent comprehensive reviews were published about the role of VLPs by generation of vaccines against alphaviruses, first and foremost to the emerging CHIKV (Gao et al. 2019; Kumar et al. 2019; Reyes-Sandoval 2019), as well as to the equine EEEV, VEEV, and WEEV (Stromberg et al. 2020). The alphavirus vectors for cancer treatment were reviewed by Zajakina et al. (2008, 2015). The place of alphaviruses by the development of the VLP production in plants was recently identified by Peyret et al. (2021).

GENOME

The alphavirus genome is unsegmented RNA of 9.7–11.8 kb. The genome of alphaviruses involves a 5′-cap and noncoding region followed by genes for the nonstructural and

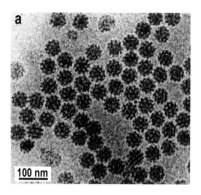

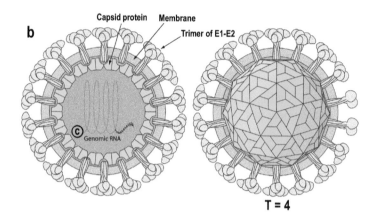

FIGURE 18.1   The portrait and schematic cartoons of the typical representatives of the *Togaviridae* family. (a) Electron cryomicrograph of Semliki Forest virus (SFV; with kind permission from Springer Science+Business Media: Nature, Cryo-electron microscopy of viruses, 308, 1984, 32–36, Adrian M, Dubochet J, Lepault J, McDowall AW.) (b) The general cartoon of the *Togaviridae* family is taken with a kind permission from the ViralZone, Swiss Institute of Bioinformatics, http://viralzone.expasy.org (Hulo C et al. 2011). (Courtesy Philippe Le Mercier.)

structural proteins and a 3′-noncoding region and a poly(A) tail. The negative-sense RNA of the replication intermediate is neither capped nor contains the poly(A) tail. A subgenomic RNA encoding the structural proteins is transcribed from the negative-sense RNA and consists of a 5′-cap and poly(A) tail. The structural proteins are translated as a polyprotein from the subgenomic mRNA. As soon as the viral capsid protein is produced, it autoproteolytically cleaves itself from the rest of the polyprotein and forms nucleocapsid cores in the cytoplasm. The E3 protein contains an ER signal sequence and translocates the polyprotein into the ER. The entire polyprotein crosses the ER membrane (going between lumen and cytoplasm) several times until host proteases cleave it into the individual structural proteins (Chen R et al. 2018).

Figure 18.2 shows a representative genomic structure of the genus *Alphavirus* members.

3D STRUCTURE

The x-ray crystal structures of the envelope glycoproteins were resolved for Sindbis virus (SINV; Li et al. 2010) and CHIKV (Voss et al. 2010), allowing 3D reconstruction by fitting into the electron cryomicroscopy (cryo-EM) patterns of the viruses in question. Figure 18.3 demonstrates the general flavivirus 3D structure obtained by electron cryomicroscopy and image reconstruction, while Figure 18.4 is intended to demonstrate how the crystal structure data were imbedded into the lower-resolution cryo-EM shell.

The 3D electron cryomicroscopy-derived structures are resolved currently for SFV, to 9 Å (Mancini et al. 2000); VEEV, to 4.4 Å (Zhang R et al. 2011); Barmah Forest virus (BFV), to 6 Å (Kostyuchenko et al. 2011); CHIKV, to 5.3 Å (Sun et al. 2013); EEEV, to 4.4 Å (Hasan et al. 2018); and RRV and Mayaro virus (MAYV), to 5.3 Å (Powell et al. 2020).

Basore et al. (2019) resolved both the CHIKV VLP and virion to 4 to 5 Å resolution by electron cryomicroscopy

reconstruction and the Mxra8, a receptor for alphaviruses, to 2.2 Å resolution by x-ray crystallography. It appeared that the Mxra8, which contains two strand-swapped Ig-like domains oriented in a unique disulfide-linked head-to-head arrangement, binds by wedging into a cleft created by two adjacent CHIKV E2-E1 heterodimers in one trimeric spike and engaging a neighboring spike. Remarkably, this study employed the mammalian cells-produced VLP preparation equivalent to the CHIKV VLPs that were evaluated in the phase I and II human clinical trials for the vaccine candidate protecting against CHIKV disease (Chang et al. 2014), as described later.

It is noteworthy that Forsell et al. (2000) have proved the model of the structural flavivirus maturation. The authors approved experimentally that the spike proteins can still direct the T = 4 geometry of the SFV virions via their horizontal interactions on the membrane surface and also transmit the corresponding regularity to the core via their vertical interactions with the capsid proteins. This suggests that the spike-mediated assembly "from without" plays an important role in alphavirus budding (Forsell et al. 2000).

SINGLE-ROUND INFECTIOUS VLPs

The alphavirus expression vector systems applicable by the sriVLP (or, in other words, suicide), also VRP, technique were engineered for SFV by Peter Liljeström and Henrik Garoff (1991), as well as for SINV (Xiong et al. 1989) and VEEV (Davis et al. 1989). Berglund et al. (1993) published a detailed prescription on the safe conditionally infectious SFV sriVLPs.

The sriVLP vector systems are based on the simultaneous expression, in mammalian cells, of a replicon carrying target gene(s) and a helper vector expressing the structural alphavirus genes for production of replication deficient sriVLPs or suicide particles. The reason for replication deficiency relates to the presence of the packaging signal

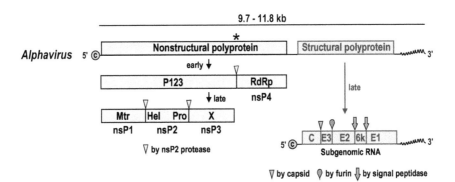

FIGURE 18.2 The genomic structure of the genus *Alphavirus* members derived from the official ICTV profile (Chen al. 2018) and from the ViralZone. The diagram shows a short 5′ noncoding region, a 3′ noncoding polyadenylated region, and the ORFs. An asterisk indicates the stop codon present in some alphaviruses that must be translationally readthrough to produce a precursor containing nsP4. Mtr, methyl transferase; Pro, protease; Hel, helicase; X, unknown function; RdRp, RNA-dependent RNA polymerase. The structural genes are colored dark pink.

in the nonstructural region of the expression vector, which prevents the packaging of the RNA from the helper vector. This and other precautions by the safe functioning of the sriVLP-packaged genomes, such as elimination of any replication proficient alphavirus and RNA recombination, are reviewed comprehensively by Lundstrom (2020).

The selected examples of the numerous sriVLP applications will be cited under specific topics devoted to specific alphavirus representatives and summarized in Table 18.1.

SEMLIKI FOREST VIRUS COMPLEX

CHIKUNGUNYA VIRUS

Natural VLPs

The true CHIKV VLPs were prepared first in mammalian cells by Akahata et al. (2010). The CHIKV structural polyprotein of capsid plus envelope (C-E) or E alone from two different CHIKV strains was inserted into a eukaryotic expression vector under the control of CMV promoter and transfected into 293T human kidney cell culture. After transfection of the cells with the C-E constructs, the E1 and E2 proteins were detected in the supernatant, suggesting the generation of the CHIKV VLPs. The examination of the purified product by electron microscopy revealed VLPs with the same morphologic appearance as the wild-type CHIKV virions. Moreover, the electron cryomicroscopy and 3D image reconstruction showed that the VLPs had an external diameter of 65 nm and a core diameter of 40 nm, where the E1-E2 glycoproteins were organized into 240 heterodimers, assembled into 80 glycoprotein spikes arranged with triangulation number T = 4 quasi-symmetry on the surface of the VLPs, in close agreement therefore with the 3D CHIKV virion structure presented in Figure 18.3. The purified CHIKV VLP vaccine candidate demonstrated full protection of rhesus monkeys against challenge with the virus (Akahata et al. 2010).

This CHIKV VLP vaccine, termed VRC-CHKVLP059–00-VP, has withstood the phase I clinical trial (Chang et al.

2014) and the randomized, placebo-controlled, double-blind, phase II trial that was conducted at six outpatient clinical research sites located in Haiti, Dominican Republic, Martinique, Guadeloupe, and Puerto Rico and enrolled a total of 400 healthy adults aged 18 through 60 years (Chen GL et al. 2020). The positive outcome of the trial included the safety and tolerability, as well as the immune response after second vaccination. The phase III trial is planned to assess clinical efficacy of the CHIKV VLP vaccine.

Noranate et al. (2014) got the CHIKV VLPs when optimizing the whole structural CHIKV region C-E3-E2–6k-E1, inserted it into a mammalian expression vector, and transfected the resulting construct into mammalian 293 cells. The spherical particles with a 50- to 60-nm diameter, which resembled the native CHIKV virions, were revealed in the cell lysate and the supernatant. The purified VLPs retained antigenicity and immunogenicity identical to those of the native virion. Theillet et al. (2019) used such 293T cells-produced CHIKV VLPs for serological detection of specific CHIKV IgM to optimize patient management.

Wang E et al. (2008) developed live virus-derived chimeric CHIKV vaccine candidates using either VEEV attenuated vaccine strain, a naturally attenuated strain of EEEV, or SINV as a backbone and the structural protein genes of CHIKV. All vaccine candidates replicated efficiently in cell cultures and were highly attenuated in mice. All of the chimeras also produced robust neutralizing antibody responses, although the VEEV and EEEV backbones appeared to offer greater immunogenicity. The vaccinated mice were fully protected against disease and viremia after CHIKV challenge (Wang E et al. 2008).

Frédéric Tangy's team elaborated a recombinant live attenuated measles virus (MV) vaccine expressing CHIKV VLPs comprising capsid and envelope structural proteins from the recent CHIKV strain La Reunion (Brandler et al. 2013). The immunization of mice susceptible to measles virus induced high titers of CHIKV antibodies that neutralized several primary isolates, as well as specific cellular immune responses that were also elicited. A single

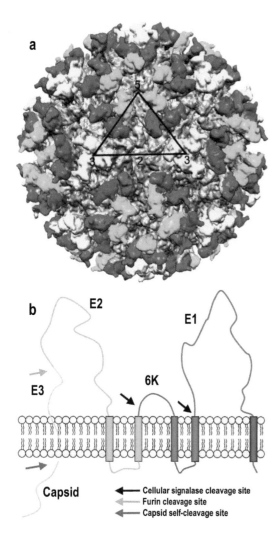

a

b

E2

E1

E3

6K

Capsid

◄── Cellular signalase cleavage site
◄── Furin cleavage site
◄── Capsid self-cleavage site

FIGURE 18.3 The structural proteins of an alphavirus. (a) The cryo-EM density of Sindbis virus showing T = 4 symmetry. The four E2 molecules in one asymmetric unit (outlined in black) are colored red, green, blue, and yellow. These give rise to one trimeric spike on each icosahedral 3-fold axis and one generally positioned spike. The E1 molecules are colored gray. (b) Threading of the Sindbis virus structural polyprotein through an endoplasmic reticulum membrane showing the position of the capsid, E3, E2, 6K, and E1 proteins. (With kind permission from Springer Science+Business Media: *Nature*, Structural changes of envelope proteins during alphavirus fusion, 468, 2010, 705–708, Li L, Jose J, Xiang Y, Kuhn RJ, Rossmann MG.)

immunization with this vaccine candidate protected all mice from a lethal CHIKV challenge. Ramsauer et al. (2015) reported successful outcome of the clinical phase I trial of this MV-based CHIKV VLP vaccine. It is worth mentioning that the live attenuated MV vector used in these studies is the well-known Schwarz strain of the measles vaccine, described in 1962 and initially introduced by the Pasteur Institute in Paris.

Wang D et al. (2011) described a recombinant CHIKV vaccine comprising a nonreplicating complex adenovirus vector encoding the structural CHIKV polyprotein cassette.

A single immunization with this vaccine completely protected mice against virus challenge.

An interesting example of a chimeric live attenuated CHIKV vaccine was generated by Erasmus et al. (2017). The authors engineered an infectious cDNA clone of EILV/CHIKV containing the structural polyprotein ORF of a 2014 CHIKV strain isolated from an infected patient from the British Virgin Islands and insect-specific alphavirus, Eilat virus (EILV), as a vector. The electron cryomicroscopy resolution to 9.85 Å of the EILV/CHIKV virions revealed no differences from the wild-type CHIKV virions. The chimeric virus mimicked the early stages of CHIKV replication in vertebrate cells from attachment and entry to viral RNA delivery yet remained completely defective for productive replication, providing a high degree of safety. A single dose of the EILV/CHIKV virions produced in mosquito cells provided complete protection in two different mouse models and in cynomolgus macaques. The EILV platform represented the first structurally native application of an insect-specific virus in preclinical vaccine development (Erasmus et al. 2017). Remarkably, the same technology of the chimeric EILV viruses was used to generate vaccine candidates against EEEV and VEEV (Erasmus et al. 2018).

Garg et al. (2020) generated a universal VLP vaccine termed CJaYZ targeting not only CHIKV but also Japanese encephalitis (JEV), yellow fever (YFV), and Zika (ZIKV) viruses from the *Flaviviridae* family, order *Amarillovirales* (see Chapter 22). For CHIKV, the vaccine included C-E3-E2-E1 genes. The stable 293T cell lines secreting VLPs containing capsid protein for all four viruses were established and adapted to grow in suspension cultures to facilitate vaccine scale up. The immunization of mice with different combinations of the capsid protein containing VLPs either as monovalent, bivalent, or tetravalent formulation resulted in generation of high levels of neutralizing antibodies. The potential tetravalent VLP vaccine candidate provided strong neutralizing antibody titers against all four viruses (Garg et al. 2020).

An example of the CHIKV DNA vaccines is worthy of mention. Tretyakova et al. (2014) used the immunization DNA (iDNA) infectious clone technology, which combined the advantages of DNA and live attenuated vaccines and was designed to initiate replication of live vaccine virus from the plasmid in vitro and in vivo. Thus, the authors presented an iDNA vaccine composed of plasmid DNA that encoded the full-length infectious genome of a live attenuated CHIKV clone downstream from a eukaryotic promoter. The vaccination of mice with a single dose of CHIKV iDNA plasmid resulted in seroconversion, induction of neutralizing antibodies, and protection from experimental challenge with a neurovirulent CHIKV. Remarkably, Hidajat et al. (2016) evaluated genetic stability of the most prospective CHIKV DNA vaccines by next-generation sequencing.

After the previously described synthesis in mammalian cells, Metz et al. (2013a) were the first who described the successful production of the CHIKV VLPs using recombinant baculoviruses. To produce the CHIKV VLPs, the recombinant baculovirus was generated by expressing the complete CHIKV S27 structural polyprotein including C, E3, E2, 6K, and E1. The insect cells-secreted CHIKV VLPs were glycosylated and fitted a size of 68 ± 14 nm, which was consistent with the reported size, 65–70 nm, of alphavirus virions. A single immunization with the non-adjuvanted CHIKV VLPs induced high titer neutralizing antibody responses and provided complete protection against viraemia and joint inflammation upon challenge with the CHIKV in an adult wild-type mouse model of CHIKV disease (Metz et al. 2013a). Metz et al. (2013b) demonstrated that the CHIKV VLPs were more immunogenic in the lethal mouse model than the glycoprotein E1 or E2 subunits produced in insect cells, as reported before (Metz et al. 2011b).

Arévalo et al. (2016, 2019) expressed the complete sequence encoding structural proteins C-E3-E2–6K-E1 of the CHIKV S27 strain, after codon-optimization, in insect cells. In parallel, the E1 and E2 genes, designed as transmembrane-truncated versions, were synthesized and cloned into the pFastBac HT vector, which added an N-terminal His$_6$ tag and tobacco etch virus (TEV) proteolytic site to each gene. The VLPs were obtained and described but not characterized by electron microscopy. Like the VLPs, the His-tagged E1 and E2 proteins were secreted into the culture media of *Sf*9 cells infected with their respective baculoviruses. The unadjuvanted CHIKV VLPs elicited immune responses that ensured full protection of adult mice against CHIKV infection. In contrast, the antibody responses elicited by the VLP-based vaccines were attenuated in aged mice, with negligible neutralizing antibody titers detected. Remarkably, the unvaccinated aged mice were resistant to CHIKV infection, while vaccination with the CHIKV VLPs exacerbated disease (Arévalo et al. 2019).

Hikke et al. (2016) used CHIKV, in parallel with SAV (see later), to develop the alphavirus core-like particle platform as an alternative to the whole enveloped VLPs. The CHIKV core-like particles were produced to high levels in insect cells by expression of the CHIKV capsid protein. The core-like particles were found in dense nuclear bodies within the infected cell nucleus. It is noteworthy that the eGFP gene was inserted at the N-terminus of the CHIKV capsid sequence and the eGFP-capsid fusion protein was visible only in a confined area within the nucleus, whereas the GFP control protein freely diffused throughout the cell (Hikke et al. 2016). However, the ability of the fusion protein to self-assemble was not addressed.

As for the VLP production in yeast cells, Saraswat et al. (2016) reported the production of the well-shaped CHIKV VLPs, sharing morphological identity to native CHIKV virions, in *Pichia pastoris*. Thus, the full sequence expressing CHIKV whole structural proteins was introduced into the expression system, and the spherical particles of approximately ~65–70 nm in diameter were revealed by electron microscopy, as well as the putative core particles with an average diameter of ~40 nm. These yeast-produced CHIKV VLPs induced both cell-mediated as well as humoral response in a balanced manner, while the generated CHIKV-specific antibodies demonstrated efficient in vitro and in vivo neutralization activity, protecting neonatal mice against challenge with two different CHIKV strains (Saraswat et al. 2016).

Chimeric VLPs

Urakami et al. (2017) developed a VLP-based vaccine platform utilizing the CHIKV VLPs. Using the previously described crystal structure of CHIKV (Voss et al. 2010), the two surface loop regions were identified within the envelope protein, where foreign antigens could be inserted without compromising VLP structure. The expression vector encoded CHIKV structural proteins C-E3-E2–6K-E1 with a short linker introduced in E2 and enabled secretion of the VLPs similar to the wild-type CHIKV VLPs into the culture supernatant, when transfected into the mammalian 293F cell culture.

The two possible insertions sites were chosen: (i) within E2 of the so-called αVLP construct carrying Ser-Gly-Gly-Gly-Gly-Ser linker between aa 206 and 207 in the E2 domain and (ii) into the mutated furin cleavage site of the so-called $_{E3}$αVLP construct carrying the Ser-Gly-Gly-Gly-Gly-Ser linker instead of the furin cleavage site. The insertion of foreign sequences resulted therefore in 240 or 480 copies of antigen displayed on the VLP surface in a highly symmetric manner thus capable of inducing strong immune responses against any inserted antigen.

The circumsporozoite protein (CSP) of the *Plasmodium falciparum* malaria parasite was used as a foreign antigen. The central NANP repeat region of CSP was chosen as the epitope to be exposed, since it was characterized as the major surface antigen displayed during the infective stage of malaria. In fact, the epitope was highly conserved among different strains of *P. falciparum* and was considered an immunodominant B cell epitope (Crompton et al. 2010). The five NANP repeats, namely the stretch GNP(NANP)$_5$NAG, was inserted into the cloning sites of the αVLP (αVLP-NANP) and $_{E3}$αVLP($_{E3}$αVLP-NANP) genes. The Dual-NANP carried the NANP repeat epitope in both insertion sites, thus displaying 480 copies of the five NANP repeats per each VLP.

After optimization of the length of the NANP repeat epitope, the malaria vaccine candidate was termed VLPM01. The latter contained the Dual-NANP inserted with 13 NANP repeats, namely the stretch GNP(NANP)$_{13}$NAG, without sacrificing the VLP status and yield. The VLPM01 induced high titer anti-NANP antibodies in mice and monkeys. The examination of the VLPM01 by electron cryomicroscopy and 3D reconstruction (Figure 18.5) showed that it had a diameter of 67 nm and a T = 4 quasi-icosahedral symmetry, which was similar to the natural CHIK VLP structure. The 3D cryo-EM reconstructed map of VLPM01 complexed with the fragment antigen-binding (Fab) region of anti-NANP repeat monoclonal antibody (clone DG2) showed that the NANP epitopes in both insertion sites were exposed on the surface of the VLPM01 particles and were

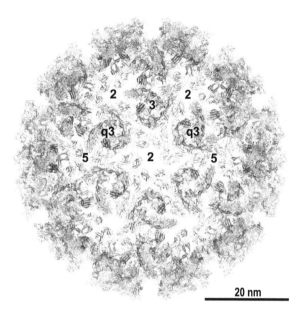

FIGURE 18.4 3D structure of Chikungunya virion: combination with cryo-EM data. The alphavirus T = 4 icosahedral surface glycoprotein shell. Atomic model of the 240 chikungunya E2-E1 heterodimers arranged as 80 spikes. E2 domain A is colored cyan, B dark green, C pink, and the β-ribbon dark purple. E1 is sandy brown for clarity. (With kind permission from Springer Science+Business Media: *Nature*, Glycoprotein organization of Chikungunya virus particles revealed by X-ray crystallography, 468, 2010, 709–712, Voss JE, Vaney MC, Duquerroy S, Vonrhein C, Girard-Blanc C, Crublet E, Thompson A, Bricogne G, Rey FA.)

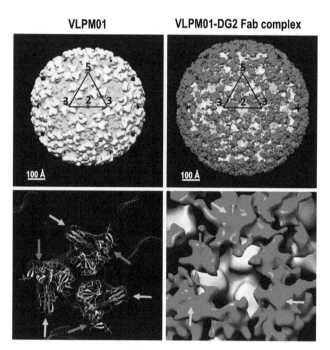

FIGURE 18.5 The electron cryomicroscopy reconstructions of VLPM01. (upper left) 3D cryo-EM map of VLPM01, viewed down an icosahedral twofold axis. The black triangle marks the boundary of an icosahedral asymmetric unit. The numbers in black show the locations of the icosahedral two-, three-, and five-fold axes around the asymmetric unit. (bottom left) Closer view of the icosahedral threefold trimeric spike in VLPM01 fitted with the crystal structure of CHIK p62-E1 (PDB ID 3N40). Gold arrows point to the insertion site at E2, and blue arrows point to the insertion site at E3. (upper right) 3D cryo-EM map of VLPM01 complexed with the DG2 Fab fragments. (bottom right) Trimeric spike on the icosahedral threefold axis of VLPM01 complexed with Fab fragments of DG2. Gold arrows point to the DG2 Fab density bound on the insertion site at E2. Blue arrows point to the DG2 Fab density bound on the insertion site at E3. (Reprinted with permission of American Society for Microbiology from Urakami A et al. *Clin Vaccine Immunol.* 2017; 24:e00090–17.)

accessible to the binding of Fab fragments. Meanwhile, another chimera, termed Dual-PyCSP, was engineered by insertion of the $G(QGPGAP)_{14}G$ sequence from CSP of rodent malaria parasite *P. yoelii* (Urakami et al. 2017).

The chimeric CSP-carrying CHIKV VLPs elicited strong immune responses against CSP in animals. The sera from immunized monkeys protected mice from malaria infection. The mice vaccinated with the *P. yoelii* CSP-containing VLPs were protected from an infectious sporozoite challenge (Urakami et al. 2017).

ROSS RIVER VIRUS

Tellinghuisen et al. (1999) expressed the nucleocapsid genes of RRV and SINV in *E. coli* and purified the capsid proteins. In the presence of single-stranded but not double-stranded nucleic acid, the proteins oligomerized in vitro into core-like particles that resembled the native SINV nucleocapsid cores (Wengler et al. 1984). Despite their similarities, RRV and SINV capsid proteins do not form mixed core-like particles.

Mukhopadhyay et al. (2002) presented the electron cryomicroscopy reconstructions of the in vitro-assembled core-like particles from RRV and WEEV capsid proteins expressed in *E. coli* and assembled with a synthetic 48-nucleotide oligomer. Remarkably, only after numerous trials and failed attempts at determining the structure of

purified core-like particles and cytoplasmic nucleocapsid cores was it concluded that extensive centrifugation steps, such as those required for sedimentation through density gradients, were detrimental and consequently were eliminated from the procedure. The cryo-EM reconstructions of nonpurified core-like particles, although only at 30 Å resolution, showed that they contained the same T = 4 quasi-symmetric arrangement of hexamers and pentamers as nucleocapsid cores found in mature virions (Mukhopadhyay et al. 2002).

Goicochea et al. (2007) used the RRV core-like particles to package gold nanoparticles by the efficient self-assembly of the alphavirus capsid around a functionalized nanoparticle core. Next, Cheng and Mukhopadhyay (2011) showed that the in-vitro-assembled RRV core-like particles can associate with alphavirus glycoproteins to form the true enveloped VLPs. Thus, the in-vitro-assembled core-like particles retained the functionality of nucleocapsid core in the absence of any host chaperones. These in-vitro-assembled cores entered the glycoprotein-expressing mammalian cells via the endocytotic pathway and generated there the enveloped VLPs. These VLPs bud from cells like native virus, were similar in size to the native virus, and could enter cells to release the contents of the core-like particle into the cytoplasm of the cell. Therefore, when packaged with biological and nonbiological cargoes, such RRV VLPs were able to infect target cells and to release the cargoes, opening novel opportunities for gene and drug delivery, as well as medical imaging (Cheng and Mukhopadhyay 2011). In fact, this was the first demonstration of how the expression in *E. coli* and mammalian cells could be combined to establish a VLP-derived gene/drug delivery platform.

SEMLIKI FOREST VIRUS

The examples of the SFV sriVLPs-based vaccine candidates are listed in Table 18.1. The SFV sriVLPs were originally used to raise protective immunity against influenza virus in mice (Zhou et al. 1994, 1995). Mossman et al. (1996) achieved protection of the immunizad pigtail macaques against lethal simian immunodeficiency virus (SIV) challenge. The novel prospective SFV sriVLP approach was reviewed soon after by Malone et al. (1997).

Particular mention should be made of the expression of the hepatitis B virus (HBV) surface antigens. Thus, Niedre-Otomere et al. (2012) constructed the SFV sriVLPs for immunization against HBV, which would optimally present broadly neutralizing epitopes of the HBV S and L proteins (see Chapter 37) in the context of strong innate immune reactions. The induction of HBV-neutralizing antibodies in mice was approved by experiments on susceptible primary hepatocyte cultures from treeshrew *Tupaia belangeri*, a technique that led to real breakthrough in the HBV studies (Köck et al. 2001; Glebe et al. 2003). Moreover, the induced response was HBV subtype-independent and S-escape-resistant (Niedre-Otomere et al. 2012).

The lentiviruses received a lot of attention for obvious reasons of the medicinal importance, because of human immunodeficiency virus (HIV) causing AIDS. For instance, the SFV sriVLPs expressing HIV-1 envelope glycoprotein (Env) were applied for immunization studies in mice in comparison to a DNA vaccine and recombinant Env gp160 protein, and the highest antibody titers were observed in animals immunized with SFV particles (Brand et al. 1998). More recently, Ajbani et al. (2017) demonstrated the immunogenicity of the SFV sriVLPs or virus-like replicon particles (VRPs) by authors' terminology, expressing the *env/gag/pol*RT genes of Indian HIV-1, clade C. The immunization of mice with recombinant sriVLPs in a homologous prime-boost protocol, either individually or in combination, elicited significant CD8[+] and CD4[+] T cells responses. Generally, the immunogenicity of the SFV-based sriVLPs was found to be higher as compared to that of RNA replicons and the sriVLP vaccine

was suggested as a promising preventive and therapeutic candidate for the control and management of HIV/AIDS (Ajbani et al. 2017).

After infectious disease, many types of cancers were targeted by the alphavirus sriVLPs in preclinical settings, as narrated in the recent comprehensive review of Lundstrom (2020). More details about the SFV sriVLP-based vaccine and therapeutic candidates are presented in Table 18.1.

By analogy with the following described properties of the VEEV sriVLPs (Thompson et al. 2006) and SINV sriV-LPs (R. Mikkelsen and R. E. Johnston, unpublished data, cited by Khalil et al. 2014b), the SFV sriVLPs demonstrated strong adjuvant activity (Hidmark et al. 2006).

EASTERN EQUINE ENCEPHALITIS COMPLEX

EASTERN EQUINE ENCEPHALITIS VIRUS

As indicated in Table 18.1, the EEEV sriVLPs carrying the EEEV replicon were used as a vaccine candidate for the EEEV infection when given individually or in combination with the analogous VEEV and WEEV vaccines possessing their individual replicons to mice or cynomolgus macaques (Reed et al. 2014). The alphavirus replicon was derived by deletion of the genes encoding viral structural proteins from full-length genomic cDNA clones and retained all of the machinery necessary for its replication and transcription once it was introduced into an appropriate cell type.

The individual replicon vaccines or the combination of all three replicon vaccines elicited strong neutralizing antibodies in mice to their respective alphavirus. The protection from either subcutaneous or aerosol challenge with VEEV, WEEV, or EEEV was demonstrated up to 12 months after vaccination in mice. The EEEV replicon and the vaccine combination vaccine elicited neutralizing antibodies against EEEV and protected against aerosol exposure to a North American variety of EEEV (Reed et al. 2014).

VENEZUELAN EQUINE ENCEPHALITIS COMPLEX

VENEZUELAN EQUINE ENCEPHALITIS VIRUS

The VEEV sriVLPs, or virus replicon particles (VRPs), were next after the SFV system that was engineered as nonpropagating viral vectors expressing high levels of an antigen protein after a single round of replication. The examples of the corresponding applications are enumerated in Table 18.1.

Pushko et al. (1997) engineered the first reliable VEEV sriVLPs that provided protection against influenza in mice. Then, the sriVLPs expressing genes from Marburg, a filo-virus related to Ebola virus, were demonstrated to provide protection against Marburg virus challenge in guinea pigs and nonhuman primates (Hevey et al. 1998). Pushko et al.

TABLE 18.1

Generation of the Single-Round Infectious VLPs (sriVLPs) by the *Alphavirus* Vectors

Vaccine Target	Gene	Testing Model	References
Eastern Equine Encephalitis Virus (EEEV)			
EEEV	EEEV replicon	Mice, cynomolgus macaques (protection)	Reed et al. 2014
Semliki Forest Virus (SFV)			
Hepatitis B virus	Protein S, preS1 (1–48)-S	Mice (virus-neutralizing B and T cell response)	Niedre-Otomere et al. 2012
Human immunodeficiency virus 1	Env gp160	Cynomolgus macaques (B and T cell response)	Berglund et al. 1997
Human immunodeficiency virus 1	Env gp160	Mice (antibody response)	Brand et al. 1998
Human immunodeficiency virus 1, clade A	HIVA	Mice (T cell reponse)	Hanke et al. 2003
Human immunodeficiency virus 1, clade C	VREC (GagPolNef; RevTatEnv)	Mice (T cell reponse)	Sundbäck et al. 2005; Gómez et al. 2007
Human immunodeficiency virus 1, clade C	Gag, Env, PolRT	Mice (T cell response)	Ajbani et al. 2015, 2017
Influenza H1N1	Nucleoprotein (NP)	Mice (CTL response)	Zhou et al. 1994, 1995
Influenza H1N1	Nucleoprotein (NP) Hemagglutinin (HA)	Mice (protection)	Berglund et al. 1999
Louping ill virus	prM-E, NS1	Sheep (protection)	Fleeton et al. 1999; Morris-Downes et al. 2001
Simian immunodeficiency virus	Env gp160	Pigtail macaques (protection)	Mossman et al. 1996
Simian immunodeficiency virus	Env, Gag-Pol, Nef, Rev, Tat	Macaques (protective by use of three different vectors)	Nilsson et al. 2001; Koopman et al. 2004; Michelini et al. 2004; Negri et al. 2004
Simian immunodeficiency virus	Env, Gag	Rhesus macaques (a boost after the VSV vector-based construct)	Gambhira et al. 2014
Brain tumors (glioblastoma)	Endostatin	Mice (tumor regression)	Yamanaka et al. 2001
Brain tumors (glioblastoma)	Interleukin-18, -12	Mice (antitumor immune response)	Yamanaka et al. 2000, 2002, 2003
Breast cancer (4T1)	Luciferase	Mice (tumor targeting)	Vasilevska et al. 2012
Breast cancer (4T1)	Interleukin-12	Mice (total cure, together with LVR01 adjuvant)	Kramer et al. 2015
Breast cancer (4T1)	Interleukin-12	Mice (tumor regression)	Chikkanna-Gowda et al. 2005
Cervical cancer (HPV)	HPV E6-E7	Mice (complete elimination of tumors)	Daemen et al. 2003
Cervical cancer (HPV)	HPV E6-E7	Mice (regression of tumor and total survival, together with a low-dose irradiation and sunitinib)	Draghiciu et al. 2014, 2015
Colon cancer (CT26)	VEGFR-2	Mice (survival, together with IL-4)	Lyons et al. 2007
Colon cancer (CT26)	Interleukin-12	Mice (necrosis of tumor cells)	Chikkanna-Gowda et al. 2005
Lung cancer (H358a)	eGFP	Mice (regression of tumor)	Murphy et al. 2000
Ovarian cancer (MOSEC)	Ovalbumin	Mice (enhanced immune response)	Zhang YQ et al. 2010
Sindbis Virus (SINV)			
Colon cancer (CT26)	LacZ	Mice (therapeutic effect)	Granot et al. 2014
Lung cancer (CT26, CL25)	LacZ	Mice (protection against tumor challenge)	Granot et al. 2014

Vaccine Target	Gene	Testing Model	References
Ovarian cancer (ES2)	IL-12	Mice (long-term survival)	Granot and Meruelo 2012
Ovarian cancer (MOSEC)	GFP, Luciferase	Mice (apoptosis)	Venticinque and Meruelo 2010
Pancreatic cancer (Pan02)		Mice (apoptosis)	Venticinque and Meruelo 2010
Venezuelan Equine Encephalitis Virus (VEEV)			
Cytomegalovirus	pp65, IE1, gB	Mice, rabbits (immune response), human (phase I trial)	Reap et al. 2007a, b; Bernstein et al. 2009
Dengue virus 1–4	Ectodomain of DENV E protein (E85)	Mice (protection)	White et al. 2007, 2013; Khalil et al. 2014a
Ebola virus	Nucleoprotein (NP); glycoprotein (GP)	Mice (protection)	Pushko et al. 2000
Ebola virus	Nucleoprotein (NP)	Mice (protection)	Wilson and Hart 2001
Ebola virus	VP24, VP30, VP35, and VP40	Mice (protection)	Wilson et al. 2001
Ebola and Sudan viruses	Glycoprotein (GP)	Cynomolgus macaques (protection)	Herbert et al. 2013
Human immunodeficiency virus 1, clade C	Gag (vaccine AVX101)	Human (phase I trial)	Wecker et al. 2012
Human papillomavirus	Oncoprotein E7	Mice (eradication of tumors)	Velders et al. 2001
Influenza H1N1	Hemagglutinin (HA)	Mice (protection)	Pushko et al. 1997
Influenza H5N1	Hemagglutinin (HA)	Chickens (protection)	Schultz-Cherry et al. 2000
Junin and Machupo arenaviruses	Glycoprotein precursors (GPC)	Guinea pigs (protection)	Johnson et al. 2020
Lassa virus	Nucleoprotein (NP)	Mice (antibody response)	Pushko et al. 1997
Lassa virus	Nucleoprotein (NP) Glycoprotein (GP)	Guinea pigs (protection)	Pushko et al. 2001
Lassa virus + Ebola virus	Nucleoprotein (NP) Glycoprotein (GP)	Guinea pigs (protection)	Pushko et al. 2001
Lassa virus	Glycoproteins (GPC)	Mice (protection)	Wang M et al. 2018
Marburg virus	Nucleoprotein (NP) Glycoprotein (GP)	Mice, cynomolgus macaques (protection)	Hevey et al. 1998
SARS-COV	Glycoprotein S	Mice (protection)	Deming et al. 2006
Simian immunodeficiency virus	Matrix-capsid region of Gag, gp160 or unanchored gp140	Rhesus macaques (protection)	Johnston et al. 2005
VEEV	VEE replicon	Mice, cynomolgus macaques (protection)	Reed et al. 2014
Bacillus anthracis	4 variants of protective antigen (PA)	Mice (protection)	Lee et al. 2003
Clostridium botulinum serotype A	Neurotoxin	Mice (protection)	Lee et al. 2001
Breast cancer (A2L2)	Truncated neu oncoprotein	Mice (regression of tumor)	Moran et al. 2007
Cervical cancer (HPV)	HPV-16 E7	Mice (prevention of tumor development)	Velders et al. 2001
Melanoma (B16)	Tyrosine related protein-2	Mice (regression of tumor)	Avogadri et al. 2010, 2014
Pancreatic cancer	Carcinoembryonic antigen (vaccine AVX701)	Mice (tumor models), human (phase I trial)	Morse et al. 2010
Prostate cancer (transgenic adenocarcinoma)	Prostate-specific membrane antigen, prostate stem cell antigen	Mice (prolonged survival)	Garcia-Hernandez et al. 2007, 2008

(Continued)

TABLE 18.1 (CONTINUED)

Generation of the Single-Round Infectious VLPs (sriVLPs) by the *Alphavirus* Vectors

Vaccine Target	Gene	Testing Model	References
Prostate cancer	Prostate-specific membrane antigen	Human (phase I trial)	Slovin et al. 2013
Prostate cancer (Du145, 22Rv1)	Prostate-specific antigen	Mice (tumor growth delay)	Riabov et al. 2015
Western Equine Encephalitis (WEEV)			
WEEV	WEEV replicon	Mice (protection), cynomolgus macaques (antibody response)	Reed et al. 2014

Notes: The targets are given in the following order: viruses, bacteria, cancer. Within each group, the alphabetical order is used.

(2000) engineered the VEEV sriVLPs that provided protection against Ebola hemorrhagic fever virus in two rodent models. Next, the sriVLPs-derived vaccines were developed for Lassa virus, and their immunogenicity and protective capability against lethal Lassa virus challenge in guinea pigs was demonstrated (Pushko et al. 2001). Moreover, the authors configured the VEEV replicons for the combined expression of vaccine-relevant genes and evaluated the protective capability of combination and bivalent vaccines against both Lassa and Ebola viruses.

Furthermore, the VEEV sriVLPs demonstrated high efficiency by the expression of genes of influenza (Schultz-Cherry et al. 2000), botulinum neurotoxin serotype A (Lee et al. 2001), Ebola virus (Wilson and Hart 2001; Wilson et al. 2001; Herbert et al. 2013), human papillomavirus (Velders et al. 2001), anthrax (Lee et al. 2003), and simian immunodeficiency virus (Johnston et al. 2005).

Next, a tetravalent VEEV-vector-based dengue (White et al. 2007, 2013; Khalil et al. 2014a) was elaborated. Then, the putative Lassa virus (LASV) vaccines were successfully developed. First, the wild-type LASV glycoprotein (GPCwt) and a noncleavable C-terminally deleted modification (ΔGPfib) were expressed from individual VEE 26S subgenomic promoters and demonstrated high immunogenicity and protectivity in mice (Wang M et al. 2018). The VEE sriVLPs were also applied for expression of GPCs of Junin (JUNV) and Machupo (MACV) viruses and elicited humoral immune responses, which correlated with complete protection against challenges with JUNV and MACV, respectively (Johnson et al. 2020).

Reed et al. (2014) achieved strong protection against VEEV challenge in mice and macaques through the previously mentioned vaccine candidate combining the sriVLPs carrying replicons of all three equine encephalitic alphaviruses, EEEV, VEEV, and WEEV.

The VEEV sriVLPs-based vaccinations against filoviruses should be highlighted. Thus, the vaccination against Sudan virus (SUDV) by a single intramuscular administration of the VEE particles expressing the SUDV GP provided complete protection in cynomolgus macaques (Herbert et al. 2013). However, the immunization did not fully protect the macaques from intramuscular challenges with Ebola virus (EBOV). When intramuscular

coimmunization with the VEE-SUDV-GP and the VEE-EBOV-GP was performed, the complete protection against challenges with both SUDV and EBOV followed. Moreover, the intramuscular immunization also resulted in complete protection against challenges with aerosolized SUDV, although two vaccinations were required to reach efficacy (Herbert et al. 2013).

The current timely coronavirus pandemic has cast a special light on the need of novel vaccines against nidoviruses. The VEEV sriVLPs were used for the expression of SARS-CoV S and N proteins from the Urbani strain (Deming et al. 2006). The immunization of mice with the CoV S-expressing particles provided complete short- and long-term protection against challenges with homologues SARS-CoV strains in both young and senescent mice. In contrast, no protection was obtained after immunization with VEE-SARS-CoV N. Related to heterologous SARS-CoV strains, the chimeric icGDO3-S virus encoding a synthetic S gene of the most genetically divergent human GDO3 strain showed strong resistance to neutralization with antisera directed against the Urbani strain. Despite that, the immunization with the VEE-SARS-CoV S sriVLPs resulted in complete short-term protection against icGDO3-S challenges in young mice but not in senescent animals.

It is reasonable to highlight the VEEV sriVLPs-based vaccines that were evaluated by human clinical trials. Thus, the cytomegalovirus (CMV) vaccine candidate (Reap et al. 2007a, b) demonstrated the successful outcome of the phase I trial (Bernstein et al. 2009). The procedure was safe, and neutralizing antibodies and multifunctional T-cell responses were generated against all three CMV antigens important for protective immunity. Then, Wecker et al. (2012) published a report on the phase I trial of the VEEV sriVLPs-based HIV vaccine AVX101 in the US and South Africa, where volunteers were subjected to subcutaneous injection of escalating doses of the particles expressing the nonmyristoylated form of the HIV-1 subtype C Gag protein. Slobin et al. (2013) performed the phase I clinical study for the VEEV sriVLPs-based vaccine against prostate cancer, which expressed prostate-specific membrane antigen, in patients with castration resistant metastatic disease. Despite the failure to generate robust immune responses and clinical benefits, the success in eliciting neutralizing antibodies indicated that dose optimization would further enhance the efficacy of the vaccine, as commented by Lundstrom (2020).

As by the SFV sriVLP vectors, many types of cancers were targeted by the VEEV sriVLPs, as described in the previously mentioned comprehensive review of Lundstrom (2020). It should be highlighted that the phase I trial study was performed successively with the VEEV sriVLPs-based vaccine against pancreatic cancer (Morse et al. 2010). More details about the VEEV sriVLP-based vaccine and therapeutic anticancer candidates are collected in Table 18.1.

By analogy with the previously described properties of the SFV sriVLPs, the VEEV sriVLPs demonstrated strong adjuvant activity (Thompson et al. 2006) and were referred to as adjuvant nVRP (Thompson et al. 2006; Tonkin et al. 2010) or GVI3000 (Steil et al. 2014). According to these studies, the adjuvant particles enhanced humoral, cellular, and mucosal immune responses when inoculated into adult mice by subcutaneous, intradermal, or intramuscular routes with a number of soluble (ovalbumin or keyhole limpet hemocyanin) or inactivated particulate antigens (norovirus, poliovirus, or influenza viruses). Khalil et al. (2014b) demonstrated the same effect in the seven-day-old neonatal mouse model, which has been reported to resemble more closely the stage of immune maturation and immune-function limitations in human neonates.

WESTERN EQUINE ENCEPHALITIS COMPLEX

SINDBIS VIRUS

As described earlier, Tellinghuisen et al. (1999) expressed the nucleocapsid genes of SINV and RRV in *E. coli*. The purified capsid proteins formed in vitro core-like particles in the presence of single-stranded but not double-stranded nucleic acid. The truncated forms of the SINV capsid protein were used to establish aa requirements for assembly. Thus, a capsid protein starting at residue 19, namely CP(19–264), was fully competent for the in vitro assembly, whereas proteins with further N-terminal truncations could not support assembly. However, a capsid protein starting at residue 32 or 81 was able to incorporate into mosaic particles in the presence of the CP(19–264) or could inhibit assembly if its molar ratio relative to CP(19–264) was greater than 1:1 (Tellinghuisen et al. 1999).

An E2-E1 recombinant protein of SINV, in which the ectodomains of E2 and E1 were connected by a flexible Strep-tag linker, was expressed in *Drosophila* Schneider 2 (S2) cells (Li et al. 2010). The size-exclusion chromatography showed that the purified protein existed in solution as trimers of the E2-E1 heterodimer over a pH range from 5.5 to 9.5. The protein was crystallized at pH 5.6, and the resultant crystal structure consisted of trimers of E2-E1 heterodimers that were remarkably similar to the trimeric spikes in the virus. The appropriate x-ray structure was used to achieve the 3D reconstruction of SINV, which is presented here in Figure 18.3.

As by the SFV sriVLP vectors, many types of cancers were targeted by the VEEV sriVLPs, as narrated in the previously mentioned review of Lundstrom (2020).

WESTERN EQUINE ENCEPHALITIS VIRUS

As described earlier, the WEEV sriVLPs were engineered by the development of a combination vaccine that would protect animals against all three encephalitic alphaviruses: EEEV, VEEV, and WEEV (Reed et al. 2014). While the individual replicon vaccines or their combination elicited strong neutralizing antibodies in mice and demonstrated good protection against aerosol challenge with EEEV and VEEV in mice and macaques, both the WEEV replicon and combination of three vaccines elicited poor neutralizing antibodies to WEEV in macaques, and the protection conferred was not as strong.

Mukhopadhyay et al. (2002) expressed the nucleocapsid gene of WEEV in *E. coli*. The purified capsid proteins formed core-like particles in the presence of a single-stranded 48-nucleotide oligomer. As described by the RRV presentation earlier, the cryo-EM reconstructions of nonpurified core-like particles showed the same T = 4 quasi-symmetry, although at low 30 Å resolution (Mukhopadhyay et al. 2002).

The E1 (Das et al. 2004) and E2 (Das et al. 2007) genes of WEEV were expressed in *E. coli* and the products appeared in inclusion bodies. The inclusion bodies were successfully solubilized, refolded, and the immunogenicity of these nonglycosylated proteins was assessed in mice. The immunization of mice with the E1 or E2 proteins generated both humoral and cell-mediated immune responses. The challenge of the E1-immunized mice with live WEEV demonstrated little or no protection, while the E2-immunized mice could be partially protected from the lethal WEEV challenge. Hu et al. (2008) expressed in *E. coli* the C-terminally His$_6$-tagged E1 and E2 genes of WEEV, and again the products appeared as inclusion bodies. As before, the refolded E1 and E2 proteins did not form any particles but were suitable as antigens in an ELISA test for detection of anti-WEEV antibodies.

In the baculovirus-driven expression system, the WEEV glycoprotein constructs encoding full-length E1, the E1 ectodomain, an E2–6K-E1 polyprotein precursor, and an artificial, secretable E2-E1 chimera were expressed (Toth et al. 2011). The nature of the WEEV construct and the timing of expression influenced both the quantity and quality of recombinant glycoprotein produced. Thus, the full-length E1 product was insoluble, irrespective of the timing of expression. Each of the other three constructs yielded soluble products, and, in these cases, the timing of expression was important, as higher protein processing efficiencies were generally obtained at earlier times of infection. The putative self-assembly of the products was not observed in this study.

SALMON PANCREATIC DISEASE VIRUS

The SAV VLPs were obtained via expression of recombinant baculoviruses encoding SAV capsid protein and glycoproteins E1 and E2 in *S. frugiperda Sf*9 insect cells (Metz et al. 2011a). However, the formation of the SAV VLPs in insect cells did not take place under standard conditions but was dependent on the level of processing of the envelope glycoprotein E2. This was only achieved when a temperature shift from 27°C to lower temperatures was applied. At 27°C, the pE2 protein, a precursor of E2, was misfolded and not processed by host furin into mature E2. Hence, E2 was detected neither on the surface of infected cells nor as VLPs in the culture fluid. However, when temperatures during protein expression were lowered, the pE2 was processed into mature E2 in a temperature-dependent manner, and VLPs were abundantly produced (Metz et al. 2011a).

This study led to further investigation of the processing and trafficking of the SAV glycoproteins E2 and E1 as a function of temperature (Hikke et al. 2014). As a result, the critical determinant for the low-temperature dependent SAV virion formation was uncovered and connected with needs for both low temperature and coexpression of E2 and E1 for proper translocation and presentation at the cell surface of insect and fish cells.

At last, Hikke et al. (2016) used SAV, in parallel with CHIKV, to develop an alternative to the whole enveloped alphavirus VLPs, namely the core-like particle platform. Thus, the authors engineered recombinant baculovirus vectors to produce high levels of alphavirus core-like particles in insect cells by expression of the SAV or CHIKV capsid proteins. The core-like particles localized in dense nuclear bodies within the infected cell nucleus and were purified through a rapid and scalable protocol involving cell lysis, sonication, and low-speed centrifugation steps. Furthermore, an immunogenic epitope from the alphavirus E2 glycoprotein, namely the E2 glycoprotein B-domain encompassing aa 158–252, was successfully fused to the N-terminus of the SAV capsid protein without disrupting the self-assembling properties. When the 96-aa E2 epitope was fused to the C-terminus of the carrier, the self-assembly was impeded and only protein aggregates were visible, caused possibly by steric hindrance. The authors proposed the novel platform of the chimeric epitope-tagged alphavirus core-like particles as a simple and perhaps more stable alternative to the true alphavirus VLPs (Hikke et al. 2016), especially taking into account the obstacle that the insect-cell derived CHIKV VLPs appeared amenable to large scale production in bioreactors (Metz and Pijlman 2011).

19 Order *Martellivirales*: *Virgaviridae*

He enters the port with a full sail.

Virgil

ESSENTIALS

The family *Virgaviridae* is a substancious and specific part of the order *Martellivirales* that was presented generally in Chapter 17.

According to the official ICTV review (Adams et al. 2017; ICTV 2020), the family *Virgaviridae* currently includes 7 genera and 59 species, where the genus *Tobamovirus* involving the famous tobacco mosaic virus (TMV) from the *Tobacco mosaic virus* species is the most representative member. TMV is one of the most thoroughly studied viruses in history, the first discovered (in 1886), the first nonbacterial pathogen isolated and named virus (in 1898), the first virus to be crystalized (in 1935), and the first visualized by electron microscopy (in 1939). TMV was the first virus for which the aa sequence of the coat protein was determined and the first plant virus for which structures and functions were known for all genes. It was the first virus for which activation of a resistance gene in a host plant was related to the molecular specificity of a viral gene product. It was not surprising that TMV became one of the first and most important subjects of the VLP nanotechnologies. In the field of plant biotechnology, it appeared as one of the most promising vectors.

The TMV particle was the first macromolecular structure shown to self-assemble in vitro, as demonstrated in the early review of Butler (1999). At last, TMV became one of the first candidates for the development of nonreplicative VLP carriers (Haynes et al. 1986)

The pioneering role of TMV in many scientific fields and applications including immunology was described in early exhaustive reviews of Harrison and Wilson (1999), Okada (1999), Van Regenmortel (1999), McCormick and Palmer (2008), as well as many others.

The *Virgaviridae* members infect a wide range of herbaceous and mono- and dicotyledonous plant species, but the host range of individual members is usually limited. All members can be transmitted experimentally by mechanical inoculation, and for those in the genus *Tobamovirus* this is the only known means of transmission (Adams et al. 2017).

Figure 19.1 presents a portrait and a cartoon of tobacco mosaic virus (TMV) from the genus *Tobamovirus*, a typical representative of the *Virgaviridae* family. The *Virgaviridae* virions are nonenveloped, rigid rods possessing a helical symmetry with a pitch of 2.3 to 2.5 nm and an axial canal. They are about 20 nm in diameter with predominant lengths that depend upon the genus. The TMV virions are 18 nm in diameter and of 300–310 nm in length. The virions can form large crystalline arrays visible by light microscopy.

In most *Virgaviridae* viruses, the capsid comprises multiple copies of a single coat protein of about 17–24 kDa. As reviewed by Adams et al. (2017), in viruses of the genera *Furovirus* and *Pomovirus*, a larger minor capsid protein is also produced by translational readthrough of the coat protein-encoding gene stop codon and can be detected at the extremity of virus particles (Cowan et al. 1997). In at least some furoviruses, a further minor coat protein of 25 kDa is initiated from a CUG codon upstream of the canonical start codon (Yang et al. 2016).

Remarkably, the only plant viruses with rod-shaped particles not included in the family are those classified in the genus *Benyvirus*, family *Benyviridae*, order *Hepelivirales*, which were presented before in Chapter 16. The benyviruses have polyadenylated RNAs and replication proteins only distantly related to those of viruses in the family *Virgaviridae* (Adams et al. 2017).

GENOME

Figure 19.2 shows the genomic structure of TMV as a typical representative of the *Virgaviridae* family. The single-stranded positive-sense RNA genome is linear, monopartite, and of 6.3–6.5 kb in size. The 3'-terminus has a tRNA-like structure, while the 5'-terminus has a methylated nucleotide cap m⁷GpppG. The virion RNA is infectious and serves as both the genome and the viral messenger RNA.

The largest ORF encodes a replication protein with conserved methyltransferase and helicase domains, an arrangement typical of alpha-like viruses. This protein is translated directly from the genomic RNA. In viruses of all genera except *Hordeivirus*, the RNA dependent RNA polymerase is expressed as the C-terminal part of this protein by readthrough of a leaky stop codon. The movement proteins and the capsid protein are expressed from separate subgenomic mRNAs (Adams et al. 2017).

GENUS *TOBAMOVIRUS*

Tobacco Mosaic Virus

3D Structure

The fine structural investigations of TMV have a strikingly long history. TMV was prepared in a fully crystalline state in the 1930s (Bawden and Pirie 1937, 1938), and the first TMV structure was determined 80 years ago by John Desmond Bernal—a pioneer of the x-ray crystallography in molecular biology—and his colleagues, using x-ray fiber diffraction method (Bernal and Fankuchen 1941a, b). The

DOI: 10.1201/b22819-24

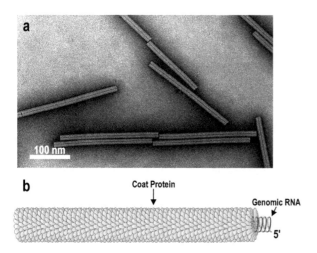

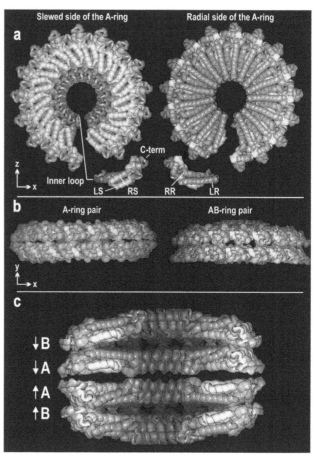

FIGURE 19.1 A portrait and schematic cartoon of tobacco mosaic virus (TMV) of the *Tomato mosaic virus* species from the genus *Tobamovirus*, a typical representative of the *Virgaviridae* family. (a) Negative contrast electron micrograph of the TMV particles stained with uranyl acetate. (Reprinted from Adams MJ et al. ICTV virus taxonomy profile: *Virgaviridae. J Gen Virol.* 2017;98:1999–2000.) (b) A cartoon of the TMV virion is taken with kind permission from the ViralZone, Swiss Institute of Bioinformatics, http://viralzone.expasy.org (Hulo C et al. 2011). (Courtesy Philippe Le Mercier.)

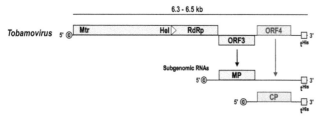

FIGURE 19.2 The genomic structure of TMV of the *Tomato mosaic virus* species from the genus *Tobamovirus*, a typical representative of the *Virgaviridae* family, combined from the official ICTV profile (Adams et al. 2017) and from the ViralZone, Swiss Institute of Bioinformatics; Hel, helicase; RdRp, RNA-dependent RNA polymerase; MP, movement protein of the 30K superfamily. The white triangular arrowhead shows the position of a suppressible stop codon that results in larger, readthrough products. The tRNA-like structure at the 3'-terminus of the genomic RNAs is shown as a gray square box. The structural gene is colored dark pink.

cautionary tale of the 3D studies on TMV was narrated in an early exhaustive review written by the Nobelist 1982 Sir Aaron Klug (1999).

The first fine TMV structure was reported more than 40 years ago (Stubbs et al. 1977), then refined to 2.9 Å by Namba et al. (1989), but the final improvements to 2.4-Å resolution were made by Bhyravbhatla et al. (1998). Figure 19.3 demonstrates this historical 3D structure of TMV. The TMV coat appeared as a compact protein of 159 aa that formed two antiparallel α-helical strands, where the N- and C-termini, as well as a surface loop between helices

FIGURE 19.3 Overview of the refined atomic model of the TMV four-layer disk aggregate. The Cα trace is drawn with fat ribbons to give the effect of a space-filling model and is color coded according to a sequence starting with yellow at the N-terminus and ending with olive green at the C-terminus. (a) Views looking perpendicular to the crystallographic twofold axis, parallel to the 17-fold noncrystallographic axis of the A-ring in the four-layer aggregate. One subunit has been pulled out of the closed-ring assembly in both figures to illustrate the packing. The loop region is in the lower radius end of the macromolecular assembly. Left: View of the right and left slewed helices and the turn of the protein chain. The small β-sheet region in the high radius end of the molecule can also be clearly seen. The right radial helix (below the slewed helices) extends into the low radius region more like a loose turn. The RNA binding region for the complete virion lies in the groove formed between the turn of the slewed helices and the right radial helix. The N-terminus can be seen tucked inside the protein chain, whereas the C-terminus extends out in the high radius end of the molecule. Right: View of the opposite side of the ring relative to figure on the left. The radial helices extend the length of the subunits, forming a much smoother surface compared to the slewed side. The left radial helix extends straight into the axial hole, but the rms fluctuations of the residues in the low radius region are very high. (b) The packing in the central AA-ring pair and the AB-ring pair of the four-layer aggregate. Left: The central A-pair of rings is related by one primary crystallographic twofold axis (along the z axis, viewing direction), and 16 other noncrystallographic 2-fold axes. This A-A ring pair is sandwiched between the two B-ring of subunits. The slewed helices (shown in a, left) of the 17 subunits overlap across the twofold

(Continued)

FIGURE 19.3 *(Continued)*
axes, sandwiching approximately two layers of solvent between them. There are two protein–protein contacts in the high radius end between the symmetry-related protein chains across the alternate dyad axes. Right: The AB-ring pair is the asymmetrical unit of the crystal (space group. P2₁2₁2), viewing along the crystal z-axis. The radial helices of the A-ring (shown in a, *right*) interact with the slewed helices of the B-ring of subunits and correspond to the surface of the B-ring in contact with the solvent on the top. The tilt in the B-ring of subunits with respect to the A-ring gives the structure a convex shape on the top. The C-terminal end of the subunit appears to be in a different conformation in both the A and B chains. There are protein–protein contacts between the A- and B-chains in the radial range of 60–80 Å. (c) The complete four-layer aggregate is shown, viewed perpendicular to the 17-fold rotational axis and the 2-fold axis in the crystal. This is a view similar to that in (b), but eight subunits from the front of the aggregate have been removed to clearly show the inner axial hole, which is filled with the solvent, the packing of the disordered segments, and the relative orientations of the A- and B-ring of subunits with respect to each other. Notice that there is considerable solvent-accessible volume between the A- and B-chains. There is a contact surface of only ~250 Å^2, with three protein–protein contacts, between the A- and B-chains compared to ~750 Å^2 and three protein–protein contacts between the A-chains across the 2-fold axes. The major overlap region across the 2-fold axes is occupied by solvent atoms. The loop region residues in the A-chain are more restricted in their movement compared to the B-chain residues. The A- and B-chains and their respective polarities are indicated on the side. By convention, the "top" of the subunit refers to the slewed helices side, and the "bottom" refers to the radial helices. Thus, the central A-pair of rings has a "top-to-top" interaction, whereas the AB-pair has a "bottom-to-top" interaction. (Reprinted from *Biophys. J.* 74, Bhyravbhatla B, Watowich SJ, Caspar DL, Refined atomic model of the four-layer aggregate of the tobacco mosaic virus coat protein at 2.4-Å resolution, 604–615, Copyright 1998, with permission from Elsevier.)

2 and 3, namely aa residues 55 and 60, were surface exposed and might therefore accommodate foreign insertions.

Culver et al. (1995) have proven that the carboxylate groups of the aa residues Glu50 and Asp77 across subunit interfaces are crucial for the virion disassembly. The site-directed mutagenesis E50Q and D77N increased virion stability, accompanied with reduced infectivity. The mixing of purified mutant coats with wild-type virions under appropriate conditions stabilized the virions, opening thus a way for the future nanotechnological ideas.

It is noteworthy that TMV was used, in parallel to the first structural essays, as a test specimen to develop techniques for high-resolution structural analysis in electron micrographs of biological assemblies with helical symmetry (Jeng et al. 1989).

More recently, Zhou et al. (2013) engineered a disulfide bond between the adjacent subunits in the nanorods by introducing a cysteine by a single T103C mutation within the inner channel of the TMV coat, which significantly improved the assembly capability and the stability of the rods in *E. coli*. This expanded the spectrum of metal fabrication within the channel to include gold nanobeads and nanorods (Zhou et al. 2015).

Going further, Zhang J et al. (2016) introduced disulfide bonds between the neighboring TMV coat disks through mutation of cysteines not only at position 103 but also at aa positions 1 or 3, which were located at the outer surface of rods. Thus, if the 103 mutation of cysteine on the inner surface promoted the assembly of TMV disks into a tube due to the S–S bonding between adjacent subunits in the longitudinal axis, the two cysteines at 1 and 3 sites on the outer surface were expected to act as "tethers" to capture disks for each other in the x–y plane. With the synergistic effects of chemical bonding of inter-cysteines and the buffer solution, including pH, temperature, and ionic strength, four unprecedentedly assembled patterns of the TMV coats, such as disk arrays, disk stacks, tube stacks, and bundles, were observed (Zhang J et al. 2016).

In the connection with the surprising variability of the TMV patterns, it is noteworthy that the TMV-like particles of differing length grown in situ on solid flat or bead supports yielded high-surface density arrays of carrier rods in nanobrush and nanostar layouts, from nanoboomerangs to tetrapods (Mueller et al. 2011; Azucena et al. 2012; Eber et al. 2013).

Expression

Bevan et al. (1985) were the first who attempted to produce the TMV coat by transformation of tobacco plants with the coat gene expressed under the cauliflower mosaic virus (CaMV) 35S promoter, but the self-assembly issue was not addressed. It soon transpired that tobacco plants transgenic for the coat were resistant to the TMV infection, and the concept of the "coat protein-mediated resistance" of plants was accepted, as reviewed early by Beachy (1999). As part of the coat-mediated resistance studies, Asurmendi et al. (2007) showed that the TMV coat subunits formed in fact higher aggregates in transgenic tobacco, although the levels of the coat produced in transgenic plants were low.

To increase the levels of the TMV coats in plants, Saunders and Lomonossoff (2017) used transient expression of the coat coding sequence in *N. benthamiana*. No detectable material accumulated when the coat gene was expressed alone, indicating that the assembly was crucial for the coat accumulation in plants. When the coat gene was expressed in the presence of RNA molecules containing the origin of assembly (OAS), the virus-like rods readily formed, the length of which was controlled by the length of the RNA. Furthermore, it was possible to fuse a nine-aa cobalt–platinum (CoPt)-binding peptide to the C-terminus of the coat without abolishing particle assembly in planta. The CoPt could be deposited on the surface of particles harboring the peptide but not on particles containing the wild-type coat (Saunders and Lomonossoff 2017). This pivotal study for the plant-based expression of the TMV coat was readily reviewed by Lomonossoff and Wege (2018).

The synthetic gene coding for the TMV coat protein was expressed in *E. coli* at the very beginning of the VLP era (Haynes et al. 1986). The expression product assembled into rod-like particles under acidic conditions in *E. coli* extracts. Thus, to induce polymerization of the TMV coat product, the sonicated bacterial lysate was dialyzed overnight at pH 5.0 and the helical rod-like structures were documented by electron microscopy.

Shire et al. (1990) cloned the coat gene of a vulgare TMV strain, expressed it in *E. coli*, and purified adequate amounts of the recombinant TMV coat for physical studies. The ability of the recombinant protein to assemble into long helical rods at low pH and subsequently depolymerize at higher pH values demonstrated that most of the coat protein expressed and purified out of *E. coli* behaved in a fashion like that of the protein isolated from virus with regard to pH dependency of self-assembly. The *E. coli*-derived coat lacked an N-terminal acetyl group and therefore carried an additional positive charge, near neutral pH, compared to the tobacco-plant-produced viral protein and changing therefore the assembly properties. Although the recombinant TMV coat was reconstituted with TMV RNA, it did so at a reduced rate compared to the reconstitution with the native TMV coat (Shire et al. 1990).

Hwang et al. (1994a) demonstrated that the former *E. coli*-made TMV coats formed in vitro at pH 5.0 stacked cylindrical aggregates that were inactive for in vitro assembly with TMV RNA and failed to be immunogold-labeled using a mouse monoclonal antibody specific for the helically assembled TMV coat. The true helical TMV-like ribonucleoprotein particles of the predicted length were formed in high yield only when chimeric transcripts containing the TMV OAS sequence were coexpressed in vivo together with the TMV coat gene. Moreover, the *E. coli* expression of a full-length cDNA clone of the TMV genome of 6.4 kb resulted in vivo in high, immunodetectable levels of the TMV coat assembly of sufficient intact genomic RNA to initiate systemic infection of susceptible tobacco plants (Hwang et al. 1994a, b).

Further development of the TMV expression models at that time progressed in two major directions: (i) by engineering of TMV-based vectors expressing foreign genes in plants as independent units, e.g., some bacterial genes (Donson et al. 1991), GFP (Casper and Holt 1996), genes encoding heavy and light chains of a colon cancer-directed monoclonal antibody (Verch et al. 1998), complete VP1 gene of foot-and-mouth disease virus (FMDV; Wigdorovitz et al. 1999), consensus sequence of hypervariable region 1 (HVR1), a potential neutralizing epitope of hepatitis C virus (HCV) genetically fused to the C-terminal of the B subunit of cholera toxin (Nemchinov et al. 2000), or core gene of porcine epidemic diarrhea virus (PEDV; Kang et al. 2004) and (ii) by means of fusion of the foreign sequences to the coat gene (Takamatsu et al. 1990; Hamamoto et al. 1993). The second direction that will be described led to the generation of the chimeric replication-competent, rather than classical noninfectious

VLPs. As will be described, many of the latter had mosaic structure.

From the very beginning, the commercial companies were involved and planned to produce and process large quantities of plant-generated viral products in the field. The development of the TMV as a commercially important vector had been reviewed at that time by Turpen (1999) and Yusibov et al. (1999), while the putative place of TMV among other viral vectors was evaluated by Porta and Lomonossoff (1998) and later by McCormick and Palmer (2008) and Smith et al. (2009).

Developing further the *E. coli* expression of the TMV coat gene, Hwang et al. (1998) found that both GroEL and DnaK chaperons had significant direct or indirect effects on the overall expression, stability, folding, and assembly of the TMV coat protein in vivo. The overproduction of GroEL or GroES alone had little effect. However, cooverexpression of GroEL and GroES resulted in a twofold increase in soluble TMV coat and a fourfold rise in the assembled TMV-like particles in vivo. Moreover, TMV coat was shown to interact directly with GroEL in vivo (Hwang et al. 1998). Later, Dedeo et al. (2010) constructed and expressed in *E. coli* a "circular permutant" of TMV coat, where the N- and C-termini were reengineered, and a short loop was placed to close the sequence gap, as described later in detail.

Bruckman et al. (2011) engineered in *E. coli* the TMV coat, to which a His$_6$ tag was added to the C-terminus and the product self-assembled into disks, hexagonally packed arrays of disks, stacked disks, helical rods, fibers, and elongated rafts. However, the His$_6$ tag significantly affected the self-assembly in comparison to the wild-type coat. Nevertheless, the His$_6$ tag interactions attributed to the alternative self-assembly of the chimeras could be controlled through ethanol and nickel-nitrilotriacetic acid (Ni-NTA) additions.

Wnęk et al. (2013) also added the C-terminal His$_6$ tag to the TMV coat expressed in *E. coli*. The chimeras did not bind TMV RNA or form disks at pH 7, but they retained the ability to self-assemble into rod-like arrays at acidic pH. The C-terminal tags in such arrays were exposed on the protein surface, allowing interaction with target species. These chimeras were utilized to create nanorods able to bind gold nanoparticles uniformly, which could be transformed into gold nanowires.

The TMV coat formed TMV-like rods spontaneously inside the cells when expressed in fission yeast *Schizosaccharomyces pombe* (Kadri 2007, cited by Mueller et al. 2010). After the wild-type TMV coat, Mueller et al. (2010) explored expression of two genetically engineered TMV coat variants in *S. pombe*. They were TMV-CP-His$_6$ containing a C-terminal His$_6$ tag and TMV-CP-E50Q with enhanced lateral coat subunit interactions. These results were compared to those with the corresponding mutant TMV coat variants prepared from plant-derived virions, which were expressed from infectious TMV constructs. The wild-type TMV and TMV-CP-E50Q yielded TMV-like rods, irrespective of the proteins' source or presence or absence of RNA. In contrast, His-tagged coat from plants produced only short rods in an inefficient manner and no rods at all when expressed in yeast (Mueller et al. 2010).

Kadri et al. (2013) compared the heterologous expression of the TMV coat protein and in vivo assembly of the rod-shaped TMV-like particles encapsidating viral or host RNA in both *E. coli* and *S. pombe*. The TMV-like particles were produced in both hosts, irrespective of whether the OAS of TMV was present. The additional plasmid providing an OAS-containing RNA was able to alter the length distribution of the TMV-like particles. The plant and yeast-expressed coats behaved similarly upon isoelectric focusing, whereas the coat expressed in bacteria migrated differently. After purification by buoyant density centrifugation, the encapsidated nucleic acids were determined to be of host origin as well as of viral origin. The OAS-containing mRNA was packaged preferentially in yeast. The majority of TMV-like particles showed the same length distribution similar to those in the absence of the OAS-containing mRNA, likely due to host RNA being primarily encapsidated (Kadri et al. 2013).

The aberrant assembly properties of the TMV coat expressed in bacteria were ascribed to the lack of acetylation of the N-terminal serine, which prevented the formation of the disk structures necessary to initiate the process (Shire et al. 1990; Wnęk et al. 2013). To alleviate the problem, Eiben et al. (2014) spiked the *E. coli*-produced TMV coat with a minimum of 20% of plant-made TMV coat, an approach that enabled efficient RNA-guided assembly of the chimeric TMV-coat-His$_6$ into particles of the expected length. The pure His$_6$-tagged coats did not form particles on their own with TMV RNA in vitro, but they were integrated into the wild-type coat-blended particles. The resulting rods formed dense monolayers with short range alignment on silicon substrates, substantially different, however, from the largely wavy patterns obtained with the wild-type TMV (Eiben et al. 2014).

Finbloom et al. (2016) produced a mutant TMV coat in *E. coli* in which two lysine residues at aa positions 53 and 68 were mutated to arginines. The modified coat was able to form stable disk structures that were subsequently modified for drug delivery purposes, as indicated later.

Going beyond the traditional bacteria and yeast expression systems, the efficient assembly of the TMV coat was observed within the hyphal cells of *Colletotrichum acutatum*, a phytopathogenic fungus (Mascia et al. 2014). In fact, the experiments were designed to develop an efficient TMV-based virus-induced gene silencing (VIGS) system for the fungus.

Lomonossoff and Wege (2018) suggested that it was likely that the assembly competent TMV coats could be generated in most eukaryotic expression systems. As the authors of the excellent review noted, the early studies on the identification of the TMV coat subgenomic mRNA showed that the coat produced in the in vitro translation system of wheat-germ extracts could assemble into rods with exogenously added TMV RNA (Hunter et al. 1976).

Chimeric Virions

As mentioned earlier, Takamatsu et al. (1990) were the first who generated chimeric TMV virions carrying foreign sequences. They reported expression of Leu-enkephalin YGGFL (Enk), a pentapeptide with opiate-like activity, as a fusion protein with the TMV coat in tobacco protoplasts by means of a TMV RNA vector. The Enk peptide with a preceding in-frame methionine was inserted just before the termination codon of the coat gene. The methionine was placed between coat and Enk sequences with an idea to isolate Enk by BrCN cleavage. In protoplasts inoculated with the Enk RNA, the coat-Enk fusion accumulated as the major protein. The virion formation was observed with one of the chimeric TMV-Enk variants, which carried in fact an extra aa sequence of SCIGAEI at its C-terminus. This observation indicated a clear benefit of the virion formation by enhancing the stability of the chimeric coat derivatives and allowing simple and rapid purification of the product from the inoculated tobacco plants in the form of virus particles (Takamatsu et al. 1990).

Hamamoto et al. (1993) were the first who generated mosaic TMV virions carrying foreign sequences over the leaky stop codon. Thus, a six base 3' context sequence, which originally permitted readthrough of the leaky UAG stop codon for the TMV 130K protein gene, was inserted between the TMV coat stop codon and the foreign angiotensin-I-converting enzyme inhibitor peptide (ACEI) encoding sequence. The ACEI was a 12-aa peptide FFVAPFPEVFGK found in the tryptic hydrolysate of milk casein, which possessed antihypertensive effects when orally administered and was described by Karaki et al. (1990). The chimeric TMV produced both an intact coat and a fused protein consisting of coat and ACEI, yielding in the mosaic virions that exposed the ACEI on their surface. This novel vector could form virus particles and spread systemically from inoculated to noninoculated leaves. The progeny virions showed the presence of not only intact coat protein but also of the coat-ACEI fusion, since the C-terminal portion of the coat protein projected outward from the particle, according to the previously described 3D structure. The production of ACEI was achieved not only in tobacco but also in fruits of tomato plants. The authors speculated that the described tomato fruits could be a dietary antihypertensive because ACEI would be released from the coat-ACEI fusion in the human intestine by trypsin digestion (Hamamoto et al. 1993).

Developing further this readthrough TMV vector, Sugiyama et al. (1995) constructed the mosaic TMV virions that presented three different kinds of epitopes, two of them from influenza virus hemagglutinin (HA)—namely aa stretches 91–108 and 139–146— and one from human immunodeficiency virus type 1 (HIV-1) envelope protein, namely the 13-aa principal neutralization determinant from the HIV-1 gp120, on the surface of the virions. Each of these mosaic TMV particles reacted with each antipeptide antiserum. The relative amount of the chimeric coat-to-wild-type coat was about 1:20 within virions, the same as in leaf extracts and corresponding to the theoretical prediction. These results indicated that the chimeric coats were incorporated into the virions as efficiently as the wild-type coat (Sugiyama et al. 1995).

Turpen et al. (1995) either inserted the B-cell epitopes of malaria CS protein into the surface loop region of the TMV coat or fused them to the C-terminus using the leaky UAG stop signal derived from the replicase protein reading frame. The epitope stretches AGDR from *P. vivax* or QGPGAP from *P. yoelii*, which were recognized by protective mAbs, were chosen. Either one or three copies of the AGDR epitope were substituted for the two aa following Pro63 within the surface loop, while the C-terminus was extended with two copies of the QGPGAP epitope. The viruses carrying the internal (AGDR)$_3$ insertion or C-terminal (QGPGAP)$_2$ addition reacted with monoclonal antibodies and were passaged for larger-scale virus purification. The tobacco plants systemically infected with each of these constructs produced high titers of genetically stable recombinant virus, enabling purification of the chimeric particles in high yield. By the C-terminal fusion, both the wild-type TMV coat and fusion protein synthesized by the leaky stop mechanism coassembled into virus particles at the predicted ratio of approximately 20:1 (Turpen et al. 1995). Generally, the loop position was found to be the least accommodating of foreign peptide insertions, which is probably due to structural constraints (McCormick and Palmer 2008).

Fujiyama et al. (2006) used the same readthrough approach to express a fragment encoding 15 aa of the poliovirus (PV) type 1 Sabin. The peptide TTHIEQKALAQGLGQ was encoded at nucleotide positions 2447–2491 of the PV genome, where the first 11 aa residues comprised the C-terminus of VP3 and the latter four aa comprised the N-terminus of VP1. The epitope was inserted downstream of the six-base 3′ context nucleotide sequence within the TMV coat gene, which allowed readthrough at the leaky UAG stop codon, thereby producing the chimeric TMV virions with both intact and chimeric coat subunits in *N. tabacum* plants. The chimeric virions induced the PV-specific antibodies in mice after intraperitoneal immunization (Fujiyama et al. 2006).

Borovsky et al. (2006) fused trypsin-modulating oostatic factor (TMOF), a mosquito decapeptide hormone, at the TMV coat C-terminus by using the readthrough leaky stop codon that facilitated expression of the chimeric coat-TMOF and wild-type coat in the same ratio of 1:20, and both proteins were coassembled within infections mosaic virions that presented TMOF on their surface. When fed to mosquito larvae, the purified mosaic virions stopped larval growth and caused death. Because of the wide host range of TMV, it was intended to use the mosaic TMV-TMOF virions as a general method to protect plants against agricultural insect pests and to control vector mosquitoes.

Bendahmane et al. (1999) performed insertions at the C-terminus of the TMV coat, namely between aa positions Ser154 and Gly155 of total 158 aa, for direct expression without any readthrough mechanism. The immunodominant

epitopes from rabies virus (RV) glycoprotein G, aa 404–418—in total three variants with differing aa residues at N-terminus and aa 253–275—and an epitope from murine hepatitis virus (MHV) glycoprotein S, aa 900–909, termed 5B19 epitope, in total four variants with different neighboring aa residues, were chosen. The selected epitopes were successfully displayed on the surface of the chimeric TMV virions, and viruses accumulated to high levels in infected leaves of *Nicotiana tabacum*. Figure 19.4 demonstrates a clear proof of the epitope exposition on the viral surface. The combinations of the epitope-neighboring charged aa residues unveiled the importance of the epitope pI and its critical effects on the successful epitope display on the surface of chimeric virions, as well as the lack of tolerance to positively charged epitopes on the TMV surface (Bendahmane et al. 1999). When mice were immunized with purified chimeric virions carrying the MHV 5B19 epitope, intranasally or subcutaneously, the MHV-neutralizing antibodies were induced. Moreover, the immunized mice

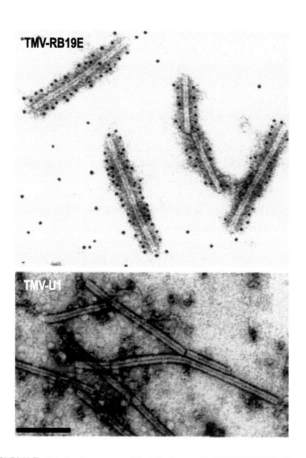

FIGURE 19.4 Immunogold labeling of TMV-RB19E (carrying the MHV 5B19 epitope 900-LLGCIGSTCA-909 with N-terminally added Arg and C-terminally added Glu) and TMV-U1 (wild-type as a negative control) using monoclonal antibody raised against the 5B19 peptide. The scale bar represents 200 nm. (Reprinted from *J Mol Biol*. 290, Bendahmane M, Koo M, Karrer E, Beachy RN, Display of epitopes on the surface of tobacco mosaic virus: impact of charge and isoelectric point of the epitope on virus-host interactions, 9–20, Copyright 1999, with permission from Elsevier.)

survived challenge with a lethal dose of MHV, whereas mice administered with wild-type TMV died ten days post challenge (Koo et al. 1999).

The first example of the chimeric TMV virions exposing a bacterial epitope, namely the so-called epitope 9–14mer covering the sequence 261-TDAYNQKLSERRAN-274 in the C-terminal portion of outer membrane protein F of *Pseudomonas aeruginosa*, was published by Staczek et al. (2000). The epitope was inserted again between aa residues Ser154 and Gly155 of the TMV coat. When mice were immunized with the chimeric TMV, the anti-peptide-9–14mer-specific antibodies were induced, and protection was achieved against challenge with wild-type *P. aeruginosa* in a mouse model of chronic pulmonary infection (Staczek et al. 2000). The attempts to express another immunodominant epitope termed 10–14mer: aa 305-NATAEGRAINRRVE-318, on the TMV coat were unsuccessful. When the chimeric TMV containing epitope 9–14mer was given together with chimeric influenza virus (see Chapter 33) carrying the 10–14mer epitope, as a combined vaccine, the immunized mice were protected against challenge with *P. aeruginosa* in the chronic pulmonary infection model at approximately the same level as provided by the individual chimeric virus vaccines (Gilleland et al. 2000). In parallel, the same epitopes were inserted into the cowpea mosaic virus (CPMV) vectors, as mentioned in Chapter 27.

Wu et al. (2003) inserted two immunodominant epitopes from VP1 of FMDV, serotype O, between the same aa residues Ser154 and Gly155 of the TMV coat. The chosen epitopes F11 and F14 contained the sequences 142—PNLRGDLQVLA—152 and 200—RHKQKIVA-PVKQTL—213, respectively. Both chimeras systemically infected the tobacco plant and produced large quantities of stable progeny viral particles assembled together with the modified coat subunits. The guinea pigs, suckling mice, and swine models demonstrated the protective effects of each of the vaccines or both together against FMDV infection.

Developing further the success with the FMDV epitopes, Jiang et al. (2006) achieved the expression of longer foreign epitopes by engineering of a new TMV-based vector in which four to six C-terminal aa residues were deleted from the coat protein subunit. The capacity of the new vector allowed expression of a peptide F25 containing two fused epitopes, F11 and F14, which failed using the original vector in tobacco. Although animal assays indicated that such expressed F25 was not as efficient as F11 in immunity, possibly due to lack of a spacer arm between the two fused epitopes, the new TMV-based vector met the requirement of expressing longer foreign peptides (Jiang et al. 2006).

Palmer et al. (2006) engineered a vaccine against both cottontail rabbit papillomavirus (CRPV) and rabbit oral papillomavirus (ROPV). The immunodominant epitope 94–122 that was mapped by Embers et al. (2002) appeared in this study as a vaccination subject. The epitope sequence comprising aa 94-VGPLDIVPEVADPGGPT-110 of the CRPV L2 protein and the homologous epitope aa

94-VGPLEVIPEAVDPAG-110 of the ROPV L2 protein were inserted between aa positions Ser155 and Gly156, of total 159 aa TMV coat sequence in this case, and two analogous constructs comprising the epitope sequence 107–122 were constructed. The rabbits receiving the CRPV L2 vaccine alone or in combination with ROPV L2 vaccines were completely protected against CRPV infections (Palmer et al. 2006).

Pogue et al. (2007) used the N-terminus of the TMV coat to fuse feline panleukopenia parvovirus (FPLV or FPV) protein VP2 epitopes MGQPDGGQPAVRNERAT or MGSDGAVQPDGGQPAV, the two principal hemagglutinating and neutralizing antibody-inducing epitopes on the surface of FPV. The corresponding chimeric virions were extracted from *Nicotiana* plants, purified, and administered to cats, resulting in partial protection against challenge with FPV.

Li Q et al. (2006) analyzed a great set of short aa stretches inserted at the C-terminus of the TMV coat and concluded that a foreign transmembrane domain within the fused peptide would result in the local lesions on the susceptible tobacco leaves. In fact, none of the TMV recombinants that systematically infected susceptible tobacco contained a transmembrane domain in the coat protein subunits. Furthermore, Li Q et al. (2007) revealed that the presence of a cysteine residue in the foreign peptides, regardless of its position and the peptide sequence, was causally related to changes in the morphology and stability of these TMV chimeras.

Frolova et al. (2010) generated a putative vaccine against breast cancer, which was targeted against the human epidermal growth factor receptor-2 (HER2/neu) breast cancer antigen. Thus, the immunodominant HER2/neu peptides of 14 or 36 aa residues, recognized by monoclonal antibody trastuzumab (or Herceptin) and called trastuzumab-binding peptides (TBPs), were added C-terminally to the TMV coat via the flexible (GGGGS)$_3$ linker. The chimeric TMV particles with exposed TBPs retained trastuzumab-binding capacity, whereas antibodies induced by them in mice did not recognize HER2/neu on the surface of human cells (Frolova et al. 2010).

Petukhova et al. (2013) elaborated a putative influenza vaccine and inserted three versions of the M2e sequence between aa residues Ser155 and Gly156. Since cysteine residues within the heterologous peptide were thought to impede assembly of chimeric particles, the variants were constructed where Cys codons 17 and 19 of the M2e epitope were substituted by codons for Ser or Ala. The chimeric viruses were capable of systemically infecting *N. benthamiana* plants, and the M2e epitopes were uniformly distributed and tightly packed on the surface of the chimeric TMV virions. The antisera raised against TMV-M2e-Ala virions in mice appeared to contain far more antibodies specific to influenza virus M2e than those specific to the TMV carrier particle, by a ratio of 5:1. In fact, the majority of the TMV coat-specific epitopes in the chimeric TMV-M2e particles were hidden from the immune system by the M2e epitopes exposed on the particle surface. The immunized mice demonstrated high resistance to lethal doses of homologous and heterologous flu strains. The biological properties of the chimeric TMV-influenza viruses were described in more detail by Petukhova et al. (2014).

McComb et al. (2015) intended to construct an epitope targeted anthrax vaccine and presented the appropriate peptides, corresponding to the protective antigen (PA) sequences of aa 152–171, 232–247, and PA 628–637, on the surface of TMV virions by the C-terminal fusion. The chimeric TMV virions carrying the 232–247 and 628–637 epitopes remained genetically conserved over the course of virus propagation, demonstrated high yield, and were purified successfully from leaf tissue. As a result, a partial toxin neutralization was observed in mice, although protective effects in cellular assays or animals were not evident (McComb et al. 2015).

Röder et al. (2017) went to the mosaic techniques of display using the insertion of the FMDV 2A sequence between the target gene and the coat, hence a ribosomal skip occurs during translation, leading to the expression of the coat fusion protein, free target protein, as well as wild-type coat (Santa Cruz et al. 1996; Donnelly et al. 2001). The FMDV 2A sequence has previously been used in the PVX vectors to avoid unfavorable limitations of the coat fusions, as described in detail in Chapter 21. Therefore, Röder et al. (2017) were the first who applied the FMDV 2A technique to the TMV coat fusions and displayed a fluorescent iLOV protein, a suitable alternative to GFP and an improved version of LOV2 (light, oxygen or voltage sensing) domain of *Arabidopsis thaliana* phototropin 2, which was constructed by Chapman et al. (2008). In parallel to the FMDV 2A-using N-terminal fusion, the iLOV sequence was genetically fused either directly or via a glycine-serine linker to the C-terminus of the TMV coat. The viruses carrying a direct fusion of iLOV and coat initially remained predominantly in the stem, but those with an additional FMDV 2A sequence achieved a rapid systemic infection. As previously observed for analogous PVX vectors (see Chapter 21), the iLOV-2A-TMV construct resulted in the expression of an iLOV-TMV fusion protein, free TMV coat, and free iLOV. The TMV virions presenting iLOV were identified by immunogold electron microscopy and were therefore suitable for the display of proteins at least as large as the 13-kDa—or 113-aa—monomeric iLOV protein, which represented a protein that usually had limitations for fusions to the TMV coat due to the unfavorable aa sequence. The densitometric analyses showed a ratio of 2:3 for iLOV-2A-TMV fusion protein to free iLOV. Generally, the applied strategy demonstrated that not only the C-terminus but also the N-terminus of the TMV coat became suitable for efficient peptide display on the mosaic particles (Röder et al. 2017).

Chimeric VLPs

Haynes et al. (1986) achieved the expression of the TMV VLPs bearing an epitope from poliovirus 3 (PV3) in *E. coli*. For this purpose, they constructed a synthetic TMV coat gene containing both a convenient arrangement of restriction enzyme cleavage sites and an optimized set of codons for efficient translation in a prokaryotic system. The poliovirus-3 VP1 octapeptide epitope QQPTTRAQ was added to the C terminus of the synthetic TMV coat gene. Interestingly, the self-assembly of the chimeric TMV coat-polio 3 monomers into typical rods and disks did not occur in the bacterial cells, but their polymerization was initiated during dialysis of bacterial lysates at pH 5.0. Such TMV coat-polio 3 particles elicited poliovirus-neutralizing antibodies following injection into rats (Haynes et al. 1986).

In plants, the chimeric TMV VLPs were obtained by the C-terminal fusion of a 13-aa epitope from murine zona pellucida ZP3 protein, between the codons for residues aa Ser154 and Gly155 of the coat molecule, with a general idea to induce antibody-mediated contraception but to avoid severe autoimmune oophoritis in mice (Fitchen et al. 1995). The ZP3 sequence included the entire B-cell epitope associated with contraception, $ZP3_{331-343}$, but lacked the critical N-terminal residue of the T-cell epitope linked to autoimmune oophoritis. The coat protein gene of a clone used to prepare infectious RNA transcripts of TMV was replaced by this modified gene after cleavage at flanking restriction enzyme recognition sites. The products of in vitro transcription from the hybrid clone were used to inoculate young tobacco leaves. The chimeric protein accumulated to high concentrations in inoculated plants. The partially purified product appeared as rod-like particles of about 16 nm in diameter, like the observed 18 nm diameter of wild-type virus. The length of the rods was highly variable but was much shorter than 300 nm, the length of wild-type TMV virions. The penetration of stain into a central channel could be seen as is observed with the wild-type virus, and striations with 2.5 nm spacing could be observed, which was also similar to the wild-type TMV particles (2.3 nm). The parenteral immunization of mice with these VLPs resulted in serum antibody recognizing both the synthetic ZP3 peptide and the authentic murine ZP3 glycoprotein (Fitchen et al. 1995).

As noted by McCormick and Palmer (2008), the paper of Fitchen et al. (1995) was of the most special importance of all, since it demonstrated that the TMV peptide-display vaccine was able to break B-cell tolerance and induce production of autoreactive antibodies in vaccinated mice. The breaking immune tolerance was a remarkable achievement in a vaccine and illustrated how TMV is at least equivalent in its ability to induce autoantibodies as other VLP platforms, such as human papillomavirus (see Chapter 7) or RNA phage (see Chapter 25) VLPs.

As mentioned earlier, Dedeo et al. (2010) constructed a "circular permutant" of TMV coat, where the N- and C-termini were reengineered to the inner pore and a short GEG loop was placed to close the sequence gap. This protein variant relocated therefore the N- and C-termini to a flexible loop that faced the center of disks and rods after assembly, as shown in Figure 19.5. This protein was produced in remarkably high yield through *E. coli* expression

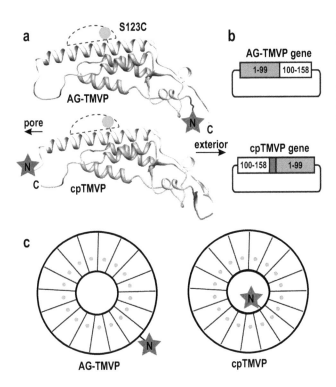

FIGURE 19.5   Design of a TMVP circular permutant that relocates termini to the pore of disks. (a) The top image shows normal AGTMVP, and the lower image shows the expected structure of a circular permutant, cpTMVP, in which the native termini are connected with a GEG linker (magenta). A flexible loop is interrupted to produce new termini. The location of the reactive cysteine 123 is shown as a green circle inside the RNA channel (dotted line) and the reactive N-terminus as a red star. (b) To create the cpTMVP gene, 5′ (blue) and 3′ (white) segments of the AG-TMVP gene were switched and connected with a short linker (magenta). (c) Self-assembly into disks and rods positioned the N- and C-termini on the exterior surfaces of AG-TMVP and the pores of cpTMVP. (Reprinted with permission from Dedeo MT et al. Nanoscale protein assemblies from a circular permutant of the tobacco mosaic virus. *Nano Lett*, 181–186. Copyright 2010, American Chemical Society.)

of the redesigned gene and self-assembled into light-harvesting rods that were much like those assembled from the wild-type protein. Moreover, the disks formed from the permutant structure were stable over a significantly wider pH range, greatly improving the practicality of this assembled form for materials applications, as will be described later. The permutant also showed the ability to coassemble with regular monomers.

Langowski et al. (2020) used the novel exposed loop of the circular permutant of the TMV coat to construct a malaria vaccine and displayed various numbers of the epitope from the C-terminal region of the circumsporozoite (CS) protein of *Plasmodium falciparum*: NPNAx3, -x4, -x5, -x7, -x10, or -x20 on the outer surface. The chimeric genes were expressed in *E. coli* and plants. The NPNAx5 variant was also displayed traditionally on the intrinsic N- or C termini of the TMV coat. However, the overlaid

predicted minimum energy structures illustrated that the N- and C-terminal display on the native TMV fold led to increased antigen flexibility compared to the loop display. The general idea of the Langowski et al. (2020) study consisted in the improvement and optimization of the hepatitis B virus (HBV) surface (HBs) protein-based *P. falciparum* CS protein repeat vaccine RTS,S/AS01, which is described in detail in Chapter 37. The RTS,S-induced NPNA-specific antibody titer and avidity have been associated with high-level protection in naïve subjects, but efficacy and longevity in target populations were relatively low. In an effort to improve upon RTS,S, a minimal repeat-only, epitope-focused, protective malaria vaccine was designed, where the repeat antigen copy number and flexibility were optimized using the TMV display platform. Comparing antigenicity of TMV displaying 3–20 copies of the NPNA repeat revealed that low copy number can reduce the abundance of low-affinity monoclonal antibody (mAb) epitopes while retaining high-affinity mAb epitopes. The TMV presentation improved titer and avidity of the repeat-specific Abs compared to a nearly full-length CS protein vaccine. The NPNAx5 antigen displayed as a loop on the TMV particle was found to be the most optimal, and its efficacy was further augmented by combination with a human-use adjuvant ALFQ that contained immune stimulators. This data was confirmed in mice and rhesus macaques where a low dose of the TMV-NPNAx5 elicited antibodies that persisted at functional levels for up to 11 months (Langowski et al. 2020). The human clinical trials are planned.

Spherical Compositions

The Joseph G. Atabekov's team generated a novel TMV-based nanoplatform, namely the so-called spherical nanoparticles (SNPs). The SNPs were achieved by two-step thermal remodeling of the native TMV virions when particles of irregular shape and varying size were generated by heating at 90°C and converted then into the SNPs by heating at 94°C (Atabekov et al. 2011). The first 90°C heating-produced irregular particles were considered therefore to be intermediate precursors of the true regular SNPs appearing after the heating at 94°C. In addition to the standard SNPs of 53 nm in diameter, which appeared to be generated by individual TMV virions, the large SNPs of 100–800 nm in diameter, were also noticed, and the size of the SNPs was dependent on the TMV concentration. Moreover, the SNPs were generated by distinct forms of the RNA-free TMV coat aggregates, as well as by individual coat subunits. Remarkably, a one-step SNP assembly appeared to occur in these cases. The authors' evidence implied that, upon thermal denaturation, the TMV coat subunits acquired a specific conformation favorable for the specific self-assembly into the SNPs (Atabekov et al. 2011).

The SNPs represented a new type of a nanoplatform where foreign protein molecules could be bound to the SNP surface. First, Atabekov et al. (2011) demonstrated that the SNPs bound GFP molecules to their surface. The fluorescence microscopy data indicated that the whole surface of all SNPs used for GFP binding was covered with fluorescent molecules.

Second, Karpova et al. (2012) generated the SNPs that contained one of the following foreign antigens: antigenic determinant A of rubella virus E1 glycoprotein, a recombinant protein containing the M2e epitope of influenza A virus protein M2, a recombinant antigen consisting of three epitopes of influenza A virus haemagglutinin, potato virus X (PVX) coat protein, the PVX coat fused with the epitope of plum pox virus (PPV) coat, as well as bovine serum albumin (BSA). The "mixed" compositions were also assembled by binding two different foreign antigens to each of the SNPs. The assembly procedure of the compositions involved a short incubation of the SNPs with a foreign protein of interest and was based on noncovalent, e.g., electrostatic and hydrophobic, bonds. Then, the SNP-protein/epitope complexes were separated from unbound protein and treated by formaldehyde to link a foreign antigen covalently to the SNPs. Remarkably, the SNP surface possessed a unique adsorption capacity, being capable of binding a diversity of various proteins. The antigenic specificity of foreign antigens was fully retained within the chimeric SNPs, whereas their immunogenicity increased significantly (Karpova et al. 2012).

In parallel, Nikitin et al. (2011) immobilized bovine serum albumin (BSA) noncovalently on the surface of SNPs coated with hydrophilic polycation poly-N-ethyl-4-vinylpyridine bromide. The synthetic polycation acted as a spacer between the negatively charged SNPs and negatively charged molecules of interest. However, such ternary SNP–polycation–protein complexes were not stable at high ionic strengths due to charge shielding effect. The ternary complexes could be stabilized by covalent binding of SNP, polycation, and a model protein or utilization of a hydrophobic polymer forming hydrophobic bonds with both SNP and protein. Thus, Nikitin et al. (2014) chose the second approach and coated the TMV SNPs with the two hydrophobic polycations based on poly-N-ethyl-vinylpyridine. The complexes consisting of the SNPs, hydrophobic polycation, and a model BSA were stable and did not aggregate in solution, particularly at high ionic strengths.

Meanwhile, Trifonova et al. (2014) demonstrated that the SNPs bound the entire isometric virions of heterogeneous nature, such as human enterovirus C or cauliflower mosaic virus. Such compositions were stabilized by covalent binding with formaldehyde. Remarkably, the proper SNP antigenic sites were hidden and masked by virions within the compositions.

Concerning the fine structure of the SNPs, Dobrov et al. (2014) found that the SNP protein differed strongly from that of the native TMV coat and was a result of transition from mainly (about 50%) α-helical structure to a structure with low content of α-helices and a significant fraction of β-sheets. Ksenofontov et al. (2019) used complementary physicochemical methods to perform a detailed structural analysis of the SNP surface and to determine the most likely exposed areas, taking therefore an important step toward fine understanding of the SNP structure.

New Materials

TMV was one of the first to receive particular attention in regard to combination of the organic and inorganic materials, and its inside surface was used to template the growth of metal oxide particles by means of template mineralization (Shenton et al. 1999). The internal and external surfaces of the TMV VLPs consisting of repeated patterns of charged aa residues, such as glutamate, aspartate, arginine, and lysine, offered a wide variety of nucleation sites for surface-controlled inorganic deposition, which, in association with the high thermal and pH stability, were exploited in the synthesis of unusual materials such as high-aspect-ratio composites and protein-confined inorganic nanowires. Thus, TMV was used as a template for reactions such as cocrystallization (CdS and PbS), oxidative hydrolysis (iron oxides), and sol-gel condensation (SiO_2). For example, the specific nucleation of CdS on the TMV surface resulted in the mineralized tubular structures, approximately 50 nm in width, which consisted of a 16-nm-thick electrondense outer crust and a 18-nm-diameter internal core. The iron oxide mineralization of TMV particles was achieved by addition of NaOH to a dispersion of virions in acidic Fe^{II}/Fe^{III} solution. The sol-gel condensation of tetraethoxysilane (TEOS) led to the formation of electron-absorbing SiO_2 rods, which were identified as TMV particles covered with a uniform 3-nm silica shell. The preferential deposition of CdS, PbS, and Fe oxides on the TMV external surface was explained by specific metal-ion binding at the numerous glutamate and aspartate surface groups. The authors planned to block the metal-ion binding sites on the external surface in order to facilitate the specific nucleation of viral-particle-encapsulated inorganic nanowires (Shenton et al. 1999). This pioneering study, together with the experiments on cowpea chlorotic mottle virus (CCMV) (see Chapter 17), was reviewed early by Trevor Douglas and Mark Young (1999) and then by Christof M. Niemeyer (2001).

Fowler et al. (2001) described the controlled hydrolysis and condensation of mixtures of TEOS and aminopropyltriethoxysilane (APTES) in a liquid crystalline gel of ordered TMV tubules arranged with a repeat distance of ca. 38 nm. The transmission electron microscopy revealed a highly ordered silica-TMV mesostructure, showing uniform lattice fringes with spacings of ca. 18 nm, consistent with a periodic array of coaligned TMV particles joined end-to-end and intercalated within a continuous framework of amorphous silica. Remarkably, the encapsulated TMV filaments were approximately 11 nm in width, which was significantly less than the external diameter of the native TMV particles of 18 nm, indicating that the protein tubules were compressed during silicification (Fowler et al. 2001).

Dujardin et al. (2003) used TMV as a template for the controlled deposition and organization of Pt, Au, or Ag nanoparticles. The chemical reduction of $[PtCl_6]^{2-}$ or $[AuCl_4]^-$ complexes at acidic pH gave rise to the specific decoration of the external surface of the wild-type TMV rods with metallic nanoparticles less than 10 nm in size. In contrast, the photochemical reduction of Ag^I salts at pH 7 resulted in nucleation and constrained growth of discrete Ag nanoparticles aligned within the 4-nm-wide internal channel. Similar experiments using the E95Q/D109N mutant of the TMV coat, which resulted in reduction of negative charge along the central cavity, confirmed that glutamic and aspartate acid groups were involved in site-specific deposition (Dujardin et al. 2003). This study had paved the way for the preparation of 1D arrays for a wide range of inorganic quantum dots by molecular engineering of the internal and external surfaces of the TMV tubules.

Knez et al. (2002) achieved the electroless deposition of metal selectively on the inner or outer TMV surface. Furthermore, Knez et al. (2003) used the central TMV channel as a template to synthesize nickel and cobalt nanowires only a few atoms in diameter, with lengths up to the micrometer range. Generally, the electroless deposition of metals was possible without virion degradation due to their stability against pH change, relatively high temperatures, and mild to relatively strong reductants. Knez et al. (2004) produced metal clusters on the nanometer length scale that were firmly attached to the virion, without destroying its geometry. To obtain a strong bond, the clusters were produced by first binding the pertaining ions to TMV and subsequently reducing them. This process was—in the case of palladium, platinum, and gold—followed by an electroless deposition of another metal from a bath. Consequently, the 3-nm nickel and cobalt wires of several 100 nm in length were synthesized. Balci et al. (2006) got copper nanowires of 3 nm in diameter and up to 150 nm in length by electroless deposition within the 4-nm-wide channel of the TMV particles. The fabrication process of the nanowires was based on sensitization of tobacco mosaic virus with Pd^{II} prior to the electroless deposition. The method was used also for nickel and cobalt nanowires deposited within the TMV channel.

Yi et al. (2005) genetically engineered cysteine onto the TMV virion surface, providing attachment sites for thiol-reactive fluorescent markers. The addition of a single Cys residue within the TMV coat was created by the insertion of a TGT codon at the third position within the coat protein gene using a PCR-based mutagenesis procedure. The resulting TMV1cys structure is shown in Figure 19.6. To pattern these viruses, the labeled virions were partially disassembled to expose 5' end RNA sequences and hybridized to virus-specific probe DNA linked to electrodeposited chitosan. The electron microscopy and ribonuclease treatments confirmed the patterned assembly of the virus templates onto the chitosan surface. Therefore, TMV nanotemplates appeared to be dimensionally assembled via nucleic acid hybridization. Royston et al. (2008) used the TMV1cys vector to vertically pattern TMV particles onto gold surfaces via gold-thiol interactions. The patterned TMV1cys virions functioned as robust templates for the reductive deposition of nickel and cobalt at room temperature via electroless

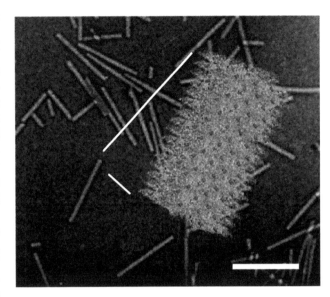

FIGURE 19.6 The visualization of the TMV1cys. Background, electron micrograph of TMV (negatively stained). Bar represents 300 nm. Foreground, structural model representing ~10% of a TMV1cys virion. Red space-filling molecules show the location of the genetically inserted cysteine residue. (Reprinted with permission from Yi H et al. Patterned assembly of genetically modified viral nanotemplates via nucleic acid hybridization. *Nano Lett*, 1931–1936. Copyright 2005, American Chemical Society.)

deposition, producing dense carpets of oriented metal-coated viral templates. The mineralized surface-assembled viruses significantly increased available surface area and enhanced electrode life and voltage output in a battery electrode system (Royston et al. 2008).

Then, Lim et al. (2010) estimated the maximum ion capacity of the TMV1cys vector for the Au^{III} and Pd^{II} ions and optimized the size of the metal clusters and their surface coverage at various metal ion loadings.

Next, Chen et al. (2010) used the metal coatings on the patterned 3D TMV1cys vector to engineer a multilayered nanoscale 3D assembled silicon anode for lithium-ion batteries, where the internal metal layer functioned as a strong current collector. Unlike the previously reported methodology by Nam et al. (2006) using the phage M13 for the synthesis of nanomaterials and relying on powder mixing and ink casting for the electrode fabrication, as described in Chapter 11, the method of Chen et al. (2010) presented direct fabrication of a nanostructured silicon electrode.

Lee SY et al. (2005, 2006) generated the TMV2cys vector that contained two additional cysteines at the position of 2 and 3 in coat protein, resulting not only in an exposure of thiols to the outer surface but also in more dense thiol array exposed per coat unit that made it possible to form a dense metal cluster layer. Thus, the authors engineered platinum conductors by the deposition of Pt clusters on the TMV coat surface. Moreover, the application of the prepared Pt cluster-deposited TMV nanotubes as conductors was presented with its electrical conductance (Lee SY et al. 2006). In fact, this was the first report on the electrical conductance of the metal cluster-coated TMV particles.

Endo et al. (2006) selected four aa residues Asn98, Gln99, Ala100, and Asn101 on the flexible loop near the interior surface and individually mutated them to a cysteine residue. The modification at C27A and the cysteines introduced to the four positions did not affect the formation of the TMV rod structures. A pyrene was then selectively introduced to the cysteine residue by using N-(1-pyrene)maleimide. The edge-to-edge length of a pyrene maleimide molecule was 1.2 nm, which was small enough to be incorporated into the inner cavity with diameter of 4 nm. Because the wild-type TMV coat had one cysteine residue at position 27, the Cys27 was first replaced by alanine, and four amino acids on the inner loop were then individually changed to cysteine for selective labeling with pyrene maleimide. Remarkably, the incorporation of a pyrene moiety at positions 99 and 100 significantly promoted the assembly of the TMV coat and stabilized the rod-like particles through a π-stacking interaction in a well-ordered way (Endo et al. 2006).

Miller et al. (2007) engineered the TMV coat mutant S123C, expressed it in *E. coli*, and constructed a light-harvesting system. The building blocks were prepared by attaching fluorescent chromophores to the Cys123. The thiol-reactive chromophores selected for this study included Oregon Green 488 maleimide as the primary donor, tetramethylrhodamine maleimide as an intermediate donor, and Alexa Fluor 594 maleimide as the acceptor. By simple adjustments in pH and ionic strength, these conjugates assembled into stacks of disks or into rods that reached hundreds of nanometers in length and ensured efficient energy transfer from large numbers of donor chromophores to a single acceptor (Miller et al. 2007). Furthermore, Miller et al. (2010) showed that the TMV rod-and-disk assemblies derived from a single stock of chromophore-labeled protein exhibited drastically different levels of energy transfer, with rods significantly outperforming disks.

Using Tyr139 of the TMV coat modified with the diazonium salt of *p*-aminoacetophenone, Holder and Francis (2007) designed rods carrying single-walled carbon nanotubes (SWCNTs), structures with rich electronic properties. The ketone groups on the TMV coats were attached to alkoxyamines on the solubilized SWCNTs through oxime formation, and the TMV-SWCNT hybrids in close contact with parallel alignment were obtained. Remarkably, in addition to the expected 300-nm rods, a significant number of the SWCNT-bound rods were oriented in a contiguous end-to-end fashion. The SWCNT-TMV conjugates retained their water solubility (Holder and Francis 2007).

Scheck et al. (2008) engineered N-terminal TMV coat mutants that could be highly reactive under pyridoxal 5′-phosphate (PLP) conditions. The idea was also to extend the N-terminus so that it would be more solvent-exposed and hopefully lead to higher conversion. The mutants were prepared with N-terminal Ala, Cys. Gly, and Ser, with and without an additional one aa spacer before the new terminus. The Ala and Gly termini gave the highest conversion to

the oxime, and Cys and Ser led to more modest yields. The linker separating the new terminal residue from the native N-terminus improved conversion, likely due to improved accessibility. The optimal conversion to the oxime product was obtained for these mutants using higher concentrations of PLP. However, even with these higher concentrations, no formation of PLP adduct was observed. The authors suggested that secondary and tertiary structure of proteins can lead to an N-terminal environment that is simply not compatible with modification by PLP (Scheck et al. 2008).

The previously described circular permutant of the TMV coat generated by Dedeo et al. (2010), as shown in Figure 19.5, has opened up new opportunities for materials application. Thus, the disks formed from the permutant structure were stable over a significantly wider pH range, greatly improving the practicality of this assembled form. The new position of the N-terminus allowed functional groups to be installed in the inner pore of the disks, affording geometries reminiscent of natural photosynthetic systems (Dedeo et al. 2010).

To broaden further putative applications, He et al. (2009) investigated behavior of the TMV rods at liquid/liquid interfaces, namely the oil/water interface. Remarkably, TMV showed different orientations at the perfluorodecalin/water interface, depending on the initial TMV concentration in the aqueous phase. Thus, at low TMV concentration, the rods oriented parallel to the interface, mediating the interfacial interactions at the greatest extent per particle. At high TMV concentrations, the rods were oriented normal to the interface, mediating the interfacial interactions and also neutralizing interrod electrostatic repulsion.

Kaur et al. (2010) used the 2D substrates coated with the TMV nanorods to study the differentiation process of bone marrow stromal cells (BMSCs) into osteoblast-like cells. The presence of the TMV nanorods significantly affected the expression levels of genes involved in osteo-differentiation and subsequent cell behavior. Thus, the early interaction of cells with TMV triggered signaling pathways and enhanced osteogenic differentiation potentials. Remarkably, the surface coating with the TMV particles was essential to accelerate differentiation from 21 to 14 days, whereas supplementing the media with virus as a solution failed to induce the similar enhanced differentiation (Sitasuwan et al. 2012). To progress from 2D to 3D models, Luckanagul et al. (2012) generated a 3D scaffold that incorporated the TMV particles without affecting their quaternary structures in the porous hydrogels. The assembly of the porous hydrogel with the TMV particles required no covalent linkages between the alginate and virus coat proteins, which simplified postassembly. As a proof of concept, mesenchymal stem cells were seeded and induced to osteogenic lineage.

Zahr and Blum (2014) published a detailed protocol of targeting the arginine residues on the TMV coat by bis(*p*-sulfonatophenyl)phenylphosphine)-passivated gold nanoparticles with high specificity to create the 22-nm rings.

Atanasova et al. (2011) reported fabrication of field-effect transistors (FETs) by TMV-directed self-assembly of in situ mineralized nanosized ZnO, a semiconducting material, building blocks, where uniform ZnO nanowires with precisely controlled coating thickness were synthesized on self-assembled TMV templates. The FETs in bottom gate/bottom contact arrangement were fabricated successfully by deposition of the TMV/ZnO composite on substrates with prestructured electrodes. Next, Atanasova et al. (2017) used TMV as a template to direct the synthesis of zinc sulfide (ZnS), and the TMV/ZnS hybrid nanowires or thin films were obtained with controllable thickness of the inorganic layer.

The astonishing role of the TMV particles in the field of novel bionanomaterials was extensively reviewed in numerous excellent papers (Young et al. 2008; Lee LA et al. 2009, 2011; Steinmetz et al. 2009; Li K et al. 2010; Rybicki 2010; Soto and Ratna 2010; Steinmetz 2010; Dedeo et al. 2011; Jutz and Böker 2011; Pokorski and Steinmetz 2011; Liu Z et al. 2012; Alonso et al. 2013; Bittner et al. 2013; Fan et al. 2013; Love et al. 2014; Capek 2015; Shukla and Steinmetz 2015; Koch et al. 2016; Wen and Steinmetz 2016; Zhang Y et al. 2016; Zhang W et al. 2017; Chu et al. 2018; Eiben et al. 2019; Balke and Zeltins 2020; Wege and Koch 2020). Gasanova et al. (2016) published a review that was dealing mostly with the TMV-based generation of putative nanovaccines. The exhaustive review of George P. Lomonossoff and Christina Wege (2018) covered systematically all extremely broad aspects of the TMV-based nanotechnological applications.

Chemical Coupling

Demir and Stowell (2002) chose the C-terminus as the point of chemical modifications and introduced a single mutation T158K into the native TMV coat. The rationale was that this single accessible lysine could be selectively coupled using facile N-hydroxy succinamide chemistry to afford a template with selectable chemospecificity. Thus, the T158K mutant virions were used for the production of a variety of chemospecific nanotubular materials, including semi-crystalline protein arrays and metallic "nanopipes" as well as nanomolecular "light sticks" (Demir and Stowell 2002).

The classical studies on the chemical coupling of VLPs were conducted by Matthew B. Francis's team. As described in Chapter 25, Hooker et al. (2004) were the first who performed the covalent modification of the interior surface of viral capsids, namely RNA phage MS2, for the putative attachment of drug cargo, site-isolated catalysts and novel nucleation sites for crystal growth. To achieve this, the authors developed an efficient four-step strategy for the interior functionalization of hollow capsid shells. This method featured a new hetero-Diels-Alder bioconjugation reaction for the attachment of olefin substrates to selectively modified tyrosine residues, in this case Tyr85 of the MS2 coat. Thus, using diazonium-coupling reactions, the Tyr85 was modified with high efficiency and selectivity.

This pioneering approach was developed further by Schlick et al. (2005) for the dual-surface modification of the TMV virions. Within the latter, the reactive Tyr139 residues appeared on the outer surface, while Glu97 and Glu106 were located on the inner surface, providing reactive handles on both the exterior and interior surfaces of the TMV tube. Accordingly, the two synthetic strategies were applied for the attachment of new functionality to either the exterior or the interior surface of the virus. The first of these was accomplished using a highly efficient diazonium coupling/oxime formation sequence, which installed more than 2,000 copies of a material component on the capsid exterior. The inner cavity of the tube was modified by attaching amines to glutamic acid side chains through a carbodiimide coupling reaction. Both of these reactions were used for a series of substrates, including biotin, chromophores, and crown ethers. Through the attachment of PEG polymers to the capsid exterior, the organic-soluble TMV rods were prepared (Schlick et al. 2005).

To display foreign antigens on the TMV scaffold by chemical coupling, as an alternative for genetic fusion to display larger foreign sequences, Smith et al. (2006) introduced a reactive lysine moiety at the externally located N-terminus of the TMV coat, resulting in the final N-terminal sequence M ADFK SYS . . . which facilitated biotinylation of the capsid. First, a model antigen GFP-streptavidin (SA) was bound to the biotinylated TMV particles, creating a GFP decorated particle, and the GFP-SA tetramer loading of 26% was obtained, corresponding to approximately 2,200 GFP moieties displayed per intact virion. Second, the N-terminal fragment of the COPV L2 protein, comprising aa 61–171 residues, was linked to streptavidin. With the TMV display, the L2 protein fragment was significantly more immunogenic than uncoupled antigens when tested in mice (Smith et al. 2006).

Mallajosyula et al. (2014) used the exposed TMV coat lysine to conjugate plant-produced hemagglutinin (HA) of influenza virus H1N1, a product with low vaccine potency because of its monomeric structure. Three chemical conjugation chemistries were tested, and all generated acceptable conjugation characteristics. Remarkably, a single dose of the TMV-HA conjugate vaccine was sufficient to generate 50% survival, or 100% survival with adjuvant, compared with 10% survival after vaccination with a commercially available H1N1 vaccine.

Next, the TMV-lysine vector of Smith et al. (2006) was used for the engineering of a multivalent subunit vaccine against tularemia (Banik et al. 2015). The potentially protective *E. coli*-derived purified proteins DnaK, OmpA, and Tul4 of *Francisella tularensisis* were conjugated to the TMV-lysine coat and tested in mice. The immunization with the TMV-conjugated proteins induced a strong humoral immune response and protected mice against respiratory challenges with very high doses of *F. tularensis* live vaccine strain. Remarkably, the TMV-monoconjugate vaccine in which all the three recombinant proteins were conjugated to a single TMV virion served as a poor vaccinogen. In contrast, the vaccine formulation that contained a multivalent blend of all the three proteins conjugated individually to TMV induced a superior protective immune response.

Developing further the chemical coupling approach, McCormick et al. (2006a) compared both chemical and N-terminal genetic fusion by testing the putative TMV-based vaccines carrying the two well-characterized CTL epitopes, namely the Ova peptide SIINFEKL and the p15e melanoma epitope KSPWFTTL. The vaccination of mice elicited measurable cellular responses and resulted in significantly improved protection from tumor challenge in both the Ova and melanoma models. Remarkably, some differences were detected by action of the chemically coupled and genetically fused products. In addition, the ability of the N-terminally modified TMV with the reactive lysine moiety for direct interaction with antigen presenting cells was measured when the TMV-lysine was labeled with Alexa 488 and incubated with spleen or bone marrow cells at different doses and for different times (McCormick et al. 2006a).

Improving the conjugated peptides, McCormick et al. (2006b) tested not only the Ova and p15e CTL epitopes but also the tyrosinase-related protein 2 (Trp2) peptides as self-antigen targets. The Ova peptide fusions to TMV were designed as bivalent formulations with peptides encoding additional T-help or cellular uptake via the integrin-receptor binding RGD peptide, showed improved vaccine potency, and protected mice from EG.7-Ova tumor challenge, which was achieved with only two doses of vaccine of ~600 ng peptide and given without adjuvant (McCormick et al. 2006b).

Niu et al. (2006, 2007a, b) exploited the outer TMV surface as template to coat conductive polymers, namely polyaniline or polypyrrole, on the 1D assembled TMV to produce composite nanofibers and macroscopic bundles of such fibers as conductive nanowires. The nanomechanical properties of the polyaniline-coated TMV were evaluated by Wang et al. (2008). In parallel, the TMV-based materials demonstrated great potential with applications in nanoelectronics and energy-harvesting devices (Kalinin et al. 2006; Tseng et al. 2006; Miller et al. 2007).

Bruckman et al. (2008) performed surface modification of the TMV virions with "click" chemistry, using Cu[I]-catalyzed azide-alkyne 1,3-dipolar cycloaddition (CuAAC) reaction, renascent of the well-known Huisgen reaction. In fact, the authors combined the CuAAC with a diazonium-coupling reaction to quantitatively functionalize tyrosine residues, namely Tyr139. As a result, the alkyne groups were quantitatively attached to the Tyr139 residues by diazonium-coupling, and a sequential CuAAC reaction with azides efficiently conjugated a wide range of compounds to the outer surface of TMV.

Bruckman et al. (2010) fabricated a thin film sensor for the detection of volatile organic compounds by deposition of oligo-aniline grafted TMV onto a glass substrate. Thus, when the oligo-aniline motifs were conjugated onto the TMV surface by a traditional diazonium coupling reaction

to tyrosine residues followed by the CuAAC reaction, the modified TMV was easily fabricated into a thin film by directly drop coating onto a glass substrate. Upon integration of the glass substrate into a prototypical device, the TMV-based thin film exhibited good sensitivity and selectivity toward ethanol and methanol vapor.

Zang et al. (2014) presented a TMV-based sensing method for the detection of 2,4,6-trinitrotoluene (TNT), a widely used explosive and also a toxic chemical for organisms from bacteria to humans. The TNT sensing unit was based on the diffusion modulation of the target molecules through the E. coli-produced suspended chimeric TMV VLPs modified with the TNT-binding peptides. The most efficient TNT-binding peptide WHWQRPLMPVSI and DNT-binding peptide HPNFSKYILHQR were found by Jaworski et al. (2008).

Moreover, the modified TMV particles were used as beneficial scaffolds to present sensor enzymes. Thus, the TMV mutant exposing a cysteine residue on outer surface enabled the coupling of bifunctional maleimide-PEG-biotin linkers, and the surface was equipped with two streptavidin-conjugated enzymes: glucose oxidase and horseradish peroxidase (Koch et al. 2015). At least 50% of the coats were decorated with a linker molecule and all thereof with active enzymes. The sensor chips combining an array of Pt electrodes loaded with glucose oxidase-modified TMV nanotubes were used for amperometric detection of glucose as a model system for the first time (Bäcker et al. 2017). Koch et al. (2018) conjugated penicillinase enzymes with streptavidin and coupled them to the TMV nanorods using biotin-linker. As a result, the field-effect penicillin biosensor was designed, which consisted of an Al-p-Si-SiO$_2$-Ta$_2$O$_5$-TMV structure and demonstrated high penicillin sensitivity in a nearly linear range from 0.1mM to 10 mM and a low detection limit of about 50 μM, e.g., for penicillin detection in bovine milk samples (Poghossian et al. 2018).

Yin et al. (2012) used the CuAAC reaction to couple the tumor-associated carbohydrate antigens (TACAs) that are popular targets for the development of putative antitumor vaccines. The TMV coat served as a carrier of a weakly immunogenic TACA, the monomeric Tn antigen. The location of Tn attachment was important. When introduced at the N terminus of TMV, Tn was immunosilent. The coupling to Tyr139 elicited strong immune responses in mice and both Tn-specific IgG and IgM antibodies were generated. Moreover, the antibodies exhibited strong reactivities toward Tn antigen displayed in its native environment, i.e., cancer cell surface, thus highlighting the potential of TMV as a promising TACA carrier by breaking the immunotolerance of Tn when coupled with the appropriate site of the TMV coat (Yin et al. 2012).

The circular permutant of the TMV coat generated by Dedeo et al. (2010), see Figure 19.5, contributed strongly to chemical coupling, in tight connection with the previously mentioned new materials applications. The permutant also showed the ability to coassemble with wild-type monomers, allowing the future generation of multicomponent rod structures that would be modified by chemical coupling on the exterior and interior surfaces, as well as in the internal RNA channel (Dedeo et al. 2010).

Bruckman et al. (2013a) used the TMV virions for the development of the first rod-shaped VLP-based magnetic resonance imaging (MRI) contrast agent. The interior (Glu97 and Glu106) and exterior (Tyr139) surfaces were selectively labeled with Gd(DOTA) molecules using a two-step bioconjugation process. The high labeling efficiency of these reactions yielded TMV rods loaded with 3417 Gd(DOTA) at interior sites or 1712 Gd(DOTA) at the exterior surface. Next, the authors applied the previously described Atabekov's et al. (2011) heating protocol to induce thermal transition of TMV rods into the TMV spherical nanoparticles (SNPs) measuring 170 nm in diameter; these TMV SNPs contained several TMV rods and thus have an increased Gd content measured on a per-particle basis. Thus, the SNPs contained over 25,000 Gd ions per SNP (Bruckman et al. 2013a). This study was reviewed early by Bruckman et al. (2013b).

Bruckman et al. (2014b) involved both rod-like TMV particles and Atabekov's SNPs, this time 54 nm-sized, in the biodistribution and pharmacokinetic studies. The availability of rods and spheres made of the same protein provided a unique scaffold to study the effect of nanoparticle shape on the in vivo fate. The PEGylated formulations were also considered for enhanced biocompatibility and all versions of nanoparticles exhibited comparable. However, the rods circulated longer than spheres, illustrating the shape effect on circulation. The PEGylation also increased circulation times. The TMV rods and spheres were cleared from circulation by macrophages in the liver and spleen, while the spheres were more rapidly cleared from tissues compared to the rods. None of the formulations induced blood clotting or hemolysis, laying the foundation for further application and tailoring of the TMV nanoparticles for biomedical applications.

The excellent and exhaustive methodological protocols for the purification and chemical conjugation of the TMV particles were published by Wen et al. (2012) and by Michael A. Bruckman and Nicole F. Steinmetz (2014).

Packaging and Delivery

As described earlier, the presence of the origin of assembly (OAS) played an important role in the RNA packaging by the TMV coat in vitro and in vivo (Hwang et al. 1994a; Kadri et al. 2013; Saunders and Lomonossoff 2017). However, it should be kept in mind that Mueller et al. (2010) found, as mentioned earlier, that both the wild-type TMV and TMV-CP-E50Q mutant yielded TMV-like rods, irrespective of the proteins' source or presence or absence of RNA.

Smith et al. (2007) introduced the TMV OAS into the RNA genome of Semliki Forest virus (SFV) and generated an SFV expression vector that was efficiently packaged in vitro by the TMV coat purified from wild-type or chimeric virions. The SFV vector was lacking the SFV structural proteins but expressing the beta-galactosidase (β-gal) reporter gene. The vector encapsidation significantly improved the humoral and cellular immune responses in mice. Furthermore, reassembly with recombinant TMV coats permitted the display of peptide epitopes on the capsid surface as either genetic fusions or through chemical conjugation to complement the immunoreactivity of the encapsidated RNA genetic payload. This SFV vector/TMV coat system provided therefore an original packaging and delivery model, as reviewed early by McCormick and Palmer (2008).

Azucena et al. (2012) solved the problem of the RNA-directed self-assembly of the TMV coat on immobilized RNA scaffolds, presenting a possibility to grow nucleoprotein nanotubes in place. The different chemistries were introduced for the site-selective, bottom-up assembly of the TMV coats after adding the corresponding coat proteins, and the dense arrays of the TMV-like particles in defined patches were completed. Schneider et al. (2016) elaborated a novel "stop-and-go" strategy for growing the TMV tubes under tight kinetic and spatial control, which combined RNA guidance and its site-specific but reversible interruption by DNA blocking elements.

Rego et al. (2013) applied the packaged RNA-driven methodology to the previously described TMV1cys vector for the programmed assembly of the viral building blocks with controlled dimensions. To do this, the DNA templates of desired lengths were synthesized, and the corresponding in vitro RNA transcripts were obtained via standard molecular biology techniques. The assembling the RNA transcripts in vitro with the TMV coat protein disks yielded viral building blocks of controlled lengths as directed by the lengths of the RNA transcripts employed. The methodology could be readily extended to the fabrication of building blocks with genetically modified coat protein subunits.

In parallel, Geiger et al. (2013) performed the similar RNA-guided self-assembly process when the TMV coat was used to establish tailored nanorod scaffolds that could be loaded not only with homogeneously distributed functionalities but with distinct molecule species grouped and ordered along the longitudinal axis. The arrangement of the resulting domains and final carrier rod length both were governed by RNA-templated two-step in vitro assembly. The two selectively addressable TMV coat mutants carrying either thiol or amino groups on the exposed surface were engineered and shown to retain reactivity toward maleimides or NHS esters, respectively, after acetic acid-based purification and reassembly to novel carrier rod types. The stepwise combination of both mutants with RNA allowed fabrication of TMV-like nanorods with a controlled total length of 300 or 330 nm, respectively, consisting of adjacent longitudinal 100–200 nm domains of differently addressable coat species. Therefore, this fine technology paved the way toward rod-shaped scaffolds with predefined, selectively reactive barcode patterns on the nanometer scale. Moreover, the presence of two domains consisting of coats with different functional groups on one artificial TMV nanoparticle gave the opportunity of using, e.g., "click reactions" to specifically couple organic molecules to only one part of the particle and thereby to synthesize Janus-type TMV particles with a hydrophilic and a hydrophobic portion (Geiger et al. (2013).

Developing further this approach, Shukla et al. (2015) prepared three TMV particle formulations varying in length but identical in width by assembling purified TMV coats with the in vitro synthesized RNA molecules of defined lengths, each containing an OAS. This yielded TMV rods with median lengths of ~300, 130, and 60 nm—corresponding to aspect ratios of 16.5, 7, and 3.5, respectively—or long, medium, and short TMV variants. Moreover, a TMV coat mutant of Geiger et al. (2013) was used, which replaced the threonine residue at position 158 with lysine. The amine group of the lysine-side chain was displayed on the particle surface to allow functionalization with fluorophores, PEG, and the RGD ligand. The stealth TMV formulations were produced by covering the particle surface with PEG to overcome immune surveillance, and targeted TMV formulations were produced by displaying the integrin-targeting cyclic peptide ligand RGD via an intervening PEG spacer. By evaluation of the biodistribution and tumor-homing of such particles, it appeared that both the aspect ratio and surface chemistry of the particles had distinct effects on in vivo behavior, including tumor localization and cancer cell versus immune cell targeting (Shukla et al. 2015).

These size-dependence and aspect ratio studies were extended by Liu X et al. (2016). In their case, the TMV nanorods with different aspect ratios were obtained by ultrasound treatment and sucrose density gradient centrifugation. By incubating with epithelial and endothelial cells, the nanorods with various aspect ratios had clearly different uptake and internalization pathways in different cell lines.

Atanasova et al. (2019) used Geiger et al.'s (2013) thiol-carrying TMV coat mutant to manipulate the surface hydrophilicity of TMV through covalent coupling of such polymer molecules as the perfluorinated (poly(pentafluorostyrene) (PFS)), the thermo-responsive poly(propylene glycol) acrylate (PPGA), and the block-copolymer polyethylene-block-poly(ethylene glycol). The covalent attachment of hydrophobic polymer molecules with proper features retained the integrity of the TMV structure, and the degree of the virus hydrophobicity was tuned by the polymer properties, opening therefore novel possibilities for the packaging and delivery applications.

The arginine mutants of Finbloom et al. (2016), which were described earlier, appeared also as good candidates by the putative drug delivery.

The numerous foreign nano-objects, such as gold nanoparticles and silver sulfide quantum dots, including also binary nanochains of these different kinds of nanoparticles, were successfully encapsulated and conveniently assembled into the highly organized 1D in the cavity of the TMV coat (Zhang J et al. 2021).

Concerning potential clinical applications, Bruckman et al. (2014a) developed the TMV-based, high aspect ratio, molecularly targeted magnetic resonance (MR) imaging contrast agent. Specifically, the vascular cell adhesion molecule (VCAM)-1, was targeted, because it is highly expressed on activated endothelial cells at atherosclerotic plaques. To achieve dual optical and MR imaging in an atherosclerotic mouse model, TMV was modified to carry near-infrared dyes and chelated Gd ions. The targeted TMV-based MR probe increased the detection limit significantly, and the injected dose of Gd ions could be further reduced 400x compared to the suggested clinical use, demonstrating the utility of targeted nanoparticle cargo delivery (Bruckman et al. 2014a).

Czapar et al. (2016) reported TMV as a delivery system for phenanthriplatin, cis-[Pt(NH$_3$)2Cl(phenanthridine)] (NO$_3$), or Phen-Pt, a cationic monofunctional DNA-binding platinum^{II} anticancer drug candidate with unusual potency and cellular response profiles. Due to a high density of negative charges lining the TMV inner channel, the particles were loaded with approximately 2000 PhenPt^{2+} cations by a one-step protocol. The Phen-Pt release from the carrier was induced by lowering the pH of the medium. This delivery system, designated PhenPt-TMV, exhibited matched efficacy in a cancer cell panel compared to free phenanthriplatin or cisplatin, owing to increased accumulation of PhenPt within the tumor tissue.

The subsequent work using TMV particles loaded with a related drug, cisplatin, showed that the TMV nanorods are an efficient way of delivering anticancer therapeutics. Thus, the TMV particles of the high aspect ratio served as a nanocarrier for cisplatin for treatment of platinum-resistant ovarian cancer cells (Franke et al. 2018). The cisplatin (cis-[Pt(NH$_3$)$_2$Cl$_2$], cisPt) is the most common drug used for intraperitoneally administered ovarian cancer therapy. The TMV-cisplatin conjugate (TMV-cisPt) was synthesized using a charge-driven reaction that, like a classic click reaction, was simple and reliable for large-scale production. Up to ~1900 cisPt were loaded per TMV-cisPt. The efficient cell uptake of the TMV- cisPt was observed when incubated with ovarian cancer cells, and the TMV-cisPt demonstrated superior cytotoxicity and DNA double-strand breakage in platinum-sensitive and platinum-resistant cancer cells when compared to free cisplatin (Franke et al. 2018).

The previously described spherical TMV nanoparticles—or SNPs—were applied as chemotherapy delivery agents targeting breast cancer (Bruckman et al. 2016). The authors probed for the first time the reactivity of SNPs toward bioconjugate reactions targeting lysine, glutamine/aspartic acid, and cysteine residues. The functionalization of SNPs using these chemistries yielded efficient payload conjugation. In addition to covalent labeling techniques, encapsulation techniques were developed, where the cargo was loaded into the SNPs during heat transition from rod to sphere. Finally, the SNP formulations, in parallel with nanorod ones, were developed, loaded with the chemotherapeutic doxorubicin, and applied for chemotherapy delivery targeting breast cancer.

At last, Shukla et al. (2021) performed the targeted delivery of drugs to cells expressing prostate-specific membrane antigen by the TMV particles. Thus, a derivative of the prostate-specific membrane antigen (PSMA) was synthesized and conjugated to the external surface of TMV. The PSMA-targeted TMV particles were subsequently loaded with the antineoplastic agent mitoxantrone (MTO) or conjugated internally with the fluorescent dye cyanine 5 (Cy5) and demonstrated ability to bind more efficiently to the surface of the PSMA⁺ cancer cells and accomplish therapeutic effect.

CUCUMBER GREEN MOTTLE MOSAIC VIRUS

The CGMMV had certain advantages over TMV as a tobamovirus-based heterologous peptide expression, since, being a virus that infects edible cucurbitaceous plants, it could potentially be taken as an edible vaccine.

The full-length clone of CGMMV was developed for the expression of hepatitis B virus (HBV) surface (HBs) protein (Ooi et al. 2006). The "a" determinant of the HBs protein (for details see Chapter 37) covering aa residues Pro111 to Thr141 was added C-terminally to the CGMMV coat. The system expressed the CGMMV coat-HBs fusion protein in parallel with a modified clone of the wild-type CGMMV coat protein at an approximate ratio of 1:1 and led to formation of mosaic virions. The assessment of the coat protein composition of the mosaic virions by SDS-PAGE analysis showed that 50% of the coat proteins were actually fused to the HBs epitope. The high content of the chimeric coat protein within the mosaic particles was explained by a suggestion that muskmelon host plant could be producing higher levels of translation nonsense suppressor tRNA, making the application of the translation readthrough signal favorable in this host. The HBs epitope was displayed on the mosaic virions and was able to induce specific anti-HBs antibodies by the in vitro cultured peripheral blood mononuclear cells (PBMCs) (Ooi et al. 2006).

Teoh et al. (2009) used the CGMMV coat to present a truncated dengue virus type 2 envelope (E) protein binding region from aa 379 to 423 (EB4). The EB4 gene was inserted at the C-terminus of the CGMMV coat, and readthrough sequences of TMV or CGMMV, CAA-UAG-CAA-UUA, or AAA-UAG-CAA-UUA were, respectively, inserted in between the coat and the EB4 genes. Only constructs with the wild-type CGMMV readthrough sequence yielded infectious viruses following infection of host plant, muskmelon *Cucumis melo* leaves. The chimeric (carrying the EB4) and wild-type CGMMV were shown to coexist in the virus population of the infected plants. The ratio of modified to unmodified coats within the mosaic virions was found to be approximately 1:1, as before in the case of the HBs-carrying mosaic particles.

TURNIP VEIN-CLEARING VIRUS

The TVCV vector played a special role by exploring of the tobamovirus capacity to accept foreign insertions. Thus, Werner et al. (2006) were the first who disproved the widely held dogma that the tobamovirus virions tolerate only short, linear epitope insertions and presented the possibility that the tobamovirus viral particles could be produced that display longer conformational epitopes. Thus, the authors displayed the 133-aa functional fragment of the *Staphylococcus aureus* protein A, which contained two of the five Ig binding domains, on the TVCV surface as a C-terminal fusion to the coat protein via a flexible 15-aa linker. The resultant chimera retained the ability to assemble into viral particles, whereas the protein A fragment remained fully functional as an immunoadsorbent. The length of the chimeric particles was similar to that of the wild-type particles, ~200–220 nm length. The high level of expression of the nanoparticles provided a very inexpensive self-assembling matrix, while the macromolecular nature of these nanoparticles allowed the design of a simple protocol for purification of mAbs with high recovery yield. The extremely dense packing of protein A on the nanoparticles, >2,100 copies per viral particle, resulted in an immunoadsorbent with extremely high binding capacity. Remarkably, both N- and C-terminal fusion proteins containing either a flexible glycine-rich linker (GGGGS)₃ or a helical linker (EAAAK)₃ were expressed at high levels and could be detected on Coomassie-stained gels as a strong band comparable to that of wild-type coat. In contrast, fusion proteins without a linker were not expressed at detectable levels. However, the coat fusion proteins could be recovered at high yield only when the protein A fragment was fused to the C-terminus of coat.

To determine whether the protein A fusions at the N-terminus of coat also assembled into viral particles, crude extracts, rather than purified samples, were analyzed by using the electron microscope. Surprisingly, only large aggregates of viral particles were observed. It appeared likely that aggregation of the N-terminal fusions is the cause for the loss of the particles during the purification procedure (Werner et al. 2006).

GENUS *FUROVIRUS*

Koenig et al. (2006) engineered a vector based on zygocactus virus X (ZVX) of the genus *Potexvirus*, family *Alphaflexiviridae*, order *Tymovirales* (see Chapter 21) and expressed in leaves of plant *Chenopodium quinoa* the coat protein genes of beet necrotic yellow vein virus (BNYVV) of the genus *Benyvirus* from the *Benyviridae* family, order *Hepelivirales* (as described in Chapter 16) and of a furovirus, namely soil-borne cereal mosaic virus (SBCMV). The SBCMV VLPs were rod-shaped, contained the typical axial canal, and were decorated strongly with SBCMV antibodies by electron microscopy investigation.

GENUS *PECLUVIRUS*

The cDNA copies of the coat gene of a pecluvirus, namely Indian peanut clump virus, strain H (IPCV-H), were introduced into plant *N. benthamiana* or *E. coli* by transformation with vectors expressing the 5'-most sequence of IPCV-H RNA-2 and the IPCV-H coat gene, respectively (Bragard et al. 2000). In both plant and bacterial cells, the IPCV coat gene was expressed and ensured formation of the rod-like VLPs by electron microscopy. In plant extracts, the smallest preponderant particle length was about 50 nm, while other abundant lengths were about 85 and about 120 nm. The commonest VLP length in bacterial extracts was about 30 nm. Many of the longer VLPs appeared to comprise aggregates of shorter particles. The lengths of the supposed "monomeric" VLPs corresponded approximately to those expected for the encapsidated coat gene transcript RNA. Both plant- and bacteria-derived VLPs contained RNA encoding the IPCV-H coat. The results showed that the encapsidation did not require the presence of the 5'-terminal untranslated sequence of the virus RNA and suggested that, if there is an "origin of assembly" motif or sequence, it lies within the coat gene. When transgenic plants expressing the IPCV-H coat gene were inoculated with IPCV-L, a strain that is serologically distinct from IPCV-H, the virus particles that accumulated contained both types of coat proteins (Bragard et al. 2000).

GENUS *POMOVIRUS*

The 21-kDa C-terminal fragment of the 67-kDa readthrough variant of the coat protein of potato mop-top virus (PMTV), a pomovirus, was cloned, fused to glutathione S-transferase, and expressed in *E. coli* (Cowan et al. 1997). An antiserum prepared against the purified fusion protein was used for the immunogold labeling of virions, which localized the readthrough protein near one extremity of some of the PMTV virions.

Hélias et al. (2003) expressed the intrinsic PMTV coat gene in *E. coli*. The molecular mass of the product was consistent with that deduced from the PMTV coat aa sequence, but the product was insoluble. It was purified by preparative gels and used for antiserum production in rabbits. In parallel, Čeřovská et al. (2003) expressed in *E. coli* the coat gene of the Czech PMTV isolate. The coat was provided in this case with His_6 tag, remained insoluble, and was used for production of antibodies. No renaturation attempts were undertaken in either study.

GENUS *TOBRAVIRUS*

The coat gene of tobacco rattle virus (TRV), a tobravirus, was integrated into the genome of *N. tabacum* and ensured production of the coat protein, the possible assembling of which was not examined (van Dun et al. 1987).

Greater use has been made of tobraviruses, especially TRV, as virus-induced gene silencing (VIGS) vectors, where insertion of a plant sequence into RNA2 of the virus induced a host-silencing response that targeted homologous plant mRNA sequences for degradation, thus reducing expression of the host gene. This particular aspect, as well as a presumed place of tobraviruses in biotechnological studies, were discussed in a special review on tobraviruses by MacFarlane (2010).

20 Order *Martellivirales*: Other Families

The part can never be well unless the whole is well.

Plato

FAMILY *CLOSTEROVIRIDAE*

The family involves 4 genera and 59 species in accordance with the current ICTV taxonomy. The genera are *Ampelovirus* with the type species *Grapevine leafroll-associated virus 3* involving Grapevine leafroll-associated virus 3 (GLRaV-3), *Closterovirus* with the *Beet yellows virus*, *Crinivirus* with *Lettuce infectious yellows virus*, and *Velarivirus* with the *Grapevine leafroll-associated virus 7* type species.

According to the latest ICTV report (Fuchs et al. 2020), the plant viruses in the family *Closteroviridae* possess long, helically constructed filamentous particles of 650–2,200 nm in length. Figure 20.1 presents a portrait of a typical representative and a schematic cartoon of the family members. Thus, the virions are filaments with a pitch of the primary helix in the range of 3.4–3.8 nm, containing about ten protein subunits per turn of the helix and showing a central hole of 3–4 nm. The virions are flexuous and have a diameter of about 12 nm and lengths ranging from 650 nm (viruses with a fragmented genome) to over 2,200 nm (viruses with a monopartite genome). The structural proteins of most members of the family consist of a major coat protein (CP) and a diverged copy of it, denoted minor coat protein (CPm), with masses ranging from 22–46 kDa (CP) and 23—80 kDa (CPm), according to species. The CPm encapsidates the 600–700 5′-terminal nucleotides of the virus RNA, at one extremity (75–100 nm) of the particle, thus forming a distinct structure, referred to with the terms "rattlesnake," "heterodimeric," or "bipolar" (Agranovsky 2016; Fuchs et al. 2020).

Figure 20.2 shows the principal genomic structure of four genera of the *Closteroviridae* family. In brief, this a large positive-sense single-stranded, mono-, bi-, or tripartite RNA genome of 13,000 to nearly 19,000 nucleotides. Regardless of whether the genome is monopartite or fragmented, the virions contain a single molecule of linear, positive-sense, single-stranded RNA. The presence of a cellular HSP70 homolog and a duplicated, diverged copy of the capsid protein, namely CPm, genes in the virus genome are hallmarks of the family (Fuchs et al. 2020).

Both CP and CPm genes of beet yellows virus (BYV) from the *Closterovirus* genus were expressed in *E. coli* by Agranovsky et al. (1994). High yields of both proteins, which were provided with the His$_6$ tags for the rapid purification needs, were achieved, but the VLP-forming ability

of the produced proteins was not addressed. Furthermore, the closterovirus coat genes were expressed in *E. coli* for lettuce infectious yellows virus (LIYV), genus *Crinivirus* (Klaassen et al. 1994); citrus tristeza virus (CTV), genus *Closterovirus* (Nikolaeva et al. 1995); sweet potato chlorotic stunt virus (SPCSV), genus *Crinivirus* (Hoyer et al. 1996); beet yellow stunt virus (BYSV), genus *Closterovirus* (Karasev et al. 1998); grapevine leafroll-associated virus 3 (GLRaV-3), genus *Ampelovirus* (Ling et al. 1997b); and cucurbit yellow stunting disorder virus (CYSDV) from the *Crinivirus* genus (Livieratos et al. 1999). No data about putative self-assembly of these proteins were presented.

In parallel, Ling et al. (1997a) expressed the coat protein gene of GLRaV-3 in *N. benthamiana* plants. Later, Vinogradova et al. (2012) expressed the coat gene of BYV in *N. benthamiana* and suspected a potential of resistance to BYV infection. Again, no data about the possible self-assembly of the synthesized coat proteins was presented.

The advances in the closterovirus research, with a special emphasis on the relationships between virus biology and vector design, were reviewed by Dolja and Koonin (2013). Recently, Qiao and Falk (2018) demonstrated successful replacement and "add a gene" strategies to develop LIYV-derived vectors for transient expression of GFP reporter in *N. benthamiana* plants.

FAMILY *ENDORNAVIRIDAE*

This family includes viruses with linear, single-stranded, positive-sense RNA genomes that range from 9.7–17.6 kb and have been reported infecting plants, fungi, and oomycetes (Valverde et al. 2019). According to this recent ICTV report, the family consists of two genera, *Alphaendornavirus* and *Betaendornavirus*, into which viruses were classified based on their genome size, host, and presence of unique domains. *Alphaendornavirus* includes 24 species whose members infect plants, fungi, and oomycetes, while the genus *Betaendornavirus* includes 7 species whose members infect ascomycete fungi.

No true virions are associated with members of this family since their genomes lack a coat protein gene and the RNA of endornaviruses is not associated with structural proteins at all. However, the endornavirus RNA genomes could be associated with pleomorphic cytoplasmic membrane vesicles. Such cytoplasmic vesicles of ~70 nm diameter were observed in the cytoplasm of broad bean *Vicia faba* infected with Vicia faba endornavirus (VfEV), as reviewed by Valverde et al. (2019).

DOI: 10.1201/b22819-25

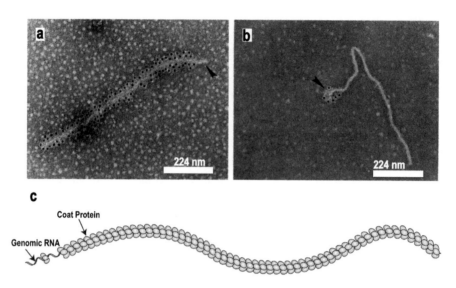

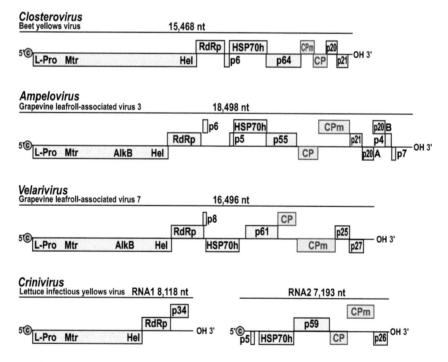

FIGURE 20.1 The portrait and schematic cartoon of the typical representatives of the *Closteroviridae* family. (a, b) The electron microscopy and immunogold labeling analysis of partially purified lettuce infectious yellows virus (LIYV) from the genus *Crinivirus*. The virion in (a) was labeled using antiserum to the LIYV coat protein. The virions in (b) were labeled using antiserum to the LIYV minor coat protein. (Reprinted with permission of Microbiology Society from Tian T et al. *J Gen Virol.* 1999;80:1111–1117). (c) The general cartoon of the *Closteroviridae* representative virions is taken with kind permission from the ViralZone, Swiss Institute of Bioinformatics, http://viralzone.expasy.org (Hulo C et al. 2011). (Courtesy Philippe Le Mercier.)

FIGURE 20.2 The genomic structure of the family *Closteroviridae* members, derived from the recent ICTV report. (Redrawn from Fuchs M et al. ICTV virus taxonomy profile: *Closteroviridae. J Gen Virol.* 2020;101:364–365.) The structural genes are colored dark pink.

FAMILY *KITAVIRIDAE*

This family involves three genera, namely *Blunervirus*, *Cilevirus*, and *Higrevirus*, covering in total 7 species. The genus *Cilevirus* with the type species *Citrus leprosis virus C* is described by Locali-Fabris et al. (2012). This genus contains mite-transmitted viruses with bacilliform particles and a bipartite ssRNA genome that is 3'-polyadenylated. The bacilliform virions are about 120–130 × 50–55 nm. The citrus leprosis virus C (CiLV-C) is an atypical virus that does not spread systemically in its plant hosts.

FAMILY *MAYOVIRIDAE*

This new family was created in 2019 and covers two genera, *Pteridovirus* and *Idaeovirus*, each involving two species. The two known pteridoviruses are Japanese holly fern mottle virus (JHFMoV; Valverde and Sabanadzovic 2009) and maize-associated pteridovirus (MaPV; Read et al. 2019). JHFMoV forms quasi-spherical particles of 30–40 nm in diameter and possesses two genomic ssRNAs of about 6.2 and 3.0 kb (Valverde and Sabanadzovic 2009).

The promising genus *Idaeovirus* represented by raspberry bushy dwarf virus (RBDV) was described by Rastgou et al. (2012). The virions are isometric, about 33 nm in diameter, and demonstrate similarity with icosahedral T = 3 virions of the *Bromoviridae* family members. As commented by Rastgou et al. (2012), the idaeovirus virions appear flattened in electron micrographs of preparations negatively stained with uranyl salts. The idaeovirus genome is bipartite, and the isometric virions encapsidate two genomic and one subgenomic positive-sense, single-stranded RNAs.

21 Order *Tymovirales*

Never give up. . . . No one knows what's going to happen next.

L. Frank Baum

ESSENTIALS

The *Tymovirales* is a great order of quite different, flexuous filamentous and isometric icosahedral, single-stranded, positive-sense RNA viruses that infect plants and fungi. The order members played a reliable role in the development of the VLP technologies. According to the current ICTV taxonomy (Adams et al. 2012b; ICTV 2020; Kreuze et al. 2020), the order *Tymovirales* is one of the three members of the class *Alsuviricetes*, together with the two other orders: *Hepelivirales* and *Martellivirales*, the subjects of the five preceding chapters. The *Alsuviricetes* class belongs to the *Kitrinoviricota* phylum from the kingdom *Orthornavirae*, realm *Riboviria*.

The order *Tymovirales* currently contains 5 families, 2 subfamilies, 24 genera, 2 subgenera, and 219 species. This order includes the two popular families: *Alphaflexiviridae* possessing filamentous virions and *Tymoviridae* possessing icosahedral virions, which both have been involved deeply in the setting up of the VLP nanotechnology field. Then there are the large *Betaflexiviridae* family with 2 subfamilies, 13 genera, and 108 species and the two lesser-known small families *Deltaflexiviridae* and *Gammaflexiviridae*, which possess each a single genus with the three species and with the single *Botrytis virus F* species, respectively. The filamentous *Alphaflexiviridae* members, in total 6 genera, 2 subgenera, and 65 species, infect plants and plant-infecting fungi (Adams et al. 2012a; Kreuze et al. 2020), while the icosahedral *Tymoviridae* members, currently 3 genera and 42 species, infect only plants (Dreher et al. 2012).

Figure 21.1 demonstrates portraits and schematic cartoons of the typical representatives of the two most interesting families. The *Alphaflexiviridae* members are looking like flexuous filaments, usually 12–13 nm (in a range 10–15 nm) in diameter and from 470–800 nm in length, depending on the genus. The viral capsid is composed of a single polypeptide ranging in size from 18–43 kDa except for members of the genus *Lolavirus*, which have two C-coterminal capsid protein variants, and members of the genus *Sclerodarnavirus*, in which no capsid protein has been identified (Kunze et al. 2020). The *Tymoviridae* is the only family within the order *Tymovirales* that has viruses with isometric particles. The virions are nonenveloped, about 30 nm in diameter, and made up of 20 hexameric and 12 pentameric subunits arranged in a T = 3 icosahedron. The RNA appears to be at least partially ordered in an icosahedral arrangement in the center of the protein shell (Dreher et al. 2012).

GENOME

Figure 21.2 shows the genomic structure of the two families, which remain the most interesting from the VLPs point of view. The *Alphaflexiviridae* members contain a single molecule of linear, positive-sense RNA of 5.5–9.0 kb, which is 5–6% by molecular mass of the virion. The RNA is typically capped at the 5′-terminus with m7G and has a polyadenylated tract at the 3′-terminus. Smaller 3′-coterminal subgenomic RNAs are encapsidated in some, but not all, members of the genus *Potexvirus*. There are five to seven genes depending upon the genus, except for members of the genus *Sclerodarnavirus*, which have a single gene (Kunze et al. 2020).

The *Tymoviridae* virions contain a single molecule of positive-sense RNA constituting 25–35% of the particle mass and ranging from 6.0–7.5 kb in length. The genomes are capped at the 5′ terminus with m7G, and most, though not all, have a tRNA-like structure at the 3′ end, which for turnip yellow mosaic virus (TYMV) and several other members accepts valine (Dreher et al. 2012).

FAMILY *ALPHAFLEXIVIRIDAE*

BAMBOO MOSAIC VIRUS

All potexviruses that are described in the following sections are assigned in 2021 to the subgenus *Mandarivirus* of the genus *Potexvirus*. The *Bamboo mosaic virus* including the popular bamboo mosaic virus (BaMV) is a type species of the genus *Potexvirus* containing preferred nanotechnology models of the *Alphaflexiviridae* family (Hsu et al. 2018). From the very beginning of the VLP era, the potexviruses were known as gene vectors to produce pharmaceutical proteins in plants, when the target gene was substituted for the virus coat gene. This was achieved for the first time with the vectors based on the potato virus X (PVX) genome (Chapman et al. 1992). The BaMV coat-deficient vectors provided the highest production efficiency in comparison with the analogous PVX- (or foxtail mosaic virus (FoMV)) based vector systems for the transient expression of human mature interferon gamma (mIFNγ) in plants *N. benthamiana* (Jiang et al. 2019). Furthermore, the yields of soluble and secreted mIFNγ were enhanced through the incorporation of various plant-derived signal peptides including the fusion of a secretion booster signal (Jiang et al. 2020). Together with high efficacy, the use of the potexvirus coat-deficient

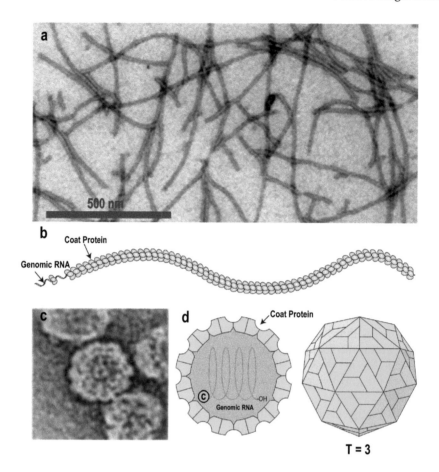

FIGURE 21.1   The portraits and schematic cartoons of the two typical representatives of the order *Tymovirales*. (a) Electron micrograph of potato virus X (PVX) from the *Potexvirus* genus, family *Alphaflexiviridae*. (From Dickmeis C et al. Production of hybrid chimeric PVX particles using a combination of TMV and PVX-based expression vectors. *Front Bioeng Biotechnol.* 2015;3:189.) (b) A cartoon of a representative of the *Alphaflexiviridae* family. (c) Electron micrograph of turnip yellow mosaic virus (TYMV) from the genus *Tymovirus* of the *Tymoviridae* family. (Reprinted from *J Mol Biol.* 24, Finch JT, Klug A, Structure of broad bean mottle virus. I. Analysis of electron micrographs comparison with turnip yellow mosaic virus and its top component, 289–302, Copyright 1967, with permission from Elsevier.) (d) A cartoon explaining the 3D structure of the *Tymoviridae* representatives. Both cartoons are taken with kind permission from the ViralZone, Swiss Institute of Bioinformatics, http://viralzone.expasy.org (Hulo C et al. 2011). (Courtesy Philippe Le Mercier.)

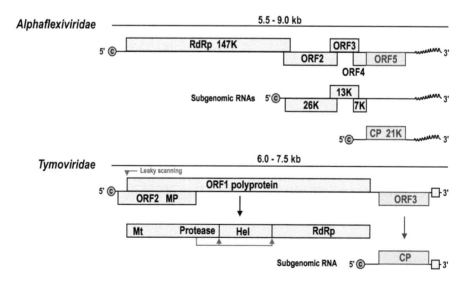

FIGURE 21.2   The genomic structure of the members of the families *Alphaflexiviridae* and *Tymoviridae*, combined from the ICTV report (Adams et al. 2012; Kreuze et al. 2020) and the ViralZone, Swiss Institute of Bioinformatics. The noncoding polyadenylated and tRNA-like 3′-termini are shown as wavy lines and gray square boxes, respectively. The structural genes are colored dark pink.

vectors corresponded to better biocontainment, since coat was necessary for the potexvirus movement (Baulcombe et al. 1995).

Nevertheless, the BaMV-based vector system was used for peptide presentation when viral coats were fused with foreign sequences and the appropriate chimeric viruses were produced. Thus, Yang CD et al. (2007) developed a new viral vaccine delivery system, where the chimeric BaMV coat carried epitope of the capsid protein VP1 of foot-and-mouth disease virus (FMDV) on the surface of the live infectious BaMV virions, but not VLPs. Namely, the chimeric BaMV coat gene was constructed by replacing the 35 N-terminal aa residues of the coat with the 37 aa residues T_{128}-N_{164} of the FMDV VP1. The chosen VP1 sequence contained therefore not only the G-H loop—the major B-cell epitope including the highly conserved Arg-Gly-Asp (RGD) tripeptide, which can bind to integrins and facilitate the internalization of FMDV into target cells—but also immunodominant T epitopes. The chimeric virus was able to infect host plants *Chenopodium quinoa* and *N. benthamiana* and to generate chimeric virions, inoculation of which in swine resulted not only in the production of anti-FMDV neutralizing antibodies but also in the protection against challenge of $10^{5.0}$ TCID$_{50}$ FMDV, an amount ten times higher than that recommended by the World Organization for Animal Health. Moreover, one inoculation was sufficient to protect the animals (Yang et al. 2007). In fact, this was the first report of a chimeric plant virus correctly expressing the FMDV VP1 epitopes to properly induce both humoral and cell-mediated immune responses. Figure 21.3 confirms the localization of the FMDV on the outer surface of the chimeric rod-shaped virions.

Furthermore, to advance the large-scale production of the FMDV-BaMV vaccines, Muthamilselvan et al. (2016)

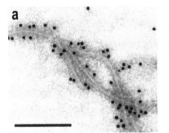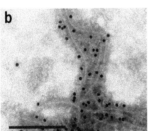

FIGURE 21.3 The immunoelectron microscopy for identification of BaMV coat protein and FMDV VP1 on the surface of virus particles. Leaf dips from *C. quinoa* infected with the chimeric virus were obtained ten days post inoculation. Grids were first incubated with leaf extract and coated with diluted anti-BaMV coat protein serum (a) or anti-FMDV VP1 serum (b) followed by gold-labeled goat antirabbit IgG complexes. The bars represent 250 nm. (With kind permission from Springer Science+Business Media: *BMC Biotechnol*, Induction of protective immunity in swine by recombinant bamboo mosaic virus expressing foot-and-mouth disease virus epitopes, 7, 2007, 62, Yang CD, Liao JT, Lai CY, Jong MH, Liang CM, Lin YL, Lin NS, Hsu YH, Liang SM.)

established transgenic cell-suspension cultures from callus derived from the transgenic *N. benthamiana* plant lines expressing different BaMV expression cassettes that encoded the chimeric virions.

Next, an epitope from VP2 protein of very virulent infectious bursal disease virus (vvIBDV) was fused to the N-terminus of the BaMV coat protein (Chen et al. 2012). The chimeric infectious virions were generated with the same BaMV vector harboring the truncated coat with the N-terminal 35-aa deletion (Yang et al. 2007) and addition of the vvIBDV VP2 P domain loop P_{BC}, namely an aa residue stretch 207-TLTAADDYQFSSQYQAGG-224 of the IBDV strain V97/TW. The chimeric virions were purified from the leaves of *C. quinoa* plants and demonstrated, after intramuscular immunization, not only specific IBDV immunogenicity but also protection of chickens against challenge with the vvIBDV, strain V263/TW (Chen et al. 2012).

Chen et al. (2017) generated the mosaic chimeric BaMV virions carrying epitopes of Japanese encephalitis virus (JEV). In this case, the JEV envelope protein domain III (EDIII) was added to the N-terminus of BaMV coat protein over an insertion of the FMDV 2A peptide to facilitate the production of both unfused and epitope-presenting coats for efficient assembly of the mosaic chimeric virions. The same BaMV coat construction with the N-terminal 35-aa deletion was used, and the 111-aa long JEV EDIII epitope was added. The cotranslational dissociation or ribosomal skip sequence 2A, namely LLNFDLLKLAGDVESNPGP, from FMDV was adopted from the strategy of Santa Cruz et al. (1996) used earlier by the PVX model (see following paragraphs). The fine mechanism of the ribosomal skip was described by Donnelly et al. (2001).

This approach allowed one to overcome the size limitations of the epitope-presentation system and provided enhanced solidity of the chimeric virions while retaining the presentation of EDIII epitopes on portions of virion surfaces. Such mosaic strategy allowed stable maintenance of the fusion construct over long-term serial passages in *C. quinoa* plants. As the most impressing result, the JEV EDIII epitope was exposed on the surface of the chimeric virions and induced effective neutralizing antibodies against JEV infection in mice (Chen et al. 2017).

At last, the same mosaic virion strategy was used to create a BaMV-based viral vector expressing recombinant proteins, collectively designated GfED, consisting of *Staphylococcus aureus* protein A domain ED (SpaED) fused to either the N- or C-terminal of an improved green florescent protein (GFP) with or without the BaMV coat protein (CP) and efficiently produced in *C. quinoa* (Kuo et al. 2018). The GfED in crude leaf extracts could specifically attach to IgG molecules of rabbits and mice, effectively labeling IgG with GFP and applicable in serological assays to easily detect plant pathogens, with results observable by the naked eye. The GfED subunits were found also within the VLPs, which were further involved in the formation of aggregates of GfED-antibody-antigen complexes with the potential for fluorescence signal enhancement (Kuo et al. 2018).

Concerning the 3D structure of BaMV, the structural studies of flexible filamentous plant viruses began more than 75 years ago but have failed, owing to the virion's extreme flexibility, as was stated by DiMaio et al. (2015). The authors imaged both the wild-type BaMV (WT) and a virion containing a deletion of 35 N-terminal residues of coat, denoted BaMV Nd35, using electron cryomicroscopy. As we know from the previous story, the replacement of the N-terminal 35 aa residues with foreign peptides from either FMDV, IBDV, or other foreign sequences did not destroy the ability of the BaMV coat to self-assemble. Meanwhile, Lan et al. (2010) directly demonstrated that up to 35 N- terminal aa of the coat can be deleted with no effect on virus replication and assembly. As a result, the structural near-atomic findings solved the mystery of how flexible virus particles maintain structural integrity as mechanical forces deform their structure (DiMaio et al. 2015). Figure 21.4 demonstrates the helical structure of the BaMV capsids.

Papaya Mosaic Virus

Natural VLPs

The VLP story of another remarkable member of the *Potexvirus* genus, papaya mosaic virus (PapMV), was started with the successful disassembly and reassembly of viral particles in vitro (Erickson et al. 1976), where the PapMV coat was prepared by the classical acetic acid degradation of purified virus (Fraenkel-Conrat 1957) and reconstituted into nucleocapsid-like structures in the absence of RNA. The aggregates ranged from 14S to 25S where 14S was a helical disk-like structure of 18–20 subunits (Erickson and Bancroft 1978; Erickson et al. 1978). These disks were the structures of two turns of the helix and were similar in architecture to the native virus particle elements

(Erickson et al. 1976, 1983). The addition of RNA (Erickson et al. 1976) or single-stranded DNA (Erickson and Bancroft 1980) to the isolated disks triggered the assembly of the long rod-shaped particles similar to the original virions.

The PapMV coat of 215 aa in length and 23 kDa of molecular mass, approximately 1,400 copies of which are assembling to form filamentous virions, seemed a useful VLP platform. Consequently, the systematic studies of the Denis Leclerc team generated the great VLP technology of the PapMV VLPs in bacteria *E. coli*. At first, Tremblay et al. (2006) achieved generation of the PapMV VLPs in *E. coli*, when the coat ORF, without N-terminal five aa residues but provided with a His$_6$ tag at the C-terminus to ease the purification process and named CPΔN5, was expressed. Twenty percent of the purified protein was found as VLPs of 50 nm in length, and 80% was found as a multimer of 450 kDa (20 subunits) arranged in a disk. Furthermore, extensive mutation analysis in the putative RNA binding domain of PapMV led to the identification of two mutants: one, K97A, that had lost its affinity for RNA and ability to self-assemble and one, E128A, that showed an improved affinity for RNA and self-assembled more efficiently into VLPs. The E128A VLPs of 150 nm in length were longer than the recombinant CPΔN5, and all protein was found as VLPs in bacteria (Tremblay et al. 2006).

Acosta-Ramírez et al. (2008) described the efficacy of a single immunization of mice with the PapMV VLPs for the efficient induction of both cellular and specific long-lasting antibody responses, without added adjuvant. Moreover, the PapMV VLPs demonstrated intrinsic adjuvanting effects in the immunization of mice with the *Salmonella enterica* serovar Typhi (*S. typhi*) outer membrane protein C when the presence of VLPs increased its protective capacity against challenge with *S. typhi*. Later, the PapMV VLPs were shown to induce innate immunity in lungs and protect mice against influenza and *Streptococcus pneumoniae* challenge (Mathieu et al. 2013). The PapMV VLPs behaved as a TLR7 agonist with strong immunostimulatory properties and significantly improved effector and memory CD8$^+$ T cell responses generated through dendritic cell vaccination increasing protection against a *Listeria monocytogenes* challenge in mice (Lebel et al. 2014).

Savard et al. (2011) used the PapMV VLPs as an adjuvant to the traditional seasonal of trivalent inactivated vaccine, which induced an antibody response toward the surface glycoproteins hemagglutin (HA) and neuraminidase (NA). The immunization of mice and ferrets with the subcutaneous injections of the adjuvanted formulation increased the magnitude and breadth of the humoral response to NP and to highly conserved regions of HA, as well as triggered a cellular mediated immune response to NP and M1 and provided the long-lasting protection of animals against challenge with a heterosubtypic influenza strain (Savard et al. 2011). Next, Rioux et al. (2014) demonstrated the efficacy of the PapMV VLPs as a potent mucosal adjuvant also by intranasal route of administration of the trivalent inactivated flu vaccine. Carignan et al. (2018) demonstrated activation

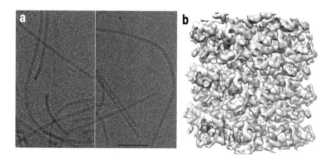

FIGURE 21.4 The BaMV helical capsid structure. (a) Micrographs of the WT (left) and Nd35 (right) capsids. Scale bar, 1,000 Å. (b) The 5.6-Å-resolution combined reconstruction, showing three full helical turns of the capsid. (With kind permission from Springer Science+Business Media: *Nat Struct Mol Biol*, The molecular basis for flexibility in the flexible filamentous plant viruses, 22, 2015, 642–644, DiMaio F, Chen CC, Yu X, Frenz B, Hsu YH, Lin NS, Egelman EH.)

of innate immunity in primary human cells by the PapMV VLPs. At last, the adjuvant effect of the PapMV VLPs by single intramuscular immunization with seasonal trivalent influenza vaccine was confirmed in phase I trial on healthy adults of 18–50 years of age (Langley et al. 2020).

The immunostimulatory potential of the PapMV VLPs was demonstrated also for cancer therapy, when their intratumor administration significantly slowed down melanoma progression and prolonged survival in the B16 syngeneic melanoma-mouse model (Lebel et al. 2016).

Chimeric VLPs

Denis et al. (2007) used the *E. coli*-derived PapMV VLPs as a carrier for an immunodominant HCV epitope derived from the envelope protein E2. Thus, the E2 sequence covering aa residues 511–530 was fused C-terminally to the 5–215 version of the PapMV coat, in front of the His_6 tag. In parallel, a monomeric form PapMVCP$_{27-215}$-E2, where the vector molecule was N-terminally truncated, was generated. While the two forms of the vaccine were both shown to be actively internalized in vitro in bone-marrow-derived antigen presenting cells, the immunogenicity in mice was strongly dependent on the antigen organization. The mice injected twice with the VLPs showed a long-lasting humoral response (more than 120 days) against both the coat and the fused HCV E2 epitope, and the antibody profile suggested a Th1/Th2 response. No immunogenicity was observed when the monomer PapMVCP$_{27-215}$-E2 was injected in mice (Denis et al. 2007).

At the same time, Leclerc et al. (2007) generated in the same expression system the stable chimeric PapMV VLPs carrying the different MHC class I HLA-A*201 epitopes at its C-terminus: a well-defined melanoma gp100 epitope 209-IMDQVPFSV-217—or influenza virus M1 matrix protein epitope 57-GILGFVFTL-65—both with five flanking aa residues at each side and again in front of the His_6 tag. As a result, this study demonstrated that the chimeric PapMV VLPs carrying standard CTL epitopes were able to mediate the proteasome-independent MHC class I cross-presentation leading to expansion of specific human T cells (Leclerc et al. 2007).

Furthermore, Leclerc's team displayed on the chimeric PapMV VLPs a classical well-studied CTL epitope (Lacasse et al. 2008). This epitope, called p33 epitope KAVYNFATM, was derived from glycoprotein of lymphocytic choriomeningitis virus (LCMV) and introduced into immunological studies by the famous Rolf M. Zinkernagel team (Pircher et al. 1989) and used as an instrument by the great Martin F. Bachmann team with the RNA phage VLP model, as described in Chapter 25. The CTL epitope was also properly processed and presented both in vitro and in vivo, when immunization of the p33-specific T-cell receptor transgenic mice with the p33-PapMV VLPs induced the activation of large numbers of specific CTLs. The immunization of nontransgenic mice with the chimeric VLPs in the absence of adjuvant developed p33-specific effector CTLs that rapidly expanded following LCMV challenge

and protected vaccinated mice against LCMV infection in a dose-dependent manner.

In parallel, Morin et al. (2007) used the PapMV VLPs for the presentation of specific peptides, not connected with the vaccine tools but rather in the manner of phage display, improving the avidity of small affinity peptides for diagnostic purposes. Thus, the authors exposed the high-affinity peptides binding to resting spores of the protist *Plasmodiophora brassicae*, an obligate parasite and a major pathogen of crucifers. The three peptides of seven aa residues each, with specific affinity to the target, were chosen and cloned at the C-terminus of the PapMV coat, generating three different high-avidity VLPs. The peptides were exposed at the surface of the VLPs, and their avidity to resting spores of *P. brassicae* was measured, where the peptide DPAPRPR showed the highest avidity. The binding avidity of the appropriate VLPs to *P. brassicae* spores was comparable to that of a polyclonal antibody and demonstrated greater specificity. The fusion of the affinity peptide to the previously described monomeric form PapMVCP$_{27-215}$ of the carrier generated a fusion protein that was unable to assemble into VLPs and did not bind resting spores. Remarkably, the avidity of the chimeric VLPs was increased by adding a glycine spacer between the C-terminus of the PapMV coat and the DPAPRPR peptide and improved even further by using a duplicated specific peptide in the fusion protein. The high-avidity VLPs possessed therefore the specificity of monoclonal antibodies but were more easily generated using the powerful selection of phage display, offering further development for the specific detection assays of economically important plant pathogens (Morin et al. 2007).

Then, Leclerc's team turned to the influenza vaccine. First, Denis et al. (2008) inserted C-terminally the traditional influenza M2e epitope, the history of which is narrated in more detail in Chapter 38, when the M2 emerged for the first time as a potential VLP-carried flu vaccine.

The chimeric PapMV VLPs carrying the M2e peptide—or N-terminus of the protein M2, positions 2–24, as always in front of the His_6 tag—induced production in mice of anti-M2e antibodies that recognized influenza-infected cells, while the M2e-PapMV discs made of 20 coat subunits were poorly immunogenic. The M2e-PapMV VLP immunization led to full protection of mice against a challenge of $4LD_{50}$ with the appropriate influenza strain.

Savard et al. (2012) generated special high affinity VLPs by adding an affinity peptide of seven aa residues directed to the nucleoprotein of influenza virus at the surface of the PapMV VLPs. The influenza nucleoprotein was expressed and purified from *E. coli* cells. The affinity peptide was selected by phage display in accordance with the previously described study of Morin et al. (2007) and added to the C-terminus of the PapMV coat. The presence of the affinity peptide increased the avidity of the chimeric PapMV VLPs to the influenza nucleoprotein and improved protection of vaccinated mice to the challenge with influenza virus (Savard et al. 2012).

Considering that some epitopes might interfere with the self-assembly of the PapMV coat when fused at the C-terminus, Rioux et al. (2012) evaluated other possible sites of insertion, using the influenza hemagglutinin epitope HA11 of 9 aa residues, namely YPYDVPDYA, as a probe.

In total, eight different sites of the HA11 insertion were evaluated, and only three sites were shown to tolerate the fusion, whereas the others led to unstable proteins. These three sites were the C-terminus and positions directly after aa 187 and aa 12, near the N-terminus. These three different recombinant proteins led to the formation of the chimeric VLPs that presented the HA11 epitope at their surface (Rioux et al. 2012).

Babin et al. (2013) demonstrated that the CTL epitope NP$_{147-155}$, namely TYQRTRALV, a H-2Kd epitope specific for Balb/C mice, derived from the influenza nucleocapsid protein, was more efficient when inserted at the N-terminus, after aa residue 12 and with five flanking aa residues at each side of the PapMV coat, than the C-terminal addition, since such chimeric VLPs were more stable at 37°C. Again, the analogous chimeric discs made of 20 subunits of the PapMV coat were less efficient for induction of the CTL response in mice. This revealed again the previously established fact that the assembly of the recombinant PapMV CP into nanoparticles was crucial to triggering an efficient CTL response (Babin et al. 2013).

After these attempts, Leclerc's team developed further the M2e-based influenza vaccine candidates. The previously described variant with the C-terminal fusion of the M2e 23 aa sequence just before the His$_6$ tag (Denis et al. 2008) was immunogenic but unstable at temperatures exceeding 30°C (Rioux et al. 2012). However, when the M2e was inserted at position 12 of the PapMV coat, the chimeric product failed to self-assemble into the VLPs (Carignan et al. 2015). To solve the assembly issue, the M2e epitope was shortened to only nine aa residues 6-EVETPIRNE-14 and inserted at the N-terminus of the PapMV coat, after position 12. This allowed assembly of highly stable and immunogenic VLPs, a single intramuscular immunization of which was sufficient to induce a potent anti-M2e humoral response capable of protecting mice against a lethal influenza challenge (Carignan et al. 2015).

The remarkable position of the highly promising PapMV VLPs in relation to other VLP candidates was reviewed thoroughly at this stage (Leclerc 2014; Lebel et al. 2015).

Later, Bolduc et al. (2018) demonstrated that an experimental vaccine formulation composed of the two types of VLPs harboring the influenza antigens NP and shortened M2e induced broad protection from a lethal influenza challenge in mice.

It is noteworthy that other laboratories expressed later their interest in the PapMV VLP platform by modeling insertion of epitopes of a member of the *Capripox* genus (Kumar et al. 2020) or by fusing 18-aa epitope of Chikungunya virus (CHIKV) envelope (Rothan and Yusof 2020).

3D Structure

To refine the PapMV VLP engineering and to predict the efficacy of coupling with a sortase (SrtA) in particular, which is described later, Thérien et al. (2017) modeled the PapMV structure based on the known structure of PapMV CP and on recent reports revealing the structure of two closely related potexviruses: pepino mosaic virus (PepMV) (Agirrezabala et al. 2015) and BaMV described earlier (DiMaio et al. 2015), as well as the structure of the truncated PapMV coat (Yang S et al. 2012). Figure 21.5 shows the results of modeling in question.

The N-terminal ends of PapMV coats were exposed therefore on the surface of the virus, while the C-termini were located in the central cavity. As a result, the fusions to the N-terminus modified the surface of PapMV nanoparticles and affected the interface of interaction between VLPs and immune cells. The C-terminal fusions caused steric hindrance in the central cavity, interfering with the VLP

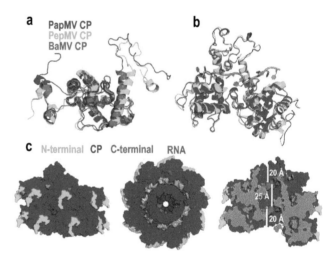

FIGURE 21.5 Modeling of the PapMV coat protein (CP) and PapMV structure. (a) The full-length structure of the PapMV CP was modeled based on the published structure of two members of the potexvirus group: BaMV CP and PepMV CP. The core region of PapMV CP (PDB 4DOX, blue; Yang S et al. 2012) superimposes well on the core region of PepMV CP (PDB 5FN1, green; Agirrezabala et al. 2015) and BaMV CP (PDB 5A2T, red; DiMaio et al. 2015). (b) Superposition of two subunits of the PapMV model (blue) with two CP subunits of the PepMV structure (green, PDB 5FN1) demonstrates the concordance between the two structures. (c) To show how each of the PapMV CP interacts with the other and with the ssRNA in the nanoparticle, the self-assembly of the 18 subunits (~2 turns) that comprise the PapMV VLPs was modeled. The CP N-terminal residues (ten first) are shown in green, the CP core in blue, the CP C-terminus in red (ten last residues), and the RNA is in orange. The last C-terminal residue of the CP is displayed in light gray in the cutaway view on the right, and the bars represent 20 Å—the distance separating CP C-terminal residues from the PapMV nanoparticle exterior at both extremities—and 32 Å, the distance between two CP C-terminal residues separated by one capsid turn. (From Thérien A et al. A versatile papaya mosaic virus (PapMV) vaccine platform based on sortase-mediated antigen coupling. *J Nanobiotechnology*. 2017;15:54.)

assembly. Considering that the C-terminal sections of two coats separated by one capsid turn define a cavity of around 25 Å, as shown in Figure 21.5c, it was anticipated that the longest C-terminal fusion allowed in the core of the VLPs would be ~17 aa if arranged as an α-helix (Thérien et al. 2017).

In parallel, Semenyuk et al. (2016) applied circular dichroism spectroscopy to compare the known CD spectra of PapMV (Lecours et al. 2006; Tremblay et al. 2006) with that of coats of Alternanthera mosaic virus (AltMV), potato aucuba mosaic virus (PAMV), and potato virus X (PVX) in a free state and in virions. Some correlation was found between specific features in the virion CD spectra and the presence of disordered N-terminal segments in the coats.

Sortase Coupling

To overcome limitations of the genetic fusion technique, Leclerc's team developed a novel approach in which peptides were fused directly to the preformed PapMV VLPs (Thérien et al. 2017). This approach was based on the use of a bacterial transpeptidase, or sortase A (SrtA), which attached the peptide directly to the nanoparticle.

Based on the PapMV structural model presented in Figure 21.5, the SrtA recognition motif LPETGG followed by the His_6 tag was added to the C-terminus of the PapMV coat. This fusion was preceded by the linker TSTTR, which allowed the SrtA motif to reach the exterior of the nanoparticles at both ends (Figure 21.5, right panel), making it available for the transpeptidation reaction.

The PapMV VLPs harboring the SrtA recognition motif allowed efficient coupling of long peptides, namely influenza M2e peptide of 26 aa residues and the HIV-1 T20 peptide of 39 aa residues. The T20 peptide was derived from the surface glycoprotein gp41 of HIV-1, since it was recognized by neutralizing monoclonal antibody 2F5 and described by Serrano et al. (2014). The engineered VLPs were capable of inducing a strong antibody response to the fused antigen. Moreover, the mice vaccinated with the M2e-PapMV VLPs were protected from infection.

Laliberté-Gagné et al. (2019) used the StrA approach to couple two full-length recombinant viral proteins, the influenza NP (which contained also shortened M2e sequence, 58 kDa in total) and the simian immunodeficiency virus (SIV) Gag, namely a construct covering the $MA_{1-15}CA-NC_{100-447}$ sequence. The coupling to the PapMV VLPs improved significantly the humoral and CTL immune response to the antigens.

Finally, in order to modulate the antigen display and their immunogenicity on the PapMV VLPs by the SrtA approach, Laliberté-Gagné et al. (2021) chose two different influenza peptides as model antigens: M2e peptide of 23 aa residues as the B-cell epitope and a peptide from NP of 23 aa residues as a source of the CTL epitope. These peptides were coupled at two different positions on the PapMV coat, the N- or the C-terminus, using the SrtA technique. The results demonstrated that coupling at the N-terminus

led to an enhanced immune response to the coupled peptide antigens as compared to coupling to the C-terminus. The difference between the two vaccine platforms was linked to the enhanced capacity of the PapMV N-terminus vaccine platform to stimulate Toll-like receptors 7 and 8. Remarkably, the strength of the immune response increased with the density of coupling at the surface of the nanoparticles (Laliberté-Gagné et al. 2021).

POTATO VIRUS X

Expression

David Baulcombe's team was the first to develop PVX as a popular vector for gene expression in plants (Chapman et al. 1992). Thus, the two viral constructs were tested. In the first, the GUS gene of *E. coli* was substituted for the viral coat protein gene, but this construct without the coat accumulated poorly in inoculated protoplasts and failed to spread from the site of infection in *N. clevelandii* plants. In the second, the foreign GUS gene was added into the viral genome coupled to a duplicated copy of the viral promoter for the coat protein mRNA. This construct also accumulated poorly in protoplasts compared to the unmodified PVX but did infect systemically and directed high level synthesis of GUS in inoculated and systemically infected tissue (Chapman et al. 1992).

Remarkably, Culver (1996) used this PVX vector to express coat protein of tobacco mosaic virus (TMV) in *N. benthamiana* and achieved cross-protection against the TMV challenge.

To express PVX coat derivatives in mammals, Massa et al. (2008) added a mutated form of the oncogenic protein E7 from human papillomavirus 16 (HPV16), the so-called E7ggg, achieved by three aa substitutions in the retinoblastoma protein (pRb)-binding, to the N-terminus of the PVX coat, directly or via a 4-aa linker. This construct was used as a DNA vaccine that proved to be able to block the growth of tumors in mice.

Morgenfeld et al. (2009) expressed similar E7-PVX chimera in tobacco chloroplasts and found that the chimera expression levels were higher than that for unfused E7, indicating that the PVX coat stabilized E7 peptide in the chloroplast stroma.

In *E. coli*, the PVX coat was expressed by Plchova et al. (2011) in the form of N- or C-terminal fusions with the E7ggg, in parallel with the plant-driven expression of the same gene fusions (see following paragraphs). Again, the E7ggg was fused to the PVX coat directly or via 4- and 15-aa linkers. The C-terminal fusions appeared more soluble than the N-terminal ones. Nevertheless, the immunoelectron microscopy revealed nice chimeric *E. coli*-produced VLPs in both cases (Plchova et al. 2011).

Generally, the superior characteristics of the PVX nanoplatform, such as ability to carry large payloads, great tumor homing, low toxicity in vivo, and superior pharmacokinetic profiles, were highlighted in the exhaustive reviews of Lico et al. (2015) and Röder et al. (2019).

Chimeric Virions

Santa Cruz et al. (1996) generated a genetically modified live PVX expressing the N-terminal fusion of 27 kDa (237 aa residues) GFP to the 25-kDa PVX coat. To overcome the steric hindrance between recombinant coats in homogeneous chimeric particles, the authors preferred generation of mosaic particles. This was achieved by introducing the ribosomal skip sequence 2A from FMDV, described earlier by the generation of the mosaic BaMV VLPs. As a result, the chimeric coat subunits were involved in mosaic virions in the presence of free unmodified coat protein subunits, and the mosaic flexuous rods demonstrated a surface overcoat of fluorescing GFP. The chimeric virions were over twice the diameter of wild-type PVX virions, accumulated as paracrystalline arrays in infected cells similar to those seen in cells, which were infected with wild-type PVX, and moved both locally and systemically in infected plants. Moreover, the assembly of mosaic virions did not reflect a unique attribute of the GFP-CP fusion. The authors referred to unpublished data that the similar fusions between the PVX coat and neomycin phosphotransferase II (31 kDa), chloramphenicol acetyltransferase (25.6 kDa), and β-galactosidase (8.5 kDa) have all resulted in the assembly and movement competent viruses (Santa Cruz et al. 1996). This FMDV 2A-based approach was used further by the chimeric derivatives of tobacco mosaic virus (TMV) of the family *Virgaviridae*, as described in Chapter 19, where PVX appeared as a prototype and control (Röder et al. 2017a).

The same strategy was used to express a single-chain antibody fragment (scFv) against the herbicide 3-(3,4-dichlorophenyl)-1,1-dimethylurea, or diuron, as a fusion to the PVX coat (Smolenska et al. 1998). The mosaic virions accumulated in inoculated *N. clevelandii* plants and assembled to give virions carrying the antibody fragment on their surface. The aim of such PVX-antibody particles would be remediation of contaminated soil and waterways. However, the recombinant virus remained infective, so careful precautions would be necessary before releasing it into the environment. This study differed from other attempts using the PVX as a vector to express scFv (Franconi et al. 1999; Roggero et al. 2001), where the scFv and coat genes were not fused and no scFv-carrying mosaic PVX particles were expected.

Moreover, Savelyeva et al. (2001) showed that the PVX coat, when delivered as a fusion gene with scFv, can promote immunity against B-cell malignancies. The expressed fusion protein was highly aggregated and promoted a CD4[+] T-cell response against the linked tumor antigen. These T cells appeared critical for protection against both lymphoma and myeloma.

Brennan et al. (1999) used the PVX vector in parallel with another great plant virus vector, namely that of cowpea mosaic virus (CPMV), described in Chapter 27. The D2 peptide derived from a *S. aureus* fibronectin-binding protein (FnBP), aa 1–38, was added N-terminally to the PVX coat, and chimeric virions induced high titers of FnBP-specific antibody in mice.

O'Brien et al. (2000) generated a fusion of the PVX coat with the protein VP6 of rotavirus, when the PVX vector was used to produce the rotavirus VLPs in plants, as described in the *Rotaviruses* paragraph of Chapter 13. This expression yielded flexuous rods containing a surface overcoat of VP6. Unexpectedly, expression of the fusion protein also yielded icosahedral VLPs, implying that presentation of VP6 on the flexuous PVX rod, followed by proteolytic cleavage, promoted its assembly into VLPs (O'Brien et al. 2000).

Marusic et al. (2001) expressed the highly conserved ELDKWA epitope from HIV-1 glycoprotein gp41, recognized by MAb 2F5, as an N-terminal fusion with the PVX coat. The resulting chimeric virions displayed the epitope on their surface and were able to elicit HIV-1-specific IgG and IgA antibodies in mice, when injected intraperitoneally or intranasally. Moreover, sera from immunized mice showed an anti-HIV-1-neutralizing activity. The chimeric coat was highly stable, retained its capability to form virions after three cycles of reinfection, and did not need therefore the mosaic technique and wild-type coats as a helper to form virions (Marusic et al. 2001). Later, the 2F5 peptide was included into a panel of 30 model peptides varying in both lengths, from 2–27 aa residues, and aa composition and added to the N-terminus of the coat of a spontaneous PVX mutant (Lico et al. 2006). The latter expressed a truncated but functional, form of the coat protein, where the aa residues 2–22 at the N-terminus were lost and substituted by a Cys residue in the correct reading frame, resulting in the size of 216 aa instead of 237 aa in the wild-type coat. The panel of peptides involved epitopes selected from melanoma-associated or HIV-1 proteins, as well as peptides, associated with celiac disease or endowed with antimicrobial activity.

Concerning involvement into the human papillomavirus (HPV) story, the PVX expression vector first was used to express the HPV16 E7 protein in the *N. benthamiana* tobacco plant as an unfused protein (Franconi et al. 2002). The PVX vector was used to express in plants the chimeric potato virus A (PVA) coat protein carrying two different epitopes from HPV16 (Cerovská et al. 2008), as described in Chapter 29.

Uhde et al. (2005) constructed chimeric PVX virions carrying two different epitopes, ep4 and ep6, as well as their tandem, from beet necrotic yellow vein virus (BNYVV), a benyvirus from the *Hepelivirales* order described in Chapter 16. The 7-aa epitopes were added N-terminally to the PVX coat and displayed therefore on the surface of PVX particles, although no wild-type coat protein subunits were present. By mixed infections with PVX vectors containing the individual ep4 and ep6 sequences, the formation of PVX particles displaying ep4 alone, ep6 alone, or both epitopes was found. Thus, it was demonstrated for the first time that PVX can be utilized to present multiple epitopes, either tandemly on every coat protein subunit or as heteromultimeric assemblies. Figure 21.6 illustrates this study.

When five serial passages through systemically infected plants were carried out with the chimeric BNYVV-PVX virions, the accumulation of several point mutations and

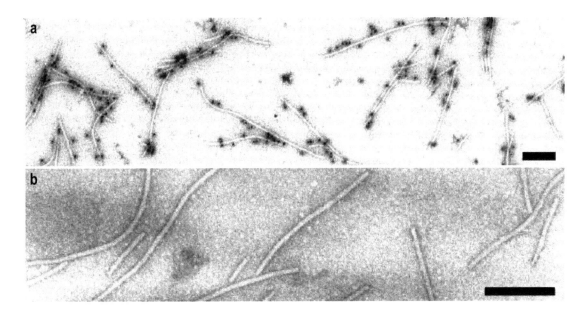

FIGURE 21.6   Immunogold labeling of PVX^epCP particles from infected *N. benthamiana* plants. (a) PVX^ep4CP decorated with mAb SCR86 specific for BNYVV epitope 4. (b) Negative control. (With kind permission from Springer Science+Business Media: *Arch Virol*, Expression of multiple foreign epitopes presented as synthetic antigens on the surface of Potato virus X particles, 166, 2005, 327–340, Uhde K, Fischer R, Commandeur U.)

deletions, predominantly affecting positively charged residues, was revealed (Uhde-Holzem et al. 2007).

Next, Uhde-Holzem et al. (2010) generated the chimeric PVX virions displaying the synthetic R9 peptide, a 27-aa consensus sequence derived from diverse variants of the hypervariable region 1 (HVR1) from the hepatitis C virus (HCV) envelope protein E2. Two different constructs were designed, with the R9 peptide expressed either as an indirect fusion via the ibosomal skip 2A of 16 aa or as a direct fusion. The systemic infection of *N. benthamiana* plants was only achieved with the construct harboring 2A element. Remarkably, the chimeric virions were recovered at yields of up to 125 mg/kg from leaf material. The immunization of mice induced specific anti-R9 IgG antibodies. Moreover, sera from patients infected chronically with HCV were found to react specifically with the chimeric virions (Uhde-Holzem et al. 2010).

In order to construct a putative tuberculosis vaccine, Zelada et al. (2006) expressed the complete ESAT-6 as a fusion protein with the 2A peptide at the N-terminus of the PVX coat. This strategy allowed the production of free coat and ESAT-6 as well as fused ESAT-2A-coat to obtain mosaic virions displaying ESAT-6 at the surface.

Marconi et al. (2006) fused two peptides chosen from glycoprotein E2, corresponding to aa 790–860 and aa 854–894, of classical swine fever virus (CSFV) of the *Flaviviridae* family of the order *Amarillovirales* (Chapter 22) to the N-terminus of the PVX coat via 2A peptide from FMDV and preceded by the His₆ tag. The relatively long, each >40 aa residues, peptide encoding sequences were correctly retained in the PVX construct after three sequential passages in *N. benthamiana* plants and were replicated with

high fidelity during PVX infection. The chimeric virions were able to induce an immune response in rabbits (Marconi et al. 2006).

Lico et al. (2009) displayed the H-2D^b-restricted CTL epitope 366-ASNENMETM-374 of influenza A virus nucleoprotein on the chimeric PVX virions. The latter activated the epitope-specific CD8⁺ T cells without adjuvant codelivery and supported therefore the idea to activate cell-mediated immune responses by the chimeric PVX virions.

Plchova et al. (2011) engineered fusions of the E7ggg both at the N- and C-termini of the PVX coat and evaluated the influence of the length of linker connecting PVX coat and E7ggg on their production. The constructs with no linker or 4-aa and 15-aa linkers were studied. The fusion proteins were successfully produced in plants, as well as in *E. coli* (see later). Although all constructs were expressed in plant cells approximately in the same amount and there were no substantial differences of expressed constructs considering the length of linkers, the authors were not able to determine if the constructs expressed in plants formed nanoparticles.

The HPV16 E7 epitope corresponding to aa residues 44–60, namely QAEPDRAHYNIVTFCCK, was used to investigate new positions for peptide presentation among seven putative surface loops of the PVX coat. To do this, the bacterial expression of 14 different PVX coats was performed, when the E7 epitope was fused with His₆ tag in different orientations in the 7 putative surface loops (Vaculik et al. 2015a).

Vaculik et al. (2015b) selected four of the tested positions to be evaluated in plants. The eight different PVX coat constructs with the E7 epitope and the His₆ tag in both orientations were expressed then transiently in *N. benthamiana* plants. Only the fusion site located after aa 3 led to systemic infection of plants and the production of recombinant proteins; however, no viral particles were detected. When the His$_6$ tag was replaced with StrepII tag, the modified virus infected plants systemically, expressed proteins assembled into viral particles, and the epitopes were located on the particle surface. It is noteworthy, however, that this novel position still belonged to the N-terminal intrinsically disordered domains of potexviruses (Solovyev and Makarov 2016; Röder et al. 2019) and was therefore not essential for particle assembly.

Next, the epitope derived from HPV-16 L2 minor capsid protein, aa 108–120, was expressed from PVX vector as N- or C-terminal fusion with the PVX coat in transgenic *N. benthamiana* plants (Cerovská et al. 2012; Hoffmeisterova et al. 2012). The N-terminal version—but not the C-terminal one—demonstrated great outcome from fresh leaf tissue, formed nanoparticles, and induced specific antibodies after immunization of mice by subcutaneous injection or tattoo administration.

Röder et al. (2017b) applied the SpyTag/SpyCatcher methodology, introduced by Zakeri et al. (2012) and described in detail in Chapter 25. Thus, the *Trichoderma reesei* endoglucanase Cel12A was covalently attached to the PVX nanoparticles. The PVX coat was modified to display the short SpyTag sequence of 14 aa residues, whereas the Cel12A enzyme of 220 aa residues was provided C-terminally with the SpyCatcher sequence of 129 aa over 10 aa of doubled G$_4$S linker. This allowed the rapid and specific irreversible attachment of the SpyCatcher fusion protein with, in this case, a ~70% coupling efficiency. The SpyTag-PVX construct therefore provided a universally applicable platform with great promise for future practice by overcoming problems of size constraints and inappropriate aa compositions that may influence the genetic engineering methods, as well as the chemical coupling methods described later.

Chimeric VLPs

In contrast to the PapMV, the PVX coat subunits have not yet been shown to assemble into filamentous VLPs in the absence of RNA either in vivo or in vitro, as reviewed by Röder et al. (2019). This was explained by the specific recognition of the virus genomic RNA by the coat, which plays a key role during the assembly of the virion, as established by Kwon et al. (2005). As a result, the final conclusion sounded that the PVX cannot assemble without its genomic RNA, and in vitro assembly did not achieve high yields of VLPs (Röder et al. 2019).

Nevertheless, some constructs cannot be regarded in frame of the chimeric virions. Thus, the fusions of the HPV epitopes to the PVX coat as a carrier occurred apart from the methodology of chimeric virions described earlier. In this case, the DNA vaccine was constructed by addition of the HPV16 E7ggg protein to the N-terminus of the PVX coat (Massa et al. 2008), as mentioned earlier.

Morgenfeld et al. (2009) expressed similar E7-PVX chimera in tobacco chloroplast and found that its expression levels were higher than that for unfused E7, indicating that the PVX coat stabilized E7 peptide in the chloroplast stroma.

Plchova et al. (2011) engineered fusions of the E7ggg both on N- and C-terminus of the PVX coat and evaluated the influence of the length of linker connecting PVX coat and E7ggg on their production. The constructs with no linker or 4-aa and 15-aa linkers were studied. The fusion proteins were successfully produced in *E. coli*, as well as in parallel in plants (as mentioned previously) and demonstrated their ability to form VLPs.

3D Structure

The potexviruses were early candidates for the advanced structural studies. Thus, Bernal and Fankuchen (1941) were the first who applied x-ray crystallography to PVX. The electron microscopy and diffraction studies have described a number of potexviruses, including PVX (Tollin et al. 1967, 1980), PapMV (Tollin et al. 1979), and narcissus mosaic virus (NMV; Tollin et al. 1975; Bancroft et al. 1980; Low et al. 1985). The architecture of all potexviruses was thought to be remarkably similar (Richardson et al. 1981).

In parallel, Kaftanova et al. (1975) performed the first successful polymerization of the PVX coat, when double-layer disks were produced and their aggregation into short, rod-like stacks was demonstrated.

Later, the analysis by fiber diffraction pattern has shown that the surface features of PVX were more flexible than those of TMV (Parker et al. 2002). Using fiber diffraction, electron cryomicroscopy, and scanning transmission electron microscopy, the low-resolution PVX structure was accomplished by Kendall et al. (2008). Nemykh et al. (2008) proposed a two-domain model explaining better the high plasticity of the PVX coat structure. Kendall et al. (2013) compared the fine structure of PVX with that of two other potexviruses: PapMV and NMV.

Generally, the flexuous rod-shaped particle of 515 × 14.5 nm comprised 1,270 coat subunits with 8.90 ± 0.01 subunits per turn, forming a 3.45 nm helical pitch (Tollin et al. 1980; Parker et al. 2002). Each coat subunit was thought to contain seven α-helices and six β-strands, with the C-terminus located inside the assembled particle and the N-terminus projected externally (Sõber et al. 1988; Baratova et al. 1992, 2004; Nemykh et al. 2008). The N-terminus therefore provided an excellent site for the presentation of recombinant peptides.

Arkhipenko et al. (2011) succeeded in the in vitro reconstitution of the PVX VLPs from viral coat and RNA including not only homologous but also foreign RNAs and showed that the morphology of particles containing heterologous RNA was identical to that of particles carrying homologous RNA. When heated to 90°C, the filamentous PVX particles formed spherical PVX VLPs (Nikitin et al. 2016), similar to TMV structures formed at higher temperatures (see Chapter 19). The average diameter of these spherical PVX particles was 48 and 77 nm at concentrations of 0.1 and 1.0 mg/ml, respectively.

At last, Grinzato et al. (2020) presented the electron cryomicroscopy structure of the PVX particle at a resolution of 2.2 Å. The well-defined density of the coat proteins and of the genomic RNA allowed a detailed analysis of protein–RNA interactions. The PVX virion is formed therefore by repeated segments made of 8.8 coat proteins, forming a left-handed helical structure. The RNA runs in an internal crevice along the virion, packaged in 5-nucleotide repeats. The resolution of the structure suggested a mechanism for the virion assembly and potentially provided a platform for the further use of PVX in nanotechnology. Figure 21.7 demonstrates the high-resolution structure and clearly illustrates why the N-terminal fusions of foreign sequences were displayed on the VLP surface, while the C-terminal fusions were not.

Chemical Coupling and Functionalization

Steinmetz et al. (2010) proclaimed PVX as a novel platform for bioconjugation protocols after the well-studied brome mosaic virus (BMV) and cowpea chlorotic mottle virus (CCMV; see Chapter 17), tobacco mosaic virus (see Chapter 19), and cowpea mosaic virus (CPMV; see Chapter 27) models. Such modifications included amine modification as well as so-called "click" chemistry or Cu^I-catalyzed azide-alkyne 1,3-dipolar cycloaddition (CuCAAC) reactions. As a result, this allowed the efficient functionalization of PVX with biotins, dyes, and polyethylene glycols (PEGs). The fluorescent-labeled and PEGylated PVX particles revealed that different fluorescent labels have a profound effect on PVX-cell interactions, therefore opening the door for chemical functionalization with targeting and therapeutic molecules (Steinmetz et al. 2010). The first success by the novel coupling and targeting model was operatively reviewed by the team leader Nicole Steinmetz (2010).

As calculated in the review of Röder et al. (2019), each PVX coat bears numerous amine and carboxylate groups among its 11 lysine, 10 aspartic acid, 10 glutamic acid, and 3 cysteine residues, although only a single lysine residue and a single cysteine residue are exposed to the solvent,

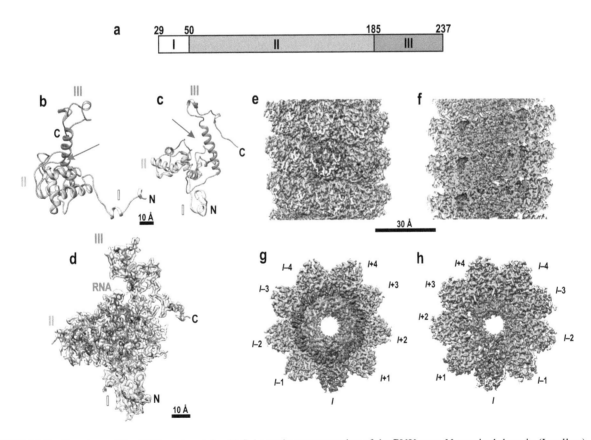

FIGURE 21.7   Structure of the PVX coat protein. (a) Schematic representation of the PVX coat, N-terminal domain (I, yellow), core domain (II, green), and C-terminal domain (III, blue). (b, c) Front (b) and side (c) views of the coat. The red arrows indicate the position of the crevice where the RNA binds. C, C-terminal end; N, N-terminal end. (d) The cryo-EM density around a coat monomer into which a single coat and a viral RNA fragment (orange sticks) have been fitted. (e, f) Surface representation of the cryo-EM 3D reconstruction: longitudinal front view (e) and cutaway view (f). (g, h) Close-up view of a section of nine consecutive CP protomers (*l*-4 to *l*+4) seen from top (g) and bottom (h). A single coat is highlighted in purple in (e, g, h). (With kind permission from Springer Science+Business Media: *Nat Chem Biol*, Atomic structure of potato virus X, the prototype of the *Alphaflexiviridae* family, 16, 2020, 564–569, Grinzato A, Kandiah E, Lico C, Betti C, Baschieri S, Zanotti G.)

making them addressable using N-hydroxysuccinimide and maleimide chemistry, respectively (Pierpoint 1974; Le et al., 2017a).

Wen et al. (2012) focused on methodical aspects of the chemical labeling of the PVX nanoparticles with fluorophores, such as Alexa Fluor 647 and polyethylene glycol (PEG), in parallel with other plant virus models: CPMV, BMV, and TMV.

Generating a potential vaccine against human epidermal growth factor receptor 2 (HER2)-positive cancers and trying to overcome immunological tolerance against HER2, Shukla et al. (2014c) conjugated the HER2 epitope P4$_{378-394}$ on the PVX particles at the solvent-exposed lysine side chains by bifunctional *N*-hydroxysuccinimide-maleimide linker (sulfo-SMCC). The carrier acted therefore as an adjuvant and improved stability and B-cell presentation of the epitope. The immunizations of the transgenic mice with the HER2-PVX nanoparticles resulted in the production of HER2-specific antibodies.

Furthermore, Jobsri et al. (2015) used nongenetic coupling of antigens to the surface of the PVX virions via biotin-streptavidin linkage and tested the ability of the PVX-based vaccine to induce antibody responses to a weak idiotypic (Id) tumor antigen. It was shown that not only was the Id-PVX conjugate vaccine superior to the DNA fusion vaccine, generated earlier by Savelyeva et al. (2001) and mentioned earlier, at induction of anti-Id antibody but it exceeded that of the gold standard Id- keyhole limpet hemocyanin conjugate vaccine (Jobsri et al. 2015).

Tumor Homing

Generally, nonspherical materials achieve better tumor homing and more efficient tumor penetration than spherical particles, as reviewed by Röder et al. (2019). Thus, Shukla et al. (2013) demonstrated enhanced tumor homing and tissue penetration of the filamentous PVX nanoparticles over icosahedral CPMV nanoparticles, due to the enhanced permeability and retention effect. Wen et al. (2013) provided more methodological details on the PVX and CPMV biodistribution and tumor-homing properties. Both models were decorated with fluorescent dyes for tracking and with PEG to reduce immunogenicity and nonspecific binding. The tumor-homing studies of the PEGylated PVX nanoparticles were carried out using avian embryo and athymic nude mouse tumor xenograft models of fibrosarcoma, squamous carcinoma, and colon and breast cancer. It was concluded that the filamentous rods demonstrated superior tumor-homing properties and enhanced transport across plasma membranes. In addition to the advantageous shape of PVX, it was also positively charged, what was regarded as a next advantage, whereas CPMV was negatively charged (Shukla et al. 2013, 2014b; Wen et al. 2013).

The non-PEGylated PVX particles were shown to adhere to red blood cells and penetrate the white pulp of the spleen (Lico et al. 2016). To increase the bioavailability and reduce the interaction with cells of the mononuclear phagocyte system (MPS), Lee et al. (2015) developed the PVX nanoparticles as PEGylated stealth filaments and evaluated the effects of PEG chain length and conformation on pharmacokinetics, biodistribution, and potential immune and inflammatory responses. The PEGylation effectively reduced immune recognition while increasing pharmacokinetic profiles. The stealth filaments showed reduced interaction with cells of the MPS and the protein: polymer hybrids were cleared from the body tissues within hours to days, indicating biodegradability and biocompatibility. Tailoring PEG chain length and conformation (brush vs. mushroom) allowed tuning of the pharmacokinetics, yielding long-circulating stealth filaments for applications in nanomedicine (Lee et al. 2015).

In parallel, Blandino et al. (2015) studied the fate and cytotoxicity of the filamentous PVX particles in hemolysis assays and early embryo assays, in parallel with the icosahedral tomato bushy stunt virus (TBSV) particles of the *Tolivirales* order (see Chapter 24). Furthermore, no evidence of apoptosis was observed when human mesenchymal stem cells were seeded onto a PVX-coated surface (Lauria et al. 2017). The PVX nanoparticles showed no sign of clinical toxicity at high concentrations and appeared promising for the targeted destruction of tumors.

Drug Delivery

Le et al. (2017b) loaded the PVX nanoparticles with doxorubicin due to the spontaneous hydrophobic interactions and π- π stacking of the planar drug molecules and polar aa residues in the surface grooves of the virus. Approximately 850–1,000 drug molecules were carried by an unmodified PVX particle, indicating that 70–80% of the coats become stably attached to the drug (Le et al. 2017b; Lee et al. 2017). The PEGylation of PVX increased its ability to carry doxorubicin, allowing the attachment of 1,000–1,500 drug molecules per particle (Le et al., 2017b).

Doxorubicin remained cytotoxic when loaded onto PVX, but its efficacy was lower than that of the free drug. However, the major drawback of this method was the need for a high molar excess of the drug and a long reaction time. No statistical differences in the tumor growth rate or survival time were observed when the PVX formulation was compared to the free drug, but the tumor volume was slightly lower in mice treated with the PVX formulation. Furthermore, the PVX formulation did not improve the treatment, but the cytotoxic efficacy was maintained (Röder et al. 2019).

Esfandiari et al. (2016) used the PVX nanoparticles to display the chemotherapeutic drug Herceptin used as a targeted therapy in HER2$^+$ breast cancer patients. The Herceptin (Trastuzumab) monoclonal antibody of 55 kDa was conjugated to the PVX by EDC/sulfo-N-hydroxysuccinimide (sulfo-NHS). The rate of cell death caused by PVX-Herceptin conjugate was significantly higher compared to that of Herceptin alone.

The binding affinity of PVX toward malignant B cells was used by the generation of a drug delivery system for non-Hodgkin's B-cell lymphomas (NHL; Shukla et al. 2020). In a metastatic mouse model of NHL, the systemically administered PVX homed to tissues harboring malignant B cells. When loaded with the chemotherapy monomethyl auristatin (MMAE), the PVX nanocarrier enabled effective delivery of MMAE to human B lymphoma cells in an NHL mouse model leading to inhibition of lymphoma growth in vivo and improved survival. Experimentally, the fluorescent PVX-Cy5 particles were synthesized by coupling NHS-Sulfo-Cy5 or Maleimide-Sulfo-Cy5 (Lumiprobe) to PVX via lysine or cysteine residues, respectively. The MMAE was conjugated to PVX via the sulfhydryl side chains on the cysteine residues using the maleimide chemistry (Shukla et al. 2020).

Imaging

As mentioned earlier, Santa Cruz et al. (1996) were the first who generated the mosaic PVX virions for optical imaging by displaying GFP on their surface.

Shukla et al. (2014a) engineered the PVX virions displaying green fluorescent protein (GFP) or mCherry as N-terminal coat protein fusions, using the FMDV 2A approach and producing a 1:3 fusion protein to coat ratio. These particles allowed the infection of *N. benthamiana* plants to be visualized clearly. The infection of plants with the recombinant GFP-PVX and mCherry-PVX particles was documented by fluorescence imaging, structural analysis, and genetic characterization to determine the stability of the chimeras and optimize the molecular farming protocols. Then, the fluorescent mCherry-PVX filaments were used as probes for optical imaging in human cancer cells and a preclinical mouse model. The cell viability assays and histological analysis following the administration of mCherry-PVX indicated the biocompatibility and rapid tissue clearance of the particles. The authors concluded that such particles could be functionalized with additional cancer-specific detection ligands to provide tools for molecular imaging, allowing the investigation of molecular signatures, disease progression/recurrence, and the efficacy of novel therapies (Shukla et al. 2014a).

To develop this approach further, Dickmeis et al. (2015) created a novel production system for the mosaic PVX particles, after the previously described 2A-based approach. The different polypeptides were displayed as the coat fusions when combinations of PVX and tobacco mosaic (TMV) expression vectors were used, each expressing different PVX coat fusions. To prove the principle of the assembly of mosaic chimeric PVX virions, the GFP and the red fluorescent protein mCherry were chosen, as well as a bimolecular fluorescence complementation (BiFC) system with split-mCherry as coat fusion proteins. The presence of assembled split-mCherry on the surface confirmed the mosaic character of the chimeric particles (Dickmeis et al. 2015).

Röder et al. (2018) fused the fluorescent protein iLOV of 113 aa residues directly to the PVX coat without impairing the assembly and systemic infection and movement of the virus. This was the largest coat fusion reported thus far, as reviewed by Röder et al. (2019).

The fluorescent dye OregonGreen 488 was conjugated to the PVX particles using both Lys-*N*-hydroxysuccinimide and Cys-maleimide chemistry, with the former achieving the best performance resulting in the modification of up to 15% of the coats (Le et al. 2017a). This may reflect the low accessibility of the Cys residue, which was thought to be located within a surface groove (Röder et al. 2019).

New Materials

The PVX particles are promising as a building block for hybrid organic–inorganic materials, as reviewed by Röder et al. (2019). Thus, the PVX particles were used to induce the deposition of silica, which could allow the development of new biomaterials with combined surface properties. The silica deposition on templates often involved the use of alkoxysilane precursors such as tetraethyl orthosilane, tetramethyl orthosilane, or (3-aminopropyl)triethoxysilane. The genetically modified PVX particles presenting the aa sequence YSDQPTQSSQRP fused to the N-terminus of the coat were able to promote mineralization with tetraethyl orthosilane at room temperature, allowing the development of hybrid materials with two or even three components designed using immunogold labeling (Van Rijn et al. 2015).

Drygin et al. (2013) reported the selective electroless deposition of platinum ions on one end of PVX particles with nucleation centers 1–2 nm in diameter.

PVX, among other plant viruses, offered new solutions for the biomedical application of biomaterials. Thus, the PVX virions displaying appropriate peptide sequences were applied as tailoring biomineralization in hydrogel-based bone tissue substitutes (Lauria et al. 2017). The PVX virions were engineered to display the RGD peptide, an integrin binding motif that promotes cell adhesion, either alone or in combination with the mineralization-inducing peptide (MIP) MSESDSSDSDSKS, revealing local high functional peptide concentrations on the virus' shell (Lauria et al. 2017). In contrast, the chemical conjugation of a cyclic RGD to coat resulted in 15% crosslinking efficiency (Shukla et al. 2015). It was hypothesized that these PVX nanoparticles would not only enhance cell adhesion mediated by RGD but also hydroxyapatite nucleation by MIP and serve therefore as an effective hydrogel-based component for bone tissue engineering, regeneration, and restoration.

Biosensing

The PVX coat was fused to the B domain of the *S. aureus* protein A to achieve efficient antibody capture and presentation on the particle surface (Uhde-Holzem et al. 2016). Thus, the protein A fragment of 60 aa residues retained its ability to immobilize antibodies when exposed on the PVX surface as a direct coat fusion. As a proof of concept, the integration of PVX[SpAB]CP particles was tested in sensing applications: PVX[SpAB]CP was immobilized on gold chips and implemented as virus sensor using quartz crystal microbalance detection.

The modified PVX particles were able to capture 300–500 antibodies per particle, which enhanced the available antibodies on the chip surface and allowed the sensitive detection of cowpea mosaic virus (CMV). In addition to sensing applications, the arrays could be used in the future to capture pollutants for cleanup or detoxification. Furthermore, when combined with medical payloads such as contrast agents or drugs, the particles could be used for molecular imaging and drug delivery (Röder et al. 2019).

The PVX nanoparticles were used to improve an ELISA for the diagnosis of primary Sjögren syndrome (Tinazzi et al. 2015). The PVX coat was genetically modified to display the immunodominant lipopeptide from lipocalin, which is involved in the pathogenesis of this autoimmune disease. The modified particles were used to coat ELISA plates for the analysis of patient serum samples and were compared to plates coated with the lipocalin peptide alone. The new ELISA demonstrated a remarkable sensitivity improvement, when compared to the old one using the synthetic peptide.

Catalysis

Carette et al. (2007) combined PVX with enzymes for the fabrication of novel biocatalysts. Thus, using the 2A approach, the authors generated the mosaic PVX particles decorated with lipase B, an enzyme derived from *Candida antarctica* of 33 kDa in size, which is known as an efficient enantiospecific catalyst for chemical hydrolysis reactions, because it specifically forms one of two possible mirror image products. The enzyme was fused N-terminally via the previously mentioned 2A stretch, and the chimeric virions were produced in *N. benthamiana* leaves. The virion-anchored enzyme of purified particles displayed catalytic activity toward the substrate *p*-nitrophenyl caproate in solution, their activity being reduced, however, by about 45 times compared with the activity of free enzyme, possibly because the coat fusion hindered substrate access to the active site. Remarkably, the authors speculated that the virus particles modified with different types of enzymes could be also obtainable, offering opportunities for the construction of defined cascade catalytic systems operating like the metabolons in nature (Carette et al. 2007).

As described earlier, Röder et al. (2017b) engineered the PVX nanoparticles displaying the *T. reesei* endoglucase Cel12A by SpyTag/SpyCatcher technique. The resulting VNP displayed ~850 enzyme molecules per PVX particle, and the retention of catalytic activity was confirmed by measuring kinetic parameters in the presence of different concentrations of 4-methylumbelliferyl-β-D-cellobioside. The affinity of the chimeric nanoparticles for the substrate was ~3.5-fold lower than that of the free enzyme, indicating that the scaffold interfered with substrate binding to some degree. However, the turnover rate k_{cat} and V_{max} of the chimeric particles were ~2.9-fold higher than that of the free enzyme, which might reflect ability of closely spaced enzymes on the 515-nm scaffold to facilitate hydrolysis (Röder et al. 2017b, 2019).

Other Potexviruses

By analogy with the well-studied PapMV VLPs, the recombinant VLPs of malva mosaic virus (MaMV), a related but distinct potexvirus, were produced in *E. coli* (Côté et al. 2008). Hanafi et al. (2010) generated in *E. coli* the chimeric MaMV VLPs that harbored the previously described CTL epitopes: HLA A*0201 epitopes derived from the gp100 melanoma antigen and influenza matrix protein M1 epitope, which were added at the C-terminal region of MaMV CP and flanked by five residues from the native antigen in the N- and C-termini of the epitope and followed by the C-terminal His$_6$ tag. Like the PapMV VLPs, the chimeric MaMV VLPs triggered a CTL response through antigenic presentation of epitopes on MHC class I. Remarkably, there was no cross-reactivity between IgG directed to the surface of PapMV and MaMV as evaluated by both ELISA and Western blot analysis, suggesting that the surfaces of both VLPs were distinct. Moreover, the PapMV and MaMV VLPs were internalized in different APCs (Hanafi et al. 2010).

The RNA-free VLPs of alternanthera mosaic virus (AltMV) were assembled in vitro into VLPs at pH 4.0 and low ionic strength, similarly to PapMV coats, when a considerable part of the AltMV VLPs stuck together, producing long bundles (Mukhamedzhanova et al. 2011). However, in contrast to PapMV, the AltMV VLPs were formed at pH 8.0 as well. Donchenko et al. (2017) resolved the 3D structures of AltMV virions and VLPs to ~13 Å by single-particle electron cryomicroscopy. Both structures possessed helical symmetry, although demonstrated some fold differences in the presence (virions) and absence of viral RNA (VLPs). This contrasted to the situation in the case of another filamentous virus, namely potato virus (PVY) of the order *Patatavirales*, as described in Chapter 29.

Recently, Thuenemann et al. (2021) achieved great efficiency of both AltMV and PapMV VLP production in *N. benthamiana* plants by use of a replicating PVX-based vector.

The near-atomic structure of the pepino mosaic virus (PepMV) virions was resolved by electron cryomicroscopy to 3.9 Å (Agirrezabala et al. 2015).

FAMILY *TYMOVIRIDAE*

Genus *Marafivirus*

Maize rayado fino virus (MRFV), a type member of the genus *Marafivirus* of 30 nm in diameter possesses T = 3 icosahedral symmetry and presents two components: empty shells and complete virions carrying the 6.3 kb genomic RNA. Both particles are composed of two serologically related, carboxy coterminal, coat proteins (CP) of apparent molecular mass 21–22 kDa (CP2) and 24–28 kDa (CP1) in a molar ratio of 3:1, respectively. CP1 contains a 37 aa N-terminal extension of CP2.

Hammond RW and Hammond J (2010) expressed both CP1 and CP2 in *E. coli*. The expression resulted in assembly of each capsid protein into VLPs, appearing in electron microscopy as stain-permeable (CP2) or stain-impermeable particles (CP1). The CP1 VLPs encapsidated bacterial 16S ribosomal RNA but not CP mRNA, while the CP2 VLPs encapsidated neither CP mRNA nor 16S ribosomal RNA. The simultaneous expression of both CP1 and CP2 in *E. coli* using a coexpression vector resulted in the assembly of VLPs, which were stain-impermeable and encapsidated CP mRNA. These results suggested that the N-terminal 37 aa residues of CP1, although not required for particle formation, may be involved in the assembly of complete virions and that the presence of both CP1 and CP2 in the particle is required for specific encapsidation of MRFV CP mRNA (Hammond RW and Hammond J 2010).

To provide an anchor for functional groups, Natilla and Hammond (2011) created three Cys-MRFV VLPs mutants by substituting several of the aa residues present on the shell of the wild-type VLPs. The resultant CP1 mutant genes were expressed in *N. benthamiana* and *N. tabacum* plants through a PVX-based vector and produced the Cys-MRFV VLPs. The thiol-selective reactivity of the Cys-MRFV VLPs was determined by reaction, under native conditions, with the thiol-specific chemical reagents fluorescein-5-maleimide and a biotinylation assay. The mutant, designated Cys 2-VLPs, displayed cysteines with a geometric arrangement of free thiolic groups. This mutant was cross-linked using NHS-PEG4-Maleimide to 17 (F) and 8 (HN) aa-long synthetic peptides, corresponding to the neutralizing epitopes of Newcastle disease virus (NDV). The resultant Cys 2-VLPs-F and Cys 2-VLPs-HN were immunoreactive with MRFV antibodies as well as with antibodies specific to F and HN. The results demonstrated that cysteine thiol groups of Cys 2-VLPs clearly bound the F and HN peptides, thereby showing the ability of the plant-produced MRFV VLPs to function as a novel platform for the multivalent display of surface ligands (Natilla and Hammond 2011). The methodological details of the MRFV engineering were presented as a step-by-step video demonstration (Natilla and Hammond 2013).

In an effort to design reliable and programmable MRFV VLPs, in which a single particle could serve dual functions for molecular display and for delivery of cargo, Natilla et al. (2015) targeted the N-terminus of CP1 for genetic modification. The N-terminus of the MRFV CP1 of 37 aa, rich in hydrophobic residues, facilitating CP-RNA interactions and nonessential for self-assembly, was predicted to have an α-helical structure. The aa substitutions were introduced in the 37 aa N-terminus by site-directed mutagenesis, and the mutant VLPs produced in plants by a potato virus X (PVX)-based vector were tested for particle stability and RNA encapsidation. All mutant CPs resulted in production of VLPs, which encapsidated nonviral RNAs, including PVX genomic and subgenomic RNAs, 18S rRNA, and cellular and viral mRNAs. In addition, the MRFV-VLPs encapsidated GFP mRNA when expressed in plant

cells. The authors concluded that the RNA packaging by the MRFV VLPs is predominantly driven by electrostatic interactions between the N-terminal 37 aa extension of CP1 and RNA and that the overall species concentration of RNA in the cellular pool may determine the abundance and species of the RNAs packaged into the VLPs. Furthermore, the RNA encapsidation was not required for the stability of VLPs. Remarkably, the VLPs formed by the MRFV CP1 were stable at temperatures up to 70°C and can be disassembled into CP monomers, which can then reassemble in vitro into complete VLPs either in the absence or presence of RNAs (Natilla et al. 2015).

To achieve mass production of VLPs, Moon et al. (2014) presented a technique for highly efficient expression of the synthetic MRFV CP2 gene in in *N. benthamiana* and production and purification of self-assembled VLPs by transient expression system using agroinfiltration. In parallel, the VLPs of brome mosaic virus (BMV) and cucumber mosaic virus (CuMV) of the *Bromoviridae* family, order *Martellivirales*, (see Chapter 17), were obtained.

Genus *Tymovirus*

Physalis Mottle Virus

Sastri et al. (1997) overexpressed the physalis mottle virus (PhMV) coat gene in *E. coli* and found excellent VLPs by electron microscopy. To establish the role of N- and C-terminal regions in the assembly, two N-terminal deletions lacking the first 11 and 26 aa residues and two C-terminal deletions lacking the last 5 and 10 aa residues were constructed and overexpressed. The variants lacking N-terminal 11 (PhCPN1) and 26 (PhCPN2) aa residues self-assembled into T = 3 capsids in vivo and were as stable as the original VLPs and encapsidated a small amount of mRNA. In contrast, both C-terminal deletion variants were present only in the insoluble fraction and could not assemble into capsids (Sastri et al. 1997).

To delineate the role of specific aa residues in the PhMV assembly, Sastri et al. (1999) expressed in *E. coli* the N-terminal 30, 34, 35, and 39 aa deletion variants, a single C-terminal (N188) deletion mutant, as well as site-specific mutants H69A, C75A, W96A, D144N, D144N-T151A, K143E, and N188A. The variant lacking 30 aa residues from the N-terminus self-assembled well, while deletions of 34, 35, and 39 aa residues resulted in the insoluble products. Surprisingly, when a foreign sequence of 41 aa, carrying a His tag and originated from vector, was added N-terminally to the PhMV coat, the chimeric VLPs assembled well into the T = 3 particles that were more compact and had a smaller diameter. In contrast to the flexible N-terminus, the deletion of even a single residue from the C terminus (N188) resulted in capsids that were unstable and disassembled to a discrete intermediate. However, the replacement of C-terminal N188 by alanine led to the formation of stable capsids (Sastri et al. 1999). These studies of the pathways and intermediates of the PhMV assembly in *E. coli* were continued by Umashankar et al. (2003), who identified six interfacial aa residues that were critical for both subunit folding and particle assembly.

At the same time, the 3D structure of the PhMV was determined by x-ray crystallography to 3.8 Å resolution (Krishna et al. 1999). Some coordinates of the related turnip yellow mosaic virus (TYMV; described later) with which PhMV coat protein shares 32% aa sequence identity were used for obtaining the initial phases. An analysis of the interfacial contacts between protein subunits indicated that the hexamers were held more strongly than pentamers, and hexamer-hexamer contacts were more extensive than pentamer-hexamer contacts. Krishna et al. (2001) resolved the 3D structure of the empty *E. coli*-produced PhMV VLPs to 3.20 Å. These two structures represented the first case, when the T = 3 virus was provided with both full and empty crystal capsid structures. The structures clearly showed that the empty shells corresponded to a "swollen state" of the virus with increased disorder of the N-terminal segments as well as some positively charged sidechains facing the interior of the virus. Figure 21.8 presents portraits of both

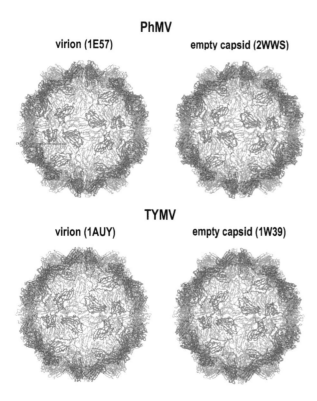

PhMV

virion (1E57)   empty capsid (2WWS)

TYMV

virion (1AUY)   empty capsid (1W39)

FIGURE 21.8 The crystal structure of the *Tymoviridae* family members, genus *Timovirus*. (Top) Physalis mottle virus (PhMV) virions (Krishna et al. 1999, 2001) and empty capsids (Krishna et al. 2001; Sagurthi et al. 2009). (Bottom) Turnip yellow mosaic virus (TYMV): virions (Canady et al. 1996) and empty capsids (van Roon et al. 2004). The 3D structures are taken from the VIPERdb (http://viperdb.scripps.edu) database (Carrillo-Tripp et al. 2009). The size of particles is in scale. The corresponding protein data bank (PDB) ID numbers are given under the appropriate virus names.

PhMV variants, compared with the analogous TYMV (see Figure 21.8 pictures).

Hema et al. (2007) started the novel era of the chimeric PhMV VLP derivatives. Thus, they substituted the N-terminus of the PhMV coat with single or tandem repeats of the B-cell epitopes of the foot-and-mouth disease virus (FMDV) nonstructural proteins (NSPs) 3B1, 3B2, 3AB, 3D, and 3ABD of lengths 48, 66, 49, 51, and 55 aa, respectively. The expression of these constructs in *E. coli* resulted in the VLPs. Summarizing, the authors were able to delete and add up to 66 aa at the N-terminus in the independent constructs without affecting the assembly process. Furthermore, the presentation of the NSP epitopes in tandem repeated with short GGS linkers gave scope to display one or more epitopes on the surface of the VLPs. The latter demonstrated expected antigenicity when used in ELISA tests detecting the FMDV NSP antibodies in field serum samples from cattle, buffalos, sheep, goats, and pigs for the differentiation of FMDV-infected animals from vaccinated animals (DIVA; Hema et al. 2007).

Chandran et al. (2009) used the PhMV VLP model to display the neutralizing epitopes of canine parvovirus (CPV) and a T-cell epitope of the fusion protein of canine distemper virus (CDV). Again, the aa substitutions were made at the N-terminus of the PhMV coat with the neutralizing epitopes and the T-cell epitope in various combinations to yield five chimeric constructs that were expressed in *E. coli* and self-assembled into excellent VLPs. The immunogenicity studies in guinea pigs and dogs resulted in the selection of variant that could be used in dogs to generate a protective immune response against diseases caused by both CPV and CDV (Chandran et al. 2009).

Shahana et al. (2015) continued the successful PhMV VLP story and achieved in *E. coli* an efficient production of the chimeric particles displaying immunodominant epitopes of Japanese encephalitis virus (JEV) envelope protein, although in this case the chimeric proteins were purified from the inclusion bodies and refolded in vitro. As a result, the chimeric protein carrying a stretch of 42 aa of the JEV epitopes at the N-terminus assembled efficiently into the VLPs that induced neutralizing antibodies in mice (Shahana et al. 2015).

Sahithi et al. (2019) replaced the N-terminus of the PhMV coat with tandem repeats of neutralizing epitopes of the infectious bursal disease virus (IBDV) protein VP2 and expressed three different chimeras in *E. coli*. The IBDV VP2 epitopes were represented by stretches of aa 207–225, 210–225, 312–324, and 329–337 in various combinations and separated by GGS linkers. All the three chimeric coat protein subunits were soluble in bacteria and self-assembled into VLPs that were applied in ELISA test as coating antigen to detect IBDV antibodies and intended as a possible vaccine candidate (Sahithi et al. 2019).

Starting a novel turn of the VLP functionalization, imaging, and drug delivery, Masarapu et al. (2017) carried out external and internal surface modification of the *E. coli*-derived PhMV VLPs with fluorophores using reactive lysine-*N*-hydroxysuccinimide ester and cysteine-maleimide chemistries, respectively. The uptake of dye-labeled particles was tested in a range of cancer cells and monitored by confocal microscopy and flow cytometry. The VLPs labeled internally on cysteine residues were taken up with high efficiency by several cancer cell lines, whereas the VLPs labeled externally on lysine residues were taken up with lower efficiency, probably reflecting differences in surface charge and the propensity to bind to the cell surface. The infusion of dye and drug molecules into the cavity of the VLPs revealed that the photosensitizer (PS), Zn-EpPor, and the drugs crystal violet, mitoxantrone (MTX), and doxorubicin (DOX) associated stably with the carrier via noncovalent interactions. The cytotoxicity of the PS-PhMV and DOX-PhMV particles was confirmed against prostate cancer, ovarian cancer, and breast cancer cell lines, respectively. Therefore, the PhMV-derived VLPs provided a new promising platform for the delivery of imaging agents and drugs, with preferential uptake into cancer cells (Masarapu et al. 2017).

Hu et al. (2019) developed the PhMV VLPs as bimodal contrast agents to achieve long circulation, specific targeting capability, and efficient delivery to tumors in vivo. After loading the internal cavity of the particles with the fluorescent dye Cy5.5 and paramagnetic Gd^{III} complexes, the outer surface was modified by PEGylation and conjugation with targeting peptides. Using this combined approach, the authors monitored a human prostate tumor model for up to ten days by near-infrared fluorescence and magnetic resonance imaging.

Hu and Steinmetz (2020b) generated a drug delivery system by loading the prodrug 6-maleimidocaproyl-hydrazone doxorubicin (DOX-EMCH) into the PhMV VLPs via a combination of chemical conjugation to cysteine residues and π–π stacking interactions with the anchored doxorubicin molecule. The DOX-EMCH prodrug featured an acid-sensitive hydrazine linker that triggered the release of doxorubicin in the slightly acidic extracellular tumor microenvironment or acidic endosomal or lysosomal compartments following cellular uptake. The VLP external surface was PEGylated to prevent nonspecific uptake and improve biocompatibility. The DOX-PhMV-PEG particles showed significantly greater efficacy in vivo compared to free doxorubicin in a breast tumor mouse model (Hu and Steinmetz 2020b).

Hu and Steinmetz (2020a) conjugated a maleimide-functionalized cisplatin prodrug containing Pt^{IV} to the internal and/or external surface of VLPs to develop a pH-sensitive drug delivery system (DDS). The internally loaded and PEGylated VLPs (Pt-PhMVCy5.5-PEG) were taken up efficiently by cancer cells where they released platinum, presumably as a reduced, DNA-reactive Pt^{II} complex, rapidly under acidic conditions in vitro. The efficacy of the VLP-based DDS was demonstrated against a panel of cancer cell lines, including cell lines resistant to platinum therapy. Furthermore, the Pt-PhMVCy5.5-PEG successfully inhibited the growth of xenograft MDA-MB-231 breast tumors

in vivo and significantly prolonged the survival of mice compared to free cisplatin and cisplatin-maleimide (Hu and Steinmetz 2020a).

At last, Hu and Steinmetz (2021) used the PhMV platform to develop the HER2-specific cancer vaccine, where the HER2-derived CH401 peptide epitopes were loaded into the interior cavity of the VLPs. The PhMV-based anti-HER2 vaccine demonstrated good efficacy in a mouse model after subcutaneous administration as an anti-HER2 cancer vaccine (Hu and Steinmetz 2021).

Tomato Blistering Mosaic Virus

Vasques et al. (2019) employed the baculovirus expression vector system (BEVS) for the production of tomato blistering mosaic virus (ToBMV) VLPs in insect *Trichoplusia ni* cells. The ToBMV gene was expressed in two variants, as a wild-type coat (CP) or a modified short N-terminal deletion (Δ2–24CP) variant, and both versions were able to self-assemble into VLPs. Therefore, the N-terminal aa residues 2–24 were shown not to be essential for the VLP self-assembly. When the major epitope of the chikungunya virus (CHIKV) envelope protein 2 (E2) of 25 aa residues, namely QRRSTKDNFNVYKATRPYLAHCPDC, was inserted instead of the intrinsic N-terminal 2–24 aa, the VLPs were produced.

Turnip Yellow Mosaic Virus

Without any doubts, turnip yellow mosaic virus (TYMV), a type member of tymoviruses, belongs to the early birds of the structural investigations. Thus, Cosslett and Markham (1948) were the first who published electron micrographs of TYMV. The Nobelist 1982 Sir Aaron Klug et al. (1957a, b) performed the first diffraction x-ray TYMV studies. At last, Canady et al. (1996) resolved the structure to 3.2 Å resolution. The crystal structure of the TYMV empty capsids was resolved to 3.75 Å (van Roon et al. 2004). Remarkably, the empty TYMV capsids were prepared from virions by a freeze–thaw method, in contrast to the empty PhMV capsids that were *E. coli*-produced recombinant VLPs, as described earlier. Such empty capsids were isolated naturally from the host plant or generated artificially by treatment under pressure (Leimkühler et al. 2001), basic environment (Kaper 1964), or repeating freeze–thaw process (Katouzian-Safadi et al. 1980). Figure 21.8 presents portraits of both full and empty TYMV variants, compared with the analogous PhMV pictures.

It is noteworthy that the crystal structure of desmodium yellow mottle virus (DYMoV) was resolved to 2.7 Å (Larson et al. 2000), a third 3D-characterized tymovirus after TYMV and PhMV.

Hayden et al. (1998) were the first who engineered chimeric TYMV virions. The authors designed live recombinants of TYMV in which selected parts of its coat were replaced with homologous regions of belladonna mottle virus (BeMV), another tymovirus, in a cDNA clone encoding the genome of TYMV. Six of ten such recombinants were fully viable, and most gave symptoms in Chinese cabbage indistinguishable from those of TYMV, although they did not always infect plants and noninfected hosts of BeMV or of other tymoviruses systemically. Remarkably, a TYMV recombinant with the N-terminal part of its coat replaced with the E71 epitope of *Plasmodium falciparum* was also viable, but others with the same region replaced with the V3 region of HIV-1 were not. The authors concluded that the N-terminus of the coat was not exposed at the surface of the virion, although it was immunogenically dominant (Hayden et al. 1998).

Qian Wang's team converted TYMV into promising building blocks for novel nanomaterials.

Thus, Barnhill et al. (2007b) employed the conventional *N*-hydroxysuccinimide-mediated amidation reaction for the chemical modification of the viral capsid. It was found that the amino groups of K32 of the flexible N-terminus made the major contribution for the reactivity of TYMV toward *N*-hydroxysuccinimide ester (NHS) reagents, where approximately 60 lysines per particle were addressed, whereas other lysines in various locations were less reactive. Therefore, in contrast to the previously referenced opinion, the TYMV coat protruded its N-terminal region housing the reactive K32, thus presenting that particular amino group in a more solvent-accessible state than the other lysines. In addition, about 90 to 120 carboxyl groups, located in the most exposed sequence, were modified with amines catalyzed with 1-(3-dimethylaminopropyl-3-ethylcarbodiimide) hydrochloride (EDC) and sulfo-NHS. TYMV was stable to a wide range of reaction conditions and maintained its integrity after the chemical conjugations. As a result, TYMV was loaded with up to 40 fluorescein or *N,N,N',N'*-tetramethylrhodamine molecules via an amidation reaction, which was equal to a local concentration of 4.6 mM of dyes surrounding the virus with no observed fluorescence quenching (Barnhill et al. 2007b).

Moreover, TYMV worked as a prototype protein scaffold for sensor development because the fluorophore was anchored via a specific ligand-receptor, namely biotin-avidin, binding (Barnhill et al. 2007a). As a result, the close proximity of orthogonal addressable amino and carboxylic groups was demonstrated by dual labeling with an NHS-activated terbium complex as donor and Alexa-488 fluorophore as acceptor by time-resolved fluoroimmuno assay.

TYMV was used as a prototype for the interfacial assembly of bionanoparticles at the oil/water (O/W) interface (Kaur et al. 2009). By some conditions, the highly ordered, hexagonal arrays of TYMV were obtained at planar O/W interfaces.

Li et al. (2009) used TYMV to synthesize bio-colloidal composite based on the noncovalent interactions with poly(4-vinylpyridine) (P4VP). TYMV particles fully covered the surface of a P4VP ball with a hexagon-like packing. The raspberry-like morphology of TYMV-P4VP colloids and the packing pattern of TYMV were revealed by numerous physical methods. The size of TYMVP4VP colloids was controlled readily by varying the mass ratio of virus and polymer.

Zan et al. (2012) developed TYMV as a scaffold for bioactive peptide display and cell culturing. The authors genetically displayed RGD motif on the coat protein of live infectious TYMV virions. The composite films composed of either wild-type TYMV or TYMV-RGD44, in combination with poly(allylamine hydrochloride) (PAH), were fabricated by a layer-by-layer adsorption of virus and PAH. The bone-marrow stem cells showed enhanced cell adhesion and spreading on TYMV-RGD44-coated substrates compared to native TYMV.

Kim et al. (2018) demonstrated a clear potential of modified TYMV as an efficient system for therapeutic cargo delivery to mammalian cells. The authors provided TYMV with Tat, a cell penetrating peptide, and demonstrated ability of such particles to enter animal cells, namely, baby hamster kidney (BHK) cells. The Tat peptide was chemically attached to the surface lysine residues of TYMV using hydrazone chemistry. The Tat conjugation was more efficient than lipofectamine in allowing TYMV to enter the cells, and the Tat-assisted transfection was associated with less loss of cell viability than lipofection. Among the cell-penetrating peptides tested (Tat, R8, Pep-1 and Pen), the R8 and Pen were also effective while Pep-1 was not. The internal space of TYMV was loaded with fluorescein dye as a model cargo, after viral RNA was removed by freezing and thawing. When the resultant empty particles were reacted with fluorescein-5-maleimide using interior sulfhydryl groups as conjugation sites, about 145 fluorescein molecules were added per particle. The fluorescein-loaded TYMV particles were conjugated with Tat and introduced into BHK cells, again with higher transfection efficiency compared to lipofection (Kim et al. 2018).

After a lengthy silence at the front of heterologous expression, the TYMV coat gene was expressed in *E. coli* and good VLPs were obtained (Powell et al. 2012). By some analogy with the PhMV coat, the authors generated the N-terminal deletion variants—namely Δ2–5, Δ2–10, and Δ2–26—and got VLPs in all cases in *E. coli* and, in parallel, infectious virus in plants. It was concluded that the TYMV virions can support productive infections despite the absence of up to 25 aa residues from the N-terminus. Nevertheless, these sequences do contribute importantly to viral fitness since all deletion mutants produced attenuated infections and less stable virions.

At last, Tan et al. (2021) produced in *E. coli* the His$_6$-tagged TYMV VLPs of 30–32 nm in size and highly stable over a wide pH range from 3.0 to 11.0 at different temperatures. The C-terminally added GSRSH$_6$ peptide was exposed on the VLP surface and could be exploited therefore as a prospective site to display functional ligands.

22 Order *Amarillovirales*

It is said that the present is pregnant with the future.

Voltaire

ESSENTIALS

The *Amarillovirales* is an order of small enveloped viruses that have an extremely bad reputation in public health, while causing serious infectious illnesses in human. According to the current taxonomy (Simmonds et al. 2012, 2017; ICTV 2020), the order *Amarillovirales* is a single member of the class *Flasuviricetes*, belonging to the *Kitrinoviricota* phylum from the kingdom *Orthornavirae*, realm *Riboviria*.

The *Amarillovirales* order currently involves one family, namely *Flaviviridae*, 4 genera, and 89 species. The genera are *Flavivirus*, *Hepacivirus*, *Pegivirus*, and *Pestivirus*. Most infect mammals and birds. Many of them are host-specific and highly pathogenic, such as hepatitis C virus (HCV) of the genus *Hepacivirus*. After HCV, a major human pathogen-causing progressive liver disease (Yu ML and Chuang 2021), the *Hepacivirus* genus involves several other viruses of unknown pathogenicity that infect horses, rodents, bats, cows, and primates (Scheel et al. 2015).

The majority of known members in the genus *Flavivirus* are arthropod borne, and many are important human pathogens such as highly dangerous dengue (DENV serotypes 1–4), Japanese encephalitis (JEV), tick-borne encephalitis (TBEV), West Nile (WNV), yellow fever (YFV), and Zika (ZIKV) viruses. Among the flaviviruses, DENV-1–4, WNV, and JEV may cause millions of infections and tens of thousands of deaths each year. Mammals and birds are the usual primary hosts, in which infections range from asymptomatic to severe or fatal hemorrhagic fever or neurological disease. Other members of the *Flavivirus* genus cause economically important diseases in domestic or wild animals, as reviewed by Simmonds et al. (2012, 2017).

Figure 22.1 presents portraits of the two typical representatives of the order and schematic cartoons of the *Flavivirus* and *Hepacivirus* genera members. The *Flaviviridae* members are enveloped, 40–60 nm virions with a single core protein (except for the *Pegivirus* genus) and two (genera *Flavivirus*, *Hepacivirus* and *Pegivirus*) or three (genus *Pestivirus*) envelope glycoproteins and possess the T = 3-like icosahedral symmetry organization.

By the genus *Flavivirus* members, the newly synthesized RNA and C protein are packaged by premembrane (prM) and E proteins to assemble into immature virions. The latter bud into the endoplasmic reticulum for glycosylation. Then, the immature virions are transported through the trans-Golgi network, where prM is cleaved by the protease furin into pr peptide and M protein, and the trimeric prM-E are rearranged to dimeric M-E heterodimers, thus forming the mature virions. The protein E and its immunoglobulin-like domain III (EDIII) containing the receptor-binding site and the major type-specific neutralization epitopes are regarded as the most promising components of a putative vaccine. By hepaciviruses, the envelope is formed by a dimer of E1 and E2 proteins.

Because of the superior medicinal importance of practically all members of the *Amarillovirales* order, it is not surprising that the goals of the VLP studies were targeted mostly to putative vaccine candidates, as well as to diagnostic tools, although to a lesser extent. The numerous thorough reviews disclose the vital role of VLPs by the generation of vaccines against flaviviruses (Krol et al. 2019; Wong et al. 2019; Araujo et al. 2020; Zhang N et al. 2020) and hepatitis C virus (Masavuli et al. 2017; Torresi 2017; Sepulveda-Crespo et al. 2020).

GENOME

Figure 22.2 shows the genomic structure of the two most representative genera of the *Flaviviridae* family. Generally, the *Flaviviridae* genomes are positive-stranded, nonsegmented RNAs of approximately 9.2–11, 8.9–10.5, 12.3–13, and 8.9–11.3 kb for members of the genera *Flavivirus*, *Hepacivirus*, *Pestivirus*, and *Pegivirus*, respectively. The genomes encode a single, long ORF flanked by 5'- and 3'-terminal noncoding regions, which form specific secondary structures required for genome replication and translation. The translational initiation of genomic RNA is cap-dependent in the case of members of the genus *Flavivirus*, whereas internal ribosome entry site elements are present in members of the other genera (Simmonds et al. 2012, 2017).

GENUS *FLAVIVIRUS*

SINGLE-ROUND INFECTIOUS VLPS

A common characteristic of the flavivirus surface proteins prM/M and E is their ability to assemble into so-called subviral particles, which are formed in variable amounts as a byproduct of flavivirus infection. These subviral particles are smaller than infectious virions, have a lower sedimentation coefficient than whole virions (70S vs. 200S), and do not contain a nucleocapsid but exhibit hemagglutination activity (Simmonds et al. 2012). The existence of such natural subviral particles triggered the idea of the engineered VLPs that will be described later. And these are the particles that will be defined throughout this chapter as the VLPs.

However, to avoid further term confusion, the specific problem of the flaviviral subgenomic replicons, so-called packaging cell lines, and single-round infectious VLPs

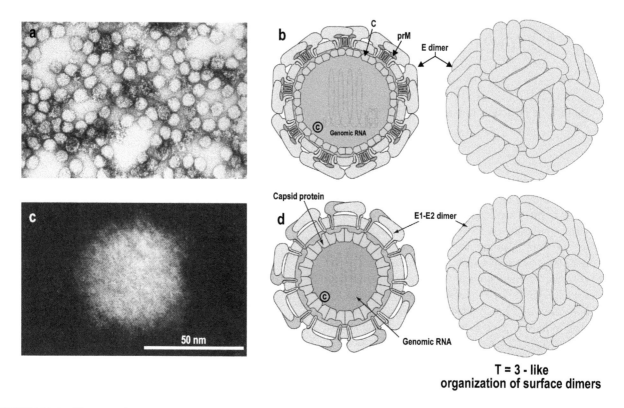

FIGURE 22.1 The portrait and a schematic cartoon of the typical representatives of the *Amarillovirales* order. (a) This 1981 transmission electron-microscopic image of yellow fever virus particles, provided by Erskine and Palmer, is taken from the CDC Public Health Image Library (PHIL). The virions are spheroidal, uniform in shape and are 40–60 nm in diameter. (b) The general cartoon of the genus *Flavivirus* members. (c) The electron micrograph of 55 to 65 nm hepatitis C virus with fine 6 nm spike-like projections from plasma sample. (Reprinted with permission of Microbiology Society from Kaito M et al. *J Gen Virol.* 1994;75:1755–1760.) (d) The cartoon of hepatitis C virus. Both cartoons are taken with kind permission from the ViralZone, Swiss Institute of Bioinformatics, http://viralzone. expasy.org (Hulo C et al. 2011). (Courtesy Philippe Le Mercier.)

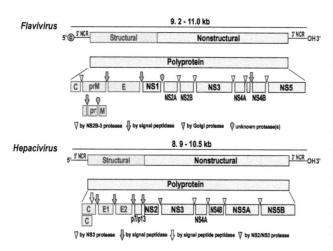

FIGURE 22.2 The genomic structure of the genera *Flavivirus* and *Hepacivirus* members from the *Flaviviridae* family, derived from the current ICTV profile (Simmonds et al. 2012, 2017). The host and viral proteases involved in cleavage of the polyprotein are indicated. The structural genes are colored dark pink.

(sriVLPs) should be clarified from the very beginning. By these three basic concepts, flaviviruses bear a remarkable

resemblance to alphaviruses of the *Togaviridae* family from the order *Martellivirales* described in Chapter 18.

Thus, for flaviviruses, the subgenomic replicons have been constructed by deleting genes for viral structural proteins (for a short, early review, see Khromykh et al. 2000). These subgenomic replicons replicated efficiently in cultured cells by virtue of functioning nonstructural proteins, but they did not produce progeny infectious viruses owing to the lack of viral structural proteins. The expression of viral structural proteins in cells harboring such replicon RNA has resulted in the secretion of particles, which have been termed VLPs by their authors. In contrast to the VLPs in our understanding, this sort of VLPs contained subgenomic replicons and induced the single-round replication by infection of susceptible cells. However, as viral structural proteins were not encoded by the subviral replicon, progeny virions were not produced, enabling safe handling under biosafety level 2 conditions. To avoid confusion with the noninfectious recombinant VLPs in our understanding, which will be described later and which are named sometimes also subviral particles by analogy with natural subviral particles typical for the flavivirus infection and mentioned earlier, the single-round infectious VLPs will be termed here *sriVLPs*.

On the other hand, the cells ensuring stable expression of the subgenomic replicons carrying structural genes, the prM and E, in the absence of the gene C, were engineered and termed packaging cells. In this case, the production of the sriVLPs appeared to be initiated by the in trans supply of the subgenomic replicons lacking structural proteins. Without the in trans supply, the packaging cell lines produced the previously mentioned subviral particles (or VLPs in our interpretation) in the same way as the traditional VLP production systems described later for each species of flavivirus.

The replicons derived from Kunjin virus (KUNV), a close relative of WNV and a member of the *West Nile virus* species, represent the most extensively characterized system of the subviral replicon subject, due to the pioneering efforts of Alexander A. Khromykh's team. The KUNV replicon system has been used successfully both to study flavivirus replication and to develop a promising new tool for heterologous gene expression (Khromykh and Westaway 1997; Khromykh et al. 1998, 2000; Varnavski and Khromykh 1999; Varnavski et al. 2000).

The first demonstration of the packaging of flavivirus RNA in trans was performed by Khromykh et al. (1998), who achieved successful packaging of KUNV replicon RNA when a Semliki Forest virus (SFV) replicon RNA expressing KUNV structural proteins was electroporated at least 12 h after electroporation of the KUNV replicon RNA. A single recombinant SFV RNA expressing the KUNV prM-E and KUNV C proteins together but under control of two separate 26S promoters was more efficient in packaging experiments than were two SFV RNAs expressing KUNV prM-E and KUNV C separately. As a result, the secreted sriVLPs were uniformly spherical with a ~35-nm diameter and contained the replication-competent KUNV RNA encapsidated by the KUNV structural proteins C, prM/M, and E (Khromykh et al. 1998).

To facilitate further applications of the KUNV replicon system in the form of the sriVLPs, Harvey et al. (2004) generated a stable BHK packaging cell line carrying the KUNV structural gene C-prM-E cassette under the control of a tetracycline-inducible promoter. Withdrawal of tetracycline from the medium resulted in production of the KUNV structural proteins that were capable of packaging transfected and self-amplified KUNV replicon RNAs into the secreted sriVLPs. The passaging of the sriVLPs on Vero cells or intracerebral injection into suckling mice illustrated the complete absence of any infectious KUNV virions. The KUNV packaging cells were also capable of packaging replicon RNAs from closely and distantly related flaviviruses, WNV and DEN-2, respectively (Harvey et al. 2004).

Scholle et al. (2004) invented a *trans*-packaging system for the WNV replicon RNAs, in which the entire WNV structural coding region encompassing C, prM, and E was expressed from the 26S subgenomic RNA promoter of a noncytopathic Sindbis virus (SINV) replicon.

Gehrke et al. (2003) constructed the first appropriate packaging cell line for TBEV. Thus, a stable CHO cell line constitutively expressing the two structural TBEV genes prM/M and E was generated and shown to efficiently export mature recombinant subviral particles (or VLPs by our terminology). When a subviral replicon lacking the prM/M and E genes was introduced into the packaging cells, the sriVLPs capable of initiating a single round of infection were released. The sedimentation analysis revealed that the sriVLPs were physically distinct from subviral particles (VLPs in our terms) and were similar to infectious virions. The sriVLPs were repeatedly passaged in the packaging cells but maintained the property of being able to initiate only a single round of infection in other cells during these passages (Gehrke et al. 2003).

Yoshii et al. (2005) developed the TBEV packaging cells using the expression of the viral C-prM-E genes in trans. The C-prM-E processing was carried out correctly by nonstructural proteins that were produced from the replicon, and replicon RNAs were incorporated into the sriVLPs without recombination between the replicon RNA and the mRNAs for the structural proteins.

Developing further this approach, Yoshii et al. (2008) applied their *trans*-packaging system to the generation of the sriVLPs with the JEV structural proteins. Although *trans*-expression of the TBEV C and JEV prM-E genes resulted in the secretion of sriVLPs, the expression of JEV C-prM-E genes did not lead to the secretion of the sriVLPs, suggesting that homologous interaction between C and nonstructural proteins or the genomic RNA was important for efficient assembly of infectious particles.

It should be particularly stressed that the sriVLPs of flaviviruses, in contrast to the subviral particles, were similar in all cases to the native virions in terms of their size and biochemical and physical features, as well as by functional characteristics for infection.

DENGUE VIRUS

Using electron cryomicroscopy and image reconstruction techniques, Zhang W et al. (2003) determined the 3D structure of mature DENV particles to a resolution of 9.5 Å. Figure 22.3 demonstrates this structure with the secondary structural disposition of the 180 proteins E and 180 proteins M in the lipid envelope. The α-helical "stem" regions of the E molecules, as well as part of the N-terminal section of the M proteins, are buried in the outer leaflet of the viral membrane. The "anchor" regions of the E and the M proteins each form antiparallel E-E and M-M transmembrane α-helices, leaving their C-termini on the exterior of the viral membrane. This structure showed in particular that the nucleocapsid core and envelope proteins did not have a direct interaction in the mature virion (Zhang W et al. 2003).

There are numerous variants of expression of the DENV structural genes in *E. coli*, which are exhaustively reviewed by Araujo et al. (2020). Remarkably, Lazo et al. (2007) synthesized in *E. coli* the DENV protein C that appeared after purification as great aggregates similar in

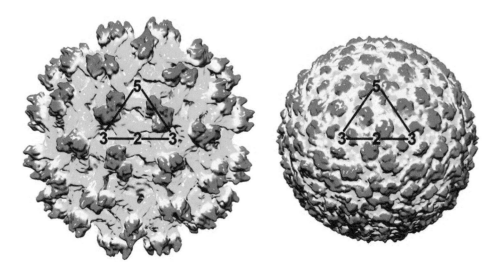

FIGURE 22.3   The 3D electron cryomicroscopy reconstruction of immature (left) and mature (right) particles of an isolate of dengue virus. (Courtesy of Richard Kuhn, Michael Rossmann, and International Committee on Taxonomy of Viruses (ICTV), https://talk. ictvonline.org/ictv-reports/ictv_online_report/positive-sense-rna-viruses/w/flaviviridae/360/genus-flavivirus.) Shown is a surface rendering of immature dengue virus at 12.5 Å resolution (left) and mature dengue virus at 10 Å resolution (right). The viruses are depicted to scale but not colored to scale. The triangles outline one icosahedral unit with the 2-, 3- and 5-fold axes of symmetry.

size to hepatitis B surface protein particles of 20 nm in diameter (see Chapter 25). Moreover, López et al. (2009) published the first report about the in vitro assembly of the recombinant *E. coli*-derived dengue capsid protein. The obtained nucleocapsid-like particles possessed a diameter of 30 nm, quite close to the diameter of the native capsid particles. These particles were formed in vitro in the presence of single-stranded oligodeoxynucleotides of 50 bases, independently of the specificity and the length of the oligonucleotides used. It is noteworthy that earlier attempts at the in vitro assembly of core-like particles using the DENV and YFV capsid proteins produced in *E. coli* have been unsuccessful (Jones et al. 2003). Lazo et al. (2010) evaluated the immunological potential of the DENV nucleocapsid-like particles.

Valdés et al. (2009) used the nucleocapsid-like particles as a carrier and designed a chimeric EDIII-capsid protein (DIIIC) of DENV serotype 2 (DENV-2) with the goal of obtaining a molecule potentially able to self-assemble and induce both humoral and cellular immunity. When mixed with oligodeoxynucleotides, the expected aggregates were obtained. These chimeric particles were immunologically evaluated in mice in comparison with nonaggregated controls. Although the humoral immune response induced by both forms of the protein was equivalent, the aggregated variant resulted in a much stronger cellular immunity (Valdés et al. 2009).

Furthermore, Suzarte et al. (2014) presented the cloning and expression in *E. coli* of the chimeric DIIIC proteins of all four dengue serotypes as a tetravalent dengue vaccine candidate with good humoral and cellular immunity. In the presence of oligonucleotides, these chimeras formed particles ranging from 50 to 55 nm in diameter. The protective immunity induced by the quadrivalent DIIIC particles

in mice and monkeys was evaluated thoroughly (Gil et al. 2015, 2017; Suzarte et al. 2015; Zuest et al. 2015; Valdés et al. 2019).

Table 22.1 establishes the numerous cases when the DENV VLPs were generated successfully from the flaviviral envelope proteins.

It was first reported in the late 1990s that the expression of a cassette encoding the DENV-1 structural proteins C-prM-E or prM-E by chromosomal integration in yeast *Pichia pastoris* led to the appearance of the correctly processed and glycosylated protein E with expected molecular mass (Sugrue et al. 1997). The E protein formed VLPs with an average diameter of 30 nm, whose morphology resembled dengue virions. Moreover, these VLPs induced a virus-neutralizing antibody response in rabbits. Liu Y et al. (2014) expressed in *P. pastoris* the prM-E genes of all four DENV serotypes, and the purified VLPs induced higher-titer antibodies compared with individual VLP dengue serotypes in mice. It is necessary to remember that the protein prM acts as a chaperone of the multifunctional protein E by the VLP self-assembly.

Meanwhile, Mani et al. (2013) expressed in *P. pastoris* the E gene without the prM gene and got VLPs ranging in diameter from 20–40 nm. Obviously, the *P. pastoris*-produced DENV E VLPs were different from infectious DENV particles whose surface is comprised of both prM and E proteins and displayed the EDIII to the immune system in radically different ways, as reviewed recently by Shukla et al. (2019). In this sense, the DENV E VLPs made in *P. pastoris* could be regarded as an alternate EDIII-displaying platform. Nevertheless, Mani et al. (2013) demonstrated that the DENV-2 E-based VLPs exposed the EDIII domain and induced high titers of neutralizing antibodies that were competent to provide significant protection when evaluated

TABLE 22.1

Formation of VLPs by the Combination of Expressed Structural Proteins Derived from the *Flavivirus* Genus Members

Serotype or Genotype	Genes	Expression System	Testing Model	References
Dengue Virus				
DENV-2	C-prM-E	Expi293TM cells	Mice	Boigard et al. 2018
DENV-2	pr/M-E	CHO-K1 cells	Mice	Konishi and Fujii 2002
DENV-1,2,3,4	prM-E	HeLa cells	Screening of human siRNA library	Wang PG et al. 2009
DENV-1,2,3,4	prM-E	Human embryonic kidney 293T cells	Mice	Zhang S et al. 2011
DENV-2	prM-E	DNA vaccine; COS-1 cells	Mice	Galula et al. 2014
DENV-1,2,3,4	prM-E	Human embryonic kidney 293 cells	NT	Metz et al. 2018 48
DENV-2	prM-E (mutations)	Human embryonic kidney 293T cells	Serodiagnosis of patients	Crill et al. 2009; Tsai et al. 2020
DENV-1,2,3,4	prM-E (mutation F108A)	FreeStyle 293F cells	Mice	Urakami et al. 2017
DENV-2	prM-E	*Spodoptera frugiperda Sf*9 insect cells	Mice	Kuwahara and Konishi 2010
DENV-2	prM-E (modified)	Mosquito C6/36 cells	NT	Charoensri et al. 2014
DENV-2	prM-E (modified)		Mice, cynomolgus macaques	Suphatrakul et al. 2015
DENV-2	prM-E; E	*Spodoptera frugiperda Sf*9 insect cells	Syncytial formation	Dai et al. 2018b
DENV-1	C-prM-E; prM-E	*Pichia pastoris*	Rabbits	Sugrue et al. 1997
DENV-2	prM-E		Mice	Liu W et al. 2010
DENV-1	prM-E		Mice	Tang et al. 2012
DENV-1,2,3,4	prM-E		Mice	Liu Y et al. 2014
DENV-1	E		Mice	Poddar et al. 2016
DENV-2	E		Mice	Mani et al. 2013
DENV-3	E		Mice	Tripathi et al. 2015
DENV-4	E		Mice	Khetarpal et al. 2017
DENV-(1+2)	E		Mice	Shukla et al. 2018
DENV-(1+2+3+4)	E		Mice	Rajpoot et al. 2018
DENV-1	C-prM-E; prM-E	Plant *Nicotiana benthamiana*	Mice	Ponndorf et al. 2020
DENV-3	prM-E	Lettuce *Lactuca sativa psbA* chloroplasts	NT	Kanagaraj et al. 2011
Japanese Encephalitis Virus				
	prM-E	Vaccinia virus HeLa cells	Mice	Mason et al. 1991; Konishi et al. 1991, 1992; Pincus et al. 1992
	prM-E	Sindbis vector BHK cells	Mice	Pugachev et al. 1995
	prM-E	cell line F (CHO-K1)	Mice	Konishi et al. 2001
	prM-E	cell line J12#26 cells (RK13)	Mice	Kojima et al. 2003; Mutoh et al. 2004
	prM-E	BJ-ME cells (BHK-21)	Mice	Hua et al. 2014
	prM-E	Lentiviral vector	Mice, pigs	de Wispelaere et al. 2015

(Continued)

TABLE 22.1 (CONTINUED)

Formation of VLPs by the Combination of Expressed Structural Proteins Derived from the Flavivirus Genus Members

Serotype or Genotype	Genes	Expression System	Testing Model	References
Genotype I	prM-E	Cell line 51–10 (CHO)	Mice, pigs	Fan et al. 2018
	prM-E	*Drosophila* cell line S2	Mice	Zhang F et al. 2007
	prM-E	*Spodoptera frugiperda Sf*9 insect cells	Mice	Kuwahara and Konishi 2010
	prM-E	*Spodoptera frugiperda Sf*9 insect cells; *Trichoplusia ni* (HighFive) cells	NT	Yamaji et al. 2012
	prM-E	*Trichoplusia ni* (HighFive) cells	Mice	Yamaji et al. 2013; Yamaji and Konishi 2016
Nakayama strain	prM-E	Silkworm pupae	Mice, rabbits	Matsuda et al. 2017
	prM-E	*Spodoptera frugiperda Sf*9 insect cells	Mice	Zhang Y et al. 2016
	prM-E; E	*Spodoptera frugiperda Sf*9 insect cells	Syncytial formation	Dai et al. 2018b
genotype V (Muar)	E	Silkworm pupae	Mice, rabbits	Nerome et al. 2018
	prM-E	*Pichia pastoris*	Mice	Zhao et al. 2017
	E		Mice	Kwon et al. 2012
Murray Valley Encephalitis Virus				
	prM-E	COS-7	Mice	Kroeger and McMinn 2002
Saint Louis Encephalitis Virus				
	prM-E	CHO-K1	ELISA diagnostics	Purdy et al. 2004
Tick-Borne Encephalitis Virus				
	C-prM-E	BHK-21 cells	Packaging of replicons	Yoshii et al. 2005
	prM-E	CHO cells	Packaging of replicons	Gehrke et al. 2003
	prM-E; E	COS-1 cells	Mice	Allison et al. 1994, 1995; Heinz et al. 1995
	prM-E	*Spodoptera frugiperda Sf*9 insect cells	ELISA diagnostics	Liu YL et al. 2005
	prM-E; E		Syncytial formation	Dai et al. 2018b
	prM-E	*Pichia pastoris*		Yun et al. 2014
West Nile Virus				
Kunjin virus	C-prM-E	BHK	Packaging of replicons	Harvey et al. 2004
	prM-E	DNA vaccine; COS-1 cells	Mice, horses	Davis et al. 2001
	prM-E	CHO-K1 cells	Mutational mapping	Crill et al. 2007
	prM-E	RK13; 293T cells	Different lengths of signal peptides upstream of the prM-E domain	Takahashi et al. 2009
	prM-E	CHO-K1 cells	Mice	Ohtaki et al. 2010
	prM-E	COS-1 cells	Mutational analysis of prM	Calvert et al. 2012
	prM-E	Herpes simplex virus 1 vector; Vero cells	Mice	Taylor et al. 2016
	prM-E	HeLa cells	Mice	Merino-Ramos et al. 2014

Serotype or Genotype	Genes	Expression System	Testing Model	References
	prM-E (mutations)	Human embryonic kidney 293T cells	Serodiagnosis of patients	Tsai et al. 2020
	C-prM-E; prM-E	*Spodoptera frugiperda Sf*9 insect cells	Mice	Qiao et al. 2004
Lineage 1	C-prM-E		ELISA diagnostics	Rebollo et al. 2018
Zika virus				
	C-prM-E; NS2B/NS3	Expi293 cells	Mice	Boigard et al. 2017
	C-prM-E; NS2B/NS3	Human embryonic kidney 293 cells	Mice	Garg et al. 2019
	C-prM-E prM-E		Mice	Garg et al. 2017
	prM-E		Mice	Espinosa et al. 2018
	prM-E		Mice	Salvo et al. 2018
	prM-E (mutations)		Serodiagnosis of patients	Tsai et al. 2020
	prM-E	HEK293SF-3F6 cells	Mice	Alvim et al. 2019
	prM-E	*Spodoptera frugiperda Sf*9 insect cells	Mice	Dai et al. 2018a
	prM-E E		Syncytial formation	Dai et al. 2018b

Notes: The data is presented in alphabetical order of viruses and then by expression systems: first mammalian cells, then insect cells, yeast, and plants; NT, not tested.

in challenge experiments in AG129 mice, a special interferon-deficient model susceptible to DENV or ZIKV infections. Furthermore, this team prepared a complete VLP set by expression in *P. pastoris* the DENV gene of other virus serotypes: 3 (Tripathi et al. 2015), 1 (Poddar et al. 2016), and 4 (Khetarpal et al. 2017).

Shukla et al. (2018) created the first bifunctional vaccine candidate, produced by co-expressing the E proteins of DENV-1 and DENV-2 in *P. pastoris*. The two E proteins coassembled into the bivalent mosaic VLPs, which preserved the serotype-specific antigenic integrity of its two component proteins and elicited predominantly EDIII-focused homotypic virus-neutralizing antibodies in mice. Moreover, the mosaic VLP-induced antibodies lacked discernible antibody-dependent enhancement (ADE) potential.

Going further, Rajpoot et al. (2018) achieved a "four-in-one" effect by coexpression of the E protein of all four DENV serotypes in *P. pastoris*, leading to their coassembly, in the absence of prM, into the tetravalent mosaic VLPs. The latter retained the serotype-specific antigenic integrity and immunogenicity of all four types of their monomeric precursors. Following a three-dose immunization schedule, the tetravalent VLPs elicited EDIII-directed antibodies in mice, which neutralized all four DENV serotypes. Importantly, the induced antibodies did not augment sublethal DENV-2 infection of the dengue-sensitive AG129 mice (Rajpoot et al. 2018). All DENV VLP production variants in *P. pastoris* are listed systematically in Table 22.1.

In mammalian cells, Konishi and Fujii (2002) were the first who generated a cell line continuously expressing subviral extracellular particles of the DENV-2. To establish the stably transfected cell line, the authors constructed a plasmid encoding the gene E and a form of prM with a pr/M cleavage site mutation designed to suppress fusion activity of the protein E. A similar strategy was implemented in the successful construction of a stably transfected cell line expressing the JEV gene E, as described later. However, the generated VLPs were not subjected to the electron microscopy analysis.

Developing further the methodology of the DENV gene expression in mammalian cells, Wang PG et al. (2009) used a codon optimization strategy to obtain the VLPs for all four DENV serotypes, while a stable HeLa cell was established for the DENV-1 prM-E. The produced E particles were mainly present in the endoplasmic reticulum and demonstrated ~20 nm in diameter and homogeneity in size and shape. The authors have used the VLP-producing cell line to develop a semiquantitative assay and screened a human siRNA library targeting genes involved in membrane trafficking, where knockdown of 23 genes resulted in a significant reduction in the VLP secretion, but 22 others increased the VLP levels in cell supernatant (Wang PG et al. 2009).

By optimizing the expression plasmids, Zhang S et al. (2011) produced VLPs of all four DENV serotypes in 293T cells. In electron microscopy, all four types of VLPs exhibited as electron-dense spherical particles of 45–55 nm in size. Therefore, the optimization of the expression plasmids and using mammalian 293T cells led to the DENV VLPs, which at last exhibited similar size and morphological features as the DENV virions.

Remarkably, the same prM and E gene combination was used by the engineering of DENV-2 DNA vaccine (Galula et al. 2014). By the transfection of COS-1 cells, the authors produced VLPs with backbones from different DENV-2 genotypes, which differed by their antigenicity and immunogenicity and by the protection level from challenge that they offered in mice. As a result, a strategy was suggested to enhance the level of VLP secretion and thereby increase the immunogenicity of VLPs by swapping the E-protein ectodomain from one genotype to the other (Galula et al. 2014).

Metz et al. (2018) devoted special attention to the VLP size of all four DENV serotypes, which were produced in the mammalian 293 cells. With particle diameters of ~ 29–34 nm, the VLPs were considerably smaller in size than natural virus particles of ~50 nm, suggesting therefore that their VLPs were assembled by 30 E-dimers in a T = 1 icosahedral lattice, different from the 90 E-dimers found on mature virus particles organized in a T = 3 symmetry and in contrast to the previously referenced findings of the DENV VLPs of 50 nm in size.

The full structural C-prM-E gene stretch was expressed in mammalian cells by Boigard et al. (2018). The authors utilized the structural proteins C-prM-E together with a modified complex of the NS2B/NS3 protease, which enhanced particle formation and yield. These VLPs of ~50 nm in diameter were produced in the Expi293TM cells and resembled native DENV virions by negative staining and immunogold labeling electron microscopy. Remarkably, it was found that the VLPs produced at lower temperature (31°C) were better recognized by a set of conformational monoclonal antibodies and differed therefore by the E protein conformation. Moreover, the mice immunized with the VLP vaccine produced at 31°C elicited the highest titer of neutralizing antibodies when compared to those elicited by equivalent doses of the vaccine produced at 37°C (Boigard et al. 2018).

Urakami et al. (2017) created the DENV-1 to -4 VLPs by coexpressing the prM and E genes but with an F108A mutation in the fusion loop structure of E to increase production of the VLPs in mammalian cells. The DENV-1–4 VLPs exhibited electron-dense 35- to 50-nm spherical particles, which were similar in size to the most previously described DENV VLPs but smaller than dengue virions, which have a diameter of around 50 nm. The immunization of mice with DENV-1–4 VLPs as individual, monovalent vaccines elicited strong neutralization activity against each DENV serotype in mice. By use as a tetravalent vaccine, the DENV-1–4 VLPs elicited high levels of neutralization activity against all four serotypes simultaneously. Moreover, the ADE effect was not observed against any serotype (Urakami et al. 2017).

A set of the mutated DENV-2 VLPs, which were produced in mammalian 293T cells, in parallel with homologous WNV and ZIKV mutants, was used to improve ELISA serodiagnosis of flavivirus infections in patients by discrimination of primary DENV, WNV, and ZIKV infections, as well as secondary DENV and ZIKV infections with a previous DENV infection (Crill et al. 2009; Lai et al. 2013; Tsai et al. 2020).

Concerning production of VLPs in insect cells, Kuwahara and Konishi (2010) produced the DENV-2 and JEV (see the following explanation) VLPs by expression of the prM and E genes in *Sf*9 cells after transfection of the latter with the appropriate RNAs. The secreted VLPs were biochemically and biophysically equivalent to the authentic antigens obtained from infected mosquito cells, but no electron-microscopy data was provided. It was emphasized, however, that the *Sf*9 cells could produce 10- to 100-fold larger amounts of the VLP product than mammalian CHO cells.

Charoensri et al. (2014) performed a broad add-on modification approach and assessed whether concurrent modifications of the prM signal peptide and E stem-anchor, replacing them with homologous cellular and viral counterparts, respectively, could lead to the enhanced VLP production in insect cells and whether additional increases would be possible with codon optimization and prM cleavage-enhancing mutation. In fact, the optimized expression cassettes appeared useful in the generation of stably expressing clones and production of the DENV VLPs for immunogenicity studies.

Continuing this approach, Suphatrakul et al. (2015) produced the DENV2 VLPs by expressing the prM-E with enhanced prM cleavage in mosquito cells. Remarkably, the two VLP preparations generated with either negligible or enhanced prM cleavage exhibited different proportions of spherical particles and tubular particles of variable lengths. The negligible-cleavage VLPs were small and spherical, ~26 nm in diameter. The majority of spherical particles in the VLPs with enhanced prM cleavage was similar to the 26-nm particles, while large spherical VLPs of 44.8 nm in diameter were rare, but the proportion of large spherical, small spherical, and tubular particles (of 19.0 × 82.8 nm by mean size) was about 1:29:47. The VLPs were moderately immunogenic in mice. In cynomolgus macaques, the VLPs with enhanced prM cleavage augmented strongly neutralizing antibody and EDIII-binding antibody responses in live attenuated virus-primed recipients, suggesting that these VLPs could be useful as the boosting antigen in prime-boost immunization (Suphatrakul et al. 2015).

Dai et al. (2018b) demonstrated that the DENV-2, as well as JEV, TBEV, and ZIKV VLPs, are produced by the baculovirus-driven expression system when only the viral E gene, without the prM, is expressed.

In plants, the first attempt to express the DENV structural genes in *N. benthamiana* plants was performed by Martínez et al. (2010), but the authors had instead chosen to concentrate on the EDIII display on the hepatitis B core (HBc) carrier (see Chapter 38).

Kanagaraj et al. (2011) expressed the DENV-3 prM-E stretch, which consisted in this case of part of C, complete prM and truncated E, in edible crop lettuce *Lactuca sativa psbA* chloroplasts. The transmission electron microscopy showed particles of ~20 nm diameter in chloroplast extracts of transplastomic lettuce, and the study was regarded as a step toward investigating an oral dengue vaccine.

At last, Ponndorf et al. (2021) performed the coexpression of DENV-1 structural proteins and a truncated version of the nonstructural proteins lacking NS5 encoding RdRp, which led to the correct assembly of the DENV VLPs in plants. The plant-produced VLPs varied in size from 25–40 nm and looked comparable to a DENV-2 VLP-positive control produced in mammalian cells, which showed a similar variation in size. The size of the VLPs and the presence of noncleaved prM suggested that the purified VLPs did not undergo the maturation process. By electron-microscopy analysis, the VLPs produced by expression of the prM-E stretch showed more impurities and fewer particles than the VLPs produced by the expression of the full structural C-prM-E gene stretch. Therefore, the prM-E expression was characterized by the authors as not successful, while the strategies used in other expression systems to improve the VLP yield did not result in increased yields in plants but, rather, increased purification difficulties. The immunogenicity assays in mice revealed that the plant-made DENV-1 VLPs led to a higher antibody response in mice compared with the DENV EDIII displayed on the bluetongue virus (BTV) core-like particles (see Chapter 13) and with the DENV EDIII subunit alone. Remarkably, the authors could not produce VLPs for the DENV serotypes 2 and 4, since no protein expression was detected for either serotype (Ponndorf et al. 2021).

Finally, it is noteworthy that a tetravalent live attenuated vaccine for dengue, namely Dengvaxia, was licensed recently in several countries, after extensive phase III clinical evaluation, as reviewed by Shukla et al. (2019). This live vaccine, developed by Sanofi Pasteur, is a mixture of four monovalent chimeric YFV vectors, each encoding the prM and E of one DENV serotype. Some other prospective live vaccines are based on attenuated DENV backbones, instead of YFV backbone, to encode prM and E of the four serotypes. The recent reviews clarify the role of the VLP-based vaccine candidates among other potential dengue vaccines (Shukla et al. 2019, 2020; Deng et al. 2020). In the context of this chapter, the optimistic expert judgment concerning the so-called four-in-one vaccine or T-mVLPs engineered in *P. pastoris* by (Rajpoot et al. 2018) and displaying all four EDIII variants is of special value.

JAPANESE ENCEPHALITIS VIRUS

The existing JEV vaccines are inactivated or live attenuated or based on the YFV backbone, as described later. Subunit vaccines were also evaluated in preclinical conditions, though they did not advance to clinical trials. No VLP vaccines have been elaborated yet.

Mason et al. (1989) expressed in *E. coli* antigenic fragments of the JEV protein E to define the boundaries of an antigenic domain that contained the binding sites for monoclonal antibodies. Two fragments of the protein E were expressed in *E. coli* by Chia et al. (2001). Wu et al. (2003) obtained a soluble recombinant protein when the EDIII domain of the protein E was fused to thioredoxin. Lin et al.

(2008) got in *E. coli* the EDIII domain corresponding to the E protein aa residues 298–399 without any fusions.

Fujita et al. (1987) were the first who expressed the JEV gene E in yeast, namely *S. cerevisiae*. The synthesized product demonstrated the expected molecular mass and induced specific immune response in mice. No evidence about its possible assembly was presented.

After 20 years, Zhao P et al. (2017) expressed the JEV prM-E gene stretch in *P. pastoris* and detected the appropriate VLPs of 30–50 nm in size. Remarkably, the JEV prM-E product was not cleaved between prM and E during this secreted expression. The purified VLPs induced efficient immune response in mice, when injected with nucleic acid as an adjuvant. Kwon et al. (2012) expressed in *P. pastoris* the E alone and demonstrated its protective effect in mice.

In mammalian cells, Mason et al. (1991) were the first who expressed different portions of the JEV genome by subcloning them in vaccinia virus vectors and expressing in BHK or Vero cells. As a result, the correct synthesis and glycosylation of the glycoprotein E, as well extracellular particles that contained fully processed forms of the M and E proteins, were detected. The diameter of particles later was estimated at 20 nm (Konishi et al. 2001). The immunization with recombinant vaccinia viruses expressing the prM and E genes protected mice from a lethal challenge by JEV (Mason et al. 1991; Konishi et al. 1991; Pincus et al. 1992).

Konishi et al. (2001) generated a cell line, termed F cells, which produced the JEV prM-E VLPs. The latter shared the biochemical properties of empty viral particles produced by JEV-infected cells of 20 nm in size. The prM in this clone contained a modification of the aa sequence at the furin cleavage site. This mutation was designed to suppress cleavage from prM to M, eliminating the fusion activity of the VLPs and allowing it to overcome the difficulty in generating cells continuously expressing particles at a high level. In fact, the engineered cell line F produced a relatively large amount of VLPs without any contamination with any infectious materials. After immunization of mice and immunological studies, the produced JEV VLPs were found useful as vaccine candidates and as diagnostic reagents in evaluating human immune responses to the JEV vaccination (Konishi et al. 2001).

Next, Kojima et al. (2003) generated a novel cell line, J12#26, after transfection of rabbit kidney-derived RK13 cells with a plasmid encoding the JEV prM-E genes. The J12#26 cells secreted VLPs into the culture medium in a huge ELISA-equivalent amount, namely, 2.5 μg per 10^4 cells.

The VLP production was stable after multiple cell passages and persisted over one year with 100% expressing cells without detectable cell fusion, apoptosis, or cell death. The mice immunized with the purified J12#26 E-antigen without adjuvant developed high titers of neutralizing antibodies for at least seven months and 100% protection against intraperitoneal challenge with JEV (Kojima et al. 2003).

Mutoh et al. (2004) adapted the J12#26 cells to serum-free medium without loss of the VLP production level, but the product in the form of almost uniform particles with an average diameter of 25 nm was easily purified with a high yield from the serum-free culture supernatants. The purified VLPs induced in mice, without adjuvant, neutralizing antibody titers as high as or higher than the licensed JEV vaccine and ensured complete protection against challenge with wild-type virus.

Okamoto et al. (2012) investigated the efficacy of a single-dose immunization by the J12#26 cells-produced JEV VLPs. Although single administration protected less than 50% of mice against lethal JEV infection, adding poly(γ-glutamic acid) nanoparticles or alum adjuvant to the JEV VLPs protected more than 90% of mice.

Later, Fan et al. (2018) constructed a CHO-heparan sulfate-deficient cell clone, termed 51–10, that stably produced JEV VLPs of genotype I (GI) and continually secreted the GI VLPs without signs of cell fusion. Such VLPs formed a homogeneously empty-particle morphology and exhibited similar antigenic activity as the GI JEV virions. The immunization of mice was satisfying, but a central aim of the study was targeted to immunization of pigs, since swine are a critical amplifying host involved in human JEV outbreaks, and vaccinating pigs is expected to suppress the viral transmission and reduce JEV infection in humans. In fact, the GI VLP-immunized swine challenged with GI or GIII viruses showed no fever, viremia, or viral RNA in their organs and demonstrated sterile protection against both JEV genotypes (Fan et al. 2018). De Wispelaere et al. (2015) engineered a lentiviral vector expressing the native prM and E, which resulted in the efficient secretion of the JEV VLPs. The immunization of mice and piglets induced high titers of anti-JEV antibodies that had efficient neutralizing activity regardless of the JEV genotype tested.

Concerning production of the JEV VLPs in insect cells, Zhang F et al. (2007) were the first who achieved this goal. The JEV VLPs were detected after expression of the prM and E genes in *Drosophila* cell line Schneider 2 (S2).

Kuwahara and Konishi (2010) produced the JEV VLPs, in parallel with the same of DENV-2, as described earlier, in *Sf*9 cells by coexpression of the corresponding prM and E genes. Then, the JEV VLPs were produced efficiently by the baculovirus-driven expression system in *Sf*9 cells (Yamaji et al. 2012) or by the stable expression in *Trichoplusia ni* (HighFive) cells (Yamaji et al. 2013; Yamaji and Konishi 2016).

Matsuda et al. (2017) performed the baculovirus-driven expression of the synthetic codon-optimized prM-E genes in silkworm pupae. Nerome et al. (2018) used the same baculovirus-driven expression in silkworm pupae, but the synthetic codon-optimized gene E of the JEV strain Muar was employed. The JEV VLPs were produced by baculovirus-driven expression system when viral E was expressed, with or without the prM protein, in insect cells (Dai et al. 2018b).

In all cases, the mice immunized with the insect cells-derived JEV VLPs developed specific antibodies, while full protection against lethal challenge of JEV was demonstrated in some studies.

It is noteworthy that the N-glycosylation of both JEV prM and E proteins that each contain a single potential N-glycosylation site was found critical for the assembly (Kim et al. 2008; Zai et al. 2013). Thus, removal of N-glycan from the prM protein resulted in a complete misfolding of the E protein and failure to form VLPs. A similar removal of N-glycan from the E protein led to a low efficiency of its folding and VLP formation. The secretion and cytotoxicity of the E protein was also markedly impaired in case the glycosylation sites in the prM or E or both proteins were removed (Zai et al. 2013).

TICK-BORNE ENCEPHALITIS VIRUS

TBEV contributed greatly to the structural 3D elucidation of flaviviruses. Thus, Rey et al. (1995) resolved the crystal structure of the TBEV protein E dimer to 2.0 Å. The idea was that the treatment of the TBEV virions with trypsin released a dimeric soluble fragment of the protein E (Heinz et al. 1991). This membrane-anchor free fragment included residues 1–395 of the 496 aa residue E polypeptide chain, and all but about 50 aa residues of its ectodomain and appeared crystallizable. The architecture of the E dimer was unusual and vastly different from that of possible homologs, such as influenza hemagglutinin, which had analogous functions. Rather than forming a spike projection, the TBEV protein E dimer extended in a direction parallel to the viral surface. Each of the monomeric subunits comprised three distinct domains I, II, and III, where the domain I was a β-barrel, from which emanated two long loops that formed the extended finger-like domain II. The domain III, which followed domains I and II in the polypeptide chain, had an immunoglobulin-like fold and bore the receptor binding site (Rey et al. 1995).

Using electron cryomicroscopy and icosahedral image reconstruction, Ferlenghi et al. (2001) resolved 3D structure of so-called subviral particles (VLPs in our terms here) to 19 Å. Then, the authors fitted the high-resolution crystal structure of the E ectodomain resolved by Rey et al. (1995) into the acquired experimental density and revealed that 30 of the E dimers were arranged in a T = 1 icosahedral lattice with outer diameter of 315 Å and allowed an approximate identification of their sites of interaction. The data allowed also one to infer the positions occupied by the protein M and to identify the positions of membrane-spanning domains. The 3D structure explained why the diameter of the subviral particles was about two-thirds of that of the whole virion. The data supported the notion that there was an arrangement of the E dimers in virions that was quasiequivalent to the one found in the subviral particles and that the presence or absence of the core determined the observed design. The diameter of virions of 50 nm would permit such quasiequivalence within a T = 3 icosahedral lattice. The only alternative, T = 4, would require a somewhat larger particle (Ferlenghi et al. 2001). Figure 22.4 demonstrates

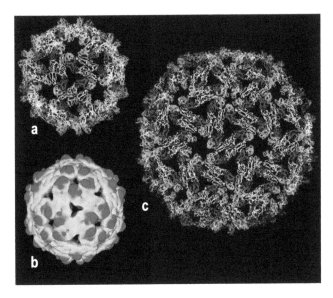

FIGURE 22.4 Proposed arrangement of the soluble E dimers in the TBEV subviral particles and virion. (a) The arrangement of the soluble E proteins in the subviral particles, generated by propagating the fit of the backbone of soluble E over the particle surface, using icosahedral symmetry. (b) Radially depth-cued surface representation of the low-pass filtered density (B = 2000 Å²) (Stewart et al. 1993; Belnap et al., 1999) derived from (a). In the subviral particles, there is additional density around the 3-fold; the authors proposed that this density corresponds to the M protein. (c) A suggested T = 3 arrangement of the E dimers in the mature virion, quasi-equivalent to the arrangement in (a). (Reprinted from *Mol Cell* 7, Ferlenghi I, Clarke M, Ruttan T, Allison SL, Schalich J, Heinz FX, Harrison SC, Rey FA, Fuller SD, Molecular organization of a recombinant subviral particle from tick-borne encephalitis virus, 593–602, Copyright 2001, with permission from Elsevier.)

the proposed arrangement of the soluble E dimers in the TBEV subviral particles and virions.

The structural investigation used the subviral particles (VLPs in our terms) isolated by method of Allison et al. (1994, 1995). Thus, Allison et al. (1994) expressed the prM-E genes in mammalian COS-1 cells. Allison et al. (1995) performed a solid quantitative study to check the requirements for secretion of soluble and particulate forms of the TBEV protein E, when full-length E and a C-terminally truncated anchor-free form were expressed in the presence and absence of prM. The formation of a heteromeric complex with prM was found to be necessary for efficient secretion of both forms of E, whereas only low levels of anchor-free E were secreted in the absence of prM. Remarkably, the prM-mediated transport function could also be provided by coexpression of prM and E from separate constructs, but a prM-to-E ratio of greater than 1:1 did not further enhance secretion. The full-length E formed stable intracellular heterodimers with prM and was secreted as a subviral particle, whereas anchor-free E was not associated with particles and formed a less stable complex with prM, suggesting that prM interacts with both the ectodomain and anchor region of the protein E (Allison et al. 1995). Furthermore, the fine details

of the intracellular assembly and secretion of subviral particles were reported by Lorenz et al. (2003).

The substantial role of TBEV by the methodology of subgenomic replicons, packaging cell lines, and sriVLPs (Gehrke et al. 2003; Yoshii et al. 2005, 2008) was translated earlier in the corresponding paragraph.

A particular interest was expressed by the isolation of the TBEV nucleocapsids and their disintegration into a capsid protein dimer by high-salt treatment, a task that remained largely unsuccessful before (Kiermayr et al. 2004). The purified capsid protein dimers were assembled in vitro into capsid-like particles when combined with in vitro transcribed viral RNA or single-stranded DNA.

Concerning expression in insect cells, Liu YL et al. (2005) used the baculovirus-driven expression system to produce VLPs (or subviral particles by the author terminology) of 30 nm in diameter and consisting of the prM and E proteins in the standard insect *Sf*9 cells.

The immunization with various soluble forms of dimeric whole or membrane anchor-free TBEV E protein failed to induce virus-neutralizing antibodies and to provide appropriate immunogenicity in mice (Heinz et al. 1995). Likewise, DNA vaccines driving intracellular expression of either whole or truncated TBEV E proteins or secretion of TBEV E dimers failed to induce virus-neutralizing antibodies and to afford robust protection against TBEV (Aberle et al. 1999).

Yun et al. (2014) produced the TBEV VLPs by expression of the pM and E genes in *P. pastoris*. This was the first report on the successful expression of the TBEV VLPs in yeast.

A rather modest role of the current TBEV VLPs in the general strategy of novel vaccines was unveiled recently in a substantial review by Kubinski et al. (2020).

West Nile Virus

The revolutionary role of WNV by the elaboration of the flavivirus subgenomic replicon expression system, packaging cell lines, and sriVLPs was described earlier.

Concerning bacterial expression of the structural genes, the WNV E protein (Wang T et al. 2001) and EDIII without fusions (Chu et al. 2007) were produced in *E. coli*. Of course, these studies did not provide for the VLP production.

Davis et al. (2001) engineered a recombinant plasmid that expressed the WNV proteins prM and E. A single intramuscular injection of the plasmid DNA induced protective immunity, preventing WNV infection in mice and horses. The plasmid-transformed COS-1 cells expressed and secreted high levels of WNV prM and E proteins into the culture medium, which were regarded likely as VLPs (extracellular subviral particles (EPs) by the authors' terminology).

Using mammalian CHO cells, Ohtaki et al. (2010) established stable cell lines continuously expressing the WNV prM-E genes. The two types of VLPs were distinguished: (i) large, 40–50 nm in size, and composed of the E and processed mature M proteins and (ii) small, 20–30 nm, consisting of the E and immature prM proteins. The large VLPs induced higher neutralizing antibody and anti-WNV IgG titers than S-VLPs. The mice vaccinated with the large VLPs showed higher protective efficacy against WNV challenge than the small VLP-immunized mice.

The mammalian cell-produced WNV VLPs, in parallel with that of DENV and ZIKV, were applied recently to distinguish three flavivirus infections (Tsai et al. 2020).

To perform immunization of birds, Merino-Ramos et al. (2014) used HeLa cells to create a cell line stably expressing the WNV prM and E proteins and constitutively secreting to the culture medium the WNV VLPs ranging from 20–40 nm in diameter and being the mean diameter of 33 ± 10 nm. Jiménez de Oya et al. (2019b) demonstrated the protective efficacy of the generated VLP-based vaccine in susceptible birds, namely magpies, and ability to control the spread of the virus, while Jiménez de Oya et al. (2019a) provided a future picture of the putative avian vaccines against WNV.

Concerning production in insect cells, Qiao et al. (2004) generated the WNV VLPs by the baculovirus-driven expression of the C-prM-E or prM-E genes. The VLPs were polymorphic in appearance and had a diameter of 40–60 nm. Both VLP variants induced WNV-specific antibodies possessing potent neutralizing activities. The immunization of mice with the prM-E VLPs induced sterilizing immunity against challenge with WNV without producing any evidence of viremia or viral RNA in the spleen or brain (Qiao et al. 2004). Later, Rebollo et al. (2018) produced the WNV VLPs by the baculovirus-directed expression of the C-prM-E genes in the same *Sf*9 cells and used these VLPs, instead of the whole virions, to perform correctly two different ELISAs for WNV diagnosis. Remarkably, the WNV VLPs found in the sucrose gradient preparation were more homogeneous in size, with a majority of structures showing a diameter of around 30 nm, whereas those found in the less purified cushion preparation were more variable in size, about 30–50 nm.

The general role of the WNV VLPs compared to other candidates during the last 20 years of progress toward a WNV was indicated by Kaiser et al. (2019).

Yellow Fever Virus

The yellow fever virus (YFV) live-attenuated vaccine termed YFV-17D was developed in 1936 using the so-called 17D strain. All YFV vaccines produced today are derived from this strain. This vaccine was produced in embryonated chicken eggs, and the techniques applied to vaccine manufacture had changed little since the 1940s, as reviewed recently by Araujo et al. (2020).

The YFV-17D vaccine was used as a vector for expressing epitopes of other flaviviruses and to construct the appropriate live vaccines, such as JEV (Monath et al. 2003), DENV (Guy et al. 2008), WNV (Monath et al. 2006), and ZIKV (Giel-Moloney et al. 2018) and of antigens from other

pathogens, such as malaria (Bonaldo et al. 2002) and HIV/ SIV (Bonaldo et al. 2010).

The YFV protein E was synthesized in Vero cells, as well as in *Spodoptera frugiperda* insect cells by the baculovirus-driven expression system (Desprès et al. 1991; Shiu et al. 1991; Barros et al. 2011), but no VLPs were reported yet. The recombinant variants of the proteins C (Jones et al. 2003) and EDIII (Volk et al. 2009) were expressed in *E. coli*. Later, the protein E was transiently produced in *Nicotiana benthamiana* as a stand-alone protein or as a fusion to the bacterial enzyme lichenase (Tottey et al. 2018). In all cases, there was no aim and chance to get VLPs.

Zika Virus

The truncated ZIKV E protein with the N-terminal (i) 90% region reserved (E90; Han et al. 2017) and (ii) 80% reserved (ED80) (Liang et al. 2018), as well as EDIII (Yang et al. 2017), were cloned and expressed in *E. coli*, with no hope of self-assembly.

The global progress in the ZIKV VLP generation is outlined in Table 22.1. The first safe, effective, and economical ZIKV VLP platform was developed in mammalian cells by different approaches. Thus, Boigard et al. (2017) coexpressed the structural C-prM-E and nonstructural NS2B/ NS3 (encoding protease NS3 with its cofactor NS2B) genes. The viral NS3 protein was truncated maintaining only its N-terminal protease domain NS3Pro, which was kept as a single transcription unit with its cofactor NS2B. The VLPs were produced in a suspension culture of mammalian cells and self-assembled into particles of an average diameter of 60 nm, which closely resembled Zika virions as shown by electron microscopy. The VLP vaccine stimulated significantly higher virus neutralizing antibody titers in mice than comparable formulations of the inactivated ZIKV vaccine (Boigard et al. 2017).

In parallel, Garg et al. (2017) started with so-called reporter virus particles (RVPs) carrying the synthetic ZIKV C-prM-E gene construct and packaging a GFP reporter-expressing WNV replicon. The ZIKV prM-E construct alone was sufficient for generating VLPs, whereas efficient VLP production from the C-prM-E construct was achieved in the presence of the WNV NS2B-3 protease, which cleaved C from prM. To facilitate the efficient VLP platform, a stable cell line expressing high levels of ZIKV prM-E proteins was generated and produced constitutively the appropriate VLPs, as well as a cell line expressing ZIKV C-prM-E proteins for RVP production. The immunization studies in mice by both C-prM-E and prM-E VLPs showed that VLPs generated higher neutralizing antibody titers than those by DNA vaccination, with the C-prM-E VLPs giving slightly higher titers than those with the prM-E VLPs (Garg et al. 2017).

Espinosa et al. (2018) engineered the ZIKV expression cassette carrying prM-E genes from the African MR766 strain but with the E ectodomain sequence from the Brazilian SPH2015 strain, with an assumption that the antibody response against a putative vaccine should be directed toward the more recent Brazilian E ectodomain. Other elements of this chimeric cassette were chosen with an aim to ensure the highest VLP production. The resulting plasmid was transiently transfected into HEK293 cells, and the ZIKV VLPs were concentrated and purified from media supernatant. The population of particles was predominantly composed of two different sizes, 35 nm and 55 nm, where approximately 80% of the VLPs were 35 nm in size. The neutralizing antibodies induced by this VLP vaccine in immunocompetent mice provided significant protection against subsequent ZIKV challenge upon transfer of antibodies to AG129 mice, a special model susceptible to ZIKV infection, as mentioned earlier.

At the same time, Salvo et al. (2018) cloned the prM and E genes of ZIKV strain H/PF/2013 with nascent signal sequence into a pCMV expression vector under the control of a cytomegalovirus (CMV) promoter and CMV polyadenylation signal. After expression in mammalian cells, the transmission electron microscopy revealed structures with similarity to ZIKV and size that ranged from 30–60 nm and an average size of ~50 nm. All vaccinated mice, including the interferon-deficient AG129 mice, survived ZIKV challenge with no morbidity or weight loss while control animals either died at nine days post challenge (AG129) or had increased viremia (BALB/c), while neutralizing antibodies were observed in all vaccinated mice (Salvo et al. 2018).

Alvim et al. (2019) also employed the prM and E to form the ZIKV VLPs, in the absence of the protein C. Most importantly, the authors demonstrated that perfusion technology using stably transfected cells constitutively expressing ZIKV VLPs was highly promising as a vaccine-manufacturing platform and brought prospects for the development of large-scale, cost-effective technologies to produce vaccines against ZIKV.

Remarkably, Urakami et al. (2017) enhanced the ZIKV VLP production in mammalian cells by introducing the same previously described for DENV mutation F108A into the ZIKV protein E.

Tsai et al. (2020) used wild-type and mutant ZIKV VLPs, which were produced in mammalian 293T cells, in parallel with homologous DENV and WNV mutants, to improve ZIKV serodiagnosis and distinguish three flavivirus infections in patients.

At last, Garg et al. (2019) solved the central methodological problem of the VLP production, namely the incorporation of the protein C into the ZIKV VLPs and formation of the complete core. To achieve this, a bicistronic vector was engineered, which expressed C-prM-E and NS2B/NS3 using an IRES sequence. This bicistronic expression cassette, in a lentiviral vector, was used to create a stable cell line that constitutively secreted the C-prM-E VLPs. By immunization of mice, the superiority of the C-prM-E VLPs over the prM-E ones in generation of neutralizing antibody response was demonstrated. The challenge of the C-prM-E VLP-immunized mice with Zika virus showed complete protection.

Moreover, Garg et al. (2020) generated a universal VLP vaccine termed CJaYZ and targeting JEV, YFV and ZIKV, as well as Chikungunya virus (CHIKV), from the *Togaviridae* family, order *Martellivirales* (see Chapter 18). For ZIKV, YFV, and JEV it included the C-prM-E genes and for CHIKV the C-E3-E2-E1 genes. The stable 293T cell lines secreting VLPs containing capsid protein for all four viruses were established and adapted to grow in suspension cultures to facilitate vaccine scale up. The immunization of mice with different combinations of the capsid protein containing VLPs either as monovalent, bivalent, or tetravalent formulation resulted in generation of high levels of neutralizing antibodies. The potential tetravalent VLP vaccine candidate provided strong neutralizing antibody titers against all four viruses (Garg et al. 2020).

The different variants of the vaccinia virus vectors-based ZIKV vaccine candidates expressing PrM and E genes were found protective in mouse models (Pérez et al. 2018; Jasperse et al. 2021).

After mammalian cells, the ZIKV VLPs were produced also by the baculovirus-driven expression system when the viral protein E was expressed, with or without the prM protein, in insect cells (Dai et al. 2018b). In both cases, the VLPs were released from the cell surface. The purified particles demonstrated size of 30–50 nm and were similar to native ZIKV virions in morphology. The immune electron microscopy indicated the exposure of E protein on the outer surface of the VLPs (Dai et al. 2018b). Dai et al. (2018a) demonstrated how the ZIKV VLPs could be quickly and easily prepared in large quantities using the baculovirus-driven expression system. Moreover, the overproduced VLPs stimulated high levels of virus neutralizing antibody titers and potent memory T-cell responses in mice.

The attempts to produce the ZIKV VLPs in plants, together with other studies on the generation of plant-based vaccine candidates against flaviviruses, were recently summarized by Peyret et al. (2021).

OTHER FLAVIVIRUSES

Kroeger and McMinn (2002) reported on the development of the mammalian expression that resulted in the secretion of VLPs of Murray Valley encephalitis virus (MVEV) upon transfection of the murine fibroblast cell line with a plasmid carrying the MVEV prM-E genes. The mice inoculated with the purified VLPs were protected from lethal challenge with the virulent prototype MVEV strain, and this protection correlated with the development of a neutralizing humoral immune response by the host.

Purdy et al. (2004) expressed in mammalian CHO cells the prM-E genes and purified the corresponding extracellular VLPs of Saint Louis encephalitis virus (SLEV). The ease of production, safety, adequate sensitivity for detection of true positives, and low cross-reactivity in ELISA screening of patient serum samples made the SLEV VLPs preferable to traditionally used antigens derived from suckling mouse brain preparations.

Holmes et al. (2005) used VLPs of JEV, WNV, SLEV, and DENV-1,2,3,4 to evaluate the cross-reactivity of seven mosquito-borne viral antigens in sera of patients including those infected with either Powassan virus (POWV) or La Crosse virus (LACV). The VLPs always performed better than the suckling mouse brain antigens in the ELISA tests and the mammalian cell lines continuously secreting these VLPs were therefore identified as a significant improvement of serodiagnostic of flavivirus infections.

As described earlier, a single inoculation of the WNV VLPs protected mice against a lethal challenge with WNV. During this previously mentioned study, Merino-Ramos et al. (2014) also evaluated the cross-reactivity of the immune response against Usutu virus (USUV), which shares multiple ecological and antigenic features with WNV. It turned out that the immunization of mice with the WNV VLPs increased specific, although low, antibody titers found upon subsequent USUV infection, promoting therefore a cross-reactive humoral response against USUV.

GENUS *HEPACIVIRUS*

HEPATITIS C VIRUS

The first HCV genome was cloned in 1989 by the staff members of the Chiron Corporation and therefore the vital problem of the mysterious non-A, non-B viral hepatitis was clarified substantially (Choo et al. 1989). The attempts to achieve the HCV VLPs quickly followed.

The long history of attempts of the Elmārs Grēns's team in late the 1990s to early 2000s to get virus- or at least nucleocapsid-like particles in *E. coli*, after expression of different gene C variants, ended with failure: the products were mostly insoluble and did not assemble (Mihailova et al. 2006; Marija Mihailova and Irina Sominskaya, a set of unpublished data). Nevertheless, Lorenzo et al. (2001) reported in *E. coli* aggregation of the protein C segment spanning aa 1–120 into VLPs with an average diameter of 30 nm. Recently, Olivera et al. (2020) reported expression in *E. coli* of the two chimeric proteins encompassing conserved and immunogenic epitopes from the HCV proteins C, E1, E2, and NS3 proteins, outside the VLP approach, and their immunogenicity in mice.

Baumert et al. (1998) were the first who achieved production of the HCV VLPs by baculovirus-driven expression system in insect cells. The expressed HCV structural proteins assembled into VLPs of 40–60 nm in diameter in large cytoplasmic cisternae. The HCV VLPs were similar to putative HCV virions by CsCl and sucrose gradient centrifugation pattern. The protein p7, a small viral polypeptide that resides between the structural and nonstructural regions of the HCV polyprotein, was not sufficient for the VLP formation. The particle-associated nucleic acids demonstrated that HCV RNAs were selectively incorporated into the particles over non-HCV transcripts (Baumert et al. 1998). The HCV VLPs were highly immunoreactive with sera from individuals infected with various HCV genotypes,

were immunogenic in mice (Baumert et al. 1999), and were used successfully as capture antigens in an ELISA to detect and quantify antibodies against HCV structural proteins in patients with acute and chronic hepatitis C (Baumert et al. 2000). Moreover, the HCV VLPs were able to bind specifically the human lymphoma and hepatoma cell lines but not mouse cell lines (Triyatni et al. 2002a; Wellnitz et al. 2002).

Lechmann et al. (2001) evaluated thoroughly the humoral and cellular immunogenicity of the insect cells-produced HCV VLPs, with or without viral p7 protein, in mice. The latter developed high titers of anti-E2 antibodies and virus-specific cellular immune response including CTL and T helper responses with γ-interferon production. The VLPs without p7 generated a higher cellular immune response with a more Th1 profile than the particles with p7. The immunization of heat-denatured particles resulted in substantially lower humoral and cellular responses, suggesting that the immunogenicity was strongly dependent on the particulate condition. Remarkably, administration of CpG oligonucleotide did not significantly enhance the immunogenicity of HCV VLPs (Lechmann et al. 2001). Murata et al. (2003) showed protection of mice by the immunization with the insect-produced HCV VLPs. Qiao et al. (2003) tried different adjuvants for the immunization. The HCV VLPs demonstrated good immunogenicity not only in mice but also in baboons (Jeong et al. 2004). Moreover, the HCV VLPs were rapidly taken up by human dendritic cells (Barth et al. 2005).

Finally, these HCV VLPs were tested in chimpanzees, the only established animal model susceptible to HCV infection (Elmowalid et al. 2007). All animals that were immunized with the HCV VLPs with or without AS01B adjuvant developed an HCV-specific immune response against core, E1, and E2. Upon challenge with an infectious HCV inoculum, one chimpanzee developed transient viremia with low HCV RNA titers, while the three other chimpanzees became infected with higher levels of viremia, but their viral levels became unquantifiable ten weeks after the challenge. Four naïve chimpanzees were infected with the same HCV inoculum, and three developed persistent infection with higher viremia. The authors commended the results as encouraging for further development of the appropriate anti-HCV vaccine (Elmowalid et al. 2007).

Coming back to the insect cells-supported expression of the HCV genes in early 2000s, Owsianka et al. (2001) noticed that such VLP formation has not been observed in mammalian cells expressing the same viral structural proteins as in the insect cells and compared the antigenic properties of the mammalian cells-produced envelope proteins with that of the insect cells-produced HCV VLPs. The latter revealed a population of particles with radii ranging from 20–44 nm, the morphology of which was consistent with that predicted by analogy to related flaviviruses. The VLPs were at least partially enveloped with spikes protruding from the surface. The presence of the HCV E2 glycoprotein on the surface of VLPs was confirmed by immunogold electron microscopy. Similarly, the E1 glycoprotein was also shown to be present on the surface of VLPs (Owsianka et al. 2001). Figure 22.5 demonstrates the good HCV VLPs produced by the baculovirus expression system.

Furthermore, the insect cells-derived VLPs were probed with numerous monoclonal antibodies against E1 and E2, and it was demonstrated definitively that the protrusions on the particles were formed by the E1 and E2 glycoproteins (Clayton et al. 2002; Triyatni et al. 2002b)

Meanwhile, Kunkel et al. (2001) expressed the HCV gene C alone in the recombinant baculovirus system and demonstrated that nucleocapsid-like structures were generated. The 124 N-terminal aa residues of the protein C were sufficient for self-assembly into nucleocapsid-like particles. The latter had a regular, spherical morphology with a modal distribution of diameters of approximately 60 nm. Their self-assembly required structured RNA molecules. Inclusion of the C-terminal domain of the protein C modified the core assembly pathway such that the resultant particles acquired an irregular outline, although these particles remained similar in size and shape to those assembled from the 124 N-terminal aa residues of the protein C.

The ability of the HCV protein C alone to assemble into particles in insect cells was confirmed by Choi et al. (2004) who repeated the baculovirus-driven expression of the structural HCV genes in the same *Sf*9 cells but constructed recombinant baculovirus expression vectors consisting of either HCV C alone, C-E1, or C-E1-E2. Surprisingly, neither C-E1 nor C-E1-E2 could assemble into VLPs, while the protein C alone was assembled into amorphous

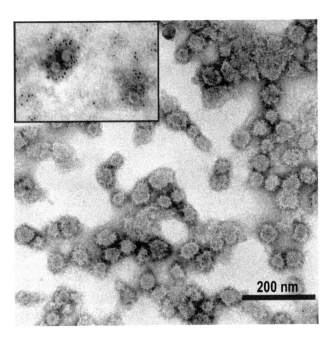

FIGURE 22.5 Electron micrograph of negative-stained HCV VLPs isolated from *Sf* cells. Inset: VLPs placed on the EM grid were incubated first with anti-E2 MAb. (Reprinted with permission of Microbiology Society from Owsianka A et al. *J Gen Virol.* 2001;82:1877–1883.)

nucleocapsid-like particles that were released into the culture medium as early as two days after infection.

Xiang et al. (2002) produced the HCV VLPs in the same *Sf*9 cells and confirmed that the structural HCV proteins underwent the appropriate processing within the particles and the assembly proceeded correctly. The authors utilized these recombinant HCV particles to evaluate sera from patients with HCV viremia—including some HCV viremic patients with negative commercial HCV immunoassay results—and to check recognition of the VLPs by cultured cells.

The HCV VLPs of 40–60 nm in diameter and similar to the HCV virions from serum samples or hepatic tissue were obtained also when the C and E1-E2 genes were inserted into the two different baculoviruses and expressed in insect cells (Zhao W et al. 2004). It was concluded that the HCV structural proteins simultaneously expressed in insect cells interacted with each other and assembled into the HCV VLPs, which were immunogenic in mice.

Girard et al. (2004) accented an effect of the 5' nontranslated region on self-assembly of the HCV VLPs in the *Sf*9 insect cells. Chapel et al. (2006) used the insect cells-produced VLPs as a model to search for potential inhibitors of the HCV morphogenesis.

The baculovirus-driven expression contributed to the fine structural investigations on HCV. In contrast to flaviviruses reach in the 3D information, the paucity of such information for HCV was oppressive. Thus, Yu X et al. (2007) presented the first electron-cryomicroscopy reconstruction of the HCV VLPs from insect cells at low resolution of 30 Å. Figure 22.6 demonstrates this 3D reconstruction.

Furthermore, Badia-Martinez et al. (2012) used electron tomography of plastic-embedded sections of insect cells to visualize the morphogenesis of recombinant HCV VLPs and provide a 3D sketch of viral assembly at the endoplasmic reticulum showing different budding stages and contiguity of buds.

The attempts to produce the HCV VLPs in yeast were made with a slight delay to that using the baculovirus expression system. Thus, Falcón et al. (1999) expressed in *P. pastoris* the gene stretch encoding the first 339 N-terminal aa of the HCV polyprotein and including the 191 aa of the C protein and 148 aa of the E1 protein. The expression led to the appearance of the core protein-formed VLPs with diameters ranging from 20–30 nm and localized along the membrane of the endoplasmic reticulum but mostly in vacuoles, either free or inside autophagic bodies. The properties of these HCV core VLPs were examined fundamentally (Acosta-Rivero et al. 2001, 2002, 2003, 2004).

The finding of Majeau et al. (2004) is noteworthy in this context, where 75 N-terminal residues of the C protein appeared sufficient to assemble and generate nucleocapsid-like particles in vitro. However, homogeneous particles of regular size and shape were observed only when the particles were produced from at least the first 79 N-terminal aa of the protein C. This small protein unit fused to the endoplasmic reticulum-anchoring domain also generated VLPs

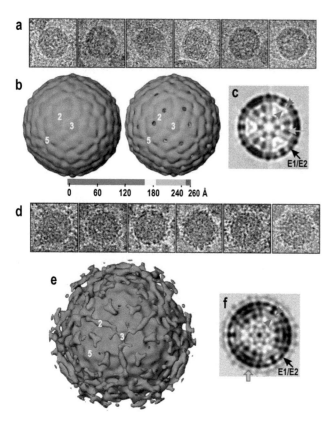

FIGURE 22.6 3D reconstructions of unlabeled and antibody-labeled 500-Å HCV VLPs. (a) Boxed-out cryoEM images of unlabeled HCV VLPs from the 500-Å size group. (b) Shaded surface representation of the 3D reconstruction of HCV VLP at 30-Å resolution, viewed along an icosahedral 3-fold axis. Densities in panels (b) and (e) are color coded according to radius (color bar). (Left) Map displayed at a contour level of one standard deviation above the mean density. (Right) The same map displayed at a slightly higher density contour level (1.1 times the standard deviation above the mean) to reveal the features with more robust density. (c) A 7-Å-thick density slice extracted from the HCV VLP reconstruction shows multilayer structural organization. The two red arrowheads indicate the lipid bilayer; the blue arrow indicates the capsid. Unlike the dengue virus, the HCV VLP lacks an intermediate density layer corresponding to the M-protein layer. The outer layer (black arrow) is attributed to the E1 and E2 proteins. (d) Boxed-out cryoEM images of antibody-labeled HCV VLPs from the same size group. (e) Shaded surface representation of the 3D reconstruction of antibody-labeled HCV VLP, viewed along an icosahedral 3-fold axis. The putative antibody densities are shown in pink. (f) A density slice extracted from the 3D reconstruction of antibody-labeled HCV VLP. The lipid bilayer and E1/E2 layer are indicated as in panel (c). The extra layer of lower density surrounding the particle is attributed to the antibodies. (Reprinted from *Virology* 367, Yu X, Qiao M, Atanasov I, Hu Z, Kato T, Liang TJ, Zhou ZH, Cryo-electron microscopy and three-dimensional reconstructions of hepatitis C virus particles, 126–134, Copyright 2007, with permission from Elsevier.)

in yeast cells (Majeau et al. 2004). A bit later, Acosta-Rivero et al. (2005) reported that the truncated variant of protein C covering the first 120 aa with a 32 aa N-terminal fusion peptide including the His$_6$ tag, purified as a monomer, formed

in vitro VLPs of 20–43 nm with an average diameter of approximately 30 nm.

More recently, Fazlalipour et al. (2015) achieved expression of the genes C, E1 and E2 in *P. pastoris*, which resulted in the generation of the HCV VLPs of 70 nm in diameter. These VLPs efficiently induced all three sorts of specific antibodies in rabbits.

In mammalian cells, the HepG2 cells produced negative-strand RNA and VLPs after transfection with RNA transcribed from a full-length HCV cDNA clone (Dash et al. 1997). The VLPs of 50–60 nm in diameter were assembled also when a transcription plasmid containing the full-length HCV genome was transfected into the HepG2 cell culture (Myung et al. 2001). Dash et al. (2001) showed later that these so-called HCV VLPs that were produced from full-length HCV clone in HepG2 cells caused infection in a chimpanzee.

The assembly of the structural HCV proteins into the noninfectious HCV VLPs was achieved for the first time in mammalian cells only when recombinant Semliki Forest virus (SFV; Blanchard et al. 2002, 2003; Ait-Goughoulte et al. 2006; Hourioux et al. 2007), vesicular stomatitis virus (VSV; Ezelle et al. 2002), or YF (Molenkamp et al. 2003) replicons expressing the appropriate genes were employed.

At the same time, Wakita et al. (2005) generated a cell culture system that reproduced the complete life cycle of HCV in vitro by using a cloned HCV genome. Using this system, Kushima et al. (2010) revealed that the disulfide-bonded dimer of protein C was formed by a single cysteine residue at aa position 128, a highly conserved residue among almost all reported isolates. The Cys128 residue was therefore responsible for the correct VLP production but had no effect on the replication of the HCV RNA genome or the several known functions of the protein C, including RNA binding ability and localization to the lipid droplet (Kushima et al. 2010).

The HCV VLPs that were produced in mammalian cells using an adenovirus-based system generated particles that were reported to resemble the native virions morphologically (Chua et al. 2012; Kumar et al. 2016). The immunization of mice with these adenovirus-derived HCV VLPs in combination with adjuvants led to significant antibody response (Chua et al. 2012). In a heterologous prime-boost strategy, immunization with recombinant adenoviruses encoding the HCV structural proteins as a final booster, following priming with the HCV VLPs, resulted in enhancement of both antibody and T-cell responses (Kumar et al. 2016).

Furthermore, the mammalian cells, namely the human hepatocyte-derived cell line Huh7, served as a real source of the potent HCV VLP vaccine. First, Earnest-Silveira et al. (2016b) produced VLPs of 40–80 nm in size, which incorporated the proteins C, E1, and E2 of HCV genotype 1a, elucidated their biochemical and morphological properties, as well as fine details of immune response in mice. Second, Earnest-Silveira et al. (2016a) presented a reliable protocol for the large-scale production of the quadrivalent HCV VLPs formed by the HCV proteins C, E1, and E2 of the genotypes 1a, 1b, 2a, or 3a after coexpression in Huh7 cell factories using a recombinant adenoviral expression system. The high efficiency of the quadrivalent HCV VLP vaccine was demonstrated in mice (Christiansen et al. 2018a, b) and pigs, the closest animal model to humans, after primates (Christiansen et al. 2019).

The availability of the quadrivalent HCV VLPs had enabled the structural studies. The atomic force microscopy (AFM) was used to define morphological and nanomechanical properties of VLPs representing four common genotypes of hepatitis C virus (Collett et al. 2019). The significant differences in size of the VLPs were observed, and particles demonstrated a wide range of elasticity. Remarkably, all VLPs were shown to be glycosylated in a manner similar to native viral particles. Concerning glycosylation of the HCV envelope, a recent exhaustive review could be recommended (Ströh and Krey 2020).

Recently, He et al. (2020) proposed a novel approach to the putative HCV vaccine through E2 optimization and nanoparticle display. The authors redesigned variable region 2 in a truncated form, termed tVR2, on E2 cores derived from genotypes 1a and 6a, resulting in improved stability and antigenicity, and they produced them in mammalian cells. The crystal structures of the three optimized E2 cores with human cross-genotype neutralizing antibodies revealed how the modified tVR2 stabilized E2 without altering key neutralizing epitopes. Then, the E2 cores were displayed on 24- and 60-meric ferritin nanoparticles. In mice, these nanoparticles elicited more effective neutralizing antibody responses than soluble E2 cores (He et al. 2020).

At last, it is worth mentioning that a *trans*-encapsidation system similar to the sriVLPs described earlier for flaviviruses was elaborated for HCV by Adair et al. (2009).

23 Order *Nodamuvirales*

The science of today is the technology of tomorrow.

Edward Teller

ESSENTIALS

The *Nodamuvirales* is a small order of nonenveloped viruses, spherical in shape and possessing icosahedral symmetry. The nodaviruses are remarkable since they played a historical role in the nurturing of the VLP field from the very beginning.

According to the current taxonomy (ICTV 2020), the order *Nodamuvirales* is a single member of the class *Magsaviricetes*, which belongs to the *Kitrinoviricota* phylum from the kingdom *Orthornavirae*, realm *Riboviria*. The order *Nodamuvirales* currently involves 2 families, 3 genera, and 11 species. The families are *Nodaviridae* and *Sinhaliviridae*, a small family consisting of a single genus with two species of honeybee pathogens. The *Nodaviridae* family includes two genera, *Alphanodavirus* and *Betanodavirus*, natural hosts of which are insects and fish, respectively (Thiéry et al. 2012; Sahul Hameed et al. 2019).

While Nodamura virus (NoV) serves as a type species of the *Alphanodavirus* genus, another alphanodavirus, namely Flock House virus (FHV), appeared to be of special importance for development of the viral nanotechnology field.

The betanodaviruses cause in marine fish "viral nervous necrosis" or "viral encephalopathy and retinopathy," which is associated with behavioral abnormalities and high mortalities. Striped jack nervous necrosis virus (SJNNV) serves as a type species of the *Betanodavirus* genus. However, another betanodaviruses, namely dragon grouper nervous necrosis virus (DGNNV) and malabaricus grouper nervous necrosis virus (MGNNV), both belonging to the *Redspotted grouper nervous necrosis virus* species, were explored preferably by the VLP generation.

Figure 23.1 presents portraits of the two special representatives of alpha- and betanodaviruses and a general schematic cartoon of the *Nodamuvirales* order members. The virions range from 25–33 nm in diameter, with or without surface projections, and they consist of 180 protein subunits arranged on an icosahedral T = 3 surface lattice. The surface projections were observed by electron microscopy in negatively stained betanodaviruses but not in alphanodaviruses. In contrast to alphanodaviruses, the maturation cleavage of the coat protein is not observed in betanodaviruses, and they contain 180 copies of a single full-length coat protein of 42 kDa. Accordingly, the image reconstruction of the MGNNV VLPs indicated that the coat protein of betanodaviruses has a two-domain structure compared to the single-domain structure of the coat protein of alphanodaviruses (see Figure 23.1).

GENOME

Figure 23.2 shows the genomic structure of the typical representatives of the order. As reviewed by the latest ICTV report (Sahul Hameed et al. 2019), the nodaviral genome is bipartite and consists of two single-stranded positive-sense RNA of 3.1 kb (RNA1) and 1.4 kb (RNA2) encoding an RNA-dependent RNA polymerase (RdRp) of about 112 kDa (983–1014 amino acids) and the coat protein precursor α, respectively. The genome translation occurs from capped genomic and subgenomic RNAs. Both RNA molecules are capped therefore at their 5'-ends, lack polyadenylation at their 3'-ends, and are encapsidated in the same virus particle. The subgenomic RNA3 is not encapsidated into virions.

GENUS *ALPHANODAVIRUS*

EXPRESSION

The alphanodavirus FHV served as one of the earliest VLP models. When insect *Spodoptera frugiperda* cells were infected with a recombinant baculovirus encoding a cDNA copy of the FHV RNA2, the coat protein α assembled into the VLP precursor particles that matured normally by autocatalytic cleavage of the protein α into polypeptide chains β of 38 kDa and γ of 5 kDa (Schneemann et al. 1993). The VLPs were morphologically indistinguishable from the authentic FHV and contained RNA derived from the coat protein message. The expression of mutants in which Asn-363 at the β-γ cleavage site of the protein α was replaced by either Asp, Thr, or Ala residues resulted in the VLPs that were cleavage defective (Schneemann et al. 1993).

The insect cells-produced FHV VLPs were incredibly important in the investigation of the specific mechanism by which the bipartite nodaviral RNA genome was selected for encapsidation in the native FHV virions (Krishna et al. 2003; Venter et al. 2005, 2007).

Gopal and Schneemann (2018) published a detailed protocol for the generation of the FHV VLPs in insect cells, including small-scale production of VLPs from *Sf*21 cells and large-scale production of them from *Trichoplusia ni* (*T. ni* or HighFive) cells.

Johnson et al. (2004) achieved the baculovirus expression system-driven generation of VLPs of Pariacoto virus (PaV), another member of the genus *Alphanodavirus*.

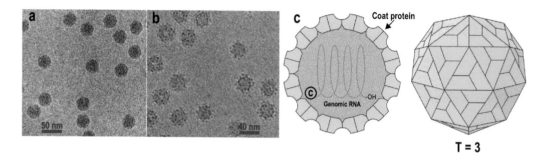

FIGURE 23.1 The portraits and a schematic cartoon of the typical representatives of the *Nodamuvirales* order. The electron cryo-micrographs of (a) Flock House virus (FHV) (Courtesy of N. Olson and T. Baker and the International Committee on Taxonomy of Viruses (ICTV), https://talk.ictvonline.org/ictv-reports/ictv_online_report/positive-sense-rna-viruses/w/nodaviridae) and (b) VLPs of malabaricus grouper nervous necrosis virus (MGNNV) produced in insect cells. (Courtesy of L. Tang and J.E. Johnson and the International Committee on Taxonomy of Viruses [ICTV], https://talk.ictvonline.org/ictv-reports/ictv_online_report/positive-sense-rna-viruses/w/nodaviridae/1002/genus-betanodavirus.) (c) The general cartoon of the *Nodamuvirales* order representative virions is taken with kind permission from the ViralZone, Swiss Institute of Bioinformatics, http://viralzone.expasy.org (Hulo C et al. 2011). (Courtesy Philippe Le Mercier.)

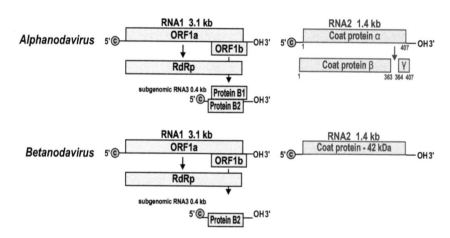

FIGURE 23.2 The genomic structure of the family *Nodaviridae* members, derived from the recent ICTV profile (Thiéry et al. 2012; Sahul Hameed et al. 2019). The structural genes are colored dark pink.

After insect cells, FHV was found to produce virions in plants (Selling et al. 1990) and yeast (Price et al. 1996) after transfection with genomic RNA, opening therefore wide-ranging applications of the FHV model.

3D STRUCTURE

The first atomic resolution structure of an insect virus determined by single crystal x-ray diffraction was published at 3.0 Å (Hosur et al. 1987) and refined later to 2.8 Å (Wery et al. 1994) for black beetle (BBV) alphanodavirus.

Figure 23.3a presents this T = 3 structure, together with other known 3D structures of nodaviruses. All known structures of the *Nodaviridae* family revealed the T = 3 icosahedral capsids assembled from 180 capsid proteins that had an eight-stranded antiparallel β-barrel topology.

The crystal structures, at 2.8 Å resolution, of the T = 3 FHV virions (Fisher et al. 1992) and mutant cleavage-defective provirion-like FHV particles obtained in baculovirus-driven expression system (Fisher et al. 1993) followed. Fisher and Johnson (1993) published the FHV crystal structure that revealed an ordered duplex RNA fragment that controlled capsid architecture by interaction with a helical protein domain of the subunit located inside the capsid shell. One of the helices that bound the RNA was part of a 44-aa polypeptide γ that was autocatalytically cleaved from the initial subunit translation product after virion assembly. Later, Tihova et al. (2004) found that the baculovirus system-expressed FHV particles encapsidated viral and heterologous RNAs having virtually identical dodecahedral structures adjacent to the coat protein, independent of sequence and length, and indicating that the FHV coat protein and not the sequence of the nucleic acid controlled the organization of the packaged RNA. Thus, the N-terminal basic segment of the coat protein known to interact with the packaged genome was not required for the RNA encapsidation (Tihova et al. 2004). The same conclusions followed, when the baculovirus system-expressed PaV

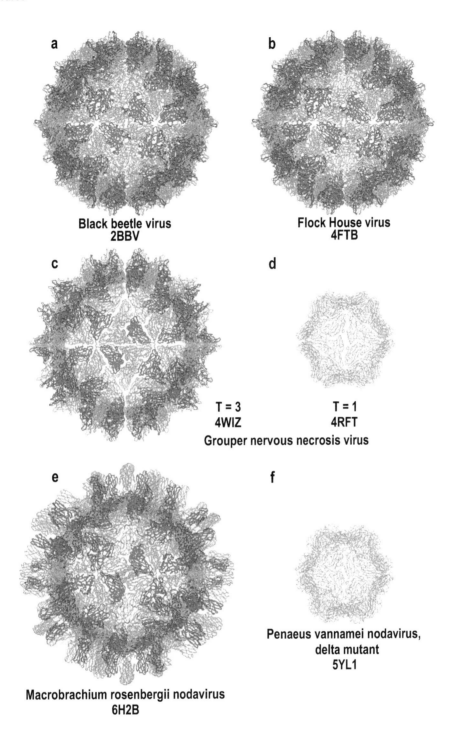

a Black beetle virus
2BBV

b Flock House virus
4FTB

c T = 3
4WIZ

d T = 1
4RFT

Grouper nervous necrosis virus

e Macrobrachium rosenbergii nodavirus
6H2B

f Penaeus vannamei nodavirus,
delta mutant
5YL1

FIGURE 23.3 The crystal structure of the *Nodaviridae* family members. (a) Black beetle virus, 2.8 Å resolution (Wery et al. 1994). (b) Flock House virus, 2.7 Å resolution (Speir et al. 2012). (c, d) Grouper nervous necrosis virus, 3.1 Å resolution, T = 3 (c) and T = 1, 3.1 Å resolution (d) (Chen NC et al. 2015); (e) Macrobrachium rosenbergii nodavirus, 3.28 Å resolution (Ho et al. 2018); (f) T = 1 subviral particle of the Penaeus vannamei nodavirus coat protein deletion mutant (delta 1–37; 251–368; Chen NC et al. 2017, 2019). The 3D structures are taken from the VIPERdb (http://viperdb.scripps.edu) database (Carrillo-Tripp et al. 2009). The size of particles is in scale. The corresponding protein data bank (PDB) ID numbers are given under the appropriate virus names.

VLPs were investigated by electron cryomicroscopy and image reconstruction at 15.4-Å resolution (Johnson et al. 2004).

Then, matrix-assisted laser desorption/ionization mass spectrometry combined with time-resolved, limited proteolysis was suggested to examine the dynamic FHV processes such as assembly, maturation, and cell entry (Bothner et al. 1998).

Figure 23.3b presents the crystal structure of the authentic FHV particle at 2.7 Å resolution (Speir et al. 2012).

The deletion of 50 N-terminal aa residues inhibited complete self-assembly of the FHV coats, while restriction of the deletion to 31 aa resulted in a heterogenous outcome of small bacilliform-like structures and irregular structures, as well as wild-type-like T = 3 particles, but not of the expected T = 1 particles (Dong et al. 1998).

The mutations at the calcium-binding sites affected FHV capsid stability and drastically reduced virus infectivity, without altering the overall architecture of the capsid. The mutations also altered the conformation of the protein γ, functioning as a membrane-disrupting agent and usually sequestered inside the capsid, by increasing its exposure under neutral pH conditions (Banerjee et al. 2010).

The crystal structure of NoV, a classical reference strain of alphanodaviruses, was determined to 3.3 Å resolution by Zlotnick et al. (1997), while Tang et al. (2001) resolved crystals of PaV at 3.0 Å.

CHIMERIC VLPs

Tisminetzky et al. (1994) were the first who proposed application of the FHV coat protein as a carrier of foreign immunological epitopes. Thus, thanks to the existing crystal structure and using molecular modeling, the authors selected four different outer loops of the FHV coat protein, which connected β-sheets of the eight-stranded β-barrel structure, to expose IGPGRAF heptapeptide, a putative neutralization domain representing aa 314–320 within the V3 loop of the gp120 protein in the HIV-1 strains MN, SC and SF2. The resulting fused genes were expressed in E. coli, but no signs of self-assembly were observed. Next, Schiappacassi et al. (1997) used the E. coli expression system to produce FHV coat protein carrying the different V3 loop core sequences, such as IQRGPGRAF (strain IIIB), KGPGRVI (RF), IGPGRTL (NY5), IGLGQAL (Z), and FGRGQAL (MAL). The fused genes were constructed by deleting aa residues 269–272 at the L2 loop of the FHV coat and replacing them with the HIV-1 epitopes. The chimeric proteins appeared in inclusion bodies and were resuspended in 8 M urea; the possible renaturation and self-assembly of them was not addressed. However, display of the different HIV-1 peptides at this site of the FHV coat protein allowed more specific serotyping of patients' sera and was suggested as a tool for the correct evaluation of the immune response against different V3 loop core sequences (Schiappacassi et al. 1997).

Next, Buratti et al. (1997) introduced three immunodominant fragments of the hepatitis C virus (HCV) core protein, namely, aa residues 1–20, 21–40, and 32–46, at the five selected loops. The chimeras gained in E. coli were not expected to self-assemble but were used successfully as diagnostic tools (Buratti et al. 1997).

Lorenzi and Burrone (1999) inserted short sequences of 7 or 10 aa residues of human IgE into one of the FHV capsid loops, purified the chimeric products from E. coli lysates in denaturing conditions, used them successfully to immunize rabbits, and established induction of specific antibody response against the short insertions. No attempts to get VLPs have been made.

In parallel, Buratti et al. (1996) used another non-VLP strategy, when they displayed the short gp41-neutralizing epitopes identified within aa residues 735–752 of the HIV-1 gp160 protein, at one of the selected FHV capsid loops. In this case, only a short fragment of the chimeric HIV-FHV genes was expressed in E. coli, in the form of a GST fusion, to solubilize products, and the VLP formation was not planned at all.

Later, an attempt to construct combined HBV/HCV vaccine was achieved, and a set of selected B-cell epitopes derived from the HBV surface protein (aa 124–147) and HCV core (aa 2–21 and 22–40) and E1 (aa 315–328) proteins was inserted at the five previously modeled sites of the FHV coat protein (Xiong et al. 2005). The resulting fused genes were expressed in E. coli, but the chimeric proteins appeared in the form of inclusion bodies and needed to be solubilized in urea. Chen Y et al. (2006) demonstrated reactivity of these epitopes with antisera from hepatitis B and C patients and induction of immune response in guinea pigs. However, no attempts to renature chimeras and get VLPs were undertaken.

In contrast to these E. coli trials, Scodeller et al. (1995) accomplished the baculovirus-driven production of the V3-loop epitope-carrying FHV VLPs in insect cells. This time, the 314-IGPGRAF-320 epitope was introduced into two positions of the FHV coat protein, first, in such a way that it replaced four original aa residues and regenerated aa 127–130 and 135 of the FHV coat protein. Second, the insertion introduced eight foreign aa residues between positions 305 and 306: seven from IGPGRAF, plus an extra E residue at the end. Third, both insertions were introduced at the same carrier protein. Unlike the E. coli expression, formation of the chimeric HIV-1-FHV VLPs was established by analysis of cell lysates in sucrose gradients, where chimeras were found as a broad sedimentation peak at the same position as authentic virions. Moreover, the maturation cleavage was observed with the chimeric constructs. Although it was strongly suggested that the chimeric proteins were assembled into VLPs, no electron microscopy was performed, and the whole-cell lysates were used as immunogens to inoculate guinea pigs. A high neutralizing capacity against the homologous strain MN, as well as heterologous strain IIIB was induced by the first variant of the V3-FHV chimera only (Scodeller et al. 1995).

Later, Manayani et al. (2007) used the baculovirus-driven expression system to display 180 copies of the high affinity, *Bacillus anthracis* protective antigen (PA)-binding von Willebrand A domain (VWA) of the ANTXR2 cellular receptor. The chimeric VLPs correctly displayed the receptor VWA domain of 181 aa in length (aa 38–218) on their surface, when inserted at the FHV coat positions 206–209 (chimera 206) or 264–268 (chimera 264), and inhibited lethal *B. anthracis* toxin action in in vitro and in vivo models of anthrax intoxication. Moreover, the VLPs complexed with the PA elicited a potent toxin-neutralizing antibody response that protected rats from anthrax lethal toxin challenge after a single immunization without adjuvant, thus representing a novel and highly effective, dually acting reagent for treatment and protection against anthrax (Manayani et al. 2007). Figure 23.4 demonstrates the corresponding electron micrographs and pseudoatomic models. The prominent peptide loops at aa positions 206 and 264, which formed trimers on the particle surface (Schneemann et al. 1998) and were selected as the insertion sites, fully met the expectations and ensured successful display of

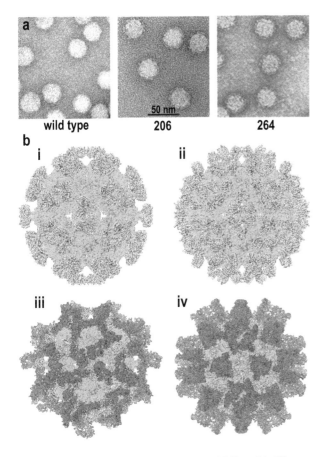

FIGURE 23.4 The VWA$_{ANTXR2}$-FHV VLPs. (a) Electron micrographs of gradient-purified wild-type and chimeric VLPs negatively stained with uranyl acetate. (b) Pseudoatomic models of VWA$_{ANTXR2}$-FHV chimeras. X-ray coordinates of FHV capsid protein (green) and ANTXR2 VWA domain (yellow) were docked into the cryoEM density of chimera 206 (i) and chimera 264 (ii). Panels show surface views of the particles in the absence of the cryoEM density maps. Note the different distributions of the ANTXR2 domains on the surfaces of the VLPs. (iii, iv) In silico model of PA$_{83}$ bound to the surface of VWA$_{ANTXR2}$-FHV chimeras. PA83 (purple) was modeled onto the surface of chimera 206 (iii) and chimera 264 (iv) using the known high resolution X-ray structure of the ANTXR2-VWA/PA63 complex as a guide (Lacy et al. 2004). Panels show surface views of the entire particles to illustrate the difference in occupancy of PA83 on the two chimeras. (Reproduced from *PLoS Pathog*, Manayani DJ, Thomas D, Dryden KA, Reddy V, Siladi ME, Marlett JM, Rainey GJ, Pique ME, Scobie HM, Yeager M, Young JA, Manchester M, Schneemann A, A viral nanoparticle with dual function as an anthrax antitoxin and vaccine, 3, 2007, 1422–1431.)

foreign peptides and entire proteins up to 20 kDa in size (Manayani et al. 2007).

At that productive time, Destito et al. (2009) published an exhaustive review presenting in a comparative way two great VLP models, FHV- and cowpea mosaic virus (CPMV)-based ones. The CPMV model is presented in Chapter 27.

Next, the great Annette Schneemann and her team demonstrated that the display of fragments from the influenza virus hemagglutinin (HA) stem region on the insect-cells produced FHV VLPs resulted in the induction of cross-reactive anti-HA antibodies (Schneemann et al. 2012). The 20-aa residue A-helix of the HA2 chain was selected as an epitope and displayed in 180 copies on the chimeric FHV VLPs. Figure 23.5 illustrates the structural outcome of this delicate structural design. As shown, the structure of helix-turn-helix protein B2 (PDB 2AZ2 and 2AZ0) was spliced first into the structure of the FHV capsid subunit in place of the 206 or 264 loops. The residues near the B2 termini in the best locations to start and end the inserts (i.e., unstructured polypeptide) while maintaining exposure and structure of the B2 helical segments were identified. To display the antigenic helix identified in the structures of the antibody CR6261-HA complexes (PDB 3GBN and 3GBM) (Ekiert et al. 2009), a portion of the long helix in B2 (residues 40–59) was then replaced with the HA2 A-helix (residues 39–58) from the human 1918 H1N1 pandemic virus (A/South Carolina/1/1918, PDB 3GBN). The A-helix was specifically oriented within the B2 helical turns so that residues in contact with the CR6261 antibody had maximum solvent exposure on the surface.

Immunologically, the A-helix was sufficient for generating an antibody response against a range of hemagglutinin subtypes in group 1, but these antibodies were unable, however, to neutralize influenza virus (Schneemann et al. 2012).

Bajaj and Banerjee (2016) used *E. coli* expression system to demonstrate the in vitro assembly of polymorphic VLPs from the FHV coat protein. In this case, the FHV coat protein was provided with the removable N-terminal His$_6$ tag separated by thrombin cleavage site, while another His$_6$ tag was introduced between positions 206 and 209 and replacing therefore aa residues 207 and 208. This site corresponded to the region exposed as a loop on the surface, capable of accommodating insertions without compromising any quaternary interactions, discovered earlier by Manayani et al. (2007) and discussed earlier. Despite the altered structure, these particles were capable of membrane disruption, like native viruses, and they were capable of incorporating and delivering foreign cargo to specific locations (Bajaj and Banerjee 2016).

GENUS *BETANODAVIRUS*

Lin CS et al. (2001) cloned and sequenced the RNA2 segment of two grouper viruses isolated from *Epinephelus malabaricus* (MGNNV) and *Epinephelus lanceolatus* (DGNNV). The sequences of the two RNAs were 99% identical, and comparison with previously sequenced

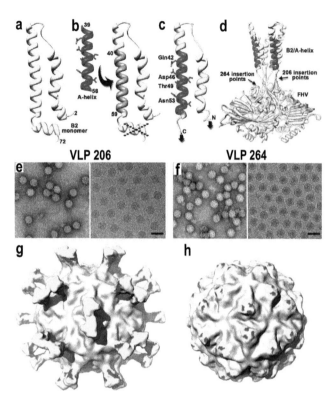

VLP 206   **VLP 264**

FIGURE 23.5 Design, construction, and structure of VLPs that multivalently display the influenza virus HA2 A-helix. (a) Monomer of the B2 protein shown as a ribbon diagram. Its small size and predominantly helical structure are valuable as a cassette for peptide substitutions. (b) The major antigenic site of the HA2 A-helix (magenta) is substituted for an equivalent length of helical residues in the long B2 helix (orange). B2 is slightly truncated to make the N and C termini meet at approximately the same position for insertion into loops. (c) The final B2/A-helix construct (yellow and magenta, respectively). The arrows denote the attachment points to the FHV capsid protein. (d) Trimer of modified FHV capsid proteins (gray) showing the complete design of the VLP with the B2/A-helix construct inserted at the 206 site. The 264 site is also marked. (e) to (h) Characterization and structure of particles produced after expression of the B2/A-helix construct inserted at the 206 (e and g) and 264 (f and h) sites of FHV. (e and f) Negative-stain (left) and cryo-EM (right) images of the 206 and 264 particles, respectively. The images show assembled icosahedral particles ~400 Å in diameter for both constructs. Scale bars, ~50 nm. (g and h) Image reconstructions of the 206 and 264 particles, respectively, at 14-Å resolution from the cryo-EM data (same color scheme as in panel d). The colored regions on the 264 particle (h) show where the inserts, if visible, would be positioned. (Reprinted with permission of American Society for Microbiology from Schneemann A et al. *J Virol.* 2012;86:11686–11697.)

RNA2 segments of such fish nodaviruses as striped jack nervous necrosis virus (SJNNV), Atlantic halibut nodavirus (AHNV), seabass nervous necrosis virus (SBNNV), and greasy grouper nervous necrosis virus (GGNNV) revealed that MGNNV and DGNNV were most closely related to GGNNV. The MGNNV coat protein was expressed in *Sf*21 cells with a recombinant baculovirus system, and VLPs were spontaneously formed. The two

types of VLPs were observed: a slower sedimenting particle was RNase-sensitive and stain-permeable, while the faster sedimenting particle survived RNase treatment and was not stain-permeable. An image reconstruction of the latter, obtained with electron cryomicroscopy data, revealed a morphology consistent with the T = 3 quasi-symmetry but with features significantly different from insect nodavirus structures at the same resolution. This was the first moderate-resolution analysis of a fish nodavirus by electron cryomicroscopy and image reconstruction (Lin CS et al. 2001).

The further electron microscopy evaluation of the MGNNV VLPs confirmed the idea that the MGNNV coat protein had two domains resembling those of tomato bushy stunt virus (see Chapter 24) and Norwalk virus (see Chapter 27), rather than the expected single-domain coat protein of insect alphanodaviruses (Tang et al. 2002).

When the insect cells-produced MGNNV VLPs or VLPs obtained after baculovirus-driven expression of the coat protein of SB2, a betanodavirus isolate from clinically affected sea bass *Dicentrarchus labrax*, were used to immunize sea bass, a strong protective immune response against experimental infection with native virus was obtained (Thiéry et al. 2006). The large-scale production and purification of the MGNNV VLPs and SB2 VLPs was performed in *Tricoplusia ni* cells. In fact, this was the first report demonstrating the potential use of VLPs to protect fish against viral infection.

Concerning bacterial expression, the partial protective immunity has been obtained in several fish species by using recombinant betanodavirus coat proteins expressed in *E. coli* but not self-assembled into VLPs (Húsgağ et al. 2001; Tanaka et al. 2001; Yuasa et al. 2002).

Lu et al. (2003) were the first who reported the generation of the betanodaviral VLPs in *E. coli* expressing the full-length ORF encoding the DGNNV coat protein. Two sizes of VLPs were observed, where the heavier particles resembled the native piscine nodavirus in size and stain permeability, and the lighter ones were approximately two-thirds of the full size. The recombinant VLPs blocked attachment of native virus to the cultured fish nerve cells, preventing therefore infection by the native virus (Lu et al. 2003).

Liu et al. (2006) used the *E. coli* bacteria-produced DGNNV VLPs to the intramuscular immunization of grouper, without any adjuvant, and demonstrated a rapid and specific humoral immune response that protected fish against virus infection.

Lai et al. (2014) produced in *E. coli* VLPs of orange-spotted grouper nervous necrosis virus (OGNNV), another representative of the *Redspotted grouper nervous necrosis virus* species of betanodaviruses, after MGNNV and DGNNV. The purified VLPs were practically indistinguishable from virus by electron microscopy, and the intramuscular vaccination with them induced humoral immune responses and activated genes associated with cellular and innate immunity against betanodavirus infection in orange-spotted grouper (Lai et al. 2014).

Lin K et al. (2016) produced the OGNNV VLPs in *E. coli* to immunize orange-spotted grouper by intramuscular injection. Remarkably, the authors were the first who used CpG adjuvant to improve outcome of fish vaccination with betanodaviral VLPs.

Chien et al. (2018) demonstrated successful oral vaccination of fish with the *E. coli* cells-produced His-tagged OGNNV VLPs that were self-assembled in vitro after purification and did not contain therefore any bacterial components.

Although production of the VLPs was not addressed specially, it is noteworthy that the cell-free protein synthesis (CFPS) system derived from crude *E. coli* cell extracts was examined to produce the His-tagged coat protein of sevenband grouper nervous necrosis virus (Kim JO et al. 2015). The remarkable efficiency of the CFPS is demonstrated in Chapter 25, in connection with RNA phage VLPs, when CFPS was shown to enhance the VLP production up to 14 times in comparison with other systems.

The bacterial expression in *E. coli* contributed strongly to the deep 3D resolution of betanodaviruses. Thus, Chen NC et al. (2015) published the crystal structure of grouper nervous necrosis virus (GNNV) in the four various forms: (i) a complete T = 3 GNNV-like particles at 3.6 Å resolution; (ii) T = 1 subviral particles of the delta-P-domain mutant at 3.1 Å; (iii) the N-terminal arginine-rich motif (N-ARM) deletion mutant at 7.0 Å; and (iv) the individual P-domain of the GNNV coat protein at 1.2 Å. The crystal structure of the T = 3 GNNV VLPs demonstrated several significant and distinct variations in capsid architecture and molecular mechanisms of capsid assembly compared to the genus *Alphanodavirus* and other RNA viruses. Various forms of the T = 3 and T = 1 particles showed that the N-ARM acted as a molecular switch. Second, the P-domain, with its DxD motif together with two bound Ca^{2+} ions, played a pivotal role in the trimerization of the GNNV coat protein and the particle assembly (Chen NC et al. 2015). Figure 23.3c, d demonstrates both GNNV T = 3 and T = 1 VLPs. Later, Wang et al. (2021) improved the 3D structure of the *E. coli*-produced DGNNV VLPs to 2.7 Å resolution by an advanced electron microscopy methodology.

Choi et al. (2013) were the first who generated the red-spotted grouper nervous necrosis virus (RGNNV) VLPs in yeast, namely, *S. cerevisiae*. The heparin chromatography-purified homogeneous nonaggregated VLPs of 25 nm in diameter elicited specific antibodies in mice. Kim HJ et al. (2016) demonstrated later that the coat protein finally underwent assembly during the chromatography and needed the presence of Ca^{2+} ions to enhance the assembly yields.

The parenteral administration of such yeast-produced RGNNV VLPs, without any adjuvants, elicited neutralizing antibody and provided the fish with full protection against RGNNV challenge, whereas oral administration provoked neutralizing antibody systemically and conferred protective immunity against virus challenge, however, only 57% of the fish survived (Wi et al. 2015). However, Kim HJ et al. (2014) demonstrated that the oral immunization

with whole *S. cerevisiae* yeast producing the RGNNV coat protein provoked a stronger humoral immune response than the purified VLPs. Consuming feed with the lysed recombinant yeast that expressed the RGNNV coat protein ensured strong protection against challenge with RGNNV (Cho et al. 2017).

Luu et al. (2017) used the dimorphic nonpathogenic yeast *Yarrowia lipolytica* as a host to express the RGNNV coat protein. When the number of integrated copies of the coat gene expression cassette was increased up to eight copies, the appropriate VLPs were detected. It is noteworthy that this was the first report on efficient expression of viral coat proteins as VLPs in *Y. lipolytica*, demonstrating its high potential as a novel VLP expression system.

At last, the famous George P. Lomonossoff's team produced VLPs of Atlantic cod nervous necrosis virus (ACNNV) by transient expression of the coat protein in *Nicotiana benthamiana* plants (Marsian et al. 2019). The correct VLPs were also produced in transgenic tobacco BY-2 cells. The high resolution electron cryomicroscopy of the obtained VLPs to 3.7 Å revealed a similar structure to those previously reported for the GNNVs and described earlier. When administered either intraperitoneally or intramuscularly to sea bass, the VLPs conferred partial protection against subsequent challenge with ACNNV (Marsian et al. 2019).

Concerning chimeric derivatives of the betanodavirus VLPs, Xie et al. (2016) not only resolved crystal structure of the *E. coli* cells-produced OGNNV VLPs at 3.9 Å but also inserted a His_6 tag into predicted insertion sites, resulting in four His-tagged VLPs at the following positions: N-terminus, Ala_{220}, Pro_{292}, and C-terminus. The His tags of the N-terminal chimera were concealed inside virions while those at Ala_{220} and C-terminus were displayed at the outer surface. As to the vector capacity, the N-terminus and Ala_{220} were able to carry short peptides, and the C-terminus even accommodated a large protein such as GFP to generate fluorescent VLPs. Therefore, the C-terminus of the betanodaviral coat protein appeared as a suitable site to accommodate foreign peptides. The His-tagged VLPs retained the same cell entry ability in the Asian sea bass cell line and provoked immune responses as strong as the original VLPs in a sea bass immunization assay (Xie et al. 2016).

UNCLASSIFIED NODAVIRUSES

SHRIMP NODAVIRUSES

Macrobrachium rosenbergii nodavirus (MrNV) has been known since 1994 and threatens the giant freshwater prawn *Macrobrachium rosenbergii* aquaculture, causing white tail disease in the prawn species that leads to 100% lethality of the infected postlarvae. Although MrNV has been proposed to be grouped into a new genus, the *Gammanodavirus* that infects prawns and would comprise MrNV and Penaeus vannamei nodavirus (PvNV) infecting shrimps (NaveenKumar et al. 2013), it is not yet classified but nevertheless clearly relates to viruses in the family *Nodaviridae* (Thiéry et al. 2012; Sahul Hameed et al. 2019).

First, the coding region of the MrNV coat protein of 371 aa residues, provided with the His tag and myc epitope sequences, was expressed in *E. coli* (Goh et al. 2011). After purification using immobilized metal affinity chromatography, the MrNV VLPs with a diameter of about 30 ± 3 nm were revealed by electron microscopy and confirmed then by dynamic light scattering (DLS). Using N-terminal and internal deletion mutagenesis, Goh et al. (2014) mapped the RNA binding region of the MrNV coat protein and found that this region was dispensable for the assembly into VLPs. Despite the first success, the *E. coli* cells-produced MrNV coat protein was characterized later as unstable, degrading, and forming heterogenous VLPs, probably due to partial degradation of the N-terminal end of the coat protein in bacteria (Goh et al. 2011).

Next, the MrNV coat protein gene, also provided with the His_6 tag, was expressed in *E. coli* by Farook et al. (2014). The product was purified by affinity chromatography and used to raise antibodies in rabbits for diagnostics but without any indications concerning self-assembly.

To overcome the problems with bacterial expression, Kueh et al. (2017) performed baculovirus-driven production of the MrNv VLPs in the *Sf9* cells. According to electron microscopy and DLS data, the recombinant protein produced by insect cells self-assembled into the highly stable, homogenous VLPs of approximately 40 nm in diameter. Therefore, the MrNV VLPs produced in *Sf9* cells were about 10 nm bigger, had a uniform morphology compared with the VLPs produced in *E. coli*, and fulfilled the expectations as the truly icosahedral T = 3 particles, although the expressed gene variant was provided with additional 25 aa residues at its N-terminal end, which contained the His_6 tag. Moreover, Kueh et al. (2017) were first who described the capability of MrNV VLPs to package both RNA and DNA of approximately 48 kb in size and suggested possible application of this phenomenon to the nucleic acid delivery and therefore a possible contribution to the field of gene therapy.

The first 3D reconstruction of the baculovirus system-produced MrNV VLPs was determined by electron cryomicroscopy at a resolution of 7 Å (Ho et al. 2017). The 3D reconstruction revealed the T = 3 icosahedral assembly that showed a striking divergence from the known structures of other nodavirus capsids. This was characterized by the presence of large dimeric blade-like spikes on the outer surface. Although the VLPs assembled in a heterologous system and in the absence of the full-length viral genome, the capsids contained density consistent with the encapsidation of RNA, which was assumed to be the cognate mRNA. Overall, the considerable morphological differences seen when comparing the structure of MrNV to other known nodaviruses supported the assertion that this virus might belong to a new genus of nodaviruses (Ho et al. 2017).

Furthermore, Ho et al. (2018) presented an atomic-resolution model of the MrNV coat protein, calculated by the electron cryomicroscopy of the insect cells-produced MrNV VLPs and a 3D image reconstruction at 3.3 Å resolution, as well as electron cryomicroscopy of MrNV virions purified from infected freshwater prawn postlarvae, which yielded altogether a 6.6 Å resolution structure and confirmed the biological relevance of the VLP structure. Thus, the MrNV coat consisted of the shell (S) and the protruding (P) domains, ranging from aa residues 1–252 and 253–371, respectively. The data revealed that, unlike other known nodavirus structures, which have been shown to assemble capsids having trimeric spikes, the MrNV coat assembled a T = 3 capsid with dimeric spikes. Moreover, Ho et al. (2018) found a number of surprising similarities between the MrNV capsid structure and that of the *Tombusviridae* family from the order *Tolivirales* (see Chapter 24), such as the extensive network of N-terminal arms, forming long-range interactions to lace together asymmetric units, a very similar fold exhibited by the protruding spike domain, stabilization of the capsid shell by three pairs of Ca^{2+} ions in each asymmetric unit. These structural similarities raised further questions concerning the taxonomic classification of MrNV and the related PvNV (Ho et al. 2018).

Chong et al. (2019) expressed in *E. coli* the recombinant domain P of the MrNV, purified it with an immobilized metal affinity chromatography and directly revealed its dimeric form with the hydrodynamic diameter of ~6 nm and 67.9% of β-sheets but without α-helical structures. This was in a good agreement with the electron cryomicroscopic analysis of MrNV, which demonstrated that the P-domain contains only β-stranded structures.

In parallel with MrNV, Chen NC et al. (2019) determined the electron cryomicroscopic structure of T = 3 and T = 1 of the PvNV VLPs, which were produced in *E. coli*, as well as crystal structures of the protrusion-domains (P-domains) of MrNV and PvNV, and the crystal structure of the ΔN-ARM-PvNV shell-domain (S-domain) in the T = 1 subviral particles. The capsid protein of PvNV revealed therefore five domains: the P-domain with a new jelly-roll structure forming cuboid-like spikes; the jelly-roll S-domain with two Ca^{2+} ions; the linker between the S- and P-domains exhibiting new cross and parallel conformations; the N-arm interacting with nucleotides organized along icosahedral 2-fold axes; and a disordered region comprising the basic N-terminal arginine-rich motif (N-ARM) interacting with RNA. The N-ARM controlled T = 3 and T = 1 assemblies (Chen NC et al. 2019). Figure 23.3 e, f demonstrates the T = 3 and T = 1 structures in question.

Selvaraj et al. (2021) markedly improved the yield of the MrNV VLPs in *E. coli* by identification of the proteases responsible for the degradation of the recombinant product and by special modulation of proteolytic activity.

The MrNV VLPs remarkably contributed to the viral nanotechnology as a novel promising platform to display foreign epitopes. First, a putative hepatitis B vaccine was constructed. Thus, Yong et al. (2015a) displayed the determinant "a" located at aa 121–149 of surface protein (HBs) of hepatitis B virus (see Chapter 37). The MrNV coat of

1–377 aa residues was provided C-terminally with the HBs "a" determinant of 49 aa residues corresponding to the HBs 111–159 aa stretch, myc epitope EQKLISEEDL and His_6 tag. The total length of the C-terminal addition reached 79 aa residues. The chimeric gene was expressed in *E. coli* and ensured generation of the appropriate VLPs. The immunization of mice with the purified chimeric VLPs induced specific antibodies against the HBs determinant "a" as well as more natural killer and cytotoxic T cells, which are vital for virus clearance (Yong et al. 2015a).

Since the *E. coli*-produced MrNV VLPs were less stable than insect cells-produced, as indicated earlier, Ninyio et al. (2020) expressed the chimeric HBs-MrNV coat gene in the *Sf*9 insect cells by baculovirus-driven expression system. The insect cells-produced chimeric VLPs were of different sizes, ranging from ~21 nm to ~55 nm, while those produced in *E. coli* were of ~30 nm in diameter. After immunization of mice, the antibody titer elicited by the *Sf*9-produced chimeric VLPs was higher than those of the *E. coli*-produced VLPs, as well as of the standard anti-hepatitis B vaccine Engerix B consisting of the recombinant yeast-produced HBs particles (see Chapter 37). In addition to the strong humoral immune response, the chimeric VLPs induced remarkable cytotoxic T lymphocyte (CTL) and natural killer cell activities in the immunized mice (Ninyio et al. 2020). Ninyio et al. (2021) characterized the HBs-MrNV VLPs by a set of biophysical and immunological methods.

Second, a prospective influenza vaccine was generated. Yong et al. (2015b) have chosen the N-terminal highly conserved 23-aa peptide ectodomain from matrix 2 (M2) protein, or peptide M2e, as a potential candidate for the development of a universal flu vaccine. More about the nature of the M2e peptide is written in the influenza Chapter 33 about its selection and adjustment as a foreign epitope of choice to be displayed on the VLPs in the RNA phage Chapter 25. Thus, one, three, or five copies of the M2e fragment were added C-terminally to the MrNV coat protein 1–377, prolonged then with the myc epitope and His_6 tag, in total 21 aa, for ease of detection and purification purposes, as in the previously described case of the chimeric HBs-MrNV vaccine. The protein domains were separated by short GGG linkers. In total, the foreign aa stretches reached 50, 102, and 154 aa residues that were added in the case of one, three, and five M2e inserted copies, respectively. The chimeric proteins formed nice VLPs of ~30 nm in diameter, when synthesized in *E. coli*, despite some N-terminal degradation of products, as mentioned also earlier, and were purified easily by immobilized metal affinity chromatography. The immunization of mice with the chimeric VLPs, in the presence of Freund's adjuvant, induced specific anti-M2e antibodies, and the titer was proportional to the copy numbers of the M2e epitope displayed on the M2e-MrNV VLPs. Furthermore, Ong et al. (2019) achieved in mice the immune responses without any adjuvants, which was induced by the chimeric VLPs carrying three M2e copies and demonstrated the protective efficacy of vaccine against influenza A virus H1N1 (A/PR/8/34)

and H3N2 (A/HK/8/68) challenges. The H3N2 and H1N1 strains were used because they were considered the most common influenza A virus subtypes that emerged during annual flu seasons.

At last, after successful generation of the promising chimeric VLP vaccine candidates, the self-assembly/disassembly capability of the MrNV coat protein was exploited to develop nano-carriers for DNA and double-stranded RNA. Thus, Jariyapong et al. (2014) expressed the His_6 tag-provided MrNV VLPs in *E. coli*, disassembled/reassembled them by successive EGTA and Ca^{2+} treatment, and packed about 2–3 molecules of 3.1-kb plasmid DNA copies per particle. These VLPs interacted with cultured insect cells and delivered loaded plasmid DNA into the cells as shown by green fluorescent protein (GFP) reporter. Next, a double-stranded RNA that was constructed to confer protection against white spot syndrome virus (WSSV), a shrimp pathogen, was packed into the MrNV VLPs, and the clear therapeutic effect was demonstrated (Jariyapong et al. 2015a, b).

Thong et al. (2019) used the thermally responsive MrNV VLPs for the targeted delivery of a cancer drug. To achieve this, folic acid (FA) was covalently conjugated to lysine residues located on the surface of the MrNV VLPs, while doxorubicin (Dox) was loaded inside the VLPs using an infusion method. This thermally responsive nanovehicle, namely FA-MrNV VLPs(Dox), released Dox in a sustained manner, and the rate of drug release increased in response to a hyperthermia temperature at 43°C. The nanocontainer enhanced the delivery of Dox to HT29 cancer cells expressing high levels of folate receptor as compared to normal cells and HepG2 cancer cells, which expressed low levels of the folate receptor. As a result, the nanocontainer increased the cytotoxicity of Dox on HT29 cells and decreased the drug's cytotoxicity on the normal and HepG2 cells (Thong et al. 2019). Recently, Somrit et al. (2020) found that the MrNV-VLPs preferentially attached to fucosylated N-glycans in the susceptible gill tissues, what could lead to the development of novel therapeutic delivery strategies.

NEMATODE NODAVIRUSES

Orsay, the first virus discovered to naturally infect *Caenorhabditis elegans* or any nematode, possesses a bipartite, positive-sense RNA genome and appears related to nodaviruses by sequence analyses, although molecular characterizations of Orsay revealed several unique features, such as the expression of a coat-δ fusion protein generated by ribosomal frameshifting and possessing no sequence homologs in GenBank, as well as the use of a noncanonical ATG-independent mechanism for translation initiation. Guo et al. (2014) were the first who overexpressed coat protein of Orsay in *E. coli* to produce recombinant VLPs and reported the crystal structure of the Orsay VLPs. Orsay capsid had the T = 3 icosahedral symmetry with 60 trimeric surface spikes. By analogy with other nodaviruses, each coat protein could be divided into three regions: an N-terminal arm that formed an extended protein interaction network at the capsid interior, an S domain with a jelly-roll, β-barrel fold forming the continuous capsid, and a P domain that formed trimeric surface protrusions. The structure of the Orsay S domain was best aligned, however, to the T = 3 plant RNA viruses but exhibited substantial differences when compared with the insect-infecting alphanodaviruses, which also lack the P domain in their coats. The Orsay P domain was remotely related to the P1 domain in calicivirus and hepatitis E virus. Removing the N-terminal arm in the truncated coat$_{42-391}$ structures produced a slightly expanded capsid with fewer nucleic acids packaged, suggesting that the arm is important for capsid stability and genome packaging. Generally, the Orsay capsid was structurally distinct from those of alphanodaviruses but suggested that the nematode-infecting viruses and betanodaviruses have a common evolutionary history (Guo et al. 2014).

24 Order *Tolivirales*

It is better to understand little than to misunderstand a lot.

Anatole France

ESSENTIALS

The *Tolivirales* is an order of nonenveloped spherical viruses with an icosahedral capsid, which infect insects and plants. According to the current taxonomy (ICTV 2020), the order *Tolivirales* is a single member of the class *Tolucaviricetes*, belonging to the *Kitrinoviricota* phylum from the kingdom *Orthornavirae*, realm *Riboviria*.

The order *Tolivirales* currently involves 2 families, 3 subfamilies, 18 genera, and 96 species. All three subfamilies belong to the large *Tombusviridae* family, which played a visible role by the development of the VLP nanotechnologies of the icosahedral T = 3 capsids. The family *Carmotetraviridae* consists of a single genus *Alphacarmotetravirus* with a single *Providence virus* species including Providence virus (PrV), an animal virus that replicates in plants or a plant virus that infects and replicates in animal cells and possesses T = 4 symmetry (Jiwaji et al. 2019).

The natural host range of the *Tombusviridae* members is relatively narrow, where viruses can either infect mono-cotyledonous or dicotyledonous plants, but no species can infect both (Domier 2012).

Figure 24.1 presents portraits of both icosahedral T = 3 and T = 4 representatives of the order *Tolivirales* and accordingly two schematic cartoons of the order members.

GENOME

Figure 24.2 shows three representative genomic structures of the three typical genera of the *Tolivirales* order. According to the official ICTV statement (Domier 2012), the *Tombusviridae* member virions, with the exception of those of the genus *Dianthovirus*, contain a single molecule of positive sense, linear ssRNA, with a size ranging from 3.7 to 4.8 kb, depending on the genus. The *Dianthovirus* virions contain two genomic RNAs of approximately 3.8 kb and 1.4 kb. The 5'-ends of the genomic RNAs are uncapped, while the 3'-ends are not polyadenylated. The dsRNAs corresponding in size to viral genomic RNA are present in infected tissues. Providence virus (PrV), a single representative of the family *Carmotetraviridae*, has a monopartite genome encoding three ORFs (Walter et al. 2010).

FAMILY *TOMBUSVIRIDAE*

3D STRUCTURE

The *Tombusviridae* family includes 3 subfamilies, 17 genera, and 95 species. The family name is derived from the type species, tomato bushy stunt virus (TBSV), a representative of the genus *Tombusvirus*. The most recent detailed ICTV summary concerning the *Tombusviridae* family was issued by Rochon et al. (2012).

TBSV, as a virus but not as VLPs, appeared as one of the pioneers among the highly resolved T = 3 viruses, when its crystal structure progressed from the 16 Å (Harrison and Jack 1975) to 5.5 Å (Winkler et al. 1977) and then to 2.9 Å (Harrison et al. 1978; Harrison 1980; Olson et al. 1983; Hopper et al. 1984) resolution. Using TBSV coordinates as a general basis, numerous crystal structures of other representatives of the *Tombusviridae* family followed, such as turnip crinkle virus (TCV) of the genus *Betacarmovirus* to 3.2 Å (Hogle et al. 1986), carnation mottle virus (CarMV) of the genus *Alphacarmovirus* to 2.9 Å (Morgunova et al. 1994), tobacco necrosis virus (TNV) of the genus *Alphanecrovirus* to 2.25 Å (Oda et al. 2000), melon necrotic spot virus (MNSV) of the genus *Gammacarmovirus* to 2.8 Å (Wada et al. 2008), panicum mosaic virus (PMV) of the genus *Panicovirus* to 2.9 Å (Makino et al. 2013), cucumber necrosis virus (CNV) of the genus *Tombusvirus* to 2.89 Å and 4.2 Å (Li et al. 2013; Sherman et al. 2017), red clover necrotic mosaic virus (RCNMV) of the genus *Dianthovirus* to 2.9 Å, and cucumber leaf spot virus (CLSV) of the genus *Aureusvirus* to 3.2 Å (Sherman et al. 2020).

Concerning the electron cryomicroscopy trials, the structure of hibiscus chlorotic ringspot virus (HCRSV), genus *Betacarmovirus*, was resolved to 12 Å (Doan et al. 2003). Then, the structure of RCNMV was resolved to 8.5 Å, and the structure was found fully consistent with other species in the *Tombusviridae* family (Sherman et al. 2006). Then, Bakker et al. (2012) used electron cryomicroscopy to resolve the 3D structures of both native and expanded forms of TCV. This allowed direct visualization of the encapsidated single-stranded RNA and coat protein N-terminal regions not seen in the high-resolution x-ray structure of the virion. The expanded form was a putative disassembly intermediate during infection, which arose from a separation of the capsid-forming domains of the coat subunits. Wang et al. (2015) obtained 4 Å-resolution structure of maize chlorotic mottle virus (MCMV), the only member of the *Machlomovirus* genus. Figure 24.3 presents the current concept concerning 3D organization of the *Tombusviridae* members.

DOI: 10.1201/b22819-29

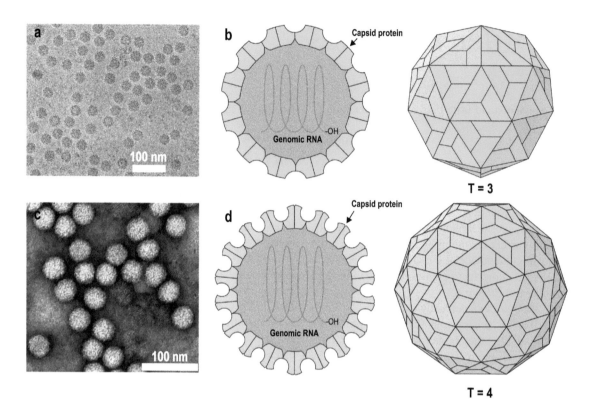

FIGURE 24.1 The portraits and schematic cartoons of the typical representatives of the order *Tolivirales*. (a) Electron cryomicrograph of tomato bushy stunt virus from the genus *Tombusvirus*, subfamily *Procedovirinae*, family *Tombusviridae* (Reprinted from *Micron*, 29, Adrian M et al. Cryo-negative staining, 145–160, Copyright 1998, with permission from Elsevier.) (b) A cartoon explaining the 3D structure of the *Tombusviridae* representatives. (c) Electron micrograph of Providence virus (PrV) from the *Carmotetraviridae* family, the diameter of the virions was estimated to be 40 nm (Reprinted from *Virology*, 306, Pringle FM et al. Providence virus: a new member of the *Tetraviridae* that infects cultured insect cells, 359–370, Copyright 2003, with permission from Elsevier.) (d) A cartoon of the *Carmotetraviridae* representatives. Both cartoons are taken with a kind permission from the ViralZone, Swiss Institute of Bioinformatics, http://viralzone.expasy.org (Hulo C et al. 2011). (Courtesy Philippe Le Mercier.)

The removal of the N-terminal 77 aa residues of the coat by trypsin digestion of purified wild-type virions of hibiscus chlorotic ringspot virus (HCRSV), a member of the genus *Carmovirus*, resulted in only T = 1 empty VLPs (Niu et al. 2014). Moreover, the authors concluded that the HCRSV coat protein was dispensable for viral RNA replication but essential for cell-to-cell movement, and virion was required for the virus systemic movement, while the N-terminal 77 aa stretch including the β-annulus domain was required by the T = 3 assembly in vitro.

The recent advances in heterologous expression in plants allowed production of sufficient quantities of VLPs for structural studies, and high resolution luteovirus structure was determined by electron cryomicroscopy (Byrne et al. 2019). Thus, this study overcame the barriers that have stood in the way of structural characterization of luteovirid capsids for many years. The high-resolution structures for the VLPs of BYDV, as well as of potato leafroll virus (PLRV), which is assigned now to the *Polerovirus* genus of the order *Sobelivirales* (see Chapter 28), were resolved at 3.0 Å and 3.4 Å, respectively. Figure 24.4 demonstrates the appropriate BYDV and PLRV structures.

CHIMERIC VLPs

The *Tombusviridae* family members served as an early platform for the epitope display by genetic fusion. Thus, in the case of the previously mentioned tomato bushy stunt virus (TBSV), the display of foreign epitopes on the capsid surface was accomplished by genetically appending a 13 aa peptide from the HIV-1 gp120 to the C-terminus of the TBSV coat via a three-aa residue linker sequence, without abolishing viral infectivity for *Nicotiana benthamiana*, and the insert was stably maintained over four serial passages (Joelson et al. 1997). These chimeric particles were generated in plants and packaged a full complement of the entire viral genome, hence were infectious. However, when 162 aa residues from the C-terminal part of gp120 were added to the TBSV coat protein, a large portion of the insert was rapidly lost on passaging of the modified virus (Oxelfelt et al. 1995). This first generation of the chimeric TBSV virions was reviewed promptly, among other pioneering models of the time based on plant viruses, by Porta and Lomonossoff (1998).

Later, another member of the *Tombusviridae* family was involved in generation of chimeric viruses. Thus,

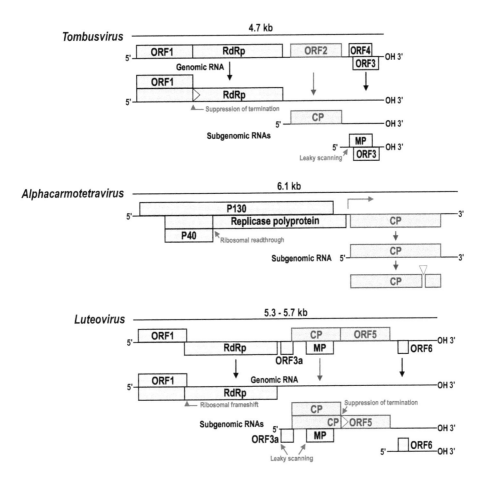

FIGURE 24.2 The typical genomic structures of the members of the *Tolivirales* order, where tomato bushy stunt (TBSV), Providence (PrV), and barley yellow dwarf (BYDV) viruses represent genomes of the genera *Tombusvirus* (family *Tombusviridae*), *Alphacarmotetravirus* (family *Carmotetraviridae*), and *Luteovirus*, respectively. The structures are redrawn from the ViralZone, Swiss Institute of Bioinformatics The structural genes are colored dark pink.

an infectious cDNA clone of tobacco necrosis virus A, a Chinese isolate (TNV-AC), was used for expression of different peptides derived from foot-and-mouth-disease virus (FMDV), serotype O, VP1 and fused C-terminally to the TNV-AC coat protein (Zhang et al. 2010). *Chenopodium amaranticolor* plants inoculated with in vitro transcripts of the chimeras developed symptoms similar to those caused by the wild-type TNV-AC. The immunogold labeling revealed that the highly expressed FMDV VP1 peptides were displayed on the virus surface and contained unmodified foreign peptides even after six successive passages in *C. amaranticolor* and three passages in *N. benthamiana*. The purified chimeric virus particles induced a strong immune response against FMDV structural protein VP1 via an intramuscular route, but when inoculated nasally the systemic and mucosal immune responses were induced in mice (Zhang et al. 2010).

Hsu et al. (2006) used the baculovirus expression system to generate VLPs that were formed by full-length and N-terminally truncated deletion mutants of the TBSV coat protein. The deletion of the majority of the coat R-domain sequence, namely residues 1–52 (NΔ52) and 1–62 (NΔ62), produced capsids similar to the wild-type T = 3 VLPs,

although the NΔ62 mutant that retained the last three aa residues of R-domain was capable of forming both the T = 3 and T = 1 particles and formation of the T = 1 capsids appeared to be preferred. There exists another mutant, NΔ72, in which R-domain, spanning residues aa 1–65, was completely removed but contained most of the β-annulus and extended arm (βA) regions exclusively formed T = 1 particles. These results suggested that as few as 3 residues 63–65 of the R-domain, which included 2 basic aa together with the arm (βA) and β-annulus regions, were sufficient for the formation of T = 3 particles. However, anywhere between 4–13 residues of the R-domain may be required for proper positioning of βA and β-annulus structural elements of the C-type subunits to facilitate an error-free assembly of the T = 3 capsids (Hsu et al. 2006).

Using efficient baculovirus-driven expression in insect cells, Kumar et al. (2009) proclaimed the TBSV coat gene as a suitable display platform and placed 180 copies of 16 aa epitope 95–110 of the ricin toxin A-chain (RTA) onto the capsid surface by fusion to the C-terminal end of the NΔ52 variant that still efficiently formed the T = 3 particles, as found earlier by Hsu et al. (2006). The expression of the chimeric RTA-TBSV coat derivative in *Sf*21 cells resulted

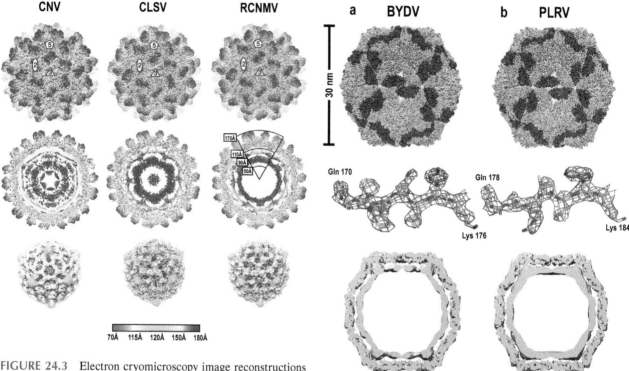

FIGURE 24.3 Electron cryomicroscopy image reconstructions of cucumber leaf spot virus (CLSV) and red clover necrotic mottle virus (RCNMV) at 3.2-Å resolution and 2.9-Å resolution, respectively, compared to the previously described 4.2-Å reconstruction of cucumber necrosis virus (CNV; Sherman et al. 2017). All panels are colored from red to blue with increasing radii. The top panels present exterior views of the particles looking down the icosahedral 3-fold (quasi-6-fold) axes. The locations of 2-fold, 3-fold, and 5-fold axes are denoted. The middle panels present thin sections of the particles showing the protein/RNA layers. The bottom panels present the inner RNA cores. (Reprinted with permission of American Society for Microbiology from Sherman MB et al. *J Virol.* 2020;94:e01439–19.)

FIGURE 24.4 The electron cryomicroscopy structures of capsids of barley yellow dwarf virus (BYDV), a current luteovirus (a) and potato leafroll virus (PLRV), a former luteovirus, now assigned to the *Polerovirus* genus, family *Solemoviridae*, of the *Sobelivirales* order (see Chapter 28) (b). Top: electron cryomicroscopy maps of whole virus capsid, colored according to the coat quasiconformers, where subunit A is blue, subunit B is green, and subunit C is red. Middle: section of representative density and molecular model for each virus. Bottom: slice through unsharpened maps, depicting density for packaged RNA and/or disordered R domain. (Reprinted from *Structure*, 27, Byrne MJ et al. Combining transient expression and cryo-EM to obtain high-resolution structures of luteovirid particles, 1761–1770.e3, Copyright 2019, with permission from Elsevier.)

in spontaneous assembly of VLPs displaying the ricin epitope. Moreover, electron cryomicroscopy and image reconstruction of the chimeric VLPs at 22 Å resolution revealed the locations and orientation of the ricin epitope exposed on the TBSV capsid surface, as shown in Figure 24.5. The injection of chimeric VLPs into mice generated antisera that detected the native ricin toxin. Remarkably, the specific TBSV coat variants, such as TBSV-NΔ52, TBSV-NΔ62 or TBSV-NΔ72, were able to generate VLPs of appropriate size, T = 1 particles of 20 nm in diameter vs. T = 3 particles of 35 nm in diameter, displaying either 60 or 180 copies of foreign epitopes, respectively (Kumar et al. 2009).

Rubino et al. (2011) expressed the coat gene of cymbidium ringspot virus (CymRSV) in *N. benthamiana* leaves and got the VLPs that had the same outward aspect and size of wild-type CymRSV and encapsidated the coat mRNA. The coat protein carrying the 10-aa Myc epitope at the C-terminus formed VLPs that were decorated by a Myc-specific antiserum and displayed therefore the Nyc epitope on the surface.

Arcangeli et al. (2014) used the capsid of artichoke mottled crinkle virus (AMCV), a member of the *Tombusvirus* genus and a close relative of TBSV, both as a carrier of immunogenic epitopes and for the delivery of anticancer molecules. Remarkably, the effects of C-terminal addition of the HIV-1 2F5 neutralizing epitope ELDKWA on the structural stability of the chimeric 2F5-AMCV VLPs were predicted and assessed by detailed inspection of the nanoparticle intersubunit interactions at atomic level, whereas the atomic structures of both wild-type and chimeric VLPs were obtained by homology modeling based on the atomic TBSV structure. Moreover, the AMCV VLPs functioned also as drug delivery vehicles able to load the chemotherapeutic drug doxorubicin (Arcangeli et al. 2014), as described later.

After the AMCV VLPs, Saunders and Lomonossoff (2015) involved a novel model, namely the previously mentioned turnip crinkle virus (TCV), of the genus

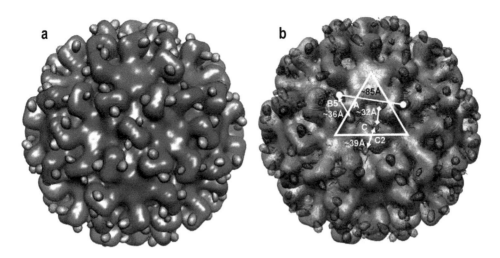

FIGURE 24.5 Electron cryomicroscopy and image reconstruction of TBSV-RTA16 at 22 Å resolution. (a) The electron density of the TBSV coat is shown in steel blue and the density corresponding to the ricin peptide is shown in gold. (b) An illustration highlighting the quality of fit of atomic structures of the TBSV coat and the ricin epitope to the electron density, which is shown as a transparent surface. Individual epitopes are exposed with a minimum separation of ~32 Å between two adjacent epitopes, identified by the line with diamonds at the end, which belong to icosahedral 2-fold-related and quasi 2-fold-related C and A-subunits respectively. The epitopes of AB(5) and CC(2) dimers are separated by 36 Å and 39 Å respectively and identified by the lines with arrows, whereas the epitopes of the A, B and C subunits that constitute the reference icosahedral asymmetric unit are separated by ~85 Å, identified with the line with spheres at the end. The central triangle (white lines) identifies the icosahedral asymmetric unit. Images were generated using the molecular graphics program *Chimera* (Pettersen et al. 2004). (Reprinted from *Virology* 388, Kumar S, Ochoa W, Singh P, Hsu C, Schneemann A, Manchester M, Olson M, Reddy V, Tomato bushy stunt virus (TBSV), a versatile platform for polyvalent display of antigenic epitopes and vaccine design, 185–190, Copyright 2009, with permission from Elsevier.)

Betacarmovirus. The transient expression of the TCV coat protein in *N. benthamiana* plants resulted in the formation of VLPs that morphologically resembled the T = 3 TCV virions but encapsidated heterogeneous cellular RNAs, rather than the specific TCV coat protein messenger RNA. The expression of an N-terminally deleted form of the coat resulted in the formation of smaller T = 1 structures that were free of RNA. The possibility of utilizing the TCV VLPs as a carrier for the presentation of foreign proteins on the particle surface was also explored by fusing the sequence of GFP to the C-terminus of the coat protein, where formation of VLPs was permitted, but the yield of chimeric VLPs was diminished compared to the yield obtained with unmodified VLPs.

Zampieri et al. (2020) used the plant expression to produce the infectious TBSV virions displaying a peptide associated with rheumatoid arthritis, in parallel with the cowpea mosaic virus (CPMV) VLPs displaying a peptide associated with another autoimmune disease, namely type 1 diabetes mellitus, as described in Chapter 27. In both cases, the prevention of autoimmune diseases was demonstrated in animal models.

The outcome of the chimeric C-terminal GFP-TCV coats, as well as of the TCV coats carrying internal insertion of the N-terminal hepatitis B core epitope MDIDPYKEFG within the P domain, between aa positions 272 and 273 was improved markedly by the generation of mosaic VLPs (Castells-Graells et al. 2018). Thus, by coexpression of one of these chimeric TCV coats with the wild-type TCV coat

protein by the coinfiltration of appropriate *Agrobacterium* suspensions, the VLP yield was enhanced through the successful formation of the mosaic VLPs.

PACKAGING AND DELIVERY

Loo et al. (2006, 2007) were the first who demonstrated the templated self-assembly of the red clover necrotic mosaic virus (RCNMV), genus *Dianthovirus*, coat protein around foreign cargo, extending their approach from gold nanoparticles (Loo et al. 2006) to magnetic nanoparticles and quantum dots (Loo et al. 2007). It was known already that the 36-nm RCNMV particle, having an inner cavity of ~17 nm, packaged its two genomic RNAs via a specific coat protein and genomic RNA interaction, where the RCNMV origin of assembly (OAS) consisted of a complex of RNA-1 with an RNA-2 stem loop previously termed a transactivator (Sit et al. 1998). A 20-nucleotide hairpin structure within the genomic RNA-2 hybridized with RNA-1 to form a bimolecular complex, which functioned as the OAS in RCNMV that selectively recruited and oriented coat subunits initiating virion assembly. The insertion or attachment of the RCNMV RNA-2 stem loop to a foreign viral RNA resulted in packaging of the foreign RNA into virions possessing a wild-type morphology. Using this basic knowledge, an oligonucleotide mimic of the OAS sequence, namely 5'-thiol deoxyuridine-modified DNA oligonucleotide analogue of the RNA-2 hairpin termed DNA-2, was attached to Au, $CoFe_2O_4$, and CdSe nanoparticles ranging from 3–15 nm, followed by addition of RNA-1 to form a synthetic OAS to direct the virion-like assembly by the RCNMV coat protein.

Figure 24.6 illustrates this approach with the following steps: (i) DNA-2 was tethered to nanoparticles using conditions that limited the oligomers bound per particle; (ii) RNA-1 interacted with DNA-2 to form the functional OAS and the sample was incubated for 10 min; (iii) the RCNMV coat was added to the mixture and self-assembly was initiated by dialyzing the sample against 50 mM Tris-HCl, pH 5.5 overnight at room temperature.

The attempts to encapsidate nanoparticles with diameters larger than 17 nm did not result in well-formed viral capsids and were consistent therefore with the presence of the ~17 nm cavity in the native RCNMV virions (Loo et al. 2006, 2007).

In parallel, Ren et al. (2006) reported hibiscus chlorotic ringspot virus (HCRSV) of the genus *Carmovirus* as a potential drug delivery platform. In this case, the whole HCRSV was propagated in kenaf (*Hibiscus rosasinensis L.*) plants, purified, and subjected to disassembly and removal of the genomic RNA. Thus, the VLPs were readily produced by destabilizing the HCRSV virions in 8 M urea or Tris buffer pH 8, in the absence of calcium ions, followed by removal of viral RNA by ultrahigh-speed centrifugation and the reassembly of the coat protein in sodium acetate buffer pH 5. The loading of foreign materials into the VLPs was dependent on electrostatic interactions. The anionic polyacids, such as polystyrenesulfonic acid and polyacrylic acid, were successfully loaded, but neutrally charged dextran molecules were not. The molecular-mass threshold for the polyacid cargo was about 13 kDa, due to the poor retention of smaller molecules, which readily diffused through the holes between the S domains present on the surface of the VLPs. These holes precluded the entry of large molecules but allowed smaller molecules to enter or exit. The polyacid-loaded VLPs had comparable size, morphology, and surface-charge density to the native HCRSV, and the amount of polyacids loaded was comparable to the mass of the native genomic materials (Ren et al. 2006).

Developing further this idea, Ren et al. (2007) employed the aid of a polyacid, when polystyrenesulfonic acid of 200 kDa was loaded simultaneously with doxorubicin into the HCRSV VLPs. The 3:1 polyacid:drug w/w ratio of mixing ensured the attraction of doxorubicin molecules to the polyacid by electrostatic forces, a net negative charge for the resultant complex, and reassembly of coats into the VLPs. Next, the folic-acid-conjugated equivalents were prepared, where the HCRSV viruses were conjugated with folic acid before they were disassembled for doxorubicin-polyacid encapsulation. Folic acid was chosen, since it was able to serve as a small molecule-ligand for targeting drug delivery systems to cancer cells (Leamon and Low 1991; Ross et al. 1994; Lu and Low 2002). As a result, the folic-acid-conjugated HCRSV cages improved the uptake and cytotoxicity of doxorubicin in the ovarian cancer cells with statistical significance (Ren et al. 2007). The fine methodology on how to reassemble the HCRSV VLPs and employ them as nanocages was published later by Wong and Ren (2018).

The first success of the *Tombusviridae* members by the development of the packaging and delivery methodologies were reviewed exhaustively at that time (Steinmetz 2010; Dedeo et al. 2011; Pokorski and Steinmetz 2011; Wen et al. 2013).

Lockney et al. (2011) engineered the first targeted RCNMV particles by conjugation of short targeting peptides, less than 16 aa residues, using the heterobifunctional chemical linker sulfosuccinimidyl-4-(N-maleimidomethyl) cyclohexane-1-carboxylate (Sulfo-SMCC). As a result, the RCNMV particles armed with a CD46-targeting peptide and loaded with doxorubicin successfully delivered the therapeutic cargo to HeLa cells.

Arcangeli et al. (2014) were the first who proposed recombinant plant-produced viral cages to produce safe vaccine vehicles and nanoparticles for drug delivery, namely the same doxorubicin. As the previously described AMCV of the genus *Tombusvirus* VLPs were prone to reversible conformational change in vitro, like the TBSV ones (Aramayo et al. 2005), small molecules were infused

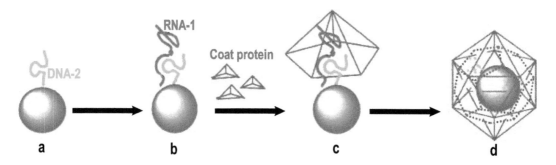

FIGURE 24.6 Schematic of nanoparticles encapsidation by RCNMV coat protein. (a) Conjugation of nanoparticle with DNA-2. (b) Addition of RNA-1 interacts with DNA-2 to form the functional origin of assembly (OAS). (c) The artificial OAS templates the assembly of coat protein. (d) Formation of virus-like particle with nanoparticle encapsidated. (Reprinted with permission from Loo L et al. Encapsidation of nanoparticles by red clover necrotic mosaic virus. *J Am Chem Soc*. 129:11111–11117. Copyright 2007 American Chemical Society.)

through the 2-nm pores in the protein shell after reversible swelling. Following these observations, the AMCV VLPs were treated with EDTA to obtain the swollen state of the VLPs. The swollen VLPs were incubated with doxorubicin molecules at a mole ratio of 1,000 molecules per capsid. After doxorubicin infusion, the pores on the VLPs were closed by addition of divalent cations and pH adjustment. The Dox-AMCV VLPs were purified from free doxorubicin by filtration, and the calculations showed that each AMCV VLP loaded an average of 61 doxorubicin molecules, while the presence of doxorubicin inside the AMCV cavity did not interfere with particle integrity (Arcangeli et al. 2014).

Alemzadeh et al. (2017) generated recombinant VLPs of Johnsongrass chlorotic stripe mosaic virus (JgCSMV), genus *Aureusvirus*, in tobacco plants and used them as the doxorubicin carrier. In this case, the JgCSMV coat protein gene was expressed in tobacco tissue, and the transformed hairy roots produced high levels of the recombinant protein that readily assembled to form empty VLPs with overall structure similar to native virions. Thus, the JgCSMV VLPs were used as a nanocontainer for loading doxorubicin, taking advantage of the reversible swelling of VLPs in vitro. Remarkably, Alemzadeh et al. (2017) infused doxorubicin into the JgCSMV-VLPs by a reversible pore-opening mechanism elaborated by Loo et al. (2008) to the infusion of dye molecules into the previously desribed RCNMV VLPs.

The previously described TCV, from the genus *Betacarmovirus*, VLPs were among the next prospective packaging candidates (Saunders and Lomonossoff 2015). The advantages of the TCV VLPs were that swollen forms of the particle could be obtained, and an in vitro assembly system was available (Sorger et al. 1986), potentially allowing the incorporation of larger molecules. Indeed, the anionic polyacids were successfully loaded into the HCRSV protein cages from a related carmovirus (Ren et al. 2006). Remarkably, as presented earlier, TCV was one of the first viruses to have its structure determined by crystallography (Hogle et al. 1986), and its swollen capsid structure was reported later by Bakker et al. (2012).

FAMILY *CARMOTETRAVIRIDAE*

Since the T = 4 tetravirus and T = 3 nodavirus capsid proteins underwent closely similar autoproteolysis to produce the N-terminal β and C-terminal, lipophilic γ polypeptides and the γ peptides and the N-termini of β also acted as molecular switches that determined their quasi-equivalent capsid structures, the crystal structure of Providence virus (PrV; Taylor et al. 2006; Speir et al. 2010) was compared first with the 3D structure of Nudaurelia capensis ω virus (NωV), a tetravirus from the *Omegatetravirus* genus of the *Alphatetraviridae* family, order *Hepelivirales*, which is described in Chapter 16. The 3D structure of PrV at 3.8 Å

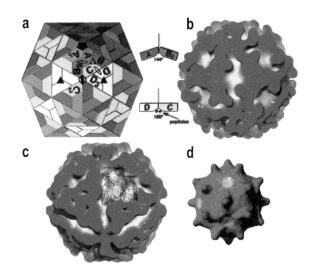

FIGURE 24.7 Electron cryomicroscopy reconstruction of Providence virus. (a) A T = 4 quasi symmetry model of the tetravirus capsids. Positions of icosahedral and quasi icosahedral rotations axes are shown as filled and unfilled geometric symbols, respectively (oval = 2-fold; triangle = 3-fold; pentagon = 5-fold; hexagon = 6-fold). The A, B, and C polygons related by a quasi-3-fold and the D polygon related to C by a quasi-2-fold define the icosahedral asymmetric unit (ABCD). Each of the polygons represents identical protein subunits but occupies slightly different geometrical (chemical) environments. Polygons with subscripts are related to those without by the icosahedral symmetry of the subscript (i.e., A to A5 by 5-fold rotation). Unlike T = 3 capsids, there is no icosahedral 2-fold dimer. Instead, the icosahedral 2-folds are coincident with quasi-6-fold arrangements of B, C, and D subunits (three sets of dimers). Looking at the arrangements of ABC and DDD subunit triangles clarifies tetravirus capsid architecture. In a clear break from quasi equivalence, ABC triangles form a bent interface with each other, and ABC-DDD triangles form a flat interface due to the insertion of subunit polypeptides at the interface. (b) Surface representation of the NωV reconstruction at ~21 Å resolution and in the same orientation as (a). Darker blue areas are at a greater radius from the particle center. The subunit Ig-like domains form large, contiguous triangular facets with curved edges around the icosahedral 3-fold axes. (c) Surface representation of PrV at ~28 Å resolution (same coloring and orientation as [a] and [b]). The subunits of one icosahedral asymmetric unit are shown as ribbons through their corresponding transparent surface. The most distinctive difference from NωV is that the triangular facets now have nearly straight edges and 3-fold-related pits (characteristic of betatetraviruses) due to a change in the orientation of the Ig-like domains. (a) Surface representation of the PrV RNA core (same orientation as [a]—[c]) after removing density corresponding to the crystal structure protein coordinates. Darker red areas are at a greater radius from the particle center. Large bulges of density extend from the core at each icosahedral 2-fold axis and make contact with the capsid protein shell. (Reprinted from *Structure* 18, Speir JA, Taylor DJ, Natarajan P, Pringle FM, Ball LA, Johnson JE, Evolution in action: N and C termini of subunits in related T = 4 viruses exchange roles as molecular switches, 700–709, Copyright 2010, with permission from Elsevier.)

revealed conserved folds and cleavage sites, but the protein termini had completely different structures and the opposite functions of those in NωV. The N termini of β formed the molecular switch in PrV, whereas γ peptides played this role in NωV. The PrV γ peptides instead interacted with packaged RNA at the particle 2-folds. Generally, the disposition of peptide termini in PrV was closely related to those in nodaviruses (Speir et al. 2010). Figure 24.7 illustrates this remarkable 3D structure comparison.

Despite repeated attempts, the expression of the PrV coat protein using recombinant baculovirus resulted in insoluble aggregates in *Sf*21 cells, even though NωV VLPs have been successfully produced in the same system (Taylor et al. 2006; Speir et al. 2010).

25 Orders *Norzivirales* and *Timlovirales*

Sometimes, said Pooh, the smallest things take up the most room in your heart.

A.A. Milne, *Winnie-The-Pooh*

Everything must be made as simple as possible. But not simpler.

Albert Einstein

ESSENTIALS

Norton D. Zinder, the famous discoverer of the first RNA phage f2, assumed that "considering its small size and large yield, the RNA phage is currently the most populous organism in the world" (Zinder 1965). However, up to the year 2021, the whole RNA phage world occupied in the ICTV taxonomy a small place of the *Leviviridae* family consisting of the two historical genera *Allolevivirus* and *Levivirus*, each of which was represented by two species: *Escherichia virus FI* and *Escherichia virus Qβ* for the *Allolevivirus* genus and *Escherichia virus BZ13* and *Escherichia virus MS2* for the *Levivirus* genus, where the famous Qβ and MS2 phages were marked as the most representative members. Before 2018, the small *Leviviridae* family was not assigned to any higher order. Then, it was placed into the great *Riboviria* realm, together with other 39 RNA virus families that were not assigned to any order at that time. Then, the *Leviviridae* family was assigned to the *Levivirales* order as the sole family, and the *Levivirales* was included as a sole order into the class *Allassoviricetes*. Nevertheless, many popular RNA phages, actively involved into the nanotechnological applications, such as so-called noncoliphages AP205, PP7, and φCb5, remained unclassified. This situation was reviewed at the *Leviviridae* stage by Pumpens (2020). Among other things, the book narrated that the then ICTV taxonomy of the RNA phages grew up from the old immunological typing of the first coliphages (Krueger RG 1969) and remained structured in a general accordance with the early serological groups that have divided the coliphages into four serogroups, namely serogroups I to IV. The serological classification scheme I to IV was confirmed later by sequencing of the first RNA coliphage genomes. As a result, the initial four serogroups were recognized as the corresponding genogroups I to IV of the RNA coliphages but not of the phages infecting other bacteria. The most popular RNA phages MS2, GA, Qβ, and SP were traditionally recognized as reference strains for the sero- and/or genogroups I, II, III, and IV, respectively. Then, the group I and II members were combined into the *Levivirus* genus, while the serogroup III and IV members formed the *Allolevivirus* genus, but the non-*E. coli* RNA phages remained unclassified by the sero/genogroup typing.

The early history, basic properties, and classification peculiarities of the RNA phages were described in detail in an original author's review (Zinder 1980) and three specialized books (Gren 1974; Zinder 1975; Pumpens 2020). It is highly symbolic that the Latin *levis* means *light (not heavy), quick, swift, fickle, dispensable, trivial, trifling.*

At last, the logic and integrity of the RNA phage unit were preserved in 2020–2021 due to the revolutionary achievements of the ICTV subcommittee *Leviviridae Study Group* guided by Colin Hill (Callanan et al. 2020a). As a great result, the *Levivirales* order was upgraded to the class *Leviviricetes*, formerly *Allassoviricetes*, which covered now the two orders, namely *Norzivirales*—which replaced the *Levivirales* order—and the novel order *Timlovirales*. The huge novel *Leviviricetes* class is expanded currently to a total of 6 families, 428 genera, and 882 species. The successful Subcommittee is to be particularly commended for the idea to name the novel taxons after the eminent actors of the RNA phage history Norton D. Zinder, Tim Loeb, Sol Spiegelman, Walter Fiers, Jan van Duin, John F. Atkins, Thomas Blumenthal, Joan A. Steitz, and other famous initiators.

According to the novel great equitable taxonomy, the famous RNA phage players are divided now among the following taxons. The order *Norzivirales* covers first the family *Fiersviridae*, which is in fact the former *Leviviridae* family and unites the most popular RNA phage MS2 of the novel *Emesvirus zinderi* species, genus *Emesvirus*; the phage GA of the *Emesvirus japonicum* species of the same *Emesvirus* genus; the phage Qβ of the *Qubevirus durum* species; and phage FI of the *Qubevirus faecium* species, both genus *Qubevirus*. Then, the *Fiersviridae* family includes the old great noncoliphages, namely the pseudomonaphages PP7 of the *Pepevirus rubrum* species, genus *Pepevirus*, and PRR1 of the species *Perrunavirus olsenii*, genus *Perrunavirus*. The family *Duinviridae* involves the nanotechnologically highly important acinetophage AP205 as a member of the *Apeevirus quebecense* species, genus *Apeevirus*. The other novel order, *Timlovirales*, includes the popular caulophage φCb5 as a member of the species *Cebevirus halophobicum*, order *Cebevirus*, from the family *Steitzviridae*. It should be kept in mind that hundreds of the novel members of the class *Leviviricetes* remain up to now the products of the modern metagenomics and never had been seen as real phages, in contrast to the "classical" RNA phages that were discovered as real infectious particles (Callanan et al. 2018, 2020b).

The first *E. coli*-infecting RNA phages, or coliphages, which traditionally played one of the central roles in viral nanotechnology, were identified in the early 1960s and termed f2 (Loeb and Zinder 1961), MS2 (Davis et al. 1961), R17 (Paranchych and Graham 1962), fr (Marvin and Hoffmann-Berling 1963), M12 (Hofschneider 1963), and

DOI: 10.1201/b22819-30

Qβ (Watanabe 1964). Later, other RNA phages that became useful as VLP carriers were described, including the *E. coli* phages SP, FI (Miyake et al. 1969), and GA (Aoi et al. 1973), the *Caulobacter crescentus* phage φCb5 (Schmidt and Stanier 1965), the *Pseudomonas aeruginosa* phage PP7 (Bradley 1966), the pseudomonaphage (or rather broad host range) P-pili-specific phage PRR1 (Olsen and Thomas 1973), and, last, the *Acinetobacter* phage AP205 (Coffi 1995; Klovins et al. 2002).

The pili-specific RNA phages, currently among the most productive and promising VLP carriers, are nonenveloped, spherical viruses with T = 3 icosahedral symmetry and diameters ranging from approximately 27–30 nm. Figure 25.1 demonstrates electron micrograph and 3D reconstruction of typical RNA phage particles located on a bacterial pili. The phage particle is composed of 178 chemically identical coat protein molecules or 89 coat dimers and one copy of maturation (or A) protein, which replaces a single coat dimer.

GENOME

The genomic MS2 RNA was the first full-length sequenced genome in the world (Fiers et al. 1975, 1976). Then, the full-length genomes were achieved for the coliphages of different geno/serogroups: fr (Adhin et al. 1990) from group I; GA (Inokuchi et al. 1986), KU1 (Groeneveld et al. 1996),

and BZ13 (Friedman et al. 2009) from group II; Qβ (Mekler 1981) and M11 and MX1 (Beekwilder et al. 1995, 1996) from group III; SP (Inokuchi et al. 1988), NL95 (Beekwilder et al. 1995, 1996), and FI (Kirs and Smith 2007; Friedman et al. 2009) from group IV.

The first full-length genome of noncoli phages, namely of the pseudomonaphage PP7 (Olsthoorn et al. 1995), demonstrated no significant nucleotide sequence identity between PP7 and the coliphages except for a few regions. The strong differences from the coliphages were found in the full-length genomes of the acinetophage AP205 (Klovins et al. 2002), the caulophage φCb5 (Kazaks et al. 2011), the pseudomonaphage LeviOr01 (Pourcel et al. 2017), the P-pili–specific phage PRR1 (Ruokoranta et al. 2006), the M-pili–specific phage M (Rumnieks and Tars 2012), and the R-plasmid–dependent phages C-1 INW-2012 and HgaI1 (Kannoly et al. 2012). These sequences evoked a strong interest to the problem of the RNA phage evolution and improvement of the classical four-group-based classification of the two historical genera of the former *Leviviridae* family.

The metagenomic sequencing opened a new method of the RNA phage search and contributed first to the modern RNA phage list by marine RNA phages EC and MB, which sequences have been assembled from the metagenomic sequencing data of marine organisms (Greninger and DeRisi 2015). In contrast with the previous genotyping data concentrated on the RNA coliphages, the EC and MB

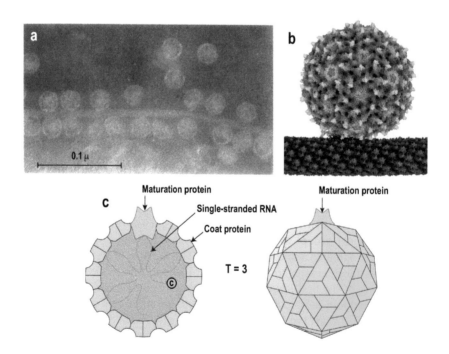

FIGURE 25.1 Virus particles of the levivirus phages. (a) An example of the classical David E. Bradley (1964) electron micrographs of RNA phages. The RNA phage ZIK/1 adsorbed on an associated filamentous type, phosphotungstate negative stain, ×360,000. (Reprinted with permission of **Elsevier** from Bradley DE. *J Ultrastruct Res.* 1964;10:385–389.) (b) A 3D model of the MS2 phage particle bound to the bacterial F pilus. The virion consists of a coat protein (teal) capsid assembled around the genomic RNA (blue) and a single copy of the maturation protein (yellow-orange), which recognizes the F pilus (gray). (Courtesy Jānis Rūmnieks.) (c) The cartoon presentation of the RNA phage virion taken with kind permission from the ViralZone, Swiss Institute of Bioinformatics, http://viralzone.expasy.org (Hulo C et al. 2011). (Courtesy Philippe Le Mercier.)

phages showed only moderate translated amino acid identity to other enterobacteria phages and appeared to constitute novel issues for the RNA phage taxonomy. The first massive metagenomic search for the RNA phage genomes was performed by Krishnamurthy et al. (2016). The main goal of the authors consisted of the idea of overcoming the limited number of known RNA phage species and, as a result, overcoming the predominant concentration on the RNA coliphages. This survey of metagenomic databases revealed 158 partial single-stranded RNA phage genome sequences belonging to 122 distinct phylotypes, 66 of which possessed a putative open reading frame predicted to be the coat gene. This study identified new dimensions of RNA phage biology, including phages with novel genome organizations, numerous open reading frames that contained novel genes with no detectable homology to the known phage genes, RNA phage presence in novel ecological niches, and the first data in support of an RNA phage infection of a Gram-positive bacterium. These results critically illuminated an unexamined dimension of molecular and ecological phage diversity and fundamentally established a necessary framework that enabled a more accurate dissection of RNA phage modulation of microbial populations.

During the same period, a great survey of RNA virus sequences from invertebrates resulted in 67 additional levi-like RNA phage genome sequences, among newly discovered 1445 RNA viruses (Shi et al. 2016). At that time, none of these levi-like genomes appeared in the NCBI taxonomy browser under *Leviviridae* family, but they were included under the *unclassified Riboviria*. At last, the great Colin Hill's team hugely succeeded by metagenomics in the enormous amplification of the known RNA phage members, from tens to over a thousand (Callanan et al. 2018, 2020b). Thus, Callanan et al. (2020b) identified 15,611 nonredundant ssRNA phage sequences, including 1015 near-complete genomes, enabling them to complete the revolutionary phylogenetic assessment of the huge but steadily overlooking the RNA phage world.

The Kaspars Tārs team in Riga performed an unprecedented similarity analysis of the coat proteins of all known levi-like RNA genomes, deciphered mostly from the metagenomic studies (Liekniņa et al. 2019) and paved the way for their use in viral nanotechnology by expression of the corresponding synthetic coat protein genes. The great set of these novel prospective vectors is presented in the Epilogue of this book.

Figure 25.2 presents detailed graphical structures of genomes of the RNA phages that were most actively involved into the VLP nanotechnology applications. The genomes are organized here by the traditional sero/geno-grouping characteristics and host specificity.

Therefore, the RNA phage genome encodes four genes, where three of them—maturation (or A) protein, coat protein, and the phage-specific replicase subunit—constitute the three-gene minimum needed to make a phage. The fourth lysis gene appeared later as a specific addition to this minimum. The strongest genomic difference of the former

Allolevivirus and *Levivirus* genera members consisted in the fact that the *Allolevivirus* genome encoded a C-terminally extended coat protein known as a minor A1 protein, which appeared as a result of ribosomal readthrough of a leaky opal UGA termination codon of the coat protein gene (Weiner and Weber 1971, 1973; Weiner et al. 1972) and was essential for the formation of viable Qβ particles in vivo (Hofstetter et al. 1974; Engelberg-Kulka et al. 1977, 1979; Skamel et al. 2014). For the first time, the A1 protein was observed by the Qβ replication in spheroplasts exposed to low concentrations of actinomycin D or rifampicin in the presence of an amino acid mixture containing radioactive leucine or histidine: this protein, which was present also in the Qβ particles, was named IIb (Garwes et al. 1969). It was calculated that the 3–10 copies of the IIb protein incorporate into the virion. After 45 years, it was established precisely that the A1 protein, or the IIb protein of Garwes et al. (1969), is incorporated as 12 copies per Qβ virion (Skamel et al. 2014). It is required for infection, but its precise function is not yet fully deciphered. A recent electron microscopy visualization of foreign epitopes carried by the A1 protein within infectious Qβ particles showed that the A1 protein molecules are occupying corners of the Qβ icosahedron (Skamel et al. 2014). The A1 protein appeared therefore as a special peculiarity of the former group III and IV phages that were forming together the *Allolevivirus* genus.

The history of the RNA phage VLP employment by viral nanotechnologies was described by Pumpens et al. (2016), while the current state of their involvement is unveiled by Peabody DS et al. (2021).

EXPRESSION OF THE COAT GENE AND NATURAL VLPs

The famous Walter Fiers team expressed for the first time the coat protein gene of the RNA phages, namely, of the phage MS2, in parallel with maturation and replicase genes, by the individual insertion of each gene into the thermoinducible expression plasmids under control of the phage λ P_L promoter (Remaut et al. 1982). The coat protein was synthesized at high efficiency, up to 20% of the total de novo protein synthesis, as early as 30 min after induction. Kastelein et al. (1983) examined the translational efficiency of the MS2 coat gene in vivo with respect to neighboring sequences by gradual shortening of the upstream region. An analogous procedure with the cloning of the Qβ genes under the same P_L promoter was performed at the same time by Karnik and Billeter (1983). However, these expression studies had no concern with possible assembly of the phage coats into VLPs, since they did not use electron microscopy or other methods to detect native status and size of the appearing coats.

For the first time, Tatyana Kozlovska from Elmārs Grēns's team demonstrated that the expression in *E. coli* of an RNA phage coat gene, in this case of the phage fr, a representative of the group I and a relative of MS2, led to the highly efficient formation of VLPs that were

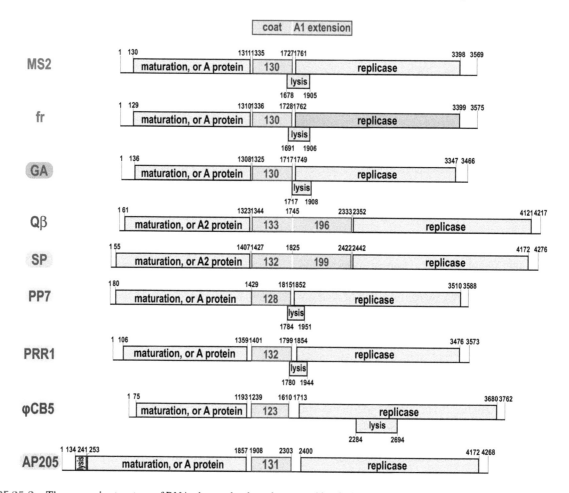

FIGURE 25.2   The genomic structure of RNA phages that have been used in viral nanotechnology applications. The specific colors of the RNA phage representative names indicate the genogroup and/or the host. Thus, the representatives of the genogroups I, II, III, and IV of coliphages are marked red, brown, blue, and violet, respectively. The pseudomonaphage PP7, pseudomonaphage or rather broad host range P-pili specific PRR1, caulophage φCB5, and acinetophage AP205 are marked green, teal, and khaki, respectively. The coat gene, including A1 extension, is colored dark pink, as also the readthrough part of the A1 protein by the former *Allolevivirus* representatives Qβ and SP. Genomes and genes are shown to scale. Amino acid length of the coats and A1 extensions is given by large numbers, but numbering of the gene location within the genome is indicated by small numbers. The genomes are collineated by the position of the coat protein gene. The sequence data is compiled mostly from the NCBI browser. It is necessary to consider that the experimentally determined lengths of the phage coat proteins are always one amino acid residue shorter than the actual proteins in the database because the N-terminal methionine is cleaved off in infected *E. coli* cells. This explains some discrepancies in the coat protein numbering in different published works.

undistinguishable in electron microscopy from the original phage (Kozlovskaia et al. 1986). This historically important electron microscopy was performed by Velta Ose. The high-level synthesis of the fr coat protein occurred under the control of the strong promoter of the *E. coli* tryptophan operon P$_{trp}$. The fr VLPs did not differ from the native fr phage also by the double radial immunodiffusion test according to Ouchterlony (1965).

Next, the coat gene of the phage JP34, an intermediate between the groups I and II, was sequenced and expressed by the Jan van Duin team (Adhin et al. 1989). Again, the question of the coat state after expression, in the form of VLPs or not, was not touched on in this study.

Then, David S. Peabody (1990) constructed a genetic two-plasmid system that placed the synthesis of a hybrid replicase-β-galactosidase enzyme under the translational control of the MS2 coat. In parallel, an analogous two-plasmid system was generated by the Peabody team for the Qβ (Lim et al. 1996) and PP7 (Lim et al. 2001) coat proteins. This pioneering system permitted the straightforward isolation of mutations that affected the repressor function of coat protein and functioned brilliantly in vivo. The capsids produced by the coat gene expression were detected in these studies by electron microscopy (Peabody DS 1990) but mostly by the simple agarose electrophoresis technique. Later, Caldeira and Peabody (2007) recommended the PP7 VLPs as a putative platform for growing viral nanotechnology. The authors tested the stability of the PP7 VLPs and specifically addressed the crosslinking of the PP7 VLPs by disulfide bonds between coat protein dimers at its 5-fold and quasi-6-fold symmetry axes, to improve thermal stability of the PP7 VLPs.

Thanks again to Tatyana Kozlovska, Grēns's team expressed in *E. coli* the coat gene of the phage Qβ, a representative of the group III, under the control of the strong *E. coli* P_{trp} promoter (Kozlovska et al. 1993). In this case, the Qβ coat gene was amplified from native Qβ RNA using a reverse-transcription-PCR technique. The amplified gene contained not only the sequence coding for the 133-aa residues of the essential Qβ coat but also the sequence encoding the additional 196-aa residues of the 329-aa readthrough protein (or protein Al) and separated from the proper coat by an opal UGA stop codon, as described previously. This obstacle allowed further progress in the elaboration of the original VLP display vectors. Moreover, the Ouchterlony double radial immunodiffusion test with antibodies against the phage Qβ and the electron microscopy evaluation performed by Velta Ose showed that, when expressed, the Qβ coat gene was responsible for the high-level synthesis and correct self-assembly of Qβ coat monomers into the VLPs indistinguishable morphologically and immunologically from the phage Qβ particles.

Table 25.1 summarizes the general data on expression of the RNA phage coat genes, with a special emphasis on the VLP production and determination of their high-resolution structure.

3D STRUCTURE

Figure 25.3 demonstrates the x-ray crystallography structures of six resolved RNA phages and/or their VLPs, which are involved in the viral nanotechnology techniques. The fine 3D structures are in fact remarkably similar, despite the marked diversity in the primary structures of their coats.

The general idea of the spatial structure of the icosahedral RNA phages was clarified in the middle 1970s. Thus, the first reliable 3D image reconstruction from electron micrographs was performed by R. Anthony Crowther et al. (1975) for the RNA phages R17 and f2. It showed that the protein subunits were located at the 2-fold positions and extended toward the quasi-3-fold positions of the T=3 icosahedral surface lattice, leaving holes on the 5-fold and 3-fold axes. The particles were reconstructed as roughly circular of diameter about 240 Å, where protein was concentrated into a relatively thin shell some 20–30 Å thick, and superposition patterns characteristic of views down 2-fold, 3-fold, and 5-fold axes of symmetry were recognized (Crowther et al. 1975).

The study of high-resolution structures began in the early 1990s with x-ray crystallography and has continued up to now. The first crystals were obtained for the phage preparations, while further investigations quite often preferred the VLPs that were obtained after expression of the RNA phage coat genes in *E. coli* or yeast. Moreover, since the active studies of the chimeric VLPs began in the mid-1990s, the determination of the 3D structure was tightly connected with the urgent problem of how to display foreign peptides on the particle surface.

The crystal structure of MS2 virion was first determined at a resolution of 3.3 Å (Valegård et al. 1986, 1990, 1991; Liljas and Valegård 1990) and then refined to the 2.8 Å resolution (Golmohammadi et al. 1993). Next, the crystal structure of the MS2 VLPs with amino acid exchanges in the FG loop was resolved (Stonehouse et al. 1996a, b). It is noteworthy from a historical point of view that the first phage MS2 crystals and the first preliminary x-ray examination data were obtained by the famous Walter Fiers lab in the 1970s (Min Jou et al. 1979). According to the 3D structure of the MS2 VLPs, the 180 coat subunits were arranged therefore in dimers as initial building blocks and formed a lattice with a T=3 triangulation number. The coat subunit consisted of a five-stranded β-sheet facing the inside of the particle and a hairpin and two α-helices on the outside. This hairpin was used further as a favorite target for the genetic insertions and/or chemical coupling of the desired foreign sequences.

A bit later, the structure of a recombinant capsid, i.e. VLPs, of the RNA phage fr was determined by x-ray crystallography at a resolution of 3.5 Å and was shown to be identical to the protein shell of the native phage fr (Bundule et al. 1993; Liljas et al. 1994). Again, this study was performed by the Lars Liljas team in Uppsala but with the active participation of the Elmārs Grēns team in Riga. Furthermore, the fr VLPs with FG loop modified by a 4-aa-long deletion were resolved (Axblom et al. 1998).

In parallel, the structures of both virions and VLPs of the phage Qβ were resolved at a resolution of 3.5 Å (Valegård et al. 1994a; Golmohammadi et al. 1996). These structures differed from the previously determined RNA phage MS2 and fr structures by the presence of stabilizing disulfide bonds on each side of the flexible FG loop, which covalently linked the coat dimers. A profound comparison with the structure of the related phage MS2 showed that, although the fold of the Qβ coat was very similar, the details of the protein–protein interactions were completely different (Golmohammadi et al. 1996).

The first MS2, Qβ, and fr structures were followed by the structure of the group II phage GA, which showed some structural differences compared to the MS2 and fr virions or VLPs, especially in the N- and C-terminal regions (Tars et al. 1997).

Then, the crystal structure of the pseudomonaphage PP7 was resolved at a resolution of 3.7 Å (Tars et al. 2000a, b). The crystal structure of another pseudomonaphage or P-pili-specific phage PRR1 was resolved to 3.5 Å and exhibited a binding site for a calcium ion close to the quasi-3-fold axis (Persson et al. 2008). Then, the crystal structure of a very distant RNA phage, the caulophage φCb5, which is assigned now to the *Timlovirales* order, was resolved to 3.6 Å, while the structure of the φCb5 VLPs was resolved to 2.9 Å (Plevka et al. 2009a). The structures appeared to be nearly identical, with some differences in the average density of RNA. Unlike other phages, the φCb5 capsids were significantly stabilized by calcium ions. The disassembly of these capsids occurred when the calcium ions were chelated

TABLE 25.1

Expression of the Coat Gene and Production of the Natural RNA Phage VLPs by Different Expression Systems

VLPs	Expression Host	References	Comments
Geno/Serogroup I			
fr	*Escherichia coli*	Kozlovskaia et al. 1986	First RNA phage VLPs detected by electron microscopy
	Saccharomyces cerevisiae	Freivalds et al. 2014	Efficient outcome of VLPs in yeast
	Pichia pastoris	Freivalds et al. 2014	Efficient outcome of VLPs in yeast
MS2	*Escherichia coli*	Remaut et al. 1982	No VLP analysis
	Escherichia coli	Kastelein et al. 1983	No VLP analysis
	Escherichia coli	Peabody DS 1990	First electron microscopy of the MS2 VLPs
	COS cells	Berkhout and Jeang 1990	First step by the development of tethering and imaging technologies; no VLPs mentioned
	Escherichia coli	Ni et al. 1995a, b	Crystal structure of the MS2 coat dimers
	Arabidopsis thaliana	Cerny RE et al. 2003	Translational repression of transgene expression; no VLP analysis
	Saccharomyces cerevisiae	Legendre and Fastrez 2005	First RNA phage VLPs in yeast; packaging of heterologous mRNAs
	Escherichia coli	Plevka et al. 2008, 2009b	Crystal structure of VLPs formed by the covalent MS2 coat dimers
	Escherichia coli	Bundy et al. 2008	The first scalable cell-free production of the RNA phage VLPs
Geno/Serogroup II			
GA	*Escherichia coli*	Ni et al. 1996	Crystal structure of the GA coat dimers
	Saccharomyces cerevisiae	Freivalds et al. 2008	First RNA phage VLPs of group II in yeast
	Pichia pastoris	Freivalds et al. 2008	Efficient outcome of VLPs
JP34	*Escherichia coli*	Adhin et al. 1989	no VLP analysis
Geno/Serogroup III			
Qβ	*Escherichia coli*	Karnik and Billeter 1983	No VLP analysis
	Escherichia coli	Kozlovska et al. 1993	First Qβ VLPs demonstrated
	Escherichia coli	Lim et al. 1996	Capsids detected by agarose gel
	Arabidopsis thaliana	Cerny RE et al. 2003	Translational repression of transgene expression; no VLP analysis
	Saccharomyces cerevisiae	Freivalds et al. 2006	First RNA phage VLPs of *Allolevivirus* genus in yeast; high level of the VLP production
	Pichia pastoris	Freivalds et al. 2006	First RNA phage VLPs in *P. pastoris*
	Escherichia coli	Bundy and Swartz 2011	First Qβ VLPs in the cell-free system
Geno/Serogroup IV			
SP	*Escherichia coli*	Priano et al. 1995	Coat expressed as a part of the A1 protein, no VLP analysis
	Saccharomyces cerevisiae	Freivalds et al. 2014	No assembly into VLPs detected
	Pichia pastoris	Freivalds et al. 2014	No assembly into VLPs detected
Acinetophage			
AP205	*Escherichia coli*	Spohn et al. 2010; Tissot et al. 2010	AP205 VLPs as a prospective platform
	Escherichia coli	Shishovs et al. 2016	Crystal/electron cryomicroscopy structure of VLPs
	Saccharomyces cerevisiae	Freivalds et al. 2014	Efficient outcome of VLPs
	Pichia pastoris	Freivalds et al. 2014	Efficient outcome of VLPs
Caulophage			
φCB5	*Escherichia coli*	Plevka et al. 2009a	Crystal structure of VLPs; stabilization by calcium ions
	Saccharomyces cerevisiae	Freivalds et al. 2014	Efficient reassembly of dimers; packaging of nanomaterial cargos
	Pichia pastoris	Freivalds et al. 2014	Efficient reassembly of dimers; packaging of nanomaterial cargos

VLPs	Expression Host	References	Comments
Pseudomonaphages			
PP7	*Escherichia coli*	Lim et al. 2001	Capsids detected by agarose gel
	Escherichia coli	Caldeira and Peabody 2007	Crosslinking of coat dimers by disulfide bonds
	Saccharomyces cerevisiae	Freivalds et al. 2014	Efficient outcome of VLPs
	Pichia pastoris	Freivalds et al. 2014	Highly efficient outcome of VLPs
PRR1	*Escherichia coli*	Persson et al. 2008	Crystal structure, stabilization of VLPs by metal ions

Note: The phages are given in alphabetical order within each specific hierarchical group.

with EDTA and/or there was a reduction in the surrounding salt concentration (Plevka et al. 2009a).

The first 3D structure of the acinetophage AP205 was achieved by electron cryomicroscopy at relatively low resolution estimated to be between 17 Å and 24 Å (van den Worm et al. 2006). After ten years of trials by Kaspars Tārs's group on the acinetophage AP205, the crystal structure of the AP205 coat dimer was solved at a resolution of 1.7 Å and then fitted into the 6.6-Å resolution electron cryomicroscopy map (Shishovs et al. 2016). Figure 25.4 presents this structure, which appeared incredibly important in the current viral nanotechnology trials. In fact, the structure of the AP205 coat dimer could be regarded as a circular permutant relative to the structures of the phage MS2 and other family members. This feature was made possible by the fact that the N-terminus of one monomer in the dimer is in close proximity to the C-terminus of the other monomer. Consequently, the AP205 structure demonstrates the N- and C-termini in the same locations as those occupied by the surface-exposed AB loops in other phage capsids. The N- and C-termini in other phage VLPs are not well exposed on the surface and are clustered around the quasi-3-fold axes. This explained the unique properties of the AP205 VLPs, which can tolerate long additions at its N- and C-termini without compromising the capsid assembly, in contrast to other RNA phage VLPs (Tissot et al. 2010).

It is noteworthy, however, that the phage AP205, due to the initiative of Kaspars Tārs's group, served as a model by the pioneering 3D studies using solid-state nuclear magnetic resonance under ultrafast magic-angle spinning (MAS; Barbet-Massin et al. 2013, 2014; Andreas et al. 2015). At last, dynamic nuclear polarization (DNP) made it possible to overcome the sensitivity limitation of the MAS NMR experiments (Jaudzems et al. 2018). Thus, the high-quality DNP-enhanced NMR spectra were obtained for the phage AP205 by combining high magnetic field (800 MHz) and fast MAS (40 kHz).

The modern atomic electron cryomicroscopy contributed strongly to the location of the single copy of the maturation protein, or A protein, or A2 protein by the former *Allolevivirus* genus. The maturation protein binds to genomic RNA and is not only absolutely necessary for the infectivity of the RNA phages by their attachment to the bacterial pili, as shown in Figure 25.1, but also could be useful in viral nanotechnology as a specific RNA-targeting tool.

As mentioned earlier, the maturation protein was found to replace a single coat dimer within the phage MS2 icosahedron (Dent et al. 2013; Koning et al. 2016). It appeared therefore that the RNA phage virion actually contains 178 coat monomers, not 180, as was stated for long time before. The actual phage MS2 structure was revealed then by electron cryomicroscopy and asymmetric reconstruction at 3.6 Å resolution (Dai et al. 2017). Gorzelnik et al. (2016) resolved the electron cryomicroscopy structures of the phage Qβ with and without symmetry applied. The icosahedral structure, at 3.7 Å resolution, resolved loops not previously seen in the published x-ray structure, whereas the asymmetric structure, at 7 Å resolution, revealed maturation (A2) protein and the genomic RNA. In fact, the A2 protein replaced one dimer of coat proteins at a 2-fold axis. Next, Cui et al. (2017) used electron cryomicroscopy to reveal structures of Qβ virions and Qβ VLPs at 4.7 and 3.3 Å resolutions, respectively. These breakthrough high-resolution electron cryomicroscopy studies that revealed the slightest details of the MS2 and Qβ interactions with the host bacteria pili were recently summarized by Gorzelnik et al. (2021).

In the pure structural connection, it is noteworthy that the interface of the SARS-CoV N protein dimer appeared as a four-stranded β-sheet, superposed by two long α-helices, i.e., this topology closely resembled that of the RNA phage capsids (Chang CK et al. 2005). The sequence alignment and secondary structure prediction suggested that other coronavirus N proteins also adopted a similar dimerization mechanism.

Recently, the phage MS2 structure was resolved by a novel approach in frame of the global single particle imaging (SPI) initiative launched in December 2014 at the Linac Coherent Light Source, SLAC National Accelerator Laboratory, USA (Sun Z et al. 2018). The all-atom molecular dynamics computer model was built for the phage MS2 particle without its genome, using high-resolution electron cryomicroscopy measurements for initial conformation (Farafonov and Nerukh 2019).

CHIMERIC VLPs

GENETIC FUSIONS

fr

The first chimeric RNA phage coats were constructed for the phage fr by the Grēns team (Borisova et al. 1987;

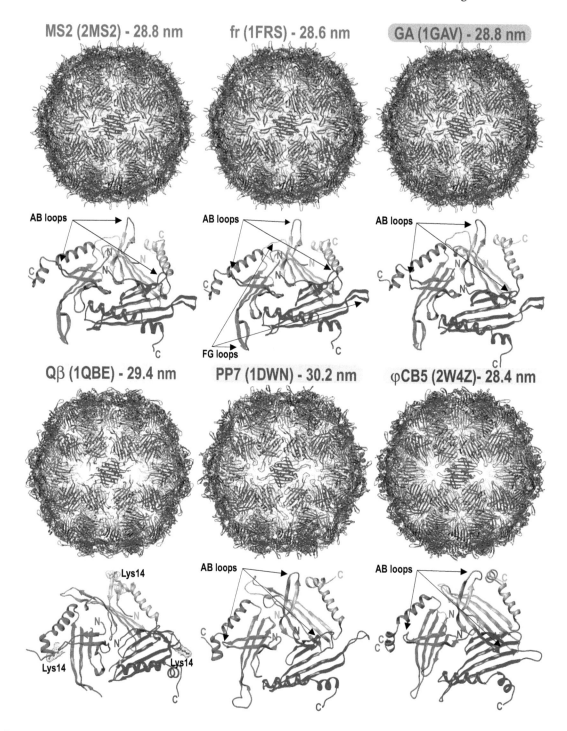

FIGURE 25.3 Crystal structures of the RNA phage VLPs. The specific colors of the RNA phage representative are the same as in Figure 25.2 and indicate the genogroup and/or the host. The protein data bank IDs are shown in parentheses and the outer diameters are indicated for each species. The coat chains A, B, and C are marked in red, green, and blue, respectively. The trimeric icosahedral asymmetric units (IAU or ASU) are located under the corresponding full-capsid 3D structures. The outer surface is oriented toward the reader. The AB loops, which are exposed on the full-capsid surfaces, are indicated within the IAUs by arrows, while the N- and C-termini are labeled by the N and C symbols. The Qβ AB loops are distinguished by Lys14 residues, which are indicated by the shaded areas. The structural data is compiled from the VIPERdb (http://viperdb.scripps.edu) database (Carrillo-Tripp et al. 2009) and is visualized using Chimera software. (Pettersen et al. 2004; adopted with permission of S. Karger AG, Basel from Pumpens P et al. *Intervirology*. 2016;59:74–110.)

Gren et al. 1987; Gren and Pumpens 1988; Kozlovskaia et al. 1988; Pushko et al. 1993) and for the phage MS2 by the Peter G. Stockley team (Mastico et al. 1993). In both cases, the main idea consisted of the presentation of the foreign immunological epitopes on the VLP surface by genetic fusion at the specially chosen sites on the VLP carrier. To identify such coat regions that could tolerate the foreign insertions in the case of the fr VLPs, the

AP205 (5LQP) - 27.6 nm

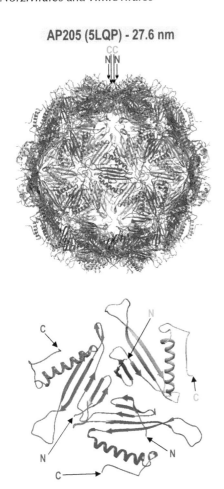

FIGURE 25.4 Crystal structure of the RNA phage AP205 that demonstrates clear differences with MS2 and other RNA phages, although the overall folds of the coat dimers are similar for both AP205 and MS2 (Shishovs et al. 2016). The coat chains A, B, and C are marked in red, green, and blue, respectively, as in Figure 25.3. From the 3D pictures, it is obvious that the AP205 coat has N- and C-termini that are located in roughly the same place as the AB loops from MS2 and other RNA phages from Figure 25.3. As a result, the N- and C-termini in the AP205 coat are well exposed on the capsid surface (blue and green, respectively, as an example). The structural data is compiled from the VIPERdb (http://viperdb.scripps.edu) database (Carrillo-Tripp et al. 2009) and is visualized using Chimera software (Pettersen et al. 2004).

synthetic oligonucleotide linkers encoding short amino acid sequences and containing convenient restriction sites were inserted into different regions of the fr coat gene (Borisova et al. 1987; Gren et al. 1987; Gren and Pumpens 1988; Kozlovskaia et al. 1988; Pushko et al. 1993). The Grēns team performed this work in parallel and with the same oligonucleotides on two putative VLP carriers, namely the fr coat and the HBc (see Chapter 38). Remarkably, this work was based on the computer predictions of the spatial structure of both VLP carriers and was conducted before the first crystal structure of an RNA phage, namely that of the homologous MS2 capsid, had been resolved (Valegård et al. 1986, 1990, 1991; Liljas and Valegård

1990; Golmohammadi et al. 1993) and the first electron cryomicroscopy (Crowther et al. 1994) and crystal (Wynne et al. 1999) structures of the HBc were achieved. The first fr chimeras that were constructed by Tatyana Kozlovska and Peter Pushko contained 2–12-aa-residue-long additions to their N- or C-termini or insertions at position 50 in the RNA binding region and were capable of self-assembly, but this was not the case at positions 97–111 in the αA-helix (Kozlovskaia et al. 1988; Pushko et al. 1993). The majority of the other fr coat mutants demonstrated reduced self-assembly capabilities and formed either coat dimers—as at the amino acid exchanges at residues 2, 10, 63, or 129—or both dimer and capsid structures, as at the residues 2 or 69 (Pushko et al. 1993).

The short epitope 31-DPAFRA-36 (or DPAFR or still DPAF), a hepatocyte-binding domain from the hepatitis B virus preS1 region, served as the first real immunological marker in the first VLP mapping studies and further in the numerous elucidations of different VLP carriers, such as, for example, the hamster polyomavirus major capsid protein VP1 (Gedvilaite et al. 2000; Zvirbliene et al. 2006). This linear epitope was recognized strongly by the classical anti-preS1 monoclonal antibody MA18/7 generated by the famous Wolfram H. Gerlich's team (Heermann et al. 1984) and mapped to the four minimal DPAF aa residues by the Grēns team (Sominskaya et al. 1992b) and then still to the three letters D, P, and F by the famous Ken Murray's team (Germaschewski and Murray 1995). When the preS1 epitope DPAFR was inserted as the standard immunological marker at positions 2, 10, and 129 of the fr coat, it appeared in all cases on the particle surface (Borisova et al. 1987). In parallel, the Grēns team used a longer immunological marker, namely the well-known and widely used V3 loop of the envelope gp120 protein of HIV-1 (Goudsmit et al. 1988), in this case, of the strain MN. This model epitope was 39 or 40 aa residues long. Unfortunately, all attempts to tolerate this 40-aa-long V3 loop into the N-terminus, at positions 10, 12, and 15 or within the FG loop, led to unassembled and mostly insoluble products (Kaspars Tārs, unpublished data).

The FG loop of the fr coat was also tested as a potential target for insertions/replacements and was initially modified by a 4-aa-long deletion (Axblom et al. 1998). The deletion variant retained the ability to form capsids, although the VLP products displayed significantly reduced thermal stability. Furthermore, the 3D structures of the mutant capsids revealed that the modified loops were disordered near the fivefold axis of symmetry and were too short to interact with each other (Axblom et al. 1998). Because of the high importance of the FG loop in capsid stability, further development of the chimeric fr VLPs based on the implementation of the FG loops was not pursued.

The fr coat vectors with the polyepitope linkers at position 2 (Borisova et al. 1987; Kozlovskaia et al. 1988; Pushko et al. 1993) were employed for the cloning of foreign sequences in all possible reading frames, with an aim of the fine mapping of linear immunological epitopes. This method did not expect an obligatory VLP formation but was managed with immunoblotting of the polyacrylamide gel distributions of the corresponding fr coat fusion libraries. Thus, the minimal length of the preS1 epitope DPAFR (Sominskaya et al. 1992b), the preS2 epitope QDPR (Sominskaya et al. 1992a), and many other hepatitis B virus-derived epitopes (Sällberg et al. 1993; Meisel et al. 1994; Sobotta et al. 2000; Sominskaya et al. 2002) was established. It is noteworthy that the fr coat epitope selection approach was used by the establishing and mapping of the structural proteins of hamster polyoma virus (Siray et al. 1999, 2000). Moreover, the fr-based expression vectors were used for the efficient expression in *E. coli* of the hamster polyoma virus VP1 gene fragments (Voronkova et al. 2007).

In parallel, in an attempt to develop chimeric VLPs as an experimental vaccine against HPV-induced tumors, the fr coat was tested as a putative VLP carrier of the relatively long stretch of the HPV16 E7 oncoprotein epitopes spanning aa residues 35–54 and 35–98 (Pumpens et al. 2002). These HPV E7 fragments were expressed in parallel on the three popular VLP carriers: fr coat and hepatitis B virus surface and core proteins. When inserted at position 2/3 of the fr coat, the E7 fragment 35–98 prevented, however, the VLP formation. Nevertheless, the purified fr-E7(35–98) induced the Th1 and Th2 subsets of T helper cells and elicited equally high antibody responses to both E7 and fr coat carrier (Pumpens et al. 2002). The fr coat also failed to form VLPs following the insertion of long HBV preS1 sequences, longer than the previously mentioned DPAFR epitope (Pumpens et al., unpublished).

Conversely, the fr coats showed unusually high capacities as VLP vectors by the addition of long segments of hamster polyomavirus VP1, a major capsid protein (Voronkova et al. 2002). Thus, the fr VLPs tolerated the N-terminal fusion of the immunodominant epitopes located at the C-terminal positions 333–384, 351–384, 351–374, and 364–384 of the hamster polyomavirus VP1. The efficient induction of the VP1-specific antibodies in rabbits and mice by immunization with chimeric VLPs harboring aa residues 333–384, 351–384, and 364–384 of the VP1 suggested the immunodominant nature of the C-terminal region of VP1 (Voronkova et al. 2002). At that time, the fr VLP studies markedly enhanced interest in using RNA phage coats as potent vaccine candidates.

Later, the purified fr VLPs carrying the hamster polyomavirus VP1 epitopes 333–384 and 351–384 were employed by the study on lymphoma outbreak in a GASH:Sal colony of Syrian golden hamsters *Mesocricetus auratus* (Muñoz et al. 2013).

MS2

As mentioned earlier, the chimeric MS2 VLPs were first generated by the Stockley's team. Thus, Mastico et al. (1993) employed for the first time the N-terminal β-hairpin exposed at the surface of the MS2 capsid, namely at the site between aa residues Gly15 and Thr16 or between positions Gly14 and Thr15 if the N-terminal methionine is not considered, as preferable for toleration of foreign insertions.

The insertion of corresponding oligonucleotides at this site allowed the production of the chimeric MS2 VLPs carrying foreign peptide sequences expressed at the central part of the hairpin. The foreign sequences for the insertion were chosen because of their known antigenic properties and resulted in several different peptides up to 24 aa residues in length. The chimeric coat proteins self-assembled into largely RNA-free chimeric VLPs in *E. coli* and were easily disassembled and reassembled in vitro. The foreign epitopes exposed on these chimeric VLPs were found to be immunogenic in mice (Mastico et al. 1993). The first short outline of this data appeared in the review papers of Hill and Stockley (1996) and Lomonossoff and Johnson (1996). The technical aspects of the constructions in question were described in a detailed methodological paper (Stockley and Mastico 2000).

The putatively protective epitope T1, spanning 24 aa residues in length and derived from the immunodominant liver stage antigen-1 (LSA-1) of the malaria parasite *Plasmodium falciparum*, must be distinguished here. It was inserted at the tip of the N-terminal β-hairpin, between positions 15 and 16, of the MS2 coat (Heal et al. 2000). The chimeric VLPs carrying the LSA-1 epitope elicited both humoral and cellular immune responses in BALB/c mice with significant upregulation of interferon-γ, a finding that corroborated naturally acquired resistance to liver stage malaria (Heal et al. 2000). This study was regarded at that time as an important step to the development of the putative LSA-1 vaccine (Kurtis et al. 2001).

In parallel, David S. Peabody (1997) has chosen so-called Flag as a standard immunological marker. The Flag octapeptide DYKDDDDK was proposed as a specific marker by Hopp et al. (1988). An antibody specific for the first four amino acid residues of this sequence was used as a detection reagent and for the affinity purification of products under mild conditions, by analogy with the previously mentioned monoclonal anti-preS1 antibody MA18/7. The authors thought that the proteins would remain unaffected by the Flag presence and would retain their biological activity, because of the small size of the peptide moiety and its hydrophilic nature. Moreover, it seemed possible to remove the Flag peptide by an enzymatic cleavage using enterokinase (Hopp et al. 1988). Peabody DS (1997) studied the consequences of the Flag insertion on the translational repressor and capsid assembly functions of the chimeric MS2 coat. When the Flag epitope was added to the N-terminus, the chimeric coat folded properly into the dimer but was defective for capsid assembly. On the other hand, a chimeric coat that was expected to display the Flag insertion as a surface loop did not fold correctly and, consequently, was proteolytically degraded. The genetic fusion of the coat dimer resulted in a protein considerably more tolerant of these structural perturbations and mostly corrected the defects accompanying the Flag peptide insertion. Thus, the genetic fusion of the subunits reversed the defects of the AB loop insertions when either half of the fused dimer contained the wild-type sequence. However, the fusion of subunits was insufficient to compensate fully for the structure defects imposed by the presence of the Flag in both AB loops of the fused dimer. It did, nevertheless, prevent degradation of the protein and allowed its accumulation within inclusion bodies. The increased resistance of the single-chain coat protein to the urea denaturation indicated that the fused dimer was substantially more stable than the wild-type one. The covalent joining of subunits was supposed to be a general strategy for the engineering of the MS2 chimeras with increased protein stability (Peabody DS 1997).

The His-tagged MS2 VLPs were constructed by the introduction of the His$_6$ linker between coat codons 15 and 16 (Cheng et al. 2006). The His$_6$ linker simplified purification of the VLP-covered RNAs, since this study was oriented mostly toward generation of the so-called armored RNA markers that will be discussed later in the *Armored RNA* paragraph.

Selby and Peterlin (1990) expressed the first chimeric MS2 coat derivative, namely the coat fused to the HIV-1 Tat, in eukaryotic cells. The product did not form VLPs (but this study paved the long way for the tethering techniques, another breakthrough) but formed non-VLP application of the RNA phages. The two others historically important non-VLP MS2 coat fusions were performed with the (i) LexA protein that started the so-called three-hybrid approach and (ii) green fluorescence protein (GFP) that gave onset to the pioneering visualization (or imaging) studies in living cells. It is remarkable here that, by the tethering studies in eukaryotic cells, the MS2 coat was provided with the HA tag, an epitope 98–106 from human influenza hemagglutinin (Kim YK et al. 2005). Then, the SNAP domain of O-6-alkylguaninalkyltranferase was fused to the MS2 coat with an idea to link the MS2-SNAP fusion protein to DNA when the latter would be labeled with the SNAP substrate benzylguanine (Paul et al. 2013).

Recently, Robinson et al. (2020) programmed the MS2 VLPs by a high-throughput technique called systematic mutagenesis and assembled particle selection (SyMAPS). The latter was applied to display 9,261 nonnative tripeptide insertions into the FG loop as a valuable position for peptide insertion and to illuminate how properties such as charge, flexibility, and hydrogen bonding could interact to preserve or disrupt capsid assembly.

Qβ

The relatively low tolerance of the fr- and MS2-based VLP vectors to long foreign insertions provoked a definite interest in the generation of so-called mosaic particles. In contrast to the homogenously formed VLPs, the mosaic particles were intended to be composed of wild-type (*helper*) and epitope-carrying (*chimeric*) subunits. This was supposed to improve the VLP capacity for the foreign insertions and was used for the first time in the case of hepatitis B virus surface antigen as a carrier (Delpeyroux et al. 1988), as described in Chapter 37.

After successful cloning of the full-length A1 gene of the phage Qβ in *E. coli* (Kozlovska et al. 1993), Tatyana

Kozlovska et al. (1996, 1997) recognized the 195-aa-residues long readthrough extension of the phage Qβ coat as a promising site for foreign insertions by the generation of the mosaic Qβ VLP particles. The readthrough A1 extension was proposed to contain elements that could protrude as the spike-like structures on the Qβ VLP surface. This assumption was achieved by the sensitive homology programs elaborated at that time by Indulis Gusārs from the Grēns team. He detected some unexpected similarities when the readthrough Qβ A1 extension was compared with the protruding preS part of the hepatitis B virus surface antigen (Kozlovska et al. 1996). The real 3D structure of the Qβ A1 readthrough domain was revealed at high resolution by x-ray crystallography after 15 years (Rumnieks and Tars 2011). The fold was found unique among all proteins in the protein data bank. However, the real fold of the protruding hepatitis B virus preS1 region remains still unresolved.

A next argument for the applicability of the readthrough extension was provided by the self-assembly capabilities of the capsids with mutually exchanged extensions of the Qβ and SP A1 proteins, which were demonstrated experimentally at that time (Priano et al. 1995).

To realize the idea of the mosaic Qβ coat/A1 VLPs carrying foreign epitopes inserted into the readthrough extension part of A1, the two possible strategies were chosen (Kozlovska et al. 1996, 1997). First, it was the expression of the original A1 gene in conditions of the enhanced UGA suppression. Second, it was the separated expression of the proper coat gene and of the full-length A1 gene after mutational changing of the UGA stop codon to UAA stop or GGA sense codons, respectively. By the second strategy, the expression of the coat and full-length A1 genes was achievable (i) from a single plasmid harboring the appropriate genes within a cassette under the control of a strong promoter, and (ii) from the two different plasmids conveying the two antibiotic resistances to the *E. coli* cells.

The preS1 epitope DPAFRA was employed as a standard foreign epitope, short enough and easy to test with well-characterized antibody MA18/7, as described earlier. The 39- or 40-aa-residues-long V3 loop of the envelope gp120 protein of HIV-1 was used as a more complicated immunological marker that would represent epitopes longer than 4–6 aa residues and possessing some definite conformational features.

Concerning the direction based on the enhanced UGA suppression, Hofstetter et al. (1974) were the first who found out that the content of A1 within the Qβ phage particles could be elevated in conditions of the UGA suppression. Following this indication, a plasmid harboring cloned gene of *opal* suppressor tRNA that recognized the UGA codon as the Trp codon (Smiley and Minion 1993) was introduced into *E. coli* K802 cells bearing the appropriate Qβ expression plasmids. As a result, the synthesis of the A1, A1 carrying the DPAFRA, or A1 carrying the V3 loop was enhanced to up to 50% of the Qβ coat synthesis. Irrespective of the method by which the wild-type Qβ helper subunits were supplied, the chimeric Qβ subunits with additions at their

C-termini, replacing the readthrough part, acquired the ability to be included into the chimeric particles in the cases when the chimeric subunits alone failed to form homogeneous particles from identical subunits (Kozlovska 1996, 1997).

Tatyana Kozlovska's group carefully mapped the potentially tolerant sites of the readthrough extension by the consistent insertion of the two previously mentioned model epitopes (Vasiljeva et al. 1998). In conditions of enhanced UGA suppression, the chimeric VLPs were always observed, although the proportion of the A1-extended to the short coats in mosaic particles dropped from 48%–14%, with an increase of the length of the A1 extension. Nevertheless, the model preS1 epitope, DPAFR in this case, was found on the surface of the mosaic Qβ particles and ensured their specific antigenicity and immunogenicity in mice (Vasiljeva et al. 1998). The antibody response to the preS1 epitope was clearly higher for the self-assembled Qβ-preS1 VLPs than for a nonassembled Qβ-preS1 fusion variant (Fehr et al. 1998). When the Qβ coat was modified to carry long hepatitis B virus preS insertions, namely full-length preS, preS1 alone, or preS2 alone, instead of the A1 extension, the mosaic Qβ particles were formed that possessed the surface-exposed preS chains, while regular VLPs did not form without the presence of the Qβ coat as a helper (Indulis Cielēns and Regīna Renhofa, unpublished data).

Further development of the idea of the mosaic Qβ coat particles was accomplished by the famous M.G. Finn's team, who performed at that time impressive studies on the functionalization of the Qβ VLPs, as described later. Concerning specifically the genetic fusions, Udit et al. (2008) coexpressed two plasmids in E. coli, where the first, carbenicillin-resistant expression vector, carried the wild-type Qβ coat gene, but the other, spectinomycin-resistant vector, encoded the chimeric Qβ coat gene with the His-tag GSGSGH$_6$ coding sequence appended to the C-terminus of the Qβ coat gene. This resulted in the polyvalent display of approximately 37 His$_6$ tags per particle and allowed immobilization of the Qβ VLPs on metal-derivatized surfaces, without any observed change in the VLP stability compared to the wild-type VLPs.

By further engineering of metal- and metallocycle-binding sites on the Qβ VLP scaffolds, the three motifs, His$_6$, His$_6$-His$_6$, and Cys-His$_6$, were compared by incorporation into the capsid via a coexpression methodology at ratios of 1.1:1, 1.1:1, and 2.3:1 for wild-type to chimeric coat protein (Udit et al. 2010b). The size-exclusion chromatography yielded elution profiles identical to wild-type particles, while Ni-NTA affinity chromatography resulted in retention times that increased according to Qβ-His$_6$ < Qβ-Cys-His$_6$ < Qβ-His$_6$-His$_6$. In addition to interacting with metal-derivatized surfaces, the Qβ-Cys-His$_6$ and Qβ-His$_6$-His$_6$ VLPs bound heme (Udit et al. 2010b).

Furthermore, instead of relying on the differential readthrough of a stop codon as in the natural virus, Brown SD et al. (2009) employed again the compatible plasmids coding for the wild-type Qβ coat and for the chimeric Qβ with the C-terminal extension of the 58 aa Z domain derived from S. aureus protein A and fused over an octapeptide spacer. As is known, the Z domain binds to the CH2-CH3 hinge region of a group of IgG subtypes and can be employed in numerous nanotechnological applications. The fusion plasmid yielded copious quantities of the Qβ-Z fusion protein, but no intact particles, consistent with the expectation that the extended subunit was incapable of assembling into a particle on its own. In contrast, when E. coli cells were transformed with both wild-type and fusion plasmids, the mosaic Qβ particles were isolated in high yields, approximately 50 mg per liter of culture, and approximately 20 Z domains were incorporated per particle. Moreover, the fused domains were accessible on the exterior surface of the particle (Brown SD et al. 2009).

Liao et al. (2017) constructed the mosaic Qβ VLPs by expressing two proteins, namely the original Qβ coat and the Qβ-GFP fusion protein, where GFP was fused to the C-terminus of the Qβ coat, linked in tandem by an internal ribosome entry site sequence in E. coli. The two proteins were expressed in roughly equal amounts, but the Qβ-GFP fusion protein was incorporated into the Qβ VLPs less well, at a ratio ~1:10 relative to the original Qβ protein.

PP7

After successful application of the MS2 VLPs as a platform for the peptide display (Peabody DS et al. 2008), David S. Peabody and Bryce Chackerian turned to the VLPs generated by the pseudomonaphage PP7. By analogy with the MS2 coat, the AB loop of the PP7 coat and of the PP7 coat single-chain dimer, between positions 11 and 12, was employed first (Caldeira et al. 2010). In this first paper, the broadly cross-type neutralizing HPV L2 epitope of 15 aa residues served as the most important model, in parallel with the Flag peptide and the V3 loop of HIV-1. The single-chain PP7 coat dimer was also highly tolerant of the random 6-, 8-, and 10-aa insertions, as in the case of the MS2 VLPs. As Caldeira et al. (2010) noted, the PP7 VLPs offered several potential advantages and improvements over the MS2 VLPs. First, the particles were more stable thermodynamically, because of the presence of the stabilizing intersubunit disulfide bonds (Caldeira and Peabody 2007). Second, the PP7 VLPs were not cross-reactive immunologically with those of the MS2 VLPs. This could be important in applications where serial administration of the VLPs would be necessary. Third, the authors anticipated that the correct folding and assembly of the PP7 VLPs might be more resistant to the destabilizing effects of peptide insertion or that the PP7 VLPs might at least show tolerance of some peptides not tolerated in the MS2 VLPs.

As the first result, the single-chain dimer of the PP7 coat demonstrated broad tolerance to random and specific peptide insertions, displayed peptides on the VLP surface, ensured high immunogenicity of the inserted epitopes, and packaged the RNA that directed their synthesis. The two prospective vaccine candidates were generated. The first specific PP7 VLP vaccine successfully induced antibodies against a peptide derived from the V3 loop of HIV-1 that was the target of neutralizing antibodies. The second vaccine induced antibodies against a broadly cross-neutralizing epitope from the minor capsid protein L2 of HPV16. It was highly important, since the neutralizing antibodies that targeted the minor capsid protein, L2, were broadly cross-neutralizing, suggesting that the neutralizing epitopes on L2 were conserved across HPV types, unlike the L1-specific neutralizing antibodies that were largely HPV type-specific. Caldeira et al. (2010) showed that the PP7 VLP vaccine displaying an L2 epitope induced antibody responses that protected mice from genital challenge with homologous (HPV16) and heterologous (HPV45) pseudovirus.

Moreover, Chackerian and Peabody's team advertised successful generation of a pan-HPV vaccine based on the PP7 VLPs displaying the broadly cross-neutralizing L2 epitope, in contrast to the classical HPV vaccines that were based on the VLPs of the major HPV capsid protein L1 (Tumban et al. 2011). It is noteworthy that Tumban et al. (2011) have removed RNA, for the first time in the history of the chimeric VLPs, from the inside of the L2-PP7 VLPs by alkaline treatment, in accordance with the procedure that was applied earlier on the nonchimeric MS2 VLPs (Hooker et al. 2004) and demonstrated its high efficiency later on the VLPs based on hepatitis B virus cores (Strods et al. 2015). Generally, the conclusive success of the PP7- and MS2-based L2 HPV vaccines was self-reviewed at that time by Tyler et al. (2014a).

The further progress in the PP7-based vaccines is presented later in the *Vaccines* paragraph. It is, however, worth emphasizing here that the PP7 coat single-chain dimer platform was provided with externally displayed cell-penetrating peptides and encapsidated GFP mRNA (Sun Yanli et al. 2016). Therefore, both display and packaging capabilities of the PP7 VLPs were involved to construct a targeted delivery vector for both peptides and mRNA.

The PP7-derived VLPs contributed the first 3D visualization of the chimeric particles obtained by the famous M.G. Finn's team (Zhao L et al. 2019). In short, the PP7 VLPs were shown to tolerate the display of sequences from 1 kDa, a cell penetrating peptide, to 14 kDa, the Fc-binding double Z-domain, on its exterior surface as C-terminal genetic fusions to the coat protein. Moreover, a single-chain dimeric construct allowed the presentation of exogenous loops between capsid monomers and the simultaneous presentation of two different peptides at different positions on the icosahedral structure. Surprisingly, the appearance of a $T=4$ structure was demonstrated for the levivirus-derived capsids for the first time, adopted unexpectedly by particles self-assembled from coat dimers (Zhao L et al. 2019).

Thus, when the constructed PP7-PP7, PP7-PP7-ZZ, and PP7- LEAEMDGAKGRL-PP7 particles were resolved by electron cryomicroscopy at ~3 Å resolution for each, it arose that all of the dimer-based structures, including the most simple PP7-PP7 dimer, formed $T=4$ capsids. This stood in contrast to the $T=3$ structure reported for the monomeric PP7 particle and confirmed for the ZZ-PP7

structure. Figure 25.5 presents these unexpected data and visualizes the extra peptides on the PP7-PP7-ZZ and PP7-LEAEMDGAKGRL-PP7 particles using low-pass filtered maps displayed at low threshold.

Concerning general plasticity of the 3D organization of the coat dimers, Asensio et al. (2016) further established that a single aa mutation S37P led to the smaller MS2 VLPs of 17 nm in diameter and switched the particle geometry from the $T=3$ to $T=1$ icosahedral symmetry. Next, de Martín Garrido et al. (2020) have assembled MS2 VLPs with nongenomic RNA containing the capsid incorporation sequence and revealed by electron cryomicroscopy not only the traditional $T=3$ but also $T=4$ and mixed capsids between these two triangulation numbers to 4 Å and 6 Å, respectively.

Continuing the 3D plasticity issue, the icosahedral Qβ VLPs were converted into rods after modification of the FG loop structure (Cielens et al. 2000). Since the Qβ VLPs were stabilized by disulfide bonds of cysteine residues 74 and 80 within the FG loop, the stability of capsids was mutationally reduced by converting the aa stretch 76-ANGSCD-81 within the FG loop into the 76-VGGVEL-81 sequence. It led to production of aberrant rod-like Qβ VLPs, along with normal icosahedral capsids. The length of the rod-like particles exceeded 4–30 times the diameter of icosahedral Qβ VLPs. The appearance of alternate VLP forms of RNA phages was further confirmed by the presence of the rod-like structures in the case of the coassembly of the phage fr and GA coats, and the rod-like structures were identified as icosahedral prolates (Rumnieks et al. 2009).

AP205

The AP205 coat demonstrated unique capabilities by the toleration of foreign insertions. An impressive collection of the chimeric AP205 VLPs was generated by the fruitful collaboration of Martin F. Bachmann and Elmārs Grēns's teams (Tissot et al. 2010). Numerous AP205-based vaccine candidates are listed in the *Vaccines* paragraph. Generally, the AP205 VLPs tolerated N- and/or C-terminally fused epitopes of up to at least 55 aa residues in length and conferred a high immunogenicity and protective ability to displayed epitopes. A set of the preS1-carrying AP205 VLPs was constructed, first of all, with the addressing/targeting/delivery but not vaccination purposes (Kalniņš et al. 2013). These data are presented in the *Packaging and Targeting* paragraph.

The His-tagged AP205-His$_6$ VLPs were further generated and compared head-to-head to the analogous HBc-His$_6$ VLPs (Gillam and Zhang 2018).

GA

The chimeric GA VLPs constructed by Regīna Renhofa's group with vaccination and addressing/targeting/delivery purposes are described in the *Vaccines* and *Packaging and Targeting* paragraphs, respectively. These derivatives formed excellent particles, as discovered by electron microscopy.

Figure 25.6 is intended to demonstrate high quality of the chimeric VLPs obtained by the gene fusion technology.

PP7-PP7-ZZ   **PP7-allop-PP7**   **PP7-allop-PP7-150loop**

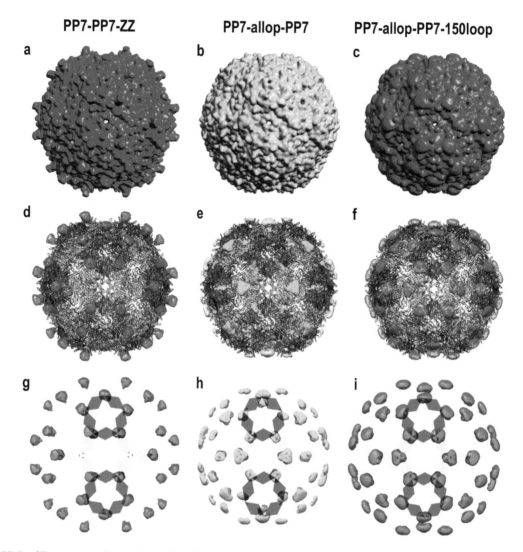

FIGURE 25.5 3D reconstruction and mapping of extra density on the surface of PP7-a-loop-PP7–or PP7-LEAEMDGAKGRL-PP7, PP7-PP7-ZZ–and PP7-a-loop-PP7-150-loop–or PP7-LEAEMDGAKGRL-PP7-NDTGHETDEN–particles. Overall surface density maps: (a) PP7-PP7-ZZ, (b) PP7-a-loop-PP7, and (c) PP7-a-loop-PP7–150-loop. Extra densities highlighted: (d) PP7-PP7-ZZ, (e) PP7-a-loop-PP7, and (f) PP7-a-loop-PP7–150-loop, comparing to PP7-PP7. (g) PP7-PP7-ZZ, (h) PP7-a-loop-PP7, and (i) PP7-a-loop-PP7–150-loop, showing the same loop insertions and C-terminal extensions as in (d–f). (Reprinted with permission from Zhao L, Kopylov M, Potter CS, Carragher B, Finn MG. Engineering the PP7 virus capsid as a peptide display platform. *ACS Nano*;13, 4443–4454. Copyright 2019 American Chemical Society.)

This picture includes derivatives of the two representative models, namely the group II phage GA and acinetophage AP205, and compares the RNA phage virions, recombinant VLPs, and chimeric VLPs carrying long foreign insertions. It is remarkable that, in these high-quality electron micrographs, the surfaces of the long insertions-carrying VLPs differ considerably from the surfaces of unmodified VLPs by the appearance of distinct knobs that are presumably formed by the inserted sequences.

PEPTIDE DISPLAY

MS2

The famous Jan van Duin's team was the first who succeeded by the generation of the peptide display on the live chimeric RNA phage, namely MS2 (van Meerten et al. 2001). This study is described later in the *Chimeric Phages* paragraph.

Following the same peptide display idea, David S. Peabody and Bryce Chackerian, together with their colleagues, proposed the immunogenic display of diverse peptides on the MS2 VLPs but not on the viable phages (Peabody DS et al. 2008). This novel approach served not only for the tailored display of the specific peptides but was also adapted to the generation of random peptide libraries, where desired binding activities could be recovered by affinity selection. The first specific peptides were represented by the V3 loop of HIV gp120 and the ECL2 loop of the HIV coreceptor, CCR5, which were inserted into a surface loop of the MS2 coat protein. Both insertions disrupted the VLP

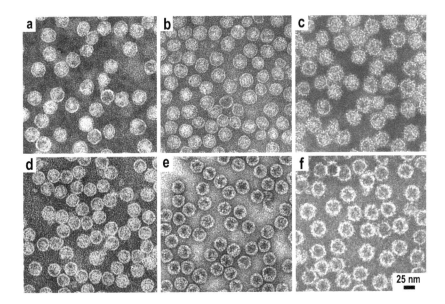

FIGURE 25.6 Electron micrographs of the chimeric RNA phage VLPs as compared with native virions and recombinant nonchimeric VLPs: (a) negatively stained AP205 virions, (b) recombinant AP205 VLPs, (c) chimeric AP205 VLPs carrying 151 aa residues of human interleukin-1β at the C-terminus, (d) GA virions, (e) recombinant GA VLPs, (f) chimeric GA VLPs carrying a 61-aa-long $Z_{HER2:342}$ affibody at the C-terminus. The VLPs were purified from the appropriate gene-expressing *E. coli* cells. The electron microscopy was performed by Velta Ose. For electron microscopy, the grids with the adsorbed particles were stained with aqueous solutions of 1% uranyl acetate (pH 4.5) or 2% phosphotungstic acid (pH 7.0) and examined with JEM-100C or JEM-1230 electron microscopes (Jeol Ltd., Tokyo, Japan) at 100 kV. Well-ordered knobs are clearly visible on the surfaces of the chimeric VLPs. (Reprinted with permission of S. Karger AG, Basel from Pumpens P, Renhofa R, Dishlers A, Kozlovska T, Ose V, Pushko P, Tars K, Grens E, Bachmann MF. *Intervirology*. 2016;59:74–110.)

assembly, apparently by interfering with the correct protein folding, but these defects were suppressed efficiently by the genetical fusion of the two coat polypeptides into a single-chain dimer, as described previously. The resulting VLPs displayed the V3 and ECL2 peptides on their surfaces, where the inserted epitopes showed the potent immunogenicity that was the hallmark of the VLP-displayed antigens. The experiments with random-sequence peptide libraries acknowledged the high tolerance of the MS2 single-chain dimer to the insertion of six, eight, and ten aa stretches. In fact, the MS2 VLPs supported the display of a wide diversity of peptides in a highly immunogenic format. Moreover, the MS2 VLPs encapsidated the mRNAs that directed their own synthesis, thus establishing the genotype/phenotype linkage necessary for the recovery of the affinity-selected sequences. Thus, the single-chain MS2 VLPs united in a single structural platform the selective power of the phage display with the high immunogenicity of the tailored VLPs (Peabody DS et al. 2008).

Furthermore, the specialized plasmid vectors were described that facilitated the construction of high-complexity random sequence peptide libraries on the MS2 VLPs and that allowed control of the stringency of affinity selection through the manipulation of display valency (Chackerian et al. 2011). The system was used to identify epitopes for several previously characterized monoclonal antibody targets and showed that the MS2 VLPs thus obtained elicited antibodies in mice whose activities mimicked those of the selected antibodies. The advantage of the MS2 VLP-based display system over the well-known filamentous phage display libraries consisted of the fact that the MS2-displayed peptides were much more immunogenic than the same epitopes displayed on the filamentous phages. The latter were usually not very immunogenic because the epitopes were presented on them at low densities that were not sufficient for efficacious B-cell activation. In contrast, the proposed MS2 VLP peptide display system combined the high immunogenicity of the MS2 VLPs with the powerful affinity selection capabilities of other phage display systems (Chackerian et al. 2011). The affinity selection in the MS2 VLP platform was performed successfully by the development of a mimotope vaccine targeting the *Staphylococcus aureus* quorum-sensing pathway (O'Rourke et al. 2014), described in detail in a special review article (O'Rourke et al. 2015).

The MS2 VLP peptide display was used for the presentation of the mimotope that mimicked a conserved epitope of the *Plasmodium falciparum* blood stage antigen AMA1 (Crossey et al. 2015b).

Then, the plasma-derived IgG from a pool of five patients with advanced ovarian cancer was subjected to iterative biopanning, using a library of the MS2 VLPs displaying diverse short random peptides (Frietze et al. 2016b). After two rounds of biopanning, the selectant population of the MS2 VLPs was analyzed by deep sequencing. One of the top 25 most abundant peptides identified, namely DISGTNTSRA, had sequence similarity to the cancer antigen 125 (CA125/MUC16), a well-known ovarian-cancer-associated antigen. The mice immunized with the chimeric MS2-DISGTNTSRA VLPs generated antibodies that cross-reacted with the purified soluble CA125 from ovarian cancer cells but not with the membrane-bound CA125, indicating that the DISGTNTSRA peptide was a CA125/MUC16 peptide mimic of the soluble CA125. Looking for anti-DISGTNTSRA, anti-CA125, and CA125 markers in preoperative ovarian cancer patient plasma, it was concluded that the deep sequence-coupled biopanning was applicable for the identification of the autoantibody responses against tumor-associated antigens in the cancer patient plasma (Frietze et al. 2016b).

Furthermore, the pathogen-specific antigen fragment library displayed on the MS2 VLPs, in combination with deep sequence-coupled biopanning, was used for the detailed mapping of the linear epitopes targeted by antibody responses to secondary Dengue virus (DENV) infection in humans (Frietze et al. 2017). Although there was considerable variation in the responses of individuals, several epitopes within the envelope glycoprotein and nonstructural protein 1 were commonly enriched. This study therefore established a novel approach for the characterization of the pathogen-specific antibody responses in human sera (Frietze et al. 2017). Furthermore, Warner et al. (2020) described refinements that expanded the number of patient samples that can be processed at one time, increasing the utility of the deep sequence-coupled biopanning technology for rapidly responding to emerging infectious diseases. The sera from primary DENV-infected patients were used as a model to pan an MS2 VLP library displaying all possible 10-aa peptides from the DENV polypeptide. The selected VLPs were identified by deep sequencing, and an impressing number of the DENV epitopes was identified (Warner et al. 2020).

Furthermore, Collar et al. (2020) employed the deep sequence-coupled biopanning technology used to map *Chlamydia trachomatis*-specific antibodies in groups of women with defined outcomes following infection with this obligate intracellular bacterium. The analysis not only yielded immunodominant epitopes that had been previously described but also identified new epitopes targeted by human antibody responses to *C. trachomatis*. Remarkably, it was established that detection of definite serum antibodies in women with urogenital *C. trachomatis* infection was not associated with protection against reinfection (Collar et al. 2020).

Using the peptide display approach, the four different functional single-chain variable fragments (scFvs) were displayed successfully on the surface of the MS2 VLPs (Lino et al. 2017). Each scFv was validated both for its presence on the surface of the MS2 VLPs and for its ability to bind its cognate antigen. This prospective paper showed clearly how the scFv libraries could be displayed in a similar manner on the VLP surface and could then be biopanned in order to discover the novel scFv targets.

The predominant epitope 131–160 of VP1 of foot-and-mouth disease virus was displayed on the top of the MS2 coat (Wang G et al. 2018). Despite the impressive length of the epitope, the chimeric protein self-assembled successfully into chimeric nanoparticles of 25–30 nm in diameter, when expressed in *E. coli*. The chimeric nanoparticles stimulated the antibody levels in mice, which were similar to that induced by the commercial synthetic peptide vaccine PepVac and were significantly more immunogenic than the control peptide. Moreover, the results from the specific interferon-γ responses and lymphocyte proliferation tests indicated that the nanoparticle-immunized mice exhibited significantly enhanced cellular immune response. This made it possible to regard the MS2-derived nanoparticles as a potential alternative vaccine for the future FMDV control (Wang G et al. 2018).

Detailed methodology of the affinity selection using the MS2 VLP display platform was published recently by Frietze et al. (2020). The products of the affinity selection by the MS2 VLP platform are listed in the *Vaccines* paragraph.

Qβ

The peptide display system was elaborated also on the Qβ platform, although on the viable Qβ phage but not on the Qβ VLPs. The live Qβ phage was employed as a display system via engineering of the A1 protein (Skamel et al. 2014). This unique system employed the high Qβ capability to mutate and adopt to the sequence changes. As an example, the chimeric Qβ phages bearing G-H loop peptide of the VP1 protein of FMDV were constructed first. The full story is placed in the *Chimeric Phages* paragraph.

CHEMICAL COUPLING

The chemical coupling of foreign oligopeptides to the surface of VLPs was developed as an alternative method to the genetic fusion of epitope-encoding sequences. For the first time, it was applied to the Qβ VLPs by the famous Martin F. Bachmann's team in Zürich (Jegerlehner et al. 2002a). They used an approach that was initially applied in their laboratory for another broadly used recombinant VLP model, namely hepatitis B virus core (Jegerlehner et al. 2002a, b, 2004). For the first time with the Qβ VLPs, the model peptide D2 of 15 aa residues with N-terminal CGG linker containing therefore a free cysteine residue at the N-terminus was coupled to a naturally exposed lysine residue on the Qβ VLP surface using the hetero-bifunctional cross-linker maleimidobenzoic acid sulfosuccinimidyl ester (Jegerlehner et al. 2002a). The modified VLPs showed efficient induction of oligopeptide-specific antibodies in mice. Moreover, it was verified experimentally that the density of epitopes displayed on the VLPs was a key parameter for efficient antibody response (Jegerlehner et al. 2002a). This chemical coupling approach initiated a novel era of therapeutic vaccination and led to the development of a panel of experimental therapeutic vaccines by the Bachmann team. Thus, Bachmann and Dyer (2004) announced the therapeutic vaccination for chronic diseases as a new field for the VLP technology and presented an impressive list of vaccine candidates to be generated. The therapeutic vaccination was proclaimed as a second vaccine revolution for the new epidemics of the twenty-first century (Dyer et al. 2006).

The fruitful combination of the chemical coupling with the VLP approach was forecasted earlier by Bachmann and Jennings (2001). In turn, the idea of the therapeutic vaccines was based on the assumption that the VLP carriers could present surface-displayed self-antigens, overcome the natural tolerance of the immune system toward self-proteins, and induce high levels of specific autoantibodies (Bachmann et al. 1993). Starting from 2002, the Qβ VLPs were the focus of attention of the Bachmann team. Figure 25.7 presents the general idea of the Qβ VLP-based chemical coupling approach to the generation of a long list of prospective vaccine candidates by this team.

Currently, the chemical coupling on the Qβ VLPs is distinguished as one of the central routes for the generation of novel vaccines (Bachmann and Zabel 2015; Mohsen et al. 2017b, 2020; El-Turabi and Bachmann 2018; Engeroff and Bachmann 2019). The full list of the chemically coupled Qβ VLP-based vaccine candidates generated by the Bachmann team is presented in the *Vaccines* paragraph.

In contrast to Bachmann's team, Bryce Chackerian and David S. Peabody used the phage Qβ—but not the Qβ VLPs—in parallel with the HPV VLPs to generate a vaccine candidate targeting the amyloid-beta (Aβ) peptide as a promising potential immunotherapy for the Alzheimer's disease patients (Chackerian et al. 2006). The bifunctional SMPH cross-linker with amine- and sulfhydryl-reactive arms was used, where the amine-reactive arm of SMPH was linked again to the surface-exposed lysines on Qβ, and then the Aβ-GGC peptide was linked to Qβ-SMPH by

virtue of the exposed sulfhydryl residue on the C-terminal cysteine residue of the peptide. The immunization with the Aβ peptide conjugated to the phage Qβ or to the HPV VLPs elicited anti-Aβ antibody responses at low doses and without the use of adjuvants. Remarkably, both HPV VLP- and Qβ phage-based Aβ vaccines induced weak or negligible T-cell responses against Aβ, while T-cell responses were largely directed against linked viral epitopes.

After classical work on the HPV VLPs and creation of the worldwide accepted HPV VLP vaccine, John T. Schiller's team, personally Bryce Chackerian, exhibited initially a purely immunological interest in the phage Qβ as a promising display system (Chackerian et al. 2008). By using the phage Qβ that was supplied by David S. Peabody in parallel with the bovine and human papillomavirus VLPs, Chackerian et al. (2008) studied the ability to distinguish between self- and foreign antigens in a B cell receptor transgenic mouse model that expressed a soluble form of hen egg lysozyme (HEL).

Furthermore, the chemical coupling of the HPV16 L2 epitopes to the phage Qβ but not to the Qβ VLPs was used in parallel to the PP7 VLP-based fusion technology, where the same peptide sequences were applied in both arms of the study (Tyler et al. 2014b). The technical data of the products is presented in the *Vaccines* paragraph.

FUNCTIONALIZATION

MS2

An alternate chemical coupling approach that could be named rather functionalization of the VLP surface was developed on the phage MS2 by David S. Peabody (2003).

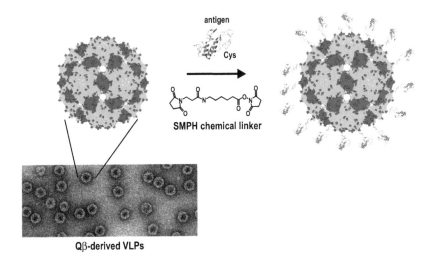

Qβ-derived VLPs

FIGURE 25.7 VLPs and the chemical coupling approach. The figure depicts a structural representation of a Qβ VLP obtained from the crystal structure of the phage Qβ. It also illustrates the modular approach to vaccine production whereby the protein antigen and VLP are separately produced and then covalently linked using a heterobifunctional chemical cross-linker such as succinimidyl-6-[(β-maleimidopropionamido) hexanoate] (SMPH). The resultant conjugate vaccine displays the target antigen in an ordered and highly repetitive fashion. Shown too is an electron-micrograph of 25–30 nm diameter icosahedral Qβ VLPs. (Bachmann MF, Jennings GT. Therapeutic vaccines for chronic diseases: Successes and technical challenges. *Philos Trans R Soc Lond B Biol Sci.* 2011;366:2815–2822. Reproduced by permission of The Royal Society of Chemistry.)

Thus, the surface cysteines were introduced onto the MS2 coat. The two cysteine residues that were present in the natural MS2 coat were internally located, while the T15C substitution displayed a reactive thiol on the VLP surface (the aa positions are given here as in the original paper, where the initial methionine was not considered). Cheng et al. (2005) performed similar site-directed mutagenesis at the codon 15 (16 by the genetic map) of the MS2 coat gene. The thiolated MS2 Cys-VLPs were then chemically modified with fluorescein-5′-maleimide, and the fluorescent nanoparticles were obtained.

In order to get dyed particles to be introduced into ecological studies, the phage MS2 (but not MS2 VLPs in this case) was labeled at that time with fluorescein-5-isothiocyanate (FITC), fluorescein, 5-(4,6-dichlorotriazinyl)amino-fluorescein (5-DTAF), or rhodamine B (Gitis et al. 2002a, b). The FITC and 5-DTAF were used for the conjugation of lysine residues. The rhodamine B and fluorescein labelings were performed by using 1-[3-(Dimethylamino)propyl]-3-ethylcarbodiimide hydrochloride (DEC) together with the dye. This procedure resulted in permanent attachment without covalent conjugation of the dye, probably due to DEC-assisted caging of the fluorescent labels in the hydrophobic environment of the dye.

In parallel, Wu M et al. (2002) set in motion another MS2 functionalization approach. They conjugated the MS2 coat to transferrin. To do this, the MS2 VLPs were reacted with a thiol-introducing reagent SATA, while transferrin was then reacted with a maleimide modification reagent SMCC, and the SATA-derivative of the coat protein was activated by hydroxylamine treatment. Finally, the two proteins were mixed and made it possible to form covalent conjugates at room temperature for 30 min. As a result, the cross-linked transferrin was found in the MS2 VLPs (Wu M et al. 2002).

Furthermore, the MS2 VLPs were converted into an efficient nanoparticle magnetic resonance imaging (MRI) contrast agent (Anderson et al. 2006). Thus, more than 500 gadolinium chelate groups were conjugated onto a viral capsid. The MS2(Gd-DTPA-ITC)m contrast agent was synthesized through the conjugation of premetalated Gd-DTPA-ITC to lysine residues.

The further strong progress in the functionalization field relates to the long-term successful activities of the Matthew B. Francis team. They were first who started to use the surface functionalization methodology not only on the outer VLP surface but also on the interiors of the MS2 VLPs by modification of tyrosine residues via a recently developed hetero-Diels-Alder bioconjugation reaction (Hooker et al. 2004). Using diazonium-coupling reactions, the tyrosine 85 was modified with high efficiency and selectivity. The virtually complete coupling was achieved in 15 min through exposure to five equivalents of a diazonium salt (Hooker et al. 2004). This pioneering approach was applied successfully for the dual-surface modification of the tobacco mosaic virus (Schlick et al. 2005).

A new protein modification reaction was based on a palladium-catalyzed allylic alkylation of tyrosine residues using π-allylpalladium complexes (Tilley and Francis 2006). This technique employed electrophilic π-allyl intermediates derived from allylic acetate and carbamate precursors and was used to modify the phage MS2 in aqueous solution at room temperature. To facilitate the detection of modified proteins, a fluorescent allyl acetate was synthesized and coupled to the phage MS2, while the tyrosine selectivity of the reaction was confirmed through trypsin digest analysis.

As described later in detail in a thorough review by Algar et al. (2011), the Francis team has specifically modified tyrosine residues native to the interior surfaces of the MS2 phage particles with nitrophenyl diazonium salts (Hooker et al. 2007, 2008; Kovacs et al. 2007; Datta et al. 2008). The nitrophenyl diazonium salts were linked to other functional groups that enabled the attachment of reporters such as organic dyes, MRI contrast agents, and radiolabels for positron emission tomography (PET). The diazonium reaction also allowed selective bioconjugation to the interior of the MS2 particles, while allowing lysine residues native to the exterior surface to be modified in orthogonal reactions. The latter included functionalization of the MS2 particles with succinimidyl esters of PEG and targeting ligands (Kovacs et al. 2007).

McFarland and Francis (2005) were the first who began to explore the possibilities of "green chemistry," resulting in new tools for the modification of the phage MS2 particles. This resulted in the application of iridium-catalyzed transfer hydrogenation by the reductive alkylation of lysine residues of the phage MS2, which contained altogether six lysine residues, at room temperature and neutral pH.

The next functionalization approach was achieved by the Francis team by oxidative coupling to aniline-containing side chains (Carrico et al. 2008; Stephanopoulos et al. 2009; Tong et al. 2009). Thus, the one-to-one oxidative coupling of aniline with N,N-diethyl-N'-acylphenylene diamine using sodium periodate was achieved for the attachment of peptides to the surface of the MS2 VLPs (Carrico et al. 2008). The aniline functionality was installed into the MS2 capsid through the incorporation of an unnatural amino acid residue, namely *para*-amino-L-phenylalanine (*p*AF), by *amber* suppression methods. To optimize the surface-accessible *p*AF incorporation, the five stop codon variants were constructed: Q6TAG, D11TAG, T15TAG, D17TAG, and T19TAG, and the appropriate suppression technique was applied, where the observable MS2 coat protein expression was found only in the presence of *p*AF. As the MS2 mutant with *p*AF at position T19 produced the highest yield and was found to be the robust oxidative coupling scaffold, it was used for all subsequent experiments. Three peptides known to target specific tissues were attached by the oxidative coupling via N,N-diethyl-N'-acylphenylene diamine, which was placed at the N-terminus of each peptide during synthesis on the solid phase. The oxidative coupling to aniline was also used to decorate the exterior surface of the MS2 particles with zinc porphyrins for photocatalysis (Stephanopoulos et al. 2009) and a DNA aptamer for cellular delivery (Tong et al. 2009). Furthermore, a highly efficient protein bioconjugation method involving addition of anilines to *o*-aminophenols in the presence of sodium periodate was used (Behrens et al. 2011). The impressive progress in the dual-surface modification of MS2 capsids and generation of novel nanomaterials was summarized at that time by Witus and Francis (2011). The detailed protocol on how to conjugate antibodies to the MS2T19*p*AF VLPs was published by ElSohly et al. (2018).

Furthermore, the Francis team mutagenically introduced cysteine at position 87 by the N87C exchange and targeted this interior residue with maleimide reagents to introduce drug molecules (Wu W et al. 2009; Stephanopoulos et al. 2010b) or imaging agents (Meldrum et al. 2010) to the interior surface of the capsids with high efficiency approaching 100%. On the basis of this success, this site was also chosen for modification with Gd^{III}-hydroxypyridonate complexes through the use of linking groups that possessed a minimal number of rotatable bonds (Garimella et al. 2011).

Seim et al. (2011) described an unprecedented oxidative coupling strategy that selectively modified tyrosine and tryptophan residues using cerium^{IV} ammonium nitrate as a one-electron oxidant. This oxidative method was found to couple electron-rich aniline derivatives directly to tyrosine and tryptophan residues. The new tyrosine and tryptophan residues, namely T15W, T19W, T15Y, and T19Y, were introduced to the external surface of the MS2 VLPs. The strategy was used to modify both native and introduced residues on proteins with polyethylene glycol and small peptides. The reaction was also used in conjunction with cysteine alkylation to doubly modify the MS2 capsids with both targeting and imaging functionalities.

Then, the modification strategy using *o*-aminophenols or *o*-catechols that were oxidized to active coupling species in situ using potassium ferricyanide was tested, among other approaches, with the aniline side chains of the *p*AF residues (Obermeyer et al. 2014b) introduced into the MS2 coat, as described earlier.

The modification of aniline-containing MS2 VLPs was performed with ortho-azidophenols under photochemical conditions (El Muslemany et al. 2014). An easily accessible and operationally simple photoinitiated reaction was proposed, which involved the photolysis of 2-azidophenols to generate iminoquinone intermediates that coupled rapidly to aniline groups. As a specific application, the reaction was adapted for the photolithographic patterning of azidophenol DNA on aniline glass substrates.

Meanwhile, ElSohly and Francis (2015) published a thorough account of the development of oxidative coupling strategies for site-selective protein modification, while ElSohly et al. (2015) presented how the synthetically modified MS2 VLPs could be used as versatile carriers by antibody-based cell targeting. Finally, ElSohly et al. (2017) identified *o*-methoxyphenols as molecules that undergo efficient oxidative couplings with anilines in the presence of periodate as oxidant. This approach was used to link epidermal growth factor to the MS2 VLPs bearing aniline moieties, therefore affording nanoscale delivery vectors

that could target a variety of cancer cell types. The most recent account on the principles and methodology of the dual-surface modification of the MS2 VLPs for the nanomaterial fabrication and delivery applications was published by Aanei and Francis (2018). The practical results of the pioneering studies of Francis's team are summarized in a table in the *Nanomaterials* paragraph.

A novel functionalization approach was achieved in the cell-free protein synthesis platform by the production of MS2 and Qβ VLPs with surface-exposed methionine analogs: azidohomoalanine and homopropargylglycine containing azide and alkyne side chains (Patel and Swartz 2011). Such VLPs could be used for one-step, direct conjugation schemes to display multiple ligands of interest.

In 2011, the phage MS2 was chosen to meet the growing requirements for the quality of bioartificial tracers for the inline measurement of virus retention in membrane processes. With this aim, the surface of the phage MS2 was modified by the grafting of enzymes, namely horseradish peroxidase (HRP; Soussan et al. 2011a, b). This tracer was thus built to enable direct detection of its induced enzymatic activity, notably by an amperometric method. The generation of the MS2 tracer was started with covalent binding of activated biotin to the lysine residues of the phage, while the neutravidin–HRP conjugates were chosen as enzymatic probes for the second labeling on biotin molecules.

The conjugated MS2 VLPs were used to study parvovirus B19, which has an extraordinary narrow tissue tropism, showing only productive infection in erythroid precursor cells in the bone marrow. The VP1u, a unique region of viral protein 1 that mediated the B19 uptake into cells of the erythroid lineage, was chemically coupled to the MS2 capsid and tested for the internalization capacity on permissive cells (Leisi et al. 2016). The MS2-VP1u bioconjugate mimicked the specific internalization of the native B19 into erythroid precursor cells, which further coincided with the restricted infection profile.

The MS2 functionalization was employed by the development of a nanoparticle system for blood–brain barrier (BBB) targeting and delivery (Curley and Cady 2018; Curley et al. 2018). To do this, the thiolated angiopep-2-FAM (AP2), a synthetic peptide that has been shown to have the greatest transcytosis efficiency across the BBB over any other targeting ligand, was conjugated to the MS2 capsid surface. The injection of the MS2-AP2 particles into the tail vein of rats was successful in allowing the particles to cross into the brain.

The MS2 VLPs were also used as a mass spectrometry signal multiplier (Yuan et al. 2019). Thus, the dibenzylcyclooctynepoly(ethylene glycol)-folate (DBCO-PEG-FA) and DOTA-Eu complex tag modified (FA-PEG)$_{69}$-MS2-(DOTA-Eu)$_{965}$ targeted the folate receptor on the human nasopharyngeal carcinoma cells and made it possible to quantify at least five cancer cells by ^{153}Eu-species unspecific isotope dilution inductively coupled plasma mass spectrometry.

Qβ

M.G. Finn's team struck the keynote of the Qβ functionalization. The practical outcome of these studies is listed later in the *Packaging and Targeting* and *Nanomaterials* paragraphs.

At the beginning, following the previously mentioned idea of Anderson et al. (2006), who worked with the phage MS2, Prasuhn et al. (2007, 2008) turned to the generation of viral MRI contrast agents but based on the Qβ VLPs, in parallel with another classical viral model, namely cowpea mosaic virus (CPMV) from the *Secoviridae* family of the *Picornavirales* order, described in Chapter 27.

The incorporation of the unnatural amino acid residue azidohomoalanine (AHA), as well as of the homopropargyl glycine, into the Qβ VLPs was described for the first time by Strable (2008). Remarkably, this pioneering labeling with azide- or alkyne-containing unnatural amino acids by expression in a methionine auxotrophic strain of *E. coli*, which was followed by the Cu^I-catalyzed cycloaddition, was performed in parallel with both Qβ VLPs and hepatitis B virus core particles, i.e. in the two most popular VLP carriers that have been applied together in numerous classical VLP studies.

In contrast to the hepatitis B virus cores, the Qβ VLPs suffered no instability and were not decomposed by the formation of triazole linkages. Therefore, the marriage of these well-known techniques of the sense-codon reassignment and bioorthogonal chemical coupling provided the capability to construct polyvalent particles displaying a wide variety of functional groups with near-perfect control of spacing (Strable et al. 2008).

Recently, Bhushan et al. (2018) proposed *S*-allylhomocysteine as an unnatural genetically encodable methionine analog for the Qβ VLPs. This analog is processed by translational cellular machinery and is also a privileged olefin cross-metathesis reaction tag in proteins. It was used for efficient Met-codon reassignment in a Met-auxotrophic strain of *E. coli*.

The next success of the Finn's team was the display of polyvalent glycan ligands for the cell-surface receptors on the Qβ VLPs, again in parallel with the CPMV model (Kaltgrad et al. 2008). The authors have immobilized glycan reaction precursors on the Qβ VLPs and allowed a "building out" from such surface to make polyvalent and precisely placed oligosaccharides. Thus, the glycans arrayed on the exterior of the Qβ VLPs were used as substrates for glycosyltransferase reactions to build di- and trisaccharides from the VLP surface. The resulting particles exhibited tight and specific associations with cognate receptors on beads and cells, in one example defeating in cis cell-surface interactions in a manner characteristic of polyvalent binding. Then, the polyvalent sugar-coated Qβ VLPs were used by the elaboration of the back-scattering interferometry method for the quantitative determination of glycan-lectin interactions, one of the most important classes of interactions in biochemistry (Kussrow et al. 2009), but Astronomo et al. (2010) tried to breach the *glycan shield* defense of HIV.

Meanwhile, Hong V et al. (2008) improved the CuAAC reaction by the use of an electrochemical potential to maintain catalysts in the active Cu^I oxidation state in the presence of air. This simple procedure efficiently achieved excellent yields of the CuAAC products without the use of potentially damaging chemical reducing agents. The electrochemically protected bioconjugations in air were performed with the Qβ VLPs that were derivatized with azide moieties at surface lysine residues, and complete derivatization of more than 600 reactive sites per particle was demonstrated (Hong V et al. 2008). The optimized CuAAC reaction was reviewed under the headline "*How to click with biomolecules*" as the most widely recognized example of the *click chemistry* applications (Hong V et al. 2009). At the same, Strable and Finn (2009) reviewed the principles of the chemical modification of viruses and virus-like particles. The principles and approach of the CuAAC reaction in connection with the Qβ VLPs, among other substrates, were reviewed further in detail (Lallana et al. 2011, 2012; Such et al. 2012; Levine et al. 2013; Kim H et al. 2015).

The CuAAC reaction was used to display the human iron-transfer protein transferrin, a high-affinity ligand for the receptors upregulated in a variety of cancers, on the Qβ VLPs (Banerjee D et al. 2010). Furthermore, the CuAAC reaction was used by the covering of the Qβ VLP surface by polymer chains, to which multiple connections were made, thereby cross-linking protein cage subunits (Manzenrieder et al. 2011). Such polymers extended in vivo circulation lifetime, diminishing nonspecific adsorption or passivating the immune response. The poly(2-oxazoline)s were chosen for the role of such polymers because of their advantageous properties of versatile controlled syntheses by means of living cationic polymerization, aqueous-phase solubility, and chemical stability that made them attractive for a variety of biomedical and materials applications.

The polymerization of oligo(ethylene glycol)-methacrylate and its azido-functionalized analog was performed directly from the outer surface of the Qβ VLPs by atom transfer radical polymerization (ATRP) (Pokorski et al. 2011a, b). The introduction of chemically reactive monomers during polymerization therefore provided a robust platform for post-synthetic modification via the CuAAC reaction. The ATRP methodology was developed further by Hovlid et al. (2014) and reviewed in an excellent minireview of Wallat et al. (2014), who specifically highlighted the concept of the *grafting-from* proteins as the new and efficient chemistry to polymerize directly from protein substrates in aqueous media, with the Qβ VLPs as the most successful story. The authors demonstrated efficient postpolymerization modification via the CuAAC reaction to introduce fluorophores, Gd-based contrast agents, and doxorubicin, a common anticancer drug. Then, Isarov et al. (2016) generated the *graft-to* protein/polymer conjugates using polynorbornene block copolymers, which were prepared via ring-opening metathesis polymerization (ROMP). The ROMP technique afforded low-dispersity polymers and allowed for the strict control over polymer molecular mass and architecture. Such polymers consisted of a large block of PEGylated monoester norbornene and were capped with a short block of norbornene dicarboxylic anhydride. This cap served as a reactive linker that facilitated attachment of the polymer to lysine residues under mildly alkaline conditions. As a result, the multivalent polynorbornene-modified Qβ VLPs were constructed.

Lee PW et al. (2017a, b, c) conjugated a set of biocompatible polymers to the Qβ VLPs, using the *grafting-to* approach with end-functionalized aminoreactive polymers. Crooke et al. (2018) described the assembly and characterization of a series of polymer conjugates of the Qβ VLPs—using (poly(oligo(ethylene glycol)methacrylate), poly(methacrylamido glucopyranose), and PEG—and investigated their ability to shield the protein from antibody recognition as a function of polymer loading density, chain length, architecture, and conjugation site.

In connection with the biocompatible polymers, the Nicole F. Steinmetz team raised a critical question of the bioinspired shielding strategies for nanoparticle drug delivery applications (Gulati et al. 2018). Since nanoparticle surface camouflage was often required to reduce immune clearance and thereby increase circulation times allowing the carriers to reach their target site, the origin of the coating remained highly important, whether of synthetic or biologic origin. To this end, polyethylene glycol has long been used, with several PEGylated products reaching clinical use. Unfortunately, the growing use of PEG in consumer products has led to an increasing prevalence of PEG-specific antibodies in the human population, which in turn has fueled the search for alternative coating strategies. Gulati et al. (2018) highlighted alternative bioinspired nanoparticle shielding strategies that may be more beneficial moving forward than PEG and other synthetic polymer coatings.

As an alternate approach to the click chemistry, Finn's team employed genetic engineering and chemical conjugation to construct the Qβ VLPs with cationic amino acid motifs on the surface (Udit et al. 2009). Thus, the application of the point mutation strategies made it possible to generate Qβ VLPs bearing the K16M, T18R, N10R, or D14R mutations. The mutants therefore provided a spectrum of particles differing in surface charge by as much as +540 units (K16M *versus* D14R). Whereas larger poly-Arg insertions, for example C-terminal Arg$_8$, did not yield intact virions, it was possible to append chemically synthesized oligo-Arg peptides to the stable Qβ wild-type and K16M platforms. These particles were applied as inhibitors of the anticoagulant action of heparin, which was a common anticlotting agent subject to clinical overdose. Moreover, the engineered cationic Qβ VLPs retained their ability to inhibit heparin at high concentrations and showed no anticlotting activity of the kind that limited the utility of antiheparin polycationic agents that were traditionally in clinical use (Udit et al. 2009). Then, Gale et al. (2011) investigated the clinical heparin reversal function of one of these active Qβ VLPs, namely the T18R mutant, in which the solvent-exposed

residue Thr18 of the coat protein was mutated to Arg, thereby giving the particle substantially greater positive surface charge. The K16M mutant served as a negative control since its positive charge was diminished relative to the wild-type structure and it did not bind or reverse heparin in vitro. The plasma specimens were collected from patients who were treated with either therapeutic heparin or with high doses of heparin while undergoing cardiac catheterization procedures. The reversal of heparin anticoagulation in these patient plasma samples was investigated in comparison to protamine. As a result, the T18R particles showed significantly more consistent heparin reversal activity, with much greater potency in terms of particle concentration versus protamine concentration (Gale et al. 2011). This pioneering work was based generally on the three successful approaches to modify the Qβ VLPs by (i) chemically appending poly-Arg peptides, (ii) point mutations to Arg on the virus capsid, and (iii) incorporation of heparin-binding peptides displayed externally on the virus surface. Each approach generated particles with good heparin antagonist activity with none of the toxic side effects of protamine that remained nevertheless the only currently FDA-approved drug for clinical use as the heparin antagonist (Udit 2013). Finally, Cheong et al. (2017) presented further strong arguments in favor of the potential clinical use of the Qβ T18R VLPs as effective, nontoxic heparin antagonists.

To further develop the heparin antagonist properties of the chimeric Qβ VLPs, Choi et al. (2018) backed off from the Qβ point mutants and the CuAAC reaction and turned to the good old two-plasmid coexpression methodology that was first described by Tatyana Kozlovska's group (Vasiljeva et al. 1998) and further developed by Brown SD et al. (2009), as narrated earlier in the *Chimeric VLPs: Genetic Fusions: Qβ* paragraph. The two-plasmid system was utilized to generate the Qβ VLPs that contained both the wild-type coat protein and a second coat protein with either a C- or N-terminal cationic peptide extension of 4–28 aa in length. The incorporation of the modified coat proteins varied from 8%–31%. The particles with the highest incorporation rate and best antiheparin activity therefore displayed the C-terminal peptide ARK$_2$A$_2$KA, which corresponded to the Cardin-Weintraub consensus sequence for binding to glycosaminoglycans.

In parallel with heparin antagonists, Mead et al. (2014) generated the sulfated phage Qβ VLPs that elicited heparin-like anticoagulant activity. The sulfated Qβ VLPs were examined thoroughly for the binding kinetics to cationic peptides (Groner et al. 2015).

The chemistry of the VLP functionalization, including both MS2 and Qβ VLPs, was reviewed by Schoonen and van Hest (2014). The methodology used to generate the hybrid Qβ particles was described in detail by Brown SD (2018).

A new direction in the Qβ VLP applications was opened by Cigler et al. (2010) by the generation of novel nanomaterials. where the organic Qβ VLPs were combined with inorganic gold nanoparticles.

The colorful Qβ VLPs were prepared by encapsulation of multiple copies of fluorescent proteins (Rhee et al. 2011). The encapsidated proteins were nearly identical in photochemical properties to monomeric analogs, were more stable toward thermal degradation, and were protected from proteolytic cleavage. It is important that the residues on the outer capsid surface could be chemically derivatized by the CuAAC reaction without affecting the fluorescence properties of the packaged proteins. Moreover, the simultaneous modification of the Qβ VLPs with the metalloporphyrin derivative for photodynamic therapy and a glycan ligand for specific targeting of cells bearing the CD22 receptor was proposed for a nonsurgical method of cancer treatment based on photosensitizer molecules that produced toxic concentrations of singlet oxygen and other reactive oxygen species upon illumination (Rhee et al. 2012).

The CuAAC click chemistry reaction allowed conjugation of C_{60}, or buckyball, possessing high electron affinity and charge transport capabilities that have made derivatives of C_{60} and carbon nanotubes particularly attractive for the next-generation photovoltaic and electrical energy storage devices. First, the grafting to the Qβ VLPs made it possible to surmount the insolubility of C_{60} in water (Steinmetz et al. 2009). Second, it led to the serious investigation of photodynamic activity of the Qβ VLPs conjugated with C_{60} (Wen et al. 2012). Moreover, the cell uptake and cell killing using white light therapy and a prostate cancer cell line was demonstrated (Wen et al. 2012).

The Finn team applied the CuAAC reaction for the conjugation of tumor-associated carbohydrate antigens (TACAs) to the Qβ VLPs (Yin et al. 2013). The TACAs were overexpressed on a variety of cancer cell surfaces and therefore presented tempting targets for the anticancer vaccine development. Since such carbohydrates were often poorly immunogenic, it was intriguing to display a very weak TACA, the monomeric Tn antigen, or GalNAc-α-O-Ser/Thr, on the Qβ VLPs. Glycan microarray analysis showed that the antibodies generated were highly selective toward Tn antigens. Furthermore, the Qβ-based VLPs elicited much higher levels of IgG antibodies than other types of VLPs, and the produced IgG antibodies reacted strongly with the native Tn antigens on human leukemia cells (Yin et al. 2013). The next imperative step to the carbohydrate-based antitumor vaccine was undertaken by Yin et al. (2016). The authors focused on the ganglioside GM2 TACA, which was overexpressed in a wide range of tumor cells. The GM2 immobilized on the Qβ VLPs through a thiourea linker elicited high titers of IgG antibodies, which recognized GM2-positive tumor cells and effectively induced cell lysis through complement-mediated cytotoxicity.

Yin et al. (2018) generated another putative cancer vaccine candidate, when conjugat mucin-1 peptides and glycopeptides, the top ranked tumor associated antigens, to the Qβ VLPs by the CuAAC reaction. The immunization of mice with these constructs led to highly potent antibody responses with IgG titers over one million, which were among the highest anti-mucin-1 IgG titers reported to date.

After putative vaccines using carbohydrate epitopes, the Qβ VLPs assisted in the elaboration of an excellent diagnostic tool of Chagas disease (Brito et al. 2016). Thus, the α-Gal antigen [Galα(1,3)Galβ(1,4)GlcNAcα] that was an immunodominant epitope displayed by infective trypomastigote forms of *Trypanosoma cruzi*, the causative agent of Chagas disease, was displayed on the Qβ VLPs by the CuAAC reaction. The Qβ VLPs displaying a high density of α-Gal were found to be a superior reagent for the ELISA-based serological diagnosis of Chagas disease and the assessment of treatment effectiveness. Next, such Qβ VLPs displaying approximately 540 α-Gal molecules were used to assess the protective effect of anti-α-Gal responses in falciparum malaria (Coelho et al. 2019).

A specific class of glycoproteins, namely fluoroglycoproteins, was generated as powerful substrates in NMR, MRI, and PET techniques (Boutureira et al. 2010). This approach exploited the incorporation of L-homopropargylglycine as nonnatural amino acid and triazole formation chemistry. The reaction of L-homopropargylglycine-modified Qβ VLPs with a fluoroglucosyl azide afforded F-glycoprotein in >95% of positions, and a particle contained therefore 180 site-selectively positioned fluoroglycans (Boutureira et al. 2010). This pioneering work was reviewed and commented on later by Boutureira and Bernardes (2015). The growing impact of the possible antiglycan vaccines was accessed recently (Polonskaya et al. 2019; Krumm and Doores 2020).

A highly specific approach for the further development of VLP vectors was achieved by the asymmetrization of the Qβ VLPs after the introduction of a single copy of the maturation protein A2, which allowed the production of VLPs with a single unique modification (Smith et al. 2012). This approach was driven by the estimations that, by bioconjugation described earlier, the theoretical maximum of the Qβ VLPs having a single molecule attached could be 37%, according to the Poisson probability distribution. The remaining VLP population would contain zero (37%) or multiple (26%) molecules attached to each VLP and would not be easily separated from the VLPs with one molecule. The *E. coli*-based cell-free protein synthesis system was employed for the coexpression of the cytotoxic A2 protein and the coat of the phage Qβ to form a nearly monodispersed population of novel VLPs. The cell-free protein synthesis allowed for direct access and optimization of both protein synthesis and VLP self-assembly. The A2 was shown to be incorporated at high efficiency, approaching a theoretical maximum of one A2 per VLP. This work demonstrated for the first time the de novo production of a novel VLP, which contained a unique site that could have the potential for future nanometric engineering applications (Smith et al. 2012).

A novel pioneering approach to the bioconjugation, brilliant fluorescent functionalization across quaternary structure of the Qβ VLPs, was elaborated by the Jeremiah J. Gassensmith's team (Chen Z et al. 2016a, 2017). Thus, the dibromomaleimide moiety was employed to break and rebridge the exposed and structurally important disulfides.

The newly functionalized Qβ particles became brightly fluorescent and could be tracked in vitro using a commercially available filter set. Therefore, this highly efficient bioconjugation reaction not only introduced a new functional handle *between* the disulfides of VLPs without compromising their thermal stability, but it was used to create a fluorescent probe (Chen Z et al. 2017).

At that time, the role of the chemical functionalization of viral architectures in order to create new technologies was reviewed by Chen Z et al. (2016b). Chen Z et al. (2018) further presented detailed methodology of how to prepare the fluorescent PEGylated Qβ VLPs by dibromomaleimide-disulfide chemistry to obtain a bright yellow fluorophore.

Moreover, the Gassensmith's team elaborated a fundamentally novel composite nanomaterial (Benjamin et al. 2018). The surface of the Qβ VLPs was equipped with natural ligands for the synthesis of small gold nanoparticles. By exploiting disulfides in the protein secondary structure and the geometry formed from the capsid quaternary structure, regularly arrayed patterns of ~6 nm gold nanoparticles across the surface of the Qβ VLP were produced. It was further shown that the entrapped genetic material held upward of 500 molecules of the anticancer drug doxorubicin without leaking and without interfering with the synthesis of the gold nanoparticles. This direct nucleation of nanoparticles on the capsid allowed for exceptional conduction of photothermal energy upon nanosecond laser irradiation. As a proof of principle, it was demonstrated that this energy was capable of rapidly releasing the drug from the capsid without heating the bulk solution, allowing for highly targeted cell killing in vitro (Benjamin et al. 2018).

Finally, Lee H et al. (2018) regulated the uptake of the Qβ VLPs in macrophage and cancer cells via a pH switch. The cellular uptake by macrophages and cancer cells of the Qβ-derived nanoparticles was inhibited by conjugating negatively charged terminal hexanoic acid moieties onto the Qβ VLP surface. When hydrazone linkers were installed between the surface of the Qβ VLPs and the terminal hexanoic acid moieties, this resulted in a pH-responsive conjugate that, in acidic conditions, released the terminal hexanoic acid moiety and allowed for the uptake of the Qβ nanoparticle. The installation of the pH switch did not change the structure-function properties of the hexanoic acid moiety and the uptake of the Qβ VLP conjugates by macrophages (Lee H et al. 2018).

The functionalization of the RNA phage capsids was exhaustively reviewed in terms of a growing novel discipline that could be classified as *physical, chemical, and synthetic virology*, with a special aim to create reprogrammed virus particles as controllable nanodevices (Chen MY et al. 2019).

AP205

In 2010, Bachmann's team involved the AP205 VLPs into the chemical coupling procedures. Following the lysine-cysteine oligopeptide coupling methodology that was described earlier in the *Chimeric VLPs: Chemical Coupling* paragraph, an experimental West Nile virus vaccine

was constructed on the AP205 VLP platform (Spohn et al. 2010). This vaccine consisted of the recombinantly expressed domain III of the WNV E glycoprotein chemically cross-linked to the AP205 VLPs. In contrast to the isolated DIII protein, which required three administrations to induce detectable antibody titers in mice, the high titers of the DIII-specific antibodies were induced after a single injection of the conjugate AP205 VLP vaccine. These antibodies were able to neutralize the virus in vitro and provided partial protection from a challenge with a lethal dose of WNV. The three injections of the vaccine induced high titers of virus-neutralizing antibodies and completely protected mice from WNV infection (Spohn et al. 2010).

At the same time, the AP205 and Qβ VLP conjugates with the model D2 peptide from outer membrane protein of *Salmonella typhi* were used by the solid immunological investigation performed by Bachmann's team on a process termed *carrier-induced epitopic suppression* (CIES) (Jegerlehner et al. 2010). The preexisting immunity against the VLP carrier proteins has been reported to inhibit the immune response against antigens conjugated to the same carrier by the CIES process. Hence understanding the phenomenon of CIES was of major importance for the further development of conjugate vaccines.

The fluorescent Alexa 647-conjugated AP205 VLPs and Alexa 488-conjugated Qβ VLPs were employed in an excellent immunological study showing that intranasal administration of VLPs resulted in the splenic B-cell responses with strong local germinal-center formation (Bessa et al. 2012).

Furthermore, Tytgat et al. (2019) developed the AP205 VLPs as a facile glycoengineering platform. The Glycoli platform forced a site-specific polypeptide glycosyltransferase together with variable glycosyltransferase modules to synthesize defined glycans, of bacterial or mammalian origin, directly onto recombinant proteins in the *E. coli* cytoplasm. The authors emphasized special capabilities of this glycoengineering platform to generate self-assembling nanomaterials bearing hundreds of copies of the glycan epitope. Thus, to decorate the AP205 VLPs with glycans, the glycosylation site was introduced at the C-terminus of the AP205 coat, and the corresponding gene was coexpressed with the asparagine (N)-glucosyltransferase of *Actinobacillus pleuropneumoniae* (ApNGT) in *E. coli*. Hexose, lactose, and common mammalian oligosaccharide epitopes 2′-fucosyllactose and 3′-sialyllactose were used to be exposed.

At last, the AP205 VLPs greatly contributed to the elaboration of the highly efficient SpyTag/SpyCatcher plug-and-display methodology (Brune et al. 2016; Thrane et al. 2016). This methodology was based on the so-called bacterial superglue phenomenon, when the isopeptide bond was spontaneously formed between a peptide and its protein couple, derived from specific domains of certain bacterial proteins (Veggiani et al. 2014; Tan et al. 2016; Bonnet et al. 2017). The two binding couples were used: SpyTag peptide/SpyCatcher protein, derived by splitting the CnaB2 domain of the fibronectin-binding protein FbaB from

Streptococcus pyogenes (Zakeri et al. 2012) and SnoopTag peptide/SnoopCatcher protein, derived by splitting the D4 domain of RrgA adhesin from *Streptococcus pneumoniae* (Veggiani et al. 2016).

Thus, the SpyTag peptide and the SpyCatcher protein spontaneously formed a highly stable amide bond between lysine and aspartic acid residues by an irreversible reaction that occurred within minutes under physiological conditions (Zakeri et al. 2012; Veggiani et al. 2014). To realize the general plug-and-display idea, the SpyTag or SpyCatcher sequences were genetically fused to the N- and/or C-terminus of the AP205 coat. After mixing the modified AP205 VLPs with the correspondingly linked peptides, the quantitative covalent coupling of the peptides to the VLPs was observed (Brune et al. 2016; Thrane et al. 2016). The practical outcome of this approach is described in the *Vaccines: Plug-and-Display* paragraph.

STABILITY

The VLP technologies inspired a novel discipline that would be named virus engineering—by analogy to gene and protein engineering—and would include the functionalization and stabilization of VLPs for different nanotechnological applications. The MS2 and Qβ VLPs, together with cowpea chlorotic mottle virus (CCMV) from the *Bromoviridae* family, order *Martellivirales* (Chapter 17), were among the main subjects of the practical recommendations that were reviewed exhaustively by Mauricio G. Mateu (2011, 2013, 2016). He regarded physical stability of the VLPs as the most critical parameter by their production, storage, and medical and/or industrial application.

As was known before, the natural RNA phage VLPs differed markedly in their stability. Thus, for example, the Qβ VLPs were more stable than the MS2 VLPs due to intersubunit cross-linking by disulfide bonds (Ashcroft et al. 2005). The mutational mapping revealed residues that were responsible for the special inter- and intramolecular contacts and therefore for the greater thermal stability of the MS2 capsids (Stonehouse and Stockley 1993).

The efforts to improve the stability of the natural RNA phage VLPs were started with the development of a methodology that allowed the screening of bacteria for the synthesis of the mutant MS2 coats with altered assembly properties (Peabody DS and Al-Bitar 2001) and the selection of the MS2 coat D11N variant that formed virions more stable than the wild-type MS2 coat (Lima et al. 2004). The introduction of interdisulfide bonds into the 5-fold axis of symmetry to crosslink the MS2 VLPs improved their thermal stability to the level of that seen in the Qβ VLPs, which possessed natural intersubunit disulfide bonds (Ashcroft et al. 2005). In contrast, the cross-linking at the threefold axis of symmetry resulted in variant coats that were unable to self-assemble (Ashcroft et al. 2005). The development of an *E. coli*-based cell-free protein synthesis system opened a direct avenue for studying the role of disulfide bond formation in the stability of the mutant MS2 VLPs in

comparison to the Qβ and HBc VLPs (Bundy and Swartz 2011). Through the construction of a set of the Qβ coat mutants, it was found that the disulfide linkages were the most important stabilizing elements in the VLPs and that interdimer interactions were less important than intradimer interactions for the Qβ VLP assembly (Fiedler et al. 2012).

The elucidation of the thermal stability of the foreign epitope-carrying VLPs, such as the chimeric MS2 VLPs formed by single-chain dimers, in comparison to the natural disulfide crosslinked PP7 VLPs (Caldeira and Peabody 2007, 2011) was of great importance for the further development of the stability principles.

The genetic fusions of two copies of the MS2 (Peabody DS et al. 2008), PP7 (Caldeira et al. 2010), and GA or Qβ (Indulis Cielēns and Arnis Strods, unpublished) coats resulted in the self-assembly competent single-chain dimer that not only increased thermodynamic stability but also considerably improved tolerance of the foreign insertions within the AB-loop, as presented earlier. The resultant correctly assembled VLPs encapsidated mostly their "own" mRNAs encoded by the coat genes and expressing the coat proteins.

The stability of the RNA phage VLPs against the attack by antibodies was achieved by the PEGylation of the VLPs (Kovacs et al. 2007). In fact, the ability of the polymer coating to block the access of polyclonal antibodies to the capsid surface was probed using a sandwich ELISA, which indicated a 90% reduction in binding. It appeared further that the stability of the viable phage MS2 was enhanced by the formulation with poly-γ-glutamic acid (PGLA), a biodegradable polymer that demonstrated remarkable potential protection of beneficial viruses (Khalil et al. 2016).

By the previously described generation of the magnetic resonance contrast agents (Hooker et al. 2007), the MS2 capsids sequestering the gadolinium-chelates on the interior surface, when attached through tyrosine residues, not only provided higher relaxivities than their exterior functionalized counterparts, which relied on lysine modification but also exhibited improved water solubility and capsid stability. Therefore, the attachment of a functional cargo to the interior surface was envisioned to minimize its influences on biodistribution, yielding significant advantages for the tissue targeting by additional groups attached to the capsid exterior (Hooker et al. 2007).

Bundy and Swartz (2011) demonstrated the ability to control the disulfide bond formation in VLPs by directly controlling the redox potential during or after production and assembly of VLPs in the open cell-free protein synthesis environment. This study used the three most popular VLP platforms: the Qβ VLPs, the VLPs of an MS2 mutant that formed disulfide bonds (Ashcroft et al. 2005), and HBc VLPs. The Qβ VLPs formed disulfide bonds at high efficiency after exposure to a reductase-free extracellular aerobic environment. Similar to the MS2 and HBc VLPs, the endoplasmic reticulum redox conditions with a glutathione redox buffer were not sufficient to stimulate complete formation of the disulfide bonds in the Qβ VLPs. By incubating with the extremely oxidizing but bioincompatible hydrogen peroxide, close to 100% of the Qβ VLP disulfide bonds were formed (Bundy and Swartz 2011).

When Finn's team engineered a set of mutations in the Qβ VLPs, it was generally concluded that the disulfide linkages were the most important stabilizing elements and that acidic conditions significantly enhanced the resistance of VLPs to thermal degradation (Fiedler et al. 2012). The interdimer interactions were found to be less important for the VLP assembly than the intradimer interactions. The special paragraphs on the general factors influencing stability of the MS2 VLPs including both physicochemical and structural observations were included in the extensive review on the MS2-based delivery platform (Fu and Li 2016).

Recently, the Francis team performed the first study on the MS2 VLPs, where the chemical modification conditions were used as a selection factor for the protein fitness and stability (Brauer et al. 2019; Maza et al. 2019).

The PP7-derived VLPs, regardless of how they were manipulated by dimerization, extension, or loop insertion, as described previously in the *Chimeric VLPs: Genetic Fusions* paragraph, reliably maintained stability to ~80°C. Appending ZZ domains to either terminus of the PP7 coat monomer or to the N-terminus of the PP7-PP7 coat dimer gave equivalently stable particles (Zhao et al. 2019).

CONTRIBUTION TO BASIC IMMUNOLOGY

The RNA phage VLPs contributed greatly to the basic immunological knowledge, due to systematic investigations performed by Martin F. Bachmann's team that grew out of the Nobelist 1996 Rolf M. Zinkernagel's school. Bachmann's team started in the early 2000s with the HBc VLPs but soon switched to the Qβ VLPs and later broadened their immunological interest also to the AP205 VLPs.

First, Bachmann's team performed deep immunological studies with the Qβ VLPs decorated by the model epitopes: p33, which was derived from the lymphocytic choriomeningitis virus (LCMV) surface glycoprotein (Bachmann et al. 2004b, 2005a; Storni et al. 2004; Schwarz K et al. 2005; Agnellini et al. 2008; Bessa et al. 2008; Keller et al. 2010a, b), OVA peptides from ovalbumin (Jegerlehner et al. 2007; Gilfillan et al. 2020), and D2 epitope derived from *Salmonella* (Jegerlehner et al. 2010). These three VLP models generated by chemical coupling have played a central role in the elucidation of the fine immunological mechanisms that governed responses to the chimeric VLPs.

Second, Bachmann's team was the first who displayed a small antigen, namely nicotine, on the Qβ VLPs (Maurer et al. 2005; Beerli et al. 2008; Cornuz et al. 2008; Lang et al. 2009). This was a novel approach that enabled the generation of strong immunological responses against nonpeptide antigens and paved the way for the development of experimental vaccines against nicotine addiction. Moreover, it led to the determination, in collaboration with the Kaspars Tārs group, of the 3D structure for the human

antibody-nicotine complex and oriented future studies to increasing the antibody affinity to nicotine (Tars et al. 2012). Later, M.G. Finn's team displayed another small antigen, a carbohydrate moiety, on the Qβ VLPs and tested them as a prospective cancer vaccine (Yin et al. 2013, 2016).

Third, Bachmann's team succeeded by the packaging of the CpG sequences into the Qβ VLPs and elaborated the allergy vaccine candidate CYT003-QbG10, which did not carry any attached epitopes but contained the encapsulated CpG sequence, so-called QbG10 (Senti et al. 2009; Klimek et al. 2011, 2013b; Beeh et al. 2013; Casale et al. 2015).

In 2002, Bachmann's team published their first self-review, which presented the two VLP models, namely HBc and Qβ, as molecular assembly systems that rendered antigens of choice highly repetitive and were able to induce efficient antibody responses in the absence of adjuvants and provide protection from viral infection and allergic reactions (Lechner et al. 2002).

Moreover, the Qβ VLPs as a standard antigen contributed markedly to the solution of basic immunological problems. Thus, the in vivo response of marginal zone and follicular B cells to the Qβ VLPs was compared (Gatto et al. 2004; Gatto and Bachmann 2005). The role of the CD21-CD35 complement receptors in the generation of the B-cell memory was elucidated by the immunization of the appropriate deficient mice with the Qβ VLPs (Gatto et al. 2005). It was shown that the early B-cell proliferation and development of B-cell memory in mice were highly antigen-dependent, whereas persisting antigen was not essential for the maintenance of B-cell and antibody memory in the late phase of the response (Gatto et al. 2007b). The heterogeneous antibody repertoire of the marginal zone B-cells was evaluated after immunization of mice with the two different VLPs, namely Qβ and AP205 (Gatto et al. 2007a).

The Qβ VLPs decorated with the p33 epitope and packaged with the CpG adjuvant were used for evaluation of the recall proliferation potential of the memory CD8[+] T cells by antiviral protection (Bachmann et al. 2005b). The p33-Qβ VLPs later played an important role by explaining VSIG4, a B7 family-related protein, as a negative regulator of the T-cell activation (Vogt et al. 2006). Bachmann et al. (2006a) compared memory CD8[+] T-cell development after infection with live LCMV or after vaccination in mice with the p33-Qβ VLPs. The p33-decorated and CpG-packaged VLPs were employed by the estimation of the differential role of IL-2R signaling for the CD8[+] T-cell responses in acute and chronic viral infections (Bachmann et al. 2007). In an extensive review, Hinton et al. (2008) explained how viruses and VLPs triggered Toll-like receptors (TLRs), which, in addition to increasing overall antibody levels, drove the switch to the IgG2a isotype that was more efficient in viral and bacterial clearance and activated complement, which in turn lowered the threshold of the B-cell receptor activation. This conclusion was critical in the VLP vaccine design, demonstrating that the safe recombinant vaccines could still remain as effective as a virus in inducing B-cell responses.

The Qβ VLPs labeled with Alexa-488 were employed for the trafficking of nanoparticles in vivo, in parallel with 20-, 500- and 1,000-nm polystyrene fluorescent nanoparticles, when injected into footpads of mice (Manolova et al. 2008). These data provided clear evidence that the particle size determined the mechanism of trafficking and that only small nanoparticles could specifically target the lymph-node-resident cells.

The Qβ and AP205 VLPs carrying single-stranded RNA of bacterial origin were used to assess the role of TLR7 signaling in driving IgA responses, both at the systemic level and in the respiratory tract mucosa of mice (Bessa et al. 2009, 2010; Bessa and Bachmann 2010). The key roles of IL-2 in inhibiting IL-21 production by CD4+ T cells and of IL-21 in negatively regulating marginal zone B cell survival and antibody production were highlighted by using the Qβ VLPs loaded with the bacterial RNA (Tortola et al. 2010).

By exhaustive investigation on the mechanisms by which Th cells promote CD8+ T-cell responses, the Qβ VLPs were used to display interleukin-15 by chemical coupling and to induce the corresponding antibodies in mice (Wiesel et al. 2010). The Qβ VLPs displaying the cytokines IL-1α and IL-1β were used to neutralize the endogenous cytokines in mice by the evaluation of how oxidative stress and metabolic danger signals converge and mutually perpetuate the chronic vascular inflammation that drives atherosclerosis (Freigang et al. 2011).

The immunization of mice with the Qβ VLPs that were packaged with the CpG oligodeoxynucleotides led to the qualified conclusion that the CpG adjuvant functioned as the TLR9 agonist and strongly promoted the T-cell-dependent antibody response (Hou et al. 2011). This effect depended on the adaptor MyD88 signaling, because the IgG response to the Qβ VLPs was almost completely ablated in Myd88-/- mice (Hou et al. 2011). Later, the Qβ VLPs carrying TLR ligands were employed by the extensive study of the fine MyD88 signaling mechanisms (Tian et al. 2018).

Link et al. (2012) performed a highly important study that demonstrated directly the major immunological difference of the Qβ VLPs and their dimeric subunits. For the first time, this study clearly identified the size and repetitive structure of the Qβ VLPs as critical factors for the efficient antigen presentation to B cells and highlighted important differences between viral particles and their subunits.

Using Qβ VLPs, Zabel et al. (2014) demonstrated rapid differentiation of the memory B cells to powerful secondary plasma cells, whereas the secondary pool of the memory B cells was to a large extent derived from naive B cells, allowing plasticity of the memory B-cell repertoire upon multiple antigenic exposures. Furthermore, Zabel et al. (2017) demonstrated that the Qβ VLP-specific memory B-cell responses exhibited a hierarchical dependence on Th cells.

Concerning the essential role of the packaged *E. coli* RNA, the Qβ VLPs were used recently to show that the RNA and the TLR7-signaling in B cells synergize for the regulation of the secondary plasma cell response (Krueger CC et al. 2019a, b).

The Qβ VLPs carrying endogenous bacterial RNA were used by the extensive studies on the indispensable role of the thioredoxin-1 (Trx1) and the glutathione (GSH)/glutaredoxin-1 (Grx1) systems for the development and functionality of the marginal zone B cells in mice (Muri et al. 2019a, b, 2021).

Using the CpG-loaded Qβ VLPs and E7 protein of HPV, Bachmann's team brought out clearly that the physical association of the VLP carrier with the vaccine protein was more critical for B- than T-cell responses, in the case when the vaccine protein possessed great size comparable with that of the VLP carrier (Gomes et al. 2017). Thus, the B-cell but not T-cell responses required antigen-linkage to the carrier and adjuvant for optimal vaccination outcome by the use of both particulate carriers and antigens of similar size. The necessity of conjugating the two entities together was avoided. This seemed to be especially important for the development of patient-specific vaccines, where VLPs could remain an attractive platform for personalized vaccines considering the convenience of production and low costs (Gomes et al. 2017).

Moreover, using the Qβ VLPs, which displayed the model LCMV p33 peptide, Mohsen et al. (2017a) demonstrated that the CpGs functioned efficiently as adjuvants when they were packaged in separate VLPs and mixed with the VLPs displaying the CTL epitopes prior to administration in vivo. This novel method generated results comparable to the standard method where the CpG adjuvants and CTL epitopes were linked to the same VLP. Thus, packaging adjuvants into the carrier VLPs eliminated the need for physical linkage to the antigenic VLPs (Mohsen et al. 2017a).

The Qβ VLPs decorated with foreign epitopes were used to generate monoclonal antibodies targeted to the epitope in question, as, for example, the Qβ VLPs carrying the major cat allergen Fel d 1 (Zha et al. 2018). Further important immunological details regarding the immunization with the Qβ VLPs were revealed in the recent papers of Bachmann's team (Gomes et al. 2019a; Krueger CC et al. 2019b).

The basic problems of the interaction of the VLPs with the innate immune system were analyzed recently in an extensive review from Bachmann's team (Mohsen et al. 2018, 2019a, b, c. 2020).

Due to David Klatzmann's vigorous activity, the AP205 and Qβ vaccine vectors were included in the impressive list of platforms by the elaboration of a methodology that used transcriptomic data in dendritic cells to predict the adaptive immune responses induced by large sets of vaccine vectors of different classes, ranging from infectious particles to VLPs and DNA (Dérian et al. 2016; Tsitoura et al. 2019).

Liao et al. (2017) developed methods to label and enrich the Qβ-VLP-specific B cells and followed these cells in immunized mice for one year.

Hong S et al. (2018) acknowledged that the antigen-specific B cells were the dominant antigen-presenting cells that initiated naive CD4+ T cell activation by immunization of mice with the Qβ VLPs.

Raso et al. (2018) used the Qβ VLPs for the identification of the regulation mechanism of the germinal center B-cell TLR signaling, mediated by α_v integrins and noncanonical autophagy. Furthermore, the Qβ VLPs were employed in a substantive study that utilized autoimmune models to assess the impact of TYK2, a JAK family member, variants on T-cell subsets and cytokine signaling and on normal and autoimmune responses in vivo (Gorman et al. 2019).

VACCINES

NATURAL VLPs

Although originally initiated by the epitope-coupling methodology (Kündig et al. 2006), the well-known therapeutic Qβ VLP-based allergy vaccine CYT003-QbG10 was elaborated finally by Bachmann's team as a natural Qβ VLP encapsulated with the TLR9 ligand, A-type CpG named QbG10. This vaccine was nominated as an efficient tool against allergy, asthma, rhinoconjunctivitis, and rhinitis (Senti et al. 2009; Klimek et al. 2011, 2013a; Beeh et al. 2013; Casale et al. 2015; Kündig et al. 2015; Bachmann and Kündig 2017).

The Qβ VLPs encapsulated with CpG demonstrated remarkable therapeutic potential against peritoneal carcinomatosis in a murine model (Miller AM et al. 2019). The robust anti-Qβ immune response likely contributed to the enhanced survival and decreased disease progression in these mice. The authors of this recent investigation concluded that the promising preclinical results suggested that the CpG-packed Qβ VLPs may have potential as an immunotherapy for the treatment of patients with peritoneal carcinomatosis and are worthy of further evaluation.

Furthermore, Lemke-Miltner et al. (2020) proclaimed the Qβ VLPs encapsulated with the A-CpG as a promising tool against B cell lymphoma. Thus, the treatment of murine A20 lymphoma enhanced survival and reduced growth of both injected and contralateral noninjected tumors in a manner dependent on both the ability of mice to generate anti-Qβ antibodies and the presence of T cells.

Finally, the biocompatible Qβ VLPs-loaded Mg-based micromotors were elaborated for the treatment of peritoneal ovarian tumors (Wang C et al. 2020). In fact, the autonomous propulsion of such Qβ VLPs-loaded Mg-micromotors in the peritoneal fluid enabled active delivery of intact immunostimulatory Qβ VLPs to the peritoneal space of ovarian-tumor-bearing mice, greatly enhancing the local distribution and retention of Qβ VLPs.

GENETIC FUSIONS

Table 25.2 presents a detailed list of the vaccine candidates that have been constructed from the RNA phage VLPs using genetic fusion methodology. Concerning the selection of optimal VLP carriers for generation of the genetically fused vaccine candidates, the AP205 VLPs have demonstrated a special capacity and tolerance to foreign insertions

(Tissot et al. 2010). Moreover, the AP205 VLPs appeared as good candidates for construction of the mosaic VLPs (Cielens et al. 2014). The exceptional capabilities of the PP7 platform, in comparison with the traditional RNA phage platforms, were revealed by Zhao L and Finn (2017). Zhao L et al. (2019) demonstrated unique properties of the chimeric PP7 VLPs, as described earlier in the *Chimeric VLPs: Genetic Fusions: PP7* paragraph.

Highlighting some prospective vaccines, the success attended first the HPV vaccine elaborated by Bryce Chackerian and David S. Peabody's team and based on the PP7 single-chain-dimer VLPs (Caldeira et al. 2010; Hunter et al. 2011; Tumban et al. 2011, 2013, 2015; Tyler et al. 2012, 2014b, c). A similar HPV vaccine candidate was then constructed by the same team on the MS2 single-chain-dimer VLPs (Tumban et al. 2012, 2015; Saboo et al. 2016; Peabody J et al. 2017). Both PP7 and MS2 VLP-based vaccines were immunogenic, but the MS2-L2 VLPs induced a broader HPV-neutralizing antibody response. This was likely because of the structural context of L2 display on the VLPs, since L2 was displayed on the AB-loop of the PP7 coat but at the N-terminus of the MS2 coat (Caldeira et al. 2010). These studies were extensively reviewed (Tyler et al. 2014a). A broader review on HPV vaccine candidates, including the RNA phage VLP-based vaccines, was published later (Jiang et al. 2016). Later, a novel MS2 VLP-based vaccine was tested and demonstrated in mice the protective power against cervicovaginal infection with HPV pseudoviruses 16, 18, 31, 33, 45, and 58 at levels similar to mice immunized with the standard Gardasil-9 vaccine (Zhai et al. 2017). The vaccine candidates were represented by the MS2 VLPs displaying the tandem HPV31/16L2 peptide 17–31 or by a mixture of VLPs displaying either the tandem peptide or the consensus peptides 69–86 from HPV L2. Moreover, the MS2-L2 VLPs were active by oral immunization and demonstrated remarkable storage potential (Zhai et al. 2019). The high potential of the RNA phage VLPs by the generation of a pan-HPV vaccine was recognized in a review dealing with the role of VLPs by the prevention of HPV-associated malignancies (Wang JW and Roden 2013). The remarkable role of the MS2 and PP7 VLPs in the development of the L2-based HPV vaccines was reviewed by Schellenbacher et al. (2017).

After prospective HPV vaccine candidates, a malaria vaccine elaborated by Chackerian and Peabody's team and based on the MS2 VLPs was also regarded as very promising (Ord et al. 2014; Crossey et al. 2015b).

A lot of work was performed on a vaccine using the predicted Zika virus epitopes (Basu et al. 2018). Because of its broad character and complexity, this study is not referenced here in the vaccine tables. Thus, the identified potential B cell epitopes on the Zika virus envelope protein, namely aa 241–259 from the domain II and aa 294–315, 317–327, 346–361, 377–388, and 421–437 from the domain III, were displayed on the three VLP platforms: MS2, PP7, and Qβ, and their immunogenicity in mice was assessed. When Zika virus epitopes could not be successfully displayed by the

TABLE 25.2

The Putative Vaccines Constructed on the RNA Phage VLPs and Viable Virions by Genetic Fusion Methodology

Vaccine Target	Source of Epitope	Epitope Length, aa	Position of Insertion or Addition	Comments and Immunological Results	References
AP205					
Acquired immunodeficiency: human immunodeficiency virus (HIV)	Co-receptor CCR5, ECL2 loops: mini-loop CRSQKEGLHYTC and full-length loop CRSQK . . . QTLKC	12 33	C-terminus	Mini-loop VLPs with GTAGGGSG but not with GSG linker. The full-length loop VLPs with both linkers. No immunological data.	Indulis Cielēns and Regīna Renhofa, unpublished
	Co-receptor CXCR4, mini-loop: CNVSEADDRYIC	12		VLPs with GTAGGGSG but not with GSG linker. No immunological data.	
	Co-receptor CXCR4, aa 1–39	39	N-terminus	Display on the VLP surface by immunological testing.	Tissot et al. 2010
	Nef protein, aa 66–100 and 132–151	55	C-terminus	Display on the VLP surface by EM.	Tissot et al. 2010
Autoimmune arthritis	Interleukin-1β, human	17 kDa	C-terminus	Display on the VLP surface by EM (Figure 25.6). No immunological data.	Juris Jansons, Irina Sominskaya, and Regīna Renhofa, unpublished
	Interleukin-1α, murine Interleukin-1β, murine and human	157 152 154	C-terminus	Mosaic particles. GSG or GSGG linkers. No immunological data.	Indulis Cielēns and Regīna Renhofa, unpublished
Cancer, prostate	Gonadotropin releasing hormone (GnRH), murine, aa 1–10	10	N-terminus C-terminus	C-terminal fusion: antibody response and inhibition of GnRH function in mice.	Tissot et al. 2010
Chikungunya virus (CHIKV)	Glycoprotein E1, the virus-neutralizing domain III (DIII); Glycoprotein E2, full-length, Domain A, Domain B, Domains A+B	84 361 131 60 298	C-terminus	Mosaic particles, by suppression of the C-terminal amber codon. GSG linker. No immunological data.	Indulis Cielēns and Regīna Renhofa, unpublished
Hepatitis B virus (HBV)	preS1, aa 21–47	27	N-terminus	SGTAGGGSGS linker more preferable than SGG. No immunological data.	
	preS1, aa 21–47, 20–58, 20–119	27 39 100	C-terminus	GTAGGGSG linker. No immunological data.	
Hepatitis C virus (HCV)	E2 protein, genotype 1α, HVR sequence, aa 384–411; a "consensus" HVR sequence	28 31	N-terminus	SGTAGGGSGS linker. No immunological data.	
Human immunodeficiency virus (HIV)	Fusion peptide FP8 AVGIGAVF	14	N-terminus of the single-chain dimer	Induction of HIV-1 neutralizing antibodies in mice.	Mogus et al. 2020
Hypertension	Angiotensin II, aa 1–8	8	N-terminus C-terminus	Display on the VLP surface by immunological testing.	Tissot et al. 2010
Influenza A virus (IAV)	M2e protein, N-terminal ectodomain, consensus sequence, aa 2–24	23	N-terminus	Anti-M2 response in mice. Protection of 100% of mice from lethal influenza infection.	Tissot et al. 2010; Schmitz et al. 2012

Vaccine Target	Source of Epitope	Epitope Length, aa	Position of Insertion or Addition	Comments and Immunological Results	References
Lymphocytic choriomeningitis virus (LCMV)	Glycoprotein, peptide p33: KAVYNFATM	9	C-terminus	No immunological data.	Indulis Cielēns and Regīna Renhofa, unpublished
Severe acute respiratory syndrome virus 2 (SARS-CoV-2)	Spike protein S, aa 437–508	72 + linker aa	C-terminus	Addition to the second copy of the covalent AP205 dimer. Virus-neutralizing response in mice.	Liu X et al. 2021
Obesity	Ghrelin, aa 24–31: GSSFLSPE	8		No immunological data.	Regīna Renhofa, unpublished
	Gastric inhibitory peptide (GIP), aa 1–15 of mature GIP: YAEGTFISDYSIAMD	15		No immunological data.	
Salmonella typhi	Outer membrane protein, D2 peptide, aa 266–280	15	N-terminus C-terminus	Antibody response in mice.	Tissot et al. 2010
Staphylococcus aureus	α-hemolysin (Hla), aa 119–131	30	C-terminus	Protection of mice.	Joyner et al. 2020
West Nile virus (WNV)	Glycoprotein E, the virus-neutralizing domain III (DIII), aa 296–406	111	C-terminus	Mosaic particles. Induction of IgG2 anti-DIII antibodies in mice.	Cielens et al. 2014

fr

Vaccine Target	Source of Epitope	Epitope Length, aa	Position of Insertion or Addition	Comments and Immunological Results	References
Hamster polyomavirus (HaPV)	VP1, aa 364–384, 351–374, 351–384, 333–384	21 24 34 52	N-terminus	Induction of anti-VP1 antibodies in rabbits and mice.	Voronkova et al. 2002

GA

Vaccine Target	Source of Epitope	Epitope Length, aa	Position of Insertion or Addition	Comments and Immunological Results	References
Human immunodeficiency virus (HIV)	Co-receptor CCR5, ECL1 loop YAAAQWDFGNTMCQ	14	FG loop	No immunological data.	Arnis Strods and Regīna Renhofa, unpublished
	Co-receptor CCR5, ECL2a loop QKEGLHYTG	9	AB loop, aa 14/15	No immunological data.	Indulis Cielēns and Regīna Renhofa, unpublished
	Co-receptor CCR5, N-terminus, aa 1–27 aa 1–31	27 31	N-terminus of the second coat copy	Mosaic particles with the GA coat as a helper were formed in both cases.	
	Co-receptor CXCR4, aa 1–40	40	N-terminus	Four variants of linkers. No immunological data.	
Influenza A virus (IAV)	M2e protein, N-terminal ectodomain consensus sequence, aa 2–24 with C-terminally added G residue	24	N-terminus C-terminus	GSGS (GSRS) and GSG linkers for N-and C-terminal insertions, respectively. No immunological data.	
	M2e protein, N-terminal ectodomain consensus sequence, aa 2–24 with C-terminally added G	24	N-terminus of the first coat copy of the coat single-chain dimer	GSG linker. No immunological data.	
Salmonella typhi	Outer membrane protein, D2 peptide, aa 266–280	15	N-terminus	No immunological data.	

(Continued)

TABLE 25.2 (CONTINUED)

The Putative Vaccines Constructed on the RNA Phage VLPs and Viable Virions by Genetic Fusion Methodology

Vaccine Target	Source of Epitope	Epitope Length, aa	Position of Insertion or Addition	Comments and Immunological Results	References
MS2					
Cancer, breast	xCT protein, ECD6 peptide LYSDPFST derived from the sixth loop	8	AB loop of the coat single-chain dimer, aa "15/14"	Antibody response and therapeutic effect in mice.	Bolli et al. 2018
Cancer, ovarian	Peptide DISGTNTSRA mimicking the cancer-associated antigen 125 (CA125/MUC16)	10	AB loop of the coat single-chain dimer, aa "15/14"	Antibodies cross-reactive with CA125 from ovarian cancer cells in mice.	Frietze et al. 2016
Flag peptide	DYKDDDDK	9	Single-chain dimer. AB loop, D12/N13	No immunological data.	Peabody DS 1997
Foot-and-mouth disease virus (FMDV)	VP1, aa 141–160 serotype O/OZK	20	AB loop	Antibody response in mice. Protection of guinea pigs and swine against FMDV challenge.	Dong et al. 2015
			AB loop	Packaging of antisense RNA against the FMDV 3D genes into VLPs carrying the 141–160 epitope presented on the surface. Protection of suckling mice (40%) and guinea pigs (85%) from FMDV challenge.	Dong et al. 2016
	VP1, aa 131–160, serotype O	30	AB loop	Chimeric VLPs more immunogenic than the control epitope tandem peptide.	Wang Guoqiang et al. 2018
Human immunodeficiency virus (HIV)	gp120 protein, V3 loop	10	AB loop of the coat single-chain dimer, aa "15/14"	First application of the single-chain dimer. Antibody response in mice. Ability to pack their own mRNAs.	Mastico et al. 1993; Stockley and Mastico 2000; Peabody DS et al. 2008
	Co-receptor CCR5, ECL2 loop	10		First application of the single-chain dimer. Antibody response in mice. Ability to pack their own mRNAs.	
Human papillomavirus (HPV)	L1 protein L2 protein	20 20	AB loop, aa "15/14"	No immunological data.	Mastico et al. 1993; Stockley and Mastico 2000
	L2 protein of HPV16, HPV18, HPV31, aa 17–31	15	N-terminus and AB loop of the coat single-chain dimer	Display of two different epitopes on the same particle. Protection of mice from genital infection with HPV pseudoviruses representing 11 diverse HPV types. Stability after storage at room temperature for 34 months. Longevity of protection: two years.	Tumban et al. 2012, 2015; Tyler et al. 2014c; Saboo et al. 2016; Peabody J et al. 2017

Vaccine Target	Source of Epitope	Epitope Length, aa	Position of Insertion or Addition	Comments and Immunological Results	References
	Tandem HPV31/16 L2 peptide, aa 17–31, consensus peptides from L2, aa 69–86 or 108–122, multivalent epitope consensus L2(108–122)/ HPV31L2(20–31)/ HPV16L2(17–31)	27 18 15 42		Protection of mice from HPV pseudoviruses 16, 18, 31, 33, 45, and 58. Oral immunization. Stability after spray-freeze drying.	Zhai et al. 2017, 2019
Hypercholesterolemia	Proprotein convertase subtilisin/kexin type 9 (PCSK9), aa 153–163, 188–200, 208–222, 368–381	11 13 17 14	N-terminus of the coat single-chain dimer	No dramatic reductions in total cholesterol in immunized mice, in comparison to chimeric Qβ VLPs (see Table 25.3).	Crossey et al. 2015a
Influenza A virus (IAV)	Hemagglutinin, epitope YPYDVPDYA	9	AB loop, aa "15/14"	No immunological data.	Mastico et al. 1993; Stockley and Mastico 2000
	M2e protein, conserved epitope EVETPIRNE	9	AB loop of the coat single-chain dimer, aa "15/14"	Purification protocol for the potential veterinary vaccine. Encapsulin nanocompartments, a novel approach to the rational vaccine design.	Lagoutte et al. 2016, 2018
Malaria	*Plasmodium falciparum*, liver stage antigen-1 (LSA-1), T1 epitope	24	AB loop, aa "15/14"	Antibody response in mice, with significant upregulation of interferon-γ.	Mastico et al. 1993; Heal et al. 2000; Stockley and Mastico 2000
	Plasmodium falciparum, RH5 protein, a peptide SAIKKPVT mimicking a linear epitope	8	AB loop of the coat single-dimer, aa "15/14"	Induction of antibodies inhibiting parasite invasion.	Ord et al. 2014
	Plasmodium falciparum, blood stage antigen AMA1 (apical membrane antigen-1)	10		Induction of murine antibodies cross-reactive with AMA1.	Crossey et al. 2015b
Staphylococcus aureus	*agr*IV quorum-sensing operon, mimotope peptides that immunologically mimic the autoinducing peptide 4 (AIP4)	6 7		Vaccine targeting the secreted autoinducing peptides of the *S. aureus agr* quorum-sensing system in mice.	O'Rourke et al. 2014, 2015
PP7					
Flag peptide	DYKDDDDK	8	AB loop of the coat and of the coat single-chain dimer, aa 11/12	Packaging of their own mRNA. Antibody response in mice.	Caldeira et al. 2010
Human immunodeficiency virus (HIV)	gp120 protein, V3 loop, peptide IQRGPGRAPV	10		Packaging of their own mRNA. Antibody response in mice.	Caldeira et al. 2010
Human papillomavirus (HPV)	L2 protein of HPV16, aa 17–31: QLYKTCKQAGTCPPD	15		Protection of mice from genital infection with HPV16 pseudovirions by intravaginal immunization.	Caldeira et al. 2010; Hunter et al. 2011

(Continued)

TABLE 25.2 (CONTINUED)

The Putative Vaccines Constructed on the RNA Phage VLPs and Viable Virions by Genetic Fusion Methodology

Vaccine Target	Source of Epitope	Epitope Length, aa	Position of Insertion or Addition	Comments and Immunological Results	References
	L2 protein of HPV1, 5, 6, 11, 16, 18, 45, 58, aa 17–31	15	AB loop of the coat single-chain dimer, aa 11/12	Protection of mice from genital infection with 8 pseudovirion types by immunization with mixture of 8 L2 VLPs. Persistence of antibodies over 18 months.	Tumban et al. 2011, 2013
	L2 protein of HPV16, aa 17–31, 35–50, 51–65, 65–79, 65–85, Consensus 65–85	21 (for 65–85 variant)		Induction of wide-range neutralizing antibodies by VLPs displaying the 65–85 consensus peptide.	Tyler et al. 2014b
	L2 protein of HPV1, HPV16, HPV18, aa 17–31	15	N-terminus and AB loop of the coat single-chain dimer	Display of two different epitopes on the same particle. Protection of mice from genital infection with pseudovirions representing 11 diverse HPV types.	Tyler et al. 2014c
Influenza A virus (IAV)	Hemagglutinin, long alpha helix, or LAH, domain, strain H1N1/09, with a GSG linker	58	N-terminus	Formation of soluble protein aggregates. Seroconversion to both group 1 and group 2 hemagglutinin antigens in mice.	Lu et al. 2018
Pregnancy	Human chorionic gonadotropin (hCG), aa 39–56, 45–55.66–81, 69–80, 111–120, 116–125, 121–130, 131–140, 136–145	8–16	AB loop of the coat single-chain dimer, aa 11/12	Inhibition of uterine weight gain by immunization of mice with VLPs displaying peptides 116–125, 126–135, and 131–140.	Caldeira et al. 2015
Prostate cancer	Prostatic acid phosphatase, 114–128	15	AB loop of the coat single-chain dimer, aa 11/12	Induction of antibodies.	Sun Yanly and Sun Yanhua 2016
Staphylococcus aureus	Accessory gene regulator operon *agr*, AIP1S peptide, YSTSDFIM, where original C in the AIP1 peptide was changed to S (red)	8	AB loop of the coat single-chain dimer, aa 11/12	Induction of murine antibodies recognizing the original AIP1 in vitro. Protection in mice upon challenge with virulent MRSA *agr* type I isolate.	Daly et al. 2017
Tuberculosis	*Mycobacterium tuberculos*, 16 kD protein, peptide 91–110	20	Single-chain coat dimer.	Stimulating antigen for the detection of interferon-γ release.	Yi et al. 2018
Zika virus (ZIKV)	Envelope protein, Domain III LEAEMDGAKGRL VEFKDAHAKRQTVVV LVDRGWGNGCGLFGKG C-terminally: NDTGHETDEN	12 15 16 10	AB loop of the coat single-chain dimer, aa 11/12, and C-terminus	High-resolution 3D structures. No immunological data.	Zhao L et al. 2019

Vaccine Target	Source of Epitope	Epitope Length, aa	Position of Insertion or Addition	Comments and Immunological Results	References
Qβ					
Foot-and-mouth disease virus (FMDV)	VP1 protein, G-H loop peptide	14	C-terminally added to the shortened A1 protein gene within the viable Qβ genome	Chimeric phage, not VLP, displaying the FMDV peptide. Induction of antibodies in guinea pigs with good affinity to both FMDV and hybrid Qβ-G-H loop.	Skamel et al. 2014
Hepatitis B virus (HBV)	preS1 epitope: 31-DPAFR-35 31-DPAFRA-36	5 6	A1 protein, C-terminal extension, aa 72/73, after aa 3, 6, 13, 19; instead of the C-terminal extension	Mosaic particles by A1-derived chimeras and Qβ coat helper via (i) suppression of leaky UGA stop codon and (ii) simultaneous expression of coat and A1-derived genes. Induction of anti-preS1 antibodies in mice.	Kozlovska et al. 1996, 1997; Vasiljeva et al. 1998
	Full-length preS1 and preS2; preS1, aa 20–47, 20–58, or 31–58	28–163	A1 protein, C-terminal extension, aa 18 or after leaky termination codon instead of the C-terminal extension	Mosaic particles by (i) suppression of terminal *amber* or *opal* codons or (ii) coexpression of the chimeric gene with the coat helper gene.	Indulis Cielēns and Regīna Renhofa, unpublished
Human immunodeficiency virus (HIV)	gp120 protein, V3 loop, aa 299–337	39	A1 protein, C-terminal extension, aa 72/73; after aa 19; instead of the C-terminal extension	Mosaic particles.	Kozlovska et al. 1996, 1997
	gp120 protein, epitopes b122a: N(410–423)-NG-(435–449)-NG-(291–341)-GSAGSAGSA-(365–392) C; OD$_{EC}$, and b122a1-b		C-terminal	Induction of neutralizing antibodies in rabbits.	Purwar et al. 2018
Paroxysmal nocturnal hemoglobinuria	C5 protein, epitope ASYKPSKEESTSGS linked with the PADRE peptide AKFVAAWTLKAAA via GSG linker and N-terminal sequence AYGG	34	C-terminal	Mosaic particles with 50 copies of extension per capsid. Protection of mice from complement-mediated intravascular hemolysis in a model of paroxysmal nocturnal hemoglobinuria.	Zhang Lingjun et al. 2017

Source: This and following tables are the extended versions of the original tables from Pumpens P et al. *Intervirology.* 2016;59:74–110, with permission of S. Karger AG, Basel, and Pumpens P. *Single-Stranded RNA Phages: From Molecular Biology to Nanotechnology.* Boca Raton—London—New York: CRC Press, 2020, with permission of Taylor & Francis Group.

Note: The data is given in alphabetical order of the (i) phage model and (ii) vaccine target for the same phage model. When the precise number of the amino acid residues within the epitope was difficult to assess, the size of the epitope was expressed in kDa. In all cases, the expression system is *E. coli*.

genetic insertion on the MS2 or PP7 VLPs due to the failure of the recombinant coat proteins to assemble into VLPs, they were displayed on the Qβ VLPs by chemical coupling. In fact, mice immunized with a mixture of VLPs displaying Zika virus epitopes elicited anti-ZIKV antibodies. However, immunized mice were not protected against a high-challenge dose of Zika virus, but sera—albeit at low titers—from immunized mice neutralized in vitro a low dose of Zika virus. Taken together, these results showed that these epitopes were B cell epitopes, and they were immunogenic when displayed on the Qβ VLP platform. The results also showed that immunization with the VLPs displaying a single B-cell epitope minimally reduced Zika virus infection, whereas immunization with a mixture of VLPs displaying a combination of the B cell epitopes neutralized Zika virus infection (Basu et al. 2018).

The MS2 (O'Rourke et al. 2014) and PP7 (Daly et al. 2017) VLPs were used by Chackerian and Peabody's team to generate promising vaccine candidates against methicillin-resistant *Staphylococcus aureus* (MRSA), a cause of the growing incidence of skin and soft tissue infections (SSTI). Compared to antibiotic-susceptible strains, the MRSA SSTI treatment failure required added interventions, with associated increases in human suffering and medical costs. Therefore, it was of high importance that the PP7 VLPs carrying a short amino acid sequence were able to demonstrate efficiency in a murine SSTI challenge model with a highly virulent MRSA isolate (Daly et al. 2017). Furthermore, Joyner et al. (2021) chose the 119-GFNGNVTGDDTGKIGGLIGAN-131 peptide, so-called linear neutralizing domain, of the pore-forming cytotoxin α-hemolysin (Hla—which is a critical factor of *Staphylococcus aureus* virulence—and generated in parallel two vaccine candidates by different strategies, namely the AP205 C-terminal fusion the Qβ chemically coupled version. As shown in Tables 25.2 and 25.3, the vaccination with either of the two VLPs protected both male and female mice from subcutaneous Hla challenge, evident by reduction in lesion size and neutrophil influx to the site of intoxication (Joyner et al. 2020). Remarkably, Mogus et al. (2020) used the same concept to compare the efficiency of the fused and coupled epitopes by the elaboration of the novel prospective HIV-1 vaccine, in the case by the MS2 and Qβ VLPs, respectively.

A new concept, the "epitope-RNA VLP vaccine," was introduced by Dong et al. (2015, 2016), who generated a foot-and-mouth disease vaccine that combined exposure of an epitope on the MS2 VLP surface with the packaged antisense RNA. This complex was expected to have the virtues of both the VLP and RNAi vaccines and was obtained by cotransformation of the two plasmids into bacteria. Thus, the antisense RNA against the *3D* genes of foot-and-mouth disease virus (FMDV) was packaged into the MS2 VLPs carrying the VP1 epitope 141–160 presented on the VLP surface. As indicated in Table 25.2, the vaccine demonstrated definite potency against FMDV infection in mice and guinea pigs (Dong et al. 2016).

Bolli et al. (2018) demonstrated inhibition of the progression of metastatic breast cancer in vivo after immunization of mice with the MS2 VLPs carrying an epitope from the cystine-glutamate antiporter protein xCT that regulates cystine intake, conversion to cysteine, and subsequent glutathione synthesis, protecting cells against oxidative and chemical insults. The MS2 VLPs carrying the xCT epitope elicited a strong antibody response against xCT including high levels of IgG2a antibody. The IgG antibodies isolated from the MS2 VLP-treated mice bound to tumorspheres, inhibited xCT function as assessed by reactive oxygen species generation, and decreased breast cancer stem cell growth and self-renewal (Bolli et al. 2018).

It is noteworthy that the MS2 VLPs carrying matrix protein 2 ectodomain (M2e) of influenza A virus were used as a positive control of the M2e display by the elaboration of a novel category of vaccines ensuring the simultaneous surface display and cargo loading of encapsulin nanocompartments (Lagoutte et al. 2018).

Some attempted to use the PP7 VLPs for the generation of chimeras that would reproduce on their surface the natural trimer of the long alpha helical (LAH) domain of the influenza virus hemagglutinin (Lu IN et al. 2018). Unfortunately, the native self-assembly did not occur, but the aggregates demonstrated good immunological capabilities. It is remarkable that the Kaspars Tārs and Andris Kazāks team succeeded by the self-assembly of analogous chimeras based on the hepatitis B core platform (Kazaks et al. 2017) as described in Chapter 38. The attempt to present the trimeric hemagglutinin on the Qβ VLPs was reported at that time by Lauster et al. (2018).

Mosaic Qβ particles were generated as a putative vaccine antagonizing the pathological effects of IL-13 in severe human diseases (Bai et al. 2018). Thus, the human IL-13 peptide was fused to the C-terminus of the Qβ coat, which was then coexpressed with the native Qβ coat and formed the mosaic Qβ-IL-13 particles.

Recently, Liu X et al. (2021) used the AP205 tandem to accommodate receptor binding motif (RBM) domain of the spike protein S of SARS-CoV-2 to construct an attractive vaccine candidate against Covid-19, as documented in Table 25.2.

CHEMICAL COUPLING

Table 25.3 presents a list of the vaccine candidates generated by the chemical coupling approach. As noted earlier, Bachmann's team moved as pioneers to the forefront of this approach, first in the field of therapeutic vaccination against noninfectious diseases. The principal validity of the chemical coupling approach was approved by a long line of experimental therapeutic vaccines. The impressive list of the latter was first proposed by Bachmann and Dyer (2004), where the chemically coupled VLPs were planned to replace the host-specific monoclonal antibodies in the treatment of acute and chronic diseases, starting with noninfectious diseases. The proposed transition from passive administration

TABLE 25.3

Vaccines Constructed on the RNA Phage VLPs or Phage Virions by Chemical Coupling Methodology

Vaccine Target	Source of Epitope	Epitope Length, aa	Coupling Site	Comments	References
AP205					
Human immunodeficiency virus (HIV)	Six peptides covering the alpha-helical regions of gp41: 3–13, 3–17, 3–20, 3–24, P1, P8, provided with a C-terminal Cys residue	12 16 19 23 40 66	Lysine residues	Induction of anti-gp41 antibodies in mice.	Pastori et al. 2012
Hypertension	ATR-AP205–001 vaccine: ATR-001 peptide AFHYESQ corresponding to an epitope of the ECL2 of human AT1R, a G protein coupling receptor, with N-terminally added cysteine	8		Induction of antibodies in mice.	Hu et al. 2017
Lymphocytic choriomeningitis virus (LCMV)	Glycoprotein, the p33 peptide KAVYNFATM	9		Induction of CTL response in mice.	Schwarz K et al. 2005
Salmonella typhi	Outer membrane protein, D2 peptide, aa 266–280: TSNGSNPSTSYGFAN with N-terminal CGG linker	15	Lysine residues	Immunization of mice with VLPs carrying 13, 56, 94, 142, and 293 peptides per VLP. Overcoming of carrier induced epitopic suppression (CIES) by high coupling density.	Jegerlehner et al. 2010
West Nile virus (WNV)	Glycoprotein E, the virus-neutralizing domain III (DIII), aa 582–696 of the WNV polyprotein precursor	115	Lysine residues	Induction of virus-neutralizing antibodies in mice. Partial protection of mice from WNV challenge.	Spohn et al. 2010
Qβ					
Acquired immunodeficiency: feline immunodeficiency virus (FIV)	Transmembrane (TM) glycoprotein, a peptide containing tryptophan-rich motif, aa 767–786: LQKWEDWVGWIGNIPQYLKG	20	Lysine residues	Induction of antibodies, reactive with the epitope but not with FIV, in cats.	Freer et al. 2004
Acquired immunodeficiency: human immunodeficiency virus (HIV)	Co-receptor CCR5, N-terminal ECL domain, circularized	20	Lysine residues	Induction of antibodies, reactive with CCR5 and inhibiting entry of pseudotype viruses in vitro, in mice and rabbits.	Huber A et al. 2008; Sommerfelt 2009
Acquired immunodeficiency: human immunodeficiency virus (HIV)/simian immunodeficiency virus (SIV)	Co-receptor CCR5, macaque, EC1, N-terminal aa MDYQVSSPTYDIDYYTSEPC; ECL2, a cyclic peptide, aa 168–177 DRSQREGLHYTG linked through an DG dipeptide spacer	20 10	Lysine residues	Induction of antibodies, reactive with CCR5 and inhibiting SIV infection in vitro, in mice and rats.	Hunter et al. 2009, 2010; van Rompay et al. 2014
Allergy	Cysteine protease, the major fecal allergen of the house dust mite *D. pteronyssinus*, Der p 1 peptide, aa 117–133: CGIYPPNANKIREALAQTHSA	21	Lysine residues	Vaccine without adjuvants was found safe and immunogenic in humans.	Lechner et al. 2002; Kündig et al. 2006

(Continued)

TABLE 25.3 (CONTINUED)

Vaccines Constructed on the RNA Phage VLPs or Phage Virions by Chemical Coupling Methodology

Vaccine Target	Source of Epitope	Epitope Length, aa	Coupling Site	Comments	References
Allergy, cat	Fel d1 protein, major cat allergen, a covalent dimer of chain 2 and chain 1 of Fel d1 spaced by a 15 aa-linker (GGGGS)x3 and added to the coding sequence for LEHHHHHHGGC at the C-terminus	23 kDa	Lysine residues	Protection against type I allergic reactions in mice.	Schmitz et al. 2009; Uermösi et al. 2010; Engeroff et al. 2018; Zha et al. 2018
Allergy, asthma and rhinitis	Two IgE peptides, different loops of the C3 domain: ADSNPRGVSAYLSRPSPGGC and YQCRVTHPHLPRALMRS	20 16	Lysine residues	Induction of antibodies in mice and nonhuman primates.	Champion et al. 2014; Akache et al. 2016; Weeratna et al. 2016
Allergy to red meat	α-1,3-galactosyl transferase, Galα3LN epitope. For conjugation, α-Gal trisaccharide and glucose were converted to their respective alkyne derivatives Each alkyne was attached by a two-step procedure in which the protein nanoparticle was first acylated with an azide-terminated N-hydroxysuccinimide ester and then addressed by copper-catalysed azide-alkyne cycloaddition	Carbohydrates are conjugated	Lysine residues	Induction of antibodies in mice.	Araujo et al. 2016
Alzheimer's disease	Aβ peptide (1–9)-GGC: DAEFRHDSGGGC	12	Lysine residues	Phage Qβ, but not Qβ VLPs. Induction of antibodies in mice without adjuvants.	Chackerian et al. (2006)
	Aβ peptide: N-terminal Aβ(1–9) or C-terminal Aβ (28–40)	12 16	Lysine residues	Qβ VLPs. Induction of antibodies in mice without adjuvants. Reduction of Aβ levels.	Li QY et al. 2010
	CAD106 vaccine: Aβ peptide 1–6 DAEFRH plus a GGC spacer	9	Lysine residues	Avoiding activation of Aβ-specific T cells and reducing the amyloid accumulation in transgenic mice. Phase II study. Favorable safety profile and acceptable antibody response in patients.	Wiessner et al. 2011; Winblad et al. 2012; Farlow et al. 2015; Vandenberghe et al. 2017
Atherosclerosis	Interleukin-1α, full length, containing aa linker at C-terminus	17 kDa	Lysine residues	Reduction of inflammatory reaction and plaque progression in mice.	Tissot et al. 2013
Autoimmune arthritis	Interleukin-1α, murine, aa 117–270 of IL-1a precursor Interleukin-1β, murine, aa 119–269 of IL-1β precursor, both provided with aa linkers at C-termini	17 kDa	Lysine residues	Induction of autoantibodies in mice. Protection of mice from inflammation and degradation of bone and cartilage in a collagen-induced arthritis model.	Spohn et al. 2008; Guler et al. 2011
Autoimmune arthritis, encephalomyelitis, and myocarditis	Interleukin 17, murine, aa 26–158, with a C-terminal linker GGGGGC	32 kDa	Lysine residues	Induction of autoantibodies in mice. Efficiency in animal models of autoimmunity.	Röhn et al. 2006; Sonderegger et al. 2006; Dallenbach et al. 2015

Vaccine Target	Source of Epitope	Epitope Length, aa	Coupling Site	Comments	References
Cancer, induction of antitumor antibodies	Tumor associated carbohydrate antigens (TACAs): monomeric Tn antigen (GalNAc-α-O-Ser/Thr) that is overexpressed on the surface of a variety of cancer cells including breast, colon, and prostate cancer and is involved in aggressive growth and lymphatic metastasis of cancers	Carbohydrates are conjugated by CuAAC	Lysine residues and N-terminus	Induction of antibodies, highly selective toward Tn antigens and reacting strongly with the native Tn antigens on human leukemia cells, in mice.	Yin et al. 2013, 2014, 2015
	Tumor-associated carbohydrate antigens (TACAs): ganglioside GM2, a tetrasaccharide, overexpressed in a wide range of tumor cells	Carbohydrates are conjugated by CuAAC	Lysine residues and N-terminus	Induction of antibodies recognizing GM2-positive tumor cells and inducing cell lysis through complement-mediated cytotoxicity.	Yin et al. 2016
	Tumor-associated mucin-1 peptides and glycopeptides. The (glyco) peptides containing 20–22 aa residues as the backbone covering one full length of the tandem repeat region with the sequence PDTRPAPGSTAPPAHGVTSA	20–22	Lysine residues and N-terminus	Induction of high titer antibodies and CTL response in mice. Killing of mucin-1 positive tumor cells.	Yin et al. 2018
	Mucin-1 peptide SAPDT*RPAP, where * denotes glycosylation	9	Lysine residues and N-terminus	Induction of high titer antibodies and CTL response in mice. Recognition of human breast cancer over normal breast tissues.	Wu X et al. 2018
Chikungunya virus (CHIKV)	E2 glycoprotein (Singapore strain), aa 2800–2818, 3025–3058, 3073–3081, 3121–3146, and 3177–3210	19, 34, 9, 26, 34	Lysine residues	Induction of high-titer antibodies in mice.	Basu et al. 2020
Chronic inflammatory disorders: rheumatoid arthritis, psoriasis, Crohn's disease	TNF-α, murine, aa 80–235 of the transmembrane form, aa 4–23 peptide: CGGSSQNSSDKPVAHVVANHQVE	156 20	Lysine residues and N-terminus	Induction of autoantibodies in mice. Protection from inflammation in a murine model of rheumatoid arthritis.	Spohn et al. 2007
Chronic inflammatory illnesses: type 2 diabetes mellitus	Interleukin-1β, murine, rhesus monkey and human, two muteins: mIL-1β (D143 K), mIL-1β (D143 K) with strongly reduced inflammatory activity.	153	Lysine residues and N-terminus	Good tolerability of vaccine in mice and nonhuman primates Phase I/II clinical trial in patients with type 2 diabetes mellitus, using the human version of the vaccine hIL1βQβ with neutralizing anti-IL-1β antibody response.	Spohn et al. 2014; Cavelti-Weder et al. 2016
	Interleukin-1β, human, a set of site-specific mutations to mimic modifications occurring during inflammatory responses and possibly leading to alterations or masking of certain cytokine sites		Lysine residues and N-terminus	Efficiency in mouse models of inflammatory bowel disease and inflammatory arthritis.	Spohn et al. 2017
	Islet amyloid polypeptide (IAPP), peptides: KCNTATCAT without s-s bond KCNTATCATGGK[Aoa] with s-s bond CGGTNVGSNTY CGGREPLNYLPL	9 11 12	Lysine residues and N-terminus	Induction of antibodies against aggregated but not monomeric IAPP in mice.	Roesti et al. 2020

(Continued)

TABLE 25.3 (CONTINUED)

Vaccines Constructed on the RNA Phage VLPs or Phage Virions by Chemical Coupling Methodology

Vaccine Target	Source of Epitope	Epitope Length, aa	Coupling Site	Comments	References
Clostridium difficile	Receptor-binding domain of the toxin B		Lysine residues and N-terminus	Induction of neutralizing antibodies in rodents.	Rebeaud and Bachmann 2012
Coronary heart disease (CHD)	Angiopoietin-like protein 3 (ANGPTL3), murine, 32-EPKSRFAMLDDVKILA-47 Angiopoietin-like protein 4 (ANGPTL4), murine, 29-QPEPPRFASWDEMNL-LAHGLLQLGH-53	20 29	Lysine residues and N-terminus	The vaccination of mice with ANGPTL3 VLPs—but not ANGPTL4 VLPs—was associated with reduced steady state levels of triglycerides.	Fowler et al. 2021
Eosinophilia	Recombinant interleukin-5; recombinant eotaxin, both murine	17 and 8 kDa	Lysine residues and N-terminus	Induction of autoantibodies in mice. Reduction of eosinophilic inflammation of the lung in an ovalbumin-based mouse model of allergic airway inflammation.	Zou et al. 2010
Hen egg lysozyme (HEL), as a model for overcoming self-tolerance	Full-length	129	Lysine residues and N-terminus	Qβ phage, but not VLPs. Induction of autoantibodies in tolerant HEL transgenic mice, without adjuvants.	Chackerian et al. 2008
Human immunodeficiency virus (HIV)	gp41 protein, cluster I, a conserved immunodominant loop connecting the heptad repeat 1 and heptad repeat 2 of the HIV-1 envelope glycoprotein	20	Lysine residues and N-terminus	Qβ phage but not VLPs. Induction of antibodies in rabbits.	Sharma et al. 2012
	gp120 protein, several conserved epitopes of broadly neutralizing antibodies		Lysine residues and N-terminus	Over 87% of seropositive patients showed specific antibody responses to the epitopes displayed on the Qβ VLPs.	Nchinda et al. 2017
	gp120 protein, variants of the b122a epitope—see Table 25.2 for the structure		Lysine residues and N-terminus	Induction of neutralizing antibodies in rabbits.	Purwar et al. 2018
	Fusion peptide FP8 AVGIGAVF	12	Lysine residues and N-terminus	Induction of HIV-1 neutralizing antibodies in mice.	Mogus et al. 2020
Human papilloma virus (HPV)	L2 protein, N-terminal HPV16 peptides: aa 34–52, 49–71, 65–85, 108–120, consensus 65–85 peptide	19 23 21 13	Lysine residues and N-terminus	Phage Qβ but not VLPs. Induction of neutralizing antibodies by the consensus 65–85 peptide-VLPs in mice.	Tyler et al. 2014b
	E7 protein; E7 peptide 49–57: RAHYNIVTFGGC	17 kDa 12	Lysine residues and N-terminus	Induction of antibodies and CTL response in mice. In contrast to antibodies, both CD4⁺ and CD8⁺ T cell responses as well as T-cell-mediated protection against tumor growth were comparable for linked and mixed antigen formulations.	Gomes et al. 2017

Vaccine Target	Source of Epitope	Epitope Length, aa	Coupling Site	Comments	References
Hypercholesterolemia	Proprotein convertase subtilisin/kexin type 9 (PCSK9), human, aa 68–76, 153–163, 207–223	9 11 17	Lysine residues and N-terminus	Significant reductions in total cholesterol, free cholesterol, phospholipids, and triglycerides in mice and macaques (see Table 25.2).	Crossey et al. 2015a
	Proprotein convertase subtilisin/kexin type 9 (PCSK9), human, aa 150–157, 161–170, 236–243, 273–281, and 577–585	8 10 8 9 9	Lysine residues and N-terminus	Induction of high titer antibodies in mice, especially with PCSK9Qβ-003 vaccine displaying the 150–157 epitope. Decrease of plasma total cholesterol and up-regulation of LDLR expression in liver.	Pan Yajie et al. 2017
Hypertension	CYT006-AngQb vaccine: Angiotensin II-derived peptide CGGDRVYIHPF where CGG is a linker	11	Lysine residues and N-terminus	Successful preclinical trials in mice and rats. Phase I and IIa clinical trials.	Bachmann et al. 2006b; Ambühl et al. 2007; Tissot et al. 2008
Hypertension, diabetic nephropathy, atherosclerosis	ATRQβ-001 vaccine: ATR-001 peptide AFHYESQ corresponding to an epitope of the ECL2 of human AT1R, a G protein coupling receptor, with N-terminally added cysteine	8	Lysine residues and N-terminus	Decrease of the blood pressure of Ang II-induced hypertensive mice and spontaneously hypertensive rats.	Chen X et al. 2013; Ding et al. 2016; Zhou et al. 2016; Pan Yajie et al. 2019
Hypertension	ATRQβ-001 vaccine: CE12 epitope CAPESEPSNSTE, human L-type calcium channel $Ca_V1.2$ α_{1C} subunit	12	Lysine residues and N-terminus	Decrease of blood pressure in hypertensive mice and rats.	Wu H et al. 2020
Inflammatory hyperalgesia: potential long-term therapy for chronic pain	Nerve growth factor (NGF), murine, aa 19–241 of pro-NGFβ and an additional 9 aa extension at the C terminus comprising a His_6 tag and GGC sequence	223	Lysine residues and N-terminus	Reduction of hyperalgesia in collagen-induced arthritis or postinjection of zymosan A, two models of inflammatory pain in mice.	Röhn et al. 2011
Influenza A virus (IAV)	M2 protein, N-terminal extracellular domain	23	Lysine residues and N-terminus	Induction of antibodies by intranasal immunization of mice. Protection from viral challenge.	Bessa et al. 2008
	M2e peptide SLLTEVETPIRNEWGC-azide coupled through the copper-free click chemistry crosslinker DBCO	16	Lysine residues and N-terminus	Preference of the VLP packing with prokaryotic RNA whenever IgG2 response is desired, while eukaryotic RNA should be employed to induce an IgG1 response.	Gomes et al. 2019b
	Hemagglutinin, globular head domain (gH1), A/California/07/2009 (H1N1) strain, aa 49–325 and C-terminal extension GGGCG	281	Lysine residues and N-terminus	Protection from mouse-adapted virus and 2009 H1N1 virus in mice. High immunogenicity and acceptable safety profile of nonadjuvanted vaccine in healthy humans.	Skibinski et al. 2013, 2018; Jegerlehner et al. 2013; Low et al. 2014

(Continued)

TABLE 25.3 (CONTINUED)

Vaccines Constructed on the RNA Phage VLPs or Phage Virions by Chemical Coupling Methodology

Vaccine Target	Source of Epitope	Epitope Length, aa	Coupling Site	Comments	References
Influenza A virus (IAV)	Two hemagglutinin stalk peptides of subtypes H1 and H3 overlapping at the C terminus	57 72	Lysine residues and N-terminus	Induction of antibodies in mice. The same protection against influenza virus infection as with a peptide vaccine.	Kiršteina 2017
Leishmaniasis	α-Gal trisaccharide epitope, *Leishmania infantum*, *Leishmania amazonensis*		Lysine residues and N-terminus	Protection against *Leishmania* challenge in a C57BL/6 α-galactosyltransferase knockout mouse model.	Moura et al. 2017
Lyme disease	modified outer surface protein CspZ of *Borrelia burgdorferi*	~25 kDa	Lysine residues and N-terminus	Induction of bactericidal antibody response in mice and clearing of spirochete infection.	Marcinkiewicz et al. 2018
Lymphocytic choriomeningitis virus (LCMV)	Glycoprotein, peptides: p33, KAVYNFATM; p13, GLNGPDIYKGVYQFKSVEFD; p33-gp61, CKSLKAVYNFATMGLNGPDIY-KGVYQFKSVEF with a GGC linker added to the C-terminus	9 20 32	Lysine residues and N-terminus	Crucial role in the studies of the CTL induction by the CpG packaging. The average presence of about two p33 peptides per subunit was found, indicating that about 360 peptides were displayed per particle.	Bachmann et al. 2004b, 2005a; Storni et al. 2004; Schwarz K et al. 2005; Bessa et al. 2008; Agnellini et al. 2008; Keller et al. 2010a, b; Mohsen et al. 2017a
Malaria	*Plasmodium falciparum*, circumsporozoite protein (CSP), an almost full-length CSP consisting of 19 NANP and 3 NVDP repeats and the majority of the N- and C-terminal regions (residues 26_{Tyr}-127_{Asp} linked to 207_{Pro}-383_{Ser})	45 kDa	Lysine residues and N-terminus	Induction of antibodies in mice and rhesus monkeys. Protection of mice against transgenic parasite challenge.	Khan et al. 2015; Phares et al. 2017
	Plasmodium falciparum, Pfs25	40 kDa	Lysine residues and N-terminus	The highest quantity of antibodies was elicited when compared to the antigen displayed by AP205 VLPs.	Leneghan et al. 2017
	Plasmodium vivax, cell-traversal protein for ookinetes and sporozoites (*P. vivax* CelTOS)	21 kDa	Lysine residues and N-terminus	No immunization regimens of four different vaccine platforms provided significant protection against a novel chimeric rodent *P. berghei* parasite *Pb-Pv*CelTOS.	Alves et al. 2017
	Plasmodium vivax, peptides from the VK210 and VK247 allelic variants 210agdr: CGGDRADGQPAGDRADGQPAGDR 210qpag: CGGAGDRADGQPAGDRADGQPAG 247gang: CGGAGNQPGANGAGNQPGANG (AGDR)₃: CGGAGDRAGDRAGDR	23, 23, 21, 15	Lysine residues and N-terminus	Induction of antibodies in mice. Partial protection against malaria challenge by the Qβ-AGDR variant.	Atcheson and Reyes-Sandoval 2020; Atcheson et al. (2020, 2021)

Vaccine Target	Source of Epitope	Epitope Length, aa	Coupling Site	Comments	References
	CSP, CIS43 mAb epitope 101-NPDPNANPNVDPNAN-115	19	Lysine residues and N-terminus	Partial protection from malaria infection in a mouse model.	Jelínková et al. 2021
Melanoma	Mel-QbG10 vaccine: Melan-A/Mart-1 A27L variant peptide CGHGHSYTTAEELAGIGILTV Packaging with QbG10 CpG: GGGGGGGGGGGGACGATCGTC-GGGGGGGGGG	20	Lysine residues and N-terminus	Promising results by phase I and IIa clinical trials on stage II-IV melanoma patients.	Speiser et al. 2010; Braun et al. 2012; Goldinger et al. 2012
	The vaccines were produced with germline epitopes, germline-multitarget vaccine (GL-MTV) or mutated epitopes (Mutated-MTV) or a combination of both (Mix-MTV) in mice transplanted with aggressive B16F10 melanoma tumors	8 9 10	Lysine residues and N-terminus	Induction of effective CD8+ T-cell responses in mice.	Mohsen et al. 2019c
Nicotine addiction	CYT002-Nic-Qb vaccine, or NIC002 (formerly known as Nicotine-Q β or Nic-Qβ): nicotine was covalently coupled to Qβ VLPs via a succinimate linker	-	Lysine residues and N-terminus	Preclinical studies in mice and successful phase I and II clinical trials. Stable vaccine formulations enabling storage.	Maurer et al. 2005; Beerli et al. 2008; Cornuz et al. 2008; Lang et al. 2009; McCluskie et al. 2015
Obesity	Qβ-GIP vaccine: gastric inhibitory peptide (GIP, also known as glucose-dependent insulinotropic polypeptide), aa 1–15 YAEGTFISDYSIAMD of mature GIP (42 aa) with C-terminally added linker GC.	15	Lysine residues and N-terminus	Induction of antibodies in mice. Reduction of body weight gain in mice fed a high-fat diet. Increased weight loss in vaccinated obese mice.	Fulurija et al. 2008
Osteoporosis	Qβ-TRANCE/RANKL vaccine: TNF-related activation-induced cytokine (TRANCE), also known as receptor activator of NF-kappaB ligand (RANKL), aa 158–316 (extracellular domain) of the mature form of murine TRANCE/RANKL with a Cys-linker and His$_6$ tag	159	Lysine residues and N-terminus	Induction of autoantibodies in mice without adjuvant. Prevention of bone loss in a mouse model of osteoporosis.	Spohn et al. 2005; Spohn and Bachmann 2007
OVA peptide	Ovalbumin, CSSAESLKISQAVHAAHAEINEAGR	25	Lysine residues and N-terminus	Determination of class switch in B cells to IgG2a by TLR9 signaling.	Jegerlehner et al. 2007
	Ovalbumin, SIINFEKL	8	Lysine residues and N-terminus	Optimization of high-avidity T-cell responses in mice.	Gilfillan et al. 2020
Parkinson's disease	Alpha-synuclein, peptides: CGGKNEEGAPQ (PD1), MDVFMKGLGGC (PD2), CGGEGYQDYEPEA (PD3)	11 11 13	Lysine residues and N-terminus	Induction of autoantibodies in mice, recognizing Lewy bodies and toxic oligomeric a-syn species. The antibodies were unable, however, to treat symptoms of Parkinson's disease in the mouse model.	Doucet et al. 2017

(Continued)

TABLE 25.3 (CONTINUED)

Vaccines Constructed on the RNA Phage VLPs or Phage Virions by Chemical Coupling Methodology

Vaccine Target	Source of Epitope	Epitope Length, aa	Coupling Site	Comments	References
Salmonella typhi	Outer membrane protein, D2 peptide, aa 266–280 TSNGSNPSTSYGFAN with N-terminal CGG linker	18	Lysine residues and N-terminus	The vaccine preparations of 13, 56, 94, 142, and 293 peptides per VLP were used. Overcoming of carrier-induced epitopic suppression (CIES) by high coupling density (see also earlier mention of AP205).	Jegerlehner et al. 2010
Salmonella enteritidis	Synthetic tetrasaccharide antigen		Lysine residues and N-terminus	Induction of antiglycan antibodies in mice. Complete protection of mice from lethal bacterial challenge.	Huo et al. 2019
Staphylococcus aureus	α-hemolysin (Hla), aa 119–131	22	Lysine residues and N-terminus	Protection of mice	Joyner et al. 2020
Streptococcus pneumoniae	Synthetic glycans, TS3, a linear repeat of a disaccharide Glcβ1–3GlcAβ1–4Glcβ1–3GlcAβ1, TS14, Galβ1–4Glcβ1–6[Galβ1–4]GlcNAcβ1, a branched tetrasaccharide		Lysine residues and N-terminus	Induction of antiglycan antibodies and long-term protective immunity in mice.	Polonskaya et al. 2017
Tauopathy	pT181, tau peptide [175]TPPAPKpTPPSSGEGGC[190], phosphorylated at Thr181 and modified with two glycine and one cysteine spacer sequence for conjugation	13	Lysine residues and N-terminus	Reduction of both soluble and insoluble species of hyperphosphorylated pTau in the hippocampus and cortex, avoiding a Th1-mediated pro-inflammatory cell response and rescuing cognitive dysfunction in a mouse model of frontotemporal dementia.	Maphis et al. 2019

Note: The order of data presentation is the same as in Table 25.2. When the precise number of the amino acid residues within the epitope was difficult to assess, the size of the epitope was expressed in kDa.

of monoclonal antibodies to active vaccination against self-antigens was a logical step in the drug development, focusing on affordable medicines and broader patient acceptance and regulatory compliance. The induction of autoantibodies was regarded as beneficial under certain physiological conditions in order to remove unwanted excess of a particular self-antigen, such as, for example, angiotensin in the case of hypertension.

To be maximally informative, Table 25.3 includes the RNA phage VLPs carrying the model epitopes p33, D2, and OVA, although these chimeric VLPs were not designed as actual vaccine candidates but were involved in the basic immunological studies, as described in the *Contribution to Basic Immunology* paragraph.

The development prospects of the chemically conjugated RNA phage VLP vaccines were assessed in review articles devoted specifically to the following diseases: Alzheimer's

disease (Chackerian 2010; Fettelschoss et al. 2014; Liu E and Ryan 2016; Sterner et al. 2016; Gallardo and Holtzman 2017; Marciani 2017; Mo et al. 2017; Bachmann et al. 2019; Kaminaka and Nozaki 2019), atherosclerosis (Govea-Alonso et al. 2017; Kobiyama et al. 2019; Amirfakhryan 2020), cancer (Grippin et al. 2017; Ong et al. 2017; Qiu et al. 2017; Wei MM et al. 2018; Mohsen et al. 2019b; Neek et al. 2019), hypercholesterolemia (Chackerian and Remaley 2016), hypertension (Ready 2005; Bachmann et al. 2006b; Brown MJ 2008; Gradman and Pinto 2008; Miller SA et al. 2008; Campbell 2009, 2012; Phisitkul 2009; Maurer and Bachmann 2010; Bairwa et al. 2014; Nakagami et al. 2014; Tamargo and Tamargo 2017; Nakagami and Morishita 2018; Azegami and Itoh 2019), influenza (Sączyńska 2014; Pushko and Tumpey 2015; Quan et al. 2016; Wong and Ross 2016; Kolpe et al. 2017), malaria (Reyes-Sandoval and Bachmann 2013; Wu Y et al. 2015; Noe et al. 2016; Draper et al. 2018; Huang WC et al. 2018), nicotine addiction (Cerny T 2005; Heading 2007; Maurer and Bachmann 2006, 2007; Didilescu 2009; Crain and Bhat 2010; Hartmann-Boyce et al. 2012; Kitchens and Foster 2012; Raupach et al. 2012; Fahim et al. 2013; Lieber and Millum 2013; Wang GB and Zhu 2013; Pentel and LeSage 2014; Collins and Janda 2016; Ilyinskii and Johnston 2016; Kalnik 2016; Kosten et al. 2017; Zhao Z et al. 2017), and obesity (Monteiro 2014; Na et al. 2014).

It should be emphasized that Bachmann's team elaborated the promising CAD106 vaccine against Alzheimer's disease (Wiessner et al. 2011). The CAD106 first-in-human study demonstrated a favorable safety profile and promising antibody response (Winblad et al. 2012). Further in-human investigations supported the CAD106 favorable tolerability profile after repeated CAD106 injections (Farlow et al. 2015; Vandenberghe et al. 2017). The CAD106 vaccine, as well as the Qβ VLP nicotine vaccine, were reviewed exhaustively in a special paper devoted to the vaccines in psychiatry (Kuppili et al. 2018).

The potential Qβ VLP-based vaccines for metabolic diseases were reviewed by Morais et al. (2014). Rynda-Apple et al. (2014) concentrated on the capability of the VLPs to induce protective immune responses in the lung.

The Qβ-IgE VLP-driven therapeutic induction of anti-IgE antibodies to trigger the human immune system was reviewed by Licari et al. (2017).

The Qβ VLP-based vaccine against the signaling molecule IL-17 was described as a safe and effective tool in the common skin disease psoriasis (Foerster and Bachmann 2015). It was specifically emphasized that the vaccination against IL-17 could be capable of replacing the costly manufacture of antibodies currently in clinical use, with huge implications for treatment availability and health economics. The prospects of an active vaccine against asthma, targeting IL-5, were outlined by Bachmann et al. (2018). The numerous anticytokine vaccine candidates including a full list of the Qβ VLP-based vaccines have been reviewed (Delavallée et al. 2008; Link and Bachmann 2010; Uyttenhove and Van Snick 2012).

More generally, the noninfectious disease vaccine candidates were thoroughly examined in a set of exhaustive reviews (Bachmann and Jennings 2010, 2011; Röhn and Bachmann 2010; Bachmann and Whitehead 2013; Chackerian and Peabody 2015; Chackerian and Frietze 2016; El-Turabi and Bachmann 2018). The remarkable advances of the VLP vectors, including the Qβ and PP7 VLPs, by mucosal immunization were reviewed by Vacher et al. (2013).

Concerning the most remarkable infectious disease vaccines, which are listed in Table 25.3, the influenza gH1-Qβ vaccine induced influenza-specific CD4+ and CD8+ T cell responses and a number of influenza-specific cytokines including antiviral IFN-γ, in both nonadjuvanted and adjuvanted formulations (Skibinski et al. 2018).

The atomic 3D structure of the previously mentioned influenza LAH epitope, when it was chemically coupled to the Qβ VLPs, was resolved by magic angle spinning (MAS) solid-state NMR (Jaudzems et al. 2021). Remarkably, the comparison of the MAS NMR fingerprints between the free and VLP-coupled forms of the antigen provided structural evidence of the conservation of its native fold upon bioconjugation to the vaccine form. In fact, this work represented the first example of an atomic-level structural analysis of a VLP-coupled protein.

An interesting effort was attempted to display a complex conformational HIV1 gp120 epitope, namely, the epitope b122a, recognized by the CD4 binding site-directed broadly neutralizing antibody b12, on the Qβ VLPs (Purwar et al. 2018). The artificially designed b122a polypeptide was generated earlier as a recombinant protein carrying a substantive part of the b12 epitope (Bhattacharyya et al. 2013). Then, this recombinant protein, in parallel with some other epitopes, was displayed on the Qβ VLPs by both fusion and coupling technologies, as presented in Tables 25.2 and 25.3.

The Chackerian team recently presented the chemically coupled Qβ VLPs carrying a short malaria epitope from the junctional region of circumsporozoite protein, as listed in Table 25.3, and demonstrated their efficiency in a mouse model (Jelínková et al. 2021).

A prospective approach for developing vaccines against bacterial infectious agents was demonstrated by the Lyme disease vaccine candidate, where outer surface protein CspZ of *Borrelia burgdorferi* was modified by eliminating ability to bind the complement regulator factor H and then coupled to the Qβ VLPs (Marcinkiewicz et al. 2018). This is the first example of the promising strategy by conjugating a bacterial antigen to a VLP and eliminating binding to the target ligand.

The Qβ VLPs were used to display high-mannose glycans that were identified on the HIV-1 envelope glycoprotein, gp120, as a target for neutralizing antibody 2G12, the first HIV-1 antiglycan neutralizing antibody described (Doores et al. 2013). This study contributed markedly to the putative HIV carbohydrate vaccine design strategies. The Qβ VLP-based glycovaccines were reviewed by Restuccia et al. (2016).

A third-generation antiglycane vaccine was generated against two *Streptococcus pneumoniae* pathogenic serotypes, where long-term protective immunity in mice was elicited with exquisite specificity (Polonskaya et al. 2017). Furthermore, the α-Gal trisaccharide epitope was identified on the surface of the protozoan parasites *Leishmania infantum* and *L. amazonensis*, the etiological agents of visceral and cutaneous leishmaniasis, respectively and displayed on the Qβ VLPs (Moura et al. 2017). The putative vaccine demonstrated high efficiency in a C57BL/6 α-galactosyltransferase knockout mouse model and protected animals against *Leishmania* challenge, eliminating the infection and proliferation of parasites in the liver and spleen as probed by qPCR. The α-Gal epitope might therefore be considered as a vaccine candidate to block human cutaneous and visceral leishmaniasis (Moura et al. 2017). A detailed protocol for the conjugation of a prototypical tumor-associated carbohydrate antigen (TACA), the Tn antigen, to the Qβ VLPs was published by Sungsuwan et al. (2017).

Recently, Alam et al. (2021) engineered another glycan-modified Qβ VLPs that evoked Th1 immune responses. It was examined how antigen structure can influence uptake and signaling from the C-type lectin DC-SIGN (dendritic cell-specific intercellular adhesion molecule-3-grabbing nonintegrin or CD209). In fact, the signaling depended on the ligand displayed on the VLP: only those particles densely functionalized with an aryl mannoside, Qβ-Man$_{540}$, elicited DC maturation and induced the expression of the proinflammatory cytokines characteristic of a Th1 response. The mice immunized with a VLP bearing the aryl mannoside and a peptide antigen (Qβ-OVA-Man$_{540}$) had antigen-specific responses, including the production of CD4$^+$ T cells producing the activating cytokines interferon-γ and tumor necrosis factor-α (Alam et al. 2021). Therefore, this study strongly highlighted the utility of the DC-targeted VLPs as vaccine vehicles that induce cellular immunity. It should be noted that the basics of the appropriate methodology were elaborated on the Qβ VLPs by Ribeiro-Viana et al. (2012).

The number of the sophisticated Qβ VLP-based vaccines is permanently growing. Thus, Mohsen et al. (2021) proposed a highly convincing strategy to engineer the Qβ VLPs chemically coupled with multineoantigens for the individualized vaccination against mammary carcinoma.

Shao et al. (2021) offered a new vaccination platform, proposing a novel scalable manufacturing approach to single-dose vaccination against HPV. The HPV16 L2 epitopes were conjugated to the Qβ VLPs that were then encapsulated into poly(lactic-coglycolic acid) (PLGA) implants, using a benchtop melt-processing system. The implants facilitated the slow and sustained release of the HPV-Qβ VLPs without the loss of nanoparticle integrity, during high temperature melt processing. The mice vaccinated with the implants generated IgG titers comparable to the traditional soluble injections and achieved protection in a pseudovirus neutralization assay. Because the melt-processing is so versatile, the technology offered the opportunity for massive upscale into any geometric form factor. Notably, microneedle patches would allow for self-administration in the absence of a healthcare professional, within the developing world (Shao et al. 2021). Ortega-Rivera et al. (2021) developed the idea of preventing cardiovascular diseases by lowering cholesterol levels in plasma and proposed a trivalent vaccine candidate targeting proprotein convertase subtilisin/kexin-9 (PCSK9), apolipoprotein B (ApoB), and cholesteryl ester transfer protein (CETP), where the Qβ VLPs displayed the corresponding antigens and were formulated as slow-release PLGA:VLP implants using hot-melt extrusion, as elaborated by Shao et al. (2021). The delivery of the trivalent vaccine candidate via the implant-produced antibodies against the cholesterol checkpoint proteins at levels comparable to a three-dose injection schedule with soluble mixtures and led to the total plasma cholesterol decrease.

The AP205 VLPs offered the first example of combination of both genetic fusion and chemical coupling for the same particle (Kirsteina et al. 2020). Thus, to generate different broadly protective influenza vaccine candidates, the 72 aa sequence encoding the 3xM2e protein was genetically fused to the C-terminus of the AP205 coat, but HA$_{stem}$ trimer was chemically coupled to the chimeric AP205 VLPs. The combination of both conserved influenza antigens into a single VLP fully protected mice from a high-dose homologous H1N1 influenza infection (Kirsteina et al. 2020). Because of the explicit complexity of this vaccine, it is not included into Table 25.3.

The major problems of immunogenicity and immunodominance in antibody responses against the epitopes displayed on the VLPs were analyzed recently by Vogel and Bachmann (2020), Chackerian and Peabody (2020), and Peabody DS et al. (2021).

PLUG-AND-DISPLAY

The plug-and-display methodology, principles of which were described earlier in the *Chimeric VLPs: Functionalization: AP205* paragraph, led to the generation of an impressive number of prospective vaccine candidates, which are listed in Table 25.4.

The first place in this list belongs, without a doubt, to the putative malaria vaccines (Brune et al. 2016; Janitzek et al. 2016; Thrane et al. 2016; Leneghan et al. 2017; Singh et al. 2017). Generally, the full-length CSP was genetically fused at the C-terminus to SpyCatcher. The CSP-SpyCatcher antigen was then covalently attached, via the SpyTag/SpyCatcher interaction, to the AP205 VLPs, which displayed one SpyTag per VLP subunit. Moreover, the malaria transmission-blocking activity was improved by introduction of the *P. falciparum* protein 48/45 into the SpyTag/SpyCatcher-mediated display on the AP205 VLPs. The potential importance of the AP205 VLPs displaying VAR2CSA epitopes was reviewed by Pehrson et al. (2017). It is noteworthy that the VAR2CSA-based vaccines were targeted against placental malaria appearing during pregnancy, since the *var2csa* gene is expressed on the surface of parasite isolates from placental tissue and by parasites selected to bind to chondroitin sulfate A (CSA).

Furthermore, three platforms were compared directly for the displaying Pfs25 target (Leneghan et al. 2017). These platforms comprised the three important routes to antigen-scaffold linkage: the plug-and-display SpyTag/SpyCatcher, chemical cross-linking, and genetic fusion. Thus, the SpyCatcher-AP205 allowed covalent conjugation of antigen to the VLPs by the formation of a spontaneous isopeptide bond between SpyCatcher and its partner SpyTag, as described earlier. It was highly important that the fusion of SpyCatcher to the VLPs and SpyTag to the antigen allowed heterologous expression of antigen and VLP in the system most suited to each and avoided the typical downfalls of antigen-fusion VLPs such as poor expression levels, poor solubility, and misfolding. The Qβ VLPs were used for the chemical crosslinking. For the

TABLE 25.4

Vaccines Constructed on the AP205 VLPs by Plug-and-Display Methodology

Vaccine Target	Source of Epitope	Epitope Size	Coupling Site	Comments	References
AP205					
Asthma/allergy	IL-5, murine	33 kDa	SpyTag or SpyCatcher: genetically fused to the N-terminus and/or C-terminus	Efficient breaking of self-tolerance in mice.	Thrane et al. 2016
Cancer	Human telomerase reverse transcriptase, the mutant Telo epitope biotin-GAHIVMVDAYKPTRE-ARPALLTSRLRFIPK	30	SpyCatcher genetically fused to the N-terminus	No immunological data.	Brune et al. 2016
	Human epidermal growth factor receptor (EGFR) from glioblastoma, fusion junction epitope LEEKKGNYVVTDHGA-HIVMVDAYKPTK—biotin	27		No immunological data.	
	Murine proteins involved in cancer (CTLA-4, PD-L1, Survivin and HER2)	kDa: 15 27 30 83	SpyTag or SpyCatcher: genetically fused to the N-terminus and/or C-terminus	Efficient breaking of self-tolerance in mice.	Thrane et al. 2016
	Full HER2 extracellular domain including subdomains I-IV, aa 23–652, genetically fused with SpyCatcher, aa 23–139, at the N-terminus—via GGS linker—and His$_6$ at the C-terminus; the SpyTag peptide sequence AHIVMVDAYKPTK was separated from the coat by a flexible linker GSGTAGGGSGS at the N-terminus and GTASGGSGGSG at the C-terminus	656	SpyTag genetically fused to both the N- and C-termini	Induction of anti-HER2 autoantibodies in mice. Reduction of spontaneous development of mammary carcinomas by 50%–100% in human HER2 transgenic mice. Inhibition of HER2-positive tumors implanted in wild-type mice.	Palladini et al. 2018
Cardiovascular disease	PCSK9, murine	84 kDa	SpyTag genetically fused to both the N- and C-termini	No immunological data.	Thrane et al. 2016
Diarrhea, Enterotoxigenic *E. coli* (ETEC)	Human heat-stable toxins STh and STh-A14T, where the SpyTag was fused to the N-terminus	4.3 kDa	SpyCatcher fused to the N-terminus	The sera of immunized mice completely neutralized the toxic activities of native STh.	Govasli et al. 2019
Human immunodeficiency virus (HIV)	The RC1 epitope that facilitates the recognition of the V3-glycan patch on the envelope protein of HIV-1		C-terminal SpyTag sequence (13 residues) was added to the RC1–4 fill epitope	Induction of anti-RC1 antibodies in mice, rabbits, and rhesus macaques.	Escolano et al. 2019
Influenza A virus (IAV)	The HA$_{stem}$ sequence #4900 (Impagliazzo et al. 2015), A/Brisbane/59/2007 (H1N1), fused at the C-terminus with a His$_6$ tag followed by a Glycin-Glycin linker and the SpyTag sequence AHIVMVDAYKPTK	~108 kDa	SpyCatcher fused to the N-terminus	Protection of mice after single dose inoculation	Thrane et al. 2020

(Continued)

TABLE 25.4 (CONTINUED)

Vaccines Constructed on the AP205 VLPs by Plug-and-Display Methodology

Vaccine Target	Source of Epitope	Epitope Size	Coupling Site	Comments	References
Malaria	*Plasmodium falciparum*, membrane protein 1 (PfEMP1) containing the complex lysine and cysteine-rich interdomain region (CIDR).	~27 kDa	SpyCatcher fused to the N-terminus	Induction of antibodies in mice after a single immunization.	Brune et al. 2016
	Plasmodium falciparum, Pfs25 protein	~27 kDa			
	Plasmodium falciparum, Pfs25	40 kDa	SpyTag-Pfs25 to the SpyCatcher-VLPs	Induction of antibodies in mice.	Leneghan et al. 2017
	Plasmodium falciparum, CSP, CIDR, VAR2CSA, and Pfs25 proteins	53, 32, 118, and 40 kDa	SpyTag or SpyCatcher: to the N-terminus and/or C-terminus	Induction of antibodies in mice.	Thrane et al. 2016; Janitzek et al. 2016
	Plasmodium falciparum, Pfs48/45 protein, region 6C. The SpyTag sequence AHIVMVDAYKPTK genetically fused along with flexible linkers GSGTAGGGSGS, N-terminus, and GGSGly, C-terminus. The DNA fragment encoding amino acids 24–139 of the SpyCatcher domain sequence	~130 kDa, ~27 kDa	SpyCatcher-R0.6C and SpyCatcher-6C to the both Spy-tagged N- and C-terminus	Induction of antibodies in mice.	Singh et al. 2017
	Plasmodium falciparum, Pfs48/45 and Pfs230 fusion proteins		SpyTag fused to the N-terminus of AP205	Induction of antibodies in mice.	Singh et al. 2019
	Plasmodium falciparum, Pfs47	58	SpyCatcher fused to N-terminus; His$_6$ to C-terminus	Induction of antibodies in mice. Strong transmission-reducing activity of the IgG.	Yenkoidiok-Douti et al. 2019
Severe acute respiratory syndrome virus 2 (SARS-CoV-2)	Spike protein S, aa 319–591	273 + SpyCatcher at the N- or C-terminus	SpyTag fused to the N-terminus of AP205	High virus-neutralizing response in mice after single immunization.	Fougeroux et al. 2021
Tuberculosis	*Mycobacterium tuberculosis*, Ag85A protein	48 kDa	SpyTag or SpyCatcher: genetically fused to the N-terminus and/or C-terminus	No immunological data.	Thrane et al. 2016

Note: The order of data presentation is the same as in Table 25.2. When the precise number of the amino acid residues within the epitope was difficult to assess, the size of the epitope was expressed in kDa.

genetic fusion, the chosen multimerizing coiled-coil protein IMX313 was a multimerization domain based on a hybrid version of the chicken complement inhibitor C4b-binding protein. It was shown before that the IMX313 when fused to various antigens increased antibody titer and avidity, as well as improved cellular immunogenicity. It was intriguing that the chemically conjugated Qβ VLPs elicited the highest quantity of antibodies, while the SpyCatcher-AP205-VLPs elicited the highest quality

anti-Pfs25 antibodies for transmission blocking upon mosquito feeding (Leneghan et al. 2017).

In parallel, high-density display of human HER2 on the surface of the AP205 VLPs allowed overcoming B cell tolerance and induced high-titer therapeutically potent anti-HER2 IgG antibodies (Palladini et al. 2018). Moreover, prophylactic vaccination with the AP205-HER2 VLPs prevented (i) tumor growth in wild-type FVB mice grafted with mammary carcinoma cells expressing human

HER2 and (ii) spontaneous development of human HER2-positive mammary carcinomas in tolerant transgenic mice. Therefore, the AP205-HER2 VLPs demonstrated definite potential to become a tool for treatment and prevention of HER2-positive cancers, while the AP205 platform appeared as a good choice for development of vaccines against other noncommunicable diseases (Palladini et al. 2018).

Recently, the AP205 plug-and-display methodology led to the promising influenza vaccine (Thrane et al. 2020) and to the two variants of the prospective SARS-CoV-2 vaccine (Fougeroux et al. 2021), as outlined in Table 25.4.

Developing this fruitful novel technique, the dual plug-and-display system was elaborated (Brune et al. 2017). Thus, a dually addressable synthetic nanoparticle was generated by engineering the IMX313 and two orthogonally reactive split proteins. The SpyCatcher protein formed an isopeptide bond with the SpyTag peptide through spontaneous amidation, while the SnoopCatcher formed an isopeptide bond with the SnoopTag peptide through transamidation. The SpyCatcher-IMX-SnoopCatcher provided a modular platform, whereby SpyTag-antigen and SnoopTag-antigen could be multimerized on opposite faces of the particle simply upon mixing.

The *stick*, *click*, and *glue* routes were compared in the exciting review published by Karl D. Brune and Mark Howarth (2018). The authors concentrated on the potential vaccine candidates that used noncovalent assembly methods, (*stick*) unnatural amino acids for bioorthogonal chemistry (*click*) and spontaneous isopeptide bond formation by SpyTag/SpyCatcher (*glue*). The impressive applications of these methods were outlined in detail and critically considered, with particular insight on the novel Tag/Catcher plug-and-display decoration (Brune and Howarth 2018). The *catching a SPY*, or using the SpyCatcher-SpyTag approach, was reviewed by Hatlem et al. (2019) and Aves et al. (2020).

The AP205 platform contributed by the generation of novel protein cages to be employed by the plug-and-display SpyTag/SpyCatcher methodology (Bruun et al. 2018). Thus, the i301 nanocage, which mimicked the structure of VLPs, was based on the 2-keto-3-deoxy-phosphogluconate aldolase from the Entner−Doudoroff pathway of the hyperthermophilic bacterium *Thermotoga maritima*. The i301 had five mutations that altered the interface between the wild-type protein trimer, promoting assembly into a higher-order dodecahedral 60-mer. The novel modular antigen was designed by fusing SpyCatcher to the N-terminus of the protein and compared with the AP205 VLPs for the ruggedness and immunogenicity of the scaffold, when different transmission-blocking and blood-stage malaria antigens were displayed (Bruun et al. 2018).

The AP205 platform was employed by the construction of the first combinatorial vaccine (Janitzek et al. 2019). Since cervical cancer and placental malaria are major public health concerns (for example, in Africa) and the target population for vaccination against both diseases, adolescent girls, would be overlapping, the authors decided to combine both vaccines by displaying the appropriate antigens on the AP205 VLPs by the plug-and-display technique. Therefore, the proof of concept for a combinatorial vaccine was demonstrated by simultaneous display of the two clinically relevant antigens, namely the HPV RG1 epitope and the placental malaria VAR2CSA antigen. The three distinct combinatorial VLPs were produced displaying one, two, or five concatenated RG1 epitopes without obstructing the VLP capacity to form. The codisplay of VAR2CSA was achieved through a split-protein Tag/Catcher interaction without hampering the vaccine stability. Vaccination with the combinatorial vaccines was able to reduce HPV infection in vivo and induced anti-VAR2CSA IgG antibodies, which inhibited binding between native VAR2CSA expressed on infected red blood cells and chondroitin sulfate A in an in vitro binding inhibition assay (Janitzek et al. 2019). This was the first successful attempt to use the AP205 plug-and-display system to make a combinatorial vaccine capable of eliciting antibodies with dual specificity. Furthermore, Janitzek et al. (2021) evaluated the AP205 platform by combination with different adjuvants and using different administration routes, while Fredsgaard et al. (2021) performed the head-to-head comparison of the modular AP205 platform with other VLP backbones and antigen conjugation systems. The SpyTag/SpyCatcher system resulted in significantly higher antigen-specific IgG titers compared to when using affinity-based conjugation (i.e., using biotin/streptavidin). Moreover, the vaccines based on the AP205 VLPs elicited significantly higher antigen-specific IgG compared to corresponding vaccines using the human papillomavirus major capsid protein, namely HPV L1 VLPs, described in Chapter 7. The AP205 VLP platform mediated induction of antigen-specific IgG with a different subclass profile (i.e., higher IgG2a and IgG2b) compared to the HPV L1 VLPs (Fredsgaard et al. 2021).

Generally, the growing success of the MS2, Qβ, and PP7 scaffolds by the generation of the novel prospective HPV vaccines was commented by the recent thorough reviews (Yadav et al. 2019; Huber B et al. 2021), as already mentioned in Chapter 7.

Li X et al. (2021) coupled the polysaccharide antigens to the surface of Qβ or AP205 VLPs using the SpyTag/SpyCatcher system. This approach enabled the uncoupled expression of polysaccharide antigens by protein-glycan coupling technologies and combined the latter with the SpyTag/SpyCatcher platform. The resulting vaccine candidates induced high-titer antibodies against bacterial lipopolysaccharide and had strong prophylactic effects against infection in a mouse model (Li X et al. 2021).

PACKAGING AND TARGETING

COAT-RNA RECOGNITION

The coat dimers of the RNA phages have been shown to recognize an RNA hairpin as an operator at the start site of the replicase gene and to form so-called repressor complex I in order to repress translation of the replicase gene, regulate the replication cycle, and perform the self-genome recognition by virion assembly (for details and references, see Pumpens 2020).

The x-ray crystallography led to a real breakthrough in understanding the protein-RNA interactions that were occurring during the RNA phage translational repression and genome encapsidation. Thus, the first crystal structure of a complex of recombinant MS2 capsids with the 19-nucleotide RNA operator was resolved at 2.7 Å (Valegård et al. 1994b, 1997; Stockley et al. 1995). Then, the crystal structures of the MS2 VLPs complexed with the RNA aptamers, which differed by their secondary structure from wild-type RNA (Convery et al. 1998; Rowsell et al. 1998; Grahn et al. 1999) or involved the presence of 2′-deoxy-2-aminopurine at the critical -10 position of the operator (Horn et al. 2004) were resolved. Further, the structure of the coat protein complexed with the operator RNA fragments were also solved by x-ray crystallography for the phages PP7 (Chao et al. 2008), PRR1 (Persson et al. 2013), and Qβ (Rumnieks and Tars 2014). Although the overall binding mode of the stem-loop to the coat was similar in all the studied cases, the details were surprisingly different among different phages. However, it should be noted here that all attempts to identify analogous coat-RNA interactions in the acinetophage AP205 and caulophage φCb5 failed until now (Kaspars Tārs, unpublished observations), suggesting that mechanisms of genome recognition and translational repression might differ significantly among the distant levivirus members. Figure 25.8 compares the different binding modes of the coat dimer-operator complexes of the phages MS2, Qβ, and PP7.

The electron cryomicroscopy studies have shown that, in addition to the operator, many other RNA sequences in the MS2 genome were able to bind to the coat dimer (Koning et al. 2003). Next, the electron cryomicroscopy structures of the phages Qβ, PP7, and AP205 were solved (van den Worm et al. 2006). Such studies have allowed the 3D visualization of the icosahedrally averaged genomic MS2 RNA at a resolution of 9 Å (Toropova et al. 2008). The direct evidence for the packaging signal-mediated assembly of the MS2 phage was presented when based on cross-linking studies of peptides and oligonucleotides at the interfaces between the capsid proteins and the genomic RNA of this phage (Rolfsson et al. 2016; Stockley et al. 2016). Remarkably, the same coat-RNA and maturation protein-RNA interfaces were identified in every viral particle.

The ability of the RNA phage coats to recognize the specific operator stem-loop—or RNA aptamer—led first to the development of an efficient VLP packaging methodology, including so-called armored RNAs, which will be described later in a special paragraph. Second, the coat-operator complexing led to the great non-VLP tethering and imaging technologies based on the mutual recognition of the coat-fused proteins to the coat-operator-tagged RNAs. Using these techniques, mRNAs that have been tagged with the operator sequence were specifically recognized by coats, which were fused to fluorescent or other functional probes. The tethering technique allowed affinity purification of

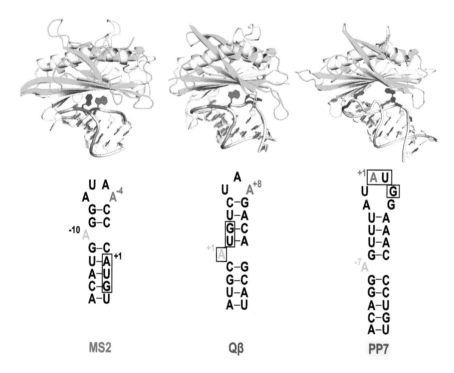

FIGURE 25.8 Different binding modes of operator stem-loop sequences to the coat dimers from MS2, Qβ, and PP7. For the operator stem-loop sequences, adenosine, which is required for the binding of loop nucleotides, is shown in red, while bulged adenosine is shown in blue. The replicase gene initiation codon is boxed. (Reprinted with permission of S. Karger AG, Basel from Pumpens P et al. *Intervirology.* 2016;59:74–110.)

the desired RNA-protein complexes. Moreover, the coat-operator tethering methodology enabled imaging of the processing, export, localization, translation, and degradation of operator-tagged mRNA in a single living cell. The tethering and imaging methodologies are exploiting mostly the coat-operator composition from phage MS2—although from the PP7—and the simultaneous application of both of them as the two-color MS2 and PP7 labeling. Nowadays, the tethering technique is applied actively in the CRISPR-Cas9 methodologies. However, these tethering and imaging approaches involved the chimeric variants of the RNA phage monomers or dimers but not their VLPs (for details and references, see Pumpens 2020) and do not act therefore as a subject of the present narrative.

Development of the Idea

The efficient encapsidation of the desired materials, together with the precise assessment of the VLP innage, are playing the central role by the development of the two major biomedicinal goals of the VLP platforms consisting, in a few words, of the immunological display and targeted drug delivery. As the materials that may need delivery in diagnostics, vaccines, and therapeutic modalities are very different and may include RNA and DNA fragments, proteins, and peptides, as well as very different low-molecular-mass molecules and inorganic nanomaterials, the requirements for the packaging are eminently wide-ranging.

The packaging idea developed from the in vitro reconstitution of RNA phages from their structural subunits, which was rather theoretical and met the challenge to produce infectious phages in a test tube and repeat and/or model therefore the natural process of life. Nowadays, as we know, the packaging is one of the most popular techniques of the VLP nanotechnology, which serves purely practical claims.

Thomas and Barbara Hohn provided the first reconstitution attempts, with a solid theoretical and experimental background, which was exhaustively presented in their comprehensive and well-illustrated review (Hohn and Hohn 1970). They were probably the first who presented systematically the ability of different nonspecific polyribonucleotides, such as plant virus TMV and TYMV RNA and even polyuridylic acid and polyvinyl sulfate, to be encapsulated into the RNA phage VLPs (Hohn 1969a; Hohn and Hohn 1970). Moreover, the TMV and TYMV RNA could be shared by two growing capsids so that a high proportion of "twinned" and "tailed" particles was synthesized (Hohn and Hohn 1970).

Nowadays, Rolfsson et al. (2010) assayed the ability of the MS2 coat to package large, defined fragments of its own genomic RNA. Thus, the efficiency of packaging into a T = 3 capsid in vitro was inversely proportional to the RNA length, implying that there was a free-energy barrier to be overcome during assembly. All the RNAs examined had greater solution persistence lengths than the internal diameter of the capsid into which they became packaged, suggesting that the protein-mediated RNA compaction must occur

during assembly. The electron cryomicroscopy structures of the capsids assembled in these experiments with the subgenomic RNAs showed a layer of RNA density beneath the coat protein shell but lacked density for the inner RNA shell seen in the wild-type virion. The inner layer was restored when full-length virion RNA was used in the assembly reaction, implying that it became ordered only when the capsid was filled, presumably because of the effects of steric and/or electrostatic repulsions. The data was consistent with mutual chaperoning of both RNA and coat protein conformations, partially explaining the ability of such viruses to assemble so rapidly and accurately (Rolfsson et al. 2010).

It is highly important, however, to remember that the early studies observed the assembly of the RNA phage VLPs in the absence of any RNA in the case of the phages fr (Herrmann et al. 1968; Hohn 1969 b; Schubert and Frank 1970 a, b, 1971; Zipper et al. 1971, 1973), f2 (Matthews 1970; Zelazo and Haschemeyer 1970; Matthews and Cole 1972 a, b, c), R17 (Samuelson and Kaesberg 1970), and MS2 (Oriel and Cleveland 1970; Rohrmann and Krueger 1970).

David S. Peabody was the first who started the practical application of the RNA operator (or aptamer) properties for the putative RNA delivery. Thus, the ability of the MS2 operator to stimulate encapsidation of the desired RNAs in vivo demonstrated for the first time the real potential of the RNA phages as potential gene delivery vectors (Pickett and Peabody 1993). In parallel, the Qβ coat residues responsible for RNA recognition were mapped (Lim et al. 1996). This was extremely helpful for the knowledge-based development of the packaging and gene transfer technologies based on the popular Qβ phage model.

Peter Stockley's team was the first to announce the cell-specific delivery by the MS2 VLPs encapsidated with the desired drug. They achieved encapsulation of the deglycosylated ricin A chain coupled to the RNA operator stem-loop and decoration of the MS2 VLPs by transferrin (Wu M et al. 1995, 1997). Such structures, called *synthetic virions* by the authors, demonstrated high toxicity to leukemia cells carrying transferrin receptor. The targeted delivery strategy was further developed by using soluble recombinant human CD4 molecules or anti-HIV antibodies as ligands (Wu M et al. 1996). Generally, this was the first example showing that the encapsidation and targeting could be an efficient way to deliver toxins and other molecules, such as antisense reagents, in a cell-specific fashion. A detailed review of these pioneering experiments was published at that time (Brown WL et al. 2002).

The first review declaring the MS2 VLPs as an element of synthetic biology in the nanotechnology field and presenting the pioneering data of the Matthew B. Francis team at that time was published by Philip Ball (2005).

Cohen and Bergkvist (2013) were the first who performed the photodynamic therapy in vitro by the porphyrin-loaded MS2 capsids that were targeted to the cancer cells by a DNA aptamer exposed on the capsid surface.

Table 25.5 summarizes results on the RNA phage VLPs as packaging, or nanocontainer, models, together with their

TABLE 25.5

RNA Phage VLPs as Models for Nanocontainer Packaging and Decoration with Addressing/Targeting/Delivery Purposes

Interior: Packaged by	Exterior: Decorated by	Addressed to	Supposed to Treat	Comments	References
AP205					
Random *E. coli* RNA, without any specific packaging	HBV, full length preS1, aa 13–119	Eukaryotic cell lines HepG2, Hek293, Jurkat, Namalwa, and BHK-21	HBV infection	The uptake took place but seemed as the cell-line unspecific and not highly efficient.	Kalniņš et al. 2013
fr					
No specific packaging	IgG-binding Z domain at the C-terminus of the coat coexpressed with native coat as a helper	Rather unspecific targeting	Potentially broad applications in diagnostics	This is a first example of the generation of mosaic VLPs carrying the IgG-binding Z domain that could be targeted to antibodies displayed on the cell surface and used in diagnostics.	Juris Jansons and Irina Sominskaya, unpublished
GA					
GA operator-specified mRNA in vivo	No decoration	No cell targeting	Broad spectrum of potential targets	Packaging of different mRNAs encoding GA coat protein, ENA-78, and GFP in vivo in yeast *Saccharomyces cerevisiae*.	Rūmnieks et al. 2008
	Cell-penetrating peptide TAT, HIV-1, 48–60, or WNV E protein DIII domain	Human peripheral blood mononuclear cells (PBMCs)	PBMC-derived failure	Production of mosaic VLPs that are packaged by mRNAs encoding IL2 or GFP was achieved in vivo in yeast *Saccharomyces cerevisiae*.	Strods et al. 2012
MS2 operator-specified mRNA in vivo	No decoration	No cell targeting	Broad spectrum of potential targets	Introduction of single (S87N, K55N, R43K) and double (S87N+K55N and S87N+R43K) aa exchanges into the GA coat allowed self-assembly and packaging of MS2 operator-carrying mRNAs in vivo in yeast *Saccharomyces cerevisiae*.	Ārgule et al. 2017
No specific packaging	Z$_{HER2:342}$ affibody, 61 aa in length	HER2 receptor on cancer cells	Diagnostics/ therapy	Ability of GA-$_{ZHER2:342}$—VLPs to recognize selectively and enter the HER2 receptor-bearing cells was demonstrated. Electron microscopy of this construction is presented in Figure 25.6.	Arnis Strods and Regīna Renhofa, unpublished
	Stromal cell-derived factor (SDF1), aa 1–41 and 1–19	CXCR4 receptor on leukocytes	Diagnostics/ therapy	Insertions at the N-terminus of the first or second single-chain dimer copy were performed. Some members of the family recognized specific cells.	
MS2					
Ricin A coupled to RNA operator; 5-fluorouridine coupled to RNA operator	Transferrin, anti-MS2 antibodies, anti-DF3 antibodies	Cells of the immune system; breast carcinoma and leukemia cells	Broad spectrum of potential targets	Classical attempt to use RNA operator as a carrier of the desired material.	Wu M et al. 1995, 1997

Interior: Packaged by	Exterior: Decorated by	Addressed to	Supposed to Treat	Comments	References
Antisense ODNs targeted to human nucleolar protein p120 mRNA	Covalent decoration with transferrin on the VLP surface	Promyelocytic leukemia cell line	Acute myelogenous leukemia	The ODNs were synthesized as covalent extensions to the 19-nt-long operator sequence.	Wu M et al. 2005
50–70 Fluorescein molecules conjugated to the interior of VLPs	180 PEG-2000 or PEG-5000 chains on the exterior of VLPs	Tumor cells	Solid tumors	An early attempt to construct an addressed nanocontainer for the potential delivery of therapeutic cargo.	Kovacs et al. 2007
Fluorescent dye conjugated to the VLP interior	A 41-nt operator-containing sequence covalently bound to the VLP surface	Tyrosine kinase receptor on the Jurkat T cells	Leukemia	Colocalization experiments using confocal microscopy indicated that the operator-labeled capsids were endocytosed and trafficked to lysosomes for degradation that could allow the targeted drug delivery of acid-labile prodrug.	Tong et al. 2009
Antisense RNA against the 5'- UTR and IRES of HCV	Cell-penetrating peptide TAT, HIV-1, 47-YGRKKRR-QRRR-57	Huh-7 cells containing an HCV reporter system	Chronic hepatitis C	The TAT peptide was conjugated chemically to VLPs. The packaged antisense RNA showed an inhibitory effect on the translation of HCV genome.	Wei B et al. 2009
Fluorescent dyes as donor chromophores	Zinc porphyrins capable of electron transfer	No cell targeting	No definite target disease specified	Specific positioning allowed energy transfer and sensitization of the porphyrin at previously unusable wavelengths, as demonstrated by the system's ability to effect a photocatalytic reduction reaction at multiple excitation wavelengths.	Stephanopoulos et al. 2009
Coupling of 180 porphyrins capable of generating cytotoxic singlet oxygen upon illumination	~20 copies of a Jurkat-specific aptamer	Jurkat leukemia T cells	Leukemia	The doubly modified VLPs were able to target and kill Jurkat cells selectively even when mixed with erythrocytes.	Stephanopoulos et al. 2010b
Porphyrin: *meso*-tetra-(4-N,N,N,-trimethylanilinium)-porphine-tetrachloride, or TMAP. ~ 250 TMAP were loaded by nucleotide-driven packaging	Cancer cells-targeting DNA aptamers via chemical conjugation	MCF-7 human breast cancer cells	Breast cancer upon photoactivation	TMAP interacted with the RNA present within the MS2 capsid. Removal of the interior RNA led to significantly lower porphyrin interaction with the MS2 capsid. The MCF-7 cells incubated with targeted, porphyrin-loaded virus capsids exhibited cell death. The strategy offered an approach for efficient targeted delivery of photoactive compounds for site-specific photodynamic cancer therapy.	Cohen et al. 2011, 2012; Cohen and Bergkvist 2013
Porphyrin	Angiopep-2	Midbrain	Intractable brain conditions, such as tinnitus	Angiopep-2 facilitates the transport of the MS2 VLPs across the blood-brain-barrier.	Apawu et al. 2018; Cacace et al. 2018
Nanoparticles, chemotherapeutic drugs, siRNA cocktails, and protein toxins	SP94 peptide that binds human hepatocellular carcinoma (HCC) cells	Hep3B cells	HCC	The targeted VLPs loaded with doxorubicin, cisplatin, and 5-fluorouracil selectively killed the HCC cell line, Hep3B. Encapsidation of a siRNA cocktail induced growth arrest and apoptosis of Hep3B cells. Loading of VLPs with ricin toxin A-chain killed the entire population of Hep3B cells.	Ashley et al. 2011

(Continued)

TABLE 25.5 (CONTINUED)

RNA Phage VLPs as Models for Nanocontainer Packaging and Decoration with Addressing/Targeting/Delivery Purposes

Interior: Packaged by	Exterior: Decorated by	Addressed to	Supposed to Treat	Comments	References
MicroRNA: pre-miR 146a	Cell-penetrating peptide TAT, HIV-1, 47-YGRKKR-RQRRR-57	Human peripheral blood mononuclear cells (PBMCs)	Systemic lupus erythematosus, osteoclastogenesis	The TAT peptide was conjugated chemically to VLPs. Restoring the loss of miR-146a was effective in eliminating the production of autoantibodies.	Pan Yang et al. 2012 a, b; Yao Y et al. 2015
RNA conjugate encompassing a siRNA and the operator sequence	Covalent attachment of human transferrin	HeLa cells	Potentially broad applications	The VLPs entered cells via receptor-mediated endocytosis and produced siRNA effects better than by traditional lipid transfection route.	Galaway and Stockley 2013
Gold nanoparticles	Alexa Fluor 488 (AF 488) labeled DNA strands	No cell targeting	No definite target disease specified	The VLP architecture by placing the dye at distances of 3, 12, and 24 bp from the surface of VLPs bearing 10-mm gold nanoparticles allowed the rapid exploration of many variables involved in metal-controlled fluorescence.	Capehart et al. 2013
Various reporters to be used by fluorescence-based flow cytometry, confocal microscopy, and mass cytometry	Antibodies using a rapid oxidative coupling strategy	Receptors on human breast cancer cell lines	Breast cancer	The broad set of conjugates with various reporters on the interior of VLPs may lead to many clinically relevant applications, including drug delivery and in vivo diagnostics.	ElSohly et al. 2015
^{64}Cu labeling	Anti-EGFR antibodies	Receptors on tumor xenografts in mice	Breast cancer	The targeting antibodies did not lead to increased uptake in vivo despite in vitro enhancements.	Aanei et al. 2016; Aanei and Francis 2018
Proteins with tags comprising anionic amino acids or DNA and gold nanoparticles with negative surface charges	No decoration	No cell targeting	No definite target disease specified	The possible application in biocatalysis, protein stabilization, vaccine development, and drug delivery.	Aanei et al. 2018a
AlexaFluor680	Antibodies specific to VCAM	VCAM1, vascular cell adhesion molecule	Atherosclerosis	The VCAM-targeted MS2 VLPs were used for the detection of plaques in the early stages of atherosclerosis development.	Aanei et al. 2018b
Doxorubicin	No decoration	Convection-enhanced delivery	Glioblastoma	The MS2 VLPs conjugated to doxorubicin, as well as tobacco mosaic virus disks, showed the best response in mice, in comparison with phage filamentous rods.	Finbloom et al. 2018
Long noncoding RNA (lncRNA): MEG3 RNA	GE11, a dodecapeptide YHWYGYTPQNVI, ligand of epidermal growth factor receptor (EGRF), chemically coupled	EGFR receptors	HCC cell line	The targeted delivery was dependent on clathrin-mediated endocytosis and MEG3 RNA suppressed tumor growth mainly via increasing the expression of p53 and its downstream gene GDF15, but decreasing the expression of MDM2.	Chang L et al. 2016

Interior: Packaged by	Exterior: Decorated by	Addressed to	Supposed to Treat	Comments	References
MicroRNA: miR-122	Cell-penetrating peptide TAT, HIV-1, 47-YGRKKR-RQRRR-57	Hep3B, HepG2, and Huh7 cells and Hep3B related animal models	Hepatocellular carcinoma	The TAT peptide was displayed on the MS2 VLPs by genetic fusion, instead of being chemically crosslinked. The MS2 VLPs displaying TAT penetrated effectively the cytomembrane and delivered miR-122. The inhibitory effect was shown on the hepatocellular carcinoma model cells.	Wang Guojing et al. 2016
Thallium^I ions from TlNO$_3$	iRGD peptide	Endotheliocytes of the tumor tissue neovasculature and certain tumor cells	Breast cancer	iRGD peptide-conjugated MS2 VLPs filled with Tl$^+$ caused cell death in two types of cultivated human breast cancer cells and effected necrosis of these tumor xenografts in mice.	Kolesanova et al. 2019
LacZ RNA fused to RNA operator	No decoration	No cell targeting	Potentially broad applications	This is a classical attempt to ensure operator-specific packaging in vivo.	Pickett and Peabody 1993
MS2 operator sequence linked to the human growth hormone (hGH) mRNA for in vivo packaging in *S. cerevisiae*	No decoration	No cell targeting	No definite target disease specified	This is a sort of application belonging to the "armored RNA" technology. Functionality of packaged mRNA was confirmed by translation of mRNAs purified from VLPs.	Legendre and Fastrez 2005
Taxol	No decoration	No cell targeting	Breast, lung, and ovarian cancers	Aa exchange N87C on the interior of VLP allowed use of sulfhydryl groups for the attachment of taxol, a potent chemotherapeutic, as a cargo.	Wu W et al. 2009
HIV-1 gag mRNAs (1544 bases) produced in *S. cerevisiae*	No decoration	No cell targeting	HIV/AIDS: as a vaccine	The HIV antigen-specific antibody responses were elicited by immunization of Balb/C mice.	Sun S et al. 2011
Alkaline phosphatase tagged with a 16 aa peptide	No decoration	No cell targeting	No definite target disease specified	This is a first attempt to encapsulate enzymes where the encapsulated enzyme had the same K(m) value and a slightly lower k(cat) value than the free enzyme.	Glasgow et al. 2012
mRNA joined to 19-nt operator/aptamer	No decoration	Macrophages	Prostate cancer	The packaged mRNA ensured strong humoral and cellular immune responses and protected mice as a therapeutic vaccine against prostate cancer.	Li J et al. 2014
No specific packaging	Tumor cell-specific peptides	Tumor cells	Solid tumors	Three peptides known to target specific tissues: (i) neuroblastoma and breast cancer cell lines, (ii) matrix metalloproteinases, (iii) kidney were chosen as attachment models.	Carrico et al. 2008
No specific packaging	Conjugation of azide- and alkyne-containing proteins (an antibody fragment and the granulocyte-macrophage colony stimulating factor), nucleic acids, and PEG	Broad spectrum of potential targeting	Tumors and other possible targets	This is a universal approach based on the inclusion of surface exposed methionine analogues (azidohomoalanine and homopropargylglycine) containing azide and alkyne side chains by cell-free synthesis technology.	Patel and Swartz 2011

(Continued)

TABLE 25.5 (CONTINUED)

RNA Phage VLPs as Models for Nanocontainer Packaging and Decoration with Addressing/Targeting/Delivery Purposes

Interior: Packaged by	Exterior: Decorated by	Addressed to	Supposed to Treat	Comments	References
No specific packaging	PEG, small peptides	Broad spectrum of potential targeting	Tumors and other possible targets	The oxidative coupling to tyrosine and tryptophan residues using cerium(IV) ammonium nitrate as an oxidant. Six MS2 viral capsids were generated with tryptophan and tyrosine residues on the exterior surface of each monomer. Two of these capsids were also expressed with a cysteine on the interior surface.	Seim et al. 2011
No specific packaging	PEG, small peptides	Broad spectrum of potential targeting	Tumors and other possible targets	A substantially more efficient version of addition of anilines to *o*-aminophenols.	Behrens et al. 2011
No specific packaging	Epidermal growth factor	Variety of cancer cell types	Tumors and other possible targets	The *o*-methoxyphenols were identified as air-stable, commercially available derivatives that undergo efficient oxidative couplings with anilines in the presence of periodate as oxidant.	ElSohly et al. 2017
PP7					
mRNA encoding GFP	Low molecular weight protamine VSRRRRRRGGRRRR	Cell-penetrating ability to mouse prostate cancer cells	Demonstration of principle	The chimeric PP7 VLPs carrying GFP mRNA penetrated the mouse prostate cancer cells RM-1 after 24 h incubation due to the displayed cell-penetrating peptide.	Sun Yanli et al. 2016
MicroRNA: pre-miR-23b	Cell-penetrating peptide TAT, HIV-1, 47-YGRKKRR-QRRR-57	Hepatoma SK-HEP-1 cells	Demonstration of principle	The TAT peptide was displayed on the PP7 VLPs by the insertion into the AB loop of the single-chain PP7 coat dimer.	Sun Yanli et al. 2017
Cytosine deaminase with Rev peptide	ZZ domain OVA1 OVA2		Potential applications in medicine and chemical manufacturing	All of the Z-domain-bearing particles recognized antibody Fc domain. An average of 15–25 cytosine deaminase molecules were packaged.	Zhao L et al. 2019
Qβ					
CpG oligodeoxynucleotides (ODNs)	With and without specific addressing	Cells of immune system	Potentially broad applications	This is a classic example of the ODN encapsulation for the broad clinical applications as prophylactic and/or therapeutic vaccines.	Bachmann et al. 2004b, 2005a; Storni et al. 2004; Schwarz K et al. 2005; Agnellini et al. 2008; Bessa et al. 2008; Senti et al. 2009; Keller et al. 2010a, b; Klimek et al. 2011, 2013b; Beeh et al. 2013; Casale et al. 2015

Interior: Packaged by	Exterior: Decorated by	Addressed to	Supposed to Treat	Comments	References
~60 Alexa Fluor 568 fluorophores per VLP	Fullerene C_{60} and PEG	HeLa cell line	Cancer	This approach overcame the insolubility of C_{60} in water and opened the door for the applications in photoactivated tumor therapy.	Steinmetz et al. 2009
Metalloporphyrin derivative for photodynamic therapy	Glycan ligand for specific targeting of cells bearing the CD22 receptor	CD22 receptor	Cancer	This approach benefited from the presence of the strong targeting function and the delivery of a high local concentration of singlet oxygen-generating payload.	Rhee et al. 2011, 2012
Aptamers embedded in a longer RNA sequence with the Qβ coat operator	No decoration	No cell targeting	Potentially broad applications in therapy	The VLPs ensured the delivery of the encapsulated aptamers that were protected from degradation and retained ability to bind their small-molecule ligands.	Lau et al. 2011; Wu Z et al. 2011
Positively charged synthetic polymer by atom transfer radical polymerization (ATRP) methodology	No decoration	No cell targeting	Potentially broad applications in therapy and imaging diagnostics	This is a robust method for removing encapsidated RNA from VLPs and the use of the empty interior space for site-specific, "graft-from" ATRP reactions.	Hovlid et al. 2014
The Rev tagged proteins: 25-kDa N-terminal aspartate dipeptidase, peptidase E, 62-kDa firefly luciferase (Luc), a thermostable mutant of Luc	No decoration	No cell targeting	No definite target disease specified	The encapsulated enzymes were stabilized against thermal degradation, protease attack, and hydrophobic adsorption.	Fiedler et al. 2010
The Rev tagged proteins: 2-deoxyribose-5-phosphate aldolase, superoxide dismutase, cytosine deaminase, purine-nucleoside phosphorylase	ZZ domain; epidermal growth factor EGF; peptide GE7, which binds the EGF receptor; peptide F56, which binds the subtype 1 of the vascular EGF receptor	No cell targeting	Potential applications in medicine and chemical manufacturing	The captured enzymes were active while inside the nanoparticle shell and were protected from environmental conditions that led to free-enzyme destruction.	Fiedler et al. 2018
Universal RNAi scaffold that could target any desired mRNA	No decoration	Specific messengers	Potentially broad applications in therapy	The dose- and time-dependent inhibition of GFP expression in human cells and of Pan-Ras expression in brain tumor cells.	Fang et al. 2017, 2018
Doxorubicin	Gold nanoparticles	No cell targeting	Cancer	The highly selective drug release and cell killing of macrophage and cancer cells in vitro exclusively within the laser path while cells outside the path, even though they were in the same culture, showed no drug release or death.	Benjamin et al. 2018
EGFP, enhanced green fluorescent protein, or CD, cytosine deaminase	IgG-binding ZZ domains	Undifferentiated cells	Cancer	After labeling with antibodies against the hPSC-specific surface glycan SSEA-5, the EGFP-containing particles were shown to specifically bind undifferentiated cells in culture, and the CD-containing particles were able to eliminate undifferentiated hPSCs with virtually no cytotoxics to differentiated cells upon treatment with the prodrug 5-fluorocytosine.	Rampoldi et al. 2018

(Continued)

TABLE 25.5 (CONTINUED)

RNA Phage VLPs as Models for Nanocontainer Packaging and Decoration with Addressing/Targeting/Delivery Purposes

Interior: Packaged by	Exterior: Decorated by	Addressed to	Supposed to Treat	Comments	References
RNAi; luciferase mRNA	Cell-penetrating peptide and apolipoprotein E peptide	Cells of murine glioblastoma models	Brain tumors: glioblastoma	The modified Qβ VLPs crossed blood–brain barrier and acted synergistically with temozolomide for promoting clinical chemotherapy.	Pang et al. 2019
No specific packaging	Copper^I-catalyzed azide-alkyne cycloaddition reaction	Variety of cells	Potentially broad applications	Bioconjugations of Qβ VLPs derivatized with azide moieties at surface lysine residues was performed. Complete derivatization of more than 600 reactive sites per particle was achieved.	Hong V et al. 2008
No specific packaging	Glycans used as substrates for glycosyltransferase reactions to build di- and trisaccharides	Cognate receptors on the appropriate beads and cells	Potentially broad applications	The elaborated methodology provided a convenient and powerful way to prepare complex carbohydrate ligands for clustered receptors.	Kaltgrad et al. 2008; Kussrow et al. 2009
No specific packaging	Cationic aa motifs by genetic engineering or chemical conjugation	Inhibiting the anticoagulant action of heparin	Clinical overdose of heparin	The polycationic motifs displayed on the mutated Qβ VLPs acted as heparin antagonists.	Udit et al. 2009; Udit 2013; Gale et al. 2011; Cheong et al. 2017
No specific packaging	Cationic aa motifs by genetic engineering	Inhibiting the anticoagulant action of heparin	Clinical overdose of heparin	The two-plasmid system was employed. The highest incorporation rate and best antiheparin activity was achieved by the mosaic particle that displayed the C-terminal peptide ARK$_2$A$_2$KA.	Choi et al. 2018
No specific packaging	IgG-binding Z domain at the C-terminus of the coat coexpressed with native coat as a helper	Rather unspecific targeting	Potentially broad applications in diagnostics	This presents generation of mosaic VLPs carrying the IgG-binding Z domain that could be targeted to antibodies displayed on the cell surface or used in diagnostics.	Brown SD et al. 2009
No specific packaging	Oligomannosides that are modeling the "glycan shield" of HIV envelope	No cell targeting	HIV/AIDS as a model	The oligomannose clusters were recognized by monoclonal anti-HIV antibody but did not induce antibodies against the HIV epitopes by immunization of rabbits.	Astronomo et al. 2010
No specific packaging	Transferrin	Transferrin receptors	Potentially broad applications in diagnostics and therapy	This approach allowed cellular internalization of chimeric VLPs through clathrin-mediated endocytosis.	Banerjee D et al. 2010
No specific packaging	DNA	No cell targeting	Potentially broad applications in nanotechnology	A noncompact lattice was created by DNA-programmed crystallization using surface-modified Qβ VLPs and gold nanoparticles, engineered to have similar effective radii.	Cigler et al. 2010

Interior: Packaged by	Exterior: Decorated by	Addressed to	Supposed to Treat	Comments	References
No specific packaging	Poly(2-oxazoline)	No cell targeting	Potentially broad applications in diagnostics and therapy	This showed that the size and content of VLP–polymer constructs could be controlled by changing polymer chain length and attachment density. The system is universal because of the convenient click chemistry applications.	Manzenrieder et al. 2011
No specific packaging	Conjugation of azide- and alkyne-containing proteins	Broad spectrum of potential targeting	Tumors and other possible targets	See previous mention; the same for MS2.	Patel and Swartz 2011
No specific packaging	Modification via the copper-catalyzed azide-alkyne cycloaddition reaction	No definite cell targeting specified	Potentially broad applications in therapy and imaging diagnostics	This is a methodology to use VLPs as multivalent macroinitiators for atom transfer radical polymerization (ATRP)	Pokorski et al. 2011a
No specific packaging	Human epidermal growth factor (EGF) as a C-terminal fusion to the Qβ coat	EGF receptor	Therapy based on interactions with cells over-expressing their cognate receptor	Mosaic particles with an approximately 5–12 EGF molecules on the VLP surface were obtained. The particles were found to be amenable to bioconjugation by standard methods as well as the high-fidelity copper-catalyzed azide-alkyne cycloaddition reaction (CuAAC).	Pokorski et al. 2011b
No specific packaging	Sulfate groups that elicit heparin-like anticoagulant activity	No cell targeting	Blood coagulation—as heparin-like drugs	Following conversion of VLP surface lysine groups to alkynes, the sulfated ligands were attached to the VLP via copper-catalyzed azide-alkyne cycloaddition (CuAAC). Three to six attachment points per coat monomer were modified via CuAAC. The sulfated VLPs were able to perturb coagulation.	Mead et al. 2014; Groner et al. 2015
No specific packaging	Polynorbornene block copolymers	No cell targeting	Potentially broad applications in therapy and imaging diagnostics	Poly(norbornene-PEG)-b-poly(norbornene anhydride) of three molecular mass: 5, 10, and 15 kDa were added to lysine residues.	Isarov et al. 2016
No specific packaging	Poly(ethylene glycol)	No cell targeting	Potentially broad applications in therapy and imaging diagnostics	The polymer conformation of poly-(ethylene glycol) was compared with those of water-soluble polyacrylate and polynorbornene when attached to the Qβ VLPs.	Lee PW et al. 2017a
No specific packaging	Poly(lactic-*co*-glycolic acid), or PLGA	No cell targeting	Potentially broad applications in therapy and imaging diagnostics	The melt-encapsulation was found to be an effective method to produce composite materials that can deliver viral nanoparticles over an extended period and elicit an immune response comparable to typical administration schedules.	Lee PW et al. 2017b, c
No specific packaging	Low molecular mass inhibitors of the GCPII	Glutamate carboxypeptidase II or GCPII	Recognition, imaging, and delivery of treatments to prostate cancer cells	The GCPII is a membrane protease overexpressed by prostate cancer cells and detected in the neovasculature of most solid tumors.	Neburkova et al. 2018

(Continued)

TABLE 25.5 (CONTINUED)

RNA Phage VLPs as Models for Nanocontainer Packaging and Decoration with Addressing/Targeting/Delivery Purposes

Interior: Packaged by	Exterior: Decorated by	Addressed to	Supposed to Treat	Comments	References
φCb5					
Gold nanoparticles, tRNA, diphteria toxin, mRNA, CpG	No decoration	No cell targeting	Potentially broad applications in therapy	The ease with which the φCb5 coat dimers can be purified and reassembled into VLPs makes them attractive for the internal packaging of nanomaterials and the chemical coupling of peptides.	Freivalds et al. 2014

Note: The data is presented in alphabetical order of the phage model. For each model, the studies are ordered chronologically but presenting first the studies describing both packaging and decoration, then data with packaging only, and then data with decoration only.

decoration with addressing/targeting/delivery tools, because of tight overlap of both the packaging and targeting goals.

HETEROLOGOUS RNAS

Legendre and Fastrez (2005) produced the MS2 VLPs in *S. cerevisiae* with an aim to package functional heterologous mRNAs containing the MS2 packaging sequence—or specific aptamer—and to adapt the MS2 VLPs therefore to the gene delivery purposes. For instance, linking the MS2 aptamer to the human growth hormone mRNA enabled the packaging of this particular mRNA into the MS2 VLPs (Legendre and Fastrez 2005). A similar approach was employed by the expression of the phage GA VLPs and packaging of the functional heterologous mRNA in vivo during expression in *S. cerevisiae* by Regīna Renhofa's group (Rūmnieks et al. 2008; Strods et al. 2012; Ārgule 2013; Strods 2015; Ārgule et al. 2017).

By the previously presented prospective studies on the immunogenic display of foreign epitopes on the MS2 VLPs, Peabody and Chackerian's team found that the covalent single-chain MS2 coat dimers not only tolerated the six-, eight-, and ten-amino acid insertions but also encapsidated the mRNAs that directed their synthesis, thus establishing the genotype/phenotype linkage necessary for the recovery of affinity-selected sequences (Peabody DS et al. 2008).

The specific siRNA was delivered by the MS2 VLPs to hepatocellular carcinoma cells (Ashley et al. 2011). In fact, it was demonstrated that the MS2 VLPs could be conjugated to peptides recognizing human hepatocellular carcinoma cells and could be loaded with vastly different types of cargo, including low molecular mass chemotherapeutic drugs, siRNA cocktails, protein toxins, and nanoparticles, resulting in the selective killing of target cells.

The MS2 VLPs were used to deliver the packaged siRNA that was targeted against the antiapoptotic factor *BCL2* into HeLa cells (Galaway and Stockley 2013). The gold-nanoshell encapsulated quantum dots were synthesized by attaching single quantum dots to hairpin RNA that directed the formation of the MS2 VLPs (Miao et al. 2010).

The Finn's team embedded a high affinity RNA aptamer against a heteroaryldihydropyrimidine structure, which represented a drug-like molecule with no cross-reactivity with mammalian or bacterial cells, as well as an aptamer against theophylline, in a longer RNA sequence that was encapsidated inside the Qβ VLPs (Lau et al. 2011). The general use of the Qβ VLP-protected aptamers for the delivery of quantum dots, gold, silver, and silica nanoparticles was reviewed by Wu Z et al. (2011).

Fang et al. (2017) demonstrated a method for production of a novel RNAi scaffold, packaged within the Qβ VLPs. The RNAi scaffold contained a functional RNA duplex with paired silencing and carrier sequences stabilized by a miR-30 stem-loop. The Qβ RNA hairpin on the 5′ end conferred affinity for the Qβ coat protein. The silencing sequences could include mature miRNAs and siRNAs and could target essentially any desired mRNA. The VLP-RNAi assembled upon coexpression of the Qβ coat and the RNAi scaffold in *E. coli*. It appeared further that the Qβ hairpin was necessary but not sufficient for the efficient packaging (Fang et al. 2018).

An extensive investigation on the microRNA delivery was published (Wang Guojing et al. 2016). This study was based on the MS2 VLPs displaying the cell-penetrating peptide TAT and aimed to treat hepatocellular carcinoma. Previously, this highly prominent team developed the MS2 VLP-based microRNA-146a delivery system when the TAT peptide was conjugated chemically to the VLPs (Wei B et al. 2009). Then, the MS2-mediated RNAi delivery was carried out in vivo when MS2 VLPs were packaged with a pre-miRNA, pre-miR-146a, and surface-decorated with HIV Tat peptides for cell penetration, while no cell-targeting ligands were used (Pan Yang et al. 2012a, b). These MS2-miR-146a VLPs effectively eliminated autoantibodies, thus ameliorating systemic lupus erythematosus progression in lupus-prone mice (Pan Yang et al. 2012b) and suppressed osteoclast differentiation (Pan Yang et al. 2012a). Upon intravenous delivery to mice, these VLPs displayed widespread biodistribution in the plasma, lung, spleen, and kidney, where high levels of mature miR-146a were detected and knockdown of known targets of miR-146a was observed. Furthermore, the MS2 VLPs were used to deliver mRNA vaccines, which successfully protected mice from prostate cancer (Li J et al. 2014). The general potential of the targeted RNAi-based anticancer therapy including the MS2 VLP-based examples were reviewed at that time by Yan et al. (2014).

The MS2 VLPs were used to encapsulate the long non-coding MEG3 lncRNA and were crosslinked with the GE11 polypeptide, proving beneficial to cancer therapy (Chang L et al. 2016). Nevertheless, the technical problems of cross-linking provoked the team to switch to the phage display technique (Wang Guojing et al. 2016). This was the first study in which the cell-penetrating peptides were displayed on the MS2 VLP surface by genetic fusion. Therefore, the MS2 VLP-based microRNA delivery system was combined for the first time with the phage surface display platform. It is necessary to mention that the HIV TAT peptide was the first reported cell penetrating peptide, since the TAT region 47–57 functioned as the clear protein transduction domain and possessed efficient transport capacity. Therefore, the TAT 47–57 peptide was displayed on the MS2 VLPs by genetic fusion instead of chemical crosslinking. As a result, the TAT peptide was connected to the N-terminus of the single-chain dimer of the MS2 coat, whereas the pre-miR122 was designed with a *pac* site to facilitate packaging. The study acknowledged not only the cell-penetrating capability of the MS2 VLPs displaying TAT peptide but also the inhibitory effect of the encapsulated miR-122 on the hepatocellular carcinoma cells (Wang Guojing et al. 2016). It seems worth mentioning that a solid article on the cell-specific delivery of mRNA and microRNA by MS2 VLPs carrying cell-penetrating peptides (Sun Yanli and Yin 2015) was retracted at that time.

Furthermore, Sun Yanli et al. (2017) inserted the TAT peptide into the AB loop of the PP7 coat single-chain dimer, while the pre-miR-23b microRNA was packaged inside the PP7 VLPs. The PP7 VLPs carrying the cell-penetrating TAT peptide and packaged microRNA were efficiently expressed in *E. coli* using the one-plasmid double expression system, penetrated hepatoma SK-HEP-1 cells, and delivered the pre-microRNA-23b, which was processed into a mature product within 24 hours and inhibited the migration of hepatoma cells by downregulating liver-intestinal cadherin.

Armored RNA

The *armored* nucleic acids were implemented as useful noninfectious, easily available reagents for quality control in the routine diagnosis of pathogenic viruses. The classical armored RNA technology is based on the MS2 VLPs, where in vivo encapsidation of a desired RNA is performed by including the MS2 operator (aptamer) sequence into the RNA molecule in question to enable its packaging. The armored RNA is therefore protected from ribonuclease digestion and can be used as stable, well-characterized controls and standards in routine clinical runs.

For the first time, the versatile armored RNA technology was elaborated by Brittan L. Pasloske and colleagues (1998) from Ambion, Inc., Austin, Texas and applied to the HIV-1 assay. Thus, the DNAs encoding the (i) MS2 coat protein, (ii) target RNA sequence represented by a 172-base consensus sequence from a portion of the HIV-1 *gag* gene, and (iii) MS2 operator sequence were cloned downstream of an inducible *lac* promoter. As the MS2 coat gene was translated, it was bound to the operator sequence at the 3′ end of the recombinant RNA, initiating the encapsidation of the recombinant RNA to produce pseudoviral particles. The capacity of the armored RNA to accept the desired foreign RNA was measured by a lambda fragment of different length and estimated as approximately 500 nucleotides. Unlike the phage MS2, which was released into the spent medium by lysing *E. coli*, the armored RNA particles were localized in the cytoplasmic fraction of the *E. coli* cells. After production and purification, the resulting HIV-1 armored RNA particles were shown to be totally resistant to degradation in human plasma and produced reproducible results in the HIV-1 assay. Furthermore, Drosten et al. (2001) employed the armored RNA in their technically reliable HIV-1 assay for high-throughput blood donor screening.

Next, the Pasloske's team generated the armored RNA control for the hepatitis C virus (HCV) genotyping assay (WalkerPeach et al. 1999), and the practical methodology of the armored RNAs was presented by DuBois et al. (1999). The MS2 VLP-based armored HCV control was then used by Forman and Valsamakis (2004).

The MS2 VLP-based armored RNAs were used further to detect enterovirus (Beld et al. 2004, Donia et al. 2005), West Nile virus (Eisler et al. 2004), severe acute respiratory syndrome coronavirus (SARS-CoV; Bressler and Nolte 2004), HCV (Konnick et al. 2005, Drexler et al. 2009), HIV, hepatitis A, C, and G, Dengue, and Norwalk virus, among others (Schaldach et al. 2006). Hietala and Crossley (2006) adopted the MS2 VLP-based armored RNA approach to the detection of the four high-consequence animal pathogens: classical swine fever virus, foot-and-mouth disease virus, vesicular stomatitis virus of New Jersey serogroup, and vesicular stomatitis virus of Indiana serogroup.

Cheng et al. (2006) prepared the first His-tagged MS2 armored RNA-carrying VLPs for RT-PCR detection of SARS-CoV, where the His_6 tag was introduced into the loop region of the MS2 coat protein and exposed on the VLP surface. Later, Zhang J et al. (2015) got ribonuclease-resistant His6-tagged MS2 VLPs carrying armored influenza RNA.

The next MS2 VLP-based armored RNAs contained the M gene of influenza H3N2 (Yu et al. 2007); the integrase region of the HIV-1 polymerase gene (Tang et al. 2007; Huang S et al. 2011) and other HIV-1 sequences (Huang J et al. 2008); the hepatitis E virus ORF3 (Zhao C et al. 2007); the rabies virus sequences (Wang YL et al. 2008); the SARS-CoV sequences (Stevenson et al. 2008); the influenza A, influenza B, and SARS-CoV sequences altogether (Yu et al. 2008); and the rotavirus NSP3 gene (Chang L et al. 2012).

Summarizing up to that time, the armored RNA contained approximately 1.7 kb of MS2 RNA encoding the maturation and coat proteins and the operator, (*pac*) site. Taking into account the MS2 genome length of 3569 nucleotides at most, 1.9 kb of the nonphage RNA sequence might be encapsulated by this method. Practically, the packaging of 500 bases of RNA has been demonstrated to be very efficient, while packaging of 1- and 1.5-kb RNAs was inefficient (Pasloske et al. 1998). Huang Q et al. (2006) used armored RNA technology to package a 1,200-nucleotide foreign RNA sequence in their RT-PCR assay for HCV, HIV-1, SARS-CoV1, and SARS-CoV2 by deleting some disposable sequences between the multiple cloning site and the transcription terminator.

Wei Y et al. (2008) generated a method for packaging long, more than 2,000 nucleotides, RNA sequences, which was referred to as armored L-RNA technology. It made it possible to pack the armored L-RNA of 2248 bases containing six gene fragments: hepatitis C virus, SARS-CoV1, SARS-CoV2, and SARS-CoV3, avian influenza virus matrix gene, and H5N1 avian influenza virus. In parallel, Wei B et al. (2008) improved the MS2 VLP-based armored RNA by increasing the number and affinity of the *pac* site in exogenous RNA and of the sequence encoding coat protein. Such one-plasmid expression system allowed the extension of the length of the armored RNA to 1891 bases and included SARS-CoV1, SARS-CoV2, SARS-CoV3, HCV, and influenza H5N1 fragments. The armored L-RNA technology with the improved *pac* site was used by the elaboration of the branched DNA assay by the HIV-1 detection (Zhan et al. 2009). This laboratory also presented protocols for the MS2 VLP-based armored RNA detection for enterovirus 71 and coxsackievirus A16 (Song et al. 2011) and avian influenza H7N9 virus (Sun Yu et al. 2013).

The MS2 VLP-based armored RNA standard was applied to measure microcystin synthetase E gene (*mcyE*) expression in toxic *Microcystis* sp., since microcystin was a secondary metabolite peptide toxin known to cause hepatotoxicosis and carcinogenicity in vertebrates (Rueckert and Cary 2009; Wood et al. 2011).

Monjure et al. (2014) were the first who applied the armored RNA technique to optimize PCR for quantification of simian immunodeficiency virus (SIV) genomic RNA in plasma of rhesus macaques. Then, the two novel control systems for RT-qPCR were established by the MS2 VLP-based armored RNAs of hantavirus and Crimean-Congo hemorrhagic fever virus (Felder and Wölfel 2014).

The armored RNA technology contributed to the efficient in vivo expression of large amounts of homogeneous RNA, when specific RNA scaffolds were extended by the addition of the MS2 operator to the original chimeric tRNA-mRNA sequence, in order to be encapsulated into the MS2 VLPs (Ponchon et al. 2013).

The armored double-stranded DNAs were encapsulated for the first time into the MS2 VLPs by Zhang Lei et al. (2015). To do this, the MS2 VLPs were dissociated into coat dimers first, and then the latter were reassembled into the VLPs in the presence of *pac* site in the RNA or DNA form. The MS2 VLPs remained indistinguishable from native MS2 capsids in size and morphology after encapsulation of the double-stranded DNA.

During the 2013–2015 Ebola virus disease humanitarian crisis, the MS2 VLP-based armored RNA assay was elaborated for differential detection of five Ebola virus species (Lu G et al. 2015), as well as for the evaluation of seven commercial Ebola virus RT-PCR detection kits (Wang Guojing et al. 2015). The MS2 armored RNA was used also in other Ebola virus kits (Dedkov et al. 2016; Grolla 2017; Shah 2017).

In 2015, the appearance of Middle East respiratory syndrome coronavirus (MERS-CoV) initiated rapid construction of the external quality assessment systems, which were based again on the MS2 VLPs encapsulating specific RNA sequences of MERS-CoV as positive specimens (Zhang Lei et al. 2016).

At that time, Mikel et al. (2015) published an extensive review on the MS2 VLP-based armored RNA with a special aim to teach how to prepare and use the armored tools by the RT-PCR and qRT-PCR detection of RNA viruses not only in clinical specimens but also in food matrices. To monitor the sample analysis, Mikel et al. (2016) generated two sorts of the MS2 VLPs with armored RNAs derived from mitochondrial DNA sequences of two extinct species, namely, thylacine *Thylacinus cynocephalus* and the moa bird *Dinornis struthoides*. Because the sequences were derived from mDNA of extinct species, its natural occurrence in the analyzed samples was highly unlikely. Next, Mikel et al. (2017, 2019) adopted His-tag system for rapid production and easy purification of MS2 VLPs. This was the first case when the single-chain version of the MS2 coat dimer containing the His-tag and capable to form intact His-tagged MS2 VLPs was supposed for the armored RNA technology.

In parallel, the MS2 VLP-based armored RNA was employed by the external quality assessment program of the RT-PCR detection of measles virus (Zhang D et al. 2015), by the rapid detection of HCV genotypes 1a, 1b, 2a, 3a, 3b, and 6a, where the appropriate armored RNAs were produced for each of six HCV genotypes (Athar et al. 2015) by the detection of hepatitis E virus (Wang S et al. 2016), a set of mosquito-borne viruses and parasites (Zhang Yingjie et al. 2017), Dengue virus serotypes (Fu et al. 2017), Zika virus (Lin et al. 2017), HCV (Zambenedetti et al. 2017), a broad set of influenza viruses (Zhang D et al. 2018), rabies virus (Dedkov et al. 2018a), filoviruses (Dedkov et al. 2018b), HIV-1 (Gholami et al. 2018a, b), and Lassa virus (Dedkov et al. 2019). Petrov et al. (2018) adapted the armored MS2 VLPs for the detection of Ebola, Marburg, Lassa, Machupo, Venezuelan encephalitis equine (VEE), Rift Valley fever, and rabies viruses. The MS2 VLP-based armored RNA technology was also employed by the p210 *BCR-ABL1* testing, as a proof of common cytogenetic abnormalities in leukemia (Fu et al. 2019). Recently, the armored MS2 technique was applied to generate standard RNA for measuring hepatitis B virus (Gao et al. 2019).

After the most popular MS2 VLPs, the Qβ VLPs were introduced for the first time in the role of the armored RNA tools by the HCV detection (Villanova et al. 2007). Then, the Qβ VLP-based armored RNA technology was applied to the norovirus detection (Zhang Q et al. 2018). At last, the armored Qβ VLPs were found to be more stable than the armored MS2 VLPs for packaging human norovirus RNA, by incubation and storage at different temperatures (Yao L et al. 2019). Furthermore, the Qβ VLPs were used to prepare armored RNA as a positive control for the advanced Covid-19 RT-PCR tests, in parallel with the VLPs of cowpea chlorotic mottle virus (CCMV), from the *Bromoviridae* family of the order *Martellivirales* (Chan et al. 2021a, b), as already mentioned in Chapter 17.

Recently, Zilberzwige-Tal et al. (2021) proposed novel in vitro and in vivo technologies to encapsulate functional ssDNA nanostructures of 200–1,500 nucleotides into the ultrastable MS2 VLPs.

CpGs

The packaging exercises in the immunological applications of the VLPs were stimulated strongly as a result of the epochal discovery that explained the role of the CpG motifs in DNA by the B-cell activation (Krieg et al. 1995). Thanks to the Martin F. Bachmann's team efforts, as demonstrated earlier in the *Contribution to Basic Immunology* and *Vaccines* paragraphs, the RNA phage VLPs were found to serve as nanocontainers that can encapsulate such specific adjuvants as the immunostimulatory CpG oligodeoxynucleotides, known as specific TLR9 ligands (Temizoz et al. 2016). Thanks to Bachmann and Grēns's team activities, the RNA phage VLPs were packaged with single-stranded or double-stranded RNA fragments, known as TLR7 and TLR3 ligands, respectively (Brencicova and Diebold 2013). Moreover, the recombinant and chimeric RNA phage VLPs contained always encapsulated bacterial RNA, which acted as an adjuvant.

The CpG oligonucleotides were studied successfully together with the model VLPs by Bachmann's team, first with the HBc VLPs (Storni et al. 2002, 2003; Schwarz K et al. 2003; Storni and Bachmann 2003). The CpG packaging into both HBc and Qβ VLPs provided with the previously mentioned LCMV gp33 epitope followed (Bachmann et al. 2004b; Storni et al. 2004). Storni et al. (2004) and showed that the packaging of CpGs into the HBc or Qβ VLPs was not only simple but could reduce the two major problems of the CpG usage, namely their unfavorable pharmacokinetics and systemic side effects, including splenomegaly. The vaccination with the CpG-loaded VLPs was able to induce high frequencies of peptide-specific CD8+ T cells. It protected from infection with recombinant vaccinia viruses and eradicated established solid fibrosarcoma tumors. It was concluded therefore for the first time that packaging CpGs into the Qβ VLPs improved both their immunogenicity and pharmacodynamics (Storni et al. 2004). This study paved the long way for the development of the CpG containing VLP vaccine candidates, especially in the allergy vaccines elaborated by Bachmann's team, as described earlier in the *Vaccines* paragraph. The Qβ VLP vaccines carrying packaged CpG were reviewed in the global context of the CpG oligodeoxynucleotide nanomedicines for the prophylaxis or treatment of cancers, infectious diseases, and allergies (Hanagata 2017; Bachmann et al. 2020; Jensen-Jarolim et al. 2020).

Meanwhile, Matthew B. Francis's team elaborated a simple and highly efficient alkali treatment method to remove the encapsidated RNA (Hooker et al. 2004) that was employed later by other VLP technologies, such as HBc VLPs, for the specific packaging (Strods et al. 2015).

Proteins

The embedding of other than RNA cargos into the RNA phage VLPs was stimulated by the rapid development of the click chemistry approaches, as described in detail in the preceding *Functionalization* paragraphs.

M.G. Finn's team was the first to achieve the in vivo packaging of proteins, which are listed in Table 25.5, into the Qβ VLPs (Fiedler et al. 2010). To facilitate this RNA-directed encapsidation, the two binding domains were introduced to the coat protein mRNA: an RNA aptamer developed by in vitro selection to bind an arginine-rich HIV-1-derived peptide Rev and the sequence of the Qβ packaging hairpin. The cargo enzymes were N-terminally tagged with the Rev peptide and inserted into another compatible group plasmid. The transformation with both plasmids and expression in *E. coli* yielded VLPs encapsidating the Rev-tagged protein. The same Rev tagging approach was applied to the four chosen enzymes that are listed in Table 25.5 (Fiedler et al. 2018). The captured enzymes were active while inside the nanoparticle shell and were protected from environmental conditions that led to the free-enzyme destruction. The technology was extended to create, via self-assembly, the VLPs that simultaneously displayed protein ligands on the exterior and contained enzymes within. The displayed ligands are listed in Table 25.5.

The two other methods of protein encapsidation were demonstrated by Matthew B. Francis's team with the MS2 capsids (Glasgow et al. 2012). First, attaching DNA oligomers to a molecule of interest and incubating it with MS2 coat protein dimers yielded reassembled capsids that packaged the tagged molecules. Second, the genetically encoded negatively charged peptide tags were employed to encapsulate alkaline phosphatase. The purified encapsulated enzyme was found to have the same K_m value and a slightly lower k_{cat} value than the free enzyme, indicating that this method of encapsidation had a minimal effect on enzyme kinetics (Glasgow et al. 2012).

Later, Francis's team presented a solid methodological account on their success by the encapsulation of proteins with tags comprising anionic amino acids or DNA and gold nanoparticles with negative surface charges into the MS2 VLPs (Aanei et al. 2018a).

OTHER CARGOS

Steinmetz et al. (2009) packaged the Qβ VLPs with fluorophores and decorated their surface with fullerene buckyballs. At the same time, the porphyrins were attached to the MS2 VLPs (Stephanopoulos et al. 2009). After this covalent bioconjugation of porphyrins, Cohen et al. (2011, 2012) employed the purified capsids of the phage MS2 as a delivery vessel of porphyrins that were bound by "nucleotide-driven packaging." These and other packaging etudes are listed in Table 25.5.

The Francis team presented the controlled integration of gold nanoparticles and organic fluorophores into the MS2 VLPs (Capehart et al. 2013). Thus, the MS2 VLPs were used to house gold particles within its interior volume. The exterior surface of each capsid was then modified with Alexa Fluor 488-labeled DNA strands. By placing the dye at distances of 3, 12, and 24 bp from the surface of capsids containing 10 nm gold nanoparticles, fluorescence intensity enhancements of 2.2, 1.2, and 1.0, respectively, were observed (Capehart et al. 2013).

Meanwhile, due to strong efforts of Andris Kazāks's group in yeast expression, Grēns's team presented a wide variety of novel VLP models, derived from the phages fr, SP, AP205, PP7, and ϕCb5, with the special aim of nanomaterial packaging (Freivalds et al. 2014). It was stressed particularly that a variety of compounds, including RNA, DNA, and gold nanoparticles, could be packaged efficiently inside the novel ϕCb5 VLPs. The ease with which the phage ϕCb5 coat protein dimers have been purified in high quantities and reassembled into VLPs made them especially attractive for the packaging of nanomaterials and the chemical coupling of peptides of interest on the surface.

The MS2 VLPs were used to elaborate a general cargo-compatible approach to encapsulate guest materials based on the apparent critical assembly concentration (CAC_{app}) of the viral nanoparticles (Li L et al. 2019). The new method drove the reassembly of the latter to encapsulate cargoes by simply concentrating an adequately diluted mixture of the VLP building blocks and cargoes to a concentration above the CAC_{app}.

TARGETING

Alongside the data presented in Table 25.5, some specific targeting approaches are highlighted here. Thus, the capsid exterior was endowed successfully with cell-specific targeting capabilities, particularly via the appendage of receptor-specific antibodies on the external surface. Given the potential of the MS2 VLPs as a drug delivery and imaging system and its success in in vitro experiments (Ashley et al. 2011; ElSohly et al. 2015), the in vivo behavior of constructs based on the MS2 scaffold warranted investigation. Thus, Aanei et al. (2016) conjugated the anti-EGFR antibodies

to the MS2 VLPs, to target them toward receptors overexpressed on breast cancer cells. Next, the epidermal growth factor itself was conjugated to the MS2 VLPs in order to target them directly to the epidermal growth factor receptor on the cancer cells (ElSohly et al. 2017).

Then, the Francis team used DNA origami as a scaffold and performed the one-dimensional arrangement of virus capsids with nanoscale precision (Stephanopoulos et al. 2010a). To do this, the interior surface of the MS2 VLPs was modified with fluorescent dyes as a model cargo, but the unnatural amino acid on the external surface was then coupled to DNA strands that were complementary to those extending from origami tiles, while the two different geometries of DNA tiles, rectangular and triangular, were used. The capsids associated with tiles of both geometries had virtually 100% efficiency under mild annealing conditions, and the location of capsid immobilization on the tile could be controlled by the position of the probe strands. The rectangular tiles and capsids could then be arranged into one-dimensional arrays by adding DNA strands linking the corners of the tiles. The resulting structures consisted of multiple capsids with even spacing of approximately 100 nm. This hierarchical self-assembly made it possible to position the MS2 particles with unprecedented control and allowed the future construction of integrated multicomponent systems from biological scaffolds using the power of rationally engineered DNA nanostructures (Stephanopoulos et al. 2010a). Furthermore, Wang D et al. (2014) presented a hierarchical assembly of plasmonic nanostructures using MS2 scaffolds on DNA origami templates.

Concerning the employment of the Qβ VLPs, Neburkova et al. (2018) involved them in a large comparative study of different VLPs, namely polymer-coated nanodiamonds, mouse polyomavirus VLPs, and polymeric poly(HPMA) nanoparticles, for their ability to display low-molecular-mass inhibitors of glutamate carboxypeptidase II (GCPII), a membrane protease that is overexpressed by prostate cancer cells. Regardless of the diversity of the investigated nanosystems, they all strongly interacted with the GCPII and effectively targeted the GCPII-expressing cells.

Finn's team demonstrated that the lung-tissue delivery of the Qβ VLPs was mediated by macrolide antibiotics (Crooke et al. 2019). It was established that azithromycin was able to direct the Qβ VLPs to the lungs in mice, with significant accumulation within 2 h of systemic injection. These results suggested that this new class of bioconjugate could serve as an effective platform for intracellular drug delivery in the context of pulmonary infections.

At last, Finn's team demonstrated how to minimize the unspecific binding of the Qβ particles to mammalian cells (Martino et al. 2021). It was found that lysine residues mediate such nonspecific interactions, presumably by virtue of protonation and interaction with anionic membrane lipid headgroups and/or complementary residues of cell surface proteins and polysaccharides. The chemical acylation of surface-exposed amines of the Qβ VLPs led to a significant reduction in the association of particles with the mammalian cells. Moreover, the single-point mutations of particular lysine residues to either glutamine, glutamic acid, tryptophan, or phenylalanine were mostly well-tolerated and formed intact capsids, but the introduction of double and triple mutants was far less forgiving. Remarkably, the introduction of glutamic acid at position 13 (K13E) led to a dramatic increase in cellular binding, whereas removal of the lysine at position 46 (K46Q) led to an equally striking reduction (Martino et al. 2021).

NANOMATERIALS

The RNA phage VLP-based applications that have been developed as novel nanomaterials are compiled in Table 25.6.

First, the use of imaging agents in combination with RNA phage VLPs has contributed to the high-resolution and noninvasive visualization of these particles, as well as to the potential treatment of diseases. Thus, the first studies on the generation of nanoparticles for magnetic resonance imaging applications and the first comparisons of interior versus exterior cargo strategies appeared in the mid-2000s (Anderson et al. 2006; Hooker et al. 2007). Second, the MS2 VLPs were loaded with positron emission tomography radiolabels (Hooker et al. 2008). Currently, major attention is focused on the use of the RNA phage-based nanoparticles as potential scaffolds for novel biomaterials and as subjects for nanoscale engineering applications involving exposure to various chemical compounds. As can be seen from Table 25.6, the MS2 phage and VLPs have played to date a leading role in the novel material and bioimaging studies.

Remarkably, Cohen et al. (2009) used the phage MS2 as a biotemplate for semiconductor nanoparticle synthesis. Thus, the synthesis of cadmium sulfide nanoparticles was demonstrated in aqueous solution with the phage MS2 as a result of the bionanofabrication approach that offered unique opportunities for nanomaterials synthesis, where choice of biotemplate could influence the size, shape, and function of the produced material. The phage MS2 was chosen as a potential template system for the bionanofabrication of nanoscale materials due to its capability to be (i) genetically modified to incorporate functional groups and sequences and (ii) devoid of the inner RNA by degradation or replacement with other oligonucleotide sequences. Remarkably, the cadmium sulfide nanoparticles possessed predicated fluorescent properties when formed in the presence of the phage MS2 and remained stable in aqueous solutions. Surprisingly, the MS2-templated CdS nanoparticles were stable at room temperature for several weeks.

Udit et al. (2010a) reviewed at that time the striking success of the MS2 VLP-based chemically tailored multivalent virus platforms that found their broad application from drug delivery to catalysis and appeared therefore as a good candidate to generate new materials.

In the context of the RNA phage VLPs as templates for nanomaterials and as carriers for vaccines, bioimaging labels or drug delivery tools, Machida and Imataka (2015)

published a comprehensive summary of the most applicable production methods for viral particles. Yeast was then evaluated as a prospective expression system to produce VLPs including MS2 and Qβ VLPs (Kim HJ and Kim HJ 2017). It is noteworthy that the stability of the phage MS2 was enhanced by the formulation with PGLA, as mentioned earlier (Khalil et al. 2016).

Hartman et al. (2018) elaborated an impressive theoretical strategy that paved the way for a basic reconstruction of the phage VLPs and generation of novel nanomaterials. This approach was termed *systematic mutation and assembled particle selection* (SyMAPS) and used in a library generation and single-step selection, in order to study the VLP self-assembly. This selection did not rely on infectivity, clinical abundance, or serum stability and therefore enabled experimental characterization of the assembly competency of all single amino acid variants of the MS2 coat protein and challenged some conventional assumptions of protein engineering. The resulting high-resolution fitness landscape (AFL) presented a fundamental roadmap to altering the MS2 coat to achieve tunable chemical and physical properties. After recapitulating the results of many previous investigations in a single experiment (Peabody DS 2003; Carrico et al. 2008; Wu W et al. 2009), Hartman et al. (2018) calculated the effect of ten physical properties on the AFL and evaluated the validity of several common protein engineering assumptions. An additional round of selection identified a previously unknown variant, the MS2 coat T71H variant, which exhibited acid-sensitive properties that were promising for the engineering controlled endosomal release of cargo by targeted drug delivery. The library of the MS2 variants can be subjected to future selections to address any number of additional engineering goals. Moreover, Hartman et al. (2018) recommended the SyMAPS as a straightforward approach that could be applied more broadly to assess the fitness landscapes of the coat proteins of clinically relevant pathogens, including HBV and HPV virions.

Jeremiah J. Gassensmith's team recently proposed a concept of the PhotoPhage, a Qβ VLP-based photothermal therapeutic agent (Shahrivarkevishahi et al. 2021). The design was based on covalent conjugation of 212 water soluble near-infrared absorbing croconium dyes to lysine residues on the VLP surface of Qβ, which turned it to a powerful NIR-absorber with photothermal efficiencies exceeding that of gold nanostructures. This PhotoPhage system generated heat upon 808 nm NIR laser radiation and caused significant cellular cytotoxicity that prevented the progression of primary tumors in mice. Moreover, the PhotoPhage acted simultaneously as an immunoadjuvant that promoted maturation of dendritic cells, triggered T lymphocyte cells, and reduced suppressive T regulatory cells, leading to effective suppression of primary tumors, reducing lung metastases, and increasing survival time (Shahrivarkevishahi et al. 2021).

The specific application of the RNA phages and their VLPs in the generation of novel nanomaterials and imaging tools remains a favorite subject for numerous review articles

TABLE 25.6

Nanomaterials Based on the RNA Phage VLPs and Virions

Contrast Agent	Comments	References
MS2		
MS2 phage was labeled with the succinimidyl ester of [Ru(2,2′-bipyridine)2(4,4′-dicarboxy-2,2′-bipyridine)]²⁺ (RuBDc), which is a very photostable probe that possesses favorable photophysical properties including long lifetime, high quantum yield, large Stokes' shift, and highly polarized emission. The RuBDc luminophore attacks lysine residues.	The intensity and anisotropy decays of RuBDc when conjugated to RNA phage MS2 were examined using frequency domain fluorometry with a high-intensity, blue light-emitting diode (LED) as the modulated light source. The results showed that RuBDc can be useful for studying rotational diffusion of biological macromolecules.	Kang and Yoon 2004
The MS2(Gd-DTPA-ITC)m contrast agent was synthesized through the conjugation of premetalated Gd-DTPA-ITC (2-(4-isothiocyanatobenzyl)-diethylenetriaminepentaacetic acid) to the lysine residues of MS2 VLPs (not phage) on the VLP exterior.	A magnetic resonance imaging (MRI) contrast agent was developed by conjugation of more than 500 Gd chelate groups onto a VLP. The high density of paramagnetic centers and slow tumbling rate of modified VLPs provided enhanced T1 relaxivities up to 7,200 mM-1s-1 per particle. A bimodal imaging agent was generated by sequential conjugation of fluorescein and Gd³⁺ chelate.	Anderson et al. 2006
In order to maximize the relaxivity of each Gd³⁺ complex attached to the phage scaffold, hydroxypyridonate (HOPO)-based contrast agents were used. The interior (Tyr85) and exterior (Lys106, Lys113, and the N-terminus) surfaces of "empty" phage were targeted independently through the appropriate choice of reagents.	The phage capsids sequestering the Gd-chelates on the interior surface (attached through tyrosine residues) not only provided higher relaxivities than their exterior functionalized counterparts (which relied on lysine modification) but also exhibited improved water solubility and capsid stability. There are strong advantages of using the internal surface for contrast agent attachment, leaving the exterior surface available for the installation of tissue targeting groups.	Hooker et al. 2007; Datta et al. 2008
The "empty" shell of phage was labeled on its inside surface with ¹⁸F-fluorobenzaldehyde through a multistep bioconjugation strategy. An aldehyde functional group was first attached to interior tyrosine residues through a diazonium coupling reaction. The aldehyde was further elaborated to an alkoxyamine functional group, which was then condensed with ¹⁸F-fluorobenzaldehyde.	This is a first example of the positron emission tomography (PET) radiolabels that have been developed on RNA phages. Relative to fluorobenzaldehyde, the fluorine-18-labeled MS2 exhibited prolonged blood circulation time and a significantly altered excretion profile.	Hooker et al. 2008
Approximately 125 xenon MRI sensor molecules were incorporated in the interior of an MS2 VLPs, conferring multivalency and other properties of the VLP to the sensor molecule.	The resulting signal amplification facilitated the detection of sensor at 0.7 pM, the lowest to that date for any molecular imaging agent used in magnetic resonance. This amplification promised the detection of chemical targets at much lower concentrations than would be possible without the VLP scaffold.	Meldrum et al. 2010
Multivalent, high-relaxivity MRI contrast agents using rigid cysteine-reactive gadolinium complexes.	Greater contrast enhancements were seen for MRI agents that were attached via rigid linkers, validating the design concept and outlining a path for future improvements of nanoscale MRI contrast agents.	Garimella et al. 2011
The PET imaging characteristics were improved by the usage of PEG chains added to MS2 VLPs. The MS2- and MS2-PEG VLPs possessing interior DOTA chelators and labeled with ⁶⁴Cu were compared by injecting intravenously into mice possessing tumor xenografts.	The biodistribution and circulation properties of the VLP-based PET imaging agents were investigated carefully, in order to realize the potential of such agents for the future use in in vivo applications.	Farkas et al. 2013

Contrast Agent	Comments	References
The VLPs were modified using an oxidative coupling reaction, conjugating ~90 copies of a fibrin targeting peptide to the exterior of each protein shell. The installation of near-infrared fluorophores on the interior surface of the capsids enabled optical detection of binding to fibrin clots. The targeted capsids bound to fibrin, exhibiting higher signal-to-background than control, nontargeted VLP-based nanoagents.	The chemically functionalized and specifically targeted VLPs were used for fibrin imaging. The modified capsids out-performed the free peptides and were shown to inhibit clot formation at effective concentrations over tenfold lower than the monomeric peptide alone. The in vitro assessment of the capsids suggests that fibrin-targeted VLPs could be used as delivery agents to thrombi for diagnostic or therapeutic applications.	Obermeyer et al. 2014a
The targeted, selective, and highly sensitive ^{129}Xe NMR nanoscale biosensors were synthesized using MS2 VLPs, Cryptophane A (CryA) molecules, and DNA aptamers. The MS2 VLPs were modified with CryA-maleimide at ~110 N87C positions on the interior surface of the capsid. This was followed by the attachment of aptamers on the exterior surface.	The biosensor showed strong binding specificity toward targeted lymphoma cell line. This work provided a strong basis for the continued exploration of targeted cancer cell imaging agents with future applications in NMR and MRI.	Jeong et al. 2016

PP7

Technetium-99m (^{99m}Tc) labeling of PP7 phage was achieved by a HYNIC bifunctional agent. Radiochemical purity higher than 90% was obtained.	The labeled PP7 phage was used as a specific tracer of *Pseudomonas aeruginosa* infection in animals.	Cardoso et al. 2016; Elena et al. 2016
Encapsulation of small-ultra red fluorescent protein (smURFP) was achieved by the PP7 VLP disassembly-reassembly.	Potential application of the construct as non-invasive in vivo imaging agents.	Herbert et al. 2020

Qβ

MRI contrast agent. The VLPs were decorated with gadolinium complexes using the Cu^I-mediated azide-alkyne cycloaddition (CuAAC) reaction. The interior surface labeling was engineered by the introduction of an azide-containing unnatural amino acid into the coat.	The circulation lifetime, plasma clearance, and distribution in major organs were studied in mice for contrast agents based on original VLPs and VLPs with mutated surface aa residues	Prasuhn et al. 2007, 2008
The FGlyco-CCHC technique, as a subset of the CuAAC reaction, was used for the generation of fluoroglycoproteins based on the Qβ VLPs as potential substrates in NMR, MRI, and PET techniques.	The reaction of L-homopropargylglycine-modified Qβ VLPs with a fluoroglucosyl azide led to occupation of more than 95% of positions with F-glycoprotein. The Qβ VLP displayed finally 180 site-selectively positioned fluoroglycans.	Boutureira et al. 2010; Boutureira and Bernardes 2015
Encapsulation of small-ultra red fluorescent protein (smURFP) was achieved by the Qβ VLP disassembly-reassembly.	Potential application of the construct as noninvasive in vivo imaging agents, the same as earlier for the PP7 VLPs.	Herbert et al. 2020

R17

The metal-ligand complex, [Ru(2,2'-bipyridine)$_2$(4,4'-dicarboxy-2,2'-bipyridine)]$^{2+}$ (RuBDc) was used as a spectroscopic probe for studying hydrodynamics of biological macromolecules.	The combination of the use of a long-lifetime Ru(II) metal-ligand complexes with blue LED excitation made it possible to perform time-resolved intensity and anisotropy decay measurements with a simpler and lower-cost instrument.	Kim MS et al. 2010

(Manchester and Singh 2006; Lee LA et al. 2009, 2011; Hooker 2010; Li K et al. 2010; Rodríguez-Carmona and Villaverde 2010; Steinmetz 2010; Chung et al. 2011; Dedeo et al. 2011; Jutz and Böker 2011; Pokorski and Steinmetz 2011; Ng et al. 2012; Bruckman et al. 2013; Farr et al. 2014; Glasgow and Tullman-Ercek 2014; Li F and Wang 2014; Capek 2015; Koudelka et al. 2015; Shukla and Steinmetz 2015; Tsvetkova and Dragnea 2015; Karimi et al. 2016; Lee EJ et al. 2016; Maassen et al. 2016; Raeeszadeh-Sarmazdeh et al. 2016; Samanta and Medintz 2016; Wen and Steinmetz 2016; Zhang Yu et al. 2016; Glidden et al. 2017; Sun H et al. 2017; Zhang W et al. 2017; Almeida-Marrero et al. 2018).

NANOREACTORS

The idea that the MS2 VLPs could be used as a nanoreactor was published first by de la Escosura et al. (2009). The authors relied on the two key studies performed by Francis's team. First, on the two orthogonal modification strategies to decorate the exterior surface of MS2 capsids with PEG chains, while installing 50–70 copies of a fluorescent dye inside as a drug cargo mimic (Kovacs et al. 2007). Second, on the solubilization of the phage MS2 in organic solvents, subsequent removal of water, and resolubilization of MS2 in a solvent of choice (Johnson et al. 2007). The resolubilized MS2 was then derivatized with stearic acid in chloroform, illustrating that the bioconjugation reactions could be performed with reagents that were completely insoluble in water. Therefore, the extended range of potential chemical modifications and the enhanced thermal stability of these MS2 cages in organic solvents opened the way for their use as organic nanoreactors (de la Escosura et al. 2009). In their review on smart nanocontainers and nanoreactors, Kim KT et al. (2010) adduced the same experimental proofs for the use of MS2 as a nanoreactor, with an additional argument that the MS2 VLPs were modified on the exterior with aptamers to direct uptake of the capsid drug carrier by specific cells, according to Tong et al. (2009). The MS2 cages were listed among the possible nanoreactor candidates also by Bode et al. (2011). Then, Cardinale et al. (2012) included both MS2 and Qβ in the list of virus scaffolds as potential enzyme nanocarriers, using the papers of Ashley et al. (2011) and Strable et al. (2008), respectively, as the experimental arguments. Cardinale et al. (2015) extended the arguments by references on the experimental studies of Wu M et al. (1995) in the case of MS2 and of Fiedler et al. (2010) and Rhee et al. (2011, 2012 in the case of Qβ. The Finn team reported packaging of the 25-kDa N-terminal aspartate dipeptidase peptidase E (PepE), 62-kDa firefly luciferase (Luc), and a thermostable mutant of luciferase (tsLuc), inside Qβ VLPs (Fiedler et al. 2010). The average number of the encapsidated cargo proteins was controlled by changing expression conditions or by removing interaction elements from the plasmids. In this way, the PepE incorporation was varied reproducibly between 2 and 18 per particle. Fewer copies of Luc proteins were packaged: 4–8 copies per particle were found for most conditions, whereas the number of the packaged tsLuc molecules varied between 2 and 11 per VLP. The encapsulated enzymes demonstrated normal functional capabilities. All production and assembly steps occurred in vivo within the bacterial cell, with indirect control of amount of packaged cargo possible by simply changing the expression media or the nature of the components of the packaging system. The VLPs were produced in high yields and were purified by a convenient standard procedure, independent of the protein packaged inside (Fiedler et al. 2010).

Furthermore, the Francis team encapsulated enzymes inside the MS2 capsids, using a new osmolyte-based method (Glasgow et al. 2012). Attaching DNA oligomers to a molecule of interest and incubating it with the MS2 coat protein dimers in vitro yielded reassembled capsids that packaged the tagged molecules. The addition of a protein-stabilizing osmolyte, trimethylamine-N-oxide, significantly increased the yields of reassembly. Second, the expressed proteins with genetically encoded negatively charged peptide tags could also induce capsid reassembly, resulting in high yields of reassembled capsids containing the protein. This second method was used to encapsulate alkaline phosphatase tagged with a 16-aa peptide. The purified encapsulated enzyme was found to have the same K_m value and a slightly lower k_{cat} value than the free enzyme, indicating that this method of encapsulation had a minimal effect on the enzyme kinetics. This method therefore provided a practical and potentially scalable way of studying the complex effects of encapsulating enzymes in protein-based compartments (Glasgow et al. 2012).

The functioning of this enzymatic nanoreactor was studied further to explore pore-structure effects on substrate and product flux during the catalyzed reaction (Glasgow et al. 2015). When the enzymes were inside capsids with pores having no charge or a charge opposite that of the substrate, the apparent K_M ($K_{M,app}$) was similar to that of the free enzyme; however, when pore and substrate charge were the same, the $K_{M,app}$ of the enzyme increased significantly. The kinetic modeling suggested this could be caused primarily by reduced product efflux, leading to inhibition of the enzyme. These experiments represented the first step in creating selective nanoreactors with the potential either to protect cargo by inhibiting entrance of interfering molecules or to enhance multistep pathways by concentrating internal intermediates. These findings also lend support to the hypothesis that protein compartments could modulate the transport of small molecules and thus influence metabolic reactions and catalysis in vitro (Glasgow et al. 2015).

M.G. Finn's team recently described an exploration of the stability of Qβ VLP-encapsidated enzymes toward a greater range of denaturing stimuli than before, including heat, organic solvents, and chaotropic agents (Das et al. 2020). The authors demonstrated the significant enhancements in enzyme stability including the ability to perform reactions at greatly accelerated rates because of greater enzyme tolerance toward elevated temperatures in case of the three enzymes, namely yeast cytosine deaminase, peptidase E, and purine nucleoside phosphorylase.

The prospective SpyTag/SpyCatcher approach was used for the construction of a catalytic nanoreactor based on the in vivo encapsulation of multiple enzymes into the MS2 VLPs (Giessen and Silver 2016a). Briefly, the sequestration of two enzymes from the indigo biosynthetic pathway in the MS2 VLPs was achieved by the SpyTag/SpyCatcher protein fusions that covalently crosslinked with the interior surface of the capsid. The functional two-enzyme indigo biosynthetic pathway was targeted to the engineered capsids, leading to a 60% increase in the indigo production in vivo. The enzyme-loaded particles were purified in their active form and showed enhanced long-term stability in vitro and demonstrated greater activity than free enzymes (Giessen and Silver 2016a). Encapsulation as a prospective strategy for the design of biological compartmentalization was reviewed by Giessen and Silver (2016b).

The success of the RNA phage-derived nanoreactors was reviewed extensively (Giessen and Silver 2016b; Koyani et al. 2017; Quin et al. 2017; Schwarz B et al. 2017; Banerjee A and Howarth 2018; Timmermans and van Hest 2018; Wilkerson et al. 2018; Ren et al. 2019).

NANOMACHINES

As mentioned earlier in the *Vaccines: Natural VLPs* paragraph, the Qβ VLPs were used to encapsulate the Mg-based micromotors that were elaborated for the active delivery of the VLPs to peritoneal space of ovarian tumors in mice (Wang C et al. 2020). The fabrication of the Qβ motors included the following steps: (i) spreading of Mg microparticles over glass slides, (ii) gold sputter over the Mg microparticles, (iii) TiO_2 atomic layer deposition, (iv) PLGA coating layer over the Mg/Au/TiO_2 micromotors, (v) Qβ VLP-loaded chitosan film over the Mg/Au/TiO_2/PLGA micromotors, and (vi) release of the resulting Qβ-motors. This approach could be regarded as the invention of the first nanomachine based on the RNA phage VLPs.

CHIMERIC PHAGES

MS2

After the numerous revolutionary topics on the gene-engineered reconstruction of the phage genome, Jan van Duin's team succeeded by the generation of a candidate of the peptide display platform on the live MS2 phage, despite the strong restrictions at the RNA genome level (van Meerten et al. 2001). To test the feasibility of this approach, van Meerten et al. (2001) took a noninfectious VLP mutant generated earlier by Elmārs Grēns's team (Pushko et al. 1993) as a prototype in which 5 aa residues, namely ASISI, were inserted at N-terminus of the coat protein of the phage fr (a close relative of MS2) and did not prevent the assembly of the fr VLPs. According to the van Meerten et al. (2001) considerations, the in vivo approach faced several additional difficulties when compared with the ex vivo studies on the VLP formation. Most important, the peptide insertion must be compatible with the formation of an infectious phage. This required the correct incorporation of the maturation protein in the now-modified virion. Furthermore, the new coat protein must still be able to act as a translational repressor by binding the replicase operator. Another problem facing peptide display on live phages was that the extra RNA encoding the peptide could come under selection pressure for various reasons. If it could not adopt a proper secondary structure, it might fall prey to *E. coli* endonucleases. The degeneracy of the genetic code allowed the authors to make various mutants, all encoding the same ASISI insert in the coat protein. Several of these mutants were unstable and suffered deletions that reduced the insert to two or three amino acids, while others underwent adaptation by base substitutions. Some mutants were fully stable. Whether or not an insert was stable appeared to be determined by the choice of the nucleic acid sequence used to encode the extra peptide. This effect was not caused by differential translation, because the coat-protein synthesis was equal in wild-type and mutants. It was concluded that the stability of the insert depended on the structure of the large RNA hairpin loop, as demonstrated by the fact that a single substitution could convert an unstable loop into a stable one (van Meerten et al. 2001).

Qβ

After 20 years from the first experiments on the mosaic Qβ particles described earlier, Skamel et al. (2014) generated the Qβ display system via engineering of the A1 protein and suggested the phage Qβ as a favorable alternative to M13 for in vitro evolution of displayed peptides and proteins due to high mutagenesis rates in the Qβ RNA replication that could better simulate the affinity maturation processes of the immune response. The authors constructed the chimeric Qβ phages bearing G-H loop peptide of the VP1 protein of foot-and-mouth disease virus (FMDV). The surface-localized FMDV VP1 G-H loop cross-reacted with the anti-FMDV monoclonal antibody and was found to decorate the corners of the Qβ icosahedral shell by electron microscopy. Thus, the A1 protein-based Qβ-display emerged as a novel framework for the rapid in vitro evolution by affinity maturation to the molecular targets (Skamel et al. 2014).

Developing this approach further, the membrane proximal external region (MPER) of HIV-1 envelope glycoprotein-41 (gp41) was selected for the presentation on the Qβ phage (Waffo et al. 2017). Thus, a fragment representing the 50-aa consensus region within the HIV-1 gp41 MPER was fused in frame with the A1 protein of the phage Qβ. The three variant MPER expression cassettes were obtained with the MPER cDNA in frame with the A1 gene, including pQβMPER, pQβMPERHis with an additional C-terminal hexa-histidine tag and QβMPERN with a C-terminal not1 site. The expression cassettes were used for the production of QβMPER recombinant phages after transformation of *E. coli* HB101 strain. The engineered Qβ phages displayed 12 molecules of MPER per phage particle on the particle surface. The antigenicity of the chimeric phages was assessed with plasma from longstanding antiretroviral naïve HIV-1-infected persons, while immunogenicity studies were done in mice. The fusion of MPER and Qβ genes was confirmed by RT-PCR followed by gel electrophoresis and sequencing. The novel recombinant QβMPER phages were proposed, first, to monitor MPER-specific immune responses in HIV-1 exposed or infected people. Second, it was concluded that the recombinant QβMPER phages could be used as immunogens for the induction of the MPER-specific immunity against HIV-1, as a component of the possible HIV-1 vaccines (Waffo et al. 2017). The detailed methodology of the Qβ phage engineering by the generation of peptide libraries for the FMDV G-H loop and the HIV-1 MPER was published by Singleton et al. (2018).

Finally, Waffo et al. (2018) generated chimeric phages QβMSP3, QβUB05, and a combined variant of the two

QβUB05-MSP3, which displayed asexual blood stage antigens UB05 and merozoite surface protein 3 (MSP3) of *Plasmodium falciparum*. Since naturally acquired immune responses to the MSP3 and UB05 were implicated in semi-immunity in populations living in malaria-endemic areas, it would be reasonable to use them by designing malaria vaccine candidates displaying the epitopes upon the surface of the phage in their native form. In fact, the chimeric phages differentially detected blood stage antigen targeting antibodies in children living in a high malaria transmission region of Cameroon. As a result, the chimeric phage QβUB05-MSP3 was validated as an appropriate antigen for tracking immunity to malaria. Furthermore, the comparative analysis of IgG responses to the QβUB05-MSP3 vaccine was performed in dual HIV-malaria infected adults living in areas differing in malaria transmission intensities (Lissom et al. 2018; Nchinda et al. 2019).

26 Order *Nidovirales*

Never laugh at live dragons.

J.R.R. Tolkien

A hospital is no place to be sick.

Samuel Goldwyn

ESSENTIALS

The *Nidovirales* is a huge and deeply structured order of the enveloped, positive-sense single-stranded RNA viruses of widely different architecture that have not experienced significant consideration from the VLP nanotechnologies thus far, because of their remarkable size and complexity. This great order currently involves 8 suborders, 14 families, 25 subfamilies, 39 genera, 65 subgenera, and 109 species. According to the modern taxonomy (ICTV 2020), the order *Nidovirales* is one of the three orders belonging to the class *Pisoniviricetes*, together with the huge *Picornavirales* order (described in the neighboring Chapter 27) and the smaller *Sobelivirales* order, presented in Chapter 28. The *Pisoniviricetes* class is belonging to the *Pisuviricota* phylum from the kingdom *Orthornavirae*, realm *Riboviria* (ICTV 2020).

Above all, the order *Nidovirales* includes the notorious family *Coronaviridae*. The coronaviruses (CoVs) have long been regarded as relatively harmless pathogens for humans. The severe respiratory tract infection outbreaks caused by severe acute respiratory syndrome (SARS) CoV and Middle East respiratory syndrome (MERS) CoV, however, have caused high pathogenicity and mortality rates in humans and attracted particular scientific interest. In fact, the coronaviruses became well-known in 2003 because of the SARS-CoV-1 outbreak, which was followed in 2012 by the MERS-CoV appearance, and at last drew total attention to the SARS-CoV-2 outbreak and global Covid-19 pandemic in 2019.

The family *Coronaviridae*, the only member of the *Cornidovirineae* suborder, includes, under the current official ICTV view (ICTV 2020), 2 subfamilies, 5 genera, 26 subgenera, and 46 species altogether. Both SARS-CoV-1 and SARS-CoV-2 belong to the *Severe acute respiratory syndrome-related coronavirus* species from the *Sarbecovirus* subgenus, genus *Betacoronavirus*, subfamily *Orthocoronavirinae* of the *Coronaviridae* family. The MERS-CoV belongs to the *Middle East respiratory syndrome-related coronavirus* species of another subgenus, namely *Merbecovirus*, within the same *Betacoronavirus* genus. Another well-studied representative of the coronaviruses, murine hepatitis virus (MHV), is a member of the *Murine coronavirus* species, belonging to the subgenus *Embecovirus* of the same *Betacoronavirus* genus.

The *Arteriviridae* family, one of the four members of the *Arnidovirineae* suborder, currently involves 6 subfamilies, 13 genera, 11 subgenera, and 23 species. Porcine reproductive and respiratory syndrome virus (PRRSV), a member of the *Betaarterivirus suid 1* species, which is the only species within the subgenus *Eurpobartevirus*, genus *Betaarterivirus*, subfamily *Variarterivirinae* of the *Arteriviridae* family, is of a special concern for the VLP technologies.

Figure 26.1 presents schematic cartoons of the order representatives, which are intended to show the previously mentioned diversity of viral architecture. As reviewed by de Groot et al. (2012), the members of the *Coronaviridae* family are roughly spherical enveloped particles, 120–160 nm in diameter, with a characteristic fringe of 15–20 nm petal-shaped surface projections termed *peplomers*. In the previously mentioned famous *Betacoronavirus* genus, a second, inner fringe of 5–7 nm surface projections is also seen. The CoV particles as studied by electron cryotomography are homogeneous in size and distinctively spherical with outer diameter 85 ± 5 nm of envelope. The envelope exhibits an unusual thickness (7.8 ± 0.7 nm), almost twice that of a typical biological membrane. The nucleocapsid is helical and tightly folded to form a compact structure that tends to closely follow the envelope. The arterivirus virions are significantly smaller than those of the other nidoviruses, spherical or egg-shaped and with a seemingly isometric core that contains the genome. The complete particles and nucleocapsids, as measured by electron cryomicroscopy, average 54 nm and 39 nm in diameter, respectively. No spikes are obvious on the arterivirus surface.

The members of the family *Coronaviridae* generally possess three or four envelope proteins. The spike protein S of corona-, toro-, and bafiniviruses are exceptionally large type I membrane glycoproteins of 1,200–1,600 aa residues, which are heavily N-glycosylated and perform the receptor-binding and membrane fusion functions. The coronavirus S proteins assemble into homotrimers, where the membrane-distal part of the peplomers comprises the receptor-binding domains and is composed of S1 subunits, but the C-terminal S2 subunits form a membrane-anchored stalk. The most abundant structural glycoprotein M has a similar triple-spanning membrane topology with a short N-terminus located on the outside of the virion and a long C-terminal endodomain, comprising an amphiphilic region and a hydrophilic tail. The pentameric integral membrane protein E, with ~20 copies per particle, exhibits ion channel and/or membrane permeabilizing activities. The highly basic, RNA-binding phosphoprotein N is involved not only in the in encapsidation and packaging of the genome but is

DOI: 10.1201/b22819-31

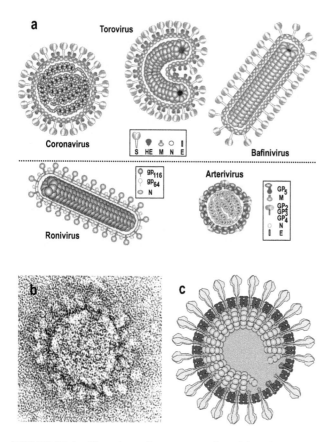

FIGURE 26.1 The schematic structures of particles of members of the order *Nidovirales*. (a) The classical classification (Reprinted from *Virus Taxonomy, Classification and Nomenclature of Viruses. Ninth Report of the International Committee on Taxonomy of Viruses*, King AMQ, Lefkowitz E, Adams MJ, Carstens EB (Eds), de Groot RJ, Cowley JA, Enjuanes L, Faaberg KS, Perlman S, Rottier PJM, Snijder EJ, Ziebuhr J, Gorbalenya AE, Order—*Nidovirales*. 785–795, Copyright 2012, with permission from Elsevier.) (b) The electron microscopic image of a negatively stained particle of SARS-CoV-2, causative agent of Covid-19 from the public CDC collection (ID 23640), provided by Cynthia S. Goldsmith and A. Tamin. (c) The modern cartoon presentation of SARS-CoV-2 taken with a kind permission from the ViralZone, Swiss Institute of Bioinformatics, http://viralzone. expasy.org (Hulo C et al. 2011). (Courtesy Philippe Le Mercier.)

involved in many other essential functions, such as RNA synthesis and translation and RNA chaperone activity.

The S protein is the major inducer of virus-neutralizing antibodies—which are directed mostly to the N-terminal half of the molecule—and a target for T cell responses. The surface-exposed N-terminus of the M protein induces virus-neutralizing antibodies, while the N protein evokes, like the S protein, protective T-cell responses (de Groot et al. 2012).

GENOME

The nidoviral genome is an infectious, linear, positive-sense single-stranded RNA molecule, which is capped and poly-adenylated. As reviewed by de Groot et al. (2012), the two groups—large and small nidoviruses—can be distinguished

by the genome size. The genomes of the large nidoviruses are well over 25 kb in length with size differences in the 5 kb range: 26.4–31.7 kb (*Coronavirus*), 28–28.5 kb (*Torovirus*), about 26.6 (*Bafinivirus*), and 26.2–26.6 kb (*Okavirus*). The small nidoviruses include the single *Arteriviridae* family with genomes of 12.7–15.7 kb in length. Therefore, the coronavirus genomes have the longest RNA virus genome known to date. Their RNA-dependent RNA polymerase is also the only one known to display a proofreading function, possibly to stabilize this long RNA sequence. Figure 26.2 presents the general structure of the SARS-CoV-2 genome, which is the most topical kind of nidovirus just now. The 5'-most two-thirds of the genome comprise the two large, partially overlapping ORFs 1a and 1b, which constitute the replicase gene and together encode a collection of enzymes that are part of the replication complex. The virion RNA functions as mRNA1 for the ORFs 1a and 1b, but the expression of the latter requires a programmed ribosomal frameshift.

FAMILY *CORONAVIRIDAE*

GENUS *ALPHACORONAVIRUS*

Subgenus *Pedacovirus*

Porcine Epidemic Diarrhea Virus

The 80–100% mortality in neonatal piglets is caused by porcine epidemic diarrhea virus (PEDV), which remains therefore the one of the most devastating viral diseases affecting swine worldwide. Wrapp and McLellan (2019) reported the electron cryomicroscopy structure of the PEDV protein S in the prefusion conformation at a resolution of 3.1 Å and revealed that the structure was substantially different from that observed in the previously determined spike structure from another representative of alphacoronavirus, namely human coronavirus NL63 (HCoV-NL63), which is described later.

Wang Cuiling et al. (2017) used the baculovirus expression system to produce the PEDV VLPs, which were composed of the spike protein S, membrane protein M, and envelope protein E, synthesized in *Sf*9 cells, and documented by electron microscopy as similar to the PEGV virions by their shape and diameter of ~150 nm. Remarkably, there was also a spike protein around the sphere. Furthermore, the baculovirus-driven PEDV VLPs displayed excellent immunogenicity in mice by induction of virus-neutralizing antibodies (Wang Cuiling et al. 2017). Hsu CW et al. (2020) engineered the PEDV VLPs containing the S, M, and E proteins, the genes of which were expressed by a novel polycistronic baculovirus expression vector. The electron microscopy demonstrated numerous VLPs, ~100 nm in diameter, displaying similar morphology to coronavirus. While pigs intramuscularly immunized with the VLPs alone were capable of eliciting systemic anti-PEDV S-specific IgG and cellular immunity, the co-administration of the PEDV VLPs with the mucosal CCL25/28 adjuvants further modulated the immune responses by enhancing

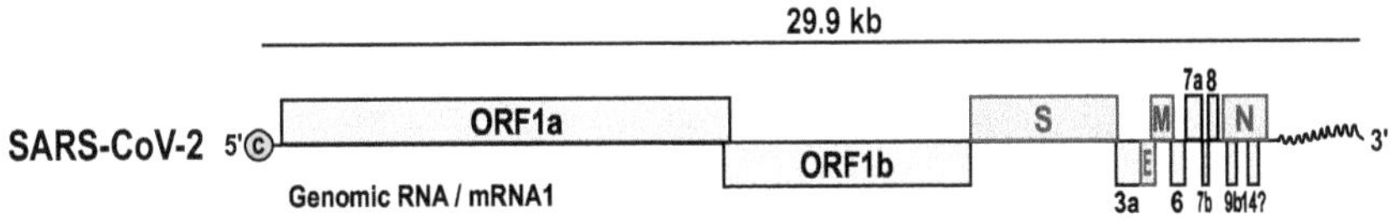

FIGURE 26.2 The genomic structure of the SARS-CoV-2 is based largely on the presentation at the ViralZone, Swiss Institute of Bioinformatics. The structural genes are colored dark pink.

systemic anti-PEDV S-specific IgG, mucosal IgA, and cellular immunity. The pigs immunized with the adjuvanted VLPs showed superior immune protection against PEDV (Hsu CW et al. 2020).

Kim et al. (2021) generated the PEDV VLPs in eukaryotic HEK293T cells by the expression of the PEDV S, E, M, and N genes. It was found that the M protein is essential for the formation of the PEDV VLPs. The formation of the latter was decreased in the presence of E protein and increased in the presence of N protein. In this study, the putative VLPs were concentrated by centrifugation through 20% sucrose cushion, but the electron microscopy was not referred. The immunization of mice with the PEDV VLPs induced Th2-dominant immune response, and both IgG and IgA antibodies were found. Moreover, these antibodies protected against virus infection in Vero cells (Kim et al. 2021).

A bit earlier, Makadiya et al. (2016) expressed the affinity tagged PEDV S1 protein, the N-terminal part of the protein S, in a secretory form in yeast *Pichia pastoris*, insect, and mammalian cells to identify the most suitable production system, where the mammalian HEK-293 T cells gave the highest yield of protein that was N-glycosylated and was the most appropriate candidate for vaccination. However, the vaccination with this S1 subunit vaccine failed to provide complete protection to suckling piglets after challenge exposure (Makadiya et al. 2016). Wang X et al. (2016) engineered the recombinant *P. pastoris* yeast expressing the PEDV S1 gene, tested the whole yeast by oral immunization of mice and piglets, and detected high levels of IgA against PEDV in piglets. Chang et al. (2020) used the silkworm *Bombyx mori* and its pupae to produce the full-length protein S, but the oral immunogen-expression strategy was not able to overcome the immunological unresponsiveness, which was possibly due to gastrointestinal specific barriers and oral tolerance. It should be mentioned here that the full-length S protein and the S1 subunit were displayed on baculovirus by Chang et al. (2018) and Hsu WT et al. (2021), as described in Chapter 6.

Subgenus *Setracovirus*

Human Coronavirus NL63

In contrast to the well-known human betacoronaviruses that are described later, the human alphacoronaviruses including human coronavirus NL63 (HCoV-NL63) cause mainly mild infections of the upper respiratory tract (van der Hoek et al. 2004), while SARS-CoV, MERS-CoV, and SARS-CoV-2 are responsible for severe disease with high

morbidity and mortality rate, as reviewed by Vellingiri et al. (2020). Moreover, the importance of the studies in human alphacoronaviruses is growing due to the established facts that the preexisting HCoV-NL63 antibody response is cross reacting with SARS-CoV-2 (Simula et al. 2020). It was suggested accordingly by the authors that the previous exposure to HCoV-NL63 epitopes would produce antibodies that could confer a protective immunity against SARS-CoV-2 and probably reduce the severity of the disease.

Naskalska et al. (2018) generated the HCoV-NL63 VLPs using the baculovirus expression system. For that, the recombinant baculoviruses coding for M, E, and S proteins of HCoV-NL63 were created and propagated in the HF insect cells. All sequences were codon-optimized for the expression in insect cells, synthesized, and subcloned to one bicistronic or two monocistronic separate donor plasmids for more flexible regulation of protein expression. Additionally, the M and E protein tagged with HA peptide and S protein fused with GFP were engineered and the expression of all tagged and untagged S, M, and E was successfully achieved. The HCoV-NL63 VLPs were properly formed, however, only when all three proteins were untagged or when M protein was HA-tagged, but the latter not only decreased the M protein synthesis but also impaired its release to the culture medium. By the coinfection of HF cells with the recombinant baculoviruses carrying separately the M + E genes and the gene S, the spherical enveloped particles, resembling the previously published coronaviral VLPs were observed. The diameter of the putative VLPs ranged from 40 nm–200 nm, and the distinctive spike projections on the outer rim of these structures were visible. The VLPs were purified from the culture media in the case of the mentioned (M + E) and S expression or from cell homogenates when the combination (M-HA + E) and S was used. The hydrodynamic particle size measurements indicated that the mean diameter of the purified VLPs was 180 nm. Next, both tagged and untagged HCoV-NL63 VLPs were effectively internalized into experimental macaque kidney epithelial cells. Furthermore, the VLPs were able to deliver cargo when the HA-tagged VLPs were decorated with anti-HA tag monoclonal antibody and selectively transduce cells expressing the ACE2 protein such as ciliated cells of the respiratory tract (Naskalska et al. 2018). This was therefore the first attempt to use the CoV VLPs as specific delivery vectors and convert them into a valuable delivery platform.

Zhang K et al. (2020) resolved the structure of the HCoV-NL63 protein S trimer to 3.4 Å by single-particle electron cryomicroscopy imaging of vitrified virions without chemical fixative. The domain arrangement appeared strikingly different from that of the SARS-CoV-2 protein S and explained their different requirements for the activating binding to the receptor.

Subgenus *Tegacovirus*

Transmissible Gastroenteritis Virus

The structural genes of transmissible gastroenteritis virus (TGEV) from the *Alphacoronavirus 1* species, which encoded nucleoprotein, integral membrane protein, and spike (peplomer) protein, were expressed in parallel by mammalian and baculovirus expression systems at the early stage of the protein engineering studies (Pulford et al. 1990). The *Sf*9 insect cells-produced spike protein S of 175 kDa was able to trimerize and was transported to the cell surface, as is the authentic TGEV S protein (Godet et al. 1991). Despite the lack of complete carbohydrate processing, the recombinant S protein exhibited antigenic properties similar to TGEV S and induced high levels of neutralizing antibodies in immunized rats. The deletion of 70 aa from the C-terminus containing the membrane anchor of the polypeptide allowed its secretion, although the oligomerization process and the antigenic profile of the anchor-free S protein was altered (Godet et al. 1991). When immunized with insect cells, which were infected with the recombinant baculoviruses encoding the protein S or its fragments, the piglets induced virus-neutralizing antibody titers (Tuboly et al. 1994). Shoup et al. (1997) reported that the parenterally administered TGEV S glycoprotein vaccines elicited virus-neutralizing antibodies to TGEV in serum and colostrum that do not fully provide active or passive immunity in swine. Furthermore, Sestak et al. (1999) examined the protective ability of the three major structural TGEV proteins S, N, and M produced by the baculovirus expression system, in the form of cell lysates, against challenge with TGEV. Later, the baculovirus-driven truncated protein S of TGEV was successfully employed by the development of a diagnostic ELISA test (López et al. 2009). It is quite clear that these expression studies did not raise the VLP issue.

The appearance of the TGEV "pseudoparticles" by the vaccinia virus-driven expression in mammalian cells was first reported by Baudoux et al. (1998). Such VLPs resembling authentic virions by size were released in the culture medium of the cells that coexpressed the TGEV genes M and E. The interferogenic activity of the purified VLPs was comparable to that of the TGEV virions, thus establishing that neither ribonucleoprotein nor spikes were required for the interferon induction. The replacement of the externally exposed, N-terminal domain of the protein M with that of bovine coronavirus (BCV) led to the production of the chimeric VLPs with no major change in interferogenicity, although the structures of the TGEV and BCV ectodomains markedly differed. Moreover, the BCV pseudoparticles were produced and demonstrated the same interferogenic activity, suggesting that the ability of coronavirus particles to induce interferon-α was a result of a specific multimeric structure (Baudoux et al. 1998).

Genus *Betacoronavirus*

Subgenus *Embecovirus*

Murine Hepatitis Virus

The early coronavirus studies were performed with murine hepatitis virus (MHV), a member of the *Murine coronavirus* species. Thus, Krijnse Locker et al. (1995) performed the vaccinia virus-driven expression of the MHV gene encoding the membrane glycoprotein M and found that the independently synthesized protein M accumulated in the Golgi apparatus in homomultimeric, detergent-insoluble structures, presumably as part of its retention mechanism. When Opstelten et al. (1995) used the vaccinia virus-driven coexpression of the MHV proteins M and S, the formation of the heteromultimeric M-S complexes was detected in the absence of other coronaviral proteins. Vennema et al. (1996) described the assembly of the MHV envelope independent of a nucleocapsid. The membrane particles containing coronaviral envelope proteins were assembled in and released from animal cells coexpressing the corresponding genes from transfected plasmids. Of the three viral membrane proteins, only two were required for particle formation, namely the membrane glycoprotein M and the small envelope protein E, while the spike protein S was dispensable but was incorporated when present. Therefore, the nucleocapsid protein N was neither required nor taken into the particles when present. The envelope vesicles formed a homogeneous population of spherical particles indistinguishable from authentic coronavirions in size (~100 nm in diameter) and shape but less dense than the virions (Vennema et al. 1996). Remarkably, Maeda et al. (1999) have found that the expression of the MHV protein E alone was sufficient for the VLP production. Godeke et al. (2000) constructed the two chimeric proteins S of MHV and an alphacoronavirus, namely feline infectious peritonitis virus (FIPV), or feline coronavirus (FCoV), a member of the *Tegacovirus* subgenus, which is described earlier. The chimeric MHV-FIPV proteins S consisted of the ectodomain of the one virus and the transmembrane and endodomain of the other. They were found to assemble only into viral particles of the species from which their C-terminal domain originated. Thus, the 64-terminal-residue sequence sufficed to draw the 1,308 (MHV)- or 1,433 (FIPV)-aa-long mature S protein into the VLPs (Godeke et al. 2000).

The individual structural MHV proteins were produced by the baculovirus-driven expression system in insect cells, namely the peplomer glycoprotein E2 (Yoden et al. 1989), the spike protein S (Parker et al. 1990; Taguchi et al. 1990; Takase-Yoden et al. 1991). These expressions resulted in the production of the glycosylated polypeptides of the expected molecular mass, which were transported to the surface of the cell but did not form any VLP structures.

Bosch et al. (2004) engineered a viable chimeric MHV virions carrying the protein S that was extended N-terminally with GFP. The efficiency of incorporation of the chimeric S-GFP protein into the virions was, however, reduced relative to that in the wild-type particles which may explain, at least in part, the reduced infectivity produced by MHV-S-GFP infection.

Walls et al. (2016) resolved the 3D structure of the MHV spike glycoprotein S trimer ectodomain to 4.0 Å by the single-particle electron cryomicroscopy. Remarkably, the structure shared a common core with the F proteins of paramyxoviruses from the order *Mononegavirales* (Chapter 31), implicating mechanistic similarities and an evolutionary connection between these viral fusion proteins. The structure was compared with the crystal structures of the human coronavirus S domains, which are reviewed later.

Subgenus *Merbecovirus*

MERS-CoV

Gao et al. (2013) performed the thorough 3D examination of the so-called fusion core of the MERS-CoV. As known, the heptad repeats HR1 and HR2 of the S protein assemble into a complex called the fusion core, which represents a key membrane fusion architecture. The HR sequences were variably truncated, then connected with a flexible aa linker and produced in *E. coli*. The recombinant protein automatically assembled into a trimer in solution, displaying a typical α-helical structure. One of these trimers was crystallized, and its structure was solved at a resolution of 1.9 Å. A canonical 6-helix bundle, like those reported for other coronaviruses, was revealed, with three HR1 helices forming the central coiled-coil core and three HR2 chains surrounding the core in the HR1 side grooves. Moreover, the complex structure of the receptor binding domain in the MERS-CoV protein S1 bound to its receptor CD26, also known as dipeptidyl peptidase 4 (DPP4), has been successfully solved by the same team (Lu G et al. 2013).

Coleman et al. (2014) produced the MERS-CoV protein S by the baculovirus expression system, by analogy with the previously published protocol for the SARS-CoV-1 (see following paragraph). The S proteins were extracted from infected insect cells pellet with a nonionic detergent and purified by column chromatography. The purified full-length MERS and SARS S proteins had an apparent molecular mass of 180 kDa and 160 kDa by SDS-PAGE respectively and formed nanoparticles of ~25 nm diameter consisting of multiple S protein molecules when observed by electron microscopy or dynamic light scattering. These S-formed VLPs, in conjunction with an adjuvant, were able to stimulate virus-neutralizing antibody responses in mice (Coleman et al. 2014). Such MERS-CoV S nanoparticles, in combination with the Matrix-M1 adjuvant, produced high-titer anti-S neutralizing antibody and protected mice from MERS-CoV infection in vivo (Coleman et al. 2017). Jung et al. (2018) developed two types of the MERS-CoV vaccines: the protein S nanoparticles of mean diameter 35 nm or ~80 nm when formulated with alum adjuvant and the recombinant adenovirus serotype 5 encoding the MERS-CoV spike gene. Both heterologous prime-boost and homologous S protein nanoparticles vaccinations provided protection from MERS-CoV challenge in mice (Jung et al. 2018).

Then, Kato et al. (2019) expressed the S protein of MERS-CoV lacking its transmembrane and cytoplasmic domains (SΔTM), which was secreted into the hemolymph of silkworm larvae using a bombyxin signal peptide and purified using affinity chromatography. The purified SΔTM formed the small nanoparticles like the full-length S protein and had the ability to bind the DPP4 receptor. These results indicate that bioactive SΔTM was expressed in silkworm larvae (Kato et al. 2019).

Wang Chong et al. (2017a, b) constructed recombinant baculovirus co-expressing the S, E, and M genes. The infection of *Sf*9 cells with this recombinant baculovirus resulted in the successful assembly of the MERS-CoV VLPs. The electron microscopy and immunoelectron microscopy results demonstrated that the MERS-CoV VLPs were structurally similar to the native virions. The VLPs not only closely resembled the SARS-CoV virions in size and particle morphology but also in virus morphogenesis. Remarkably, the S and M proteins alone cannot assemble into the VLPs, as assessed by electron microscopy. The rhesus macaques inoculated with the MERS-CoV VLPs and Alum adjuvant induced a T-helper 1 cell (Th1)-mediated immunity and virus-neutralizing antibodies including specific IgG antibodies against the receptor binding domain (Wang Chong et al. 2017a, b).

After the previously described SΔTM nanoparticles, Kato et al. (2019) produced in the insect *Bm*5 cells the true MERS-CoV VLPs when coexpressed with the S, M, and E genes. Surprisingly, the S protein was not displayed on the VLPs even though both E and M proteins were secreted into the culture supernatant. The S protein-displaying nanovesicles with diameters of ~100–200 nm were prepared by surfactant treatment and mechanical extrusion using S protein- or three structural protein-expressing *Bm*5 cells, as confirmed by immune electron microscopy (Kato et al. 2019).

Generally, the MERS vaccine candidates were comprehensively reviewed (Excler et al. 2015; Wang L et al. 2015; Du et al. 2016; Yong et al. 2019; Hashemzadeh et al. 2020).

Subgenus *Sarbecovirus*

SARS-CoV-1

The phylogenetic analysis of the SARS-CoV-1 showed that it was neither a mutant nor a recombinant of previously characterized CoVs and formed a new, distinct group within the genus (Eickmann et al. 2003; Marra et al. 2003; Rota et al. 2003; Stadler et al. 2003; Gibbs et al. 2004). As it was reviewed in brief by Ingallinella et al. (2004), the spike protein S, which was displayed in ~200 copies on the viral membrane as a trimer, was divided into two subdomains of similar size: S1 and S2, where S1 formed the globular portion of the spike, which mediated binding to host cell receptors, but the more conserved S2 formed the membrane-anchored stalk region and mediated viral-host cell fusion. The cleavage between S1 and S2 was not an absolute requirement for the fusion of CoV, and the available data suggested that the S protein of SARS-CoV-1 was not cleaved into subunits. The S2 contained two predicted amphipathic α-helical regions, with a 4,3 heptad repeat (HR), characteristic of coiled coils and a common feature of Type I viral fusion proteins. As a result, Ingallinella et al. (2004) showed that the peptides including the heptad repeats (HRs), namely the entire HR1 (residues 889–972) and HR2 (residues 1142–1185) regions of the SARS-CoV-1 S2 protein, formed a trimeric, proteolytically stable, α-helical core complex. Furthermore, Supekar et al. (2004) crystallized this complex and resolved its structure to 1.6 Å. Duquerroy et al. (2005) resolved the crystal structure of the HR1/HR2 complex of the SARS-CoV-1 to 2.2 Å and complemented therefore the previous 3D studies on the SARS-CoV-1 protein S (Superkar et al. 2004; Xu Y et al. 2004). Li F et al. (2005) determined the crystal structure at 2.9 Å resolution of the receptor-binding domain (RBD) bound with the peptidase domain of human angiotensin-converting enzyme 2 (ACE2) as the receptor. Beniac et al. (2006) demonstrated the structure of the SARS-CoV-1 virions by the electron microscopy and 3D image reconstruction at the 16 Å resolution. The virions averaged 1,185 Å in diameter, including the spikes, with a 36-Å-thick lipid bilayer envelope that was approximately spherical and had an average diameter of 865 Å. The envelope-anchored spike had a structure about 180 Å in diameter, with three distinct protuberances or domains 50 Å thick on each subunit of the trimer (similar in appearance to the blades of a propeller) and a thin stalk connecting the spike to the viral envelope. The "blades" were twisted at an angle of ~30° to the axis of symmetry and were almost certainly composed of the S1 domain of the SARS-CoV-1 S polypeptide (Beniac et al. 2006). In parallel, the single-particle image analysis was applied to compare the 3D structure of the SARS-CoV-1 virions with that of MHV and FCoV (Neuman et al. 2006a, b).

Ho et al. (2004) used the baculovirus expression system to produce the SARS-CoV-1 VLPs in insect *Sf*21 cells. The VLPs were assembled by the coinfection of the cells with recombinant baculoviruses, which separately expressed the S, M, and E genes. The M and E proteins were sufficient for the efficient formation of the VLPs, while the S protein was able to incorporate into the VLPs also. The VLPs of 110 nm in diameter bore resemblance to the authentic virions when examined by electron microscopy (Ho et al. 2004). Mortola and Roy (2004) generated the recombinant baculoviruses that expressed not only the S, M, and E but also the nucleocapsid N gene, either individually or simultaneously, in *Sf*9 cells. As expected, the coinfections of insect cells with two recombinant viruses demonstrated that M and E could assemble readily to form the smooth-surfaced VLPs, but in the absence of the E protein the M and S did not assemble into the VLPs. Moreover, the simultaneous high-level expression of S, E, and M by a single recombinant virus allowed the very efficient assembly and release of the VLPs that appeared as morphological mimics of the virion particles. When the *Sf*9 cells were coinfected with the triple recombinant

baculovirus together with the single recombinant virus expressing the N gene, the examination of the purified VLPs released into the media still had only S, M, and E proteins but did not exhibit the presence of the N protein, despite the fact that the N protein was synthesized in the infected *Sf*9 cells (Mortola and Roy 2004).

Later, Khan et al. (2007) presented methodology of how to obtain the insect cells-derived SARS-CoV-1 VLPs that consisted ot the M and E proteins. Lu X et al. (2007) not only got the SARS-CoV-1 VLPs from the S, M, and E proteins but also demonstrated their high immunogenicity in mice and protective immunity against the infection of the S protein-pseudotyped murine leukemia virus. In fact, this was the first study describing the immunogenicity and protective potency of the SARS CoV-1 VLPs. Bai et al. (2008) generated two recombinant baculoviruses that expressed the S gene of SARS-like coronavirus isolated from bats and the E and M genes from SARS-CoV-1, respectively. The immunological properties of such mixed VLPs were thoroughly evaluated in mice. Lu B et al. (2010) turned to the mucosal immunization of mice with the SARS-CoV-1 VLPs and found that both intranasal and intraperitoneal immunizations, with or without the CpG adjuvant, elicited both systemic and mucosal immune responses.

Liu Y V et al. (2011) produced in insect *Sf*9 cells high levels of the mixed SARS-CoV-1 VLPs containing the SARS protein S and the influenza M1 protein. These mixed VLPs had a similar size and morphology to the wild-type SARS-CoV-1 and demonstrated promising results in the mouse lethal challenge model. Moreover, it was found in this study that the purified SARS protein S formed ~25 nm diameter particles consisting of multiple S protein molecules when observed by dynamic light scattering and electron microscopy (Liu YV et al. 2011). As narrated earlier in the MERS paragraph in detail, Coleman et al. (2014) concentrated on this novel type of the putative VLPs and purified them for both MERS and SARS-1 CoVs. These S-formed VLPs, in combination with an adjuvant, were able to stimulate the appropriate virus-neutralizing antibody responses in mice (Coleman et al. 2014).

In mammalian expression systems, Huang Y et al. (2004) expressed the codon-optimized SARS-CoV-1 genes S, M, N, and E in human 293 renal epithelial cells and found that the M and N proteins were necessary and sufficient to self-assemble. The addition of the gene S facilitated budding of the VLPs that contained a corona-like halo resembling SARS-CoV-1 when examined by electron microscopy. The S, M, and N—but not E—proteins of the SARS-CoV-1 were found therefore necessary and sufficient for the VLP assembly (Huang Y et al. 2004). Hsieh et al. (2005) expressed the four His-tagged SARS-CoV-1 genes S, M, N, and E in Vero E6 cells and concluded that that both E and M expressed alone could be released as the self-assembled particles and that the E and M proteins were likely to form the VLPs when they were coexpressed, while the protein N had an essential role in the packaging of SARS-CoV-1 RNA.

To solve the problem of the minimal SARS-CoV-1 self-assembly requirements, either M and E proteins or M and N proteins, Siu et al. (2008) demonstrated that both E and N proteins must be coexpressed with M protein for the efficient production and release of the VLPs by transfected Vero E6 cells. This suggested that the mechanism of the SARS-CoV-1 assembly differed from that of other studied coronaviruses, which only required M and E proteins for the VLP formation. When coexpressed, the native trimeric protein S was incorporated onto VLPs. Remarkably, when a fluorescent protein tag was added to the C-terminal end of N or S protein (but not M protein) the chimeric viral proteins assembled within the VLPs and allowed visualization of the VLP production and trafficking in living cells by imaging technologies (Siu et al. 2008). In parallel, Nakauchi et al. (2008) confirmed the fact that the four proteins S, M, N, and E were necessary to form the correct SARS-CoV-1 VLPs, when synthesized in 293T cells. Moreover, a specific interaction between the N and M proteins was demonstrated, suggesting that the protein N binds directly to the protein M to be incorporated into the VLPs (Nakauchi et al. 2008).

In contrast to these data, Tseng YT et al. (2010) found the self-assembly and release of the SARS-CoV-1 protein M in medium of a variety of cells such as 293T, HeLa, or Vero-E6, in the form of membrane-enveloped vesicles with densities lower than those of VLPs formed by M plus N. Although the M self-assembly involved both N- and C-terminal regions along the M sequence, the N-terminal 50 aa residues containing the first transmembrane domain were sufficient for the M self-association (Tseng YT et al. 2010).

Lokugamage et al. (2008) produced in 293T or CHO cells the mixed VLPs that consisted of the SARS-CoV-1 protein S and the MHV proteins M, E, and N. The immunization experiments of mice showed that such mixed VLPs could be an effective vaccine strategy against SARS-1 infection. These mixed VLPs were included into the study of Tseng CT et al. (2012), who evaluated four different vaccine candidates in mice. The authors concluded, however, that all tested variants given any of the vaccines led to occurrence of Th2-type immunopathology after challenge of mice, suggesting hypersensitivity to SARS-CoV-1 components, and a definite caution was recommended in proceeding to application of a SARS-CoV-1 vaccine in humans (Tseng CT et al. 2012).

As to the yeast expression system, Cao et al. (2005) expressed at high level the SARS-CoV-1 gene N in *P. pastoris*. The produced protein N displayed the expected β-sheet secondary structure in solution, but no data about its putative self-assembly were presented. Chen WH et al. (2020b) developed the vaccine candidate that was based on a high-yielding, *P. pastoris*-derived receptor-binding domain, so-called RBD219-N1, of the SARS-CoV-1 protein S. When formulated with Alhydrogel, the RBD219-N1 not only induced high levels of neutralizing antibodies against a clinical SARS-CoV-1 isolate but also fully protected mice

from the lethal SARS-CoV-1 challenge and was superior on these indicators to the whole protein S.

Concerning the bacterial expression, Chen J et al. (2005) demonstrated that the RBD-based SARS-CoV-1 vaccines could also be produced in *E. coli*, without counting obviously on the VLP production.

SARS-CoV-2

The genome of the modern age SARS-CoV-2 encodes the same four structural proteins: the spike S, membrane M, envelope E, and nucleocapsid N, as other coronaviruses. However, the ~600-kDa trimeric S protein, which is one of the largest known class I fusion proteins, is heavily glycosylated with 66 N-linked glycans and each S protomer comprises the S1 and S2 subunits with a single transmembrane anchor (Walls et al. 2020; Watanabe et al. 2020; Wrapp et al. 2020). As currently reviewed by the numerous op-ed articles, the most vaccine-related of which are listed at the end of this paragraph, the S protein binds to the cellular surface receptor angiotensin-converting enzyme 2 (ACE2) through the receptor binding domain (RBD), and the cleavage of S1/S2 by furin-like protease is required for the activation of the protein S. The receptor binding sites are exposed only when the RBDs adopt an "up" conformation. The appropriate "RBD down," "one RBD up," and "two-RBD up" conformations have been observed in the recombinantly expressed S proteins of SARS-CoV-2 (Henderson et al. 2020; Walls et al. 2020; Wrapp et al. 2020). Casalino et al. (2020) unveiled the essential structural role of N-glycans at sites N165 and N234 in modulating the conformational dynamics of the spike's RBD. The first largest electron cryotomography dataset of the SARS-CoV-2 was published by Yao et al. (2020), where the native structures of the S proteins in pre- and postfusion conformations were determined to average resolutions of 8.7–11 Å. The full-length model of the glycosylated SARS-CoV-2 S protein, both in the open and closed states, was presented. Moreover, the native conformations of the ribonucleoproteins and their higher-order assemblies were revealed, including the global architecture of the SARS-CoV-2 and the issue of how the virus packs its ~30-kb-long single-segmented RNA into the ~80-nm-diameter lumen (Yao et al. 2020). The general outcome of this structural study is shown in Figure 26.3. A few months later, Ke et al. (2020) imaged by electron cryotomography the intact SARS-CoV-2 virions and determined the high-resolution structure, conformational flexibility, and distribution of the S trimers in situ on the virion surface. Figure 26.4 presents a portrait that was based on both studies and appeared at the popular PDB-101 portal intended to promote exploration in the world of proteins and nucleic acids.

Fujita et al. (2020) applied the baculovirus-silkworm system to produce the ectodomain of the SARS-CoV-2 protein S. Dai et al. (2020) used both baculovirus and mammalian cell expression systems to produce single-chain dimers of the RBD tandem repeats from the protein S of MERS-CoV, SARS-CoV-1, and SARS-CoV-2 spanning aa

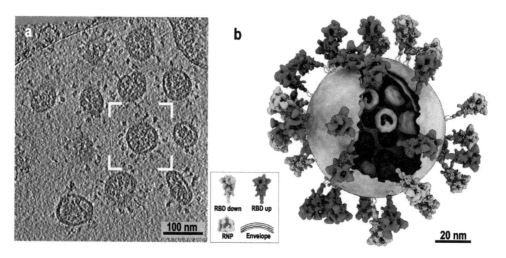

FIGURE 26.3 The molecular architecture of the SARS-CoV-2. (a) A representative tomogram slice (5 Å thick) showing pleomorphic SARS-CoV-2 virions. (b) A representative virus boxed in (a) is reconstructed by projecting the prefusion S in the "RBD down" conformation (salmon) and "one RBD up" conformation (red), the lipid envelope (gray), and RNPs (yellow) onto their refined coordinates. RNPs away from the envelope are hidden for clarity. The unsolved stem regions of the spikes are sketched from a predicted model (https://zhanglab.ccmb.med.umich.edu/COVID-19). RBD, receptor binding domain; RNP, ribonucleoprotein. (Reprinted from *Cell*. 183, Yao H, Song Y, Chen Y, Wu N, Xu J, Sun C, Zhang J, Weng T, Zhang Z, Wu Z, Cheng L, Shi D, Lu X, Lei J, Crispin M, Shi Y, Li L, Li S, Molecular architecture of the SARS-CoV-2 virus, 730–738.e13, Copyright 2020, with permission from Elsevier.)

FIGURE 26.4 SARS-CoV-2 and neutralizing antibodies. This painting shows a cross section through SARS-CoV-2 surrounded by blood plasma, with neutralizing antibodies in bright yellow. The painting was commissioned for the cover of a special COVID-19 issue of *Nature*, presented 20 August 2020, and is currently in the collection of the Cultural Programs of the National Academy of Sciences. It incorporates information from two electron cryomicroscopy studies that explore the shape and distribution of spikes and the nucleoprotein (Ke et al. 2020; Yao et al. 2020). (Illustration by David S. Goodsell, RCSB PhgfSsvcfg65rotein Data Bank; doi: 10.2210/rcsb_pdb/goodsell-gallery-019.)

367–606, 306–527, and 319–537, respectively. For each construct, signal peptide sequence 1–17 of MERS-CoV S protein was added to the N-terminus for protein secretion, and the His$_6$ tag was added to the C-terminus to facilitate further purification processes. The RBD-single-chain dimers of MERS-CoV and SARS-CoV-2 were developed for pilot scale production in GMP grade manufacturing at high yields in an industry-standard Chinese hamster ovary (CHO) cell system (Dai et al. 2020).

Li T et al. (2020) employed the baculovirus expression system to evaluate four versions of the S protein, namely RBD, S1 subunit, the wild-type S ectodomain, and the prefusion trimer-stabilized form, where RBD appeared as a monomer in solution, whereas the other three variants associated as homotrimers with substantial glycosylation. Goepfert et al. (2021) performed the first ongoing in-human trial of the baculovirus-driven and insect cells-produced SARS-CoV-2 preS dTM vaccine, which was based on a stabilized prefusion protein S and developed by Sanofi Pasteur. It is obvious therefore that none of the baculovirus-driven vaccine candidates was developed on the VLP scaffold. Yang et al. (2020) constructed a subunit vaccine composed of the aa residues 319–545 of the SARS-CoV-2 RBD and produced it through the baculovirus expression system. The preclinical study reported that the vaccine could protect the nonhuman primates from the SARS-CoV-2 infection with little toxicity (Yang et al. 2020).

The first SARS-CoV-2 protein S-based vaccine that demonstrated immunogenicity in humans was reported by Keech et al. (2020). This vaccine used the full-length, i.e., including the transmembrane domain, wild-type SARS-CoV-2 protein S and was produced in *Sf*9 cells.

Xu R et al. (2020) were the first ones to report the generation of the SARS-CoV-2 VLPs. The mammalian cells were selected for the expression due to the correct protein glycosylation patterns. Among the four SARS-CoV-2 structural proteins, the expression of the membrane protein M and small envelope protein E genes were necessary and sufficient for the efficient formation and release of the SARS-CoV-2 VLPs. Moreover, the corona-like structure presented in the SARS-CoV-2 VLPs from Vero E6 cells was more stable and unified, as compared to those from HEK-293T cells. Remarkably, the M protein was able to easily release into media independent of other structural proteins, while the E protein could also be secreted when expressed alone but to a much smaller extent compared to M, indicating that the M was an essential driver of VLP formation. The efficient formation of the S-containing VLPs could be driven by the coexpression of any other proteins, but the presence of M resulted in the cleavage of the membrane fusion subunit of the protein S. The gained VLPs displayed typical corona-like structure when evaluated by electron microscopy. Compared to the SARS-CoV-2 VLPs constructed from Vero E6 cells, the VLPs formed by HEK-293T cells were more variable both in shape and size. Specifically, the average diameter of the SARS-CoV-2 VLPs from HEK-293T cells fell around 90.33 ± 32.45 nm, whereas those assembled in Vero E6 cells were smaller, showing about 71.02 ± 21.98 nm. Apart from the diameter, the length of spike protein from these two systems was remarkably distinctive, exhibiting 8.33 ± 2.55 nm in the SARS-CoV-2 VLPs from HEK-293T cells and 10.72 ± 3.253 nm in Vero E6 systems (Xu R et al. 2020). Plescia et al. (2020) also assessed the four structural proteins from SARS-CoV-2 for their ability to form the VLPs from human HEK293 cells and provided methods and resources of producing, purifying, fluorescent, and APEX-2 labeling of the SARS-CoV-2 VLPs for the evaluation of mechanisms of viral budding and entry as well as assessment of drug inhibitors under BSL-2—but not under BSL-3 conditions. Boson et al. (2021) concentrated on the specific functions and processing of the structural proteins by their expression in Vero E6 and 293T cells. Swann et al. (2020) reported assembly of the SARS-CoV-2 VLPs by coexpressing the genes S, M, and E in mammalian 293T cells. The assembled SARS-CoV-2 VLPs possessed the protein S spikes on particle exterior, and the particles ranged in shape from spherical to elongated with a characteristic size of 129 ± 32 nm. Moreover, the authors tested the structural integrity of the SARS-CoV-2 VLPs attached to dry glass using atomic force microscopy (Swann et al. 2020).

Kumar B et al. (2021) generated the SARS-CoV-2 VLPs in 293T cells by concomitant synthesis of three viral membrane proteins S, E, and M, along with the cytoplasmic nucleocapsid N, and the VLP production and secretion was found to be highly dependent on the protein N. The N proteins from related betacoronaviruses variably substituted for the homologous SARS-CoV-2 N, and the chimeric betacoronavirus N proteins effectively supported the VLP production if they contained the SARS-CoV-2 N C-terminal domains (CTD), which were established as critical features of the VLP assembly. Second, the authors produced the VLPs with Nluc peptide fragments appended to E, M, or N proteins, with each subsequently inoculated into target cells expressing complementary Nluc fragments. The complementation into the functional Nluc was used therefore to assess virus-cell entry by mimicking the natural SARS-CoV-2 assembly and cell entry processes. The VLP system utilized an 11 aa Nluc fragment called HiBiT, which was appended to the CoV structural proteins and served as a marker for virus assembly, secretion from VLP-producing cells, and subsequent VLP entry into target cells. The HiBiT marker was identified by protein complementation with a larger Nluc fragment called LgBiT, with resulting Nluc detection by luminometry (Schwinn et al. 2018). Remarkably, the elaborated SARS-CoV2 VLP systems facilitated studies of virus assembly and entry in the BSL-1 laboratory settings (Kumar B et al. 2021).

Lu J et al. (2020) presented an mRNA vaccine candidate that was formulated from a cocktail of mRNAs encoding the S, M, and E genes and ensured therefore the formation of the SARS-CoV-2 VLPs. The latter were detected in culture medium of the transfected HEK 293A cells. The mice receiving the VLP vaccine had the strongest immune response and developed significantly higher titers of the S-specific antibody than mice receiving a vaccine consisting of the S-encoding mRNA only (Lu J et al. 2020).

Ju et al. (2021) developed the BSL-2 cell culture system for production of the transcription- and replication-competent SARS-CoV-2 VLPs (trVLPs). This trVLP expressed the reporter gene GFP that replaced the gene N. The complete viral life cycle was achieved in the cells ectopically expressing the SARS-CoV-1 or SARS-CoV-2 proteins N but not the MERS-CoV N. This system led to the development of the efficient SARS-CoV-2 reverse genetics tool to dissect the virus life cycle under the BSL-2 condition. Suprewicz et al. (2021) used this system to evaluate binding of vimentin by the trVLPs.

Concerning the yeast expression, Arora et al. (2020) were the first ones to report the production of the SARS-CoV-2 VLPs by the coexpression of the genes S, M, and E in *Saccharomyces cerevisiae*. The electron microscopy images clearly outlined the presence of the corona-like morphology and uniform size distribution of the VLPs. The size of the VLPs of 90–124 nm, with a mean of 103 $\pm$ 17 nm, was determined by the dynamic light scattering (DLS) technique. Based on this expression, Mazumder et al. (2020) developed the PRAK-03202 vaccine candidate, which induced the SARS CoV-2 specific neutralizing antibodies in BALB/c mice. The PBMCs from convalescent patients, when exposed to the PRAK-03202, showed lymphocyte proliferation and elevated IFN-γ levels, supporting therefore the clinical development and testing of the PRAK-03202 for use in humans.

Developing further the previously mentioned success of the RBD-based *P. pastoris*-derived SARS-CoV-1 vaccine candidate, the homologous non-VLP RBD-based SARS-CoV-2 vaccine RBD219-N1C1 was elaborated (Chen WH et al. 2021; Pollet et al. 2021).

As a specific tool of the yeast expression system to be mentioned, Thi Nhu Thao et al. (2020) used the *S. cerevisiae*-based transformation-associated recombination (TAR) cloning technique to maintain the SARS CoV-2 genome as a yeast artificial chromosome and T7 RNA polymerase that was then used to generate infectious RNA to rescue viable virus. This approach allowed engineering and generation of chemically synthesized clones of SARS-CoV-2 in only a week after receipt of the synthetic DNA fragments.

At last, Ward et al. (2021) expressed the full-length gene S glycoprotein of SARS-CoV-2 in *Nicotiana benthamiana* plants. The S protein was modified with R667G, R668S, and R670S substitutions at the S1/S2 cleavage site to increase stability and K971P and V972P substitutions to stabilize the protein in prefusion conformation. The signal peptide was replaced with a plant gene signal peptide, and the transmembrane domain (TM) and cytoplasmic tail 220 (CT) of the protein S was also replaced with TM/CT from Influenza H5 A/Indonesia/5/2005 to increase the VLP assembly and budding. The self-assembled VLPs bearing S protein trimers were isolated from the plant matrix and subsequently purified using a process similar to that described for the influenza vaccine candidates (Chapter 33). As a result, the trimeric spike glycoproteins were displayed at the surface of the VLPs, so-called Co-VLPs, which mimicked the shape and the size of the virions. The candidate CoVLP vaccine administered alone or with AS03 or CpG1018 adjuvants was evaluated in a Phase 1 trial in healthy adults. It was concluded that the CoVLPs+AS03 demonstrated a good benefit/risk ratio and supported the transition of this formulation to studies in expanded populations and to efficacy evaluations (Ward et al. 2021).

Generally, the prospective ways to develop the SARS-CoV-2 vaccines, including the putative employment of the VLP approach, as well as the promising vaccine candidates together with the acting companies, were promptly followed and reviewed in the papers published in 2020: March (Chen WH et al. 2020a; Zhang J et al. 2020), April (Callaway 2020), May (Uddin et al. 2020), June (Conte et al. 2020), July (Chauhan et al. 2020; He et al. 2020; Sheikhshahrokh et al. 2020; Speiser and Bachmann 2020), August (Ng et al. 2020), September (Haq et al. 2020; Pandey et al. 2020; Rabaan et al. 2020), October (Dong et al. 2020; Kaur and Gupta 2020; Krammer 2020), November (Cardoso et al. 2020; Poland et al. 2020; Varghese et al. 2020), and December (Zhao et al. 2020) and in 2021: January (Karpiński et al. 2021; Ura et al. 2021), February (Forni et al. 2021); March (Funk et al. 2021; Huang L et al. 2021; Shahcheraghi et al. 2021), April (Peacock et al. 2021), May (Raoufi et al. 2021),

June (Kumar M et al. 2021), July (Peyret et al. 2021), and August (Fathizadeh et al. 2021).

Special attention should be focused on the reviews that addressed the role of the putative nanotechnologies and nanomaterials by the fight against the SARS-CoV-2 infection and Covid-19 pandemic (Gurunathan et al. 2020; Shin et al. 2020; Sportelli et al. 2020; Weiss et al. 2020; Mousavi et al. 2021; Saravanan et al. 2021).

Finally, it should be mentioned that the SpyCatcher multimerization was used to engineer the prospective SARS-CoV-2 vaccine candidate (Rahikainen et al. 2021; Tan et al. 2021). More data on how the SpyTag/SpyCatcher technology works is presented in Chapter 25 by the detailed description of the VLPs based on the RNA phages, the members of the orders *Norzivirales and Timlovirales*. The SARS-CoV-2 vaccine was based on the display of the protein S RBD on a synthetic VLP platform, SpyCatcher003-mi3, using SpyTag/SpyCatcher technology. To create the RBD-SpyVLP vaccine candidate, the SpyTag stretch AHIVMVDAYKPTK) was fused between the signal sequence from influenza H7 HA (A/HongKong/125/2017) and the N-terminus of the monomeric RBD aa 331–529, NITN . . . GPKK6 (SpyTag-RBD) and the glycoprotein was expressed in mammalian cells Expi293F. The purified SpyTag-RBD was then conjugated to the SpyCatcher003-mi3 VLPs produced in *E. coli* (Bruun et al. 2018) to generate the SpyTag-RBD:SpyCatcher003-mi3 (RBD-SpyVLPs) immunogen. In fact, the SpyTag-RBD was efficiently conjugated to the SpyCatcher003-mi3 VLPs, with 93% display efficiency, which corresponded to an average of 56 RBDs per VLP. The low doses of the RBD-SpyVLPs in a prime-boost regimen induced a strong neutralizing antibody response in mice and pigs that was superior to convalescent human sera. The anti-RBD-SpyVLP antibodies recognized the RBD key epitopes, reducing the likelihood of selecting neutralization-escape mutants. Moreover, the RBD-SpyVLPs were thermostable and could be lyophilized without losing immunogenicity, to facilitate global distribution and reduce cold-chain dependence (Tan et al. 2021).

GENUS *GAMMACORONAVIRUS*

Subgenus *Igacovirus*

Infectious Bronchitis Virus

Infectious bronchitis virus (IBV) from the *Avian coronavirus* species is a highly infectious avian pathogen that causes the associated disease avian infectious bronchitis; affects the respiratory tract, gut, kidney, and reproductive systems of chickens; and remains responsible therefore for substantial economic loss within the poultry industry. Corse and Machamer (2000) were the first ones to demonstrate that the coexpression of the IBV genes M and E by the vaccinia virus expression system resulted in the release of the IBV VLPs into the supernatant of BHK-21 cells. The IBV protein E was released into the supernatant regardless of whether the protein M was present in transfected cells, while the protein

M was not released into the supernatant unless the protein E was present. These results suggested therefore that the E was required for the release of these proteins from cells. Next, Corse and Machamer (2003) examined the ability of the mutant and chimeric E and M proteins to form the IBV VLPs. The M proteins that were missing portions of their cytoplasmic tails or transmembrane regions were not able to support VLP formation, regardless of their ability to be crosslinked to the protein E. The interactions between the E and M proteins and the membrane bilayer were likely to play an important role in the VLP formation and virus budding (Corse and Machamer 2003).

Liu G et al. (2013) reported that the transfection of *Sf*9 cells with a single recombinant baculovirus encoding the IBV proteins M and S resulted in the assembly of the appropriate IBV VLPs with a diameter of about 100 nm. In fact, this was the first report for coronaviruses that the S protein with the M protein alone were able to be assembled into the VLPs. The generated IBV VLPs induced humoral immune responses in mice and chickens in a level comparable to that of the inactivated IBV vaccine, and more importantly the IBV VLPs elicited significantly higher cellular immune responses than the inactivated IBV vaccine (Liu G et al. 2013).

Xu PW et al. (2016) engineered the IBV VLPs by the coinfection of *Sf*9 cells with three recombinant baculoviruses separately encoding M, E, or the recombinant S (rS) genes. The rS gene was obtained by fusion of the S1 and TM/CT fragments of the gene S, and the rS protein was sufficiently flexible to retain the ability to participate in the VLPs. The size and morphology of the IBV VLPs were similar to the authentic IBV virions. The IBV VLPs elicited virus-neutralizing antibodies in chickens via subcutaneous inoculation (Xu PW et al. 2016).

Zhang Y et al. (2021) were the first who produced in *Sf*9 cells the three sorts of the IBV VLPs carrying first the three structural proteins S, M, and E, as well as the VLPs carrying the S and M proteins and the VLPs carrying the M and E proteins. Among the three VLP variants, the SME-VLPs stimulated the strongest humoral, cellular, and mucosal immune responses in chickens and provided effective protection against viral challenge.

Lv et al. (2014) developed the first chimeric VLPs as an IBV vaccine by the baculovirus-driven expression in *Sf*9 cells. The chimeric VLPs were composed of matrix 1 protein from avian influenza H5N1 virus, a member of the order *Articulavirales* (Chapter 33), and a fusion protein neuraminidase (NA)/spike 1 (S1) that was generated by fusing the IBV protein S1 to the cytoplasmic and transmembrane domains of the protein NA of avian influenza H5N1 virus. The chimeric VLPs elicited significantly higher S1-specific and virus-neutralizing antibody responses in intramuscularly immunized mice and chickens than inactivated IBV viruses (Lv et al. 2014).

Wu et al. (2019) engineered the next variant of the chimeric IBV VLPs by combination with Newcastle disease virus (NDV), a member of the *Mononegavirales* order

(Chapter 31), another pathogen seriously affecting the poultry industry. Thus, the IBV protein S1 and the ectodomain of the NDV protein F were separately linked with the transmembrane and C-terminal domain of the IBV protein S, composing the rS and rF chimeras. The novel chimeric IBV-NDV VLPs containing the rS, rF, and the IBV protein M were produced by the baculovirus-driven expression in *Sf*9 cells and demonstrated similar morphology to the natural IBV virions. The immunization of chickens with the VLPs efficiently induced humoral and cellular immune responses and provided 100% protection against IBV or NDV virulent challenge from death (Wu et al. 2019).

It should be mentioned here that Chen HW et al. (2016) engineered the synthetic IBV VLPs (sVLPs) by incubating 100-nm gold nanoparticles in a solution containing an optimized concentration the protein S, achieving therefore the size similarity to the native virions. The vaccination of mice and chickens with the sVLPs showed enhanced lymphatic antigen delivery, stronger antibody titers, increased splenic T-cell response, and reduced infection-associated symptoms in an avian model of coronavirus infection (Chen HW et al. 2016).

Liang et al. (2020) integrated the NanoLuc Binary Technology to the reverse genetics of IBV. The 11-aa HiBiT tag was inserted to the S or M protein, and the chimeric IBVs were produced. The replication of the HiBiT-tagged IBV allowed continuos monitoring in an infected chicken embryo, and the chimeric virus was genetically stable for at least 20 passages (Liang et al. 2020). This was therefore the first example of the integration of the HiBiT tagging system, a novel reporter system, with the coronavirus reverse genetics, facilitating future studies of replication and pathogenesis of the CoVs.

FAMILY *ARTERIVIRIDAE*

PORCINE REPRODUCTIVE AND RESPIRATORY SYNDROME VIRUS

As reviewed by de Groot et al. (2012), the structural proteins of arteriviruses are apparently unrelated to those of the other members of the order *Nidovirales*. In porcine reproductive and respiratory syndrome virus (PRRSV), a single member the *Betaarterivirus suid 1* species from the subgenus *Eurpobartevirus*, genus *Betaarterivirus*, subfamily *Variarterivirinae*, and equine arteritis virus (EAV), a single member of the *Alphaarterivirus equid* species from the genus *Alphaarterivirus*, subfamily *Equarterivirinae*, six envelope proteins have been identified, each essential for virion infectivity, while a single protein species, N, forms the nucleocapsid. As shown in Figure 26.1, the nonglycosylated membrane protein M structurally resembles the M protein of coronaviruses. It forms a disulfide-linked heterodimer with the major glycoprotein GP5, which is also a putative triple-spanning membrane protein. The viral glycoproteins GP2, GP3, and GP4 are minor virion components and form heterotrimers. The envelope protein E is small, hydrophobic, and nonglycosylated and believed to function as an ion-channel protein.

PRRSV is the causative agent of a disease called porcine reproductive and respiratory syndrome, also known as blue-ear pig disease, leading to the reproductive failure in pigs worldwide. The first fine structure of the structural PRRSV proteins was resolved to 2.6 Å for the protein N of 123 aa residues, which was overexpressed in *E. coli*, crystallized, and existed in crystals as a tight dimer forming a four-stranded β-sheet floor superposed by two long α-helices and flanked by two N- and two C-terminal α-helices (Doan and Dokland 2003a, b). The structure of the protein N represented therefore a new class of viral capsid-forming proteins, distinctly different from those of other enveloped viruses but reminiscent of the coat protein of the RNA phage MS2 from the *Norzivirales* order (Chapter 25). Figure 26.5 demonstrates the domain organization and crystal structure of the PRRSV protein N dimer. Later, Spilman et al. (2009) carried out the electron cryomicroscopy and tomographic reconstruction of the PRRSV virions grown in mammalian cells. The virions were pleomorphic, round to egg-shaped particles ranging from about 50–65 nm, with an average diameter of 58 nm. The particles displayed a smooth outer surface with only a few protruding features, presumably corresponding to the envelope protein complexes. The overall organization of the virus was therefore similar to that of coronaviruses like SARS-CoV-1, except for the absence of the prominent spikes characteristic of the coronaviruses.

The virions contained a double-layered, hollow core with an average diameter of 39 nm, which was separated from the envelope by a 2–3-nm gap. The analysis of the 3D structure suggested that the core was composed of a double-layered helical chain of nucleocapsid proteins bundled into a hollow ball (Spilman et al. 2009).

The major structural PRRSV proteins were produced early by the baculovirus expression system without any ambitions to engineer VLPs (Kreutz and Mengeling 1997; Plana Duran et al. 1997). Groot Bramel-Verheije et al. (2000) engineered the chimeric PRRSV virions that carried an epitope of the hemagglutinin (HA) protein of human influenza A virus, which was added at the 5' end and at the 3' end of the gene N. Later, PRRSV served as a model virus for the genomic cDNA expression and recombination into baculovirus (Zheng et al. 2010). Thus, the infectious PRRSV particles were generated and secreted by expression of the full-length cloned genome from the modified baculovirus vector in *Sf*9 cells. The infectious PRRSV particles were also produced in mammalian cell cultures inoculated with the appropriate chimeric baculovirus.

Nam et al. (2013) were the first ones to compose the PRRSV VLPs from the GP5 and M proteins by the baculovirus-driven expression in *Sf*9 cells. The size and morphology of the VLPs closely resembled those of the PRRSV virions, while the intramuscular immunization of mice demonstrated the capacity of the VLPs to induce both neutralizing antibodies and cellular response.

A bit earlier, Wang W et al. (2012) engineered the first chimeric PRRSV VLPs as a putative vaccine candidate. The VLPs were composed of the matrix protein M1 from H1N1 influenza virus and a fusion protein, denoted as NA/GP5, containing the cytoplasmic and transmembrane domains of the H1N1 virus neuraminidase (NA) and PRRSV GP5 protein. The NA/GP5 fusion was incorporated efficiently into the chimeric VLPs along with the influenza virus M1 protein.

The intramuscular immunization of mice with the chimeric VLPs stimulated humoral and cellular responses to the PRRSV protein GP5 (Wang W et al. 2012).

Binjawadagi et al. (2016) used the baculovirus expression system to generate two types of the PRRSV VLPs, containing either two GP5-M or five GP5-GP4-GP3-GP2a-M structural proteins. To improve their immunogenicity, the authors entrapped them together in PLGA nanoparticles and coadministered them intranasally with a potent *Mycobacterium tuberculosis* whole cell lysate adjuvant to pigs. Nevertheless, the PRRSV VLPs were found poorly immunogenic and did not induce adequate protective immunity (Binjawadagi et al. 2016). In parallel, García Durán et al. (2016) generated the PRRSV VLPs consisting of the two major proteins M and GP5 and of the six proteins M, GP2, GP3, GP4, GP5, and E–ensuring therefore the complete structural pattern–and produced them in *Sf*9 cells. The electron microscopy showed that the two types of the VLPs were indistinguishable and similar in shape and size to the native PRRSV virions.

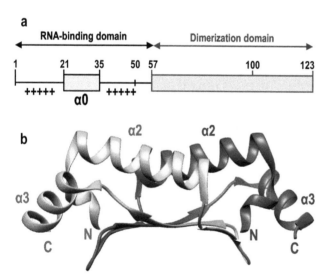

FIGURE 26.5 The crystal structure of the PRRSV protein N. (a) Domain organization of the 123-aa N protein. The C-terminal dimerization domain is shown as a pink box, while the predicted α-helix in the N-terminal domain (α0) is shown as a gray box. The regions of positive charges are indicated by (+) signs. (b) The ribbon representation of the crystal structure of the N dimer including aa residues 57–123, PDB ID 1P65 (Doan and Dokland 2003b). The N- and C-termini are indicated, as are the α2 and α3 helices. (Redrawn with permission from Microbiology Society from Spilman MS et al. *J Gen Virol.* 2009;90:527–535.)

Van Noort et al. (2017) obtained the insect cells-produced PRRSV VLPs consisting of the M, N, E, and GP5 proteins with the size of ~50 nm. However, the intranasal immunizations of pigs with the PRRSV VLPs and VLPs plus the 2′, 3′-cGAMP VacciGrade™ adjuvant exacerbated viremia. Overall, the PRRSV VLPs and PRRSV VLPs plus the adjuvant failed to provide protection against PRRSV challenge. It was concluded that the different dose of VLPs and/or alternative route of vaccination such as intramuscular injection should be explored in the future studies to fully assess the feasibility of such a vaccine platform for the PRRSV control and prevention (Van Noort et al. 2017).

Concerning the plant expression system, Thuenemann et al. (2013) invented the expression of the major structural PRRSV genes M and GP5 in plants based on the data on their expression in mammalian cells (Jiang et al. 2007; Cruz et al. 2010). The data agreed with the presence of conformational epitopes, relevant for protection, in the disulfide-linked GP5-M heterodimer, the major component of the PRRSV envelope (Dokland 2010). When the full-length sequences of both genes were expressed in *N. benthamiana*, it was not possible to detect the expression of either of the proteins. When some partially deleted versions of GP5 and M proteins were designed, the production of vesicles exposing the PRRSV antigens in a VLP-like structure was observed (Thuenemann et al. 2013).

Uribe-Campero et al. (2015) reported for the first time the generation and purification of the PRRSV VLPs by expressing GP5, M, and N genes in *Nicotiana silvestris* plants. The highly purified VLPs were identified by electron microscopy with a size of 60–70 nm and demonstrated therefore the appropriate size of the PRRSV virions. The VLPs were immunogenic in mice (Uribe-Campero et al. 2015).

The comprehensive review by Dhakal and Renukaradhya (2019) could be recommended to evaluate the current status of the putative VLP vaccine candidates against porcine reproductive and respiratory syndrome responsible for, e.g., over $1 billion loss per year through direct and indirect costs in the U.S. swine industry (Holtkamp et al. 2013).

EQUINE ARTERITIS VIRUS

Early research on the structural EAV genes started with a series of recombinant fusion glutathione-S-transferase proteins with the fragments encoded by the EAV gene N and produced in *E. coli* (Chirnside et al. 1995). Although a good correlation existed between virus neutralizing antibody and recombinant N_{1-69} positive values in post-infection sera, all the recombinant N proteins failed to induce any virus neutralizing response in immunized rabbits. The segments of the EAV genome encoding the predicted ORFs of the M and N genes were expressed in *E. coli* as maltose-binding fusion proteins by Kheyar et al. (1997). This study recommended the recombinant N and M proteins for the serodetection of the EAV-infected animals. Balasuriya et al. (2000) used the RNA replicon particles derived from a vaccine strain of Venezuelan equine encephalitis virus (VEE) from the family *Togaviridae*, order *Martellivirales* (Chapter 18), were used as a vector for the expression of the major EAV envelope proteins (G_L and M), both individually and in heterodimer form (G_L/M), in mammalian cells. All mice developed antibodies against the recombinant proteins with which they were immunized, but only the mice inoculated with the replicon particles expressing the G_L/M heterodimer developed antibodies that neutralized EAV.

Later, Kabatek and Veit (2012) described synthesis of the ectodomains of the EAV proteins GP2b, GP3, and GP4 in *E. coli* and their refolding and oligomerization from inclusion bodies. After extraction of the inclusion bodies and further dialysis, the Gst- (but not His-) tagged proteins refolded into a soluble conformation. However, when dialyzed together with the Gst-GP3 or with Gst-GP4, the His-GP2b and His-GP4 remained soluble and were recovered as oligomers are by affinity chromatography (Kabatek and Veit 2012). Unfortunately, no electron microscopy data was presented in this study.

27 Order *Picornavirales*

Everything is theoretically impossible, until it is done.

Robert A. Heinlein

ESSENTIALS

The *Picornavirales* is an order of icosahedral T = 3 and pseudoT = 3 (pT = 3) single-stranded positive-sense RNA viruses that have been shown to be excellent models of both genetic fusions and chemical coupling, as well as the beloved new material scaffolds. This huge order currently involves 8 families, 1 subfamily, 103 genera, 3 subgenera, and 323 species. According to the modern taxonomy (ICTV 2020), the order *Picornavirales* is one of the three orders belonging to the class *Pisoniviricetes*, together with the currently famous *Nidovirales* order and the small order *Sobelivirales*, which are described in the neighboring Chapter 26 and Chapter 28, respectively. The *Pisoniviricetes* class is belonging to the *Pisuviricota* phylum from the kingdom *Orthornavirae*, realm *Riboviria* (Zell et al. 2017).

Above all, the order *Picornavirales* includes the three great families *Caliciviridae*, *Picornaviridae*, and *Secoviridae*, as well as the less popular family *Dicistroviridae*, which was nevertheless used by the VLP nanotechnologies.

The family *Caliciviridae* joined the order *Picornavirales* in 2019 and includes, under the current official ICTV view (Vinjé et al. 2019; ICTV 2020), 11 genera and 13 species. The genus *Lagovirus* that includes the well-studied rabbit hemorrhagic disease virus (RHDV) causing a highly contagious disease in rabbits and is thus an economically important pathogen for commercial rabbit production, as well as European brown hare syndrome virus (EBHSV) and the genus *Norovirus* with Norwalk virus (NoV), are especially essential for the VLP nanotechnologies. Some representatives of the genera *Sapovirus* and *Vesivirus* also played a remarkable role in this field. Generally, the *Caliciviridae* members infect mammals, birds, and fish (Vinjé et al. 2019). The most clinically important representatives are human noroviruses, which remain a leading cause of acute gastroenteritis in humans (Green 1997).

The impressive *Picornaviridae* family currently includes 68 genera and 158 species, where the following genera are of primary interest for the VLP techniques: *Aphthovirus* with foot-and-mouth disease virus (FMDV); *Cardiovirus* with Mengo virus (MV) and porcine encephalomyocarditis virus (EMCV); *Enterovirus* with coxsackievirus B (CVB3), enterovirus 71 (EV71), and poliovirus (PV); and *Hepatovirus* with hepatitis A virus (HAV; Zell et al. 2017; ICTV 2020). Most of the known picornaviruses infect mammals and birds, but some have also been detected in reptiles, amphibians, and fish. Many picornaviruses are important human and veterinary pathogens and may cause diseases of the central nervous system, heart, liver, skin, gastrointestinal tract, or upper respiratory tract. Most picornaviruses are transmitted by the fecal-oral or respiratory routes (Zell et al. 2017).

The *Secoviridae* family is the source of the famous cowpea mosaic virus (CPMV) from the genus *Comovirus* of the *Comovirinae* subfamily, which appeared as one of the first and most favorite targets of the VLP strategies and appears now as a highly promising anticancer agent. The family *Secoviridae* includes 1 subfamily, 8 genera, 3 subfamilies, and 86 species (Thompson et al. 2017; ICTV 2020). The family members infect plants, mainly dicots, and are transmitted predominantly by insects or nematodes, although some seed transmission was demonstrated (Thompson et al. 2017).

Figure 27.1 presents typical portraits of the order representatives and schematic cartoons of members of the important families. The *Caliciviridae* virions possess icosahedral T = 3 symmetry and diameter of 27–40 nm, where smaller empty, so-called 23-nm particles with T = 1 symmetry were observed after the mature 38-nm virions. The T = 3 capsid is composed of 90 dimers of the major structural protein VP1. The *Picornaviridae* virions are 30–32 nm in diameter, possess T = 1 or pT = 3 symmetry, and consist of 60 identical protomers, each comprising four VP1-VP4 capsid proteins, in other words protein 1A and paralogous 1B, 1C, and 1D or three capsid proteins when 1AB remains uncleaved. The mature capsid proteins 1B, 1C, and 1D and the uncleaved 1AB possess a core structure of an eight-stranded β-barrel, also known as a "jelly roll" (Jiang P et al. 2014). The same structure is typical for the *Secoviridae* virions that are also of similar 25–30 nm in diameter. The capsid is built by large (CPL or simply L) and small (CPS or S) capsid proteins. Many virus preparations contain empty virus particles. In the case of viruses, such as CPMV, with a bipartite genome, the two RNAs are encapsidated in separate virions. Thus, the B-particle containing one molecule of RNA1 and the M-particle containing one molecule of RNA2 were identified (Thompson et al. 2017).

GENOME

Figure 27.2 shows the genomic structure of the *Picornavirales* families and genera, which are most involved into the 3D and VLP studies. In brief, the *Caliciviridae* members possess monopartite RNA genome of 7.4–8.3 kb, with a 5′-terminal genome-linked virus protein VPg of 10–15 kDa and 3′-terminal poly(A) tract. The genome is organized into either two or three major ORFs, while a further ORF4 of murine norovirus (MNV) encodes virulence factor VF1 (Vinjé et al. 2019). The *Picornaviridae* genomic RNA,

DOI: 10.1201/b22819-32

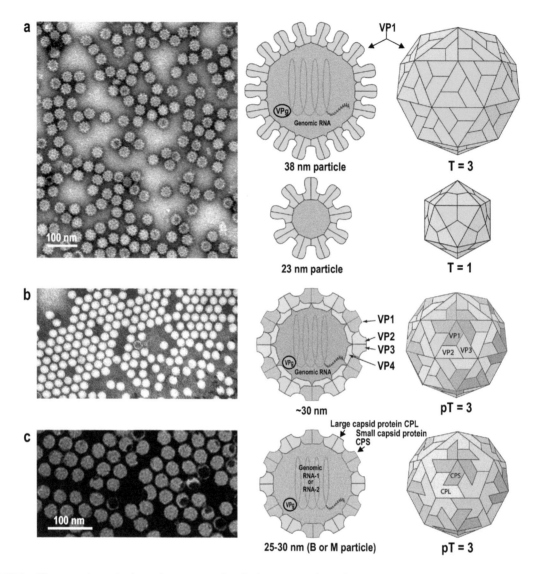

FIGURE 27.1 The portraits and schematic cartoons of typical representatives of the order *Picornavirales*. (a) Family *Caliciviridae*. The negatively stained micrograph of rabbit hemorrhagic disease virus (RHDV). (Reprinted from Wang X et al. *PLOS Pathog.* 2013; 9:e1003132.) (b) Family *Picornaviridae*. The negatively stained micrograph of poliovirus (PV) by Fred Murphy and Sylvia Whitfield (1975) is from the public CDC Image Library (PHIL), ID 1875. (c) Family *Secoviridae*. The negatively stained micrograph of cowpea mosaic virus (CPMV). (Reprinted from Thompson JR et al. ICTV virus taxonomy profile: *Secoviridae*. *J Gen Microbiol.* 2017;98:529–531.) The cartoons of the *Picornavirales* representatives are redrawn with kind permission from the ViralZone, Swiss Institute of Bioinformatics, http://viralzone.expasy.org (Hulo C et al. 2011). (Courtesy Philippe Le Mercier.)

ranging in size from 6.7 kb to 10.1 kb, commonly contains a single large ORF coding for a polyprotein, whereas a typical picornavirus genome encodes three or four capsid proteins and at least seven nonstructural proteins. The IRES element and the 2A genome region may have been exchanged between the genera. There may be more than one protein produced from the L or 2A genome regions, as well as multiple copies of 3B in the picornavirus genome (Zell et al. 2017). The *Secoviridae* genomic RNA of total 9.0–13.7 kb size could be mono- or bipartite RNA, as mentioned earlier for CPMV. The genomic RNAs are covalently linked to a small viral genome-linked protein VPg of 2–4 kDa at their 5' end and have a 3'-terminal poly(A) tract.

Each RNA encodes, in most of the cases, a single polyprotein (Thompson et al. 2017).

FAMILY *CALICIVIRIDAE*

GENUS *LAGOVIRUS*

Rabbit Hemorrhagic Disease Virus

Expression

Laurent et al. (1994) were the first who expressed the RHDV capsid gene encoding protein VP1—or VP60—of about 60 kDa, the unique component of viral capsid, by the baculovirus expression system. The used RHDV isolate

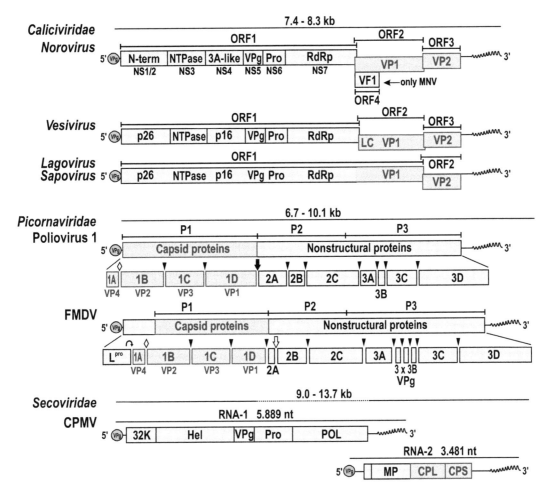

FIGURE 27.2 The genomic structure of the most representative families, genera, and viruses of the order *Picornavirales*. The data are collected from the ICTV reports (Thompson et al. 2017; Zell et al. 2017; Vinjé et al. 2019) and from the ViralZone, Swiss Institute of Bioinformatics, http://viralzone.expasy.org (Hulo et al. 2011). The structural genes are colored dark pink. For the *Picornaviridae*, the cleavages of polyprotein, which are facilitated by 2A^{pro} of the enteroviruses (black arrow) or by an NPG↓P motif at the C-terminus of 2A of foot-and-mouth disease virus (FMDV; white arrow), release the P1 and P1–2A proteins, respectively. The leader proteinase L^{pro} releases itself from the polyprotein by cleavage at its own C-terminus. The P2 and P3 polypeptides are precursors of the nonstructural proteins necessary for genome replication. Further polyprotein processing is mediated by 3C^{pro} (cleavage sites indicated by arrow heads). Processing of 1AB, the precursor of 1A and 1B, is thought to be autocatalytic and occurs in empty capsids or at virion maturation (white diamond) (Zell et al. 2017).

was collected from a wild rabbit in France. The recombinant VP60 self-assembled to form VLPs that did not need any other viral component and were easily purified from the supernatant of infected *Sf*9 cells. The RHDV VLPs were structurally and immunologically indistinguishable from the RHDV virions. The intramuscular vaccination of rabbits with the VLPs conferred complete protection in 15 days, while this protection was found to be effective from the fifth day after VLP injection and was accompanied by a strong humoral response (Laurent et al. 1994).

A bit later, the self-assembly and protective immunogenicity of the VP60 of a Czechoslovakian RHDV strain V-351 were demonstrated. By electron microscopy evaluation, the VLPs were similar in size and morphology to the native RHDV virions. The immune response to the intramuscularly injected VLPs completely protected rabbits following challenge with the virulent RHDV. In hemagglutination

assays, the VLPs bound to human red blood cells like the native virions (Nagesha et al. 1995). Sibilia et al. (1995) obtained nice RHDV VLPs when they expressed the gene in insect cells either as an individual protein source from a mRNA analogous to the viral subgenomic RNA or as part of a polyprotein that included the viral 3C-like protease and the RNA polymerase. Moreover, the authors clearly showed that the RHDV VLPs did not package RNA and remained empty.

It is noteworthy that Marin et al. (1995) did not detect VLPs when they used the baculovirus system to express the VP60 gene of a Spanish RHDV field isolate AST/89, although infection of *Sf*9 cells with the appropriate vector resulted in the production of high yields of the VP60 protein. The latter was able nevertheless to elicit a protective response in rabbits against a nasal challenge with lethal RHDV doses. Meanwhile, the VP60 gene of the same

Spanish AST/89 isolate was expressed in *E. coli*, and the product was found within the insoluble fraction of bacteria (Boga et al. 1994). Then, Plana-Duran et al. (1996) not only got the RHDV VLPs when they expressed the VP60 gene of another Spanish isolate, Olot/89, in insect cells but also successfully protected rabbits by oral administration at doses as low as 3 μg. Later, Bárcena et al. (2004) demonstrated the excellent VLPs derived from the AST/89 strain and produced by the baculovirus expression system. Next, Pérez-Filgueira et al. (2007) generated a low-cost source of the VLPs of the RHDV strain AST/89 by baculovirus-driven expression in *Trichoplusia ni* insect larvae. In parallel, Gromadzka et al. (2006) demonstrated the RHDV VLPs after baculovirus-driven expression of a Polish strain SGM that differed markedly from that of the reference strain. The RHDV strain SGM VLPs induced RHDV-specific antibodies in rabbits and guinea pigs, but the rabbits immunized with the VLPs were fully protected against challenge with a virulent RHDV. Young SL et al. (2006) combined the immunogenicity of the *Sf*21 insect cells-produced RHDV VLPs with the CpG adjuvant by transcutaneous immunization, resulting in the enhanced T-cell responses to the VLPs.

Later, the codon-optimization of the VP60 gene led to increased expression in *Sf*9 cells when the resulting supernatant was directly used as a low-cost vaccine antigen without the need for concentration or purification (Gao J et al. 2013). López-Vidal et al. (2015) used so-called TB expression cassette to improve the baculovirus-driven production of the RHDV VLPs, in parallel with the same strategy for the VLPs of porcine circovirus type 2 virus (PCV2) of the *Piccovirales* order described in Chapter 9. Zheng X et al. (2016) developed efficient production of the RHDV VLPs in silkworm pupae.

Bárcena et al. (2015) were the first who published the baculovirus-driven production of VLPs of RHDV2, a new highly contagious variant of RHDV, that was discovered in 2010 in northwestern France but further spread through Europe. Later, the baculovirus system-produced RHDV2 VLPs were presented by other groups (Miao et al. 2019; Müller et al. 2019). The mosaic bivalent RHDV1-RHDV2 VLPs were generated by simultaneous expression of both VP60 genes in *Sf*9 cells (Qi et al. 2020). Yang DK et al. (2015) used the baculovirus system to express the VP60 gene of another novel highly pathogenic RHDV strain that circulated in the Korean rabbit population since 2007 and had a devastating effect on the rabbit industry in Korea. All novel vaccines were fully protective.

In mammalian cells, the VP60 gene was expressed by recombinant vaccinia (Bertagnoli et al. 1996b), myxoma (Bertagnoli et al. 1996a; Bárcena et al. 2000; Torres et al. 2000), or canarypox (Fischer et al. 1997) viruses. Although the problem of the possible VLP formation was not touched in these studies, the inoculations of recombinant viruses by different routes including oral ones allowed protection of rabbits against a challenge with virulent RHDV. Later, Yuan et al. (2013) performed protective DNA vaccination of rabbits with the gene encoding VP60 and demonstrated

appearance of the native virion-like RHDV VLPs when the gene was expressed in rabbit kidney epithelial cell culture.

In yeast *Saccharomyces cerevisiae*, Boga et al. (1997) demonstrated formation of the RHDV VLPs similar in size and appearance to native virions. The subcutaneous vaccination of rabbits with a single dose of this antigen in the absence of adjuvants conferred complete protection against the hemorrhagic disease.

Later, the RHDV gene VP60 of the Spanish isolate AST/89 (Farnós et al. 2005) or of a Chinese isolate (Yan et al. 2005a) was expressed in yeast *Pichia pastoris*. While Yan et al. (2005a) reported formation of the virion-like VLPs, the product obtained by Farnós et al. (2005) was insoluble and associated with the yeast membranous system. Nevertheless, this product was expressed at high levels, was found N-glycosylated, and protected rabbits by subcutaneous or oral administration against challenge with 100 LD_{50} of homologous virus (Farnós et al. 2005). Moreover, this *P. pastoris*-produced RHDV product induced a strong humoral and cell-mediated immune response following intranasal immunization in mice (Farnós et al. 2006). Although oral vaccination employing the cell debris fraction or the whole transformed yeast appeared as an effective immunization method, the use of a formulation to be administered by the parenteral route inexorably implied the laborious solubilization process of the antigen. For this reason, Farnós et al. (2009) succeeded by obtaining of a soluble VP60 variant at high levels in *P. pastoris*, although the soluble product formed aggregates of 30 nm in diameter rather than assembled into the ordered VLPs. The scale-up of the *P. pastoris*-produced particles was performed (Fernández et al. 2013). Fernandez et al. (2015) markedly improved secretion and assembly of the particles in high-cell-density yeast fermentations. Moreover, a technology was used, which employed recombinant nonspecific BPTI-Kunitz-type protease inhibitor (rShPI-1A) isolated from the sea anemone *Stichodactyla helianthus*.

In plants, Castañón et al. (1999) expressed the VP60 gene in transgenic potato and demonstrated full protection of rabbits after subcutaneous/intramuscular immunization with leaf extracts from the gene-expressing plants. Fernández-Fernández et al. (2001) reported protection of rabbits by immunization with the VP60 protein synthesized in *Nicotiana clevelandii* plants with a potyvirus-based vector. Martín-Alonso et al. (2003) achieved production of the VP60 protein in transgenic tubers of potato plants. However, no one of these reports touched the problem of the possible VP60 self-assembly. Thus, Gil et al. (2006) were the first who observed the true RHDV VLPs in tissue sections of *Arabidopsis* plants. The mice fed with leaves of the transgenic plants expressing VP60 were primed to a subimmunogenic baculovirus-derived vaccine single dose.

At last, in bacteria, the VP60 gene was inserted into an expression vector containing a small ubiquitin-like modifier (SUMO) tag that promoted the soluble expression of heterologous proteins in *E. coli* cells. After expression and purification of the His-SUMO-VP60 and cleavage of the SUMO tag, the RHDV VP60 protein was self-assembled into the VLPs with a similar shape, but with a smaller size compared with authentic RHDV virions, namely with diameters ranging from 25–30 nm against approximately 32–44 nm diameters of the natural capsids.

Nevertheless, the bacteria-produced VLPs ensured full protectivity of rabbits against lethal virus challenge (Guo et al. 2016).

3D Structure

It was structurally intriguing that RHDV possessed a single capsid protein, VP60 (or VP1 as it was termed later) unlike most animal viruses. For the first time, the structure of the insect-cells produced RHDV VLPs was determined by electron cryomicroscopy and 3D reconstruction at low 32 Å resolution in complex with a neutralizing monoclonal antibody (Thouvenin et al. 1997). The atomic coordinates of a Fab fragment were fitted to the VLPs, which had a T = 3 icosahedral lattice consisting of a hollow spherical shell with 90 protruding arches with the mAb occupation of 50%.

Nagesha et al. (1999) reported that deletions of 30 aa at the N terminus of VP60, which were replaced by a heterologous six-residue epitope, gave rise to smaller particles of ~27 nm in comparison with the native virion-like VLPs of ~40 nm in diameter, when the corresponding genes were expressed by the baculovirus system. Later, Laurent et al. (2002) showed that the first 42 aa residues were not essential for assembly of the 38-nm capsids. However, a truncated VP60 with an internal deletion of 62 aa residues at the N-terminus (between residues 31 and 93) assembled into the 27-nm particles. As described later, these deletions were compensated in some cases by insertion of foreign sequences.

Bárcena et al. (2004) established the mechanism that allowed RHDV coat protein to switch among quasi-equivalent conformational states to achieve the appropriate curvature for the formation of a closed shell by the same baculovirus-driven expression of a set of the corresponding mutants. Thus, the quasi-equivalent interactions between the coat subunits were ensured by the N-terminal region of a subset of subunits, which faced the inner surface of the capsid shell. The mutant coat protein lacking this N-terminal sequence assembled into T = 1 capsids (Figure 27.1). The fact that an insertion mutant with an addition of 12 heterologous aa residues assembled into the virion-like T = 3 particles indicated that the aa residues at the most N-terminal region are disordered because this heterologous epitope did not interfere with the switching mechanism. The polymorphism of the RHDV T = 3 resembled therefore the similar appearance by the T = 3 plant viruses. Moreover, the similar design of RHDV and NoV capsids was emphasized in this study (Bárcena et al. 2004).

Using the electron cryomicroscopy approach, Hu Z et al. (2010) resolved the 3D structure of the native RHDV virions to 11 Å and compared the obtained structure with those of NoV from the genus *Norovirus* and San Miguel sea lion virus (SMSV) belonging to the species *Vesicular exanthema of swine virus* from the genus *Vesivirus* (see following sections). In parallel, Katpally et al. (2010) improved the electron cryomicroscopy resolution of RHDV to ~8 Å. The 3D structure of murine norovirus 1 (MNV-1) was also determined in this study (see following sections).

At last, Wang Xue et al. (2013) reported the electron cryomicroscopic reconstruction of wild-type RHDV at 6.5 Å resolution and the crystal structures of the shell (S) and protruding (P) domains of the VP60, expressed by the baculovirus system, each at 2.0 Å resolution. Figure 27.3 demonstrates this comprehensive 3D structure. Therefore, the authors built a complete atomic model of the RHDV capsid, where VP60 had a conserved S domain and a specific P2 sub-domain that differed from those found in other caliciviruses. The N-terminal arm domain of VP60 folded back onto its cognate S domain. The sequence alignments of VP60 from six groups of RHDV isolates revealed seven regions of high variation that could be mapped onto the surface of the P2 sub-domain and suggested three putative

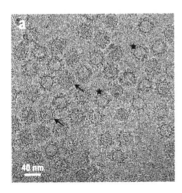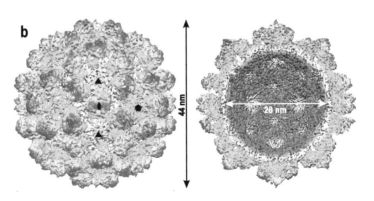

FIGURE 27.3 Electron cryomicroscopy and 3D image reconstruction of RHDV. (a) CryoEM micrograph of purified RHDV. The black arrows and stars point to RHDV particles with differing amounts of internal density. (b) Reconstructed cryoEM map of the RHDV virion, color-coded by radius. Icosahedral 2-, 3- and 5-fold axes are indicated by black symbols, and AB and CC capsomers are identified. On the right, the closest half of the density map has been removed to reveal internal features in the RHDV density map. The contour threshold of the cryoEM map here was set to 3.3σ above the mean. (Reprinted from Wang X et al. *PLOS Pathog.* 2013;9:e1003132.)

pockets might be responsible for binding to histo-blood group antigens.

In parallel, Luque et al. (2012) designed several insertion mutants to validate the VP60 (or VP1 in this paper) pseudoatomic model, placed foreign epitopes at the N- or C-terminal end as well as in an exposed loop on the capsid surface, and expressed the chimeric genes by the baculovirus expression system. A set of T- and B-cell epitopes of various lengths derived from viral and eukaryotic origins was employed. Whereas most insertions were well tolerated, VP1 with a feline calicivirus (FCV) capsid protein-neutralizing epitope at the N terminus assembled into mixtures of T = 3 and larger T = 4 capsids (Luque et al. 2012). Figure 27.4 shows the outfit of this novel class of the RHDV-derived capsids.

Chimeric VLPs

As mentioned earlier by the expression in baculovirus system, when the N-terminal 30-aa residues of RHDV coat protein were substituted by a well-characterized six-residue epitope, namely an epitope from the bluetongue virus (BTV) capsid protein VP7 (Btag), the fusion protein retained its ability to self-assemble into VLPs (Nagesha et al. 1999). However, the size of these particles was only 27 nm, compared to the 40-nm VLPs derived from the original coat. When the Btag was fused to the C-terminus of the RHDV capsid protein without deletion, the fusion proteins formed VLPs of 40 nm in size and retained their antigenicity, but the Btag antigenicity appeared to be low in this construct (Nagesha et al. 1999).

Continuing the story of the chimeric RHDV VLPs obtained by the baculovirus-driven expression, El Mehdaoui

et al. (2000) fused the N-terminally truncated VP60 protein lacking 42 aa residues with sequences that were expected to be sufficient to confer DNA packaging and gene transfer properties to the chimeric VLPs. Therefore, each of the two putative DNA-binding sequences of major L1 and minor L2 capsid proteins of human papillomavirus type 16 (HPV-16) was fused at the N-terminus of the truncated VP60 protein. The two recombinant chimeric proteins expressed in insect cells self-assembled into VLPs similar in size and appearance to the authentic RHDV virions.

As mentioned earlier, Bárcena et al. (2004) demonstrated that insertion of a foreign epitope at the N- or C-terminal regions did not alter the ability of VP60 to form VLPs similar to authentic virions. The 11-aa linear epitope from the transmissible gastroenteritis virus (TGEV) nucleoprotein, which was recognized by monoclonal antibody DA3, was used in this study.

Crisci et al. (2009) inserted the CD8+ T cell epitope OVA corresponding to the aa 257–SIINFEKL–264 stretch from chicken ovalbumin at two different locations: (i) the N-terminus, predicted to be facing to the inner core of the VLPs and (ii) a novel insertion site predicted to be located within an exposed loop. Both constructions correctly assembled into VLPs and induced specific cellular responses mediated by cytotoxic and memory T cells by the immunization of with the chimeric VLPs without any adjuvant. More importantly, the immunization with chimeric VLPs was able to resolve an infection by a recombinant vaccinia virus expressing OVA protein (Crisci et al. 2009). The longer OVA variant GSQLESIINFEKLTEGS was used by Luque et al. (2012) for the insertion into the predicted exposed loop in the P domain of the VP60 protein. Later,

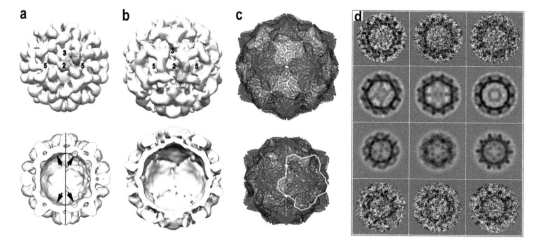

FIGURE 27.4   NT42 T = 3 and T = 4 capsid structures. (a) Surface-shaded representations of the outer (top) and inner (bottom) surfaces of the NT42 T = 3 capsid, viewed along an icosahedral 2-fold axis at 20-Å resolution. A, B, and C subunits are shown as ribbons. Inner surfaces of NT42 (bottom left) and wt (bottom right) T = 3 capsid show the differences at the 3-fold axis (arrows indicate the sectioned 3-fold axis). (b) Outer (top) and inner (bottom) surfaces of the NT42 T = 4 capsid, viewed along an icosahedral 2-fold axis at 25-Å resolution. The icosahedral asymmetric unit is shown (A, blue; B, red; C, green; D, yellow). (c) S domains in the T = 4 (top) and T = 3 (bottom) icosahedral shells. Interacting surfaces between A, B, C, and D β-barrels are quasi-equivalent. (d) Images of T = 4 (row 1) and T = 3 (row 4) capsids taken directly from original cryomicrographs compared to the projected views (T = 4, row 2; T = 3, row 3) of the 3D reconstruction in the corresponding orientation. Selected capsids are oriented close to a 2-fold (column 1), 3-fold (column 2), and 5-fold (column 3) symmetry axis. (Reprinted with permission of ASM from Luque D et al. *J Virol.* 2012;86:6470–6480.)

the same OVA epitope was introduced at the four locations: (i) at the N-terminus, (ii) replacing aa positions 2–14, (iii) replacing aa positions 196–207, and (iv) replacing aa positions 217–228 (Chen M et al. 2014). All variants were expressed by the baculovirus expression system, formed VLPs, and were immunogenic in mice.

Crisci et al. (2012b) generated the chimeric RHDV VLPs containing at the N-terminus of the VP60 protein a well-known T epitope of FMDV, namely AAIEFFEGMVHDSIK, located at aa positions 21–35 of the FMDV nonstructural protein 3A and efficiently recognized by lymphocytes from infected pigs. As a result, the chimeric 3A-RHDV VLPs were able to induce immune response in pigs after intranasal or intramuscular inoculation. Luque et al. (2012) generated the excellent T = 3 RHDV VLPs bearing 42-aa insertion with a B and a T cell epitope derived from FMDV at the C-terminus.

As described earlier and illustrated in Figure 27.4, the RHDV VLPs tended to form both T = 3 and T = 4 particles when the VP60 protein was provided with the N-terminal addition of a virus-neutralizing FCV epitope, namely, 42-aa stretch [GSGNDITTANQYDAADIIRN]$_2$GS bearing two copies of a B-cell epitope from the FCV coat, within the so-called NT42 construct (Luque et al. 2012). In parallel, the same FCV epitope stretch was inserted into the predicted internal loop in the P domain. All chimeric products were evaluated by electron cryomicroscopy (Luque et al. 2012). The evident progress of the RHDV VLPs against the global VLP background was reviewed at that time by Crisci et al. (2012a).

Moreno et al. (2016) used two foreign B-cell epitopes: a novel neutralizing FCV B-cell epitope of 22 aa residues and the well-characterized M2e stretch from the extracellular domain of the influenza A virus M2 protein. The chimeric RHDV VLPs were generated in insect cells by insertion of the foreign B-cell epitopes at three different locations within VP60 protein: (i) between aa residues 2 and 3 facing the inner core, (ii) between aa 306 and 307 within loop L1 at the tip of the P2 subdomain, the most surface-exposed region, and (iii) at the C-terminal end of VP60 protein, which faces the outer surface of the viral capsid at the pentameric and hexameric cup-shaped depressions. Thus, the chosen positions involved therefore different levels of surface accessibility. Moreover, the epitopes were inserted in different copy numbers per site. By immunization of mice, the chimeric RHDV VLPs elicited potent protective humoral responses against displayed foreign B-cell epitopes, demonstrated by both in vitro neutralization and in vivo protection against a lethal challenge (Moreno et al. 2016). Figure 27.5 presents the chimeric RHDV VLPs carrying different numbers of the FCV epitope copies at the different VP60 insertion sites.

Donaldson et al. (2017) generated a therapeutic cancer vaccine and evaluated it in a murine model of colorectal cancer. Thus, the epitopes derived from the colorectal cancer tumor-associated antigens topoisomerase IIα and survivin were added to the N-terminus of the VP60 protein

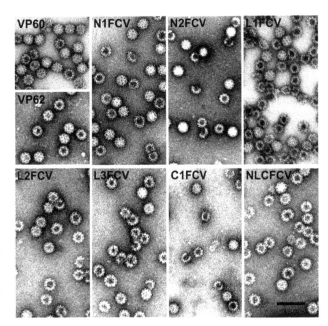

FIGURE 27.5 Analysis of the VP60-related VLPs by electron microscopy. Negatively stained purified particles corresponding to RHDV VP60 and FCV VP62 (top, left), and the indicated VP60 chimeric mutants were analyzed by electron microscopy: N1FCV and N2FCV are carrying one or two copies of the FCV epitope, respectively, at the N-terminus; L1FCV, L2FCV, and L3FCV are carrying one, two, or three copies of the FCV epitope within the L1 loop; C1FCV carries the FCV epitope at the C-terminus; NLCFCV bears one copy of the FCV epitope in each of the three insertion sites. Scale bar = 100 nm. (Reprinted from Moreno N et al. *Sci Rep.* 2016;6:31844.)

via GGS linker. The length of the inserted epitopes was of 8 (survivin), 9 (topIIα), and 20 (both epitopes connected via ALL linker) aa residues, and they did not prevent the self-assembly of the insect *Sf* 21 cells-produced chimeric VP60 molecules. Moreover, the chimeric VLPs formed a composite particle in the presence of CpGs. Although overall survival of mice bearing MC38-OVA tumors was markedly improved by the vaccination with all three VLP variants, the multitarget VLPs carrying both epitopes significantly prolonged the vaccine-induced remission period in comparison to each mono-therapy.

Recently, Rangel et al. (2021) engineered the chimeric RHDV VLPs displaying the FMDV epitopes by the baculovirus expression system and demonstrated induction of virus-neutralizing antibodies and partial protection in pigs.

In yeast *P. pastoris*, Yan et al. (2005b) were the first who got chimeric VLPs by adding the His$_6$ tag to the N-terminus of VP60 protein and demonstrated assembly of the latter into the VLPs, which were morphologically and antigenically similar to native RHD virions but did not package RNA.

Chemical Coupling

Peacey et al. (2007) were the first who intended to overcome limitations imposed by established genetic antigen incorporation procedures leading to the chimeric genetically fused RHDV VLPs and subjected the *Sf*21 insect cells-produced VLPs to chemical coupling. The VLPs formed by authentic VP60, as well as chimeric VP60 carrying N-terminal fusions of short peptides, including the haemagglutinin helper T cell epitope (HAT), were used as models for the chemical coupling. The direct labeling of the VLPs with biotin was achieved by incubating with sulfo-NHS-biotin. To couple short peptides to the VLPs, the heterobifunctional linker sulfosuccinimidyl 4-(*N*-maleimidomethyl) cyclohexane-1-carboxylate (sulfo-SMCC) was used. To be coupled to the VLPs, the larger proteins, such as eGFP or ovalbumin, were first treated with *N*-succinimidyl S-acetylthioacetate (SATA) followed by addition of hydroxylamine, and thiol-modified proteins were added to the maleimide-activated VLPs. The administration of the conjugated ovalbumin-VLPs induced high titer ovalbumin-specific antibody in mice, demonstrating that the immune stimulatory properties of the capsid were conferred to a conjugated foreign antigen. The VLPs facilitated delivery of the conjugated ovalbumin to dendritic cells, eliciting proliferative responses in naïve TCR transgenic T helper cells that were at least tenfold greater than ovalbumin antigen delivered alone (Peacey et al. 2007).

Peacey et al. (2008) demonstrated that the vaccination of mice with the RHDV VLPs that were chemically conjugated with ovalbumin or with both ovalbumin-derived CD4 (OTII) and CD8 (OTI) epitopes delayed the growth of the aggressive B16.OVA melanoma in mice. Neither VLP. OTI nor VLP.OTII alone were capable of inhibiting tumor growth. Win et al. (2011) translated the observation of the VLP cross-presentation reported by Peacey et al. (2008) into a more relevant human system and subsequently characterized the pathways used by dendritic cells to endocytose the VLPs coupled to model antigens, process them into peptides, and cross-present them to antigen-specific CD8[+] T cells. Furthermore, the authors explored the potential to utilize human tumor lysates as a source of undefined antigen to couple to the VLPs to induce tumor-specific CD8[+] T-cell responses in vitro (Win et al. 2011). Thus, Win et al. (2012) conjugated the Mel888 melanoma lysates to the RHDV VLPs. The chimeric VLPs were able to induce specific immune responses toward tumor cells while negating the inhibitory effects of lysates delivered alone.

McKee et al. (2012) noncovalently combined the RHDV VLPs with a galactose-containing adjuvant α-galactosylceramide (α-GalCer), a glycolipid that has been shown to specifically stimulate invariant natural killer-like T cells, forming therefore a composite particle for codelivery of antigen and adjuvant to the same antigen-presenting cell. The vaccination of mice with the α-GalCer-VLP complex led to the generation of antigen-specific T cells that protected prophylactically against subcutaneous tumor challenge and was more effective at generating antitumor immune responses than either component individually (McKee et al. 2012).

The next possible anticancer vaccine candidate was generated by incorporation of the so-called gp33 epitope SAVYNFATM, a MHC-class I peptide, derived from lymphocytic choriomeningitis virus (LCMV), into the RHDV VLPs, when the epitope was coupled genetically or chemically (Li Kunyu et al. 2013). Both VLPs, when delivered as prophylactic vaccines, inhibited the growth of Lewis's lung carcinoma tumors expressing gp33 in mice to a similar degree.

Jemon et al. (2013) designed a vaccine against HPV16 positive tumors. Thus, the universal T helper cell epitope PADRE of the sequence AKFVAAWTLKAAA was genetically fused to the N-terminus of the VP60, while the MHC I-restricted peptide 48–EVYDFAFRDL–57 from the HPV16 E6 was chemically coupled to the *Sf*21 insect cells-produced RHDV VLPs. To do this, the E6 peptide was provided with N-terminal cysteine, whereas a biotin marker was added to the C-terminus of the peptide to readily detect it. Remarkably, there was an E6-independent reduction in tumor area in mice vaccinated with the RHDV-VLPs in a therapeutic tumor-challenge model, which was attributed to a nonspecific immune stimulatory effect of the RHDV-VLPs. When the modified E6-RHDV-VLPs carrying the PADRE epitope were used, the tumor growth was further delayed in the E6 peptide-dependent manner (Jemon et al. 2013).

To enhance antigen uptake through mannose receptors, leading to improved immune responses, Al-Barwani et al. (2014) conjugated monomannoside and dimannoside to the insect cells-produced RHDV VLPs, providing approximately 270 mannose groups on the surface of each VLP. Such chimeric VLPs exhibited significantly enhanced binding and internalization by murine dendritic cells, macrophages, and B cells as well as human dendritic cells and macrophages. Moreover, fluorescently labeled VLPs conjugated to the mannoside or dimannoside were used to determine the effect of the VLP mannosylation on the VLP internalization and processing by key antigen-presenting cells. Kramer et al. (2019) conjugated mono- and dimannosides to the surface of the genetically constructed gp100-RHDV VLPs, which carried up to three copies of gp100$_{25-33}$ epitope KVPRNQDWL, a melanoma-associated antigen, containing proteasome cleavable linkers to target the correct processing of the epitope. The mannosides were intended therefore to utilize a second pathway of internalization, mannose receptor mediated, to further augment antigens internalized by phagocytosis/macropinocytosis. It was demonstrated that the mannosylation of the chimeric RHDV VLPs translated into enhanced uptake of the VLPs and superior survival of mice in a melanoma tumor trial.

Concerning application of the yeast *P. pastoris*-produced RHDV VLPs, Fernández et al. (2013) chemically coupled to them the 24-aa linear protective B-cell epitope KEDYRYAISSTNEIGLLGAEGLTC from the envelope protein E2 of classical swine fever virus (CSFV), a pestivirus from the *Amarillovirales* order described in Chapter 22.

The conjugates significantly enhanced the peptide-specific antibody response in vaccinated pigs.

Packaging and Delivery

As described earlier, El Mehdaoui et al. (2000) were the first ones to attempt an idea to package foreign DNA into the RHDV VLPs by addition of the corresponding HPV sequences at the N-terminus of the coat. In fact, the two chimeric VLPs had acquired the ability to bind DNA and package a plasmid. However, only the chimeric VLPs containing the DNA packaging signal of the L1 protein demonstrated efficient gene transfer into Cos-7 cells at a rate similar to that observed with the papillomavirus L1 VLPs. It was possible to transfect only a very limited number of RK13 rabbit cells with the chimeric RHDV capsids containing the L2-binding sequence. The chimeric RHDV VLPs containing the L1-binding sequence transferred genes into rabbit and hare cells at a higher rate than did the HPV-16 L1 VLPs. Unfortunately, no gene transfer was observed in human cell lines. Thus, it was shown for the first time that the insertion of a DNA packaging sequence into a VLP that was not able to encapsidate DNA transformed this capsid into a VLP that could be used as a gene transfer vector, opening the way to designing new vectors with different cell tropisms (El Mehdaoui et al. 2000).

As described earlier, the chemical conjugation of the RHDV VLPs with the appropriate ligands stimulated specific targeting to dendritic cells (Win et al. 2011, 2012). In addition, Donaldson et al. (2017) described formation of the composite RHDV VLPs carrying CpGs. The appearance of the CpG-carrying composite VLPs was proven by dialysis of solutions containing VLPs and CpGs with 1 MDa tubing. When CpGs were dialyzed alone with this tubing they were completely dialyzed out of solution; however, in the presence of the VLPs they remained. Therefore, the presence of the VLPs in solution provided some attractive force, impairing the ability of CpGs to defuse out of solution. The amount of CpGs associated with the VLPs following dialysis was calculated at around 20 μg CpGs per 1 mg of the VLPs. The amount of CpGs was approximately doubled when associated with the chimeric RHDV VLPs carrying the L1 DNA-binding site of HPV16 but was adversely affected by insertion of a positively charged octaarginine sequence. As a result, the combination of the VLPs and CpGs delivered together was more immunogenic than when administered separately and could be regarded as a form of the CpG packaging by the VLPs.

European Brown Hare Syndrome Virus

Laurent et al. (1997) expressed in the baculovirus system the capsid protein of a French isolate of European brown hare syndrome virus (EBHSV), a member of the second of the two species of the *Lagovirus* genus, classified as the *European brown hare syndrome virus* species. The nice VLPs that were produced in *Sf*9 insect cells were indistinguishable from the infectious EBHSV virions and displayed morphological characteristics similar to those described earlier for the insect cells-produced RHDV VLPs. The cross-protection experiments showed that vaccination with the EBHSV VLPs did not protect rabbits against an RHDV challenge. A set of MAbs was raised against the EBHSV VLPs and used together with anti-RHDV and anti-EBHSV MAbs produced against native viruses to study the antigenic relationships between the two caliciviruses. Finally, a classification of RHDV and EBHSV as two serotypes of a single serogroup was proposed (Laurent et al. 1997).

To investigate the folding of the lagovirus capsid protein, Laurent et al. (2002) constructed chimeric genes encoding fusion proteins derived from the N- and C-terminal parts of the RHDV capsid protein and the EBHSV capsid protein. Thus, the RE protein consisted of the 300-aa N-terminal region of the RHDV capsid protein fused to the last 281 aa residues of the C-terminal region of the EBHSV capsid protein. The ER protein consisted of the 297-aa N-terminal region of the EBHSV capsid protein fused to the C-terminal region of the RHDV capsid protein (last 279 amino acid residues). The chimeric genes encoding these two proteins were expressed by in vitro transcription and translation in rabbit reticulocyte lysates. The obtained mapping data allowed to conclude that the C-terminal part of the capsid protein is accessible to the exterior whereas the N-terminal domain of the protein constituted the internal shell domain of the particle (Laurent et al. 2002). As follows from the previously described matter, this statement was actively used by the generation of the chimeric RHDV VLPs.

GENUS *NOROVIRUS*

Norwalk Virus

Expression

The 56.6-kDa capsid protein of Norwalk virus (NoV), a member of the only species, namely *Norwalk virus*, of the *Norovirus* genus, was produced by the baculovirus-driven expression and demonstrated an ability to self-assemble into empty VLPs similar to native virions in size and appearance (Jiang et al. 1992). These particles induced high levels of Norwalk virus-specific serum antibody in laboratory animals following parenteral inoculation. The early availability of large amounts of the nice NoV VLPs allowed the development of rapid, sensitive, and reliable tests for the diagnosis of Norwalk virus infection as well as the implementation of structural studies. Concerning diagnostics, the new NoV VLPs-based tests were developed (Jiang et al. 1992; Parker et al. 1993; Treanor et al. 1993; Graham et al. 1994), and new insights about the molecular characterization and epidemiology of human caliciviruses have come from application of the new tests (Gray et al. 1993, 1994; Lew et al. 1994; Numata et al. 1994; Parker et al. 1994). Further studies in mice as well as in volunteers demonstrated that the NoV VLPs are a promising mucosal vaccine for NoV infections (Ball et al. 1996, 1998; Guerrero et al. 2001).

In a similar way to the reference NoV VLPs, without requiring any other viral components, the native virion-like VLPs were produced in insect cells after expression of the capsid genes from numerous members of the *Norovirus* genus: a Mexican strain of Snow Mountain (Jiang et al. 1995), Lordsdale (Dingle et al. 1995), Toronto (Leite et al. 1996), Hawaii (Green et al. 1997), Grimsby (Hale et al. 1999), and Burwash Landing, White River, and Florida (Belliot et al. 2001) viruses. Surprisingly, unlike baculovirus-driven expression, the Hawaii-strain VLPs were not observed in mammalian cells (Pletneva et al. 1998).

The phase I trials on healthy volunteers found the NoV VLP vaccine candidate to be safe and immunogenic when administered orally without adjuvants (Ball et al. 1999). El-Kamary et al. (2010) conducted two phase I double-blind, controlled studies of the VLPs derived from norovirus GI.1 genotype adjuvanted with monophosphoryl lipid A (MPL) and the mucoadherent chitosan, where the majority of vaccine recipients developed virus-specific serum antibodies. For the sake of clarity, the explanation of the modern norovirus genotype designation GI.1 is available in the recently updated classification of norovirus genogroups and genotypes by Chhabra et al. (2019). Moreover, Atmar et al. (2011) established that the NoV VLP vaccine provided protection against illness and infection after challenge with a homologous virus. It is worth mentioning that the systemic vaccination with the baculovirus-driven NoV VLPs recalled human IgA responses at higher magnitudes than IgG responses (Onodera et al. 2019). In this study, the NoV VLPs of the genotypes GI.1, GII.4, as well as of mouse norovirus MNV-S7, were produced in the baculovirus expression system.

In plants, Mason HS et al. (1996) gained the nice 38-nm NoV VLPs in transgenic tobacco and potato plants. The NoV VLPs from tobacco leaves and potato tubers were identical to the insect cell-derived VLPs by structural and immunological characteristics and were orally immunogenic in mice. Moreover, when potato tubers expressing the NoV capsid gene were fed directly to mice, they developed specific immune response (Mason HS et al. 1996). The transgenic NoV VLPs-containing potatoes induced immunological response in human volunteers as well (Tacket et al. 2000), while Tacket et al. (2003) detected humoral, mucosal, and cellular immune responses to the oral NoV VLPs in volunteers during clinical trial.

Furthermore, Huang et al. (2005) presented the successful production of the perfectly assembled 38-nm virion-size icosahedral T = 3 NoV VLPs, similar to those produced in insect cells, in tomato fruits. However, the tomato material contained more 23-nm VLPs, presumably T = 1 particles, with relatively little indication of the 38-nm VLPs. It is remarkable that in the same study the hepatitis B virus surface (HBs) particles were produced in *N. benthamiana* plants, as described in Chapter 37. Using a plant-optimized NoV capsid gene, the expression of the latter in tomato and potato plants increased fourfold (Zhang X et al. 2006). To overcome long generation time and modest level of antigen

accumulation, the NoV VLPs were produced by an efficient tobacco mosaic virus (TMV)-derived transient expression system using leaves of *N. benthamiana* (Santi et al. 2008).

After the prototype NoV variants, other representatives of the Norwalk virus species were involved into the VLP production in plants, such as, for example, Narita 104 virus of the norovirus genogroup II (Mathew et al. 2014).

At last, Diamos and Mason HS (2018) elaborated a plant-based recombinant protein expression system based on agroinfiltration of a replicating vector derived from the geminivirus bean yellow dwarf virus (BeYDV) and achieved and produced in *N. benthamiana* the NoV GII.4 VLPs at >1 mg/g leaf fresh weight, over three times the highest level ever reported in plant-based systems.

In yeast *P. pastoris*, Xia et al. (2007) expressed the capsid gene of the NoV genotype II.4 and demonstrated VLPs like those expressed in other expression systems. The oral administration of raw material from the yeast cell lysates without any adjuvant resulted in systemic and mucosal immune responses in mice. Tomé-Amat et al. (2014) secreted the NoV VP1 protein from *P. pastoris* cells and obtained the appropriate fully assembled VLPs of 40 nm in diameter from the culture supernatant.

In bacteria, Tan et al. (2004) demonstrated that the *E. coli*-expressed NoV capsid proteins maintained the same antigenicity and receptor binding specificity as that of the baculovirus-expressed capsid, although the *E. coli*-expressed proteins did not form VLPs. Huo et al. (2018) cloned the VP1 coding sequence, with or without N-terminal deletions (N26 and N38, 26 and 38 aa deleted from the N-terminus, respectively), of a previously isolated Sydney 2012-like strain into a cold shock *E. coli* expression vector. The electron microscopy observation indicated the in vivo assembly of VLPs with two sizes in accordance with those observed in *Sf*9 cells. The immunization of mice with the purified VLPs derived from the N38 variant demonstrated higher IgG antibody titers and blocking antibody titers when compared with the full-length capsid protein assembled VLPs from recombinant baculovirus expression system.

Hwang et al. (2021) expressed the NoV VP1 gene in *E. coli* and self-assembled the capsid protein into VLPs by a specific chaperna-based assembly mechanism. Thus, an tRNA-interacting domain (tRID) was genetically fused with the VP1 capsid protein, as a tRNA docking tag, in the bacterial host to transduce chaperna function for de novo viral antigen folding. The tRID/tRNA removal prompted the in vitro assembly of monomeric antigens into the VLPs that elicited robust protective immune responses after immunization (Hwang et al. 2021).

3D Structure

The structures of the first two caliciviruses were determined by electron cryomicroscopy and image reconstruction, i.e. baculovirus-expressed NoV VLPs (Prasad et al. 1994b) and primate calicivirus (Prasad et al. 1994a). The VLP structures revealed a diameter of 38 nm and exhibited a T = 3 icosahedral lattice composed of 90 arch-like capsomers formed by dimers of the capsid protein at 22 Å resolution. These structures showed distinct similarity to the single-stranded RNA plant viruses such as tomato bushy stunt virus (TBSV) and turnip crinkle virus (TCV) from the *Tolivirales* order, which is described in Chapter 24. The smaller form of the NoV VLPs found in insect cells, 23 nm in diameter, appeared to be a T = 1 symmetry variant of capsid self-assembly, consisting of 60 polypeptide molecules (White LJ et al. 1997).

Next, a high-resolution structure of the T = 3 NoV VLPs assembled from 180 copies of capsid protein with a classical eight-stranded β-sandwich motif was determined by x-ray crystallography at 3.4 Å (Prasad et al. 1999). The capsid protein folded into two principal domains, a shell S domain and a protruding P domain, which contained two subdomains, P1 and P2. The S domain was formed by the N-terminal 225 residues, and residues 50–225 folded into an eight-stranded antiparallel β sandwich structure, a folding commonly seen in many viral capsids. The P domains of neighboring subunits, related by the local and icosahedral twofold axes, interacted with each other to form a prominent protrusion. The residues beyond 225 formed the P1 and P2 domains (Prasad et al. 1999).

Bertolotti-Ciarlet et al. (2002) identified functional domains of the NoV capsid protein involved in assembly when analyzing the ability of a set of deletion mutants of the capsid protein to assemble into VLPs after expression in insect cells. The deletion of the N-terminal 20 residues, suggested by the x-ray structure to be involved in a switching mechanism during assembly, did not affect the ability of the mutant capsid protein to self-assemble into 38-nm VLPs with the T = 3 icosahedral symmetry. Further deletions in the N-terminal region affected particle assembly. The deletions in the C-terminal regions of the P domain, involved in the interactions between the P and S domains, did not block the assembly process, but they affected the size and stability of the particles. The mutants carrying three internal deletion mutations in the P domain, involved in maintaining dimeric interactions, produced significantly larger 45-nm particles, albeit in low yields. The complete removal of the protruding domain resulted in the formation of smooth particles with a diameter that was slightly smaller than the 30-nm diameter expected from the NoV structure (Bertolotti-Ciarlet et al. 2002).

Katpally et al. (2008) determined the electron cryomicroscopy structure of MNV-1 to ~12-Å resolution and found that, compared to the NoV VLPs (Prasad et al. 1999) and SMSV virions (Chen R et al. 2003, 2006), the protruding domains were rotated by ~40° in a clockwise fashion and lifted up by ~16 Å. To better understand the MNV-1 conformation, the resolution of infectious MNV-1 was further improved to ~8 Å, in parallel with the similar reconstruction of the RHDV VLPs (Katpally et al. 2010), as mentioned earlier.

Although structural analysis of the norovirus VLPs showed that the capsid had a T = 3 icosahedral symmetry

and was composed of 180 copies of VP1 that were folded into three quasi-equivalent subunits A, B, and C, the VLP structures of the two norovirus GII.4 genetic variants that have been identified in 1974 and 2012 and termed CHDC-1974 and NSW-2012, respectively, were determined by electron cryomicroscopy as T = 4 structures (Devant et al. 2019). Surprisingly, more than 95% of these GII.4 VLPs were larger than virions, and 3D reconstruction showed that these VLPs exhibited the T = 4 icosahedral symmetry. The T = 4 particles were assembled from 240 copies of VP1 that adopted four quasi-equivalent conformations A, B, C, and D and formed two distinct dimers, A/B and C/D. The VLPs consisted therefore of 60 A/B dimers and 60 C/D dimers, with B, C, and D subunits located at the 2-fold axis and the A subunits at the 5-fold axis. The protruding domains were elevated ~21 Å off the capsid shell, which was ~7 Å more than in the previously studied GII.10 T = 3 VLPs. A small cavity and flap-like structure at the icosahedral 2-fold axis disrupted the contiguous T = 4 shell. The data showed that the VP1 sequences detected several decades apart folded into T = 4 VLPs that shared common structural features not observed in the T = 3 VLPs. Overall, these results indicated that the NSW-2012 VP1 could form both T = 3 and T = 4 particles, but only ~5% of the population were T = 3 (Devant et al. 2019). Figure 27.6 demonstrates the T = 3 and T = 4 of the NSW-2012 VLPs at resolutions 7.3 Å and 15 Å, respectively.

Recently, Snowden et al. (2020) involved reverse genetics to study dynamics in the MNV capsid by high-resolution electron cryomicroscopy. The analysis revealed that the P domain dimers were independently mobile elements (i.e., they moved in a manner that was not coordinated with other P domain dimers) with the ability to sample a wide conformational space whilst maintaining infectivity (Snowden et al. 2020).

Chimeric VLPs

Koho et al. (2015) presented a nanocarrier platform based on the modified NoV VLPs and noncovalent chemical conjugation. The VLPs were modified by adding a C-terminal polyhistidine tag, which projected out of the VLP surface, produced by the baculovirus expression system. The norovirus genotype GII.4 utilized in this study has been previously expressed as unmodified VLP by Allen et al. (2009) and Koho et al. (2012). The chimeric NoV VLPs were generated by only the His-tagged capsid proteins or in the form of mosaic particles containing both the His-tagged and the original wild-type NoV capsid protein. The polyhistidine tag was first utilized in VLP purification and later employed to attach different cargo molecules noncovalently on the VLP surface via tris-nitrilotriacetic acid (trisNTA) adaptors. The surface-modified NoV VLP nanocarriers were implemented first in delivering a conjugated fluorescent dye as a model molecule into human cells. Generally, this technology provided a universal nanoparticle tool for the adaptation of vaccine delivery or targeting, for increasing and fine-tuning vaccine immunogenicity or bioavailability

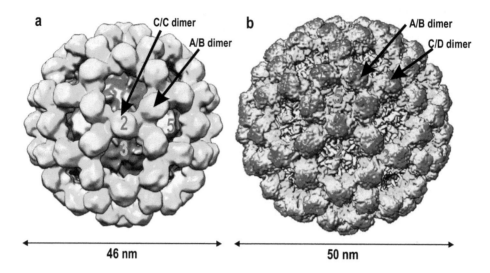

FIGURE 27.6 Electron cryomicroscopy reconstruction of norovirus VLPs. (a) NSW-2012 T = 3 VLPs. (b) NSW-2012 T = 4 VLPs. (Reprinted from *Antiviral Res.* 168, Devant JM, Hofhaus G, Bhella D, Hansman GS, Heterologous expression of human norovirus GII.4 VP1 leads to assembly of T = 4 virus-like particles, 175–182, Copyright 2019, with permission from Elsevier.)

via a displayable molecule, and for use as surface building blocks (Koho et al. 2015).

The next step by the development of the NoV VLP-based modular vaccine platform was made by Lampinen et al. (2021). The authors displayed SpyTags on the NoV VLPs and covalently decorated the particle surface with SpyCatcher conjugation technology. The highly prospective SpyTag/SpyCatcher methodology, which was introduced by Zakeri et al. (2012), is described in more detail in Chapter 25. With this modular system, the SpyCatcher-fused antigen was produced and purified separately from the NoV SpyTag-VLPs and then decorated the latter via isopeptide bonds forming spontaneously between SpyTag and SpyCatcher. In this study, two conserved influenza antigens were prepared as SpyCatcher fusions to present them onto the NoV VLPs. Thus, the latter were decorated with the ectodomain of influenza M2 ion channel protein (M2e) and a minimized stem-fragment of hemagglutinin glycoprotein 2 (HA2), both protein fragments being highly conserved across different influenza strains from a long time span. In mouse immunization experiments, the decorated NoV VLPs raised high titers of IgG antibodies against HA2 and SpyCatcher proteins. Remarkably, the presentation of HA2 on the NoV VLPs showed a trend toward higher antibody titers compared to soluble SpyCatcher-HA2 alone (Lampinen et al. 2021).

P Particles

Tan and Jiang (2005) started a special story of the NoV P particles. First, it appeared that the isolated P domain containing the hinge formed a dimer and bound to histo-blood group antigen (HBGA) receptors with a low affinity (Tan et al. 2004). Next, they found that the P domain without the hinge, when expressed in *E. coli*, formed a small particle with a significantly increased receptor binding affinity. The

glutathione S-transferase (GST)-P fusion proteins were synthesized in *E. coli*, and the P proteins were released from GST by thrombin cleavage. An end-linked oligopeptide containing one or more cysteines such as CNGRC promoted the P-particle formation by forming intermolecular disulfide bridges. The binding sensitivity of the P particle to HBGAs was enhanced >700-fold compared to the P dimer, which was comparable to that of VLPs. This binding enhancement was observed in the P particles of both norovirus GI and GII strains. The P particle was estimated to have molecular mass of ~830 kDa and contain 12 P dimers, in which the P2 subdomain built up the outer layer, while the P1 subdomain formed the internal core. By electron microscopy, the P particles revealed ring- or pentagon-shaped structures with a diameter of ~5 nm, indicating that they were closed, spherical particles with a cavity inside. The P particle was predicted to be a T = 1 icosahedron (Tan and Jiang 2005). The highly conserved arginine cluster R at the C-terminus of the P domain was critical for the P particle formation, while replacement of the R cluster with histidines resulted in low efficiency, but replacement with alanines led to complete loss of the self-assembly (Tan et al. 2006). Tan et al. (2008) produced the P particles not only in *E. coli* but also in yeast *P. pastoris*. The authors analyzed the P particle structure with and without hinge and described the symmetry as octahedral one. The dimeric packing of the proteins in the P particles was similar to that in the NoV capsid, in which the P2 subdomain with the receptor-binding interface was located at the outermost surface of the P particle. The P particles were immunogenic and revealed similar antigenic and HBGA-binding profiles with their parental VLPs, which were produced in parallel by the baculovirus expression system (Tan et al. 2008). Next, a new type of subviral particle, the small P particles, were obtained through a further modification, either an addition of the

Flag tag DYKDDDDK or a change of the arginine cluster, at the C-terminus of the cysteine-containing P domain. The electron cryomicroscopy showed that the small P particles were tetrahedrons formed by 6 P dimers or 12 P monomers that were half the size of the P particles. The fitting of the crystal structure of the P domain into the cryo-EM density map of the particle indicated similar conformations of the P dimers as those in the original P particles (Tan et al. 2011a).

The P particles became a successful platform for the presentation of foreign antigens. Thus, a small His tag peptide of 7 aa residues and a large rotavirus protein VP8 of 159 aa were inserted into one of the three surface P loops (Tan et al. 2011b). Neither insertion affected P particle formation, while both antigens were presented well on the P particle surface. The immune-enhancement effect of the P particle was demonstrated by significantly increased antibody titers induced by the P particle-presented antigens compared to the titers induced by free antigens. Moreover, the level of protection against rotavirus shedding was significantly higher in mice immunized with the VP8-chimeric P particles than those of mice immunized with the free VP8 antigen. The P particle-VP8 chimeras appeared as a dual vaccine candidate against both rotavirus and norovirus (Tan et al. 2011b). Tan et al. (2011c) presented insertion of a number of small (5 aa) to large (238 aa) antigens into these loops without affecting P particle formation and production. Among others. the protection against influenza virus and rotavirus challenges was demonstrated in mice after immunization with the chimeric P particles carrying influenza M2e and rotavirus VP8 antigens, compared to free M2e and VP8 antigens, respectively.

The P particles were thoroughly studied by their immunological properties (Tamminen et al. 2012), heat inactivation (Li D et al. 2012), and mass spectroscopy parameters (Bereszczak et al. 2012). The P particles appeared at that time as the true vaccine candidates against norovirus, rotavirus, and influenza viruses (Tan and Jiang 2012a). The deep investigation of the outstanding immunological and vaccine properties of the P particles was continued in detail (Tan and Jiang 2012b; Fang H et al. 2013; Kocher et al. 2014; Su et al. 2015).

Wang L et al. (2014) generated a dual vaccine candidate against norovirus and hepatitis E virus (HEV) from the *Hepelivirales* order (Chapter 16). The dimeric P domains of NoV and HEV were fused together, designated as NoV P(-)-HEV P, which was then linked with the dimeric glutathione-S-transferase (GST). After expression and purification in *E. coli*, the GST-NoV P(-)-HEV P fusion protein assembled into polyvalent complexes with a mean size of 1.8 μm, while the NoV P(-)-HEV P formed oligomers ranging from 100 to 420 kDa. Both GST-NoV P(-)-HEV P and NoV P(-)-HEV P complexes induced in mice significantly higher antibody titers to NoV P(-) and HEV P, respectively, than those induced by a mixture of the NoV P(-) and HEV P dimers.

The P particles were used as a platform to display the gp41 membrane proximal external region (MPER) of HIV-1 and induced MPER-specific antibody responses in immunized guinea pigs (Zang et al. 2014; Yu et al. 2015). The Alzheimer's disease (AD) immunogen Aβ1–6 was inserted into the three loops of the P particle to generate an AD protein vaccine (Fu et al. 2015, 2017; Li Y et al. 2016).

The putative influenza vaccines were generated using the P particles. Thus, the selected B cell epitopes from hemagglutinin (HA) were used to construct chimeric trivalent HA-P particles active against H1N1, H3N2, and B influenza (Gong X et al. 2016). The P particles carrying M2e of an emerging swine influenza virus and highly conserved two each of H1N1 peptides of pandemic 2009 and classical human influenza viruses were entrapped into biodegradable polylactic-co-glycolic acid (PLGA) nanoparticles (Hiremath et al. 2016). The vaccine induced the virus specific T cell response in the lungs and reduced the challenged heterologous virus load in the airways of the vaccinated pigs. Remarkably, the rotavirus inner capsid VP6 VLPs exhibited a notable improving impact on the immune responses in mice, which was induced by the P particles carrying the M2e sequence (Heinimäki et al. 2020).

Beyond the direct application to construct vaccines, the P particles were used as an in vitro model to assess the interactions of noroviruses with probiotics (Rubio-del-Campo et al. 2014), for the identification of sialic acid-containing glycosphingolipids (gangliosides) as ligands for NoVs (Han L et al. 2014) and for the anchoring on the surface of *E. coli* (Niu et al. 2015).

It is noteworthy that Chen YL et al. (2018) came back recently to the production of the P particles in *P. pastoris* and succeeded by the establishment of a high purification scheme (Chen YL et al. 2020).

The main challenges facing the development of a NoV vaccine, including the P particle-based candidates, were reviewed by Lucero et al. (2018) and Mattison et al. (2018). The general success of the P particles was thoroughly reassessed in the excellent reviews of Tan and Jiang (2015, 2019).

New Materials

When Douglas and Young M (1999) performed their experiments on cowpea chlorotic mottle virus (CCMV), as described in Chapter 17, they used in parallel NoV as a model system for nanophase crystal growth. The authors emphasized that NoV formed particles virtually identical to those observed for CCMV under similar conditions, although the NoV particles were not known to swell. The mineralization reaction was shown with vanadate, molybdate, and tungstate within the CCMV, while the same mineralization reaction was observed with tungstate in the case of NoV.

GENUS *SAPOVIRUS*

The expression of the capsid protein of Sapporo virus (SaV) in insect cells resulted in the self-assembly of VLPs that had a morphology similar to that of the native virus with distinct cup-like depressions on the surface of the particles, referred to as a "Star of David" appearance (Numata et al. 1997). Later, Kitamoto et al. (2002) generated the Sapporo virus VLPs by the baculovirus expression system for mapping of monoclonal antibodies.

Jiang et al. (1999) described expression of the viral capsid protein of the two Sapporo-like virus (SLV) strains, Hou/90 and Hou/86, in the baculovirus system. The expressed capsid protein self-assembled into the nice VLPs of 30–35 nm in diameter, round, and with ten spikes on the edge of the particles. The addition of the MEG tri-peptide at the N-terminus did not prevent self-assembly, while addition of His$_6$ tag to the N-terminus of capsid protein blocked VLP formation.

Hansman et al. (2005a, c) reported the baculovirus-driven expression and self-assembly of capsid proteins from the genotypes GI, GII, and GV of sapoviruses, where constructs began exactly from the predicted VP1 start AUG codon. The VLPs, with diameters of 41–48 nm, were morphologically similar to those of native sapoviruses, whereas a fraction of the GV VLPs was smaller, with diameters of 26–31 nm and spikes on the outline. The ELISA showed that the GI VLPs were antigenically distinct from GII and GV VLPs (Hansman et al. 2005c).

When the constructs containing N- and C-terminal-deleted sapovirus protein VP1 were expressed in the baculovirus system, the VLPs, including both small and native-size, self-assembled only by proteins derived from N-terminally deleted VP1 constructs that began 49 nucleotides downstream. These results were similar to those reported for the RHDV N- and C-terminally deleted VP1 expression studies but were distinct from those reported for the NoV N- and C-terminally deleted VP1 data (Hansman et al. 2005a).

GENUS VESIVIRUS

Feline Calicivirus

A distinguishing feature of the vesivirus genome, in contrast to that in other caliciviruses, is that it encodes a capsid protein precursor (73–78 kDa) that is proteolytically processed by the viral protease to yield a mature capsid protein of ~60 kDa (Desselberger 2019), which is significantly larger in size than the capsid protein (~55 kDa) of human caliciviruses.

FCV, a member of the *Vesivirus* genus, provided a tractable model since it was able to propagate in cell culture. The expression of the FCV capsid gene in cultured feline cells resulted in the VLPs that appeared uniform in morphology and size of 35 nm and were indistinguishable from intact calicivirus virions (Geissler et al. 1999). A bit earlier, DeSilver et al. (1997) expressed the precursor capsid protein gene of FCV in the baculovirus system and demonstrated the appropriate VLPs. Then, Di Martino et al. (2007) produced in insect cells the VLPs of the FCV strain F9. The FCV VLPs were morphologically and antigenically

similar to the native virions and elicited in rabbits the virus-neutralizing antibodies against clinical FCV strains.

Bhella et al. (2008) determined the 3D structure of the FCV-receptor complex by electron cryomicroscopy and image reconstruction to 18 Å-resolution. Furthermore, Ossiboff et al. (2010) determined the crystal structure of FCV at 3.6-Å resolution.

As described earlier, Moreno et al. (2016) produced the nice FCV VLPs by the expression of the protein VP62 gene in the baculovirus system, along with the RHDV-derived VP60 chimeric mutants. These FCV VLPs that are shown in Figure 27.5 were used by comparison of the immunogenicity induced by the FCV epitope incorporated into the RHDV chimeric VLPs with that elicited by the same epitope in its natural context.

Conley et al. (2017) reported the structure of the FCV strain F9 to 7 Å resolution, in parallel with the 3D structure of a vesivirus 2117, an adventitious agent that was identified in 2009 as a contaminant of Chinese hamster ovary cells propagated in bioreactors at a pharmaceutical manufacturing plant belonging to Genzyme and caused significant economic losses. By further electron cryomicroscopy studies on vesivirus 2117, Sutherland et al. (2021) showed that the outer face of the dimeric capsomers, which contains the receptor binding site and major immunodominant epitopes in all caliciviruses studied thus far, is quite different from that of FCV. This is a consequence of a 22 aa insertion in the sequence of the FCV capsid protein that forms a "cantilevered arm," which plays an important role in both receptor engagement and undergoes structural rearrangements thought to be important for genome delivery to the cytosol (Sutherland et al. 2021).

San Miguel Sea Lion Virus

The *Vesicular exanthema of swine virus* species from the *Vesivirus* genus includes the well-studied San Miguel sea lion virus (SMSV). Since SMSV was successfully propagated in vitro, in contrast to many caliciviruses, the x-ray crystallographic structure of virions of the SMSV serotype 4 was resolved to 3.2-Å resolution by molecular replacement techniques (Chen R et al. 2003, 2006) by using a 19-Å cryo-EM map of SMSV (Chen R et al. 2004) as an initial phasing model. The crystal structure of SMSV showed significant and distinct variations, some of which were important for host specificity and antigenic diversity.

McClenahan et al. (2010) engineered the sequences encoding the major VP1 and minor VP2 capsid proteins from two marine vesivirus isolates, namely Steller sea lion viruses V810 and V1415, for expression of the VLPs in the baculovirus expression system. The resulting VLPs were morphologically similar to native vesivirus virions. The purified VLPs were probed in immunoblots with pooled antisera specific for nine SMSV types, and a predominant protein of ~60 kDa was detected. The VLPs were used as antigens to develop an ELISA for the detection of serum antibodies to marine vesiviruses in animals from two free-ranging populations of Steller sea lion in Alaska (McClenahan et al. 2010).

FAMILY *DICISTROVIRIDAE*

The capsid structures of the two dicistroviruses, cricket paralysis virus (CrPV) of the *Cripavirus* genus (Tate et al. 1999) and triatoma virus (TrV) of the *Triatovirus* genus (Squires et al. 2013), were resolved to 2.4 Å and 2.5 Å, respectively. These capsids were quite similar naked T = 3 icosahedrons built by 60 copies of each of the three major structural proteins VP1–3, all folded with a jelly-roll core. Agirre et al. (2013) comprised the electron cryomicroscopy reconstruction of the TrV virions and native empty particles at ~15 and ~19 Å resolution, respectively. The structure of black queen cell virus (BQCV), another representative of the *Triatovirus* genus, was determined at the 3.4 Å resolution by Spurny et al. (2017). Then, the electron cryomicroscopy structures were resolved for the two representatives of the *Aparavirus* genus, namely Israeli acute bee paralysis virus (IAPV) to 3.2 Å (Mullapudi et al. 2016, 2017) and mud crab dicistrovirus (MCDV) to 3.3 Å (Gao Y et al. 2019).

Sánchez-Eugenia et al. (2015) described the baculovirus-driven expression of the two ORFs, nonstructural NS and structural P1, encoded by the TrV genome. TrV infects blood-sucking insects belonging to the *Triatominae* subfamily that act as vectors for the transmission of *Trypanosoma cruzi*, the etiological agent of the Chagas disease. The proteolytic processing of the P1 polyprotein was strictly dependent upon the coexpression of the NS polyprotein, and that NS/P1 coexpression led to assembly of the VLPs that exhibited a morphology and a protein composition akin to the natural TrV empty virions. Remarkably, the unprocessed P1 polypeptide assembled into quasispherical structures conspicuously larger than the VLPs produced in NS/P1-coexpressing cells, likely representing a previously undescribed morphogenetic intermediate. This intermediate has not been found in members of the related *Picornaviridae* family currently used as a model for dicistrovirus studies, thus suggesting the existence of major differences in the assembly pathways of these two virus groups (Sánchez-Eugenia et al. 2015). Sánchez-Eugenia et al. (2016) used x-ray crystallography to solve the atomic structure of the TrV empty particle, which appeared after RNA release. It was observed that the overall shape of the capsid and of the three individual proteins was maintained in comparison with the mature virion.

FAMILY *PICORNAVIRIDAE*

Genus *Aphthovirus*

Foot-and-Mouth Disease Virus

The crystal structure of foot-and-mouth disease virus (FDMV) was determined at the 2.9 Å resolution by Acharya et al. (1989). The capsid was composed of 60 copies each of four proteins, VP1–VP4. The proteins VP1–3 were partly exposed on the capsid surface, according to

the icosahedral pT = 3 symmetry, while VP4 was internal and had an N-terminal myristic acid. Each of the proteins VP1–3 contained an eight-stranded β-barrel very similar in tertiary structure and distribution within the virus particle to the capsid proteins of the small RNA-containing T = 3 plant viruses. The crystal structures of picornaviruses have been already determined at that time to enteroviruses and rhinoviruses of the *Enterovirus* genus and Mengo virus of the *Cardiovirus* genus (see following sections). The FMDV structure showed similarities with the resolved picornaviruses but also demonstrated several unique features. The characteristic canyon or "pit" found in other picornaviruses was absent. The most immunogenic portion of the capsid, which acted as a potent peptide vaccine, formed a disordered protrusion on the virus surface (Acharya et al. 1989). The crystal structure of FMDV is presented in Figure 27.7, together with other most studied representatives of the *Picornaviridae* family. More recently, Kotecha et al. (2017) used electron cryomicroscopy to visualize the cell-entry mechanism by attachment of FMDV to an integrin receptor, generally αvβ6, via a conserved arginine-glycine-aspartic acid (RGD) motif in the exposed, antigenic, GH loop of the VP1 protein.

In mammalian cells, the empty FMDV VLPs were obtained following the use of such expression systems as vaccinia virus (Abrams et al. 1995) and adenovirus (Mayr et al. 1999, 2001; Mason PW et al. 2003).

In the baculovirus system, Roosien et al. (1990) were the first who expressed the capsid precursor P1–2A, with and without the proteases L and 3C, of the FMDV serotypes O1K and A10 in *Sf* cells. The processed procapsid proteins were produced in cells infected with recombinant baculoviruses, when L and 3C were present in the constructs, indicating that these FMDV proteases were active in insect cells. The analysis of baculovirus-expressed products in sucrose gradients showed that a fraction of the capsid proteins was present in an aggregated form, migrating at 70S and possibly resembling the empty FMDV VLPs (Roosien et al. 1990). Oem et al. (2007) produced the pentamer-like structures after baculovirus-driven expression of the genes for the FMDV VP1 and 3C proteins from individual promoters.

At last, Cao et al. (2009) constructed a recombinant baculovirus that simultaneously expressed the genes for the P1–2A and 3C proteins of FMDV, serotype Asia I, from individual promoters. The capsid proteins expressed in HighFive insect cells were processed by viral 3C protease and formed the empty VLPs that were similar to the authentic so-called 75S empty capsids of FMDV in terms of their size of 30 nm in diameter, shape, and sedimentation velocity. The immunoelectron microscopy revealed some small-sized particles of about 10 nm besides the virion-like empty VLPs. The latter induced high levels of FMDV-neutralizing antibodies in guinea pigs (Cao et al. 2009). Furthermore, Cao et al. (2010) optimized codon usage to express the VP1–2A-VP3 and VP0 genes of the FMDV serotype O from individual promoters in *Sf*9 cells. The expressed VP1–2A-VP3 was autocatalytically cleaved into the individual proteins,

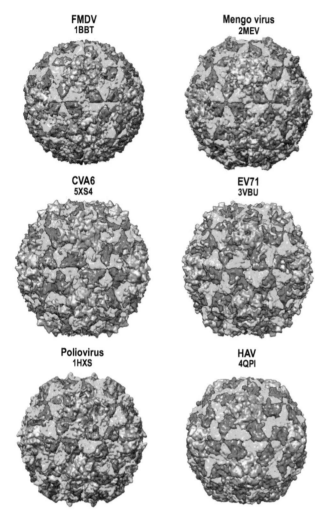

FMDV
1BBT

Mengo virus
2MEV

CVA6
5XS4

EV71
3VBU

Poliovirus
1HXS

HAV
4QPI

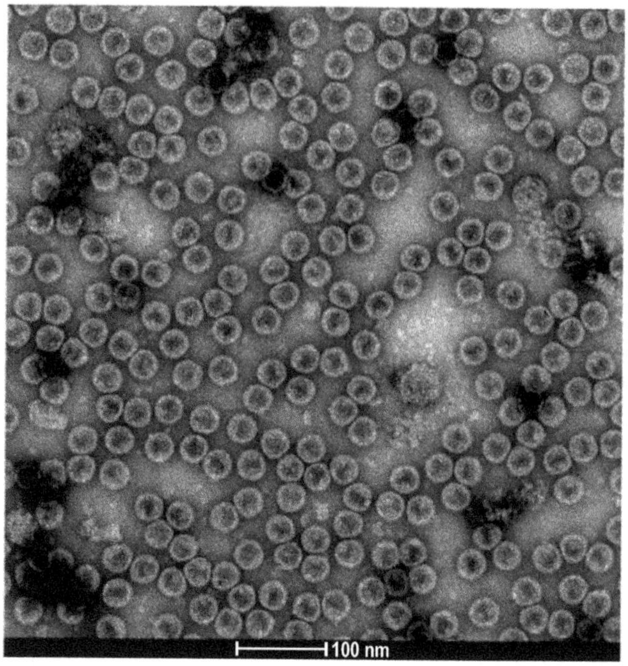

FIGURE 27.8 Analysis of the empty FMDV, serotype A22, VLPs produced in *Sf*9 cells. The peak fractions from the sucrose gradient were analyzed by electron microscopy. (Reprinted from Porta C, Xu X, Loureiro S, Paramasivam S, Ren J, Al-Khalil T, Burman A, Jackson T, Belsham GJ, Curry S, Lomonossoff GP, Parida S, Paton D, Li Y, Wilsden G, Ferris N, Owens R, Kotecha A, Fry E, Stuart DI, Charleston B, Jones IM, Efficient production of foot-and-mouth disease virus empty capsids in insect cells following down regulation of 3C protease activity. *J Virol Methods.* 2013;187:406–412.)

FIGURE 27.7 The crystal structure of the *Picornaviridae* family members: foot-and-mouth disease virus (FMDV), genus *Aphthovirus*, 2.60 Å resolution, outer diameter 304 Å (Acharya et al. 1989; Fry et al. 1993); Mengo virus, genus *Cardiovirus*, 3.00 Å, 326 Å (Krishnaswamy and Rossmann 1990); Coxsackievirus A6 (CVA6) VLPs, genus *Enterovirus*, 3.10 Å, 340 Å (Xu et al. 2017); enterovirus type 71 (EV71), genus *Enterovirus*, 4.00 Å, 334 Å (Wang Xiangxi et al. 2012); poliovirus, genus *Enterovirus*, 2.20 Å, 324 Å (Hogle et al. 1985; Filman et al. 1989); hepatitis A virus (HAV), genus *Hepatovirus*, 3.01 Å, 312 Å (Wang Xiangxi et al. 2015). The 3D structures are taken from the VIPERdb (http://viperdb.scripps.edu) database (Carrillo-Tripp et al. 2009). The size of particles is to scale. All structures possess the icosahedral pseudoT = 3 (pT = 3) symmetry. The corresponding protein data bank (PDB) ID numbers are given under the appropriate virus names.

VP1–2A and VP3, and the proteins VP0, VP3, and VP1–2A were self-assembled into VLPs resembling the authentic FMDV virions.

Porta et al. (2013) presented the highly efficient baculovirus-driven production of the beautiful (by the electron microscopy analysis) empty VLPs of FMDV, serotype A22, with the P1–2A-3C cassette that contained a modified variant of the 3C enzyme. Figure 27.8 demonstrates these VLPs. The intermediate levels of the latter resulted

in efficient processing of the P1–2A precursor into the structural proteins VP0, VP3, and VP1, which assembled into the empty FMDV VLPs. Similarly, Bhat et al. (2013) produced the FMDV, serotype O, VLPs by the P1–2A and 3C (wild-type and mutant) gene expression by baculovirus under polyhedrin promoter, since the wild 3C caused toxicity to insect cells. The immunogenicity and protective efficacy of the FMDV VLPs was demonstrated in guinea pigs in terms of humoral immune response and protection against virulent virus challenge (Bhat et al. 2013). Kumar et al. (2016) elaborated a highly efficient protocol to produce the FMDV, serotype O, VLPs in Eri silkworm (*Samia cynthia ricini*) larvae. The polyprotein P1–2A-3C-encoding gene with the mutated 3C was used in this study that paved the way to the large scale production of the FMDV VLPs for the putative vaccine or diagnostic use.

In *E. coli*, Lewis SA et al. (1991) expressed the VP1 and 3C genes of the FMDV serotype A12, resulting in efficient synthesis and processing of the structural protein precursor and assembly into the 70S empty capsids, while the electron microscopy data were not presented. This material reacted with neutralizing monoclonal antibodies recognizing only conformational epitopes and elicited a significant neutralizing antibody response in vaccinated guinea pigs.

Later, the soluble forms of the FMDV VP1 and VP3 proteins were successfully produced in *E. coli* as fusions with SUMO (*small ubiquitin-related modifier*) protein, and SUMO was removed from the fusion proteins by SUMO protease (Lee CD et al. 2008). Moreover, Lee CD et al. (2009) expressed SUMO fusion proteins of the three FMDV capsid proteins VP0, VP1, and VP3 in *E. coli*. These three fusion proteins formed the ternary VP0-VP1-VP3 complex. The proteolytic removal of SUMO moieties from the ternary complexes resulted in VLPs with size and shape resembling the authentic FMDV virions, as visualized by electron microscopy (Lee CD et al. 2009). The FMDV VLPs produced by a SUMO fusion protein system in *E. coli* induced potent protective immune responses in guinea pigs, swine, and cattle (Guo HC et al. 2013). Recently, Guo M et al. (2021) improved the heat resistance of this vaccine via biomineralization. Thus, four simple biomimetic mineralization methods with the use of calcium phosphate were applied, and the biomineralized VLPs were stored at temperatures of 25°C for eight days and 37°C for four days, while the animal experiments showed that the biomineralization had no effect on the immunogenicity of VLPs.

In yeast, Balamurugan et al. (2005) expressed the P1–2A genome stretch of all the four serotypes of FMDV (O, Asia 1, A22, and C), which included VP1–3, partial VP4, and 2A in each case, in *P. pastoris*. It was expected that the P1 polyprotein would fold in a way maintaining discontinuous epitopes involved in virus neutralization. In fact, the partially purified P1 protein demonstrated some protective efficacy after challenge of the vaccinated guinea pigs with different adjuvant formulations.

In plants, Carrillo et al. (1998) expressed the sequence encoding the FMDV VP1 in *Arabidopsis thaliana*. The mice immunized with leaf plant extracts were protected against challenge with virulent FMDV. In fact, this was the first study showing protection against a viral disease by immunization with an antigen expressed in a transgenic plant (Carrillo et al. 1998, 2001; Wigdorovitz et al. 1999a, b). Furthermore, Dus Santos et al. (2005) reported production of the P1 polyprotein and 3C protease of FMDV in transgenic alfalfa plants. Although no definitive data has been produced, the electron microscopy analysis showed the presence of the spherical VLPs of 30 nm in diameter in the plant expressing the P1–3C stretch. The product ensured complete protection of mice against the experimental challenge with the virulent virus. Then, the foliar extracts from transgenic tomato plants expressing the polyprotein P1–2A and protease 3C of FMDV elicited a protective response in guinea pigs (Pan et al. 2008).

The further development of the plant-based FMDV vaccines was achieved during the work undertaken by the EC FP7 Plant Production of Vaccines (PLAPROVA) consortium, which investigated the use of transient expression technologies to produce VLPs of both human and veterinary pathogens—including FMDV—and was summarized by Thuenemann et al. (2013). Thus, the cleavage

sites within the P1 polyprotein were modified to be recognized by the CPMV 24K protease, rather than the native FMDV 3C protease, since this was less toxic than 3C when expressed in plants (Saunders et al. 2009). In an attempt to express the empty FMDV capsids in plants, the P1 precursor and the CPMV 24K protease were coinfiltrated on separate constructs into *N. benthamiana* leaves. The modified P1 polyprotein carrying the recognition sites for 24K protease was expressed and processed into shorter proteins if the CPMV 24K protease was coexpressed. However, no VLP-like particles were observed by electron microscopy of protein samples isolated from agroinfiltrated plant leaves. This suggested that processing of the P1 polyprotein in plant cells and/or some steps of assembly of capsids differed from that occurring in animal cells. Alternatively, any empty capsids that did form might be unstable in plants, a plausible scenario given the acid-instability of FMDV capsids (Thuenemann et al. 2013).

The problem of the acid-instability was solved recently by Xie et al. (2019) for the VLPs of the FMDV serotype O that appeared as more sensitive than those of other serotypes. The novel FMDV mutants with increased acid resistance were isolated using mammalian BHK-21 cells cultured under low-pH conditions. The aa substitutions Q25R, K41E, and N85A in the VP1 capsid protein and K154Q in the VP3 capsid protein were detected in all six mutants. Based on these aa replacements, the empty FMDV, serotype O, VLPs were produced in insect cells and demonstrated resistance to the acid-induced dissociation of the capsid into pentameric subunits that were produced in insect cells. These VLPs induced strong neutralizing antibodies in guinea pigs and protected all the guinea pigs from FMDV challenge (Xie et al. 2019).

At last, the cell-free expression system of rabbit reticulocyte lysates was specifically used to study the precursor processing and pentamer assembly stages of the FMDV capsid assembly (Goodwin et al. 2009; Newman et al. 2018, 2021).

GENUS *CARDIOVIRUS*

Mengo Virus

Mengo encephalomyelitis virus, a member of the *Cardiovirus A* species, is characterized by an extremely wide host range, which includes primates. The crystal structure of Mengo virus was determined at the 3.0 Å resolution by Luo et al. (1987). The structure of Mengo virus was substantially different from the structures of rhinoviruses and polioviruses. Although the organization of the major capsid proteins VP1, VP2, and VP3 of Mengo virus was essentially the same as in rhino- and polioviruses, the large insertions and deletions, mostly in VP1, radically altered the surface features. In particular, the putative receptor binding "canyon" of human rhinovirus 14 (HRV14) became a deep "pit" in Mengo virus because of polypeptide insertions in VP1 that filled part of the canyon. The minor capsid peptide, VP4, was completely internal in Mengo virus (Luo et al. 1987).

The crystal structure of Mengo virus was further refined by Krishnaswamy and Rossmann (1990). This structure is presented in Figure 27.7. The pH-dependent structural changes of Mengo virus related to its host receptor attachment site and disassembly were described by Kim S et al. (1990).

Altmeyer et al. (1994) engineered the first chimeric infectious Mengo virus expressing the sequence encoding 147 aa of the gp120 glycoprotein from HIV-1, strain MN, within the RNA genome of a stably attenuated Mengo virus strain. The HIV-1 stretch was fused to the N-terminus of a short, nonstructural Mengo virus leader polypeptide L of 67 aa and was recognized by a gp120 V3 loop-specific monoclonal antibody on the virion surface. The immunization of mice or cynomolgus monkeys with the chimeric HIV-Mengo virus induced both humoral and cellular immune response. Furthermore, Altmeyer et al. (1995) inserted at the same place the well-characterized CTL epitope of the lymphocytic choriomeningitis virus (LCMV) nucleoprotein, namely the 119-PQASGVYMG-127 stretch within the cloned 117–130 fragment. Van der Ryst et al. (1998) engineered chimeric Mengo viruses expressing either a large region (aa 65–206) of the HIV1 *nef* gene product or CTL epitope regions from the SIV Gag (aa 182–190), Nef (aa 155–178) and Pol (aa 587–601) gene products. The heterologous antigens were expressed either as fusion proteins with the L polypeptide or in cleaved form through autocatalytic cleavage by the FMDV 2A protein. However, the chimeric Mengo viruses were characterized as rather weak immunogens. Moreover, it was concluded that the expression of certain heterologous sequences as fusion proteins with L can result in the loss of ability to infect normally susceptible animals (Van der Ryst et al. 1998).

Porcine Encephalomyocarditis Virus

Porcine encephalomyocarditis virus (EMCV) is another well-known member of the *Cardiovirus A* species. Jeoung et al. (2010) constructed a plasmid containing the P1–2A and 3C genes of the EMCV-K3 viral strain. The EMCV VLPs were successfully assembled in mammalian 293FT cells and were identified as particles of about 30–40 nm by electron microscopy. The immunization of mice with the plasmid DNA induced high levels of neutralizing antibody and led to a significant protection ratio after challenge with wild-type EMCV (Jeoung et al. 2010). Furthermore, Jeoung et al. (2011) successfully generated the EMCV VLPs of the same 30–40 nm size by the baculovirus expression system in *Sf*9 cells. The high neutralizing antibody titers were observed following double immunization of pigs. Finally, the ECMV VLP vaccine with an alum adjuvant was recommended as a safe candidate for protecting against EMCV-induced reproductive failure in pig farms (Jeoung et al. 2012).

GENUS *ENTEROVIRUS*

Coxsackievirus

The group of six serotypes of Coxsackievirus (CVB1-CVB6) are the representatives of the *Enterovirus B* species

and pathogenic enteroviruses that trigger illness ranging from gastrointestinal distress to full-fledged pericarditis and myocarditis. The crystal structure of CVB3 was resolved to 3.5 Å by Muckelbauer et al. (1995). The β-sandwich structure of the viral capsid proteins VP1, VP2, and VP3 was conserved between CVB3 and other picornaviruses. The structural differences between CVB3 and other enteroviruses and rhinoviruses were located primarily on the viral surface. The hydrophobic pocket of the VP1 β-sandwich was occupied by a pocket factor, modeled as a C16 fatty acid. An additional study has shown that the pocket factor could be displaced by an antiviral compound. Myristate was observed covalently linked to the N terminus of VP4, as described earlier in the case of FMDV.

Reimann et al. (1991) constructed a chimeric CVB4 virus. Thus, five aa of the putative BC loop of the structural protein VP1 of CVB4 were inserted into the corresponding loop of CVB3 by site-directed mutagenesis of infectious recombinant CVB3 cDNA. The chimeric cDNA could induce an infectious cycle upon transfection of permissive host cells. The resulting chimeric virus CVB3/4 was neutralized and precipitated by CVB4 and CVB3 serotype-specific polyclonal antisera, demonstrating that it unified antigenic properties of both coxsackievirus serotypes. The chimera elicited antibodies in rabbits, which could neutralize the two coxsackievirus serotypes CVB3 and CVB4.

Zhang L et al. (2012) constructed the baculoviruses carrying either the intact, entire coding region of CVB3 or the four individual coding regions for virus proteins VP1–4. The electron microscopy revealed the presence of the CVB3 VLPs in the appropriate sucrose gradient purification fractions. The mice vaccinated with the CVB3 VLPs developed antibodies after the first boost. Moreover, the vaccinated animals were protected from myocarditis when subsequently challenged with a cardiovirulent CVB3 strain. Remarkably, the vaccination with VLPs produced from the complete CVB3 coding region gave a greater immune response and afforded better protection than with the VLPs from the quadruple expression vector (Zhang L et al. 2012).

Xu L et al. (2017) isolated in mammalian cells two stable particles of Coxsackievirus A6 (CVA6), a representative of the *Enterovirus A* species, that has recently emerged as a major cause of hand-foot-and-mouth disease (HFMD) in children worldwide. These particles were: one empty CVA6 procapsid and the other infectious CVA6 A-particle. The capsid structures of CVA6 procapsid and A-particle at near atomic ~3 Å resolution both closely resembled those of the uncoating intermediates of other enteroviruses. This structure of the A-particle is presented in Figure 27.7. In addition, structural and functional studies of the CVA6 A-particle complexed with the Fab of a neutralizing antibody identified four surface loops on VP1 that could be targeted for vaccine design (Xu L et al. 2017).

In parallel, the CVA6 VLPs were produced in insect cells (Shen et al. 2016) and yeast *P. pastoris* (Zhou et al. 2016). In both cases, the CVA6 VLPs efficiently induced specific antibodies that protected mice against lethal viral challenges. Furthermore, Chen J et al. (2018) resolved for the first time the atomic structure of the CVA6 VLPs to 3.0 Å by electron cryomicroscopy and located the two CVA6-specific conserved linear B-cell epitopes of the VP1 protein on the outer VLP surface.

Liu Q et al. (2012) produced the CVA16 VLPs in insect cells by coexpression of the P1 and 3CD genes of CVA16 using recombinant baculoviruses. The CVA16 VLPs consisted of processed VP0, VP1, and VP3 and were present as ~30 nm spherical particles. The passive immunization with anti-VLP sera conferred protection against lethal CVA16 challenge in neonate mice, indicating therefore a humoral mechanism of protection.

In yeast, Zhao H et al. (2013) produced the CVA16 VLPs by co-expressing P1 and 3CD of CVA16 in *S. cerevisiae*. These VLPs exhibited similarity in both protein composition and morphology as empty particles from the CVA16-infected cells. As before, the passive immunization with anti-VLPs sera conferred full protection against lethal CVA16 challenge in neonate mice.

Enterovirus Type 71

The reliable atomic models of enterovirus type 71 (EV71), a member of the *Enterovirus A* species and the major agent of HFMD in children that can cause severe central nervous system disease and death, were provided by Wang Xiangxi et al. (2012) at resolutions of 2.3, 2.6, 2.9, and 3.8 Å respectively for two independent structure determinations each for full virus and empty particles. The structure of the empty EV71 particle is presented in Figure 27.7. Shingler et al. (2013) used electron cryomicroscopy to reconstruct these two capsid states and provided remarkable insight into the mechanics of genome release.

By the baculovirus-driven expression, Hu YC et al. (2003) were the first who reported production of the crystalline VLPs morphologically resembling the authentic EV71 aggregates after coinfection of the insect *Sf*9 cells with two recombinant baculoviruses encoding the polyprotein P1 and protease 3CD of EV71. Then, this previous work was continued by further constructing a new recombinant baculovirus simultaneously expressing P1 and 3CD. The new baculovirus was used for single infection of the *Sf*9 cells, coexpression of P1 and 3CD, and purification of VLPs that were indistinguishable from the authentic virions in size of 25–27 nm, morphology, and composition of surface epitopes (Chung YC et al. 2006). These EV71 VLPs elicited potent immune responses and protected mice against lethal virus challenge (Chung YC et al. 2008). Chung CY et al. (2010) optimized production and achieved in *Sf*9 cells cultured in the bioreactor the best extracellular VLP yield of ~64.3 mg/L, representing a ~43-fold increase over the yield attained using the old process in the spinner flasks. The macaque monkeys developed both specific humoral and cellular immune responses to the EV71 VLPs (Lin YL et al. 2012). The role of the EV71 VLPs in comparison with other vaccine candidates was reviewed early by Chong et al. (2012).

Lin SY et al. (2015) performed next regular improvements of the design by the baculovirus-driven expression of the EV71 VLPs and achieved high level extracellular yields of the product in HighFive cells. Zhao D et al. (2015) improved the purification of the insect cells-produced EV71 VLPs by high-performance liquid chromatography (HPLC), while the diameter of purified EV71 VLPs was analyzed not only by electron microscopy but also by dynamic light scattering.

Somasundaram et al. (2016) presented obstacles for the enhanced production of both EV71 and CVA16 VLPs, comparing the VLP yields in *Sf*9 and HighFive cells. The authors used high-resolution asymmetric flow field-flow fractionation couple with multiangle light scattering (AF4-MALS) for the first time to characterize the EV71 and CVA16 VLPs, displaying an average root mean square radius of 15 ± 1 nm and 15.3 ± 5.8 nm, respectively.

Gong M et al. (2014) not only generated the secreted versions of the EV71 and CVA16 VLPs using the baculovirus expression system but also reconstructed the 3D structures of both VLPs by electron cryomicroscopy at 5.2-Å and 5.5-Å resolutions, respectively. The obtained structures of the EV71 and CVA16 VLPs highly resembled the crystal structures for EV71 natural empty particles (Wang Xiangxi et al. 2012) and CVA16 135S-like expanded particles (Ren et al. 2013), respectively. Moreover, Ku et al. (2013, 2014) reported the immunogenicity and high protective efficacy of a bivalent HFMD vaccine obtained by combination of the insect cells-produced EV71 and CVA16 VLPs.

Zhang Wei et al. (2018) proposed a tetravalent VLP vaccine by combination of the insect cells-produced EV71, CVA6, CVA10, and CVA16 VLPs. The tetravalent vaccine elicited in mice antigen-specific and long-lasting serum antibody responses comparable to those elicited by its corresponding monovalent vaccines. Moreover, the tetravalent vaccine immune sera strongly neutralized EV71, CVA16, CVA10, and CVA6 strains with neutralization titers similar to those of their monovalent counterparts, indicating a good compatibility among the four antigens in the combination vaccine. The passively transferred tetravalent vaccine-immunized sera conferred efficient protection against single or mixed infections with EV71, CVA16, CVA10, and CVA6 viruses in mice, whereas the monovalent vaccines could only protect mice against homotypic virus infections but not heterotypic challenges (Zhang Wei et al. 2018).

The baculovirus-driven production of the EV71 VLPs by expression of the P1 and 3CD genes in *Sf*9, *Sf*21, or High-Five insect cells was further optimized by Kim HJ et al. (2019).

In yeast, Li HY et al. (2013) produced the EV71 VLPs in *S. cerevisiae* by coexpressing P1 and 3CD genes. The yeast-produced VLPs exhibited similar morphology and protein composition as the empty EV71 particles from the EV71-infected cells. The immunization of mice elicited neutralizing antibodies against EV71 and potent cellular immune response. Moreover, the VLP-induced immune sera conferred protection against the lethal EV71 challenge in neonate mice.

Wang Xiaowen et al. (2016) produced the EV71 VLPs in *S. cerevisiae* by coexpression of the genes encoding four structural proteins VP1, VP2, VP3, and VP4. The neonatal mice model demonstrated that the VLP immunization conferred protection to suckling mice against the lethal viral challenge.

Zhang C et al. (2015) produced the EV71 VLPs in yeast *P. pastoris* by coexpressing the EV71 P1 and 3CD genes, when high levels up to 4.9% of total soluble protein were achieved. More importantly, the maternal immunization with the yeast-produced VLPs protected neonatal mice in both intraperitoneal and oral challenge experiments. Yang Z et al. (2020) expressed the codon-optimized P1 and 3C genes of EV71 in *P. pastoris* and presented a high yield and simple manufacturing process of the EV71 VLPs, with the expression level reaching 270 mg/L. The produced EV71 VLPs consisted of processed VP0, VP1, and VP3 within ~35-nm spherical particles. The immunization of mice provided effective protection against lethal challenge in both maternally transferred antibody and passive transfer protection mouse models.

Zhao H et al. (2015) engineered the first chimeric EV71-based VLPs, in order to produce bivalent vaccine against HFMD in *S. cerevisiae*. Thus, the neutralizing epitope SP70 within the capsid protein VP1 of EV71 was replaced with that of CVA16. In fact, the replacement occurred in the GH loop of VP1, where only four aa residues K215L, E217A, K218N, and E221D were replaced. The structural modeling revealed that the replaced CVA16-SP70 epitope was well exposed on the surface of the chimeric VLPs. The latter elicited robust Th1/Th2 dependent immune responses against EV-A71 and CVA16 in mice, while passive immunization with antichimeric VLP sera conferred full protection against lethal challenge of both EV71 and CVA16 infection in neonatal mice. To elucidate the structural basis of the VLPs produced from yeast as vaccine candidates, Lyu et al. (2015a) determined the crystal structures of the chimeric VLPs, in parallel with the original EV71 VLPs. Both structures shared similarity with that of the naturally occurring empty particle (Lyu et al. 2015a) and showed that both the linear and conformational neutralization epitopes identified in EV71 were structurally preserved on both VLPs.

In mammalian cells, Tsou et al. (2015) engineered a recombinant adenovirus-vectored vaccine with the EV71 P1 and 3CD genes inserted into the E1/E3-deleted adenoviral genome. The EV71 VLPs were produced in HEK-293A cells. The VP0, VP1, and VP3 were expressed from the P1 gene after correct digestion by 3CD protease. The antiviral immunity against EV71 was clearly demonstrated in mice vaccinated with the chimeric adenovirus.

Lin YJ et al. (2015) produced the EV71 VLPs in enoki mushrooms *Flammulina velutipes*. Polycistronic expression vectors harboring the glyceraldehyde-3-phosphodehydrogenase promoter to codrive the EV71 P1 and 3C genes using the 2A peptide of porcine teschovirus-1 were constructed and introduced into *Flammulina velutipes* via *Agrobacterium tumefaciens*-mediated transformation.

The P1 and 3C genes were integrated into the chromosomal DNA through a single insertion, and their resulting mRNAs were transcribed. The resulting EV71 VLPs were composed of the four subunit proteins digested from the P1 polyprotein by 3C protease. The 3D reconstruction of the EV71 VLPs was performed to confirm their similarity to the EV71 virions (Lin YJ et al. 2015).

A novel type of the *E. coli*-produced EV71 VLPs was invented by Xue et al. (2017). The authors combined the peptides P_{70-249}, $P_{324-443}$, and $P_{746-876}$ from the VP1-VP3 proteins into one fusion protein. The expressed fusion protein finally assembled into spheric VLPs of 25–40 nm in diameter and proved efficient in induction of humoral and cell-mediated immunity in mice (Xue et al. 2017). Moreover, the fusion protein-based EV71 VLPs demonstrated an immunoprotective role in neonatal mice through maternal antibodies and active immunization, which remarkably alleviated EV71-mediated damage (Liu J et al. 2020).

Enterovirus D68

The enterovirus D68 (EV-D 68) belongs to the *Enterovirus D* species and is suspected to play a definite role in acute flaccid myelitis (AFM), an uncommon but serious neurological condition, mostly in young children. Zheng Q et al. (2019) resolved atomic structures of the EV D68 major phases throughout its life cycle, including mature, cell-entry intermediate (A-particle), and empty forms and of virus complex with two monoclonal antibodies, defining therefore distinct mechanisms of viral neutralization.

Dai et al. (2018) successfully generated the EV-D68 VLPs in the insect *Sf*9 cells infected with a recombinant baculovirus coexpressing the P1 precursor and 3CD protease of EV-D68. The VLPs were composed of VP0, VP1, and VP3 capsid proteins derived from precursor P1 and were visualized as spherical particles of ~30 nm in diameter. The passive transfer of anti-VLP sera completely protected neonatal recipient mice from lethal EV-D68 infection. Moreover, the maternal immunization with these VLPs provided full protection against lethal EV-D68 challenge in suckling mice.

At the same time, Zhang C et al. (2018) coexpressed the P1 precursor and 3CD protease of EV-D68 in *P. pastoris* yeast, which resulted in the generation of the EV-D68 VLPs composed of the processed VP0, VP1, and VP3 proteins and visualized as ~30 nm particles. As earlier with the insect cells-produced VLPs, the yeast-produced EV-D68 VLPs ensured full protection of mice by active and passive immunization.

Poliovirus

The 3D structure of poliovirus (PV) type 1 Mahoney, a member of the *Enterovirus C* species and the causative agent of poliomyelitis, was resolved by x-ray crystallography to 2.9 Å resolution by Hogle et al. (1985). Each of the three major capsid proteins VP1, VP2, and VP3 contained a "core" consisting of an eight-stranded antiparallel β-barrel with two flanking helices. The arrangement of β-strands and helices was structurally similar and topologically identical to the folding pattern of the capsid proteins of the icosahedral plant viruses, as in the case of other picornaviruses. In each of the major capsid proteins, the "connecting loops" and N- and C-terminal extensions were structurally dissimilar. The packing of the subunit "cores" to form the virion shell was reminiscent of the packing in the T = 3 plant viruses but was significantly different in detail. Several of the "connecting loops" and C-terminal strands formed prominent radial projections, which were the antigenic sites of the virion (Hogle et al. (1985). This structure is presented in Figure 27.7.

It is noteworthy that the naturally occurring empty PV VLPs have been discovered very early besides the mature virions (Hummeler et al. 1962).

A set of different PV vectors was engineered with an aim to express foreign sequences by chimeric PV virions. First, the chimeric PV viruses were used for the display of foreign epitopes that were inserted into one of the capsid proteins. Thus, Burke et al. (1988) constructed a PV chimera carrying a defined region of type 3 inserted into type 1 and demonstrating the composite antigenicity and immunogenicity in small animals and primates. The authors proposed these techniques for modifications of the poliovirus type 1 Sabin strain not only to the development of new improved type 2 and type 3 polio vaccines but also to vaccines against other picornaviruses, such as hepatitis A virus. Martin A et al. (1988) engineered a poliovirus type 2 antigenic site on a type 1 capsid. Murray MG et al. (1988) replaced six aa residues in antigenic site I of the Mahoney strain with a sequence specific for the Lansing strain. In parallel, Kohara et al. (1988) replaced the sequence encoding the antigenic determinants in viral capsid proteins of the Sabin 1 genome by the corresponding sequences of the type 2 and type 3 genome, respectively. Minor et al. (1991) presented PV chimeras of type 1 and type 3, which involved antigenic sites of the PV types 2, 3, and 4. Moreover, Evans et al. (1989) constructed the PV Sabin type 1 chimera carrying an epitope from the transmembrane glycoprotein (gp41) of HIV-1. To facilitate the production of this chimera, a cassette vector was engineered by introduction of unique restriction sites that flanked the region encoding aa residues 91–102 of the VP1 protein. This region of VP1 was known to elicit neutralizing antibodies and on the 3D map of PV was seen to form a distinct surface projection at the pentameric apex of the icosahedral virion, according to the previously mentioned pioneering paper of Hogle et al. (1985). The antibodies raised by this chimera in rabbits neutralized a wide range of American and African HIV-1 isolates and also inhibited virus-induced cell fusion. Crabbe et al. (1990) replaced the antigenic site 1 of VP1 by the epitopes from HIV-1 and came to the conclusion, after the appropriate calculations, that a viable virus will only be formed when antigen chimeras modified at the antigenic site would have a loop occupying a similar volume in space to that occupied by the original antigenic site. In addition, the modified loop must fit with the peptide bond angles and distances at the top of the beta-barrel of VP1. Altmeyer et al. (1991)

constructed a chimera, in which the neutralization antigenic site IA of PV type 1 strain Mahoney was replaced by neutralization immunogenic site IA of HRV14. Dedieu et al. (1992) got a number of viable HIV-1-PV chimeras. Lemon et al. (1992) inserted peptide sequences from hepatitis A virus (HAV) capsid proteins into the B-C loop of VP1 of the Sabin strain type 1 poliovirus (PV-1) and got a number of viable chimeras.

Second, the dicistronic PV RNAs were constructed by duplicating the 5' noncoding region of the PV genomic RNA, namely the internal ribosomal entry site (IRES), in which foreign polypeptides were expressed by using one IRES, and essential viral proteins were produced by using the other IRES (Alexander et al. 1994; Lu et al. 1995).

Third, the PV minireplicons were constructed in which PV structural protein genes were replaced by foreign sequences (Choi et al. 1991; Ansardi et al. 1995; Porter et al. 1995).

To overcome limitations of these three strategies, including the small size of the tolerated insert, genetic instability of the inserted sequences, and a requirement for helper virus for viral propagation, Andino et al. (1994) constructed PV chimeras, where exogenous peptides of up to 400 aa residues and an artificial cleavage site for viral protease 3C were fused to N-terminus of the viral polyprotein P1. The extended chimeric polyprotein was produced in infected cells and proteolytically processed into the complete array of viral proteins plus the foreign peptide, which was excluded from mature virions. The recombinants retained exogenous sequences through successive rounds of replication in culture and in vivo (Andino et al. 1994). Moreover, the foreign sequences were inserted at different positions of the P1 and separated by artificial 3C or 2A protease cleavage sites. The 3C or 2A proteases accurately recognized and cleaved the inserted synthetic proteolytic sites, freeing the exogenous protein sequences from the rest of the PV polyprotein. In this manner, all of the poliovirus proteins were correctly produced, and normal viral replication proceeded. Thus, Tang et al. (1997) described the construction of a PV chimeric virus carrying SIV Env, Gag, and Nef antigenic sequences, and the SIV aa stretches were effectively produced in the PV chimera-infected cells. The infection of susceptible mice with the PV chimeras elicited humoral immune responses to SIV proteins in a dose-dependent manner. The administration of a mix of chimeras carrying five different SIV antigens elicited antibodies that recognized each of the SIV proteins (Tang et al. 1997).

Concerning production of the empty PV VLPs in mammalian cells, Ansardi et al. (1991) used a vaccinia vector to express the P1 polyprotein stretch of PV type 1 Mahoney in HeLa cell culture. The coinfection of the P1-expressing vaccinia virus together with a second recombinant vaccinia virus that expressed the PV proteinase 3CD resulted in the correct processing of the P1 to yield the three individual capsid proteins VP0, VP3, and VP1. The examination of the appropriate sucrose gradient fractions by electron microscopy revealed the empty PV VLPs of ~27 nm in diameter

(Ansardi et al. 1991). In parallel, Jore et al. (1991) observed formation of similar PV particles by in vitro translation of subgenomic PV RNAs in rabbit reticulocyte lysates.

Recently, Viktorova et al. (2018) demonstrated a different design of a PV vaccine based on the in situ production of VLPs. The PV genes P1 and protease CD were expressed from a Newcastle disease virus (NDV) vector, a negative-strand RNA virus from the *Mononegavirales* order (Chapter 31) with mucosal tropism. In this system, the PV VLPs were produced in the cells of vaccine recipients and were presented to their immune systems in the context of active replication of NDV, which served as a natural adjuvant. The intranasal administration of the vectored vaccine to guinea pigs induced strong neutralizing systemic and mucosal antibody responses.

As to the baculovirus-driven expression, Urakawa et al. (1989) expressed the complete coding region, nucleotide residues 743 to 7363, of the P3/Leon/37 strain of PV type 3 within a baculovirus vector in the *S. frugiperda* insect cells. The infected insect cells produced the structural PV proteins VP0, VP1, and VP3, which were recovered from extracts of the infected cells as the empty VLPs, approximately 27 nm in diameter, demonstrating the typical icosahedral morphology of the previously mentioned empty PV particles. The immunization of mice with the PV VLPs induced virus-neutralizing antibodies. Another recombinant baculovirus, which contained the majority of the PV structural region, made an unprocessed precursor to the PV structural proteins. It was concluded that the PV-encoded proteases were responsible for the processing of the viral polyprotein (Urakawa et al. 1989).

Bräutigam et al. (1993) individually cloned the VP0, VP3, and VP1 genes of the PV type 3, strain P3/Leon/37, in recombinant baculoviruses, expressed them in *Sf* cells, and isolated the PV VLPs that corresponded in size of 27 nm in diameter, appearance, and antigenicity to those expected for the PV virion. However, the yields of the particles were low when compared to those derived from a construct that expressed the complete coding region of poliovirus type 3, indicating that procapsid synthesis from the polyprotein P1 was a more favored route (Bräutigam et al. 1993).

In yeast, Jore et al. (1994) transferred first the PV sequence encoding protease 3CD to *S. cerevisiae*, as an inducible transcription unit. Further, by simultaneous induction of both P1 and 3CD expression, the cell extracts revealed the presence of the PV VLPs that resembled the authentic empty PV particles. However, using a competition immunoprecipitation assay, Rombaut et al. (1994) showed that these empty particles were so-called H-antigenic, meaning that these particles behaved as being heat-denatured, did not possess any neutralizing determinant, and were unable to induce neutralizing and protective antibodies. In contrast, Rombaut and Jore (1997) demonstrated that the natural, or N-antigenic, empty PV VLPs can be synthesized in the same *S. cerevisiae* yeast expression system, when pirodavir, a capsid-binding compound, was added to the yeast cells. The N-antigenic empty VLPs that were purified by

immunoaffinity chromatography with monoclonal antibodies induced virus-neutralizing antibodies in mice (Rombaut and Jore 1997). Later, Wang XW et al. (2013) expressed in *S. cerevisiae* the synthetic optimized genes P1 and 3CD of PV type 1 and obtained the expected PV VLPs.

The obvious fact that the genome-free empty PV VLPs were unstable and readily changed antigenicity to a form not suitable as a vaccine remained the major problem for a long time. To overcome this uncertainty, the stable PV capsid variants were selected. Thus, Adeyemi et al. (2017) demonstrated that the empty VLPs of type 1 PV can be stabilized by selecting heat-resistant viruses, and the selected mutants could be applied as candidates to synthesize stable VLPs as the genome-free PV vaccines. Fox et al. (2017) reported the genetic manipulation of the virus to generate stable empty capsids for all three PV serotypes, and the obtained particles were shown to be extremely stable and to generate high levels of protective antibodies in animal models.

The stabilized PV VLPs were successfully expressed in plants. Thus, Marsian et al. (2017) developed in *N. benthamiana* the synthetically produced stabilized virus-like particle (sVLP)-based PV vaccine with D antigenicity, without the drawbacks of current vaccines. In parallel, the VLPs of wild-type PV type 3 and the sVLPs of the stabilized PV3 mutant SktSC8, described by Fox et al. (2017), were produced in *N. benthamiana*. Figure 27.9 demonstrates both traditional VLPs and the revolutionary plant-made sVLPs. Such plant-produced sVLPs retained the native antigenic conformation and the repetitive structure of the original virus particle but lacked infectious genomic material. The mice carrying the gene for the human PV receptor were protected from wild-type PV when immunized with the plant-made PV sVLPs. The structural analysis of the stabilized mutant at 3.6 Å resolution by electron cryomicroscopy and single-particle reconstruction revealed a structure almost indistinguishable from the wild-type PV3 and showed that it adopted a native, D, antigenic conformation. Moreover, the analysis of the structure revealed the fine mechanism of action of the stabilizing mutations (Marsian et al. 2017).

Following this successful introduction of the PV particles with the enhanced thermostability, Xu Y et al. (2019) produced the sVLPs of PV1, PV2, and PV3 in large quantities by using the baculovirus expression system and *Sf*9 cells. To do this, the individual expression of the appropriate gene variants encoding the PV proteins VP1, VP0, and VP3–2A was performed, which allowed coassembly of the latter into the sVLPs. The mice immunized with the PV sVLPs generated strong PV-neutralizing antibodies.

Furthermore, Sherry et al. (2020) developed a highly efficient methodology to produce the PV VLPs in *P. pastoris*, in preliminary experiments of which the thermostable mutant sVLPs previously proposed and used by Fox and others (Fox et al. 2017; Marsian et al. 2017; Xu Y et al. 2019) were self-assembled. Therefore, the *Pichia* expression system was concluded to be able to efficiently produce the D antigen and contribute therefore to the creation of a VLP vaccine not only for a polio-free world but also as a model

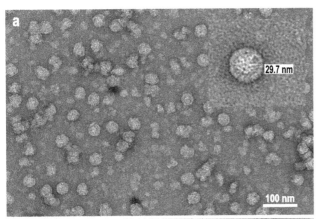

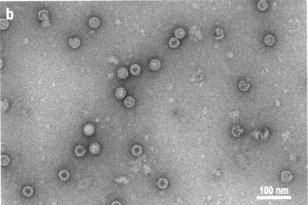

FIGURE 27.9 Electron microscope analysis of plant-produced PV VLPs. (a) *N. benthamiana* produced wt PV3 VLPs. (b) PV3 SktSC8 sVLPs visualized by negative staining and electron microscopy. The inset in (a) represents a higher magnification image of a particle within the main field. (Reprinted from Marsian J et al. *Nat Commun.* 2017;8:245.)

system for enterovirus VLP vaccine production (Sherry et al. 2020).

Remarkably, Bahar et al. (2021) proposed an alternate way to get the good sVLPs. The authors presented an efficient mammalian expression strategy producing good yields of wild-type PV VLPs for all three serotypes and the sVLP variant for PV3. In brief, the modified vaccinia virus Ankara (MVA) expression system in BHK-21 cells was used for the coexpression of the P1 and 3CD sequences. Whilst the wild-type VLPs were predominantly in the nonnative C-antigenic form, the PV3 sVLPs adopted the native D-antigenic conformation eliciting neutralizing antibody titers and were indistinguishable from the natural empty PV particles by electron cryomicroscopy with a similar stabilizing lipidic pocket-factor in the VP1 β-barrel. This factor was not available in alternative expression systems, which would require synthetic pocket-binding factors (Bahar et al. 2021).

Human Rhinovirus 14

The x-ray crystal structure of human rhinovirus 14 (HRV14), a member of the *Rhinovirus B* species and the causative

agent of common cold, was resolved to 3.0 Å by Rossmann et al. (1985). Together with the previously described poliovirus virions, they were the first structurally resolved representatives of the great *Picornaviridae* family. Both the tertiary fold of the VP1, VP2, and VP3 polypeptide chains and their pT = 3 quaternary organization within the HRV14 capsid were very similar to the earlier resolved T = 3 structures of the two RNA plant viruses, namely tomato bushy stunt virus (TBSV) of the *Tolivirales* order described in Chapter 24 and southern bean mosaic virus (SBMV) of the *Sobelivirales* order (Chapter 28). The β-barrels in HRV14, like those in SBMV, were wedge-shaped, with the thin end pointing toward the 5- or 3-fold (quasi-6-fold) axes (Rossmann et al. 1985). Furthermore, Edward Arnold and Michael G. Rossmann (1990) refined the structure and identified immunogenic regions, as well as the hydrophobic pocket in VP1, which was the locus of binding for the so-called WIN agents, discovered initially by the Sterling-Winthrop Research Institute and used against HRV.

Gail Ferstandig Arnold and her team presented the design and construction of the chimeric live HRV14 that contained immunogenic regions from other pathogens as part of their surface coat proteins (Arnold GF et al. 1994). The short segments encoding the PV3 Sabin VP1 and VP2 proteins, the influenza hemagglutinin (HA) glycoprotein, and the HIV-1 gp120 surface and gp41 transmembrane glycoproteins were inserted into a full-length clone of HRV14 at regions corresponding to the neutralizing immunogenic sites IA (NIm-IA) and II (NIm-II). Of 12 chimeric constructs described, 3 produced viable virus. An HRV14 chimeric virus containing five aa of influenza HA had wild-type HRV14 growth characteristics and was neutralized by anti-HA antisera. However, antisera raised in two guinea pigs against the HA-HRV14 chimera did not show significant neutralization of relevant strains of influenza (Arnold GF et al. 1994). To increase the likelihood of recovering viable HRV14 chimeras displaying the transplanted HIV-1 V3 loop sequences in conformations that mimic that of HIV, Resnick et al. (1994) used random systematic mutagenesis to produce libraries of chimeric HRV14 in which the transplanted epitope from HIV-1 was flanked by one or more randomized aa residues. This allowed the HIV epitope to be accommodated into the HRV14 coat proteins in many conformations, some of which should result in the production of viable, immunogenic hybrids. Using this approach, a library containing the sequence XXIGPGRAXX, where X could be any of the 20 aa, was generated. A subset of chimeras was identified that reacted with neutralizing anti-HIV-1 V3 loop antibody preparations, indicating that the antigenicity of the epitopes had been transplanted. Another chimeric virus library was designed to reflect the natural diversity of the V3 loop by incorporating aa residues at frequencies similar to those found among naturally occurring isolates of HIV-1 (Resnick et al. (1994). Smith AD et al. (1994) studied the sequence XXIGPGRAXX inserted at the NIm-II of HRV14 between VP2 residues 159 and 160. Twenty-five unique chimeric viruses were isolated, and a nonrandom aa distribution that may reflect structural requirements for viability was observed at the randomized positions. Remarkably, 15 of 25 chimeras were neutralized by one or more of a panel of four anti-HIV-1 V3 loop antibodies (Smith AD et al. 1994). Resnick et al. (1995) performed immunoselection of the HRV14 library by the neutralizing V3 loop-directed MAbs and found five chimeras, where insertions did not match those of any known isolate of HIV-1. Nonetheless, all five chimeras were neutralized by specific HIV-1 antibodies and were able to elicit the production of antibodies that bound V3 loop peptides from diverse HIV-1 isolates.

The exciting story of the live HIV-1:HRV14 chimeras was reviewed early on (Arnold GF et al. 1996; Smith AD et al. 1997). It should be emphasized that this applied random systematic mutagenesis methodology was of utmost importance at that time for the further progress in rapid transplantation of foreign sequences into target proteins to produce libraries containing members with the desired properties. However, the number of foreign aa residues that were displayed on the surfaces of picornaviruses was limited, and the protein epitopes were quite small, typically consisting of only up to two tenths of aa residues.

Developing the live HRV14 approach, Smith AD et al. (1998) identified more HIV-1:HRV14 chimeras that carried the V3 loop sequence IGPGRAFYTTKN of HIV-1, strain MN, and were able to elicit neutralizing antibody responses in guinea pigs, while three of them elicited HIV neutralization titers that exceeded those of all but a small number of previously described HIV immunogens. Furthermore, Zhang A et al. (1999) optimized the HIV-1 V3 insertion HIGPGRAF on the VP2 capsid protein when the HIV sequence was flanked by (i) a Cys residue that could form a disulfide bond and (ii) randomized aa in either of two arrangements to generate numerous presentations of the Cys-Cys loop. The presence of the disulfide bond in the chimeric virus was verified by proteolytic digestion with and without a reducing agent. At last, Ding et al. (2002) reported the 2.7-Å resolution structure of a chimeric rhinovirus, MN-III-2, which displayed part of the HIV-1 gp120 V3 loop and elicited HIV-neutralizing antibodies. The MN-III-2 virus was engineered by Smith AD et al. (1998) and exposed 12 aa of the gp120 V3 loop of HIV-1$_{MN}$, name, IGPGRAFYTTKN, flanked on the N-terminal side by the randomized linker ADT. The insertion was made between Ala159 and Asn160 of the VP2 puff of the neutralizing immunogenic site II (Nlm-II). Due to the symmetry of the viral capsid, the insertion was displayed in 60 copies on the surface of HRV14.

Later, Arnold GF et al. (2009) chose another popular HIV-1 epitope, namely, the well-conserved ELDKWA epitope of the membrane-proximal external region (MPER) of HIV-1 protein gp41. The epitope was inserted at the same previously described site of the HRV14 protein VP2 by the traditional combinatorial library approach. The optimal location of the ELDKWA epitope onto the HRV14 surface loop was achieved by connection via linkers of variable lengths and sequences, and the appropriate viruses were selected by recognition with a specific neutralizing MAb (Arnold GF et al. 2009). Moreover, the molecular in silico modeling was employed to design an optimal structure (Lapelosa et al. 2009, 2010). In summary, guided by x-ray crystallography, molecular modeling, combinatorial chemistry, and powerful selection techniques, six combinatorial libraries of the chimeric HRV14 displaying the MPER epitopes and connected to an immunogenic surface loop of HRV14 via linkers of varying lengths and sequences were designed and produced (Yi et al. 2013). Although not all libraries led to viable chimeric viruses, the combinatorial approach allowed examination of large numbers of MPER-displaying chimeras. Among the chimeras were five that elicited antibodies capable of significantly neutralizing HIV-1, in one case leading to neutralization of all six subtypes tested.

Although not directly connected with the VLP matter, the oncolytic property of PVSRIPO, a recombinant PV-HRV chimera, should be mentioned. The PVSRIPO chimera, where the internal ribosomal entry site (IRES) of PV was substituted with the IRES from HRV2, has shown promise for glioblastoma treatment. The impact of the PVSRIPO therapy is substantially reviewed (Goetz and Gromeier 2010; Goetz et al. 2011; Brown and Gromeier 2015; Denniston et al. 2016; Brown et al. 2017; Desjardins et al. 2018; Gromeier and Nair 2018; Iorgulescu et al. 2018; Carpenter et al. 2021; Gromeier et al. 2021).

Swine Vesicular Disease Virus

The preliminary x-ray data to 3.6 Å resolution of swine vesicular disease virus (SVDV), strain JX/78, a member of the *Enterovirus B* species, which shared some antigenic properties with Coxsackievirus B5 (CVB5) and could be a recently evolved genetic sublineage of this important human pathogen, were published first by Lin W et al. (2002). Then, three different crystal forms of SVDV, isolate SPA/2/'93, were obtained by Jimenez-Clavero et al. (2003). Fry et al. (2003) presented the 3.0-Å crystal structure of highly virulent strain UK/27/72 of SVDV, which revealed the expected similarity in core structure to those of other picornaviruses, showing most similarity to the closest available structure to CBV5, that of CBV3. Remarkably, the authors mapped the aa substitutions that might have occurred during the supposed adaptation of SVDV to a new host.

Ko et al. (2005) performed the baculovirus-driven production of the SVDV, strain UKG/27/72, VLPs by simultaneous expression of the P1 and 3CD genes under different promoters in insect *Sf*9 cells. The antigenic differences between recombinant VLPs and SVDV virions were not statistically significant, indicating that the VLPs could be used in the place of the SVDV antigen in ELISA kits (Ko et al. 2005). In fact, this was the first report of the production and diagnostic application of the SVDV VLPs. Later, Xu W et al. (2017) optimized the baculovirus-driven production of the SVDV VLPs in *Sf*9 and HighFive cells and applied them in the competitive and isotype-specific

ELISA. Following the methodology of Xu W et al. (2017), Yang M et al. (2020) generated the SVDV VLPs of the same strain UK 27/72, performed large-scale expression in *Sf*9 cells, and developed and evaluated the isotype ELISA for anti-SVDV detection.

GENUS *HEPATOVIRUS*

Hepatitis A Virus

The high-resolution electron cryomicroscopy pattern of hepatitis A virus, a member of the *Hepatovirus A* species and the causative agent of a type of the liver inflammation, was visualized by Wang Xiangxi et al. (2017). Earlier, Wang Xiangxi et al. (2015) published the high-resolution x-ray structures for the mature HAV and its empty particle. The structures of the two particles were indistinguishable, apart from some disorder on the inside of the empty particle. The full virus contained the small viral protein VP4, whereas the empty particle harbored only the uncleaved precursor VP0. This structure is presented in Figure 27.7. The HAV particle contained no pocket factor and was able to withstand remarkably high temperature and low pH, and empty particles were even more robust than full particles (Wang Xiangxi et al. 2015).

Wychowski et al. (1990) were the first ones to generate a chimeric HAV derivative by replacing the HAV capsid VP4 gene with the poliovirus (PV) VP4 gene within a full-length infectious clone of HAV cDNA in an attempt to produce viable chimeric viruses. Although the viable chimeric viruses were not detected, the in vitro translation was successfully performed in rabbit reticulocyte lysate. However, the putative appearance of the chimeric virions was not addressed in this study.

Beneduce et al. (2002) engineered the first chimeric HAV capsids, where nonstructural, but morphogenesis-involved protein 2A, being a part of the capsid protein precursor P1–2A, was used as a target. It was known before that a small quantity of the VP1–2A is present in the infectious HAV virions. Thus, the foreign antigen was planned to be exposed on the particle surface by introducing extraneous sequences within the context of the 2A gene of HAV. For that purpose, the aa residues 44–53 of protein 2A were replaced by a foreign antigenic epitope, namely the popular 7-aa stretch ELDKWAS of the HIV-1 envelope protein gp41, also called 2F5 epitope, and the chimeric empty capsids of 70S were found after vaccinia virus MVA-T7-mediated expression of the cDNA in mammalian COS7 and Huh-T7 cells (Beneduce et al. 2002). Extending this work, Kusov et al. (2007) reported that the chimeric HAV-gp41 virus replicated in HAV-susceptible cells as well as in nonhuman primates. The infected marmosets developed both an anti-HAV and anti-2F5 epitope immune response. Furthermore, an HIV-neutralizing antibody response was elicited in guinea pigs immunized with the chimeric HAV-gp41 virions.

Although not directly relevant in the sense of the chimeric VLPs, it should be mentioned that Xiang et al. (2017) developed a combination vaccine against hepatitis A and E infections, where a HEV neutralization epitope HEp148 located at aa 459–606 of the HEV capsid protein was inserted into the 2A/2B junction of the HAV genome, in frame with the rest of the HAV polyprotein. The recombinant virus expressed the HEp148 protein in a partially dimerized state in HAV-susceptible cells. The immunization with the HAV-HEp148 virus induced a strong HAV- and HEV-specific immune response in mice, demonstrating a novel approach to the development of a combined hepatitis A and E vaccine (Xiang et al. 2017).

Concerning bacterial expression of the structural HAV genes, Ostermayr et al. (1987) produced the HAV VP1 as a fusion of β-galactosidase. Johnston et al. (1988) fused the VP1 to the *E. coli* TrpE coding sequences under the control of the strong tryptophan promoter. Gauss-Müller et al. (1990) expressed six overlapping genomic regions of the HAV VP1 and VP3 as β-galactosidase- fusions, where chimeric proteins were poorly soluble. Ross and Anderson (1991) produced in *E. coli* all four structural HAV proteins as the β-galactosidase fusions and found that the rabbit antisera against them were unable to neutralize viral infectivity or react with HAV by radioimmunoassay. Years later, Jang et al. (2014) expressed His-tagged versions of the HAV VP1 in *E. coli* and proposed them as a useful source for developing HAV subunit vaccine candidates. da Silva Junior et al. (2017) thoroughly purified the *E. coli*-produced VP1 and established an in-house ELISA for the detection of IgM antibodies in sera from HAV-positive patients. Meanwhile, Li Chuanfeng et al. (2013) significantly increased the outcome of such HAV-derived antigens when the His-tagged and codon-optimized VP1 gene version of duck HAV were expressed in *E. coli*. Intriguingly, the HAV VP1-P2a stretch, which was fused with the fragment of the *E. coli* flagellin, was expressed not only in *E. coli* but also in food-grade lactic acid bacterium *Lactococcus lactis* (Berlec et al. 2013). The resulting antigen-displaying bacteria induced specific humoral immune response when orally administered to mice. The *L. lactis* cells carrying the VP1 gene of duck HAV were used for oral vaccination of mice and ducklings (Song et al. 2019).

Concerning appearance of the possible HAV-derived self-assembled structure in bacterial cells, Pintó et al. (2002) expressed in *E. coli* the full-length HAV polyprotein and found some particulate structures. Further analysis unveiled the synthesis of 14S pentamers and 70S empty capsids by expressing the viral genome for periods of time longer than 4 h in *E. coli* (Sánchez et al. 2003). The 14S HAV pentamers were able to self-assemble into the 70S capsids in vitro. Remarkably, the antibodies induced by these structures recognized and neutralized HAV (Sánchez et al. 2003). Unfortunately, no electron microscopy data was presented.

A specific sort of the HAV-derived VLPs was generated by Kui et al. (2009). Thus, the gene fragment encoding HAV VP1, aa 24–171, and the HEV ORF2 gene fragment encoding aa 431–615 were spliced together via a 9-peptide linker. The fusion protein of 342 aa residues was produced in *E.*

coli and specifically reacted with human hepatitis A- and E-positive sera. After renaturation of the fusion protein the spheric aggregates of 10–20 nm were found and evaluated by electron microscopy. The chimeric HAV-HEV particles induced strong anti-HAV and anti-HEV humoral immune responses in mice (Kui et al. 2009).

As to the expression in insect cells, Harmon et al. (1988) were the first ones to produce the structural HAV proteins, namely VP1, by the baculovirus-driven system. The protein was found predominantly in the cytoplasm of infected insect cells, probably as an insoluble aggregate. Stapleton et al. (1991) expressed the entire HAV ORF and identified products of the same molecular mass as HAV proteins, although several larger HAV proteins were detected in cells infected with the recombinant baculovirus, suggesting that posttranslational processing was inefficient or retarded in *Sf* cells. Lee JM et al. (2009) described the secretory expression and immunogenicity of the HAV VP1 from stably transformed *Drosophila melanogaster* S2 (Schneider 2) cells.

The previously described ideas of the HAV 2A usage were applied by Ou et al. (2013) to the baculovirus-driven expression system, where a novel therapeutic antihypertension angiotensin II vaccine was produced in *Sf*9 cells. This vaccine presented four successive repeated Ang IIs as the functional epitope on the surface of the HAV VLPs. The Ang IIs epitope was inserted into the 2A region of the modified HAV genome and, as a result, 240 copies of Ang II were transferred onto the surface of the HAV VLPs, significantly increasing the antigenicity and immunogenicity of Ang II. The vaccine demonstrated good immunogenicity and good effect on reduction of blood pressure in the spontaneous hypertensive rat model (Ou et al. 2013).

The efficient baculovirus-driven production of the duck HAV VLPs was described by Wang A et al. (2018). The structural polyprotein precursor gene P1 and the protease gene 3CD were expressed in insect cells and the recombinant structural proteins spontaneously assembled into the VLPs, which were composed with the three structural proteins. The vaccination with the VLPs induced high humoral immune response and provided strong protection of ducklings.

In mammalian cells, Winokur et al. (1991) engineered recombinant vaccinia viruses that expressed the HAV polyprotein and the P1 structural region. The HeLa cell lysates demonstrated that the polyprotein was cleaved into immunoreactive proteins that comigrated with the HAV capsid proteins VP0 and VP1 and self-assembled into the 70S and 15S particles, similar to those of the HAV empty capsids and pentamers, respectively. The 70S particles were identified by immune electron microscopy as the appropriate HAV VLPs. The Pl construct produced, however, a 90-kDa protein, which showed no evidence of posttranslational processing (Winokur et al. 1991). At the same time, Karayiannis et al. (1991) engineered the vaccinia virus that contained an HAV cDNA fragment encoding structural polypeptides VP4, VP2, and VP3 and the N-terminus of

VP1. Zhu et al. (1994) reported production of the nice HAV VLPs when the recombinant vaccinia virus encoded the entire HAV polyprotein.

In plants, Chung HY et al. (2011) fused the HAV VP1 to the immunoglobin Fc fragment and expressed the chimera in tobacco leaves using a vector based on beet curly top virus (BCTV). The fusion of heterologous peptides with Fc was undertaken to provide stability and facilitate the purification process in a plant system. The purified chimeric protein elicited production of specific antibodies after intraperitoneal immunization of mice.

FAMILY *SECOVIRIDAE*

GENUS *COMOVIRUS*

Cowpea Mosaic Virus
3D Structure

The well-known cowpea mosaic virus (CPMV) is a 31-nm, icosahedral plant virus that grows in the common cowpea plant *Vigna unguiculata*. As mentioned earlier, CPMV has a bipartite RNA genome with each RNA molecule RNA1 and RNA2 encapsidated in a separate particle. The CPMV capsids are composed of 60 copies each of a large L of 42 kDa and small S of 24 kDa capsid protein to form the pseudoT = 3 icosahedral particle. As stated by George P. Lomonossoff (2008) in the *Encyclopedia of Virology*, the discovery that the CPMV particles contained equimolar amounts of two different polypeptides suggested that the capsids had an architecture more similar to the animal picornaviruses than to other plant viruses of known structure. This provided an early clue as to the common origins of plant and animal viruses (Lomonossoff 2008). Figure 27.10a illustrates the organization of the CPMV capsid as a space-filling drawing.

Historically, Crowther et al. (1974) published the first 3D image reconstruction from electron micrographs of CPMV and showed that the virus possesses icosahedral symmetry. The model for the structure supposed the 12 pentamers at the 5-fold positions and 20 trimers at the 3-fold positions. White JM and Johnson (1980) reported a hexagonal crystal form of CPMV and proposed a packing model based on the particle size, crystal density measurements, and steric limitations. Johnson and Hollingshead (1981) supported evidence for the proposed crystal structure of CPMV using both electron microscopy and x-ray diffraction and performed data collection to 15 Å resolution. Remarkably, this paper invented comparison of the 3D structures of CPMV and southern bean mosaic virus (SBMV) from the order *Sobelivirales* described in Chapter 28, which were identical in the *c*-axis projection but quite different three dimensionally and found a new interpretation of the data for the SBMV crystals. The structural data of comoviruses with an accent on the similarities between comoviruses and picornaviruses were extensively reviewed at that time by Lomonossoff and Johnson (1991) who broke down barriers dividing the studies on "structure" and "molecular genetics" by providing

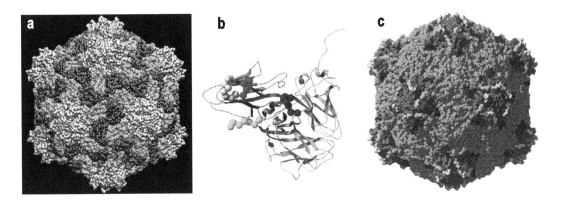

FIGURE 27.10 The 3D structure of CPMV. (a) A space-filling model of the CPMV capsid showing the L subunit in dark green and light green spheres and the S subunit shown in gray spheres. (b) Asymmetric unit (one L and one S subunit) of the CPMV capsid showing locations of insertion sites on the capsid proteins. On the S subunit, the βB-βC loop is in highlighted in yellow, the C'-C" loop in blue, the S-subunit C-terminus in green, and on the L subunit the βE-βF loop in purple. (c) Space-filling model of the whole CPMV capsid showing the locations of the same insertion sites on the capsid surface. There are 60 copies of each of these sites per particle. (With kind permission from Springer Science+Business Media: *Curr Top Microbiol Immunol.*, Biomedical nanotechnology using virus-based nanoparticles, 327, 2009, 95–122, Destito G, Schneemann A, Manchester M.)

as complete a description as was presently possible of the function, mode of expression, and structure of the coat proteins of comoviruses.

At last, Lin T et al. (1999) refined the crystal structure of CPMV to 2.8 Å resolution. It is this structure that is depicted in Figure 27.10. The roughly 300-Å capsid was similar to the picornavirus capsid displaying a pseudoT = 3 surface lattice. The three β-sandwich domains adopt two orientations, one with the long axis radial and the other two with the long axes tangential in reference to the capsid sphere. The structural peculiarities of CPMV were exhaustively reviewed and illustrated by Destito et al. (2009).

Expression

Nida et al. (1992) were the first to express the 60K coat protein precursor of CPMV in *N. tabacum* plants. The precursor neither underwent autoproteolysis to generate the mature viral coat proteins nor assembled into VLPs, suggesting that the precursor processing is required for virus assembly. Wellink et al. (1996) introduced the coding regions for the CPMV coat proteins L and S introduced separately into a transient plant expression vector, and significant expression of either capsid protein was observed only in the cowpea protoplasts transfected simultaneously with both constructs. The immune electron microscopy revealed the presence of the virion-like VLPs in extracts of these protoplasts. Moreover, an extract of protoplasts transfected with both constructs together with RNA1 was able to initiate a new infection, showing that the two capsid proteins of CPMV can form functional particles containing RNA1 and that the 60-kDa capsid precursor is not essential for this process.

When the regions of the CPMV RNA2 that encode the L and S coat proteins were expressed together but from separate promoters in the same construct in the insect *Sf*21 cells using baculovirus vectors, the empty VLPs were formed, whose morphology closely resembled that of the native CPMV virions (Shanks and Lomonossoff 2000). These L- and S-expressing stretches were the same as described earlier and previously used for the transient expression of the individual CPMV coat proteins in protoplasts by Wellink et al. (1996). No VLPs were formed when the individual L and S proteins were expressed. These results confirmed the former idea that the 60 kDa L–S fusion is not an obligate intermediate in the virion assembly pathway.

To determine whether the CPMV VLPs could be formed when the mature L and S proteins were produced by proteolytic processing in trans, Saunders et al. (2009) examined the processing of the RNA2-encoded precursors by the RNA1-encoded 24K proteinase in insect cells, and the VLPs were efficiently produced when the L and S proteins were released from either full-length RNA2 polyproteins or from VP60 by the in trans processing. However, while processing and VLP formation from the full-length RNA2 polyproteins required the simultaneous presence of both the 32K cofactor and the 24K proteinase, the processing from VP60 required just the 24K proteinase and was efficient in generating VLPs. The expression of the VP60 and 24K proteinase constructs in plants gave rise to VLPs, suggesting that the results obtained in insect cells were relevant to the in planta situation. Thus, the processing of VP60 by the 24K proteinase led to the formation of the empty VLPs in both insect cells and plants (Saunders et al. 2009). Hesketh et al. (2015) determined the electron cryomicroscopy reconstructions for the wild-type CPMV and for the empty *N. benthamiana*-produced VLPs, to 3.4 Å and 3.0 Å resolution, respectively and built de novo atomic models of their capsids. These new structures revealed the C-terminal region of the small coat protein subunit, which was essential for virus assembly and which was missing from previously determined crystal structures, as well as residues that bind to the viral genome. These observations allowed the development of a new model for genome encapsidation and capsid assembly (Hesketh et al. 2015). Moreover, Huynh et al. (2016) reported the crystal structure of the empty CPMV VLPs determined using x-ray crystallography at 2.3 Å resolution and compared it with the previously described electron cryomicroscopy structure, as well as with the virion crystal structures. Although the x-ray and electron cryomicroscopy structures of the empty VLPs were mostly similar, there existed significant differences at the C-terminus of the small subunit S. The intact C-terminus of the S subunit played a critical role in enabling the efficient assembly of the CPMV virions and the empty VLPs but underwent proteolysis after particle formation. It was also identified that the C-termini of the S subunits underwent proteolytic cleavages at multiple sites instead of a single cleavage site as previously observed (Huynh et al. 2016). Then, Hesketh et al. (2017) used electron cryomicroscopy to refine the atomic models of the wild-type CPMV containing RNA2 and of the naturally formed empty CPMV capsids. The resolution of these structures was sufficient to visualize large aa residues and to identify the essential aa side groups involved in genome encapsidation.

Chemical Coupling and New Materials

This pioneering nanotechnological story was developed first by the famous M.G. Finn team. Thus, Wang Q et al. (2002c) reported the first chemical derivatization of the native CPMV virions and the crystal structure of a derivatized particle as well as the preparation of site-specific mutations that allowed the attachment of fluorescent dyes and gold clusters through maleimide linkers. The paper invented the CPMV virions to be exploited as addressable nanoblocks imbued with a variety of the appropriate chemical and physical properties. Wang Q et al. (2002a) identified up to four lysine residues per asymmetric unit, which could be addressed by lysine-selective species; described the chemical reactivity of these lysine residues; and demonstrated that the native CPMV virion is a starting material for the diverse nanotechnological applications based on the CPMV chemical and physical properties. This powerful CPMV platform was especially impressive in combination with the engineered cysteine reactivity when the cysteine residues were introduced onto the CPMV surface using site-directed mutagenesis and were proven useful for the positioning of dye molecules (Wang Q et al. 2002b). The putative modification and insertion sites on the CPMV surface are shown in Figure 27.10b and c. Wang Q et al. (2003) decorated the displayed cysteine residues by bromoacetamide reagents and showed that the cysteine residues on the wild-type capsid were at least sixfold less reactive than those displayed on the inserted loop of the mutant. Most importantly, the authors employed the blue fluorescent antibody as a convenient probe of the location of its stilbene hapten on a nanoparticle scaffold and demonstrated that this interaction was inhibited by the site-directed attachment of a PEG chain, starting therefore the virion-displayed antigens and new materials approach. It should be noted that the

used electrophilic reagents were quite hydrophobic, demonstrating that the virus survived the attachment of such species and the use of aqueous-organic cosolvent mixtures (Wang Q et al. 2003). The new functions to the CPMV virions were brought by covering with PEG (Raja et al. 2003b) or decoration with sugar molecules (Raja et al. 2003a). The latter carbohydrates-carrying CPMV particles were studied for their binding behavior with both carbohydrate-binding protein and cells (Raja et al. 2003a).

In parallel, Fang J et al. (2002) reported a fascinating phenomenon, namely the formation of complex patterns by the drying of the concentrated CPMV droplets on surfaces. The CPMV virions self-organized into parallel and orthogonal lines, forming fingerlike patterns on freshly cleaved mica and cross-like patterns on acid-treated mica. The atomic force microscopy showed that the parallel and orthogonal lines had a uniform width and thickness (Fang J et al. 2002). Smith JC et al. (2003) used the dip pen nanolithography (DPN) to nanopattern the cysteine mutant CPMV virion monolayers. Then, a model consisting of the cysteine mutant CPMV virions and gold substrates together with functionalized alkanethiols as the linkers was employed to engineer new nanobuilding blocks for material fabrication (Cheung et al. 2003). Strable et al. (2004) attached the complementary 20-mer oligonucleotides composed of a repeating GC-rich 3-base pair sequence to the CPMV scaffolds either at lysine residues or at the genetically engineered cysteines. Remarkably, there was little or no disassembly of the virions during the coupling process. Falkner et al. (2005) used the CPMV crystals as nanocomposite scaffolds when cross-linked with glutaraldehyde and filled with metal, platinum, and palladium, getting putative new material candidates to be engaged as sensors or in x-ray optical systems. Medintz et al. (2005) decorated the CPMV virions with luminescent quantum dots, while Portney et al. (2008) developed further the CPMV-based quantum dots. Lewis JD et al. (2006) conjugated the CPMV virions with fluorophores A555 (Alexa Fluor 555) and A488 (Alexa Fluor 488) at different dye-to-capsid molar ratios (70 and 120, respectively) and used then for the in vivo vascular imaging in mice and chick embryos. Later, Brunel et al. (2010) reached the maximum number of fluorophores, and a fluorescent PEGylated peptide, in this case, was attached to the CPMV virion and limited to 200 using an efficient hydrazone ligation chemistry. Sapsford et al. (2006) coupled to the CPMV virions not only fluorophores but also functional antibodies. Destito et al. (2007) demonstrated that the CPMV virions conjugated with the folic acid-PEG moiety recognized in vitro the tumor cells bearing the folate receptor. Li T et al. (2008) generated the raspberry-like assemblies of virus–polymer nanocomposites by a controlled assembly based on the noncovalent interactions between the CPMV particles and poly(4-vinyl-pyrindine) as a polymer.

Prasuhn et al. (2007) generated the CPMV-based MRI contrast agents, in parallel with the RNA phage Qβ VLPs of the class *Leviviricetes* (Chapter 25). Then, the polyvalent glycan ligands for the cell-surface receptors were displayed on the CPMV virions, again in parallel with the Qβ VLPs (Kaltgrad et al. 2008). Astronomo et al. (2010) tried to breach the glycan shield defense of HIV and coupled the corresponding oligomannosides to the surface lysines on the CPMV and Qβ VLPs, with more success with the latter. Strable and Finn (2009) reviewed these studies together with basic principles of the chemical modification of viruses and virus-like particles by the CuAAC reaction. These studies are described in more detail in Chapter 25.

Nicole F. Steinmetz et al. (2006a, b, c, 2008a, b, 2009a, c) developed the idea of the CPMV-based nanobuilding blocks applied to various materials and employed by many different applications to the extremely high theoretical and experimental standard. Thus, the engineered CPMV derivatives were used as a template for fabrication of biotin-streptavidin complexes (Steinmetz et al. 2006a), carboxylate groups (Steinmetz et al. 2006b), organometallic complexes (Steinmetz et al. 2006c), quartz crystals (Steinmetz et al. 2008a), polyelectrolytes (Steinmetz et al. 2008b), buckyballs, C60 fullerenes (Steinmetz et al. 2009a), silica nanocomplexes (Steinmetz et al. 2009c), and PEGylated derivatives (Steinmetz and Manchester 2009).

Later, the fluorescent imaging of human prostate tumor xenografts on the chicken chorioallantoic membrane animal model was achieved by using CPMV decorated with NIR Alexa Flour 647 and functionalized with peptide bombesin to target gastrin-releasing peptide receptors (Steinmetz et al. 2011a). Moreover, the CPMV nanoparticles targeted surface vimentin on cancer cells (Koudelka et al. 2009; Steinmetz et al. 2011b, c).

Aljabali et al. (2011b) coupled specific peptides to the CPMV virions, which promoted further templated mineralization of particles. Such peptides were previously identified through an evolutionary screening process using phage display technologies, and specifically directed mineralization by CoPt (CNAGDHANC; peptide-CoPt), FePt (HNKHLPSTQPLA; peptideFePt), and ZnS (CNNPMHQNC; peptideZnS) were selected (Mao et al. 2004; Reiss et al. 2004). The subsequent mineralization of the peptide-CPMV conjugates produced monodisperse nanoparticles of ~32 nm diameter coated with, for example, cobalt–platinum, iron–platinum or zinc sulfide, which cannot be readily prepared by other methods. This route was particularly attractive as it avoided the need to genetically engineer the protein surface of the virus to provide chimeras for templated-mineralization (Aljabali et al. 2011b). Nevertheless, the previously described production of the plant-produced empty CPMV VLPs opened a new possibility to the internal mineralization of the particle (Sainsbury et al. 2011). Thus, the presence of the C-terminal 24-aa peptide of the S protein was found to inhibit the internal mineralization, an effect that was eliminated by enzymatic removal of this region. Remarkably, the substitution of this region with His_6 tag generated stable particles and facilitated external mineralization by cobalt. Therefore, the problem of the consistent internal and external mineralization of CPMV was solved (Sainsbury et al. 2011).

Aljabali et al. (2011a) described the use of the polyelectrolyte surface-modified CPMV virions for the templated synthesis of gold nanoparticles. Thus, a cationic polyelectrolyte, poly(allylamine) hydrochloride, was electrostatically bound to the external surface of the virus capsid. It promoted the adsorption of anionic gold complexes, which were then easily reduced under mild conditions to form a metallic gold layer. The gold surface was further modified with thiol reagents. In contrast, reaction of the polyelectrolyte-modified CPMV with preformed gold nanoparticles resulted in the self-assembly of large, hexagonally packed, tessellated spheres (Aljabali et al. 2011a).

Blum et al. (2011) engineered biosensors, where the CPMV virions were used as a scaffold to control the positions of gold nanoparticles. The nanoparticles were then interconnected by thiol-terminated conjugated organic molecules, resulting in a 3D conductive network. The biotin molecules were attached to the virus scaffold using linkers to act as molecular receptors. As a result, the binding of avidin to the biotin receptors on the self-assembled nanosensors caused a significant change in the network conductance that was dependent on the charge of the avidin protein (Blum et al. 2011).

Patil et al. (2012) used the CPMV virions to elaborate a synthetic strategy for the nanoscale engineering of capsid surfaces as a new class of nanostructured biohybrid materials. In fact, the polymer-surfactant/CPMV constructs were produced as highly concentrated solvent-free viscoelastic liquids or soft solids that can be readily stored, transported, and used as infective agents. Furthermore, the surface-engineered virions could be redissolved in aqueous media or dissolved in a range of organic solvents and easily sprayed as aerosols for potential delivery and application on synthetic or biological substrates (Patil et al. 2012).

Tiu et al. (2016) were the first ones to assemble viral nanoparticles using a nanopatterned conducting polymer array. The poly(pyrrole-co-pyrrole-3-carboxylic acid) was selected in this study as the conducting polymer, since polypyrroles were well known for biocompatibility, excellent electrical properties, energy storage capacity, and mechanical flexibility. The pyrrole monomer selected had a carboxylic acid moiety, which had the capability to interact with CPMV electrostatically in a pH lower than its isoelectric point, between 3.4 and 4.5. The authors achieved the formation of the hierarchical CPMV nanoparticle assemblies on colloidal-patterned, conducting polymer arrays using a protocol combining colloidal lithography, electrochemical polymerization, and electrostatic adsorption (Tiu et al. 2016).

Packaging and Delivery

Rae et al. (2005) were the first ones to inspect the fate of the plant-produced CPMV virions by the systemic trafficking in mice via the oral route, suggesting that the CPMV derivatives could be used as edible or mucosally delivered vaccines or therapeutics. First, the CPMV virions were shown to be stable under simulated gastric conditions in vitro. Second, the pattern of localization of the CPMV virions to mouse tissues following oral or intravenous dosing was then determined, where CPMV was found in a wide variety of tissues throughout the body, including the spleen, kidney, liver, lung, stomach, small intestine, lymph nodes, brain, and bone marrow. Thus, the stability of the CPMV virions in the gastrointestinal tract followed by their systemic dissemination supported their use as orally bioavailable nanoparticles (Rae et al. 2005).

Wu et al. (2005) developed the fluorescent CPMV probe for near-infrared fluorescence and demonstrated that the high loading of local dye concentration over 30 dye molecules per particle in this probe were sufficient to fulfill the methodology needs without significant fluorescence quenching. The successful tomographic fluorescence imaging was demonstrated in a tissue-like phantom (Wu et al. 2005). As mentioned earlier, Lewis JD et al. (2006) assessed the utility of the fluorescent CPMV particles for the intravital imaging by injecting and visualizing the vasculature of mouse embryos and shell-free chick embryos. The authors conducted also the vascular mapping studies to visualize the extent of angiogenesis induced by tumor implants on the chick chorioallantoic membrane (Lewis JD et al. 2006).

Martin BD et al. (2006) published an excellent study that described a new aspect of the use of CPMV as a scaffold and bright tag for the cargo capture and transport by *Drosophila* kinesin-driven microtubules in standard gliding assays. The capture occurred through both NeutrAvidin (NA)-biotin and antibody (IgG)-antigen interactions. The microtubules were derivatized with rabbit antichicken IgG or biotin, and the virus was conjugated with chicken IgG or NA. Since CPMV was a very effective scaffold for fluorescent dye molecules and was able to accommodate up to 60–120 dyes per virus depending on the mutant used, the fluorescent CPMV appeared to be superior to fluorescent polystyrene spheres of the same size, as both a reporter tag and a scaffold for the microtubule-transported cargo proteins, because of its negligible nonspecific adsorption and superior brightness (Martin BD et al. 2006).

Therefore, the uptake of the CPMV virions was successfully studied in cells of the immune system (Gonzalez et al. 2009), allowing the visualization of inflammation (Shriver et al. 2009). The previously mentioned decoration of the CPMV virions with PEG polymers inhibited the binding of the viral capsid to a variety of cell types (Lewis JD et al. 2006; Steinmetz and Manchester 2009; Brunel et al. 2010), resulting in a marked increase in plasma half-life (Singh et al. 2007) and allowing the visualization of blood flow in living organisms (Lewis JD et al. 2006). At that time, Leong et al. (2010) published an intersectoral and thorough study on the intravital imaging of embryonic and tumor neovasculature using the CPMV virions. All aspects of the synthesis, purification, and fluorescent labeling of the CPMV virions, along with their use for the imaging of vascular structure and for the intravital vascular mapping in developmental and tumor angiogenesis models, were unveiled. It was emphasized that the dye-labeled CPMV

virions could be synthesized and purified in a single day, and imaging studies could be conducted over hours, days, or weeks, depending on the application (Leong et al. 2010).

Wen et al. (2012) presented fine methodology of the propagation of the CPMV virions in *V. ungiuculata* plants and detailed purification and chemical labeling protocols, with a focus on the labeling of VNPs with fluorophores, e.g., Alexa Fluor 647 and PEG. The data was presented along with that on other advanced plant virus models, such as brome mosaic virus (BMV) of the family *Bromoviridae*, tobacco mosaic virus (TMV) of the family *Virgaviridae* (both from the order *Martellivirales*), and potato virus (PVX) from the order *Tymovirales*, described in Chapters 17, 19, and 21, respectively.

Hovlid et al. (2012) modified the CPMV surface to display the integrin-binding RGD oligopeptide sequence derived from human adenovirus type 2 (HAdV-2). Concurrently, wild-type CPMV was modified via NHS acylation and Cu^I-catalyzed azide-alkyne cycloaddition (CuAAC) chemistry to attach the RGD peptide. Both types of particles showed strong and selective affinity for several different cancer cell lines that expressed RGD-binding integrin receptors (Hovlid et al. 2012).

Aljabali et al. (2013) performed the covalent modification of the CPMV virions with the chemotherapeutic drug doxorubicin (DOX). The DOX molecules, 80 per particle, were covalently bound to the external surface carboxylates of the CPMV virion, and the CPMV conjugate was targeted to the endolysosomal compartment of the cells, in which the proteinaceous drug carrier was degraded and the drug released. The two chemical ligation strategies were evaluated: (i) covalent modification via a stable amide bond and (ii) a labile disulfide bridge. When DOX was conjugated via stable amide bond to the CPMV carrier system, the formulation induced time-delayed but enhanced toxicity to HeLa cells compared to a free drug (Aljabali et al. 2013). In fact, this study was the first demonstrating the utility of the CPMV virions as a real drug delivery vehicle.

Yildiz et al. (2013) reported a noncovalent infusion technique that facilitated efficient cargo loading. The authors achieved infusion and retention of 130–155 fluorescent dye molecules per CPMV using DAPI (4′,6-diamidino-2-phenylindole dihydrochloride), propidium iodide (3,8-diamino-5-[3-(diethylmethy-lammonio)propyl]-6-phenylphenanthridinium diiodide), and acridine orange (3,6-bis(dimethylamino) acridinium chloride), as well as 140 copies of therapeutic payload proflavine (acridine-3,6-diamine hydrochloride). The loading was performed through interaction of the cargo with the virion-encapsidated RNA molecules. Remarkably, the loading mechanism was RNA-specific, since empty RNA-free CPMV VLPs were not loaded. The cargo-infused CPMV capsids remained chemically active, and the surface lysine residues were covalently modified with dyes, leading to the development of the dual-functional CPMV carriers. The delivery of cargo was demonstrated to a panel of cancer cells, such as cervical, breast, and colon, while the CPMV carriers entered cells via the surface marker vimentin and were targeted to the endolysosome, where the carrier was degraded, and the cargo was released, allowing imaging and/or cell killing (Yildiz et al. 2013).

In contrast to the CPMV virions-based applications, Aljabali et al. (2010) stressed the great role of the plant-produced empty CPMV VLPs as a packaging unit. Until the previously described study of Saunders et al. (2009), CPMV had not been used to encapsulate materials as it had been exceedingly difficult to obtain the empty RNA-free particles, as these comprised only 5–10% of particles produced during an infection. Some attempts to inactivate or eliminate the viral RNAs either by irradiation with ultraviolet light or chemically did not satisfy the biotechnological needs. Aljabali et al. (2010) solved the problem of how to encapsulate cobalt or iron oxide, the possible agents of the targeted magnetic-field hyperthermia therapy, within the CPMV VLPs by environmentally benign processes. Since the external CPMV surface remained amenable to chemical modification, the way to the CPMV-driven targeted delivery of cargos or therapeutics was opened.

At last, the CPMV VLPs were targeted to treat deep vein thrombosis (occlusive venous clot formation), which remains a leading cause of death worldwide (Park et al. 2021). Thus, the peptide ligands were selected that recognized the myeloid related protein 14 (MRP-14, also known as S100A9) and formulated as nanoparticles by using CPMV and tobacco mosaic virus (TMV). The intravascular delivery of the MRP-14-targeted nanoparticles in a murine model of disease resulted in enhanced accumulation in the thrombi and reduced thrombus size (Park et al. 2021).

Chimeric Virions

The famous George P. Lomonossoff team was the first to engineer the chimeric CPMV viruses, the structure of which was based on the knowledge of the atomic resolution 3D structure coupled with the availability of the appropriate infectious cDNA clones. It should be noted also that CPMV was the first plant virus to be developed as a system for the display of foreign peptides. Thus, Usha et al. (1993) showed that it was possible to insert a foreign sequence, namely an epitope derived from VP1 of foot-and-mouth disease virus (FMDV), in the S protein without abolishing the ability of the chimeric CPMV virus to multiply and form capsids in whole plants. The chimeras were designed so that the FMDV epitope was expressed either as an insertion or as a replacement for part of the wild-type βB-βC loop sequence, comprising residues 20–27 of the S protein. In the first construct, the FMDV epitope comprising aa 136–160 of VP1 were inserted. In the second construct, the FMDV epitope encoding VP1 aa 141–160 was inserted in place of the wild-type loop sequence. In the first construct the size of the βB-βC loop was increased by 27 aa, while in the second construct the net increase was only 14 aa residues. While RNA from both chimeras was able to replicate in cowpea protoplasts only the first construct containing the FMDV epitope as an insertion was able to direct capsid formation

and infect whole cowpea plants. The infection, however, differed from a normal CPMV infection in that the lesions produced on the inoculated leaves were small, the infection did not spread systematically through the plant, and the yield of virus particles was lower than with a wild-type CPMV infection (Usha et al. 1993).

Using the information gained from the FMDV insertions, Porta et al. (1994) generated the CPMV chimeric viruses expressing epitopes derived from human rhinovirus 14 (HRV-14) and human immunodeficiency virus 1 (HIV-1) in a vector, where no CPMV-specific sequences were removed. Therefore, the following epitopes were inserted: aa 85–98 of HRV-14 VP1, aa 731–752 from gp41 of HIV-1, the so-called Kennedy epitope by Kennedy et al. (1986), as well as aa 141–159 of FMDV VP1. The chimeric viruses possessed the antigenic properties of the inserted sequences, and, in the case of the HRV-14-derived construct, the immunogenicity of the inserted epitope was shown in rabbits (Porta et al. 1994). The successful development of these CPMV chimeras was described by Porta et al. (1996) in detail. Moreover, Lin T et al. (1996) resolved the crystal structure of the CPMV/HRV14 chimera to 2.8 Å. The immunogenicity of the CPMV/HIV-1 chimera and induction of HIV-1-neutralizing antibodies by parenteral immunization of mice were demonstrated by McLain et al. (1995, 1996). Durrani et al. (1998) showed the efficiency of intranasal immunization of mice with the CPMV/HIV-1 chimera, while oral immunization was less efficient.

Next, Dalsgaard et al. (1997) reported the insertion of a 17-aa epitope DGAVQPDGGQPAVRNER from the N-terminal part of the VP2 capsid protein of mink enteritis virus (MEV), corresponding to aa residues 3–19 from VP2 of canine parvovirus (CPV) and the successful expression of the epitope on the surface of the chimeric CPMV when propagated in cowpea. The efficacy of the chimeric virus was established by the demonstration that one subcutaneous injection of 1 mg of the chimera in mink conferred protection against clinical disease and virtually abolished shedding of virus after challenge with virulent MEV. Remarkably, the epitope used occurred in three different virus species—MEV, CPV, and feline panleukopenia virus (FPLV) from the family *Parvoviridae*, order *Piccovirales* (Chapter 9), and thus the same vaccine could be used in three economically important viral hosts—mink, dogs, and cats, respectively (Dalsgaard et al. 1997). In fact, Langeveld et al. (2001) found that the CPMV chimeras carrying the same 17-aa epitope corresponding to aa residues 3–19 from VP2 of the CPV, inserted between aa 22 and 23 of the CPMV S protein, provided complete protection of dogs from clinical disease and CPV shedding in vivo. Furthermore, the fine character of immune response (Nicholas et al. 2002) and the effect of priming/booster immunization protocols (Nicholas et al. 2003) on immune response to CPV in mice were characterized.

Brennan et al. (1999c) expressed in tandem two peptides, 10 and 18, in total 34 aa residues, from the outer-membrane protein F of *Pseudomonas aeruginosa* on each of the two coat proteins of CPMV. The chimeric viruses expressing the peptides on the S and L proteins were used to immunize mice and demonstrated promising results for the development of a protective vaccine against *P. aeruginosa*. The ability of the chimeras to generate protective immunity against *P. aeruginosa* in mice was fully established by Brennan et al. (1999b). Gilleland et al. (2000) evaluated the CPMV chimeras carrying a third peptide of 14 aa residues from the F protein, so-called peptide 9, in various combinations with the 10 and 18 peptides when inserted into the L or S subunits, altogether 9 combinations, using in parallel TMV and influenza A virus (IAV) vectors, as described in Chapters 19 and 33, respectively.

Brennan et al. (1999d) expressed the D2 peptide derived from a *Staphylococcus aureus* fibronectin-binding protein on the surface of the CPMV virions (aa 1–30 of D2) and, in parallel, on the rod-shaped potato virus X (PVX) from the order *Tymovirales*, in this case aa 1–38 of D2, as described before in Chapter 21. The immunization of mice and rats showed that the D2 peptide was highly immunogenic when expressed on the two different plant viruses. Furthermore, the nasal as well as oral administration of the CPMV-D2 chimera induced systemic and mucosal immune responses in mice (Brennan et al. 1999a).

At that time, the factors impressing the fate of the CPMV chimeras were studied. Taylor et al. (1999) evaluated many insertions and concentrated on the position-dependent processing of the inserted epitopes. A chimera expressing a rotavirus epitope, though viable, rapidly lost most of the insert. Although many peptides gave an authentic immune response, this was not the case for the NIm-1A epitope from HRV-14. The crystallography revealed significant differences between the structure of NIm-1A on CPMV compared with its native configuration. Next, the influence of peptide structure on immunogenicity was investigated by constructing a series of the CPMV chimeras expressing the 14-aa NIm-1A epitope from HRV-14 at different positions on the capsid surface (Taylor et al. 2000). The biochemical and crystallographic analysis of a CPMV/HRV chimera expressing the NIm-1A epitope inserted into the βC'-βC" loop of the S protein revealed that, although the inserted peptide was free at its C-terminus, it adopted a conformation distinct from that previously found when a similarly cleaved peptide was expressed in the βB-βC loop of the S protein. The adjustment of the site of insertion within the βB-βC loop resulted in the isolation of a chimera in which cleavage at the C-terminus of the epitope was much reduced. The crystallographic analysis confirmed that in this case the epitope was presented as a closed loop. Polyclonal antisera raised against the CPMV/HRV chimera presenting the NIm-1A epitope as a closed loop had a significantly enhanced ability to bind to intact HRV-14 particles compared with antisera raised against chimeras presenting the same sequence as peptides with free C-termini (Taylor et al. 2000). Porta et al. (2003) addressed the growth properties of a series of chimeras with inserts that differed both in size and isoelectric point (pI) and examined whether the

presence of a heterologous peptide affected the virus host range, its ability to be transmitted by beetle vectors, and its level of transmission through seed. The results indicated that both length and pI of the insert had profound effects on the growth of chimeras, while the foreign insertions did not alter the virus host range or increase the rate of transmission by beetles or through seed or change the insect vector specificity. However, the presence of a foreign peptide, if anything, decreased the ability of the virus to spread in the environment (Porta et al. 2003).

Khor et al. (2002) created an antiviral against measles virus (MV) by displaying a peptide known to inhibit MV infection. This peptide sequence corresponded to a portion of the MV binding site on the human MV receptor CD46. The CPMV-CD46 chimera efficiently inhibited MV infection of HeLa cells in vitro, while wild-type CPMV did not. Furthermore, the CPMV-CD46 chimera protected mice from mortality induced by an intracranial challenge with MV. The CD46 peptide presented in the context of CPMV was up to 100-fold more effective than the soluble CD46 peptide at inhibiting MV infection in vitro (Khor et al. 2002). In fact, this study represented the first utilization of a plant virus chimera as an antiviral agent.

Yasawardene et al. (2003) attempted to construct a malaria vaccine by the insertion of the 19-aa B cell epitope VTHESYQELVKKLEALEDA, termed P109, from the merozoite surface antigen-1 (MSA-1) of the malaria parasite *Plasmodium falciparum* into the βB-βC loop of the CPMV protein S. The immunization of rabbits with the chimera yielded antibodies that, however, did not react with the native antigen on merozoite.

To prepare an experimental anthrax vaccine, Phelps et al. (2007) displayed the 25-aa peptide PA1: SNSRKKRST-SAGPTVPDRDNDGIPD from the *Bacillus anthracis* protective antigen on the surface loop of CPMV protein L, between aa 98 and 99. Special attention was devoted to the effective protocol for the inactivation of the chimeric virus propagated in cowpea plants.

Chatterji et al. (2005) engineered five different His-tag mutants of CPMV by genetically introducing the His$_6$ stretch at various locations on the virion, namely at the loops βB-βC, βC'-βC" and at the C-terminus, wild-type (aa 213) or truncated (aa 189), all above protein S, as well as βE-βF of the protein L. The mutant particles showed differential affinity for binding nickel, and their electrostatic properties were controlled as a function of the protonation state of the exposed histidine sequence. The specific addressability of the His$_6$ tag was corroborated by the selective modification of the histidine sequence with nanogold cross-linked to the Ni-NTA moiety (Chatterji et al. 2005).

Chimeric VLPs

The story of how the chimeric CPMV VLPs were produced in cowpea plants was told in detail by Montague et al. (2011). First, the C-terminus of the VP60 precursor was modified by replacing the terminal 24 aa of the S protein with the His$_6$ tag. Previous attempts to introduce modifications at this position using the infection route resulted in the recovery of only very low levels of particles (Cañizares et al. 2004). As a result, the capsid assembly was not affected by the replacement of the naturally occurring C-terminal 24 aa of the S protein with the His_6 tag, demonstrating the more general possibility to replace the C-terminal region of the S protein with a heterologous sequence when the VLP technology is used. Moreover, the incorporation of the His-tag opened further possibilities for the construction and purification of the chimeric CPMV VLPs (Montague et al. 2011).

Next, a whole protein, namely green fluorescent protein (GFP), was fused to the C-terminus of the wild-type S coat protein of the VP60 precursor (Montague et al. 2011). The plants exhibited GFP fluorescence three days post infiltration, suggesting correct folding of the GFP expressed from the construct. The purification results showed that, while a portion of GFP was still attached to the VLPs, some was cleaved off during purification, probably as a result of cleavage, which occurred at the last 24 aa of the S protein. The presence of the VP60-GFP also suggested that the GFP may interfere with the correct processing of the VP60. Nonetheless, it was excluded that the display of whole proteins on the surface of the CPMV VLPs could be possible (Montague et al. 2011).

Zampieri et al. (2020) used the CPMV VLPs, in parallel with the chimeric virions of tomato bushy stunt virus (TBSV), order *Tolivirales* (Chapter 24), for the prevention and treatment of autoimmune diseases. The immunodominant peptide associated with type 1 diabetes mellitus (T1D), so-called p524, was genetically engineered on the CPMV VLP surface. Using a mouse model, the authors showed that the chimeric CPMV VLPs prevented autoimmune diabetes, and this effect was based on a strictly peptide-related mechanism in which the virus nanoparticle acted both as a peptide scaffold and as an adjuvant, showing an overlapping mechanism of action (Zampieri et al. 2020).

Cancer Vaccines

As mentioned earlier, the CPMV virions conjugated with the folic-acid-PEG moiety recognized in vitro the folate receptor on the tumor cells (Destito et al. 2007). The CPMV virions were recognized also by gastrin-releasing peptide receptors (Steinmetz et al. 2011a) and vimentin on cancer cells (Koudelka et al. 2009; Steinmetz et al. 2011b, c). The fine methodology of imaging cancer by the CPMV-derived particles was presented by Cho et al. (2014).

Miermont et al. (2008) employed the CPMV virions for the development of the carbohydrate-based antitumor vaccines. The Tn antigen GalNAc-α-O-Ser/Thr was chosen as the model antigen to be conjugated to CPMV capsid. The Tn antigen is a tumor associated carbohydrate antigen (TACA) overexpressed on the surface of a variety of cancer cell surfaces including breast, colon, and prostate cancer, rendering it an excellent immunotherapy target. The Tn was derivatized with either a maleimide or a bromoacetamide moiety that was conjugated selectively to the previously described cysteine mutant of CPMV. The antibodies generated in the immunized mice were able to recognize Tn antigens presented in their native conformations on the surfaces of breast cancer cells (Miermont et al. 2008).

Lizotte et al. (2016) were the first ones to directly investigate the efficacy of the CPMV VLPs, not virions in this case and without any foreign components, as an immunotherapy in models of metastatic lung melanoma and other cancers and found it had striking efficacy in mouse models as an in situ vaccination reagent by simple inhalation. The inhaled CPMV VLPs were rapidly taken up by and activated neutrophils in the tumor microenvironment as an important part of the antitumor immune response. The CPMV VLPs also exhibited clear treatment efficacy and systemic antitumor immunity in ovarian, colon, and breast tumor models in multiple anatomic locations. It was important that the CPMV VLPs were stable, nontoxic, and modifiable with drugs and antigens, and their nanomanufacture was highly scalable (Lizotte et al. 2016). Furthermore, Czapar et al. (2018) developed slow-release assemblies of the CPMV VLPs making use of charged dendrimers and electrostatic self-assembly protocols. Patel et al. (2018) combined the CPMV VLPs with radiation therapy in a preclinical syngeneic mouse model of ovarian carcinoma. The CPMV presented a unique and promising hypofractionated radiation adjuvant that led to increased antitumor cytotoxic and immune signaling (Duval et al. 2020). Kerstetter-Fogle et al. (2019) demonstrated the potent efficiency of the CPMV VLPs against intracranial glioma, while Wang C et al. (2019) showed its efficiency in a mouse ovarian tumor model. Cai et al. (2019) successfully combined the CPMV VLP therapy with low doses of cyclophosphamide, while Wang C and Steinmetz (2020) suggested to combine the CPMV therapy with selected checkpoint-targeting antibodies.

The solid attempts were undertaken to show the unique character of the antitumor activity of the CPMV VLPs. First, Murray AA et al. (2018) tried the in situ vaccination with the CPMV VLPs versus the tobacco mosaic virus (TMV) nanoparticles against melanoma and demonstrated that the CPMV ones elicited potent antitumor immunity, while the TMV ones only slowed tumor growth and increased survival time, however, at significantly lower potency compared to that of the CPMV VLPs. The differential potency of CPMV vs. TMV was explained with differences in immune activation, where CPMV stimulated an antitumor response through recruitment of monocytes into the tumor microenvironment, establishing signaling through the IFN-γ pathway, which also led to recruitment of tumor-infiltrated neutrophils and natural killer cells (Murray AA et al. 2018). Moreover, Shukla et al. (2020b) evaluated, in parallel, the popular icosahedral ~30 nm VLP models: cowpea chlorotic mottle virus (CCMV), family *Bromoviridae*, order *Martellivirales* (Chapter 17); physalis mottle virus (PhMV), order *Tymovirales* (Chapter 21), RNA phage Qβ, class *Leviviricetes* (Chapter 25); sesbania mosaic virus (SeMV), order *Sobelivirales* (Chapter 28); and hepatitis B virus (HBV) cores (HBc), order *Blubervirales*

(Chapter 38). The established tumor models of melanoma, ovarian cancer, and colon cancer, as well as ex vivo and in vitro assays, were used to probe the immunological properties and potential to elicit antitumor immunity of these viruses and/or their VLPs. All tests revealed unique features of CPMV that made it an inherently stronger immune stimulant, and the CPMV in situ vaccine outperformed all listed candidates (Shukla et al. 2020b).

The technological and scientific aspects of the CPMV success story were not neglected. Zheng Y et al. (2019) involved freeze-drying to produce the efficacious CPMV VLPs with the nearly eliminated traces of encapsulated RNA. Then, Shukla et al. (2020c) compared the empty CPMV VLPs and the complete CMPV virions carrying genomic RNA in an aggressive ovarian tumor mouse model and established that both significantly increased the survival of tumor-bearing mice and showed promising antitumor efficacy. However, they demonstrated distinct yet overlapping immunostimulatory effects due to the presence of virus RNA in wild-type particles, indicating their suitability for different immunotherapeutic strategies. Specifically, the RNA-containing CPMV particles were uniquely able to boost populations of potent antigen-presenting cells, such as tumor-infiltrating neutrophils and activated dendritic cells. As a result, the parallel development of both CPMV and CPMV VLPs as immunotherapeutic vaccine platforms with tailored responses was postulated (Shukla et al. 2020c). To develop the CPMV virions as the safe and potent therapy agent, Chariou et al. (2021) reported inactivation of CPMV using UV light and chemical inactivation using β-propiolactone or formalin.

At last, Stump et al. (2021) showed that the CPMV code-livered with irradiated ovarian cancer cells constituted an effective prophylactic vaccine against a syngeneic model of ovarian cancer in mice. Following two vaccinations, 72% of vaccinated mice rejected tumor challenges, and all those mice survived subsequent rechallenges, demonstrating immunologic memory formation. Therefore, this study supported remission-stage vaccines using irradiated patient tumor tissue as a promising option for treating ovarian cancer, and validated CPMV as an antitumor vaccine adjuvant for that purpose (Stump et al. 2021).

Reviews

The historical role, biotechnological importance, and broad variety of applications that have employed the CPMV packaging and display technologies including vaccines, antiviral therapeutics, nanoblock chemistry, and materials science urgently needed a growing number of exhaustive reviews. The earliest classical reviews (Porta et al. 1994; Lomonossoff and Johnson 1995; Johnson et al. 1997; Cañizares et al. 2005a, b) were devoted to the CPMV model as a novel high-yielding system for the presentation of foreign peptides. Then, a solid number of reviews delt with the CPMV platform and its application for the diagnostic imaging (Manchester and Singh 2006; Li Kai et al. 2010; Steinmetz 2010; Bruckman et al. 2013; Tsvetkova and Dragnea 2015; Zhang Wenjing et al. 2017). Steinmetz and

Evans (2007) accented the role of the CPMV model as a source of the advanced building blocks in bionanoscience and indicated their potential for future application. The CMPV, alone or along with other models, was reviewed in the sense of the novel biotemplates for materials and their use in nanotechnology (Steinmetz and Evans 2007; Aniagyei et al. 2008; Young M et al. 2008; Destito et al. 2009; Lee LA et al. 2009, 2011; Steinmetz et al. 2009a, b; Strable and Finn 2009; Soto and Ratna 2010; Dedeo et al. 2011; Jutz and Böker 2011; Montague et al. 2011; Pokorski and Steinmetz 2011; Liu Z et al. 2012; Bittner et al. 2013; Wen et al. 2013; Glasgow and Tullman-Ercek 2014; Li F and Wang 2014; Lomonossoff and Evans 2014; Maassen et al. 2016; Wen and Steinmetz 2016; Zhang Y et al. 2016). The more specific reviews unveiled the role of the CPMV model in the engineering of the Gd-loaded nanoparticles by MRI (Bruckman et al. 2013), more generally, as PET and MRI contrast agents (Shukla and Steinmetz 2015), in photonics and plasmonics (Wen et al. 2015), for the production of novel metal nanomaterials (Capek 2014; Love et al. 2014), by the putative application as versatile nanomachines (Koudelka et al. 2015).

Sainsbury et al. (2014) unveiled the fine methodology concerning production and purification of the CPMV VLPs. The role of the CPMV model on the background of other virus-based nanocarriers for drug delivery was touched on in reviews by Ma et al. (2012) and Lee EJ et al. (2016). Meshcheriakova et al. (2017) concentrated on the combination of the high-resolution electron cryomicroscopy and mutagenesis to further develop the CPMV model for bionanotechnology. The recent review demonstrated the growing CPMV potential for the diagnostic and therapeutic applications (Shukla et al. 2020a).

Bean Pod Mottle Virus

The crystal structure of bean pod mottle virus (BPMV), a close relative of CPMV, was resolved to 3.0 Å by Chen ZG et al. (1989). As in the case of CPMV, the tertiary and quaternary structures of the BPMV capsid proteins were found similar to those observed in animal picornaviruses, supporting the close relation between plant comoviruses and animal picornaviruses, which was also established by previous biological studies. As in the CPMV capsid, the two capsid proteins of BPMV were folded into three antiparallel β-barrel structures, where the small protein formed one barrel and the large protein formed two barrels that were covalently connected to form a single polypeptide. The 60 copies of each protein type in the virus generated 180 β-barrel domains that were arranged in the pseudoT = 3 manner.

OTHER GENERA

Genus *Cheravirus*
Apple Latent Spherical Virus

The apple latent spherical virus (ALSV) has isometric virus particles about 25 nm in diameter that contain two ssRNA species RNA1 and RNA2 and three capsid proteins Vp25, Vp20, and Vp24 (Li Chunjiang et al. 2000, 2004). This virus was originally isolated from an apple tree in Japan, but the virus was able to infect a broad range of herbaceous plants such as tomato, cucurbits, lettuce, spinach, okra, chard, zucchini, soybean, *Arabidopsis thaliana*, and tobacco without causing any obvious symptoms, as reviewed by Li Chunjiang et al. (2014). The ALSV vectors were constructed for the expression of foreign genes in plants (Li Chunjiang et al. 2004). The ALSV model was recognized therefore as a tool to produce edible vaccines in fruits and vegetables by infecting them with an ALSV presenting the epitope sequences of pathogenic viruses on ALSV particles. Thus, Li Chunjiang et al. (2014) engineered the viable ALSV vectors, where foreign peptides could be fused to either of two C-terminal regions of Vp20 at aa positions between G171 and P172 or between P172 and L173 or to the C-terminus, position T192, of Vp24. An ALSV vector presenting the 20-aa epitope sequences of the coat protein of zucchini yellow mosaic virus (ZYMV) systemically infected host plants, and the epitope sequence was stably maintained in the chimeric ALSV for more than ten serial passages and for at least six months. The purified chimeric ALSV virions induced a ZYMV-specific immune response in rabbits. Then, the ALSV vector was used for expression of an epitope from the protein VP1 of foot-and-mouth disease virus (FMDV; Li Chunjiang et al. 2014).

Genus *Nepovirus*

Grapevine Fanleaf Virus

The grapevine fanleaf virus (GFLV) causes a severe degeneration of grapevines. The crystal structure of the wild-type GFLV strain was resolved to 2.7 Å by Schellenberger et al. (2010, 2011) and demonstrated a structure of approximately 30 nm in diameter, comprising 60 identical coat subunits of 56 kDa, 504 aa, arranged in a pseudoT = 3 symmetry. The GFLV particles were observed in transgenic plants expressing the GFLV coat coding sequence, suggesting that the GFLV coat was able to self-assemble into the VLPs (Barbier et al. 1997, cited by Belval et al. 2016). After 20 years. Belval et al. (2016) showed that the GFLV coat is a highly versatile protein of biotechnological interest, compatible with the simultaneous encapsulation and exposure of large proteins through the genetic fusion to the CP N- or C-terminal end. Thus, the production of the spherical empty VLPs exposing fluorescent proteins at either their outer surface or inner cavity was demonstrated. Both N- and C-terminal ends of the GFLV coat allowed the genetic fusion of proteins as large as 27 kDa and the *N. benthamiana* plant-based production of the nucleic acid-free VLPs. Remarkably, the expression of N- or C-terminal coat fusions resulted in the production of VLPs with recombinant proteins exposed to either the inner cavity or the outer surface, respectively, while coexpression of both fusion proteins led to the formation of the mosaic VLPs, although rather inefficiently (Belval et al. 2016).

Yazdani et al. (2019) studied the feasibility of using GFLV VLPs as a scaffold for HPV L2 epitope, aa 17–31, presentation. To achieve this goal, the GFLV VLPs displaying the HPV L2 epitope as well as the natural GFLV VLPs were expressed in both *E. coli* and *P. pastoris*. The epitope sequence was genetically inserted in the αB-αB" domain C of the GFLV coat. The highest expression yield was obtained in *E. coli*. Using this system, the VLP formation required a denaturation-refolding step, whereas VLPs with lower production yield were directly formed using *P. pastoris*, as confirmed by immune electron microscopy. Since the GFLV-L2 VLPs were found to interact with the HPV L2 antibody, it was assumed that the inserted epitope was located at the VLP surface with its proper ternary structure (Yazdani et al. 2019).

28 Order *Sobelivirales*

If we wait for the moment when everything, absolutely everything is ready, we shall never begin.

Ivan Turgenev, *Virgin Soil*

ESSENTIALS

The small *Sobelivirales* order unites icosahedral spherical T = 3 and bacilliform T = 1 single-stranded positive-sense RNA viruses that have played a substantial role in the structural evaluation of the plant viruses. The order currently involves 3 families, 6 genera, and 61 species. According to the modern taxonomy (ICTV 2020), the order *Sobemovirales* is one of the three orders belonging to the class *Pisoniviricetes*, together with the famous *Nidovirales* and huge *Picornavirales* orders described in the neighboring Chapter 26 and Chapter 27, respectively. The *Pisoniviricetes* class belongs to the *Pisuviricota* phylum from the kingdom *Orthornavirae*, realm *Riboviria*.

The order *Sobelivirales* involves the most studied family *Solemoviridae* of plant viruses represented by 4 genera and 59 species (Sõmera et al. 2021) together with the small *Avernaviridae* family with a single genus *Dinornavirus* and a single species *Heterocapsa circularisquama RNA virus 01*, which infects the harmful bloom-forming dinoflagellate, *H. circularisquama* and which is assumed to be the major natural agent controlling the host population. The members of both families are spherical and possess T = 3 symmetry, while the single species *Mushroom bacilliform virus* of the single genus *Barnavirus* from the third family, namely *Barnaviridae*, is bacilliform and possesses T = 1 symmetry. This virus infects the cultivated white button mushroom *Agaricus bisporus*, and, although it is sometimes associated with severe La France disease, it is not the causal agent.

The family *Solemoviridae* includes four genera, the most important of which for our goals are *Polerovirus* and *Sobemovirus*. The latter, involving 20 species, played a substantial historical role by the 3D characterization of the icosahedral T = 3 plant viruses.

Figure 28.1 presents typical portraits of both T = 3 and T = 1 representatives of the *Sobelivirales* order. The T = 3 virions built by 180 monomers are 26–32 nm in diameter for the sobemoviruses and 34 nm for the single polemovirus (Sõmera et al. 2021). The coat protein monomers are chemically identical but can adopt three conformations A, B, and C. The A subunits form 12 pentamers at 5-fold axes, whereas three of each B and C subunits assemble into 20 hexamers at 3-fold axes. The assembly of the T = 3 capsid of sobemoviruses is controlled by the N-termini, known as the random domain of C subunits, which are partly ordered and inserted between the interacting sides of the subunits. The N-terminal arms of A and B subunits are completely disordered. The C-terminal shell domain forms a canonical single jelly-roll β-sandwich fold (Rossmann et al. 1983). The bacilliform T = 1 virions of the barnavirus are typically 19 × 50 nm but range between 18 and 20 nm in width and 48 and 53 nm in length (Revill 2012).

GENOME

Figure 28.2 shows the genomic structure of the two families of the *Sobelivirales* order. The genome of the *Solemoviridae* family comprises a polycistronic, positive-sense, single-stranded RNA molecule of 4–4.6 kb. The genome organization is conserved, with 4–5 ORFs. The protein VPg is covalently attached to the 5′-terminus of viral and subviral RNAs, interacts with the translation initiation complex, and regulates the activity of viral protease and viral RNA-directed RNA polymerase (RdRP), expressed by means of ribosomal frameshifting. The two variants of the polyprotein are translated via a ribosomal leaky scanning mechanism from genomic RNA, and they undergo proteolytic processing at conserved cleavage sites between the domains. The coat protein (CP) is translated from the subgenomic RNA. The 3′-terminus is nonpolyadenylated but has a stable stem-loop or tRNA-like structure (Sõmera et al. 2021).

The genome of the single barnavirus is a positive-sense ssRNA of 4,009 nucleotides in size, which carries the linked VPg and appears to lack a poly(A) tail (Revill 2012).

The *Polerovirus* and *Enamovirus* member virions, the members of the former *Luteoviridae* family, contain a single molecule of infectious, linear, positive-sense ssRNA, the size of which is fairly uniform ranging from 5.6–6.0 kb. The genomic RNAs do not have a 3'-terminal poly(A) tract but a small protein VPg is covalently linked to the 5' end of the genomic RNAs of poleroviruses and the one enamovirus.

FAMILY *SOLEMOVIRIDAE*

GENUS *SOBEMOVIRUS*

3D Structure

The first x-ray diffraction studies of southern bean mosaic virus (SBMV), cowpea strain, were performed by Johnson et al. (1974). Futhermore, Johnson et al. (1976) resolved the crystal structure of SBMV to 22.5 Å. Rayment et al. (1978) subsequently extended the resolution to 11 Å using a single isomorphous heavy atom derivative. These results were an essential step in the determination of the 5.0 Å resolution electron density map of SBMV, which was achieved by Suck et al. (1978). Meanwhile, Hsu et al. (1976) detected

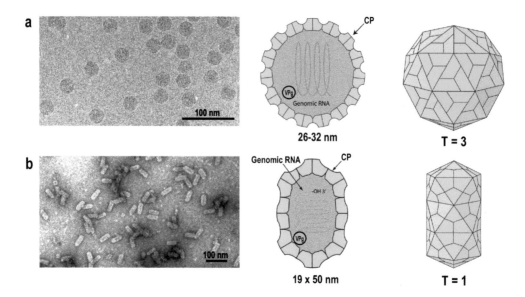

FIGURE 28.1 The portraits and schematic cartoons of the representatives of the order *Sobelivirales*. (a) The electron cryomicrograph of southern bean mosaic virus, cowpea strain. (Reprinted from *J Mol Biol*. 303, Opalka N, Tihova M, Brugidou C, Kumar A, Beachy RN, Fauquet CM, Yeager M, Structure of native and expanded sobemoviruses by electron cryo-microscopy and image reconstruction, 197–211, Copyright 2000, with permission from Elsevier.) (b) Mushroom bacilliform virus, genus *Barnavirus*, family *Barnaviridae*. The electron micrograph is taken from the ICTV report. (Reprinted from *Virus Taxonomy, Classification and Nomenclature of Viruses. Ninth Report of the International Committee on Taxonomy of Viruses*, King AMQ, Lefkowitz E, Adams MJ, Carstens EB (Eds), Revill PA, Family—*Barnaviridae*. 961–964, Copyright 2012, with permission from Elsevier.) The cartoons of the *Sobelivirales* representatives are redrawn with kind permission from the ViralZone, Swiss Institute of Bioinformatics, http://viralzone.expasy.org (Hulo C et al. 2011). (Courtesy Philippe Le Mercier.)

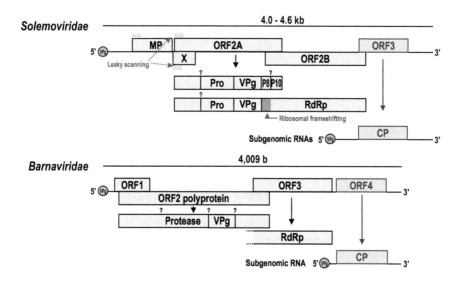

FIGURE 28.2 The genomic structure of the members of the order *Sobelivirales*, taken from the ViralZone, Swiss Institute of Bioinformatics (Hulo et al. 2011). The structural genes are colored dark pink.

stabilizing effect of divalent metal ions, Mg^{2+} and Ca^{2+}, on the SBMV virions.

At last, Abad-Zapatero et al. (1980) published the classical x-ray crystal structure of SBMV at 2.8 Å resolution and concluded that the SBMV coat protein subunit had a fold similar to the shell domain of tomato bushy stunt virus (TBSV), a member of the order *Tolivirales* (Chapter 24). However, the protruding domain of TBSV was absent in SBMV. The

tertiary structure observed in both viruses was found to be particularly suitable for the formation of the protein coat in the spherical plant RNA viruses. The SBMV structure was finally refined by Silva and Rossmann (1987). The pioneering SBMV structure, together with other currently available 3D structures of sobemovirus, is displayed in Figure 28.3.

When the N-terminal 61 aa residues of the SBMV coat protein were removed by mild trypsinolysis, resulting in

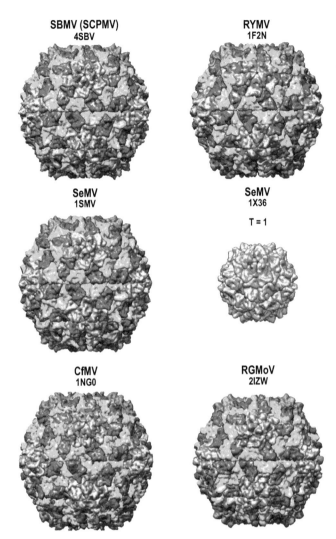

SBMV (SCPMV)
4SBV

RYMV
1F2N

SeMV
1SMV

SeMV
1X36

T = 1

CfMV
1NG0

RGMoV
2IZW

FIGURE 28.3 The crystal structure of the representatives of the genus *Sobemovirus* from the *Solemoviridae* family: southern bean mosaic virus (SBMV), 2.80 Å resolution, outer diameter 320 Å (Abad-Zapatero et al. 1980; Silva and Rossmann 1987); rice yellow mottle virus (RYMV), 2.80 Å, 318 Å (Qu et al. 2000); sesbania mosaic virus (SeMV), T = 3, 3.00 Å, 320 Å (Bhuvaneshwari et al. 1995); sesbania mosaic virus (SeMV), N-terminal deletion mutant, T = 1, 2.70 Å, 192 Å (Sangita et al. 2005); cocksfoot mottle virus (CfMV), 2.70 Å, 320 Å (Tars et al. 2003); Ryegrass mottle virus (RGMoV), 2.90 Å, 312 Å (Plevka et al. 2007). The 3D structures are taken from the VIPERdb (http://viperdb.scripps.edu) database (Carrillo-Tripp et al. 2009). All particles are T = 3 except for the SeMV structure, ID 1X36, which is T = 1. The size of particles is to scale. The corresponding protein data bank (PDB) ID numbers are given under the appropriate virus names.

a 22 kDa fragment P22, the cleaved protein assembled, in the absence of nucleic acid, into the spherical T = 1 particles of 17.5 nm in diameter (Erickson and Rossmann 1982). These particles were formed between pH 4 and 7 at low ionic strength and also between pH 8 and 11 when Ca^{2+} ions were present, while Mg^{2+} ions had no effect on the assembly of these spheres. The T = 1 particles were crystallized with Ca^{2+} ions (Erickson and Rossmann 1982),

and their 3D structure was resolved to 6 Å (Erickson et al. 1985).

Savithri and Erickson (1983) reassembled the SBMV VLPs from isolated coat protein and RNA components in vitro at low ionic strength. The SBMV RNA was separated into low—0.3–0.6 x 10^6 Da—and high—1.0–1.4 x 10^6 Da—molecular mass fractions. The assembly with high mass RNA resulted in T = 3 particles at pH 7 and 9. Low mass RNA assembled with coat protein into T = 1 particles at pH 5 and 7 and into T = 3 particles at pH 9. The formation of T = 3 particles at pH 9 and of T = 1 particles at pH 7 required the presence of Ca^{2+} and Mg^{2+} ions. The proteolytic digestion of the basic N-terminal arm of the coat protein in the absence of RNA indicated that the arm-RNA interaction was an early event in the assembly of an initiation complex for both types of particles. The effects of pH and divalent cations on SBMV assembly suggested that the charge configuration of carboxyl group clusters in the putative initiation complex regulated further subunit interactions and, hence, the mode T = 3 vs T = 1 of assembly (Savithri and Erickson 1983).

Rossmann et al. (1983) performed a detailed structural comparisons of SBMV with satellite tobacco necrosis virus (STNV) belonging to the unassigned genus *Albetovirus* (Chapter 30). Although there was no clear indication of homologous aa sequences between SBMV and STNV, the conservation of some functional groups was clearly evident. The surface domains of SBMV and the previously mentioned TBSV of the order *Tolivirales* were more like each other than like STNV (Rossmann et al. 1983).

It is notable to remember that the previously presented data were assigned to the so-called cowpea strain SBMV that was further defined as southern cowpea mosaic virus (SCPMV; Hull et al. 2000).

The crystal structure of Sesbania mosaic virus (SeMV) infecting *Sesbania grandiflora* plants in Andhra Pradesh, India, commonly known as vegetable hummingbird or West Indian pea, was resolved first to 4.7 Å by Subramanya et al. (1993). Bhuvaneshwari et al. (1995) improved resolution to 3.0 Å. Despite the overall similarity between SeMV and SBMV (or SCPMV by modern nomenclature) in the nature of the polypeptide fold, these viruses showed a number of differences in intermolecular interactions. The polar interactions at the quasi-threefold axis were substantially less in SeMV, and positively charged residues on the RNA-facing side of the protein and in the N-terminal arm were not particularly well conserved. Bhuvaneshwari et al. (1995) concluded therefore that the protein-RNA interactions could likely be different between the two viruses. Murthy et al. (1997) summarized fine details of the studies on the 3D structure of SeMV virions. Moreover, the authors presented the in vitro self-assembly of SeMV coat protein with SeMV RNA at pH 5, 7, and 9 into VLPs that were identical to native virions, in contrast to the previously mentioned assembly of SBMV, which was strongly pH-dependent.

The crystal structure of rice yellow mottle virus (RYMV), a major pathogen that dramatically reduces rice production in many African countries, was determined and refined to 2.8 Å resolution (Qu et al. 2000). The RYMV capsid structure was similar to those of other sobemoviruses. When compared with these viruses, however, the βA arm of the RYMV C subunit, which is a molecular switch that regulates quasi-equivalent subunit interactions as mentioned earlier, was swapped with the 2-fold-related βA arm to a similar, noncovalent bonding environment. This exchange of identical structural elements across a symmetry axis was categorized as 3D domain swapping and produced long-range interactions throughout the icosahedral surface lattice. The molecular switch was found therefore to modulate the stability of the viral capsids. The biochemical analysis supported the notion that the 3D domain swapping increased the stability of RYMV (Qu et al. 2000).

The native structures of SCPMV and RYMV were examined by electron cryomicroscopy and image reconstruction at 25 Å resolution (Opalka et al. 2000). While having a similar topography, the surface of RYMV was comparatively smooth. The presence of divalent cations led to swelling and fracturing of the SCPMV virions, whereas the expanded form of RYMV was stable (Opalka et al. 2000).

Tars et al. (2003) resolved the x-ray crystal structure of cocksfoot mottle virus (CfMV) to 2.7 Å. The CfMV coat protein had a jelly-roll β-sandwich fold and its conformation was very similar to that of other sobemoviruses and tobacco necrosis virus (TNV) of the previously mentioned *Tolivirales* order (Chapter 24). The N-terminal arm of one of the three quasiequivalent subunits was partly ordered and followed the same path in the capsid as the arm in RYMV. In other sobemoviruses, the ordered arm followed a different path, but in both cases the arms from three subunits met and formed a similar structure at a threefold axis. A comparison of the structures and sequences of sobemoviruses showed that the only conserved parts of the protein–protein interfaces were those that formed binding sites for Ca^{2+} ions (Tars et al. 2003).

At last, Plevka et al. (2007) resolved the x-ray crystal structure of ryegrass mottle virus (RGMoV) to 2.9 Å. Again, the coat protein had a canonical jelly-roll β-sandwich fold. However, in comparison to other sobemoviruses, the RGMoV coat protein was missing several residues in two of the loop regions. The first loop contributed to contacts between subunits around the quasi-threefold symmetry axis. The other loop that was smaller in the RGMoV structure contained a helix that participated in stabilization of the β-annulus in other sobemoviruses. The loss of interaction between the RGMoV loop and the β-annulus was compensated for by additional interactions between the N-terminal arms. As a consequence of these differences, the diameter of the RGMoV virion was 8 Å smaller than that of the other sobemoviruses (Plevka et al. 2007).

Expression

Andris Zeltiņš and his group in Riga performed before the 2000s some successful attempts to get VLPs of CfMV and RYMV in *E. coli*. Unfortunately, these promising data were never published.

A bit later, Tamm and Truve (2000) published the expression of the CfMV coat gene in *E. coli*, although without any indication to the putative self-assembly.

Lokesh et al. (2002) found the efficient self-assembly of SeMV coat into the VLPs encapsidating 23S rRNA or coat mRNA in *E. coli*. Thus, the SeMV coat protein lacking 22 aa from the N-terminus assembled into stable T = 3 capsids that appeared similar to the SeMV virions, indicating that these N-terminal 22 aa were dispensable for the T = 3 assembly. The two distinct capsids, T = 1 and pseudoT = 2 (pT = 2), were observed when the N-terminal 36 aa encompassing the arginine-rich motif of the N-terminal arm was removed. Only T = 1 particles were observed upon deletion of 65 aa from the N-terminus, which also included the sequence element for the β-annulus. These results clearly revealed that the N-terminal arm acted as a molecular switch in regulating T = 3 assembly. The formation of stable pT = 2 particles showed that pentamers of the AB dimers could nucleate assembly at icosahedral 5-folds. The capsids assembled from the N-terminally truncated proteins also encapsidated 23S rRNA and coat mRNA, suggesting the presence of sites outside the N-terminal 65 aa that could be involved in the RNA-protein interactions (Lokesh et al. 2002).

Sangita et al. (2002, 2004, 2005a, b) presented the structural fate of the N-terminally truncated mutants of the SeMV coat protein that lacked 36 or 65 residues from the N terminus and were produced in *E. coli*. Both produced similar T = 1 capsids of approximate diameter of 20 nm. In contrast to the wild-type particles, the T = 1 VLPs contained only 60 copies of the truncated protein coats. Though the 36-aa truncation mutant had the β-annulus segment, it did not form a T = 3 capsid, presumably because it lacked an arginine-rich motif found close to the N-terminus. The T = 1 VLPs of both 36-aa and 65-aa truncation mutants retained many key features of the T = 3 quaternary structure. The Ca^{2+} binding geometries at the coat interfaces of these two particles were also nearly identical. When the conserved aspartate residues that coordinated the calcium, D146 and D149, in the 65-aa deletant, were mutated to asparagine, the subunits assembled into the T = 1 VLPs but failed to bind Ca^{2+} ions. It was suggested that, although Ca^{2+} ion binding contributed substantially to the stability of the T = 1 VLPs, it was not mandatory for their assembly. In contrast, the presence of a large fraction of the N-terminal arm including sequences that preceded the β-annulus and the conserved D149 appeared to be indispensable for the error-free assembly of the T = 3 VLPs (Sangita et al. 2004). Furthermore, Sangita et al. (2005b) showed that deletion of the N-terminal 31 aa resulted in the formation of T = 1 capsids and resolved the corresponding x-ray crystal structure. The structure of the N-terminally truncated T = 1 VLP variant consisting of the SeMV coats lacking 31 aa is shown in Figure 28.3.

In a further study, it was shown that replacement of the arginines of N-terminus by glutamates resulted in the formation of empty T = 3 particles suggesting the importance of arginine residues for RNA encapsidation (Satheshkumar et al. 2005). Furthermore, to examine the importance of the β-annulus as a molecular switch that determines the T = 3 capsid assembly, Satheshkumar et al. (2005) engineered a deletion mutant of the SeMV coat, in which aa 48–59 involved in the formation of the β-annulus were deleted, retaining the rest of the residues in the N-terminal segment. When expressed in *E. coli*, the deleted SeMV protein assembled into the T = 3 VLPs of sizes close to that of the wild-type virions. Pappachan et al. (2008) purified these VLPs, crystallized them, and determined their 3D structure by x-ray crystallography at 3.6 Å resolution. The mutant VLPs closely resembled the native virions. However, surprisingly, the structure revealed that the assembly of the VLPs has proceeded without the formation of the β-annulus. It was concluded therefore that the β-annulus was not essential for the T = 3 capsid assembly and might be formed as a consequence of the particle assembly (Pappachan et al. 2008). In fact, this was the first structural demonstration that the VLP morphology of sobemoviruses with and without the β-annulus could be closely similar.

Chimeric VLPs

After establishing the previously described fact that the SeMV coat consists of a disordered N- terminal R-domain and an ordered S-domain, Gulati et al. (2016) replaced the R-domain with unrelated polypeptides of similar lengths: the B-domain of *Staphylococcus aureus* protein A (SpA) of 58 aa residues covering three helical segments that bind IgG and two SeMV-encoded polypeptides, one of 74 aa and chosen as an intrinsically disordered polypeptide and another of 96 aa having ATPase activity and predicted to have an ordered structure. The chimeric proteins formed T = 3 or larger VLPs, instead of T = 1 particles that were obtained after removal of the R-domain but could not be crystallized. The presence of metal ions during purification resulted in a large number of heterogeneous nucleoprotein complexes. Remarkably, the chimera NΔ65-B where R-domain was replaced with the B-domain of SpA could also be purified in a dimeric form. Its crystal structure revealed the T = 1 VLPs devoid of metal ions and having disordered B-domain (Gulati et al. 2016). The SeMV VLPs carrying the B-domain of SpA efficiently delivered three different monoclonal antibodies, which targeted abrin, intracellular tubulin, and HER2 receptor inside the cells (Abraham et al. 2016). Such a mode of delivery was much more effective than antibodies treatment alone.

The in vivo behavior of the SeMV VLPs and the clinical impact following its delivery via the oral or intravenous route was studied by Vishnu Vardhan et al. (2016), and no toxic effects were observed in mice administered with high SeMV doses. Then, Vishnu Vardhan et al. (2019) conjugated SeMV with fluorophores by reactive lysine-N-hydroxysuccinimide ester and cysteine-maleimide chemistries and used them for imaging. Remarkably, SeMV had a natural preference for entry into breast cancer cells, although they could also enter various other cell lines. The fluorescence

of SeMV particles labeled via the cysteines with Cy5.5 dye was found to be more stable and was detectable with greater sensitivity than that of particles labeled via the lysines with Alexa Fluor, providing thus a new platform technology that could be used to develop in vivo imaging and targeted drug delivery for cancer diagnosis and therapy (Vishnu Vardhan et al. 2019).

OTHER GENERA

The members of the *Polerovirus* genus, which are icosahedral plant viruses, attracted the special attention by a definite similarity to some RNA phages (see Chapter 25) due to a mechanism of coat protein readthrough for 300 additional aa residues, with subsequent insertion of the extended coats into the luteovirus capsids as minor components (Bahner et al. 1990). Thus, the coat protein of potato leafroll virus (PLRV) from the *Polerovirus* genus existed in two forms: (i) a 23 kDa protein, the product of the coat protein gene, and (ii) a 78 kDa protein, the product of the coat protein gene and an additional ORF expressed by readthrough of the coat protein gene stop codon. The same readthrough structure of the ORFs 3 and 5 appears also by the representatives of the *Enamovirus* genus. As reviewed by the ICTV (Domier 2012), the readthrough protein may be associated with aphid transmission and/or virus particle stability.

By close analogy with the corresponding RNA phages, this readthrough coat protein of PLRV was chosen to be exploited as a putative carrier. First, the fusion was performed with the full-length green fluorescent protein (GFP), when the latter was incorporated as a fusion with the readthrough protein into the live PRLV particles in plants (Nurkiyanova et al. 2000). Later, a fluorescent viral clone of another polerovirus, namely, turnip yellows virus (TuYV), was engineered by introducing the enhanced green fluorescent protein (eGFP) sequence into the nonstructural domain sequence of the readthrough protein (Boissinot et al. 2017). The resulting chimeric virus was infectious in several plant species.

The PLRV coat protein gene was expressed in transgenic potato plants, and, although accumulation of the coat was not detected, some level of protection was conferred against the PLRV multiplication (van der Wilk et al. 1991).

The first expressions of the coats of luteoviruses by the taxonomy of that time, PRLV (López et al. 1994) and soybean dwarf virus (Smith et al. 1993, 1998), were performed in *E. coli* in the form of fusions proteins and without any desire to generate true VLPs.

Tian et al. (1995) observed the first recombinant polerovirus VLPs, when both coat and readthrough protein genes of BWYV, another polerovirus, were expressed in *Bombyx mori* larvae by the baculovirus-driven expression system.

The PRLV coat protein, without the C-terminal addition, was synthesized by the baculovirus-driven expression in insect cells and formed good VLPs, although an N-terminal addition of a long MHHHHHHGDDDDKDAMG tag was performed (Lamb et al. 1996). Sułuja et al. (2005) expressed the coat of a Polish isolate of PLRV, N-terminally His-tagged with the same sequence as earlier but in the form of two variants of VLPs, one with the C-terminus of the coat corresponding to the wild-type protein and the second with a clathrin binding domain, namely aa LLDLD, attached to the end of the coat. These VLPs protected some small nucleic acids against DNase I and RNase A treatment, suggesting therefore a possible nucleic-acid-packaging usage of the insect cells-produced PLRV VLPs.

The encapsidation of baculovirus mRNAs was confirmed by the baculovirus-driven expression of the coat protein gene of pea enation mosaic virus 1 (PEMV1), a representative of the genus *Enamovirus*, with and without His tag (Sivakumar et al. 2009). The *Sf*21 cell line provided the highest level of the coat expression of the cell lines tested and resulted in production of the good VLPs. However, the readthrough protein was not detected in recombinant baculovirus-infected cells by Western blot. The baculovirus-expressed VLPs purified on a nickel column were of variable size of 13–30 nm and contained coat mRNA, as well as other baculovirus mRNAs, indicating nonspecific RNA encapsidation in the absence of viral RNA replication (Sivakumar et al. 2009).

At last, Bolus et al. (2020) used the baculovirus-driven system to express the coat gene of Miscanthus yellow fleck virus (MYFV), a new polerovirus infecting *Miscanthus sinensis*, a grass used for sugarcane breeding and bioenergy production. The purified VLPs 28 nm in diameter were found to conform to icosahedral shapes, characteristic of other poleroviruses.

Concerning 3D structure, the recalcitrance of luteovirids to crystallization was likely due in part to the highly disordered readthrough domain projecting outward from the capsid surface, which might not uniformly adopt an ordered conformation, as concluded by Chavez et al. (2012). The computational modeling of the readthrough domain had also proved difficult, as it did not bear homology to any protein for which a crystal structure was available, and de novo modeling was also negatively affected by the high degree of disorder predicted in this domain (Chavez et al. 2012). Therefore, the structure of poleroviruses LPRV (Terradot et al. 2001; Lee et al. 2005) and BWYV (Brault et al. 2003) coats was examined first by immunological and site-directed mutagenesis. Later, the protein interaction reporter (PIR) technology was employed, a strategy that used chemical cross-linking and high resolution mass spectrometry, to discover topological features of the PLRV (Chavez et al. 2012; DeBlasio et al. 2015) and TuYV (Alexander et al. 2017) coat and readthrough proteins that were required for the diverse biological functions of the polerovirus virions.

As mentioned in Chapter 24, high-resolution structures of poleroviruses were determined by electron cryomicroscopy (Byrne et al. 2019). The high-resolution structures for the VLPs of PLRV and barley yellow dwarf virus (BYDV), a current luteovirus BYDV from the order *Tolivirales* (Chapter 24), were resolved at 3.0 Å and 3.4 Å, respectively.

The enamovirus PEMV1 proved recalcitrant to forming ordered VLPs, an issue previously encountered during expression of the PEMV1 coat in insect cells (Sivakumar et al. 2009). The PEMV1 electron cryomicroscopy structure was not determined, but, given the high sequence homology between BYDV, PLRV, and PEMV1, the homology modeling was applied to generate the PEMV1 structure (Byrne et al. 2019). The BYDV and PLRV structures were shown earlier in Figure 24.4.

It is noteworthy that Skurat et al. (2017) generated a chimeric virus as a more efficient source of the PLRV coat. A binary vector was constructed, which carried chimeric cDNA, where the coat protein gene of the tobacco mosaic virus was substituted for the PLRV coat gene. The latter packed RNAs from the helical TMV in spherical virions, and the yield of isolated chimera was about three orders higher than the yield of the native PLRV.

FAMILY *ALVERNAVIRIDAE*

The 3D reconstruction of Heterocapsa circularisquama RNA virus (HcRNAV) was performed by electron cryomicroscopy at 18 Å resolution and provided the first structural information on this marine virus family (Miller et al. 2011). The HcRNAV virions had a diameter of 34 nm and T = 3 symmetry. The 180 quasi-equivalent monomers had an unusual arrangement in that each monomer contributed to a "bump" on the surface of the protein. Though the coat protein probably had the classic "jelly-roll" β-sandwich fold, this was a new packing arrangement and was distantly related to the other positive-sense ssRNA virus capsid proteins. The handedness of the structure was determined by a novel method involving high-resolution scanning electron microscopy of the negatively stained viruses and secondary electron detection (Miller et al. 2011).

Wada et al. (2011) expressed the optimized de novo synthesized coat gene of HcRNAV in *E. coli*. The bacterially produced protein, which was purified after a procedure involving denaturation and refolding, successfully formed VLPs that significantly resembled native HcRNAV virions. The purified denatured protein was used as an antigen to immunize rabbits, and the resulting antiserum was shown to be strongly reactive to not only the bacterially expressed recombinant protein but also to the native HcRNAV coat protein (Wada et al. 2011).

Kang et al. (2015) engineered the target-specific HcRNAV VLPs for the delivery of algicidal compounds to the harmful algae. Thus, the coat gene of HcRNAV34 was expressed in *E. coli*, and the corresponding VLPs were self-assembled in vitro. Next, the algicidal compound, thiazolidinedione 49 (TD49), was encapsidated into the HcRNAV34 VLPs for specific delivery to *H. circularisquama*. Consequently, the HcRNAV34 VLPs demonstrated the same host selectivity as naturally occurring HcRNAV34 virions, while TD49-encapsidated VLPs showed a more potent target-specific algicidal effect than TD49 alone (Kang et al. 2015).

29 Order *Patatavirales*

Science is magic that works.

Kurt Vonnegut

ESSENTIALS

The *Patatavirales* is an order of a single family, where the family *Potyviridae* is the largest family of RNA plant viruses. There are 12 genera covering 235 species altogether, distinguished by the host range, genomic features, and phylogeny of the member viruses. Some members cause serious disease epidemics in cultivated plants. According to the modern taxonomy (ICTV 2020), the order *Patatavirales* is one of the two orders, together with the *Astravirales* (Chapter 30), belonging to the class *Stelpaviricetes*. The latter is a part of the *Pisuviricota* phylum from the kingdom *Orthornavirae*, realm *Riboviria*.

Figure 29.1 presents two schematic cartoons of the order members and an electron microscopy of a typical representative. As summarized by Wylie et al. (2017), the nonenveloped flexuous filamentous virions are 680–900 nm long and 11–20 nm in diameter, possess helical symmetry with a pitch of about 3.4 nm, and are formed by a single coat protein of 28.5–47 kDa. The particles of some viruses appear longer in the presence of divalent cations than in the presence of EDTA. The virions are moderately immunogenic, and there are serological relationships among many members. The N- and C-terminal residues are positioned on the exterior of the virion. The mild trypsin treatment removes N- and C-terminal segments, leaving a trypsin-resistant core of about 24 kDa. All potyvirus coats display significant aa sequence identity in the trypsin-resistant core, but little identity in their N and C-terminal segments (Wylie et al. 2017). As recently reviewed by Martínez-Turiño and García (2020), the virions may contain, in addition to coats, the three other viral proteins: VPg, HC-Pro and CI, which have been found associated with one end of some potyviral particles.

GENOME

Figure 29.2 shows the genomic structure of the *Potyviridae* family members. As summarized by Wylie et al. (2017), the genomes range from 8.2–11.3 kb, with an average size of 9.7 kb. The most genomes are monopartite, but those of members of the genus *Bymovirus* are bipartite. The genomes have a VPg of about 24 kDa, which is covalently linked to the 5′- end, while the 3′-terminus is polyadenylated. The encoded large polyprotein is self-cleaved into a set of functional proteins, and the gene order is generally conserved throughout the family.

The members of the *Potyviridae* family are related to the order *Picornavirales* members (Chapter 27) in their genome structure, in particular by the presence of a single ORF translated into a polyprotein and by the block of the replication genes: helicase, VPg, 3C-like protease and polymerase. However, they differ from picornaviruses in virion morphology, the size of VPg, and the type of helicase (Wylie et al. 2017).

As recently reviewed by Martínez-Turiño and García (2020), the flexous filamentous particles of plant viruses in general and potyviruses in particular are comparatively much less exploited than the spherical particles, although their use is increasing. The putative advantages, especially for biomedical applications, conferred by the flexuous rod-shaped structures compared to other types of architectures, have been presented by Röder et al. (2019) in the case of potato virus X (PVX) and were discussed accordingly in Chapter 21. As suggested by Martínez-Turiño and García (2020), an additional advantage of potyvirus-based nanotechnology tools, not shared by other flexuous viruses such as PVX, is the possibility of using the internal channel of RNA-free VLPs not only to display peptides but also to carry drug cargos or imaging molecules. Until now, the primary use of potyviral capsids has been as antigen presentation systems, but they are increasingly being used for other purposes, such as in the assembly of improved enzyme systems (Martínez-Turiño and García 2020).

FAMILY *POTYVIRIDAE*

GENUS *MACLURAVIRUS*

When the coat gene of cardamom mosaic virus (CdMV), provided with the upstream His$_6$ tag, was expressed in *E. coli*, formation of filamentous aggregates was observed by immuno-gold electron microscopy (Jacob and Usha 2002). After purification of the bacterially produced CdMV coat protein by Ni-NTA affinity chromatography under denaturing conditions, the product aggregated irreversibly upon renaturation at concentrations above 0.07 mg/ml. The filamentous aggregates were of 100–150 nm in length and did not contain any traces of the coat protein mRNA. The introduction of mutations into the core region of the CdMV coat by deleting either Arg or Asp residues, which have been postulated earlier by Jacquet et al. (1998a) to be involved into the assembly of potyviruses, led to formation of inclusion bodies, which were soluble only in buffers with 8 M urea, as in the case of the wild-type coat. The renaturation of the mutant coats, as well as wild-type coats, by dialysis after the affinity chromatography resulted in the aggregation. It was assumed that these deletions had no role in

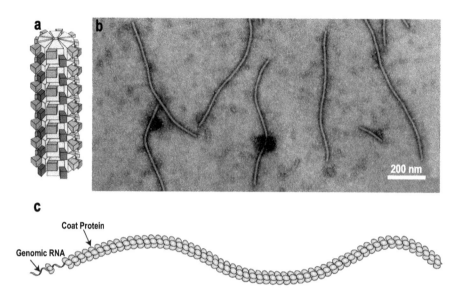

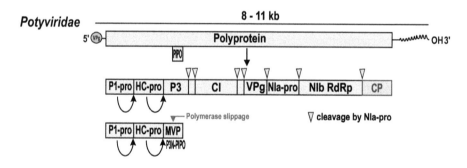

FIGURE 29.1   The portrait and schematic cartoon of the typical representatives of the *Potyviridae* family. (a) Schematic diagram of a potyvirus particle. The N-terminus (~30 aa; large rectangle) and C-terminus (~19 aa; small rectangle) of the coat protein is exposed on the surface of the intact virus particle. (Reprinted from *Adv Virus Res*. 36, Shukla DD, Ward CW, Structure of potyvirus coat proteins and its application in the taxonomy of the potyvirus group, 273–314, Copyright 1989, with permission from Elsevier.) (b) Negative-contrast electron micrograph of particles of an isolate of plum pox virus (PPV), stained with 1% PTA, pH 6.0. (Courtesy of I.M. Roberts and International Committee on Taxonomy of Viruses [ICTV], https://talk.ictvonline.org/ictv-reports/ictv_online_report/posi tive-sense-rna-viruses/w/potyviridae.) (c) The general cartoon of the *Potyviridae* representative virions is taken with kind permission from the ViralZone, Swiss Institute of Bioinformatics, http://viralzone.expasy.org (Hulo C et al. 2011). (Courtesy Philippe Le Mercier.)

FIGURE 29.2   The genomic structure of the members of the *Potyviridae* family, taken from the ViralZone, Swiss Institute of Bioinformatics (Hulo et al. 2011). VPg, viral protein genome-linked; P1-Pro, protein 1 protease; HC-Pro, helper component protease; P3, protein 3; PIPO, pretty interesting *Potyviridae* ORF; 6K, six kilodalton peptide; CI, cytoplasmic inclusion; NIa-Pro, nuclear inclusion A protease; NIb, nuclear inclusion B RNA-dependent RNA polymerase; CP, coat protein. The interpretation of the protein names is from Wylie et al. (2017). The structural gene is colored dark pink.

inducing the self-aggregation property of the CdMV coat protein (Jacob and Usha 2002).

Damodharan et al. (2013) were the first who employed the CdMV coat to display foreign epitopes and invented production of the appropriate chimeric VLPs. Thus, the N- and C-termini of the coat protein were engineered with the so-called Kennedy peptide and the 2F5 and 4E10 epitopes of the glycoprotein gp41 of human immunodeficiency virus 1 (HIV-1). The Kennedy peptide is located in the cytoplasmic tail of gp41, corresponds to aa sequence 735–752 of the precursor envelope glycoprotein, and contains three epitopes: 734-PDRPEG-739, 740-IEEE7–43, and 746-ERDRD-750, where the last one was strongly characterized as a virus-neutralizing (Cleveland et al. 2003). The highly conserved membrane proximal external region (MPER) comprises the last 24 C-terminal aa of the gp41 ectodomain and contains epitopes recognized by three HIV-1 broadly neutralizing monoclonal antibodies, namely 2F5, 4E10, and Z13, where the 2F5 epitope has been mapped to the motif ELDKWA (Muster et al. 1993). The monoclonal antibody 4E10 recognizes a contiguous epitope at the C terminus of the 2F5 binding region (Stiegler et al. 2001). Both 2F5 and 4E10 epitopes were identified as virus-neutralizing. As a result, Damodharan et al. (2013) engineered 16 constructs carrying various combinations of the Kennedy peptide, 2F5 and 4E10 epitopes at the N- and C-termini of the full-length

CdMV coat of 308 aa residues. The chimeric proteins derived from the 13 constructs were refolded and purified in the same manner as the wild-type coat protein, expecting therefore the formation of filamentous VLPs (Damodharan et al. 2013). However, although the purified proteins demonstrated good HIV-related immunological characteristics, no electron microscopy or other arguments of the VLP presence were demonstrated in this paper. Later, Kumar et al. (2016) performed molecular modeling and in-silico engineering of the CdMV coat protein for the presentation of immunogenic epitopes of the LipL32 protein of *Leptospira*. The structure of the putative chimeric proteins carrying the mapped LipL32 epitopes was predicted for insertions at N-, C-, and both of the termini of the CdMV coat.

Vijayanandraj et al. (2013) used large cardamom chirke virus (LCCV), another representative of the *Macluravirus* genus, which causes so-called chirke disease, an important constraint in large cardamom production in India. To improve the immune diagnostics of the disease, the authors expressed the His-tagged LCCV coat gene in *E. coli*. The resulting protein was purified in denaturing conditions followed by renaturation, which resulted in high yield and good quality of the protein for the diagnostic purposes (Vijayanandraj et al. 2013). Again, no data about the structural VLP state of the recombinant LCCV coat was reported.

Genus *Poacevirus*

The full-length coat gene of sugarcane streak mosaic virus (SCSMV) encoding 283 aa residues was expressed in *E. coli*, with the N-terminal addition of foreign 36 aa residues from vector along with the His$_6$ tag (Hema et al. 2003). The product was soluble, underwent purification by affinity chromatography on Ni-NTA column, and was used to immunize rabbits. Although the recombinant SCSMV coat satisfied the ELISA needs, no data about its putative assembly or aggregation state were presented. Later, Hema et al. (2008) confirmed by electron microscopy that this *E. coli*-produced protein actually formed the potyvirus-like VLPs, which carried the encapsidated coat mRNA.

Tatineni et al. (2013) used triticum mosaic virus (TriMV), a member of the reference species of the genus *Poacevirus*. The TriMV coat gene was expressed in *E. coli*. The 294 aa residues, or 37.0 kDa, of the TriMV coat were N-terminally provided with a foreign stretch of 8 kDa consisting of the vector-derived His-tag, S-Tag, and enterokinase sequences. The mostly soluble product was purified and used to immunize rabbits, while no data about its aggregation state was disclosed.

Genus *Potyvirus*

Johnsongrass Mosaic Virus

Jagadish et al. (1991) were the first ones to invent a potyvirus, namely Johnsongrass mosaic virus (JGMV), as a putative VLP carrier of foreign epitopes. First, the full-length

JGMV coat gene was expressed in *E. coli* and in yeast *Saccharomyces cerevisiae*, and in both cases the coat protein assembled to form the potyvirus-like VLPs. The latter were heterogeneous in length with a stacked-ring appearance and closely resembled the JGMV virions in their flexuous morphology and width. Remarkably, in *E. coli* the full-length protein JGMV coat should contain an additional 16 aa stretch at its N-terminus, with 12 generated from the multiple cloning sites present in the vector and four from the C-terminal region of the NIb gene. In the yeast construct, the synthesis of the full-length unfused coat occurred. After two sucrose gradient centrifugations, the products of both *E. coli* and yeast appeared as VLPs by electron microscopy. This is demonstrated in Figure 29.3, where the VLPs are in particular heavily decorated with specific antibodies. The mutant forms of the JGMV coat produced by site-directed mutagenesis of the highly conserved residues RQ at positions 194/195 to DL abolished particle formation, even though the levels of the mutant coat protein produced in *E. coli* remained high (Jagadish et al. 1991). Therefore, the efficient *E. coli* expression system allowed further mapping of the JGMV coat and identification of key aa residues required for assembly (Jagadish et al. 1993b). Later, Saini and Vrati (2003b) markedly improved the production efficiency of the JGMV VLPs in *E. coli*, at least 16-fold, by relieving the initiation codon from the stable hairpin-loop structure after appropriate nucleotide substitutions.

The versatile nature of the JGMV model was further confirmed by high-level production of the correct JGMV VLPs during the baculovirus-driven expression in the insect *Sf*9 cells (Edwards et al. 1994) and during expression in mammalian cells by the recombinant vaccinia virus expression system (Hammond et al. 1998). It should be noted, however, that the first signs of this versatility appeared earlier when the purified potyvirus coats were reassembled into the typical VLPs in vitro, in the absence of viral RNA (McDonald et al. 1976; McDonald and Bancroft 1977).

The first chimeric JGMV VLPs were produced in *E. coli* after insertion of an octapeptide epitope SNTFINNA from the merozoite surface antigen (MSA2) of the malaria parasite *Plasmodium falciparum* and a decapeptide hormone QHWSYGLRPG, namely luteinizing hormone releasing hormone (LHRH), at the N- or at both the N- and C-termini of the JGMV coat (Jagadish et al. 1993a). Moreover, a full-length protein, Sj26-glutathione S-transferase of 26 kDa from the helminth parasite *Schistosoma japonicum*, was introduced into the JGMV coat, replacing the N-terminal 62 aa stretch (Jagadish et al. 1993a). The electron microscopy of ultrathin sections of *E. coli* revealed the appearance, within the cytoplasm, of parallel strands sometimes extending the length of the cell and stringing cells together, with "threads" of the JGMV VLPs appearing to connect individual bacterial cells. The JGMV VLPs bearing the 26-kD antigen Sj26 were shorter and wider and elicited an efficient Sj26-specific immune response after administration to mice. These studies led to conclusion that the potyvirus coat can accommodate peptides or even large antigens

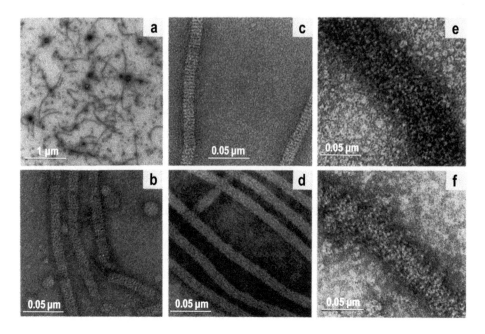

FIGURE 29.3 Electron micrographs of the JGMV PVLPs from *E. coli* (a, b and e) and *S. cerevisiae* (c). JGMV particles purified from plants are shown in (d). The VLPs from *E. coli* and the VLPs [reassembled following formic acid (60%) denaturation of JGMV followed by dialysis against 10 mM-phosphate buffer pH 7.2 containing 100 mM-NaCl at 4 °C] decorated with specific antibodies are shown in (e) and (f), respectively. (Reprinted with permission of Microbiology Society from Jagadish MN et al. *J Gen Virol*. 1991;72:1543–1550.)

and is able to present antigens on the surface of the JGMV VLPs (Jagadish et al. 1993a). Generally, the JGMV vectors allowed for not only the addition of short peptides to their N- or C-termini but also for the fusion of large antigens to the N terminus or replacing most of the N- or C-terminal exposed regions (Jagadish et al. 1996). Moreover, the chimeric JGMV VLPs were highly immunogenic in mice and rabbits even in the absence of any adjuvant (Jagadish et al. 1993a, 1996). Using their efficiently improved *E. coli* expression system, Saini and Vrati (2003a) engineered the chimeric JGMV VLPs carrying on their surface the virus-neutralizing epitopes from the envelope (E) protein of Japanese encephalitis virus (JEV). The four peptides from different locations within the JEV protein E were chosen and fused to the C-terminus of the JGMV coat. The fusion protein formed VLPs that were purified by sucrose gradient centrifugation. The immunization of mice with the chimeric VLPs containing JEV peptide sequences induced antipeptide and anti-JEV antibodies. The JGMV VLPs carrying the 27-aa peptide containing aa 373 to 399 from JEV protein E induced virus-neutralizing antibodies, without any adjuvant, and the immunized mice showed significant protection against a lethal JEV challenge. Next, Choudhury et al. (2009) performed an attempt to engineer a contraception vaccine and investigated the potential of the JGMV VLPs to present gamete epitopes, namely mouse zona pellucida (ZP3) peptide QAQIHGPR and spermatozoan specific YLP$_{12}$ peptide YLPVGGLRRIGG. Moreover, the fusion peptide comprised of both YLP$_{12}$ and ZP3 epitopes separated by a diglycine spacer (YLP$_{12}$-ZP3) was positioned onto the VLPs. The immunization of female mice with the

VLPs presenting the YLP$_{12}$-ZP3 fusion peptide and a physical mixture of the VLPs presenting either YLP$_{12}$ or ZP3 epitope, administered in saline without any adjuvant, led to generation of specific antibody responses and a significant reduction in litters born per mice (Choudhury et al. 2009).

In parallel, the chimeric JGMV VLPs were produced not only in *E. coli* but also in mammalian cells (Hammond et al. 1998). The vaccinia virus insertion vectors contained the JGMV coat gene with N-terminal insertion of the previously mentioned decapeptide of LHRH or with both the N-terminal insertion of the octapeptide from the MSA2 of *P. falciparum* and the C-terminal insertion of LHRH. The infection of cells with such vaccinia virus chimeras resulted in the formation of the appropriate VLPs. This approach made the vaccinia vectors suitable for the putative delivery of the potyvirus-like VLPs displaying vaccine antigens in vivo without the need for particle purification and/or inclusion of adjuvant (Hammond et al. 1998).

Lettuce Mosaic Virus

The lettuce mosaic virus (LMV) was chosen together with other potyvirus, potato virus A (PVA), described later, to establish a new experimental approach allowing the single particle resolution imaging of immune complexes on virus particles by combined atomic force-electrochemical microscopy (AFM-SECM; Nault et al. 2015). In fact, when operated in a molecule touching (Mt) mode and combined with redox-immunomarking, this technique enabled the in situ mapping of the distribution of proteins on individual virus particles and made localization of individual viral proteins possible. The acquisition of a topography image allowed the isolated virus particles to be identified and structurally characterized, while the simultaneous acquisition of a current image allowed the sought protein, marked by redox-antibodies, to be selectively located (Nault et al. 2015). The fine methodology of this novel approach was presented in the corresponding laboratory manual published by Anne et al. (2018).

Although this is not connected directly with the VLP approach, it is interesting to note here that the chimeric LMV virions tagged with the GFP gene were engineered and used to test the LMV resistance in lettuce (Candresse et al. 2002). The GFP was inserted in frame into the viral polyprotein and was expressed as a translational fusion to the viral helper component protein HC-Pro.

Papaya Ringspot Virus

Chatchen et al. (2006) expressed the coat gene of papaya ringspot virus (PRSV) in *E. coli* and simultaneously applied it to the engineering of chimeric molecules. The 15-aa epitope from the VP2 of canine parvovirus (CPV) was either inserted into the PRSV coat gene at the 5' or 3' ends, at the both 5' and 3' ends, or substituted into the 3'-end of the PRSV coat gene. After lysis of bacteria, both of the native and chimeric coats were present in the pellet and remained insoluble. The protein aggregates were dissociated into the soluble fraction by dissolving with sodium-phosphate buffer pH 9 and purified by Ni^{2+} ions affinity column. The yield of the native and chimeric coats after affinity column purification was approximately 80–200 μg/l of bacterial cultures. All chimeric coats stimulated immune response, while the chimeras carrying the CPV epitope at the C-terminus and at both N- and C-termini elicited ten times higher specific antisera in immunized mice compared with the other two chimeras that contained the CPV epitope at the N-terminus or as a substitution at the C-terminus. No data about the structural state of the chimeras as putative VLPs or aggregates was presented (Chatchen et al. 2006).

For the first time in the case the PRSV coat, Guerrero-Rodríguez et al. (2014) demonstrated the nice PRSV VLPs by electron microscopy after expression of the corresponding gene in *E. coli*. Remarkably, although the coat protein lacked a histidine tag, its purification by Ni^{2+} affinity chromatography was achieved. Further, the VLPs were enriched by polyethylene glycol precipitation and diafiltration. The obtained PRSV VLPs were chemically coupled to a model antigen, namely GFP, and the adjuvant effect of the VLPs was demonstrated in mice. The immunization resulted in a significant increase in anti-GFP IgG response, particularly IgG1 class, suggesting that the PRSV VLPs were able to recruit cells of the immune response to the site of injection. The PRSV VLP platform in *E. coli* was used to generate a chimeric vaccine candidate against porcine circovirus type 2 (PCV2), a virus that is described in Chapter 10. Thus, Aguilera et al. (2017) inserted the protective PCV-2 epitope 19-FNLKDPPLNP-28 into the three different regions of the PRSV coat, namely the N- and C-termini and between aa residues 18 and 19 into the predicted antigenic region. Remarkably, this was the first report that showed formation of the chimeric VLPs using PRSV as epitope-presentation scaffold. It was found that the PCV2 epitope localization strongly influenced the VLP length, and all three chimeric VLPs induced high levels of anti-PCV2 antibodies in mice. In parallel, Cárdenas-Vargas et al. (2016) used the PRSV VLPs as an adjuvant for the synthetic 34-mer peptide 87-NSENGTCYPGDFID YEELREQLSSVSSFEKFEIF-120 from the HA1 region of the hemagglutinin (HA) of the influenza virus A/South Carolina/1/18 (AH1 N1). The peptide has been described previously to induce reactive and neutralizing sera in pigs against an influenza virus (Vergara-Alert et al. 2012). As a result, the PRSV VLPs provided the influenza HA peptide with high B- and T-cell response in mice (Cárdenas-Vargas et al. 2016).

Pepper Vein Banding Virus

The pepper vein banding virus (PVBV) was assigned recently to the genus *Potyvirus* as chilli veinal mottle virus-pepper vein banding virus (ChiVMV-PVB), a member of the species *Chilli veinal mottle virus*. Joseph and Savithri (2000) expressed the PVBV coat gene in *E. coli* and demonstrated appearance of the filamentous potyvirus-like PVBV VLPs, which encapsidated the coat protein mRNA. The particles were heterogeneous in length as compared to the purified PVBV virions. The additional 44 aa at the N-terminus of the coat, including the His tag, which appeared due to the applied cloning strategy, did not affect the assembly of the resulting coat protein into the PVBV VLPs. This result conformed the previously published results discussed earlier on the display of foreign epitopes at the N-terminus of the JGMV coat with further assembly of the appropriate JGMV VLPs. When the PVBV coat was expressed in another vector lacking additional 41 aa and without the His tag, the product also assembled into VLPs as was apparent from electron microscopy (Joseph and Savithri 2000). Then, Anindya and Savithri (2003) analyzed the N- and C-terminal deletion mutants of the PVBV coat and found the N-terminal 53 and C-terminal 23 aa residues to be crucial for the intersubunit interactions involved in the initiation of virus assembly. However, these segments were surface exposed in the ring-like intermediate and dispensable for further interactions that resulted in the formation of the PVBV VLPs.

Continuing later the story of the *E. coli*-produced PVBV VLPs, Sabharwal et al. (2019) elucidated the endocytic uptake pathway for the flexuous rod-shaped particles and their subsequent intracellular trafficking into epithelial HeLa and HepG2 cells. The kinetics of internalization in HeLa cells demonstrated the biodegradable and biocompatible nature of the PVBV VLPs. Furthermore, vimentin and Hsp60 were identified as the cell-surface proteins required for the binding/internalization of the PVBV VLPs in HeLa and HepG2 cells, respectively. Sabharwal et al. (2020) developed the chimeric PVBV VLPs and used them as a delivery vehicle for antibodies against intracellular antigens as well as for the future applications in immunodiagnostics. The chimeric PVBV particles were generated by genetically engineering the B domain of *Staphylococcus aureus* protein A (SpA) at the N-terminus of the PVBV coat. The chimeric VLPs purified by sucrose density gradient centrifugation had ~440-fold higher affinity toward IgG antibody when compared to the natural SpA. Interestingly, the unassembled chimeric coat with the B-domain at the N-terminus purified by Ni-NTA chromatography as a monomer demonstrated ~45-fold higher affinity toward antibodies compared to the SpA. As a result, the chimeric PVBV VLPs were able to bind and deliver antibodies against both intracellular (α-tubulin) and surface-exposed antigens (CD20). In contrast to the chimeric VLPs, the unassembled chimeric monomer did not enter mammalian cells. Thus, for the first time, the VLP assembly was proven to be essential for internalization of the product (Sabharwal et al. 2020).

Plum Pox Virus

Concomitant with the development of the *E. coli*-based JGMV expression model, the coat gene of plum pox virus (PPV) was expressed in transgenic *Nicotiana* plants (Ravelonandro et al. 1992; Wypijewski et al. 1995). The transgenic plums transformed with the PPV coat gene displayed a resistance to the sharka disease (Jacquet et al. 1998b). The two modified PPV coat variants were tested, where the first one contained a deletion of the nucleotides encoding for the DAG aa triplet involved in the virus aphid-transmission but the second one lacked the first 420 nucleotides of the PPV coat gene. However, the described resistance against PPV was in fact RNA but not coat protein-mediated and was classified as a "sense suppression" or homology-dependent resistance (Jacquet et al. 1998b).

The first expression of the PPV coat gene in *E. coli* was reported by Mattanovich et al. (1989), while the first PPV VLPs were obtained in *E. coli* with the full-length coat and with the coat lacking the previously mentioned DAG triplet involved in aphid transmission by Jacquet et al. (1998a). Moreover, no VLPs were observed with the coat lacking R220, Q221, or D264, the aa residues known to be essential for the assembly of other potyvirus coats. The transgenic *Nicotiana benthamiana* lines expressing the different PPV coat constructs were infected with a non-aphid-transmissible strain of zucchini yellow mosaic virus (ZYMV-NAT). The aphid transmission assays performed with these plants demonstrated that the developed strategies provided an effective means of minimizing the biological risks associated with heteroencapsidation (Jacquet et al. 1998a).

Developing the PPV model in the form of the live chimeric infectious viruses rather than VLPs, the N-terminal part of the PPV coat was chosen as the site for expression of foreign antigenic peptides (Fernández-Fernández et al. 1998). First, the modifications in this site were engineered to avoid the capability of natural transmission by aphids of the resulting PPV vector. As the first practical attempt, the different forms of an antigenic peptide of 15 aa residues from the VP2 capsid protein of canine parvovirus (CPV)

were expressed in a single form or in a tandem repetition. Both chimeras were able to infect *Nicotiana clevelandii* plants with characteristics similar to the wild-type virus. The chimeras remained genetically stable after several plant passages. Moreover, mice and rabbits immunized with the chimeric virions developed CPV-specific antibodies, which showed sufficient neutralizing activity (Fernández-Fernández et al. 1998). Furthermore, Fernández-Fernández et al. (2002) improved the PPV vector that was used for the display of the CPV epitopes. After a PEPSCAN analysis and identification of the PPV coat immunogenic regions, the more efficient antigen presentation vectors were designed. As predicted by the PEPSCAN analysis, a small displacement of the insertion site in a previously constructed vector turned the derived chimeras into more efficient immunogens. The vectors expressing foreign peptides at different positions within the highly immunogenic region of aa residues 43 to 52 within the N-terminal domain of the PPV coat were the most effective at inducing specific antibody responses against the foreign sequence (Fernández-Fernández et al. 2002).

Meanwhile, Varrelmann et al. (2000) constructed different mutants of an infectious full-length PPV clone: three mutants with mutations of the assembly motifs RQ and DF in the coat and two coat chimeras with exchanges in the coat core region of zucchini yellow mosaic virus (ZYMV) and potato virus Y (PVY). The assembly mutants were restricted to single infected cells, whereas the PPV chimeras were able to produce systemic infections in *N. benthamiana* plants (Varrelmann et al. 2000).

Potato Virus A

Čeřovská et al. (2002) were the first to express the coat gene of potato virus A (PVA) in *E. coli*. The bacterially produced PVA coat protein was purified by centrifugation in CsCl density gradient or on a sucrose cushion, and the production of the PVA VLPs was proved by electron microscopy. The PVA VLPs were used for preparation of a mouse antiserum used further in ELISA. Pokorna et al. (2005) linked a short sequence coding for the human papilloma virus 16 (HPV16) oncoprotein E7 peptide, aa 44–60, containing immunodominant epitopes for B and T cells to the C-terminus of the PVA coat or to its deleted form with a short C-terminal deletion of 5 aa residues. Both chimeric proteins, just like the original PVA coat, spontaneously assembled into the VLPs in both *E. coli* and mammalian 293T cells, while both chimeric genes used as a DNA vaccine protected mice against the development of the induced tumors, also when the vaccination was performed after tumor cell administration. Furthermore, the expression of the PVA-based HPV vaccine was completed in *Nicotiana* plants using viral potato virus X (PVX)-based expression vector (Čeřovská et al. 2008). In this case, the PVA coat carried the two different HPV16 epitopes, where the 14-aa epitope LVEETSFIDAGAPG derived from minor capsid protein L2 was added to the N-terminus, while the 17-aa E7 epitope remained at the C-terminus. The plant-produced

VLPs were purified and documented by electron microscopy. Hoffmeisterová et al. (2008) presented the expression of the PVA-HPV16 chimera in an edible plant, namely turnip *Brassica rapa*, cv. Rapa, which is suitable for human consumption in its raw state. Therefore, this expression was planned as a possible contribution to the putative oral vaccination purposes.

At last, Besong-Ndika et al. (2016) used PVA virions as enzyme-nanocarriers by a noncovalent approach for adhesion of enzymes to virus particles. This approach made use of z33, a peptide derived from the B-domain of *Staphylococcus aureus* protein A, which binds to the Fc domain of many immunoglobulins. The authors demonstrated that, with specific antibodies addressed against the viral coats, an 87% coverage of z33-tagged protein can be achieved on the potyvirus particles. The enzymes 4-coumarate coenzyme A ligase (4CL2) and stilbene synthase (STS), which catalyze consecutive steps in the resveratrol synthetic pathway, were modified to carry the N-terminal z33 peptide and a C-terminal His_6 tag produced in *E. coli*, and assembled readily into macromolecular complexes with the PVA virions and anti-PVA coat protein antibodies. The immobilized enzymes were able to synthesize resveratrol, presenting therefore a bottom-up approach to immobilize active enzyme cascades onto virus-based nanocarriers as the putative nanodevices (Besong-Ndika et al. 2016).

Potato Virus Y

Stram et al. (1993) expressed the full-length coat gene of potato virus Y (PVY) in *E. coli* and revealed the presence of the appropriate VLPs in the cell extracts by immune electron microscopy. Later, Folwarczna et al. (2008) expressed the PVY coat gene in *E. coli*, purified the product, and used it to induce polyclonal antibodies in rabbits. The PVY coat was obtained in a soluble fraction, but its putative VLP state was not addressed.

Andris Zeltiņš's group in Riga not only expressed the PVY coat gene in *E. coli* and demonstrated the corresponding VLPs, also when the coat carried the $(G_4S)_3$ linker at the N-terminus, but also successfully generated a set of the chimeric PVY-derived VLPs (Kalnciema et al. 2012). Thus, the PVY coat with the N-terminal insertion of the 22-aa preS1 epitope from the surface protein (Chapter 37) of hepatitis B virus epitope (HBV) or the whole protein rubredoxin of 54 aa retained its ability to form the potyvirus-like filamentous particles, whereas adding a foreign sequence to the C-terminus of the PVY coat generated mostly unstructured protein aggregates. Figure 29.4 demonstrates the chimeric PVY VLPs. Thus, the preparations contained flexible filamentous VLPs with a diameter of approximately 11 nm, which resembled morphologically native PVY virions. The measurement of the length of the VLPs demonstrated that there was a whole spectrum of particles measuring between 30 and 2,200 nm, although the VLPs of 400–800 nm were predominant. Surprisingly, the chimeric coats formed VLPs that were longer than natural PVY virions. It followed therefore that the PVY VLP

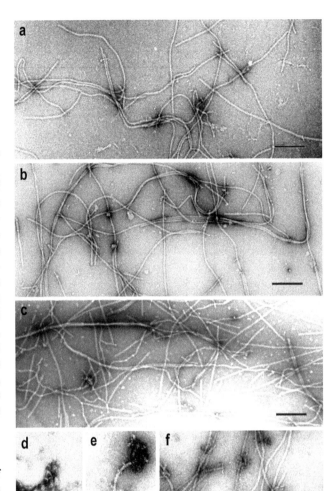

FIGURE 29.4 Electron micrographs of PVY coat-derived VLPs. (a) Purified VLPs formed from PVY coat after expression of the PVY coat gene. (b) PVY VLPs with N-terminal GS-linker NG_4S. (c) PVY-preS1 VLPs displaying the HBV preS1 epitope, aa 20–41. (d) PVY coat with C-terminal GS-linker CG_4S. (e) N-terminally truncated PVY coat. (f) PVY VLPs displaying rubredoxin. Bar 100 nm (a, b, c, f), 50 nm (d, e). (With kind permission from Springer Science+Business Media: *Mol Biotechnol*, Potato virus Y-like particles as a new carrier for the presentation of foreign protein stretches, 52, 2012, 129–139, Kalnciema I, Skrastina D, Ose V, Pumpens P, Zeltins A.)

carrier accommodated well a foreign protein sequence of up to 71 aa residues in length, located it on the VLP surface, and ensured bacterial production in preparative amounts. However, after inserting the GFP gene at the same N-terminal position, the chimera was not able to form VLPs. Remarkably, the obtained PVY VLPs were stable in physiological conditions but sensitive to EDTA, high salt, and extreme pH. When mice were immunized with chimeric PVY VLPs carrying the preS1 epitope, the strong anti-preS1 immune response was found, even in the absence of any adjuvants (Kalnciema et al. 2012).

Meanwhile, the near-atomic 3D structure of both PVY virions and VLPs was resolved by electron cryomicroscopy (Kežar et al. 2019). The general PVY virion structure is presented in Figure 29.5. It was resolved to 3.4 Å and comprised the diameter of 130 Å with a left-handed helical arrangement of coats assembled around viral RNA, with 8.8 coats per turn, confirming architectural conservation between genetically unrelated potyviruses and potexviruses (Chapter 21). The extended N- and C-terminal regions were devoid of secondary structure elements. In contrast, the globular core subdomain contained seven α-helices and one β-hairpin and closely resembled the corresponding regions in the potyvirus watermelon mosaic virus (WMV) described later.

To determine how viral RNA affects PVY virion architecture, the RNA-free VLPs were produced in parallel in *E.* *coli*. While the absence of viral RNA did not substantially affect the diameter of self-assembled VLP filaments, the lengths of VLP filaments varied from 30 nm to 3.5 μm, as opposed to the approximately 730-nm-long PVY virions. The 3D structure of the PVY VLPs at 4.1-Å resolution revealed stacked-ring architecture. This nonhelical assembly of filamentous flexuous VLPs, as already indicated by negative-stain transmission electron microscopy micrographs, was unique compared to the potexviral *Alternanthera mosaic virus* (AltMV) VLPs, which resembled the helical architecture of an authentic virus (Chapter 21). This structure of the PVY VLPs is demonstrated in Figure 29.6. The PVY VLPs are assembled therefore from the octameric coat rings of 125-Å diameter, stacked along the longitudinal axis of filaments devoid of nucleic acid. The two neighboring rings are 42.7 Å apart with a -31.8° twist angle between them. By the

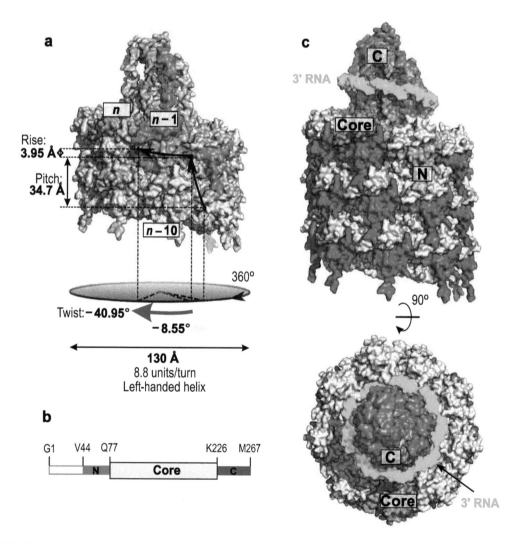

FIGURE 29.5 Structural features of potato virus Y. (a) Two turns of the PVY filament, surface representation. Colored units: CP^n (pink), CP^{n-1} (blue), and CP^{n-10} (yellow). (b) A linear representation of the CP (N-terminal region, red; core, gray; C-terminal region, blue). The empty box at the N-terminus depicts a structurally disordered part. (c) Side and top views of the PVY filament (three turns of CPs and four turns of ssRNA), surface representation; the colors are as in (b). The position of one CP unit (pink) is shown in the filament. (Adapted from Kežar A, Kavčič L, Polák M, Nováček J, Gutiérrez-Aguirre I, Žnidarič MT, Coll A, Stare K, Gruden K, Ravnikar M, Pahovnik D, Žagar E, Merzel F, Anderluh G, Podobnik M. Structural basis for the multitasking nature of the potato virus Y coat protein. *Sci Adv.* 2019;5:eaaw3808.)

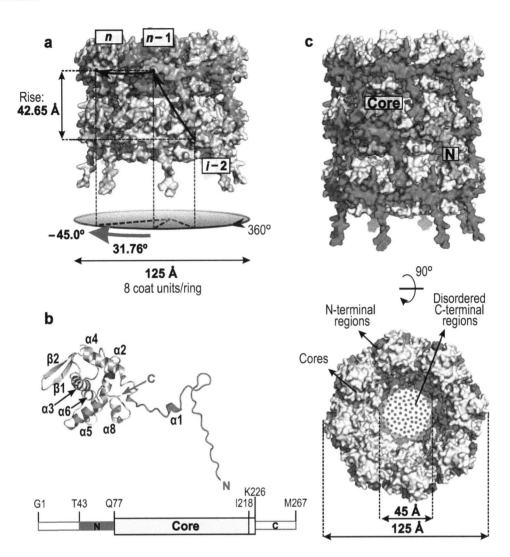

FIGURE 29.6 Structural features of the PVY VLPs. (a) Surface representation of two rings in a VLP filament with colored units CP^n (pink), CP^{n-1} (blue), and CP^{i-2} (yellow). (b) Structure of the VLP coat (N-terminal region, red; core, gray). Bottom: alinear representation of the coat, using the same colors as in the model above. Dashed boxes at N- and C-terminal regions depict structurally disordered parts. (c) Side and top views of the VLP filament; the colors are as in (b). Position of one coat unit (pink) is shown in the filament. (Adapted from Kežar A, Kavčič L, Polák M, Nováček J, Gutiérrez-Aguirre I, Žnidarič MT, Coll A, Stare K, Gruden K, Ravnikar M, Pahovnik D, Žagar E, Merzel F, Anderluh G, Podobnik M. Structural basis for the multitasking nature of the potato virus Y coat protein. *Sci Adv.* 2019;5:eaaw3808.)

structural analysis of the large set of the shortened PVY coats, it was established that the deletion of 49 N-terminal aa residues prevented filament assembly, resulting in the formation of separated octameric rings, while deletion of 69 N-terminal aa residues resulted in the monomeric coat units. On the other hand, the deletion of up to 60 C-terminal aa residues did not affect filament assembly but rather led to formation of hollow flexuous nanotubes (Kežar et al. 2019).

At last, Wi et al. (2020) reported bacterial production and in vitro self-assembly of the PVY nanofilaments for CO_2 capture with immobilized enzyme at a high density in a spatially oriented and ordered fashion along the surface of the VLPs. The appropriate filamentous VLPs were successfully formed by genetic fusion of carbonic anhydrase (hmCA) of 65.2 kDa from the marine bacterium *Hydrogenovibrio*

marinus to the N-terminus of the PVY coat of 31.1 kDa via the flexible (GGGGS)$_2$ linker. The instability of the VLPs against proteolytic degradation was circumvented by the periplasmic export of the fusion protein. The truncated form of the PVY coat coexpressed by internal translation was crucial for the successful formation of long filamentous VLPs by alleviating steric hindrance by mosaic assembly. The diameter of VLPs from the chimeric VLPs was up to 21 nm, which was much thicker than the 12 nm of the bare PVY VLPs, indicating that the bulky enzyme was successfully displayed on the VLP surface. Moreover, the chimeric VLPs were heterogeneous in diameter within a single particle, which appeared to be the result of mosaic assembly from the mixture of full-length and truncated forms of the PVY coats. It is worth noting that the displayed size of the enzyme was even larger than the size of the PVY coat and was one of the largest reported proteins genetically fused to plant viral coats. The VLPs showed CO_2 hydration activity comparable to that of free enzyme owing to the oriented and ordered immobilization of enzyme on the surface of nanofilaments. In conclusion, this fast and economic bottom-up fabrication of highly active nanobiocatalyst potentially allowed the nanofilaments to be efficiently used and recovered in putative biocatalytic and biosensor systems (Wi et al. 2020).

Soybean Mosaic Virus

Liu et al. (1993) were the first who expressed the full-length coat gene of soybean mosaic virus (SMV) in *E. coli*, while the appearance of VLPs was not reported. Kendall et al. (2008) used fiber diffraction, electron cryomicroscopy, microscopy, and scanning transmission electron microscopy to determine the symmetry of SMV, in parallel with the structure of a potexvirus, namely PVX (Chapter 21), as a reference. As a result, the low-resolution structure of both viruses was obtained. It was concluded that these viruses, and, by implication, most or all flexible filamentous plant viruses, share a common coat protein fold and helical symmetry, with slightly fewer than nine subunits per helical turn (Kendall et al. 2008).

Sugarcane Mosaic Virus

The sugarcane mosaic virus (SCMV) coat was expressed in *E. coli* by Smith et al. (1995) in the form of an N-terminal fusion with the *MalE* maltose binding protein (MBP), while the viral coat was separated by the protease factor X_a cleavage site. The fusion protein was extracted from the bacterial lysate by amylose resin column affinity chromatography and the two domains separated by the factor X_a proteolysis. The SCMV coat protein was then purified from the maltose binding protein by ion exchange chromatography in denaturing conditions, namely in buffer-containing 6 M urea. The purified SGMV induced high titers of specific antibodies in rabbits. However, a putative ability of the bacterially produced SCMV coat to form VLPs was not addressed in this study.

Tobacco Etch Virus

Although not directly connected with the VLP story, it should be noted that the first potyvirus-based vector system for the foreign gene expression in planta was elaborated on tobacco etch virus (TEV) by Dolja et al. (1992), after the efficient vector systems elaborated earlier in the 1980s and based on brome mosaic virus (BMV), order *Martellivirales*, family *Bromoviridae* (Chapter 17); tobacco mosaic virus (TMV), order *Martellivirales*, family *Virgaviridae* (Chapter 19); and potato virus X (PVX), order *Tymovirales* (Chapter 21). After TEV, other potyviruses followed as subjects by the elaboration of the in planta expression vectors, namely plum pox virus (PPV; Guo et al. 1998), lettuce mosaic virus (LMV; German-Retana et al. 2000), clover yellow vein virus (ClYVV; Masuta et al. 2000), pea

seed-borne mosaic virus (PSbMV; Johansen et al. 2001), zucchini yellow mosaic virus (ZYMV; Arazi et al. 2001b, 2002; Hsu et al. 2004), turnip mosaic potyvirus (TuMV; Beauchemin et al. 2005; Chen et al. 2007; Touriño et al. 2008), and potato virus A (Kelloniemi et al. 2008). From the general point of development of the advanced nanotechnological applications, the well-known and universally used TEV protease (Kapust and Waugh 2000) is also worth mentioning here.

Voloudakis et al. (2004) expressed the TEV coat gene in *E. coli*. The TEV coats from which 28, 63, or 112 aa residues were deleted from the N-terminus polymerized into the potyvirus-like VLPs, but these structures were more rigid and progressively smaller in diameter than those produced by the full-length TEV coat. The coat lacking 175 N-terminal aa failed to assemble into VLPs. It was concluded that the coat fragment containing aa stretch 131 to 206 was necessary and sufficient to assemble the TEV VLPs in *E. coli*.

Manuel-Cabrera et al. (2012) applied the TEV virions as a model for chemical coupling. Using a biotin-tagged molecule that reacted specifically with amino groups, it was found that the TEV coat possessed amino groups on its surface available for coupling to other molecules via crosslinkers. Moreover, the authors evaluated the immune response when the TEV virions were administered alone intraperitoneally to mice and concluded that the model possesses definite potential to be employed as a vaccine adjuvant when chemically coupled to antigens of choice. Next, Manuel-Cabrera et al. (2016) moved from the virions to VLPs and expressed the TEV coat gene in *E. coli* in two versions, namely as a wild-type variant and as a C-terminally His$_6$-tagged version. Although both versions were expressed in the soluble fraction of *E. coli* lysates, only the His-TEV coat self-assembled into the micrometric flexuous filamentous VLPs—some of them reaching 2-μm length—enabled high yields, and facilitated purification of the TEV VLPs. To evaluate the adjuvanting capacity of the His-TEV VLPs, they were administered to mice together with the chimeric protein comprising GP3 and GP4 epitopes, and GP5 and M ectodomains of porcine reproductive and respiratory syndrome virus (PRRSV) from the order *Nidovirales* (Chapter 26), as well as thioredoxin as a fusion partner. In fact, the His-TEV VLPs changed the IgG2/IgG1 ratio against the PRRSV chimeric protein, when compared to the potent IgG1 response induced by the protein alone, suggesting therefore that the adjuvant could modulate the immune response against a soluble nonassembled antigen.

Zapata-Cuellar et al. (2021) used the TEV VLPs for the noncovalent coupling of full antigens, where coiled-coil heterodimerization motifs were employed. To achieve this, the complementary heterodimerization motifs were added to the TEV coat and GFP, the proteins were expressed and purified separately, and, finally, TEV VLPs were coupled to GFP by mixing, avoiding therefore translational fusions and chemical coupling and simplifying the whole process dramatically.

Turnip Mosaic Virus

Sánchez et al. (2013) were the first ones to exploit the chimeric virions derived from turnip mosaic potyvirus (TuMV) as nanoscaffolds for peptide presentation by the infection of plants with the appropriate chimeric viruses. For this purpose, the 20-aa peptide derived from human vascular endothelial growth factor receptor 3 (VEGFR-3) was fused to the N-terminus of the TuMV coat between the first and second aa and expressed in the previously mentioned TuMV-derived vector described earlier by Touriño et al. (2008) in *Arabidopsis thaliana* or Indian mustard *Brassica juncea*. The chimeric virions that held the VEGFR-3 peptide were purified and used to immunize mice, whose sera showed log increases of antibodies against the VEGFR-3 peptide when compared with mice immunized with the peptide alone. Then, the purified chimeric virions also showed log increases in their ability to detect VEGFR-3 antibodies in sera, when used as reagents in the appropriate ELISA assays (Sánchez et al. 2013). The fine methodology of how to generate the chimeric TuMV virions presenting a foreign peptide, using a VEGFR-3 peptide as an example, produce them in plants, and purify them was published by Sánchez and Ponz (2018). Nonetheless, some peptides hindered the ability of the virus to infect host plants. Thus, González-Gamboa et al. (2017) found that the fusion of the 14-aa peptide derived from the human thrombin receptor (TR), spanning aa 42–55, inhibited TuMV infectivity. To avoid this problem, the authors moved from the chimeric virions to the chimeric VLPs. The latter were produced in *N. benthamiana* plants by transient high-level expression of wild-type or chimeric coat protein—without virus replication—and observed by electron microscopy. The overall architecture of the chimeric TuMV-TR VLPs retained characteristics of a long flexuous potyvirus, 15–20 nm in width, and was like that of the wild-type TuMV VLPs. However, both wild-type and TR-carrying VLPs were more heterogeneous in length, displaying particles as small as 100 nm, up to 2000 nm, in contrast to the monodisperse appearance of TuMV virions observed earlier by Sánchez et al. (2013). Also, both sorts of the VLPs were empty, and their thermal stability appeared to be lower in comparison with virions (González-Gamboa et al. 2017).

Yuste-Calvo et al. (2019b) engineered the TuMV VLPs with the N-terminal addition of an epitope, described previously to be involved in inflammation processes and autoimmune diseases such as multiple sclerosis, namely the chaperonin Hsp60 peptide, which comprised aa stretch 301–320 of human Hsp60, which is conserved also in mice. The TuMV VLPs were produced in *N. benthamiana*, reached about 700 nm in length by the electron microscopy data, and allowed the quantitative detection of anti-Hsp60 autoantibodies in the in vivo model of intestinal inflammation induced by dextran sodium sulfate (Yuste-Calvo et al. 2019b).

Cuenca et al. (2016) used the TuMV virions as scaffolds for chemical coupling. The lipase B of *Candida antarctica* was conjugated onto amino groups of the external viral surface by glutaraldehyde as a conjugating agent. The appropriate TuMV nanonets were formed, with large enzyme aggregates deposited. The enzyme remained active in the nanoimmobilized form, even gaining an increased relative specific activity, as compared to the nonimmobilized form. Yuste-Calvo et al. (2019a) elaborated the general strategy of the TuMV chemical modification and evaluated the capacity to conjugate different molecules to the active natural residues, obtaining the multiderivatized VLPs by combination with the genetic fusions. This approach is illustrated in Figure 29.7. The modification strategy was based on the recent 3D structure obtained by electron cryomicroscopy for both the TuMV virions and VLPs to final resolutions 5 and 8 Å, respectively (Cuesta et al. 2019). In contrast to the previously described situation in PVY, the TuMV VLPs produced in *N. benthamiana* plants conserved the helical architecture of the virion. However, the absence of the single-stranded RNA precluded the interaction between coat subunits mediated by the N-terminal arm (Cuesta et al. 2019).

Watermelon Mosaic Virus

A bit earlier than for the previously described TuMV virions and VLPs, the electron cryomicroscopy was resolved to 4.0 Å for the watermelon mosaic virus (WNV) virions from infected squash plants (Zamora et al. 2017). It should be noted that WMV is a potyvirus with one of the widest host ranges and is present worldwide, infecting more than 170 different plant species from 26 families, including cucurbits and legumes. The WMV atomic model showed the conserved helical fold typical for other flexible filamentous plant viruses, including a universally conserved RNA binding pocket, which could be regarded as a potential target for antiviral compounds. Moreover, the authors concluded that this conserved fold of the coat protein is widely distributed in eukaryotic viruses and is also shared by nucleoproteins of enveloped viruses with segmented negative-sense single-stranded RNA genomes, including influenza viruses from the order *Articulavirales* (Chapter 33).

Zucchini Yellow Mosaic Virus

A set of chimeric virions were engineered for zucchini yellow mosaic virus (ZYMV) as a live scaffold (Arazi et al. 2001a). First, the N-terminal truncated versions of the coat by 8, 13, and 33 aa residues demonstrated systemic infectivity and produced symptoms similar to those of the natural virus. The tagging of these deletion mutants with either human 16-aa c-Myc peptide ASEQKLISEEDLGS or His$_6$ tag maintained viral infectivity, as the addition of these peptides to the intact ZYMV coat. In contrast to the c-Myc epitope fusion, addition of the 16-aa immunogenic epitope from foot-and-mouth disease virus (FMDV) did not permit systemic infection. However, the fusion of the c-Myc peptide to the N-terminus of the FMDV peptide restored the capability of the virus to spread systemically. All fused peptides were exposed on the virion surface, masking the

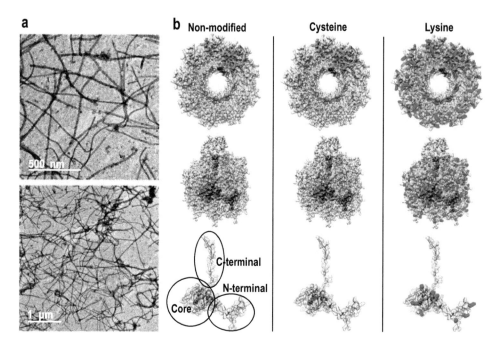

FIGURE 29.7 TuMV structural aspects. (a) TEM images of purified virion particles (upper panel) or VLPs (bottom panel) with no genetic or chemical modifications (wild type). (b) Structural TuMV model based on the WMV structure: two assembled particle representations (3 turns, monomer in gray) and one of the CP monomer (bottom). The monomer representation allows us to distinguish the core region and both amino and carboxy domains. These representations are repeated marking the cysteine residue (blue) and the lysines (red), three of them at the represented region of the amino domain (Lys69, Lys74, and Lys82; other 12 are not visible) exposed on the viral nanoparticle surface. (Reprinted from Yuste-Calvo C, González-Gamboa I, Pacios LF, Sánchez F, Ponz F. Structure-based multifunctionalization of flexuous elongated viral nanoparticles. *ACS Omega.* 2019a;4:5019–5028.)

natural immunogenic determinants of the ZYMV coat (Arazi et al. 2001a).

Pille et al. (2013) developed a novel approach allowing the noncovalent assembly of proteins on the ZYMV virion scaffolds. The antibody-binding peptide z33 was genetically fused to the monomeric yellow fluorescent protein and 4-coumarate:CoA-ligase 2. This z33 tag allowed their patterning on the surface of the ZYMV virions by means of specific antibodies directed against the viral coat. The coverage efficiency was about 87%, while the fluorescence and enzymatic activity were fully retained after assembly. The authors regarded the principle using combination of the scaffold-specific antibody and z33-fusion proteins as an efficient solution to a wide variety of proteins/enzymes and antigenic scaffolds to support coupling for creating, for example, functional biochips with optical or catalytic properties (Pille et al. 2013).

30 Other Positive Single-Stranded RNA Viruses

The one thing that matters is the effort.

Antoine de Saint-Exupery

ORDER *STELLAVIRALES*

FAMILY *ASTROVIRIDAE*

According to the current taxonomy (ICTV 2020), the order *Stellavirales* is one of the two members of the class *Stelpaviricetes*, together with the *Patatavirales* order described in the neighboring Chapter 29. The *Stelpaviricetes* class belongs to the *Pisuviricota* phylum from the kingdom *Orthornavirae*, realm *Riboviria*. The small *Stellavirales* order currently involves a single family *Astroviridae* with two genera *Avastrovirus* and *Mamastrovirus* and 22 species altogether. Human astrovirus (HAstV) is an important cause of acute gastroenteritis.

Figure 30.1 presents a portrait of typical representative and schematic cartoons of an astrovirus. The virions are spherical nonenveloped T = 3 icosahedrons of 28–30 nm in diameter, according to the most recent ICTV report (Bosch et al. 2012). A distinctive five- or six-pointed star is discernible on the surface of about 10% of virions. The virions derived from cell culture are up to 41 nm in diameter, with well-defined surface spikes. The surface projections are small, and the surface appears rough, where spikes are protruding from the 30 vertices. As shown in Figure 30.2, the astrovirus genome is arranged in three ORFs: ORF1a and ORF1b at the 5' end encoding the nonstructural proteins and ORF2 at the 3' end encoding the structural proteins (Bosch et al. 2012). The VP90 capsid precursor protein undergoes C-terminal cleavages by host caspases to generate VP70 during virus maturation as 180 copies of VP70 per particle. The infectious particles are generated by further cleavages of VP70 by extracellular proteases resulting in three structural proteins, VP34, VP27, and VP25 (Dryden et al. 2012).

The electron microscopy of negatively stained astrovirus showed pentagonal and hexagonal contours, possibly corresponding to projections of icosahedral particles, with well-defined spikes/projections protruding from the viral surface (Risco et al. 1995). Later, a low-resolution electron cryomicroscopy image of astrovirus revealed a rippled, solid capsid shell of 33 nm in diameter that was decorated with 30 dimeric spikes extending 5 nm from the surface (Méndez and Arias 2007). The structure suggested that VP34 built up the continuous capsid shell, while VP27/29 and VP25/26 formed the dimeric projections seen in the reconstruction image. Dong et al. (2011) reported the crystal structure

of the HAstV-8 dimeric surface spike resolved to 1.8-Å resolution. The overall structure of each spike/projection domain had a unique three-layered β-sandwich fold, with a core six-stranded β-barrel structure that was also found in the capsid protrusions of hepatitis E virus (HEV) from the *Hepelivirales* order (Chapter 16), suggesting a close phylogenetic relationship between these two viruses. Like in HEV, the astrovirus projection domain formed stable dimers in both solution and crystal. Based on the HEV capsid model, the homology modeling was performed, resulting in the T = 3 astrovirus capsid model with features remarkably similar to those observed in the electron cryomicroscopy reconstruction. This structure is shown in Figure 30.3. Moreover, Toh et al. (2016) solved the crystal structure of VP9071–415 (aa 71 to 415 of VP90) of HAstV-8 at 2.15 Å resolution. The VP9071–415 encompassed the conserved N-terminal domain of VP90 but lacked the hypervariable domain, which formed the capsid surface spikes. The structure of the VP9071–415 was comprised of two domains: an S domain, which adopted the typical jelly-roll β-barrel fold, and a P1 domain, which formed the previously mentioned squashed β-barrel that was observed in the HEV capsid structure. The fitting of the VP9071–415 structure into the electron cryomicroscopy maps of HAstV produced an atomic model for a continuous, T = 3 icosahedral capsid shell (Toh et al. 2016).

York et al. (2015) reported the crystal structures of the two main structural domains of the HAstV-1 coat: the core domain at 2.60 Å resolution and the spike domain at 0.95 Å resolution. The fitting of these structures into the 25 Å-resolution electron cryomicroscopy density maps of HAstV allowed to characterize the molecular features on the surfaces of immature and mature T = 3 HAstV-1 particles. The latter is shown in Figure 30.4. The mapping of conserved aa residues onto the HAstV-1 capsid core and spike domains in the context of the immature and mature HAstV particles revealed dramatic changes to the exposure of conserved residues during virus maturation. Remarkably, the antibodies raised against mature HAstV were reactive to both the HAstV-1 capsid core and spike domains, revealing therefore that the core domain was antigenic (York et al. 2015). Furthermore, Bogdanoff et al. (2017) reported the structure of the HAstV-1 capsid spike domain bound to a neutralizing monoclonal antibody.

DuBois et al. (2013) used limited proteolysis to isolate a stable fragment of the turkey astrovirus 2 (TAstV-2) capsid corresponding to the surface-exposed spike domain and determined its crystal structure to 1.5-Å resolution. Surprisingly, the overall TAstV-2 capsid spike structure

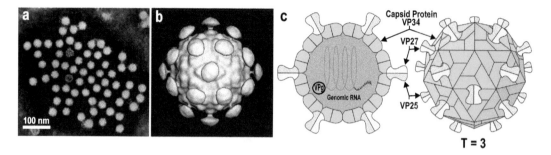

FIGURE 30.1 The portraits and schematic cartoons of the representatives of the order *Astrovirales*. (a) The negative contrast electron micrograph of virions of human astrovirus from a stool specimen. (b) The surface-shaded view of human astrovirus 1 (HAstV-1) reconstructed from 367 images recorded by electron cryomicroscopy and calculated at 30 Å resolution. The map displays a diameter of 440 Å and T = 3 icosahedral symmetry. (Courtesy of M. Yeager and International Committee on Taxonomy of Viruses (ICTV), https://talk. ictvonline.org/ictv-reports/ictv_9th_report/positive-sense-rna-viruses-2011/w/posrna_viruses/248/astroviridae-figures.) (c) The general cartoon of astroviruses taken with kind permission from the ViralZone, Swiss Institute of Bioinformatics, http://viralzone.expasy.org (Hulo C et al. 2011). (Courtesy Philippe Le Mercier.)

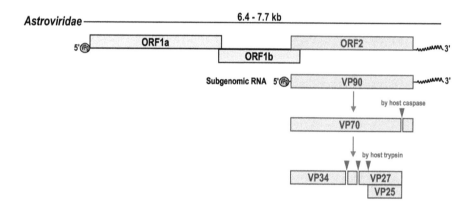

FIGURE 30.2 The genomic structure of the members of the *Astroviridae* family, taken from the ViralZone, Swiss Institute of Bioinformatics (Hulo et al. 2011). The structural genes are colored dark pink.

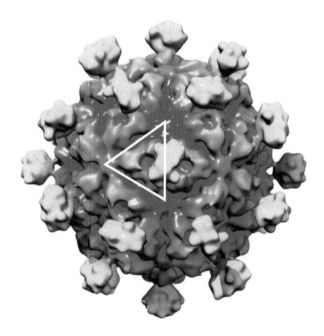

FIGURE 30.3 The T = 3 astrovirus capsid model rendered at 15-Å resolution. The white lines highlight one asymmetric unit. (From Dong J et al. Crystal structure of the human astrovirus capsid spike. *Proc Natl Acad Sci U S A*. 108:12681–12686, Copyright 2011 National Academy of Sciences, U.S.A.)

was unique, with only distant structural similarities to the HAstV-8 capsid spike and other viral capsid spikes. Arias and DuBois (2017) reviewed the current knowledge on the astrovirus capsids, combining the structural, biochemical, and virological data.

At last, Aguilar-Hernández et al. (2018) showed that the mature infectious particles were composed by only two proteins; VP34, which formed the core domain of the virus, and VP27, which constituted the 30 dimeric spikes present on the virus surface. The authors found that, during the transition of immature (90 spikes) to mature (30 spikes) virus particles, which occurred during trypsin activation, the viral protein VP25, which most likely formed the 60 spikes that were lost during maturation, detached from the virion.

Dalton et al. (2003) were the first who generated vaccinia virus recombinant expressing the structural 87-kDa polyprotein of HAstV. The formation of VLPs was observed in rhesus monkey kidney cells in the absence of other astrovirus proteins and genomic RNA. The purified trypsin-activated VLPs strongly resembled the complete astrovirus particles.

The expression of the complete ORF2 of HAstV-1 in the baculovirus system also led to the formation of VLPs of

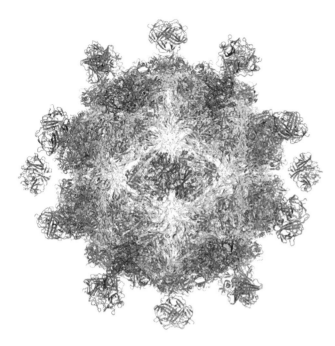

FIGURE 30.4 The model of the mature T = 3 HAstV-1 virion. (Reprinted with permission of ASM from York RL, Yousefi PA, Bogdanoff W, Haile S, Tripathi S, DuBois RM. *J Virol.* 2015;90:2254–2263.)

around 38 nm (Caballero et al. 2004). The same kind of VLPs were obtained either with the expression of a truncated form of ORF2 lacking the first 70 aa residues or with the same truncated form in which those 70 aa were replaced by GFP. All three kinds of VLPs were equally recognized by an anti-HAstV-1 polyclonal antibody and by monoclonal antibodies, indicating a nonessential role of those aa neither in the capsid assembly nor in the antigen structure. A second type of structure consisting of 16-nm ring-like units was observed in all of the cases, mostly after disassembling the 38-nm VLPs by EDTA. The removal of the EDTA and the addition of Mg^{2+} ions promoted the reassembly of the 38-nm VLPs. The 16-nm ring-like structures were suspected to be T = 1 icosahedrons (Caballero et al. 2004). Lee et al. (2013) expressed the capsid gene of an 11672 isolate of chicken astrovirus (CAstV) by the baculovirus expression system, but the CAstV VLPs were not visualized in this study.

Royuela (2010) expressed the sequence corresponding to the HAstV-2 VP25 (VP26 in the paper) in *E. coli*, where the problem of assembly was not addressed. Then, the C-terminus of the duck astrovirus (DAstV) ORF2 was expressed in *E. coli*, and the product was used in ELISA (Wang et al. 2011).

At last, a trivalent vaccine candidate should be mentioned, which was elaborated by Xia et al. (2016a, b) against astrovirus, HEV, and norovirus from the *Picornavirales* order (Chapter 27) but did not form true VLPs. This trivalent subunit vaccine was engineered by fusion together of the dimeric protruding domains of the three viruses and expression of the chimeric gene in *E. coli*. The resulting

fusion protein did not form very large aggregates as was previously expected, but it did form most likely tetramers with molecular weights of ~320 kDa, providing nevertheless the enhanced immunogenicity in mice (Xia et al. 2016a).

ORDER *OURLIVIRALES*

FAMILY *BOTOURMIAVIRIDAE*

The order *Ourlivirales* is the only member of the class *Miaviricetes*. The latter belongs to the *Lenarviricota* phylum from the kingdom *Orthornavirae*, realm *Riboviria*. The small *Ourlivirales* order currently involves a single family *Botourmiaviridae* with 6 genera and 35 species altogether. As stated by the recent ICTV report (Ayllón et al. 2020), the members of the family *Botourmiaviridae* infect plants and filamentous fungi. The members of the genus *Ourmiavirus* are plant viruses and possess bacilliform virions of 18 nm × 30–62 nm size, which are composed of a single 23.8 kDa coat protein. The bacilliform virions of ourmiaviruses constitute a series of particles with conical ends (apparently hemi-icosahedra) and cylindrical bodies 18 nm in diameter. The bodies of the particles are composed of a series of double disks, the most common particle having two disks (particle length 30 nm), a second common particle having three disks (particle length 37 nm), with rarer particles having four disks (particle length 45.5 nm) or six disks (particle length 62 nm). The members of the other genera are nonencapsidated fungal viruses (Ayllón et al. 2020).

ORDER *WOLFRAMVIRALES*

FAMILY *NARNAVIRIDAE*

The order *Wolframvirales* is the only member of the class *Amabiliviricetes*, which belongs to the *Lenarviricota* phylum, together with three other classes inluding the class *Leviviricetes*, the host of the RNA phages, from the kingdom *Orthornavirae*, realm *Riboviria*. Therefore, the narnaviruses are interesting as the far relatives of RNA phages from the orders *Norzivirales* and *Timlovirales*, the favorite VLP model (Chapter 25). The extremally small *Wolframvirales* order currently involves the single family *Narnaviridae* with the single genus *Narnavirus* of two species, *Saccharomyces 20S RNA narnavirus* and *Saccharomyces 23S RNA narnavirus*. The narnaviruses are especially interesting since they contain the simplest genomes of any RNA virus, ranging from 2.3–3.6 kb and encoding only a single polypeptide that has an RNA-dependent RNA polymerase domain (Hillman and Cai 2013).

ORDER *CRYPPAVIRALES*

FAMILY *MITOVIRIDAE*

The members of the *Cryppavirales* order are the closest relatives of narnaviruses and have been included earlier in the *Narnaviridae* family (Hillman and Cai 2013).

The *Cryppavirales* order is the only member of the class *Howeltoviricetes*, which belongs to the *Lenarviricota* phylum, together with three other classes including the class *Leviviricetes*, the host of the orders *Norzivirales* and *Timlovirales*, and the previously desribed classes *Amabiliviricetes* with the *Wolframvirales* order and *Miaviricetes* with the *Ourlivirales* order. All these classes belong the kingdom *Orthornavirae*, realm *Riboviria*. The extremally small *Cryppavirales* order currently involves the single family *Mitoviridae* with the single genus *Mitovirus* with the five mitovirus species. In contrast to narnaviruses that have been found in the yeast *Saccharomyces cerevisiae* and in the oomycete *Phytophthora infestans* and are confined to the cytosol, the mitoviruses have been found only in filamentous fungi and were located in mitochondria. None identified thus far encoded a capsid protein, but their genomes are entrapped within lipid vesicles (Hillman and Cai 2013).

FAMILY *PERMUTOTETRAVIRIDAE*

GENUS *ALPHAPERMUTOTETRAVIRUS*

The novel *Permutotetraviridae* family is not assigned to any order but is governed directly by the *Orthornavirae* kingdom and involves a single genus *Alphapermutotetravirus* with two species, *Euprosterna elaeasa virus* and *Thosea asigna virus*, the former members of the *Betatetravirus* genus from the former *Tetraviridae* family and similar therefore by their 3D structure to the *Alphatetraviridae* family members, order *Hepelivirales* (Chapter 16). The Thosea asigna virus (TaV) infects larvae of *Setothosea asigna*, the major defoliating pests of oil and coconut palms. TaV is also well known because of its 2A sequence that is widely used, along with the foot-and-mouth-disease virus (FMDV) 2A sequence, in multicistronic vectors by protein engineering. The quite short 2A sequences of 54–60 nucleotides are *cis*-acting hydrolase elements that mediate a ribosomal skip between 2A-linked genes, resulting in stoichiometric protein production that can improve transgene expression and function, as explained for example by Arber et al. (2013).

As follows from the family name, the *Permutotetraviridae* virions exhibit T = 4 icosahedral shell quasi-symmetry, and they are nonenveloped, roughly spherical, and about 40 nm in diameter (Dorrington et al. 2012). The advanced electron microscopy analysis of TaV was performed by Pringle et al. (1999). The TaV virions were characterized as the true T = 4 particles of 38 nm in diameter, more closely related by phylogenetic analysis of the capsid proteins to Nudaurelia capensis β virus (NβV) than to Nudaurelia capensis ω virus (NωV) described in detail earlier by the *Hepelivirales* order (Chapter 16).

Pringle et al. (2001) addressed the TaV capsid precursor protein, which is cleaved twice to generate three proteins. Two of the proteins, L (58.3 kDa) and S (6.8 kDa), were incorporated into the TaV virion. The third, nonstructural protein, produced from the N terminus of the precursor protein, was up to 17 kDa in size and was of unknown function. The production of

VLPs by the baculovirus expression system was used to analyze the capsid processing strategy employed by TaV. The VLPs were formed in both the presence and absence of the 17 kDa N-terminal region of the capsid precursor. The VLPs were not formed, however, when the L and S regions were expressed from separate promoters, indicating that cleavage between the L and S capsid proteins was an essential part of the TaV capsid assembly. The expression of the TaV 17 kDa protein in bacteria did not produce intracellular tubules similar to those formed by bacterial expression of the p17 protein from Helicoverpa armigera stunt virus (HaSV) of the *Hepelivirales* order, as mentioned in Chapter 16.

SATELLITE VIRUSES

ESSENTIALS

The following paragraphs are dealing with four novel genera and one family of satellite viruses that were invented into viral taxonomy by the opportune initiative of Krupovic et al. (2016) on the satellite classification in order to overcome the unsubstantiated separation of satellite viruses from the remainder of the viral world. Thus, the four new genera—*Albetovirus*, *Aumaivirus*, *Papanivirus*, and *Virtovirus*—were created for positive-sense single-stranded plus-RNA satellite viruses that infect plants and the family *Sarthroviridae*, including the genus *Macronovirus*, for the plus-RNA satellite viruses that infect arthropods. These classification units are directly governed by the realm *Riboviria*, without any intermediate steps.

The satellite viruses encode structural proteins required for the formation of infectious particles but depend on helper viruses for completing their replication cycles. This is especially intriguing, since many RNA viruses such as endornaviruses, hypoviruses, narnaviruses, and umbraviruses do not form virions. Krupovic et al. (2016) reasonably argued that all nucleic-acid-containing nonorganismal entities that encode their own capsid proteins are to be classified within proper viral taxa, regardless of whether or not they depend on another virus for replication.

All of the plant satellite viruses are packed into small capsids exhibiting T = 1 icosahedral symmetry, as reviewed early on by Ban et al. (1995). The corresponding virions are constructed from 60 copies of the coat protein, which adopts the classical jelly-roll topology. Based on sequence similarity, Krupovic et al. (2016) classified the plant satellite viruses into the four groups and proposed the four genera that are described in the following sections.

GENUS *ALBETOVIRUS*

This genus involves three species of satellite viruses, *Tobacco albetovirus 1, 2,* and *3*. These satellite virus particles are found in plant hosts in association with taxonomically diverse helper viruses. The T = 1 isometric particles are about 17 nm in diameter. The capsid consists of 60 copies of a single protein of 17–24 kDa, which is encoded by the positive-sense ssRNA satellite virus genome as the only ORF. In brief, the linear STNV genome consists of 1,239 nucleotides and encodes a single coat protein of 195 aa, which is necessary and sufficient for virion formation. The coats of the three characterized serotypes, STNV-1 (or STNV), STNV-2, and STNV-C, are ~50–63% identical in sequence, as reviewed by Krupovic et al. (2016).

The satellite tobacco necrosis virus (STNV) from the *Tobacco albetovirus 1* species played a special historical role by the development of the x-ray crystallography of viruses. The first crystallographic analysis of STNV was performed by Crowfoot and Schmidt (1945), who used very large single crystals obtained in August 1944 by N.W. Pirie of Rothamsted. After 20 years, Fridborg et al. (1965) reported extensive purification, crystallization, electron microscopy, and preliminary x-ray diffraction analysis of STNV. Liljas et al. (1982) resolved the 3D structure of STNV to 3.7 Å. Rossmann et al. (1983) compared the 3D structure of STNV with those of tomato bushy stunt virus (TBSV) from the *Tombusviridae* family of the order *Tolivirales* (Chapter 24), southern bean mosaic virus (SBMV), family *Solemoviridae*, order *Sobelivirales* (Chapter 28), and alfalfa mosaic virus, family *Bromoviridae*, order *Martellivirales* (Chapter 17) and demonstrated remarkable, although unexpected, similarity. The polypeptide folds of these viruses had greatest similarity in the β-sheet region of the eight-stranded antiparallel β-barrel, while the largest differences occurred in the connecting segments. Then, Jones and Liljas (1984) refined the structure of STNV to 2.5 Å. This structure, together with those of other satellite viruses, is shown in Figure 30.5. The model located three different metal ion sites in the protein shell, linking the protein subunits together and occupied probably by Ca^{2+} ions. It should be noted that Saldin et al. (2011) used STNV as a model to approve an advanced idea

of reconstruction of an icosahedral virion from single-particle scattering experiments by free electron lasers (FELs). Hosseinizadeh et al. (2014) recovered 3D of STNV to high resolution by the FEL approach and reconstructed the final structure to atomic level. Martín-Bravo et al. (2020) used the full-atom STNV results (Dykeman and Sankey 2008) by the developing of a minimal coarse-grained model for the low-frequency normal mode analysis of icosahedral viral capsids.

The well-known Walter Fiers team expressed the coat gene of STNV in *E. coli*, but fast degradation of the bacterially synthesized STNV coat protein was observed (van Emmelo et al. 1984). Later, Lane et al. (2011) expressed the codon-optimized gene of STNV in *E. coli* and demonstrated the in vivo assembly of the coat into VLPs closely resembling the T = 1 virions. The x-ray crystal structure of the VLP has been solved and refined at 1.4 Å resolution and shown to be very similar to that of wild-type STNV, as demonstrated in Figure 30.6. The VLPs packaged the recombinant mRNA transcript and could be disassembled and reassembled using different buffer conditions. An additional low-resolution x-ray crystal structure determination revealed well-ordered RNA fragments lodged near the inside surface of the capsid, close to basic clusters formed by the N-terminal helices that projected into the interior of the particle. The RNA consisted of multiple copies of a 3-bp helical stem, with a single unpaired base at the 3′ end, and probably consisted of a number of short stem-loops where the loop region was disordered. The arrangement of the RNA is shown in Figure 30.7.

Moreover, Bunka et al. (2011) established conditions for the in vitro disassembly and reassembly of the STNV VLPs. While the in vivo assembly was dependent on the presence of the N-terminal region coat, the in vitro assembly required RNA. The authors selected the suitable RNA aptamers, one of which matched the STNV-1 genome in 16 out of 25 nucleotide positions, including across a statistically

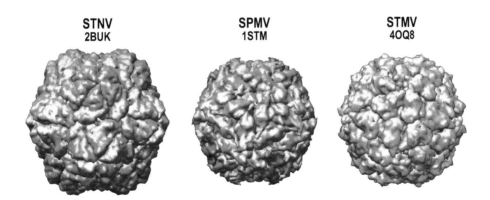

STNV
2BUK

SPMV
1STM

STMV
4OQ8

FIGURE 30.5 The crystal structure of the satellite viruses: satellite tobacco necrosis virus (STNV), genus *Albetovirus*, 2.45 Å resolution, outer diameter 196 Å (Jones and Liljas 1984); satellite panicum mosaic virus (SPMV), genus *Papanivirus*, 1.90 Å, 170 Å (Ban and McPherson 1995); satellite tobacco mosaic virus (STMV), genus *Virtovirus*, 1.45 Å, 174 Å (Larson et al. 2014). The 3D structures are taken from the VIPERdb (http://viperdb.scripps.edu) database (Carrillo-Tripp et al. 2009). The size of particles is to scale. All structures possess the icosahedral T = 1 symmetry. The corresponding protein data bank (PDB) ID numbers are given under the appropriate virus names.

a

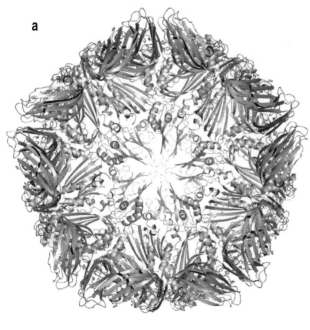

b

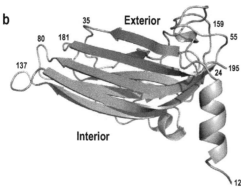

FIGURE 30.6 The 3D structure of the STNV VLPs. (a) Secondary structure representation of the refined STNV capsid, viewed along the 5-fold icosahedral axis. The "front" of the particle is cut away to show the interior. The coat subunits are colored gold, expect for the N-terminal helices that are in light gold. The clusters of three helices form protrusions from the interior surface of the capsids (figures were prepared with PyMOL: www.pymol.org/). (b) The coat subunit showing the N-terminal helix (light gold) protruding away from the domain. The loop at 137 makes contact to other subunits at the 5-fold vertices of the capsid, while the helix forms a cluster at the 3-fold axes with two neighboring subunits. (Reprinted from *J Mol Biol.* 413, Lane SW, Dennis CA, Lane CL, Trinh CH, Rizkallah PJ, Stockley PG, Phillips SE, Construction and crystal structure of recombinant STNV capsids, 41–50, Copyright 2011, with permission from Elsevier.)

significant 10/10 stretch. This ten-base region folded into a stem-loop displaying the motif ACAA and was able to bind the STNV coat. These results strongly contributed to the evaluation of the STNV assembly mechanism based on kinetically driven folding of the RNA using stem-loops that displayed versions of this motif (Dykeman et al. 2013; Ford et al. 2013; Patel et al. 2015, 2017).

Developing further the encapsidation studies, Kotta-Loizou et al. (2019) developed an in planta packaging assay based on the transient expression of the STNV-1 coat and assessed the ability of the resulting VLPs to encapsidate the mutant STNV-1 RNAs expected to have different encapsidation potential based on the in vitro studies. The results revealed, however, that >90% of the encapsidated RNAs were host-derived, although there was some selectivity of packaging for the STNV-1 RNA and certain host RNAs.

Genus *Aumaivirus*

The only species *Maize aumaivirus 1* is involved. The satellite maize white line mosaic virus (SMWLMV), a member of this species (Zhang et al. 1991), depends on maize white line mosaic virus (MWLMV) from the *Tombusviridae* family, order *Tolivirales* (Chapter 24). As reviewed by Krupovic et al. (2016), the ssRNA genome of SMWLMV is 1,168 nucleotides in length, encodes one capsid protein, and produces virions of 17 nm in diameter, but there is only limited sequence similarity between the SMWLMV capsid protein and corresponding proteins of the STNV-like viruses.

Genus *Papanivirus*

The only species *Panicum papanivirus 1* is involved. The most studied member of this species, namely satellite panicum mosaic virus (SPMV), is completely dependent for replication and systemic spread in plants on panicum mosaic virus (PMV), a member of the species *Panicum mosaic virus*, genus *Panicovirus*, family *Tombusviridae*, order *Tolivirales* (Chapter 24). As reviewed by Krupovic et al. (2016), the 826 nt-long ssRNA genome of SPMV contains two ORFs; only one of them, which encodes coat protein, was found to be expressed in the in vitro translation assays. The sequence of the SPMV coat is not appreciably similar to those of STNV-like viruses, only below 15% identity, while the x-ray structure resolution to 1.9 Å revealed the same jelly-roll fold similar to that of the STNV coat (Ban and McPherson 1995; Ban et al. 1995). This structure is shown in Figure 30.5. Furthermore, the two other viruses encoding SPMV-like coats were reported. The first one, satellite St. Augustine decline virus (SSADV), is associated with the St. Augustine decline strain of PMV, is 95% identical to SPMV over the entire genome length, and can be considered a different strain of SPMV. The second putative satellite virus, satellite grapevine virus (SGVV), was discovered by deep sequencing of total intracellular RNA from grapevine. However, neither the viral particles nor the associated helper virus have been characterized, as reviewed by Krupovic et al. (2016).

Qiu and Scholthof (2004) expressed the SPMV coat from a potato virus X (PVX) gene vector in *N. benthamiana* and concluded that, after its encapsidation function, the coat demonstrates pathogenicity in both host and nonhost plants, as well as could be interfering with suppression of gene

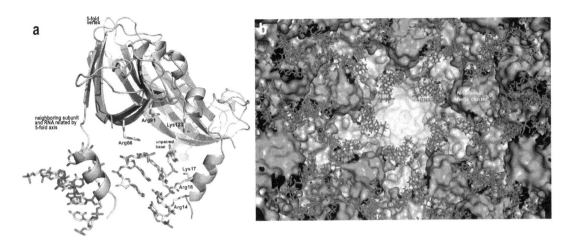

FIGURE 30.7 The packaging of RNA by the STNV VLPs. (a) The coat subunit with its associated RNA fragment. Three basic residues on the N-terminal helix (Arg14, Lys17 and Arg18) can make good hydrogen bonds to the three phosphate groups of the RNA duplex in a 1:1 relationship. Three basic residues on the internal surface of a neighboring subunit (Arg66, Arg91 and Lys123) can make contact to phosphates in a loop linking the two strands of the duplex. The neighboring subunit is related by a 5-fold axis and is shown with its associated RNA fragment. (b) The view of the inner surface of the capsid. Looking along a 5-fold axis from the center of the particle, the RNA fragments form a ring around the 5-fold, with each associated with one N-terminal helix cluster. (Reprinted from *J Mol Biol*. 413, Lane SW, Dennis CA, Lane CL, Trinh CH, Rizkallah PJ, Stockley PG, Phillips SE, Construction and crystal structure of recombinant STNV capsids, 41–50, Copyright 2011, with permission from Elsevier.)

silencing. Moreover, the coat protein gene of SPMV was used as a model by a novel primer design method for site-directed fragment deletion, insertion, and substitution (Qi and Scholthof 2008). As a result, a series of deletions, insertions, and substitutions were introduced into the SPMV coat gene.

Genus *VIRTOVIRUS*

The only species *Tobacco virtovirus 1* is involved. A representative of this species, satellite tobacco mosaic virus (STMV) was isolated from tree tobacco *Nicotiana glauca* and was naturally associated with and dependent on tobacco mild green mosaic virus (TMGMV), a member of the species *Tobacco mild green mosaic virus*, genus *Tobamovirus*, family *Virgaviridae*, order *Martellivirales* (Chapter 24). As reviewed by Krupovic et al. (2016), STMV can adapt and replicate in many plant hosts, e.g., tobacco, pepper, tomato, in association with other tobamoviruses, including tobacco mosaic virus (TMV). Thus far, STMV is the only known satellite virus that uses rod-shaped viruses as helpers. The STMV genome is a linear ssRNA molecule of 1,059 nt that contains two ORFs, one of which encodes the STMV coat that does not have identifiable homologs in sequence databases. The 3D structure showed that the STMV capsid possesses the same jelly-roll fold similar to those of STNV and SPMV, as follows from Figure 30.5, and suggests therefore the evolutionary relation of the three satellite viruses. The STMV capsid was subjected to thorough x-ray crystallography analysis (Koszelak et al. 1989; Larson et al. 1993a, b; Larson et al. 1998). At last, the x-ray structure was refined

to the 1.4 Å resolution (Larson et al. 2014). Remarkably, the STMV virion appeared as the first all-atom model for the complete structure of any virus, demonstrating 100% of the atoms and using the the complete protein and RNA sequences (Zeng et al. 2012). It should be noted that the STMV capsids served as one of the models to grow crystals in microgravity during the International Microgravity Laboratory-2 (IML-2) mission in July of 1994, using the European Space Agency's Advanced Protein Crystallization Facility (APCF; Koszelak et al. 1995). In fact, the cubic STMV crystals more than 30 times the volume of crystals grown in the laboratory were produced in microgravity. The x-ray diffraction analysis demonstrated that the STMV crystals diffracted to significantly higher resolution and had superior diffraction properties (Koszelak et al. 1995).

Family *SARTHROVIRIDAE*

This independent family is represented by a single genus involving a single species *Macrobrachium satellite virus 1*. As described in the recent ICTV report (Sahul Hameed et al. 2018), the extra small virus (XSV) from this species is a satellite virus of Macrobrachium rosenbergii nodavirus (MrNV), an unclassified virus related to members of the family *Nodaviridae*, order *Nodamuvirales*, and described in detail in Chapter 23. Both viruses have isometric, spherical virions, infect giant freshwater prawns, and together cause white tail disease (WTD), which is responsible for mass mortalities and important economic losses in hatcheries and farms. The XSV virions are spherical, about 15 nm in diameter, icosahedral in shape, and serologically unrelated to those of MrNV. The two overlapping polypeptides of about 17 kDa and 16 kDa were found in the XSV virions. The XSV genome is a linear positive-sense RNA molecule of 796 nucleotides, which, unlike other satellite viruses, contains a short poly(A) tail of 15–20 nucleotides at the 3'-end (Sahul Hameed et al. 2018).

Neethi et al. (2012) expressed the His-tagged 640-bp gene fragment encoding the capsid protein of XSV in *E. coli*. Although no self-assembly was observed, the recombinant protein induced in mice polyclonal antibodies, which reacted specifically with the recombinant protein and XSV in WTD-infected tissues. The *E. coli*-produced His-tagged nonassembled XSV capsid protein was used as an ELISA reagent (Naveen Kumar et al. 2020) and as an oral vaccine against WTD in freshwater prawn (NaveenKumar et al. 2021).

Section V

Negative Single-Stranded RNA Viruses

31 Order *Mononegavirales*

Big Brother is watching you.

George Orwell

ESSENTIALS

The order *Mononegavirales* is the second largest among other negative single-stranded RNA virus orders after the *Bunyavirales* order (see Chapter 32) by number of families, currently 11, according to the recent detailed taxonomy (ICTV 2020; Kuhn et al. 2020). This is the major order in the class *Monjiviricetes* of the subphylum *Haploviricotina*, phylum *Negarnaviricota*, kingdom *Orthornavirae*, realm *Riboviria*. The other order of the *Monjiviricetes* class, namely *Jingchuvirales*, consists of the only family *Chuviridae* with the only *Mivirus* genus of 30 insect-specific species.

The order *Mononegavirales* includes therefore in total 11 families, 7 subfamilies, 89 genera, and 422 species. The order is of great medicinal and veterinary importance. There are such well-known pathogens as Borna disease virus (BoDV) of the *Mammalian 1 orthobornavirus* species from the *Orthobornavirus* genus, family *Bornaviridae*; members of the *Ebolavirus* and *Marburgvirus* genera, family *Filoviridae*; measles virus (MV) of the *Measles morbillivirus* species and peste des petits ruminants virus (PPRV) of the *Small ruminant morbillivirus* species, both *Morbillivirus* genus; mumps virus (MuV) of the *Mumps orthorubulavirus* species, *Orthorubulavirus* genus, Menangle virus of the *Menangle pararubulavirus* species and Tioman virus of the *Tioman pararubulavirus* species, both *Pararubulavirus* genus; Nipah virus of the *Nipah henipavirus* species, *Henipavirus* genus; Newcastle disease virus (NDV) of the *Avian orthoavulavirus 1* species from the *Orthoavulavirus* genus and Sendai virus (SenV) of the *Murine respirovirus* species, as well as human parainfluenza viruses 1 and 3 belonging the *Human respirovirus 1* and *3* species, respectively, of the *Respirovirus* genus, all belonging to different subfamilies of the great *Paramyxoviridae* family; human respiratory syncytial virus (hRSV) of the *Human orthopneumovirus* species of the *Orthopneumovirus* genus and human metapneumovirus (HMPV) of the *Human metapneumovirus* species, genus *Metapneumovirus*, family *Pneumoviridae*; rabies virus (RABV) of the *Rabies lyssavirus* species, genus *Lyssavirus*; and vesicular stomatitis virus (VSV) of the *Indiana vesiculovirus* species from the *Vesiculovirus* genus, family *Rhabdoviridae*.

Figure 31.1 presents typical portraits of the *Mononegavirales* order representatives. The virions are mostly spherical, with diameters from 70–150 nm. Nevertheless, the filamentous, circular, U- or 6-shaped (the *Filoviridae* family) or bullet-shaped and bacilliform, 180 nm long and 75 nm wide (the *Rhabdoviridae* family), are common. The family *Mymonaviridae* demonstrates filamentous, potentially enveloped ribonucleocapsids with a diameter of 20 nm and 300–400 nm in length. The virions of the minor families *Lispiviridae* and *Xinmoviridae* are not observed yet, but they could presumably be enveloped.

GENOME

The *Mononegavirales* order members have linear monopartite nonsegmented RNA genome of negative polarity. The genomic RNAs possess inverse-complementary 3' and 5' termini; they are not covalently linked to a protein and remain noninfectious.

Figure 31.2 demonstrates the genomic structure of the most popular families of the order. The size of the genomes varies in wide range from 8–19 kb. The genomes demonstrate the characteristic gene order 3'-UTR—core protein genes—envelope protein genes—RNA-dependent RNA polymerase gene—5'-UTR (3'-N-P-M-G-L-5'), with some exceptions. The infectious helical ribonucleocapsids are enveloped into virions.

Therefore, as stated in Figure 31.1, irrespective of the genomic structure, the virions of the *Mononegavirales* order members can be very similar to these of arenaviruses and orthobunyaviruses of the *Bunyavirales* order (see Chapter 32) in the case of spherical paramyxoviruses, pneumoviruses, bornaviruses, and nyamiviruses but are more distinct in the case of bullet-shaped rhabdoviruses and obviously distinct in the case of long filamentous filoviruses. We shall follow this order of ribonucleocapsid shapes in the description of the VLP applications of the *Mononegavirales* order members.

The general problems of the VLP generation by the representatives of the order and other single-stranded negative-strand ssRNA viruses were reviewed earlier, for example by Diederich et al. (2015) and Pushko and Pumpens (2015).

FAMILY *PARAMYXOVIRIDAE*

According to the latest ICTV issues (Rima et al. 2019), the *Paramyxoviridae* family unites 4 subfamilies, 17 genera, and 78 species. A look of a typical paramyxovirus is shown in Figure 31.1c. The members of the family infect mammals and birds or, in some cases, reptiles and fish. Many viruses are host-specific, and several, such as measles (MV), mumps (MuV), Nipah (NiV), Hendra (HeV), and several parainfluenza viruses, are pathogenic for humans. The virus transmission is horizontal, mainly through direct contact and airborne routes; no vectors are known.

The structural proteins are nucleocapsid protein N, matrix protein M, and envelope proteins: fusion protein F

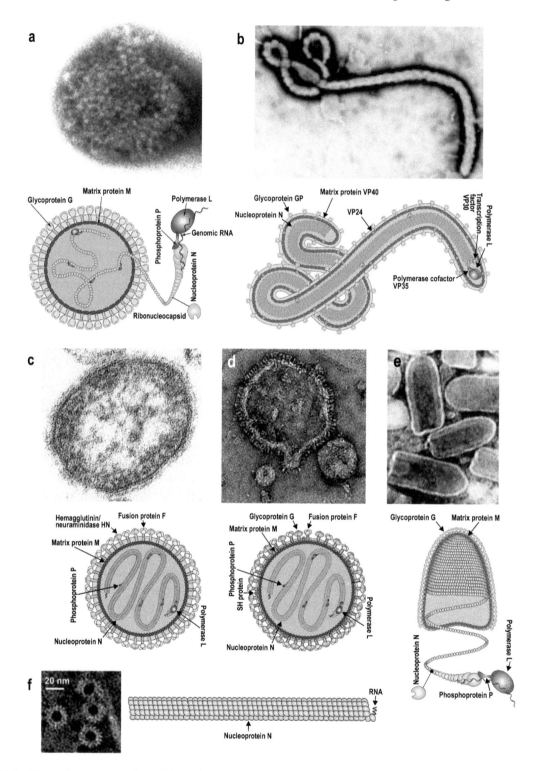

FIGURE 31.1 The major representatives of the order *Mononegavirales*. (a) families *Bornaviridae*, *Nyamiviridae*, and *Artoviridae*. The electron micrograph of Borna disease virus of *mammalian 1 orthobornavirus* species, by M. Eickmann is taken from the ICTV website. Diameter of ~100 nm; (b) family *Filoviridae*. The filamentous virions are ~790 nm long for Marburg virus and 970 nm long for Ebolavirus, with diameter of ~80 nm. The electron micrograph of an Ebolavirus is from the CDC Public Health Image Library (PHIL); (c) *Paramyxoviridae*, the electron micrograph is from the CDC PHIL; (d) families *Pneumoviridae* and *Sunviridae*. The electron micrograph of human respiratory syncytial virus (hRSV), species *human orthopneumovirus*, by Kyle Dent, Neil Ranson, and John Barr, University of Leeds, is reproduced with permission from Rima et al. (2017); (e) family *Rhabdoviridae*, the electron micrograph is from the CDC PHIL; (f) family *Mymonaviridae*. The electron micrograph demonstrates rings from nucleoprotein monomers of Sclerotinia sclerotimonavirus (Jiāng et al. 2019). The virions are filamentous, 25–50 nm in diameter, ~1,000 nm in length, and may be enveloped by a membrane, while the nucleocapsids released from virions are single, left-handed, helical structures that, when tightly coiled, have a diameter of 20–22 nm and a length of 200–2,000 nm. All cartoons of virions are reproduced with kind permission from the ViralZone, Swiss Institute of Bioinformatics (Hulo C et al. 2011). (Courtesy Philippe Le Mercier.)

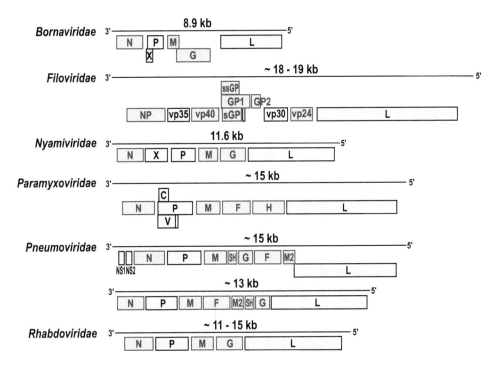

FIGURE 31.2 The genomic structure of the major families of the order *Mononegavirales*. The structural genes are colored dark pink. The sequence data is compiled from the NCBI browser. In the case of the *Pneumoviridae* family, the genetic maps of representatives of the genera *Orthopneumovirus* (mentioned earlier) and *Metapneumovirus* (mentioned later) are shown.

and receptor binding protein (RBP) designated variably as haemagglutinin–neuraminidase (HN), haemagglutinin (H), or glycoprotein (G).

Of four subfamilies belonging to the *Paramyxoviridae* family, the smallest *Metaparamyxovirinae* subfamily consists of a single species as a member of a single genus, namely *Synodovirus*, which was not addressed yet by the VLP technique.

SUBFAMILY *AVULAVIRINAE*

The subfamily consists of 3 genera with 22 species. Newcastle disease virus (NDV) of the *Avian orthoavulavirus 1* species, *Orthoavulavirus* genus, causes highly contagious disease of birds, which is transmissible to humans. Transmission occurs by exposure to fecal and other excretions from infected birds and through contact with contaminated food, water, equipment, and clothing.

The avulaviral VLPs were the first produced in standard expression systems among representatives of paramyxoviruses. Thus, this early report described the expression of the NDV HN protein in a baculovirus-driven insect cell model and presented clear electron microscopic observation of NDV-like particles (Nagy et al. 1991).

The electron cryotomography disclosed general features of the NDV virions (Battisti et al. 2012). The authors combined the electron cryotomography with x-ray crystallography and clearly revealed dimers of the matrix protein M that assembled into pseudotetrameric arrays generating the membrane curvature necessary for virus self-assembly,

anchoring the glycoproteins in the gaps between the matrix proteins, and associating the helical nucleocapsids, but about 90% of virions lacked matrix arrays (Battisti et al. 2012).

The high-resolution crystallographic structure of the fusion protein F was obtained for NDV by Chen et al. (2001).

The high-level production of the nucleocapsid-like, or so-called herringbone-like, particles of approximately 20–22 nm in diameter and variable length from 50–100 nm was detected after expression of the NDV protein N in insect cells and *E. coli* (Errington and Emmerson 1997).

The matrix protein M of NDV was expressed in its native form in *S. cerevisiae* and also in *E. coli*, after refolding in vitro (Iram et al. 2014).

The crystal structure of the NDV hemagglutinin–neuraminidase HN and its stalk region was resolved by Crennell et al. (2000) and Yuan et al. (2011).

The expression results are summarized in Table 31.1. The coexpression of all possible combinations of N, M, F, and HN of NDV was performed in avian cells and resulted in the release of VLPs with densities and efficiencies of release similar to those of authentic virions (Pantua et al. 2007). Expression of the protein M alone—but not the N, F, or HN proteins individually—resulted in the efficient VLP release, while expression of all different combinations of proteins in the absence of the protein M did not result in VLP release (Pantua et al. 2007).

The NDV VLPs directed the attachment to cell surfaces and induced immune response in mice that was comparable to the response to equivalent amounts of inactivated NDV vaccine virus (McGinnes et al. 2010).

Concerning construction of chimeric VLPs, the first examples of the insertion of epitopes into avulaviral structural proteins are related to the protein N as a scaffold of the nucleocapsid-like, or ringlike, or herringbone-like, particles. First, chimeric N proteins of NDV were constructed in which the antigenic regions of the NDV HN and F proteins, myc epitope, and a His$_6$ tag were linked to the C-terminus of the N monomer, expressed in *E. coli* and self-assembled into ringlike structures (Rabu et al. 2002). Second, six fragments of the protein VP1 of enterovirus EV71 were fused to the C-terminal end of the full-length or truncated protein N of NDV with further expression of chimeric proteins in *E. coli* (Sivasamugham et al. 2006). The chimeric proteins self-assembled into intact ringlike structures and induced a strong immune response against the complete EV71 VP1 in rabbits (Sivasamugham et al. 2006), mice (Ch'ng et al. 2011), and hamsters (Ch'ng et al. 2012).

Avulaviruses possess a long and successful history of generation of viral vectors for the expression of foreign proteins, including foreign glycoproteins and/or chimeric glycoproteins consisting of foreign glycoprotein fragments inserted into authentic glycoproteins and acting as vaccine and gene therapy candidates. Thus, chimeric NDV and human parainfluenza virus 3 (Deng et al. 1997) and NDV and avian paramyxovirus (Peeters et al. 2001) glycoproteins HN could be mentioned among the earliest examples of the vaccine and virus delivery approach based on the construction of mixed envelopes.

The NDV virus expressing the hemagglutinin H5 protected chickens against avian H5N1 influenza, as well as NDV challenge (Nayak et al. 2009). Then, NDV-based chicken infectious laryngotracheitis virus vaccine was constructed (Zhao et al. 2014).

The authentic VLP approach was realized by the construction of a chimeric hRSV vaccine candidate based on the NDV VLPs formed by the proteins M and N and carrying a chimeric protein that contained the cytoplasmic and transmembrane domains of the NDV HN protein and the ectodomain of the hRSV protein G (Murawski et al. 2010). Further, the NDV-based hRSV vaccine candidate was improved by addition of the ectodomains of both hRSV F and G proteins fused to the transmembrane and cytoplasmic domains of NDV F and HN proteins, respectively (McGinnes et al. 2011). The vaccine candidate was purified from avian cells and demonstrated complete protection of mice from hRSV replication in lungs (McGinnes et al. 2011) and efficient stimulation of long-lived hRSV-specific, T-cell-dependent secretion of neutralizing antibodies and hRSV-specific memory response (Schmidt MR et al. 2012). Next, an NDV-based influenza vaccine candidate was constructed by the expression of VLPs composed of the proteins M1 and HA from avian influenza virus (AIV) and a chimeric protein containing the cytoplasmic and transmembrane domains of AIV neuraminidase protein NA and the ectodomain of the NDV protein HN (Shen et al. 2013). A single immunization of chickens with the chimeric VLPs induced both AIV H5- and NDV-specific antibodies and conferred

complete protection against NDV challenge (Shen et al. 2013). Next, the NDV VLPs containing the NDV protein F along with the influenza virus matrix protein M1 were produced in insect cells and protected chickens against lethal NDV challenge (Park JK et al. 2014). Furthermore, the dendritic cell-binding peptide-decorated chimeric VLPs containing NDV HN and AIV HA were developed (Xu X et al. 2020).

Moreover, the VLPs based on NDV M and N proteins were engineered by fusing the gp350 ectodomain—or receptor binding domain RBD—of Epstein-Barr virus (EBV) with the NDV F protein (Ogembo et al. 2015). These EBV-NDV VLPs, efficiently produced in CHO cells, were similar in diameter and shape to EBV virions and elicited a long-lasting neutralizing antibody response in mice.

As mentioned in Chapter 2, the NDV VLP platform was successfully used to generate vaccines against Epstein-Barr virus (EBV) infection (Perez EM et al. 2017) and for the prevention of infection and its associated malignancies by Kaposi sarcoma-associated herpesvirus (KSHV; Barasa et al. 2017). At last, Yang Y et al. (2021) used the NDV VLP platform to display the SARS-CoV-2 spikes and achieve potent virus-neutralizing responses in mice.

It is also noteworthy that the NDV-based constructions were used as oncolytic viruses (Zamarin and Palese 2012).

Subfamily Orthoparamyxovirinae

The subfamily consists of 8 genera with 34 species. Regarding public healthcare, the most important and dangerous members of the orthoparamyxoviruses are Nipah virus of the *Henipavirus* genus, measles virus of the *Morbillivirus* genus, and human parainfluenza viruses (PIVs) 1 and 3 belonging to the *Respirovirus* genus. The orthoparamyxoviruses remain a serious burden of animal and poultry husbandry. Thus, peste des petits ruminants virus (PPRV) of the *Morbillivirus* genus causes disease affecting goats and sheep with up to 80% mortality rate in acute cases, and members of the *Atlantic salmon aquaparamyxovirus* species of the *Aquaparamyxovirus* genus are one of the causes of proliferative gill inflammation of salmons that leads to considerable losses in fishery. Sendai virus of the *Respirovirus* genus is responsible for a highly transmissible respiratory tract infection in mice and occasionally in pigs. Importantly, the rinderpest, or cattle plague, caused by rinderpest virus (RPV) of the *Rinderpest morbillivirus* species, genus *Morbillivirus*, was officially proclaimed by the UN Food and Agriculture Organization as fully eradicated, making it the second eliminated disease after smallpox in world history (*A world without rinderpest* 2014).

Concerning 3D structure of orthoparamyxoviruses, the electron cryotomography at low resolution has revealed general features of the virions of Sendai (Loney et al. 2009) and measles (Liljeroos et al. 2011) viruses.

The helical structure of the nucleocapsids formed by the protein N was discovered for measles virus at a relatively low resolution, not higher than 12 Å (Bhella et al.

2004; Schoehn et al. 2004). Next, NMR was employed to improve resolution of the measles virus nucleocapsid structure (Barbet-Massin et al. 2014). At last, electron cryomicroscopy and single particle-based helical image analysis were used to determine the structure of the helical nucleocapsid formed by the folded domain of the measles virus nucleoprotein encapsidating an RNA at a resolution of 4.3 Å (Gutsche et al. 2015).

The high-resolution crystallographic structure of the fusion protein F was obtained for measles virus (Zhu et al. 2003) and Nipah and Hendra viruses (Lou et al. 2006). The crystal structure of the hemagglutinin–neuraminidase HN and its stalk region was resolved for the hemagglutinin H of measles virus (Hashiguchi et al. 2007) and its complex with the CD46 receptor (Santiago et al. 2010), the attachment glycoprotein G of Nipah and Hendra viruses (Bowden et al. 2008; Xu K et al. 2008, 2013). Furthermore, the crystal structure of the Nipah virus multimeric phosphoprotein P that tethers the viral polymerase to the nucleocapsid was resolved (Bruhn et al. 2014).

The structure of unassembled protein N in complex with its viral chaperone was resolved for Nipah virus (Yabukarski et al. 2014). Recently, the electron cryomicroscopy structure of the Nipah nucleocapsid protein-RNA assembly was resolved at near-atomic resolution (Ker et al. 2021).

Concerning production of orthoparamyxoviral VLPs in standard expression systems, production of VLPs was confirmed in many expression systems very soon for the protein N, as indicated in Table 31.1. First, production of nucleocapsid-like structures was shown by the vaccinia virus-driven expression of the measles virus protein N in avian and mammalian cells (Spehner et al. 1991). Second, high-level production of the herringbone-like particles was detected after baculovirus-driven expression in insect cells for the protein N of measles virus (Fooks et al. 1993), Nipah virus (Eshaghi et al. 2005), and dolphin morbillivirus (Grant et al. 2010). Third, the same herringbone-like particles were formed also by the expression in *E. coli* in the case of the protein N of measles virus (Warnes et al. 1995), RPV and PPRV (Mitra-Kaushik et al. 2001), and Nipah virus (Tan et al. 2004). Fourth, the most impressive results on the expression of the protein N in the form of the herringbone-like structures were achieved in yeast by the Kęstutis Sasnauskas team. So, both *S. cerevisiae* and *P. pastoris* were used for the expression of the protein N of measles virus (Slibinskas et al. 2004). Furthermore, the proteins N of Sendai virus (Juozapaitis et al. 2005), Hendra and Nipah henipaviruses (Juozapaitis et al. 2007b), and human respiroviruses 1 and 3 or PIVs 1 and 3 (Juozapaitis et al. 2008) were also expressed in yeast.

The first evidence on the determinative role of the matrix protein M in the assembly of virus-like and not of the nucleocapsid-like structures came from the experiments in mammalian cells on mutants of Sendai virus belonging to the species *Murine respirovirus* of the *Respirovirus* genus (Stricker et al. 1994). These studies resulted in the unveiling of an orderly mechanism of the protein M-dependent self-assembly of paramyxoviruses (Stricker et al. 1994; Takimoto and Portner 2004). The expression of the Nipah virus protein M was shown to be sufficient for the production of VLPs in mammalian cells (Ciancanelli et al. 2006) and in *E. coli* (Subramanian et al. 2009).

The coexpression of the proteins M and N of human PIV type 1 in mammalian cells showed that the M protein alone can induce the budding of VLPs and that the protein N can assemble into intracellular nucleocapsid-like structures (Coronel et al. 1999). Furthermore, the coexpression of the proteins M, N, F, and HN of Sendai virus in mammalian cells resulted in the generation of VLPs that had morphology and density similar to those of authentic virus particles and allowed investigation of the roles played by individual proteins in the self-assembly and budding (Sugahara et al. 2004; Gosselin-Grenet et al. 2010).

The central role of the protein M in the self-assembly of N, F, and G (analogue of HN) was demonstrated also in the vaccinia virus-driven expression of Nipah virus proteins in mammalian cells (Patch et al. 2007). The VLPs were detected after coexpression of the proteins M and F of measles virus (Pohl et al. 2007). The same vaccine potential was confirmed also in the case of the Nipah VLPs composed of the proteins M, G, and F (Walpita et al. 2011).

Furthermore, the production of the PPRV VLPs was achieved by baculovirus-driven coexpression of the proteins M and N (Liu et al. 2014a); M, N, and H (Liu et al. 2014b); M, H, and F (Li W et al. 2014); and M, H, F, and N (Yan et al. 2019) in insect cells. The efficient simultaneous baculovirus-driven expression of the three M, F, and H (Yan et al. 2020) or four M, F, H, and N (Yan et al. 2019) structural PPRV proteins resulted in excellent VLPs that induced both humoral and cell-mediated immune responses in mice and goats.

Concerning construction of chimeric VLPs, a fusion of the CS protein from *Plasmodium berghei* to the protein N of measles virus was expressed in *P. pastoris* (Jacob et al. 2014). The chimeric protein generated highly multimeric but heterogenic NP bearing the CS protein on the surface that ensured significant reduction of parasitemia after immunization of mice with whole heat-inactivated yeast cells and following challenge with a high dose of parasites (Jacob et al. 2014).

By analogy with the previously described avulaviruses, the orthoparamyxoviruses contributed to the generation of live attenuated viral vectors expressing foreign glycoproteins and/or chimeric glycoproteins and acting as vaccine and gene therapy candidates (for a review, see Nakanishi and Otsu 2012; Le Bayon et al. 2013), as well as oncolytic viruses (Galanis 2010; Hudacek et al. 2013). The mutually exchanged RPV and PPRV glycoproteins F and H were among the first examples of the vaccines based on the construction of mixed envelopes (Das SC et al. 2000). Further, human PIV-vectored Ebola virus (EBOV) vaccine was constructed by the introduction of the EBOV glycoprotein GP (Yang L et al. 2008) and demonstrated high protective efficiency in guinea pigs by intranasal inoculation (Bukreyev

TABLE 31.1

Formation of VLPs by the Combination of Expressed Structural Proteins Derived from the *Mononegavirales* Order Members

Genus	Strain	Expression System									
		Mammalian		Insect		Plant		Yeast		Bacteria	
		NO	YES	NO	YES	NO	YES	NO	YES	NO	YES
Family *Bornaviridae*											
Orthobornavirus	Borna disease virus										M
Family *Filoviridae*											
Cuevavirus	Lloviu virus		NP+GP+VP40								
Ebolavirus	Zaire Ebolavirus		GP+VP40		GP+VP40	GP				GP	
			NP+GP+VP40		NP+GP+VP40						
			GP+VP40		NP+GP+VP40+ VP24						
	Sudan virus				GP+VP40						
Marburgvirus	Marburg virus		GP+VP40		NP+GP+VP40						
			NP+GP+VP40		GP+VP40						
Family *Paramyxoviridae*											
Orthoavulavirus	Newcastle disease virus	N+F+ HN	M		HN?				M		N
			M+N+F+HN								M
Henipavirus	Nipah virus	N+F+ G	M		N				N		N
			M+F+G								M
			M+N+F+G								
	Hendra virus								N		
Morbillivirus	Cetacean (dolphin) morbillivirus				N						
	Measles virus		N		N				N		N
			M								
			M+F								
	Peste des petits ruminants virus				M+N						N
					M+N+H						
					M+H+F						
					M+H+F+N						
	Rinderpest virus										N
Respirovirus	Human parainfluenza virus 1		M						N		
			M+N								
	Human parainfluenza virus 3								N		
	Sendai virus		M						N		
			M+N+F+HN								

Genus	Strain	Mammalian NO	Mammalian YES	Insect NO	Insect YES	Plant NO	Plant YES	Yeast NO	Yeast YES	Bacteria NO	Bacteria YES
Orthorubulavirus	Human parainfluenza virus 2							N			
	Mumps virus	M						N		N	
		M+N+F									
Pararubulavirus	Tioman virus							N			
Family *Pneumoviridae*											
Metapneumovirus	Human metapneumovirus	F	M+G+F					N	M		
			G								
			M+F								
Orthopneumovirus	Bovine respiratory syncytial virus			F					N		
	Human respiratory syncytial virus	F	M+F		N	G		N	M		N
		M		F	Influenza M1+G					F	
					Influenza M1+F						
Family *Rhabdoviridae*											
Ephemerovirus	Bovine ephemeral fever virus			G							
Lyssavirus	European bat lyssavirus 1							N			
	European bat lyssavirus 2							N			
	Lagos bat lyssavirus										M?
	Mokola lyssavirus										M?
	Thailand dog virus										M?
	Rabies virus		G	G	N	G	N?	N	N		
					N+M				G		
Novirhabdovirus	Infectious hematopoietic necrosis virus								N		
Vesiculovirus	Chandipura virus	G							G		
	Vesicular stomatitis virus		M	N	M				N		
				G							

Notes: The viruses are given in alphabetical order within each specific hierarchical group. References are given in the text.

et al. 2009). Next, the promising bivalent human PIV5-based hRSV vaccine was constructed (Phan et al. 2014). At last, An D et al. (2021) developed a reasonable PIV 5-based Covid-19 vaccine candidate and demonstrate protection of the hACE2 mice and ferrets against SARS-CoV-2 challenge after a single-dose mucosal immunization.

The paramyxoviruses demonstrated numerous examples of pseudotyping on traditional pseudotype carriers. First, Moloney MLV was pseudotyped with the glycoprotein F of Sendai virus (Spiegel et al. 1998) and the proteins H and F of measles virus (Voelkel et al. 2012). Second, lentivirus vectors were pseudotyped by the glycoproteins of Sendai virus (Kobayashi et al. 2003), human PIV3 (Jung et al. 2004), measles virus (Funke et al. 2009), Nipah and Hendra henipaviruses (Khetawat and Broder 2010), and tupaia paramyxovirus (Enkirch et al. 2013). Third, pseudotyping of VSV vectors was performed by the glycoproteins of Nipah virus (Kaku et al. 2009), and an advanced high-throughput serum neutralization assay for Nipah virus antibodies was constructed (Kaku et al. 2012). Fourth, the protein H of RPV was incorporated into extracellular baculovirus during expression in insect cells and ensured induction of neutralizing antibody response in cattle (Sinnathamby et al. 2001).

The measles virus platform was used to generate vaccine candidates against HBV (Singh M et al. 1999).

SUBFAMILY RUBULAVIRINAE

The subfamily consists of 2 genera with 18 species. The most dangerous members of the rubulaviruses are human parainfluenza viruses 2 and 4 (PIVs) and mumps virus belonging to the *Orthorubulavirus* genus and Menangle and Tioman viruses of the *Pararubulavirus* genus.

The herringbone-like particles were formed by the expression of the protein N of mumps virus in *E. coli* (Cox R et al. 2009).

The numerous herringbone-like structures of the rubulaviral protein N were achieved in yeast by Kęstutis Sasnauskas's team. So, both *S. cerevisiae* and *P. pastoris* were used for the expression of the gene N in the case of mumps virus (Samuel et al. 2002; Slibinskas et al. 2003). Then, the protein N of Menangle (Juozapaitis et al. 2007a) and Tioman (Petraityte et al. 2009) viruses was synthesized in yeast *S. cerevisiae*. Later, *S. cerevisiae* was used for the expression of gene N of human parainfluenza virus 2 (Bulavaitė et al. 2016).

The VLPs were detected after coexpression of the proteins M, N, and F of mumps virus (Li M et al. 2009).

The x-ray crystal structure of the nucleoprotein N-RNA complex from PIV5 belonging to the species *Mammalian orthorubulavirus 5* was resolved to 3.11 Å (Alayyoubi et al. 2015).

PNEUMOVIRUSES

The *Pneumoviridae* family consists of 2 genera with 5 species. This taxon formerly was a subfamily within the *Paramyxoviridae* but was reclassified as a family in 2016 (Rima et al. 2017). Members of genus *Orthopneumovirus* infect mammals, while members of *Metapneumovirus* are specific for mammals or birds. Some viruses are specific and pathogenic for humans such as hRSV of the *Orthopneumovirus* genus and human metapneumovirus (HMPV) of the *Metapneumovirus* genus. The hRSV and HMPV remain among the leading causes of childhood hospitalization and a major health burden worldwide. There are no known vectors for pneumoviruses, and transmission is thought to be primarily by aerosol droplets and contact (Rima et al. 2017).

As the paramyxoviruses, the pneumoviruses possess enveloped spherical pleomorphic virions of about 150 nm in diameter, similar to one shown in Figure 31.1d. The structural proteins are the NP (or N), the matrix protein M, and the envelope proteins: fusion protein F and glycoprotein G.

The first expression of the hRSV structural genes was achieved in mammalian cells by the generation of a rescue system of the virus particles entirely from cloned complementary DNAs (Teng and Collins 1998).

The protein N of hRSV formed typical herringbone-like structures by expression in insect cells (Méric et al. 1994) and yeast *S. cerevisiae* (Juozapaitis et al. 2006), as well as the protein N of HMPV in *S. cerevisiae* (Petraitytė-Burneikienė et al. 2011).

The expression of the hRSV protein N in *E. coli* led to formation of ringlike structures (Roux et al. 2008) that protected mice against hRSV challenge and provided cross protective immunity in calves against a viral challenge with bovine respiratory syncytial virus (bRSV) (Riffault et al. 2010). However, attempts to prepare similar ringlike structures by expression of the protein N from bRSV were unsuccessful (Riffault et al. 2010). Further, the safety and efficacy of the mucosal vaccine based on the *E. coli*-derived hRSV protein N rings was demonstrated in the mice neonates (Remot et al. 2012). The *E.-coli*-derived ringlike structures allowed generation of the first chimeric pneumoviral particles by the chimeric hRSV protein N carrying GFP added to its N-terminus (Roux et al. 2008).

The expression of the hRSV protein F in insect cells led to the production of the 40-nm nanoparticles composed of multiple hRSV F oligomers arranged in the form of rosettes that were able to induce neutralizing antibodies (Smith G et al. 2012). The Phase I trial of such hRSV F nanoparticle vaccine candidates in healthy adults demonstrated good toleration and induction of hRSV-neutralizing antibodies (Glenn et al. 2013). Another protein F-derived vaccine candidate was based on the trimers of a truncated secreted version of the hRSV protein F that was expressed in mammalian cells and demonstrated protective efficacy in mice after formulation with the CpG adjuvant (Garlapati et al. 2012).

The crystal structure of the hRSV nucleocapsid-like protein–RNA complex was determined at about 3 Å resolution (Tawar et al. 2009; El Omari et al. 2011; Bakker et al. 2013), and general hRSV architecture, which indicated important structural differences between the *Pneumoviridae* and *Paramyxoviridae* families, was elucidated by electron cryotomography (Liljeroos et al. 2013). The crystal structure of the protein M of GMPV in its native dimeric state was resolved at 2.8 Å resolution (Leyrat et al. 2014).

The fine elucidation of functional hRSV self-assembly revealed absolute requirement for the protein M (Mitra et al. 2012). The coexpression of the HMPV proteins M, G, and F in mammalian cells led to formation of VLPs with a similar morphology to the filamentous virus morphology that was observed in HMPV-infected cells; unexpectedly, the protein G only was able to form VLPs in the absence of the other virus proteins (Loo et al. 2013). Contrary to the idea of a dominant role of the protein G in the pneumoviral self-assembly, the proteins M and F were found sufficient to form HMPV VLPs by expression in mammalian cells (Cox RG et al. 2014). The two doses of such VLPs conferred complete protection against HMPV replication in the lungs of mice (Cox RG et al. 2014).

The additional advance in the pneumoviral VLP approaches was achieved in insect cells by efficient baculovirus vector-driven coexpression of the hRSV proteins F or G together with the influenza matrix protein M1 (Quan et al. 2011). Intramuscular vaccination of mice with such VLPs provided effective protection against hRSV infection (Quan et al. 2011), but the vaccine candidate was improved by mixing both F and G VLPs (Lee S et al. 2014) or by coadministration with DNA vaccines (Hwang et al. 2014; Ko et al. 2015). The direct comparison of the combined VLP and DNA vaccines with the formalin-inactivated hRSV vaccine in mice revealed clear advantages of the VLP vaccine (Lee JS et al. 2014; Kim KH et al. 2015).

Later, the influenza M1 protein was used as a core in the VLPs containing tandem repeat of the RSV protein G (Kim AR et al. 2018).

As with the paramyxoviruses described earlier, the pneumoviruses were used in a long line of the mixed infectious virus constructions with the envelope exchange. For example, the hRSV envelope proteins G and F were expressed on the surface of such chimeric viruses based on recombinant bovine/human PIV type 3 (Schmidt AC et al. 2001), Sendai virus (Takimoto et al. 2004; Zimmer et al. 2005; Zhan et al. 2007), NDV (Martinez-Sobrido et al. 2006), measles virus (Mok et al. 2012), and influenza virus (Bian et al. 2014; Fonseca et al. 2014; Zhang P et al. 2014).

The pseudotyping with the hRSV envelope proteins G and F was performed on VSV (Kahn et al. 1999) and with the HMPV proteins G and F on MLV (Lévy et al. 2013). The major immunogenic domains of the hRSV proteins G and F were displayed on the surface of *Lactococcus lactis* by fusion to the appropriate bacterial anchors (Lim et al. 2010) and paved a way to a protective and safe vaccine candidate (Rigter et al. 2013).

As mentioned earlier, the ectodomains of hRSV F and G proteins were fused to the transmembrane and cytoplasmic domains of NDV F and HN proteins—respectively—and provided complete protection of mice from hRSV replication in lungs (McGinnes et al. 2011; Schmidt MR et al. 2012).

The neutralizing epitope of hRSV from the glycoprotein F played an important role for providing the proof of principle data for epitope-focused vaccine design (Correia et al. 2014). This epitope was selected since it was targeted by the licensed, prophylactic neutralizing antibody palivizumab (also known as Synagis, or pali) and an affinity-matured variant, motavizumab, or mota (for details and references, see Correia et al. 2014).

Finally, the hRSV protein F played a pioneering role in the development of modern bionanomaterials as putative vaccines by display on a gold nanorod (Stone et al. 2013).

BORNAVIRUSES AND NYAMIVIRUSES

The two other families encompassing spherical viruses were involved in only a few VLP studies. The *Bornaviridae* family contains 3 genera with 11 species, while the *Nyamivirdae* family lists 6 genera with 12 species (Dietzgen et al. 2017; ICTV 2020). Figure 31.1a presents a typical example of these spherical viruses.

Borna disease viruses 1 and 2 (BoDV-1 and BoDV-2) causing Borna disease in mammals are members of the species *Mammalian 1 orthobornavirus* in the genus *Orthobornavirus*. The BoDVs are the rare animal RNA viruses that can establish a persistent infection in the host cell nucleus. Moreover, there are DNA sequences derived from the mRNAs of ancient bornaviruses in the genomes of vertebrates, including humans, designated as endogenous Borna-like elements, but BDV does not integrate into host's genome (for references, see Horie et al. 2013). However, the way the bornavirus participates in the evolution of host genomes remains unclear. BoDVs are neurotropic agents and infect a wide variety of mammalian species from rodents to birds and readily establish a long-lasting, persistent noncytolytic infection in brain cells. Moreover, BoDVs are involved in the pathogenesis of some human psychiatric disorders.

Nyamanini virus (NYMV) and Midway virus of the *Nyavirus* genus, typical members who provided the family with its name, are tick-borne agents that infect land birds and seabirds, respectively (Kuhn et al. 2013).

The expression of the BoDV protein M in *E. coli* resulted in the production of tetramers with the tendency to assemble into high-molecular-mass lattice-like complexes (Kraus et al. 2005). The crystal structure of the BoDV protein M tetramer exhibited structural similarity to the N-terminal domain of the Ebola virus protein M (VP40; Neumann et al. 2009). Moreover, it was found that two distinct proteins of NYMV serve a matrix protein function as previously described for members of the *Filoviridae* family (Herrel et al. 2012). The BoDV protein N synthesized in *E. coli* revealed a planar tetrameric structure by crystallographic analysis at 1.76 Å resolution (Rudolph et al. 2003).

Pseudotyping by the BoDV protein G was performed on the replication-competent VSV (Perez M et al. 2007), but no epitopes from bornaviruses or nyamiviruses were used for the expression on VLP and/or non-VLP carriers yet.

The use of a reverse genetic approach allowed identification of the BoDV proteins required for packaging of BoDV RNA analogues into infectious VLPs and establishment of the BoDV model as a vector for the cloning of foreign genes (Perez M and de la Torre 2005), first of all, for the specific expression in the central nervous system (Daito et al. 2011).

RHABDOVIRUSES

The *Rhabdoviridae* family, which is the largest family of the *Mononegavirales* order, contains bullet-shaped and bacilliform viruses (shown in Figure 31.1e) and consists of 3 subfamilies, 40 genera with 246 species (Walker et al. 2018). The most popular representatives of the *Rhabdoviridae* family are rabies virus (RABV) of the *Lyssavirus* genus and vesicular stomatitis virus (VSV) of the species *Indiana vesiculovirus* from the genus *Vesiculovirus*. The VSV is one of the classical objects and subjects of modern virology, immunology, and gene therapy. Infectious hematopoietic necrosis virus (IHNV) of the *Salmonid novirhabdovirus* species, genus *Novirhabdovirus*, remains a global burden of fishery, since it causes the disease known as infectious hematopoietic necrosis in salmonid fish such as trout and salmon.

The structural proteins of rhabdoviruses are the NP (or N), the matrix protein M, and the glycoprotein G. The loosely coiled strands of varying length formed by the protein N of rabies virus have been detected by electron microscopy at the very beginning of the genetic engineering era (Schneider et al. 1973).

The first expression of the structural genes of VSV (Pattnaik and Wertz 1991) and rabies virus (Conzelmann and Schnell 1994) was achieved in mammalian cells by the generation of a rescue system of the virus particles entirely from cloned complementary DNAs. Moreover, it was shown that the matrix protein M of VSV has intrinsic VLP-forming and budding activity when expressed alone in mammalian cells (Justice et al. 1995). Furthermore, the vaccinia virus-free expression and recovery of the VSV single-round infectious particles, so-called sriVLPs, was markedly improved (Harty et al. 2001, Ito et al. 2003).

Although not always observed by electron microscopy, the structures similar to herringbone-like particles described earlier for other negative-stranded viruses might have been produced by the baculovirus-driven expression of the protein N from rabies virus (Préhaud et al. 1990; Reid-Sanden et al. 1990; Fu et al. 1991) and VSV (Katz et al. 1995) in insect cells. The VSV protein M synthesized in insect cells induced formation of vesicles, which were released from the cell surface in the form of liposomes (Li Y et al. 1993). The baculovirus vector itself or recombinants expressing VSV glycoprotein G or nucleocapsid protein N did not produce the formation of vesicles.

Further evaluation confirmed the presence of the ring-like structures in the insect cell-derived preparations of the protein N from rabies virus (Pinto et al. 1994; Iseni et al. 1998). Such preparations demonstrated immunogenicity (Hooper et al. 1994) and conferred protective immunity in mice against challenge with lethal doses of rabies virus (Yin et al. 2013).

The expression of the rhabdoviral protein N in *E. coli* did not reveal any nucleocapsid-like structures in the case of IHNV (Oberg et al. 1991), VSV (Das T and Banerjee 1993), and rabies virus (Goto et al. 1995). However, coexpression of the VSV protein N and phosphoprotein P resulted in the formation of the native N–P complex (Gupta and Banerjee 1997).

The typical nucleocapsid-like particles were obtained by the expression of the protein N from rabies virus and two other lyssaviruses: European bat lyssavirus 1 and 2 in *S. cerevisiae* (Kucinskaite et al. 2007). The high-level expression of the protein N from rabies virus was achieved in transgenic tomato and tobacco plants and partial protection of mice with plant-derived material against viral challenge was presented, however, without any indications about the structural status of the produced protein N (Perea Arango et al. 2008).

Like the protein N, the rhabdoviral glycoprotein G was expressed in mammalian cells very early in the case of rabies (Kieny et al. 1984; Burger et al. 1991) and Chandipura (Masters et al. 1989) viruses. Next, the glycoprotein G of rabies virus (Prehaud et al. 1989), IHNV (Koener and Leong 1990), and bovine ephemeral fever virus (Johal et al. 2008) was expressed in insect cells, but no data on its self-assembly was presented. The insect cell–derived glycoprotein G of rabies virus conferred protection against a lethal challenge with rabies virus to mice (Prehaud et al. 1989) and to raccoons by oral field vaccination (Fu et al. 1993), but the IHNV glycoprotein G provided very limited protection in rainbow trout (Cain et al. 1999). Noticeably, according to two investigations, expression in mammalian cells of the glycoprotein G from VSV (Rolls et al. 1994) and rabies virus (Fontana et al. 2014) may lead to the appearance of some sort of virus-like particles. Targovnik et al. (2019) developed an easy one-step metal ion affinity chromatography process to purify a high amount of the ectodomain of rabies glycoprotein G from insect cells, however without any indications to self-assembly into VLPs.

The expression of the glycoprotein G of INHV (Verjan et al. 2008) and rabies virus (Singh A et al. 2012) in *E. coli* did not result in any self-assembled protein G products. The same was true for the expression of the glycoprotein G of rabies virus in yeast (Klepfer et al. 1993; Sakamoto et al. 1999) and plants: tomatoes (McGarvey et al. 1995) and carrot (Ashraf et al. 2005).

The pivotal role of the matrix protein M in the rhabdoviral self-assembly was identified from the very beginning in the case of VSV (Lyles et al. 1996) and rabies virus (Mebatsion et al. 1999). Bacterial expression of the protein M from three lyssaviruses, Lagos bat, Mokola, and Thailand dog, allowed high-quality purification for further crystallization and elucidation of their 3D structure (Assenberg et al. 2008).

The crystal structure was determined for the protein N from rabies virus (Albertini et al. 2006, 2007) and VSV (Green and Luo 2009; Leyrat et al. 2011), for the glycoprotein G from VSV (Roche et al. 2007) and Chandipura virus (Baquero et al. 2012), and for the protein M from VSV (Gaudier et al. 2002; Graham et al. 2008) and lyssaviruses mentioned earlier (Assenberg et al. 2008; Graham et al. 2008). Remarkably, the structures of the protein M from VSV and Lagos bat lyssavirus both shared a common fold despite sharing no identifiable sequence homology (Graham et al. 2008).

Finally, an electron cryomicroscopy model of VSV was constructed and showed that each virion contains two nested, left-handed helices: an outer helix of matrix protein M and an inner helix of NP N and RNA (Ge et al. 2010).

The use of rhabdoviral structural proteins as putative carriers of foreign sequences includes fusion of GFP to the protein N that was efficiently expressed in mammalian cells and incorporated into infectious rabies virions (Koser et al. 2004). Moreover, the fusion protein induced a strong humoral immune response against GFP in mice (Koser et al. 2004).

The rhabdoviruses are widely used objects and subjects of the pseudotyping technology, predominantly VSV, which is one of the most common pseudotype objects functioning as a support for foreign envelope proteins, as well as subjects providing the protein G for the envelopment of retroviral and lentiviral cores and participating in many efficient gene therapy vectors. As an early review on the VSV pseudotype potential, we would recommend a protocol book chapter (Lo and Yee 2007) and restrict our consideration to some examples of the rhabdoviral contribution to the pseudotyping approaches: retargeting of the VSV-pseudotyped lentiviral vectors by polymer nanomaterials (Liang et al. 2013) and targeting of rabies virus glycoprotein-pseudotyped lentiviral vectors to assess nerve recovery in nerve injury models (Wei et al. 2014) and pseudotyping of VSV by rabies glycoprotein to study transport among neurons in vivo (Beier et al. 2013) and glycoproteins of highly pathogenic AIVs (Zimmer et al. 2014).

It is noteworthy that rhabdoviruses can be used as viral vectors for the delivery of VLPs, for example, of human norovirus VLPs (Ma and Li 2011).

At last, self-propagating, infectious, virus-like vesicles (VLVs) were generated when an alphavirus RNA replicon expressed the VSV glycoprotein G as the only structural protein (Rose et al. 2014). The authors claimed that the VLVs arose from membrane-enveloped RNA replication factories, or spherules, containing VSV G protein that were largely trapped on the cell surface and suggested a basic mechanism of propagation used by primitive RNA viruses lacking capsid proteins. This VLV platform was used to express the middle surface envelope glycoprotein (MHBs) of hepatitis B virus (Reynolds et al. 2015). A single immunization with VLV-MHBs protected mice from HBV hydrodynamic challenge, and this protection correlated with the elicitation of a CD8 T cell recall response. In contrast to MHBs, a VLV expressing HBV core (HBc) protein neither induced a CD8 T-cell response in mice nor protected against challenge (Reynolds et al. 2015).

Generally, the VSV platform was used to generate promising vaccine candidates against EBOV (Marzi et al. 2011, 2015), EBOV and MARV (Geisbert and Feldmann 2011), HBV (Cobleigh et al. 2013), hepatitis C virus (HCV; An HY et al. 2013), Chikungunya virus (CHIKV) and Zika virus (ZIKV; Chattopadhyay et al. 2013, 2018), Lassa virus (Safronetz et al. 2015), and finally against Covid-19 (Hennrich et al. 2021; Yahalom-Ronen et al. 2021). The rabies virus platform was used to construct vaccines against anthrax (Smith ME et al. 2006) and a multivalent vaccine

against Dengue 2 virus and HBV (Harahap-Carrillo et al. 2015). A detailed protocol for the construction of rabies viral vectors was published by Osakada and Callaway (2013).

FILOVIRUSES

The *Filoviridae* family contains variously shaped, often filamentous viruses (shown in Figure 31.1b) and consists of 6 genera with 11 species (Kuhn et al. 2019). Several filovirus representatives, such as various variants of EBOV of the genus *Ebolavirus* and Marburg virus (MARV) of the genus *Marburgvirus*, are pathogenic for humans and highly virulent. The natural hosts for filoviruses could be bats, as for Ebola- and Marburgviruses, whereas others may infect fish.

The filoviral structural proteins are the NP, the glycoprotein G1,2 (GP), the matrix protein M (VP40), and the protein VP24 that is not homologous to genes of other *Mononegavirales* representatives and participates in the nucleocapsid assembly, together with the NP and VP35 protein, a cofactor of the viral RNA polymerase complex. Regarding the filovirus evolutionary history, the evidence of their origin assuming EBOV and cuevavirus divergence from marburgviruses since the early Miocene was provided (Taylor et al. 2014). The largest outbreak of EBOV disease spreading through Guinea, Liberia, Sierra Leone, and Nigeria in 2014 discovered novel virus variants (Gire et al. 2014) and stimulated the search for novel vaccine candidates against filoviruses (Cohen 2014). Notably, like bornaviruses, the filovirus genome fragments are found integrated into vertebrate genomes (Belyi et al. 2010).

The first success in the construction of filovirus vaccines arose from the expression of MARV genes encoding NP, GP, VP40, VP35, or VP24 by RNA replicon based upon VEEV, when NP or GP immunization induced protection of guinea pigs against viral challenge, while GP and simultaneous NP and GP immunization ensured full protection of nonhuman primates, namely cynomolgus macaques (Hevey et al. 1998). Next, VEEV vector-driven expression of EBOV genes encoding GP or both NP and GP resulted in full protection of mice and guinea pigs against lethal EBOV infection, while NP immunization protected mice but not guinea pigs (Pushko et al. 2000). The dual expression of bivalent alphavirus particles that carried glycoprotein genes of both EBOV and Lassa virus (LASV) protected guinea pigs against challenges with EBOV and LASV (Pushko et al. 2001). However, further evaluation led to conclusions that the disease observed in primates differed from that in rodents, suggesting that rodent models of EBOV may not predict the efficacy of candidate vaccines in primates and that protection of primates may require different mechanisms (Geisbert et al. 2002).

In parallel, the development of the filoviral vaccine strategy focused on the VLP approach, as indicated in Table 31.1. The EBOV and MARV VLPs composed of the glycoprotein GP and matrix protein VP40 and resembling the distinctively filamentous infectious virions were generated by the expression in mammalian cells (Bavari et al. 2002). The vaccination of rodents with EBOV (Warfield et al. 2003) or MARV (Swenson et al. 2004) VLPs conferred induction of the immune response and protected mice against EBOV (Warfield et al. 2003) or guinea pigs against MARV (Warfield et al. 2004) lethal challenge. Further, EBOV and MARV VLPs were shown to be more effective stimulators of human dendritic cells than the respective viruses (Bosio et al. 2004). A pan-filovirus hybrid VLP vaccine candidate based on EBOV and MARV proteins GP and VP40 demonstrated that only GP was required and sufficient to protect against a homologous filovirus challenge (Swenson et al. 2005). Moreover, the EBOV and MARV VLPs produced in mammalian cells fully protected nonhuman primates (cynomolgus macaques) from the lethal viral challenge (Warfield et al. 2007b). Finally, it was concluded that the expression of the matrix protein VP40 alone is sufficient for VLP production, but addition of other filovirus proteins increases the efficiency of VLP production in mammalian cells and results in the case of the coexpression of GP and VP40 in the promising vaccine candidate (for a full list of references, see review by Warfield et al. 2005 and Warfield and Aman 2011). Later, the VLPs consisting of NP, GP, and NP40 of lloviu cuevavirus of the genus *Cuevavirus* were prepared (Maruyama et al. 2014). The modified vaccinia Ankara (MVA) vaccine platform was used to generate the EBOV (Schweneker et al. 2017) and MARV (Malherbe et al. 2020) VLPs and demonstrate their protective effect in animals. Similarly, the MARV GP+NP40 VLPs were produced by the pox virus-based platform (Lázaro-Frías et al. 2018).

The same GP and VP40-based VLPs as in the mammalian cells were produced by the baculovirus-driven expression in insect cells (Ye et al. 2006). Next, the EBOV and MARV VLPs similar to those produced in mammalian cells were generated by the baculovirus-driven coexpression of the proteins GP, VP40, and NP, and their protective efficiency against a lethal viral challenge in mice was demonstrated (Warfield et al. 2007a). Weiwei et al. (2017) got the MARV VLPs by coexpression of the GP and VP40 in *Sf*9 cells and demonstrated their excellent immunogenicity in rhesus macaques. The insect *Sf*9 cell-produced EBOV GP–VP40 VLPs conferred full protection but showed strong dose dependence in immunization of mice (Sun et al. 2009). Furthermore, the efficiency of the insect cell-produced EBOV VLP vaccine (Warfield et al. 2007a) was confirmed by the first trial on captive chimpanzees (Warfield et al. 2014).

The nice recombinant EBOV VLPs were produced efficiently in *Drosophila melanogaster* Schneider 2 (S2) cells by simultaneous expression of the GP, NP, VP24, and VP40 genes (Park EM et al. 2018).

After traditional Zaire EBOV, the VLPs of another representative of the *Ebolavirus* genus, namely Sudan virus (SUDV), were generated by baculovirus-driven coexpression of GP and VP40 genes in insect *Sf*9 cells (Wu et al. 2020). The filamentous SUDV VLPs demonstrated excellent

immunogenicity in mice and horses and represented therefore a promising approach for vaccine development against SUDV infection.

The large volume of biophysical data on the conformational stability of insect cell-produced EBOV and MARV VLPs characterized by various spectroscopic techniques over a wide pH and temperature range was generated, in order to select optimized solution conditions for further vaccine formulation and long-term storage (Hu et al. 2011). No attempts to produce the filoviral VLPs in bacteria or plants were found, while the situation by the plant expression was recently described in detail by Peyret et al. (2021).

Regarding high-resolution elucidation of spatial structure of filoviral proteins, the crystal structures of the EBOV GP ectodomain (Weissenhorn et al. 1998; Malashkevich et al. 1999), VP40 (Dessen et al. 2000a, b; Gomis-Rüth et al. 2003), GP (Lee JE et al. 2008), VP35 (Leung et al. 2009, 2010), and the MARV GP ectodomain (Koellhoffer et al. 2012) and VP24 (Zhang AP et al. 2014) were published.

The crystal structure of the EBOV VP40 revealed its topological distinction from all other known viral matrix proteins, consisting of two domains with unique folds connected by a flexible linker (Dessen et al. 2000b). Later, the crystal structure of a disk-shaped octameric form of VP40 composed of the four antiparallel homodimers of the N-terminal domain was shown (Gomis-Rüth et al. 2003).

The successful attempts to view the complete picture of filoviral components and their morphogenesis (Noda et al. 2005; Bharat et al. 2011, 2012), as well as the budding process (Welsch et al. 2010), were undertaken by electron cryotomography. A system of so-called infectious VLPs that carry a minigenome consisting of the negative-sense copy of the GFP (Watanabe et al. 2004) or luciferase (Wenigenrath et al. 2010) gene was constructed for EBOV and MARV, respectively. The reverse genetics approach is also fully adapted to filoviruses (Hoenen et al. 2011). It is worthwhile to mention the recombinant MARV with the eGFP gene that was inserted between the second and third genes, encoding VP35 and VP40, respectively, and conferred expression of the eGFP gene from an additional transcription unit (Schmidt KM et al. 2011).

The filoviruses also have a long and successful history of the use of their glycoproteins as pseudotyping agents. First, the pseudotyped lentiviral vectors were constructed for the delivery of lentiviral vectors to the cells exposing receptors for the filoviruses, for example, folate receptor alpha, a glycosylphosphatidylinositol-linked surface protein on the apical surface of airway epithelia (Kobinger et al. 2001; Medina et al. 2003; Sinn et al. 2003). Another retroviral vector, namely the MLV vector, was used to construct cross protective filoviral vaccine candidates based on the retro-VLPs that were pseudotyped by the EBOV GP lacking mucin-like domain or on the appropriate DNA plasmids, or so-called plasmo-retro VLPs able to direct production of retro-VLPs (Ou et al. 2012).

Second, the EBOV GP was used to pseudotype the VSV vector, and the VSV-based vaccine was shown to protect rhesus monkeys against an EBOV challenge (Geisbert et al. 2008b). Further, complete protection of cynomolgus macaques by Zaire EBOV or MARV VLPs against a challenge with the respective virus was demonstrated (Geisbert et al. 2008a). Furthermore, complete protection was reported also against a challenge with Sudan EBOV of rhesus monkeys immunized with VSV-based vaccine displaying the Sudan EBOV GP (Geisbert et al. 2008c). A single immunization with VSV-based vaccine expressing the Zaire EBOV GP, but not Côte d'Ivoire EBOV GP, provided some cross protection of cynomolgus macaques against a challenge with Bundibugyo EBOV that was found as a new EBOV species following an outbreak in Uganda in 2007 (Falzarano et al. 2011). Advanced research on the VSV-based vaccines against both EBOV and MARV by improving cross protection (Geisbert and Feldmann 2011; Marzi et al. 2011) and avoiding neurovirulence in nonhuman primates (Mire et al. 2012) was performed. The GP of the novel lloviu cuevavirus was used to pseudotype the VSV vector (Maruyama et al. 2013).

Third, the EBOV GP was used to pseudotype paramyxovirus vectors of human PIV (Bukreyev et al. 2009) and NDV (Wen et al. 2013).

32 Order *Bunyavirales*

The whole is more than the sum of its parts.

Aristotle

ESSENTIALS

The order *Bunyavirales* is the largest among other negative single-stranded RNA virus orders by number of families, currently 12, according to the recent detailed taxonomy (Abudurexiti et al. 2019). This is the only order in the class *Ellioviricetes* of the subphylum *Polyploviricotina*, phylum *Negarnaviricota*, kingdom *Orthornavirae*, realm *Riboviria*.

The huge *Bunyavirales* order includes therefore in total 12 families, 4 subfamilies, 54 genera, and 477 species altogether. Moreover, the order is of great medicinal importance. Although members of the order are generally found in arthropods or rodents, some of them occasionally infect humans, such as highly dangerous representatives of the families *Arenaviridae* (Lassa, Tacaribe, and Junin viruses of the *Mammarenavirus* genus, as well as many other mammarenaviruses), *Hantaviridae* (Hantaan, Puumala, and other orthohantaviruses of the *Orthohantavirus* genus), *Nairoviridae* (Crimean-Congo hemorrhagic fever virus of the *Orthonairovirus* genus), *Peribunyaviridae* (numerous orthobunyaviruses including Bunyamwera virus of the *Orthobunyavirus* genus), and *Phenuiviridae* (Rift Valley fever and other phleboviruses of the *Phlebovirus* genus). Some of the *Bunyavirales* order representatives infect plants, such as the members of the *Fimoviridae* and *Tospoviridae* families.

In contrast to the previously mentioned densely inhabited families of the order, the families *Cruliviridae*, *Leishbuviridae*, *Mypoviridae*, and *Wupedeviridae* are represented by a single species as a member of a single genus.

Figure 32.1 presents typical portraits of the *Bunyavirales* representatives. The virions are typically 80–120 nm in diameter. The members of the *Arenaviridae* family may demonstrate, however, a diameter from 40–200 nm with trimeric surface spikes (Radoshitzky et al. 2019). The glycoproteins at the surface of the envelope are arranged on an icosahedral lattice with $T = 12$ symmetry, at least in the case of Rift Valley fever phlebovirus (Huiskonen et al. 2009; Sherman et al. 2009).

GENOME

The *Bunyavirales* order members have segmented RNA genomes. Figure 32.2 demonstrates the genomic structure of the most popular families of the order.

The genome of the *Arenaviridae* family consists of two single-stranded RNAs, denoted Small (S) and Large (L) and encoding four viral proteins in an ambisense coding strategy. The S-segment RNA is approximately 3.5 kb and encodes the viral nucleocapsid protein N and envelope glycoprotein GP. The L-segment RNA is approximately 7.5 kb and encodes the viral RNA-dependent RNA-polymerase L and a small RING-domain containing matrix protein Z.

The *Fimoviridae* family of plant-infecting emaraviruses contains tetrapartite genome.

The other ten families of the *Bunyavirales* order have tripartite genomes consisting of large (L), medium (M), and small (S) RNA segments (however, the segments of the *Mypaviridae* genome are denoted 1, 2, and 3). These RNA segments are single stranded and exist in a helical formation within the virion. The L segment encodes the polymerase L, while the M segment encodes the viral glycoproteins, and the S segment encodes the nucleocapsid protein N. In some cases, the S and M segments are ambisense.

ARENAVIRUSES

According to the latest ICTV issues (Radoshitzky et al. 2019), the *Arenaviridae* family unites 4 genera with 54 species, where the genus *Mammarenavirus* contains 40 species including the most familiar and dangerous arenavirus strains, such as Lassa virus (LASV) from the traditional Old World arenavirus group and Guanarito virus (GTOV), Junin virus (JUNV), Machupo virus (MACV), Sabia virus (SABV), and Whitewater Arroyo virus (WWAV) from the traditional New World virus group, as well as Lujo virus (LUJV), which may cause severe hemorrhagic fever syndromes resulting in significant mortality. First, the observed severe cases of disease have introduced the LASV as a reference arenavirus strain to the scientific community (Buckley et al. 1970). Second, one of the reference strains that was the most-studied representative of the family by the theoretical virological and immunological investigations, namely, lymphocytic choriomeningitis virus (LCMV), may cause not only influenza-like syndromes but also severe aseptic meningitis. LCMV exists in both geographic areas but is regarded rather as an Old World virus.

The S-segment of the arenaviral genome encodes the glycoprotein precursor GPC, which is cleaved further into three parts: attachment glycoprotein GP1, fusion transmembrane glycoprotein GP2, and a stable signal peptide SSP, and, in ambisense, the nucleoprotein NP; the L-segment encodes the matrix protein Z and, in negative sense, the multifunctional protein L (for references see Olschläger and Flatz 2013).

The two RNA segments are embedded into the nucleocapsid that interacts with the Z protein under the plasma membrane and buds, releasing the virion covered with surface glycoprotein spikes. The Z protein forms homooligomers

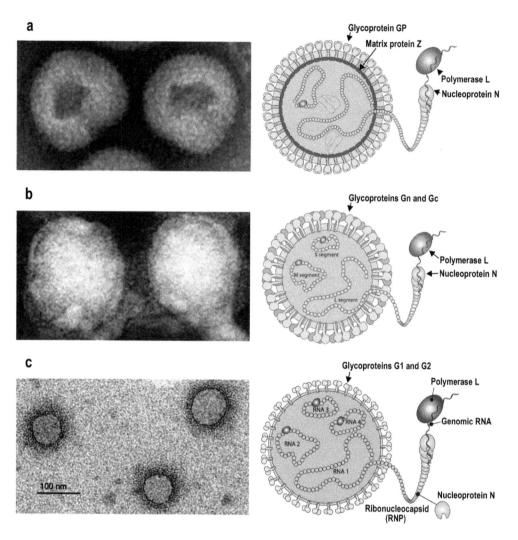

FIGURE 32.1 Typical representatives of the members of the order *Bunyavirales*. (a) family *Arenaviridae*; (b) families *Cruliviridae*, *Hantaviridae*, *Leishbuviridae*, *Mypoviridae*, *Nairoviridae*, *Peribunyaviridae*, *Phasmaviridae*, *Phenuiviridae*, *Tospoviridae*, and *Wupeviridae*; (c) family *Fimoviridae*. The electron micrographs of the virions are from the CDC Public Health Image Library (PHIL). The virions are 80–120 nm in diameter. The glycoproteins at the surface of the envelope are arranged on an icosahedral lattice, with T = 12 symmetry. The immunosorbent electron micrograph of virions of European mountain ash ringspot-associated virus of the *Fimoviridae* family is taken from the International Committee on Taxonomy of Viruses (ICTV) report, courtesy Inga Ludenberg, https://talk.ictvonline.org/ictv-reports/ictv_9th_report/negative-sense-rna-viruses-2011/w/negrna_viruses/214/emaravirus-figures, the bar represents 100 nm. The virion cartoons are reproduced with kind permission from the ViralZone, Swiss Institute of Bioinformatics (Hulo C. et al. 2011). (Courtesy Philippe Le Mercier.)

that represent structural scaffold of the virion. The life cycle of the arenaviruses is restricted to the cell cytoplasm. The arenavirus virions contain sand-looking particles that are responsible for the name of the family (*arena* means sand in Latin) and are nothing else than ribosomes acquired from the host cells.

Expression

The attempts to express the nearly full-length or fragmented structural LASV proteins N (Barber et al. 1987; ter Meulen et al. 2000; Branco et al. 2008) and glycoproteins GP1 and GP2 (Branco et al. 2008) in *E. coli* led to assembly defective products that were useful, however, in

the induction of the human T-cell proliferative response (ter Meulen et al. 2000) and by the ELISA diagnostics (Branco et al. 2008). In order to construct a putative live mucosal vaccine, the complete LASV NP sequence was cloned in a recombinant *aroA* attenuated *Salmonella typhimurium*, and the presence of the nonassembled NP in whole cell extracts was detected, with some cross-protective efficiency of the recombinant *Salmonella* against LCMV infection, which was demonstrated in mice (Djavani et al. 2000, 2001). Similar to the bacterial system, baculovirus-driven expression of LASV NP (Barber et al. 1990; Saijo et al. 2007) and GPC (Hummel et al. 1992) in insect cells did not result in assembly competent products but warranted production of valuable diagnostic reagents.

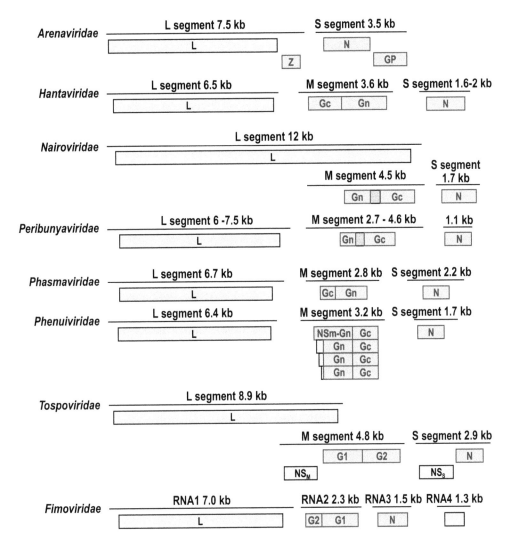

FIGURE 32.2   The genomic structure of the major families of the order *Bunyavirales*. The structural genes are colored dark pink. The sequence data is compiled from the NCBI browser.

The homologous eukaryotic vaccinia virus-driven expression of LASV NP resulted in the production of the corresponding nonassembled protein that was able to induce protection in guinea pigs against lethal challenge with LASV (Clegg and Lloyd 1987; Morrison et al. 1989). The same was true for the vaccinia virus-driven expression of LASV GPC that demonstrated in infected cells the presence of GPC precursor and GP1 and GP2 posttranslational cleavage products that did not show any signs of assembly but nevertheless acted as an efficient vaccine against the LASV lethal challenge in guinea pigs (Auperin et al. 1988; Morrison et al. 1989). Moreover, vaccinia virus-driven expression of GPC protected rhesus monkeys against the LASV challenge (Fisher-Hoch et al. 1989). The simultaneous vaccinia virus-driven expression of NP and GPC resulted in the synthesis of authentic proteins in cells with respect to electrophoretic mobility, glycosylation, and posttranslational cleavage—but again—without any signs of self-assembly (Morrison et al. 1991). In general, the early work on protective properties of the LASV and LCMV proteins was highly important in the

sense of immunological theory, since these studies clearly demonstrated the importance of the strong cellular immune response against arenavirus proteins in the absence of measurable neutralizing antibodies, which protected experimental animals from clinical disease but not from infection (for more details, see a review by Meulen 1999).

The comprehension of general mechanisms of arenaviral self-assembly and the ways to involve arenaviruses in the VLP methodology evolved with understanding of the role played by the protein Z of LASV and LCMV and its ability to self-assemble in vitro (Kentsis et al. 2002) and in mammalian cells (Perez et al. 2003; Strecker et al. 2003). In fact, the small RING finger protein Z acts as a matrix protein that is a driving force for the viral assembly and budding. The protein Z self-assembles in the absence of any other viral proteins and is sufficient for the release of enveloped protein Z-based VLPs (for a review, see Urata and Yasuda 2012). Although protein Z demonstrates strong variability among different arenaviral species, all arenaviruses encode and use it to initialize self-assembly.

3D Structure of the VLPs

Electron microscopy showed that the protein Z-induced VLPs did not significantly differ in their morphology and size from LASV particles (Eichler et al. 2004). The protein Z recruits NP to cellular membranes where virus assembly takes place, but mutation of two proline-rich domains PTAP and PPXY within the protein Z drastically reduces the release of VLPs (Eichler et al. 2004). The fine 3D structure of the LASV protein Z was determined by triple-resonance NMR techniques (Volpon et al. 2008) and refined by the homology models and replica exchange molecular dynamics (May et al. 2010).

The crystal structure of the full-length LASV NP of Josiah (Qi et al. 2010) and AV (Brunotte et al. 2011) strains was solved at a resolution of 1.80 and 2.45 Å, respectively. Separately, crystal structures of functional LASV NP domains were determined. First, the structure of N-terminal domain in complex with single-stranded RNA was shown, suggesting the likely assembly by which viral ribonucleoprotein complexes are organized (Hastie et al. 2011b). Second, the structure of the immunosuppressive C-terminal portion that possesses exonuclease activity with strict specificity for double-stranded RNA substrates was demonstrated (Hastie et al. 2011a). At last, the crystal structure of the homologous C-terminal NP portion was determined for the JUNV (Zhang Yinjie et al. 2013), LCMV (West et al. 2014), and Tacaribe arenavirus (Jiang Xue et al. 2013).

The crystal structures were presented also for the attachment glycoprotein GP1 of MACV at 1.7 Å resolution (Bowden et al. 2009) and for the recombinant ectodomain of the LCMV transmembrane glycoprotein GP2 at 1.8 Å resolution (Igonet et al. 2011). Direct interaction of the protein Z with GPC, first of all, with SSC, was shown for LASV and LCMV by confocal microscopy and coimmunoprecipitation assays (Capul et al. 2007).

The combination of all available structural information into an integrated arenaviral high-resolution structure with its publication in the VIPER-dB structural database at http://viperdb.scripps.edu/index.php (Carrillo-Tripp et al. 2009) remains a problem of the near future.

Putative VLP Carriers

Generation of Arenaviral VLPs

Preparation and broad applications of arenavirus VLPs that are based on the protein Z self-assembly were hampered by the lack of assembled VLPs after the expression of arenavirus, namely JUNV, protein Z in bacterial and baculovirus-driven insect expression systems, the two broadly used systems for VLP production (Goñi et al. 2010). General complexity of the arenaviral structural organization prevented therefore simple solutions with bacterial or baculovirus-driven expression. It is not excluded that the localization in membranes and strong need for correct myristoylation of the protein Z were the reason for this failure (Strecker et al. 2006).

Nevertheless, a gradual progress in the elucidation of fine molecular mechanisms of self-assembly and budding of arenaviruses by the dissection of the whole processes into consecutive steps offered the prospect of the arenaviral VLP preparation at least in mammalian cells. A number of arenavirus VLPs have been prepared, as revealed in Table 32.1. First, obstacles of homotypic LCMV and heterotypic (LCMV, LASV, and MACV) self-association of their NPs were documented (Ortiz-Riaño et al. 2012). It was shown that LASV NP forms trimers upon expression in mammalian cells (Lennartz et al. 2013). The contribution of certain domains (Levingston Macleod et al. 2011) and mapped amino acid residues (D'Antuono et al. 2014) to self-assembly of TCRV NP was demonstrated.

Second, the ability of the protein Z to self-assemble into VLPs with the NP was shown by coexpression of the protein Z and NP of Mopeia virus, a close relative of the pathogenic LASV, when highly selective incorporation of the NP into protein Z-induced VLPs was observed (Shtanko et al. 2010). The self-assembly of the NP and protein Z is complicated, however, by participation of a host protein, namely an ESCRT-associated protein ALIX/AIP1 (Shtanko et al. 2011).

Third, the ability of the protein Z to self-assemble into VLPs with the GPs was demonstrated (Schlie et al. 2010).

Fourth, all three structural arenaviral components, protein Z, NP, and GPs, were found to self-assemble into VLPs after expression of the respective genes of JUNV and TCRV in mammalian cells (Casabona et al. 2009). Efficient multimilligram generation of LASV VLPs structurally and morphologically similar to native LASV virions but lacking replicative functions, was achieved (Branco et al. 2010). The LASV VLPs demonstrated typical pleomorphic distribution in size and shape by electron microscopy analysis and induced antibody response against individual viral proteins in mice in the absence of adjuvants (Branco et al. 2010).

The first application of the arenaviral chimeric VLPs was shown by the preparation of the JUNV Z-based fusion protein with eGFP (García CC et al. 2009). Further, both the ability of JUNV Z-eGFP fusion protein to generate VLPs in the absence of any other viral protein and the capacity of the Z protein to support fusions at its C-terminus, without impairing budding activity, were shown in a standard mammalian cell line 293T (Borio et al. 2012).

Mareze et al. (2016) generated a chimeric JUNV-based vaccine against dengue virus (DENV) by fusing the coding sequences of the JUNV protein Z to those of two cryptic DENV peptides that were conserved on the envelope protein of all serotypes of DENV. The VLP formation was observed by transmission electron microscopy. The chimeric VLPs were not only immunogenic in mice but also protective in vitro against infection of cells with DENV (Mareze et al. 2016).

Expression by VLPVs

The single-cycle virus-like-particles vehicles or vectors (VLPVs) are employing encapsulated alphavirus replicons

TABLE 32.1

Formation of VLPs by the Combination of Expressed Structural Proteins Derived from the *Bunyavirales* Order Members

Genus	Strain	Mammalian		Insect		Plant		Yeast		Bacteria	
		NO	YES	NO	YES	NO	YES	NO	YES	NO	YES
Family *Arenaviridae*											
Mammarenavirus	Junin virus		Z+NP+GP	Z						Z	
	Lassa virus	NP	Z	NP						NP	
		GPC	Z+GP	GPC						GPC	
		NP+GPC	Z+NP+GP								
	LCMV		Z								
	Mopeia virus		Z+NP								
	Tacaribe virus		Z+NP+GP								
Family *Hantaviridae*											
Orthohantavirus	Andes virus		Gn+Gc								
	Dobrava-Belgrade virus							N			
	Hantaan virus		N+Gn+Gc	N+Gn+Gc	N			Gc		N	
	Puumala virus		Gn+Gc			N		N			
Family *Nairoviridae*											
Orthonairovirus	Crimean-Congo hemorrhagic fever virus	N		N			Gn Gc				
Family *Peribunyaviridae*											
Orthobunyavirus	California encephalitis virus (snowshoe hare bunyavirus)			N							
	Schmallenberg virus							N		N	
Family *Phenuiviridae*											
Phlebovirus	Rift Valley fever phlebovirus	Gn+Gc	N+Gn+Gc		N+Gc	N+ Gn	Gn chimeric			N	
					N+Gn+ Gc						
					Gn+Gc						
	Human severe fever with thrombocytopenia syndrome phlebovirus									N	
Family *Tospoviridae*											
Orthotospovirus	Tomato spotted wilt tospovirus					N				N	

Notes: The viruses are given in alphabetical order within each specific hierarchical group. References are given in the text.

expressing foreign gene(s) of interests, as reviewed by Pushko and Tretyakova (2014). The alphavirus RNA replicon technology provided a reasonable compromise, in terms of safety and immunogenicity, between live-attenuated and inactivated vaccines. The vaccine particles prepared by this technology are not able to spread beyond the initially infected cells but can efficiently deliver and transduce the gene(s) of interest into target cells (e.g., dendritic cells, DCs). This vaccine technology was based usually on an attenuated strain of Venezuelan equine encephalitis virus (VEEV) (Pushko et al. 1997), and numerous alphavirus replicon-based vaccine candidates reached preclinical and clinical development (Pushko and Tretyakova 2014).

Thus, the VEEV-based VLPVs expressing LASV NP resulted in a highly effective production of a foreign gene (Pushko et al. 1997). Next, expression of the LASV NP and GPC proteins and high protective efficiency of the generated vaccine constructs was demonstrated in guinea pigs (Pushko et al. 2001). Moreover, dual expression of glycoprotein genes of both Ebola virus (EBOV) and LASV was achieved by using bivalent alphavirus particles (Pushko et al. 2001), as described in the *Filoviruses* paragraph of Chapter 31.

Furthermore, this technology was used to design VLPVs expressing LASV GP with enhanced immunogenicity and bivalent VLPVs expressing cross-reactive GP of Junin virus (JUNV) and Machupo virus (MACV), causative agents of Argentinian and Bolivian hemorrhagic fever (Carrion et al. 2012).

In the case of LASV, the bicistronic RNA replicons encoding wild-type LASV GPC (GPCwt) and C-terminally deleted, noncleavable modified glycoprotein (ΔGPfib) were encapsidated into the VLPV particles using VEEV capsid and glycoproteins provided in trans. In transduced cells, the VLPVs induced simultaneous expression of LASV GPCwt and ΔGPfib from 26S alphavirus promoters (Wang M et al. 2018). The LASV ΔGPfib was predominantly expressed as trimers, accumulated in the endoplasmic reticulum, induced ER stress, and apoptosis promoting antigen cross-priming (Wang M et al. 2018).

In the case of the bivalent JUNV and MACV vaccine, the bicistronic RNA replicons expressing multiple GPCs derived from both pathogenic arenaviruses were generated and proof of concept immunogenicity and efficacy studies performed (Johnson et al. 2020). However, the assembly of the expressed LASV proteins was not studied.

After alphaviral vectors, the yellow fever vaccine was used for the expression of LASV GPC or GP1 and GP2 (Bredenbeek et al. 2006) and demonstrated proper processing of GPC and protection against fatal LASV challenge in guinea pigs but not sterilizing immunity, unlike immunization with live reassortant LASV vaccine ML29 (Jiang Xiaohong et al. 2011).

Later, Kainulainen et al. (2018) presented a vaccine candidate that combined the scalability and efficacy benefits of a live vaccine with the safety benefits of single-cycle replication. The system consisted of Lassa virus replicon particles devoid of the virus essential glycoprotein gene and a cell line that expressed the glycoprotein products, enabling efficient vaccine propagation. The guinea pigs vaccinated with these particles were protected against fever, weight loss, and lethality after infection with Lassa virus (Kainulainen et al. 2018). At last, Salvato et al. (2019) constructed and demonstrated preclinical efficacy of a LASV vaccine candidate that was produced by the Modified Vaccinia Ankara (MVA) vector and resulted in the VLPs formed by GPC and Z proteins from the prototype Josiah strain lineage IV. A single intramuscular dose of the vaccine led to full protection of mice challenged with a lethal dose of virus, delivered directly into the brain (Salvato et al. 2019).

Exchange of Arenaviral Envelopes or Pseudotyped Arenaviruses

Although the substitution of arenaviral envelopes with foreign ones does not fit precisely into the traditional concept of the VLPs, we decided to include these approaches into the section describing arenaviruses as potential VLP-based epitope carriers. First, a chimeric LCMV variant, where the glycoprotein G of vesicular stomatitis virus (VSV) was used to substitute the GP of LCMV, was constructed (Pinschewer et al. 2003). This allowed identification of arenaviral GP as a serious Achilles heel hampering generation of live attenuated arenavirus vaccines by reverse genetic engineering (Bergthaler et al. 2006). The rLCMV/VSV-G chimeras provided strong immunological background for further development of GP exchange vaccines for combating arenaviral hemorrhagic fevers (Pinschewer et al. 2010). Further, placing LASV GP on the backbone of LCMV resulted in a chimeric virus that displayed high tropism for dendritic cells following in vitro or in vivo infection of mice (Lee AM et al. 2013). Introduction of point mutations and module exchanges into LASV GP on the pseudotyped rLCMV allowed further mapping of immunological and functional units of arenaviral GPs (Sommerstein et al. 2014). Another example of a chimeric virus is the Mopeia-Lassa reassortant ML29 vaccine virus, which contained L and Z genes from Mopeia virus, while NP and GP genes were derived from LASV, as reviewed substantially by Carrion et al. (2012).

PUTATIVE VLP SUBJECTS

Pseudotyping of Non-Arenaviral VLPs by Arenaviral GPs

The pseudotyping history started with the generation of replication-competent VSV vectors expressing the glycoproteins of LASV and other viral hemorrhagic fever agents such as Ebola and Marburg viruses (Garbutt et al. 2004). Further, pseudotyping of murine leukemia retroviral vectors by LASV, LCMV, JUNV, and MACV GPs not only allowed identification of arenaviral receptors on the cells but also explored the role played by the arenaviral GPs in viral entry in the absence of other arenavirus proteins (Reignier et al. 2006). An efficient model for the fast and quantifiable detection of neutralizing antibodies in human and animal sera was constructed on the basis of LASV GP-pseudotyped murine leukemia retroviral cores (Cosset et al. 2009). A lentiviral pseudotyping model was used for the studies on LCMV (Dylla et al. 2011) and a nonpathogenic Pichinde virus (PICV; Kumar et al. 2012).

The pseudotyping of rVSV with a set of arenaviral GPs including LASV, JUNV, SABV, MACV, Chapare virus (CHAPV), and a novel highly pathogenic arenavirus LUJV with production of the respective pseudotyped viruses in mammalian 293T cells was performed (Tani et al. 2014).

Remarkably, pseudotyping of VSV by LCMV GP contributed to the clinical development of the VSV as an anticancer agent for the VSV-based oncolytic virotherapy by preventing neurotoxicity of the pseudotyped rVSV, when pseudotyped with the nonneurotropic LCMV GPs (Muik et al. 2011). Furthermore, the involvement of the rVSV pseudotyped with LCMV GPs solved the problem of antivector immunity induced by vaccine and/or gene therapy vector and preventing boosts with the same vector (Tober et al. 2014). The implementation of LCMV-pseudotyped rVSV allowed the boost in mice receiving VSV vectors encoding ovalbumin as a model antigen (Tober et al. 2014).

Arenaviruses as a Source of Epitopes

A unique contribution of arenaviruses to the VLP methodology consisted in the discovery and applications of widely used classic model epitopes. First of all, this is the cytotoxic T cell (CTL) epitope with a basic sequence KAVYNFATC derived originally from the leader peptide of LCMV GP1 (Gairin et al. 1995; Hombach et al. 1995) and known under the names gp33–41, gp33, p33, or finally as a gold standard epitope of vaccinology and gene therapy (Desjardins et al. 2009). The immunodominant character and fine mechanisms of the gp33–41 presentation remained a subject of basic immunological investigations for more than 15 years after its discovery (Aichele et al. 1997; Hudrisier et al. 1997; Gallimore et al. 1998; Ludewig et al. 1998; Kotturi et al. 2008; Bunztman et al. 2012). Crystal structure of the complex formed by the gp33–41 epitope with the murine MHC class I molecules was resolved (Achour et al. 2002; Sandalova et al. 2005).

The gp33–41 epitope was used as a model epitope in the following VLP models: hepatitis B virus core (HBc; Ruedl et al. 2002, 2004; Storni et al. 2002; Storni and Bachmann 2003; Marsland et al. 2005), RNA phages Qβ (Ruedl et al. 2004; Bachmann et al. 2005; Schwarz et al. 2005; Agnellini et al. 2008; Bessa et al. 2008; Keller et al. 2010) and AP205 (Schwarz et al. 2005), murine leukemia virus (MLV) (Derdak et al. 2006; Desjardins et al. 2009), papaya mosaic virus (Lacasse et al. 2008), hamster polyomavirus (HaPyV; Mazeike et al. 2012), and rabbit hemorrhagic disease virus (Li K et al. 2013), as described in detail in the appropriate chapters.

Aside from VLPs but in the context of the potential vaccine applications, the gp33–41 epitope was used to label nonstructural protein NS1 of influenza virus A/PR/8/34,

which appeared to be a promising protective antigen for the design of novel modified live influenza virus vaccines (Mueller et al. 2010). Another application was the expression of the gp33–41 by the pseudogenome encapsidated into human papillomavirus VLPs as a potential gene therapy tool (Shi et al. 2001).

Another important LCMV CTL epitope covering amino acid residues 118–132 of the LCMV NP was inserted into the VP2 coat protein of the porcine parvovirus as a VLP model and assured protection of mice against lethal challenge with LCMV (Sedlik et al. 1997, 1999, 2000; Casal 1999; Rueda et al. 1999). Further, the NP 118–132 epitope was used as an immunological marker for a very interesting and unusual VLP carrier, namely tubule-forming nonstructuralprotein NS1 of bluetongue virus VLP model (Ghosh et al. 2002). Chimeric NS1-based tubules protected the immunized mice against challenge with a lethal LCMV dose (Ghosh et al. 2002). This story is described in Chapter 13.

After these highly successful CTL epitopes, LASV, LCMV, and other arenaviruses continued to be a source of potential CTL epitopes for further use in the VLP studies (see, e.g., Botten et al. 2006, 2010)

Finally, one study involved replacement of human immunodeficiency virus envelope protein domains, signal peptide, transmembrane, and cytoplasmic tail with the analogous elements of other viral envelopes including those of LASV GPs in an attempt to enhance incorporation of chimeric products into VLPs (Wang BZ et al. 2007).

HANTAVIRUSES

According to the latest ICTV issues (Laenen et al. 2019; Kuhn et al. 2020), the *Hantaviridae* family contains 4 subfamilies, 7 genera, and 53 species. Similar to arenaviruses, hantaviruses infect rodents, but they are transmitted by rodent feces.

Regarding structural proteins, as illustrated in Figure 32.2, the M RNA encodes a single open reading frame that is processed into the viral glycoproteins Gn and Gc, which are embedded into the lipid bilayer envelope of the virion, but the S RNA segment encodes the protein N, which, in addition to encapsidation of the genomic RNAs, functions as an RNA chaperone (for references see a review of Soldan and González-Scarano 2014). Unlike arenaviruses, the hantaviruses do not encode a matrix protein as a possible analogue of the protein Z that played the central role in the generation of arenaviral VLPs. It was speculated that the function of matrix protein in this case is carried out by the cytoplasmic tail of glycoprotein Gn that interacts with the nucleocapsid protein and/or genomic RNA (Strandin et al. 2013).

Concerning the 3D structure of hantaviral proteins, the presence of coiled-coil motifs and di- and trimerization with further multimerization was experimentally confirmed first for the hantaviral protein N (Alfadhli et al. 2001; Kaukinen et al. 2001). Next, direct interaction of the hantaviral protein N with the envelope glycoproteins Gn and Gc during intracellular trafficking was demonstrated (Wang H et al. 2010; Shimizu et al. 2013).

The electron cryomicroscopy and tomography studies showed that the structure of hantaviruses (Huiskonen et al. 2010; Battisti et al. 2011) demonstrated a unique paradigm and differed from the T = 12 symmetrical structure proposed for the phleboviruses (see later mention).

The crystal structure of the ectodomain Gn of Puumala hantavirus (PUUV) of the *Puumala orthohantavirus* species, genus *Orthohantavirus*, subfamily *Mammantavirinae*, expressed in mammalian cells was determined to 2.3 Å resolution using the single-wavelength anomalous diffraction method (Li S et al. 2016). Furthermore, using electron cryotomography of related apathogenic Tula orthohantavirus (TULV), the structure of the envelope glycoprotein spike complex was resolved to 16 Å (Li S et al. 2016).

The 3D structure of the ectodomain of the PUUV Gc glycoprotein, expressed in inset *Sf*9 cells, was solved using x-ray crystallography at pH 6.0 and 8.0 to 1.8 Å and 2.3 Å resolutions, respectively, where both structures revealed a class II membrane fusion protein fold in its post-fusion trimeric conformation (Willensky et al. 2016).

The VLP-forming ability of the hantaviral structural proteins is summarized in Table 32.1. The early attempts to produce hantaviral structural proteins N, Gn, and Gc in standard expression systems did not result in the generation of VLPs. This was observed by the vaccinia virus-driven expression of the glycoproteins Gn and Gc of Hantaan virus (HTNV; Pensiero et al. 1988), as well as by the expression of the protein N of HTNV (Schmaljohn et al. 1988) in mammalian cells. Second, no VLPs were detected by the baculovirus-driven expression of the proteins N, Gn, and Gc of HTNV in insect cells (Schmaljohn et al. 1990; Yoshimatsu et al. 1993). Third, no VLPs were detected after expression of the protein N of two hantaviruses in *E. coli* (Gött et al. 1991). Fourth, expression of the protein N of PUUV in transgenic tobacco and potato plants did not lead to any self-assembled products (Kehm et al. 2001). Fifth, the same was true for the expression of the glycoprotein Gc of HTNV in yeast *Pichia pastoris* (Ha et al. 2001) and of the protein N of PUUV (Dargeviciute et al. 2002), Dobrava (Geldmacher et al. 2004), or other (Razanskiene et al. 2004) orthohantavirus strains in *Saccharomyces cerevisiae*.

Nevertheless, further studies demonstrated great progress by the hantaviral VLP generation. The formation of hantaviral nucleocapsid-like structures, similar to authentic ribonucleoproteins appearing by detergent disruption of virions, was reported for the first time by the baculovirus-driven expression of the HTNV protein N in insect cells (Betenbaugh et al. 1995). Moreover, the first VLPs similar to the original HTNV virions were observed during simultaneous expression of the protein N and envelope glycoproteins Gn and Gc by the vaccinia virus-driven system in mammalian Vero or BHK cells but not by the baculovirus-driven expression in insect *Sf*9 cells (Betenbaugh et al. 1995). Later, coexpression of the HTNV protein N together with glycoproteins Gn and Gc in Chinese hamster ovary

(CHO) cells led to classical VLPs that stimulated highly specific antibody response against HTNV virus N protein and glycoproteins, comparable to that induced by commercial inactivated bivalent hantaviruses vaccine but with a higher level of specific cellular response to N protein than that of the inactivated vaccine (Li C et al. 2010).

The expression of Gn and Gc envelope glycoproteins of Andes orthohantavirus (ANDV) and PUUV in mammalian 293FT cells led to their self-assembly into pleomorphic VLPs, which did not require the protein N and were released to cell supernatants (Acuña et al. 2014). Moreover, a Gc endodomain deletion mutant did not abrogate VLP formation (Acuña et al. 2014), while acidification triggered rearrangement of Gc into a stable post-fusion homotrimer (Acuña et al. 2015).

As to the chimeric derivatives of hantaviruses, the fusion of GFP to the PUUV protein N expressed in *S. cerevisiae* did not result in the nucleocapsid-like particles (Antoniukas et al. 2006).

Concerning pseudotyping of nonhantaviral models by hantaviral GPs, first, Moloney MLV vector was used for the generation of particles pseudotyped by HTNV (Ma et al. 1999). Second, the VSV pseudotype bearing hantavirus envelope glycoproteins was produced and used in a neutralization test as a substitute for native hantaviruses (Ogino et al. 2003; Higa et al. 2012) and as an alternative vaccine in mice (Lee BH et al. 2006). Third, the lentiviral gene therapy vectors were pseudotyped with hantavirus envelopes in order to improve the transduction efficiency into vascular smooth muscle and endothelial cells (Qian et al. 2006). The recombinant pseudotyped lentivirus containing the HTNV glycoproteins was used also as a potential vaccine candidate (Yu et al. 2013).

As to hantaviruses as a source of epitopes, after determination of the N-terminal sequence of the PUUV as a major antigenic domain of the whole virus (Elgh et al. 1996; Gött et al. 1997), extensive attempts were initiated by Rainer G. Ulrich in 1997 (Koletzki et al. 1997) and were continued for a long time (Zvirbliene et al. 2014) in order to expose the N-terminal N domains on widely used hepatitis B core and hamster polyoma VLP vectors, as described in Chapters 38 and 8, respectively.

NAIROVIRUSES

According to the latest ICTV issues (Garrison et al. 2020), the *Nairoviridae* family contains 7 genera and 47 species. These viruses are maintained in arthropods or transmitted by ticks among mammals, birds, or bats. The most important nairovirus with public-health impact is Crimean-Congo hemorrhagic fever virus (CCHFV), which is tick-borne and endemic in much of Asia, Africa, and Southern and Eastern Europe. The most significant nairovirus with veterinary importance is Nairobi sheep disease virus (NSDV), also known as Ganjam virus, which is also tick-borne and causes lethal hemorrhagic gastroenteritis in small ruminants in Africa and India.

As shown in Figure 32.2, the nairoviral structural proteins generally comply with that of previously described hantaviruses.

The early attempts to produce structural CCHFV proteins N, Gn, and Gc in homologous and heterologous expression systems did not result in the generation of VLPs. This was observed by the expression of the protein N in mammalian cells (Garcia S et al. 2006). No VLPs were detected also by the baculovirus-driven expression of the CCHFV protein N (Saijo et al. 2002; Dowall et al. 2012).

At last, the genome-free nucleocapsid-like particles formed by the protein N and similar to those generated for HTNV in mammalian and insect cells (Betenbaugh et al. 1995) were prepared for CCHFV by using a baculovirus-insect cell expression system (Zhou ZR et al. 2011). Ghiasi et al. (2011) expressed the CCHFV glycoproteins Gn/Gc in transgenic tobacco, and feeding the leaves and roots to mice elicited specific IgA and IgG production, but no attempt to demonstrate VLPs was made in this study.

The crystal structure of the CCHFV protein N was determined (Carter et al. 2012a; Wang Y et al. 2012), Notably, structural alignment of the CCHFV protein N with other bunyaviral N proteins revealed that the closest CCHFV relative is the LASV arenavirus (Carter et al. 2012b).

As to high-resolution structures of the nairoviral envelope glycoproteins, the first one was the structure of the C-terminal cytoplasmic tail of the CCHFV glycoprotein Gn determined by NMR (Estrada and De Guzman 2011).

PERIBUNYAVIRUSES

According to the latest ICTV issues (Hughes et al. 2020), the *Peribunyaviridae* family contains 4 genera and 112 species. Typical for the family are members of the *Orthobunyavirus* genus, first, Bunyamwera virus (BUNV). Most orthobunyaviruses are transmitted by mosquitoes, and human infections occur through blood feeding by a vector arthropod. The orthobunyaviruses are associated with a broad spectrum of human disease, including encephalitides (La Crosse virus or LACV), febrile illnesses (BUNV), and viral haemorrhagic fever (Ngari virus, NRIV, Garissa variant). The adverse veterinary outcomes include fetal abnormalities and abortion storms among livestock (Cache Valley virus, CVV, and Schmallenberg virus, SBV; Hughes et al. 2020).

No VLPs were detected by the baculovirus-driven expression of the protein N of snowshoe hare orthobunyavirus in insect cells (Urakawa et al. 1988) or by the expression of the protein N of SBV in *E. coli* (Zhang Yongning et al. 2013) or *S. cerevisiae* (Lazutka et al. 2014).

The x-ray crystallographic structure at 3.3 Å resolution was determined for the BUNV protein N expressed in *E. coli* (Rodgers et al. 2006). The crystal structure of the protein N of SBV was resolved by Dong et al. (2013). The first complete electron cryotomography structure of BUNV was published by Bowden et al. (2013).

The first expression of the structural genes of BUNV in mammalian cells was achieved by the generation of a rescue system of the virus particles entirely from cloned complementary DNAs (Bridgen and Elliott 1996).

The idea that the glycoproteins Gn and Gc are the major viral components required for the formation of VLPs was confirmed by a so-called infectious VLP model including a viral minigenome (Shi et al. 2006, 2007, 2009). It was shown in this model that the cytoplasmic tails of both glycoproteins Gn and Gc contain specific information necessary for efficient virus particle generation.

The replacement of the N-terminal half of the glycoprotein Gc ectodomain of BUNV with the sequences of fluorescent proteins, either eGFP or mCherry, led to the generation of chimeric infectious viruses that allowed visualization of different stages of the infection cycle (Shi et al. 2010).

An example of the pseudotyping approach was represented by preparation of a recombinant virus from La Crosse (LACV) and Jamestown Canyon (JCV) viruses, both belonging to the California encephalitis virus serogroup of orthobunyaviruses (Bennett et al. 2012). As a result, LACV expressing the attachment/fusion glycoproteins of JCV demonstrated protection in mice against lethal challenge with both viruses and against challenge with JCV in rhesus monkeys (Bennett et al. 2012).

Next, Moloney MLV vector was used for the generation of particles pseudotyped by either LACV or HTNV (Ma et al. 1999).

PHENUIVIRUSES

According to the latest ICTV issues (Kuhn et al. 2020), the *Phenuiviridae* family contains 20 genera and 137 species. The most famous member is Rift Valley fever phlebovirus (RVFV) of the genus *Phlebovirus.*

The presence of coiled-coil motifs and di- and trimerization with further multimerization was experimentally confirmed for the protein N of RVFV (Le May et al. 2005), after the same was found for tomato spotted wilt tospovirus (TSWV; Uhrig et al. 1999; see later mention) and hantavirus (Alfadhli et al. 2001; Kaukinen et al. 2001). Direct interaction of the protein N with the envelope glycoproteins Gn and Gc during intracellular trafficking was demonstrated in the case of Uukuniemi phlebovirus (Overby et al. 2007a), in parallel with TSWV (Snippe et al. 2007; Ribeiro et al. 2009) and hantaviruses (Wang H et al. 2010; Shimizu et al. 2013).

The 1.6 Å crystal structure of the RVFV hexameric ring-shaped protein N forming a functional RNA binding site and expressed in *E. coli* was determined (Ferron et al. 2011). This structure explained the switch from an intra- to an intermolecular interaction mode of the N-terminal arm as a general principle that underlies multimerization and RNA encapsidation by bunyaviruses. Next, the crystal structure of the protein N of an emerging phlebovirus, namely human severe fever with thrombocytopenia syndrome phlebovirus (SFTSV), was resolved (Zhou H et al. 2013). Then, the crystal structure of the RVFV glycoprotein Gc followed and demonstrated an important but unanticipated evolutionary link between bunyavirus and flavivirus envelopes (Dessau and Modis 2013; Rusu et al. 2012), namely, the gained structure was similar to that described earlier for the alphavirus E1-E2 proteins.

The first complete bunyaviral 3D structures were resolved by electron cryotomography for RVFV (Freiberg et al. 2008) and Uukuniemi phlebovirus (Overby et al. 2008). These structures suggested the presence of an icosahedral quasi-symmetry with T = 12 triangulation for the glycoprotein protrusions (or spikes) in the case of the most regular particles. Intriguingly, this was the first appearance of the T = 12 icosahedral symmetry in the viral world. As mentioned earlier, further electron cryomicroscopy and tomography studies showed that the structure of hantaviruses (Huiskonen et al. 2010; Battisti et al. 2011) differed from the T = 12 symmetrical structure proposed for the phleboviruses.

At the very beginning of the VLP era, the vaccinia virus-driven expression of the RVFV glycoproteins Gn and Gc did not result in VLPs (Kakach et al. 1988). No VLPs were detected by the baculovirus-driven expression of the RVFV glycoproteins Gn and Gc in insect cells either (Schmaljohn et al. 1989; Takehara et al. 1990) or after expression of the RVFV protein N in *E. coli* (Jansen van Vuren et al. 2007). No data on production of VLPs was reported in efforts for the expression of the SFTSV protein N in *E. coli* (Jiao et al. 2012).

Later, the enormous potential of insect cells to produce excellent RVFV VLPs by the baculovirus-driven expression of the three structural proteins N, Gn, and Gc or two proteins N and Gc was clearly demonstrated by the famous Polly Roy's team (Liu et al. 2008). Both efficient insect and mammalian cell expression systems were developed to produce high-quality RVFV VLPs consisting of the proteins N, Gn, and Gc (Mandell et al. 2010b). Further, capability of the RVFV glycoproteins Gn and Gc alone, without the protein N, to self-assemble into VLPs was shown in a *Drosophila* insect cell expression system and used to generate an RVFV vaccine candidate that fully protected mice from a lethal challenge with RVFV (de Boer et al. 2010).

A large-scale process was generated for the baculovirus-driven production of RVFV VLPs, carrying N, Gn, and Gc proteins, in insect cells (Li Y et al. 2016, 2020).

The idea that the glycoproteins Gn and Gc are the only viral components required for the formation of VLPs was developed on Uukuniemi phlebovirus in a so-called infectious VLP model (Overby et al. 2006). It was shown in this model that the cytoplasmic tails of both glycoproteins Gn and Gc contain specific information necessary for efficient virus particle generation (Overby et al. 2007b). The same conclusions followed from the experiments on the role of the glycoproteins Gn and Gc from Bunyamwera orthobunyavirus, as described earlier.

In addition to the purely structural gene expression, the infectious Uukuniemi phlebovirus VLP model (Overby et al. 2006, 2007b) included a minigenome encoding reporter genes within the VLPs and resembled more a gene therapy vector rather than the classic genome-free VLP approach. The same idea of a minigenome (or minireplicon according to authors' terminology) approach was realized in the case of RVFV, where transcriptionally active RVFV nucleocapsids were formed by expression of the recombinant polymerase L and the protein N from the minireplicon packaged into VLPs by additional expression of viral glycoproteins (Habjan et al. 2009). Such infectious VLPs resembled authentic RVFV virions, were able to infect new cells, and conferred protection against lethal RVFV challenge in mice after three subsequent immunizations (Näslund et al. 2009). Replacement of the protein N-encoding gene with the reporter gene within the RVFV minireplicon improved the safety and efficiency of the proposed vaccine against lethal RVFV challenge in mice (Pichlmair et al. 2010).

The same minigenome (or minireplicon) idea was realized in the so-called nonspreading RVFV vaccine produced in a mammalian cell line (Kortekaas et al. 2011). The nonspreading vaccine that was improved by the expression of the glycoprotein Gn from the encapsidated minigenome demonstrated sterile protection against lethal RVFV challenge in lambs after a single intramuscular vaccination (Oreshkova et al. 2013).

A specific approach similar to the pseudotyping concept was used by the generation of mosaic (or chimeric according to the authors' terminology) RVFV VLPs that contained the Gag protein of Moloney MLV together with the RVFV proteins N, Gn, and Gc and demonstrated efficient protection of mice and rats against lethal RVFV challenge (Mandell et al. 2010a).

Concerning expression in plants, the RVFV protein N and a truncated version of Gn were produced in transgenic *Arabidopsis*, and the proteins were shown to be immunogenic when the plants were fed to mice (Kalbina et al. 2016). At last, the success of the influenza VLP production in plants led to the production of the RVFV VLPs in *N. benthamiana* via transient expression (Mbewana et al. 2019). Notably, the transmembrane domain of the RVFV Gn protein was replaced with that of HA from an avian influenza H5N1 strain, which allowed VLPs of 49–60 nm to be produced and purified, and these were shown to elicit a specific antibody response in mice, without any adjuvant, demonstrating therefore the first chimeric RVFV VLPs and the first demonstration of a detectable yield of RVFV Gn in plants (Mbewana et al. 2019). Generally, the production of the putative vaccines based on the *Bunyavirales* members in plants was systematically reviewed by Peyret et al. (2021), on the solid background of similar efforts in other enveloped viruses.

TOSPOVIRUSES

According to the latest ICTV issues (Kuhn et al. 2020), the *Tospoviridae* family contains 1 genus and 26 species. The representative member is TSWV of the only genus *Orthotospovirus*. The members of the *Tospoviridae* family infect plants and have become one of the limiting factors for vegetable crops such as tomato, pepper, and lettuce (for a review, see Turina et al. 2012).

As becomes clear from Figure 32.2, the genome of tospoviruses resembles that of the previously described *Hantaviridae*, *Nairoviridae*, *Perivunyaviridae*, and *Phenuiviridae* families.

No VLPs were detected after expression of the protein N of TSWV in *E. coli* (Richmond et al. 1998). Moreover, the homologous expression of the protein N of TSWV or other tospoviruses in plants did not lead to any self-assembled products (Chen et al. 2005).

Concerning chimeric derivatives of the structural TSWV proteins, the fusions of the protein N with fluorescent proteins did not result in the nucleocapsid-like particles (or VLPs) in mammalian cells (Snippe et al. 2007) or plants (Lacorte et al. 2007).

The presence of coiled-coil motifs and di- and trimerization with further multimerization was experimentally confirmed first for the protein N of TSWV (Uhrig et al. 1999), in parallel with that of hanta- and phleboviruses, as described earlier. Direct interaction of the protein N with the envelope glycoproteins Gn and Gc during intracellular trafficking was found also in the case of TSWV (Snippe et al. 2007; Ribeiro et al. 2009). The crystal structure of the protein N of TSWV was resolved by Komoda et al. (2013).

33 Order *Articulavirales*

The most amazing combinations can result if you shuffle the pack enough.

Mikhail Bulgakov, *The Master and Margarita*

The best thing about getting a flu shot is that you never again need to wash your hands. That's how I see it.

Chuck Palahniuk

ESSENTIALS

The order *Articulavirales* is the smallest among other negative single-stranded RNA virus orders by number of families and species. According to the latest ICTV issues (ICTV 2020), the order *Articulavirales* consists of 2 families, 8 genera, and 10 species. The family *Orthomyxoviridae* is well known world-wide since this is the home of influenza viruses. The minor *Amnoonviridae* family consists of a sole genus with a sole species, namely *Tilapia tilapinevirus*, that infects both wild and aquacultured populations of tilapia fish, farming of which goes back to Ancient Egypt.

According to the current new taxonomy, the order *Articulavirales* is the only member of the class *Insthoviricetes*, which, together with the class *Ellioviricetes* (with the only *Bunyavirales* order, see Chapter 32) forms the *Polyploviricotina* subphylum of the phylum *Negarnaviricota*, kingdom *Orthornavirae*, realm *Riboviria*.

The order is of extremely high medical importance, because of the continuous healthcare concern associated with influenza infections. There are four types of influenza virus, namely types A, B, C, and D belonging to the genera *Alphainfluenzavirus*, *Betainfluenzavirus*, *Gammainfluenzavirus*, and *Deltainfluenzavirus*, respectively. The influenza A and B viruses are clinically relevant for humans. Influenza A viruses are responsible for annual epidemics and all known influenza pandemics. The influenza B virus is primarily a human pathogen, which is not associated with an animal reservoir. It causes similar symptoms and disease as the influenza A virus. However, the frequency of the severe cases of influenza B infections appears to be significantly lower than that of influenza A. As to the influenza C, well-defined outbreaks have rarely been detected in humans, and the virus is rarely associated with severe syndromes. Most people have antibody to the influenza C virus by early adulthood, and influenza C is not included in the current influenza vaccine formulations. For more information and references, reviews of Pushko et al. (2008), Pushko and Tumpey (2015), and Pushko and Tretyakova (2020) are recommended.

Figure 33.1 presents a typical image of a member of the species *Influenza A virus* of the *Alphainfluenzavirus* genus. The virions are pleomorphic, mostly spherical in shape, and 80–120 nm in diameter. Filamentous particles can also be found, with 80–120 nm diameter and up to 2,000 nm in length.

GENOME

Figure 33.2 shows the negative-sense, single-stranded RNA genome of the family *Orthomyxoviridae* that contains 8 segments coding for 11 proteins and is encapsidated by nucleoprotein NP. The segment size ranges from 890–2,341 nucleotides with the total genome size of 13.5 Kb.

The genome encodes following structural proteins: hemagglutinin (HA), neuraminidase (NA), matrix (M1), and M2 proteins. The outer envelope of the virion particle is made of a lipid bilayer that contains glycoprotein spikes of two types, HA (~14 nm long trimer) and NA (~6 nm long tetramer). The HA and NA represent the two major glycoprotein antigens on the surface of influenza virions. Depending on antigenic characteristics of the HA and NA, influenza A (but not B) viruses are divided into antigenic subtypes—or serotypes—based on the antibody response to the viral HA (H) and NA (N). For influenza A viruses, there are currently 18 different HA serotypes (H1–H18) and 11 different serotypes of NA (N1–N11). The serotypes that have been found infectious in humans are: H1N1 (caused Spanish Flu in 1918 and Swine Flu in 2009), H1N2, H2N2 (caused Asian Flu in 1957), H3N2 (caused Hong Kong Flu in 1968), H5N1 (caused Bird Flu in 2004), H5N8, H6N1, H7N2, H7N3, H7N7, H7N9 (avian flu), H9N2 (avian flu), H10N7, and H10N8.

Some subtypes of the influenza A viruses presented a pandemic threat, the worst of which has been the Spanish flu pandemic, which circled the globe in 1918. The first pandemic of the twenty-first century was caused by the swine-origin 2009 H1N1 virus. An outbreak of human infection caused by an avian-origin H7N9 virus emerged in eastern China in the spring of 2013 and immediately raised pandemic concerns. In addition to H7N9 virus, avian-origin influenza viruses of the H5N1 subtype possess high pandemic potential. After the H7N9 and H5N1 subtypes, the H9N2 and H6N1 viruses also represent human pathogens with a pandemic potential (for references see Pushko and Tumpey 2015).

The HA protein is normally synthesized in the infected cells as HA_0 precursor that is cleaved posttranslationally by cellular proteases into HA_1 and HA_2. The cleavage exposes hydrophobic N-terminus of HA_2, which then mediates fusion between viral envelope and endosomal membrane. There is a direct link between cleavage and virulence of avian influenza viruses (Horimoto and Kawaoka 2001).

DOI: 10.1201/b22819-39

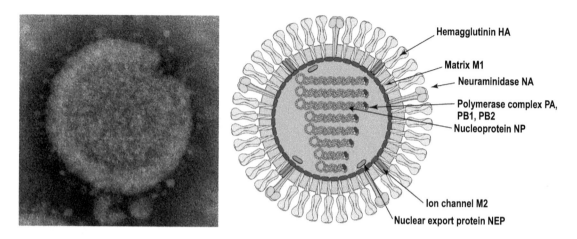

FIGURE 33.1 Electron micrograph and schematic cartoon of a typical representative of the order *Articulavirales*. The electron micrograph of an influenza virus is taken from the CDC Public Health Image Library (PHIL). The cartoon of a representative of the *Orthomyxoviridae* family is reproduced with kind permission from the ViralZone, Swiss Institute of Bioinformatics, http://viralzone. expasy.org (Hulo C et al. 2011). (Courtesy Philippe Le Mercier.)

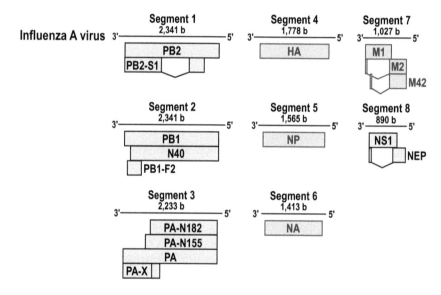

FIGURE 33.2 The genomic structure of a typical representative of the order *Articulavirales*, influenza A virus A/Puerto Rico/8/34 (H1N1). The structural genes are colored dark pink. The sequence data are from the classical sequence determination (Winter and Fields 1980, 1982; Fields et al. 1981; Hall and Air 1981; Van Rompuy et al. 1981; Winter et al. 1981; Fields and Winter 1982).

The NA has a sialidase activity that removes sialic acid moieties from the HA, NA, and host cell surfaces, thus facilitating the release of progeny virions from infected cells.

In addition to HA and NA, the outer lipoprotein envelope of the virus particle also contains several molecules of M2, an ion channel protein. The M2 protein is a minor component of the envelope that has been implicated in ion channel activity and efficient uncoating of incoming viruses during the infection cycle.

The inner side of the envelope is lined by the matrix protein M1, which is the most abundant structural protein of the influenza A virion. The M1 and M2 proteins are encoded within the same RNA segment 7, from which M1 and M2 are generated by using alternatively spliced RNA.

As to the minor *Amnoonviridae* family, the RNA genome of the *Tilapia tilapinevirus* species is segmented in ten segments encoding ten proteins and total genome's size of 10.3 kb, while each of the ten segments ranges in size from 465 to 1,641 nucleotides.

EXPRESSION OF THE STRUCTURAL GENES

Table 33.1 summarizes the available data on the formation of the influenza VLPs in different expression systems.

MAMMALIAN CELLS

Concerning studies on homologous expression of the influenza structural genes, it started with introduction of the

TABLE 33.1

Formation of VLPs by the Combination of Expressed Structural Proteins Derived from the *Alphainfluenzavirus* Genus Members

Subtype	Strain	Genes	Expression System	Testing Model	References
Genus: Alphainfluenzavirus species: Influenza A virus					
H1N1	A/WSN/33	HA, NA, M1	Human HeLa cells	NT	Ali et al. 2000
H1N1	A/WSN/33; A/Puerto Rico 8/34	NP, HA, NA, M1, M2, *PB2, PB1, PA, NS2*	Human embryonic kidney 293T cells	Mice	Neumann et al. 2000; Watanabe et al. 2002
H1N1	A/Puerto Rico/8/34	HA, NA, HIV-1 Gag-GFP	Human embryonic kidney 293SF cells	NT	Venereo-Sánchez et al. 2016, 2017, 2019
H1N1	A/California/04/09 (NA) A/New York/312/2001 (M1)	NA, M1	Human embryonic kidney 293T cells	Mice	Easterbrook et al. 2012
H1N1	A/ Gansu/Chenguan/1129/07; A/California/04/2009	NA	Human embryonic kidney 293T cells	NT	Lai JCC et al. 2010
H1N1 (COBRA)	COBRA (HA); A/mallard/ Alberta/24/01 H7N3 (NA)	HA, NA, HIV-1 Gag	Human embryonic kidney 293T cells	Mice,	Carter et al. 2016
H1N1	A/Puerto Rico 8/34	HA, M1	*Spodoptera frugiperda Sf*9 insect cells	Mice	Quan et al. 2007, 2012, 2013b
H1N1	A/California/04/09	HA, M1	*Spodoptera frugiperda Sf*9 insect cells	Mice	Quan et al. 2010
H1N1	A/South Carolina/1/18 HA); A/New York/312/01 (M1)	HA, M1	*Spodoptera frugiperda Sf*9 insect cells	Mice	Schwartzman et al. 2015; McCrow et al. 2018
H1N1	A/South Carolina/ 1/1918 (HA); A/Brevig_Mission/ 1/1918 (NA)	HA, NA, M1, M2	Spodoptera frugiperda *Sf*9 insect cells	Mice	Matassov et al. 2007
H1N1	reconstructed 1918 virus (Tumpey et al. 2005)	HA, NA, M1	*Spodoptera frugiperda Sf*9 insect cells	Mice, ferrets	Perrone et al. 2009
H1N1	A/California/04/09	HA, NA, M1	*Spodoptera frugiperda Sf*9 insect cells	Mice, ferrets	Pushko et al. 2010
H1N1	A/California/04/09 (HA, NA) A/Indoneisa/05/05 H5N1 (M1)	HA, NA, M1	*Spodoptera frugiperda Sf*9 insect cells	Human phase 2 trial	López-Macías et al. 2011
H1N1	A/California/04/09	HA, NA, M1	*Spodoptera frugiperda Sf*9 insect cells	Pigs	Pyo et al. 2012
H1N1	A/Puerto Rico 8/34	HA, M1, flagellin	*Spodoptera frugiperda Sf*9 insect cells	Mice	Wang BZ et al. 2008
H1N1	A/Puerto Rico/8/34	HA, NA, MLV Gag	*Spodoptera frugiperda Sf*9 insect cells	Mice, ferrets	Haynes et al. 2009
H1N1	A/WSN/33	M1, M2	*Spodoptera frugiperda Sf*9 insect cells	Mice	Song JM et al. 2011
H1N1	A/WSN/33	M1	*Spodoptera frugiperda Sf*9 insect cells	Mice	Song JM et al. 2011
H1N1	A/California/04/09	NA	*Spodoptera frugiperda Sf*9 insect cells	Mice	Kim KH et al. 2019
H1N1	A/Puerto Rico/8/34	HA, Gag	*Trichoplusia ni (T. ni)-* derived insect cell line *Tnms42*	Mice	Klausberger et al. 2020

(Continued)

TABLE 33.1 (CONTINUED)

Formation of VLPs by the Combination of Expressed Structural Proteins Derived from the *Alphainfluenzavirus* Genus Members

Subtype	Strain	Genes	Expression System	Testing Model	References
H1N1	A/New Caledonia/20/99	HA, M1	Plant *Nicotiana benthamiana*	Mice	D'Aoust et al. 2008
H1N1	A/New Caledonia/20/99	HA	Plant *Nicotiana benthamiana*	Mice	D'Aoust et al. 2008, 2009, 2010
H1N1	A - California - 04 - 09 (swine flu)	HA	Plant *Nicotiana benthamiana*	Mice	D'Aoust et al. 2010; Landry et al. 2014; Makarkov et al. 2017
H1N1 H3N2 Influenza B (quadrivalent)	A/California/07/09 H1N1; A/Victoria/361/11 H3N2; B/Brisbane/60/08; B/Wisconsin/1/10	HA	Plant *Nicotiana benthamiana*	Human Phase 2 trial	Pillet et al. 2016
H3N2	A/Victoria/3/75	NP, HA, NA, M1, M2, *PB2*, *PB1*, *PA*, *NS1*, *NS2*	Mammalian COS-1 cells	NT	Mena et al. 1996; Gómez-Puertas et al. 1999, 2000
H3N2	A/Victoria/3/75	HA, NA, M1	Mammalian COS-1 cells	NT	Gómez-Puertas et al. 2000
H3N2	A/Victoria/3/75	HA, M1	Mammalian COS-1 cells	NT	Gómez-Puertas et al. 2000
H3N2	A/Taiwan/083/06	HA, NA, M1, M2	Mammalian Vero cells	Mice	Wu et al. 2010
H3N2	A/Hong Kong/4801/14	HA, NA, M1	Mammalian CHO-K1 or Vero or 293T cell lines	Mice, Human in vitro model	Buffin et al. 2019
H3N2	A/Udorn/72	HA, NA	Human embryonic kidney 293T cells or HeLa cells	NT	Chen et al. 2007
H3N2 (COBRA)	COBRA (HA); A/mallard/ Alberta/24/01 H7N3 (NA)	HA, NA, HIV-1 Gag	Human embryonic kidney 293T cells	Mice, ferrets	Wong et al. 2017
H3N2	A/Udorn/72	HA, NA, M1, M2	*Spodoptera frugiperda Sf*9 insect cells	NT	Latham and Galarza 2001
H3N2	A/Udorn/72	HA, M1	*Spodoptera frugiperda Sf*9 insect cells	Mice	Galarza et al. 2005
H3N2	A/Udorn/72	M1	*Spodoptera frugiperda Sf*9 insect cells	NT	Latham and Galarza 2001
H3N2	A/Fujian/411/02	HA, NA, M1	*Spodoptera frugiperda Sf*9 insect cells	Mice, ferrets	Bright et al. 2007
H3N2	A/Hong Kong/68	HA, NA, MLV Gag	*Spodoptera frugiperda Sf*9 insect cells	Mice	Haynes et al. 2009
H3N2	A/Brisbane/10/07	NA, M1	*Spodoptera frugiperda Sf*9 insect cells	Ferrets	Smith GE et al. 2017
H3N2	A/Brisbane/10/07	HA, NA, M1	*Spodoptera frugiperda Sf*9 insect cells	Ferrets	Smith GE et al. 2017
H3N8	A/pintail/Ohio/339/1987 (HA); A/New York/312/01 (M1)	HA, M1	*Spodoptera frugiperda Sf*9 insect cells	Mice	Schwartzman et al. 2015
H5N1	A/Thailand/KAN-1/04	HA, NA, M2, MLV Gag	Human embryonic kidney 293T cells	Mice	Szécsi et al. 2006
H5N1	A/ Hanoi/30408/05	HA, NA, M1, M2	Mammalian Vero cells	Mice	Wu et al. 2010
H5N1	A/Vietnam/1203/04	HA	Mammalian Vero cells	Mice	Schmeisser et al. 2012

Subtype	Strain	Genes	Expression System	Testing Model	References
H5N1	A/Vietnam/1203/04 (HA) A/New York/312/2001 H1N1 (M1)	NA, M1	Human embryonic kidney 293T cells	Mice	Easterbrook et al. 2012
H5N1	A/Cambodia/JP52a/05	NA	Human embryonic kidney 293T cells	NT	Lai JCC et al. 2010
H5N1 (COBRA)	COBRA (HA); A/Thailand/1(KAN-1)/04 (NA); A/Puerto Rico/8/34 (M1)	HA, NA, M1	Human embryonic kidney 293T cells	Mice, ferrets, cynomolgus macaques	Giles and Ross 2011; Giles et al. 2012
H5N1	A/Thailand/1(KAN-1)/04 (HA); A/Viet Nam/1203/04 H5N1 (NA); A/WSN/33 H1N1 (M1, M2)	HA, NA, M1, M2	*Spodoptera frugiperda Sf*9 insect cells	NT	Pan et al. 2010
H5N1	A/Viet Nam/1203/04 (clade 1) and A/Indonesia/05/05 (clade 2)	HA, NA, M1	*Spodoptera frugiperda Sf*9 insect cells	Mice, ferrets, Human phase I/II trials	Bright et al. 2008; Crevar and Ross 2008; Mahmood et al. 2008; Khurana et al. 2011
H5N1	A/chicken/Hubei/489/04	HA, NA, M1	*Spodoptera frugiperda Sf*9 insect cells	Mice	Tao et al. 2009
H5N1	A/Viet Nam/1203/04	HAΔ (multibasic amino acid deletion), M1	*Spodoptera frugiperda Sf*9 insect cells	Mice	Song JM et al. 2010a
H5N1	A/chicken/Korea/Gimje/08	HA, M1	*Spodoptera frugiperda Sf*9 insect cells	Chickens	Choi et al. 2013
	A/chicken/Korea/Gimje/08	HA	*Spodoptera frugiperda Sf*9 insect cells	Chickens	Choi et al. 2013
H5N1	A/mallard/Maryland/802/07 (HA); A/New York/312/01 (M1)	HA, M1	*Spodoptera frugiperda Sf*9 insect cells	Mice	Schwartzman et al. 2015
H5N1	A/Indonesia/05/05	NA, M1	*Spodoptera frugiperda Sf*9 insect cells	Ferrets	Smith GE et al. 2017
H5N1	A/Indonesia/05/05	HA, NA, M1	*Spodoptera frugiperda Sf*9 insect cells	Ferrets	Smith GE et al. 2017
H5N1	A/Viet Nam/1203/04	HA, NA, M1	*Spodoptera frugiperda Sf*9 insect cells	Mice	Kang SM et al. 2009c
H5N1	A/Indonesia/05/05	HA, NA, M1	*Spodoptera frugiperda Sf*9 insect cells	Mice	Smith GE et al. 2013
H5N1	A/Vietnam/1203/04 or A/Indonesia/5/05	HA, NA, MLV Gag	*Spodoptera frugiperda Sf*9 insect cells	Mice, ferrets	Haynes et al. 2009
H5N1	A/chicken/Germany/14; A/chicken/West Java/Subang/29/07; A/chicken/Egypt/121/2012	HA, NA, BIV Gag	*Spodoptera frugiperda Sf*9 insect cells	Chickens	Kapczynski et al. 2016
H5N1	A/chicken/West Java Sbg/29/07 (HA); A/chicken/Egypt/121/12 (NA); A/turkey/Oregon/71 H7N3 (HA); A/turkey/Wisconsin/1/1966 H9N2 (HA);	HA, NA, BIV Gag	*Spodoptera frugiperda Sf*9 insect cells	Chickens	Pushko et al. 2017

(Continued)

TABLE 33.1 (CONTINUED)

Formation of VLPs by the Combination of Expressed Structural Proteins Derived from the Alphainfluenzavirus Genus Members

Subtype	Strain	Genes	Expression System	Testing Model	References
H5N1	A/tufted duck/ Fukushima/16/11	HA	Silkworm pupae	Chickens, mice	Nerone et al. 2015, 2017
H5N1	A/ Indonesia/5/05	HA, NA, M1, M2, APRIL/ HA$_{tm}$	*Spodoptera frugiperda Sf*9 insect cells	Mice	Hong et al. 2019
H5N1	A/ Indonesia/5/05	HA, NA, M1, M2, BAFF/ HA$_{tm}$	*Spodoptera frugiperda Sf*9 insect cells	Mice	Hong et al. 2019
H5N1	A/Viet Nam/1203/2004 (HA); A/PR/8/34 (M1)	HA, M1, M2e5x	*Spodoptera frugiperda Sf*9 insect cells	Mice	Kang HJ et al. 2019a
H5N1	A/ Thailand/1(KAN-1)/04 (HA); A/Viet Nam/1203/04 (NA); A/WSN/33 H1N1 (M1, M2)	HA, NA, M1, M2	*Spodoptera frugiperda Sf*9 insect cells	Mice	Hong et al. 2019
H5N1	A/tufted duck/ Fukushima/16/11	HA	Eri silkworm pupae		Maegawa et al. 2018
H5N1	A/ Indonesia/5/05	HA	Plant *Nicotiana benthamiana*	Mice, ferrets, Phase I clinical trial,	D'Aoust et al. 2008, 2009, 2010; Landry et al. 2010, 2014; Makarkov et al. 2017
H5N3	A/Duck/ France/02166/02 (HA, NA) H7N1 A/Chicken/ Italy/1067/99 (M1)	HA, NA, M1	*Spodoptera frugiperda Sf*9 insect cells	Ducks	Prel et al. 2008
H7N1	A/Chicken/ FPV/Rostock/34	HA, NA, M2, MLV Gag	Human embryonic kidney 293T cells	Mice	Szécsi et al. 2006
H7N3	A/chicken/Jalisco/CPA1/12	HA, NA, M1	*Spodoptera frugiperda Sf*9 insect cells	Mice	Smith GE et al. 2013
H7N3	A/Environment/ Maryland/261/2006 (HA); A/ New York/312/01 (M1)	HA, M1	*Spodoptera frugiperda Sf*9 insect cells	Mice	Schwartzman et al. 2015
H7N7	A/duck/Korea/A76/10	HA	Silkworm pupae	Chickens, mice	Nerone et al. 2017
H7N7	A/duck/Korea/A76/10	HA	Eri silkworm pupae	NT	Maegawa et al. 2018
H7N9	A/Wuxi/1/13	HA, NA, M1	Human embryonic kidney 293T cells	Mice	Zhang L et al. 2015
H7N9	A/Anhui/1/13	HA	*Spodoptera frugiperda Sf*9 insect cells	NT	Buckland et al. 2014
H7N9	A/Anhui/1/13	HA	*Spodoptera frugiperda Sf*9 insect cells	Mice, ferrets	Pushko et al. 2015
H7N9	A/Anhui/1/13 (HA, NA); A/ Indonesia/05/05 H5N1 (M1)	HA, NA, M1	*Spodoptera frugiperda Sf*9 insect cells	Mice; Human Phase 1 trial	Fries et al. 2013; Smith GE et al. 2013
H7N9	A/Shanghai/2/13; A/Puerto Rico/8/34 H1N1 (NA, M1)	HA, NA, M1	*Spodoptera frugiperda Sf*9 insect cells		Li X et al. 2015
H7N9	A/Anhui/1/13	HA, NA, M1	*Spodoptera frugiperda Sf*9 insect cells	Mice, ferrets	Liu YV et al. 2015

Subtype	Strain	Genes	Expression System	Testing Model	References
H7N9	A/Zhejiang/DTID-ZJU01/13 (HA); A/PR/8/34 H1N1 (NA, M1)	HA, NA, M1	*Spodoptera frugiperda Sf*9 insect cells	Mice	Xu et al. 2016
H7N9	A/Taiwan/S02076/13	HA, NA, M1	*Spodoptera frugiperda Sf*9 insect cells	Mice, chickens	Hu et al. 2017
H7N9	A/ Taiwan/1/13	HA, NA, M1	*Spodoptera frugiperda Sf*9 insect cells; HighFive cells	NT	Lai CC et al. 2019
H7N9	A/Shanghai/2/13	HA, NA, M1	*Spodoptera frugiperda Sf*9 insect cells	Mice	Ren et al. 2018
H7N9	A/Chicken/Guangdong/53/14; A/swine/Guangdong/01/1998 H3N2 (HA-TM)	HA or HA-TM (H7/H3), M1	*Spodoptera frugiperda Sf*9 insect cells	Mice	Qin et al. 2018
H7N9	A/Anhui/1/13 or A/Shanghai/1/13 (HA); A/Udorn/307/72 (M1)	HA, M1	Insect *Trichoplusia ni* cell line HighFive (BTI-TN-5B1–4)	Mice	Klausberger et al. 2014
H7N9	A/Shanghai/1/2013	HA	Silkworm pupae	Chickens, mice	Nerone et al. 2017
H7N9	A/Hangzhou/1/13	HA	Plant *Nicotiana benthamiana*	Mice, ferrets	Pillet et al. 2015
H9N2	A/HongKong/1073/99	HA, NA, M1	*Spodoptera frugiperda Sf*9 insect cells	Mice, rats, ferrets	Pushko et al. 2005, 2007
Genus: *Betainfluenzavirus* species: *Influenza B virus*					
	B/Phuket/3073/2013	HA, NA, M1	Mammalian CHO-K1 or Vero, or 293T cell lines	Mice, Human in vitro model	Buffin et al. 2019
	B/Florida 04/06; B/Brisbane 60/2008	HA, NA, M1	*Spodoptera frugiperda Sf*9 insect cells	NT	Gavrilov et al. 2011

Notes: The viruses are given in order of their subtypes and expression systems: first mammalian cells, then insect cells and plants; NT, not tested.

reverse genetics system for influenza viruses (Luytjes et al. 1989; Jin et al. 1994, 1997; Zhang J et al. 2000). Mena et al. (1996) generated a system in which a synthetic influenza A virus-like chloramphenicol acetyltransferase (CAT) RNA could be encapsidated, replicated, and packaged into VLPs in mammalian cells expressing all virus-encoded polypeptides from plasmids. Further, this rescue system was optimized by a factor of ~ 50–100-fold and used for the study of the role of proteins in formation of VLPs (Gómez-Puertas et al. 1999). Next, Gómez-Puertas et al. (2000) demonstrated that the matrix protein M1 was the only essential viral component for the VLP formation, while the viral ribonucleoproteins were not required for the particles. Moreover, when M1 was expressed alone in mammalian COS-1 cells, it assembled into VLPs that were released in the culture medium, whereas the two virus glycoproteins HA and NA were unable to form VLPs in the absence of M1 (Gómez-Puertas et al. 2000).

Contrary to these early findings in which matrix protein M1 was considered to be a real basis of VLPs, Chen BJ et al. (2007) generated the two-protein influenza VLPs, when HA treated with exogenous NA or coexpressed with viral NA was released from mammalian cells independently of M1. Such VLPs did not include M1, whereas incorporation of M1 into VLPs required HA expression. Notably, when M1 was omitted from VLPs, the latter were similar morphologically to those of wild-type VLPs or viruses (Chen BJ et al. 2007).

Moreover, Lai JCC et al. (2010) found that the expression of NA alone was sufficient to generate and release VLPs of the H5N1 and H1N1 subtypes in human embryonic kidney 293T cells. The biochemical and functional characterization of the NA-containing VLPs demonstrated that they were morphologically similar to influenza virions. Later, the N1 NA-alone VLPs derived from the 2009 pandemic H1N1 influenza virus were generated by baculovirus-driven expression in *Spodoptera frugiperda Sf*9 cells, and the N1 VLP-immunized mice developed cross-protective immunity against antigenically different influenza viruses (Kim KH et al. 2019).

The generation of the H5N1 HA-alone VLPs in mammalian cells was demonstrated when the HA gene was expressed by use of modified vaccinia virus Ankara (MVA) vectors (Schmeisser et al. 2012). However, the coexpression of the NA gene increased VLP production.

Nevertheless, in another study, when the NP and M1 proteins were synthesized in mammalian BHK-21 cells using Semliki Forest virus replicons, both M1 and NP were engaged in extensive homooligomerization reactions soon after synthesis, but there was no detectable heterooligomerization taking place between the two viral proteins or between these and host proteins (Zhao et al. 1998).

In parallel, other expression systems were established for generating infectious influenza virus entirely from cDNAs in mammalian cells (Fodor et al. 1999; Neumann et al. 1999). The plasmid-driven system was used for production of replication-incompetent influenza NS2 gene-knockout VLPs in mammalian cells (Neumann et al. 2000; Watanabe et al. 2002).

Generally, the production of VLPs containing HA in mammalian cells required coexpression of NA or exogenously added NA for the effective release of VLPs into culture media (Ali et al. 2000; Gómez-Puertas et al. 2000; Chen BJ et al. 2007), whereas in insect cells the VLPs containing HA were produced in the absence of NA expression (Latham and Galarza 2001; Guo L et al. 2003; Galarza et al. 2005; Quan et al. 2007). In contrast to mammalian cells, insect cells did not add sialic acids to N-glycans during the posttranslational modification (Lanford et al. 1989), which explained why VLPs containing HA could be released from the insect cell surfaces without requiring cleavage of sialic acid by the neuraminidase.

Insect Cells and Insects

The further development of the influenza VLPs was enhanced by the baculovirus-driven expression system. The first influenza VLPs, potentially applicable as vaccine candidates, were generated by Latham and Galarza (2001). Thus, the quadruple baculovirus recombinant that simultaneously expressed in Sf9 cells the HA, NA, M1, and M2 genes of influenza virus A/Udorn/72 of the H3N2 subtype ensured formation of VLPs. The latter demonstrated surface projections that closely resembled those of wild-type influenza virus. The VLPs were observed in cases when a combination of single recombinants included the M1 protein.

The investigation of the minimal number of structural proteins necessary for VLP assembly and release using single gene baculovirus recombinants showed that the expression of M1 protein alone led to the release of vesicular particles, which in gradient centrifugation analysis migrated in a similar pattern to that of the VLPs (Latham and Galarza 2001). Furthermore, the HA gene within the quadruple recombinant was replaced either by a gene encoding the G protein of vesicular stomatitis virus (VSV) or by a hybrid gene containing the cytoplasmic tail and transmembrane domain of the HA and the ectodomain of the G protein. Each of these constructs was able to drive the assembly and release of VLPs (Latham and Galarza 2001).

Furthermore, the VLPs formed by M1 and HA and produced in insect cells demonstrated great immunogenicity and protective efficacy following immunization of mice by either intranasal instillation or intramuscular injection (Galarza et al. 2005). The intranasal instillation of VLPs elicited antibody titers that were higher than those induced by either intramuscular inoculation of VLPs or intranasal inoculation with two sublethal doses of the challenge influenza virus of control group. The antibody responses were enhanced when the VLP vaccine was formulated with IL12 as an adjuvant. All mice were challenged with 5 LD_{50} of a mouse-adapted influenza A/Hong Kong/68 (H3N2) virus and demonstrated full protection (Galarza et al. 2005). Similarly, the two-protein HA and M1 VLPs were generated successfully in Sf9 cells and induced detectable protective immune responses in mice for the subtype H1N1 (Quan et al. 2007).

In parallel, Peter Pushko et al. (2005, 2007) generated the VLPs of avian influenza virus A/HongKong/1073/99 of the H9N2 subtype upon infection of insect Sf9 cells with recombinant baculoviruses expressing the three structural genes of influenza encoding the HA, NA, and M1 proteins. The three proteins were self-assembled and released into the culture medium as VLPs. The electron microscopic examination of negatively stained samples revealed the presence of the H9N2 VLPs with a diameter of approximately 80–120 nm, which showed surface spikes characteristic of influenza HA protein on virions and were associated frequently as groups. The VLP size variation observed was similar to previous observations of pleomorphic influenza virions. Moreover, the VLPs exhibited functional characteristics of influenza virus including hemagglutination and neuraminidase activities. Figure 33.3 demonstrates an example of the purified influenza H9N2 VLPs.

In mice, the VLPs elicited serum antibodies specific for influenza A/Hong Kong/1073/99 (H9N2) virus and inhibited replication of the influenza virus after challenge (Pushko et al. 2005). Although in insect cells HA_0 was not processed into HA_1 and HA_2 polypeptides, the antibody elicited with VLPs in mice efficiently recognized the HA_1 and HA_2 proteins derived from the wild-type influenza virus. Furthermore, the H9N2 VLPs were evaluated not only in mice and rats but also in ferrets, the most suitable animal model for preclinical evaluation of human influenza vaccine candidates. The adjuvants, such as Novasome—a nonphospholipid liposome nanoparticle—or alum adjuvants were used to enhance immunogenicity and protection efficacy of the VLPs, after challenge with influenza A/Hong Kong/1073/99 (H9N2) virus (Pushko et al. 2007).

Then, the three-component HA, NA, and M1 VLPs were produced by the baculovirus-driven expression system in insect cells for the influenza subtypes H3N2 (Bright et al. 2007), H5N1 (Bright et al. 2008), and H1N1 of the 1918 influenza A virus (Matassov et al. 2007; Perrone et al. 2009). The latter success suggested feasibility to make a safe and immunogenic vaccine to protect against the extremely virulent 1918 virus, using the VLP technology. The mucosal and traditional parenteral administrations of H1N1 VLPs were compared for the ability to protect against the reconstructed 1918 virus and the highly pathogenic avian H5N1 virus isolated from a fatal human case (Perrone et al. 2009). Thus, the mice that received two intranasal immunizations of H1N1 VLPs were largely protected against a lethal challenge with both the 1918 virus and the H5N1 virus.

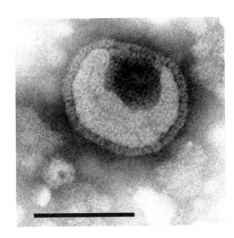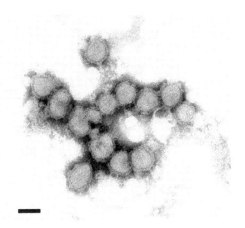

FIGURE 33.3 Negative staining electron microscopy of influenza H9N2 VLPs comprised of the HA, NA, and M1 proteins. Bars represent 100 nm. (Reprinted from *Vaccine*. 25, Pushko P, Tumpey TM, Van Hoeven N, Belser JA, Robinson R, Nathan M, Smith G, Wright DC, Bright RA. Evaluation of influenza virus-like particles and Novasome adjuvant as candidate vaccine for avian influenza, 4283–4290, Copyright 2007, with permission from Elsevier.)

The H3N2 VLPs induced broader immune responses than the whole virion inactivated influenza virus or recombinant HA vaccines (Bright et al. 2007). In the case of the H5N1 subtype, the HA, NA, and M1 of both clade 1 and clade 2 H5N1 isolates were used (see Table 33.1), and the H5 proteins were engineered not to contain polybasic cleavage site. Nevertheless, such cleavage-deficient HA proteins were capable of efficiently assembling into VLPs in *Sf*9 insect cells (Bright et al. 2008). The mice vaccinated with such VLPs were protected against challenge regardless if the H5N1 clade was homologous or heterologous to the vaccine. Moreover, the H5N1 VLPs were more effective influenza vaccines than recombinant non-VLP HA, so-called rH5, immunogens (Bright et al. 2008).

Generally, in the experiments that involved *Sf*9 cells and recombinant baculoviruses, two strategies have been used for expression of influenza VLPs. In one strategy, influenza structural genes were coexpressed from a single recombinant baculovirus (Latham and Galarza 2001; Pushko et al. 2005, 2007; Bright et al. 2007). The other strategy involved coinfection of the *Sf*9 cells with the two recombinant baculoviruses, one expressing the HA gene, whereas the other recombinant baculovirus expressed the M1 gene (Galarza et al. 2005; Quan et al. 2007). Usually, the influenza virus genes were codon-optimized for high-level expression in *Sf*9 cells and synthesized chemically (see e.g. Smith GE et al. 2017).

The presence of additional influenza proteins, such as NA and M2, in the VLPs provided additional benefits including induction of broader immune responses, especially in outbred populations, because the NA and M2 proteins contain additional antigenic epitopes (Bui et al. 2007). However, coexpression of several structural influenza genes in the cells during manufacturing process reduced the overall levels of expression of VLPs due to promoter dilution effect or because of excessive metabolic burden (Roldão et al. 2007). Furthermore, over-expression of the M2 gene dramatically

decreased the yields of VLPs (Gómez-Puertas et al. 1999). The authors hypothesized that the overproduction of M2, an ion channel protein, inhibited intracellular transport and drastically reduced accumulation of coexpressed HA and hence reduced the yield of VLPs.

Therefore, most of the basic approaches described earlier for expressing influenza VLPs included M1 protein, because M1 was the most abundant protein in the virion and because it has been shown initially to be the driving force of influenza virus budding (Gómez-Puertas et al. 2000). By the baculovirus-driven expression in insect cells, the influenza VLPs that contained the HA protein only were reported first by Choi et al. (2013). Pushko et al. (2015) generated in insect cells the H7 HA subviral particles of ~20 nm in diameter composed of approximately 10 HA_0 molecules, while no significant quantities of free monomeric HA_0 were observed in preparations. Such particles protected mice and ferrets from challenge with H7N9 influenza virus (Pushko et al. 2015).

Historically, the first baculovirus-driven expression of the avian H5 and H7 subtype HA gene in insect cells deserves special mention (Crawford et al. 1999). The purified HA proteins from insect cell cultures retained hemagglutination activity and formed rosettes in solution, indicating proper folding. Moreover, these proteins functioned as an effective subunit vaccine in chickens (Crawford et al. 1999). At the same time, the famous Walter Fiers lab fused the head part of the NA gene deprived of the transmembrane NA part to a tetramerizing leucine zipper domain of the yeast transcription factor GCN4, and the resulting product appeared as enzymatically active tetrameric neuraminidase (Fiers et al. 2001). The protective immunity induced by this engineered neuraminidase, however, remained fairly strain-specific.

After successful application of the insect *Sf*9 cells, the large-scale production of the influenza VLPs was achieved in silkworm pupae (Nerome et al. 2015, 2017; Maegawa et al. 2018). Thus, Nerome et al. (2015) efficiently produced

H5 HA VLPs with approximately 14-nm long spikes, similar to an authentic influenza HA spike, after expression of a single HA gene. The quantitative analysis of the yield of the avian H5 and H7 HA VLPs produced in silkworm pupae revealed a yield of 0.40 million HA units per pupa—corresponding to 1832 µg protein—and a million HA units, respectively, when the codon-optimized DNA was used, while the electron microscopy demonstrated the presence of large VLPs and small HA particles (Nerone et al. 2017). Furthermore, the dramatic overexpression of the H5 HA VLPs in Eri silkworm pupae was achieved (Maegawa et al. 2018).

At last, Correia et al. (2020) used the stepwise adaptive laboratory evolution (ALE) to adapt the insect HighFive cells to grow at a neutral culture pH 7.0 and markedly improved, up to fourfold, the production of HA-displaying VLPs and demonstrated the positive effect at the bioreactor scale. Moreover, the storage of the influenza HA-VLPs was improved by a trehalose-glycerol natural deep eutectic solvent system (Correia et al. 2021).

PLANTS

First, the H1 and H5 influenza VLPs consisting of M1 and HA proteins were produced by agroinfiltration of *Nicotiana benthamiana* plants (D'Aoust et al. 2008). The immunization of mice with the H5 VLPs conferred complete protection against a lethal influenza virus challenge. These results showed, for the first time, that plants could produce the enveloped influenza VLPs budding from the plasma membrane. Later, the assembly of HA into VLPs upon expression in *N. benthamiana* was demonstrated for several other HAs from type A influenza—including H2, H3, H6, and H9—and from type B influenza (D'Aoust et al. 2009). Finally, the authors assessed the speed of the platform with pandemic A-H1N1 swine flu outbreak in 2009 (D'Aoust et al. 2010). In support of this claim, Pillet et al. (2015) reported rapid generation of the H7 HA VLPs in plant, although without any indication on the structure and size of the HA-only VLPs. Later, immunological properties (Landry et al. 2014; Makarkov et al. 2017) as well as the structural composition (Le Mauff et al. 2015) of the tobacco plant-produced HA-only VLPs were evaluated in detail. In total, the industrial and clinical development of the influenza VLP-based vaccine was performed by Medicago, a Canadian company. First, the monovalent HA VLPs were generated against pandemic strains such as H5N1 (Landry et al. 2010; Pillet et al. 2018) and H7N9 (Pillet et al. 2015). Then, the quadrivalent HA VLP formulation for seasonal flu was generated, which successfully completed phase 1 (Pillet et al. 2016), phase 2 (Pillet et al. 2019), and phase 3 clinical trials (Ward et al. 2020). As noticed by Peyret et al. (2021), while Medicago is undoubtedly the world leader in plant-produced influenza VLPs, the other groups also demonstrated production of the HA VLPs in plants. Thus, Ed Rybicki (2014) demonstrated H5N1 HA VLP extraction from apoplastic washes of *N. benthamiana* leaves, and

Smith T et al. (2020) demonstrated the potential of plant-produced H6N2 HA as a veterinary vaccine for poultry.

The revolutionary strategy of the prospective VLP-based influenza vaccine engineering in heterologous expression systems was reviewed operationally by Pushko et al. (2008), Kang SM et al. (2009a, b), D'Aoust et al. (2010), Rybicki (2014), and Quan et al. (2016). Generally, the production of the influenza vaccines in plants, on the background of that of other enveloped viruses, was systematically reviewed by Peyret et al. (2021).

COMPARATIVE ANALYSIS

Buffin et al. (2019) emphasized some definite advantages of the mammalian expression systems, such as flexibility and the same glycosylation patterns as in a human virus. The authors generated the influenza A and B VLPs of HA, NA, and M1 proteins in CHO-K1, Vero, or 293 T cell lines using transient transfection. After production in a bioreactor and extensive purification, the mammalian VLPs closely emulated the exterior of authentic virus particles in terms of both antigen presentation and biological properties. The two VLPs produced contained more NA proteins on their surface with a HA:NA ratio around 1:1 than influenza viruses, which presented a HA:NA ratio of around 4:1. The successful immunogenicity studies were performed in mice and in a human in vitro model, the so called MIMIC system (Buffin et al. 2019).

When the yields of the influenza VLPs were compared directly between insect *Sf*9 cells and mammalian HEK293 suspension cells, Thompson et al. (2015) found that the *Sf*9 cells produced approximately 35 times more VLPs than HEK293 cells. Moreover, the VLPs purified from the *Sf*9 cells also showed more homogeneity, higher HA titer, and a stronger safety advantage, since the HEK293-derived product was contaminated with exosomes and microvesicles. In contrast, the *Sf*9 cells were maintained usually in a serum- and animal product-free medium. Remarkably, the yield and quality of the *Sf*9-produced influenza VLPs were enhanced by addition of the M2 inhibitor amantadine, a popular substrate of the early anti-influenza drugs (Zak et al. 2019). Nevertheless, the challenge for the baculovirus-driven VLP expression remained in the difficulties of separating the produced VLPs from the baculovirus expression vector during vaccine manufacturing.

NATURAL VLPs AS VACCINES

Currently licensed popular inactivated influenza vaccines are composed of formalin-treated whole virus or detergent-split viral components. The vaccine strains are selected based on epidemiologic and antigenic considerations of circulating human strains and their anticipated prevalence during the coming year. To obtain high-yield vaccine seed viruses, the chosen strains are adapted to grow in embryonated eggs, or reassortant viruses are generated containing HA and NA genes of current strains and genes for internal proteins of influenza A/Puerto Rico/8/34 (H1N1) virus, which confer high growth capacity in eggs (Robertson et al. 1992).

These live virus strains are cold adapted, temperature sensitive, and attenuated so as not to produce influenza-like illness by limiting their replication to only the upper respiratory tract in humans. The reassortant strains developed by serial passage at sequentially lower temperatures acquire attenuated phenotypes because of multiple mutations in gene segments that encode viral internal proteins (Murphy and Coelingh 2002).

As stated earlier in the *Expression of the Structural Genes* paragraph, the influenza VLPs induced protective immunity against infection in animal models (Galarza et al. 2005; Pushko et al. 2005, 2007; Bright et al. 2007, 2008; Matassov et al. 2007; Quan et al. 2007; Crevar and Ross 2008; Mahmood et al. 2008) and paved the way to the development of the safe influenza VLP vaccines. The good manufacturing process of the typical baculovirus-driven HA + NA + M1 VLP vaccine, namely H7N9 VLPs, was described in detail by Hahn et al. (2013) and Lai CC et al. (2019).

In 2013, the first egg-free recombinant trivalent influenza vaccine FluBlok based on baculovirus expression system-produced HA (Cox and Hollister 2009) received FDA approval (Woo et al. 2017). In 2017, the trivalent version was replaced by the quadrivalent version. For the 2020–2021 influenza season, the FluBlok vaccine was formulated to contain 180 μg HA per 0.5 mL dose, with 45 μg HA of each of the following four influenza virus strains: A/Hawaii/70/2019 (H1N1), A/Minnesota/41/2019 (an A/Hong Kong/45/2019-like virus) (H3N2), B/Washington/02/2019, and B/Phuket/3073/2013. Each of the four HAs is expressed in the *Sf*9 cell line using a baculovirus vector, namely Autographa californica nuclear polyhedrosis virus vector, extracted from the cells with Triton X-100 and further purified by column chromatography. Although not defined originally as a VLP vaccine (Cox and Hollister 2009), the Flublok vaccine has demonstrated the typical influenza HA-only particles (Buckland et al. 2014). In fact, such a purified HA formed rosette-shaped nanoparticles of approximately 30–50 nm in diameter (Holtz et al. 2014). The similar subviral particles consisting of approximately three to four trimers of the full-length HA were isolated from the *Sf*9 cells treated with mild detergent (Pushko et al. 2015).

Landry et al. (2010) published results of the phase I clinical trial with the plant-derived H5 HA-only VLP vaccine candidate generated originally by D'Aoust et al. (2010), as described earlier. The vaccine candidate demonstrated a good safety profile and promised immunogenicity in humans.

Fries et al. (2013) described the two phase I clinical trials of the experimental H7N9 VLP vaccines consisting of the HA and NA of A/Anhui/1/13 (H7N9) with the M1 of A/Indonesia/5/05 (H5N1) with and without the saponin-based ISCOMATRIX adjuvant. By day 35, after immunizations

at day 0 and day 21, the seroconversion and HA-inhibition (HAI) reciprocal antibody titers of 40 or more (a value of clinical benefit) were detected in 5.7% and 15.6% of participants receiving 15 μg and 45 μg of HA, respectively, without adjuvant. In contrast, three of the four treatment groups receiving VLPs with adjuvant had HAI seroconversion and reciprocal titers of 40 or more in more than 60% of participants. The significant increases in N9 NA-inhibiting antibodies occurred in up to 71.9% of recipients of the vaccine without adjuvant, 92.0% of recipients of vaccine with 30 units of adjuvant, and 97.2% of recipients of vaccine with 60 units of adjuvant. Remarkably, the unadjuvanted VLP vaccines generated primarily antibodies targeting the C-terminus of the HA_1 domain, predicted to be mostly buried on the native HA spikes, while the adjuvanted VLP vaccine generated antibodies against large epitopes in the HA_1 spanning the receptor binding domain, promoting therefore higher quality antibody immune response against avian influenza in naïve humans (Chung et al. 2015). It is necessary to emphasize also that the vaccines that were studied were released for human use within three months after the availability of the HA and NA sequences (Fries et al. 2013).

Smith GE et al. (2017) generated in baculovirus-driven expression system the VLP vaccines that were constructed from the N1 NA of A/Indonesia/05/2005 (clade 2.1.3.2) H5N1 virus or control N2 NA of A/Brisbane/10/2007 H3N2 virus, both together with M1. Therefore, the N1 vaccines included two VLP groups, namely N1 + M1 and H3 + N1 + M1, which were directly compared to control N2 VLPs (H3 + N2 + M1). The N1 + M1 VLPs were designed to study the effect of N1 without HA interference, while the hybrid H3 + N1 + M1 VLPs were designed to serve as a control for the effect of heterosubtypic H3 and conserved M1. The VLPs expressing homologous H5 HA (H5 + N1 + M1) representing a monovalent pandemic vaccine were also generated and used as the positive control vaccine for optimal protection. The ferrets vaccinated with the influenza N1 NA VLPs elicited high-titer serum NA-inhibition antibody and were protected from lethal challenge with A/Indonesia/05/05 virus (Smith GE et al. 2017).

The elaboration of a set of the VLP vaccines against influenza of the H7N9 subtype is particularly noteworthy. The cases of the H7N9 human infections caused by an avian-origin H7N9 virus emerged in eastern China in March 2013, immediately raised pandemic concerns and provoked rapid development of recombinant vaccines including VLPs. The list of the latter is given in Table 33.1 and reviewed extensively by Pushko and Tretyakova (2020). It is worthwhile to mention in this context the work of Schwartzman et al. (2015), who immunized mice with a cocktail of H1 + H3 + H5 + H7 VLPs, all elaborated on the HA + M1 basis.

By analogy, Luo et al. (2018) proposed sequential immunizations with baculovirus-driven heterosubtypic HA + M1 VLPs harboring H1, H8, and H13 from the HA phylogenetic group 1; H3, H4, and H10 from the HA phylogenetic group 2; or in various combinations. In this case, the immunized mice were fully protected when challenged with lethal doses of heterosubtypic viruses from either phylogenetic group.

Khurana et al. (2011) performed a phase I/II in humans vaccinated with H5N1 A/Indonesia/05/05 (clade 2.1) VLP vaccine manufactured in *Sf*9 insect cells and comprised of the HA, NA, and M1 proteins (Pushko et al. 2005; Bright et al. 2007, 2008). These findings represented the first report describing the quality of the antibody responses in humans following avian influenza VLP immunization.

López-Macías et al. (2011) reported a two-stage, phase II, randomized, double-blind, placebo-controlled study conducted in 4,563 healthy adults, 18–64 years of age, during the H1N1 2009 pandemic in Mexico. The VLP vaccine consisting of the HA, NA, and M1 proteins was safe and well-tolerated and induced robust HA inhibition immune responses after a single vaccination, with high rates of seroprotection (López-Macías et al. 2011).

Next, Pillet et al. (2016) presented the results from a phase I/II randomized clinical trial conducted to assess the safety, tolerability, and immunogenicity of a nonadjuvanted plant-derived quadrivalent VLP influenza vaccine (for the vaccine components see Table 33.1) in healthy adults. The plant-derived VLP vaccine was well-tolerated and elicited sustained, cross-reactive humoral and cell-mediated immune responses. Furthermore, it was shown that the VLP vaccine elicited a balanced immune response even in very old mice with comorbidities (Hodgins et al. 2019). At last, Ward et al. (2020) reported the two phase III efficacy studies of the plant-derived quadrivalent VLP vaccine, one in adults aged 18–64 years (the 18–64 study) and one in older people aged 65 years and older (the 65-plus study). These efficacy studies that were the first large-scale studies of any plant-derived human vaccine showed that the quadrivalent VLP vaccine can provide substantial protection against respiratory illness and influenza-like illness caused by influenza viruses in adults. It should be noted that Ward et al. (2021) announced a plant-produced SARS-CoV-2 vaccine, where the transmembrane domain TM and cytoplasmic tail CT of the protein S was also replaced with the TM/CT from Influenza H5 A/Indonesia/5/2005 to achieve the VLP assembly and budding. The self-assembled VLPs bearing S protein trimers were isolated from the plant matrix, purified, and evaluated in the phase I trial (Ward et al. 2021).

In order to overcome the high variability of influenza HA, in particular at the level of the head, Giles and Ross (2011) invented the generation of computationally optimized broadly reactive antigens (COBRAs) for the influenza HA. Thus, the COBRA HA H5N1 VLP vaccine elicited broadly reactive antibodies and demonstrated high protective efficiency against highly pathogenic avian influenza virus challenge in mice and ferrets (Giles and Ross 2011), as well as in cynomolgus macaques (Giles et al. 2012). In total, the COBRA H5 designs were shown to elicit increased breadth of antibody responses to 25 H5N1 strains that were isolated over a 12-year span. Then, Carter et al. (2016) used the COBRA approach for the H1N1 influenza virus. In this case, the generated COBRA HA VLPs elicited antibody responses to a panel of 17 H1N1 viruses isolated over almost a 100-year span of time and assured universal, broadly reactive, protective response against seasonal and pandemic H1N1 isolates in mice (Carter et al. 2016; Sautto et al. 2018). Sautto et al. (2020) explained the mechanism by which the COBRA HA-based VLPs elicit broad protective immunity against a diverse panel of influenza viruses. Wong et al. (2017) performed the COBRA approach for the H3N2 influenza viruses, where 17 prototype H3N2 COBRA HA proteins were screened in mice and ferrets for the elicitation of antibodies with HA inhibition activity against human seasonal H3N2 viruses that were isolated over the last 48 years.

At last, El-Husseiny et al. (2021) presented the insect cells-produced VLPs of the H5N1 virus of clade 2.2.1.2 and demonstrated full protection of pathogen-free chickens against virus challenge.

3D STRUCTURE

First, the x-ray crystal structure of the N-terminal portion of the M1 protein of the influenza A/PR/8/34 (H1N1) virus was determined at 2.08 Å resolution (Sha and Luo 1997). This structure revealed that residues 2–158 of the 258-residue-long M1 fold into two domains, the N-terminal and the middle M ones, each of which contained four α-helices. The protein formed a dimer. The highly positively charged region on the dimer surface was positioned to bind RNA while the hydrophobic surface opposite the RNA binding region was involved in interactions with the membrane.

The first electron cryotomograms of influenza virus, namely of the A/Aichi/68 (H3N2) strain, were published by Harris et al. (2006). The tomograms distinguished two kinds of glycoprotein spikes, HA and NA, in the viral envelope, which were formed by trimers of HA and tetramers of NA. A spherical influenza virion of average diameter 120 nm possessed ~375 spikes. The tomograms also resolved the matrix protein layer lining the envelope and depicted internal configurations of ribonucleoprotein complexes. They also revealed the stems that linked the glycoprotein ectodomains to the membrane and interactions among the glycoproteins, the matrix, and the ribonucleoprotein. Moreover, some virions had substantial gaps in their M1 layer, while others appeared to lack a M1 layer entirely, supporting the idea of an alternative budding pathway in which M1 is minimally involved (Harris et al. 2006).

The first electron cryomicroscopy image of the influenza VLPs was published by Pushko et al. (2008) for the purified H5N1 VLPs that contained engineered H5 protein, as described earlier. This image is presented in Figure 33.4.

In parallel, the 3D structures for the HA (Yamada et al. 2006) and NA (Russell et al. 2006) have been determined with high resolution using x-ray crystallography.

McCraw et al. (2018) reported the electron cryomicroscopy analysis of the recombinant H1 HA VLPs of the 1918 pandemic influenza virus. Figure 33.5 demonstrates the general structural schematics of this analysis. According

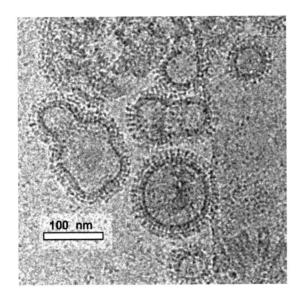

FIGURE 33.4 Electron cryomicroscopy of the H5N1 influenza VLPs containing engineered H5 protein. The VLPs were generated from influenza A/Indonesia/5/05 (H5N1) HA, NA, and M1 proteins. The HA protein was engineered to lack the multibasic cleavage site. Bar represents 100 nm. (Reprinted from Bright RA et al. *PLOS ONE* 2008;3:e1501.)

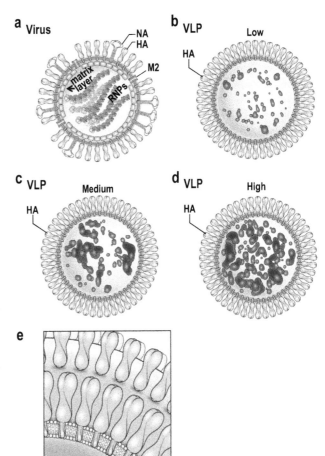

FIGURE 33.5 Structural schematics for the comparison of the molecular organization of influenza virus and VLPs. (a) Schematic of an influenza virus particle. (b, c, d) Schematics of VLPs containing low, medium, and high amounts of internal components, respectively. HA is light blue with the membrane in gray. Internal components of VLPs are brick red. For the virus, additional components are viral glycoproteins, which are NA (yellow), M2 (light purple), matrix layer (green) with genomic RNP filaments inside with polymerase proteins (yellow) at one end. For clarity only one row of surface glycoproteins is shown. (e) Schematic of HA on the surface of VLPs illustrating neighboring HA molecules that are beside, in front, and behind each other on the surface. (Reprinted from McCraw DM et al. *Sci Rep* 2018; 8:10342.)

to the authors' conclusion, the VLPs were dominated by spherical morphologies that were similar in sizes to influenza virions. The membrane-embedded HA spike components were projecting from the VLP surfaces, whereas some internal component material was present within the VLPs (McCraw et al. 2018).

When the influenza VLPs generated in mammalian cells from HA, NA, and HIV-1 Gag fused to GFP were characterized by nano-LC-MS/MS analysis to identify the possible protein cargo, nucleolin appeared as the most abundant cellular protein present in VLPs (Venereo-Sánchez et al. 2019). It is worth mentioning that the LC/MS/MS technique was used for the first time by the characterization of the influenza H3N2 and H5N1 VLPs that were generated in Vero cells and consisted of the four proteins: HA, NA, M1, and M2 (Wu et al. 2010). Now, the high-resolution LC-MS methods allow precise quantitation of both HA and NA protein concentrations in influenza VLP vaccine candidates (Guo J et al. 2020).

For purification of influenza VLPs, a nitrocellulose membrane-based filtration system (Park and Song 2017) and a cascade of ultrafiltration and diafiltration steps, followed by a sterile filtration step (Geisler and Jarvis 2018; Carvalho et al. 2019), were used successfully. The bioprocessing of influenza VLPs was recently reviewed (Durous et al. 2019).

HYBRID VLPs: REASSORTANCE AND PSEUDOTYPING

The configurations of influenza VLPs have included a heterologous M1 derived from an unrelated influenza virus strain (Prel et al. 2008) or even a retrovirus Gag protein in place of M1. Thus, the coexpression of murine leukemia virus (MLV) Gag and influenza HA, NA, and M2 in mammalian 293T cells led to VLPs that mimicked the properties of the viral surface of the two highly pathogenic avian influenza viruses of either H7N1 or H5N1 antigenic subtype (Szécsi et al. 2006). Then, the involvement of the MLV Gag by baculovirus-driven expression resulted in retrovirus-like particles containing HA of the H5N1 subtype, in that these pseudotyped particles appeared as consistent 100-nm spheres (Haynes et al. 2009). A Gag-to-HA ratio was noted to be approximately 3–4:1 and to exhibit hemagglutination specific activity, which were like those found in the influenza virus. Therefore, while there may exist specific interactions between cytoplasmic domains of viral envelope proteins and matrix or core components, such virus specific interactions were not required for efficient release of a self-sufficient budding protein Gag (Haynes et al. 2009).

A novel platform of hybrid influenza VLPs containing HA subtypes derived from the three distinct strains was generated by Pushko et al. (2011). This recombinant VLP design resulted in the expression of three HA subtypes colocalized within a VLP, as demonstrated on Figure P.4 in the *Prologue* chapter. The experimental triple-HA VLPs containing HA proteins derived from seasonal H1N1, H3N2, and type B influenza viruses were immunogenic and protected ferrets from challenge with all three seasonal viruses.

Similarly, VLPs containing HA subtypes derived from H5N1, H7N2, and H2N2 viruses protected ferrets from three potentially pandemic influenza viruses (Pushko et al. 2011). Furthermore, the hybrid VLPs that colocalized HA proteins derived from H5N1, H7N2, and H9N2 viruses, as well as NA and M1 derived from influenza PR8 strain, were generated (Tretyakova et al. 2013). The triple-HA VLPs exhibited hemagglutination and NA activities and morphologically resembled influenza virions. The intranasal vaccination of ferrets with the VLPs resulted in induction of serum antibody responses and protection against experimental challenges with H5N1, H7N2, and H9N2 viruses (Tretyakova et al. 2013). A seasonal trivalent VLP vaccine composed of the influenza A H1N1 and H3N2 and influenza B VLPs elicited antigen-specific immune responses in mice and ferrets (Ross et al. 2009). In this study, the ferrets vaccinated with the highest dose of the VLP vaccine and then challenged with the homologous H3N2 virus had the lowest titers of replicating virus in nasal washes and showed no signs of disease.

This highly prospective hybrid VLP technology represented a novel strategy for the rapid development of trivalent seasonal and pandemic influenza vaccines and was reviewed thoroughly by Pushko et al. (2013).

Furthermore, Kapczynski et al. (2016) presented a recombinant VLP containing HA proteins derived from three distinct clades of H5N1 viruses as an experimental, broadly protective H5 avian influenza vaccine. The vaccination with such VLPs protected chickens from H5N1 and H5N8 influenza viruses. Next, Pushko et al. (2017) prepared triple-subtype VLPs that colocalized H5, H7, and H9 antigens derived from H5N1, H7N3 and H9N2 viruses, together with N1 NA and Gag protein of bovine immunodeficiency virus (BIV). The immune responses to H5, H7, and H9 antigens and protective efficacy of VLPs after heterologous virus challenges were demonstrated in chickens (Pushko et al. 2017).

By using BIV Gag in place of M1, quadri-subtype VLPs were prepared, which coexpressed within the VLP the four HA subtypes derived from avian-origin H5N1, H7N9, H9N2 and H10N8 viruses (Tretyakova et al. 2016). Such quadrivalent VLPs showed hemagglutination and neuraminidase activities, reacted with specific antisera, and elicited antibodies in ferrets to the homologous H5, H7, H9, and H10 antigens (Tretyakova et al. 2016). The antiserum was also evaluated for cross-reaction with multiple clades of H5N1 virus, and cross-reactivity has been confirmed. The level of immune response suggested protection against multiple influenza subtypes, which was experimentally confirmed by challenge with H10 IAV, and ferrets were protected from the challenge with H10 virus (Pushko et al. 2016).

Overall, such multisubtype, mosaic VLPs that colocalized distinct HA subtypes in the envelope showed broader protection range against different influenza viruses (Pushko et al. 2016, 2017, 2018). Multisubtype mosaic VLPs combined advantages of conserved HA epitope and blended VLPs, as VLPs contained both subtype-specific head epitopes and the conserved stem epitopes, as demonstrated in a recent review by Pushko and Tretyakova (2020).

Recently, a VLP preparation consisting of HIV-1 Gag-VLPs pseudotyped with the HA was expressed using the novel *Trichoplusia ni (T.ni)*-derived insect cell line *Tnms*42 and tested successfully to assess the sole contribution of anti-HA immunity in limiting post-influenza secondary *Staphylococcus aureus* bacterial infection, morbidity, and mortality in a situation of a vaccine match and mismatch (Klausberger et al. 2020). The authors demonstrated that matched anti-HA immunity elicited by the VLP preparation might suffice to prevent morbidity and mortality caused by lethal secondary bacterial infection. Moreover, a fast chromatography step purification method together with the LC-MS quality control was developed for the chimeric HIV-1 Gag influenza-HA VLPs produced in *Tnms*42 insect cells (Reiter et al. 2020).

It is worth mentioning in this connection that the reassortant influenza viruses have been generated long before by coinfection of embryonated chicken eggs with influenza type A and type B viruses (Gotlieb and Hirst 1954). More recently, the live attenuated influenza viruses containing HA from A/PR8(H1), A/HK68 (H3), or A/VN (H5) strains in the backbone of the B/Yamagata/88 virus were reported (Hai et al. 2011). The reassortment phenomenon is routinely used in licensed influenza vaccines. In the reassortant vaccine virus, the vaccine-relevant HA and NA glycoproteins are embedded into the envelope, whereas other viral proteins are derived from type-specific donor strains, such as the A/PR/8/34 (H1N1) or B/Ann Arbor/1/66 viruses (Chen Z et al. 2008).

The phenotypically mixed HA/SHIV VLPs were generated and produced in baculovirus expression system by Guo L et al. (2003). The authors expected that the sialic acid-binding activity of the HA protein would provide a mechanism for targeting of the chimeric VLPs to mucosal surfaces.

The pseudotyped VLPs consisting of an influenza M1 core and respiratory syncytial virus (RSV) proteins F or G on the surface were constructed (Quan et al. 2011). This study demonstrated clearly that the VLP vaccination could provide effective protection against RSV infection in mice and the combined VLPs containing RSV-F and/or RSV-G were suggested as potential vaccine candidates against RSV.

Other pseudotyped influenza VLPs were constructed with SARS-CoV antigens (Liu YV et al. 2011). The baculovirus-driven VLPs were based on the influenza matrix M1 and carried the SARS spike (S) protein. These chimeric SARS VLPs have a similar size and morphology to the wild-type SARS-CoV. The SARS VLP vaccine completely protected mice from death when administered intramuscularly or intranasally in the absence of any adjuvant (Liu YV et al. 2011).

CHIMERIC VLPs AS INFLUENZA VACCINES

The chimeric influenza VLPs were generated, which carried a membrane-anchored form of the Toll-like receptor 5 (TLR5) ligand flagellin, the major proinflammatory determinant of enteropathogenic *Salmonella*, found to be glycosylated and expressed on cell surfaces (Wang BZ et al. 2008). Such chimeric influenza VLPs contained A/PR8/34 (H1N1) HA and M1 and the modified flagellin. The membrane-anchored flagellin was encoded by the gene that was generated by fusing a signal peptide from honeybee mellitin and the transmembrane and cytoplasmic tail regions from the influenza A virus PR8 HA to the 5' and 3' termini of the flagellin gene in frame, respectively. The flagellin-containing VLPs elicited higher specific IgG responses than standard HA and M1 VLPs, indicating the adjuvant effect of flagellin. Moreover, the enhanced IgG2a and IgG2b but not IgG1 responses were observed with flagellin-containing VLPs, illuminating the activation of Th1 class immunity. When immunized mice were challenged with homologous live PR8 virus, complete protection was observed for both the standard and chimeric VLP groups, whereas chimeric VLPs carrying flagellin as a membrane-anchored adjuvant induced enhanced cross-protective heterosubtypic immune responses (Wang BZ et al. 2008, 2010; Ko et al. 2019).

Wei et al. (2011) fabricated the chimeric influenza VLPs using M2 fusion protein, as it was illustrated by the M2 fusions with the enhanced green florescence protein (eGFP), flagellin, and profilin. Such influenza VLPs that were molecularly fabricated using M2 fusion proteins were found to have the similar particle sizes and morphologies and their biological functions, such as cell binding and the activation and maturation of dendritic cells, in vitro. The flagellin- and profilin-fabricated VLPs elicited more potent neutralizing antibody responses in mice against the homologous and the heterologous H5N1 viruses compared to the nonfabricated VLPs (Wei et al. 2011).

It should also be mentioned that Pan et al. (2010) were the first who provided influenza VLPs, in this case consisting of HA, NA, M1, and M2 proteins and produced in *Sf*9 cells, with eGFP fused to NA, as a tool for the influenza virus tracking.

The immunogenicity of HA + M1-containing influenza VLPs was increased significantly by a membrane-anchored incorporation of GM-CSF (granulocyte macrophage colony stimulating factor), an adjuvant molecule (Kang SM and Compans, unpublished data, cited by Kang SM et al. 2009a). The engineered GM-CSF was provided with a membrane-anchoring domain via glycosylphosphatidylinositol (GPI) and incorporated into the budding influenza VLPs.

Later, Liu WC et al. (2016) generated the three types of chimeric VLPs with incorporation of FliC, GM-CSF, or both GM-CSF and FliC. As a result, the immunizations with the chimeric FliC VLPs and GM-CSF/FliC H5N1 VLPs elicited more potent and broadly neutralizing antibodies and neuraminidase-inhibiting antibodies in murine sera and induced higher numbers of hemagglutinin-specific antibody-secreting cells and germinal center B cell subsets in splenocytes. The immunization with the chimeric GM-CSF H5N1 VLPs induced stronger Th1 and Th2 cellular responses. The chimeric GM-CSF/FliC H5N1 VLP constructs were further obtained to include H7 or H1H7 bi- or tri-subtype (Liu WC et al. 2016).

Hong et al. (2019) reported to obtain the highly immunogenic influenza VLPs by molecular incorporation with B-cell-activating factor (BAFF) or proliferation-inducing ligand (APRIL). Since BAFF and APRIL act as homotrimers to interact with their receptors, the authors engineered the chimeric proteins by direct fusion of BAFF or APRIL to the transmembrane anchored domain of the H5 HA gene. Therefore, the resulting VLPs contained combination of the viral HA, NA, M1, and M2 proteins, with addition of the chimeric APRIL/HA$_{tm}$ or BAFF-HA$_{tm}$. The results showed that immunizations with the HA-transmembrane anchored BAFF- or APRIL-VLPs elicited neutralizing antibodies in mice. Moreover, the authors obtained multisubtype H5+H7 BAFF-VLPs and H1+H5+H7 BAFF-VLPs and demonstrated that these multisubtype BAFF-VLPs were able to induce the production of neutralizing antibodies against multiple HA subtypes (Hong et al. 2019).

Wang BZ et al. (2012) developed the flagellin- and M2-enhanced influenza VLP vaccines with broad protective efficacy using the flagellin and M2e fusion. Thus, the authors designed a membrane-anchored fusion protein by replacing the hyperimmunogenic region of flagellin with four repeats of M2e and fusing it to a membrane anchor from HA. The fusion protein was incorporated into influenza virus M1-based VLPs. These VLPs retained TLR5 agonist activity comparable to that of soluble flagellin, while the mice immunized with the chimeric VLPs by either intramuscular or intranasal immunization showed high levels of systemic M2-specific antibody responses (Wang BZ et al. 2012).

A novel class of the chimeric influenza VLPs was started by Sang-Moo Kang's lab (Song JM et al. 2011; Kim MC et al. 2013a, b, 2014a, b; Song BM et al. 2016; Kim KH et al. 2018; Lee YT et al. 2018). First, Song JM et al. (2011) generated the VLP vaccine that contained the protein M1 together with the highly conserved M2 protein in a membrane-anchored form, the so-called M2 VLPs. Next, the authors introduced a tandem repeat of the extracellular domain of M2—or M2e—named now M2e5x, in order to increase the density and variation of M2e epitopes. The M2e5x was composed of heterologous M2e sequences including conserved sequences derived from human, swine, and avian origin influenza A viruses (Kim MC et al. 2013a, b, 2014a, b). Also, a domain, named general control nondepressible 4, which was known to stabilize oligomer formation, was linked to the C-terminal part of the M2e5x tandem repeat. The signal peptide from the honeybee protein melittin was added to the N-terminus of the M2e5x gene for the efficient expression on the insect cell surfaces, thus enhancing incorporation into the chimeric VLPs. Finally, the transmembrane and cytoplasmic tail domains were replaced with those derived from HA of A/PR/8/34 virus to increase the incorporation into the VLPs. Such a chimeric M2e5x gene was expressed then in the insect $Sf9$ cells in a membrane-anchored form and efficiently incorporated into the matrix protein M1-based VLPs at a several 100-fold higher level than that on influenza virions. The intramuscular immunization of mice with the M2e5x VLP vaccines led to cross-protection regardless of influenza virus subtypes in the absence of any adjuvant (Kim MC et al. 2013a, b, 2014a, b).

However, vaccination with the M2e5x VLPs alone was unable to protect chickens from highly pathogenic avian influenza virus (HPAIV) infection, resulting in no protection (Song BM et al. 2016). To provide an improved strategy of vaccination inducing enhanced protection against avian influenza virus, Kang HJ et al. (2019a) combined the M2e5x VLPs with the H5 HA VLP vaccine. These studies paved the way to a novel vaccination strategy by enhancing the cross-protective efficacy of live attenuated influenza virus vaccines by supplemented vaccination with M2e5x VLPs (Lee YT et al. 2019).

It is remarkable that the M2e5x VLPs, as well as H5 HA VLPs, contributed strongly to the elaboration of the prospective microneedle immunization technology (Kim YC et al. 2010; Song JM et al. 2010a, b; Quan et al. 2013a; Kim MC et al. 2015).

Qin et al. (2018) proposed a strategy of using chimeric HA proteins within the VLPs. Thus, the authors evaluated the biochemical characteristics and immunogenicity of the H7 VLPs composed of M1 and chimeric HA-TM protein, transmembrane domain TM of which was replaced by that from H3N2 subtype. The H7 VLPs-TM could assemble and release into the supernatant of $Sf9$ cells and demonstrated similar morphological characteristics as the wild-type H7 VLPs. Moreover, the H7 VLPs-TM provided better protection against homologous and heterologous H7N9 virus challenge in mice.

CHIMERIC VLPs AS A CARRIER FOR FOREIGN EPITOPES

The story of the chimeric VLPs was started with the generation of a chimeric influenza virus. Thus, Muster et al. (1994) inserted the highly conserved epitope ELDKWA from the ectodomain of HIV-1 gp41 into the loop of antigenic site B of the influenza virus HA. The resulting chimeric virus was able to elicit ELDKWA-specific immunoglobulins G and A in antisera of mice. Moreover, the distantly related human immunodeficiency virus type 1 isolates MN, RF, and IIIB were neutralized by these antisera (Muster et al. 1994).

Next Gilleland et al. (2000) inserted an 11-aa portion AEGRAINRRVE of so-called epitope 10 (aa 305–318) of outer membrane protein F of *Pseudomonas aeruginosa* into the HA antigenic site B, in parallel with the generation of analogous chimeric derivatives of tobacco mosaic virus (TMV) and cowpea mosaic virus (CPMV), as described in Chapters 19 and 27, respectively. This chimeric influenza virus protected against challenge with *P. aeruginosa* in the mouse model of chronic pulmonary infection. When the chimeric influenza virus containing epitope 10 and the chimeric TMV containing epitope 9 (aa 261-TDAYNQKLSERRAN-274 in the mature F protein were given together as a combined vaccine, the immunized mice produced antibodies directed toward both epitopes 9 and 10. The combined vaccine afforded protection against challenge with *P. aeruginosa* in the chronic pulmonary infection model at approximately the same level of efficacy as provided by the individual chimeric virus vaccines. These results proved in principle that a combined chimeric viral vaccine presenting both epitopes 9 and 10 of protein F has vaccine potential warranting continued development into a vaccine for use in humans (Gilleland et al. 2000).

Then, the 90- or 140-aa fragments of *Bacillus anthracis* protective antigen (BPA) were inserted at the C-terminal flank of the HA signal peptide and expressed as the HA_1 subunit (Li ZN et al. 2005). The chimeric proteins could be cleaved into the HA_1 and HA_2 subunits by trypsin and could be also incorporated into recombinant influenza viruses suggesting that viral envelope can tolerate foreign inserts without precluding assembly. The inserted BPA domains were maintained in the HA gene segments following several passages in Madin-Darby canine kidney (MDCK) cell cultures or embryonated chicken eggs. The immunization of mice with either recombinant viruses or with DNA plasmids that expressed the chimeric BPA/HA proteins induced antibody responses against both the HA and BPA components of the protein (Li ZN et al. 2005). Although VLPs were not generated in this study, these experiments suggested that similar modifications of HA protein with foreign epitopes may be compatible with the formation of chimeric influenza VLPs.

The chimeric influenza PR8/NA-F85–93 virus was generated, where the CTL epitope from the hRSV protein F located at aa positions 85–93 was inserted into the influenza NA stalk (De Baets et al. 2013). The PR8/NA-F85–93 virus induced in mice the specific CTL response after a single intranasal immunization and provided a significant reduction in the lung viral load upon a subsequent challenge with hRSV (De Baets et al. 2013).

Martina et al. (2014) engineered the chimeric influenza virus, strain A/PR/8/34, carrying the major immunogenic domain DIII of the glycoprotein E (gE) protein of West Nile virus (WNV), when the DIII stretch was fused to the

N-terminal region of NA, as described earlier for chimeric influenza virus expressing GFP (Rimmelzwaan et al. 2007). The one subcutaneous immunization provided full protection of vaccinated mice against WNV challenge, and the protection was mediated by antibodies and CD4+ T cells. Moreover, the vaccinated mice developed protective influenza virus-specific antibody titers, suggesting the influenza virus as an attractive platform for the development of bivalent WNV-influenza vaccines (Martina et al. 2014).

The early secretory antigenic target 6 protein (ESAT-6) epitope from *Mycobacterium tuberculosis*, a potent T-cell epitope of 20 aa residues in length, was engineered into the antigenic region B of the HA from strain A/New Caledonia/20/99 H1N1 (Krammer et al. 2010). The VLPs consisting of M1 and ESAT6-HA or HA were produced in insect cells and subjected to immunization studies in mice. The high serum antibody titers detected against the ESAT-6 clearly demonstrated the feasibility of influenza A VLPs serving as an efficient platform for the presentation of valuable foreign immune epitopes. As mentioned earlier, the influenza VLPs could incorporate the VSV protein G (Latham and Galarza 2001).

The influenza VLPs were used to generate a vaccine candidate against porcine reproductive and respiratory syndrome virus (PRRSV), where HA and M1 proteins from the H3N2 influenza virus and the PRRSV GP5 protein were fused to the cytoplasmic and transmembrane domains of the NA protein, and both incorporated into the chimeric VLPs (Xue et al. 2014). The latter elicited in mice serum antibodies specific for both PRRSV GP5 and the H3N2 HA protein, and they stimulated cellular immune responses compared to the responses to equivalent amounts of inactivated viruses. Then, a vaccine was generated against human clonorchiasis caused by the infection of *Clonorchis sinensis*, a liver fluke belonging to the class *Trematoda*, which is one of the major health problems in Southeast Asia (Lee DH et al. 2017). Thus, the coexpression of M1 as a core together with the gene of *C. sinensis* tegumental protein 22.3 kDa in insect cells led to formation of the regular VLPs that elicited *C. sinensis*-specific B and CD4+/CD8+ T cell responses, resulting in protection against worm infection in a rat model.

A putative hepatitis B VLP vaccine was generated by fusion of hepatitis B virus (HBV) preS with HA and coexpression of the fused protein together with M1 of the A/sw/Spain/53207/04 virus as a core in HEK 293T cells (Cai et al. 2018). The full-length preS sequence of the HBV subtype *adw*, 163 aa residues in length, was added to the transmembrane domain and cytoplasmic tail of the HA, aa 521–566, as the scaffold. The resulting preS-influenza VLPs stimulated the preS-specific CD8+ and CD4+ T cell responses in Balb/c mice and HBV transgenic mice. Moreover, the immunization with the preS-influenza VLPs provided protection against hydrodynamic transfection of HBV DNA in mice and suggested their development into promising prophylactic and therapeutic hepatitis B vaccines (Cai et al. 2018).

The influenza VLPs were used to construct a novel vaccine against toxoplasmosis, a disease of major medical and veterinary importance, which is caused by *Toxoplasma gondii*, an obligate intracellular protozoan parasite and remains widespread throughout the world. It causes congenital disease and abortion in humans and domestic animals. Lee SH et al. (2017, 2019b) generated multiantigen VLPs in insect cells coinfected with recombinant baculoviruses presenting IMC, ROP18, and MIC8 of *T. gondii* together with M1 as a core protein. In parallel, the authors generated the three combination VLPs carrying separately the IMC, ROP18, or MIC8 proteins, each together with M1. Compared to the combination VLPs, the multiantigen VLPs showed significantly higher levels of CD4+ T cell in mice, resulting in significant reduction on parasite burden and better survival of mice (Lee SH et al. 2019a, b). When the ROP4 and ROP13 of the rhoptry organelle proteins (ROPs) secreted by *T. gondii* were used to generate VLPs with the M1 as a core, the protection of mice against *T. gondii* infection was enhanced (Kang HJ et al. 2019c), and the ROP13-M1 VLPs were suggested as a potential vaccine candidate against *T. gondii* infection (Kang HJ et al. 2019b).

The baculovirus-driven M1-based influenza VLPs were used as a core to generate a malaria vaccine candidate (Lee SH et al. 2020). The VLPs carried the merozoite surface protein 9 (MSP-9), one of the antigens located within the merozoite surface and parasitophorous vacuole of the infected erythrocytes, from *Plasmodium berghei*. The VLP administration in a mouse model resulted in a significantly prolonged survival time of infected animals.

INFLUENZA AS A GENETIC VIRUS VECTOR

By elaboration of molecular engineering technologies, a concept of using influenza as a genetic vector was applied to the development of vaccine candidates for several pathogens (García-Sastre and Palese 1995; Palese et al. 1997). Both propagation-competent and propagation-incompetent influenza virus vectors have been described. Strategies for the construction of propagation-competent influenza vectors included the insertion of foreign antigenic epitopes into influenza virus glycoproteins, as described earlier (Li ZN et al. 2005), rescue of bicistronic genes into infectious viruses, and the expression of polyproteins. The influenza virus vectors have been obtained, which expressed both B- and T-cell epitopes from different pathogens. These constructs were shown to induce in vaccinated animals systemic and local antibody responses, and cytotoxic T-cell responses against the expressed antigenic epitopes (García-Sastre and Palese 1995).

For example, vaccination of mice with recombinant influenza and vaccinia viruses expressing antigens from *Plasmodium yoelii* resulted in a significant protective immune response against malaria in this model. Mice immunized with recombinant influenza viruses expressing human immunodeficiency virus (HIV) epitopes generated long-lasting HIV-specific serum antibody response as well as secretory IgA in the nasal, vaginal, and intestinal mucosa (Palese et al. 1997).

The propagation-incompetent influenza vectors were produced by Mena et al. (1996) by expressing all 10 influenza-encoded proteins, as described earlier. As with other propagation-incompetent virus vectors, the advantage of such vectors was that, when injected in a susceptible host, they generated immune response almost exclusively against the foreign antigen expressed from the vector, whereas low, if any, immune response was generated against the structural proteins of the vector itself (Pushko et al. 1997).

PACKAGING AND TARGETING

When Mena et al. (1996) expressed all ten influenza virus-encoded proteins in the COS-1 cells, as mentioned earlier, the transfected CAT RNAs were rescued into influenza VLPs that were budded into the supernatant fluids. The released VLPs not only resembled influenza virions but also transferred the encapsidated CAT RNA to the model MDCK cell cultures. Such VLPs required trypsin treatment to deliver the RNA to fresh cells and could be neutralized by a monoclonal antibody specific for the influenza A virus HA. These data indicated that influenza VLPs were capable of encapsidating a synthetic RNA, which could be delivered to fresh cells for expression of foreign genes of interest. For other virus vectors, it has been shown that such "vector VLPs" encapsidated nucleic acids by utilizing the ability of viral structural proteins to recognize specific encapsidation signals within the nucleic acid sequences.

When Wang BZ et al. (2008) generated their flagellin-carrying influenza VLPs, they checked ability of the pre-existing antiflagellin antibodies to enhance the targeting of the chimeric VLPs to antigen-presenting cells by the Fc portion of the VLP-bound antiflagellin IgG. However, they did not observe significant differences in responses with or without preexisting antiflagellin immunity.

Pan et al. (2010) were the first who followed the uptake of the chimeric NA-eGFP-incorporated VLPs by Jurkat cells and therefore shed some light on the mechanism of virus-mediated membrane fusion and its inhibition.

NANOMATERIALS

The influenza vaccines, both classical and recombinant, possess a long history of combination with different adjuvants,

application of which could be regarded as a starting point of many nanotechnological issues. For the general overview of the influenza adjuvanting history with a specific interest in the nanoparticle usage, the appropriate reviews are recommendable (Deng and Wang BZ 2018; Tregoning et al. 2018; Wang Y et al. 2018; Jazayeri and Poh 2019).

The ferritin-based influenza nanoparticles are described in Chapter 39.

Concerning polymer applications, the influenza antigens were entrapped into PLGA (Galloway et al. 2013; Herrmann et al. 2015; Hiremath et al. 2016; Watkins et al. 2017) or chitosan (Amidi et al. 2007; Chowdhury et al. 2017; Dhakal et al. 2018b) nanoparticles.

The desolvation-driven technology was used to generate nanoparticles from full-length HA protein, HA stalk antigens, immune stimulating protein molecules, and influenza internal nucleoprotein (NP) or polypeptide of known T cell epitopes of NP (Wang L et al. 2014, 2017; Chang et al. 2017; Deng et al. 2017, 2018). Thus, Deng et al. (2018) generated a double-layered protein nanoparticle was generated by desolvating tetrameric M2e peptides into the protein nanoparticle cores and coating these cores by crosslinking headless HAs. The resulting double-layered protein nanoparticles induced complete cross-protection against viruses of the same HA phylogenetic groups, while a cocktail of two representative layered HA stalk nanoparticles from both HA phylogenetic groups induced protection against divergent viruses spanning the whole spectrum of type A influenza.

Dhakal et al. (2018a) used liposome nanoparticles as a carrier to incorporate ten peptides representing the highly conserved T and B cell epitopes and delivered with monosodium urate crystal as adjuvant. Intranasal immunization of pigs with liposomal delivery of conserved peptide vaccines improved specific cellular and mucosal humoral immune responses (Dhakal et al. 2018a).

Concerning metal nanoparticles, the M2e peptide of influenza virus was conjugated to gold nanoparticles and codelivered with CpG oligodeoxynucleotide as an adjuvant (Tao et al. 2014, 2017; Tao and Gill 2015). The gold nanoparticles also conjugated influenza HA and adjuvant protein flagellin and triggered strong immune responses conferring heterologous protection in mice (Wang C et al. 2017, 2018).

34 Other Negative Single-Stranded RNA Viruses

By the pricking of my thumbs,
Something wicked this way comes.

William Shakespeare, *Macbeth*

ESSENTIALS

According to the current taxonomy (ICTV 2020), the remaining four orders of the negative-sense single-stranded RNA viruses are members of the *Haploviricotina* subphylum, phylum *Negarnaviricota*, from the kingdom *Orthornavirae*, realm *Riboviria*, just as in the order *Mononegavirales* described in Chapter 31 and in contrast to the orders *Bunyavirales* (Chapter 32) and *Articulavirales* (Chapter 33), which both form together the other subphylum, namely *Polyploviricotina*, of the *Negarnaviricota* phylum. The difference between both subphyla of the latter is clearly evidenced by the subphyla names.

The three orders *Muvirales, Serpentovirales*, and *Goujianvirales* are single members of the classes *Chunqiuviricetes, Milneviricetes*, and *Yunchangviricetes*, respectively, while the fourth order *Jingchuvirales* forms the *Monjiviricetes* class together with the huge and incredibly important *Mononegavirales* order, a subject of Chapter 31. All these four classes form together the previously mentioned *Haploviricotina* subphylum.

This advanced taxonomy was established by the well-justified proposals of the responsible ICTV members (Wolf et al. 2017; Koonin et al. 2019).

A specific story is connected with the genus *Deltavirus*, which is described here, since the RNA genome of its member, hepatitis delta virus (HDV), is single-stranded and negative-sensed. The genus was independent and not governed by any higher taxonomic unit until 2021, but currently it is subjected to the novel *Kolmioviridae* family of the freshly established realm *Ribozyviria*. Urban et al. (2021) published a comprehensive review on the present state of the HDV virology, immunology, and newest treatment approaches.

ORDER *MUVIRALES*

FAMILY *QINVIRIDAE*

The single family of the small *Muvirales* order involves the single *Yingvirus* genus that consists of 8 species. The member viruses have bisegmented genomes and infect invertebrates. The story of the appropriate names is interesting and is presented here as provided by the corresponding ICTV proposal (Wolf et al. 2017). The order *Muvirales* is after Mù (穆) who was the Duke of Qín, the ancient state during

the Spring and Autumn Period. The family *Qinviridae* is after the ancient Chinese Qín (秦) State during the Spring and Autumn Period. The genus *Yingvirus* is after the ancestral name of Mù, the Duke of the Qín State. The class *Chunqiuviricetes* is after Chūnqiū Shídài (春秋时代), namely the Spring and Autumn Period, in which Qín and Yuè were states.

ORDER *SERPENTOVIRALES*

FAMILY *ASPIVIRIDAE*

The single family *Aspiviridae*, former *Ophioviridae*, of the small *Serpentovirales* order involves the single *Ophiovirus* genus that includes 7 species. The members of the family are filamentous plant viruses, with single-stranded negative-sense—and possibly ambi-sense—RNA genomes of 11.3–12.5 kb divided into three to four segments and coding for up to seven proteins (García et al. 2017). The virions are naked filamentous nucleocapsids about 3 nm in diameter, nonenveloped, in the shape of kinked, probably internally coiled circles of at least two different contour lengths, 700 nm and about 2,000 nm. The open form circles can collapse to form pseudo-linear duplex structures of 9–10 nm in diameter. The virions are built by coat protein subunits of 43–50 kDa, varying in mass according to species and isolate (García et al. 2017).

The full-length coat gene of citrus psorosis virus (CPsV), which is a putative causal agent of psorosis, a widespread damaging disease of citrus, was cloned in *E. coli* (Salem et al. 2018). The product of 48 kDa carrying the His$_6$ tag was purified using batch chromatography under denaturing conditions and used to produce antibodies for the ELISA tests in mice. The ability of the recombinant coat to assemble was not addressed. The current insights on CPsV as a type member of ophioviruses were recently presented by Belabess et al. (2020).

ORDER *GOUJIANVIRALES*

FAMILY *YUEVIRIDAE*

The single family of the extra small *Goujianvirales* order involves the single *Yuyuevirus* genus with only 2 species. The yueviruses were identified by the systematic redefining of the invertebrate RNA virosphere by Shi et al. (2016). Again, the story of the corresponding virus names is intriguing. The order *Goujianvirales* is after the king of Yuè State, the ancient state during the Spring and Autumn Period. The

DOI: 10.1201/b22819-40

Yueviridae family is after the Yuè State, while the genus *Yuyuevirus* name is a synonym for Yuè (Wolf et al. 2017).

ORDER *JINGCHUVIRALES*

The populous *Jingchuvirales* order currently contains five families. The historically first family *Chuviridae* involves 14 genera with the historically first *Mivirus* genus and 36 species altogether. Remarkably, the order *Jingchuvirales* name is a synonym for the ancient Chinese Chǔ (楚) State during the Spring and Autumn Period, encompassing the middle and lower reaches of the Yangzi River. The family *Chuviridae* name is after the Chǔ State, and Mǐ (芈) for the genus *Mivirus* is the ancestral name of King Zhuang of the Chǔ State (Wolf et al. 2017).

Li et al. (2015) were the first who systematized a number of the arthropods-infecting chuviruses found in the geographic location Chǔ in China. Remarkably, these newly discovered chuviruses fell between the phylogenetic diversity of segmented and the unsegmented viruses. Although monophyletic, the chuviruses displayed a wide variety of genome organizations including unsegmented, bisegmented, and a circular form, each of which appeared multiple times in the phylogeny. The chuviruses had unique and variable orders of genes: the linear chuvirus genomes began with the glycoprotein gene G, followed by the nucleoprotein gene N, and then the polymerase gene L, whereas the majority of circular chuviruses were most likely arranged in the order L-(G)-N (Li et al. 2015).

GENUS *DELTAVIRUS*

This is a genus of the novel *Kolmioviridae* family from the newly established realm *Ribozyviria*, which involves 8 genera and 15 species altogether, the logical descendants of the former single species *Hepatitis delta virus*, with up to 40% nucleotide variation between isolates of the single member, namely hepatitis delta virus (HDV). Currently, there are eight clades of HDV with different geographical distributions (Magnius et al. 2018). The HDV genome is similar to that of viroids with circular RNA but encodes a structural protein, which forms a ribonucleocapsid (Bonino et al. 1986), thereby adhering to the definition of HDV as a satellite virus. Moreover, HDV requires the envelope of the helper virus, namely hepatitis B virus (HBV) from the family *Hepadnaviridae*, order *Bluberivirales* (Chapter 37), to produce infectious particles of 36–43 nm in diameter, without visible surface projections. The woodchuck hepatitis virus (WHV), a close relative of HBV, can also act as a helper in laboratory infection of woodchucks (Ponzetto et al. 1984). Figure 34.1 presents a cartoon of HDV, taken from the recent ICTV report (Magnius et al. 2018). This report describes HDV as a unique human hepatotropic pathogen with ~1.7 kb circular negative-sense RNA genome encoding a protein, hepatitis delta antigen (HDAg), which occurs in two forms, small S-HDAg (p24) and large L-HDAg (p27), both with unique structural and regulatory functions. The

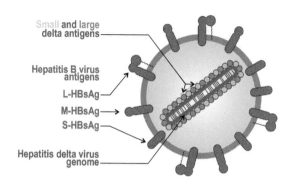

FIGURE 34.1 Schematic representation of a particle of hepatitis delta virus (HDV). (Redrawn with permission of Microbiology Society from Magnius L et al. *J Gen Virol*. 2018;99:1565–1566.)

HDV virion has an inner ribonucleoprotein comprising approximately 70 copies of variable amounts of S-HDAg and L-HDAg in close association with HDV RNA (Gudima et al. 2007). The L-HDAg differs from S-HDAg by a 19 aa C-terminal extension to the 195 aa of the S-HDAg (Chao et al. 1990). The S-HDAg is essential for HDV replication, whereas L-HDAg is essential for packaging and in some situations may inhibit replication (Taylor 2015). Remarkably, the L-HDAg must be present for delta antigen-containing particles to be released, while the S-HDAg is packaged if present in the cell but is not sufficient for particle formation. The HDV nucleocapsids can be released by treatment of virions with nonionic detergent and dithiothreitol. As stated by Magnius et al. (2018), the nucleocapsid symmetry has not been determined yet.

It should be noted, however, that Chang WS et al. (2019) described some HDV-like viruses in metagenomic samples from birds and snakes, which might use other helper viruses for generating infectious virion particles. This idea is strongly supported by recent studies showing that human HDV can replicate with the assistance of helper viruses other than HBV, such as vesiculoviruses, flaviviruses, and hepaciviruses (Perez-Vargas et al. 2019). In particular, hepatitis C virus (HCV) from the family *Flaviviridae*, order *Amarillovirales* (Chapter 22), was able to propagate HDV infection in the liver of coinfected humanized mice for several months (Perez-Vargas et al. 2019). Anyway, HDV remains the only known circular RNA virus, and it could be regarded as potentially originating from the human genome (Salehi-Ashtiani et al. 2006).

Recently, Szirovicza et al. (2020) cloned the snake deltavirus (SDeV) genome in a mammalian expression plasmid and initiated SDeV replication in cultured snake and mammalian cell lines. By superinfecting persistently SDeV-infected cells with reptarenaviruses and hartmaniviruses or by transfecting their surface proteins, the production of infectious SDeV particles was induced. Therefore, deltaviruses can likely use a multitude of helper viruses or even viral glycoproteins to form infectious particles.

To express the HDV gene in mammalian cells, Macnaughton et al. (1990) inserted it into a Rous sarcoma virus (RSV) expression vector and directed the synthesis of the recombinant HDAg in a continuous hepatoma HepG2 cell line. The product sedimented in rate zonal centrifugation with a value of 50S, close to that of particulate HBV surface antigen (HBsAg) of 22 nm in diameter, consisted of a single polypeptide of 24 kDa corresponding to the S-HDAg, and was serologically equivalent to the natural liver-derived HDAg. Gowans et al. (1990) used the recombinant HDAg in diagnostic assays for HDV antibody. Karayiannis et al. (1993) used the vaccinia virus-directed production of HDAg for immunization of woodchucks. Eckart et al. (1993) produced the two variants of HDAg, p24 and p27 by the vaccinia virus expression system and both did not provide protective immunity conferred by the recombinant vaccinia virus. In contrast, the repeated immunization of woodchucks with purified, recombinant p24 HDAg did confer significant protection against HDV challenge in some of the vaccinees (Ponzetto et al. 1993). The outcome of the HDV immunization experiments in woodchucks was thoroughly reviewed by Fiedler and Roggendorf (2001).

Chang FL et al. (1991) expressed both HDAg forms in hepatoma Huh-7 cell line and concluded that the L-HDAg is crucial for the HDV assembly, as mentioned earlier. Wang CJ et al. (1991) cotransfected Huh-7 cells with plasmids each encoding HDV cDNA and HBsAg gene and concluded that only the HBsAg was able to help in the assembly of HDV-like particles that were of 36–38 nm in diameter. Ryu et al. (1992) found that the HDAg can be packaged by surface proteins of either HBV or WHV. Among the three cocarboxy-terminal coat proteins of WHV, the smallest form was sufficient to package the HDV genome; even in the absence of HDV RNA, the HDAg was packaged by this WHV surface protein. Also, of the two coamino-terminal forms of the HDAg, only the L-HDAg was essential for packaging (Ryu et al. 1992). Later, Gudima et al. (2007) achieved efficient assembly of the HDV particles by cotransfection of Huh7 cells with two plasmids: one to provide expression of the large, medium, and small envelope proteins of HBV and another to initiate replication of the HDV RNA genome. Forouhar Kalkhoran et al. (2013) expressed the L-HDAg in Huh7 and HEK 293 cell lines.

Wang KC et al. (2005) used baculovirus vectors encoding L-HDAg and HBsAg to transfect hepatoma Huh-7 cells and demonstrated production of the true HBsAg-enveloped HDV VLPs. After this first success, Chen et al. (2005) performed the cotransduction of BHK cells with the baculovirus vectors and got VLPs resembling authentic virions in size and appearance. The production process was transferred to a bioreactor, in which the transduction efficiency was up to 90% and ensured high yields of the VLPs (Chen et al. 2005). Then, Chiang et al. (2006) employed the recombinant baculovirus transduced BHK cells to produce His-tagged L-HDAg. However, the L-HDAg alone was localized within the nucleus, in contrast to the secreted combination of both HBsAg and HDAg.

The first attempt to generate a chimeric HDAg derivative in the VLP form was performed by Lee et al. (1995). To discern whether the last 19 C-terminal aa of the L-HDAg solely constituted the signal for packaging with HBsAg, the authors constructed the LHDAg deletion mutants and tested their abilities to be packaged with HBsAg in cotransfection experiments. The results suggested that only the last 19 C-terminal aa of LHDAg were required for packaging and contained a packaging signal. When c-H-ras protein was provided with the 19 aa in question, it was cosecreted with HBsAg in the cotransfection experiment. Ward et al. (2001) continued this story and fused several C-terminal truncations of the HCV core protein to varying lengths of the N-terminally truncated L-HDAg. These constructs were analyzed for their ability to be expressed, and the particles secreted in the presence of HBsAg after transfection of Huh-7 cells were tested. However, the secretion efficiency of the HCV core-HDAg chimeric proteins was generally poor. The constructs containing full-length HDAg appeared to be more stable than truncated versions, and the length of the inserted protein was restricted to around 40 aa residues. Consequently, a polyepitope containing a B-cell epitope from human papillomavirus (HPV 16) and multiple T-cell epitopes from the HCV polyprotein was used to create the construct L-HDAg-polyB. This chimeric protein was shown to be reliant on the coexpression of HBsAg for secretion into the cell culture fluid and was secreted more efficiently than the previous HCV core-HDAg constructs. The resulting chimeric L-HDAg-polyB VLPs had similar biophysical properties to the HBsAg-enveloped L-HDAg VLPs (Ward et al. 2001).

Guilhot et al. (1994) produced transgenic mice that expressed the S-HDAg and L-HDAg in hepatocyte nuclei at levels equal to those observed during natural HDV infection, and no biological or histopathological evidence of liver disease was detectable during 18 months of observation, suggesting that neither of both forms was directly cytopathic to the hepatocyte in vivo. Hu et al. (1996) produced S-HDAg and L-HDAg by the in vitro transcription and translation in rabbit reticulocyte lysate.

Wang KS et al. (1987) were the first who reported the synthesis of the delta antigen in E. coli. Then, a chimeric protein was produced that carried 64 aa, a fragment of aa 13–76 of HDAg, fused to the N-terminal 98-aa fragment of polymerase of the RNA phage MS2, family Fiersviridae, order Norzivirales (Chapter 25), which was one of the most popular N-terminal fusion elements by bacterial expression of foreign genes at that time (Puig and Fields 1989; Saldanha et al. 1990). However, the immunization of woodchucks with the fusion protein did not protect against HDV infection (Karayiannis et al. 1990). Calogero et al. (1993) presented a purification procedure that allowed high yields of recombinant HDAg consisting of 8 aa of the N-terminus of β-galactosidase, HDAg sequence truncated at aa 204 and a trailer of 49 aa encoded by tet gene, of more than 1 mg of recombinant protein from 1 liter of E. coli cells. At last, Sheu and Lai

(2000) developed conditions for expressing S-HDAg and L-HDAg in *E. coli* as soluble proteins without any fusions and achieved large-scale purification of them to homogeneity. Later, Ding et al. (2014) produced the S-HDAg in high yield due to the optimization of the gene codons according to the codon preference of *E. coli* and using host cells with appropriate cell density. Under optimal expression conditions, the S-HDAg protein expression yielded 30 mg/l or 20% of the total cell mass in the soluble form, which could be regarded as one of the highest among any proteins expressed in *E. coli*. This procedure was described in detail by Lu et al. (2016). In parallel, Tunitskaya et al. (2016) expressed the His-tagged S-HDAg in *E. coli* and presented its purification and application to produce polyclonal anti-S-HDAg antibodies in rabbits.

Kos et al. (1991) were the first who reported the baculovirus-directed high level expression of HDAg in *Sf*9 cells. The recombinant HDAg located in the nucleus of the cells was purified in a denatured form by SDS-PAGE and induced an immune response in chimpanzees. Hwang et al. (1992) expressed both forms of HDAg in *Sf*9 cells and purified them to near homogeneity. The HDAg expressed in insect cells was antigenically and biochemically similar to that from mammalian cells. The L-HDAg was more heavily phosphorylated than the S-HDAg. Furthermore, the L-HDAg, but not S-HDAg, was isoprenylated. Both the large and small HDAg had similar RNA-binding properties and formed two distinct RNA-protein complexes with HDV RNA (Hwang et al. 1992). Then, Karayiannis et al. (1993) reported the baculovirus expression system producing HDAg, which was used then to immunize woodchucks, in parallel with the previously mentioned vaccinia virus-directed HDAg. Again, the immunization of woodchucks with the HDAg expressed by vaccinia or baculovirus did not elicit a humoral immune response. Hourioux et al. (1998) analyzed the binding capacity of the baculovirus-expressed S-HDAg and L-HDAg to synthetic peptides specific for the HBV envelope.

As to the expression in yeast, Wu et al. (1997) coexpressed L-HDAg and small HBsAg in *Saccharomyces cerevisiae*, and the assembly of VLPs from both proteins was demonstrated. Figure 34.2 demonstrates the obtained VLPs, in comparison with those produced by the expression of both genes in mammalian cells. The mice immunized with yeast-derived HDV-like particles simultaneously acquired antibodies against HBsAg and HDAg, indicating that both viral proteins were antigenic. Cartwright et al. (2017) expressed the His-tagged S-HDAg gene in *Pichia pastoris*, and the protein was purified by a Ni²⁺ affinity column.

To study the oligomerization of HDAg, Rozzelle et al. (1995) synthesized a 50-aa peptide quadrin that was based on the first third of HDAg, namely HDAg-(12–60)-Tyr = quadrin-(1–50). The quadrin was predominantly α-helical and contained a thermally stable α-helical coiled

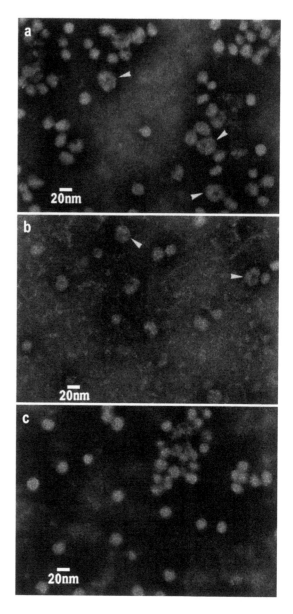

FIGURE 34.2 Electron microscopic examination of HBsAg-containing particles derived from yeast. (a) Electron micrograph of particles purified by immunoaffinity chromatography from yeast cells cotransformed with the HBsAg and L-HDAg encoding plasmids. (b) Culture supernatant of Huh-7 cells cotransfected with the HBsAg and L-HDAg encoding plasmids. (c) Yeast cells cotransformed with the HBsAg and S-HDAg encoding plasmids pGPD/S and pGAD/S-HDAg. Representative particles with a diameter larger than 30 nm are indicated by white arrowheads. (Reprinted from *Virology*. 236, Wu HL, Chen PJ, Mu JJ, Chi WK, Kao TL, Hwang LH, Chen DS, Assembly of hepatitis delta virus-like empty particles in yeast, 374–381, Copyright 1997, with permission from Elsevier.)

coil. It reacted with human anti-HDAg antibodies and interfered with the self-association of natural HDAg into oligomers. The x-ray crystallographic structure of quadrin was determined, where eight quadrin monomers were assembled

FIGURE 34.3 The 400-residue quadrin octamer. The four dimers are shown in light and dark gray for contrast. (Saderholm MJ et al., Characterization of deltoid, a chimeric protein containing the oligomerization site of hepatitis delta antigen. *Biopolymers.* 2007.88.764–773. Copyright Wiley-VCH Verlag GmbH & Co. KGaA. Reproduced with permission.)

into a square octamer (Zuccola et al. 1998). Figure 34.3 presents the unusual structure of this interesting HDV-derived peptide. Furthermore, Saderholm et al. (2007) designed and synthesized so-called deltoid with the following structure: CH₃CO-[Cys23]HDAg(12–27)-seryl-tRNA synthetase(59–65)-[Cys42]HDAg(34–60)-Tyr-NH₂, a chimeric protein that structurally resembled one end of the quadrin dimer and contained a single oligomerization site of HDV. In contrast to quadrin, the deltoid monomer/dimer system provided a simple model for studying the isologous association of two oligomerization sites to form a quadrin-like corner.

Section VI

Single-Stranded RNA Viruses
Using Reverse Transcription

35 Order *Ortervirales*

Why try to pursue what is completed?

Mikhail Bulgakov, *The Master and Margarita*

ESSENTIALS

The order *Ortervirales*, where *Orter-* is an inversion of *retro*, brings together most of the reverse-transcribing viruses. First, there are all accepted species of single-stranded RNA viruses that replicate through a DNA intermediate and belong to the Baltimore class VI, namely retroviruses as the members of the great family *Retroviridae*, which is divided into two subfamilies: *Orthoretrovirinae* and *Spumaretrovirinae*, and the families *Belpaoviridae*, *Metaviridae*, *Pseudoviridae*, better known as Bel/Pao, Ty3/Gypsy, and Ty1/Copia retrotransposons, respectively (Krupovic et al. 2018; Llorens et al. 2021). Second, the order *Ortervirales* includes the *Caulimoviridae* family of double-stranded DNA viruses that replicate through an RNA intermediate and belong therefore to the Baltimore class VII, in contrast to the other previously mentioned order members. For this reason, the family *Caulimoviridae* is separately described in Chapter 36 within Section VII of our book.

The current combination of the order members was justified by the structure of their polymerase/reverse transcriptase proteins that are sharing a common origin. The polymerase proteins are similar in structure and include aspartic protease (retroviral aspartyl protease) and an integrase belonging to the DDE recombinase superfamily. They also share similar capsid and nucleocapsid proteins/domains (Krupovic and Koonin 2017). Moreover, belpaoviruses, metaviruses, pseudoviruses, and retroviruses have some other important features in common (Krupovic et al. 2018).

Therefore, the order *Ortervirales* unites all reverse-transcribing viruses except the famous family *Hepadnaviridae* that forms an independent order *Blubervirales* and is a subject of Chapters 37 and 38 within Section VII.

The Baltimore class VI single-stranded RNA members of the order *Ortervirales* frequently integrate into the host genomes as part of their replication cycles. In contrast, the Baltimore class VII members of the families *Caulimoviridae* and *Hepadnaviridae*, often referred to as pararetroviruses (Hull and Will 1989), package circular dsDNA genomes and do not actively integrate into host chromosomes. The caulimoviruses also share some features with belpaoviruses, metaviruses, pseudoviruses, and retroviruses such a homologous aspartate protease. On the other hand, the hepadnaviruses of the order *Blubervirales* in the Baltimore group VII appear to be more distantly related to the order *Ortervirales*.

The belpaoviruses, metaviruses, pseudoviruses, retroviruses, and caulimoviruses share not only homologous proteins involved in genome replication and polyprotein processing but also the two principal protein components of the virions, namely the capsid and nucleocapsid proteins/domains (Krupovic and Koonin 2017; Koonin et al. 2018). However, the nucleocapsid domain appears to be absent in spumaretroviruses of the family *Retroviridae*. Furthermore, the similarities between belpaoviruses, metaviruses, pseudoviruses, retroviruses, as well as caulimoviruses extend to the mechanism of replication priming. All these viruses utilize host tRNA molecules as primers for genome replication by reverse transcription (Menéndez-Arias et al. 2017), in contrast to hepadnaviruses of the order *Blubervirales*, which are using a specific protein priming mechanism mediated by the polymerase terminal protein domain (for references and more information see Krupovic et al. 2018).

Altogether, the order *Ortervirales* belongs to the class *Revtraviricetes* of the phylum *Artverviricota*, kingdom *Pararnavirae*, realm *Riboviria*. According to the latest ICTV issues (Krupovic et al. 2018; ICTV 2020), the order *Ortervirales* consists of 5 families, 2 subfamilies, 28 genera, and 238 species. The reverse-transcribing viruses are widespread in animals, plants, algae, and fungi, and this broad distribution suggests the ancient origin(s) of these viruses, possibly concomitant with the emergence of eukaryotes (Krupovic et al. 2018). From the point of view of public health, the order *Ortervirales* is important because of such prominent human pathogens as human immunodeficiency viruses 1 and 2 (HIV-1; HIV-2) of the subfamily *Orthoretrovirinae* from the family *Retroviridae*.

Figure 35.1 presents a characteristic scheme of the popular member of the order *Ortervirales*, together with a portrait of its most well-known representative, HIV-1.

GENOME

Figure 35.2 presents typical genomic structures of all families of the order *Ortervirales*, except the *Caulimoviridae* family described separately in Chapter 36. Typically, the genomes of the presented family members share long terminal repeats (LTRs) and follow a layout of 5'—*gag—pol—env*—3' in the RNA genome, where *gag* and *pol* encode polyproteins managing the capsid and replication, respectively. The *pol* region encodes enzymes necessary for viral replication, such as reverse transcriptase, protease, and integrase. In contrast to the *Retroviridae* and *Metaviridae* members, the *Belpaoviridae* and *Pseudoviridae* transposons are nonenveloped and do not carry the *env* gene but the *gag* and *pol* genes only.

Generally, Krupovic and Koonin (2017) assumed that belpaoviruses, caulimoviruses, metaviruses, pseudoviruses, and retroviruses have evolved from a common viral ancestor, rather than from distinct capsid-less retrotransposons.

DOI: 10.1201/b22819-42

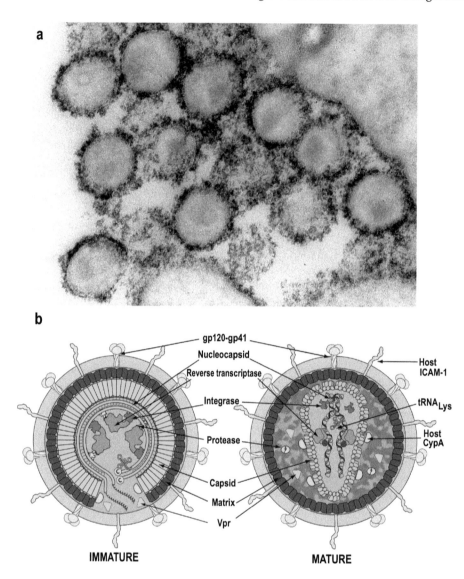

FIGURE 35.1 Electron micrograph and schematic cartoon of a representative of the order *Ortervirales*. (a) Human immunodeficiency virus-1 (HIV-1) particles that had been stained with ruthenium red, in order to show surface glycoprotein knobs. Ruthenium red stains the sugar portions of the surface glycoprotein molecules. The electron micrograph of Dr. Edwin P. Ewing, Jr (1985) is taken from the CDC Public Health Image Library (PHIL). (b) The cartoon of a representative of the *Retroviridae* family is reproduced with kind permission from the ViralZone, Swiss Institute of Bioinformatics (Hulo C et al. 2011). (Courtesy Philippe Le Mercier.)

It is noteworthy that the *Retroviridae* members package dimers of the single-stranded, positive-sense, linear RNA genomes. In addition to the *gag*, *pol* and *env* genes, these viruses may have accessory genes, which are located between *pol* and *env*, downstream from the env, or within the *env* gene. Some retroviruses may carry genes called oncogenes or *onc* genes, which are known for their ability to cause tumors in animals and transform cells in a culture into an oncogenic state.

FAMILY *PSEUDOVIRIDAE*

As summarized by the recent ICTV report (Llorens et al. 2021), the *Pseudoviridae* family combines LTR retrotransposons of the so-called Ty1/Copia group and involves currently 3 genera and 34 species altogether. The *Pseudoviridae* family members replicate via natural VLPs that are not infectious like normal virions but nevertheless can assure the pseudoviral lifecycle. However, unlike most other viruses, they do not have an extracellular phase. The *Pseudovirus* genus covers 20 species including the *Saccharomyces cerevisiae Ty1 virus* species, which played a unique role by the development of the VLP technologies. The *Hemivirus* genus involves, among other, two popular species: *Drosophila melanogaster copia virus* and *Saccharomyces cerevisiae Ty5 virus*. while the *Sirevirus* genus could be represented by the *Glycine max SIRE1 virus* species.

The yeast Ty1 retrotransposon became a leading candidate for the development of the VLP technologies due to

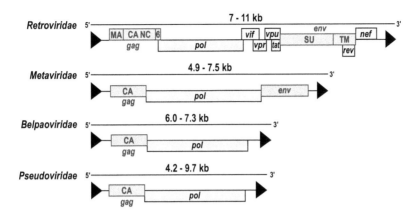

FIGURE 35.2 The genomic structure of the families belonging to the order *Ortervirales*. The structural genes are colored dark pink. The long terminal repeats are shown as black triangles. Abbreviations: 6, 6-kDa protein; CA, capsid protein; *gag*, group-specific antigen gene; *env*, envelope genes; INT, MA, matrix protein; NC, nucleocapsid; *nef, tat, rev, vif, vpr*, and *vpu*, genes that express regulatory proteins via spliced mRNAs; *pol*, polymerase gene; SU, surface glycoprotein; TM, transmembrane glycoprotein. The genomic structures were derived basically from Krupovic et al. (2018).

the classical studies of Alan and Susan Kingsman and their group. The eukaryotic Ty1 elements that are dispersed in the *S. cerevisiae* genome were found to show structural and functional similarities not only to prokaryotic transposons but also to retroviral proviruses a bit more than 35 years ago. These similarities consisted in the production of a Gag-Pol analogue by a specific frameshifting event (Mellor et al. 1985b), which fused two out-of-frame open reading frames TyA and TyB (Mellor et al. 1985c; Wilson et al. 1986), and the expression of the VLPs (Garfinkel et al. 1985). The TyA and TyB, overlapping by either 38 or 44 nucleotides, were found to produce 50-kDa p1 proteins and 190-kDa p3 proteins as a TyA:TyB fusion protein, which are processed and assembled into VLPs that contain Ty1 RNA and reverse-transcriptase activity (Adams et al. 1987b; Kingsman AJ et al. 1987).

The major structural core proteins of the Ty1 VLPs were thus generated by proteolytic cleavage of the p1, by analogy to the Gag precursor of retroviruses, by TyB-encoded protease. As a result, the wild-type Ty1 VLPs contained two proteins, p1 and p2. The structural protein p1 (49 kD, 441 aa) is the product of TyA, which is also known as p58 because of its apparent molecular mass on sodium dodecylsulfate polyacrylamide gels (Adams et al. 1987b) and is cleaved at the C-terminus by a protease encoded by TyB, producing the 45-kDa p2 protein, also known as p54. The cleavage takes place in the intermediate p1 particles, but the p2 protein is the main component of mature functional particles, which comprises 300 subunits and a 5-kb genomic RNA (Burns et al. 1992).

The p1 protein contains sufficient information for the assembly of pre-VLPs, which do not require the presence of either Ty1 protease or reverse transcriptase (Adams et al. 1987b). The expression of a truncated TyA gene encoding the first 381 aa of the p1 sequence led to the formation of p1–381 particles that were more uniform in diameter of 30–40 nm and morphology than wild-type particles (Burns

et al. 1992; Palmer et al. 1997). Like the wild-type VLPs, the p1–381 particles contained approximately 300 subunits (Brookman et al. 1995b).

The Ty1 VLPs were found to form open structures, unable to protect the encapsulated RNA, which distinguished them from typical viral capsids (Burns et al. 1992). The electron cryomicroscopy examination of the Ty1 VLPs showed that they were highly polydisperse in their radius distribution, and many of them formed an icosahedral T-number series (Palmer et al. 1997). The 3D reconstruction to 38-Å resolution from micrographs of T = 3 and T = 4 shells revealed that the single structural protein p1 encoded by the TyA gene was assembled into spiky shells from trimer-clustered units (Palmer et al. 1997). Moreover, the length of the C-terminal region of the p1 was found to dictate the T-number—and thus the size—of the assembled particles (Al-Khayat et al. 1999). The particles that were assembled from the full-length immature protein p1–440, mature processed protein p1–408 named p2, and its minimal self-assembly competent variant p1–346 demonstrated the prevalence of T = 9/T = 7, T = 7/T = 4, and T = 4/T = 3 shells, respectively (Al-Khayat et al. 1999). Figure 35.3 demonstrates the 3D reconstruction of the T-series of the Ty1 VLPs.

Intriguingly, the p1 product self-assembled well in *E. coli*, an organism that lacks endogenous retrotransposons, into spherical particles similar but not identical to the Ty1 VLPs from yeast (Luschnig et al. 1995). Years later, the chimeric Ty1 VLPs formed by the truncated p1 protein, (p1–379) and carrying the two well-characterized epitopes from beet necrotic yellow vein virus (BNYVV) were achieved not only in *E. coli* but also in yeast *Pichia pastoris*, reaching great expression levels of 20 and 65 mg/l, respectively (Uhde-Holzem et al. 2010). The epitopes were placed near the externally located N-terminus and at the internally located C-terminus of the p1 protein. Table 35.1 summarizes the gained data on the formation of the retroviral VLPs in different expression systems.

a   b   c

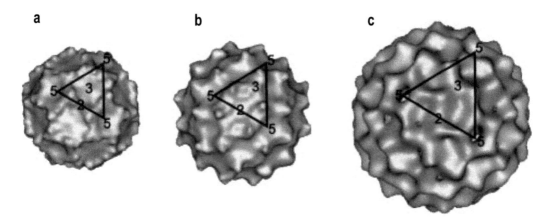

FIGURE 35.3   T-number series of Ty structures. 3D reconstruction of p1–346 T = 3, p1–408 T = 4 and a preliminary four particle T = 7 reconstruction were calculated to (a) 25 Å, (b) 35 Å, and (c) 60 Å resolution, respectively. The icosahedral symmetry axes are labeled, and the triangles enclose the icosahedral faces in the three maps. The reconstructions are viewed along the 2-fold symmetry axis and are surface contoured to enclose the calculated molecular volume for each map. (Reprinted from *J Mol Biol.*, 292, Al-Khayat et al. Yeast Ty retrotransposons assemble into virus-like particles whose T-numbers depend on the C-terminal length of the capsid protein, 65–73, Copyright 1999, with permission from Elsevier.)

The wild-type Ty1 VLPs are 15–39 nm in diameter, have an ovoid to spheroid morphology, and display electron-dense structures with a central lumen (Mellor et al. 1985a; Müller et al. 1987). They may appear also as paracrystalline inclusions in the cytoplasm of the yeast cell (Müller et al. 1987).

An attempt to map the p1 regions necessary for the formation of the Ty1 VLPs showed that the border of dispensability lay between aa positions 340 and 363 from the C terminus and between aa 40 and 71 from the N terminus, but region 62–114 was identified as indispensable for self-assembly (Brookman et al. 1995a). Further mapping revealed three structural determinants of the Ty1 VLPs and explained how point mutations can lead to increasing the diameter of the particles as much as eightfold (Martin-Rendon et al. 1996). Later, a two-hybrid system was used to specify the p1 regions critical for Ty1 VLP self-assembly, where the large internal deletion 147–233 was described unexpectedly as dispensable for monomer interactions (Brachmann and Boeke 1997).

From the first experiments, the chimeric Ty1 VLPs were found to be capable of stimulating antibody response, T-cell proliferative responses, and CTL responses, as reviewed by Kingsman AJ et al. (1995). The intrinsic epitopes of the Ty1 VLPs were also mapped by generating a panel of monoclonal and polyclonal antibodies against the p1 particle-forming protein (Brookman et al. 1995b). As a result, two N-terminal regions of the p1 protein were mapped as projecting from the surface of the shell, but the C-terminus was seen to be buried within the particle core and not available for antibody binding.

Originally, the idea of a new vaccine strategy using the Ty1 VLPs emerged in 1987 (Adams et al. 1987a, Kingsman SM and Kingsman AJ 1988), and the development of the Ty1 VLP technologies was explained in comprehensive

reviews (Adams et al. 1994a, b; Burns et al. 1994; Kingsman AJ et al. 1995). The immunodeficiency viruses HIV-1 and SIV were chosen as the first targets for introduction into the Ty1 VLP vectors. In order to construct the first HIV:Ty1 VLPs, two HIV-1 gp120 segments comprising aa 130–342 and 274–466 were added C-terminally after aa position 381 of the p1 (Adams et al. 1987a).

The exposition of gp120 segments on the Ty1 VLPs developed from the two long (approximately 200 aa gp120) stretches mentioned earlier (Adams et al. 1987a) to the V3 loop of 40 aa, which was added to the same position 381 of the p1 (Griffiths et al. 1991). In this study, immunogenicity of V3:Ty1 VLPs was tested for the first time in a nontoxic formulation licensed for use in humans, namely aluminum hydroxide, and induction of high-titer-neutralizing antibodies and an HIV-specific T-cell proliferative response was achieved (Griffiths et al. 1991; Harris et al. 1992). Then, mosaic V3:Ty1 VLPs were generated, which carried various V3 loops of different HIV isolates on the same particle, and such VLPs were able to induce V3-specific CTL response in BALB/c mice in the absence of any adjuvant or lipid vehicle (Layton et al. 1993). After comparison with recombinant gp120 and gp160 and a 40-mer V3 peptide, the particulate nature of the chimeric Ty1 VLPs was found to be crucial for the efficient uptake into antigen-presenting cells with subsequent access to the MHC class I processing pathway (Layton et al. 1993). Peifang et al. (1994) studied the V3:Ty1 VLPs for immunogenicity and antigenicity in vitro by using p1-specific human CD4+ T cell lines and clones. Interestingly, the intact particulate structure of the chimeric VLPs was strictly necessary for the CTL induction, since the presence of adjuvants in the V3:Ty VLP formulations suppressed CTL response (Harris et al. 1996).

The next object of presentation on the Ty1 VLPs, the HIV-1 p24 protein, was added first to position 381 of the p1

TABLE 35.1

Expression of the Capsid Genes from the Members of the Order *Ortervirales* and Production of the Corresponding VLPs by Different Expression Systems

VLPs	Expression Host	References	Comments
Metavirus			
Ty3	*Saccharomyces cerevisiae*	Kuznetsov et al. 2005	The Ty3 VLPs are of T = 7 symmetry.
	Saccharomyces cerevisiae	Dodonova et al. 2019	The fine structure of the Ty3 VLPs.
	E. coli	Larsen et al. 2008	The Ty3 VLPs formed by Gag3 or capsid domain CA.
Pseudovirus			
Ty1	*Saccharomyces cerevisiae*	Adams et al. 1987a	First chimeric Ty1 VLPs.
	Pichia pastoris	Uhde-Holzem et al. 2010	Chimeric p1–379 VLPs.
	E. coli	Luschnig et al. 1995	First Ty1 VLPs in bacteria.
	E. coli	Uhde-Holzem et al. 2010	First chimeric p1–379 VLPs in bacteria.
Retroviridae: Alpharetrovirus			
Rous sarcoma virus	African green monkey kidney cells CV-1	Wills et al. 1989	Self-assembly of the Pr76^{gag}, a polyprotein precursor of viral capsid.
	African green monkey kidney cells COS-1	Weldon et al. 1990	The first chimeric retroviral capsid structures.
	Trichoplusia ni (T. ni)-derived insect cell line *Tn-5B1–4;* silkworm larvae	Deo et al. 2011	The large-scale application of the RSV-Gag-577 VLPs.
	Silkworm larvae	Tsuji et al. 2011	The synthesis of the RSV-Gag-701 VLPs.
	E. coli	Campbell and Vogt 1997	Spontaneous formation of the Gag VLPs in vitro.
Retroviridae: Betaretrovirus			
Mason-Pfizer monkey virus	*E. coli*	Kliková et al. 1995; Rumlová-Kliková et al. 1999, 2000	Both the CA and the NC domains were necessary for the assembly of macromolecular sheets, while the CA N-terminus defined assembly of the spherical capsids.
Retroviridae: Deltaretrovirus			
Bovine leukemia virus	African green monkey kidney cells CV-1	Hertig et al. 1994	Production of the Gag VLPs in the absence of the *env* products.
	Spodoptera frugiperda (Sf9) cells	Kakker et al. 1999	Baculovirus-driven production of the Gag VLPs in insect cells.
Retroviridae: Gammaretrovirus			
Feline leukemia virus	*Spodoptera frugiperda (Sf9)* cells	Thomsen et al. 1992	Baculovirus-driven production of the Gag/Env VLPs in insect cells.
Murine leukemia virus	Human embryonic kidney 293FT cells	Nikles et al. 2005	The first combined retrovirus VLPs (carrying murine prion protein).
Retroviridae: Lentivirus			
Bovine immunodeficiency virus	*Spodoptera frugiperda (Sf9)* cells	Rasmussen et al. 1990	Baculovirus-driven production of the Gag VLPs in insect cells.
Feline immunodeficiency virus	*Spodoptera frugiperda (Sf9)* cells	Morikawa et al. 1991	Baculovirus-driven production of the Gag VLPs in insect cells.

(Continued)

TABLE 35.1 (CONTINUED)

Expression of the Capsid Genes from the Members of the Order Ortervirales and Production of the Corresponding VLPs by Different Expression Systems

VLPs	Expression Host	References	Comments
Human immunodeficiency virus 1	African green monkey kidney cells CV-1	Karacostas et al. 1989; Haffar et al. 1990	Production of the Gag VLPs by a vaccinia virus expression vector.
	African green monkey kidney cells CMT3 COS	Smith et al. 1990, 1993	Production of the Gag VLPs under SV40 promoter.
	Human alveolar basal epithelial A549 cells	Vernon et al. 1991	Production of the Gag VLPs by an adenovirus expression vector.
	Murine A9 and NIH3T3 cells	Diaz-Griffero et al. 2008	Production of the Gag VLPs in mouse cells.
	Spodoptera frugiperda (*Sf*9) cells	Gheysen et al. 1989; Wagner et al. 1991, 1992	Baculovirus-driven production of the Gag VLPs in insect cells.
	Trichoplusia ni- derived High-Five insect cell line	Tagliamonte et al. 2011	Baculovirus-driven production of the Gag VLPs presenting trimeric HIV-1 gp140 spikes.
	Spodoptera frugiperda (*Sf*9) and HighFive cells	Fernandes et al. 2020	Adaptive laboratory evolution (ALE) to improve the Gag VLP production.
	E. coli	Gross et al. 1997, 1998	The in vitro assembly of purified capsid proteins.
	Nicotiana benthamiana plants	Kessans et al. 2013, 2016	Generation and budding of the VLPs consisting of full-length p55^{Gag} and a deconstructed variant of gp41.
Human immunodeficiency virus 2	African green monkey kidney cells CV-1	Voss et al. 1992	Production of the Gag VLPs by a vaccinia virus expression vector
	Spodoptera frugiperda (*Sf*9) cells	Luo et al. 1990	Baculovirus-driven production of the Gag VLPs in insect cells.
Simian immunodeficiency virus	Primary chicken embryo dermal cells; African green monkey kidney cells BSC-40	Jenkins et al. 1991	Production of the Env-carrying Gag VLPs by a fowlpox expression vector.
	African green monkey kidney cells CV-1 or BSC-40	González et al. 1993; Shen et al. 1993	Production of the Env-carrying Gag VLPs by a vaccinia virus expression vector.
	Spodoptera frugiperda (*Sf*9) cells	Delchambre et al. 1989	Baculovirus-driven production of the Gag VLPs in insect cells.
		Yamshchikov et al. 1995	Baculovirus-driven production of the VLPs from the Gag and Env.

Notes: The viruses are given in alphabetical order within each specific hierarchical group.

protein after the recognition sequence for the blood coagulation factor Xa (Gilmour et al. 1989). The introduction of the protease cleavage site between the yeast carrier moiety and the protein of interest was thus adapted to facilitate the release of the latter after purification in particulate form.

The Ty1 VLPs containing all or parts of the HIV-1 proteins p24, Nef, gp41, and HIV-2 gp36 were constructed, purified, and used to develop a rapid immunoassay to detect and differentiate between HIV-1 and HIV-2 antibodies in a single test (Gilmour et al. 1990). The p17/p24:Ty1-VLPs (for simplicity, p24:Ty1-VLPs) were chosen for further immunogenicity studies as a potent vaccine candidate. The first

immunization of cynomolgus macaques with the p24:Ty1 VLPs led to the induction of a cellular response against highly conserved Th epitopes (Mills et al. 1990).

The putative p24:Ty1-VLP vaccine produced by British Biotech Pharmaceuticals Limited was safe in HIV-seronegative volunteers and induced both cellular and humoral immunity to HIV-1 Gag p17 and p24 components, but no Gag-specific CTL responses were detected (Martin et al. 1993). In the subsequent Phase I study, the p24:Ty1 VLPs elicited substantial humoral and T-cell proliferative responses, although again no p24-specific CTL responses were found (Weber et al. 1995).

In 1993, the p24:Ty1 VLP vaccine was tested in the double-blind placebo-controlled multicenter therapeutic phase II vaccination trial, with a goal to augment the immune response to p24 in the p24 antibody-positive asymptomatic HIV-1-infected subjects (Veenstra et al. 1996). Although the p24:Ty1 VLP immunizations were well tolerated, and the CD4 changes of patients were encouraging, no significant changes in humoral or cellular responses or viral load were found (Veenstra et al. 1996; Peters et al. 1997). The p24:Ty1 VLP vaccine induced marginal Gag-specific proliferative responses in the limited numbers of HIV-1-seropositive individuals, with some showing transient elevation of the HIV-1 viral load (Klein et al. 1997).

The vaccination with the VLP vaccine plus zidovudine did not significantly alter either antibody or proliferative responses to p24 or the CD4$^+$ cell number, immune activation, or viral load over 12 months (Kelleher et al. 1998); it augmented the HIV-specific CTL activity (Benson et al. 1999).

Finishing the phase II trial, Lindenburg et al. (2002) evaluated progression to AIDS, death, and a CD4$^+$ cell count ≤200 cells/mm^3 of 56 individuals who participated in the trial and compared the rate of CD4$^+$ cell count decline between the vaccinated and nonvaccinated groups. As a result, no difference between vaccine and placebo groups was found in progression to death, AIDS, and the CD4$^+$ cell count. Thus, the therapeutic vaccination with p24:Ty1 VLPs had unfortunately no long-term effect on progression to clinical endpoints or on immunological deterioration in HIV-1 disease progression (Lindenburg et al. 2002).

After exposition of the HIV proteins, a large Gag segment of the related SIV was added to the same position 381 of the p1 protein in such a way that the chimera contained aa 116–363 of the SIV Gag, namely 20 C-terminal residues of p17 contiguous with full-length p27 (Burns et al. 1990). This model vaccine was developed especially with a view to designing human anti-HIV vaccines for the prevention of homosexual transmission of HIV infection (Mills et al. 1991).

Thus, the p27:Ty1 VLPs were used for ororectal mucosal immunization of macaques and found to be capable of inducing not only T-cell proliferative responses but also rectal secretory IgA antibodies (Lehner et al. 1993, 1994a, b, 1995). Moreover, mucosal immunization with the p27:Ty1 VLPs elicited SIV Gag-specific CTLs in the regional lymph nodes as well as in the spleen and PBMC of macaques (Klavinskis et al. 1996).

The progress in the induction of CTL responses led to the conclusion that the Ty1 VLPs could be a potent means of inducing CTL responses to a variety of viral CTL epitopes, including influenza virus nucleoprotein, Sendai virus, and vesicular stomatitis virus (VSV) nucleoproteins, and the V3 loop of HIV gp120, which was described earlier (Layton et al. 1996). In particular, five variants of the CTL epitope 325–332 from the protein N of Sendai virus were cloned within the Ty1 VLPs. The ability of Ty1 VLPs carrying two different epitopes or the ability of a mixture of two different

Ty1 VLP chimeras to prime CTL responses in two different strains of mice or against two epitopes in the same individual was shown for the first time (Layton et al. 1996).

The Ty1 VLP technology was used to study Gag and Env products of an ovine/caprine lentivirus, namely, maedi-visna virus (MVV). First, the major MVV core protein p25 was fused, by analogy to the HIV p24, at the Factor Xa cleavage site after the p1 moiety (Reyburn et al. 1992). Then, the overlapping segments of the envelope glycoprotein gp135 of the MVV were exposed on the appropriate Ty1 VLPs (Carey et al. 1993). These MVV:Ty1 VLPs were used to produce antibodies and map epitopes and thus to develop diagnostic assays. In the same fashion, the Ty1 VLPs were used to construct fusions with the rubella envelope glycoprotein E1 and to map the binding sites of the appropriate monoclonal antibodies (Lindqvist et al. 1994).

The Ty1 VLP-driven technologies ensured an important breakthrough in herpesvirus vaccine development, namely human herpesvirus 3 or varicella-zoster virus (VZV). First, the overlapping segments of the envelope glycoprotein G (gE) of the VZV were exposed on the Ty1 VLPs and helped to map immunodominant and neutralizing epitopes on the gE (Fowler et al. 1995). Second, the intrinsic immunogenicity of chimeric gE:Ty1 VLPs carrying aa 1–134 or 101–161, which contained the immunodominant sequences, has been proven (Garcia-Valcarcel et al. 1997b). The results demonstrated that the gE:Ty1 VLPs can prime potent humoral and cellular anti-VZV responses in small animals and warrant further studies as a potential vaccine. Next, a segment of 300 aa of the VZV assembly protein (AP) was inserted into the Ty1 VLPs, which made possible the examination of the genetic organization and proteolytic maturation of VZV assemblin (Garcia-Valcarcel et al. 1997a). Finally, both gE- and AP:Ty1 VLPs were found to be capable of being recognized by lymphocytes of varicella and zoster patients, which highlighted their potential use as a candidate vaccine (Welsh et al. 1999).

Of particular value were the Ty1 VLP-initiated studies on malaria vaccine candidates. The Ty1 VLPs carrying a well-defined CTL epitope 252-SYIPSAEKI-260 from the circumsporozoite (CS) protein of *Plasmodium berghei* were found to induce a remarkable CTL response, on par with a lipid-tailed peptide of this same sequence, in mice (Allsopp et al. 1996). Moreover, these CTLs were able to recognize a naturally processed antigen expressed by a recombinant vaccinia virus. A string of up to 15 defined CTL epitopes from *Plasmodium* species added to the Ty1 VLPs primed protective CTL responses in mice following a single administration without adjuvant, and effective processing of epitopes from the string was demonstrated in vitro and in vivo and was not affected by flanking sequences (Gilbert et al. 1997).

Then, the authors proposed a novel prime/boost combination of CS:Ty-VLPs and modified vaccinia virus Ankara (MVA) containing a single *P. berghei* CTL-epitope among a string of 8–15 human *P. falciparum* CTL epitopes, which provided high-level protection against malaria in mice and was suggested as a candidate for vaccination against human *P. falciparum* malaria (Plebanski et al. 1998). The immunization scheme was found to be particularly important: the use of the Ty1 VLPs followed by MVA was critical for protection, but the reciprocal sequence of immunization or homologous boosting was not protective (Plebanski et al. 1998). The authors confirmed their success, showing that complete protection against malaria in mice can be achieved by priming with one type of vaccine (Ty1 VLPs or DNA-based) and boosting with another, namely MVA (Gilbert et al. 1999). Oliveira-Ferreira et al. (2000) found that the chimeric Ty1 VLPs carrying the CTL epitope SYVPSAEQI of the circumsporozoite (CS) protein of *P. yoelii* protected 62% of mice against a subsequent live *P. yoelii* sporozoite challenge. when the VLP immunization was boosted with the entire CS protein or irradiated sporozoites.

Some nontraditional objects for the insertion into the Ty1 VLPs were represented first by the group-I (Der p1) allergen of *Dermatophagoides pteronyssinus*, a house dust mite. The T helper (Th) and CTL epitopes of the Der p1 allergen were predicted, synthesized, exposed on the Ty1 VLPs, and found capable not only of inducing Th and CTL responses in mice (Harris et al. 1997), but also of suppressing allergen-specific proliferation when mice were treated with Der p1:Ty-VLPs after sensitizing them to the Der p1 (Hirschberg et al. 1999).

Regarding the exposition of subjects of nonviral origin on the Ty1 VLPs, the expression and purification of human interferon-α 2 must be mentioned first (Malim et al. 1987). Then, a line of ovine cytokines—interleukins 1α and β (Fiskerstrand et al. 1992), tumor necrosis factor α (Green et al. 1993), and interleukin 2 (Bujdoso et al. 1995)—was purified as chimeric Ty1 VLPs and released from the particles by cleavage with the Factor Xa enzyme.

An interesting use of the Ty1 VLPs was supposed by the insertion of two nucleases into the particles, which retained ability to degrade encapsidated RNA (Natsoulis and Boeke 1991). This strategy was referred to as capsid targeted viral inactivation and was proposed as useful for interfering with the replication of retroviruses.

The Ty1 VLPs also assured the search, characterization, and purification of potent mitochondrial genes—reverse transcriptases from mitochondrial introns of different eukaryotic organisms (Fassbender et al. 1994) and mitochondrial tyrosyl-tRNA synthetase (Pöggeler et al. 1996).

All previously mentioned studies used the C-terminal additions to the Ty1 p1–381 protein. Marchenko et al. (2003) reported the first N-terminal p1–381 fusion of peptides or proteins. Thus, the authors expressed peptides from the *Aspergillus fumigatus* Asp f 2 protein at the C- and/or N-termini of p1–380 and showed that the foreign insertions did not interfere with the VLP self-assembly and were exposed at the VLP surface. Moreover, Marchenko et al. (2003) generated mixed particles, which coexpressed the two hybrid p1 proteins with different heterologous protein fragments at the C-terminus. To do this, the yeast cells were transfected with a mixture of two recombinant DNA encoding Asp f 2 peptide and green fluorescent protein (GFP), and both Asp f 2 peptide and GFP were expressed within the same particle.

As mentioned earlier, Uhde-Holzem et al. (2010) generated the p1–379 sequence with added BNYVV epitopes 4 (RTPPGQ) and 6 (SANVRRD) at either the N-terminus or C-terminus or both and demonstrated great production of the chimeric Ty1-based VLPs in *E. coli* and *P. pastoris*.

Concerning internal positions of the p1 protein, Richardson and colleagues were the first to show that the HIV-1 V3 loops triggered a humoral immune response when inserted at an internal position of the p1 close to the N-terminus (cited in Brookman et al. 1995b and Uhde-Holzem et al. 2010). This indicated that the HIV-1 V3 loops, when fused close to the N-terminus, were presented on the Ty1 VLP surface.

FAMILY *BELPAOVIRIDAE*

The family involves one genus, namely *Semotivirus*, of 11 species and generally is referred to as Bel/Pao transposons (de la Chaux and Wagner A 2011; Llorens et al. 2011). The Bel/Pao transposons are a widespread kind of transposable element present in eukaryotic genomes, more abundant than the Ty1/Copia elements and second only to the Ty3/Gypsy elements. They occur in multiple phyla, including basal metazoan phyla, suggesting that Bel/Pao elements arose early in animal evolution, but they do not occur in mammals (de la Chaux and Wagner A 2011).

The typical semotivirus species are *Ascaris lumbricoides Tas virus*, *Drosophila melanogaster Bel virus*, and *Bombyx mori Pao virus*.

FAMILY *METAVIRIDAE*

The members of the family *Metaviridae* are often referred to as Ty3/gypsy retrotransposons. The representatives of the family that includes 2 genera and 31 species were found in all eukaryotes known and studied (Roth 2000; Llorens et al. 2008, 2011, 2020).

The Ty3 VLPs were observed in yeast cells expressing the Ty3 transposon (Hansen and Sandmeyer 1990). The Ty3 VLPs were approximately 50–60 nm in diameter (Hansen et al. 1992), which was comparable to the size of the Ty1 VLPs described earlier. The Ty3 capsid was formed by a Gag3 protein comprised of capsid (CA), spacer (SP3) and nucleocapsid (NC) domains (Sandmeyer et al. 2002).

Using atomic force microscopy, Kuznetsov et al. (2005) imaged the Ty3 VLPs from *S. cerevisiae* cells producing wild-type and protease and reverse transcriptase mutant Ty3. The wild-type VLPs were in the range of 25 to 52 nm in diameter, with particles in the 42- to 52-nm diameter range consistent with T = 7 symmetry. Both classes of mutant VLPs fell into a narrower range of 44 to 53 nm in diameter and appeared to be consistent with T = 7 icosahedral symmetry. Generally, the Ty3 VLPs did not undergo major external rearrangements during proteolytic maturation (Kuznetsov et al. 2005).

Larsen et al. (2008) reported that the expression of the Ty3 Gag3 or capsid domain in *E. coli* was sufficient for the Ty3 VLP assembly. The VLPs assembled from Gag3 were similar in size to immature particles from yeast and contained nucleic acid. However, the VLPs assembled from the CA domain were variable in size and displayed much less organization than native particles. The authors concluded that the VLP assembly was driven through interactions among capsid subunits in the particle, but the nucleocapsid domain, likely in association with RNA, conferred order upon this process, and the organization of the Ty3 VLPs in *E. coli* was greatly enhanced therefore by inclusion of the NC domain (Larsen et al. 2008).

The fine capsid morphology of the Ty3 VLPs was achieved by 3D and 2D electron cryomicroscopy of the immature and mature Ty3 particles within yeast cells (Dodonova et al. 2019). For this reason, the expression of the wild-type or protease mutant Ty3 was induced in yeast cells from which endogenous copies of Ty3 had been deleted. This study confirmed first the previously mentioned fact that, in contrast to retroviruses, the Ty3 VLPs did not change size or shape upon maturation. Both immature and mature particles have the same external diameter of ~42 nm, corresponding to a true diameter of ~50 nm. Figure 35.4 demonstrates the 3D reconstruction of the purified immature Ty3 particles resolved at 7.5 Å and revealing the T = 9—but not T = 7 as earlier—symmetry.

FAMILY *RETROVIRIDAE*

This is a type of viruses with an obligatory integration step, when a copy of viral RNA genome is inserted into the DNA of a host cell, thus changing the genome of the infected cell. The new DNA is incorporated into the host cell genome by an integrase enzyme, at which point the retroviral DNA is referred to as a provirus. The host cell then transcribes and translates the viral genes along with own genes. The family *Retroviridae* includes 2 subfamilies: *Orthoretrovirinae* and *Spumaretrovirinae*, containing in total 11 genera and 68 species.

SUBFAMILY *ORTHORETROVIRINAE*

This subfamily consists of six genera, five of which unite so-called oncogenic retroviruses, but one genus, namely *Lentivirus*, consists of so-called slow retroviruses. There are 49 retrovirus species in total.

Oncogenic Retroviruses

There are five genera including such medically and scientifically relevant species as *Rous sarcoma virus* from the genus *Alpharetrovirus*, *Mason-Pfizer monkey virus* of the genus *Betaretrovirus*, *Bovine leukemia virus*, *Primate T-lymphotropic virus 1*, and *Primate T-lymphotropic virus 2* of the genus *Deltaretrovirus*, and *Moloney murine*

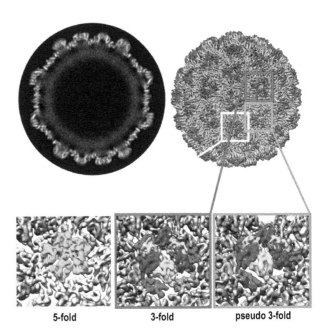

FIGURE 35.4 The electron cryomicroscopy reconstruction of a T = 9 PR- Ty3 particle. (Left) Slice through the center of the PR- Ty3 particle reconstruction resolved at 7.5 Å. The outer layer with distinct α-helical densities is the capsid (CA) layer, and the fainter blurred layer underneath likely represents spacer, nucleocapsid (NC), and the associated genome. (Right) 3D reconstruction of a PR- Ty3 particle colored radially from green (low radius) to blue (high radius). (Insets) Close-up views of the 5-fold (yellow box), 3-fold (red box), and pseudo-3-fold (orange box) positions. Within the positions with different symmetries, the densities are colored to indicate the different domains of CA: 5-fold CA- N-terminal domain (NTD) is colored green, CA-C-terminal domain (CTD) is colored yellow, 3-fold and pseudo-3-fold CA-NTD are colored cyan/blue in conformation A/B, and CA-CTD is colored orange/red in conformation A/B. (Reprinted from Dodonova et al. Structure of the Ty3/Gypsy retrotransposon capsid and the evolution of retroviruses. *Proc Natl Acad Sci U S A.* 2019;116:10048–10057.)

sarcoma virus and *Murine leukemia virus* of the genus *Gammaretrovirus*, 39 species in total.

In the 1990s, numerous studies were undertaken on the construction and expression in mammalian cells of the artificial recombinants of the Gag proteins of such prototype oncogenic retroviruses as Rous sarcoma virus (RSV) of the genus *Alpharetrovirus* and murine leukemia virus (MLV) of the genus *Gammaretrovirus* with other oncoretroviruses and lentiviruses; for example, there are pioneering studies of Deminie and Emerman (1993), Berkowitz et al. (1995), Campbell and Vogt (1995), Zhang Y and Barklis (1995), Bowzard et al. (1998), Bennett and Wills (1999).

The François-Loic Cosset team generated the first MLV Gag VLPs pseudotyped with a foreign envelope, namely hepatitis C virus envelope proteins E1 and E2 (Bartosch et al. 2003a, b). Moreover, the authors packaged a GFP marker gene within these HCV pseudoparticles, allowing thus reliable and fast determination of infectivity mediated by the HCV glycoproteins (Bartosch et al. 2003). Later, the MLV Gag VLPs carrying the HCV E1/E2 or E1 alone and devoid of any viral genome or viral enzymes, such as reverse transcriptase and integrase, were tested in a putative HCV vaccine (Garrone et al. 2011). Huret et al. (2013) developed the real MLV Gag VLP platform to display HCV antigens. This remarkable approach was reviewed by the authors (Bellier and Klatzmann 2013; Bellier et al. 2015).

Nikles et al. (2005) used MLV to set up the prion protein on the retrovirus VLPs. Thus, the authors incorporated the two variants of the cellular form of the mouse prion protein (PrP), the full-length mouse PrP209 or the C-terminal 111 aa PrP111 fused to the transmembrane domain of the platelet-derived growth factor receptor, into the MLV VLP membrane. The mammalian cells-derived VLPs displayed the PrP, proved to be highly immunogenic in PrP-deficient mice, and, even more importantly, evoked native PrP-specific antibody responses in wild-type mice (Nikles et al. 2005).

The similar approach was used to display the VSV glycoprotein on the MLV Gag VLPs produced in mammalian cells (Bach et al. 2007). In contrast to the UV-inactivated VSV, the MLV-Gag-VSV VLPs were highly immunogenic in mice and induced VSV-neutralizing IgM responses that switched to the IgG subclass.

Then, a system was elaborated for the predictable decoration of the MLV Gag VLPs, synthesized in mammalian cells, with functionally active molecules (Derdak et al. 2006; Kueng et al. 2007; Leb et al. 2009). This system circumvented the need to modify viral proteins themselves. By this methodology, the molecules to be exposed were fused at their C-terminus to the glycosyl phosphatidyl inositol (GPI) acceptor sequence of human Fcγ receptor III. Thus, Derdak et al. (2006) decorated the MLV VLPs with T cell receptor/CD3 ligands. Furthermore, Kueng et al. (2007) used the GPI-anchoring system for the decoration of the MLV Gag VLPs with functionally active cytokines. The human interleukins -2 and -4, human granulocyte-macrophage colony-stimulating factor (GM-CSF), and murine interleukin-2 were used as model cytokines The biologically active cytokine molecules equipped with GPI anchors were efficiently displayed on the MLV Gag VLPs, as demonstrated in vitro and in vivo, by their generation in the mammalian producer 293 cells (Kueng et al. 2007). Such MLV Gag VLPs decorated with lipid-modified cytokines were intended to generate a novel approach of the therapeutic vaccination among other putative applications of viral nanotechnology (see Chapter 25 for more data and discussion on this subject).

Leb et al. (2009) succeeded by modulation of allergen-specific T-lymphocyte function by the MLV VLPs decorated with HLA class II/allergen-peptide complexes and regarded the VLP product as a promising tool for specific immunotherapy of allergic diseases.

The MLV Gag was used as a scaffold by the generation of influenza vaccines, where it was used to replace the influenza matrix M1 protein. This specific sort of the MLV Gag applications is described in Chapter 33.

Kirchmeier et al. (2014) constructed a prophylactic vaccine to prevent the congenital transmission of human cytomegalovirus (HCMV). Thus, the MLV VLPs were used to express either full-length CMV protein gB or the full extracellular domain of CMV gB fused with the transmembrane and cytoplasmic domains from VSV protein G. Schädler et al. (2019) were the first to use the MLV VLPs as a scaffold for the vaccine against infectious laryngotracheitis virus (ILTV), a herpesvirus and member of the *Gallid alphaherpesvirus 1* species. Recently, the MLV platform was employed to produce VLPs pseudotyped with the short (ΔS) version of the spike protein S from severe acute respiratory syndrome coronavirus-2 (SARS-CoV-2; Roy et al. 2021).

Concerning progress on the RSV VLPs, Wills et al. (1989) succeeded in the assembly of the RSV Gag protein by the expression of the Pr76^{gag}, a polyprotein precursor, gene in mammalian cells. Second, the first successful attempt to construct chimeric particles was realized by addition of a long foreign sequence, the iso-1-cytochrome C (CYC1), from *Saccharomyces cerevisiae*, to the C-terminally truncated Pr76^{Gag}, in which the chimeric VLPs revealed a morphology similar to that of the immature type-C retrovirions (Weldon et al. 1990).

Furthermore, the conditions were found under which the RSV Gag VLPs were formed spontaneously in vitro from the Gag protein fragments purified after expression in *E. coli* (Campbell and Vogt 1997). The major self-assembly characteristics of the *E. coli*-produced RSV Gag VLPs were revealed further by Joshi and Vogt (2000), Yu et al. (2001), and Johnson et al. (2002).

Later, the large-scale production of the Gag VLPs of a short RSV-Gag-577 version carrying 577 aa residues was established in insect cells and silkworm larvae (Deo et al. 2011). Tsuji et al. (2011) achieved synthesis of the longer RSV-Gag-701 variant in silkworm larvae. Due to the efficient baculovirus-driven expression, the combined retrovirus VLPs were generated: the RSV-Gag-577 were coexpressed in silkworm larvae with human prorenin receptor (hPRR), and the hPRR was displayed on the surface of RSV the VLPs, which was detected by immunoelectron microscopy (Tsuji et al. 2011).

For a long time, bovine leukemia virus (BLV), a member of the genus *Deltaretrovirus*, served only as a pioneering source of epitopes for insertion into such carrier vectors as HBc (see Chapter 38), until its Gag was chosen as a potent VLP carrier candidate (Kakker et al. 1999). The baculovirus-driven expression of the BLV Pr44^{gag} gene encoding the Gag precursor protein in insect cells resulted in the self-assembly and release of VLPs. The recombinant baculoviruses expressing matrix (MA) or capsid-nucleocapsid (CA-NC) proteins of BLV were generated, but neither of these domains was capable, however, of assembling into particulate structures. Moreover, chimeras exchanging MA and CA-NC proteins of BLV and other leukemia viruses (e.g., human T-cell leukemia virus type I) or such an evolutionarily divergent retrovirus group as lentiviruses (e.g., simian immunodeficiency virus) assembled efficiently and budded as VLPs (Kakker et al. 1999). In parallel, the formation of VLPs in insect cells was achieved for human lymphotropic T-cell virus type II (HTLV-II) (Takahashi et al. 1999). However, it is noteworthy that Hertig et al. (1994) were the first who showed that the expression of the BLV *gag-pro* or *gag-pro-pol* genes by recombinant vaccinia virus led to production of the BLV Gag VLPs. Therefore, the assembly of the Gag VLPs occurred in the absence of *env* products and other BLV-specific proteins, and the budding and particle maturation did not require copackaging of genomic viral RNA. The BLV VLPs exhibited morphology similar to that of BLV, and their diameters ranged from 80 to 115 nm in diameter (Hertig et al. 1994)

The Gag precursor of the feline leukemia virus (FeLV) belonging to another genus of the *Orthoretrovirinae* subfamily, namely *Gammaretrovirus*, was expressed in insect cells and was shown to form VLPs and release them from the cell (Thomsen et al. 1992). Moreover, the authors found that the coexpression of the FeLV Gag with the FeLV envelope glycoprotein gp85 resulted in the formation and budding of Gag/gp85 particles.

The first high-resolution data on the oncoretroviral capsid structure was achieved by NMR, which described the 3D solution structure of the capsid protein (CA) from the human lymphotropic T-cell virus type I (HTLV-I; Khorasanizadeh et al. 1999) and RSV (Campos-Olivas et al. 2000). Then, the 3D reconstructions and high-resolution capsid structures were determined for the alpharetrovirus RSV (Kingston et al. 2000; Mayo et al. 2002; Keller et al. 2013), the betaretrovirus Mason–Pfizer monkey virus (Macek et al. 2009; Bharat et al. 2012), and the gammaretrovirus MLV (McDermott et al. 2000; Zuber and Barklis 2000; Zuber et al. 2000; Ganser et al. 2003; Mortuza et al. 2004, 2008; Qu et al. 2018).

Lentiviruses

The *Lentivirus*, covering the slow retroviruses, is a one of the six genera of the subfamily *Orthoretrovirinae* and includes 10 species, among them the well-known and greatly dangerous species *Human immunodeficiency virus 1*, *Human immunodeficiency virus 2* (the cause of the AIDS disease), as well as the *Simian immunodeficiency virus* species.

The HIV-1 and HIV-2 viruses have attracted much attention regarding the construction of promising vaccines against AIDS. Other closely related immunodeficiency viruses were developed as AIDS models. The Gag proteins turned out to be necessary and sufficient for the VLP self-assembly and budding, as in other retroviruses (for a review, see paper of Freed 1998). The Gag gene encodes a Pr55Gag precursor consisting, like Gag of leukemia and other retroviruses, of the domains matrix (MA), capsid (CA), nucleocapsid (NC), and p6, which are separated by the viral proteinase inside the nascent virion, leading to the morphological maturation of an infectious virus. In the mature virus, the CA forms a capsid surrounding the ribonucleoprotein core consisting of

the NC and the genomic RNA, but a lipid bilayer containing envelope proteins surrounds the Gag capsids.

The crucial lentiviral VLP studies were opened by the baculovirus-driven expression of the Gag precursor of SIV (Delchambre et al. 1989; González et al. 1993; Yamshchikov et al. 1995), HIV-1 (Gheysen et al. 1989; Wagner R et al. 1991, 1992; Hughes et al. 1993), HIV-2 (Luo et al. 1990), bovine immunodeficiency virus (BIV; Rasmussen et al. 1990), and feline immunodeficiency virus (FIV; Morikawa et al. 1991) in insect cells. The expression of the SIV Pr57^{Gag} (Delchambre et al. 1989) and the HIV-1 Pr55^{Gag} (Gheysen et al. 1989) precursors were the first to demonstrate that the unprocessed Gag product can spontaneously assemble into 100- to 120-nm VLPs, indistinguishable from immature lentivirus particles, in the absence of other viral proteins, but myristylation of the N-terminal glycine guarantees its budding and extracellular release.

The inclusion of the *pol* gene and expression of viral protease activity in the system resulted in efficient processing of the Gag precursor to p17, p24, p9, and p6 proteins and abolishing particle formation (Gheysen et al. 1989; Morikawa et al. 1991; Wagner R et al. 1992). Moreover, the SIV matrix (MA) domain of the Gag polyprotein (p17 Gag) self-assembled into 100-nm VLPs, which were released into the culture medium. The coexpression of HIV Pr55 and gp160 resulted in the apparent incorporation of gp160 into the Gag VLPs during the budding process (Tobin et al. 1996b).

Later, Tagliamonte et al. (2011) established a modern insect cell expression system that allowed high-level production of the HIV-1 Gag VLPs presenting trimeric HIV-1 gp140 spikes, when both genes were constitutively expressed in stable double transfected High-Five insect cell line. This system allowed efficient generation of the HIV-1 VLPs exposing different HIV-1 glycoproteins (Visciano et al. 2011). Recently, Fernandes et al. (2020) used a special adaptive laboratory evolution (ALE) approach to improve the production of the HIV Gag VLPs in stable *Sf*9 and HighFive cell lines.

While the baculovirus-driven expression system was the most efficient, the self-assembly of the Gag precursor VLPs was also observed in mammalian cells using vaccinia (Karacostas et al. 1989; Voss et al. 1992), SV40 (Smith et al. 1990, 1993), and adenovirus (Vernon et al. 1991) expression vectors.

In mammalian cells, the coexpression of SIV MA and Env proteins resulted in the incorporation of gp120 and gp41 proteins into the recombinant p17-made particles (González et al. 1993). Furthermore, the two domains critical for VLP formation were located in the central hydrophobic α-helix of the SIV MA (González et al. 1996). The SIV MA—but not HIV-1 MA—was found later to be sufficient for VLP formation and budding in eukaryotic cells as well (Giddings et al. 1998).

The domains of the HIV Gag precursor were expressed in and purified from *E. coli*, where in vitro assembly of CA yielded tubular structures with a diameter of approximately

55 nm and heterogenous length (Gross et al. 1997), while extending CA by 5-aa residues was sufficient to convert the assembly phenotype into spherical particles (Gross et al. 1998). Depending on the CA gene composition and expression characteristics, the bacterially derived HIV-1 VLPs formed tubes, spheres, and cones in vitro (Ehrlich et al. 1992, 2001; Campbell and Vogt 1995, 1997; Gross et al. 1998; Ganser et al. 1999; von Schwedler et al. 1998).

The two-step self-assembly of the bacterially expressed HIV Gag protein lacking only the C-terminal p6 was modeled in vitro and resulted in the particles equivalent to the Gag VLPs obtained from the Gag-expressing cells and similar to authentic immature HIV particles (Morikawa et al. 1999). Finally, three stages of the Gag VLP formation were described (Morikawa et al. 2000). The MA domain was found to be responsible for the trimerization of the assembly competent Gag; the p2 domain, located at the CA–NC junction and responsible for further multimerization and a change in Gag assembly morphology from tubes to spheres; and finally the N-terminal myristoylation, responsible for converting these multimers into Gag VLPs (Morikawa et al. 2000).

Surprisingly, at that time there was no clear evidence regarding the symmetry of the Gag VLPs. The image processing of the immature VLPs from insect cells (Nermut et al. 1998) confirmed the so-called fullerene-like model for the prebudding HIV-1 Gag assemblies (Nermut et al. 1994) and revealed the presence of threefold symmetry and a hexagonal network of rings with a resolution of 29 Å in the VLPs and greater than 25 Å in the membrane-associated assemblies. This data was consistent with the hypothesis that the immature HIV particles possessed icosahedral symmetry with the triangulation number T = 63 (Nermut et al. 1994, 1998). On the other hand, electron cryomicroscopy of the VLPs from insect cells and the immature HIV-1 particles from human lymphocytes showed no evidence of icosahedral symmetry at all (Fuller et al. 1997).

The 3D structure of the HIV-1 MA protein was determined at high resolution by the NMR and found to be mostly α-helical and forming an icosahedral shell, which is associated with the inner membrane of the mature virus (Massiah et al. 1994). The dimers of the HIV-1 CA protein were studied by x-ray diffraction (Momany et al. 1996) and by multidimensional heteroNMR spectroscopy for the N-terminal (aa residues 1–151; Gitti et al. 1996) and C-terminal (aa residues 146–231; Gamble et al. 1997) parts of the HIV-1 CA.

Unlike the structure of the previously characterized viral coats at that time, the N-terminus of the HIV CA showed predominant α-helical organization, composed of seven α-helices, two β-hairpins, and an exposed partially ordered loop. The C-terminal CA domain, which contained a 20-aa major homology region (MHR) and was essential for capsid dimerization and viral assembly, consisted of four α-helices and an extended N-terminal strand (Gamble et al. 1997). This data allowed for the identification of the side of the dimer that would be on the exterior of the VLPs (Momany

et al. 1996). The 3D structure of the HIV-1 NC bound to the genomic RNA packaging signal was determined by heteroNMR (De Guzman et al. 1998), and it showed tight binding of the NC zinc fingers to the RNA. It is worthy of note that the NC zinc fingers were the first structures determined at atomic resolution for a retroviral zinc finger-like complex (Summers et al. 1990). The structure of the HIV-1 proteins was reviewed at that early time by Darlix et al. (1995) and Roques et al. (1997).

The electron cryomicroscopy and image reconstructions of the HIV-1 CA tubes was performed by Li et al. (2000). All tubes were composed of hexameric rings of CA arranged with approximate local p6 lattice symmetry. The HIV-1 capsid followed the principles of a fullerene cone, in which the body of the cone is composed of curved hexagonal arrays of CA rings.

Later, the high-resolution HIV-1 CA structures were achieved (Pornillos et al. 2009; Wagner JM et al. 2016). The modern structure and understanding of retroviral capsid were summarized by Perilla and Gronenborn (2016). Figure 35.5 demonstrates the current view of the immature and mature HIV-1 capsid cores.

From the very first expression experiments in insect cells, the recombinant lentivirus VLPs demonstrated high immunogenicity in experimental animals (mice, rabbits) and seemed to represent a noninfectious and attractive candidate for a basic vaccine component (Luo et al. 1990; Rasmussen et al. 1990; Wagner R et al. 1992). Further, a correlation of aa changes in the CA domain with the HIV-1 escape from CTL surveillance (Zhang WH et al. 1994) favored the idea that the CTL epitopes were capable of controlling HIV replication in long-term nonprogressors and that the HIV-specific CD8-positive CTL response could be

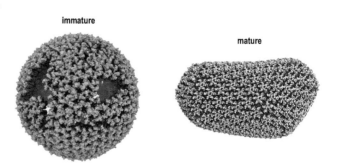

FIGURE 35.5 The mature and immature retroviral capsid cores. (Left) The computational model of the immature HIV-1 VLPs is derived from the electron cryomicroscopy (Schur et al. 2015). The lattice in the model is composed entirely of hexamers and contains multiple gaps to permit the spherical morphology. (Right) The mature HIV-1 capsid core, composed of hexamers and pentamers. The pseudo-atomic model of the mature HIV-1 capsid core contains 286 hexamers (brown) and 12 pentamers (green) (Zhao et al. 2013). (Reprinted from *Trends Biochem Sci.* 41, Perilla JR, Gronenborn AM, Molecular architecture of the retroviral capsid, 410–420, Copyright 2016, with permission from Elsevier.)

induced by the recombinant Gag VLPs. This approach was exhaustively reviewed by Wagner R et al. (1999).

The deletion and mutagenesis mapping of the Gag was followed by the construction of its chimeric derivatives. Thus, the importance among lentiviruses (HIV-1, HIV-2, SIV, and FIV) of the conserved C-terminal part of the CA domain (aa 341–346 and 350–352) for the VLP self-assembly was proven by the appropriate deletions within the HIV-1 protein (von Poblotzki et al. 1993). The deletion of aa 99–154 overlapping the MA-CA cleavage site completely abolished the capacity of the HIV-1 Gag polyprotein to form VLPs, but deletions of aa 211–241 within the CA and aa 436–471 within the p6 portion had no effect on the assembly, ultrastructure, biophysical properties, or yields of mutant VLPs (Wagner R et al. 1994). On the other hand, other authors found the aa 210–241 deletion to have a significant influence on the Gag self-assembly, as well as on deletions 277–306 (MHR), 307–333 (within the CA), 358–374 (CA spacer peptide 2 (sp2) junction), and 375–426 (sp2-NC junction and most of the NC; Carrière et al. 1995). The substitution of aa 341–352 within the C-terminal portion of the HIV-1 CA for alanines led to an inhibitory effect on its capacity to form VLPs, but it completely blocked the assembly and release of HIV in a replication-competent system (Kattenbeck et al. 1997).

The mapping data on the HIV-2 Gag showed, however, that the C-terminal p12 region of the Gag precursor protein and the zinc finger domain were dispensable for the VLP assembly, but the presence of at least one of the three proline residues located between aa positions 372 and 377 of HIV-2 was required (Luo et al. 1994). Therefore, the C-terminally truncated HIV-2 Gag protein of 376 aa in length—but not the protein of 372 aa in length—was found to be self-assembly competent (Zhang Y and Barklis 1997). The thorough examination of the effects of the NC and p6 mutations showed that the p6 domain appeared to affect not virus release efficiency but specificity of genomic HIV-1 RNA encapsidation. The NC domain was responsible, however, not only for the specificity and efficiency of RNA encapsidation but also for particle assembly and release (Zhang Y and Barklis 1997). The replacement of both NC and p6 domains by foreign polypeptides able to form interprotein contacts permitted the efficient assembly and release of particles, although no packaged RNA was detected (Zhang Y and Barklis 1997; Zhang Y et al. 1998). Remarkably, a new function of particle size control was attributed to the HIV-1 p6 after analysis of a large collection of the Gag mutants (Garnier et al. 1998).

An attempt to construct a minimal HIV-1 Gag capable of VLP assembly and release was undertaken by the combination of an assembly competent MA deletion mutant with progressive C-terminal truncations of the Gag (Wang CT et al. 1998). The smallest HIV *gag* gene product still capable of VLP formation was a 28-kDa protein consisting of a few MA amino acids and the CA-sp2 domain (Wang CT et al. 1998).

The potential of a Gag precursor as a carrier for the presentation of foreign epitopes was first documented by the insertion of the neutralizing domain V3 or the neutralizing V3 plus CD4 binding domains of the Env protein into the C-terminally truncated 41-kDa HIV-2 *gag* gene with further expression in baculovirus-infected insect cells (Luo et al. 1992). It appeared that insertion of the *env* gene sequences into the C terminus—but not into the middle of the carrier—allowed for the production of VLPs, which were slightly larger than the initial Gag VLPs and similar to the mature HIV particles.

The immunization of rabbits with these chimeric VLPs elicited high titers of neutralizing antibodies. Furthermore, the Gag-V3 VLPs showed an ability to elicit a strong anti-V3 CTL response in immunized mice (Griffiths et al. 1993; Wagner R et al. 1993). As suitable sites for insertion, the replacement of two domains of a predicted high surface probability (211–241 and 436–471)—but not of the aa 99–154—with foreign epitopes (V3 loop; CD4-binding-domain; Nef CTL epitope) or fusion of the previously mentioned sequences to the C terminus of the Pr55Gag were suggested in the vaccinia expression system (Wagner R et al. 1994a). In general, these chimeric Gag VLPs carrying T-helper and CTL epitopes from the HIV-reading frames other than Gag demonstrated good capabilities to ensure a cell-mediated immunity in addition to a humoral immune response (Wagner R et al. 1994b). Although antisera exhibited only weak neutralizing activity, the immunized mice developed a strong CTL response against a CTL epitope within the V3 domain, and the magnitude of this response was not influenced by the position of the V3 domain within the Pr55Gag carrier moiety (Wagner R et al. 1996b). Moreover, the Pr55Gag-V3 polypeptide retained high CTL-priming activity even after denaturation to the monomeric state (Schirmbeck et al. 1995).

In a similar way, the precise epitope-against-epitope replacement of a mapped Pr55Gag epitope 196–228—or 196–226—by the consensus V3 and/or the CD4-binding domains led to VLP production in and release from baculovirus-infected insect cells (Brand et al. 1995; Truong et al. 1996). Although the inserted V3 was still antigenic, it was not exposed at the surface of the particles according to the immune electron microscopy data (Brand et al. 1995) and did not elicit neutralizing antibodies either (Truong et al. 1996).

To regulate the specificity of the expected CTL response, the two types of the Gag VLP vaccines were constructed: type 1 by replacing defined Gag domains by selected HIV-1 CTL epitopes and type 2 by stable anchoring derivatives of the HIV-1 Env protein on the surface of the VLPs (Wagner R et al. 1996a). In complete absence of adjuvants, the type 1 and type 2 VLPs stimulated specific CTL in response, not only in mice but also in macaques. Unlike the type 1 VLPs, which generated the HIV-specific CTL responses in the absence of Env-specific antibodies, the type 2 VLPs induced both arms of the immune system, including reasonable levels of neutralizing antibodies (Wagner R et al. 1996a).

Further attempts at improving the V3-carrying Gag VLPs were turned to the extreme multiplication of different

V3 variants and other gp160 epitopes included into the chimeric particles. Thus, the HIV-2 Gag chimeras with four neutralizing epitopes from HIV-1 gp160; three tandem copies of the consensus V3 domain, which were derived from 245 different isolates of HIV-1; or four V3 domains from the most prevalent strains were constructed and expressed in insect cells (Luo et al. 1998). All three constructs appeared as spherical particles similar to the immature HIV but slightly larger than the Gag VLPs and induced neutralizing antibodies and strong anti-V3 CTL activity. The authors also found that the deletion of up to 143 aa at the C terminus of the HIV-2 Gag, leaving 376 aa at the N terminus of the protein, did not affect the VLP formation (Kang et al. 1999). Therefore, the C-terminal p12 region of the HIV-2 Gag precursor protein and zinc finger domains were dispensable for the Gag VLP assembly, but the presence of at least one of the three prolines at aa positions 373, 375, or 377 was obligatory.

Osterrieder et al. (1995) succeeded by the generation of a putative equine herpesvirus type 1 (EHV-1) glycoprotein gp14 vaccine, where the EHV-1 gp14 was included into the HIV-1 Gag VLPs, when synthesized in insect cells by coexpression of both corresponding genes in insect cells.

In principle, another approach to constructing chimeric VLPs was possible. This involved a specific entrapping of the chimeric Env protein derivatives into the nonchimeric Gag VLPs. The first attempts in this direction were made by multiplying V3 domains by their insertion at various positions within the initial Env protein (Rovinski et al. 1992). To improve such entrapping and obtain the Gag VLPs containing high quantities of the Env products, the glycoprotein gp120 was covalently linked at different C-terminal positions to a transmembrane (TM) domain from the Epstein-Barr virus (EBV) major Env glycoprotein gp220/350 (Deml et al. 1997a).

When coexpressed with the Pr55Gag in insect cells, such Env-TM chimeras incorporated more efficiently into the VLPs than authentic Env proteins. The immunization studies carried out with such immunologically expanded VLPs elicited a consistent anti-Pr55Gag as well as an anti-Env antibody response, including neutralizing antibodies, and the remarkable induction of CTLs in the complete absence of additional adjuvants (Deml et al. 1997b).

Later, the TM sequences, as well as signal peptide (SP) and cytoplasmic tail (CT) domains of HIV-1 Env, were replaced with those of other viral or cellular proteins individually or in combination (Wang BZ et al. 2007). The substitution of the HIV TM-CT with sequences derived from the mouse mammary tumor virus (MMTV) envelope glycoprotein, influenza virus hemagglutinin, or baculovirus (BV) gp64—but not from Lassa fever virus glycoprotein—was found to enhance Env incorporation into the VLPs. The highest level of the Env incorporation into the VLPs was observed in chimeric constructs containing the MMTV and BV gp64 TM-CT domains, in which the Gag/Env molar ratios were estimated to be 4:1 and 5:1, respectively. The electron microscopy revealed that the VLPs

with the chimeric HIV Env were similar to HIV-1 virions in morphology and size and contained a prominent layer of Env spikes on their surfaces (Wang BZ et al. 2007). Quan et al. (2007) generated the HIV-1 Gag VLPs that exposed modified SHIV envelope proteins. Earlier, Yao et al. (2000) produced the SHIV VLPs that were built up from the SIV Gag and HIV-1 Env synthesized simultaneously in insect cells. Such SHIV VLPs were used further to incorporate the CD40 ligand (Zhang R et al. 2010).

Next, the HIV-1 VLPs were generated, which carried the Flt3 ligand (FL; a dendritic cell growth factor) and were targeted therefore to dendritic cells (Sailaja et al. 2007). To do this, the human FL-encoding gene was modified to contain the sequences coding for the HIV TM with partial extracellular and cytoplasmic tail domains and a signal peptide from tissue plasminogen activator. The *Sf*9 insect cells coinfected with baculovirus vectors expressing HIV-1 Gag, Env, or FL released VLPs containing HIV Gag + Env or VLPs containing Gag + Env + FL in the culture supernatants (Sailaja et al. 2007).

Following the GPI-anchoring technique elaborated for the MLV Gag VLPs and described earlier, Skountzou et al. (2007) generated the SIV VLPs with the GPI-anchored GM-CSF or CD40L and showed that chimeras carrying GM-CSF induced SIV Env-specific antibodies as well as neutralizing activity at significantly higher levels than those induced by standard SIV VLPs. Moreover, the mice immunized with the chimeric SIV VLPs containing either GM-CSF or CD40L showed significantly increased $CD4^+$- and $CD8^+$-T-cell responses to SIV Env, compared to the standard SIV VLPs.

Di Bonito et al. (2009) used Nef as an anchor to incorporate human papillomavirus 16 (HPV-16) E7 protein into the HIV-1 Gag VLPs. In fact, the inoculation in mice with the VLPs carrying the HPV-16 E7 protein fused to Nef led to an anti-E7 $CD8^+$ T cell response much stronger than that elicited by the E7 recombinant protein inoculated with incomplete Freund's adjuvant and developed a protective immune response against tumors induced by E7 expressing tumor cells.

The next approaches at constructing chimeric VLPs involved the combination of both genes, *env* and *gag*, in the form of in-frame (or frameshift) Gag-Env fusions (Tobin et al. 1996a). Neither simple coinfection of insect cells with two Env- and Gag-expressing vectors nor the in-frame fusion strategies produced large quantities of structurally stable chimeric VLPs. The frameshift fusion method utilized the retroviral Gag-Pol ribosomal frameshift mechanism for the coexpression of the Gag and Gag-frameshift-Env fusion proteins, and reliable quantities of the VLPs containing both the Gag and Env epitopes were produced in this case (Tobin et al. 1996a).

This strategy was used further to translate the C-terminal 65% of the gp120 as chimeric Gag-Pol fusion protein, after insertion of the Env fragment immediately downstream of the Gag stop codon (Tobin et al. 1997). The mice inoculated with such VLPs developed CTL to both HIV-1 Gag and Env epitopes yet humoral immune responses only to Gag epitopes.

Nevertheless, despite HIV-1-specific neutralizing and CTL responses, both types of the HIV Pr55Gag VLPs, constructed either by replacing Gag sequences by the V3 loop and a linear portion of the CD4 binding domain or by stable anchoring of modified gp120 on the surface of particles, failed to protect macaques against simian human immunodeficiency virus (SHIV) infection after intramuscular immunization (Wagner R et al. 1998). A clearly accelerated reduction of the plasma viremia as compared to control animals was achieved after priming animals with the corresponding set of recombinant Semliki Forest viruses (Notka et al. 1999).

In the 2000s, the embedding of the native HIV Env molecules into the native HIV Gag VLPs synthesized in mammalian cells returned to the stage (McBurney et al. 2007). This approach allowed direct comparison of the breadth of cell-mediated immune responses elicited by consensus and polyvalent Env vaccines, when diverse Env variants used to be exposed on the HIV VLPs (McBurney and Ross 2009; Tong et al. 2012).

Chapman et al. (2020) generated a vaccine of the HIV-1 Gag Pr55 VLPs carrying chimeric envelope glycoproteins on the surface. In this case, the native HIV envelope signal sequence was replaced with the human tissue plasminogen activator (TPA) leader sequence, and two chimeras were made. In the first chimera, the HIV gp120 and gp41 ectodomain were retained, and the TM and CT were replaced with that of the influenza hemagglutinin protein HA$_2$. In the second chimera, the entire HIV gp41 was replaced with the influenza hemagglutinin HA$_2$, which is the analogous transmembrane subunit of the glycoprotein. DNA and modified vaccinia Ankara (MVA) vaccines expressing these chimeras were constructed. All the vaccines constructed were shown to express the Gag VLPs containing Env or Env chimeras (Chapman et al. 2020).

Next, a few facts of introduction of non-HIV sequences into the Gag carriers are worthy of consideration. The pseudorabies virus glycoprotein gD expressed on a separate vector was incorporated into the Gag VLPs in insect cells (Garnier et al. 1995). Further studies supported the idea that the incorporation of the surface glycoproteins into the retroviral particles was not a specific process and that many heterologous viral and cellular glycoproteins could be incorporated as long as they did not have long cytoplasmic C-terminal regions. Thus, the incorporation of the wild-type human epidermal growth factor receptor (EGFR) and of its C-terminally truncated variant into the Gag VLPs was achieved successfully (Henriksson et al. 1999). In agreement with the aforementioned concept, the C-terminal variant of the EGFR, with only 7 C-terminal aa, was incorporated into the VLPs more efficiently than the nonmodified EGFR, with 542 C-terminal cytoplasmic aa. Moreover, the Gag-β-galactosidase fusions were capable of being incorporated into VLPs in the presence of the intact HIV Gag protein and formed therefore mosaic particles (Henriksson et al. 1999).

Steel et al. (2010) generated a promising influenza vaccine candidate involving the conserved hemagglutinin (HA) stalk domain. While transient expression of the headless HA constructs alone in 293T cells was not found to result in VLP production, cotransfection with the HIV Gag-based construct did lead to the production of headless HA-containing particles. Specifically, when one of the headless HA constructs was coexpressed with a Gag-eGFP (enhanced green fluorescent protein) fusion protein, the appropriate VLPs were released into the cell culture medium (Steel et al. 2010).

Cervera et al. (2013) not only optimized high yield production of the HIV-1 Gag VLPs in mammalian 293T cells but also generated a Gag-GFP fusion construct that self-assembled into fluorescent VLPs. Thus, the great majority of the Gag-GFP present in cell culture supernatants was shown to be correctly assembled into VLPs of the expected size and morphology consistent with immature HIV-1 particles. Recently, Lavado-García et al. (2021) performed an analytical study to clarify the current variability of Gag stoichiometry in the HIV-1 VLPs depending on the cell-based production platform, directly determining the number of Gag molecules per VLP in each case. An average of 3617 ± 17 monomers per VLP was obtained for HEK293 cell line, substantially varying between platforms, including mammalian and insect cells.

Finally, Kessans et al. (2013) achieved in *Nicotiana benthamiana* plants the assembly and budding of the HIV-1 VLPs consisting of full-length p55^{Gag} and a deconstructed variant of gp41 (dgp41)—comprised of its membrane proximal external region (MPER), TM and CT domains. The coexpression of the plant-optimized recombinant genes encoding these proteins was achieved through the novel combination of traditional stable nuclear transformation and a tobamovirus-based transient overexpression system (Kessans et al. 2013). This plant-produced HIV-1 VLP vaccine induced reliable antibody titers against both the Gag and gp41 and Gag-specific CD4 and CD8 T-cell responses in mice (Kessans et al. 2016). The generation of the HIV VLPs in plants was reviewed on the background of other VLP platforms by Peyret et al. (2021).

Nowadays, the HIV-1 VLPs were used construct putative vaccines against enterovirus 71 (Wang X et al. 2018), influenza (Fernandes et al. 2021), and SARS-CoV-2 (Miyakawa et al. 2021).

SUBFAMILY *SPUMARETROVIRINAE*

In contrast to other retroviruses, the subfamily *Spumaretrovirinae* members—or foamy viruses—contains both ssRNA and dsDNA in extracellular particles, and reverse transcription occurs during virus assembly and disassembly (Krupovic et al. 2018).

Effantin et al. (2016) published the first electron cryo-microscopy and tomography ultrastructural data on a prototype foamy virus (PFV) obtained from human cells. The mature PFV particles had a distinct morphology with a capsid of constant dimension as well as a less ordered shell of density between the capsid and the membrane likely formed by the Gag N-terminal domain and the cytoplasmic part of the Env leader peptide. The viral membrane contained trimeric Env glycoproteins partly arranged in interlocked hexagonal assemblies. The electron cryomicroscopy allowed one to obtain a 9 Å resolution map of the glycoprotein in its prefusion state, which revealed extensive trimer interactions by the receptor binding subunit at the top of the spike and three central helices derived from the fusion protein subunit. Thus, the mature PFV particles were quite different from the mature particles of other retroviruses, which consist of a polyhedral core of CA containing the NC/RNA complex, as well as from their immature forms. These differences were related to differences in posttranslational processing, since Gag from most retroviruses is cleaved after the release of immature virions from the cell at various positions, which triggers the reorganization of the MA, CA, and NC domains inside the virus. However, for PFV, there was only one primary cleavage of Gag by the viral protease occurring in the course of virus assembly in infected cells (for details, references, and discussion see Effantin et al. 2016).

PACKAGING

Concerning packaging potential of the HIV-1 and HIV-2 VLPs, the first idea was examined by Wu et al. (1995). They proposed to target foreign proteins to HIV particles via fusion with Vpr and Vpx, since the latter were packaged into virions through virus type-specific interactions with the Gag polyprotein precursor. The *vpr* and *vpx* ORFs were fused in frame with genes encoding the bacterial staphylococcal nuclease (SN)—an enzymatically inactive mutant of SN (SN*)—and chloramphenicol acetyltransferase (CAT). The transient expression in a T7-based vaccinia virus system demonstrated the synthesis of appropriately sized Vpr1-SN/SN* and Vpx2-SN/SN* fusion proteins, which, when coexpressed with their cognate p55^{Gag} protein, were efficiently incorporated into VLPs, while remaining enzymatically active.

Later, Peretti et al. (2005) used HIV-1 Nef to incorporate proteins of even up to 30 kDa upon C-terminal fusion with them. This was demonstrated first by assessing the intracellular fluorescence of cells challenged with the HIV-1 VLPs pseudotyped with the VSV glycoprotein and incorporating Nef fused to GFP. Furthermore, the biologic activity of

products delivered by Nef-based VLPs was demonstrated by tagging Nef with the herpes simplex virus-1 thymidine kinase (HSV-1 TK). Both cell lines and primary human macrophages challenged with the VSV-G Nef7/TK VLPs died after five to seven days of treatment with ganciclovir (Peretti et al. 2005). This allowed consideration of the Nef-based VLPs as promising platforms for protein delivery, and a bit later Peretti et al. (2006) demonstrated how to kill HIV-1-infected cells with the chimeric VLPs. The methodology of this approach was described in detail by Muratori et al. (2010). Then, Sistigu et al. (2011) extended these investigations by testing the antigenic and immunogenic properties of the Nef-based VLPs incorporating much larger heterologous products, i.e., human hepatitis C virus (HCV) NS3 and influenza virus NP proteins, which were composed of 630 and 498 amino acids, respectively.

Meanwhile, Voelkel et al. (2010) reported that they were able to deliver the protein GFP and functional Flp recombinase to cells using another method than HIV-1 scaffold, namely MLV particles.

At last, Kaczmarczyk et al. (2011) involved Gag of avian sarcoma leukosis viruses (ASLVs) as a background for a safe and efficient system to deliver functional proteins into cells. This system was safe both because it was based on an avian retrovirus—which cannot replicate in human cells—and because the construct used to express the VLP proteins did not contain the Pol gene, and, as a consequence, the VLPs did not contain either reverse transcriptase or integrase. The system was efficient because the VLPs were composed entirely of the Gag fusion protein, and a single VLP delivered therefore 2,000–5,000 copies of the Gag fusion protein into a transduced cell. The authors generated VLPs that contained Gag-Cre recombinase, Gag-Fcy::Fur, and Gag-human caspase-8 as a proof of concept and demonstrated that the encapsidated proteins were active in recipient cells. Notably, the protein Fcy converts 5-fluorocytosine into 5-fluorouracil (5-FU), a highly cytotoxic compound routinely used in cancer chemotherapy. The protein Fur converts 5-FU to 5-F UMP, an irreversible inhibitor of thymidylate synthase, which blocks DNA synthesis.

In addition to intracellular delivery, Kaczmarczyk et al. (2011) created VLPs that displayed protein ligands on their surface. Thus, a truncated form of influenza neuraminidase (NA) or hemmagglutinin (HA) was fused to murine IFN-γ and human TNF-related apoptosis-inducing ligand (TRAIL). The presence of the protein fusions on the surface of VLPs activated the appropriate cellular receptors on cells.

Section VII

Double-Stranded DNA Viruses
Using Reverse Transcription

36 Family *Caulimoviridae*

The only thing that should surprise us is that there are still some things that can surprise us.

Francois de La Rochefoucauld

ESSENTIALS

The *Caulimoviridae* is a family of nonenveloped viruses with a single-core capsid protein that are infecting plants. The family combines all plant viruses with a double-stranded DNA genome that have a reverse transcribing phase in their lifecycle. According to modern classification, the family *Caulimoviridae* is a member of the order *Ortervirales* (Krupovic et al. 2018; ICTV 2020). The family is separated here from its native order *Ortervirales* that is described within the section VI in Chapter 35 because it belongs, together with the order *Blubervirales*, to the Baltimore class VII.

In contrast to the Baltimore class VI single-stranded RNA members of the order *Ortervirales*, which frequently integrate into the host genomes, the Baltimore class VII members of the family *Caulimoviridae*, as well of hepadnaviruses of the order *Blubervirales* described in Chapters 37 and 38, often referred together to as pararetroviruses (Hull and Will 1989), do not actively integrate into host chromosomes, while their episomal replication cycles do not involve an integration phase. However, the capture of pararetroviral DNA in host genomes, presumably by illegitimate recombination, is commonplace, particularly in plants, giving rise to the corresponding endogenous elements (Feschotte and Gilbert 2012; Diop et al. 2018; Krupovic et al. 2018).

By analogy with the typical members of the order *Ortervirales*, the cauliviruses also possess a homologous aspartate protease domain in their polymerase polyprotein (see Krupovic et al. 2018 for references) but lack an integrase and long terminal repeats (LTRs). However, RT-based phylogenies consistently classify these plant-infecting viruses as a sister clade of the metaviruses (compare the genome structures in the corresponding figures of Chapter 35 and here). As a result, Krupovic et al. (2018) proclaimed that the encapsidation of a DNA genome and not of a RNA genome is a homoplasious trait and, therefore, is not a reliable criterion for the current ICTV classification.

The family *Caulimoviridae* consists currently of 11 genera and 94 species. Some viruses cause economically important diseases of tropical and subtropical crops (Teycheney et al. 2020). The transmission occurs through insect vectors (aphids, mealybugs, leafhoppers, lace bugs) and grafting. The activation of infectious endogenous viral elements occurs in *Musa balbisiana*, *Petunia hybrida*, and *Nicotiana edwardsonii*. However, most endogenous cauliviruses are not infectious (Teycheney et al. 2020). The cauliflower mosaic virus-Cabb (V00141) (CaMV) from the *Cauliflower mosaic virus* species of the genus *Caulimovirus* is a typical representative of the family *Caulimoviridae*.

Figure 36.1 presents a typical cartoon of a member of the family *Caulimoviridae*, as well as electron microscopy portraits and 3D reconstruction of the CaMV particle. Generally, the virions of the family members are either isometric of 45–52 nm in diameter or, in the case of members of the genera *Badnavirus* and *Tungrovirus*, bacilliform of 30 nm × 60–900 nm (Teycheney et al. 2020).

As summarized at the ICTV home page, the CaMV virions are 52 nm in diameter with an icosahedral T = 7 symmetry. The C-terminus of the capsid protein, which has a nucleic acid-binding motif, is embedded inside the pores surrounding the capsomers and traverses the capsid protein layers to reach the genomic DNA. Remarkably, the CaMV capsid protein is glycosylated.

The virion-associated proteins (VAPs) are incorporated into the virion as a triskelion structure that cements three hexavalent or pentavalent capsomers together. The N-terminus of the VAP is facing out of the CaMV capsid. It contains two coiled-coil motifs with opposite handedness, forming dimers by antiparallel interaction with adjacent capsid molecules to create a network around the virus particle (Plisson et al. 2005). The VAP facilitates interaction of the virion with host and other viral proteins—i.e., the movement protein (MP) and the aphid transmission factor (ATF)—but is not essential for the virion formation.

GENOME

The *Caulimoviridae* genome is represented by a single molecule of noncovalently closed circular dsDNA of 7.1–9.8 kb with discontinuities at specific sites in the negative-sense (one) and positive-sense strand (one to three). The genomes do not carry LTRs and may contain one to eight ORFs encoding five to six conserved protein domains, depending on the genus. Figure 36.2 demonstrates the genomic structure of the CaMV as the most popular representative of the family.

Generally, the genome organization of the family members is dependent upon the genus and is one of the main characteristics that distinguish the genera from each other (Geering and Hull 2012).

3D STRUCTURE

The virions are either isometric or bacilliform depending on the genus. There is no envelope. The first 3D investigation of the CaMV particle was performed by neutron small-angle scattering in buffers containing various amounts of

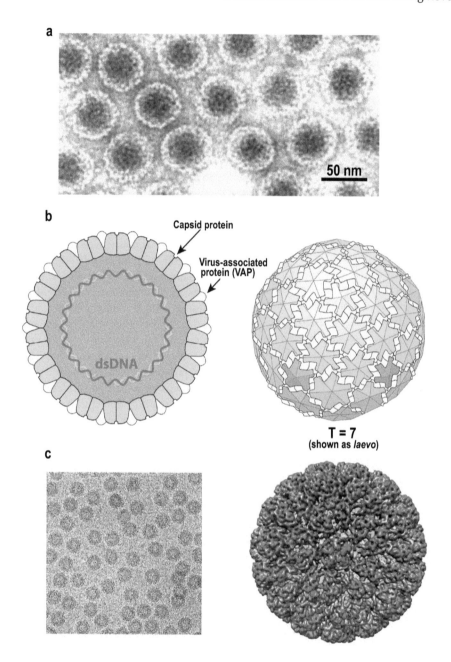

FIGURE 36.1 Electron micrograph and schematic cartoon of a typical representative of the family *Caulimoviridae*. (a) The electron micrograph of cauliflower mosaic virus, a type member of the *Caulimovirus* genus in the *Caulimoviridae* family. (From Haas M, Bureau M, Geldreich A, Yot P, Keller M: Cauliflower mosaic virus: still in the news. *Mol Plant Pathol.* 2002.3.419–429. Copyright Wiley-VCH Verlag GmbH & Co. KGaA. Reproduced with permission.) (b) The cartoon of a representative of the *Caulimoviridae* family is reproduced with kind permission from the ViralZone, Swiss Institute of Bioinformatics (Hulo C et al. 2011). (Courtesy Philippe Le Mercier.) (c) Electron micrograph of virions and tridimensional reconstruction of the cauliflower mosaic virus particle. (Images courtesy of Patrick Bron and Andrew D.W. Geering and the International Committee on Taxonomy of Viruses (ICTV), https://talk. ictvonline.org/ictv-reports/ictv_online_report/reverse-transcribing-dna-and-rna-viruses/w/caulimoviridae.)

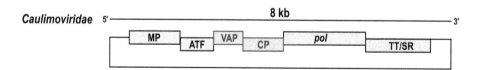

FIGURE 36.2 The genomic structure of the order *Caulimoviridae* family represented by the genome of cauliflower mosaic virus. The structural genes are colored dark pink. Abbreviations: ATF, aphid transmission factor; CP, capsid protein; MP, movement protein; pol, polymerase gene; TT/SR, translation trans-activator/suppressor of RNA interference; VAP, virion-associated protein. The circular character of the genome is displayed by the lower connecting bracket.

D_2O, and the capsid symmetry was described as probably possessing a T = 7 icosahedral organization (Kruse et al. 1987).

Next, Cheng et al. (1992) published the first electron cryo-microscopy and 3D image reconstruction of the CaMV particle. According to this study, the virus particle appeared as spherical, with a maximum diameter of 53.8 nm, composed of three concentric layers of solvent-excluded density that surrounded a large, solvent-filled cavity of ~27 nm. The outermost layer contained 72 capsomeric morphological units, with 12 pentavalent pentamers and 60 hexavalent hexamers for a total of 420 subunits (37–42 kDa each) arranged with T = 7 icosahedral symmetry. Therefore, CaMV appeared as the first example of a T = 7 virus that obeyed the rules of stoichiometry proposed for isometric viruses by Caspar and Klug (1962). The current 3D reconstruction of the CaMV particle is presented in Figure 36.2.

EXPRESSION

The CaMV capsid protein gene was expressed in *E. coli* (Chapdelaine and Hohn 1998). The authors found that folded sheets or large tubular structures appeared by the in vitro renaturation of the CaMV protein pIV derivatives, whereas only the central core of the pIV protein was required to initiate assembly. Namely, the region between aa positions 77 and 332 was sufficient for in vitro self-aggregation, while C-terminal deletion to aa position 265 still allowed dimerization but prevented further aggregation.

37 Order *Bluebervirales*: Surface Protein

My tastes are simple: I am easily satisfied with the best.

Winston S. Churchill

ESSENTIALS

The order *Bluebervirales* contains the sole family *Hepadnaviridae*, DNA viruses that are placed within the realm *Riboviria* because of homology between the hepadnavirus reverse transcriptase and the RNA-directed RNA polymerase of RNA viruses (Koonin et al. 2020). According to the current ICTV taxonomy report (Magnius et al. 2020) and current global review (Glebe et al. 2021), the *Hepadnaviridae* family includes 18 officially registered species divided into 5 genera depending on the viral host: *Avihepadnavirus* infecting birds, 3 species altogether including well-studied duck hepatitis B virus or DHBV; *Herpetohepadnavirus* infecting reptiles and frogs with one species; *Metahepadnavirus* infecting fresh and salt water teleosts, one species; *Orthohepadnavirus* infecting mammals and demonstrating a narrow host range for members of each 12 virus species; and *Parahepadnavirus* infecting freshwater teleosts, one species. The representative *Orthohepadnavirus* genus includes the hepadnaviral type representatives: hepatitis B virus (HBV), which infects humans and primates, as well as well-studied woodchuck (WHV) and ground squirrel (GSHV) hepatitis viruses. The hepadnaviruses are hepatotropic, and infections may be transient or persistent.

A new clade of teleost viruses has been identified by search in the NCBI databases (Lauber et al. 2017). These shared many features with members of the *Hepadnaviridae* and were therefore considered a sister family to this family They were designated as nackednaviruses, a name given due to their absence of an envelope, which is a fundamental difference to hepadnaviruses. The provisional taxon designation *Nudnaviridae* was proposed for this family by the ICTV (Magnius et al. 2020).

The first electron microscopy visualization of the HBV virions was published by Dane et al. (1970) and launched the informal Dane particle name for them. The typical widely used electron micrograph of HBV as a *Hepadnaviridae* type species, together with the corresponding virion map, are presented in Figure 37.1. Generally, the hepadnaviruses are small spherical, occasionally pleomorphic, enveloped particles of 42–50 nm diameter.

Figure 37.2 demonstrates HBV virions how they were resolved later by electron cryomicroscopy at 14–16 Å (Dryden et al. 2006). In parallel, the excellent electron cryomicrographs of HBV virions were published by Seitz et al.

(2007), who concentrated especially on the specific interactions of the outer envelope with the inner nucleocapsid.

GENOME

The hepadnaviral genome is represented by circular relaxed partially double-stranded DNA of approximately 3.0–3.4 kb in length, which is delivered to the nucleus of the host cell and converted into a covalently closed circular DNA molecule. The first full-length sequences of cloned HBV genomes were published for the genotype A, subtype *adw* (Valenzuela et al. 1981; Ono et al. 1983) and the genotype D, subtype *ayw* (Galibert et al. 1979; Bichko et al. 1985), which were followed later by enormous number of hepadnaviral sequences, including the metagenomic ones lately. Overall, the HBV genome is represented by 10 main genotypes, A–J, which differ by more than 8% at the nucleotide level (Kramvis 2014). In addition to the genotype classification, HBV is characterized by subtypes—or serotypes—based on the reactivity against HBs proteins. All genotypes have a common reactivity against the major antigenic determinant "a" but further express two mutually exclusive allelic antigenic determinants "d" or "y" and "w" or "r."

Figure 37.3 demonstrates genomic maps of the hepadnavirus species that were most regularly involved in the viral nanotechnology applications.

The partially dsDNA in the hepadnaviral genome is held in a circular conformation by base pairing in a cohesive overlap between the 5′-ends of the two DNA strands (Summers et al. 1975). One strand (negative-sense, i.e. complementary to the viral mRNAs) is full-length, whereas the positive-sense strand varies in length. The 3′-end of the positive strand terminates at a variable position in different molecules, creating a single-stranded gap that may account for as much as 60% of the HBV genome.

The hepadnaviral genome has three open reading frames (ORFs): precore/core (preC/C), polymerase (P), env or surface (preS/S), and, regarding orthohepadnaviruses, an additional X ORF. The common features of all members of the *Hepadnaviridae* family are the expression of the corresponding three major sets of proteins: the two structural—preC/C and preS/S—and one functional—polymerase—and replication by reverse transcription within immature nucleocapsids in the cytoplasm of infected hepatocytes. The structural surface gene preS/S is the subject of the present chapter, while the core gene preC/C is described in Chapter 38.

The history and peculiarities in the designations of the HBV genes and their products, as well as their antigenic

DOI: 10.1201/b22819-45

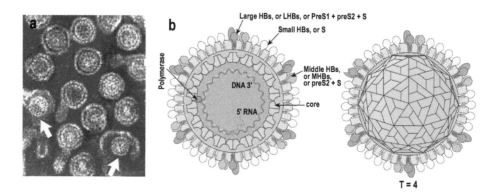

FIGURE 37.1 Virus particles of hepatitis B virus (HBV), a major representative of the *Hepadnaviridae* family. (a) A classical photograph by Linda M. Stannard (1985). (Copyright University of Cape Town (2021), published with a kind permission.) The virions are 42 nm in diameter and possess an isometric nucleocapsid or core of 27 nm in diameter, surrounded by an outer coat approximately 4 nm thick. The envelope protein, referred generally as HBsAg, is represented by the three forms, the major small HBs protein and the two minor proteins, namely, middle and large HBs proteins, which are N-terminal extensions of the small HBs with the appropriate preS1 and preS2 sequences. The surface antigen is generally produced in vast excess and is found in the blood of infected individuals in the form of filamentous and spherical particles. The filamentous particles vary in length and have a mean diameter of about 22 nm. (b) A cartoon of the HBV virion taken with kind permission from the ViralZone, Swiss Institute of Bioinformatics, http://viralzone.expasy.org (Hulo C et al. 2011). (Courtesy Philippe Le Mercier.)

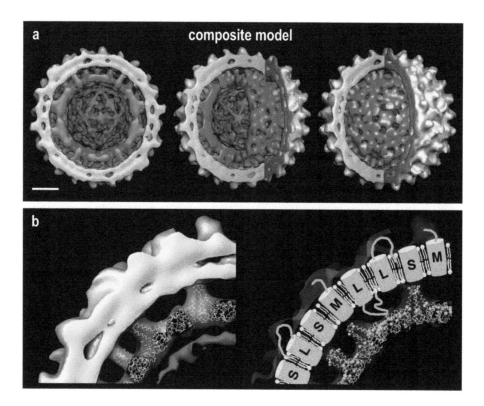

FIGURE 37.2 Model and cartoon of HBV virions depicted from electron cryomicroscopy data. (a) Cut away views of a composite model of the HBV virion comprised of a T = 4 icosahedral capsid with 120 spikes and an outer envelope with protein projections spaced ~60 Å apart. Views are cross-sections (left) and two cut aways. (b) X-ray crystal structure of recombinant capsid docked into the electron cryomicroscopy density map of the virion capsid (left). The tips of the core spikes are in close apposition but do not penetrate the envelope. Additional details and cartoon of interpretation (right). The surface protein projections are ascribed to HBs and are designated as large (L), medium (M), and small (S) according to the three start codons of the HBs open reading frame. Residues in the capsid tip are colored according to charge and hydrophobicity (negative, red; positive, blue; hydrophobic, gold; and hydrophilic, gray) and show a mixed distribution of charge. Residues denoted by green spheres have been implicated in viral envelopment via interaction with the L protein (Ponsel and Bruss 2003). HBs (yellow) modeled as S box with M and L loops. Note that ~50% of the L molecules have an interior loop that is predicted to be disordered, which interacts with specific residues in HBc (green spheres). Bar, 100 Å. (Reprinted from *Mol Cell.* 22, Dryden KA, Wieland SF, Whitten-Bauer C, Gerin JL, Chisari FV, Yeager M, Native hepatitis B virions and capsids visualized by electron cryomicroscopy, 843–850, Copyright 2006, with permission from Elsevier.)

specificities, were thoroughly explained by Wolfram H. Gerlich et al. (2020).

SURFACE GENE

The surface—or preS/S—gene encodes the three envelope proteins, L (LHBs), M (MHBs), and S (HBs), which form a C-terminal nested set, HBs being the smallest, MHBs the next largest, and LHBs the largest. The extra domains in the MHBs and LHBs proteins are named preS2 and preS1, respectively. The virions and empty subviral particles may contain two or three such surface proteins, with a common C-terminus but distinct N-termini due to different sites of translation initiation, known now collectively as the HBs proteins or hepatitis B surface antigen (HBsAg) in an immunological sense (Heermann et al. 1984; reviewed in detail by Heermann and Gerlich 1991; Bruss 2007; Gerlich 2013, 2015; Magnius et al. 2020; Seitz et al. 2020).

The first information about the HBsAg appeared when Baruch Samuel Blumberg—the 1976 Nobelist—and colleagues discovered that a serum antigen, previously identified in their lab and found to be common in leukemia patients treated via blood transfusion, was actually the component of the HBV virion (Blumberg et al. 1967, 1971; Blumberg 1977). This serum antigen, known as Australia antigen because of its identification in an Australian aborigine, was found as 22-nm rods and spheres in the blood of infected individuals. The latter particles were subviral and comprised the three previously mentioned HBV envelope proteins, as reviewed in detail by Heermann and Gerlich (1991), Seeger and Mason WS (2015), and Seeger et al. (2020).

The virions of orthohepadnaviruses contain SHBs, MHBs, and LHBs proteins that are myristoylated at the N-terminus. More than one form of each of the aforementioned proteins occurs due to alternative patterns of glycosylation. Moreover, hepadnaviruses induce the overproduction of surface proteins that are secreted into the blood as pleomorphic particles. For HBV, these are 17–22-nm spherical particles and filaments, known as 22-nm particles, as mentioned earlier. For HBV and WHV, the virions and filaments are enriched in the LHBs protein. The empty spheric 22-nm particles consist predominantly of SHBs protein. The DHBV virions contain only LHBs and SHBs proteins, which are distributed evenly between particle types.

As shown in Figure 37.3, the open reading frame preS/S encodes a total of 389 (almost all genotypes) or 400 (genotype A) aa residues. All HBs proteins contain therefore the essential S sequence of 226 aa, whereas the M protein has a 55-aa preS2 extension at the N-terminus, and the L protein has a preS1 N-terminal extension of 108 or 119 aa (depending on the HBV genotypes) in addition to the preS2 (Heermann and Gerlich 1991). The additional N-terminal 11 aa of the preS1 of the HBV genotype A seem to be nonessential for any structural and immunological features of the HBs proteins. The proportion of S to M to L (S:M:L) proteins in the HBV particle envelope is about 7:2:1 (Heermann et al. 1984).

EXPRESSION OF THE SURFACE GENE

As mentioned earlier, the HBs proteins possessed the unique property of self-assembling with cellular lipids into

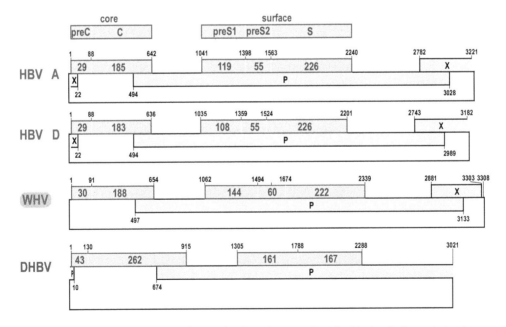

FIGURE 37.3 The genomic structure of hepadnaviruses that have been used preferably in viral nanotechnology applications. The specific colors of the representative names indicate the genus, where HBV is represented by two popular genotypes: A and D. The core and surface genes as sources of nanoparticles are colored dark pink. Amino acid length of the core and surface proteins encoded by the two structural genes is given by large numbers, but numbering of the gene location within the genome is indicated by small numbers. The genomes are collineated by the position of the precore initiation codon of the core gene, while circular character of the genome is displayed by the lower connecting bracket. (The sequence data is compiled from the NCBI browser.)

noninfectious empty envelope 22-nm particles, spherical or tubular, of about 22 nm in diameter, which were secreted in large excess from hepatocytes during hepatitis B infection and consisted mostly of the SHBs protein, in contrast to the HBV virions. Table 37.1 shows that the same self-assembling property was retained by the homologous and heterologous expression of the preS/S gene in popular expression systems, first by the homologous expression in mammalian cells lines. Thus, the eukaryotic cell lines showed good capacity of producing 22-nm particles after expression of S and M—but not L—genes within the appropriate expression cassettes (Christman et al. 1982; Gough and Murray 1982; Wang et al. 1982; Laub et al. 1983; Michel et al. 1984).

The synthesis of the HBs proteins in mammalian cells was used as a standard gene to prove the usefulness of the most of powerful eukaryotic vector systems elaborated at that time and based on regulatory elements taken from SV40 (Liu CC et al. 1982), retrovirus (Stratowa et al. 1982), vaccinia virus (Smith GL et al. 1983), herpes simplex virus (Shih et al. 1984), papillomavirus (Denniston et al. 1984), adenovirus (Davis AR et al. 1985), and varicella zoster virus (Shiraki et al. 1991; Kamiyama et al. 2000).

Very early on, the expression of the gene S was achieved in transgenic mice, and the appropriate animal models of the chronic HBsAg carrier state were generated (Babinet et al. 1985; Chisari et al. 1985).

Thereafter, reasonable levels of HBsAg synthesis have been achieved in baculovirus vector-driven insect cells (Kang et al. 1987), in silkworm larvae (Higashihashi et al. 1991), and in transgenic plants such as tobacco (Mason HS et al. 1992; Thanavala et al. 1995), potato (Ehsani et al. 1997; Richter et al. 2000; Kong et al. 2001; Smith ML et al. 2003; Shulga et al. 2004), lettuce, and lupin (Kapusta et al. 1999, 2001; Pniewski et al. 2006). The S gene was also expressed in stably transfected *Drosophila melanogaster* Schneider-2 cells under an inducible *Drosophila* metallothionein promoter (Deml et al. 1999).

Nevertheless, the results of the greatest practical utility were achieved by the successful expression of S, M, and L genes in yeast *Saccharomyces cerevisiae*, where the expression products self-assembled into particles with a widely estimated diameter of 20 ± 4 nm (Valenzuela et al. 1982; Dehoux et al. 1986; Itoh et al. 1986; Kniskern et al. 1988; Kobayashi et al. 1988; Heijtink et al. 2000). It was important that in yeast *Saccharomyces cerevisiae*, in contrast to the eukaryotic cell lines, the expression of the L gene ensured formation of spherical particles exclusively consisting of the entire L protein (Kuroda et al. 1992; Yamada et al. 2001). Such particles contained from about 110 HBs molecules and demonstrated a broad range of diameter between 50 and 500 nm. The further history of the highly efficient synthesis of the 22-nm particles in yeasts *Pichia pastoris* and *Hansenula polymorpha* is documented in Table 37.1. Generally, the yeast expression paved the way to the classical hepatitis B vaccines that are described later.

When the SHBs gene was coexpressed together with the HBc gene, the virion-like particles were formed in yeast (Shiosaki et al. 1991) and insect cells (Takehara et al. 1988).

The HBV gene S was one of the first viral structural genes expressed in bacteria (Burrell et al. 1979; Charnay et al. 1980; Edman et al. 1981; MacKay et al. 1981). Despite serious efforts mounted to generate "bacterial" HBs in *E. coli*, the self-assembly of the latter failed, possibly because specific lipids and/or folding and budding mechanisms were missing in bacteria (Fujisawa et al. 1983; Pumpens et al. 1983, 1984, 1985, 1992; Smirnov et al. 1983; Kozlovskaia et al. 1984; Kozlovskaia and Pumpens 1986). After more than 30 years, Li H et al. (2017) solved the problem when removing the transmembrane domain 1 and the C-terminal long hydrophobic sequence of the SHBs gene. The SHBs ΔNC fragment spanning the transmembrane domain 2 to N-terminal half of the transmembrane 3, aa 75–172, was successfully expressed in *E. coli*, after some N-terminal optimization (mΔNC). The mΔNC product formed VLPs with 28.5 ± 6.2 diameter and molecular mass of ~670 kDa, which reacted with anti-HBs antibody, indicating therefore surface location of the "a" determinant on the ΔNC particles. Taken together, the *E. coli*-derived mΔNC particles are good candidates to serve as a novel VLP platform ready to substitute eukaryotic cell-derived HBs particles for versatile applications (Li H et al. 2017).

Remarkably, the experiments on the HBs expression and purification were never stopped or exhausted. Thus, Reuschel et al. (2019) reported a study on the comparative purification and characterization of the HBs VLPs produced by recombinant vaccinia viruses in human hepatoma cells and human primary hepatocytes and purified by four different protocols. The pure HBs VLPs consisted of both the nonglycosylated p25 and the glycosylated gp27 forms and assembled into typical 22-nm particles.

3D AND ANTIGENIC STRUCTURE

Major structural and immunological features of the HBs monomeric subunit are shown in Figure 37.4. A linear presentation of the HBs protein with location of the putative α-helices and major B-cell and CTL epitopes including the major immunodominant B-cell epitope "a" (a) is illustrated with localization of the same crucial elements on the putative 3D map of the HBs monomer (b).

As mentioned earlier, the spherical 22-nm particles from human blood consisted mainly of the S protein, in contrast to the HBV virions, which contained the L and M proteins. To simulate the possible 3D structure of the 22-nm particles and HBs monomers, many alternative mathematical and experimental approaches were employed. First, computer-aided topological models were employed (Stirk et al. 1992; Berting et al. 1995; Sonveaux et al. 1995). A summarizing picture of such modeling is presented in Figure 37.4b. These models were confirmed by the identification of membrane-associated regions protected against proteolysis (Sonveaux et al. 1994) by circular dichroism

TABLE 37.1

Expression of the *Hepadnaviridae* Surface Gene and Production of the HB VLPs by Different Expression Systems

VLPs	Expression Host	References	Comments
Avihepadnavirus			
Duck hepatitis B virus genus	*E. coli*	Schlicht et al. 1987	Fusion with the MS2 polymerase. No particles.
	Ducks	Triyatni et al. 1998	DNA vaccine.
	Chicken hepatoma cell line LMH	Grgacic et al. 2000	Expression in homologous chicken liver cells
	Hansenula polymorpha (syn. *Pichia angusta*)	Wetzel et al. 2018	Elaboration of an efficient and versatile VLP platform.
Goose Hepatitis B virus (SGHBV)	Chicken hepatoma cell line LMH	Greco N et al. 2014	Expression of the gene S of the only known hepadnavirus that packages capsids containing ssDNA.
Orthohepadnavirus			
Hepatitis B virus	*Escherichia coli*	Burrell et al. 1979; Charnay et al. 1980; Edman et al. 1981; MacKay et al. 1981; Pumpens et al. 1983	First expression of the SHBs gene in bacteria. No 22-nm particles.
		Fujisawa et al. 1983; Kozlovskaia et al. 1984; Pumpens et al. 1984	First direct expression of the entire SHBs sequence in bacteria. No 22-nm particles.
		Li H et al. 2017	Expression of a SHBs deletion variant, which resulted in 28-nm VLPs.
	Aspergillus niger	Plüddemann and Van Zyl 2003; James et al. 2007, 2012	First production of the 22-nm particles in filamentous fungi.
	Hansenula polymorpha	Brocke et al. 2005; Huang Y et al. 2007; Seo et al. 2008	Production of the 22-nm particles in *H. polymorpha*.
	Pichia pastoris	Bo et al. 2005; Han et al. 2006; Ottone et al. 2007; Liu R et al. 2009	First production of the 22-nm particles in *P. pastoris*.
	Saccharomyces cerevisiae	Valenzuela et al. 1982; Miyanohara et al. 1983; McAleer et al. 1984; Dehoux et al. 1986; Itoh et al. 1986	First production of the 22-nm SHBs particles in yeast.
		Valenzuela et al. 1985a	First production of the 22-nm MHBs particles in yeast.
		Shiosaki et al. 1991	First production of HBV virion-like particles in yeast.
	Leishmania tarentolae	Czarnota et al. 2016	Production of the 22-nm particles in trypanosomes.
	Drosophila melanogaster Schneider-2 cells	Deml et al. 1999; Jorge et al. 2008	First production of the 22-nm particles in the *Drosophila* cells.
	Spodoptera frugiperda	Kang et al. 1987	First production of the 22-nm particles in insect cells.
	Silkworm larvae	Higashihashi et al. 1991	First production of the 22-nm particles in silkworm larvae.
	Arachis hypogaea	Chen HY et al. 2002; Zhang et al. 2005; Zhu JG et al. 2006	First production of the 22-nm particles in peanut.
	Glycine max L. Merr. cv Williams 82	Smith ML et al. 2002	First production of the 22-nm particles in soybean.

(Continued)

TABLE 37.1 (CONTINUED)

Expression of the *Hepadnaviridae* Surface Gene and Production of the HB VLPs by Different Expression Systems

VLPs	Expression Host	References	Comments
	Daucus carota subsp. *Sativus*	Zhao XY et al. 2002; Imani et al. 2002	First production of the 22-nm particles in carrots.
	Lactuca sativa	Kapusta et al. 1999, 2001; Marcondes and Hansen 2008	First production of the 22-nm particles in lettuce.
	Laminaria japonica	Jiang et al. 2002	First production of the 22-nm particles in kelp.
	Lotus corniculatus	He HX et al. 2007	First production of the 22-nm particles in crowfoot.
	Lupinus luteus	Kapusta et al. 1999, 2001; Pniewski et al. 2006	First production of the 22-nm particles in lupine.
	Lycopersicon esculentum	Zhao CH et al. 2000; Hao et al. 2007; Salyaev et al. 2007; Srinivas et al. 2008	First production of the 22-nm particles in tomato fruits.
	Malus pumila cultivar Aikanokaoli	Lou et al. 2005	First production of the 22-nm particles in apples.
	Musa acuminate	Mason HS et al. 2002; Sunil Kumar et al. 2005; Hu et al. 2008	First production of the 22-nm particles in banana fruits.
	Nicotiana tabacum	Mason HS et al. 1992; Liu YL et al. 1994; Thanavala et al. 1995	First production of the 22-nm particles in plants.
	Physalis ixocarpa Brot	Gao et al. 2003	First production of the 22-nm particles in cherry tomatillo.
	Solanum tuberosum	Domansky et al. 1995; Ehsani et al. 1997; Richter et al. 2000; Kong et al. 2001	First production of the 22-nm particles in potato.
	Zea mays	Hayden et al. 2012; Shah et al. 2015	A commercially feasible oral subunit vaccine production system in maize.
	Mammalian cells	Christman et al. 1982; Gough and Murray 1982; Liu CC et al. 1982; Stratowa et al. 1982; Wang et al. 1982; Carloni et al. 1984; Aprelikova et al. 1987; Chen ZH et al. 1991	First synthesis of the 22-nm particles in the homologous expression system.
	Live animals	Davis HL et al. 1993, 1994, 1995, 1996, 1997; Michel et al. 1995; Schirmbeck et al. 1995	First examples of the DNA immunization.
	Live animals	Babinet et al. 1985; Chisari et al. 1985	First transgenic mice producing the 22-nm particles.
Woodchuck hepatitis virus	Woodchucks	Lu et al. 1999	DNA vaccine.

Notes: The viruses are given in alphabetical order within each specific hierarchical group.

studies (Antoni et al. 1994) and small-angle neutron scattering (Sato et al. 1995). By the latter method, recombinant yeast-produced HBs protein was characterized as a spherical particle with a 29 nm diameter, where lipids and carbohydrates formed a spherical core with a 24 nm diameter and the S protein on the surface. The examination of yeast-produced HBs by high-resolution negative staining electron microscopy (EM) revealed a spherical to slightly ovoid character of particles with an average diameter of 27.5 nm, consisting of 4-nm subunits with a minute central pore. The ice-embedding EM gave a diameter value of 23.7 nm and a 7–8-nm-thick cortex surrounding an electron translucent core (Yamaguchi et al. 1998) that corresponded well to the small-angle neutron scattering data. The natural

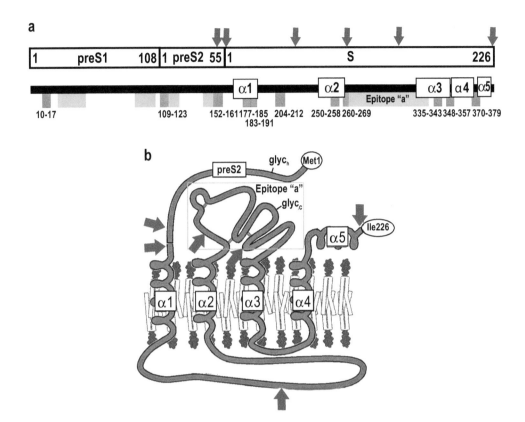

FIGURE 37.4 Structurally and immunologically relevant motifs of the HBs protein. (a) A linear presentation of the S protein with the preS1 and preS2 domains. The location of the transmembrane α-helices and major immunodominant epitopes (B cell, blue; CTL, orange) is presented. (b) Putative topological model of the MHBs (preS2-S) protein, the lipid bilayer of the ER membrane is symbolized by the lipid molecules. The top of the figure corresponds to the lumen of the ER or, after multimerization and budding of the HBs protein, to the surface of the particle. The bottom of the figure corresponds to the cytosol or to the interior of the particle. The major immunodominant region "a" is boxed. Cysteine bridges and positions of complex biantennary glycoside ($glyc_c$) within the "a" epitope and hybrid-type glycoside ($glyc_h$) on the preS2 segment are shown. The most appropriate targets for foreign insertions are depicted by blue arrows on the linear and spatial presentations of the HBs protein. (Courtesy of Wolfram H. Gerlich.) (The figure was published earlier with permission of Wolfram H. Gerlich and Bruce Boschek by Pumpens P, Ulrich R, Sasnauskas K, Kazaks A, Ose V, Grens E. In: *Medicinal Protein Engineering.* Khudyakov Y (Ed). CRC Press, Taylor & Francis Group, Boca Raton London, New York, 2008, pp. 205–248.)

human blood-derived HBs particles, examined by the same EM methods, were found to be smaller in size (Yamaguchi et al. 1998). Briefly, according to the proposed folding of L, M, and S proteins, they consisted of four membrane-spanning helices that were assembled into a highly hydrophobic complex where access to the water environment was not allowed. The proteins projected their N- and C-termini (including preS1 and preS2 sequences of L and M proteins) as well as the second hydrophilic region of protein S, bearing the major immunodominant B-cell epitope "a" (aa residues 111–149 of the protein S corresponding to aa 274–312 of the whole sequence in Figure 37.4a) to the outer surface of HBs particles.

The HBs epitope "a" of the three surface proteins L, M, and S was of special importance since it induced virus-neutralizing humoral responses and conferred the protection as demonstrated by hepatitis B vaccines. Thus, the exact 3D structure of the determinant "a," which still remained unknown, is crucial since the vast majority of induced anti-HBs antibodies recognize conformational epitopes in this

region. Moreover, the mutations in the "a" epitope led to the appearance of the HBV escape variants and failure of both diagnostics and traditional hepatitis B vaccines.

The first electron cryomicroscopy reconstruction of the natural spherical HBs particles at approximately 12 Å resolution (Gilbert et al. 2005) is shown in Figure 37.5a, b, in comparison with electron micrographs of human blood- and yeast-derived 22-nm particles. This revolutionary paper proposed that the HBs particles may possess octahedral symmetry. According to this reconstruction, the HBs particles were built of the HBs dimers as building blocks. However, it is important to recall that the authors used 22-nm particles of the SHBs that was synthesized in a transgenic mouse and could be therefore more similar to the recombinant yeast-produced than to the native human blood-derived HBs particles.

A bit later, the well-known R. Anthony Crowther team published the excellent cryomicrographs of the human blood-derived 22-nm particles shown in Figure 37.6 but concentrated on the 3D structure of the 22-nm tubes (Short

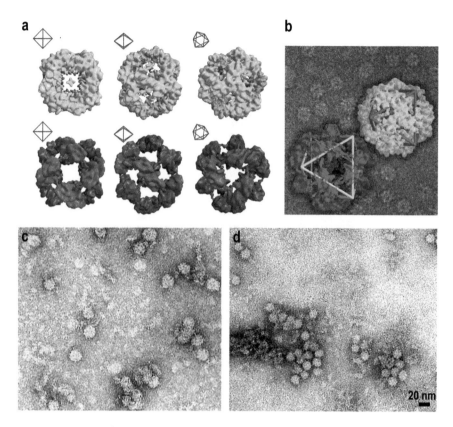

FIGURE 37.5 Structural characteristics of the HBs VLPs. (a) Small and large S particle reconstructions. Surface-rendered views of S particle reconstructions. S particles occur in two sizes, brought about by conformational switching. The smaller of the two is shown in cyan in the upper row, the larger in red in the lower row. The insets show an octahedral skeleton in the same orientations at the actual structures, to indicate the relationship of the subjects shown for the density maps to the symmetry of the actual object. (From Gilbert, R.J. et al., *Proc. Natl. Acad. Sci. U S A*, 102, 14783, 2005.) (b) Views of the small (cyan) and large (red) S particle structures, viewed along their threefold axes. An octahedral skeleton in the opposing color is shown within each semitransparent rendered surface. The background shows a typical cryo-EM field of view such as that which afforded the data used to generate these 3D maps. (Courtesy of Robert Gilbert and David J. Rowlands.) (c) Electron microscopy of the HBs particles from blood of a human carrier. (d) Electron microscopy of the HBs particles from *S. cerevisiae* yeast. Scale bar, 20 Å. (Reprinted with permission of Taylor & Francis Group from Pumpens P, Ulrich R, Sasnauskas K, Kazaks A, Ose V, Grens E. In: *Medicinal Protein Engineering*. Khudyakov Y (Ed). CRC Press, Taylor & Francis Group, Boca Raton London, New York, 2008, pp. 205–248.)

et al. 2009). The authors found that the helical tubes had a diameter of approximately 250 Å—with spike-like features projecting from the membrane—and proposed a model for the packing arrangement of surface protein dimers in the tubes.

Recently, Cao et al. (2019) determined the 3D structure of the human blood-derived 22-nm particles by electron cryomicroscopy at the resolution of ~30 Å (cryo-EM). The authors suggested that the spherical 22-nm particles are irregularly organized, where spike-like features are arranged in a crystalline-like pattern on the surface, as shown in Figure 37.7. Moreover, the HBs protein in the native 22-nm particles folded as protrusions on the surface—as those on the tubular 22-nm particles and Dane particles, or HBV virions—but was largely different from that in the recombinant octahedral subviral particles. Remarkably, these results suggested a universal folding shape of the HBs protein within the native HBV viral and subviral particles (Cao et al. 2019).

Although the knowledge of the folding and molecular architecture of the HBs protein remained insufficient for the reliable prediction of preferential insertion sites for the needs of viral nanotechnology, the available data provided some hints for replacing surface-exposed protein domains by foreign sequences without destabilizing the self-assembly of the HBs particles. Thus, the direct mutational analysis of the HBs proteins revealed domains indispensable for self-assembly (Bruss and Ganem 1991; Machein et al. 1992; Prange et al. 1995a). The role of intra- and intermolecular disulfide bonds within the hydrophilic regions of S protein, which were also responsible for its antigenic properties, was tested experimentally by site-directed in vitro mutagenesis substituting alanine for cysteine (Antoni et al. 1994; Mangold et al. 1995, 1997). Moreover, mutagenesis of the HBs led to the important conclusion that mutated, assembly noncompetent polypeptides can be rescued into mosaic particles in the presence of native helpers (Bruss and Ganem 1991).

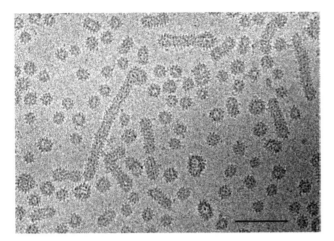

FIGURE 37.6 Cryomicrograph of an HBsAg preparation showing a variety of isometric and tubular particles. HBsAg was isolated from the blood of HBV carriers. The scale bar represents 1000 Å. (Reprinted from *J Mol Biol*. 390, Short JM, Chen S, Roseman AM, Butler PJ, Crowther RA, Structure of hepatitis B surface antigen from subviral tubes determined by electron cryomicroscopy, 135–141, Copyright 2009, with permission from Elsevier.)

The most preferable sites for the foreign insertions were localized in preS2 and S parts of the HBs protein. These sites are depicted in Figure 37.6, both on linear (a) and putative 3D (b) maps. The most recent dimeric topology of the SHBs molecule is presented in Figure 37.8.

The initial steps how the HBs grew to become one of the most prospective VLP carriers were described in detail by Kozlovskaia and Pumpens (1986), Pumpens and Grens (2002), and Pumpens et al. (2008). The fascinating story of the HBs as a recombinant hepatitis B vaccine was published by Huzair and Sturdy (2017). Nowadays, an exhaustive review on the HBs as a VLP carrier has been written by the Hans J. Netter team (Ho et al. 2020).

VACCINES

NATURAL VACCINES

SHBs Vaccines

The classical and widely used prophylactic hepatitis B vaccine is nothing else than the recombinant yeast-produced non-chimeric HBs VLPs, or 22-nm particles. This was the first vaccine produced by gene technology, which demonstrated clear and revolutionary success in the prevention and eradication of hepatitis B infection. Nevertheless, the hepatitis B problems that remain today include the inability to achieve a complete cure for chronic HBV infections, the recognition of occult HBV infections, their potential reactivation, and the incomplete protection against escape mutants and heterologous HBV genotypes by the HBsAg vaccine. As will be explained in Chapter 38, the generation of promising therapeutic anti-HBV vaccines based on or including HBc activity must be a current goal.

Nevertheless, the first prominent success of the VLP protein engineering was reached by the construction of the human hepatitis B vaccines based on the recombinant HBs protein. The yeast-derived HBs vaccine remained for long time not only the first but also the only available gene-engineered vaccine in the world.

Although human-plasma-derived and recombinant yeast-derived HBs proteins possessed many similar properties, it was necessary to keep in mind that differences did exist. Besides glycosylation, which was clearly different in yeast, such differences were seen not only in the size of particles but also in both epitope presentation and anti-HBs antibody-binding properties (Heijtink et al. 2000). Nevertheless, the pioneering high-yield production of the HBs protein in yeast resulted, first, in construction of two widely available and safe human vaccine products, which have been used since 1987 and are still leading the hepatitis B vaccine market: Engerix-B (developed by SmithKline Beecham, now produced by GlaxoSmithKline Biologicals; FDA 1988; Greenberg et al. 1996; Adkins and Wagstaff 1998; Keating and Noble 2003) and H-B-Vax II, or Recombivax HB (Merck, Sharp & Dohme, now Merck & Co, Inc.; FDA 1987), both of which were based on nonglycosylated protein S-derived particles and produced in yeast *Saccharomyces cerevisiae*.

The next popular *S. cerevisiae* S-antigen based vaccine of this type was Euvax B (Teles et al. 2001), which was developed by the Aventis Pasteur and produced by the LG Chem Ltd. Then, FENDrix was developed by the GlaxoSmithKline Biologicals and licensed in Europe (Kundi 2007). The FENDrix vaccine contained the same *S. cerevisiae*-produced SHBs protein but was formulated with a novel adjuvant system: aluminum phosphate and 3-O-desacyl-4-monophosphoryl lipid A. The vaccine possessed superior immunogenicity and induced higher antibody concentrations that reached protective levels in a faster fashion and were retained for a longer period. The FENDrix vaccine was intended for use in adults with renal insufficiency including prehemodialysis and hemodialysis patients (Nevens et al. 2006; Beran 2008).

Some other registered and commercially successful hepatitis B vaccines using yeast-derived protein SHBs were produced by original technologies based on the employment of highly productive species of yeast other than *S. cerevisiae*. First, methylotrophic yeast *Hansenula polymorpha* was used to produce the SHBs protein. The Brazilian vaccine ButaNG was developed using H. polymorpha by the N.G. Biotecnologia Ltda and produced by the Instituto Butantan (Costa et al. 1997; Ioshimoto et al. 1999). Hepavax-Gene vaccine was also developed using *H. polymorpha* by Rhein Biotech (Hieu et al. 2002), now belonging to the Dynavax company (United States) and produced by the Korean Green Cross. A new low-cost *H. polymorpha*-produced vaccine GeneVac B was registered in India (Vijayakumar et al. 2004; Kulkarni et al. 2006).

Second, yeast *Pichia pastoris* was used for another set of hepatitis B vaccines. They were Cuban SHBs-based vaccine

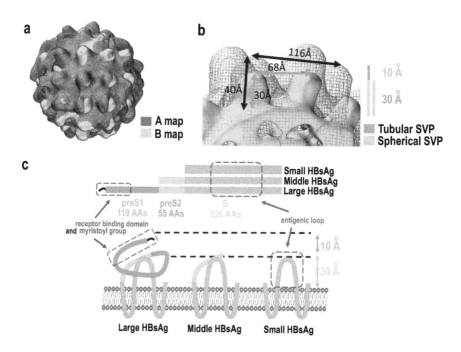

FIGURE 37.7 Comparison of the density maps of spherical and tubular subviral particles (SVPs). (a) The superposition of two EM density maps of spherical SVP shows a slight difference in size, but the local spike-like features have almost identical arrangement. (b) The superposition of B map of spherical SVP and the map of tubular SVP (Short et al. 2009, see Figure 37.6). The mesh map indicates the tubular SVP, while the solid map indicates the spherical SVP. The protrusions in spherical SVP are ~30 Å in height, about 10 Å shorter than that in tubular SVP. The distance between two tips of protrusions in spherical SVP is measured ~68 Å, apparently smaller than that in tubular SVP (~116 Å). (c) Illustration of S (yellow), preS2 (green), and preS1 (blue) domains in S/M/L HBsAg and an assembly model of the S/M/L HBsAg on the surface of HBV particles. Black line indicates the myristoyl group at the amino-terminus of large HBsAg. The preS1 (blue), preS2 (green), and S (yellow) regions are colored accordingly in the model. The exposed "antigenic loop" is indicated by a red dash line box, and the 50 amino acid residues in N-terminus of preS1 which are involved in HBV functional receptor binding are indicated by a blue dash line box (Reprinted from *Virus Res.* 259, Cao J, Zhang J, Lu Y, Luo S, Zhang J, Zhu P, Cryo-EM structure of native spherical subviral particles isolated from HBV carriers, 90–96, Copyright 2019, with permission from Elsevier.)

Heberbiovac HB elaborated by the Centro de Ingenieria Genetica Y Biotecnologia and produced by the Heberbiotec SA (Cuba) company (Galbán García et al. 1992) and its analogue, which was produced in India under the trademark Enivac HB (Kaur and Mani 2000; Estévez et al. 2007). A low-cost *P. pastoris*-produced vaccine Shanvac B was registered in India (Abraham et al. 1999).

The pioneering character of the SHBs-based developments was supported by the next revolutionary achievement, namely by one of the first practical employments of the immunostimulatory sequences (ISS)—or so-called CpGs—which targeted toll-like receptors (TLRs; Schmidt 2006; Sung and Lik-Yuen 2006). Heplisav-B (FDA 2020), a hepatitis B prophylactic and therapeutic vaccine, was developed by the California-based Dynavax company with a strong emphasis on the recently acquired Rhein Biotech company (Barry and Cooper 2007; Malhame 2007; Cooper and Mackie 2011). Under the terms of the agreement with Coley Pharmaceutical, the Dynavax received a nonexclusive license under Coley's ISS oligonucleotide patent estate for the commercialization of the Heplisav-B vaccine. The principle behind Coley's ISS drug targeting TLR9 is to use synthetic forms of cytosine-phosphate-guanine (CpG) to bind and activate the receptor. In turn,

that especially triggered a Th1 immune response resulting in the activation of T cells. The Heplisav-B vaccine combined therefore the ISS (or CpGs) with the HBs and was designed to significantly enhance the level, speed, and longevity of protection against hepatitis B. In several clinical studies, the Heplisav-B vaccine induced a protective level of antibodies against hepatitis B faster and with fewer doses than conventional hepatitis B vaccines. Additionally, Heplisav-B has provided 100% seroprotection in all subjects who have received the full regimen, including those who belong to the so-called difficult-to-immunize patients group.

The ability of the added CpG as an adjuvant to enhance immunogenicity of Engerix-B vaccine has been approved before in a phase I/II clinical study for the first clinical evaluation of the safety, tolerability, and immunogenicity of CpG when added to this commercial HBV vaccine (Cooper et al. 2004, 2005; Siegrist et al. 2004; Payette et al. 2006). Strikingly, when the patients were immunized with the Engerix-B vaccine together with CpGs, the protective levels of anti-HBs IgG were developed within just two weeks after application of the priming vaccine dose. A trend toward higher rates of specific cytotoxic T lymphocyte (CTL) responses has been also reported.

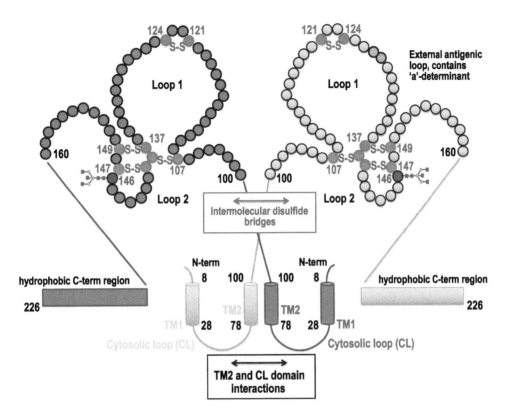

FIGURE 37.8 Proposed model for the formation of the HBs homodimer based on Suffner et al. (2018). Dimerization of two HBs monomers represented in purple and blue is facilitated by interactions of the transmembrane domains (TM2), cytosolic loops (CL) and intermolecular disulfide bridges (SS) between cysteine residues (red circles) in the external loop region. The facultative N-glycosylation site at position N146 is indicated (dark blue). The orientation of the hydrophobic C-terminal region is illustrated in a simplified form. (Reprinted from Ho JK, Jeevan-Raj B, Netter HJ. Hepatitis B virus (HBV) subviral particles as protective vaccines and vaccine platforms. *Viruses.* 2020;12:126.)

In preclinical animal studies, many other immunostimulatory adjuvants were suggested at that time to improve the efficacy of the S protein-based vaccination, first to improve cytotoxic response: PLGA (poly(D,L-lactic-co-glycolic acid) microspheres (Jaganathan et al. 2004; Feng et al. 2006), muramyl dipeptide C, or MDP-C (Yang et al. 2005) and so-called nanodecoy systems, i.e., spherical ceramic cores formed by self-assembling hydroxyapatite and cellobiose (Goyal et al. 2006).

The recombinant SHBs-based vaccines underwent permanent improvements including combination with other popular vaccines (hepatitis A, diphtheria, tetanus, acellular pertussis, inactivated poliovirus, *Haemophilus influenzae* type b) into multitargeted formulations (for more details see WHO recommendations). For example, the yeast-derived SHBs vaccines were combined with poxvirus-based vaccine (Hutchings et al. 2005) or coadministered with recombinant human papilloma virus (HPV) cervical cancer vaccines (Wheeler et al. 2008; Leroux-Roels et al. 2011) that are described in Chapter 7.

Edible Vaccines

The plant-derived SHBs protein appeared in the next pioneering step as a real edible vaccine candidate. This plant-derived antigen showed a strong immunogenicity in mice

(Richter et al. 2000; Kong et al. 2001; Kostrzak et al. 2009; Kapusta et al. 2010; Pniewski et al. 2011; Rukavtsova et al. 2011; Hayden et al. 2012) and humans (Kapusta et al. 1999, 2001).

The vegetable and fruit crops would be ideal host systems to produce an oral hepatitis B vaccine. Thus, the considerable levels of the SHBs protein were achieved in banana, carrots, and cherry tomatilla. However, nonfood/feed crops such as tobacco, potato (uncooked), lettuce, and lupin, as well as in vitro systems such as plant cell cultures and hairy roots, have many advantages, which may be exploited for the future production of hepatitis B vaccines, as reviewed at that time by Kumar et al. (2007) and Pniewski (2012).

For the first time, a double-blind placebo-controlled clinical trial was performed on the SHBs protein produced in potatoes and delivered orally to previously vaccinated individuals (Thanavala et al. 2005). The potatoes accumulated HBs protein at approximately 8.5 μg/g of potato tuber, and doses of 100 g of tuber were administered by ingestion. After volunteers ate uncooked potatoes, serum anti-HBs titers increased definitely, and the first plant-derived orally delivered hepatitis B vaccine was suggested by the authors as a viable component of a global immunization program.

An interesting alternative variant of an edible vaccine did not use HBs VLPs but was based on encapsulation of a HBs peptide representing aa residues 127–145 of the immuno-dominant epitope "a" in PLGA microparticles (Rajkannan et al. 2006).

Unfortunately, most of the published data related to the edible hepatitis B vaccine candidates did not present a reference control of the purified SHBs particles. Thus, the highly purified SHBs protein, namely, a bulk of the ButaNG vaccine described earlier, or the whole yeast cells containing SHBs protein as biomass after fermentation at the ButaNG manufacturing did not elicit any detectable anti-HBs antibodies in mice multifed with such preparations, according to personal communication of Nikolai Granovski, a "father" of the ButaNG vaccine, in 2007.

Combination with the HBc VLPs

One of the most attractive approaches for the development of a multitargeted VLP-based hepatitis B vaccine consisted in a simultaneous exploitation of advantages of both chimeric HBs and HBc VLPs while minimizing their drawbacks. This idea was realized consequently by Cuban scientists as the therapeutic nasal HBc/HBs vaccine HeberNasvac, later simply NASVAC. The history and distinctive features of this highly promising vaccine are presented in the *Vaccines: Natural VLPs* paragraph of Chapter 38.

Granovski et al. (2017) expressed in *Hansenula polymorpha* three HBV genes: HBc, SHBs, and chimeric SHBs carrying the major preS1 epitope 21–47 of 26 aa residues and the three repetitions of the preS2 epitope 131–145 of 14 aa residues representing the three different HBV genotypes.

The immunization tests showed that a combination of low doses of the highly purified antigens was effective in eliciting specific anti-HBs, anti-preS1, anti-preS2, and anti-HBc antibodies in mice.

Concerning the pivotal role of both recombinant HBs and HBc VLPs by the elaboration of vaccination strategies, an early classical review of the famous Ken Murray's team is highly recommended (Murray et al. 1989).

PreS-Containing Vaccines

Many attempts have been undertaken to improve the hepatitis B vaccine by fusion of the preS elements to the HBs. The first attempts to enrich HBs vaccines with the preS2 (Machida et al. 1987) and preS1 (Neurath et al. 1989) peptides were realized by covalent binding of the appropriate synthetic peptides to the HBs (see later discussion). Then a chimeric protein, which carried the preS1 region aa 21–47 at the C-terminus of the S protein (at aa position 223), was expressed by a recombinant vaccinia virus and secreted from several mammalian cell lines (Xu X et al. 1994). The HBV envelope was rearranged also by fusing part or all of the preS1 region to either the N- or C-terminus of the S protein (Prange et al. 1995b). Fusion of the first 42 residues of preS1 to either site allowed efficient secretion of the modified particles and rendered the linked sequence accessible at the surface of the particle. In opposite, fusion of the preS1 sequences to the C-terminus of the M protein completely blocked

secretion. Although all these particles displayed preS1, preS2, and S-protein antigenicity, high titers of preS1- and preS2- (but not S-specific) antibodies were induced. Another group constructed three fusion proteins containing preS2 (aa 120–146) and preS1 (aa 21–47) epitopes at the N-terminus and truncated C-terminus of the S protein and expressed them using a recombinant vaccinia virus system (Hui et al. 1999a). The fusion proteins were efficiently secreted in particulate form, which displayed S, preS1, and preS2 antigenicity and elicited strong antibody responses against S, preS1, and preS2 in mice.

A yeast-derived HBV vaccine containing preS1 (aa 12–52) and preS2 (aa 133–145) sequences within a rearranged, so-called L* protein, in addition to the S protein, was constructed by Leroux-Roels et al. (1997a, b). The vaccine particles consisted of S and L* at a ratio of 7:3.

GenHevac B was the first approved preS-, namely preS2-containing hepatitis B vaccine. It was generated in 1993 by the Pasteur Mérieux Connaught (France) via expression of full-length nonreconstructed M and S genes. GenVac B was produced in mammalian cells and contained both M and S proteins. This vaccine showed good efficacy for therapeutic vaccination in chronic hepatitis B patients, with or without combination with interferon-α (Senturk et al. 2002, 2009).

The next commercial preS-containing vaccine was the Bio-Hep-B vaccine, which was produced by Biotechnology General Ltd (Israel) via expression of M and L genes in mammalian cells. It contained therefore not only the M and S but also some L protein molecules. The Bio-Hep-B vaccine showed improved efficacy in a specific group of patients with end-stage renal disease (ESRD), who were badly responding to classical S vaccines (Weinstein et al. 2004). The Bio-Hep-B vaccine was highly efficient in newborns after only two injections and elicited multivalent response against L, M, and S proteins (Madaliński et al. 2004).

The Sci-B-Vac vaccine was based on the complete HBs gene encoding S, M, and L—including native promoter, enhancer, and poly(A) signal—and produced in CHO cell line, which contained more than 100 HBs coding copies per cell (Gerlich et al. 1990; Shouval et al. 1991, 1994). The Sci-B-Vac vaccine was developed by the SciGene company (Hong Kong; Yap et al. 1992). Thus, it contained major S and minor M and L proteins, by analogue with the Bio-Hep-B vaccine. Sci-B-Vac showed high therapeutic efficacy when applied to chronic hepatitis B patients after liver transplantation receiving additional lamivudine prophylaxis. The effect was due to its high and multivalent (S, preS1, preS2) immunogenicity, which was a key factor to prevent hepatitis B recurrence due to emergence of S escape mutants (Lo et al. 2007; Hellström et al. 2009; Sylvan et al. 2009). In general, the Sci-B-Vac demonstrated an excellent safety record in clinical studies, when more than 50% of vaccinees developed earlier seroprotection against HBV, in comparison with yeast-derived hepatitis B vaccines (Shouval et al. 2015).

A similar Hepimmune vaccine was developed by the Berna Biotech company (Switzerland). It was also produced in mammalian cells and contained S, M, and L proteins. A trial test in non- and low responders demonstrated enhanced antibody responses to the Hepimmune vaccine in comparison to conventional S-based vaccines (Rendi-Wagner et al. 2006). It appeared also to be a good candidate for short-time immunization protocols, which were developed for vaccination of living liver donors before transplantation of the liver to chronic hepatitis B patients (Schumann et al. 2007). The time to achieve hepatitis B immunity in those liver donors was usually short (one to two months). For this reason, the authors established a short-time immunization protocol (four injections in two-week intervals) with the Hepimmune and showed efficient cellular and humoral immune responses.

Further development of the preS inclusion strategy by the Medeva plc company led to the generation of an efficient preS1- and preS2-containing vaccine Hepagene (or Hepacare or Hep B-3), which was the first vaccine of this type to have a pan-European license. Unlike preS-containing vaccines described earlier, the Hepagene vaccine was generated by selection of desired preS epitopes and their combination within the recombinant genes. The Hepagene vaccine contained the preS1 sequence spanning aa 20–47, the complete preS2, and the complete S region of two different subtypes, *adw* and *ayw*. It was produced in the mouse C127I clonal cell line after transfection of the cells with genes encoding the three antigens. Because these genes were expressed in mammalian cells, the Hepagene components were glycosylated and therefore resembled the native viral proteins more closely. The Hepagene vaccine was shown to stimulate stronger and more rapid cellular and humoral immune responses than S-based vaccines and to circumvent anti-HBs nonresponsiveness (Jones CD et al. 1998, 1999; McDermott et al. 1998; Pride et al. 1998; Waters et al. 1998). Clinical trials in both Europe and the United States have clearly demonstrated that the Hepagene vaccine was highly immunogenic for both B- and T-cells and induced higher anti-S "a" determinant antibody titer than the S-based vaccines (Page et al. 2001; Zuckerman and Zuckerman 2002). After the joining of the Medeva to the Celltech company, the Hepagene vaccine was further evaluated successfully by Celltech and PowderJect companies utilizing PowderJect's needleless injection technology (Jones T 2002).

Concerning edible preS-carrying vaccines, the gene M was expressed in potato (Joung et al. 2004), in parallel with the expression of the gene M in *N. benthamiana* leaves (Huang Z et al. 2005). The transgenic potato tubers were then applied to mice, and the latter mounted immune responses against the anti-S and anti-preS2 antibodies that were sustained for the whole period without decrease (Youm et al. 2007). Next, the gene M was expressed in the transgenic tomato plants (Salyaev et al. 2007, 2010).

Furthermore, the L protein gene was expressed in tomato fruits (Lou et al. 2007). As detected by electron microscopy, the L gene expression led to efficient synthesis of the HBs capable of assembling not only in the capsomers but also in VLPs.

Qian B et al. (2008) expressed in transgenic rice a chimeric gene that encoded a fusion protein consisting of aa 21–47 of the preS1 sequence fused to the truncated C-terminus of the S protein. The chimeric gene was specifically expressed in rice seeds, with the highest expression level being about 31.5 ng/g dry weight grain, while the chimeric protein was observed to form VLPs as in the 22-nm particles and induced immune response against both the S and preS1 epitopes in mice.

Both M and L genes were expressed in tobacco and lettuce (Pniewski et al. 2012). The products displayed a common S domain and characteristic preS2 and preS1 domains and were assembled into VLPs.

It should be noted that the Rudolf Valenta's team constructed and clinically evaluated a hypoallergenic vaccine for grass pollen allergy, BM32, which was based on fusion proteins consisting of peptides from the IgE binding sites of the major grass pollen allergens fused to the HBV preS (Cornelius et al. 2016; Tulaeva et al. 2020). The potential of this "dual use" vaccine, although not connected to the VLP methodology, to protect against infection with HBV was professionally reviewed by Wolfram H. Gerlich (2020).

DNA Vaccines

The SHBs served as one of the first DNA vaccine models (Davis HL et al. 1993, 1994, 1995, 1996, 1997; Michel et al. 1995; Schirmbeck et al. 1995), which was based on a simple conversion of DNA expression vectors for HBs-based VLPs to the corresponding DNA vaccines. Although no DNA vaccine candidate was approved, DNA-based vaccination was demonstrated to improve anti-HBs immune responses in animal systems. In DNA vaccination, HBs proteins were synthesized from antigen-encoding expression plasmid DNA in the in vivo transfected cells of the vaccine recipients in their native conformation with correct posttranslational modifications. This ensured the integrity of antibody-defined epitopes and supported the generation of protective virus-neutralizing antibodies. Furthermore, DNA vaccination was an exceptionally potent strategy to stimulate CTL responses because antigenic peptides were efficiently generated by endogenous processing of intracellular protein antigens and thereby presented by the MHC class I pathway. These key features invented DNA-based immunization as an attractive strategy not only for prophylactic but also for therapeutic vaccination against hepatitis B (Michel and Loirat 2001; Schirmbeck and Reimann 2001; Geissler et al. 2004; Michel and Mancini-Bourgine 2005).

Since viral persistence by chronic hepatitis B was related to poor HBV-specific T-cell responses, DNA vaccination as a strong T-cell inducer was thought to be an adequate measure. A phase I clinical trial was performed in chronic HBV carriers to investigate whether the SHBs DNA vaccination could restore T-cell responsiveness (Mancini-Bourgine et al. 2004). This study provided evidence that DNA vaccination can activate T-cell responses in some chronic HBV carriers who do not respond to current antiviral therapies. A subsequent phase I clinical trial confirmed the safety and immunological efficacy of HBs DNA vaccination and demonstrated restoration or activation of T-cell responses in chronic HBV carriers (Mancini-Bourgine et al. 2006).

The safety and efficacy of the first powdered HBs DNA vaccine was demonstrated in a clinical study (Roberts et al. 2005). This clinical study evaluated for the first time the performance of a single-use disposable, commercial prototype device for particle-mediated epidermal delivery of DNA vaccine. However, patients were previously immunized with a licensed protein-based HBV vaccine but received a single boost vaccination of the DNA vaccine.

Although the initial expectations of the performance of DNA vaccination in chronic hepatitis B patients were optimistic, a therapeutic HBs DNA immunization approach in a chimpanzee chronic HBV carrier with a high viral load failed to control HBV viremia (Shata et al. 2006).

The obvious difficulties in DNA vaccination pressed investigators to apply many advanced immunization strategies, with the use of different kinds of adjuvants and prime/boost regimens. Thus, primary immunization with a DNA construct encoding interleukin-12 and boosting with canarypox vectors expressing all HBs genes six months thereafter was tried but did not lead to reduction in viral load in the chimpanzee model (Shata et al. 2006).

Promising results were obtained in a murine model by vaccine regimens using combinations of plasmid DNA encoding the gene M, poxvirus (modified vaccinia virus Ankara, MVA) encoding the same gene, and the SHBs protein (Engerix-B vaccine) to induce strong HBs-specific cellular and humoral responses (Hutchings et al. 2005). Boosting of the humoral immune response to HBs DNA vaccine was achieved by coadministration of prothymosin α as an adjuvant (Jin et al. 2005).

Next, an optimization of the HBs DNA vaccine plasmid vector was tried by modification of the content of immunostimulatory CpG motifs (Payette et al. 2006). However, two doses of DNA vaccine did not generate any detectable anti-HBs antibody response in either of two chimpanzees in this study.

The progressive oral and intranasal immunization schemes were under investigation for the putative HBs DNA vaccines. Single-dose oral immunization with biodegradable PLGA (DL-lactide-coglycolic acid) microparticles carrying HBs-encoding DNA led to the induction of a long-lasting and stable mucosal and systemic immune response in mice (He XW et al. 2005). In contrast, naked DNA vaccines given by intramuscular injection induced only systemic cellular and humoral responses to HBs, which were much lower than the responses elicited by oral administration of DNA encapsulated in PLGA microparticles at equivalent doses. Intranasal HBs DNA vaccination using a cationic emulsion as a mucosal gene carrier dramatically enhanced S gene expression in both nasal and lung tissue and increased the cellular and humoral response in mice

(Kim et al. 2006). In contrast, very weak humoral and cellular immunities were observed following immunization with naked DNA.

To model the vaccination of chronic carriers, a modified SHBs with a preS1 aa 21–47 peptide fused to the C terminus, at aa 223, of S protein was shown to induce strong humoral and CTL responses in transgenic mice after a single DNA injection (Hui et al. 1999b).

The easiest way to exploit the previously mentioned combination of both HBs and HBc proteins was with DNA vaccination, which allowed for a simultaneous expression and unlimited modifications of both genes. The examples of this approach are presented in the *Vaccines: Natural VLPs* paragraph of Chapter 38.

The woodchucks immunized with a DNA plasmid encoding the MHBs protein developed WHsAg-specific proliferative response of peripheral mononuclear blood cells but no measurable anti-WHs response (Lu et al. 1999). Nevertheless, a rapid anti-WHs response developed during the second week after virus challenge, and neither woodchuck developed any signs of WHV infection.

The genetic vaccination of ducks with DNA plasmids encoding preS/S and S proteins was effective to prime a protective anti-DHBs antibody response (Triyatni et al. 1998).

CHIMERIC VACCINES

Genetic Fusions

Table 37.2 presents a list of the chimeric HBs-based vaccine candidates according to the insertion sites, namely, N-terminus, internal, or C-terminus, of the SHBs or MHBs molecule.

Despite differences among human- and yeast-derived HBs, the high-level expression in yeast stimulated numerous attempts to employ the HBs VLPs for protein engineering studies. Chronologically, the first example of the chimeric HBs-derived VLPs was the insertion of a long segment from herpes simplex virus 1 glycoprotein gD into the N-terminal part of the MHBs sequence and the expression of such chimeric gene in yeast by Pablo Valenzuela et al. (1985b). With this revolutionary paper, the authors were among the first who invented VLPs, in this case HBs VLPs, as a "matrix for the presentation of other antigens" and "developed a new concept of hybrid HBsAg particles to improve epitope presentation of "foreign" antigens" (Valenzuela et al. 1985b). In fact, the excellent electron micrographs approved generation of the typical 22-nm particles formed by the chimeric MHBs-gD molecules.

At virtually the same time, a VP1 segment spanning aa 93–103, which included the linear part of the neutralization epitope from poliovirus type 1 (PV-1) or 2 (PV-2), was inserted between positions 113 and 114 or between positions 50 and 51 at the sites located within the two major hydrophilic domains of the S protein (Delpeyroux et al. 1986, 1987, 1988a, 1990). The recombinant genes were expressed in mouse L cells from the SV-40 early promoter

TABLE 37.2

The Putative Vaccines Constructed on the *Hepadnaviridae* Surface Proteins by Genetic Fusion Methodology

Vaccine Target	Source of Epitope	Position of Insertion or Addition	Expression System	Major Immunological Activity	References
DHBs					
Bovine viral diarrhea virus (BVDV)	E2, aa 1–344 aa 1–196 (mosaic VLPs)	N-terminus N- or C-terminus	*H. polymorpha*	Identity of foreign antigen	Wetzel et al. 2018
Classical swine fever virus (CSFV)	E2, aa 1–337 aa 1–184 aa 1–102 (mosaic VLPs)	N-terminus	*H. polymorpha*	Display of foreign antigen	Wetzel et al. 2018
Feline leukemia virus (FeLV)	env, aa 1–412 aa 1–132 (mosaic VLPs)	N-terminus	*H. polymorpha*	Not examined	Wetzel et al. 2018
West Nile virus (WNV)	E, domain III, aa 293–454 (mosaic VLPs)	N-terminus	*H. polymorpha*	Display of foreign antigen	Wetzel et al. 2018
Plasmodium falciparum	Pfs230, aa 443–1132 aa 542–736 Pfs25, aa 23–193	N-terminus	*H. polymorpha*	Display of foreign antigen; B cells (rabbits); Transmission-reducing activity	Chan et al. 2019; Wetzel et al. 2019
HBs N-terminal insertions					
Dengue virus	DENV-2, E protein, aa 1–395	↓1	*P. pastoris*	B cells (mice)	Bisht et al. 2001, 2002
	DENV-2, E protein, DIII, aa 297–400 (mosaic particles)	↓1	*P. pastoris*	ND	Khetarpal et al. 2013
	DENV-1, -3, -4, and -2, Eprotein, DIII, ~104 aa each (mosaic particles)	↓1	*P. pastoris*	B cells (mice, macaques); neutralizing activity	Ramasamy et al. 2018
	DENV-2, E protein, DIII, aa 292–393	↓1	Vero cells	B cells (mice); neutralizing activity	Harahap-Carrillo et al. 2015
Hepatitis C virus (HCV)	Nucleocapsid, aa 1–58	N-terminus of M or S	DNA vaccine	B and T cells (mice)	Major et al. 1995
Herpes simplex virus 1	Glycoprotein gD, 300 aa of total	preS2 7↓8	*S. cerevisiae*	ND	Valenzuela et al. 1985b
Human immunodeficiency virus 1 (HIV-1)	gp120, C-terminal part, aa 384–467	polylinker instead of preS2	Chinese hamster ovary cells	B and T cells (rabbits, rhesus monkeys)	Michel et al. 1988, 1990
	gp120, V3 loop, aa 308–331 (mosaic particles)	polylinker instead of preS2	Chinese hamster ovary cells	B, T, and CTL response (rabbits, rhesus monkeys)	Schlienger et al. 1992; Mancini et al. 1994
	Polyepitope polHIV-1.opt, 103 aa, plus V3 loop, 35 aa	preS2	DNA vaccine	B and T cells, CTLs (mice)	Michel et al. 2007
	Polyepitope polHIV-1.opt, permutation of epitopes	preS2	DNA vaccine	Order of epitopes is strategic to enhance immune responses toward HBs VLPs	Cervantes Gonzalez et al. 2009
	Polyepitope polHIV-1.opt	preS2	*Nicotiana tabacum, Arabidopsis thaliana*	CTL activation by oral administration (mice)	Greco R et al. 2007; Guetard et al. 2008

(Continued)

TABLE 37.2 (CONTINUED)

The Putative Vaccines Constructed on the *Hepadnaviridae* Surface Proteins by Genetic Fusion Methodology

Vaccine Target	Source of Epitope	Position of Insertion or Addition	Expression System	Major Immunological Activity	References
	gp41, membrane proximal external region (MPER), aa 656–683	2↓3	HEK293 cells	ND	Phogat et al. 2008
	gp120, MN subtype, V3 loop, 49 aa residues	Polylinker instead of preS2	DNA vaccine	B cell and CTL response (mice)	Fomsgaard et al. 1998; Bryder et al. 1999
	gp120, Lai subtype, V3 loop, 24 aa residues	Polylinker instead of preS2	DNA vaccine	B cell and CTL response (mice, rhesus monkeys)	Le Borgne et al. 1998
	TBI, Env/Gag polyepitope, 146 aa	N-terminus	*Lycopersicon esculentum*	B cells (mice fed by tomatoes)	Pozdniakov et al. 2003; Nesterov et al. 2004; Shchelkunov et al. 2004, 2005, 2006
Simian immunodeficiency virus (SIV)	gp140, V2 loop, aa 176–189 (mosaic particles)	Polylinker instead of preS2	Chinese hamster ovary cells	B cells, no protection (rhesus monkeys)	Schlienger et al. 1994
SIV/HIV	Polyepitopes from SIV Gag or HIV Env, up to 120 aa	Polylinker instead of preS2	DNA vaccine	B cells and CTLs (mice)	Marsac et al. 2005
Plasmodium falciparum	Circumsporozoite (CS) protein, repetitive NANP epitope, one or 15 repeats (mosaic particles: R16-HBsAg vaccine)	↓42 of preS2	*S. cerevisiae*	B cells (mice, rabbits)	Rutgers et al. 1988
	Circumsporozoite (CS) protein, aa 210–398 (mosaic particles: RTS,S vaccine)	↓4 of preS2	*S. cerevisiae*	phase I trial	Gordon et al. 1995 (further trials and numerous references in the text)
	Circumsporozoite (CS) protein, 184 aa (uniform particles: R21 vaccine)	N-terminus	*P. pastoris*	protection (mice); Phase I/IIa clinical trials	Collins et al. 2017; Draper et al. 2018
	Pfs16, 86 aa (mosaic particles)	↓4 of preS2	*S. cerevisiae*	B cells (mice, rabbits)	Moelans et al. 1995
	Pfs25 (Plug-and-display technique)		*P. pastoris*	B cells (mice)	Marini et al. 2019
Plasmodium vivax	Merozoite surface protein 1 (MSP-1), C-terminus, 111 aa	N-terminus	DNA vaccine	T cells (mice)	de Oliveira et al. 1999
	Circumsporozoite (CS), synthetic protein VMP001, (mosaic particles: CSV-S,S vaccine)	↓4 of preS2	*S. cerevisiae*	B and T cells (rhesus monkeys)	Vanloubbeeck et al. 2013
	Circumsporozoite (CS) protein, VK210/247 alleles (uniform particles: Rv21 vaccine)	N-terminus	*P. pastoris*	protection (mice)	Salman et al.
HBs internal insertions					
Hepatitis C virus (HCV)	E2, HVR, aa 384–411; hydrophilic regions: 384–432; 474–493; 520–543; 583–618; 637–663	113↓114	HeLa cells	ND	Lee et al. 1996

Vaccine Target	Source of Epitope	Position of Insertion or Addition	Expression System	Major Immunological Activity	References
	E2, HVR1 and downstream E2 sequence, from 19 to 79 aa	127↓129	HuH7 cells	B cells (mice); block of the HCVpp entry	Netter et al. 2001, 2003; Vietheer et al. 2007
	E2 or E1, full-length (mosaic particles)	Replacing the N-terminal transmembrane domain	Chinese hamster ovary cells	B cells (rabbits); neutralization of HCVpp	Patient et al. 2009; Beaumont et al. 2013
	E1/E2 epitopes ITGHRMAWDMMMNWS; QLINTNGSWHIN; GVPTYSWGENETD; HVR1 mimotope ETYVSGGSAARNAYGL-TSLFTVGPAQK	127↓128	HEK293T cells	B cells (mice); neutralization of HCVpp and HCVcc	Wei et al. 2018
	E2, aa 412–425	127↓128	*Leishmania tarentolae*	B cells (mice)	Czarnota et al. 2016
Human immunodeficiency virus 1 (HIV-1)	gp41, membrane proximal external region (MPER), aa 656–683	125↓128	HEK293 cells	ND	Phogat et al. 2008
	gp41, MAb 2F5 epitope, aa 662–667 of gp160, ELDKWA	113↓114	*P. pastoris*	B cells (mice)	Eckhart et al. 1996
Human papilloma virus (HPV) 16	E7, aa 35–54	139↓142	*S. cerevisiae*	B and T cells (mice)	Pumpens et al. 2002
	E7, aa 35–98	139↓174	*S. cerevisiae*	ND	Pumpens et al. 2002
Poliovirus 1 and 2	VP1, aa 93–103	50↓51 113↓114	mouse L cells	B cells (mice, rabbits)	Delpeyroux et al. 1986, 1987, 1988a, 1990
Helicobacter pylori	*katA* gene, C terminus	127↓129	HuH7 cells	B and T cells (mice)	Kotiw et al. 2012
Plasmodium falciparum	gp190, a precursor of the major merozoite surface antigen, epitopes of 32, 36, or 61 aa	113↓157	CV1 cells	B cells (mice, rabbits)	von Brunn et al. 1991
	Circumsporozoite (CS) protein, repetitive NANP epitope, 4 or 9 repeats	127↓128	HEK293F cells	B cells (mice)	Kingston et al. 2019

HBs C-terminal insertions

Vaccine Target	Source of Epitope	Position of Insertion or Addition	Expression System	Major Immunological Activity	References
Hepatitis C virus (HCV)	CTL polyepitopes	226↓	DNA vaccine	CTL response (mice)	Memarnejadian et al. 2009; Arashkia et al. 2010
Human immunodeficiency virus 1 (HIV-1)	gp120, aa 42–125 plus aa 205–503	226↓	CV-1 cells	ND	Berkower et al. 2004
	gp120, aa 42–125 plus aa 205–503	226↓	*Sf*9 cells	B cells (rabbits)	Berkower et al. 2004
	gp41, membrane proximal external region (MPER), aa 656–683, with different additions	226↓	HEK293 cells; insect Hi5 or Sf8 cells; DNA vaccines	B cells (mice, rabbits)	Phogat et al. 2008
	gp120, MN subtype, V3 loop, 49 aa residues	C-terminal	DNA vaccine	B cell and CTL response (mice)	Bryder et al. 1999

(Continued)

TABLE 37.2 (CONTINUED)

The Putative Vaccines Constructed on the *Hepadnaviridae* Surface Proteins by Genetic Fusion Methodology

Vaccine Target	Source of Epitope	Position of Insertion or Addition	Expression System	Major Immunological Activity	References
Human papilloma virus (HPV) 16	E7, aa 1–98 aa 36–98	226↓	*P. pastoris*	B cell and T cells (mice)	Báez-Astúa et al. 2005
	E7, aa 11–20 plus 82–90 (combined with N-terminal CCL19/MIP-3β and C-terminal IL-2 insertions)	226↓	*P. pastoris*; DNA vaccine	Therapeutic anticancer effect (mice)	Juarez et al. 2012
Chlamydia trachomatis	Major outer membrane protein (MOMP), polyepitope of T- and B-cell epitope-rich peptides	226↓	DNA vaccine	protection (mice), stronger than by the N-terminal addition of MOMP	Zhu S et al. 2014
Alzheimer's disease	Aβ, aa 1–11 plus HBc T helper epitope, aa 70–85	226↓	DNA vaccine	B cells (mice); therapeutic effect in old mice	Olkhanud et al. 2012
Promotion of growth	Somatostatin, one or two copies	226↓	CHO cells; DNA vaccine	B cells (rats or rabbits); increase in growth rate	Xu WZ et al. 1994; Dai et al. 2008; Liang et al. 2008

Note: The data ia given in alphabetical order of the vaccine target in following order: viruses, bacteria, noninfectious diseases. The chimeras that were found self-assembly competent are included primarily.

and enabled secretion of chimeric 22-nm particles. When the PV-1 epitope was inserted into both internal positions of the same carrier, the correct 22-nm particles were not only formed but also demonstrated the higher PV-specific antigenicity and immunogenicity (Delpeyroux et al. 1990). Moreover, a longer fragment, namely a 32-aa fragment from diphtheria toxin spanning aa residues 201–230 of the *tox*228 gene, was inserted between aa 113 and 114 (Delpeyroux et al. 1987, 1988a). Generally, the yields of secreted 22-nm particles were dependent on the site of insertion, but the position 113/114 appeared more preferable by immunological reasons.

The poliovirus model allowed generation of the first mosaic SHBs-based particles (Delpeyroux et al. 1988b). Thus, cotransfection with the two different plasmids carrying either modified or unmodified S genes led to the formation of mosaic particles presenting HBs-PV-VP1 fusion polypeptides on the helper HBs. Surprisingly, the mosaic particles induced much higher titers of neutralizing antibodies to poliovirus than did the chimeric ones. This study thus initiated the promising idea of designing multivalent VLPs carrying various peptide sequences or presenting several heterologous epitopes of interest on the surface of the same VLP carrier molecule (Delpeyroux et al. 1988b).

Human immunodeficiency virus served as the next source of epitopes for the chimeric HBs vaccine candidates. The idea was particularly attractive because of an overlap between populations at risk for HBV and HIV infections and stimulated therefore the sound hope to develop a bivalent hepatitis B/HIV vaccine (Michel et al. 1988, 1990;

Filatov et al. 1992; Schlienger et al. 1992; Eckhart et al. 1996).

First, the famous Pierre Tiollais team inserted an 84-aa-residue-long C-terminal fragment of HIV-1 gp120 into the preS2 part of the protein M and produced the fusion protein in eukaryotic cells (Michel et al. 1988). Immunization with the chimeric particles allowed not only for the generation of a humoral response in rabbits (Michel et al. 1988) but also the induction of neutralizing antibodies and proliferative T-cell and CTL responses in rhesus monkeys to both parts of the hybrid particle, i.e., HIV and HBs protein (Michel et al. 1990).

Second, the V3 loop (a putative virus-neutralizing epitope) of HIV-1 gp120 was inserted at the same site of the preS2 (Schlienger et al. 1992). It was detected this time that two in-frame initiation codons within the constructed gene allowed the expression of both the chimeric HBs-V3 gene and the natural SHBs gene, leading therefore to formation of mosaic VLPs. Immunization of rhesus monkeys with these VLPs induced generation of proliferative T-cell responses and, in some cases, neutralizing antibodies and antibody-dependent cellular cytotoxicity (Schlienger et al. 1992).

The same Tiollais team attempted to repeat this work with the V2 region of the simian immunodeficiency virus (SIV) gp140 protein (Schlienger et al. 1994). However, vaccinated macaques were not protected against the homologous SIV challenge, despite having consistent SIV-neutralizing antibody titers.

Later, Michel et al. (2007) optimized their expression system, inserted an HIV-1 polyepitope of up to 138 aa residues (which included also the V3 loop, instead of the preS2), and succeeded with efficiently secreted recombinant VLPs. In parallel, the authors performed DNA immunization of transgenic mice, which resulted in the recovery of humoral response against the carrier and enhanced levels of HIV-1 specific CD8+ T lymphocyte activation.

After the N-terminal insertions, the long region of the HIV-1 gp120 was added to the C-terminus of the SHBs and demonstrated ability to induce specific neutralizing antibodies to the fused HIV-1 epitopes (Berkower et al. 2004). Phogat et al. (2008) compared insertions of the membrane proximal external region (MPER) of HIV-1 gp41 into the N-, and C-, and internal region 125/128 of the SHBs. The schematic representation of these insertions is presented in Figure 37.9. As a result, the MPER appended at the C-terminus of the SHBs protein was recognized by specific antibodies with the highest affinity compared to the positioning of MPER at N-terminus or the extracellular loop of the carrier. Remarkably, the chimeric HBs-HIV genes (Berkower et al. 2004; Phogat et al. 2008) were adopted for the VLP purification not only from mammalian but also from insect cells, as well as were used as DNA vaccines.

In contrast to the previously listed HBs-HIV chimeras, which were produced in mammalian or insect cell lines, the following trials were performed in yeast. Much attention was given to the V3 epitope of the gp120, which not only contained T-, CTL-, and B-cell epitopes including those inducing neutralizing antibodies but also participated in coreceptor interaction and guided cell tropism of the virus. First, in early 1990s, the well-known "yeast expressing" Kęstutis Sasnauskas team inserted the HIV-1 V3 loop, subtype MN, at the internal position 139 of the SHBs molecule and got reliable synthesis of chimeras in yeast *S. cerevisiae* (Razanskas et al. 1996). Further, the SHBs served as a carrier for the popular virus-neutralizing gp41 epitope that was inserted at the internal site 113, and the gained chimera was expressed in *P. pastoris* (Eckhart et al. 1996). Thus, the chimeras harboring the 6-aa sequence ELDKWA from the gp41 protein induced specific immune response in mice, but the antibodies failed to neutralize HIV-1 in vitro, although they recognized recombinant HIV-1 gp160 protein (Eckhart et al. 1996).

When chimeric HBs-HIV genes were used for DNA immunization without dwelling on the synthesis and purification of the corresponding chimeric proteins, the HIV-1 V3 loop was inserted instead of the preS2 part of the protein M (Fomsgaard et al. 1998; Le Borgne et al. 1998) or at the C-terminal positions of the protein S (Bryder et al. 1999). These DNA vaccines elicited rapid and high B-cell and CTL responses in mice and rhesus monkeys.

Generally, the DNA vaccines exploited the potent immunogenicity of the HBs protein in the 1990s–2000s to deliver foreign insertions carrying CTL epitopes and to induce strong CTL response, as described specifically by Woo et al. (2006), Haigh et al. (2007, 2010), and reevaluated later by Chen D et al. (2010). The authors generated the DNA vaccine where polyepitope of the well-recognized CTL

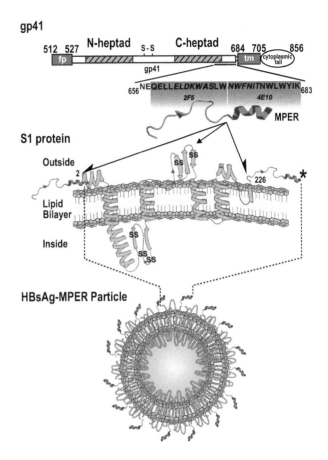

FIGURE 37.9 Schematic representation of the HIV-1 gp41, the MPER, the hepatitis B S1 protein and the HBsAg-MPER recombinant particles. The top bar diagram displays the overall organization of the HIV-1 gp41 transmembrane glycoprotein, which contains the MPER between amino acids 656—683 (fusion peptide (fp); N-heptad (heptad repeat 1); SS (cysteine-linked immunodominant region); C-heptad (heptad repeat 2); transmembrane region (tm)). In the following expanded section, the MPER residues are shown. The residues in contact with 2F5 or 4E10 in the crystal structures are "boxed" in red or purple, respectively, and the core residues defined in the original peptide mapping of the epitopes are italicized (Cardoso et al. 2005; Ofek et al. 2004). From the crystal structures of peptide with antibody, the 2F5 peptide is depicted as an extended loop (in red), and the 4E10 epitope as an alpha helix (in purple; Cardoso et al., 2005; Ofek et al., 2004). In the next panel, the MPER is schematically shown fused to either the N- or C-terminus of the hepatitis B surface antigen S1 membrane-spanning protein as indicated by the ½ arrowheads. The S1 protein is in blue, and the predicted alpha helices and beta strands are shown (from Mahoney and Kane, 1999). The center full arrowhead indicates the approximate position where the MPER was inserted into the S1 extracellular loop. An "SS" marks the approximate positions of the five S1 protein cysteine residues involved in intermolecular dimerization and oligomeric particle formation. The black asterisk indicates where additional hydrophobic C-terminal residues were added to the MPER in selected constructs. Directly following is a schematic diagram of an HBsAg-MPER particle containing an array of a (theoretical) 180 S1-C-terminal MPER fusion sequences per particle (not to scale). (Reprinted from *Virology*. 373, Phogat S, Svehla K, Tang M, Spadaccini A, Muller J, Mascola J, Berkower I, Wyatt R, Analysis of the human immunodeficiency virus type 1 gp41 membrane proximal external region arrayed on hepatitis B surface antigen particles, 72–84, Copyright 2008, with permission from Elsevier.)

epitopes from human papilloma virus (HPV16) E7 protein, respiratory syncytial virus (hRSV) F protein, influenza A matrix, human metapneumovirus (hMPV) N and SH proteins, Epstein-Barr virus (EBV) LMP1 protein, and ovalbumin was added to the C-terminus of the S gene. Moreover, the artificial gene S was generated where the aforementioned foreign CTL epitopes replaced the original HBs CTL epitopes at their individual locations. The immune response to the CTL epitopes correlated with disease protection in murine models of infectious and neoplastic disease. It was particularly important that the efficacy of the DNA vaccines to elicit the CTL response to an inserted foreign epitope was not diminished in the presence of a preexisting anti-HBs antibody (Chen D et al. 2010).

A similar idea but of replacing the major immunodominant region, including the "a" epitope and spanning aa residues 140–173, for the foreign epitopes within the HBsΔ vector was applied earlier for the presentation of a 35–98 aa segment of the human papillomavirus oncoprotein E7 (Pumpens et al. 2002). Remarkably, these HBsΔ "replacement" particles revealed only fields of the aggregated protein on electron micrographs, while the full-length HBs carrying a 35–54 aa segment of the HPV16 E7 protein inserted between aa residues 139 and 142 was capable of forming 22-nm VLPs, which did not differ morphologically from those formed by the natural HBs protein in *S. cerevisiae*, as shown in Figure 37.10.

Later, Juarez et al. (2012) generated a novel prospective therapeutic vaccine against HPV. The chimeric HBs

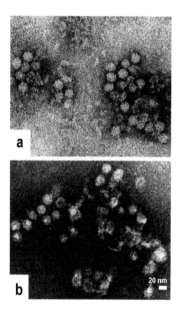

FIGURE 37.10 Electron micrographs of the negatively stained SHBs VLP preparations purified from *S. cerevisiae*. (a) Natural SHBs particles, (b) HBs particles carrying E7 region spanning aa residues 35–54. Bar, 20 nm. (Reprinted with permission of S. Karger AG, Basel from Pumpens P, Razanskas R, Pushko P, Renhof R, Gusars I, Skrastina D, Ose V, Borisova G, Sominskaya I, Petrovskis I, Jansons J, Sasnauskas K. *Intervirology*. 2002;45: 24–32.)

particles were composed of the SHBs molecules flanked their N-terminus by chemokine CC ligand 19/macrophage inflammatory protein-3β (CCL19/MIP-3β), and at the C-terminus by interleukin 2 (IL-2) and an artificial HPV16 E7 polytope encompassing aa 11–20 and 82–90. Both CCL19 and IL-2 conserved their functionality within the chimeric particles. This protein mounted specific T cell responses against E7 after immunization of transgenic mice with the corresponding plasmid, prevented the development of tumors after implantation of the appropriate tumor cell line, and demonstrated therefore definite potential of therapeutic value in the treatment of patients suffering from cervical precancer or cancer lesions caused by the high-risk HPVs (Juarez et al. 2012).

Future improvement of the HBs as a carrier of the immune response-improving epitopes might also be achieved by the addition of, for example, a well-known Th epitope derived from tetanus toxoid (Chengalvala et al. 1999). In fact, such chimeric 22-nm particles produced in a recombinant adenovirus expression system showed a several-fold enhancement of the anti-HBs response in mice relative to native HBsAg and were suggested for further exploitation as a new class of highly immunogenic HBs carriers.

As in the case of HIV, the HCV epitopes caused a particular interest because of practical attractiveness of the putative combined HBV/HCV vaccines. Thus, both DNA-based immunization and vaccinia virus expression approaches were tested first. In the plasmid DNA vector, the 58-aa-long N-terminal fragment of the HCV nucleocapsid was fused N-terminally to the M or S proteins (Major et al. 1995). In the vaccinia virus expression system, the five hydrophilic domains of the HCV E2 envelope as well as the hypervariable region (HVR) of the E2 protein were presented on the HBsAg surface by insertion at the internal site 113 (Lee et al. 1996).

In the next study, chimeric HBs-HVR1 VLPs carrying the HCV E2 HVR1 domain within the "a" determinant region of HBs elicited HVR1-specific antibodies in mice (Netter et al. 2001). Remarkably, the resulting anti-HVR1 antibody titers were found similar in mice with or without preexisting anti-HBs, suggesting therefore the successful application of the putative bivalent HBV/HCV vaccine in HBV carriers or vaccinees, i.e. individuals already immunized with the standard HBs vaccine (Netter et al. 2003). Moreover, the anti-(HBs-HVR1 VLP) antisera were able to immunoprecipitate native HCV envelope complexes (E1E2) containing homologous or heterologous HVR1 sequences and prevented entry of the HCV E1E2 pseudotyped HIV-1 particles (HCVpp) into HuH-7 target cells (Vietheer et al. 2007). These HBs-HCV VLP studies led to a strategy that looked for locations allowing the use of shared residues between the wild-type subunit sequence and the foreign insert and facilitating therefore the knowledge-based design of chimeric VLPs carrying multiple epitopes (Cheong et al. 2009a).

Improving further the putative HBs-HCV vaccine, the Philippe Roingeard's team turned to the mosaic particles, where natural HBs subunits supported assembly of the HBs molecules carrying full-length HCV E1 or E2 sequences (Patient et al. 2009; Beaumont and Roingeard 2013, 2015; Beaumont et al. 2013). To do this, the latter were inserted into the HBs molecule instead of its N-terminal transmembrane domain. Later, Wei et al. (2018) inserted three conserved HCV-neutralizing epitopes and one HVR1 mimotope into the external hydrophilic loop of the SHBs to self-assemble into subviral particles in mammalian cells, without the SHBs helper in this case.

Concerning hepatitis E, i.e. a candidate for the bivalent HBV/HEV vaccine, Li HZ et al. (2004) produced chimeric HBs-HEV VLPs in *P. pastoris*, where the only identified neutralization epitope of HEV capsid spanning aa 551–607 was coupled to the HBs molecule. Shrivastava et al. (2009) preferred, however, a combination vaccine where both HBs and recombinant HEV neutralizing region of ORF2 were entrapped into liposomes as DNAs, proteins, or DNA + protein, and the two last formats induced excellent immune response to both the components in mice.

The chimeric HBs VLPs carrying influenza CTL activity were generated by Hans J. Netter's team (Cheong et al. 2009b). For this purpose, the SHBs CTL at various sites were substituted with a conserved CTL epitope derived from the influenza matrix protein. The self-assembly depended on the insertion site. Immunizations of transgenic mice with the chimeric VLPs induced anti-influenza CTL responses proving that the inserted foreign epitope was correctly processed and cross-presented. The chimeric VLPs in the absence of adjuvant were able to induce memory T-cell responses, which were recalled by influenza virus infections in the mouse model system (Cheong et al. 2009b).

The first putative edible bivalent HBV/HIV vaccine was based on the SHBs carrying Gag and Env epitopes and was accomplished in tomato plants (Pozdniakov et al. 2003; Shchelkunov et al. 2004). The HIV-1 Gag and Env epitopes were generated in turn as an artificial four α-helix bundle TBI (or *T* and *B* cell epitopes containing *i*mmunogen) gene (Eroshkin et al. 1995). As a result, the mice fed with the recombinant tomato fruits demonstrated HBV- and HIV-specific antibodies in the serum (Shchelkunov et al. 2005, 2006). These and other plant-based hepatitis B vaccines were reviewed exhaustively at that time by Shchelkunov and Shchelkunova (2010), while a complete overview specifically on the expression of HBs in transgenic plants was published by Guan et al. (2010).

By the generation of a putative Dengue vaccine to be expressed in *P. pastoris*, Bisht et al. (2001) noticed the greater yield of the chimeric product when the Dengue virus type 2 envelope (E) protein was added to the N-terminus of the SHBs molecule in comparison with the C-terminal addition. Moreover, the N-terminal chimeras formed nice 22-nm particles, despite the 395-aa-residues-long Dengue virus E protein addition (Bisht et al. 2002). Recently, Ramasamy et al. (2018) designed and created a recombinant *d*engue *s*ubunit *v*irus-like particle *t*etravalent (DSV4) vaccine candidate based on the SHBs VLP platform, using the yeast *P. pastoris* expression system. Fusing of the EDIIIs of the four DENV serotypes via flexible peptide linkers led to the EDIII-based tetravalent antigen that was added N-terminally to the SHBs molecule and produced within the mosaic VLPs with the natural SHBs as a helper.

And finally, this is the greater success of the HBs as a carrier with the well-known malaria vaccine RTS,S, or RTS,S/AS01, or RTS,S/AS02A, or Mosquirix by trade name. The RTS,S is the world's first licensed malaria vaccine and also the first vaccine licensed for use against a human parasitic disease of any kind. The vaccine was generated originally in 1984 by scientists working at SmithKline Beecham Biologicals (now GlaxoSmithKline Vaccines) laboratories in Belgium but further developed through a collaboration between the GSK and the Walter Reed Army Institute of Research (WRAIR; Heppner et al. 2005) and has been funded in part by the PATH, a global nonprofit health organization based in Seattle; the Malaria Vaccine Initiative; and the Bill and Melinda Gates Foundation.

Thus, first, Rutgers et al. (1988) generated two variants of the fusion proteins, where 1 or 15 tetrapeptide NANP repeats from the circumsporozoite protein (CSP) of *P. falciparum* were added before the aa position 42 of the preS2 part of the MHBs protein. In both hybrid proteins one tetrapeptide repeat NVDP was also present resulting from the fusion of the CSP sequences NV to the preS2 sequences DP. The synthesis of the chimeric proteins and generation of the 22-nm-like mosaic particles, which were formed by the chimeras together with the natural SHBs protein translated from its intrinsic ATG codon, was achieved in *S. cerevisiae*. The initial construct with the 15 NANP and one NVDP, referred to as R16-HBsAg, was formulated with alum and proved safe and immunogenic in 20 hepatitis B-seronegative adults at the University of Nijmegen (Vreden et al. 1991).

In 1987, following the recognition of the potential role of the cell-mediated immunity and of nonrepeat sequence of CSP, a new CSP-HBsAg fusion protein was developed at GSK, which incorporated the CSP C-terminal flanking region of 189 aa residues, known to contain B- and T-cell epitopes, into a chimeric gene expressed in *S. cerevisiae* (Gordon et al. 1995). This construct was named "RTS" to indicate the presence of the CSP repeat region (R), T-cell epitopes (T), and the HBs antigen (S). The presence of the coexpressed SHBs (of 230 aa in this case, i.e. 226 aa together with four additional N-terminal aa residues) within the mosaic particles in a ratio of RTS to S of 1:4 led to the name "RTS,S." A process compatible with GMP manufacturing of RTS,S was then developed at GSK (cited by Heppner et al. 2005).

The first clinical trial of RTS,S evaluated two formulations, RTS,S/MPL (monophosphoryl lipid A)/alum and RTS,S/alum, developed at GSK for safety and efficacy in malaria naive adults in 1992 at the WRAIR (Gordon et al. 1995). Next, one formulation, RTS,S/AS02A, consisting of RTS,S in an oil-in-water emulsion with MPL and QS21 demonstrated great antibody response (Garçon et al. 2003). In 1996, volunteers immunized with either RTS,S/AS03 (an oil-in-water emulsion without additional adjuvants) or RTS,S/AS02A in a Phase 2A trial at WRAIR developed the highest antirepeat antibody responses ever induced by a CS-based vaccine, but RTS,S/AS02A induced a higher T-cell response as measured by IFN-γ production (Sun et al. 2003). While the RTS,S/AS03 formulation was marginally protective (two of seven), the RTS,S/AS02A protected an unprecedented six of seven volunteers (Stoute et al. 1997).

The RTS,S/AS02A, RTS,S/AS01E, and other adjuvanted variants were subjected to numerous phase I/IIa,b/III trials, especially in African countries (Gordon et al. 1995; Stoute et al. 1997, 2006; Kester et al. 2001, 2007, 2009; Bojang et al. 2005, 2009; Enosse et al. 2006; Macete et al. 2007a, b; Abdulla et al. 2008; Sacarlal et al. 2008; Ballou 2009; Bojang et al. 2009; Lell et al. 2009; Owusu-Agyei et al. 2009; Polhemus et al. 2009; Waitumbi et al. 2009; Agnandji et al. 2010; Casares et al. 2010; Asante et al. 2011; Wilby et al. 2012; Campo et al. 2014; Foquet et al. 2014; RTS,S Clinical Trials Partnership 2015; Olotu et al. 2016; Regules et al. 2016; Adepoju 2019; Schuerman 2019; Kurtovic et al. 2020; Valéa et al. 2020). The RTS,S vaccine under the Mosquirix name was approved for use the European Medicines Agency's Committee for Medicinal Products for Human Use on 24 July 2015 as the first malaria vaccine and with a special recommendation "for use in areas where malaria is regularly found, for the active immunization of children aged 6 weeks to 17 months against malaria caused by the *Plasmodium falciparum* parasite, and against hepatitis B" (www.ema.europa.eu/en/news/first-malaria-vaccine-receives-positive-scientific-opinion-ema). According to the WHO recommendations, the vaccinations by the ministries of health of Malawi, Ghana, and Kenya began in April and September 2019 and will target 360,000 children per year in areas where vaccination would have the highest impact. Based on the lessons learned, the RTS,S vaccine team generated the CSV-S,S vaccine against *P. vivax* (Vanloubbeeck et al. 2013).

To develop a next-generation RTS,S-like vaccine, called R21, Collins et al. (2017) greatly increased the proportion of CS-polypeptides, since the R21 particles were formed from a single HBs-CS fusion protein, without a helper, in contrast to the RTS,S vaccine. The R21 vaccine used the shortened CS version of 184 aa, in contrast to the RTS,S with the 198 aa long CS version spanning aa 210–398. Then, the R21 vaccine was produced in *Pichia pastoris*, and purification protocol was optimized. As a result, the R21 vaccine induced a sterile protection in mice against a challenge with transgenic sporozoites, while anti-HBs response was compromised, possibly because the high content of the CS-polypeptide blocked access to the "a"-determinant. The R21 was subjected to evaluation in Phase I/IIa clinical trials by the University of Oxford (Draper et al. 2018; ClinicalTrials.gov Identifier: NCT02925403; NCT03947190).

Furthermore, Salman et al. (2017) generated the Rv21 vaccine, an analogue of the R21 vaccine but targeted against *P. vivax*. The Rv21 vaccine was tested in novel *P.*

berghei-PvCSP challenge model in mice and entered a path toward clinical trials.

In parallel with the famous RTS,S vaccine, some computer-predicted B- and T-cell epitopes from the major merozoite surface antigen gp190 sequence were introduced by replacing the proper antigenic determinants of the SHBs protein, with further successful production of the chimeric VLPs in mammalian cells via vaccinia virus expression (von Brunn et al. 1991). Such replacement of 43 aa residues of the major hydrophilic domain of the SHBs by the three gp190-derived sequences ranging in size between 32 and 61aa did not impair the formation of 22 nm-like particles.

Then, a sexual stage/sporozoite-specific antigen Pfs16 was fused to the HBs protein and synthesized in yeast cells in the form of mosaic particles in the presence of the helper HBs (Moelans et al. 1995).

It is noteworthy that the merozoite surface protein 1 gene of *P. vivax* was also suggested as a DNA vaccine candidate when fused to the SHBs as a carrier (de Oliveira et al. 1999).

Recently, Kingston et al. (2019) proposed an interesting alternative to the RTS,S vaccine by retaining the specific malaria antigen but changing the site of the epitope insertion within the HBs molecule and expression platform. Thus, the authors inserted the four or nine NANP repeats into the SHBs immunodominant region between aa residues 127 and 128 and achieved large-scale production of the chimeric nonmosaic HBs-malaria VLPs in human embryonic kidney cells but not in yeast. An N-terminal myc-tag sequence was included into the chimeric molecules to allow detection of the VLPs independent of the SHBs backbone. As a result, the VLPs with NANP$_9$ repeats induced antibodies with the ability to activate complement, which could contribute to the inactivation of invading sporozoites. It is highly important that the expression of the chimeric VLPs in mammalian cell lines allowed additional manipulations of the HBs backbone to further enhance immunogenicity, as it was established earlier by the Hans J. Netter' team (Cheong et al. 2012; Hyakumura et al. 2015).

At last, Shanmugam et al. (2019) produced in *P. pastoris* the HBs VLP-based vaccine against Zika virus (ZIKV). This vaccine included two proteins, ZS and S, in a genetically predetermined ratio of 1:4, where ZS was an in-frame fusion of ZIKV envelope domain III with the HBs, and S was the unfused HBs. Both proteins coassembled into the VLPs that were immunogenic in mice, elicited ZIKV-neutralizing antibodies, and did not enhance a sublethal DENV-2 challenge in the specialized mice.

DHBs as a Carrier

Wetzel et al. (2018) demonstrated a versatile and cost-effective platform for the large-scale development of novel mosaic VLPs. This fundamental investigation was based on the previous studies protected by patents (Anderson and Grgacic 2004; Grgacic et al. 2006). Namely, the DHBs was chosen as the original VLP scaffold, and the industrially applied and safe yeast *Hansenula polymorpha* appeared as the highly efficient expression host.

Although the DHBV virions are of comparable size (42–50 nm in diameter) with the HBV virions, the naturally occurring DHBs VLPs are described as 35–60 nm particles (Mason WS et al. 1980), and the ratio of the large to the small DHBV surface proteins within the VLP is identical, approximately 1:4 (Schlicht et al. 1987; Grgacic and Anderson 2005), to that found in the virions' envelope (Mason WS et al. 2012). This is in definite contrast to the natural HBs 22-nm VLPs that are smaller and enriched with the SHBs molecules. Moreover, the DHBs VLPs are lacking the immunodominant region that could be equivalent to the "a" determinant of the SHBs (Schlicht et al. 1987; Yuasa et al. 1991).

One of the main goals of this impressing DHBs platform project was the development of VLPs suitable for the application in the veterinary sector, namely compatible with the *differentiating infected from vaccinated animals* (DIVA) strategy and independent of antibiotic resistance genes during all stages of development and production. As the result, eight different, large molecular mass antigens of up to 412 aa derived from four animal infecting viruses were genetically fused to the DHBs, and recombinant production strains were isolated. The mosaic VLPs were engineered by genetic fusion of foreign antigens to either the N- or C-terminus of the DHBs. Coexpression of the fusion proteins with the VLP-forming DHBs protein as a helper allowed the formation of the mosaic VLPs in all studied cases. Such mosaic VLPs were up to 95% pure and displayed up to 33% fusion protein.

Furthermore, a prospective malaria vaccine candidate, under the Metavax name, was generated on this versatile DHBs platform (Chan et al. 2019; Wetzel et al. 2019). The production process was highly efficient. Thus, up to 100 mg mosaic VLPs were isolated from 100 g dry cell mass with a maximum protein purity of 90%. The mosaic VLPs contained 70% DHBs helper to 30% fusion protein. Remarkably, the mosaic VLPs displaying the transmission stage proteins Pfs230 and Pfs25 effectively induced antibodies to the specific antigens with minimal induction of antibodies to the DHBs carrier. Antibodies to Pfs230 recognized native protein on the surface of gametocytes, and antibodies to both Pfs230 and Pfs25 demonstrated transmission-reducing activity in standard membrane feeding assays. Generally, this study strongly contributed to tomorrow's vaccine generation technique by the record-breaking presentation of leading malaria transmission blocking antigens of up to 80 kDa on the surface of mosaic VLPs.

Chemical Coupling

Machida et al. (1987) were the first who used chemical coupling to supplement natural 22-nm HBs particles with the additional preS2 antigenicity and improve therefore their antiviral immunogenicity. The synthetic 19-aa preS2 peptide spanning aa 14–32, a hydrophilic area of the pre2 sequence, was conjugated to the natural HBs particles from human blood, rich or poor in the preS2, with 1-ethyl-3-(3-dimethylaminopropyl)carbodiimide (EDAC). The HBs

particles supplemented with the preS2 peptide raised antibody to the preS2 region product in mice, which were found to agglutinate HBV virions in immune electron microscopy (Machida et al. 1987).

Neurath et al. (1989) chose the other path by their coupling methodology. Considering the fact that the chemical coupling could have deleterious effect on the antigenicity and immunogenicity of the SHBs protein (Machida et al. 1987), the authors took advantage of the hydrophobic surface of the 22-nm particles, which strongly interacted with alkyl or amino alkyl chains containing at least nine CH_2 groups. As a result, the "hybrid antigens" were prepared, which contained the preS1 peptide or the HIV-1 gp120 peptide, which were attached to the 22-nm particles through a hydrophobic tail, i.e. long-chain fatty acids or mercaptans covalently linked to the synthetic peptides. The combination of the myristylated preS1(12–47) peptide with the natural 22-nm particles containing the S and preS2 regions resulted in an immunogen that elicited a broad spectrum of protective antibodies in mice and rabbits. The combination of the 22-nm particles with the HIV-1 gp120 V3 loop, aa 306–338, resulted in an immunogen eliciting anti-HIV-l in the immunized animals.

The idea of the HIV-1 V3 loop-coupled HBs vaccine was developed further by Iglesias et al. (2005). The authors prepared cocktails of so-called multiple antigen peptides (MAPs), which included eight B cell epitopes comprising the central 15 aa from the gp120 V3 loop from different HIV-1 isolates and a T helper epitope selected comprising aa 830–844 from the tetanus toxin. The MAPs had a 2:1 B to T cell epitopes ratio and coupled with EDAC to the succinic anhydride-activated recombinant HBs particles. Immunization of mice demonstrated improved levels of cross-reactivity against the variable V3 loop epitopes. Furthermore, when compared to the traditional carriers such as keyhole limpet hemocyanin (KLH) or recombinant meningococcal protein P64k, the HBs particles were more advantageous by their immunogenicity in mice, to say nothing of their ability to generate bivalent vaccines (Cruz et al. 2009).

Peng et al. (2007) developed an IgE peptide-based vaccine, in which small peptides of 8–16 aa each corresponding to human IgE receptor-binding sites were coupled to the HBs carrier. The glutaraldehyde method was used to couple the peptides to the commercial HBs prepared from human blood. As a result, the in vitro effect of the antibodies induced by the chimeric chemically coupled HBs-based vaccine was investigated in blocking the binding of human IgE to its high-affinity receptor FcεRI and the cross-reactivity of the antibodies with IgE of other species.

A capsular polysaccharide molecule was first coupled to the HBs VLPs by Qian W et al. (2020) to develop a new pneumococcal conjugated vaccine. Thus, the vaccine consisting of pneumonia type 33 F capsular polysaccharide (Pn33Fps) conjugated to the SHBs with EDAC induced strong specific T-cell responses against both antigens in vivo in immunized mice.

Plug-and-Display

For the decoration of VLPs, the modern plug-and-display (or superglue) platform was applied first for the RNA phages in 2016, as described in the *Chimeric VLPs: Plug-and-Display* paragraph of Chapter 25. In short, the plug-and-display decoration of VLPs uses the ability of a peptide termed SpyTag and a protein termed SpyCatcher to form spontaneous covalent isopeptide bonds between lysine and aspartic acid residues under physiological conditions (Zakeri et al. 2012).

In the case of the HBs platform, the SHBs molecule was fused to the SpyCatcher protein, so that the antigen of interest, linked to the SpyTag peptide, could be easily displayed on the resulting chimeric HBs VLPs (Marini et al. 2019). The SpyCatcher-HBs VLPs were decorated with Pfs25-SpyTag, where Pfs25 was a 25 kDa *Plasmodium falciparum* protein that is expressed on the surface of zygotes and ookinetes in the mosquito midgut.

In fact, Marini et al. (2019) reproduced their previous work on the transmission-blocking malaria vaccine by the decoration of the RNA phage AP205 VLPs with the Pfs25 protein (Leneghan et al. 2017), which is described in Chapter 25. The HBs VLPs seemed definitely preferable because of the well-established safety profile in numerous preclinical and clinical trials, which was not the case for the RNA phages. Moreover, the natural and recombinant HBs VLPs have been safely used in humans for decades as an effective hepatitis B vaccine, and the chimeric malaria vaccine Mosquirix was approved by the European Medicines Agency, as stated earlier.

The SpyCatcher-HBs VLPs appeared as stable vaccine carriers, and their production was easily scalable in yeast *Pichia pastoris*. As little as 10% of the VLPs decorated by the antigen induced a higher and more efficient antibody response compared to the soluble protein, with 50 and 90% of SpyCatcher-HBs conjugated to Pfs25-SpyTag further enhanced the antigen-specific antibody response. Moreover, in contrast to the RNA phage Qβ and AP205 VLP platforms, preexisting antibodies directed toward the carrier did not interfere with the specific immune response in mice (Marini et al. 2019).

Dalvie et al. (2021) used the SpyTag/SpyCatcher methodology to engineer a HBs VLP-based SARS-CoV-2 vaccine. The SpyTag peptide was fused onto the SARS-CoV-2 receptor binding domain (RBD) of protein S and this fusion protein was manufactured in an engineered strain of yeast *Komagataella phaffii* (in other words, *Pichia pastoris*), while the HBs-SpyCatcher was produced in yeast *Hansenula polymorpha*. Thus, the RBD-SpyTag and HBs-SpyCatcher VLPs were each purified separately and then conjugated in a GMP process to produce the RBD-HBs VLPs. The latter were formulated with two adjuvants: Alum or Alum combined with CpG1018—a potent commercial TLR9 agonizing adjuvant known to elicit Th1-like responses. Remarkably, the addition of the CpG1018 improved the cellular response while reducing the humoral response in cynomolgus macaques (Dalvie et al. 2021).

As reviewed by Zhao J et al. (2020), this vaccine is safe and immunogenic and can be made in mass production. This first VLP vaccine against Covid-19 is developed by SpyBiotech/Serum Institute of India and entered phase I/II trials in September 2020 in Australia.

PACKAGING AND TARGETING

The Shun'ichi Kuroda team generated the unique drug delivery platform using the HBs VLPs, so-called bionanocapsules (BNCs). The principles of the BNCs are exhaustively reviewed by the authors (Yamada T et al. 2004; Yu et al. 2006; Kasuya and Kuroda 2009; Yamada M et al. 2012; Kuroda 2014; Somiya and Kuroda 2015, 2020a, b).

The BNCs are mimicking the early mechanism of HBV infection and using the LHBs molecules. After many unsuccessful attempts to express the full L gene sequence, Kuroda's team found in 1992 that the addition of an N-terminal signal peptide could overcome the inhibition of the LHBs protein synthesis and accomplished, for the first time, the overexpression of the LHBs particles in yeast *S. cerevisiae* (Kuroda et al. 1992). The LHBs VLPs that got the BNC name appeared as ~100-nm-sized spherical hollow particles that consisted of a liposome embedded with approximately 110 protein molecules (Yamada T et al. 2001). The BNCs are readily purified now by heat treatment, affinity column chromatography, and gel filtration, after disruption of yeast cells with glass beads (Jung et al. 2011).

These BNCs were able to specifically attach to and then internalize into human hepatic cells by implementing the early mechanism of infection by HBV virions. This was due to the functional domains in the preS1 region, namely the fusogenic domain, aa 9–24, responsible for the low pH-dependent membrane fusion of the BNCs and the heparin-binding domain, aa 30–42, with a strong affinity to heparin as compared to that of known heparin-binding peptides, such as vitronectin and gp120 of HIV-1. The heparin-binding domain recognized heparan sulfate proteoglycan (HSPG) at the cell surface of human hepatic cells. The authors precisely mapped both critical membrane fusion (Somiya et al. 2015a) and heparin binding, i.e. receptor recognition (Liu Q et al. 2018) activities. The BNCs were able to enter the human liver cells through either clathrin-dependent endocytosis or macropinocytosis at the same rate as HBV virions (Yamada M et al. 2012).

The first approval of the ability of the BNCs to deliver their payload specifically to human hepatic cells in vitro and in vivo, in a mouse xenograft model, with high transfection efficiency, after having genes and drugs loaded into their cavity by electroporation, was demonstrated by Yamada T et al. (2003). In this revolutionary paper, the BNCs transferred the GFP gene or a fluorescent dye, which resulted in observable fluorescence only in human hepatocellular carcinomas but not in other human carcinomas or in mouse tissues. When the gene encoding human clotting factor IX was transferred into the xenograft model, the factor IX was produced at levels relevant to the treatment of hemophilia B (Yamada T et al. 2003).

Yu et al. (2005) performed the fusion of eGFP at the C-terminus of the LHBs and found that some truncation of the C-terminus was required to obtain sufficient expression levels in Cos7 cells. The chimeric BNCs delivered eGFP to human hepatoma cells, displaying the eGFP moiety outside or enclosing it inside. Furthermore, Shishido et al. (2006) developed a new BNC production system using the stably transfected *Trichoplusia ni* insect cell line. The BNCs were efficiently secreted by the overexpression of the LHBs protein, which was fused to the secretion signal peptide. Remarkably, the BNC production was maintained for at least 75 days.

The first clear demonstration that nanoparticles could be used for delivery of therapeutic genes with antitumor activity into human liver tumors was achieved by Iwasaki et al. (2007). Thus, systemic injection of the BNCs containing the herpes simplex virus thymidine kinase gene (HSV-tk), followed by the administration of ganciclovir, inhibited the growth of human hepatic tumor in a rat xenograft model.

The authors steadily improved the BNC quality. Thus, their stability was enhanced when aggregation was avoided by replacing 8 unessential cysteine residues of total 14 cysteines (Nagaoka et al. 2007). Then, the lower immunogenic BNCs were generated by inserting two point mutations in the LHBs protein, which were found in HBV escape mutants (Jung et al. 2018).

Since electroporation often induced unexpected fusion of BNCs, the latter were found to form a stable complex with liposomes, facilitating the incorporation of various therapeutic materials, e.g., >45-kb plasmid, 100-nm polystyrene beads, into the BNCs (Jung et al. 2008). Such BNC–liposome complexes delivered the incorporated materials to the specific site in vitro and in vivo with the same or greater efficiency as BNCs alone. Since abandonment of electroporation allowed greater control over the manufacturing protocol, these BNC-liposome complexes, or virosomes, are recognized as a second-generation BNC technology. Such virosomes were filled with anticancer drug doxorubicin (Kasuya et al. 2009). The efficiency of the doxorubicin encapsulation was optimized, and the virosomes exhibited effective antitumor growth activity in xenograft mice harboring target cell-derived tumors (Liu Q et al. 2015). The virosomes were filled efficiently also with siRNA (Somiya et al. 2015b).

To target cells other than hepatocytes, the specificity of the BNCs was altered by genetic modifications. Varieties of specificity-altered BNCs were produced by deleting the preS region having specificity for hepatocytes and inserting binding molecules targeting other cells (Kasuya et al. 2008, 2009).

The specificity of the BNCs was altered further by insertion of biotin-acceptor peptide (BAP), which was efficiently biotinylated using biotin ligase BirA from *E. coli* (Shishido et al. 2009a). Using streptavidin as a linker, the biotinylated BNCs displayed various biotinylated ligands that were otherwise difficult to fuse with BNCs, such as antibodies, synthetic peptides, and functional molecules. The BAP-fused BNC served as a versatile carrier for drug delivery to a variety of target cells. They were efficiently biotinylated, effectively displayed streptavidin, and were internalized into target cells via biotinylated nanobody displayed on the BNC surface (Shishido et al. 2009a).

To be maximally universal, most of the preS1 and preS2 regions, aa 51–153, were replaced with the tandem form of the IgG Fc-binding Z domain from *S. aureus* protein A (Tsutsui et al. 2007; Iijima et al. 2010). Such hybrid ZZ-BNCs were conjugated then with a specific antibody against human epidermal growth factor receptor (anti-EGFR) recognizing receptor EGFRvIII and efficiently delivered to glioma cells but not to normal glial cells. Similarly, in vivo, the ZZ-BNCs displaying anti-CD11c antibodies accumulated in CD11c expressing dendritic cells after intravenous injection (Matsuo et al. 2012).

Next, the BNCs displaying the ZZ domain but packaging a model protein, namely, eGFP, inside the particles were engineered in insect cell expression systems (Kurata et al. 2008). The eGFP was genetically fused to the C-terminus of the ZZ-BNC molecule and incorporated therefore into the BNCs. When the ZZ domain presented anti-EGFR antibodies, the chimeric BNCs ensured specific delivery of the packaged eGFP to HeLa cells, by recognition of the EGFR. Further, Iijima et al. (2019) clearly demonstrated that the therapeutic efficacy of the anti-EGFR drug was enhanced by coupling to the ZZ-BNCs.

Remarkably, the ZZ-BNCs adsorbed onto the gold surface of the sensor chip of the quartz crystal microbalance markedly enhanced the sensitivity and antigen-binding capacity of the chip (Iijima et al. 2011). Moreover, using high-speed atomic force microscopy, Iijima et al. (2012) showed that the ZZ-BNCs tethered a maximum of 60 IgG molecules, displayed the Fv regions outwardly in an oriented-immobilization manner, and enabled the Fv regions to undergo rotational Brownian motion.

Furthermore, the affibody molecules, which comprised a new class of affinity ligands derived from the ZZ domain and could bind a range of different proteins, e.g. insulin, HER2, and EGFR, were used as a substitute for antibodies, while an arginine-rich peptide was displayed on the BNCs to permit the delivery into various types of cells (Shishido et al. 2009b). Thus, the Z_{HER2} affibody displaying BNCs recognized specific HER2-expressing cells such as breast cancer and ovarian cancer cells (Shishido et al. 2010). The Z_{EGFR} dimer-displaying BNCs effectively recognized the EGFR-expressing cells and delivered drugs to the cytosol (Nishimura et al. 2017).

To confer the ability of endosomal escape to the Z_{HER2}-BNCs, Nishimura et al. (2014) displayed a pH-sensitive fusogenic GALA peptide consisting of 30 aa residues on the surface of the particles. The GALA-displaying Z_{HER2}-BNCs purified from yeast uneventfully formed a particle structure and ensured endosomal escape after endocytic uptake and released the inclusions to the cytoplasm without the cell toxicity.

The ZZ-BNCs were evaluated as a vaccine carrier with the E protein DIII domain of Japanese encephalitis virus (JEV; Miyata T et al. 2013). The lysine-rich ZZ moiety exposed on the surface of the ZZ-BNCs was used for chemical conjugation with the JEV DIII antigen, which had been expressed and purified from *E. coli*. Immunization of mice with the ZZ-BNC:D3 augmented antibody response against JEV and increased protection against lethal JEV infection. Matsuo et al. (2018) improved the immunogenic activity of the ZZ-BNC:D3 by specific targeting to dendritic cells when anti-CD11c antibody was exposed on the ZZ domain and demonstrated that such complicated BNCs elicited strong immunity in mice.

The ZZ-BNCs were modified with a transactivator of transcription (TAT) peptide, in order to induce strong cell adhesion after local administration into the uterine cavity (Koizumi et al. 2019). As a result, the gene transfer using the TAT-ZZ-BNCs was approximately 5,000- or 18-fold more efficient than the introduction of the same dose of naked DNAs or the use of the cationic liposomes, respectively. The TAT-ZZ-BNCs were rapidly eliminated from the uterus and had no effect on the pregnancy rate, litter size, or fetal growth and appeared therefore as a useful gene delivery system for uterine endometrial therapy via local uterine injection (Koizumi et al. 2019).

Since the Z domain limited the animal species and subtypes of IgGs that can be displayed on the ZZ-BNCs, Tatematsu et al. (2016) introduced into the BNCs an Ig κ light chain-binding B1 domain of *Finegoldia magna* protein L (protein-L B1 domain) and an Ig Fc-binding C2 domain of *Streptococcus species* protein G (protein-G C2 domain) to produce LG-BNCs. The LL-BNCs were constructed in a similar way using a tandem form of the protein-L B1 domain. Both LG-BNCs and LL-BNCs displayed rat IgGs, mouse IgG1, human IgG3, and human IgM—all of which were not binding to the ZZ-BNCs—and accumulated in target cells in an antibody specificity-dependent manner. Thus, when these BNCs displayed anti-CD11c IgG or anti-EGFR IgG, both of which cannot bind to the Z domain, they could bind to and then enter their respective target cells (Tatematsu et al. 2016). Li H et al. (2018, 2020) introduced the tandem form of G and L domains, as a GL domain, instead of the ZZ domain and generated both GL-BNCs and GL-virosomes, which functioned as promising macrophage-specific and phagocytosis-inducing drug delivery nanocarriers.

Nishimura et al. (2012) adopted the BNCs for the specific packaging of proteins, when the latter were preencapsulated during the course of nanoparticle formation. Briefly, because of the process of the BNC formation in a budding manner on the endoplasmic reticulum (ER) membrane, the association of target proteins to the ER membrane with lipidation sequences (ER membrane localization sequences) directly generated the protein-encapsulating BNCs in collaboration with coexpression of the LHBs proteins. Thus, the membrane-localized proteins were automatically enveloped into the BNCs during the budding event.

It is noteworthy that the BNC technology provoked construction of other nanocontainers targeted to liver cells by the preS peptides, such as phosphorylcholine-based amphiphilic block copolymer micelles as a biocompatible, drug delivery carrier for treatment of human hepatocellular carcinoma with paclitaxel (Miyata R et al. 2009).

NANOMATERIALS

The HBs was involved in the generation of complex nanoparticles because of urgent vaccination needs. The use of alum, as a traditional vaccine carrier, had serious limitations. The alum-based vaccines were limited to parenteral administration, required a tight control of the storage conditions, and lacked the capacity to stimulate cell-mediated immunity. It was becoming clear that the antigen-loaded nanocarriers may facilitate antigen presentation to T cells, when their resemblance to virus in terms of size, <100 nm, is optimal and could target antigens directly to the dendritic cells in the lymph nodes (Manolova et al. 2008; Bachmann and Jennings 2010). Therefore, the search for the appropriate polymeric carriers became the first priority for developing needle-free vaccination methods, such as oral, nasal, or aerosol technologies, with high T-cell efficiency.

The SHBs vaccine was coated onto biodegradable particles formed by poly(D,L-lactide-*co*-glycolide; or PLGA; Shi et al. 2002; Feng et al. 2006; Gupta et al. 2007; Paolicelli et al. 2010; Giri et al. 2011); poly(lactic acid) (PLA) and polyethyleneglycol (PEG) copolymers (Jain et al. 2009, 2010); chitosan, a linear polysaccharide derived from chitin by deacetylation (Prego et al. 2010; Premaletha et al. 2012; Farhadian et al. 2015); and lipopolysaccharide derived alginate coated chitosan (Saraf et al. 2020).

Correia-Pinto et al. (2015) generated nanovaccines (<50nm) by multienveloping of the HBs particles with cationic polymers protamine (PR) and polyarginine (PARG) as a first layer, and then, as a next layer, with anionic polymers dextran sulfate (DS) and alginate (ALG), and the immunostimulant polynucleotide poly (I:C) (pIC). The resulting multienveloped nanoparticles (HB:PR; HB:PR:DS; HB:PARG:ALG; HB:PARG:pIC) were shown to be internalized by macrophages.

Another important direction of the nanomaterial development was connected with the possible visualization tools. Thus, Huang Z and Mason HS (2004) described an *Agrobacterium*-mediated transient assay that provided enough antigen-expressing material at two days post-transfection to evaluate antigen conformation. It was highly important in the line of possible nanomaterials that the authors used in their studies the SHBs as a model antigen and the green fluorescent protein (GFP) as a model fusion partner. As a result, the transient synthesis of VLPs occurred with the SHBs and with an SHBs fusion with GFP at the N-terminus (GFP:HBs) but not with the SHBs fusion with GFP at the C-terminus (HBs:GFP). The GFP:HBs demonstrated the functional "a" determinant and formed VLPs, similar to the yeast-derived vaccine HBs VLPs. Thus, it was feasible to modify the SHBs with an N-terminal fusion of up to 239 aa without altering its major antigenic properties. Moreover, when *N. benthamiana* leaves were cotransfected simultaneously with plasmids encoding HBs and GFP:HBs, the latter cointegrated into the same VLPs with the original natural HBs.

In parallel, Lambert et al. (2004) had reached the same conclusion when they tried both N- and C-terminal additions of GFP to the SHBs molecule. The N-terminally fused GFP:HBs was cointegrated into the envelope, giving rise to fluorescent particles, while a dual location of GFP, inside and outside the envelope, was observed. The GFP:HBs was also efficiently packaged into the viral envelope, and these GFP-tagged virions retained the capacity for attachment to HBV receptor-positive cells in vitro. These two studies definitely paved the way for future applications of the nanomaterials based on the fluorescent HB derivatives.

38 Order *Blubervirales*: Core Protein

How far that little candle throws his beams! So shines a good deed in a weary world.

William Shakespeare

If you want a thing done well, do it yourself.

Napoleon I

CORE GENE

The structure of the pleiotropic core (C) gene of HBV is depicted in Figure 38.1, as a general prototype of the core genes of *the Hepadnaviridae* family members. Thus, the core gene encodes five well-studied and functionally relevant core-related polypeptides, two of which are major. The first one is p21, also called HBcAg, Cp, or HBc monomeric unit (185 and 183 aa in length for the *adw* and *ayw* subtypes, or A and D genotypes, respectively), which serves as a structural constituent by the HBV nucleocapsid formation and is a target of cloning and expression for the production of the HBc VLPs. The second, p17 (HBeAg) is a nonstructural but immunologically important HBe protein of 159 aa residues (from position -10 of the preC precursor to position 149 of the HBc protein). The p17 protein is mostly collinear to p21 and is formed as a result of both N- and C-terminal cleavage of the preC precursor protein p25 of 214 or 212 aa residues (for the *adw* and *ayw* subtypes, respectively). The third, namely p22, HBe precursor protein appears after N-terminal cleavage of the p25 precursor. The fourth protein, p21.5, was identified as the p22 HBe precursor protein but lacked C-terminal amino acid residue (Wang Yi-Ling et al. 2015). The fifth protein, p22cr, was found in HBV DNA-negative Dane particles and identified as the p25 precursor without C-terminal domain (Kimura et al. 2005). All five peptides are summarized by diagnostics under the term *core-related proteins.*

The HBc protein together with polymerase is translated from terminally redundant pregenomic RNA (pgRNA). The preC protein is translated from a precore RNA that is collinear with the pgRNA, except for a short extension at the 5'-end that permits translation from the precore AUG (for review, see Seeger and Mason 2015). The p21.5 protein is translated from a 2.2 kb singly-spliced RNA, the most abundant spliced HBV RNA that is widely expressed among all HBV genotypes and is strongly associated with hepatopathology during HBV infection (Wang Yi-Ling et al. 2015).

The HBc protein undergoes self-assembly and acts in the HBV life cycle as an icosahedral scaffold of the HBV nucleocapsid, which carries genomic HBV DNA, polymerase and possibly cellular protein kinase. In addition to the HBV nucleocapsid building function, the core-related proteins, mostly in the HBc form, participate in every step of the HBV life cycle and regulation by (i) synthesizing dsDNA through the specific recognition of pgRNA and self-assembly on the pgRNA complex together with HBV polymerase, (ii) complexing HBV DNA and affecting its epigenetics through preferential binding to CpG islands, (iii) signaling the completion of reverse transcription and maturation of nucleocapsids carrying partially double-stranded relaxed circular DNA (rcDNA), (iv) recognizing specific sites on the HBV envelope proteins, thereby enveloping and further supporting the egress of HBV virions, (v) regulating envelopment via the 3D structure and content of the nucleocapsids, since DNA- but not RNA-containing nucleocapsids undergo envelopment, (vi) performing highly specific phosphorylation, when hyper- and hypo-phosphorylation refer to the DNA- and RNA-containing particles, respectively, and (vii) providing nuclear localization signals that target nucleocapsids to the cell nucleus after infection and form an HBV mini-chromosome carrying HBV covalently closed circular DNA (cccDNA) that acts as a template for the synthesis of HBV mRNAs.

Due to the crucial structural and regulatory roles of the HBc protein, its self-assembly is now widely used as a model to investigate the general principles of virus assembly and as a specific target for highly promising anti-HBV drugs. For more detail on the multifunctional behavior of the HBc protein, see the excellent review articles (Gerlich 2013; Seeger and Mason 2015; Zlotnick et al. 2013, 2015; Selzer and Zlotnick 2015).

Despite highly regulated envelopment mechanisms, HBV particles with "empty" nucleocapsids can accumulate in the sera of HBV-infected patients to high as 90% of secreted particles, and the option for their envelopment remains unknown (Ning et al. 2011; Luckenbaugh et al. 2015).

The HBc protein consists of two clearly separated domains: the N-terminal self-assembly (SA) domain at aa residues 1–140 and the RNA-binding protamine-like arginine-rich C-terminal domain (CTD) at aa 150–183 (Birnbaum and Nassal 1990). The domains are separated by a hinge peptide 141-STLPETTVV-149 that can perform morphogenic functions (Seifer and Standring 1994). The SA domain of the HBc protein possesses a set of variable and conserved stretches that correspond to immunological B-cell epitopes and structural elements, respectively, whereas the CTD and hinge peptide are the most conserved HBc regions and do not demonstrate any immunological importance (for a more detailed review see Pumpens et al. 1995, 2008; Pumpens and Grens 1999, 2001, 2016; Chain and Myers 2005). The CTD is necessary for the reverse

DOI: 10.1201/b22819-46

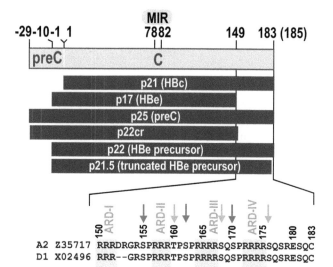

FIGURE 38.1 The structure of the pleiotropic core (C) gene of hepatitis B virus as a prototype of the *Orthohepadnavirus* genus core genes with five possible polypeptide products and functional structure of the CTD for the two subgenotypes A2 and D1 of the popular HBV genotypes A and D, respectively. The major immunodominant region (MIR) is localized. The aa sequence of the CTD is depicted, where the NLS-forming arginine-rich domain (ARD) aa stretches I—IV are displayed on a green background. The classical and novel phosphorylation sites are marked by red and pink arrows, respectively. The numbering of the CTD aa residues corresponds to the genotype D, or subtype *ayw*, sequence, which is 183 aa in length.

transcription of pgRNA and for the formation of rcDNA. The HBc capsids that are self-assembled from the HBc proteins lacking the CTD, so-called the HBcΔ variants, do not encapsidate RNA (Borisova et al. 1988; Gallina et al. 1989; Nassal 1992). As follows from the structure depicted in Figure 38.1, the HBe protein lacks the CTD of the HBc molecule.

EXPRESSION OF THE CORE GENE

For the first time, the core gene was expressed in *E. coli* in the late 1970s (Burrell et al. 1979; Pasek et al. 1979). Almost immediately, the HBc VLPs, also named core-like particles (CLPs), gained broad acceptance as one of the most powerful and prospective models for the novel VLP approach. Since then, hundreds of labs worldwide have expressed the core gene of HBV. Table 38.1 summarizes the general data on expression of the *Hepadnaviridae* core genes, with a special emphasis on the VLP production and determination of their high-resolution structure. Thus, Elmārs Grēns's team presented the first technologically sound protocol of the expression and purification conditions for *E. coli*, with clear identification of the HBc VLP structure by electron microscopy evaluation (Borisova et al. 1984, 1985, 1987, 1988, 1989; Pumpens et al. 1985, 1986; Gren et al. 1987; Gren and Pumpens 1988) on the basis of an original HBV genome of the *ayw* subtype, genotype D, termed HBV320 (Pumpens

et al. 1981; Bichko et al. 1985) and retaining its popularity up to now. Later, this HBV core gene was expressed to high yields and purity in *Pichia pastoris* (Freivalds et al. 2011). It is historically interesting that the expression of the full-length preC protein was described in one of the early papers of the Grēns team (Borisova et al. 1985).

The HBc protein demonstrated the unique capability to self-assemble in nearly all attempted homologous and heterologous expression systems. This is one of the reasons why the HBc VLPs were successful over the past 40 years and still remain among the most used VLP carriers, as argued earlier in detail by Pumpens and Grens (2002) and Pushko et al. (2013). The expression of the HBc protein was covered in numerous review articles (Gren and Pumpen 1988; Milich et al. 1995; Pumpens et al. 1995, 2008; Borisova et al. 1996, 1997; Schödel et al. 1996a, b, c; Ulrich et al. 1998; Murray and Shiau 1999; Pumpens and Grens 1999, 2001, 2002; Whitacre et al. 2009, 2015a; Pushko et al. 2013; Roose et al. 2013; Zeltins 2013). The more recent special reviews written by the pioneers of HBc VLP theory and praxis are also highly recommended (Gerlich 2015; Kolb et al. 2015a; Peyret et al. 2015b; Sällberg 2015; Whitacre et al. 2015a; Pumpens and Grens 2016).

The correct folding of the HBc monomer and the formation of authentic HBc VLPs have been documented in such heterologous platforms as bacteria *E. coli* (Burrell et al. 1979; Pasek et al. 1979; Edman et al. 1981, 1992; Stahl et al. 1982; Borisova et al. 1984, 1985; Pumpens et al. 1985; Uy et al. 1986; Lanford et al. 1987; Nassal 1988; Khudyakov et al. 1991; Zheng J et al. 1992; Berzins et al. 1994; Maassen et al. 1994), *Acetobacter methanolicus* (Schröder et al. 1991), *Bacillus subtilis* (Hardy et al. 1981), *Mycobacterium smegmatis* (Yue et al. 2007), and *Salmonella typhimurium* (Schödel et al. 1990a); yeast *S. cerevisiae* (Kniskern et al. 1986; Miyanohara et al. 1986; Imamura et al. 1988; Shiosaki et al. 1991) and *P. pastoris* (Rolland et al. 2001; Watelet et al. 2002; Freivalds et al. 2011); insect *Spodoptera frugiperda* cells (Lanford et al. 1988, Takehara et al. 1988; Hilditch et al. 1990; Lanford and Notvall 1990; Seifer et al. 1998); and plants *Nicotiana tabacum* (Tsuda Shinya et al. 1998), *Nicotiana benthamiana* (Huang Z et al. 2006; Mechtcheriakova et al. 2006, 2008), and *Vigna unguiculata* (Mechtcheriakova et al. 2006). Moreover, the HBc gene was coexpressed with the HBs gene, in order to form virion-like particles, in yeast (Shiosaki et al. 1991) and insect cells (Takehara et al. 1988). It should be noted that the production of the HBc VLPs in *P. pastoris* was recently used as a model by the elaboration of modern sensors to evaluate biomass concentration in a bioreactor (Grigs et al. 2021).

In parallel, the self-assembly of the HBc protein was demonstrated in mammalian cells of different origin. Thus, for example, the HBc gene was expressed successfully in human HeLa (Hirschman et al. 1980) and 293 (Jean-Jean et al. 1989) cell lines, stable rat and mouse cell lines (Gough and Murray 1982; Gough 1983; Roossinck et al. 1986;), mouse NIH 3T3 fibroblasts (McLachlan et al. 1987), African Green Monkey (*Cercopithecus aethiops*) kidney (AGMK)

TABLE 38.1

Expression of the *Hepadnaviridae* Core Gene and Production of Hepadnaviral Core VLPs by Different Expression Systems

VLPs	Expression Host	References	Comments
Avihepadnavirus			
Duck hepatitis B virus genus (DHBV)	*Escherichia coli*	Schlicht and Schaller 1989	No VLP analysis. The product is not phosphorylated.
	Escherichia coli	Zheng J et al. 1992	Diameter 25–28 nm by electron microscopy. No images are presented.
	human hepatoma HepG2 cell line	Schlicht and Schaller 1989	No VLP analysis. The product is phosphorylated.
	Escherichia coli	Kenney et al. 1995	First electron cryomicroscopy resolution.
	Escherichia coli	Nassal et al. 2007	Fine structural model.
Goose Hepatitis B Virus (SGHBV)	Chicken hepatoma cell line LMH	Greco N et al. 2014	Expression of the gene C of the only known hepadnavirus that packages capsids containing ssDNA.
Heron hepatitis B virus (HHBV)	*Escherichia coli*	Nassal et al. 2007	Fine structural model.
Orthohepadnavirus			
Arctic ground squirrel hepatitis virus (AGSHV)	*Escherichia coli*	Billaud et al. 2005b	
Ground squirrel hepatitis virus (GSHV)	*Escherichia coli*	Billaud et al. 2005b	
Hepatitis B virus (HBV)	*Escherichia coli*	Burrell et al. 1979; Pasek et al. 1979; Edman et al. 1981; Stahl et al. 1982; Borisova et al. 1984	First gene-engineered hepadnaviral cores. No VLP analysis.
		Cohen and Richmond 1982; Uy et al. 1986; Borisova et al. 1987, 1988; Nassal 1988; Khudyakov et al. 1991	First electron micrographs of the gene-engineered hepadnaviral core VLPs in bacteria.
		Borisova et al. 1988; Gallina et al. 1989; Inada et al. 1989; Birnbaum and Nassal 1990; Bundule et al. 1990	First expression of the HBcΔ gene.
		Bundy et al. 2008; Bundy and Swartz 2011	First scalable cell-free production of the hepadnaviral core, namely, HBcΔ.
	Acetobacter methanolicus	Schröder et al. 1991	Electron microscopy of the HBc VLPs.
	Bacillus subtilis	Hardy et al. 1981	No VLP analysis.
	Mycobacterium smegmatis	Yue et al. 2007	Fused HBcΔ(1–155)-preS1(1–55) gene. No VLP analysis.
	Salmonella typhimurium Salmonella dublin	Schödel et al. 1990a	Fused HBcΔ(1–155)-preS2(133–143) gene. No VLP analysis.
	Saccharomyces cerevisiae	Kniskern et al. 1986; Miyanohara et al. 1986	First production and electron microscopy of the gene-engineered hepadnaviral core VLPs in yeast.
		Shiosaki et al. 1991	First production of HBV virion-like particles in yeast.

(Continued)

TABLE 38.1 (CONTINUED)

Expression of the *Hepadnaviridae* Core Gene and Production of Hepadnaviral Core VLPs by Different Expression Systems

VLPs	Expression Host	References	Comments
	Pichia pastoris	Rolland et al. 2001; Watelet et al. 2002; Chen H 2007	First production of the hepadnaviral core VLPs in *P. pastoris.*
		Freivalds et al. 2011	Efficient outcome of the HBc VLPs in yeast.
		Spice et al. 2020	First production in a yeast cell-free system.
	Hansenula polymorpha	Granovski et al. 2017	High-level production and high-yield purification.
	Spodoptera frugiperda	Lanford et al. 1987a; Takehara et al. 1988; Hilditch et al. 1990; Lanford and Notvall 1990	First production of the hepadnaviral core VLPs in insect cells.
		Seifer et al. 1998	Generation of replication-competent HBc in insect cells.
	Nicotiana tabacum	Tsuda et al. 1998	First production of the hepadnaviral core VLPs in plants.
	Nicotiana benthamiana	Huang Z et al. 2006	High immunogenicity in mice by oral or nasal administration.
	Vigna unguiculata, Nicotiana benthamiana	Mechtcheriakova et al. 2006, 2008; Sainsbury and Lomonossoff 2008; Huang Z et al. 2009; Thuenemann et al. 2013	The use of viral vectors to produce HBc VLPs in plants.
	Xenopus oocytes	Standring et al. 1988, Hatton et al. 1992; Zhou S and Standring 1992a, b; Zhou S et al. 1992; Seifer et al. 1993	Synthesis of HBc and HBe proteins. No electron microscopy.
	Xenopus oocytes	Chang et al. 1994	Phenotypic mixing between the *Orthohepadnavirus* cores by assembly. No VLP analysis.
	Rabbit reticulocyte lysates	Weimer et al. 1987	First production of the HBc VLPs in vitro.
		Ludgate et al. 2016; Liu K and Hu 2018	A facile and ingenious system to study self-assembly.
	Wheat germ extract	Lingappa et al. 1994, 2005	Efficient production of HBc VLPs in a heterologous eukaryotic system.
	Mammalian cells	Townsend et al. 1997; Sällberg et al. 1998; Wild et al. 1998	First examples of the DNA immunization.
Roundleaf bat hepatitis B virus (RLBHBV)	*Escherichia coli*	Zhang TY et al. 2020	First expression of bat hepadnaviral core VLPs.
Woodchuck hepatitis virus (WHV)	*Escherichia coli*	Roos et al. 1989; Zheng J et al. 1992; Schödel et al. 1993a; Menne et al. 1997; Billaud et al. 2005a, b	First expression of an orthohepadnaviral core gene, other than as the HBV core.
		Billaud et al. 2005a, b; Zhang Z et al. 2006	Expression of the WHcΔ genes.
Other species			
African cichlid nackednavirus (ACNDV)*	*Escherichia coli*	Lauber et al. 2017	First expression of a nackednaviral core.

Notes: The viruses are given in alphabetical order within each specific hierarchical group.
*Not present in the ICTV list.

cells (Will et al. 1984), *Xenopus* oocytes (Standring et al. 1988), as well as by vaccinia virus-driven production of both surface and core proteins in different cells (Kunke et al. 1993). Furthermore, the HBc VLPs were produced in vitro by rabbit reticulocyte lysates (Weimer et al. 1987). Recently, Boudewijns et al. (2021) proposed yellow fever vaccine YF17D (see Chapter 22) as a live attenuated vector to express the HBc gene, and it constituted therefore a novel therapeutic HBc vaccine. In fact, this vaccine induced strong polyfunctional cytotoxic T cell responses in mice.

The HBcΔ VLPs self-assembled from the HBcΔ protein lacking the CTD, the aa sequence of which is shown in Figure 38.1, were produced in *E. coli* (Borisova et al. 1988; Gallina et al. 1989; Inada et al. 1989; Birnbaum and Nassal 1990; Bundule et al. 1990; Sominskaya et al. 2013). Then, the highly efficient production of the HBcΔ VLPs was achieved by a bacterial cell-free protein synthesis system, in parallel with RNA phage MS2 and Qβ VLPs (Bundy et al. 2008; Bundy and Swartz 2011). Up to now, the novel cultivation and purification methodologies have been steadily elaborated to improve and standardize the HBc VLP production (Hillebrandt et al. 2021; Zhang B et al. 2021; Zhang Yao et al. 2021).

One noticeable fact to be addressed is that the HBe protein ranging from aa -10–149 acquired capacity to efficiently self-assemble into VLPs, when expressed in *E. coli*, after extending to the HBe(-10)151 with 2 arginine residues or to the HBe(-10)154 with an RRRGR stretch (Watts et al. 2020).

The DNA-driven in vivo expression and immunization could be regarded as a special way of efficient production of the HBc VLPs (Townsend et al. 1997; Sällberg et al. 1998; Wild et al. 1998; Shao et al. 2003; Xing et al. 2005).

Recently, a special attention was devoted to the achievement and comparison of the HBc VLPs from different HBV genotypes. Thus, Andris Dišlers's group in Riga produced in *E. coli* and purified and characterized extensively the excellent HBc VLPs of the HBV genotypes A, B, C, D, E, F, and G, where the best VLP yields were found in case of the HBV genotypes D and G.

After human HBc protein, the core genes were expressed in *E. coli* for other members of the *Orthohepadnavirus* genus: woodchuck hepatitis virus (WHV) core—or WHc (Zheng J et al. 1992; Schödel et al. 1993a; Menne et al. 1997; Billaud et al. 2005a, b)—ground squirrel hepatitis virus (GSHV) core—or GSHc (Billaud et al. 2005b)—and arctic squirrel hepatitis virus (ASHV) core—or ASHc (Billaud et al. 2005b). The WHc core was also expressed by the DNA immunization (Lu M et al. 1999; García-Navarro et al. 2001; Siegel et al. 2001; Wang J et al. 2007).

The core gene of duck hepatitis B virus (DHBV), a representative of the *Avihepadnavirus* genus, was expressed in *E. coli* (Schlicht and Schaller 1989; Zheng J et al. 1992) and human hepatoma HepG2 cell line (Schlicht and Schaller 1989). It should be noted that the *Orthohepadnavirus* but not *Avihepadnavirus* cores are practically identical and demonstrate 85–87% similarity at the aa level. Moreover,

the HBV, WHV, and GSHV, but not DHBV core proteins are able to crossoligomerize into mixed particles, when expressed in *Xenopus* oocytes (Chang et al. 1994).

Finally, there should be the most complete list of hepadnaviral core proteins subjected to phylogenetic analysis (Revill et al. 2020). This impressive list includes also *Hepadnaviridae*-like nackednaviruses, the cores of which are described later in the corresponding paragraph.

3D STRUCTURE

Figure 38.2 demonstrates the x-ray crystallography structures of the core VLPs of the two most studied *Orthohepadnavirus* genus representatives, HBV and WHV, which are involved now actively in the viral nanotechnology applications. Both fine 3D structures are in fact remarkably similar in accordance with the high similarity of the aa sequences. Figure 38.3 is intended to show first that the basic structural unit of both the HBc and HBe proteins is a dimer (for review and references, see Zlotnick et al. 2013, 2015).

The fine spatial structure of the icosahedral HBc capsids was first resolved for recombinant *E. coli*-produced HBc particles by cryoelectron microscopy (Crowther et al. 1994; Böttcher et al. 1997; Conway et al. 1997) and later by x-ray crystallography (Wynne et al. 1999). The HBc particles appeared as two isomorphs with triangulation numbers T=4 and T=3, 35 and 32 nm in diameter, consisting of 120 and 90 HBc dimeric units, or 240 and 180 HBc monomers, respectively. The actual percentage of the "small" (T=3) particles in human sera was approximately 5% of the total nucleocapsids (for references see Zlotnick et al. 2015).

Although DHBV, a representative of the *Avihepadnavirus* genus, demonstrated significant difference of the DHBV core (DHBc) from the *Orthohepadnavirus* cores in both length and aa content, electron cryomicroscopy revealed similar dimer-clustered T=3 and T=4 icosahedral organizations (Kenney et al. 1995). Fine structural model of the DHBc, together with heron hepatitis B virus core (HHBc), was generated by Nassal et al. (2007). Recently, Makbul et al. (2020) presented high-resolution structures of several DHBV VLPs determined by electron cryo-microscopy. As for HBV, the DHBV VLPs—or core-like particles (CLPs) by the authors' terminology—consisted of a dimeric α-helical framework with protruding spikes at the dimer interface. A fundamental new feature was a ~ 45 aa proline-rich extension in each monomer replacing the tip of the spikes in the HBc. In vitro, the folding of the extension took months, implying a catalyzed process in vivo. The DHBc variants lacking a folding-proficient extension produced regular "core-like particles" in *E. coli* but failed to form stable nucleocapsids in hepatoma cells. It was proposed that the extension domain acts as a conformational switch with differential response options during viral infection (Makbul et al. 2020).

After the 3D structure of the dominant T=4 particles was resolved by X-ray (Wynne et al. 1999; Bourne et al.

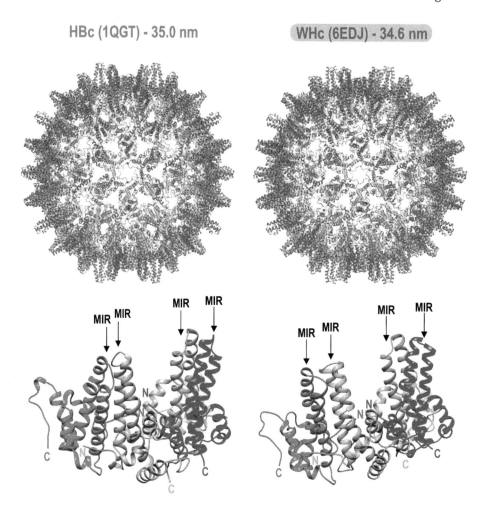

FIGURE 38.2 Crystal structures of the HBc and WHc VLPs. The protein data bank IDs are shown in parentheses, and the outer diameters are indicated for each species. The monomer chains A, B, C, and D are marked in red, green, blue, and orange, respectively. The tetrameric icosahedral asymmetric units (IAU, or ASU) are located under the corresponding full-capsid 3D structures. The MIR loops, which are exposed on the full-capsid surfaces, are indicated within the IAUs by arrows, while the N- and C-termini are labeled by the N and C symbols. The structural data are compiled from the VIPERdb (http://viperdb.scripps.edu) database (Carrillo-Tripp et al. 2009) and are visualized using Chimera software (Pettersen et al. 2004).

2006), the quasi-atomic pattern of the T=3 isomorph was reconstructed by docking the dimers of the T=4 crystal structure (Roseman et al. 2012), as demonstrated in Figure 38.3. Recently, Wu W et al. (2020) compared the two capsids by electron cryomicroscopy at 3.5 Å resolution and found that the chains have different conformations and potential energies, with the T=3 C chain having the lowest. Remarkably, the dimorphism of HBc VLPs was regulatable by controlling the rate of capsid nucleation using cations such as K$^+$ or Ca^{2+}: a quick addition of highly concentrated monovalent and/or multivalent counter-cations resulted in a morphism transition from a thermodynamically more stable, T=4 capsid-dominant state (>80% of total capsids) to a new state containing ~1: 1 amounts of T=3 and T=4 capsids (Sun Xinyu et al. 2018).

Therefore, the HBc monomer is formed by five alpha helices, where the helix 3-helix 4 hairpin dimerizes with a second monomer to form a central helical bundle. The 3D structure reveals the organization of the SA domain but not of the CTD. The structure of the CTD was also not resolved by recent high 3.5 Å resolution electron cryomicroscopy (Yu et al. 2013). It was concluded therefore that the CTD functions as a spatially flexible region of the rigid HBc nucleocapsid structure (Yu et al. 2013). Nevertheless, Böttcher and Nassal (2018) recently succeeded to resolve at 2.6- to 2.8 Å some residues downstream of Pro144 that preceded the phosphorylation sites. As stated, these residues packed against the neighboring subunits and increased the interdimer contact suggesting that the CTD would playing an important role in capsid stabilization and provide a much larger interaction interface than previously observed.

The crystal structure of the HBe protein revealed a radical reorientation of the dimeric interface, preventing the self-assembly of dimers (DiMattia et al. 2013). Whereas the HBe monomer retained the same helical structure, the changes in the dimer interface were critical for the alternate structure and activity of HBe, as demonstrated in Figure 38.3. The self-assembly competent HBc protein

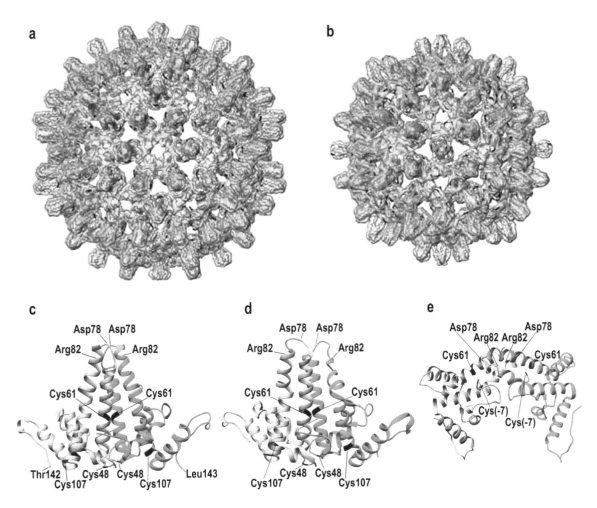

FIGURE 38.3 The 3D structures of the core-related proteins. (a) T = 4 particle, subtype adw (PDB ID: 1QCT) (Wynne et al. 1999); (b) T = 3 shell: electron microscope map with dimers from the crystal structure fitted, subtype adw (Roseman et al. 2012); (c) the HBc dimer from the 1QCT structure; (d) the "free" HBc dimer from a self-assembly incompetent HBc variant bearing a Y132A mutation (PDB ID: 3KXS; Packianathan et al. 2010); (e) the HBeAg dimer (PDB ID: 3V6Z; DiMattia et al. 2013). Two HBc or HBe monomer chains are shown by different colors: tan (chain A) and orange (chain B). The cysteine residues are marked black. The central MIR stretch 78-DPASR-82 is marked green, and its border Asp78 and Arg82 residues are indicated on the map. (Reprinted with permission of SpringerNature from Pumpens P. and Grens E. *Mol. Biol.* 2016;50:489–509.)

can form a Cys61-Cys61 disulfide bridge connecting two monomers, although this connection was not obligatory for self-assembly. The HBe structure was stabilized by a novel disulfide bridge between Cys(-7) of the preC precursor and Cys61 and was therefore blocked from self-assembly. As a result, the major HBc epitope at the tip of the dimer spike was completely disrupted (for a detailed explanation, see the original self-review of Zlotnick et al. 2013). However, when the Cys(-7)-Cys61 disulfide bond was reduced, the resulting protein readily assembled into morphologically normal capsids (DiMattia et al. 2013).

Both p22 and p21.5 proteins were able to form homodimers that interacted with HBc dimers and formed hybrid self-assembly structures. However, the p21.5 protein, in contrast to p22, was unable to self-assemble into capsids and interfered with the self-assembly of the natural HBc protein during nucleocapsid formation (Wang Yi-Ling et al. 2015).

Remarkably, the native HBc nucleocapsids were purified and resolved by electron cryomicroscopy at 14–16 Å more than ten years later than their gene-engineered counterparts (Dryden et al. 2006). Furthermore, the 14.5 Å resolution electron cryomicroscopy of the natural HBV nucleocapsids allowed detection of the viral polymerase, a distinct, donut-like dense structure that pinned the pgRNA to the capsid inner surface (Wang JC et al. 2014). A novel crystal structure with improved resolution revealed the complex of an inhibitor of HBV replication with the HBc protein, a favorite target for current anti-hepatitis-B drug discovery (Klumpp et al. 2015a).

In contrast to the HBc protein, only half of the WHc protein assembled into T=4 capsids composed of 120 dimers. The other half of the WHc protein exhibited a broad distribution of larger species that extended to more than 210 dimers with "spiral-like" morphologies (Pierson et al. 2016).

It is important for the nanotechnological applications that the SA domain is necessary and sufficient for the formation of dimers with further self-assembly into icosahedral HBc capsids, whilst the CTD is totally dispensable for self-assembly (Borisova et al. 1988; Gallina et al. 1989; Nassal 1992). Thus, the HBc protein tolerated C-terminal truncation up to aa 140 while maintaining self-assembly competence by forming capsids that were morphologically indistinguishable from natural nucleocapsids (for references and discussion see Sominskaya et al. 2013; Zlotnick et al. 2015). Direct electron cryomicroscopy analysis confirmed the principal identity of the HBcΔ VLPs formed by monomers of different C-terminal length (Liu S et al. 2010).

The in vitro self-assembly of HBc starts with a slow nucleation step, namely formation of a trimer of dimers, followed by a rapid elongation phase, and remained allosterically regulated (for qualified analysis and more references see Zlotnick et al. 2015).

The HBc protein undergoes a conformational change to become assembly active and to participate in self-assembly (for arguments and references see Zlotnick and Fane 2011). To count and size the self-assembly intermediates and readymade capsids in real time, the resistive-pulse sensing technique was used with single particle sensitivity and at biologically relevant concentrations (Harms et al. 2015). The aa positions that are critical for self-assembly are depicted in Figure 38.4.

The minor 3D-structure fluctuations of the nucleocapsids are functionally important: the finding that the mature DNA-containing HBc particles differ by their outer surface from immature RNA-containing particles (Roseman et al. 2005) needs further development.

The current high resolution structural studies are progressing in the four methodological directions. First, the novel HBc VLPs are resolved by electron cryomicroscopy. Thus, the 3D HBc structure of an unusual HBV, genotype G, was resolved at 14 Å (Cotelesage et al. 2011). This HBc possessed a unique 36-bp insertion downstream of the core gene start and therefore a 12 aa insertion at the N-terminal end of the HBc, and two stop codons in the preC region that prevented the expression of HBe (Stuyver et al. 2000; Kato et al. 2002). Then, the complexes of HBc VLPs with specific antibodies were resolved by electron cryomicroscopy (Kandiah et al. 2012; Roseman et al. 2012; Wu W et al. 2013).

Second, the hydrogen-deuterium exchange coupled to mass spectrometry (HDX-MS) (Bereszczak et al. 2013) and the native mass spectrometry (MS) and gas phase electrophoretic mobility molecular analysis (GEMMA) (Bereszczak et al. 2014) were applied to resolve HBc-antibody complexes. These alternative methods were less time consuming than electron cryomicroscopy and still sufficient for the preliminary characterization of virus-antibody complexes.

Third, the HBc VLPs were studied by x-ray crystallography. So, the crystal structure of some HBc VLP candidates carrying foreign insertions on their surface was resolved (Tan WS et al. 2007; Kikuchi et al. 2013a, b). Of special structural importance are the crystal structures of HBc VLPs complexed with different drugs that interfere with self-assembly (Bourne et al. 2006; Katen et al. 2013; Klumpp et al. 2015a).

Fourth, the solid-state NMR spectroscopy at 100 and 150 kHz magic-angle spinning (MAS) was applied to the high-resolution 3D elucidation of the HBc VLPs and their

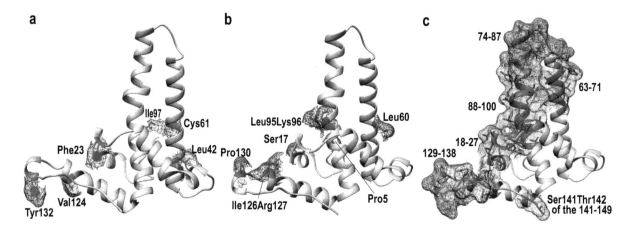

FIGURE 38.4 Schematic presentation of the functionally important amino acid residues and sequence stretches mapped onto the HBc monomer, chain A (PDB ID: 1QCT; Wynne et al. 1999). (a) Amino acid residues affecting self-assembly (red); (b) amino acid residues with mutations that may support the intracellular capsid formation but that inhibit the envelopment process and further secretion of virions (red); (c) immunological B-cell (green) and CTL (red) epitopes. The data is compiled from Zlotnick et al. (2015) for the structural features and from other reviews (Pumpens et al. 1995, 2008; Pumpens and Grens 1999, 2001) for the immunological epitopes and visualized by a Chimera software (Pettersen et al. 2004). The selected functionally important elements are provided with atoms/bonds and transparent surfaces, while the basic structures are presented by ribbons. (Reprinted with permission of SpringerNature from Pumpens P. and Grens E. *Mol. Biol.* 2016;50:489–509.)

assembly modulation (Lecoq et al. 2018a, b, 2019; Wang S et al. 2019; Schleedorn et al. 2020). Recently, using solid-state NMR, Lecoq et al. (2021) identified a stable alternative conformation of the HBc protein, where the structural variations focused on the hydrophobic pocket of the HBc, a hot spot in capsid-envelope interactions. This structural switch was triggered by specific, high-affinity binding of a pocket factor, and the conformational change induced by the binding was reminiscent of a maturation signal, leading formulation of the "synergistic double interaction" hypothesis, which explained the regulation of capsid envelopment and indicated a concept for therapeutic interference with HBV envelopment (Lecoq et al. 2021).

Until quite recently, the HBc particles remained the subject of the scrutinous structural studies of the fine assembly/disassembly processes (Chevreuil et al. 2020; Patterson A et al. 2020; Zhao Z et al. 2021) and in silico modeling (Klijn et al. 2019; Vormittag et al. 2020a, b; Pérez-Segura et al. 2021). The structure of the manifold and beautiful HBc particles was recently reviewed by Bettina Böttcher (2021).

STRUCTURAL FATE OF THE CTD

The early HBc studies pointed to the possible alternate localization of the CTD both inside (Machida et al. 1989) and outside (Bundule et al. 1990; Vanlandschoot et al. 2005) the nucleocapsid. The cryoelectron microscopy data supported the interior localization of the CTD (Zlotnick et al. 1997; Wang JC et al. 2012). It is only logical that the CTD must be regarded as a flexible structure, which justifies the inability to determine the distinct tertiary structure of the CTD by high-resolution structural methods and explains the current lack of a precise CTD atomic model.

The CTD is highly conserved and contains nucleic acid-binding sites (Hatton et al. 1992) that are organized in four arginine-rich blocks and correspond to the nuclear localization signal (NLS; Yeh and Ou 1991; Liao and Ou 1995; Haryanto et al. 2012), as shown in Figure 38.1. The NLS permanently appears on the capsid exterior (Kann et al. 1999; Chen C et al. 2011; Ludgate et al. 2011). According to theoretical explanations, a significant portion of the CTD appears on the exterior of the RNA-containing immature nucleocapsid, whilst the CTD is buried within the DNA-containing mature nucleocapsid (Meng et al. 2011).

The activity of the CTD in native HBV nucleocapsids is regulated by phosphorylation (for the analysis of hypotheses describing the role of phosphorylation in viral replication and references, see the exhaustive review of Zlotnick et al. 2015). The phosphorylation affects not only the transient exposure of the CTD but also the stability of empty $T = 4$ capsids formed by full-length HBc protein and is therefore responsible for the crosstalk between CTDs within a capsid (Selzer et al. 2015). The serine residues 155, 162, and 170 (numerated traditionally as for the D genotype, *ayw* subtype) within the CTD are common phosphoacceptor sites, as shown in Figure 38.1. The additional phosphoacceptor sites Thr162, Ser170, and Ser178 in subtype *adw* (corresponding

to the Thr160, Ser168, and Ser176 positions in ayw subtype) were identified later (Jung et al. 2014).

The phosphorylated CTD is an active CTD form for the pgRNA encapsidation, reverse transcription, and viral transport and may also function as a nucleic acid chaperone (Lewellyn and Loeb 2011a, b). Recently, this idea was confirmed experimentally (Chu TH et al. 2014).

It is noteworthy that the native self-assembly of HBc capsids is facilitated by heat shock protein 90 (Hsp90) that binds HBc dimers, catalyzes their self-assembly, and reduces the degree of capsid dissociation at various temperatures (Shim et al. 2011). The Hsp90-driven capsid formation is facilitated by reactive oxygen species (ROS; Kim YS et al. 2015). Moreover, the sequestering of the CTD within the capsid interior is performed by serine arginine protein kinase (SRPK) that binds to the CTD and acts as a self-assembly chaperone (Chen C et al. 2011). Other kinases that are directly connected with the HBc self-assembly include protein kinase C (PKC) (Kann et al. 1993) and a 46 kDa core-associated kinase (CAK) that is distinct from PKC and stems from the ribosome-associated protein fraction of the cytoplasm (Kau and Ting 1998). The kinases CAK (Kau and Ting 1998) and cyclin-dependent kinase 2 (Ludgate et al. 2012) are packaged within the nucleocapsids.

The NLS phosphorylation appeared to be necessary for the CTD interaction with the nuclear pore (Haryanto et al. 2012). Further, the interaction of HBc dimers with nucleophosmin (B23), a cell nucleolar protein, was identified (Lee SJ et al. 2009) and proven to be the self-assembly promoting activity (Jeong et al. 2014). The role of the CTD phosphorylation as a fine regulator of the interaction between HBc and nucleic acid was recently reviewed by Philippe Roingeard's team (de Rocquigny et al. 2020).

The most important targets of the HBc protein during HBV replication were the nuclei of infected cells, as the HBc protein shuttled between the nucleus and cytoplasm. The importin alpha and beta complexes, or importin alpha alone, were regarded therefore as potential transporters of capsids to the nuclear pores (Kann et al. 2007), where capsids interacted with nucleoporin 153 and dissociated during nuclear entry (Rabe et al. 2009; Schmitz et al. 2010).

The native HBc protein interacted with virus-suppressing proteins such as Np95/ICBP90-like RING finger protein (NIRF), a novel E3 ubiquitin ligase (Qian et al. 2012, 2015), and cytidine deaminase APOBEC3A (Zhao D et al. 2010; Lucifora et al. 2014), and packed them into the HBc nucleocapsids. Some other proteins, such as the MxA protein, an interferon-inducible cytoplasmic dynamin-like GTPase that possesses anti-HBV activity, might interfere with the HBc self-assembly (Li N et al. 2012). Finally, the proper self-assembly and release of the HBc capsids depended on functional Rab33B, a GTPase participating in autophagosome formation via interaction with the Atg5-Atg12/Atg16L1 complex (Döring and Prange 2015).

At last, a recent thorough study demonstrated that the HBcΔ(1–140) lacking the CTD cannot self-assemble in human hepatoma Huh7 cell line (Rat et al. 2020).

ANTIGENIC STRUCTURE AND IMMUNOLOGICAL FATE

The general antigenic structure and location of the major immunological epitopes on the HBc monomer are presented in Figure 38.4. Unlike the structural topics, no ground-breaking findings have resulted from the antigenic studies of core-derived proteins since the more than 20-year-old review by Pumpens and Grens (1999). Nevertheless, an exhaustive study on the molecular basis for the high degree of antigenic cross-reactivity between HBc and HBe proteins must be mentioned (Watts et al. 2010). This study used a classical approach for the discrimination of both activities (Milich et al. 1988).

The B-cell epitopes c (HBc epitope) and e1 (HBe epitope 1) were localized on the major immunodominant region (MIR) of the HBc protein, around the protruding region 78–82 on the tip of the spike (Salfeld et al. 1989; Sällberg et al. 1991a). The epitope e2 (HBe epitope 2) was localized on the other surface-exposed region of the HBc protein, around positions 129–132 (Sällberg et al. 1991b, 1993). These epitopes remain dominant B-cell epitopes of both major core-derived proteins (for more details, see reviews Pumpens and Grens 1999, 2001; Pumpens et al. 2008). The scanning of these epitopes by monoclonal antibodies helped to localize the exposed and buried regions of the HBc capsids (Pushko et al. 1994).

The important CTL epitope 18–27 (Bertoletti et al. 1993) was studied as a potential vaccine component, and many others also remain on the current list. More recent studies on the CTL epitopes 64–72 (Liu Q et al. 2012) and 141–149 (Sun L et al. 2014) deserve special consideration. Then, an exhaustive study of 20 potential CTL epitopes from chronic hepatitis B patients was performed (Brinck-Jensen et al. 2015). The HBc-related Th-cell epitopes have been found along the whole HBc molecule, apart from the CTD (for a more detailed list of epitopes, see Pumpens and Grens 1999, 2001; Pumpens et al. 2008).

More than a dozen reviews would be necessary to present the immunogenicity of the core-derived proteins, a basic, highly important, but complicated topic in the HBV pathogenesis. We are limiting ourselves here to basic citations of the most fundamental paper on both the T-cell-independent and T-cell-dependent nature of the HBc capsids (Milich and McLachlan 1986), an exhaustive presentation of the interaction of the HBc with the innate immune system (Lee BO et al. 2009), and a study that extended the T-cell independent behavior of the HBc VLPs onto foreign epitopes exposed on their surface (Fehr et al. 1998).

NATURAL VLPs

The natural nonchimeric HBc and WHc VLPs produced in *E. coli* were tested as possible vaccine candidates against infection with the corresponding viruses in humans and woodchucks, respectively. Moreover, the HBc VLPs were examined as a vaccine against WHV infection in woodchucks. The story of the natural hepadnaviral VLPs as vaccines is narrated later in the *Vaccines: Natural VLPs* paragraph.

CHIMERIC VLPs

GENETIC FUSIONS

HBc

The HBc protein was first presented in the role of a VLP carrier for the exposition of foreign immunological epitopes and other sequences at a vaccine conference in 1986 (see Newton et al. 1987), published in 1987 (Borisova et al. 1987; Clarke et al. 1987), and at last nominated in 1988 by Elmārs Grēns's team as a universal VLP platform (Gren and Pumpens 1988). Since then, the HBc VLPs have been used in thousands of nanobiotechnological efforts to expose immunological epitopes and/or cell-targeting signals and to package poly- and oligonucleotides as genes and immune stimulatory sequences (for global reviews in comparison with other VLP carriers see Pushko et al. 2013, Zeltins 2013). Figure 38.5 presents typical examples of the natural and chimeric HBc-derived VLPs.

To identify the HBc regions indifferent to foreign insertions and still ensuring the correct self-assembly of the modified HBc monomers into chimeric VLPs, the HBc molecule was subjected to scanning by a short epitope as an immunological marker. As in the case of the RNA phage VLPs (see Chapter 25), the short HBV preS1 epitope 31-DPAFRA-36 (or DPAFR or DPAF) was used. This epitope was necessary and sufficient (Sominskaya et al. 1992) to be recognized by the monoclonal anti-preS1 antibody MA18/7 (Heermann et al. 1984). As a result, three suitable regions were found, namely, N-terminus, MIR, and C-terminus (Berzins et al. 1994; Borisova et al. 1999; Lachmann et al. 1999).

Concerning the composition of the HBc vectors, a set of methodologies was applied after the full-length HBc gene as an original platform. First, the previously described C-terminally truncated HBc variants, the HBcΔ vectors, mostly HBc1–144 or HBc1–149 by their composition, appeared necessary and sufficient to serve as icosahedral scaffolds for the correct self-assembly of the chimeric VLPs (see the most recent exhaustive study of Sominskaya et al. (2013) for arguments and discussion).

Second, the mosaic particle technology was applied to the HBc/HBcΔ VLPs. The generation of mosaic particles was developed by the introduction of a linker containing translational stop codons UGA or UAG between sequences encoding an HBcΔ vector and a foreign protein sequence and production of desired VLPs in the appropriate suppressive *E. coli* strains (Koletzki et al. 1997, 1999; Ulrich et al. 1997, 1999; Kazaks et al. 2002). The mosaic strategy was applied also by the incorporation of modified HBc molecules carrying internal deletions (Kazaks et al. 2003) or long internal insertions (Loktev et al. 1996; Kazaks et al. 2004) within the MIR. In the fortunate cases, chimeric monomers formed hetero- and/or homodimers with natural

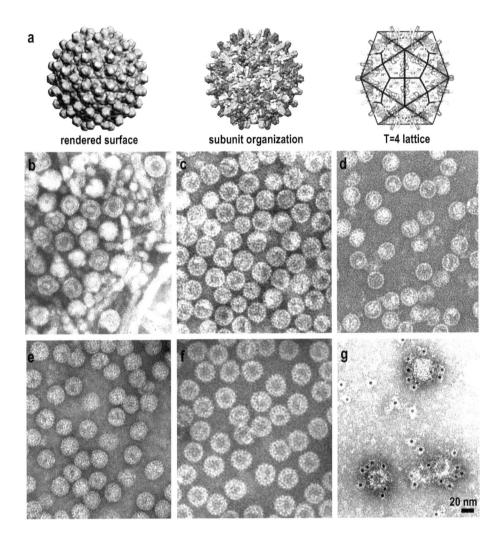

FIGURE 38.5 Structural characteristics of the VLPs on the basis of the HBc particles. (a) X-ray structure of the HBc particle according to the VIPER presentation directed by Vijay Reddy. (b) HBV preparation from the blood of an infected patient, envelopes and internal cores of HBV virions (so-called Dane particles) are visible, as well as filamentous forms of the 22 nm HBs particles, (c–g), electron microscopy of the *E. coli*-expressed HBc VLPs: (c) full-length HBc particles without any insertions; (d) C-terminally truncated HBc particles without any insertions; (e) C-terminally truncated HBc particles with long C-terminal insertion; (f) full-length HBc particles with long internal (into the MIR) insertion; (g) immunogold mapping of the C-terminally truncated HBc particles with internal insertion (into the MIR) with specific monoclonal antibodies against inserted epitope. (Reprinted with permission of Taylor & Francis Group from Pumpens P, Ulrich R, Sasnauskas K, Kazaks A, Ose V, Grens E. Construction of novel vaccines on the basis of the virus-like particles: Hepatitis B virus proteins as vaccine carriers. In: *Medicinal Protein Engineering*. Khudyakov Y (Ed). CRC Press/Taylor & Francis Group, Boca Raton, 2008, pp. 205–248.)

HBc monomers by simultaneous expression in *E. coli* and assembled into perfect VLPs.

Third, the highly ingenious SplitCore technology was elaborated by the Michael Nassal's team (Skamel et al. 2006; Nassal et al. 2008; Walker et al. 2008, 2011; Kolb et al. 2015a; Heger-Stevic et al. 2018). The SplitCore was based on the ability of two parts of the HBc molecule, coreN and coreC, to efficiently form correct HBc VLPs during expression in *E. coli*. The SplitCore tolerated numerous long fusions (more than 300 aa) to coreN and/or coreC and paved the way to the generation of complex triple-layered VLPs (Walker et al. 2011; Lange et al. 2014; Kolb et al. 2015a, b). The highly attractive SplitCore idea inspired other

authors to present arrays of the receptor-contacting epitopes of human IgE on the surface of HBc VLPs (Baltabekova et al. 2015). The SplitCore methodology was used further to elaborate an intein-mediated trans-splicing technique, where split inteinC (intC) was added to the C-terminus of the split HBc N-core (Wang Z et al. 2021). While the split HBc with the insertion of intC at the C-terminus of N-core (designated as HBc N-intC-C) existed in inclusion bodies, the introduction of a soluble tag, gb1, to the intC C-terminus remarkably improved the solubility of recombinant protein (named HBc N-intC-gb1-C). The newly designed recombinant spontaneously assembled into the VLPs and were endowed efficiently, coupling two different model antigens onto HBc N-intC-gb1-C VLPs. The model antigens delivered by the intein-driven HBc VLP scaffold induced a dramatically enhanced antigen-specific immune response (Wang Z et al. 2021).

Fourth, the HBcΔ was used as a carrier of substrate-binding domain (SBD) of the bacterial molecular chaperone DnaK, structure of which is similar to GFP (Wang XJ et al. 2009, 2010). Addition of the peptide motif NRLLLTG, which is recognized by the DnaK, allowed decoration of the HBcΔ-SBD VLPs with peptide vaccines and paved the way for generation of highly efficient peptide-based vaccines. In particular, Wang XJ et al. (2009, 2010) succeeded in a vaccine to treat gonadotropin-releasing hormone-dependent diseases.

Fifth, a novel and highly prospective class of HBc vectors was based on a genetically fused HBc dimer, the so-called "tandem core" elaborated by David J. Rowlands's team (Shepherd et al. 2013; Holmes et al. 2015; Peyret et al. 2015a, b; Stephen et al. 2018). The tandem technology was useful in both plant and bacterial expression systems and can be applied to the display of natively folded proteins on the surface of HBc particles, either through direct fusion or noncovalent attachment via a nanobody (Peyret et al. 2015a). Next, Kaspars Tārs and Andris Kazāks's team succeeded by the self-assembly of HBc dimers by generation of a prospective influenza vaccine (Kazaks et al. 2017).

Sixth, a sortase-mediated site-specific tagging of antigens onto HBc VLPs was proposed (Tang S et al. 2016). In this strategy, the HBc molecule was split into two parts, aa 1–79 N-core and aa 80–183 C-core, by analogy with the SplitCore technique. The conserved LPETGG motif was fused to the C-terminus of the N-core to be exposed on the HBc VLPs surface and recognized by the transpeptidase, sortase A. The latter breaks the bond between threonine and glycine residues and links the cysteine 184 of sortase A to LPET. An acylated intermediate is formed at the "docking" step. Sortase A is then released from the VLP scaffold by peptides or proteins that harbor the N-terminal oligoglycines nucleophilic attack (Tian and Eriksson 2011). Hence, oligoglycine-based antigens are tagged on the N-LPETGG-C VLP surface at the "coupling" step. After docking and coupling, N-terminal oligoglycine-based antigens are conjugated to the surface of the N-LPETGG-C VLPs. Using this approach, Tang S et al. (2016) linked a protein from HCMV and two epitopes SP70 from VP2 of enterovirus 71, 15 aa with additional three glycines at the N-terminus, to the HBc VLPs.

Seventh, insertion of the strep tag into the MIR allowed the coupling of foreign antigens, produced as streptavidin fusion proteins, on the surface of the capsids, and that allowed the use of the capsid as a universal antigen carrier (Akhras et al. 2017). Recently, the strep-tag added at the C-terminus of the HBcΔ vectors was employed for the purification needs (Aston-Deaville et al. 2020) but could be used also for the addition of desired peptides. Then, Zahn et al. (2020) fused the preS1/preS2 domain to monomeric streptavidin and produced the cargo-loaded HBc VLPs. The in

vivo imaging of immunized mice demonstrated membrane permeability of cargo-loaded carriers and spread of antigens over the whole organism, which led to the destruction of the HBV-positive cells and induction of HBV-specific neutralizing antibodies. The membrane permeability of these carriers allowed for needle-free application of the antigen-loaded VLPs via oral or transdermal vaccination (Zahn et al. 2020).

At last, the plug-and-display decoration of VLPs, which is based on the ability of a peptide termed SpyTag and a protein termed SpyCatcher to form spontaneous covalent isopeptide bonds, was applied to the HBc-derived VLPs (Ji M et al. 2020a, b). The principle of this revolutionary approach is described in the *Chimeric VLPs: Functionalization: AP205* paragraph of Chapter 25 devoted to RNA phages. After the RNA phages, the plug-and-display technique was applied to the HBs VLPs, as described in the *Chimeric Vaccines: Plug-and-Display* paragraph of Chapter 37.

In the case of the HBc VLPs, Ji M et al. (2020a) constructed an appropriate HBc VLP platform by inserting SpyCatcher into the MIR of HBcΔ, while the SpyTag conjugated epitopes, such as linear, cyclic, and phosphorylated peptides including Aβ(1–6), Aβ(1–15), cAβ(1–7), cEP1, cEP2 from β-amyloid monomer or oligomers, T294, pTau396–404, and pTau422 from tau proteins, were glued onto the HBc-SpyCatcher VLPs. The chimeric VLPs efficiently elicited specific Th2-type immune responses and demonstrated clear therapeutic effect in transgenic mice. This approach paves the way not only for efficient Alzheimer's disease treatment but, more generally, for the development of personalized vaccines. In parallel, Ji M et al. (2020b) used the HBc-SpyCatcher VLPs to bind the SpyTag conjugated with OVA epitope peptides. The vaccination of mice with these VLPs inhibited tumor growth in both prophylactic and treatment ways in the OVA tumor-bearing animals (Ji M et al. 2020b).

Peyret et al. (2020) investigated the conjugation of the tandem HBc (tHBc) VLPs and the model antigen GFP in vivo in *N. benthamiana* and found that it was optimal to display SpyCatcher on the tHBcAg VLPs and SpyTag on the binding partner. To test transferability of the GFP results to other antigens, the authors successfully conjugated the tHBc VLPs to the HIV capsid protein P24 and demonstrated an efficient strategy that can lead to time and cost saving post-translational, covalent conjugation of recombinant proteins in plants (Peyret et al. 2020).

Insertion Sites

As mentioned earlier, the N-terminus, MIR, and C-terminus of HBc molecules were mapped originally as appropriate sites for the introduction of foreign polypeptides (Berzins et al. 1994; Borisova et al. 1999; Lachmann et al. 1999). For ease of use, Table 38.2 classifies the most complete list of the generated chimeric HBc vaccine candidates so far according to the (i) insertion site—namely N-terminus, MIR, or C-terminus—and (ii) status of the HBc vector, the full-length HBc, or the C-terminally truncated HBcΔ.

TABLE 38.2

The Putative Vaccines Constructed on the *Hepadnaviridae* Cores by Genetic Fusion Methodology

Vaccine Target	Source of Epitope	Position of Insertion or Addition	Expression System and Major Immunological Activity	References
HBc N-terminal insertions (full-length HBc vectors)				
Epstein-Barr virus (EBV)	EBNA1, 407-HPVGEADYFEY-417	↓1	*E. coli*	Zhang B et al. 2021
Feline leukemia virus (FeLV)	FeLV-A/Glasgow-1, gp70, aa 137–153	1↓2	*E. coli*; B cells (guinea pigs)	Clarke et al. 1990
Foot-and-mouth disease virus (FMDV)	VP1, serotype O₁. aa 142–160	↓—6	Vaccinia virus-infected HeLa cells; *S. cerevisiae;* B cells (guinea pigs)	Clarke et al. 1987; Newton et al. 1987; Beesley et al. 1990
Hepatitis B virus (HBV)	preS1, aa 12–47	↓—4	*E. coli*; B cells (mice)	Schödel et al. 1992, 1993b
	preS1, aa 27–53	↓—4	*E. coli*; B cells (mice)	Schödel et al. 1993b
	preS1, aa 27–53	1↓2	*E. coli*; B cells (mice)	Schödel et al. 1992, 1993b
Hepatitis C virus (HCV)	Core, aa 10–53	↓1	*E. coli*	Zhang B et al. 2021
Human immunodeficiency virus 1 (HIV-1)	gp120, aa 303–327 p24, aa 288–304	↓—6	*E. coli*	Moriarty et al. 1990
Human rhinovirus 2 (HRV2)	VP2, aa 156–170	1↓2	*E. coli*; B cells (guinea pigs)	Clarke et al. 1990; Francis et al. 1990; Brown et al. 1991
Influenza A	M2, aa 2–24	1↓5	*E. coli*; B cells (mice); protection (mice)	Neirynck et al. 1999; Fiers et al. 2004, 2009; De Filette et al. 2005, 2006a
	M2, aa 1–23	↓1	*E. coli*; B cells (mice)	Jegerlehner et al. 2002b
	M2e, aa 1–23, provided with the Cwh signal peptide[1]	↓1	*L. casei*; secretion of the HBc-M2	Themsakul et al. 2016
Influenza A, avian	M2e, A/Duck/ Potsdam/1402–6/1986 (H5N2)	↓1	*N. benthamiana* plants; B cells (mice); protection (mice)	Ravin et al. 2012
Nipah virus (NiV)	Nucleocapsid NP, aa 401–532	↓1	*E. coli*	Yap et al. 2012
Poliovirus 1	VP1, aa 93–103	1↓2	*E. coli*; B cells (guinea pigs)	Clarke et al. 1990; Brown et al. 1991
Simian immunodeficiency virus (SIV)	env, aa 170–189 aa 655–675	↓—6	*E. coli*; B cells (guinea pigs)	Yon et al. 1992
Bordetella pertussis	Pertactin (P. 69), aa 537–566	↓—6	*S. cerevisiae*; B cells (mice)	Charles et al. 1991
Chlamydia trachomatis	Serovar E, major outer membrane protein, multiepitope (MOMPm), aa 370–387		*E. coli*; protection (mice)	Jiang et al. 2015
Human chorionic gonadotropin	hCG, subunit β, aa 109–145	↓—6	*S. cerevisiae*; B cells (guinea pigs)	Beesley et al. 1990
HBc N-terminal insertions (C-terminally truncated HBcΔ vectors)				
Hantavirus	Protein N, aa 1–45	↓1	*E. coli;* B cells (bank voles)	Ulrich et al. 1999; Koletzki et al. 2000; Krüger et al. 2001
Hepatitis B virus (HBV)	preS1, aa 31–35	↓2	*E. coli*	Lachmann et al. 1999
	preS1, aa 31–36 aa 94–105 (94–105)₃ preS2, aa 133–143	↓4	*E. coli*; B cells (rabbits, mice)	Kalinina et al. 1995; Makeeva et al. 1995

(Continued)

TABLE 38.2 (CONTINUED)

The Putative Vaccines Constructed on the *Hepadnaviridae* Cores by Genetic Fusion Methodology

Vaccine Target	Source of Epitope	Position of Insertion or Addition	Expression System and Major Immunological Activity	References
Human immunodeficiency virus 1 (HIV-1)	gp41, aa 593–604 p34 pol, aa 940–949 p17, aa 99–115	↓4	*E. coli*; B cells (rabbits)	Isaguliants et al. 1996a, b
Influenza A	M2, aa 2–24 (2–24)$_2$ (2–24)$_3$	1↓5	*E. coli*; protection (mice); Phase I trial	De Filette et al. 2005, 2006a, b, 2008; Fiers et al. 2009; Ibañez et al. 2013
	M2e, aa 1–24[1] (1–24)$_3$[1]	↓1	*E. coli*; B cells (mice)	Sun Xincheng et al. 2015
HBc internal insertions (full-length HBc vectors)				
Coxsackievirus A10	VP2, neutralizing epitope P28, aa 136–150	76↓84	*E. coli*; virus-neutralizing antibodies and passive immunization (mice)	Dai et al. 2019
Enterovirus 71	VP1, neutralizing epitopes SP55, aa 163–177 SP70, aa 208–222	76↓84	*E. coli*; virus-neutralizing antibodies and passive immunization (mice)	Ye et al. 2014
	VP1, SP55, aa 163–177; SP70, aa 208–222 (Sortase technique)	79↓80	*E. coli*; protection (mice)	Tang S et al. 2016
Foot-and-mouth disease virus (FMDV)	VP1, aa 135–160	81↓82	*E. coli*; B cells (guinea pigs)	Chambers et al. 1996
	VP4, 1-GAGQSSPATGSQNQS-15	MIR	*E. coli*; B cells (mice)	Swanson et al. 2021
Hepatitis B virus (HBV)	preS1, aa 12–60 + 89–119	78↓79	*E. coli*; B and T cells (mice)	Skrastina et al. 2008
	preS1, aa 1–42[1]	79↓80	*E. coli*; B and T cells (mice)	Malik et al. 2012
	preS2, aa 133–143	82↓83	*E. coli*	Schödel et al. 1990b
	HBs, aa 137–147 aa 27–37	81↓82	*S. typhimurium*; *E. coli*; B and T cells (mice)	Karpenko and Il'ichev 1998; Karpenko et al. 2000a, b; Veremeiko et al. 2007
Hepatitis C virus (HCV)	T epitope, aa 35–44	79↓80	*E. coli*; T cells (mice)	Chen Jia-Yu and Li 2006
Hepatitis E virus (HEV)	ORF2, aa 551–607	78↓83	*N. benthamiana;* recognized by positive swine serum	Zahmanova et al. 2021
Human cytomegalovirus (HCMV)	glycoprotein B (gB), AD-4 domain, aa 112–132 and 344–438	78↓79	*E. coli*; B cells (mice)	Tang S et al. 2016
Human immunodeficiency virus 1 (HIV-1)	gp120, V3 loop, HBX2 V 3 loop, MN	83↓84	*S. cerevisiae*; T cells (mice and guinea pigs)	Takeda et al. 2004
	env, broadly neutralizing epitopes VRC01, 2F5, 4E10	78↓84 78↓88	*E. coli*; B cells (mice)	Karpenko et al. 2005; Rudometov et al. 2018, 2019
Human papilloma virus (HPV)	E7, aa 10–14	81↓82	*E. coli*; B cells (mice)	Tindle et al. 1994
	E7, aa 35–54		*E. coli, S. typhimurium*; B and T cells (mice)	Tindle et al. 1994; Londoño et al. 1996
	E7, aa 10–14 + 35–54		*E. coli*	Tindle et al. 1994
	E7, aa 10–14 + 86–93 E7, aa 10–14 + 82–90 + 86–93		*E. coli, S. typhimurium*; B cells, no CTL (mice)	Street et al. 1999

Vaccine Target	Source of Epitope	Position of Insertion or Addition	Expression System and Major Immunological Activity	References
Human T-cell leukemia virus 1 (HTLV-1)	Tax, aa 8–22 including the Tax11–19 epitope recognized by HLA-A*0201	MIR	*P. pastoris*; CTL response (mice)	Kozako et al. 2009
Infectious bursal disease virus (IBDV)	VP2, 5EPIS, five-mimotope polypeptide	79↓80	*E. coli*; protection (chickens)	Wang Yong-shan et al. 2012
Influenza A	M2, aa 2–24	75↓85	*E. coli*; protection (mice)	Fiers et al. 2004
	M2, aa 2–24	78↓79	*S. cereviciae*; protection (mice); ineffective (rhesus monkeys)	Fu et al. 2009
Nipah virus (NiV)	Nucleocapsid NP, aa 401–532	78↓81	*E. coli*	Yap et al. 2012
Simian immunodeficiency virus (SIV)	env, aa 121–147 aa 655–675 aa 738–763	81↓82	*E. coli*; B cells (guinea pigs)	Yon et al. 1992
Venezuelan equine encephalomyelitis (VEE)	E2, aa 233–240	81↓82	*E. coli*	Loktev et al. 1996
Borrelia burgdorferi	OspA, aa 18–273 (SplitCore technique)	78↓81	*E. coli*; B cells (mice)	Walker et al. 2011
	Salp15, aa 20–135 Iric-1, aa 20–134 *Ixodes scapularis*, tHRF, aa 1–173 (SplitCore technique)	79↓81	*E. coli*; B cells (mice); killing of bacteria in vitro	Kolb et al. 2015b
	BBK32, aa 130–166 aa 160–175 aa 153–175	78↓79	*E. coli*	Ranka et al. 2013
Chlamydia trachomatis	serovar E, major outer membrane protein, multiepitope (MOMPm), aa 370–387	74↓82	*E. coli*; protection (mice)	Jiang et al. 2015
Plasmodium falciparum	CSP, full-length CS27IVC, 319 aa (SplitCore technique)	78↓81	*E. coli*	Walker et al. 2011
Theileria annulata	SPAG-1, aa 785–892	81↓82	*E. coli*; B and T cells (calves)	Boulter et al. 1995, 1999
Theileria parva	Sporozoite surface antigen p67, aa 572–651	78↓79	*E. coli*; B and T cells (calves)	Lacasta et al. 2021
Allergy	*Chenopodium album* polcalcin (Che a 3)—derived peptide, aa 45–86	77↓78	*E. coli*; B and T cells (mice)	Sani et al. 2021
Atherosclerosis	Cholesteryl ester transfer protein C-terminal fragment (CETPC), aa 451–476, where aa 461–476 is a B epitope	80↓81	DNA vaccine; B cells (rabbits)	Mao et al. 2006
Cancer	Vascular endothelial growth factor (VEGF), murine, bevacizumab epitope, 13 aa	80↓81	DNA vaccine; survival (mice)	Kyutoku et al. 2013
	Ovalbumin, CTL epitope SIINFEKL, aa 257–264	78↓82	*E. coli*; preventive and therapeutic effect (mice)	Shan et al. 2019
	EndophilinB2-Sp, murine, tumor-associated neoepitope Db/Sp$_{244–252/R251H}$	78↓82	DNA vaccine; priming of CD8$^+$ T cells (mice)	Stifter et al. 2020

(Continued)

TABLE 38.2 (CONTINUED)

The Putative Vaccines Constructed on the *Hepadnaviridae* Cores by Genetic Fusion Methodology

Vaccine Target	Source of Epitope	Position of Insertion or Addition	Expression System and Major Immunological Activity	References
Hepatic fibrosis	Transforming growth factor-beta1 (TGF-β1), human, aa 198–225	MIR?	*E. coli*; B cells (mice); attenuation of hepatic fibrogenesis (rat model)	Ji H et al. 2017
Hypertension	Angiotensin II: DRVYIHPF	80↓81	DNA vaccine; therapeutic effect (rats)	Koriyama et al. 2015
Systemic lupus erythematosus	Interleukin 17A, murine, aa 65–72 aa 110–116 human, aa 62–69 aa 107–113	80↓81	DNA vaccine; therapeutic effect (mice)	Koriyama et al. 2020
HBc internal insertions (full-length HBc vectors) + N-terminal insertions				
Human rhinovirus 2 (HRV2) Poliovirus 1 (PV1)	HRV2, VP2, aa 156–170 (internal) Poliovirus, VP1, aa 93–103 (N-terminal)	80↓81 1↓2	*E. coli*; B cells (guinea pigs)	Brown et al. 1991; Clarke et al. 1991
Influenza A (IAV)	M2e, aa 2–24 (internal) M2e, aa 1–24 plus B subunit of heat labile enterotoxin (LTB)	80↓81 1↓2	*E. coli*; mucosal immunogenicity (mice)	Zhang GG et al. 2009
HBc internal insertions (C-terminally truncated HBcΔ vectors)				
Dengue virus 2 (DENV2)	envelope domain III, 104 aa	75↓81	*E. coli, P. pastoris*; B cells (mice)	Arora et al. 2012, 2013
Enterovirus 71 (EV71)	VP4, 20 N-terminal aa	78↓79	*E. coli*; B cells (mice); virus neutralization	Zhao M et al. 2013
	VP2, aa 141–155	78↓83	*E. coli*; passive protection (mice)	Xu Longfa et al. 2014
	VP1, epitope SP70, aa 208–222 VP2, aa 141–155 both VP1 and VP2 epitopes	78↓82	*E. coli*; cross-protection against EV71 and A16 (mice)	Xu Longfa et al. 2015
Enterovirus 71 (EV71) and Coxsackievirus A16 (CVA16)	VP1, epitope SP70, aa 208–222 VP1, epitope PEP91, aa 271–285	74↓82	*E. coli*; cross-protection (mice)	Huo et al. 2017
Enterovirus 71 (EV71) and varizella zoster virus (VZV)	EV71, VP1, aa 208–222, VP2, aa 141–155; VZV, gE, aa 121–135	78↓82	*E. coli*; B cells (mice); cross-neutralization, also against CVA16	Wu Y et al. 2017
Epstein-Barr virus (EBV)	gp350, receptor binding domain, peptides P1 (aa 16–29), P2 (aa 142–161), and P3 (aa 282–301) in different order, separated by linkers	78↓82	*E. coli*; B cells (mice); virus neutralization	Zhang X et al. 2020
Foot-and-mouth disease virus (FMDV)	Different combinations of multiepitopes: [VP1(141–160)—VP4(21–40)]; [VP1(141–160)—VP4(21–40)—VP1(141–160)]	74↓79 74↓81 74↓83	*E. coli*; B cells (mice); virus neutralization	Zhang YL et al. 2007
	VP1, aa 140–160	80↓81	*Nicotiana tabacum*; protection (mice)	Huang Y et al. 2005
	VP1, complete, 213 aa	76↓83	HeLa cells; DNA vaccine; B and T cells (mice); virus neutralizing	Jin Huali et al. 2007

Vaccine Target	Source of Epitope	Position of Insertion or Addition	Expression System and Major Immunological Activity	References
Hantavirus	Protein N, aa 1–45	78↓79	*E. coli*; protection (bank voles)	Ulrich et al. 1998, 1999; Koletzki et al. 2000
	Protein N, aa 1–120 (1–120)₂	78↓79	*E. coli*; B cells (mice)	Koletzki et al. 1999; Geldmacher et al. 2004, 2005
Hepatitis B virus (HBV)	preS1, aa 31–35	73↓82	*E. coli*; B cells (mice)	Borschukova et al. 1997; Borisova et al. 1999
	preS1, aa 31–36	78↓79	*E. coli*; B and T cells (mice)	Borschukova et al. 1997; Borisova et al. 1999
	preS1, aa 31–35	78↓79	*E. coli*	Borschukova et al. 1997; Borisova et al. 1999; Lachmann et al. 1999
	preS1, aa 31–35	78↓82	*E. coli*; B and T cells (mice)	Borschukova et al. 1997; Borisova et al. 1999
	preS1, aa 31–35	78↓86	*E. coli*; B and T cells (mice)	Borschukova et al. 1997; Borisova et al. 1999
	preS1, aa 31–35	78↓89	*E. coli*; B and T cells (mice)	Borschukova et al. 1997; Borisova et al. 1999
	preS1, aa 31–35	78↓94	*E. coli*; B cells (mice)	Borschukova et al. 1997; Borisova et al. 1999
	preS1, aa 31–36 aa 94–105	81↓82	*E. coli*; B cells (mice, rabbits)	Makeeva et al. 1995
	preS1, aa 32–58 (32–58)₂ (32–58)₃	78↓82	*E. coli*; B cells (mice)	Yang HJ et al. 2005
	preS1, aa 31–58	78↓79	*E. coli*; B and T cells (mice)	Sominskaya et al. 2010
	preS1, aa 31–58	78↓79	*E. coli*	Su et al. 2013b
	preS1, aa 12–60 + 89–119	78↓79	*E. coli* B and T cells (mice)	Skrastina et al. 2008
	HBs, aa 111–149	78↓82	*E. coli*; B and T cells (mice)	Borisova et al. 1993
	HBs, aa 139–148	78↓79	*E. coli*; B cells (rabbits)	Chen SY et al. 2010
	HBs, aa 119–152	78↓79	*E. coli*; B and T cells (mice)	Su et al. 2013a
	HBs, aa 113–135	78↓82	*E. coli*; B and T cells (mice)	Zhang TY et al. 2020
Hepatitis C virus (HCV)	Core, aa 1–98[1]	78↓79	*E. coli*	Mihailova et al. 2006
	E2, hypervariable region I, 27 aa (SplitCore technique)	78↓81	*E. coli*; B cells (mice)	Lange et al. 2014
Hepatitis E virus (HEV)	ORF2, aa 613–654[1]	72↓89	Insect cells	Touze et al. 1999
	Peptides HPTLLRI and SILPYPY mimicking the neutralization epitope	78↓83	*E. coli*	Gu et al. 2004
	ORF2, L1/L2/L3 region, aa 394–458	78↓83	*E. coli*; B cells (mice); protection (rhesus monkeys)	Tang ZM et al. 2015
Human immunodeficiency virus 1 (HIV-1)	gp120, aa 107–131	78↓79	*E. coli*	Borisova et al. 1996, 1997
	gp120, aa 303–327	78↓81	*E. coli*; B cells (mice)	von Brunn et al. 1993a, b

(Continued)

TABLE 38.2 (CONTINUED)

The Putative Vaccines Constructed on the *Hepadnaviridae* Cores by Genetic Fusion Methodology

Vaccine Target	Source of Epitope	Position of Insertion or Addition	Expression System and Major Immunological Activity	References
	gp120, aa 299–338	78↓81	*E. coli*; B cells (mice)	von Brunn et al. 1993a, b
	gp120, aa 306–328	78↓81	*E. coli*; B cells (mice)	von Brunn et al. 1993a, b
	A set of T helper cell and CTL epitopes including the H-2D^d-restricted CTL epitope RGPGRAFVTI	80↓81	DNA vaccine; T cells (mice, rhesus macaques)	Fuller et al. 2007
Human papilloma virus (HPV)	E7, aa 49–57, RAHYNIVTF	78↓79	*E. coli*; suppression of tumor growth (mice)	Chu X et al. 2016
Influenza A	M2, aa 2–24	75↓85	*E. coli*; protection (mice)	Fiers et al. 2004, 2009; De Filette et al. 2005, 2006a
	M2e, "swine," "human," or "avian," aa 2–24 (2–24)$_2$ (2–24)$_4$	78↓81	*E. coli;* protection (mice)	Kotlyarov et al. 2010; Ravin et al. 2015; Tsybalova et al.
	M2e, aa 1–24[1] (1–24)$_3$[1]	78↓81	*E. coli*; B cells (mice)	Sun Xincheng et al. 2015
	M2e, aa (2–24)$_3$; M2e, aa (2–24)$_3$ + nucleoprotein N, aa 418–426	74↓86	*E. coli;* protection (mice)	Gao et al. 2013
	H7N9 strain, hemagglutinin (HA), stalk region, long alpha helix (LAH), aa 76–130	74↓86	*E. coli;* protection (mice)	Chen S et al. 2015; Zheng D et al. 2016
Porcine epidemic diarrhea virus (PEDV)	Spike protein, aa 744–752, 756–771, 1371–1377; membrane protein, aa 195–200; individually and a multiepitope	79↓80	*E. coli*; B cells (mice); virus neutralization	Gillam et al. 2018
	Spike protein, 748-YSNIGVCK-755	79↓80	*E. coli;* protection (piglets)	Lu Yi et al. 2020a
Porcine reproductive and respiratory syndrome virus (PRRSV)	GP5, multiepitope of one B cell and two T-cell epitopes	78↓79	*E. coli*; virus blocking assay	Murthy et al. 2015
	GP3, two epitopes; GP5, two epitopes, different combinations	78↓79	*E. coli*; virus-neutralizing response (pigs)	Lu Yi et al. (2020b)
Respiratory syncytial virus (RSV)	G protein, aa 144–204	78↓79	*E. coli;* protection (mice)	Qiao et al. 2016
Rubella virus (RV)	E1, aa 214–285 aa 214–240 aa 245–285	79↓78	*E. coli*; B cells (mice)	Skrastina et al. 2013
Simian immunodeficiency virus (SIV)	A large set of T helper cell and CTL epitopes	80↓81	DNA vaccine; T cells (mice, rhesus macaques)	Fuller et al. 2007
Varicella-zoster virus (VZV)	Glycoprotein E (gE), aa 121–135	78↓82	*E. coli*; B cells (mice); virus neutralization	Zhu et al. 2016; Wu Y et al. 2017
Bacillus anthracis	Protective antigen domain 2, 2β2–2β3 loop, aa 302–325[2]	78↓79	*E. coli*; protection (guinea pigs)	Yin et al. 2008
	Protective antigen domain 2, 2β2–2β3 loop, aa 302–325	78↓79 78↓82	*E. coli*; protection (mice)	Yin et al. 2014

Vaccine Target	Source of Epitope	Position of Insertion or Addition	Expression System and Major Immunological Activity	References
	Anthrax protective antigen domain 2, aa 1–128[1]	78↓81	*E. coli, Nicotiana tabacum*; B cells (mice)	Bandurska et al. 2008
Borrelia burgdorferi	OspA, aa 18–273	78↓81	*E. coli*; B cells (mice)	Nassal et al. 2005, 2008
	OspA, aa 18–273 (SplitCore technique)	78↓81	*E. coli*; B cells (mice)	Walker et al. 2008, 2011
	OspC$_{a/b}$, aa 23–211	78↓81	*E. coli*; B cells (mice)	Skamel et al. 2006; Nassal et al. 2008
Mycobacterium tuberculosis	ESAT-6, full-length	78↓79	*E. coli*; B and T cells (mice)	Yin et al. 2011
Neisseria meningitidis	Surface protein A, aa 1–174	78↓79	*E. coli*; protection (mice)	Hou et al. 2019
	Adhesin NadA, extracellular domain, aa 26–309	78↓81	*E. coli*; B and T cells (mice); bactericidal activity	Aston-Deaville et al. 2020
Plasmodium berghei	CS, (DP$_4$NPN)$_2$	75↓81	*S. typhimurium*; B and T cells (mice)	Schödel et al. 1994b
Plasmodium falciparum	CS, (NANP)$_4$	75↓81	*S. typhimurium*; B and T cells (mice)	Schödel et al. 1994b, c
	CS, (NANP)$_4$	75↓81 77↓78 78↓79	*E. coli*; B and T cells (mice)	Milich et al. 2001
	CS, (NANP)$_4$NVDP	78↓79	*E. coli*; protection (mice)	Birkett et al. 2001, 2002; Hutchings et al. 2007
Plasmodium vivax	CS, (AGDRADGQP)$_3$	78↓79	*E. coli*; B and T cells (mice)	Almeida et al. 2014
Plasmodium yoelii	CS, aa 139–162, (QGPGAP)$_4$	75↓81	*S. typhimurium*; B and T cells (mice)	Schödel et al. 1997
	CS, full length CS27IVC, 319 aa (SplitCore technique)	78↓81	*E. coli*; B cells (mice)	Walker et al. 2011
Toxoplasma gondii	ROP2-SAG1 (RS) multiepitope	74↓82	*E. coli*; B and T cells (mice)	Wang W et al. 2017
Allergy, IgE-mediated	Human membrane-bound IgE, CεmX peptide, 52 aa, or its fragments	78↓81	*E. coli*; B cells (mice)	Lin et al. 2012
Alzheimer's disease	β-Amyloid peptide, 2 fragments of aa 1–15 connected with 2 lysine residues	71↓88	*E. coli*; protection (mice, rats)	Feng et al. 2011a, b, 2012, 2013; Jin Hui et al. 2014
	tau$_{294–305}$ epitope	78↓79	*E. coli*; B cells and therapeutic effect (transgenic mouse model)	Ji M et al. 2018
	SpyCatcher, aa 1–139 SpyTag-conjugated peptides are listed in the text	78↓79	*E. coli*; B cells and therapeutic effect (transgenic mouse model)	Ji M et al. 2020
Asthma	TNFα, 4 peptides	78↓79	*E. coli*; B cells (mice, asthma model)	Ma AG et al. 2007
	Interleukin 4, murine, aa 2–15: DKNHLREIIGILNE	78↓79	*E. coli*; B cells (mice, asthma model)	Ma Y et al. 2007b
	Interleukin 13, murine, aa 12–26: LTLKELIEELSNITQ	78↓79	*E. coli*; B cells (mice, asthma model)	Ma Y et al. 2007a, c
	Interleukin 10, murine, 5 peptides	78↓79	*E. coli*; B cells (mice), enhancing IL-10 activity	Zhou G et al. 2010

(Continued)

TABLE 38.2 (CONTINUED)

The Putative Vaccines Constructed on the *Hepadnaviridae* Cores by Genetic Fusion Methodology

Vaccine Target	Source of Epitope	Position of Insertion or Addition	Expression System and Major Immunological Activity	References
	Interleukin 33, aa 109–266	78↓79	*E. coli;* suppressing IL-33 pathological roles (mice, asthma model)	Long et al. 2014
Cancer	EGFRvIII, fusion junction peptide Pep-3: LEEKKGNYVVTDH	77↓89	*E. coli;* B cells (mice)	Duan et al. 2007, 2009
	Claudin-18 isoform 2 (CLDN18.2), aa 32–41	78↓81	*E. coli;* antibodies killing cells (mice, rabbits); protection (mice)	Klamp et al. 2011
	Z_{HER2} affibody, ~7 kDa	78↓81	*E. coli;* specific targeting of cancer cells (mice)	Nishimura et al. 2013; Suffian et al. 2017
Colitis	Transforming growth factor-beta1 (TGF-β1), murine, aa 66–74	78↓79	*E. coli;* B cells (mice, model of colitis)	Ma Y et al. 2010
	IL-12/IL-23 p40 subunit, murine, aa 38–46; 53–71; 160–177	78↓79	*E. coli;* B cells (mice, model of colitis)	Guan et al. 2009, 2011, 2012a
	Interleukin 23 p19 subunit, murine. aa 139–147: EDHPRETQQ	78↓79	*E. coli;* B cells (mice, model of colitis)	Guan et al. 2013
	Interleukin 17, murine, aa 32–42; 81–91	78↓79	*E. coli;* B cells (mice, model of colitis)	Guan et al. 2012b
Gonadotropin-releasing hormone (GnRH)-dependent diseases	DnaK, substrate-binding domain, aa 394–504	80↓81	*E. coli;* B cells (mice)	Wang XJ et al. 2009, 2010
Hepatic fibrosis	Transforming growth factor-beta1 (TGF-β1), murine	71↓89	*E. coli;* B cells (mice)	Guo YH et al. 2005a, b
	Connective tissue growth factor (CTGF), aa 138–159	78↓81	*E. coli;* protection (mice)	Li S et al. 2016
Hepatocellular carcinoma	Polyepitope: HBx(92–100)-Th epitope-HBx(140–148)-HBx(52–60)-HBx(115–123)	78↓79	*E. coli;* B and T cells (mice)	Ding et al. 2009
Human chorionic gonadotropin	hCG, subunit β, 37-CTP, aa 109–145	75↓83	*E. coli;* B cells (mice)	Li G et al. 1996
Hypertension	CE12-CQ10-CE12-CQ10, where CE12 is CAPESEPSNSTE, an epitope of the E3 region of domain IV of the human L-type calcium channel $Ca_V1.2$ $α_{1C}$ subunit; CQ10 = ATR-001 is CAFHYESQ, an epitope of the second extracellular loop of human AT_1 receptors (see also ATRQβ-001 vaccine in Table 24.3 of Chapter 24)	78↓81	*E. coli;* therapeutic effect (mice, rats)	Wu H et al. 2020
Melanoma	MAGE-3, aa 168–176: EVDPIGHLY	75↓85	*E. coli*	Kazaks et al. 2008

Vaccine Target	Source of Epitope	Position of Insertion or Addition	Expression System and Major Immunological Activity	References
HBc internal insertions (C-terminally truncated HBcΔ vectors) + N-terminal insertions				
Hepatitis B virus (HBV)	preS1, aa 31–36 (internal) preS2, aa 133–143 (N-terminal)[1]	81↓82 ↓4	*E. coli*	Makeeva et al. 1995
HBc internal insertions (C-terminally truncated HBcΔ vectors) + C-terminal insertions				
Enterovirus 71 + Coxsackievirus A16	E71, VP1, epitope SP70, aa 208–222 + A16, VP1, epitope PEP91, aa 271–285 (internal) E71, VP2, epitope A3, aa 248–263 (C-terminal)	74↓82 144↓	*E. coli;* protection (mice)	Huo et al. 2017
Hepatitis B virus (HBV)	preS1, aa 27–53 (internal) preS2, aa 133–143 (C-terminal)	75↓81 156↓	*E. coli, S. typhimurium, S. dublin, S. typhi*; B cells (mice, rabbits)	Schödel et al. 1992, 1993b, 1994a, b; Tacket et al. 1997
Hepatitis B virus (HBV) Hepatitis C virus (HCV)	preS1, 31–58 (internal) Core, 1–60 (C-terminal)	78↓79 144↓	*E. coli*; B and T cells (mice)	Sominskaya et al. 2010
Respiratory syncytial virus (RSV)	G protein, aa 144–204 (internal) M2, aa 82–90 (C-terminal)	78↓79 149↓	*E. coli;* protection (mice)	Qiao et al. 2016
Neisseria meningitidis	Adhesin NadA, extracellular domain, aa 26–309 (internal) Factor H binding protein, C-terminal domain, aa 157–274 (C-terminal)	78↓81 148↓	*E. coli*; B and T cells (mice); bactericidal activity	Aston-Deaville et al. 2020
Plasmodium falciparum	CS, (NANP)₄ (internal) + CS, aa 326–345 (C-terminal) CS, (NANP)₄NVDP (internal) + CS, aa 326–345 (C-terminal)	78↓79 149↓	*E. coli*; B and T cells (mice, human)	Milich et al. 2001; Sällberg et al. 2002; Birkett et al. 2001, 2002; Nardin et al. 2004; Langermans et al. 2005; Oliveira et al. 2005; Walther et al. 2005; Gregson et al. 2008
Taenia solium cysticercosis	KETc1 and KETc12 + GK-1	MIR 149↓	*E. coli*, DNA vaccine; B cells (mice, pigs)	Wu L et al. 2005
HBc internal insertions (C-terminally truncated HBcΔ vectors) + internal Cys48↓Ser49 insertion				
Cancer	Affibody, EGRF receptor I (internal MIR) mCardinal FP (internal Cys48↓Ser49)	78↓79 48↓49	*E. coli*; tumor targeting and imaging (mice)	Kim SE et al. 2017
HBc internal insertions (C-terminally truncated HBcΔ vectors) + internal Cys48↓Ser49 insertion + N-terminal insertion				
Cancer	Affibody, EGRF receptor I (internal MIR) mCardinal FP (internal Cys48↓Ser49) mCardinal FP (N-terminal)	78↓79 48↓49 ↓1	*E. coli*; tumor targeting and imaging (mice)	Kim SE et al. 2017
HBc internal insertions (C-terminally truncated HBcΔ vectors): tandem with full-length HBc				
Influenza A	H3N2 strain, hemagglutinin (HA), stalk region, long alpha helix (LAH)	77↓78	*Pichia pastoris*; protection (mice)	Kazaks et al. 2017
	Tandiflu1 VLP (multiple antigens: LAH and M2e₃ on each of tandem members)	77↓78	*E. coli*; protection (mice)	Ramirez et al. 2018
Dengue virus	cEDIII, 103 aa (on the full-length HBc monomer)	78↓79	*N. benthamiana*; B cells (mice)	Pang et al. 2019

(Continued)

TABLE 38.2 (CONTINUED)

The Putative Vaccines Constructed on the *Hepadnaviridae* Cores by Genetic Fusion Methodology

Vaccine Target	Source of Epitope	Position of Insertion or Addition	Expression System and Major Immunological Activity	References
Human papillomavirus (HPV)	E2, aa 14–122	77↓78	*N. benthamiana*; B cells (mice); neutralization of pseudovirions	Diamos et al. 2019
Zika virus (ZIKV)	Envelope, DIII, aa 301–406	77↓78	*N. benthamiana*; B cells (mice); neutralization of ZIKV	Diamos et al. 2020
HBc C-terminal insertions (full-length HBc vectors)				
Hepatitis B virus (HBV)	preS1, aa 31–35	144↓145	*E. coli*	Lachmann et al. 1999
	preS1, aa 31–80	144↓145	*E. coli*; B and T cells (mice)	Borisova et al. 1989, 1990; Zakis et al. 1992
	preS1, aa 80–118	144↓145	*E. coli*	Borisova et al. 1989, 1990; Zakis et al. 1992
	preS2, aa 118–173	144↓145	*E. coli*; B and T cells (mice)	Borisova et al. 1989, 1990; Zakis et al. 1992
	preS1, aa 31–34	183↓	*E. coli*	Lachmann et al. 1999; Berzins et al. 1994
Human immunodeficiency virus 1 (HIV-1)	gp41, aa 589–640	144↓145	*E. coli*	Borisova et al. 1989; Ulrich et al. 1990, 1993
Murine cytomegalovirus (MCMV)	immediate early 1 protein pp89, CTL epitope, aa 168–176	179↓180	Mammalian cells; DNA vaccine; CTL response and protection (mice)	Del Val et al. 1991 de Andrade et al. 2007
Simian immunodeficiency virus (SIV)	env, aa 170–189 aa 324–339 aa 594–616 aa 655–675	183↓	*E. coli*; no B cells (guinea pigs)	Yon et al. 1992
Chlamydia trachomatis	Serovar E, major outer membrane protein, multiepitope (MOMPm), aa 370–387		*E. coli*; protection (mice)	Jiang et al. 2015
HBc C-terminal insertions (C-terminally truncated HBcΔ vectors)				
Bovine leukemia virus (BLV)	gp51, aa 89–137	144↓	*E. coli*	Borisova et al. 1989; Ulrich et al. 1991, 1993
Foot-and-mouth disease virus (FMDV)	VP1, aa 200–213 + 131–160	144↓	*E. coli*; B cells (rabbits)	Nekrasova et al. 1997
Hantavirus	Protein N, aa 1–45	144↓	*E. coli*	Koletzki et al. 2000
	Protein N, aa 38–82	144↓	*E. coli*; B cells (bank voles)	Ulrich et al. 1998, 1999
	Protein N, aa 75–119	144↓	*E. coli*; B cells (bank voles)	Ulrich et al. 1998, 1999; Koletzki et al. 2000
	Protein N, aa 1–114[3]	144↓	*E. coli*; B cells (bank voles)	Koletzki et al. 1997; Ulrich et al. 1999
	Protein N, aa 1–120[3]	144↓	*E. coli*	Koletzki et al. 1999
Hepatitis B virus (HBV)	preS1, aa 31–35	144↓	*E. coli*	Lachmann et al. 1999
	preS1, aa 12–31	144↓	*E. coli*; B cells (rabbits)	Stahl and Murray 1989
	preS1, aa 12–47	144↓	*E. coli*; B cells (rabbits)	Stahl and Murray 1989
	preS1, aa 31–79	144↓	*E. coli*; B and T cells (mice)	Borisova et al. 1989, 1990; Zakis et al. 1992

Vaccine Target	Source of Epitope	Position of Insertion or Addition	Expression System and Major Immunological Activity	References
	preS1, aa 1–42	144↓	*E. coli*	Zhao Yangqing and Zhan 2000, 2002
	preS1, aa 3–55	155↓	*E. coli*; B and T cells (mice)	Chen X et al. 2004
	preS2, aa 118–173	144↓	*E. coli*; B and T cells (mice)	Borisova et al. 1989, 1990; Zakis et al. 1992
	preS2, aa 124–174	144↓	*E. coli*; B and T cells (mice)	Borisova et al. 1989, 1990; Zakis et al. 1992
	preS2, aa 120–145	144↓	*E. coli*; B cells (rabbits)	Stahl and Murray 1989
	preS2, aa 133–143	156↓	*E. coli, S. typhimurium, S. dublin, S. typhi*; B cells (mice)	Schödel et al. 1990a, b, 1992, 1993b
	HBs, aa 111–156	144↓	*E. coli*; B and T cells (rabbits)	Stahl and Murray 1989; Shiau and Murray 1997; Murray and Shiau 1999
	HBs, aa 111–165	144↓	*E. coli*; B and T cells (rabbits)	Stahl and Murray 1989; Shiau and Murray 1997; Murray and Shiau 1999
	preS1, aa 1–20 + preS2, aa 1–26 + HBs, aa 111–156	144↓	*E. coli*	Murray and Shiau 1999
Hepatitis C virus (HCV)	core, aa 6–77	144↓	*E. coli*	Claeys et al. 1992
	core, aa 6–143	144↓	*E. coli*	Claeys et al. 1992
	NS3, aa 1359–1449	144↓	*E. coli*	Claeys et al. 1995
	NS3, aa 1460–1532	144↓	*E. coli*	Claeys et al. 1995
	core, aa 39–75; aa 1–91 aa 1–180 (1–180)$_2$ (1–180)$_3$ (1–180)$_4$	149↓	*E. coli*	Yoshikawa et al. 1993
	Core, aa 1–60	144↓	*E. coli*; B and T cells (mice)	Sominskaya et al. 2010
	Core, aa 1–98 aa 1–98 + RGD (N-term) aa 1–98 + SPRRR aa 1–151 aa 1–173 NS3, aa 327–482 aa 202–326[1]	144↓	*E. coli*; B and T cells (mice)	Mihailova et al. 2006
HBV + HCV	preS1, aa 1–20 + preS2, aa 1–26 + core, aa 1–98	144↓	*E. coli*; B cells (rabbits, mice)	Wu CL et al. 1999
Hepatitis E virus (HEV)	ORF2, aa 613–654	146↓	Insect cells	Touze et al. 1999
Human cytomegalovirus (HCMV)	gp58, aa 599–644	144↓	*E. coli*; B cells (rabbits)	Tarar et al. 1996
Human immunodeficiency virus 1 (HIV-1)	gp120, aa 299–338	144↓	*E. coli*; B cells (mice)	Borisova et al. 1997
	gp120, aa 306–328	144↓	*E. coli*; B cells (rabbits, mice)	Borisova et al. 1997; Grene et al. 1997

(Continued)

TABLE 38.2 (CONTINUED)

The Putative Vaccines Constructed on the *Hepadnaviridae* Cores by Genetic Fusion Methodology

Vaccine Target	Source of Epitope	Position of Insertion or Addition	Expression System and Major Immunological Activity	References
	gp120, aa 303–327	154↓	*E. coli*; B cells (mice)	von Brunn et al. 1993a, b
	gp41, aa 616–632 gp41, aa 667–680 gp41, aa 728–751	144↓	*E. coli*; B cells (rabbits)	Stahl and Murray 1989
	gp41, aa 589–640	144↓	*E. coli*	Borisova et al. 1989
	Gag, aa 121–210	144↓	*E. coli*; B cells (mice)	Ulrich et al. 1992; Grene et al. 1997
	Nef, aa 113–130	144↓	*E. coli*	Ulrich et al. 1993
	Env, gp120 plus ectodomain of gp41[1] gp120[1]	149↓	Mammalian cells; CD4 binding	Patterson LJ et al. 2001; Berkower et al. 2004
Human papilloma virus	E7, aa 35–98	144↓	*E. coli*; B and T cells (mice)	Pumpens et al. 2002
Simian immunodeficiency virus (SIV)	A large set of T helper cell and CTL epitopes	144↓	DNA vaccine; T cells (mice, rhesus macaques)	Fuller et al. 2007
West Nile virus (WNV)	Envelope, DIII, aa 296–415	149↓	*N. benthamiana;* B cells (mice)	He et al. 2021
Zika virus (ZIKV)	Envelope, DIII, aa 303–403	149↓	*N. benthamiana;* B cells (mice); neutralization of ZIKV	Yang M et al. 2017
	Envelope, DIII, aa 301–406	149↓	*N. benthamiana;* B cells (mice); neutralization of ZIKV	Diamos et al. 2020
Neisseria meningitidis	Factor H binding protein, C-terminal domain, aa 157–274	148↓	*E. coli;* B and T cells (mice)	Aston-Deaville et al. 2020
Plasmodium falciparum	CS, aa 326–345	149↓	*E. coli*; B and T cells (mice, human)	Birkett et al. 2001, 2002
Porphyromonas gingivalis	Rgp-1, aa 865–911	157↓	*E. coli*; B cells (rabbits, mice)	Dawson and Macrina 1999
Hepatocellular carcinoma	MAGE-1, aa 278–286 MAGE-3, aa 271–279 AFP1, aa 158–166 AFP2, aa 542–550	144↓	*E. coli*; B and T cells (mice)	Zhang Yan et al. 2007
Human chorionic gonadotropin	hCG, subunit β, 37-CTP, aa 109–145	154↓	*E. coli*	Li G et al. 1996
	Polyepitope: HBx(92–100)-Th epitope-HBx(140–148)-HBx(52–60)-HBx(115–123)	144↓	*E. coli*; B and T cells (mice)	Ding et al. 2009
Melanoma	MAGE-3, aa 163–181	161↓ 168↓ 176↓	*E. coli*	Kazaks et al. 2008

RBHBc internal insertions (C-terminally truncated RBHBcΔ vectors)

Hepatitis B virus	HBs, aa 113–135	78↓82	*E. coli*; eradication of DNA in carrier mice; B cells (cynomolgus monkeys)	Zhang TY et al. 2020

WHc internal insertions (full-length WHc vectors)

Hepatitis B virus	preS1, 8 neutralizing B cell epitopes	78↓79	*E. coli*; passive protection, circumvention of immune tolerance (mice)	Whitacre et al. 2020

Vaccine Target	Source of Epitope	Position of Insertion or Addition	Expression System and Major Immunological Activity	References
Respiratory syncytial virus (RSV)	Fusion (F) protein, palivizumab helix-loop-helix epitope, aa 254–277	78↓79	*E. coli*; protection (mice)	Schickli et al. 2015
Zika virus (ZIKV)	E protein: EDIII, aa 578–713; CD Loop, aa 628–656; Fusion Loop, aa 388–400	76↓83	*P. pastoris*; WHc-CD Loop VLPs elicited protective immune response in murine model	Cimica et al. 2020
Plasmodium falciparum		78↓79	*E. coli*; protection (mice)	Whitacre et al. 2015b
WHc internal insertions (full-length WHc vectors) + C-terminal				
Plasmodium falciparum *Plasmodium vivax*	CS, repeat NANP- and NVDP-based B-cell epitopes, other B-cell epitopes (internal) CS, CD4/CD8 T cell epitopes (C-terminal)	78↓79 187↓	*E. coli*; B and T cells (rabbits); protection (mice)	Whitacre et al. 2015a, b
WHc internal insertions (C-terminally truncated WHcΔ vectors)				
Influenza	M2e, MSLLTEVETPTRN-GWECSASDSSD	76↓77	*S. typhimurium*; B cells (mice)	Ameiss et al. 2010
Plasmodium falciparum	CSP, NANPNVDP(NANP)₃ CSP, full length CS27IVC, 319 aa (SplitCore technique)	78↓81	*E. coli*; B cells (mice)	Walker et al. 2011

Note: The data is given in alphabetical order of the vaccine target in following order: viruses, bacteria, noninfectious diseases.

[1] No VLPs were formed.

[2] The chimeric protein contained 14 amino acid residues coming from vector and added to the N-terminus of the HBc derivative.

[3] Chimeras self-assemble only as mosaic particles in the presence of wild-type HBc.

As follows from the Table 38.2 data, the unique flexibility of the HBc monomer/dimer in their ability to maintain self-assembly properties led to fantastically high capacities for foreign insertions into the three specified sites of the HBc molecule. The most immunogenic, by definition, MIR tolerated exposure on the outer VLP surface of such long foreign sequences as the 41 aa-long immunodominant region "a" of the HBs molecule (Borisova et al. 1993), the two tandem copies of the entire 120 aa immunoprotective region of the hantavirus nucleocapsid protein (Koletzki et al. 1999; Geldmacher et al. 2004), and the full-length 238 aa green fluorescent protein (GFP) (Kratz et al. 1999). The HBc VLPs carrying MIR-tolerated GFP insertion were subjected to a comprehensive electron cryomicroscopy evaluation that allowed one to establish an enormous flexibility and robustness as a whole, as well as within the subunits (Böttcher et al. 2006). Thus, the spikes of the latter were able to rotate by as much as 40° against the distal interdimer contact sites, and the likely hinge for the swiveling movement was the conserved Gly111 residue at the inner surface of the capsid.

To increase more the HBc MIR vector capacity, it was attempted to shorten the MIR sequence and partially replace it with novel foreign sequences (Borisova et al. 1996).

Recently, great capacity of more than 300 aa residues in the MIR was achieved and more, in combination with a long C-terminal insertion (Aston-Deaville et al. 2020).

The N-terminal insertions that guaranteed the outer exposure of the foreign sequences appeared as the first HBc VLP-based vaccine candidates, namely against FMDV (Clarke et al. 1987; Newton et al. 1987). The N-terminal insertions showed remarkable insertion capacities of up to 120 aa in some cases (Geldmacher et al. 2004, 2005) and played a pioneering role in the VLP vaccine ideology by the generation of an acceptable influenza vaccine (see original paper of Neirynck et al. (1999) and the most recent review by Deng et al. (2015)).

The discovery (Stuyver et al. 2000; Kato et al. 2002) and structural characterization (Cotelesage et al. 2011) of the natural N-terminally extended HBc in the HBV, genotype G, paved the way for novel ideas and applications for the N-terminal foreign additions with a guaranteed exposure on the external VLP surface, as supposed by Andris Dišlers's group (Petrovskis et al. 2021). It should be noted that the uncommon HBV genotype G naturally coinfects patients together with the other HBV genotypes, in particular, A and E (Basic et al. 2021).

Furthermore, the tendency of the N-terminally inserted 37-aa long extension to form an ordered trimeric spike on the HBc VLP surface was confirmed by electron cryomicroscopy (McGonigle et al. 2015). The N-terminal insertions got therefore a special structural advantage for the correct folding of some conformational epitopes.

In contrast to the MIR and N-terminus, C-terminal insertions stemming from the spatial CTD organization within the HBc nucleocapsid did not automatically promise high exposure on the outer surface and remarkable antigenicity but rather the opposite. Although some efforts demonstrated the enormous capacity of the C-terminal vectors reaching up to 559 aa and even 741 aa HCV sequences (Yoshikawa et al. 1993) and possible outer exposure of the inserted sequences in some special cases (Grene et al. 1997), the overall praxis demonstrated the internal localization of the inserted aa stretches, as described by Borisova et al. (1989, 1990, 1997), Ulrich et al. (1991), and Schödel et al. (1992) and reviewed by Ulrich et al. (1998) and Pumpens and Grens (2001). To turn the C-terminal insertions out of the inner region of the HBc VLPs, a special class of HBc VLP carriers, the so-called HBcG vectors, in which the arginine residues of the CTD are fully or partially replaced by glycine residues (Dishlers et al. 2015), was constructed. The elimination of the positively charged CTD in the HBcG carriers prevented the encapsidation of bacterial RNA by cultivation in *E. coli*, allowed the exposure of a C-terminally inserted model epitope (the major epitope of the HBV preS1 sequence) onto the outer surface of the HBcG-derived VLPs, and substantially improved the immunogenicity of the inserted epitope in experimental animals (Dishlers et al. 2015). The fact that the substitution of the arginine motifs at the C-terminus with noncharged residues may induce a conformational change that would allow C-terminal insertions to become more exposed and more immunogenic on the VLP surface was first demonstrated for the WHc VLPs by David R. Milich's team (Schickli et al. 2015; Whitacre et al. 2015a, b).

The representative model epitopes: MHC class I-restricted peptide p33, namely the 33-KAVYNFATM-41 stretch (Pircher et al. 1989) and MHC class II-restricted peptide p13 GLNGPDIYKGVYQFKSVEFD (Oxenius et al. 1995) from the lymphocytic choriomeningitis virus (LCMV) glycoprotein were fused to the C-terminus of the full-length HBc via LLL or RSSGMY linkers—respectively—and used in the fine immunological studies of the Bachmann's team (Storni et al. 2002, 2003; Ruedl et al. 2002, 2005; Schwarz et al. 2003, 2005; Storni and Bachmann 2004).

The three insertion sites were provided first with restriction site linkers for the insertion of foreign sequences (Borisova et al. 1987; Gren and Pumpens 1988) and short N-terminal tags (Böttcher et al. 1998a). Then, to facilitate purification, the HBcΔ variant encoding the HBc1–156 stretch was C-terminally supplied with an additional glycine and a six histidine tag His_6 (Wizemann and von Brunn 1999). A set of the N-terminally His_6-tagged HBc VLP vectors was constructed by Yap WB et al. (2009, 2010). Gillam and Zhang (2018) generated the N-terminally His_6- tagged

HBcΔ VLPs in parallel with the RNA phage AP205 His_6-tagged VLPs. The application of His_6 tags had become customary by the construction of chimeric HBc VLPs. Thus, many of subjects in Table 38.2 were provided with N- or C-terminal His_6 tags, which is not indicated there for brevity.

Schumacher et al. (2018) demonstrated the clear stabilizing effect of the C-terminal His_6 tag on the chimeric HBcΔ VLPs with model MIR insertions between aa positions 78 and 81, which was illustrated by excellent electron cryomicrographs and reconstructions.

WHc and Others

Apart from the previously described HBc vectors, a clever and prospective vector system was generated by the famous David R. Milich's team for the WHc as a prospective universal carrier and used successfully for the generation of highly promising vaccine candidates (Billaud et al. 2005a, b, 2007; Whitacre et al. 2009, 2015a, b; Ameiss et al. 2010). As mentioned earlier, the vectors with noncharged residues instead of the natural C-terminal polyarginines were adopted to expose the C-terminal insertions on the WHc VLP surface (Whitacre et al. 2015a, b).

The remarkable success of the WHc VLPs as carriers was recently continued by the generation of a prospective anti-RSV vaccine candidate via the exposure of an epitope recognized by the neutralizing monoclonal anti-RSV antibody palivizumab on the WHc VLP surface (Schickli et al. 2015).

As stated earlier in the *3D Structure* paragraph, the HBc and WHc are in fact remarkably similar, but the nonhuman hepadnaviral cores have definite advantages over the HBc for the construction of medicinally important products such as vaccines (Whitacre et al. 2009). According to the Milich's team idea, the WHc is not burdened by the problem of preexisting immunity and, more importantly, avoids immune tolerance in the human population, since WHc is minimally cross-reactive with HBc at the T-cell level (Billaud et al. 2005b, 2007).

Recently, Zhang TY et al. (2020) involved a novel vector, namely roundleaf bat hepatitis B virus core (RBHBc), to generate a promising hepatitis B vaccine.

FUNCTIONALIZATION AND CHEMICAL COUPLING

The Martin F. Bachmann's team started the chemical coupling studies in the early 2000s with the HBc VLPs but soon switched to the RNA phage Qβ and AP205 VLPs, as described in the *Chimeric VLPs: Chemical coupling* of Chapter 25. At that time, the Bachmann's team published their first self-review, which presented the HBc VLP model in detail, as a molecular assembly system for chemical coupling, which rendered antigens of choice highly repetitive and was able to induce efficient antibody responses in the absence of adjuvants and provide protection from viral infection and allergic reactions, while RNA phage VLPs remained at that time in the background (Lechner et al. 2002). Figure 38.6 is taken from this breakthrough review and demonstrates how the hetero-bifunctional cross-linker

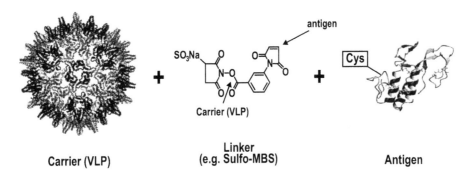

Carrier (VLP)　　　　　Linker　　　　　Antigen
　　　　　　　　　　(e.g. Sulfo-MBS)

FIGURE 38.6 HBc VLPs and the chemical coupling approach. Schematic representation of a modular system to attach an antigen containing a free cysteine (Cys) to a VLP (HBc) using a bifunctional cross-linker. Sulfo-MBS = Sulfo-m-maleimidobenzoyl-N-hydroxysuccinimide ester. (Reprinted with permission of S. Karger AG, Basel from Lechner F et al. *Intervirology.* 2002;45:212–217.)

sulfo-m-maleimidobenzoyl-N-hydroxysuccinimide ester (Sulfo-MBS) was used to couple model peptides with N-terminal CGG linker containing a free cysteine residue at the N-terminus to lysine residue exposed at the MIR on the surface of the HBcΔ VLPs. Therefore, in contrast to the RNA phage VLPs, the HBc model needed functionalization of the surface before the chemical coupling. Thus, this lysine was introduced into the HBcΔ MIR by replacing the original P79 and A80 by the peptide GGKGG, while both cysteine residues C48 and C107 aa of the HBc molecule were mutated to serines. The functionalized product was called HBcAg (1–149)-lys-2cys-Mut (Jegerlehner et al. 2002b). The first antigen models were a model peptide FLAG (CGGDYKDDK) of 9 aa residues, a 66 aa domain from GRA2, a protein derived from *Toxoplasma gondii*, M2(1–23) of total 27 aa residues, and TNFα of 19 aa with C or GGC additions at the N- or C-terminus, respectively, and a full-length protein, phospholipase A2 (PLA2), the major allergen in bee venom, of 134 aa in length (Jegerlehner et al. 2002b). However, immunization of mice with the HBc-M2 vaccine resulted in much weaker protection than that achieved by vaccination with UV-inactivated influenza virus (Jegerlehner et al. 2004). Later, the M2 was conjugated to the HBc VLPs, in parallel with the yeast-produced genetic fusions, by Fu et al. (2009).

Then, Bachmann's team coupled the D2 peptide of 15 aa, derived from the outer membrane protein of *Salmonella typhi*, plus CGG at the N-terminus, to the previously mentioned HBcAg (1–149)-lys-2cys-Mut VLPs and used as an immunological model in the same way as on the RNA phage VLPs (Jegerlehner et al. 2002a). Kündig et al. (2006) coupled Der p1 peptide, derived from a cysteine protease, the major fecal allergen of the house dust mite *D. pteronyssinus* and recognized as a possible antiallergic agent was coupled to the HBc VLPs, as a control to the classical study on the Der p1-coupled Qβ VLPs (see Chapter 25).

After 13 years, a set of special HBc vectors for chemical coupling and oligo-/polynucleotide packaging (see following discussion) was invented by Regīna Renhofa's group (Strods et al. 2015). Figure 38.7 demonstrates the introduced lysine residues—together with the intrinsic aspartic and glutamic acid residues—and how they are exposed on the tips of the HBc spikes and are ready for chemical coupling of

the chosen peptide and/or nucleic acid sequences. Together with complete removal of the internal residual RNA from bacteria- and yeast-derived HBc VLPs by alkaline hydrolysis (see following discussion), the proposed HBc VLP functionalization ensured a standard and easy protocol for the development of versatile HBc VLP-based vaccine and gene therapy applications.

Meanwhile, a novel "nanoglue" approach was invented to the functionalization and chemical coupling of the HBc VLPs (Lee KW et al. 2012b). The authors displayed the cell-internalizing peptides (CIPs) as delivery agents via the GSLLGRMKGA decapeptide, which interacted specifically at the tips of the spikes and was named "nanoglue" for this reason. To present CIPs on the nanoparticles, the CIP was cross-linked to the nanoglue, but the latter was cross-linked to the D78 of the HBcΔ VLP tips by using 1-ethyl-3-(3-dimethylaminopropyl)carbodiimide hydrochloride (EDC) and N-hydroxysulfosuccinimide (Sulfo-NHS).

In parallel, the nanoglue was used as a "binding tag" of IL2 (Blokhina et al. 2013b) and M2e (Blokhina et al. 2013a) but confined to noncovalent protein–protein interactions and avoided chemical coupling.

The history of the nanoglue peptide is highly instructive. First, Dyson and Murray (1995) found that small peptides with the motif LLGRMK recognize the HBc MIR, block the interaction of core particles with the HBs, and interfere therefore with the virion assembly. Second, Böttcher et al. (1998b) recognized the decapeptide GSLLGRMKGA as having the greatest affinity to the HBc molecule. Recently, Bettina Böttcher's team demonstrated by electron cryomicroscopy that the binding occurs at the tips of the spikes at the dimer interface and the stretch SLLGRM was mapped as the core binding motif (Makbul et al. 2021). Remarkably, this shortened peptide bound only to one of the two spikes in the asymmetric unit of the capsid and induced a much smaller conformational change.

The tandem HBc technology (Shepherd et al. 2013; Holmes et al. 2015; Peyret et al. 2015a, b; Kazaks et al. 2017), which is described earlier, could be regarded as one of the most promising functionalization approaches.

In parallel to the Qβ VLPs, as described in the Chapter 25, the incorporation of the unnatural amino acid

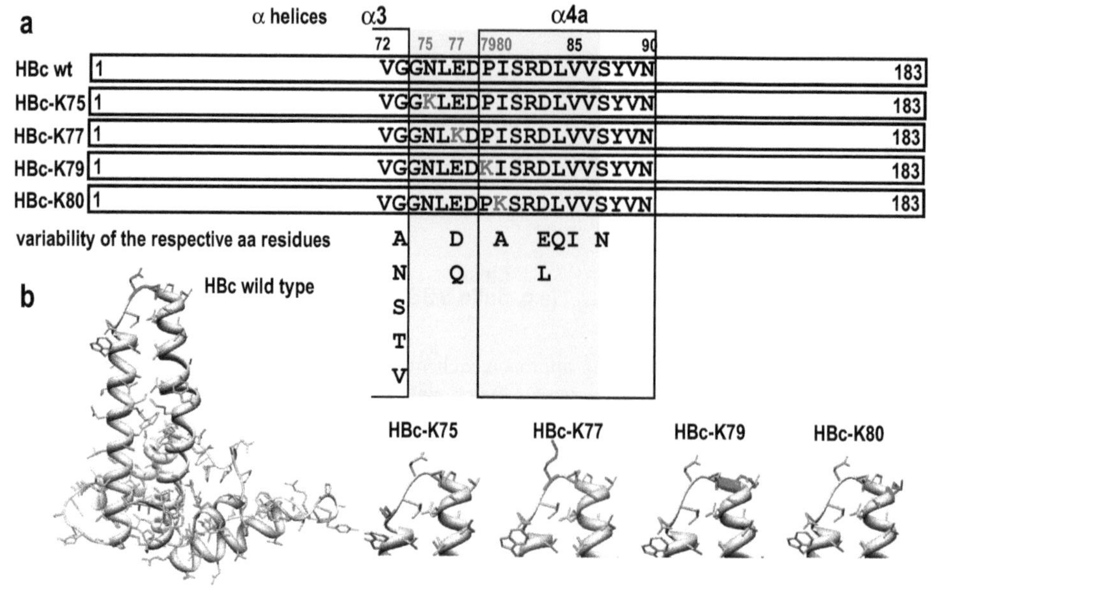

FIGURE 38.7 General structure of the initial wt HBc molecule from the HBV320 genome, genotype D1, subtype *ayw2*, GenBank accession number X02496 (Bichko et al. 1985) and four variants exposing lysine residues at the tips of the spikes. (a) Primary structure of the central part of the HBc molecule, with alternative naturally occurring aa residues. (From Pumpens and Grens 2001; Chain and Myers 2005.) (b) 3D maps for the initial HBc monomer with Glu77 and Asp78 marked red and for the tips of the spikes of four lysine-exposing HBc variants with inserted lysine residues marked red. The maps are based on the crystal structure of recombinant HBc VLPs produced in bacteria. (Reprinted from Strods et al. *Sci Rep*. 2015;5:11639.)

residue azidohomoalanine (AHA), an analogue of methionine displaying an azide. as well as of the homopropargyl glycine was invented by Strable et al. (2008). This labeling with azide- or alkyne-containing unnatural aa residues at position Met66 occurred by the HBcΔ expression in a methionine auxotrophic strain of *E. coli*, which was followed by the Cu^I^-catalyzed cycloaddition. As stated earlier, this pioneering labeling was performed with both HBc and Qβ VLPs, i.e. with the two most popular VLP carriers that have been applied together in numerous classical VLP studies. However, the HBc VLPs were decomposed by the formation of more than 120 triazole linkages per capsid in a location-dependent manner, whereas the Qβ VLPs suffered no such instability (Strable et al. 2008).

Further development of the functionalization of the HBc VLPs was possible due to the *E. coli* cell-free protein synthesis (CFPS) technique. The CFPS excluded possible cell-driven problems and allowed rapid screening of the "improved" or partially "artificial" versions of the HBc VLPs. As mentioned earlier, James R. Swartz's team succeeded in the high-level cell-free production of the HBcΔ VLPs (Bundy et al. 2008; Bundy and Swartz 2011). Next, this team used the CFPS system to display unnatural aa residues with alkyne or azide functional group into HBc molecule for the direct protein–protein coupling of antigens to VLPs using Cu^I^-catalyzed azide-alkyne cycloaddition reactions (Lu Yuan et al. 2013). Thus, an azide moiety was exposed on the surface of the HBcΔ VLPs when a methionine codon was introduced to encode residue 76 for

incorporation of the unnatural aa residue, namely AHA. In parallel, the M66 codon was replaced with a serine codon to avoid introduction of the AHA residue at this site. As a result, Lu Yuan et al. (2013) decorated the functionalized HBcΔ VLPs with flagellin that was also synthesized by the CFPS system.

Moreover, Lu Yuan et al. (2015) applied the CFPS to rapidly produce and screen HBc protein variants that still self-assemble into VLPs. This was the first study that focused on the sequence plasticity of VLP monomers and opened a general approach for functional improvement of natural VLPs. Thus, to improve stability of the HBcΔ VLPs, artificial covalent disulfide bridges were introduced throughout the VLP, by replacing D29 and R127 residues with cysteine residues, and negative charges on the HBc VLP surface were then reduced to improve surface conjugation. The solubility and assembly as well as surface conjugation were greatly improved by transplanting a rare spike region onto the common shell structure. The newly stabilized and extensively modified HBc VLPs had almost no immunogenicity in mice, demonstrating great promise for medical applications (Lu Yuan et al. 2015).

The previously mentioned providing of the HBcΔ VLPs with the N-terminal His_6 tags appeared in fact as one of the great functionalization approaches. Thus, the well-known M.G. Finn's team presented the first reported example of the use of the hexahistidine peptide sequences to anchor transition metal complexes for polyvalent display and mimicking of natural systems (Prasuhn et al. 2008). As

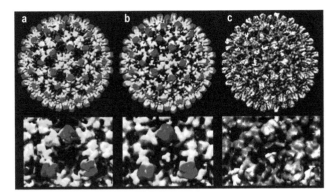

FIGURE 38.8 Structure of His$_6$-tagged HBcΔ VLPs. (a and b) Electron cryomicroscopic reconstruction of (a) wild-type HBc (resolution 9.2 Å) and (b) HBcΔ(His$_6$)$_{240}$ (resolution 8.3 Å). (c) Difference map, showing in yellow the density observed in HBΔ(His$_6$)$_{240}$ but not in wild-type HBc, overlaid on the wild-type HBc density in white. The yellow material therefore presumably represents the His$_6$ tags inserted at the N terminus of each protein subunit, three of which come together at the 3-fold symmetry axis. Close-up views down a 3-fold axis are shown below the whole-particle images. (Reprinted with permission of Cell Press from Prasuhn et al. *Chem Biol.* 2008;15:513–519.)

shown in Figure 38.8, the addition of 240 N-terminal His$_6$ tags gave rise to a particle with 80 sites of high local density of histidine side chains. Iron protoporphyrin IX was found to bind tightly at each of these sites, making a polyvalent system of well-defined spacing between metalloporphyrin complexes. The spectroscopic and redox properties of the resulting particle were consistent with the presence of 80 site-isolated bis(histidine)-bound heme centers, comprising a polyvalent b-type cytochrome mimic (Prasuhn et al. 2008).

For the simple detection reasons, the HBc-derived VLPs were labeled with popular diagnostic agents, such as Alexa Fluor™ 488 or technetium-99m (Suffian et al. 2017).

In order to reduce the HBc antigenicity, which could hamper clinical applications, the HBc VLPs were shielded with an amine-functionalized hydrophilic biodegradable polymer, poly(2-ethyl-2-oxazoline) (PEtOx-NH$_2$) was synthesized using the living cationic ring-opening polymerization (CROP) technique and covalently conjugated to the protruding spikes of the HBc VLPs via carboxyl groups (Fam et al. 2019)

At last, Tsuda Shugo et al. (2018) reported the synthesis of the HBcΔ, or Cp149-NH$_2$ by authors' terminology, where a prospective easy-to-attach/detach solubilizing tag-aided chemical approach was used to handle the aggregative core peptides.

After HBc VLPs, the DHBc VLPs were also used as a carrier by chemical coupling (Gathuru et al. 2005). Thus, mucin-1 was coupled to the DHBc VLPs and potency of the latter as an immunological carrier, like classical KLH (keyhole limpet hemocyanin), was demonstrated.

VACCINES

NATURAL VLPs

The natural HBc VLPs are still regarded as a possible prophylactic and/or therapeutic vaccine or at least as a crucial component of both possible hepatitis B vaccines in question. The real trials to protect chimpanzees with the HBc VLPs as a prophylactic hepatitis B vaccine, although not fully convincing, were performed more than 30 years ago by the Ken Murray's team as the first HBc vaccination pioneers (Murray et al. 1984, 1987; Iwarson et al. 1985).

The next step to a core-driven *Hepadnaviridae* vaccine was taken by the Michael Roggendorf's team, who demonstrated the WHc VLPs as a necessary and sufficient agent to protect woodchucks against WHV infection (Roos et al. 1989; Schödel et al. 1993a; Menne et al. 1997). Woodchucks were protected against WHV infection not only with the WHc but also with the HBc VLPs (Schödel et al. 1993a). Moreover, DNA immunization of woodchucks with plasmids expressing both WHc and WHs efficiently suppressed WHV infection (Lu M et al. 1999). The effect of the WHc DNA vaccine was enhanced by coadministration of γ-interferon (Siegel et al. 2001) or interleukin-12 (García-Navarro et al. 2001). Furthermore, Wang J et al. (2007) showed that the bicistronic WHc and γ-interferon vaccine protected animals from WHV infection but did not elicit sterilizing antiviral immunity. Comprehensive reviews on the natural WHc vaccines were published at that time by Lu M and Roggendorf (2001), Billaud et al. (2007), and Roggendorf et al. (2007, 2010, 2015). Later, Roggendorf's team observed different immune response features by immunization of mice with DNA or protein prototypes of the WHc vaccine (Zhang E et al. 2015).

Remarkably, the whole-cells expressing the DHBc after transfection with the appropriate DNA plasmid functioned as a therapeutic vaccine that was able to resolve chronic DHBV infection by induction of a DHBc-specific immune response (Miller et al. 2006).

The success of the WHc VLPs as a vaccine against WHV in woodchucks stimulated further interest to the HBc VLPs as a vaccine against HBV infection in humans. Thus, high mucosal immunogenicity including cellular response to the HBc VLPs was demonstrated in mice after nasal administration (Lobaina et al. 2003, 2006).

The high mucosal immunogenicity after oral and nasal administration was found also when plant-derived HBc VLPs were tested in mice (Huang Z et al. 2006). Furthermore, the HBc VLPs from tobacco (Pyrski et al. 2017) and lettuce (Pyrski et al. 2019) were presented as real therapeutic vaccine candidates.

By another approach, the attenuated *Salmonella* strains expressing the HBc protein were used for nasal immunization of mice, instead of the purified HBc VLPs (Nardelli-Haefliger et al. 2001; Stratford et al. 2005). Moreover, empty bacterial cell envelopes, namely *E. coli* ghosts carrying HBcΔ VLPs anchored either in the inner or on the outer membrane, induced significant immune responses by subcutaneous immunization of mice (Jechlinger et al. 2005). In parallel, the purified HBcΔ as a putative therapeutic hepatitis B vaccine raised a genuine interest from the well-known Rhein Biotech company.

The adjuvating effect of the HBc VLPs by the simultaneous immunization of mice with other proteins including HBs, because of RNA bound to the arginine-rich CTD of the HBc, was demonstrated by Reinhold Schirmbeck's team (Riedl et al. 2002a, b). Later, this team clearly demonstrated that particle-bound mammalian RNA functioned as TLR7 ligand and induced a Th1-biased humoral immunity in B6 but not in TLR7$^{-/-}$ mice by protein and DNA vaccination (Krieger et al. 2018, 2019). Moreover, prevention of specific phosphorylation in cationic domains, either by exchanging the serine residues S155, S162 and S170 with alanines or by exchanging the entire cationic domain with other cationic domains, enhanced the encapsidation of RNA into mutant core particles. Therefore, both endogenous bacterial and mammalian RNAs functioned as a natural adjuvant facilitating priming of Th1-biased immune responses by protein or DNA-based immunization, respectively.

The high-level immunogenicity and adjuvanting effect of the HBc VLPs led to idea of combination with the traditional HBs vaccine. Thus, the therapeutic nasal HBc/HBs vaccine HeberNasvac, later simply NASVAC, was designed in the Cuba's laboratories (Palenzuela et al. 2002; Aguilar et al. 2004; Lobaina et al. 2005, 2010, 2015; Betancourt et al. 2007; Lobaina Mato et al. 2016; Lopez et al. 2017; Bourgine et al. 2018; Fernández et al. 2018). A preliminary study showed that two different HBc variants, one of which contained eight residual aa in the C-terminal region, did not differ in their adjuvant effect on the HBs protein (Lobaina et al. 2005, 2006). It is necessary to emphasize that the phase I clinical trial performed by Betancourt et al. (2007) was the first demonstration of safety and immunogenicity for a nasal vaccine candidate comprising both HBs and HBc antigens.

Later, the HBc suitability as a candidate for a therapeutic vaccine was corroborated in Japan and Bangladesh (Akbar et al. 2012, 2013, 2016) and provoked collaborative studies of both teams (Al-Mahtab et al. 2013, 2018). The NASVAC vaccine underwent phase IIb/III clinical trials—the last one a phase III study by Al-Mahtab et al. (2018), for chronically infected hepatitis patient—and was produced by the ABIVAX company. Recently, an exhaustive study in a *Tupaia* model delt with possible formulations of this intranasal therapeutic vaccine (Sanada et al. 2019).

It is remarkable that the HBc/HBs combination was used as an adjuvant in the Teravac-HIV-1 vaccine, Cuba's anti-HIV vaccine based on a HIV multiepitopic recombinant protein (Iglesias et al. 2006, 2008; García-Díaz et al. 2013).

The Ulrike Protzer team, together with the Rhein Biotech company, contributed strongly to the HBc usage in a therapeutic vaccination approach (Buchmann et al. 2013). These activities resulted in the therapeutic TherVac vaccine that involved immunization with both HBc and HBs

VLPs (Backes et al. 2016). As a result, the complex vaccination schemes succeeded on the appropriate mouse models (Kosinska et al. 2019; Michler et al. 2020).

A therapeutic vaccine DV-601 that also combined HBs and HBc underwent phase I clinical trials (Spellman and Martin 2011).

The scientific validation of the therapeutic HBc together with HBs vaccination was offered in the numerous reviews (Michel and Mancini-Bourgine 2005; Kutscher et al. 2012; Barnes 2015; Elvidge 2015; Akbar et al. 2016; Lobaina and Michel 2017; Lobaina Mato 2019; Ezzikouri et al. 2020; Meng Z et al. 2020).

In parallel, a combination of the prospective combined HBcAg/HBsAg vaccine with immune stimulating oligonucleotide CpG adjuvant was presented (Li Jianqiang et al. 2015).

As in the case of the HBs immunization described in the *DNA Vaccines* paragraph of Chapter 37, numerous HBc DNA vaccine candidates were elaborated (Townsend et al. 1997; Sällberg et al. 1998; Wild et al. 1998; Shao et al. 2003; Xing et al. 2005). The DNA immunization seemed especially promising for the previously evaluated HBc and HBs combination. Thus, the HBs and HBc genes were coexpressed using a synthetic bidirectional promoter under the control of the tetracycline-inactivated transactivator. A genetic construct containing this expression unit was transfected into mammalian cells and was also injected into mice (Kwissa et al. 2000). Coadminstration of both genes as the HBc/HBs DNA vaccine to mice induced an immune response against both antigens. No interference between immunoresponses to both antigens were observed (Musacchio et al. 2001). The first trial of therapeutic immunization with the DNA vaccine expressing both HBs and HBc genes in chronically infected chimpanzees was successful in an animal having a relatively low viral load but was less successful in an animal with a high viral load (Shata et al. 2006).

The ideas based on the superiority of the DNA vaccines over protein-based ones continue to exist (Chen M et al. 2016). Thus, INO-1800, a DNA vaccine (Obeng-Adjei et al. 2012, 2013) encoding two HBcAg and HBsAg variants of the A and C HBV genotypes, was studied under phase I clinical trial (cited from Elvidge 2015). The DNA vaccine HB-110 (Yoon SK et al. 2015), which included the HBc plasmid, also underwent a phase I clinical trial by the Ichor Medical Systems company.

Brass et al. (2015) developed a therapeutic DNA vaccine candidate encompassing the codon-optimized HBc and IL-12 expressing plasmids, which were delivered by targeted high-pressure injection combined with in vivo electroporation. The fine mechanisms of the HBc action by the DNA vaccination were unveiled by Matti Sällberg (2015).

GENETIC FUSIONS

As mentioned earlier, Table 38.2 summarizes the most complete records on the generation of chimeric HBc VLPs by genetic fusion, primarily with an aim to construct immunologically active substances, first and foremost, vaccines.

The first chimeric HBc-derived vaccine candidate that was elaborated in 1980s by David J. Rowlands's team against foot-and-mouth disease virus (Clarke et al. 1987; Newton et al. 1987) was in fact one of the first attempts to invent chimeric VLPs as real vaccines, in parallel with two other attempts of the middle 1980s: HBs (see Chapter 37) and tobacco mosaic virus (see Chapter 19).

From the next numerous attempts, which were performed by genetic fusion and are collected in Table 38.2, the idea of a combined therapeutic and prophylactic vaccine against hepatitis B is worth mentioning first. From the very beginning, the potentially virus-neutralizing preS and S epitopes were regarded as the best candidates to be exposed on the HBc VLPs for this purpose. The appropriate HBc VLPs carrying preS or S epitopes were generated first by Elmārs Grēns's team (Borisova et al. 1989, 1990, 1993, 1999; Borschukova et al. 1997; Fehr et al. 1998; Kazaks et al. 2004; Skrastina et al. 2008) and Ken Murray's (Stahl and Murray 1989; Shiau and Murray 1997; Murray and Shiau 1999) teams and by Florian Schödel and colleagues (Schödel et al. 1990b, 1991, 1992, 1993b, 1994a, b). In the late 1990s, the British company Medeva started an ambitious *Hepacore* project headed by Mark Page and performed in tight collaboration with Elmārs Grēns's team, which aimed at the construction of a real therapeutic preS1-carrying vaccine based on the HBc VLP vector. A healthy volunteer study using chimeric HBc VLPs exposing the preS1 sequence 20–47 inserted into the MIR was planned for 2000 but never occurred after joining the Medeva to the Celltech company. A set of the HBc VLPs carrying different fragments of the preS1 were used to map the immunodominant stretches of the preS1 attachment site (Bremer et al. 2011). Later, Matti Sällberg's team added additional reasons for the HBc-preS vaccine (Malik et al. 2012). Moreover, Dishlers et al. (2015) opened novel biotechnological opportunities for the high-quality purification, standardization, and quality control of the genetically fused HBc-preS versions. The idea of a therapeutic vaccine combining both HBc and preS was supported by Wolfram H. Gerlich (2017).

At last, Whitacre et al. (2020) generated a real therapeutic vaccine candidate, when exposed neutralizing preS1 epitopes on the full-length WHc VLPs. The immunization of immune tolerant HBV transgenic mice with the WHc-preS1 VLPs elicited high levels of neutralizing anti-preS1 antibodies, passive transfer of which into human-liver chimeric mice prevented acute infection and cleared serum HBV from mice previously infected with HBV in a model of chronic hepatitis B. At the T cell level, the WHc-preS1 VLPs elicited HBc-specific CD4+ Th and CD8+ CTL responses. This study made a real contribution toward the long-awaited therapeutic hepatitis B vaccine.

Meanwhile, the old idea of chimeric HBc VLPs with the major HBs epitope exposed by genetic fusion (Borisova et al. 1993) was reactivated by Su et al. (2013) and seems to deserve further evaluation. Moreover, Mobini et al. (2020)

performed computer modeling and succeeded in the generation of the HBc dimer-based VLPs that carried the preS1 aa residues 1–50 in the first monomer and the HBs "a" determinant 118–150 in the second monomer. The stereochemistry and geometry of the final VLPs were investigated by the following molecular dynamics simulation (Mobini et al. 2020).

Furthermore, a possible combination of the therapeutic potency against HBV and HCV within the same HBcΔ VLPs (Mihailova et al. 2006; Sominskaya et al. 2010) remains a realistic goal in the very near future.

Concerning nonconformist HBc-based vaccine candidates against hepatitis B, the whole yeast vaccine GS-4774 Tarmogen, where GS is for Gilead Sciences, could be mentioned here, although it does not belong exactly to the HBc VLP-based vaccines. It consists of heat-inactivated yeast cells that express HBV antigens including a fusion protein of HBcAg and HBsAg, a fusion protein expressing the most highly conserved regions of HBxAg, HBsAg, and HBcAg, namely, X-S-Core or GI-13020/GS-4774 (Gaggar et al. 2014; King et al. 2014). The immunogenicity of the GS-4774 vaccine was found in murine and human ex vivo T-cell models, as a proof of concept justifying assessment in human clinical trials for the treatment of chronic hepatitis B (Boni et al. 2018).

After hepatitis B, influenza appeared as the next promising goal of the HBc-based VLP vaccines, when the M2 epitope was inserted at the N-terminus of the HBc molecule. (Neirynck et al. 1999; Fiers et al. 2004, 2009; De Filette et al. 2005, 2006a, b, 2008; Schotsaert et al. 2009, 2016; Ibañez et al. 2013; Kolpe et al. 2017). The WHc VLPs was used as a basis of a prospective influenza vaccine candidate (Ameiss et al. 2010). Then, as mentioned earlier, the tandem HBc technology was used to expose the conserved alpha-helix (LAH) region of hemagglutinin (HA) within the HBc MIR and guaranteed effective immune responses and strong protection in mice (Kazaks et al. 2017).

Malaria was always in the spotlight of the HBc VLP-based vaccines (Milich et al. 2001; Birkett et al. 2001, 2002; Sällberg et al. 2002; Nardin et al. 2004; Langermans et al. 2005; Oliveira et al. 2005; Walther et al. 2005; Gregson et al. 2008). After longitudinal studies of the most suitable malaria vaccine, the David R. Milich's team elaborated a prospective candidate that was based on the WHc vectors and systematically explored dozens of VLP variants with simultaneous insertions of epitopes at the internal (for B cell epitopes) and C-terminal site (for T cell epitopes; Whitacre et al. 2015a, b). For the first time in the VLP history, the chimeric WHc VLPs elicited sterile immunity to blood stage malaria in a mouse model and provided the strong basis for a bivalent *P. falciparum/P. vivax* malaria vaccine.

Rainer G. Ulrich and his colleagues generated a set of putative vaccine candidates against hantavirus infection (Ulrich et al. 1998, 1999; Koletzki et al. 1997, 1999, 2000; Krüger et al. 2001; Geldmacher et al. 2004, 2005). The previously mentioned ingenious SplitCore methodology has led to a prospective vaccine against Lyme disease caused by *Borrelia burgdorferi* (Walker et al. 2008, 2011).

Concerning tuberculosis, Dhanasooraj et al. (2013, 2016) inserted culture filtrate protein 10 (CFP-10) of *Mycobacterium tuberculosis*, an important vaccine candidate, into the HBcΔ MIR. The authors compared the pure protein, a mixture of antigens, and the chimeric HBcΔ-CFP10 VLPs by immunization of mice and found that the chimeric VLPs generated the stronger antigen-specific immune response in a Th1-dependent manner.

As for vaccines against noninfectious diseases, a prospective antihypertension vaccine candidate deserves special attention (Wu H et al. 2020). Remarkably, this genetically fused HBcΔ-based candidate was generated and tested in parallel with the chemically coupled ATRQβ-001 based on the RNA phage Qβ VLPs (see Chapter 25), and the HBcAg-CE12-CQ10 vaccine provided a novel and promising therapeutic approach for hypertension. Mention should be made here of the studies where preventive immunization of mice with the appropriate chimeric HBc VLPs suppressed growth of tumors (Chu X et al. 2016).

Even more than in the case with the natural HBc and WHc DNA expression vectors and DNA vaccines described earlier in the *Expression of the Core Gene* and *Vaccines: Natural VLPs* paragraphs, respectively, the DNA vectors were beneficial for the chimeric HBc-derived fused genes, which is not surprising, since an obvious advantage of DNA vaccination consisted in the elimination of always critical self-assembly and low-production barriers typical for the gene engineering at the protein level. For example, a genetic HBcΔ-preS1 fusion that failed to assemble into VLPs, when expressed in *M. smegmatis*, served as a modest DNA vaccine (Yue et al. 2007).

Chemical Coupling

As demonstrated earlier in the *Chimeric VLPs: Functionalization and Chemical Coupling* paragraph, Martin F. Bachmann's team (Jegerlehner et al. 2002b) outlined the prospects of the chemically coupled HBc vaccines in three directions, against (i) influenza (M2 peptide), (ii) allergy (PLA2 peptide), and (iii) chronic noninfectious diseases (a peptide derived from TNFα). The last idea that relied on the ability of highly repetitive epitopes exposed on a VLP carrier to break B-cell tolerance and induce self-specific antibody responses was developed further by generation of numerous therapeutic vaccines on the RNA phage VLP vectors (see Chapter 25). Nevertheless, it is important to remember the HBc as the first model that demonstrated the ability of the coupled peptides to induce strong antibody response to self-specific peptides in the absence of adjuvants. Moreover, these antibodies were cross-reactive with native TNFα. This finding suggested a potential for vaccination with VLPs to block the function of self-molecules, such as TNFα, in order to treat certain chronic diseases, such as arthritis, colitis, and asthma.

At the same time, Jegerlehner et al. (2002a) performed the chemical coupling of the model D2 peptide in parallel with the HBc and RNA phage Qβ VLPs and passed down the initiative therefore from the HBc to the Qβ as a leading carrier.

The work on the chemically coupled HBc VLP vaccines was continued methodically only with the putative influenza vaccine. Jegerlehner et al. (2002b) compared the immunogenicity in mice of both genetically fused and chemically coupled HBc-M2 VLPs and found that the M2-fusion induced the weaker immune response. The authors explained this phenomenon by poor accessibility of the genetically fused M2 to B cells, since they were partly buried within the particle while the coupled M2 peptide was maximally exposed on the tip of the HBc MIR. Moreover, mice vaccinated with M2 peptides coupled to VLPs were completely protected from lethal challenge with 10×LD$_{50}$ doses of influenza virus, when mice vaccinated with M2 peptide fused to VLPs were not protected. At last, viral titers in the lungs of mice vaccinated with M2 coupled to VLPs were very low whereas mice vaccinated either with M2 peptide or control mice exhibited high viral titers in the lung seven days after challenge with LD$_{50}$ doses of influenza virus (Jegerlehner et al. 2002b).

However, further detailed evaluation of the HBc-M2 vaccine led to the conclusion that the induced anti-M2 antibodies neither bound efficiently to the free virus nor neutralized virus infection but bound to M2 protein expressed on the surface of virus-infected cells (Jegerlehner et al. 2004). It was supposed that protection was mediated via antibody-dependent, cell-mediated cytotoxicity. The absence of neutralizing antibodies resulted in much weaker protection than that achieved by vaccination with UV-inactivated influenza virus. Thus, whereas neutralizing antibodies completely eliminated signs of disease even at high viral challenge doses, M2-specific antibodies did not prevent infection but merely reduced disease at low challenge doses. The M2-specific antibodies failed to protect from high challenge doses, as vaccinated mice underwent lethal infection under these conditions. The authors concluded that the protection mediated by the HBc-M2 VLPs would be insufficient during the yearly epidemics—for which full protection was desirable—and overall was clearly inferior to protection achieved by immunization with classical inactivated viral preparations (Jegerlehner et al. 2004).

After ten years, Blokhina et al. (2013b) used the nanoglue as a "binding tag" of M2e protein. These HBc-M2 VLPs appeared to be highly immunogenic and protected immunized mice against a lethal influenza challenge. However, when challenged with 5×LD$_{50}$ of mouse-adapted influenza strains, the immunization did not prevent morbidity but significantly reduced it relative to the control. It is necessary to mention, however, that the immune response of the HBc-M2 complex was enhanced by supplementing with murine IL2 that carried the same nanoglue peptide and was able to associate therefore with the HB carrier, as was defined before (Blokhina et al. 2013a). Later, Blokhina and Ravin (2018) turned, however, to the genetic fusion of the influenza HA2 sequence to the N-terminus and of the M2e peptide to the MIR of the HBc protein. At last, Swanson et al.

(2021) compared the HBc VLPs carrying the genetically fused and glued epitopes, namely an FMDV epitope from the VP4 protein. Remarkably, the VLPs with the VP4 epitope inserted into the MIR induced VP4-specific antibodies in mice, while the VLPs with the same peptides attached to the spikes did not (Swanson et al. 2021).

As just mentioned earlier and illustrated in Figure 38.7, Regīna Renhofa's group generated a set of universal HBc vectors with a special aim to be employed for the chemical coupling of peptides (Strods et al. 2015).

Furthermore, the chemical coupling was critical, when conjugating the polysaccharides to VLPs was found necessary. Thus, Xu Lingling et al. (2019) generated the first HBc-based polysaccharide conjugate vaccine candidate, namely against meningitis caused by *Neisseria meningitidis*. The meningococcal group C polysaccharides were conjugated to the full-length HBc VLPs by heterobifunctional polyethylene glycol (PEG) linkers of different length (2, 5 and 10 kDa). The immunization with the conjugate vaccines generated about a tenfold increase in polysaccharide-specific IgG titers and induced a shift to a Th1 cellular immune type response, as assessed by the increased IgG2a subclass production.

Bayliss et al. (2020) employed tandem HBc VLPs to expose capsular polysaccharide of *Burkholderia thailandensis* by development of the melioidosis vaccine. To use the tandem HBc VLPs for chemical conjugation to polysaccharide antigens, six lysine residues flanked on either side by three aspartic acid residues were inserted into the MIRs. The chimeric tandem HBc VLPs were produced then in *N. benthamiana* leaves, and capsular polysaccharide was oxidized and conjugated to carrier proteins by reductive amination. The conjugate vaccine was able to protect mice against intraperitoneal *B. pseudomallei* challenge of multiple median lethal dose (Bayliss et al. 2020).

These two studies pave the way for the active development of the HBc VLP-based polysaccharide conjugate vaccines effective in eliciting long-lasting and strong cellular immune response.

Cheng et al. (2020) combined chemical coupling with the genetic fusion by the engineering of the "hybrid" HBc VLPs that demonstrated therapeutic effect against melanoma in model mice. The "hybrid" VLPs were composed of the two *E. coli*-produced monomers carrying each the gp100 epitope KVPRNQDWL or the OVA epitope 257-SIINFEKL-264 inserted between aa 78 and 79 of the MIR, His$_6$ tag at the C-terminus and conjugated then with fluorescent dyes (Cyanine5.5, Cy 5.5) or the corresponding fluorescence quenchers, namely Black Hole Quencher 3 (BHQ-3), respectively. In fact, the fluorescence was quenched when the chimeric monomers reassembled into the complex "hybrid" VLPs (Cheng et al. 2020). Another efficient tumor vaccine based on the HBc VLPs carrying the same OVA epitope was recently prepared by combination with mesoporous silica nanoparticles as an adjuvant, where both components were capsulated into microneedles together (Guo Q et al. 2021).

PACKAGING

This incredibly important route of nanotechnological HBc applications is based on the HBc VLP ability to package nucleic acids, proteins, polypeptides, low molecular mass drugs, and inorganic nanoparticles, where only the problem of RNA and DNA packaging could be regarded nowadays as adequately solved.

It is essential to remember here that the natural HBc packaging with the intrinsic RNA relates to phosphorylation of the nucleocapsids, which was shown necessary and sufficient for the pgRNA encapsidation (Gazina et al. 2000). Thus, the pgRNA is packaged by the phosphorylated cores, while dephosphorylation of the cores enables synthesis of genomic dsDNA (Basagoudanavar et al. 2007) and triggers maturation, i.e. envelopment of cores and secretion of virions (Mabit and Schaller 2000; Perlman et al. 2005). As mentioned earlier, the mature DNA-containing HBc particles demonstrate significant differences in structure versus immature RNA-containing particles (Roseman et al. 2005). In opposite to RNA, DNA appears therefore as a poor substrate for encapsidation and the dsDNA-filled HBc particles are spring-loaded (Zlotnick et al. 1997). Moreover, it is noteworthy that a significant portion of the CTD is exposed at the surface of the immature RNA-containing HBc particles, whereas the CTD is mostly confined within mature DNA-containing HBc particles (Meng D et al. 2011).

The full-length CTD-carrying HBc VLPs from *E. coli* cells package heterologous RNA in amounts comparable with the pgRNA content (Zlotnick et al. 1997; Porterfield et al. 2010; Sominskaya et al. 2013). The yeast-produced CTD-carrying HBc VLPs reveal packaging of approximately 3,000–3,200 nt (Strods et al. 2015). The bacteria-produced HBc VLPs demonstrate a higher level of packaging by at least 300 nt, which could be correlated with at least 240 phosphogroups on the inner HBc VLP surface because of phosphorylation of the HBc protein in yeast, while bacterial HBc VLPs remain unphosphorylated (Freivalds et al. 2011).

This packaged heterologous RNA may function as an adjuvant and stimulate the T cell specific response in immunized animals (Zlotnick et al. 1997; Riedl et al. 2002a, b; Vanlandschoot et al. 2005; Broos et al. 2007; Porterfield et al. 2010; Sominskaya et al. 2013; Strods et al. 2015).

Like natural HBc within viral nucleocapsids during the recognition of pgRNA, recombinant HBc VLPs prefer single-stranded RNA for packaging, whereas the elimination of the CTD domain prevents the RNA packaging in *E. coli* cells (Borisova et al. 1988; Gallina et al. 1989; Nassal 1992). In eukaryotic cells in vivo, the encapsidation of short hairpin RNAs by HBc nucleocapsids resulted in the construction of the HBV "Trojan horse" vector that targeted hepatocytes (Shlomai et al. 2009).

In early experiments, HBc demonstrated a definite affinity not only to RNA, but also to DNA in vitro (Hatton et al. 1992). In the later trials, the HBc demonstrated maximal affinity for the pgRNA and random ssRNA in total

(Newman et al. 2009; Porterfield et al. 2010), to a lesser extent for ssDNA and minimally for dsDNA (Newman et al. 2009; Dhason et al. 2012), irrespective of phosphorylation.

The Bachmann team were the first who demonstrated that the immunogenicity of chimeric HBc VLPs was enhanced by mixing with the CpG oligonucleotides as TLR9 ligands (Storni et al. 2002, 2003; Schwarz K et al. 2003; Storni and Bachmann 2003). Moreover, it appeared that the packaging of CpGs into VLPs, HBc, or RNA phage Qβ was the simplest and most attractive method to avoid the two major CpG adjuvanting problems (Storni and Bachmann 2004; Storni et al. 2004), as described in the *Packaging and Targeting: CpGs* paragraph of Chapter 25. In short, the CpGs packaged into VLPs were first resistant to DNase I digestion, enhancing their stability. Second, in contrast to free CpGs, the packaging CpGs prevented splenomegaly in mice, without affecting their immunostimulatory capacity. Later, successful encapsidation of CpGs was achieved by other groups (Song et al. 2007, 2010; Kazaks et al. 2008).

In parallel, the HBc-driven packaging was described for short oligodeoxynucleotides, other than CpG (Cooper and Shaul 2005).

The controlled encapsidation and quality control of bacteria- and yeast-derived HBc VLPs were hindered by the presence of random internal RNA of host origin. Strods et al. (2015) were the first who achieved full deprivation of both bacteria- and yeast-derived natural and chimeric HBc VLPs from the contaminating encapsidated RNAs during simple, rapid, and highly efficacious alkaline treatment at pH 12. In fact, the alkali treatment method to remove the encapsidated RNA was used first by the Matthew B. Francis team on the natural RNA phage MS2 VLPs (Hooker et al. 2004). Then, the alkaline treatment was applied for purification of the chimeric RNA phage PP7 VLPs carrying human papillomavirus epitopes (Tumban et al. 2013). These studies are described in Chapter 25.

Interestingly, the incorporated RNA could be removed by a 17-kD nuclease, which was fused to the position 155 of the HBcΔ protein (Beterams et al. 2000). The packaged nuclease retained enzymatic activity, and the chimeric protein was able to form mosaic particles with the wild-type HBc protein.

As to methodology of packaging, traditional attempts to prepare full-length HBc particles for in vitro packaging by desired molecules were based on the two approaches: (i) non-dissociating osmotic shock (Kann and Gerlich 1994) and (ii) full HBc dissociation after micrococcal nuclease (Newman et al. 2009) or guanidine chloride (Porterfield et al. 2010) treatment.

Regīna Renhofa's group elaborated a simple RNA and DNA packaging methodology (Strods et al. 2015). Thus, the empty HBc particles, after alkali treatment to remove random RNA, encapsidated both RNA and DNA by the following two alternate approaches: (i) the restoration of HBc VLPs after 7 M urea treatment and (ii) direct contact of nucleic acids with empty HBc VLPs. Remarkably, the encapsidation of bacterial rRNA and some specific mRNAs was accompanied by the cleavage of the packaged RNA. This phenomenon suggested that HBc itself may possess ribonucleolytic activity that could prevent the encapsidation of unspecific RNAs in vivo (Strods et al. 2015).

According to Strods et al. (2015), the HBc VLPs protected the long double-stranded DNAs, just exceeding the length of the HBV genome, from DNase attack by forming VLP chains that covered the DNA. This protection was mediated by a strong excess of HBc VLPs compared with the DNA. When the VLP to DNA ratio was approximately equimolar, HBc VLPs packaged dsDNA fragments of approximately 1200 bp in length. Nevertheless, the most efficient packaging was achieved for single-stranded CpGs during the strong oligonucleotide excess over the HBc VLPs. No significant differences were observed in the ability of both bacteria- and yeast-produced HBc VLPs to reassemble and package desired RNAs or DNAs.

After RNA and DNA, other polyanions (poly-glutamic acid and polyacrylic acid) but not low molecular mass anions (inositol triphosphate) or polycations (polylysine and polyethylenimine) were packaged into the HBc VLPs (Newman et al. 2009). Moreover, the HBcΔ VLPs were able to package magnetic Fe_3O_4 nanoparticles with cationic coating (Renhofa et al. 2011).

Later, the encapsidation of magnetic nanoparticles was described for HBcΔ VLPs carrying His tags (Shen et al. 2015). Continuing this approach, Zhang Q et al. (2018) generated the multifunctional nanotheranostic Fe_3O_4-MTX@ HBc core-shell particles. The magnetic nanoparticles were packaged into the engineered VLPs through the affinity of His tags for the methotrexate (MTX)-$Ni2^+$ chelate. The magnetic resonance imaging (MRI) results showed that the Fe_3O_4-MTX@HBc VLPs were reliable T_2-type MRI contrast agents for tumor imaging and could act as a promising theranostic platform for multimodal cancer treatment.

Malyutin et al. (2015) packaged into HBc VLPs the hydrophobic FeO/Fe_3O_4 nanoparticles that were prepared by thermal decomposition of iron oleate and coated with poly-(maleic acid-*alt*-octadecene) modified with PEG tails of different lengths and grafting densities. Remarkably, the packaged HBc VLPs allowed the PEG tails to extend through the capsid, in opposition to brome mosaic virus (BMV) VLPs, which preferentially entrapped the tails in the interior (see also Chapter 17).

At last, Rybka et al. (2019) used full-length plant-produced HBc VLPs to encapsulate superparamagnetic iron oxide nanoparticles (SPIONs), which are widely utilized by magnetic bioseparation, magnetic hyperthermia, targeted drug delivery, and in diagnostics as MRI contrast agents. The dihexadecyl phosphate (DHP) and PL-PEG-COOH were chosen for the SPION functionalization. Both compounds were successfully used in a previous study regarding brome mosaic virus (BMV) VLPs, as described in Chapter 17.

The gold nanoparticles, together with CpG, were encapsulated into full-length HBc VLPs (Wang Yarun et al.

2016). Remarkably, the packaging of gold nanoparticles increased CD4$^+$ and CD8$^+$ T cell numbers and interferon-γ, when compared to the HBc VLPs adjuvanted with conventional Freund's adjuvant and enhanced therefore immunogenicity of CpG and VLPs on both humoral and cellular immune pathways.

Regīna Renhofa from Elmārs Grēns's team and Yury Dekhtyar from Riga Technical University tried to package silica nanoparticles into the HBc VLPs but were confined by the exposure of the latter on silica. As a result, silica nanoparticles were recommended as a novel, efficient adjuvant to promote the immunological activity of the HBc VLPs (Dekhtyar et al. 2012; Skrastina et al. 2014).

In parallel, Regīna Renhofa succeeded also in the packaging of proteins into the HBc VLPs, results of which remain still unpublished. Meanwhile, Lee KW and Tan (2008) succeeded in packaging of GFP into the HBcΔ VLPs after dissociation with denaturing agents and reassociation in the presence of GFP.

Finally, Andris Dišlers's group from Elmārs Grēns's team succeeded by the efficient packaging of ssRNA and dsRNA fragments of different length into the HBc VLPs derived from the HBV genotypes D and G (Petrovskis et al. 2021). This was possible because of the well-elaborated protocol for the VLP dissociation and reassociation that always maintained the native HBc structure. Remarkably, the dsRNA of ~400 nt used in this work was the functional ingredient of a popular immunomodulatory drug Larifan, developed by Guna Feldmane in Riga and described in an RNA phage book (Pumpens 2020).

Concerning encapsulation of proteins, Wei et al. (2020) generated a biomimetic influenza vaccine by displaying the M2e on the exterior of HBc VLPs through genetic fusion and by thermal encapsulation of a conserved internal nucleoprotein (NP) antigen peptide inside the VLPs. For comparison, another non-biomimetic dual-antigen vaccine with interior M2e/exterior NP, and other four VLP-based single-antigen vaccines with NP or M2e either being encapsulated inside or genetically displayed outside the VLP were also constructed. The dual-antigen VLP influenza vaccines elicited both NP and M2e-specific antibodies, which were stronger than those elicited by the single-antigen vaccines. Most importantly, after a lethal challenge of H1N1 virus, the biomimetic dual-antigen vaccine conferred the mice 100% protection without noticeable body weight loss in the absence of any adjuvant (Wei et al. 2020).

TARGETING AND DELIVERY

The controlled and targeted delivery of the filled HBc nanocontainers remains the most urgent problem for the near future. Indeed, there are no satisfactory described mechanisms for the native uptake of the recombinant HBc VLPs in vivo and/or in vitro by different types of cells. The only exception is the observation that the HBc VLPs possess a natural target, namely B cells. These function as antigen presenting cells (APCs) for the HBc VLPs both in mice

(Milich et al. 1997) and humans (Cao et al. 2001; Lazdina et al. 2001, 2003). Furthermore, HBc nucleocapsids possess a highly specific nuclear targeting ability, which may allow for the highly specific delivery of packaged DNA to the cell nucleus (for review, see Kann and Gerlich 1997). Employing B cells as APCs for HBc protein explains its enhanced immunogenicity in mice and humans and the contribution of the HBc antigen in the induction and/or maintenance of chronic HBV infection (Milich et al. 1997). Therefore, specific HBc nanocontainer-driven gene and/or drug therapy against B cells could be possible without any additional decoration of the HBc VLPs by cell targeting sequences.

Providing HBc VLPs with specific cell targeting or addressing sequences seemed, however, the reasonable way to ensure the specific delivery of the HBc-carried nanocargo to the desired cells. As in the case of the first chimeric HBc VLPs, the first artificial HBc targeting attempts were performed by David J. Rowlands's team, and the RGD loop of the VP1 protein of foot-and-mouth disease virus (FMDV) was inserted into the HBc MIR and served as the first cell-targeting sequence employed (Chambers et al. 1996; Sharma et al. 1997; Choi et al. 2011). Then, the HBc-RGD VLPs were used to treat hepatocellular carcinoma (HCC) by the targeted delivery of the Pokemon siRNA to HCC xenografts in mice (Kong et al. 2015). Recently, Wang Yunlong et al. (2020) used the MIR-inserted RGD to target the resulting HBc VLPs that have been filled with the near-infrared dye indocyanine green (ICG) without any chemical modifications and achieve therefore the visual diagnosis of tumors.

After RGD, a translocation motif (TLM) of protein transduction domain (PTD) was used as an instrument (Brandenburg et al. 2005). The α-helical TLM peptide PLSSIFSRIGDP was derived from HBV preS2 and inserted into N-terminus of full-length HBc or HBcΔ. As a result, the authors designed cell-permeable TLM-PTD VLPs harboring marker genes and enabling an efficient gene transfer into primary hepatocytes. Furthermore, the HBcΔ-TLM VLPs carrying tandem TLM at the N-terminus were provided with the strep-tag at the MIR between aa residues 78 and 83, loaded with various antigens, and used to deliver cargo into the cell cytoplasm (Akhras et al. 2017). As mentinoned earlier, the strep tag allowed the coupling of foreign antigens, produced as streptavidin fusion peroteins, and that ensured the use of the cell-permeable VLPs as a universal antigen carrier. Using such cell-permeable VLPs loaded with ovalbumin as a model antigen, activation of antigen presenting cells and ovalbumin-specific CD8$^+$ T-cells, which correlated with enhanced specific killing activity, was found.

The HBV preS1 sequence that participates in the natural recognition of hepatocyte receptors remains the most intensively studied candidate for the targeted delivery of HBc transported nanocargos to hepatocytes (for references see reviews Pumpens and Grens 2001, 2002, 2016; Pumpens et al. 2008; Pushko et al. 2013). A first modest attempt to demonstrate the affinity of the preS1 bearing HBc VLPs to the appropriate eukaryotic cells was performed by Regīna Renhofa's group (Kalniņš et al. 2013). Another group presented a tricky method to display the preS1 sequence on the HBc VLP surface and to deliver such VLPs to liver cells (Lee KW et al. 2012a).

Ranka et al. (2013) described targeting of the HBc VLPs to fibronectin by exposing fragments of BBK32, a multifunctional surface lipoprotein, of *Borrelia burgdorferi*, the causative agent of Lyme disease.

The human epidermal growth factor receptor-related 2 (HER2)-expressing breast cancer cells were targeted with the Z_{HER2} affibody-decorated HBcΔ VLPs, where the Z_{HER2} affibody peptide was inserted into the HBcΔ MIR. (Nishimura et al. 2013; Suffian et al. 2017). The HBcΔ-Z_{HER2} VLPs retained ability for specific recognition of the HER2-expressing breast cancer cells, exhibited different binding amounts in accordance with the HER2 expression levels, and ensured uptake by the cells. When functional siRNA was encapsulated within the HBcΔ-Z_{HER2} particles following disassembly and reassembly, the effect of reducing the solid tumor mass was exhibited in an intraperitoneal tumor model following intraperitoneal injection in mice (Suffian et al. 2018).

A universal method to display cell targeting or cell internalizing peptides on the tips of HBc spikes was presented as the previously mentioned "nanoglue" technique (Lee KW et al. 2012b).

At last, Regīna Renhofa's group combined successfully the targeting problem with the problem of packaging and generated a universal chemical-coupling based methodology of targeted delivery of the HBc VLPs packaged with desired RNAs, DNAs, proteins, or magnetic nanoparticles (Renhofa et al. 2015a, b; Strods et al. 2015). The authors regarded decoration of the HBc VLPs with the preS1 peptides and further targeting to hepatocytes as the first example of the practical application.

The targeting and immunological behavior of the chimeric HBc VLPs can be improved in many ways, including

artificial ubiquitination (Chen JH et al. 2011) and the display of flagellin from *Salmonella typhimurium*, a TLR5 activator (Lu Yuan et al. 2013; Zhao Yiwen et al. 2020), of the M1, an engineered form of the streptococcal superantigen SMEZ2 (McIntosh et al. 2014), and of the B domain of staphylococcal protein A (Kim HJ et al. 2016).

When constructing novel variants of the HBc carriers, it is necessary to consider the natural mutational variability of the HBc molecule (Homs et al. 2011) to generate more prospective VLPs in the structural and immunological sense. For example, a R154G replacement is capable of decreasing HBc binding to heparan sulphate proteoglycans, thereby avoiding the nonspecific uptake of HBc VLPs by eukaryotic cells (Suffian et al. 2015). This is highly important for specific cell targeting and nanocargo delivery.

The targeting of the HBcΔ VLPs was combined ingeniously with the packaging providing not only delivery of the cargos but also imaging of the surrounding. Thus, Kim SE et al. (2017) encapsulated a far-red fluorescent protein mCardinal FP into the HBcΔ VLPs and provided the latter with cancer cell receptor-binding affibodies recognizing human epidermal growth factor (EGRF) receptor. The genetic insertion of the FP was performed into a novel particular site between Cys48 and Ser49 of HBc subunit and led to the formation of an internal layer of FPs upon the self-assembly of the modified subunits. Then, the additional mCardinal FP was added to the N-terminus of the double-modified HBcΔ VLPs and exposed on the VLP surface, which resulted in the triple modification and FP-double-layering of the cell-targeted VLPs. Such triple gene-engineered and FP-double-layered VLPs reached more than 40 nm in diameter and effectively detected tumor in live mice with enhanced tumor targeting and imaging efficiency and with far less accumulation in the liver, compared to a conventional fluorescent dye (Kim SE et al. 2017). These monster HBc-derived VLPs are depicted in Figure 38.9.

In parallel, Chen Jiang-Yan et al. (2018) have provided a reasonable outline how to select long flexible linkers

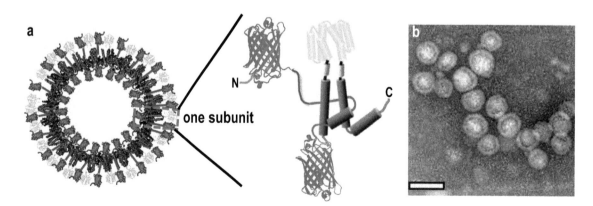

FIGURE 38.9 Genetic encapsulation and assembly to produce fluorescently engineered HBV capsids with cancer targeting capability. (a) Construction of double-layered fluorescent protein nanoparticle with cancer targeting capability, GFPs are blue, but targeting domain is yellow. (b) Electron micrograph of the triple-engineered particle, diameter 42.9 ± 6.3 nm. (Adapted from Kim SE, Jo SD, Kwon KC, Won YY, Lee J. Genetic assembly of double-layered fluorescent protein nanoparticles for cancer targeting and imaging. *Adv Sci (Weinh)*. 2017;4:1600471.)

between the HBc (of HBV genotype G, which just had extra 12 aa insertion at the N terminus) and fluorescent proteins—when the latter were to be fused at the N-terminus of HBc—but to separate the functions of both proteins. The best efficiency in rescuing the replication of an HBV replicon deficient in the HBc expression was achieved for GFP and RFP N-terminal fusions with Gly- and Ser-containing linkers of 47 aa and 186 aa in length, respectively. Nevertheless, both fusion proteins failed to support the formation of the relaxed circular DNA (Chen Jiang-Yan et al. 2018).

Shan et al. (2018b) inserted the tumor-targeting peptide RGD accompanied with glycine-rich linkers into the MIR between aa residues 78 and 81 and packaged the RGD-HBcΔ VLPs with doxorubicin (DOX). The packaging of DOX was confined by hydrophobic lipophilic HCV peptide NS5A$_{1-31}$ added at the C-terminus of the RGD-HBcΔ vector, while supplemental addition of His$_6$ tag allowed the nano-formulation to escape from the endo/lysosome. The treatment with such DOX-loaded particles showed a significant inhibition of tumor growth in mice.

Next, Shan et al. (2018a) fabricated the tumor-targeting RGD-HBc VLPs for cancer optotheranostics. Thus, for image-guided cancer phototherapy, indocyanine green was loaded into the full-length RGD-HBc VLPs via a disassembly/reassembly pathway and electrostatic attraction with high efficiency. The original CTDs were retained in order to encapsulate the negatively charged drug. As a result, the RGD-HBc VLPs significantly improved body retention, aqueous stability, and target specificity of the drug and promoted more accurate and sensitive imaging of the tumor.

Furthermore, Zhang Q et al. (2019) encapsulated a natural medicine quercetin (QR) into the RGD-exposing HBc VLPs for imaging and targeted treatment of hepatic fibrosis. The HBc-RGD/QR VLPs exhibited a rather high selectivity toward activated hepatic stellate cells via the binding affinity with integrin $\alpha_\nu\beta_3$. Once encapsulating quercetin-gadolinium complex and/or labeled with the near infrared fluorescence probes, the HBc-RGD/QR-Gd VLPs showed great potential as fluorescent and MRI contrast and drug delivery agents for hepatic fibrosis in vivo in a mouse model.

In parallel, Biabanikhankahdani et al. (2016) loaded DOX together with polyacrylic acid inside the HBcΔ VLPs, while folic acid was conjugated on the surface of the particles via a nanoglue to increase the specificity and efficacy of the drug delivery system. Then, Biabanikhankahdani et al. (2017) introduced a novel approach to display the DOX on the HBcΔ VLPs, via nitrilotriacetic acid conjugation, by exploiting the His$_6$ tag exposed on the surface of the particles. The His tags served as pH responsive nanojoints, which released DOX from VLPs in a controlled manner. This approach did not involve a time-consuming drug packaging step and allowed any drug that binds to the His tags to be displayed easily on the surface of VLPs by a simple add-and-display step. At last, Biabanikhankahdani et al. (2018) settled on the dual conjugated drug delivery HBcΔ VLP-based system, where DOX was coupled covalently to the external surface of the particles via carboxylate groups,

while folic acid was conjugated to lysine residues of the HBcΔ and targeted the VLPs to cancer cells over-expressing folic acid receptor. As a result, such dual bioconjugated VLPs increased the accumulation and uptake of DOX in the human cervical and colorectal cancer cell lines compared with free DOX, resulting in enhanced cytotoxicity of DOX toward these cells. Akwiditya et al. (2020) used folic acid to address the HBc VLPs carrying the encapsidated plasmid encoding a short hairpin RNA (shRNA) that targeted the anti-apoptotic *Bcl*-2 gene.

Zhao J et al. (2020) exposed brain target TGN peptide at the MIR between the aa residues 78 and 81, with two glycine-rich linkers, in the full-length HBc vector, packaged the TGN-HBc VLPs with epileptic drug phenytoin via disassembly/assembly, and targeted the filled nanocages to brains in mice. The 12-aa TGN peptide, namely, TGNYKALHPHNG, was obtained earlier by rounds of in vivo phage display screening from a 12-mer peptide library for brain targeting (Li Jingwei et al. 2011). As a result, the TGN-HBc-based nanocages specifically and efficiently targeted the brain tissue by 2.4-fold and increased the antiepileptic efficiency of phenytoin about 100-fold in pilocarpine induced models of epilepsy.

At last, Yang J et al. (2020) proposed dual-targeting delivery system with the primary brain targeting peptide TGN for blood–brain barrier penetration and tumor vascular preferred ligand RGD for glioblastoma targeting. Chemo- and gene-therapeutic agents of paclitaxel and siRNA were copackaged inside the VLPs. The latter delivered efficiently the packaged agents to invasive glioblastoma tumor sites. Thus, the combination of chemo and gene therapies demonstrated synergistic antitumor effects in mice through enhancing necrosis and apoptosis, as well as being able to inhibit tumor invasion with minimal cytotoxicity.

Gan et al. (2018) identified a cell penetrating peptide with the sequence NRPDSAQFWLHH from a phage displayed peptide library, which targeted the human squamous carcinoma A431 cells through an interaction with the epidermal growth factor receptor (EGFR). When displayed on the HBcΔ VLPs via the nanoglue, this peptide ensured successful delivery of the nanoparticles into the A431 cells. Next, Gan et al. (2020) showed how the cell-penetrating HBc VLPs targeted delivery of 5-fluorouracil-1-acetic acid (5-FA), a less toxic 5-fluoruracil derivative, to cancer cells expressing EGFR. Remarkably, the cytotoxicity of 5-FA increased significantly after being conjugated on the VLPs that internalized cancer cells and killed them in an EGFR-dependent manner.

Chen R et al. (2021) adapted the chimeric HBc VLPs for a subtle way of fighting obesity. The authors generated the adipose-targeting HBc VLPs that contained a traceable photosensitizer, namely zinc phthalocyanine tetrasulfonate (ZnPcS$_4$) and a browning agent (rosiglitazone) that allowed simultaneous photodynamic and browning treatments, with photoacoustic molecular imaging. After intravenous injection in obese mice, the VLPs bound specifically to white adipose tissues—especially those rich in blood supply—and drove adipose reduction thanks to the synergy of $ZnPcS_4$ photodynamics and rosiglitazone browning. Using photoacoustic molecular imaging, the changes induced by the treatment were monitored by Chen R et al. (2021).

A TARGET FOR ANTIVIRALS

This is currently an urgent issue, as the specific blocking of the HBV nucleocapsid self-assembly in vivo with low probability to induce escape mutations would be a much more efficient anti-HBV therapy than the blocking of replication by nucleos(t)ide analogues, resulting frequently in viral relapse after the cessation of therapy and the rapid emergence of resistance.

The potential effects of antivirals belonging to different classes of low molecular mass chemical drugs may have different phenotypic effects, including misdirected assembly, fast assembly, failure of capsids to package RNA, and interference with establishment of new covalently closed circular DNA (cccDNA). For a detailed analysis of the listed points, a qualified prediction of the antiviral future, and exhaustive references, see exhaustive review by Zlotnick et al. (2015). The small synthetic molecules targeting HBc self-assembly were reviewed at that time by Klumpp and Crépin (2014) and Liu N et al. (2015). From the recent medicinally prospective candidates, the highly promising sulfamoylbenzamide derivative NVR 3–778 from the Novira company (Gane et al. 2014; Flores et al. 2015; Klumpp et al. 2015b, 2018; Lam et al. 2015, 2019) deserves special attention. The heteroaryldihydropyrimidines Bay 41–409 (Stray and Zlotnick 2006), GLS4 (Wu G et al. 2013), and HAP_R01 (Ko et al. 2019) were tested as putative HBc protein allosteric modulators. Remarkably, DBT1, a dibenzothiazepine, accelerated HBc assembly, induced formation of aberrant particles, and paradoxically caused preformed HBc to dissociate (Schlicksup et al. 2020).

The antiviral field grew rapidly and employed a highly qualified in silico docking technique for the selection of potential drugs (Hayakawa et al. 2015). Generally, the putative attacks against the HBc particle by small molecules were reviewed recently (Ko et al. 2017; Yang L and Lu 2018; Kuduk et al. 2021; Senaweera et al. 2021).

It seems that the near future of hepatitis B treatment will consist therefore of a combination of novel antinucleocapsid self-assembly drugs with classical nucleos(t)ide analogues and interferon.

It is an unexpected and intriguing finding that the HBc peptide 147–183 representing the arginine rich domain (ARD) of the HBc molecule possesses broad spectrum anti-Gram positive and anti-Gram negative microbial activity (including some multidrug resistant microorganisms) at micromolar concentrations (Chen HL et al. 2013, 2016). The antimicrobial activity is specific and becomes apparent by membrane permeabilisation or DNA binding. The sequences HBc153–176 or HBc147–167 were necessary and sufficient for antimicrobial activity against *Pseudomonas aeruginosa* and *Klebsiella peumoniae*. The introduction of

such promising antimicrobial peptides would be currently important, as overall microbial resistance to antibiotics is rapidly growing.

NANOMATERIALS

As mentioned briefly earlier, Michael Nassal's team was the first who succeeded in the unexpectedly plain and effective expression of the full-length 238-aa green fluorescent protein (GFP) inserted at the HBcΔ MIR (Kratz et al. 1999). The chimeric HBcΔ-GFP VLPs (CLPs, or core-like particles, according to the authors' original nomenclature) were fluorescent and demonstrated therefore both functionality and external localization of the GFP insertion. Moreover, the chimeric particles elicited a potent humoral response against native GFP in rabbits. In this early report, the authors predicted that the HBcΔ-GFP VLPs "might be used to follow the intracellular trafficking of authentic HBV core protein and also to trace the fate of capsid-like particles after administration into a complete organism" (Kratz et al. 1999). In fact, these HBcΔ-GFP VLPs were used to investigate the interaction of HBc with nuclear pore complexes in permeabilized HeLa cells (Lill et al. 2006).

Then, an impressing set of the HBcΔ-GFP VLPs was constructed, which differed by linkers surrounding the GFP insertion, as well as carried truncated GFP variant (Böttcher et al. 2006). These structures were monitored by electron cryomicroscopy and image reconstruction, one example of which is presented in Figure 38.10.

Developing further their approach, Vogel et al. (2005a) presented an in vitro approach that allowed coassembly of natural HBcΔ subunits with an assembly-competent and an assembly deficient core-GFP fusion, namely, enhanced green fluorescent protein (eGFP) inserted into the MIR and fused to aa 148—respectively—and by mosaic VLP formation from two core fusions with different GFP derivatives

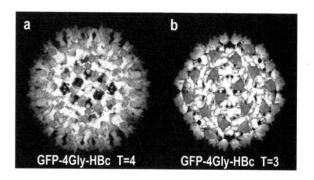

FIGURE 38.10 Organization of the outer GFP-derived shells in GFP-4Gly-HBc CLPs. The color-coding of the surface is chosen according to radii; surface in the core protein region is shown in white, surface at larger radii (GFP region) is shown in orange. (a) Large GFP-4Gly-HBc CLPs; (b) small GFP-4Gly-HBc CLPs. (Reprinted from *J Mol Biol*. 356, Böttcher B, Vogel M, Ploss M, Nassal M, High plasticity of the hepatitis B virus capsid revealed by conformational stress, 812–822, Copyright 2006, with permission from Elsevier.)

that exerted fluorescence resonance energy transfer (FRET) in an assembly status dependent way.

Then, Vogel et al. (2005b) generated HBcΔ VLPs carrying monomeric yellow, cyan, and red fluorescent proteins (mYFP, mCFP and mRFP1) at the MIR but failed by insertion of red fluorescent proteins DsRed1, DsRed2, and HcRed, which also folded into the GFP-typical β-can structure but demonstrated oligomerization tendency. Later, Yoo et al. (2012) succeeded in vivo by the construction of the HBcΔ VLPs carrying DsRed or eGFP at the MIR. Moreover, the authors performed tests in animals and found that their VLPs stably emitted high-level fluorescence inside mice for a prolonged period. It paved the way for the application of the fluorescent HBc-based VLPs not only as an in vitro fluorescent reporter but also as a noncytotoxic tool for in vivo optical imaging with targeted delivery.

Park et al. (2009) replaced the $P_{79}A_{80}$ in the HBcΔ MIR with the tandem repeated B domains of staphylococcal protein A, a specific target for the Fc domain of immunoglobulin G and exposed it therefore on the VLP surface. In parallel, the His_6 tag was added to the N-terminus of the VLPs to provide the latter with a strong affinity for nickel. Thus, a 3D assay system was developed by combining the chimeric nanoparticles with a nickel nanohair structure or porous membrane, then adding antibodies to specifically capture protein markers. This methodology was adjusted to the detection of troponin I in patients (Park et al. 2009) but remains open to form similar highly sensitive diagnostic assays for a variety of other protein markers. Later, Kwon et al. (2017) developed on this background a notably advanced one-step immunoassay based on accurate, rapid, and label-free self-enhancement of immunoassay signals, which was achieved by tightly coupling the 3D bioprobe-based sensitive assay with the coordinated assembly of gold nanoparticles in an assay solution.

The HBc derivatives—but not VLPs in this case—were used as biosensors (Lim et al. 2017, 2020). The authors reported the synthesis of bioconjugated polyacrylamide-based hydrogel that responded to the presence of HBc by a weight change in the hydrogel. This hydrogel contained the affinity crosslinks formed by the biospecific coupling between the bound dimer of HBc mutant Y132A—which was unable to self-assemble into VLPs—and anti-HBc antibody. In the presence of free HBc, the immobilized HBc-antibody complexes in the hydrogel were disrupted, triggering the swelling of hydrogel. The latter corresponded to the concentration of free HBc, and the biosensing property of the hydrogel remained highly specific.

Furthermore, the HBcΔ VLPs (but not a dimer) were used as a biosensor. Thus, Abd Muain et al. (2018) immobilized the HBcΔ VLPs onto the gold nanoparticles-decorated reduced graphene oxide (rGO-en-AuNPs) nanocomposite and used the antigen-functionalized surface to sense the presence of anti-HBc antibodies.

The HBc VLPs were loaded into poly(d,l-lactic-acid-co-glycolic acid) (PLGA) nanoparticles with or without monophospholipid A (MPLA), a Th1-favoring immunomodulator.

These particles were around 300 nm in diameter, spherical in shape, had approximately 50% HBc VLP encapsulation efficiency, and markedly improved the profile of Th1 immune response with IFN-γ production, stimulating thus development of therapeutic vaccine candidates (Chong et al. 2005).

The HBc VLPs were used to generate inorganic nanostructures by biomineralization, with copper sulfide for instance. Thus, Jia et al. (2019) presented a straightforward method for efficiently synthesizing CuS nanoparticles inside the inner cavity of the HBcΔ VLPs via the affinity of histidine tags to copper ions. The monodispersed and well-defined HBcΔ-CuS VLPs have been yielded with high efficiency and used as a potential photothermal agent for

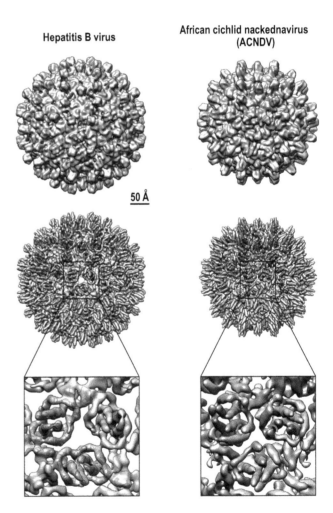

FIGURE 38.11 Capsid ultrastructure of nackednaviruses. Comparison of the capsid structure of HBV (T=4) (Yu et al. 2013) and ACNDV (T=3). Cryoelectron microscopy maps low-pass filtered at 12 Å (top row) and 8 Å (middle row). Bottom row: zoomed view onto a local (pseudo-)3-fold axis. Additional α+ helices in ACNDV highlighted by red arcs. (Reprinted from *Cell Host Microbe*. Lauber C, Seitz S, Mattei S, Suh A, Beck J, Herstein J, Börold J, Salzburger W, Kaderali L, Briggs JAG, Bartenschlager R, Deciphering the origin and evolution of hepatitis B viruses by means of a family of non-enveloped fish viruses. 2017;22:387–399.e6.)

cancer therapy due to their excellent photothermal conversion efficiency, high cellular uptake, and biocompatibility (Jia et al. 2019).

NACKEDNAVIRUSES

Lauber et al. (2017) discovered a diversified family of fish viruses that lack the envelope protein gene but otherwise exhibit key characteristics of hepadnaviruses including genome replication via proteinprimed reverse-transcription and utilization of structurally related capsids. Phylogenetic reconstruction indicated that these two virus families separated more than 400 million years ago, before the rise of tetrapods. The authors showed that HBV-like viruses are of ancient origin—descending from nonenveloped progenitors in fishes—and designated them nackednaviruses, or putative *Nackednaviridae* family members, Swabian German for "naked DNA viruses."

The nackednaviral core proteins showed little sequence similarity with those of hepadnaviruses, while secondary structure predictions revealed the conserved arrangement of α-helices characteristic for the HBc, as well as an additional short α+ helix at the extreme N-terminus. The authors expressed core proteins of the African cichlid nackednavirus (ACNDV) in *E. coli*, purified VLPs, and performed electron cryomicroscopy. As follows from Figure 38.11, the 3D particle reconstruction showed T=3 icosahedral symmetry, where the overall fold of the ACNDV core protein was similar to that of HBc. In contrast to HBc, at the local (pseudo-)3-fold axes the holes in the particle shell were plugged by the additional N-terminal helices, which might aid in protecting the genomes of the nonenveloped viruses against environmental damage (Lauber et al. 2017). Moreover, by full analogy with the HBc, it was found that the 174 aa long ACDNV core allowed truncation of 28 C-terminal aa residues, a nucleic acid-binding domain, and formed the ACDNV coreΔ(1–146) VLPs.

This revolutionary finding of "fossil" hepadna-like viruses may promote unexpected solutions in the hepadnaviral VLP-based nanotechnology.

Epilogue

IMPROVING VIRUS TAXONOMY

What's in a name? That which we call a rose by any other name would smell as sweet.

William Shakespeare, *Romeo and Juliet*

Disputing with the great William Shakespeare, there is, in fact, strong need for the continuous improvement of virus taxonomy, especially now, when old traditional collections of viruses are overwhelmed with the new and rapidly growing metagenomics data. As a result, we need many more names than before. This point is exemplified by the recent rearrangement of the class *Leviviricetes* by Colin Hill's team, as described in Chapter 25. Remarkably, the required hundreds of novel names forced the researchers to use special scripts, intended to slightly alter the spelling of chosen terms (Callanan et al. 2020a). Earlier, the urgent need for the permanent cataloguing of viruses was convincingly justified by Jens H. Kuhn et al. (2019). As the authors wisely concluded, "the gain is worth the pain."

For our specific purpose, the improvements by the ICTV (2020) that were completed and are planned in the near future by Eugene V. Koonin and his colleagues (2020, 2021) provided encouragement to us to present the whole VLP field in close relationship with the systematic classification of the virus world—as explained previously in the *Prologue* to this book—and place it in the context with the modern comprehensive hierarchical taxonomy of viruses. This allowed one to replace alphabetical or other ways of VLP descriptions and to add to the VLPs the spirit of unity with viruses, while also emphasizing differences between the VLPs and viruses. Moreover, the obvious success of the enthusiasts of the novel taxonomy, which started with the creation of the realm *Riboviria* representing the RNA virus world (Walker et al. 2019, 2020), provided an inspiration and made this book possible. After the *Riboviria*, the novel *Ribozyviria* realm was described, which covered the mysterious *Deltavirus* genus including the well-known hepatitis D (delta) virus (Hepojoki et al. 2020), as described in Chapter 34. Then, the evolution of the virus hierarchy also reached the DNA viruses, namely double-stranded DNA viruses of the Baltimore class I and the novel realms *Varidnaviria* (Woo et al. 2021) and *Adnaviria* (Krupovic et al. 2021) were described, which let us to systematically arrange the first five chapters of the book, including Chapter 5 devoted to the mysterious "other dsDNA viruses." It should be emphasized that the three crucial realms were established at a time when we finished draft compilation of the VLP data for this book and started to look for the most relevant structural organization of the available material. We believe that the whole VLP field can be better understood in the context of the continuously improving

hierarchy of the viral world that was presented in Figure P.6 of the *Prologue*. Moreover, the hierarchical organization of the chapters will encourage the reader to focus on the similarities and differences of the VLPs in the context of the neighboring virus taxons. It is our sincere hope that this first broad and systematic presentation of the VLPs and VLPs scaffolds will result in new ideas and lead to VLP candidates with novel and unexpected characteristics.

COMPARATIVE DESCRIPTION OF THE EXISTING VLP SCAFFOLDS

Write injuries in dust, benefits in marble.

Benjamin Franklin

This is illustrated in Figure E.1. Although the estimation is approximate and takes into account the number of infectious and/or noninfectious diseases that were targeted by the VLPs of any particular chapter (but not the number of different epitopes and constructs), it demonstrates the general tendency. As expected, the hepatitis B core (HBc) and RNA phage VLPs that were among the pioneers of the protein engineering of VLPs remain the absolute leaders. It is remarkable that there are teams that contributed in parallel to both these VLP scaffolds. They are teams headed by Peter H. Hofschneider in Munich, Elmārs Grēns in Riga, Martin Bachmann in Zurich/Oxford/Bern, and Michael Nassal in Freiburg. D.H.L. Bishop, a successful collaborator of Sol Spiegelman, strongly contributed to the RNA phages at their early stages and further participated in the HBc VLP studies. There are some pioneering (Borisova et al. 1987; Gren and Pumpens 1988) and more recent (Bundy et al. 2008; Strable et al. 2008; Bundy and Swartz 2011; Wu et al. 2020) studies that intended the head-to-head comparison of both RNA phage and HBc scaffolds. The authors of this book also worked on both RNA phage and HBc VLPs. This in part explains why Chapters 25 and 38 are more detailed than other chapters and provide the complete review of all works with both these VLP types. Then, Chapter 33 should be also highlighted as a special target of our research. Finally, we reviewed available literature including our own experimental contributions to VLPs described in Chapters 7, 8, 18, 22, 26, 31, 32, and 37.

Regarding the sources of the epitopes used in VLPs, the leadership clearly belongs to influenza A virus (IAV), which was targeted by the VLPs described in 23 chapters out of 38. The human immunodeficiency virus (HIV) epitopes are described in 20 chapters, while the malaria and foot-and-mouth disease virus (FMDV) epitopes are described in 18 and 15 chapters, respectively. The human papilloma virus (HPV) and hepatitis B virus (HBV)

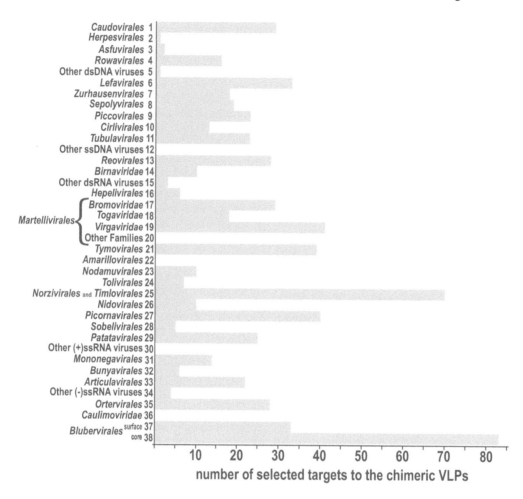

FIGURE E.1 A rough estimation of the number of different viral and nonviral infections that have been targeted by the engineered chimeric VLPs of viruses that were presented in the book chapters.

epitopes appeared in 13 chapters each. The three pathogenic coronaviruses SARS, MERS, and SARS-CoV-2 are described in 11 chapters. The epitopes of HCV, anthrax, and different forms of cancer were targeted in 10, poliovirus (PV) in 8, and Alzheimer' disease in 7 chapters each. It should be emphasized that the envelope proteins of HBV, as well as lymphocytic choriomeningitis virus (LCMV), served as sources of the widely used and practical model epitopes, namely the preS1 peptide DPAFR and the gp33 epitope, respectively.

LOOKING FOR NOVEL SCAFFOLDS

If you don't like someone's story, write your own.

Chinua Achebe

In addition to viral taxonomy, we paid a lot of attention to the VLP architecture and structural characteristics of the viral sources of the VLPs including the chimeric VLPs. The amazing virus architecture always fascinated biologists, physicists, or even mathematicians. As proof, elegant works were recently published by structural classicists David P. Wilson (2020), Antonio Šiber (2020), John E. Johnson and

Arthur J. Olson (2021), and Pierre-Philippe Dechant and Reidun Twarock (2021).

The challenges of novel VLP scaffolds in modern times can be exemplified by the current status in the previously mentioned RNA phages, described in Chapter 25. Metagenomic studies identified hundreds of novel levivirus genomes (Krishnamurthy et al. 2016; Shi et al. 2016; Callanan et al. 2018, 2020a, b; Greninger 2018). Many metagenomically predicted coat genes were expressed from the very different levi-like RNA genomes in *E. coli* in Kaspars Tārs lab (Liekniņa et al. 2019). These authors performed the BLAST analysis that revealed approximately 14 distinct levi-like RNA phage coat types, which was a considerable increase from the three coat protein types known before. In total, the authors have chosen 110 coat sequences from the metagenomic data to cover all coat groups to express the candidate genes in *E. coli*. The vast majority of the coats were produced in the expected high yields and only in a few cases was there no detectable expression. However, only about 60% of the coat proteins were at least partially soluble, while the rest were found in inclusion bodies. In an effort to mitigate the issue, the insoluble proteins were expressed in *E. coli* at 15°C, which allowed 85% of the previously insoluble coats to become at least partially soluble, and only six remained in inclusion bodies at the lower temperature. In total, 80, or approximately 72%, of the soluble coats assembled into VLPs as confirmed by electron microscopy. In most cases, the VLP morphology resembled that of the previously characterized RNA phage VLPs with an apparent spherical shape 28–30 nm in diameter that corresponded to a T = 3 icosahedral particle, as expected for the classical RNA phage virions. However, notable deviations from the standard particle size and shape were not uncommon. Some VLPs were noticeably larger, reaching 35–40 nm in diameter, which could correspond to a T = 4 icosahedral particle. The two coats assembled into small particles approximately 18 nm in diameter with a presumed T = 1 icosahedral symmetry. A sizeable proportion of one of the coats appeared to have an elongated shape, while two coats demonstrated a mixture of T = 3 and T = 1 particles. The authors addressed the possible disulfide bonds, thermal stability, and potential RNA-protein interactions as the parameters crucial for the further nanotechnological applications (Liekniņa et al. 2019).

This study demonstrated conclusively for the first time that the environmental viral sequences uncovered in metagenomic studies can be successfully employed to prepare VLPs in a laboratory setting for potential nanotechnological applications. The 80 novel VLP platforms presented opportunity for VLP applications. Thus, Liekniņa et al. (2020) attached a model peptide to the N- and C-termini of the novel coat proteins. As a model peptide, a triple repeat of 23 N-terminal residues of the ectodomain of the influenza M2 protein was used, the structure that was actively used in the development of the influenza vaccines based on the VLP approach and described in numerous chapters of this book. After examining 43 novel phage coat proteins for the ability to form the chimeric VLPs, 10 new promising candidates were selected for further vaccine design, 5 of which were tolerant to insertions at both the N- and C-termini. Furthermore, it was demonstrated that most of the chimeric VLPs had retained antigenic properties as concluded from their reactivity with anti-M2 antibodies (Liekniņa et al. 2020).

Developing further the theoretical basis for the novel scaffolds, Rūmnieks et al. (2020) determined the crystal structure of 22 previously unknown metagenomically detected RNA phage capsids that covered nine distinct coat protein types. These structures shown in Figure E.2 were determined at <4 Å resolution and revealed substantial deviations from the previously known RNA phage coat fold, which is shown in classical figures in Chapter 25. Within the sample set of 22 VLPs, only in 10 cases was the coat structure consistent with the canonical MS2 coat fold with the N-terminal hairpin, five-stranded β-sheet, and two C-terminal α-helices, while notable deviations were observed for the other coats. Moreover, the novel structures uncovered an unusual prolate particle shape and revealed a previously unseen dsRNA binding mode. This data markedly expanded the ideas for further nanotechnological

applications (Rūmnieks et al. 2020). However, the most amazing thing, in our opinion, is that the fusions of the foreign epitopes and 3D reconstructions were performed in this case with the VLPs that were derived from the viruses nobody had ever seen and the actual hosts of which remained unknown. In fact, this is the first situation when the VLP scaffolds appeared and were resolved ahead of the corresponding viruses and their virions.

Speaking about another well-known VLP scaffold, HBc, the nonhuman, rodent HBV-like viruses were actively developed by the David R. Milich's team (Billaud et al. 2007) into the generation of the prospective vaccine candidates, as described in Chapter 38. Moreover, the exploration of still unassigned nackednaviruses, the nonenveloped hepadnavirus-like fish viruses, by Ralf Bartenschlager's team (Lauber et al. 2017) created new opportunities for the research on this currently well-studied class of VLPs. Nevertheless, despite many obstacles during more than 30 years of exploration, the structural HBV proteins HBc and HBs remain favorite actors of many research programs aimed at the generation of vaccine candidates against malaria, influenza, hepatitis C virus (HCV), and other human and animal diseases, as noted by Wolfram H. Gerlich (2015).

EXPLORING NOVEL ISSUES

The only way to discover the limits of the possible is to go beyond them into the impossible.

Arthur C. Clarke
Technology and the Future

The rapid progress of the VLP nanotechnology has led to numerous innovative solutions and combinations. For example, Yang et al. (2021) recently contributed to the outcome problem of the popular VLPs by the optimization of conditions for the assembly and purification of six representative VLPs, namely that formed by the hepatitis B virus core antigen (HBc), phages Qβ, MS2, and P22, cowpea chlorotic mottle virus (CCMV), and tobacco mosaic virus (TMV), and expressed in *E. coli*.

After the technological improvements, novel applications are being developed, including by using classic RNA phage VLPs. Thus, Liu X et al. (2019) described experiments to combine the RNA phage VLP approach with exosomes to generate novel cell-targeting and drug delivery devices. Exosomes represent a naturally secreted nanoparticle family that carry RNA and protein cargos and are currently being actively developed. Liu X et al. (2019) repurposed the exosomes for targeting peptide screening. To do this, the signal peptide region of Lamp2b, a membrane protein on the exosomes, was fused in the N-terminus with ten aa-long random peptides, while the C-terminus of Lamp2b was fused to the MS2 coat protein. Then, the whole Lamp2b-MS2 coat open reading frame was further engineered to harbor a 3′UTR sequence of MS2 RNA. The resultant exosomes from engineered Lamp2b-MS2-coat expressing cells displayed the ten aa peptides on the outside while containing

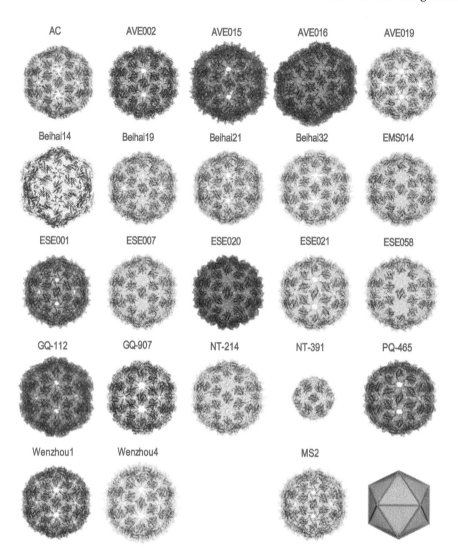

FIGURE E.2 Structure of the novel ssRNA phage VLPs. Coat protein dimers are shown in cartoon representation inside semitransparent VLP molecular surfaces and are differently colored as per different coat protein similarity groups. The back sides of the particles are clipped for clarity. All particles are shown on the same scale and in an orientation corresponding to the regular icosahedron on bottom right. Structure of the ssRNA bacteriophage MS2 is included for comparison. (From Rūmnieks J et al. Three-dimensional structure of 22 uncultured ssRNA bacteriophages: Flexibility of the coat protein fold and variations in particle shapes. *Sci Adv.* 2020;6:eabc0023.) (The high-resolution image is a kind gift of Jānis Rūmnieks.)

the genetic information inside. In proof-of-principle experiments, the authors showed that the exosomes with different peptides could be preferentially distributed to different tissues. Furthermore, the target sequences for different tissues were enriched by some rounds of selection. This approach was termed *exosome display* and proposed to be used for displaying and screening targeted peptides for the cells outside the capillary with condense barriers, like neurons in the brain (Liu X et al. 2019).

The cell-type-specific aptamer-functionalized VLPs could be an important tool for targeted disease therapy (Zhou and Rossi 2014), reviving the more than 100-year-old Paul Ehrlich's idea of a *magic bullet*. This concept presumed that a therapeutic agent would only kill the specific cells it targeted. The VLPs provided with the specific aptamers

could be considered therefore as promising prototypes of such magic bullets or, in other words, targeted therapies.

Next, the success of VLPs inspired a novel field of VLPs formed by proteins of nonviral origin, such as ferritin, heat shock protein, lumazine synthase, encapsulin, aldolase, and others, recently reviewed by Nguyen et al. (2021), as well as by purely synthetic artificial proteins, as described, for example, by Arin Ghasparian and John A. Robinson (2015). Remarkably, the previously mentioned recent study of Yang et al. (2021) optimized the outcome of the nonviral ferritin and encapsulin VLPs produced in *E. coli*, in parallel to that of the six popular viral VLPs.

Finally, the bacterial microcompartment (BMC) particles consisting of a protein shell and an encapsulated enzymatic core could also be mentioned in this regard. Thus,

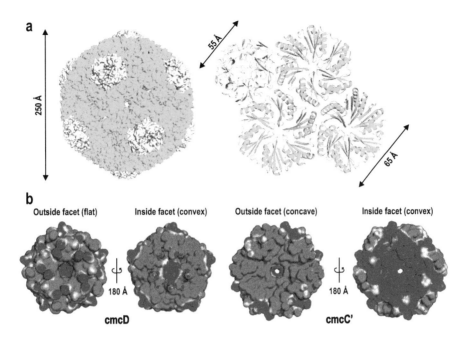

FIGURE E.3 The electron cryomicroscopy structure of pT = 4 quasi-icosahedral bacterial microcompartment (BMC) shell-derived particles (BDPs) and its penatameric and hexameric components. (a) Surface model of pT = 4 quasi-icosahedral BDP particle, displayed on the left side. A ribbon model of a cmcD pentamer and three cmcC′ hexamers is displayed on the right side. The pentameric cmcD protein is colored in yellow and hexameric cmcC′ is colored in green. Note that the 5-fold symmetry axis is located at the center of cmcD pentamer and the 3-fold axis is located in the middle between three cmcC′ hexamers. (b) The electrostatic surface potential of pentameric cmcD and hexameric cmcC′. Note the pores in the centers of pentamers and hexamers. The surface contour levels were set to −1 kT/e (red) and +1 kT/e (blue). (From Kalnins G et al. Encapsulation mechanisms and structural studies of GRM2 bacterial microcompartment particles. *Nat Commun.* 2020;11:388.)

Kalnins et al. (2020) reported the isolation and expression of the BMC structural genes from the *Klebsiella pneumoniae* GRM2 locus and resolved an electron cryomicroscopy structure of a pT = 4 quasi-symmetric icosahedral shell particle to 3.3 Å. Figure E.3 is intended to show the clear parallels by the structural organization of viral and nonviral VLPs.

The rapidly growing approaches that employ the nonviral and/or artificial VLPs deserve, however, special consideration and could be a subject of another book or even several books.

PANDEMIC CHALLENGES AND NEW DEVELOPMENTS

> The reasonable man adapts himself to the world; the unreasonable one persists in trying to adapt the world to himself. Therefore, all progress depends on the unreasonable man.

George Bernard Shaw, *Man and Superman*

The Covid-19 pandemic ruthlessly challenged the capabilities of rapid vaccine development and production around the world. Among other approaches, VLPs were also used for vaccine development, and 7 of 38 chapters in this guide show VLP applications to generate experimental SARS-CoV-2 VLPs. In addition, five and four chapters,

respectively, describe vaccine development efforts for the previous pathogenic coronaviruses SARS-CoV-1 and MERS. The race in vaccine development resulted in predictable victory of the novel mRNA vaccines and virus vector approaches driven by the international teams and commercial companies, which demonstrated outstanding scientific and technological potential. This is the time for the VLP engineers to ask "where we stand" and what are the "challenges ahead," as Forni and Mantovani (2021) did in the title of their recent review article. The efforts of the scientific groups, university laboratories, and biotech companies are all needed, since they possess unique knowledge and scientific potential to develop new ideas including the use of the various classes of the VLPs. For example, Rothen et al. (2021) demonstrated that the cucumber mosaic virus platform CuMV$_{TT}$ carrying the chemically coupled receptor binding domain (RBD) of the spike protein S of SARS-CoV-2 and described in Chapter 17 behaved as an excellent vaccine prototype. The vaccine was self-adjuvated with prokaryotic ssRNA, a well-known TLR7/8 agonist, which was packaged during expression and assembly in the bacterial *E. coli* system and demonstrated efficient antibody response in mice after intranasal administration (Rothen et al. 2021).

Since this and other VLP platforms are excellent packagers of RNA, the combination of the VLP methodology with the mRNA vaccination seems to be one of the logical and realistic outcomes. Such VLP vectors are overlapping with virus vector approaches, especially replication-defective virus vectors. We learned earlier in this book that the VLPs can be successfully administered by nasal and/or oral routes and induce mucosal immunity, the ways potentially preferable for vaccination against Covid-19. As mentioned throughout the book, there are many other advantages that support the idea of the VLP vaccine development. No doubt, Covid-19 came to stay, and the success in generating innovative vaccines will be a real practical test for the applicability and efficacy of the current VLP technologies. What is essential, in the words of Ralph Waldo Emerson, "Do not go where the path may lead, go instead where there is no path and leave a trail."

Name Index

Subject Index

Looking at this world, one cannot help wondering!
Kozma Petrovich Prutkov, *Fruits of Reflection* (1853–1854)

Beware of false knowledge; it is more dangerous than ignorance.
George Bernard Shaw

A

Acanthamoeba polyphaga marseillevirus (APMaV), 52, *see also Marseillevirus marseillevirus*
Acanthamoeba polyphaga mimivirus, species, *see* Acanthamoeba polyphaga mimivirus (APMV)
Acanthamoeba polyphaga mimivirus (APMV), 50–52, *see also Acanthamoeba polyphaga mimivirus*
Acheta domesticus densovirus (AdDNV), 123, *see also Blattodean pefuambidensovirus 1*
Acidianus rod-shaped virus 1 (ARV1) 48, *see also Itarudivirus ARV1*
Acidianus two-tailed virus, species, *see* Acidianus two-tailed virus (ATV)
Acidianus two-tailed virus (ATV), 60–61, *see also Acidianus two-tailed virus*
Acholeplasma virus L2 (AVL2) 60
Acholeplasma virus L2, species, 62
Ackermannviridae, family, 13
Adeno-associated dependoparvovirus A, species, *see* adeno-associated virus (AAV), types 1–4 and 6–13
Adeno-associated dependoparvovirus B, species, *see* adeno-associated virus 5 (AAV5)
adeno-associated virus (AAV), types 1–4 and 6–13, 110, *see also Adeno-associated dependoparvovirus A*
adeno-associated virus 5 (AAV5), 110–112, 114, *see also Adeno-associated dependoparvovirus B*
Adenoviridae, family, 37, 39
Adintoviridae, family, 56
Adnaviria, realm, 8–9, 47, 62, 73, 93, 603
Aedes albopictus C6/36 cell densovirus (C6/36DNV), 124, *see also Dipteran brevihamaparvovirus 1*
Aeropyrum coil-shaped virus (ACV), 150–151
Aeropyrum pernix bacilliform virus 1 (APBV1), 47, 60–61
African cichlid nackednavirus (ACNDV), 566, 601–602
African green monkey polyomavirus (AGMPyV or LPV), 106, *see also* lymphotropic polyomavirus (LPV)
African horse sickness virus, species, *see* African horse sickness virus (AHSV)
African horse sickness virus (AHSV), 155, 157, 163–164, *see also African horse sickness virus*
African swine fever virus, species, *see* African swine fever virus (ASFV)
African swine fever virus (ASFV), 33–36, *see also African swine fever virus*

Albetovirus, genus, 435, 454–456
Alefpapillomavirus, genus, 73
Alefpapillomavirus 1, species, 73
Aleutian mink disease parvovirus (ADV), 121, *see also Carnivore amdoparvovirus 1*
alfalfa mosaic virus (AMV), 207–211, 455
Alfamovirus, genus, 207–208
Algavirales, order, 7, 50–51, 53
Alloherpesviridae, family, 25
Allolevivirus, former genus, 315, 317–318, 320–321
Alphaarterivirus, genus, 393
Alphaarterivirus equid, species, *see* equine arteritis virus (EAV)
Alphabaculovirus, genus, 65–66, 68
Alphacarmotetravirus, genus, 307, 309
Alphachrysovirus, genus, 187–188
Alphacoronavirus, genus, 384–386
Alphacoronavirus 1, species, *see* transmissible gastroenteritis virus (TGEV)
Alphaendornavirus, genus, 257
Alphafusellovirus, genus, 61
Alphaherpesvirinae, subfamily, 25–28
Alphaflexiviridae, family, 255, 261–262, 271
Alphaglobulovirus, genus, 62
Alphaglobulovirus PSV, species, *see* Pyrobaculum spherical virus (PSV)
Alphaglobulovirus TTSV1, species, *see* Thermoproteus tenax spherical virus 1 (TTSV1)
Alphaguttavirus, genus, 61
Alphainfluenzavirus, genus, 487, 489–490, 492
Alphalipothrixvirus, genus, 47
Alphanodavirus, genus, 297–300
Alphaovalivirus, genus, 62
Alphapapillomavirus, genus, 73–85
Alphapapillomavirus 4, species, *see* human papillomavirus 57 (HPV57)
Alphapapillomavirus 7, species, *see* human papillomavirus 18 (HPV18)
Alphapapillomavirus 9, species, *see* human papillomavirus 16 (HPV16); human papillomavirus 31 (HPV31)
Alphapapillomavirus 10, species, *see* human papillomavirus 6 (HPV6); human papillomavirus 11 (HPV11)
Alphapartitivirus, genus, 185
Alphapolyomavirus, genus, 93, 94–101, 106
Alphaportoglobovirus, genus, 62
Alpharetrovirus, genus, 521
Alphasatellitidae, family, 150
Alphasphaerolipovirus, genus, 59
Alphatectivirus, genus, 57–58

Alphatetraviridae, family, 195, 203–204, 313, 454
Alphatorquevirus, genus, 150
Alphatristromavirus, genus, 49
Alphatristromavirus PFV1, species, *see* Pyrobaculum filamentous virus 1 (PFV1)
Alphatristromavirus PFV2, species, *see* Pyrobaculum filamentous virus 2 (PFV2)
Alphaturrivirus, genus, 56
Alsuviricetes, class, 7, 195, 207, 261
Alternanthera mosaic virus (AltMV), 267, 274, 446
Alvernaviridae, family, 438
Amabiliviricetes, class, 7, 453–454
Amalgaviridae, family, 185
Amalgavirus, genus, 185
Amarillovirales, order, 8–9, 221, 227, 230, 269, 279–295, 404, 506, 604
Amdoparvovirus, genus, 121
Amnoonviridae, family, 487–488
Ampelovirus, genus, 257–258
Ampullaviridae, family, 59–60
Andes orthohantavirus (ANDV), 482
Anelloviridae, family, 125–126, 149
Anser anser polyomavirus 1, species, *see* goose hemorrhagic polyomavirus (GHPyV)
Anseriform dependoparvovirus 1, species, *see* goose parvovirus (GPV); muscovy duck parvovirus (MDPV)
Antheraea mylitta cytoplasmic polyhedrosis virus (AmCPV), 171
Anulavirus, genus, 207–208
Aparavirus, genus, 411
Apeevirus, genus, 315
Apeevirus quebecense, species, *see* phage, AP205
Aphthovirus, genus, 397, 411–412
apple latent spherical virus (ALSV), 430–431
Aquabirnavirus, genus, 173–174
Aquaparamyxovirus, genus, 464
Arctic ground squirrel hepatitis virus (AGSHV), 565
Arenaviridae, family, 475–477, 479
Arfiviricetes, class, 125, 148–149
Arnidovirineae, suborder, 383
Arteriviridae, family, 383–384, 393–395
Articulavirales, order, 4, 6, 8–9, 207, 393, 449, 487–504, 505, 604
artichoke mottled crinkle virus (AMCV), 310, 312–313
Artoviridae, family, 462
Artverviricota, phylum, 7, 513
Ascaris lumbricoides Tas virus, species, 520
Ascoviridae, family, 52

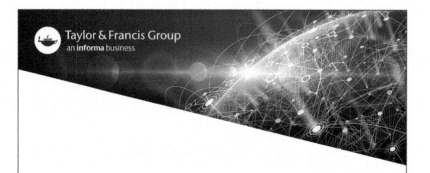

9 781032 246734